FOURTH EDITION

Modern Hydronic Heating & *Cooling*

for residential and light commercial buildings

John Siegenthaler, P.E.
Associate Professor Emeritus
Mohawk Valley Community College,
Utica, New York

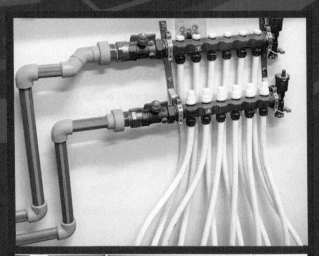

Australia • Brazil • Canada • Mexico • Singapore • United Kingdom • United States

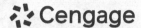

***Modern Hydronic Heating and Cooling: For Residential and Light Commercial Buildings,* 4e**
John Siegenthaler, P.E.

SVP, Course Solutions Product: Erin Joyner

VP, Product Management, Learning Experiences: Thais Alencar

Product Director: Jason Fremder

Product Manager: Laruen Whalen

Product Assistant: Bridget Duffy

Learning Designer: Elizabeth Berry

Content Manager: Sangeetha Vijay, Lumina Datamatics, Inc.

Digital Delivery Quality Partner: Elizabeth Cranston

Director, Product Marketing: Neena Bali

IP Analyst: Ashley Maynard

IP Project Manager: Anjali Kambli

Designer: Felicia Bennett

Production Service: Lumina Datamatics, Inc.

Cover Image Source: Pixacon/Shutterstock.com

Interior Image Source: Africa Studio/Shutterstock.com; Pavel Vaschenkov/Shutterstock.com

© 2023, 2012 Cengage Learning, Inc. ALL RIGHTS RESERVED.

WCN: 01-100-414

No part of this work covered by the copyright herein may be reproduced or distributed in any form or by any means, except as permitted by U.S. copyright law, without the prior written permission of the copyright owner.

For product information and technology assistance, contact us at
**Cengage Customer & Sales Support, 1-800-354-9706
or support.cengage.com.**

For permission to use material from this text or product, submit all requests online at **www.copyright.com**.

Library of Congress Control Number: 2022932545

ISBN: 978-1-337-90491-9

Cengage
5191 Natorp Boulevard
Mason, OH 45040
USA

Cengage is a leading provider of customized learning solutions. Our employees reside in nearly 40 different countries and serve digital learners in 165 countries around the world. Find your local representative at: **www.cengage.com.**

To learn more about Cengage platforms and services, register or access your online learning solution, or purchase materials for your course, visit **www.cengage.com.**

Printed Number: 9   Print Year: 2024
Printed in Mexico

Contents

	Preface	ix
Chapter 1	**Fundamental Concepts**	**1**
	Objectives	1
	1.1 What Is a Hydronic Heating System?	1
	1.2 Benefits of Hydronic Heating	2
	1.3 Heat and Heat Transfer	8
	1.4 Four Basic Hydronic Subsystems	13
	1.5 The Importance of System Design	17
	Summary	18
	Key Terms	18
	Questions and Exercises	19
Chapter 2	**Space Heating and Domestic Water-Heating Loads**	**20**
	Objectives	20
	2.1 Introduction	20
	2.2 Design Space Heating Load	21
	2.3 Conduction Heat Losses	22
	2.4 Foundation Heat Loss	26
	2.5 Infiltration Heat Loss	33
	2.6 Example of Complete Heating Load Estimate	36
	2.7 Computer-Aided Heating Load Calculations	40
	2.8 Estimating Annual Heating Energy Use	41
	2.9 Estimating Domestic Water-Heating Loads	47
	Summary	49
	Key Terms	49
	Questions and Exercises	50
Chapter 3	**Hydronic Heat Sources**	**52**
	Objectives	52
	3.1 Introduction	52
	3.2 Classification of Hydronic Heat Sources	53
	3.3 Gas- and Oil-Fired Boiler Designs	54
	3.4 Conventional Versus Condensing Boilers	59
	3.5 Domestic Hot Water Tanks as Hydronic Heat Sources	67
	3.6 Power Venting Exhaust Systems	69
	3.7 Combustion Air Requirements	71
	3.8 Boiler Heating Capacity	74
	3.9 Efficiency of Gas- and Oil-Fired Boilers	75
	3.10 Multiple Boiler Systems	80
	3.11 Electric Boilers	89
	3.12 Electric Thermal Storage Systems	91
	3.13 Hydronic Heat Pump Fundamentals	93
	3.14 Water-to-Water Heat Pumps	98
	3.15 Air-to-Water Heat Pumps	106
	3.16 System Design Considerations for Hydronic Heat Pumps	114

	3.17 Solar Thermal Systems	116
	3.18 Biomass Boiler Fundamentals	124
	3.19 Cordwood Gasification Boilers	125
	3.20 Pellet Boilers	131
	Summary	136
	Key Terms	136
	Questions and Exercises	137
	References	138

Chapter 4 — Properties of Water — 139

	Objectives	139
4.1	Introduction	139
4.2	Sensible Heat Versus Latent Heat	140
4.3	Specific Heat and Heat Capacity	140
4.4	Density	141
4.5	Sensible Heat Quantity Equation	142
4.6	Sensible Heat Rate Equation	143
4.7	Vapor Pressure and Boiling Point	145
4.8	Viscosity	146
4.9	Dissolved Air in Water	147
4.10	Incompressibility	148
4.11	Water Quality	149
	Summary	154
	Key Terms	154
	Questions and Exercises	155

Chapter 5 — Piping, Fittings, and Valves — 157

	Objectives	157
5.1	Introduction	157
5.2	Piping Materials	158
5.3	Common Pipe Fittings	176
5.4	Specialized Fittings for Hydronic Systems	179
5.5	Thermal Expansion of Piping	182
5.6	Common Valves	185
5.7	Specialty Valves for Hydronic Applications	189
5.8	Schematic Symbols for Piping Components	213
5.9	Tips on Piping Installation	214
	Summary	219
	Key Terms	219
	Questions and Exercises	220

Chapter 6 — Fluid Flow in Piping — 221

	Objectives	221
6.1	Introduction	221
6.2	Basic Concepts of Fluid Mechanics. What is a Fluid?	222
6.3	Analyzing Fluid Flow in Smooth Pipes	231
6.4	Hydraulic Resistance of Fittings, Valves, and Other Devices	236
6.5	The System Head Loss Curve	239
6.6	Piping Components Represented as Series Resistors	240
6.7	Parallel Hydraulic Resistances	243
6.8	Reducing Complex Piping Systems	247
6.9	Software-Based Circuit Analysis	249

	6.10 Pipe Sizing Considerations	250
	Summary	254
	Key Terms	254
	Questions and exercises	255

Chapter 7 Hydronic Circulators — 257

	Objectives	257
	7.1 Introduction	257
	7.2 Circulators for Hydronic Systems	258
	7.3 Placement of the Circulator Within the System	263
	7.4 Circulator Performance	266
	7.5 Variable-Speed/ECM-Powered (VS/ECM) Circulators	275
	7.6 Analytical Methods for Circulator Performance	283
	7.7 Circulator Efficiency	285
	7.8 Operating Cost of a Circulator	288
	7.9 Cavitation	292
	7.10 Special Purpose Circulators	297
	7.11 Selecting a Circulator	302
	Summary	304
	Key Terms	304
	Questions and Exercises	304

Chapter 8 Heat Emitters — 307

	Objectives	307
	8.1 Introduction	307
	8.2 Classification of Heat Emitters	308
	8.3 Finned-Tube Baseboard Convectors	308
	8.4 Thermal Ratings and Performance of Finned-Tube Baseboard	312
	8.5 Sizing Finned-Tube Baseboard in Series Circuits	320
	8.6 Hydronic Fan-Coils and Air Handlers	327
	8.7 Thermal Performance of Fan-Coils	334
	8.8 Panel Radiators	340
	8.9 Other Hydronic Heat Emitters	354
	8.10 Head Loss of Heat Emitters	357
	8.11 Heat Loss from Copper Tubing	359
	8.12 Thermal Equilibrium	361
	Summary	366
	Key Terms	366
	Questions and Exercises	367

Chapter 9 Control Strategies, Components, and Systems — 368

	Objectives	368
	9.1 Introduction	368
	9.2 Closed-Loop Control System Fundamentals	369
	9.3 Controlling the Output of Heat Sources	377
	9.4 Controlling Heat Output from Heat Emitters	382
	9.5 Outdoor Reset Control	384
	9.6 Switches, Relays, and Ladder Diagrams	398
	9.7 Basic Hydronic System Control Hardware	406
	9.8 Basic Boiler Control Hardware	426
	9.9 Mixing Strategies and Hardware	436
	9.10 Control System Design Principles	457
	9.11 Example of a Modern Control System	459

vi   Contents

	9.12 Communicating Control Systems	462
	Summary	465
	Key Terms	466
	Questions and Exercises	467

Chapter 10   Hydronic Radiant Panel Heating   469

Objectives   469
10.1   Introduction   469
10.2   What Is Radiant Heating?   470
10.3   What Is a Hydronic Radiant Panel?   471
10.4   A Brief History of Radiant Panel Heating   471
10.5   Benefits of Radiant Panel Heating   473
10.6   Physiology of Radiant Panel Heating   474
10.7   Methods of Hydronic Radiant Panel Heating   476
10.8   Slab-on-Grade Radiant Floors   477
10.9   Concrete Thin-Slab Radiant Floors   487
10.10 Poured Gypsum Thin-Slab Radiant Floors   493
10.11 Above-Floor Tube and Plate Systems   497
10.12 Below-Floor Tube and Plate Systems   504
10.13 Extended Surface Suspended Tube Systems   508
10.14 Plateless Staple-Up Systems   510
10.15 Prefab Subfloor/Underlayment Panels   511
10.16 Radiant Wall Panels   513
10.17 Radiant Ceiling Panels   518
10.18 Tube Placement Considerations (Floor Panels)   521
10.19 Radiant Floor Panel Circuit Sizing Procedure   530
10.20 System Piping and Temperature Control Options   548
Summary   561
Key Terms   561
Questions and exercises   562

Chapter 11   Distribution Piping Systems   563

Objectives   563
11.1   Introduction   563
11.2   Zoning Considerations   564
11.3   System Equilibrium   567
11.4   The Concept of Iterative Design   568
11.5   Single Series Circuits   569
11.6   Single Circuit/Multizone (One-Pipe) Systems   572
11.7   Multicirculator Systems and Hydraulic Separation   575
11.8   Multizone Systems—Using Circulators   585
11.9   Multizone Systems Using Zone Valves   593
11.10 Parallel Direct-Return Systems   599
11.11 Parallel Reverse-Return Systems   603
11.12 Home-Run Distribution Systems   606
11.13 Primary/Secondary Systems   610
11.14 Distribution Efficiency   622
11.15 Hybrid Distribution Systems   625
Summary   632
Key Terms   633
Questions and Exercises   633

Chapter 12 — Expansion Tanks — 635

Objectives — 635
12.1 Introduction — 635
12.2 Standard Expansion Tanks — 636
12.3 Diaphragm-Type Expansion Tanks — 639
12.4 Estimating System Volume — 645
12.5 The Expansion Tank Sizer Software Module — 646
12.6 Point of No Pressure Change — 647
Summary — 651
Key Terms — 651
Questions and Exercises — 652

Chapter 13 — Air & Dirt Removal & Water Quality Adjustment — 654

Objectives — 654
13.1 Introduction — 654
13.2 Problems Created by Entrapped Air — 655
13.3 Types of Entrapped Air — 657
13.4 Air Removal Devices — 659
13.5 Correcting Chronic Air Problems — 664
13.6 Filling and Purging a System — 668
13.7 Make-Up Water Systems — 672
13.8 Dirt & Magnetic Particle Separation — 673
13.9 Internal Cleaning of Hydronic Systems — 677
13.10 Water Quality in Hydronic Systems — 678
Summary — 686
Key Terms — 686
Questions and Exercises — 687

Chapter 14 — Auxiliary Loads and Specialized Topics — 688

Objectives — 688
14.1 Introduction — 688
14.2 Liquid-to-Liquid Heat Exchangers — 689
14.3 Domestic Water Heating — 709
14.4 Intermittent Garage Heating — 724
14.5 Pool Heating — 726
14.6 Hydronic Snow Melting — 730
14.7 Buffer Tanks — 743
14.8 Minitube Distribution Systems — 749
14.9 Heat Metering — 753
14.10 Introduction to Balancing — 763
Summary — 776
Key Terms — 776
Questions and Exercises — 777

Chapter 15 — Fundamentals of Hydronic Cooling — 779

Objectives — 779
15.1 Introduction — 779
15.2 Benefits of Chilled Water Cooling — 780
15.3 Thermodynamic Concepts Associated with Cooling Systems — 784
15.4 Chilled-Water Sources — 791
15.5 Chilled-Water Terminal Units — 800

15.6	Chilled-Water Piping Insulation	810
15.7	Example Systems	813
	Summary	821
	Key Terms	821
	Exercises	822

Appendix A Schematic Symbols — 823
Appendix B R-Values of Common Building Materials — 827
Appendix C Useful Conversion Factors and Data — 829
Glossary — 831
Index — 857

Preface

This book, like its preceding editions, focusses on the design and installation of state-of-the-art hydronic heating systems for residential and light commercial buildings. It was written to provide comprehensive, contemporary, and unbiased information for heating technology students, as well as heating professionals.

I believe that residential and light commercial buildings deserve better comfort systems than they often get. After all, even smaller heating systems affect the comfort and well being of countless people over many years. The information in this book is for those wanting to design and install superior heating systems for such buildings. Systems that are efficient, reliable, and deliver unsurpassed comfort.

For decades, much of the available information on designing hydronic systems has been aimed at engineers, and intended for use in larger buildings. It is often impractical and imprudent to scale such systems down for use in smaller buildings.

Manufacturer's information, while accurately representing products, seldom leads a heating professional through the entire design process.

Most heating systems for smaller buildings are designed by the same firm that performs the installation. Over the last four decades, I've had opportunity to see both ends of the design and installation spectrum. From systems that are "textbook examples" of proper design and consummate craftsmanship, to multithousand dollar assemblies of hardware that will never deliver what is expected of them.

One of the consistent differences between these successes and failures has been a commitment by the designer/installer to go beyond joining pipes and turning wrenches, to learn the fundamental principles at work in every hydronic system. Without this commitment installers can quickly slide into design stagnation, where they approach every project with the same method regardless of its appropriateness, or just rely on hardware suppliers to sketch out a system that they can assemble. They often fail to capitalize on the versatility of modern hydronics technology and the many profitable opportunities it offers.

Since the third edition was released in 2011, there have been many new developments in the hardware and design concepts that now represent cutting-edge hydronics technology. This edition reflects those changes. It emphasizes simplicity wherever possible, and conservative use of distribution energy to move heat through hydronic distribution systems.

Interest in renewable energy, and decarbonization has also grown exponentially since the release of the previous edition. A thorough examination of renewable energy heating systems reveals that *hydronics technology is the "glue" that holds nearly all these systems together.* Although heat sources such as solar thermal collectors, geothermal heat pumps, air-to-water heat pumps, and solid fuel boilers are the "engines" of such systems, hydronics technology is the "drivetrain" that delivers the heat when and where it's needed. Without proper hydronic detailing none of these renewable energy heat sources will perform up to their full potential.

Current global trends aimed at phasing out fossil fuels, and replacing them with renewably-sourced electricity, will profoundly impact the future of hydronics technology. Heat pumps, in particular, will continue to gain market share against fossil fuel boilers as heat sources for residential and light commercial systems.

The vast majority of heat pump capable of supplying warm water for hydronic heating can also supply chilled water for cooling. This is a "game changer" when it comes to the potential for hydronic-based systems to supply the *year-round* comfort requirements of smaller buildings. It eliminate a shortfall that has constrained the potential growth of hydronics technology for decades - the complexity and cost of installing a completely separate cooling system in a building that uses hydronic heating.

This edition recognizes the emerging market for hydronic cooling in smaller buildings. To that end it includes a new chapter that covers the fundamentals of small scale chilled water cooling.

Other additions to this fourth edition include:

- Broader applications for variable speed circulators
- Discussion of new piping materials and joining methods
- Application of air-to-water heat pumps
- Improving water quality in hydronic systems
- Increased use of modern air, dirt, and magnetic particle separator technology
- Additional buffer tank piping options
- Use of panel radiators at lower water temperatures
- Extended analysis and applications for plate heat exchangers

Organization

The initial chapters of this book acquaint readers with the fundamental physical processes involved in hydronic heating. Topics such as basic heat transfer, heating load calculations, and properties of fluids are discussed. A solid understanding of these basics is essential for both design and troubleshooting.

Later chapters use the fundamental principles for overall system design. Readers are referred to relevant sections of earlier chapters during the design process to reinforce the importance of these principles. Example systems at the end of some of the later chapters show complete system piping and control wiring diagrams.

Chapter 1 provides a summary of hydronic heating, and the basic concepts involved. It emphasizes comfort as the ultimate goal of the heating professional. It encourages an attitude of craftsmanship and professionalism as the reader continues through the text.

Chapter 2 covers heating load estimating calculations. Experience indicates that such calculations can be a stumbling block for students, as well as for those in the trade, who want to jump right into design and layout without first determining what the system needs to provide. Without proper load information, any type of heating system can fail to deliver the desired comfort. A complete method for determining design heating loads is presented.

Chapter 3 surveys a wide spectrum of hydronic heat sources. These include conventional gas and oil-fired boilers, as well as contemporary devices such as geothermal heat pumps, air-to-water heat pumps, modulating/condensing boilers, pellet-fueled boilers, cordwood gasification boilers, and solar thermal collectors. Emphasis is placed on the need to match the balance of the system to the operating requirements of these heat sources.

Chapter 4 describes, in simple terms, the physical properties of water including specific heat, density, viscosity, incompressibility and solubility of air. Two very important equations that relate heat and heat flow to water temperature and flow rate are introduced. It also covers water quality and the use of demineralized water.

Chapter 5 is a "show and tell" chapter covering the proper use of tubing, fittings, and valves in hydronic systems. The use of copper tubing as well as materials such as PEX and PEX-AL-PEX is discussed. Common valves are illustrated and their proper use is emphasized. Several specialty fittings and valves are also discussed.

Chapter 6 remains as the key analytical chapter of the text. It defines vital parameters associated with the hydraulic performance of any hydronic system. It also introduces a method for calculating the head loss of a fluid as it flows through a piping system. This method uses the concept of *hydraulic resistance* to build an analogy between fluid flow in piping circuits and the principles of current, voltage, and resistance in electrical circuits. This method can be applied to both simple and complex piping circuits.

Chapter 7 presents qualitative and quantitative information on circulators for hydronic systems. The pump curve is introduced, and uses to properly match a circulator to a piping system. The intersection of the pump curve of a candidate circulator, and system head loss curve of the piping system, reveals the operating flow rate of the circulator/piping system combination. Quantitative and graphical methods are shown for finding this point. The chapter also places strong emphasis on what circulator cavitation is, and how to avoid it. The fourth edition presents the latest information on variable speed pressure-regulated circulators with electronically commutated motors. These circulators represent a major shift in capability and energy efficiency, and are quickly becoming the "new normal" in residential and light commercial hydronic systems, It is essential that heating professional understand how to apply them.

Chapter 8 surveys several types of hydronic heat emitters including finned-tube baseboard convectors, fan-coils, panel radiators, and radiant baseboards. The advantages and disadvantages of each type are discussed. Performance and sizing information is also given. The fourth edition presents new methods for analyzing the performance of panel radiators at lower water temperatures, allowing them to be successfully used with contemporary hydronic heat sources.

Chapter 9 is a major chapter dealing with control components and systems for hydronic heating. Control systems for multiload/multitemperature systems are discussed. Ladder diagrams are used as a framework on which to design such control systems. The continued growth of wireless and Internet-based control technology is reflected in this fourth edition.

Chapter 10 is devoted entirely to hydronic radiant panel heating. This edition includes coverage of radiant floors, walls and ceilings including new materials and thermographic images of operating panels.

Chapter 11 covers several types of distribution piping configurations including series loop, one-pipe diverter tee, two-pipe reverse return, homerun, and primary/secondary systems. The analytical methods first introduced in Chapters 6 and 7 are reference through step by step design procedures for a wide range of distribution systems. The fundamental concepts of hydraulic separation and distribution efficiency are emphasized.

Chapters 12 and **13** deal with the specialized topics of expansion tanks and air removal. Both survey the latest types of hardware, and show how to properly select it. The fourth edition includes new material on dirt and magnetic particle separation technology.

Chapter 14 covers several miscellaneous topics. Auxiliary loads such as indirect water heating, pool/spa heating, and intermittent garage heating are discussed. Extensive material on the design and application of hydronic snow and ice melting (SIM) systems is also presented. Unique concepts for incorporating buffer tanks are presented. The fourth edition provides expanded coverage of flat plate heat exchangers, and heat metering technology.

Chapter 15 is entirely new to the fourth edition. It covers the fundamentals of chilled water cooling and how it can be applied in residential and light commercial buildings. Topics include the thermodynamic properties of air, basic psychometrics, performance of chillers, importance of pipe insulation, discussion of hardware suitable for chilled water cooling, and a range of example systems.

Design Assistance Software

This edition references several software products for expediting calculations and documenting hydronic system designs. These softwares include:

- Hydronics Design Studio
- Heat Load Pro
- HydroSketch

The **Hydronics Design Studio** is built around the analytical methods presented in this book. It allows rapid simulation of flow in user-defined piping systems including those with multiple parallel branch circuits. It also includes modules for room heat loss estimating, expansion tank sizing, and detailed simulation and sizing of series-loop baseboard systems. Several examples of how the software can be used to evaluate long numerical calculations are given.

Heat load Pro is a subset of the Hydronics Design Studio. It was created as a simple, low-cost tool for calculating building heating loads, and estimating the operating costs associated with different fuel types and heat sources.

HydroSketch is a cloud-based tool for quickly constructing piping and electrical schematic diagrams. It includes many of the component symbols given in Appendix A and used throughout this book. HydroSketch uses a simple "drag and drop" interface to quickly place component symbols on a drawing canvas and connect them into circuits.

More information on these software products is available at www.hydronicpros.com

Other Features of the Fourth Edition

- Updated photographs and illustrations show more installation details and proper use of the latest hydronic hardware.
- A full description of all variables, along with the required units, is given whenever an equation is introduced. All quantities are expressed in the (IP) units commonly used in much of the North American HVAC trade.
- Many example calculations are organized as a situation statement, solution procedure, and discussion of the results.
- A summary of key terms is given at the end of each chapter. An extended glossary is provided for quick reference to concise descriptions of key terms. Readers are encouraged to test their understanding of each chapter by briefly describing each of these key terms.
- Additional questions and exercises have been developed for several chapters.

Previous editions of this text have been used by the author as the basis of a one semester course on hydronic heating. This course is offered in the second year of a two-year Associate Degree program in Air Conditioning Technology at Mohawk Valley Community College in Utica, NY.

Acknowledgments

My appreciation is extended to the many manufacturers who have supplied images and information for this edition. Their cooperation, and willingness to accommodate what have often been tight schedules has been truly outstanding.

Finally, I wish to thank the Cengage Learning production team, and especially Vanessa Myers, Jennifer Alverson, Jennifer Ziegler, and Kimberly Klotz for their diligence and patience as the fourth edition has taken form.

Dedication

This edition is dedicated to Mario Restive, an accomplished engineer, professional associate at Mohawk Valley Community College, and long time friend. Mario worked diligently to transform ideas and algorithms for the previously mentioned software tools into thousand of lines finished code. Without his skill this key companion to the text would not exist. His loyal friendship over the last 40 years is deeply appreciated.

John Siegenthaler, P.E.

Chapter 1
Fundamental Concepts

Objectives

After studying this chapter, you should be able to:
- Describe the advantages of hydronic heating.
- Define heat and describe how it is measured.
- Describe three methods by which heat travels.
- Explain thermal equilibrium within a hydronic heating system.
- Define four basic hydronic subsystems.
- Explain the differences between a radiator and a convector.
- Explain the differences between an open-loop and a closed-loop hydronic system.
- Summarize the basic components of a hydronic heating system and explain how they operate.
- Explain the concept of distribution efficiency and use it to compare different distribution systems.

1.1 What Is a Hydronic Heating System?

Hydronic heating systems use water (or water-based solutions) to move thermal energy from where it is produced to where it is needed. The water within the system is neither the source of the heat nor its destination, only its "conveyor belt." **Thermal energy** is absorbed by the water at a **heat source**, conveyed by the water through the **distribution system**, and finally released into a heated space by a **heat emitter**. Ideally, the same water remains in the system year after year.

Hydronic systems are not limited to heating. They can also be used to convey cooling effect (e.g., the lack of heat) from a source of chilled water to cooling emitters located in one or more locations within a building.

Water has many characteristics that make it ideal for heating and cooling applications. It is readily available, nontoxic, nonflammable, and has one of the highest heat storage abilities of any material known to man. All three states of water (solid, liquid, and vapor) are used for various building heating and cooling applications. The modern hydronic systems discussed in this book make use of the liquid state only.

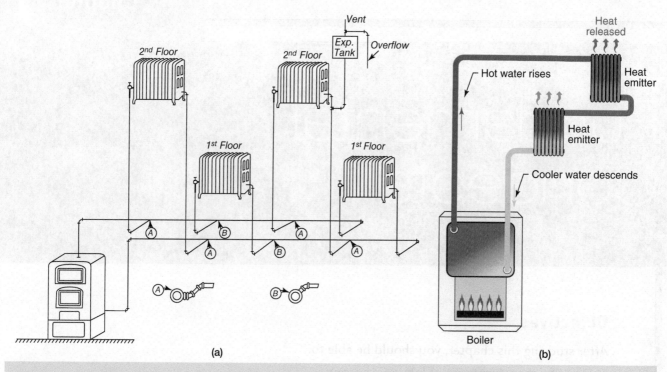

Figure 1-1 (a) An early (1930s vintage) hydronic heating system without a circulator. (b) Flow occurs due to differences in density between hot water and cooler water.

The practical temperature range for water in residential and light commercial buildings is from about 32 to 250 °F.

At the upper end of this range, the water is maintained in a liquid state by system pressurization. The lower end of the range can be extended well below 32 °F by the addition of antifreeze. Such a solution is called **brine**. It would be used in special applications such as hydronic snowmelting or the earth loop of a geothermal heat pump system.

Early hydronic systems, as depicted in Figure 1-1a and 1-1b, relied on the **buoyancy** of hot water to move water between the boiler and the heat emitters. Because of its lower density, hot water would rise upward from a boiler through supply pipes into heat emitters. After releasing heat, the now slightly denser water flows downward back to the boiler, as shown in Figure 1-1b.

These early hydronic heating systems required careful pipe sizing and installation since buoyancy-driven flows are weak, and their designs were significantly limited in comparison to what can be done with modern methods. The emergence of electrically powered **circulators** made it possible to move water at higher flow rates through much more elaborate piping systems.

Modern hydronics technology enables heat to be delivered precisely when and where it is needed. Hundreds of system configurations are possible, each capable of meeting the exact comfort requirements of its owner. Some may be as simple as a boiler serving a single piping circuit through several series-connected heat emitters. Others may use two or more hydronic heat sources operated in stages, releasing their heat through a wide assortment of heat emitters. Those same heat sources could also provide the building's **domestic hot water**. They might even heat the swimming pool or melt snow as it falls on the driveway. If the heat source is a reversible heat pump, it could also supply chilled water for cooling. Well-designed and properly installed hydronic systems provide unparalleled versatility, unsurpassed comfort, and fuel efficiency for the life of the building.

1.2 Benefits of Hydronic Heating

This section discusses several benefits of hydronic heating. Among these are

- Comfort
- Energy savings
- Design flexibility
- Clean operation
- Quiet operation
- Noninvasive installation
- High distribution efficiency

Comfort

Contrary to common belief, heating systems are not created to heat buildings. Instead, they are created for sustaining human thermal comfort within buildings. Think about it: Does a window, concrete block, or insulation batt really "care" whether its temperature is 40 or 70 °F? Of course not. However, the selection and placement of these materials can have a profound impact on human thermal comfort, or lack thereof.

Providing comfort should be the primary objective of any heating system designer or installer. Unfortunately, this objective is too often compromised by other factors, the most common of which is cost. Even small residential heating systems affect the health, productivity, and general well-being of several people for many years. It makes sense to plan and install them accordingly.

The average North American building owner spends little time thinking about the consequences of the heating system they select. Many view such systems as a necessary but uninteresting part of a building. When construction budgets are tightened, it is often the heating system that is compromised to save money for other, more impressive amenities.

Heating professionals should take the time to discuss the full range of benefits of hydronic systems, including superior comfort, as well as price with their clients. Often people who have lived with uncomfortable heating systems do not realize what they have been missing. In retrospect, many would welcome the opportunity to live or work in truly comfortable buildings and would willingly spend more money, if necessary, to do so.

Maintaining comfort is not a matter of supplying heat to the body. Instead, it is a matter of controlling how the body loses heat. *When interior conditions allow heat to leave a person's body at the same rate at which it is generated, that person feels comfortable.* If heat is released faster or slower than the rate at which it is produced, some degree of discomfort is experienced.

A normal adult engaged in light activity generates heat at a rate of approximately 400 **British thermal units** per hour (Btu/h). Figure 1-2 shows the various processes by which the body of a person at rest releases heat to a typical indoor environment.

Note that a large percentage of the body's heat loss comes from **thermal radiation** to surrounding surfaces. Most people will not be comfortable in a room containing several cool surfaces such as large windows or cold floors, even if the room's air temperature is 70 °F. Remaining heat loss occurs through a combination of **convection**, **evaporation**, respiration, and a small amount of **conduction**. The latter occurs through surfaces in direct contact with the body.

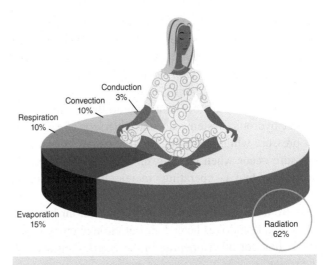

Figure 1-2 | Processes the body uses to release heat to a typical indoor environment. *Courtesy of Robert Bean.*

The body can adjust these heat loss processes, within certain ranges, to adapt to different interior environments. For example, if air temperature around the body increases, convection heat loss will be suppressed. The body responds by increasing evaporation heat loss (perspiration) and increasing skin temperature to increase radiative heat loss.

Properly designed hydronic systems influence both the air temperature and surface temperatures of rooms to maintain optimal comfort. Modern controls can maintain room air temperature to within ±1 °F of the desired **setpoint temperature**. Heat emitters such as radiant floors or radiant ceilings raise the average surface temperature of rooms. Since the human body is especially responsive to radiant heat loss, these heat emitters significantly enhance comfort. Comfortable humidity levels are also easier to maintain in hydronically heated buildings.

Several factors such as activity level, age, and general health determine the comfortable environment for a given individual. When several people are living or working in a common environment, any one of them might feel too hot, too cold, or just right. Heating systems that allow various "zones" of a building to be maintained at different temperatures can better adapt to the comfort needs of several individuals. This is called **zoning**. Although both forced-air and hydronic heating systems can be zoned, the latter is usually much simpler and easier to control.

Energy Savings

Ideally, a building's rate of heat loss would not be affected by how that heat is replaced. However, experience has shown that otherwise identical buildings can have significantly different rates of heat loss based on the types of heating systems installed. *Buildings with hydronic heating systems have consistently shown lower heating energy use than equivalent structures with forced-air heating systems.*

A number of factors contribute to this finding. One is that hydronic systems do not adversely affect room air pressure while operating. Small changes in room air pressure occur when the blower of a ducted forced-air heating system is operating. Increased room air pressure is often created by a lack of adequate return air flow from the rooms back to the furnace or air handler. This condition drives heated air out through every small crack, hole, or other opening in the exterior surfaces of the room.

One study that compared several hundred homes, some with ducted forced-air systems and others with hydronic baseboard convectors, found that air leakage rates averaged 26% higher and energy usage averaged 40% greater in homes with forced-air heating.

Another factor affecting building energy use is air temperature **stratification** (e.g., the tendency of warm air to rise toward the ceiling while cool air settles to the floor). In extreme situations, the difference in air temperature from floor to ceiling can exceed 20 °F. Stratification tends to be worsened by high ceilings, poor air distribution, and heating systems that supply air into rooms at high temperatures. Maintaining comfortable air temperatures in the occupied areas of rooms plagued with a high degree of temperature stratification leads to significantly higher air temperatures near the ceiling as shown in Figure 1-3. This, in turn, increases heat loss through the ceiling.

Hydronic systems that transfer the majority of their heat by thermal radiation reduce air temperature stratification and thus reduce heat loss through ceilings. Comfort can often be maintained at lower air temperatures when a space is radiantly heated. This leads to further energy savings. Zoned hydronic systems provide the potential for unoccupied rooms to be kept at lower

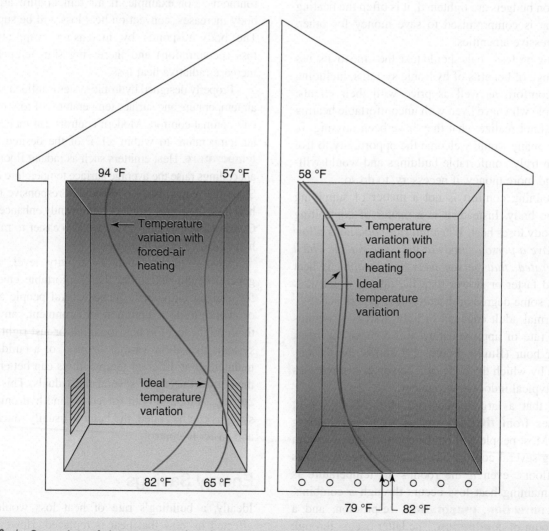

Figure 1-3 | Comparison of air temperatures from floor to ceiling for forced-air heating (left) and radiant floor heating (right).

temperatures, which lowers heat loss and reduces fuel consumption. Some hydronic systems also automatically reduce the water temperature in their distribution piping as the outdoor temperature increases. This reduces heat loss from piping and increases heat source efficiency.

The electrical energy consumption of the circulator(s) used in a well-planned modern hydronic system can be a small fraction (often less than 10%) of the electrical energy required by a blower in a forced-air heating system of equal capacity. This saving is often overlooked by those who only consider the energy use associated with *producing* heat or chilled water for cooling. This difference in energy use by various heating or cooling distribution systems can be quantified using the concept of **distribution efficiency**, which is discussed later in this chapter.

Design Flexibility

Modern hydronics technology offers virtually unlimited potential to accommodate the comfort needs, usage, aesthetic tastes, and budget constraints of almost any building. A single system can be designed to supply **space heating**, domestic hot water, and specialty **loads** such as pool heating or **snowmelting** (see Figure 1-4). Such **"multiload" systems** reduce installation costs because redundant components such as multiple heat sources, exhaust systems, electrical hookups, safety devices, and fuel supply components are eliminated. They also tend to improve the heat source efficiency and thus reduce fuel usage relative to systems where each load is served by its own dedicated heat source.

The space heating needs of some buildings are best served through the use of different heat emitters. For example, it is possible for hydronic radiant floor heating to be used in the basement and first floor of a house while the second floor rooms are heated using panel radiators or fin-tube baseboard. Modern hydronics technology makes it easy to combine different heat emitters into the same system.

The wide variety of hydronic heat sources now available also allows systems to easily adapt to special circumstances or opportunities such as "time-of-use" electrical rates, on-site renewable energy availability, or waste heat recovery.

Clean Operation

A common complaint about forced-air heating is its propensity to move dust and other airborne particles such as pollen and smoke throughout a building. In buildings where air-filtering equipment is either of low quality or is poorly maintained, dust streaks around ceiling diffusers, as seen in Figure 1-4a, are often evident. Eventually duct systems, such as that shown in Figure 1-5, require internal cleaning to remove dust and dirt that have accumulated over several years of operation.

In contrast, few hydronic systems involve forced-air circulation. Those that do create room air circulation rather than whole building air circulation. This reduces the dispersal of airborne particles and microorganisms, which is a major benefit in situations where air cleanliness is imperative, such as for people with allergies or respiratory illness, or in health care facilities and laboratories. When floor heating is used in entry areas,

Figure 1-4 | The hydronic system that heats this house also melts snow on the walkway. *Courtesy of G. Todd.*

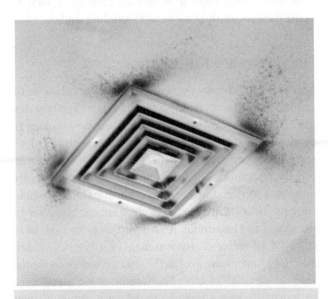

Figure 1-4a | Dust streaks surrounding a ceiling diffuser indicate poor air filtering in a ducted forced-air distribution system. *Courtesy of John Siegenthaler*

Figure 1-5 | Dirt and dust accumulation in ducting. *Courtesy of L. Drake.*

Figure 1-5a | Ducting enclosed in a framed soffit within living space. *Courtesy of Angela McCormick.*

the floor can dry rapidly to reduce tracking water and dirt further into the building.

Quiet Operation

Most people want their home to be a quiet refuge from the pace and *noise* of modern life. They don't want to hear sounds emanating from their heating and cooling systems. A properly designed and installed hydronic system can operate with virtually no detectable sound levels in the occupied areas of a home. Modern systems that use constant circulation and variable water temperature control minimize expansion noises that can occur when high-temperature water is injected directly into a room temperature heat emitter. These characteristics make hydronic heating ideal in sound-sensitive areas such as home theaters, reading rooms, or recording studios.

Noninvasive Installation

Consider the difficulty encountered when ducts have to be concealed from sight within a typical house. The best that can be done in many situations is to encase the ducting in exposed soffits, as shown in Figure 1-5a. Such situations often lead to compromises in duct sizing and/or placement.

By comparison, hydronic heating systems are easily integrated into the structure of most small buildings without compromising their structure or the aesthetic character of the space. The underlying reason for this is the high **heat capacity of water**. *A given volume of water can absorb over 3,400 times more heat as the same volume of air for the same temperature change.* The volume of water that must be moved through a building to deliver a certain amount of heat is only about 0.03% that of air, assuming that the air and water undergo the same temperature change. This greatly reduces the size of the distribution "conduit."

For example, a 3/4-inch-diameter tube carrying water at 6.0 gpm around a hydronic system operating with a 20°F temperature drop transports as much heat as a 14-inch by 8-inch duct carrying 130°F air at 1,000 feet/min. Figure 1-7 depicts these two options side by side.

Notching into floor joists to accommodate the 14-inch by 8-inch duct would destroy their structural integrity. By contrast, smaller tubing is easily routed through the framing, especially if it happens to be one of several flexible tube products now available.

If the distribution system is insulated, which is now a code requirement in many areas, considerably less material is required to insulate the tubing compared to the ducting. When insulated with the same material, the heat loss of the 14-inch by 8-inch duct is almost 10 times greater than that of the 3/4-inch tube.

Hydronic systems using small flexible tubing are much easier to retrofit into existing buildings in comparison to ducting. The tubing can be routed through open or closed framing cavities much like electrical cable, as seen in Figure 1-8.

Figure 1-6 | Hydronic systems operate with virtually no noise in occupied space. *Courtesy of John Siegenthaler.*

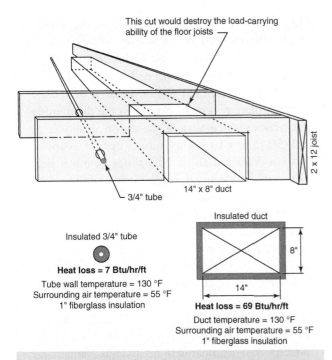

Figure 1-8 | Small-diameter flexible hydronic tubing routed through framing cavities. *Courtesy of John Siegenthaler*

Figure 1-7 | A 3/4-inch-diameter tube carrying water can transport heat at the same rate as a 14-inch by 8-inch duct in a forced-air system. If the water in the tube and air in the duct are of the same temperature, there is almost 10 times more heat loss from the duct due to its larger surface area.

One modern strategy is to route 3/8-inch or 1/2-inch flexible tubing from a central **manifold station** to a heat emitter in each room. This "homerun" approach allows the option of maintaining different temperatures in each room. The concept is shown in Figure 1-9.

For buildings where utility space is minimal, small wall hung boilers using sealed combustion systems can often be mounted in a closet. In many cases, these compact boilers supply the building's domestic hot water as well as its heat. The entire system might occupy less than 10 square feet of floor area.

Distribution Efficiency

The electrical power required to move heat or cooling effect through a building is an important consideration when trying to reduce a building's overall energy use. It's possible to compare the relative electrical energy use of various heating and cooling distribution systems using the concept of distribution efficiency, which is defined by Equation 1.1.

Equation 1.1:

$$n_d = \frac{q}{p_e},$$

where,
- n_d = distribution efficiency (Btu/h/watt)
- q = rate of thermal energy delivery by distribution system (Btu/h)
- p_e = electrical power required to operate distribution system (watt)

Figure 1-9 | A homerun distribution system uses supply and return tubing from each heat emitter to a central manifold station. *Courtesy of Caleffi North America.*

Example 1.1

A hydronic heating distribution system has three circulators, each requiring 45 watts of electrical power input when operating at peak capacity. Under this condition the distribution system delivers 160,000 Btu/h of thermal energy to the building. Determine the distribution efficiency of this system.

Solution:

Since both the rate of thermal energy delivery and the associated electrical power input to the distribution system are known, the distribution efficiency is easily calculated:

$$n_d = \frac{q}{p_e} = \frac{160,000 \text{ Btu/h}}{3(45 \text{ watt})} = 1,185 \frac{\text{Btu/h}}{\text{watt}}.$$

Discussion:

Neither the type of heat source that produced the 160,000 Btu/h of heat output, nor the efficiency of that heat source, is relevant when determining distribution efficiency.

The calculated value of 1,185 Btu/h/watt can be interpreted as follows: This hydronic distribution system can transfer 1,185 Btu/h of heat from where it is produced to where it's needed in the building for each watt of electrical power required to operate the distribution system. Still, the value of 1,185 Btu/h/watt is somewhat meaningless without something to compare it to. This will be done in Example 1.2.

Example 1.2

A ducted forced-air heating system uses a blower to deliver 84,000 Btu/h to a building. The blower requires 740 watts of electrical power input when operating. What is the distribution efficiency of this system?

Solution:

Again the calculation is straightforward:

$$n_d = \frac{q}{p_e} = \frac{84,000 \text{ Btu/h}}{(740 \text{ watt})} = 113.5 \frac{\text{Btu/h}}{\text{watt}}.$$

Discussion:

In this case the distribution system only delivers 113.5 Btu/h to where it's needed in the building for each watt of electrical power required to operate the distribution system. A ratio of the distribution efficiencies calculated in these two examples is 1,185/113.5 = 10.4. This implies that the hydronic system delivered about 10.4 times as much heat per watt of electrical power input compared to the forced-air system. It could also be interpreted as the hydronic system delivering a given amount of heat using about 9.6%, of the electrical power required by the forced-air system. In either case the hydronic distribution system described in Example 1.1 holds a clear advantage over the ducted forced-air distribution system described in Example 1.2. That advantage could result in thousands of dollars of electrical energy savings over the life of these heating systems. Bear in mind that this comparison was for two specific systems, and that the results should not be generalized to all hydronic versus ducted forced-air system comparisons.

Using state-of-the-art design techniques and modern hardware, it is possible to construct hydronic systems having distribution efficiencies over 3,000 Btu/h/watt. This distinct benefit has the potential to significantly reduce electrical energy use in a wide range of buildings using all types of hydronic heating and cooling sources.

1.3 Heat and Heat Transfer

Before attempting to design any type of heating system, it is crucial to understand the entity being manipulated: heat.

What we commonly call heat can also be described as energy in thermal form. Other forms of energy such as electrical, chemical, mechanical, and nuclear energy can be converted into heat through various processes and devices. All heating systems consist of devices that convert one form of energy into another.

Heat is our perception of atomic vibrations within a material. Our means of expressing the intensity of these vibrations is called temperature. The more intense the vibrations, the greater the temperature of the material and the greater its heat content. Any material above absolute zero temperature (−460 °F) contains some amount of heat.

There are several units for expressing a quantity of heat. In North America, the most commonly used unit of heat is the British thermal unit (Btu). A Btu is defined as the amount of heat required to raise 1 lb. of water by 1 °F.

Heat always moves from an area of higher temperature to an area of lower temperature. In hydronic systems, this occurs at several locations. For systems using a boiler as their heat source, the heat moves from the hot gases in the combustion chamber to the cooler metal walls of the boiler's heat exchanger. The heat continues to move through the metal walls of the heat exchanger into the cooler water within. After being transported to a heat emitter on a flowing "conveyor belt" of water, the heat passes through the metal walls of the heat emitter into the still cooler air and objects of a room. Finally, heat moves through the exposed surfaces of a room into the outside air. Figure 1-10 depicts this heat transfer.

In every instance, heat moved from an area of higher temperature to an area of lower temperature. *Without this temperature difference, there would be no heat transfer.*

In this book, the *rate* of heat transfer is expressed in British thermal units per hour, abbreviated as Btu/h, or Btuh. It is very important to distinguish between the *quantity* of heat present in an object (measured in Btu) and the *rate* at which heat moves in or out of the object (measured in Btu/h). These terms are often misquoted by people, including those in the heating and cooling profession.

The *rate* of heat transfer from one location to another is governed by several factors. One is the temperature difference between where the heat is and where it is going. *Temperature difference is the driving force that causes heat to move.* Without a temperature difference between two locations, there can be no heat transfer. The greater the temperature difference, the faster heat flows. In most instances, the rate of heat transfer through a material is directly proportional to the temperature difference across the material. Thus, if the temperature difference across a material were doubled, the rate of heat transfer through the material would also double.

Another factor affecting the rate of heat transfer is the type of material through which the heat moves. Some materials, such as copper and aluminum, allow

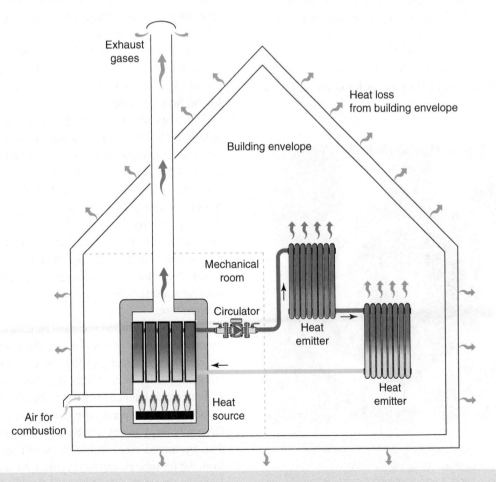

Figure 1-10 | Heat movement within a hydronic heating system and the building it serves.

heat to move through very quickly. Other materials, such as polyurethane foam, greatly inhibit the rate of heat transfer.

Three Modes of Heat Transfer

Thus far, two important principles of heat transfer have been discussed. First, heat moves from an area of higher temperature to an area of lower temperature. Second, the rate of heat transfer depends on temperature difference and the type of material. To gain a more detailed understanding of how the thermal components of a hydronic system work, we need to classify heat transfer into three modes: conduction, convection, and thermal radiation.

Conduction is the type of heat transfer that occurs through solid materials. Recall that heat has already been described in terms of atomic vibrations. Heat transfer by conduction is a dispersal of these vibrations from a source of heat, out across trillions of atoms that are bonded together to form a solid material. The index that denotes how well a material transfers heat is called its **thermal conductivity**. The higher a material's thermal conductivity, the faster heat can pass through it, all other conditions being equal. Heat passing though the walls of a coffee cup, and then to hands wrapped around that cup, as shown in Figure 1-10a is an example of conduction heat transfer. Heat moving from the hot inner surface of a pipe to its cooler outer surface is another example.

Figure 1-10a — Heat passes through the wall of the coffee cup by conduction.
Africa Studio/Shutterstock.com

The rate of heat transfer by conduction is directly proportional to both the temperature difference across the material and its thermal conductivity. It is inversely proportional to the thickness of a material. Thus, if one were to double the thickness of a material while maintaining the same temperature difference between its sides, the rate of heat transfer through the material would be cut in half.

In some locations within hydronic systems, designers try to enhance conduction. For example, the higher the thermal conductivity of a flooring material installed over a heated floor slab, the faster heat can pass upward through it and into the room. In contrast, the slower the rate of conduction through the insulation of a hot water storage tank, the better it retains heat. Equations for calculating heat flow by conduction are presented in Chapter 2.

Convection heat transfer occurs when a fluid at some temperature moves along a surface at a different temperature. The term "fluid" can refer to a gas (such as air), or a liquid (such as water).

Consider the example of water at 100°F flowing along a surface maintained at 150°F. The cooler water molecules contacting the warmer surface absorb heat from that surface. These molecules are churned about as the water moves along. The heated molecules are constantly swept away from the surface into the bulk of the water stream and replaced by cooler molecules. One can think of the heat as being "scrubbed" off the surface by the flowing water.

The speed of the fluid moving over the surface greatly affects the rate of convective heat transfer. We've all experienced the increased "wind chill" effect as cool air is blown across our skin rather than lying relatively stagnant against it. Although the air temperature may not be extremely cold, the speed at which it moves over our skin greatly increases the rate of convective heat loss. Although it may feel like the air is very cold, we are actually sensing the *rate* of heat loss from our skin rather than the air temperature. To achieve the same cooling sensation in calm air requires a much lower air temperature.

When fluid motion is caused by either a circulator (for water) or a blower or fan (for air), the resulting heat transfer is more specifically called **forced convection**. When buoyancy differences within the fluid cause it to move without any influence from a circulator or blower, the heat transfer is more specifically called **natural convection**. Generally, heat moves much slower by natural convection than by forced convection. The warm air currents rising from the fin-tube element in Figure 1-11 are an example of natural convection. The heat transferred to the air pushed along by the blower is an example of forced convection.

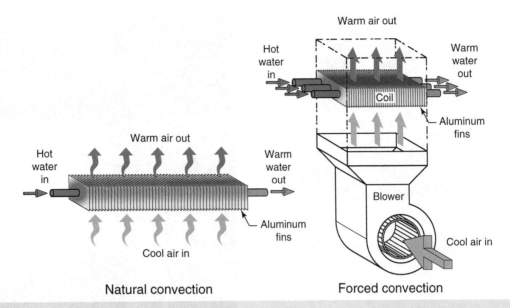

Figure 1-11 | Two types of convective heat transfer.

Some hydronic heat emitters, such as a finned-tube baseboard, are designed to release the majority of their heat output to the surrounding air by natural convection. Such devices are appropriately called **convectors**. Heat emitters that use fans or blowers to force air through a heat exchanger are usually called **fan coils** or **air handlers**.

Thermal radiation is probably the least understood mode of heat transfer. Just like visible light, thermal radiation is **electromagnetic energy**. It travels outward from its source in straight lines, at the speed of light (186,000 miles/s), and cannot bend around corners, although it can be reflected by some surfaces. Unlike conduction or convection, thermal radiation needs no material to transfer heat from one location to another.

Consider a person sitting a few feet away from a campfire on a cold winter day, as shown in Figure 1-11a. If pointed toward the fire, the person's face probably feels warm, even though the surrounding air is cold. This sensation is the result of thermal radiation emitted by the fire traveling through the air and being absorbed by their exposed skin. The air between the fire and the person's face is not heated as the thermal radiation passes through it. Likewise, thermal radiation emitted from the warm surface of a heat emitter passes through the air in a room without directly heating that air. When the thermal radiation strikes another surface in the room, most of it is absorbed. At *that instant*, the energy carried by the thermal radiation becomes heat.

The main difference between thermal radiation and visible light is the wavelength of the radiation. Anyone who has watched molten metal cool has noticed how the bright orange color eventually fades to duller shades of red, until finally the metal's surface no longer glows. As the surface of the metal cools below about 970 °F, our eyes can no longer detect *visible* light from the surface. Our skin, however, still senses that the surface is very hot. Though unseen, thermal radiation in the nonvisible infrared portion of the electromagnetic spectrum is still being strongly emitted by the metal's surface.

Any surface continually emits thermal radiation to any cooler surface within sight of it. The surface of a heat emitter that is warmer than our skin or clothing surfaces

Figure 1-11a | Thermal radiation passes from the fire to skin surfaces without warming the air between. *Pavel Vaschenkov/Shutterstock.com*

Figure 1-11b — Comfort is reduced due to radiant heat loss to cold glass surfaces even though interior air temperature is acceptable. *Courtesy of Caleffi North America.*

transfers heat to us by thermal radiation. Likewise, our skin and clothing give off thermal radiation to any surrounding surfaces at lower temperatures.

The term **mean radiant temperature** describes the area-weighted average temperature of all surfaces within a room. As the mean radiant temperature of a room increases, the air temperature required to maintain comfort decreases. As the temperature of the room's surfaces increases, the heat released from the body by thermal radiation decreases, so the amount released by convection must increase to keep the total rate of heat release constant. The converse is also true. This is why a person in a room with an air temperature of 70°F may still feel cool if surrounded by cold surfaces, such as large, undraped windows, on a cold winter day, as shown in Figure 1-11b.

As thermal radiation strikes an opaque surface, part of it is absorbed as heat and part is reflected away from the surface. The percentage of incoming radiation that is absorbed or reflected is determined by the optical characteristics of the surface and the wavelength of the radiation. Most interior building surfaces absorb the majority of thermal radiation that strikes them. The small percentage that is reflected typically strikes another surface where most of it will be absorbed, and so on. Very little, if any, thermal radiation emitted by warm surfaces in a room escapes from the room.

Although the human eye cannot see thermal radiation, there are devices that can detect it and display an image that uses colors to represent different surface temperatures. Such an image is called an **infrared thermograph**. Figure 1-12a shows an infrared-detecting camera pointed at a tiled floor. The visible floor surface gives no evidence that it is being heated. Figure 1-12b is an infrared thermograph of the same floor area produced by the infrared camera. The bright colors show areas of different surface temperatures. From the shape of these color patterns, it is evident that a

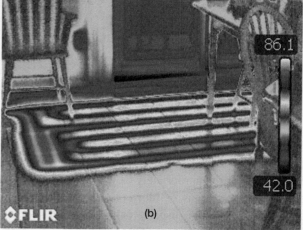

Figure 1-12 — (a) An infrared camera aimed at a floor. *Courtesy of John Siegenthaler* (b) Infrared thermograph of same floor reveals heat from tubing installed below floor. *Courtesy of John Siegenthaler*

source of heat is installed under the floor; in this case, it is tubing carrying heated water. The color spectrum on the far right of the image gives the range of surface temperatures in the image. In this case they range from the relatively cool sill of the patio door (about 42.0°F) to the warmest floor areas directly above the embedded tubing (about 86°F). Infrared thermography is a powerful tool in diagnosing the thermal characteristics of buildings as well as mechanical systems. It can also be used to locate heated objects embedded behind surfaces as Figure 1-12b demonstrates.

The rate at which thermal radiation transfers heat between two surfaces depends upon their temperatures, an optical property of each surface called emissivity, and the angle between the surfaces.

Hydronic heat emitters deliver a portion of their heat output to the room by convection and the rest by radiation. The percentages delivered by each mode depend on many factors such as surface orientation, shape, type of finish, air flow rate past the surface, and surface temperature.

When a heat emitter transfers over 50% of its heat output by radiation, it is called a **radiant panel**. Heated floors, walls, and ceilings are all examples of radiant panels. They are discussed in detail in Chapter 10.

Thermal Equilibrium

If a material is not gaining or losing heat and remains in a single physical state (solid, liquid, or gas), its temperature does not change. This is also true if the material happens to be gaining heat from one object while simultaneously releasing heat to another object at the same rate. These principles have many practical applications in hydronic heating.

For example, if you observed the operation of a hydronic heating system for an hour and found no change in the temperature of the water leaving the boiler, although it was firing continuously, what could you conclude? Answer: Since there is no change in the water's temperature, it did not undergo any net gain or loss of heat. The rate that the boiler injects heat into the water stream passing through it is the same as the rate the heat emitters remove heat from the water stream (see Figure 1-13).

Under these conditions, the system is in thermal equilibrium. It is important to understand that *all hydronic heating systems inherently attempt to find a condition of* **thermal equilibrium** *and remain in operation at that condition*. The goal of the system designer is to ensure that thermal equilibrium is established at a condition that maintains comfort in the building and does not adversely affect the operation, safety, or longevity of the system's components.

1.4 Four Basic Hydronic Subsystems

In a hydronic heating system, water is heated by a heat source and conveyed by means of a distribution system to heat emitters where it is released to the building. A control system regulates these elements in an attempt to keep the rate of heat delivery very close to the rate of building heat loss. The overall hydronic system thus consists of four interrelated subsystems:

1. Heat source
2. Distribution system
3. Heat emitters
4. Control system

This section shows how these subsystems are connected to form a simple hydronic heating system. The discussion begins with the concept of a simple piping loop. Several essential components are described and then situated in the loop. Finally, a composite schematic shows all components in their preferred locations relative to each other. After reading this section, you will have a basic understanding of what the essential components of a hydronic system are and what they do. The details of proper component selection and sizing are covered in later chapters.

Basic Hydronic Circuit

The simplest hydronic system can be described as a loop or piping circuit. If the circuit is sealed off from the atmosphere at all locations (as is true for most modern hydronic systems), it is called a **closed-loop system**. If the circuit is open to the atmosphere at any point, it is called an **open-loop system**. Figure 1-14 shows the

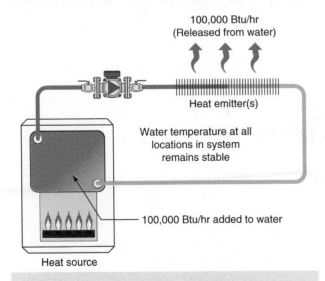

Figure 1-13 | When the rates of heat input and heat release are equal, the system is in thermal equilibrium and fluid temperatures at all locations in the system are stable.

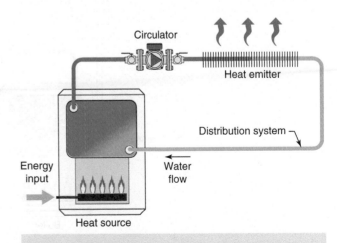

Figure 1-14 | The basic hydronic heating circuit.

simplest form of a piping circuit. The basic components are represented by schematic symbols.

When there is a demand for heat, water flow in the circuit is established by the circulator. The water carries heat from the heat source to the heat emitter where the heat is released into the space. Think of the flowing water as a "conveyor belt" for the heat.

In an ideal application, the rate of heat production by the heat source would exactly match the rate of heat dissipation by the heat emitters. This heat flow would also match the rate of heat loss from the building. Unfortunately, in real systems such ideal conditions seldom exist.

For example, it is fairly common to select a boiler with a heating capacity slightly greater than the building's rate of heat loss on the coldest day of the year. During a milder day, a boiler selected by this method could deliver heat to the building much faster than the building loses heat. Continuous operation of the boiler under such circumstances would quickly overheat the building. Obviously, some method of controlling the system's heat delivery is needed.

Temperature Controls

Figure 1-15 adds two simple control devices, the **room thermostat** and the **temperature-limiting controller**, to the basic system. The room thermostat determines when the building requires heat based on its **setpoint temperature** and the current indoor air temperature.

When the indoor air temperature drops slightly below the thermostat's setpoint, the switch contacts inside the thermostat close. This in turn signals other electrical circuits in the system to turn on the circulator and "enable" the heat source to produce heat.

The temperature-limiting controller ensures that the water temperature within the heat source remains within a predetermined range while the demand for heat is present. It does so by turning the heat-producing components of the heat source on and off as needed. Like the room thermostat, the temperature-limiting controller is a temperature-operated switch. For example, assume that the temperature-limiting controller on a typical gas- or oil-fired boiler is set for 160°F with a 10°F **differential**. If the water temperature inside the boiler drops to 150°F (160°F setpoint − 10°F differential), a switch contact in the temperature-limiting controller closes to operate the burner. The burner remains on until the water in the boiler reaches its upper limit temperature of 160°F or the heat demand is no longer present (e.g., the room thermostat is "satisfied").

Together, the room thermostat and temperature-limiting controller provide the necessary start/stop signals to the heat source and circulator to reasonably match the heat output to the heating requirement of the building. Such components have been used in millions of hydronic heating systems for several decades. These devices, as well as several more sophisticated controllers, are discussed in Chapter 9.

Expansion Tank

As water is heated, it expands. This increase in volume is an extremely powerful but predictable characteristic that must be accommodated in any type of closed-loop hydronic system. Figure 1-16 shows an **expansion tank**

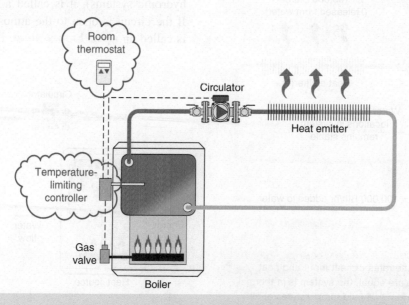

Figure 1-15 | Adding a room thermostat and heat source temperature-limiting controller to the system.

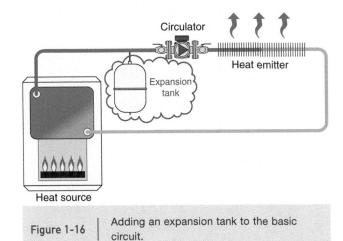

Figure 1-16 | Adding an expansion tank to the basic circuit.

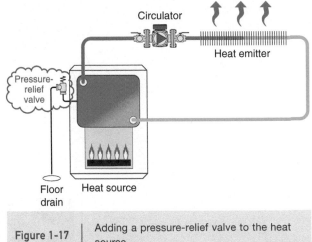

Figure 1-17 | Adding a pressure-relief valve to the heat source.

added to the basic system. The tank contains a captive volume of air. As the heated water expands, it pushes into the tank and slightly compresses the captive air volume. As a result, the system pressure rises slightly. As the water cools, its volume decreases, allowing the compressed air to expand, and the system pressure returns to its original value. This process repeats itself each time the system heats up and cools off.

Most modern hydronic systems use diaphragm-type expansion tanks. Such tanks contain their captive air in a sealed chamber. Older hydronic systems often used expansion tanks without diaphragms. Such tanks had to be considerably larger than modern diaphragm-type tanks. They also had to be mounted higher than the heat source.

The sizing and placement of the system's expansion tank are crucial to proper system operation. Both are discussed in Chapter 12.

Pressure-Relief Valve

Consider the fate of a closed-loop hydronic system in which a defective controller fails to turn off the heat source after its upper temperature limit has been reached. As the water gets hotter and hotter, system pressure steadily increases due to the water's expansion. This pressure could eventually exceed the pressure rating of the weakest component in the system. Most residential system components have pressure ratings of at least 60 pounds per square inch (psi) and may withstand two or more times that pressure before bursting. The consequences of a system component bursting at such high pressures and temperatures could be devastating. For this reason, all closed-loop hydronic systems must be protected by a **pressure-relief valve**. This is a universal requirement of all mechanical codes in North America.

Pressure-relief valves are designed and labeled to open at a specific pressure. Residential and light commercial systems typically have pressure-relief valves rated to open at 30 psi pressure. In systems with boilers, the pressure-relief valve is almost always attached directly to the boiler, or mounted very close to the boiler with nothing that could impede its operation between it and the boiler. A pressure-relief valve has been added to the heat source in Figure 1-17. Pressure-relief valves are discussed in more detail in Chapter 5.

Make-Up Water System

Most closed-loop hydronic systems experience minor water losses over time due to evaporation from valve packings, pump seals, air vents, and other components. These losses are normal and must be replaced to maintain adequate system pressure. The common method for replacing the water is through a **make-up water system** consisting of a **pressure-reducing valve**, **backflow preventer**, pressure gauge, and shutoff valves.

Because the pressure in a municipal water main or private water system is often higher than the pressure-relief valve setting in a hydronic system, such water sources cannot be directly connected to the loop. A pressure-reducing valve, also known as a **feed water valve**, is used to reduce and maintain a constant minimum pressure in the system. This valve allows water into the system whenever the pressure on the outlet side of the valve drops below the valve's pressure setting. This often occurs as air is vented from the system at start-up or after servicing. Most pressure-reducing valves have an adjustable pressure setting. Determining the proper setting is covered in Chapter 12.

The backflow preventer does just what its name implies. It stops any water that has entered the system from returning and possibly contaminating the potable

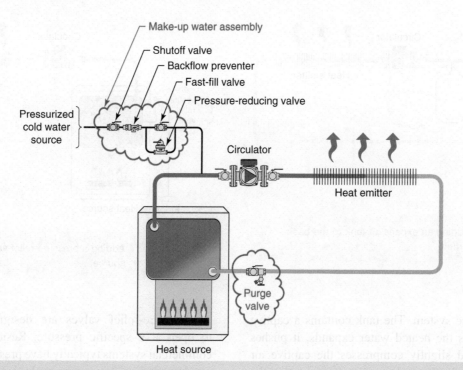

Figure 1-18 | Adding a make-up assembly.

water supply system. Most municipal codes require such a device on any heating system connected to a public water supply. Even in the absence of code requirements, installing a backflow preventer is a wise decision, especially if any antifreeze, corrosion inhibitors, or other chemicals are, or ever might be, used in the system.

The shutoff valves are installed to allow the system to be isolated from its water source and to allow components between the shutoff valves to be isolated if they need to be serviced. An optional "fast-fill" valve is sometimes installed in parallel with the pressure-reducing valve so water can be rapidly added to the system. This is particularly beneficial when filling larger systems. The components that constitute the make-up water system are shown in Figure 1-18.

Also shown in the lower portion of Figure 1-18 is a **purging valve** that allows most of the air in the system to be eliminated as it is filled with water. Purging valves consist of a ball valve, which is in line with the distribution piping, and a side-mounted drain port. To **purge** the system, the inline ball valve is closed, and the drain port is opened. As water first enters the system through the make-up water assembly, air is expelled through the drain port. Eventually, most of the air originally in the system will have been expelled through the drain port of the purging valve.

Flow-Check Valve

Another component commonly used in older hydronic systems is a **flow-check valve**. This valve contains a weighted plug that sits over the orifice within the valve. Some minimum differential pressure is required to lift this plug off the valve's orifice and thus enable flow through the valve. That minimum differential pressure will only be present when the circulator for that circuit is operating.

Another valve, known as a **spring-loaded check valve**, is sometimes used in place of a flow-check valve. This practice is common in modern systems. Both types of valves are discussed in more detail in Chapter 5. Either of these valves can serve one or two purposes depending upon the system it is installed in.

In single-loop systems, the flow-check valve prevents hot water in the boiler from slowly circulating through the distribution system when the circulator is off. Whenever a device containing heated water is part of an unblocked piping path having some vertical displacement, the potential for such flow exists. This flow is driven by the difference in density of hot water in the heat source relative to that of the cooler water in the distribution system. If not prevented, such **thermosiphoning** allows heat to be dissipated by slow but persistent flow through a piping circuit in an uncontrolled manner, often ending up where it is not desired. Figure 1-19 shows the

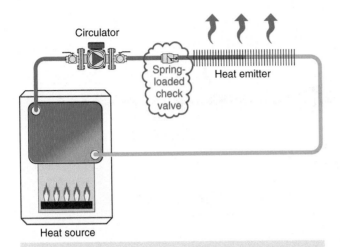

Figure 1-19 | Adding a flow-check valve to the basic circuit.

preferred placement of a flow-check valve (or spring-loaded check valve) in a single-loop system.

In multizone systems using circulators, a flow-check valve is installed in each zone circuit to eliminate thermosiphoning as well as flow reversal through inactive zones.

Air Separator

An **air separator** is designed to separate air from water and eject it from the system. Modern air separators create regions of reduced pressure as water passes through them. The reduced pressure causes molecules of oxygen, nitrogen, and other gases dissolved in the water to form bubbles. Once formed, these bubbles are guided upward into a collection chamber where an automatic air vent expels them from the system.

The process of separating air from water is enhanced when the water is heated. For best results, the air separator should be located where fluid temperatures are highest—in the supply pipe from the heat source. An air separator is shown in Figure 1-20. Air elimination is covered in detail in Chapter 13.

1.5 The Importance of System Design

Figure 1-21 is a composite drawing showing all the components previously discussed in their proper positions relative to each other. By assembling these components, we have built a simple hydronic heating system. It must be emphasized, however, that just because all the components are present does not guarantee that the system will function properly. Combining these components is not simply a matter of choosing a favorite product for each and connecting them as shown. Major subsystems such as the heat emitters and heat source have temperature and flow requirements that must be properly matched if they are to function together as a system. The objective is to achieve a stable, dependable, affordable, and efficient overall system. Failure to respect the operating characteristics of all components, and how they interact, will result in installations that underheat, overheat, waste energy, or otherwise disappoint their owners. The chapters that follow present detailed information on all the basic components just discussed. This information is essential in planning systems that deliver optimal performance.

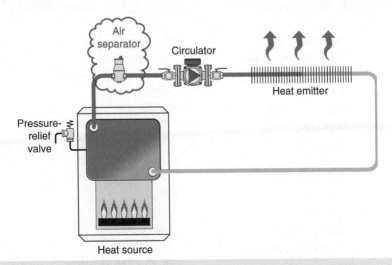

Figure 1-20 | Adding an air separator to the basic circuit.

18 Chapter 1 Fundamental Concepts

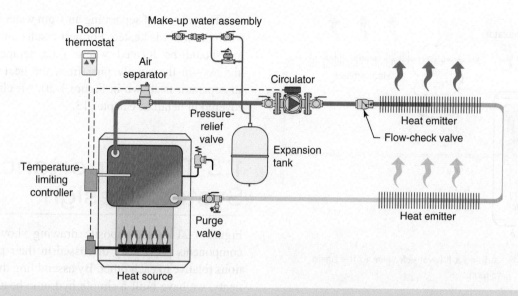

Figure 1-21 | Composite drawing showing all the basic hydronic circuit components.

Summary

The primary objective of any heating system is to provide the best possible comfort for the building occupants. Other important objectives include making efficient use of fuel and electricity, providing quiet and dependable operation, and maximizing the service life of all system components.

Hydronic heating systems offer tremendous potential for attaining all these objectives. The key to turning potential into reality is a solid understanding of the processes and components introduced in this chapter. A lack of understanding of these basics is often the cause of improper system design or difficulty in diagnosing the cause of poor system performance. Before continuing into more detailed aspects of design and installation, take time to test your understanding of these all-important fundamentals.

Key Terms

air handlers
air separator
air vent
backflow preventer
brine
British thermal units
buoyancy
circulators
closed-loop system
conduction
convection
convectors
differential
distribution efficiency
distribution system
domestic hot water
electromagnetic energy
evaporation
expansion tank
fan coils
feed water valve
flow-check valve
forced convection
heat capacity of water
heat emitter
heat source
infrared thermograph
loads
make-up water system
manifold station
mean radiant temperature
multiload systems
natural convection
open-loop system
pressure-reducing valve
pressure-relief valve
purge
purging valve
radiant panel
room thermostat
setpoint temperature
snowmelting
space heating
spring-loaded check valve
stratification
temperature-limiting controller
thermal conductivity
thermal energy
thermal equilibrium
thermal radiation
thermosiphoning
zoning

Questions and Exercises

1. Why does ducting in forced-air heating systems have to be so much larger than the tubing in a hydronic heating system of equal heating capacity?

2. Why is it better to surround a person with warm surfaces as opposed to just warm air?

3. What type of heat transfer creates the wind chill effect we experience during winter?

4. A certain block of material conducts heat at a rate of 100 Btu/h. One side is maintained at 80 °F and the other at 70 °F. Describe what happens to the rate of heat transfer when the:
   a. 80 °F side is raised to 130 °F
   b. thickness of the block is doubled
   c. 80 °F side is raised to 130 °F, and the thickness is cut in half

5. At approximately what temperature does a surface that is cooling from a high temperature stop emitting visible radiation?

6. How does thermal radiation differ from visible light? In what ways are they similar?

7. Describe thermal equilibrium within a hydronic system.

8. What is the function of a feed water valve in a hydronic system?

9. What is the difference between an open-loop system and a closed-loop system?

10. Are there any closed-loop hydronic heating systems on which a pressure-relief valve is not required? Why?

11. List two types of hydronic heat emitters other than finned-tube baseboard convectors.

12. What is the function of a flow-check valve in a single-loop system?

13. What is a "brine"? In what type of hydronic heating application would it be used?

14. Why is it necessary to have a backflow preventer in the make-up water line?

15. What are the customary units for heat and heat transfer rate in the North America?

16. What is a "zoned" hydronic system?

17. What made water flow through early hydronic systems before circulators were used?

18. What causes air bubbles to form within an air separator?

19. How are the expansion tanks used in modern hydronic systems different from those used in older systems?

20. What are some common ways small amounts of water leak out of a closed-loop hydronic system?

Chapter 2

Space Heating and Domestic Water-Heating Loads

Objectives

After studying this chapter, you should be able to:
- Describe what a design heating load is and why it is important to heating system design.
- Explain the difference between room heating loads and building heating loads.
- Determine the thermal envelope of a building.
- Calculate the effective total *R*-value of a building surface.
- Describe how the unit *U*-value for windows and doors is determined and how it is used.
- Estimate infiltration heat loss using the air change method.
- Determine the heat loss of foundations and slab floors.
- Explain what degree days are and how they are used.
- Estimate the annual space heating energy usage of a building.
- Estimate the daily energy used for domestic water heating.

2.1 Introduction

The design of any space heating or domestic water-heating system must start with an estimation of the **thermal load**. The care given to this step will directly affect system cost, efficiency, and most importantly, customer satisfaction.

It is natural for those learning to design hydronic heating systems to want to "dive into" a discussion of the hardware involved. However, this is like selecting the foundation for a building without knowing how much weight that foundation must support. It is pointless to start designing any heating system for a building without knowing the rate at which that building requires heat. To that end, this chapter lays out the basics for determining both space heating and domestic water-heating loads. Using the information in this chapter, designers can accurately estimate these loads and then optimize their system designs to meet them.

The majority of this chapter deals with estimating space heating loads. Simple mathematical methods for **conduction heat loss** and **air infiltration heat loss** are introduced, as are methods of estimating heat losses from slab on grade floors and basements. These methods are then applied to an example house on a room-by-room basis. Software-based approaches to heating load estimation are also discussed. The concept of **degree days** is introduced and used as a way to extrapolate the design heating load into an estimate of annual heating energy use for the building in a specific climate.

Information for estimating daily and peak hourly domestic water-heating loads is also presented. Domestic water-heating load information will be combined with space heating loads and used to design combisystems in later chapters.

2.2 Design Space Heating Load

It is important to understand what a design space heating load is before attempting to calculate it. *The design heating load of a building is an estimate of the rate at which a building loses heat during the near-minimum outdoor temperature.* This definition contains a number of key words that need further explanation.

First, it must be emphasized that a heating load is a calculated *estimate* of the rate of heat loss of a room or building. Because of the hundreds of construction details and thermal imperfections in an object as complex as a building, even a simple house, it is simply not possible to determine its design heating to the nearest Btu/h. Some factors that add uncertainty to the calculations are:

- Imperfect installation of insulation materials
- Variability in *R*-value of insulation materials
- Shrinkage of building materials, leading to greater air leakage
- Complex heat transfer paths at wall corners and other intersecting surfaces
- Effect of wind direction on building air leakage
- Overall construction quality
- Traffic into and out of the building

Second, the heating load is a *rate* of heat flow from the building to the outside air. It is often misstated, even among heating professionals, as a number of Btus rather than a *rate of flow* in Btus per hour (Btu/h). This is like stating the speed of a car in miles rather than miles per hour. Recall from the discussion of thermal equilibrium in Chapter 1 that when the rate of heat flow into a system matches the rate of heat flow out of a system, the temperature of that system remains constant. In the case of a building, *when the rate of heat input to the building equals the rate the building loses heat, the indoor temperature remains constant.* This is also true for each room within the building.

Finally, the design heating load is estimated assuming the outside temperature is *near* its minimum value. This temperature, called the **97.5 percent design dry bulb temperature**, is not the absolute minimum temperature for the location. It is the temperature that the outside air is at or above during 97.5% of the year. Although outside temperatures do occasionally drop below the 97.5 percent design dry bulb temperature, the duration of these low temperature excursions is short enough that most buildings can "coast" through them using heat stored in their thermal mass. Use of the 97.5 percent design temperature for heating system design helps prevent oversizing of the heat source.

Building Heating Load Versus Room Heating Load

One method of calculating the design heating load of a building yields a single number that represents the design heating load of the entire building. This value is called the **building design heating load**, and it is useful for selecting the heating output of the heat source as well as estimating annual heating energy use. However, this single number is not sufficient for designing the heat distribution system. It does not tell the designer how to properly proportion the heat output of the heat source to each room. Even when the building heating load is properly determined, failure to properly distribute the total heat output can lead to overheating in one room and underheating in another.

One should not assume that individual **room design heating loads** are proportional to room floor area. For example, a small room with a large window area could have a greater heating load than a much larger room with smaller windows.

The proper approach is to perform a design heating load estimate calculation for each room. Then, once these individual room heating loads are determined, add together to obtain the building design heating load.

Why Bother with Heat Load Calculations?

Would an electrician select wiring without knowing the amperage it must carry? Would a structural engineer select steel beams for a bridge without knowing the forces they must carry? Obviously, the answer to these questions is no. Yet, many so-called heating professionals routinely select equipment for heating systems with only a guess as to the building's heating load.

Statements attempting to justify this approach range from "I haven't got time for doing all those calculations," to "I'd rather be well oversized than get called back on the coldest day of the winter," to "I did a house like this one a couple of years ago, it can't be much different." Instead of listing excuses, let's look at some of the consequences of not properly estimating the building heating load.

- If the estimated heat loss is too low, the building will be uncomfortable during cold weather (a condition few homeowners will tolerate). An expensive callback will likely result.
- Grossly *overestimated* heating loads lead to systems that needlessly increase installation cost. Oversized systems may also cost more to operate due to lower heat source efficiency. This reduced efficiency can waste *thousands* of dollars worth of fuel over the life of the system.

Most building owners never know if their heating systems are oversized and simply accept the fuel usage of these systems as normal. Most trust the heating professional they hire to properly size and select equipment in their best interest. Heating professionals who deserve that trust adhere to the following principle:

Before attempting to design a space heating system of any sort, always calculate the design heating load of each room in the building using credible methods and data.

The remainder of this chapter shows you how to do this.

2.3 Conduction Heat Losses

Conduction is the process by which heat moves through a solid material whenever a temperature difference exists across that material. The rate of conduction heat transfer depends on the **thermal conductivity** of the material as well as the temperature difference across it. The relationship between these quantities is given in Equation 2.1.

Equation 2.1:

$$Q = A\left(\frac{k}{\Delta x}\right)(\Delta T)$$

where,
- Q = rate of heat transfer through the material (Btu/h)
- k = thermal conductivity of the material (Btu/T·h·ft)
- Δx = thickness of the material in the direction of heat flow (ft)
- ΔT = temperature difference across the material (°F)
- A = area across which heat flows (ft²)

Example 2.1

Determine the rate of conduction heat transfer through the 6-inch-thick wood panel shown in Figure 2-1.

Solution:

As shown in Figure 2-1, wood has a thermal conductivity of 0.1 Btu/°F·h·ft. Substituting the data into Equation 2.1:

$$Q = A\left(\frac{k}{\Delta x}\right)(\Delta T)$$

$$Q = 200 \text{ ft}^2\left(\frac{0.1 \text{ Btu/°F·h·ft}}{0.5 \text{ ft}}\right)(70°F - 10°F)$$

$$= 2,400 \text{ Btu/h}$$

Discussion:

Notice how the units of the inserted data cancel out so that the resulting units are Btu/h. Thermal conductivity data for various materials may be stated in units other than those given below Equation 2.1. If so, be sure to convert that conductivity value to the stated units, or the results of the calculation will be invalid.

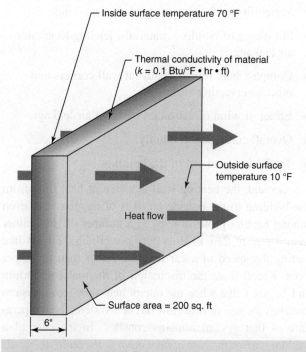

Figure 2-1 | Heat flow through a material by conduction.

Thermal Resistance of a Material

Since building designers are usually interested in *reducing* the rate of heat flow from a building, it is more convenient to think of a material's *resistance* to heat flow rather than its thermal conductivity. An object's **thermal resistance** (also referred to as its **R-value**) can be defined as its thickness in the direction of heat flow divided by its thermal conductivity.

Equation 2.2:

$$R\text{-value} = \frac{\text{thickness}}{\text{thermal conductivity}} = \frac{\Delta x}{k}$$

The greater the thermal resistance of an object, the slower heat passes through it when a given temperature differential is maintained across it. Equations 2.1 and 2.2 can be combined to yield an equation that is convenient for estimating heating loads:

Equation 2.3:

$$Q = \left(\frac{A}{R}\right)(\Delta T)$$

where,
 Q = rate of heat transfer through the material (Btu/h)
 ΔT = temperature difference across the material (°F)
 R = R-value of the material (°F·ft²·h/Btu)
 A = area across which heat flows (ft²)

Some interesting facts can be demonstrated with this equation. First, *the rate of heat transfer through a given object is directly proportional to the temperature difference (ΔT) maintained across it.* If this temperature difference were doubled, the rate of heat transfer would also double.

Second, *the rate of heat transfer through an object is inversely proportional to its R-value.* For example, if its R-value were doubled, the rate of heat transfer through the object would be halved (assuming the temperature difference across the object remained constant).

Finally, the larger the object's surface area, the greater the rate of heat transfer through it. For example, a 4-foot by 4-foot window with an area of 16 ft² would transfer heat twice as fast as a 2-foot by 4-foot window with an area of 8 ft².

Example 2.2

Determine the rate of heat transfer through the 6-inch-thick wood panel shown in Figure 2-2.

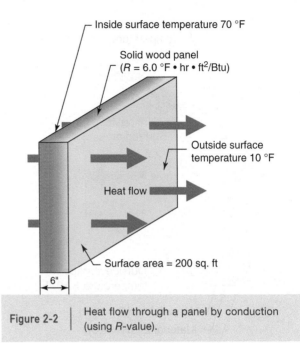

Figure 2-2 | Heat flow through a panel by conduction (using *R*-value).

Solution:

Substituting the relevant data into Equation 2.3:

$$Q = \left(\frac{A}{R}\right)(\Delta T) = \left(\frac{200}{6}\right)(70 - 10) = 2{,}000 \text{ Btu/h}$$

The R-value of a material is also directly proportional to its thickness. If a 1-inch-thick panel of extruded polystyrene insulation has an *R*-value of 5.4, a 2-inch-thick panel would have an *R*-value of 10.8, and a 1/2-inch-thick piece would have an *R*-value of 2.7. This relationship is very useful when the *R*-value of one thickness of a material is known, and the *R*-value of a different thickness needs to be determined.

The *R*-value of a material is slightly dependent on the material's temperature. As the temperature of the material is lowered, its *R*-value increases slightly. However, for most building materials, the change in *R*-value is small over the temperature ranges the material typically experiences and thus may be assumed to remain constant.

Total *R*-Value of an Assembly

Since the walls, floors, and ceilings of buildings are rarely constructed of a single material, it is often necessary to determine the total *R*-value of an assembly made up of several materials in contact with each other. This is done by adding up the *R*-values of the individual

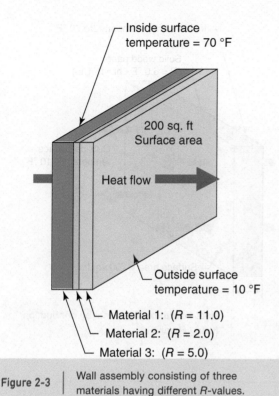

Figure 2-3 | Wall assembly consisting of three materials having different R-values.

materials. For example, the total R-value of the assembly shown in Figure 2-3 is:

$$R_{total} = 11.0 + 2.0 + 5.0 = 18.0$$

The heat flow across the panel can now be calculated using the **total R-value** substituted into Equation 2.3:

$$Q = \left(\frac{A}{R}\right)(\Delta T) = \left(\frac{200}{18}\right)(70 - 10) = 667 \text{ Btu/h}$$

The R-values of many common building materials can be found in Appendix B. To find the R-values of other materials, consult manufacturer's literature or the current edition of the *ASHRAE Handbook of Fundamentals*.

Air Film Resistances

In addition to the thermal resistances of the solid materials, there are two thermal resistances, called the **inside air film** resistance and **outside air film** resistance, which affect the rate of heat transfer through a panel separating inside and outside spaces. These thermal resistances represent the insulating effects of thin layers of air that cling to all surfaces. The R-values of these air films are dependent on surface orientation, air movement along the surface, and reflective qualities of the surface. Values for these resistances can also be found in Appendix B. The R-values of air films include the combined effect of conduction, convection, and radiation heat transfer, but are stated as a conduction-type thermal resistance for simplicity.

Example 2.3 shows how the total R-value of the wall assembly shown in Figure 2-3 is affected by the inclusion of the air film resistances. Notice that the temperatures used to determine the ΔT in this example are the indoor and outdoor *air temperatures*, rather than the *surface temperatures* used in the previous examples.

Example 2.3

Determine the rate of heat transfer through the assembly shown in Figure 2-4. Assume still air on the inside of the wall and 15-mph wind outside. See Appendix B for values of the air film resistances.

Solution:

The total R-value of the assembly is again found by adding the R-values of all materials, including the R-values of the inside and outside air films:

$$R_{total} = 0.17 + 11.0 + 2.0 + 5.0 + 0.68 = 18.85$$

The rate of heat transfer through the assembly can once again be found using Equation 2.3:

$$Q = \left(\frac{A}{R}\right)(\Delta T) = \left(\frac{200}{18.85}\right)(70 - 10) = 637 \text{ Btu/h}$$

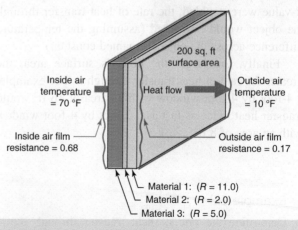

Figure 2-4 | Air film resistances add R-value to a wall assembly.

> **Discussion:**
>
> Including the R-value of the inside and outside air films caused a slight decrease in the rate of heat transfer through the panel. For the panel construction shown in Figure 2-4, the decrease in the rate of heat transfer was about 4.5%. The greater the total R-value of the materials making up the panel, the smaller the effect of the air film resistances. However, the R-value of the air films should always be included when determining the total R-value of an assembly.

Effect of Framing Members

The rate of heat flow through assemblies made of several materials stacked together like a sandwich can now be calculated. In such situations, and when edge effects are considered insignificant, the heat flow through the assembly is uniform over each square foot of surface area. However, few buildings are constructed with walls, ceilings, and other surfaces made of simple stacked layers of materials. This is particularly true of wood-framed buildings with wall studs, ceiling joists, window headers, and other framing members. Wooden framing members that span across the insulation cavity of a wall tend to reduce its **effective total R-value**. This is because the thermal resistance of wood, about 1.0 per inch of thickness, is usually less than the thermal resistance of insulation materials it displaces. Thus, the rate of conduction through framing materials is usually faster than through the surrounding insulation materials. This effect can be detected with infrared thermography, as shown in Figure 2-5. In this case, the wall studs as well as top and bottom plates are readily visible to the infrared camera because they create "stripes" of higher-surface temperature along the outside wall surface on a cold winter night.

It is possible to adjust the R-value of an assembly to include the thermal effects of framing. This requires that the assembly's total R-value be calculated both between the framing and at the framing. The resulting total R-values are then weighted according to the percentage of solid framing in the assembly. Equation 2.4 can be used to calculate an "effective total R-value," which can then be applied to the entire panel area.

Equation 2.4:

$$R_{\text{effective}} = \frac{(R_i)(R_f)}{p(R_i - R_f) + R_f}$$

where,

$R_{\text{effective}}$ = effective total R-value of panel (°F·ft²·h/Btu)
P = percentage of panel occupied by framing (decimal %)
R_f = R-value of panel at framing (°F·ft²·h/Btu)
R_i = R-value of panel at insulation cavities (°F·ft²·h/Btu)

Although the percentage of the panel occupied by framing could be calculated for every construction assembly, such calculations can be tedious and often make insignificant changes in the results. Typical wood-framed walls will have between 10 and 15% of the wall area as solid framing across the insulation cavity. The lower end of this range would be appropriate for 24 inches on-center framing, with insulated headers over windows and doors. The upper end of this range is typical of walls with 16 inches on-center framing and solid headers.

To illustrate these calculations, the effective total R-value of the wall assembly shown in Figure 2-6 will be calculated assuming that 15% of the wall is solid framing.

Notice that the effective R-value of this wall (R-22.9) is about 13.6% lower than the R-value through the insulation cavity (R-26.5). If the person estimating the heat loss simply ignored the effect of framing, the estimated heat loss for the wall would be 13.6% low. It follows that a heating system designed according to this underestimated heat loss might not be able to maintain comfort on a design load day.

To obtain the total heat loss of a room, the effective total R-value of each different **exposed surface** within the room must be determined using a procedure similar to that shown in Figure 2-6. Fortunately, many buildings have the same type of construction for most exposed

Figure 2-5 Infrared thermograph of framed wall reveals "stripes" of higher outside surface temperatures due to lower R-value of studs relative to cavity insulation.

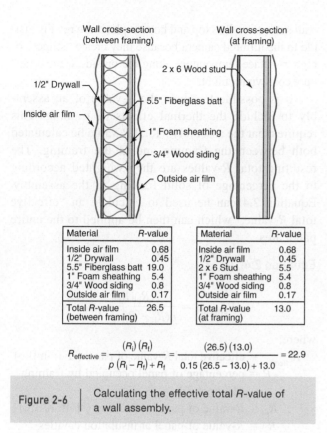

Figure 2-6 | Calculating the effective total R-value of a wall assembly.

$$R_{\text{effective}} = \frac{(R_i)(R_f)}{p(R_i - R_f) + R_f} = \frac{(26.5)(13.0)}{0.15(26.5 - 13.0) + 13.0} = 22.9$$

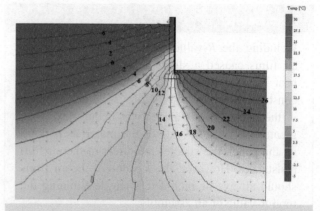

Figure 2-7a | Simulated heat flow through a basement wall and uninsulated heated slab floor. *Courtesy of Beaver Plastics LTD.*

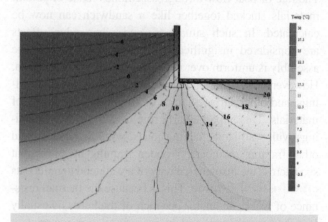

Figure 2-7b | Simulated heat flow through a basement wall and heated slab floor with 2 inches of expanded polystyrene insulation below floor slab. *Courtesy of Beaver Plastics LTD.*

walls, ceilings, and so on. In such cases, the effective R-value of the assembly need only be calculated once.

It is good practice to make a sketch or CAD drawing of the cross-section of each exposed assembly (i.e., wall, ceiling, floor), similar to that shown in Figure 2-6. The R-value of each material can then be determined from Appendix B and the effective total R-value calculated. These sheets or CAD drawing files showing the cross-sections of the assemblies and their effective total R-values can be saved and referenced on other buildings with the same construction.

2.4 Foundation Heat Loss

Heat flow from a basement or slab-on-grade foundation is determined by complex interactions between the building, the surrounding soil, insulation materials (if present), and the air temperature above grade.

Figure 2-7a shows a **finite element analysis** computer simulation of heat flow from a residential basement during February, in a cold winter climate. The basement has a width of approximately 26 feet. Only half of this width is shown because heat flow is assumed to be symmetrical about a vertical centerline through the basement. The floor slab is heated by embedded tubing and is assumed to be covered by a ¾-inch-thick hardwood flooring. *There is no under-slab insulation.* The basement walls are insulated with 2 inches of expanded polystyrene.

The contour lines with temperatures labeled in °C are called **isotherms**. They represent locations having the same temperature. Heat flow is perpendicular to the isotherms in all locations. In this case, heat flows downward from the slab, passes under the footing, then passes outward and upward toward the cold exposed soil surface. In this case, almost half of the purchased energy used to heat the slab floor is being driven downward into the soil.

Figure 2-7b shows the same situation, but with 2 inches of expanded polystyrene insulation installed under the floor slab. Notice that soil temperatures below the slab are significantly lower, because less heat is being driven downward. It follows that heat loss from this basement will be significantly less. For the cold northern climate assumed in this simulation, which represented an average between the typical weather conditions in Toronto, Ontario, and Calgary, Alberta, the use of 2-inch expanded polystyrene insulation under the heated slab reduced heat loss by approximately 50%. The energy savings over a complete heating season were estimated at 8,945 Btu/ft² of floor area.

The results of complex computer simulations of foundation heat loss have been used to develop simpler calculation procedures that can be used for load estimating. One such method is useful for estimating the heat loss of partially buried basement walls, the other for estimating the heat loss from floor slabs.

Heat Loss Through Basement Walls

The heat loss of a basement wall is significantly affected by the height of soil against the outside of the wall. Soil adds thermal resistance between the wall and the outside air and thus helps reduce heat loss.

Figure 2-8 illustrates the theoretical heat flow paths through a uniformly insulated basement wall and adjacent soil. Notice that the heat flow paths from the inside basement air to the outside air are longer for the lower portions of the wall. As the path length increases, so does the total thermal resistance of the soil along that path.

One method of approximating the heat loss through basement walls is based on dividing the wall into horizontal strips based on the height of finish grade. The upper strip includes all wall area exposed above grade as well as the wall area to a depth of 2 feet below grade. The implicit assumption is that by mid-winter, shallow soils will be at approximately the same temperature as the outside air. The middle strip includes the wall area from 2 feet to 5 feet below grade. The lower strip includes all wall area deeper than 5 feet below grade. Figure 2-9 illustrates these areas.

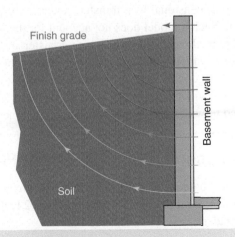

Figure 2-8 | Theoretical heat flow paths through a uniformly insulated basement wall.

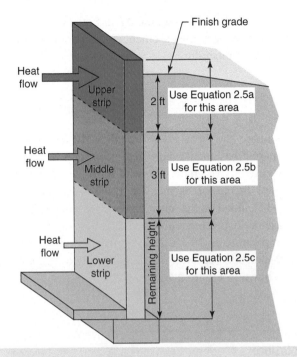

Figure 2-9 | Dividing the basement wall into three horizontal strips to estimate heat loss.

Use Equation 2.5a to calculate the effective R-value of basement walls for all above grade areas as well as areas down to 2 feet below grade.

Equation 2.5a:

$$R_{\text{effective}} = R_{\text{wall}} + R_{\text{insulation}}$$

Use Equation 2.5b to calculate the effective R-value of basement walls for areas from 2 feet to 5 feet below grade.

Equation 2.5b:

$$R_{\text{effective}} = 7.9 + 1.12(R_{\text{insulation}})$$

Use Equation 2.5c to calculate the effective R-value of basement walls for areas deeper than 5 feet below grade.

Equation 2.5c:

$$R_{\text{effective}} = 11.3 + 1.13(R_{\text{insulation}})$$

The parameter ($R_{\text{insulation}}$) in Equations 2.5a through 2.5c represents the R-value of any insulation *added* to the foundation wall over the particular strip for which the effective R-value is being determined. The foundation wall itself is assumed to be a masonry or concrete wall between 8 and 12 inches thick.

Once the R-values of each wall strip have been determined, use Equation 2.3 to calculate the heat loss of each area. In each case, the ΔT in Equation 2.3 is the difference between the basement air temperature and the outside air temperature.

Example 2.4

Estimate the heat loss through a 10-inch-thick concrete basement wall 9 feet deep with 6 inches exposed above grade. The wall is 40 feet long and has 2 inches of extruded polystyrene insulation (R-11) added on the outside over its full depth. The basement air temperature is 70 °F. The outside air temperature is 10 °F. Assume the R-value of the 10-inch concrete wall is 1.0. A drawing of the wall is shown in Figure 2-10.

Solution:

For the upper wall strip, use Equation 2.5a to determine the effective R-value. Then use Equation 2.3 to calculate the heat loss.

$$R = 1.0 + 11 = 12$$
$$A = 40\,\text{ft}\,(2.5\,\text{ft}) = 100\,\text{ft}^2$$
$$\Delta T = (70\,°\text{F} - 10\,°\text{F}) = 60\,°\text{F}$$

$$Q = \left(\frac{A}{R}\right)(\Delta T) = \left(\frac{100}{12}\right)(60) = 500\ \text{Btu/h}$$

For the middle wall strip, use Equation 2.5b to determine the effective R-value. Then use Equation 2.3 to calculate the heat loss.

$$R = 7.9 + 1.12(R_{added}) = 7.9 + 1.12(11) = 20.2$$
$$A = 40\,\text{ft}(3\,\text{ft}) = 120\,\text{ft}^2$$
$$\Delta T = (70\,°\text{F} - 10\,°\text{F}) = 60\,°\text{F}$$

$$Q = \left(\frac{A}{R}\right)(\Delta T) = \left(\frac{120}{20.2}\right)(60) = 356\ \text{Btu/h}$$

For the lower wall strip, use Equation 2.5c to calculate the R-value. Then use Equation 2.3 to calculate the heat loss.

$$R = 11.3 + 1.13(R_{added}) = 11.3 + 1.13(11) = 23.7$$
$$A + 40\,\text{ft}(3.5\,\text{ft}) = 140\,\text{ft}^2$$
$$\Delta T = (70\,°\text{F} - 10\,°\text{F}) = 60\,°\text{F}$$

$$Q = \left(\frac{A}{R}\right)(\Delta T) = \left(\frac{140}{23.7}\right)(60) = 354\ \text{Btu/h}$$

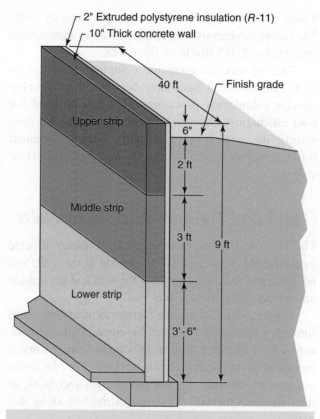

Figure 2-10 | Basement wall used in Example 2.4.

The total heat loss for this wall is the sum of the heat losses from each wall strip:

$$Q_{total} = 500 + 356 + 354 = 1{,}210\ \text{Btu/h}$$

Discussion:

Note that the same ΔT term (the difference between the basement air temperature and the outside air temperature) was used in all three calculations. It should also be noted that these results are based on a two-dimensional heat transfer model of the situation. Such a model does not account for the more complex three-dimensional heat transfer at wall corners, which will slightly increase basement heat loss. In general, narrow basements with several corners will have greater rates of heat loss compared to wider basements with minimal corners.

Heat Loss Through Basement Floors

Equation 2.6a can be used to estimate downward heat loss for basement floor slabs that are at least 2 feet below finish grade:

Equation 2.6a:

$$Q_{bslab} = \left(\frac{A}{R_{bslab}}\right)\Delta T$$

where,

Q_{bslab} = rate of heat loss through the basement floor slab (Btu/h)
A = floor area (ft²)
R_{bslab} = effective R-value of slab *(designated as R_i in Equation 2.6b, or as R_u in Equation 2.6c)*
ΔT = a. (basement air temperature minus outside air temperature) (°F) *for unheated slabs*
b. (average slab temperature minus outside air temperature) (°F) *for heated slabs*

The effective R-value of the basement slab (R_{bslab}) is found using either Equation 2.6b or 2.6c. Use Equation 2.6b if the slab is *insulated with at least R-3 (°F·h·ft²/Btu) underside insulation*. Use Equation 2.6c if the slab is *uninsulated*. Equations 2.6b and 2.6c are based on the width of the shortest side of the floor slab.

Equation 2.6b:
$$R_i = 20.625 + 1.6063(w)$$

Equation 2.6c:
$$R_u = 14.31 + 1.1083(w)$$

where,
R_i = effective R-value for *insulated* slab (°F·h·ft²/Btu)
R_u = effective R-value for *uninsulated* slab (°F·h·ft²/Btu)
w = width of *shortest* side of floor slab (ft)

Example 2.5
Determine the rate of heat loss from an unheated basement floor slab measuring 30 feet by 50 feet, when the outside temperature is 10 °F, and the basement air temperature is 65 °F. Assume the basement floor has no underside insulation.

Solution:
Since the slab has no underside insulation, Equation 2.6c will be used to calculate the effective total R-value.

$$R_u = 14.31 + 1.1083(w)$$
$$= 14.31 + 1.1083(30) = 47.56$$

The heat loss from the slab can now be determined using Equation 2.6a.

$$Q_{bslab} = \left(\frac{A}{R_{bslab}}\right)\Delta T = \frac{30 \times 50}{47.56}(65 - 10) = 1{,}735 \text{ Btu/h}$$

Discussion:
The effective R-value of the slab determined using either Equation 2.6b or Equation 2.6c includes the effects of the slab, underside insulation (if present), and several feet of soil in the path of heat flow from the basement air to the outside air. Thus, the effective R-value in relatively high, even for an uninsulated slab.

Heat Loss Through Slab-on-Grade Floors

Many residential and commercial buildings have concrete slab-on-grade floors rather than crawl spaces or full basements. Heat flows from the floor slab downward to the surrounding earth and outward through any exposed edges of the slab. Of these two paths, outward heat flow through the slab edge tends to dominate. This is because the outer edge of the slab is exposed to outside air or to soil at approximately the same temperature as the outside air. Soil under the slab and several feet in from the perimeter tends to stabilize at somewhat higher temperatures once the building is maintained at normal comfort conditions. Hence, downward heat losses are relatively small from interior areas of the floor slab. Areas of the floor close to the exposed edges have higher rates of heat loss. This often justifies the use of higher R-value insulation near the perimeter of the slab compared to under interior areas.

As with basement walls, the heat flow paths from a slab to the soil and outside air are complex and vary with foundation geometry, soil conditions, and the insulation materials used. Figure 2-11 shows isotherms and arrows indicating the direction of heat flow for a heated slab on grade with continuous underside and edge insulation.

The method used to estimate the heat loss from a slab-on-grade floor depends on the amount of insulation (if any) used near the perimeter of the slab, the orientation of that insulation, and the type of soil present at the perimeter of the slab. Dense soils with significant water content will conduct heat better than lighter/drier soils and thus increase rates of heat loss.

Two scenarios will be considered: One in which no underslab or edge insulation is used, and another where a rigid insulating panel having an R-value between 4 and 16 (°F·h·ft²/Btu) is installed between the edge of the slab and the foundation as well as under the outer four feet of the slab. Both scenarios are shown in Figure 2-12.

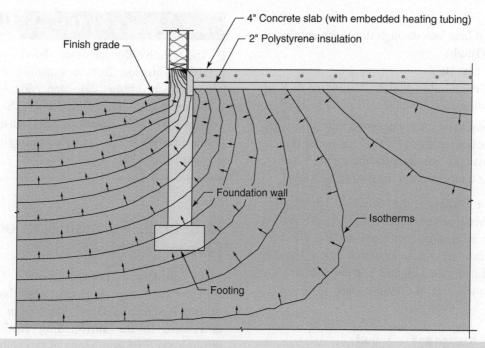

Figure 2-11 | Isotherms and arrows indicating the direction of heat flow for a heated slab-on-grade foundation.

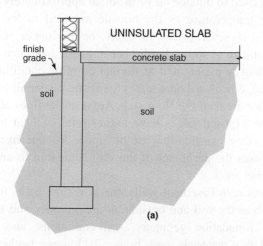

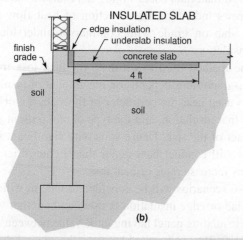

Figure 2-12 | (a) Uninsulated floor slab. (b) Insulated floor slab.

The heat loss from an *uninsulated* slab can be estimated using Equation 2.7a.

Equation 2.7a:

$$Q_{\text{slab}} = c_0 L (\Delta T)$$

where,

Q_{slab} = rate of heat loss through the slab edge and interior floor area (Btu/h)

L = exposed edge length of the floor slab (ft)

ΔT = a. difference between inside and outside air temperature (°F) for unheated slabs

b. difference between average slab temperature and outside air temperature (°F) for heated slabs

c_0 = a number based on soil type (see table in Figure 2-13)

Heavy/damp soil	Heavy/dry soil (or) light/damp soil	Light/dry soil
$c_0 = 1.358$	$c_0 = 1.18$	$c_0 = 0.989$

Figure 2-13 | Values of c_0 for Equation 2.7a (uninsulated slabs).

Example 2.6

Determine the rate of heat loss from the shaded portion of an unheated and uninsulated floor slab shown in Figure 2-14. The soil outside the slab is classified as heavy and damp. The inside air temperature is 70 °F and the outside air temperature is 10 °F.

Solution:

Only two sides of the room's slab are exposed to the outside. The total linear footage of exposed slab edge is 20 + 25 = 45 feet. The value of c_0 for heavy/damp soil in Figure 2-13 is 1.358. Substituting the data into Equation 2.7a yields:

$$Q_{slab} = c_0 L(\Delta T) = 1.358(45)(70 - 10)$$
$$= 3{,}670 \text{ Btu/h}$$

Discussion:

It should also be stressed that unlike many of the equations used for estimating heat loss, Equation 2.7a is based on the *length* of the exposed perimeter of the slab and not the slab's area.

If the slab contained embedded tubing for heating, it would typically be at temperatures higher than room air temperature during the heating season. This would increase the rate of downward and outward heat loss. *It is suggested that the air temperature in Equation 2.7a be replaced by a temperature that is 25 °F higher than room air temperature when estimating the heat loss of a heated floor slab under design load conditions.*

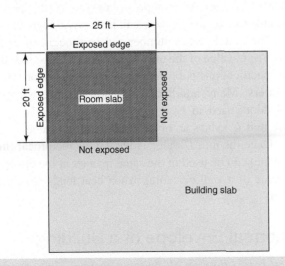

Figure 2-14 | Room with exposed slab edge used in Example 2.6.

If the slab-on-grade floor is *insulated* as shown in Figure 2-12b, with insulation having an R-value between 4 and 16 (°F·h·ft²/Btu), the heat loss can be determined using Equation 2.7b.

Equation 2.7b:

$$Q_{slab} = L(\Delta T)\left[b_0 + b_1 R_{edge} + b_2(R_{edge})^2\right]$$

where,

Q_{slab} = rate of heat loss through the slab edge and interior floor area (Btu/h)

L = exposed edge length of the floor slab (ft)

ΔT = (a. difference between inside and outside air temperature (°F) for unheated slabs)/(b. difference between the average slab temperature and outside air temperature (°F) for heated slabs)

R_{edge} = R-value of the edge and underslab insulation (°F·h·ft²/Btu) (4 ≤ R_{edge} < 16)

b_0, b_1, b_2 = constants based on soil type (see Figure 2-15)

Example 2.7

Determine the rate of heat loss from the shaded portion of the unheated floor slab shown in Figure 2-14, *assuming 2 inches of extruded polystyrene insulation has been placed at the edge of the slab and under the outer 4 feet of the slab*. The soil outside the slab remains classified as heavy and damp. The inside air temperature is 70 °F and the outside air temperature is 10 °F.

Solution:

Using the data in Appendix B, the R-value of 2 inches of extruded polystyrene insulation is 2 × (5.4) = 10.8 (°F·h·ft²/Btu). Substituting the values of b_0, b_1, and b_2

Heavy/damp soil	$b_0 = 0.755$	$b_1 = -0.0425$	$b_2 = -0.00126$
Heavy/dry soil (or) Light/damp soil	$b_0 = 0.595$	$b_1 = -0.0360$	$b_2 = -0.00108$
Light/dry soil	$b_0 = 0.411$	$b_1 = -0.0293$	$b_2 = -0.0009$

Figure 2-15 | Values of b_0, b_1, and b_2 for Equation 2.7b (insulated slabs).

for heavy/damp soil along with the remaining data into Equation 2.7b yields the following:

$$Q_{slab} = 45(70 - 10)[0.755 + (-0.0425)10.8 + (0.00126)(10.8)^2]$$
$$= 1,196 \text{ Btu/h}$$

> **Discussion:**
>
> Comparing the results of Examples 2.6 and 2.7 shows that there is a 67% decrease in heat loss from the slab when 2 inches of extruded polystyrene insulation is installed. Insulating slab floors in heated buildings is essential to good performance. The estimated heat loss includes loss through the exposed edge of the slab as well as downward heat flow from the shaded interior area. Equation 2.7b should only be used when the R-value of the under-slab insulation is between 4 and 16 (°F·h·ft²/Btu).

Designers and building owners should consider that there is most likely only one opportunity to properly insulate a concrete slab-on-grade floor. Although "retrofitting" underslab insulation is not impossible, it is extremely disruptive and very expensive. From a practical standpoint, such retrofitting is very unlikely to happen. Thus, the choice of underslab insulation will most likely have to suffice for the life of the building. It is imperative to "get it right" the first time. The author recommends a *minimum* of 2-inch-thick extruded polystyrene insulation (R — 10.8 °F•h•ft²/Btu) under *all* slab-on-grade floor areas in heated spaces. If the building program emphasizes high energy efficiency, even more underslab insulation (3 to 4 inches of extruded polystyrene) is likely justified, especially in cold northern climates. This insulation should be placed under all floor areas with exception of footings for load bearing columns or walls. Extruded polystyrene insulation is available with compressive stress ratings of 15 to 100 psi to accommodate slabs ranging from interior residential floors to heavy vehicle maintenance garages and aircraft hangers.

Heat Loss Through Windows, Doors, and Skylights

The windows, doors, and skylights in a room can represent a large percentage of that room's total heat loss. The construction of windows and skylights can range from old, single glass assemblies with metal frames to modern multi-pane assemblies filled with Argon gas to suppress convection, low-E (low emissivity) coatings to suppress radiation heat transfer, and insulated polymer frames to reduce edge conduction losses. Doors can range from solid wood to foam-core insulated panels with steel or fiberglass claddings.

The thermal resistance of a typical window, door, or skylight unit varies depending upon *where* on the cross-section of the unit the thermal resistance is evaluated. For example, a wooden window frame will provide higher thermal resistance than will two panes of glass with an air space in between. Similarly, the thermal resistance at the edge of a multi-pane glazing assembly will be less than that measured at the center of the glazing unit. This is the result of greater conduction through the edge spacers that separate the panes versus the air space at the center of the unit.

To provide a uniform and simplified approach to dealing with these effects, manufacturers, working cooperatively through the National Fenestration Rating Council (NFRC), have developed a standard for determining the **unit U-value** of a window, door, or skylight unit.

The R-value is the reciprocal of U-value. If either the U-value or R-value is known, the other can be easily determined.

$$R = \frac{1}{U}$$

The lower the unit U-value of a window, door, or skylight, the lower its heat loss, all other conditions being equal.

The methods used in the standard NFRC 100 *Procedures for Determining Fenestration Product U-Factor* account for the thermal resistance at the window frame as well as at the edge and center of the glazing unit. These thermal resistances are then "area weighted" based on the shape of the unit. An example of how this would apply to a simple rectangular window unit is shown in Figure 2-16.

The unit U-value number can be thought of as the *effective U*-value of the unit. Values for it can be found in the specifications published by window and door manufacturers. Many building codes now require windows and doors used in new construction or remodeling to have unit U-values at, or below, a specified value.

Once the unit U-value is known, its reciprocal (unit R-value) can be used in the same manner at the effective R-value of a wall or ceiling when heat loads are calculated.

Thermal Envelope of a Building

Heat is lost through all building surfaces that separate heated space from unheated space. Together, these surfaces (walls, windows, ceilings, doors, foundation, etc.)

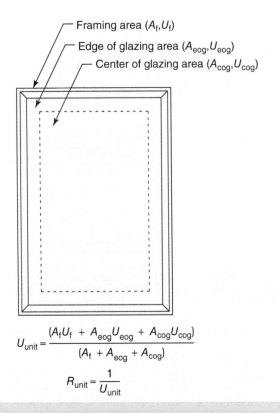

$$U_{unit} = \frac{(A_f U_f + A_{eog} U_{eog} + A_{cog} U_{cog})}{(A_f + A_{eog} + A_{cog})}$$

$$R_{unit} = \frac{1}{U_{unit}}$$

Figure 2-16 | Areas and formula for determining unit *U*-value of a rectangular window unit.

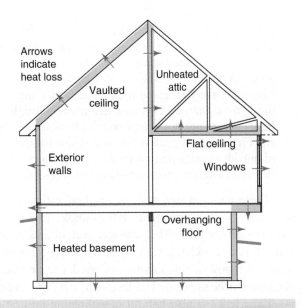

Figure 2-17 | A portion of the thermal envelope of a house (shown shaded).

are called the **thermal envelope** of a building. *It is only necessary to calculate heat flow though the surfaces that constitute this thermal envelope.* Walls, ceilings, floors, or other surfaces that separate one heated space from another should not be included in these calculations.

2.5 Infiltration Heat Loss

Figure 2-17 illustrates a part of the thermal envelope for a house.

One can also envision each room in a building as a "compartment." The thermal envelope of that compartment would consist of all surfaces that *separate heated space from unheated space*. Heat loss calculations would only be performed for these surfaces.

In addition to conduction losses, heat is also carried out of buildings by uncontrolled air leakage. This is called **infiltration heat loss**. The faster air leaks into and out of a building, the faster heat is carried away by that air. The thermal envelope of an average building contains hundreds of imperfections through which air can pass. Some leakage points, such as fireplace flues, exhaust hoods, and visible cracks around windows and doors, are obvious. Others are small and out of sight, but when taken together represent significant amounts of leakage area. These include cracks where walls meet the floor deck, air leakage through electrical outlets, small gaps where pipes pass through floors or ceilings, and many other small imperfections in the thermal envelope. Some of these may be the result of structural or aesthetic detailing of the building.

It would be impossible to assess the location and magnitude of all the air leakage paths in a typical building. Instead, an estimate of air leakage can be made based on the air sealing quality of the building. This approach is known as the **air change method** of estimating infiltration heat loss. It relies on a somewhat subjective classification of air sealing quality as well as experience. The designer chooses a rate at which the interior air volume of the building (or an individual room) is exchanged with outside air. For example: 0.5 air change per hour means that half of the entire volume of heated air in a space is replaced with outside air once each hour.

The infiltration rates shown in Figures 2-18a and 2-18b are suggested in *ACCA Manual J* (8th edition). They are categorized based on single- or two-story constructions. Use Figure 2-18a for single-story construction and Figure 2-18b for two-story construction.

The descriptions for air sealing quality (e.g., tight, semi-loose) are as follows:

- **Tight:** All structural joints and cracks are sealed using *meticulous* work along with air infiltration barriers, caulks, tapings, or packings. Window and doors are rated at less than 0.25 CFM per foot of

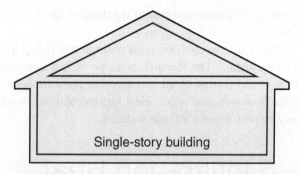

Single-story Floor area (ft²)	900 or less	901–1,500	1,500–2,000	2,000–3,000	3,001 or more
Tight	0.21	0.16	0.14	0.11	0.10
Semi-tight	0.41	0.31	0.26	0.22	0.19
Average	0.61	0.45	0.38	0.32	0.28
Semi-loose	0.95	0.70	0.59	0.49	0.43
Loose	1.29	0.94	0.80	0.66	0.58

Figure 2-18a | Air change rates for single-story construction.

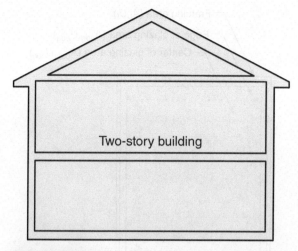

Two-story Floor area (ft²)	900 or less	901–1,500	1,500–2,000	2,000–3,000	3,001 or more
Tight	0.27	0.20	0.18	0.15	0.13
Semi-tight	0.53	0.39	0.34	0.28	0.25
Average	0.79	0.58	0.50	0.41	0.37
Semi-loose	1.23	0.90	0.77	0.63	0.56
Loose	1.67	1.22	1.04	0.85	0.75

Figure 2-18b | Air change rates for two-story construction.

crack at 25-mph wind speed. All exhaust fans are equipped with backdraft dampers. The building either does not have recessed light fixtures piercing through thermal envelope surfaces, or has recessed lights with negligible air leakage. The house does not have powerful (150 CFM or greater) exhausting range hoods. The house either does not have combustion-type heat sources within conditioned space, or any combustion heat sources present are direct vented. If fireplaces are present, they are provided with their own combustion air and have tight-fitting glass doors.

- **Semi-Tight:** Conditions range between tight and average.

- **Average:** All structural joints and cracks are sealed using *average* work along with air infiltration barriers, caulks, tapings, or packings. Window and doors are rated at 0.25 to 0.50 CFM per foot of crack at 25-mph wind speed. All exhaust fans are equipped with backdraft dampers. The building either does not have recessed light fixtures piercing through the thermal envelope or has recessed lights with negligible air leakage. The house does not have powerful (150 CFM or greater) exhausting range hoods. The house either does not have combustion-type heat sources within conditioned space or any combustion heat sources present are direct vented. If fireplaces are present, they are provided with their own combustion air and have tight fitting glass doors.

- **Semi-Loose:** Conditions range between average and loose.

- **Loose:** There is little or no effort to seals joints or cracks in the building envelope. Windows and doors are either not rated or are rated for more than 50 CFM per foot of crack at 25-mph wind speed. There are many recessed light fixtures piercing through the thermal envelope that are not airtight. Exhaust fans are not equipped with backdraft dampers. Fireplaces, if present, do not have a source of outside combustion air or tight glass doors. Powerful kitchen exhaust blowers are present.

These air exchange rates are at best estimates and they vary over a wide range. For example, in a single-story building with 1,800 square feet of floor area, the air leakage rate associated with "loose" quality air sealing results in almost six times more heat loss due to air leakage compared to the same size building with "tight" air sealing.

Experience is also helpful in estimating infiltration rates. A visit to the building while it is under construction helps give the heating system designer a feel for the air sealing quality being achieved. An inspection visit to an existing building, though not as revealing, can still help the designer assess air sealing quality. Things to look for or consider include:

- Visible cracks around windows and doors
- Continuous spray foam insulation versus batt insulation

- Poorly installed or maintained weather stripping on doors and windows
- Slight air motion detected near electrical fixtures
- Smoke that disappears upward into ceiling lighting fixtures
- Deterioration of paint on the downwind side of building
- Poor-quality duct-system design (especially a lack of proper return air ducting)
- Interior humidity levels (drier buildings generally indicate more air leakage)
- Presence or lack of flue damper on fossil fuel heating system
- Windy versus sheltered building site
- Storm doors and entry vestibules help reduce air infiltration
- The presence of fireplaces (especially those of older masonry construction)
- Casement and awning windows generally have less air leakage than double hung or sliding windows
- Cold air leaking in low in the building generally indicates warm air leaking out higher in the building

Figure 2-18c | A blower door test being performed during house construction. *Courtesy of © John P. Gordon, AIA.*

Blower Door Testing

The airtightness of an existing or partially constructed building can often be assessed through blower door testing. A **blower door** consists of an adjustable frame covered by an airtight fabric that expands to form an airtight seal around the perimeter of an exterior door opening, as shown in Figure 2-18c. A variable speed fan mounts through the fabric panel.

When the fan in the door panel is operated, a negative air pressure is created within the building. This causes outside air to move into the building (e.g., infiltrate) through any possible leakage paths. Infiltrating air can be detected by a **smoke pencil** held near suspected leakage paths such as the perimeter of doors and windows, electrical junction boxes or pipe penetrations, or seams and cracks in surfaces. The smoke pencil emits an inert smoke that responds to the slightest air movement. In many cases, the detected leakage paths can be repaired using sealants, tapes, or similar measures. This is especially true when the building is under construction. Ideally, a blower door test is performed when the thermal envelope is completed, but the wall, ceiling, and floor finishes have not yet been installed. This allows good access to detect and repair leakage paths. The blower door can then be used to assess the degree to which the leakage paths have been reduced or eliminated.

Some blower doors are equipped with instruments that precisely measure the differential pressure between outside and inside the building. The overall air tightness of the building is determined by increasing the fan speed until this differential pressure reaches 50 Pascals, which is approximately 0.2 inches of water column pressure. This creates a condition that approximates a 20-mph wind striking all surfaces of the building simultaneously. The air changes per hour leakage rate at 50 Pascals differential pressure can then be determined based on the fan speed and house volume. An average house without special air leakage detailing could have upward of 15 air changes per hour at a 50 Pascal differential pressure. An **EnergyStar-certified** house requires an air leakage rate less than or equal to four air changes per house at 50 Pascals. An ultra-tight house built to the German **PassivHaus standard** must have a tested air leakage rate not exceeding 0.6 air changes per hour at 50 Pascals differential pressure.

Estimating Infiltration Heat Loss

Air leakage rates can be converted into rates of heat loss using Equation 2.8:

Equation 2.8:

$$Q_i = 0.018(n)(v)(\Delta T)$$

where,
- Q_i = estimated rate of heat loss due to air infiltration (Btu/h)
- n = number of air changes per hour, estimated based on air sealing quality (1/h)
- v = interior volume of the heated space (room or entire building) (ft^3) 0.018 = heat capacity of air (Btu/ft^3/°F)
- ΔT = inside air temperature minus the outside air temperature (°F)

Example 2.7
Determine the rate of heat loss by air infiltration for a 30-foot by 56-foot single-story building with 8-foot ceiling height and average quality air sealing. The inside temperature is 70 °F. The outside temperature is 10 °F.

Solution:
The floor area of this building is 30 × 56 = 1,680 ft². Referring to Figure 2-18a for a 1,680-square-foot *single-story* building with average air sealing quality, the suggested air leakage rate is 0.38 air changes per hour. Substituting this value along with the other data into Equation 2.8 yields:

$$\begin{aligned} Q_i &= 0.018(n)(n)(\Delta T) \\ &= 0.018(0.38)(30 \times 56 \times 8)(70 - 10) \\ &= 5{,}516 \text{ Btu/h} \end{aligned}$$

Discussion:
As the conduction heat losses of a building are reduced, air infiltration becomes a larger percentage of total heat loss. When higher insulation levels are specified for a building, greater efforts at reducing air infiltration should also be undertaken. These include the use of high-quality windows and doors, close attention to caulking, sealed infiltration barriers, and so on.

2.6 Example of Complete Heating Load Estimate

Finding the total design heat loss of a room is simply a matter of calculating both the conduction heat loss and the infiltration heat loss and then adding them together. When this is done on a room-by-room basis, the resulting numbers can be used to size the heat emitters as the hydronic distribution system is designed. The sum of the room heat losses—the total building design heat loss—helps determine the required heat output of the heat source.

This section demonstrates a complete heat loss estimate for a small house having the floor plan shown in Figure 2-19. Simple tables show the results of using the equations present earlier in this chapter. Other assumed values for the calculations are as follows:

- Unit *R*-value for all windows: *R*-3.0
- Unit *R*-value for exterior door: *R*-5
- *R*-value of foundation edge insulation: *R*-10
- Rate of air infiltration in vestibule and utility room: 1.0 air change/h
- Rate of air infiltration in all other rooms: 0.5 air change/h
- Outdoor design temperature: −5 °F
- Desired indoor temperature: 70 °F
- Window height: 4 feet
- Exterior door height: 6 feet 8 inches
- Wall height: 8 feet
- Light/dry soil under slab

Determining the Total *R*-Value of Thermal Envelope Surfaces

Walls: The wall cross-section is shown in Figure 2-20. It is made up of 2 × 6 wood framing spaced 24 inches on center with *R*-21 fiberglass batt insulation in the stud cavities. The inside finish is 1/2-inch drywall. The outside of the wall is sheathed with 1/2-inch plywood, and covered with a Tyvek® infiltration barrier and vinyl siding.

2.6 Example of Complete Heating Load Estimate

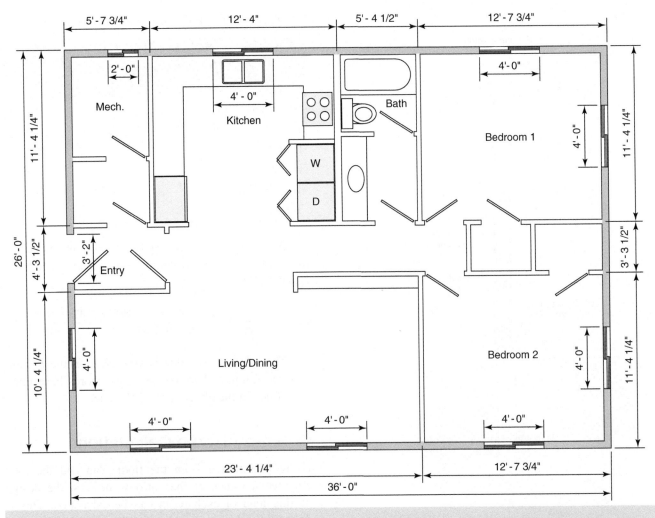

Figure 2-19 | Floor plan of example house for which heating load will be estimated.

The R-values of the materials are as follows:

Material	R-value between framing	R-value at framing
Inside air film	0.68	0.68
1/2-inch drywall	0.45	0.45
Stud cavity	21.0	5.5
1/2-inch plywood sheathing	0.62	0.62
TYVEK® infiltration barrier	~0	~0
Vinyl siding	0.61	0.61
Outside air film	0.17	0.17
Total	23.5	8.03

Assuming 15% of wall is solid framing, the effective total R-value of the wall is determined using Equation 2.4:

$$R_{\text{effective}} = \frac{(R_i)(R_f)}{p(R - R_f) + R_f} = \frac{(23.53)(8.03)}{0.15(23.53 - 8.03) + 8.03} = 18.3$$

Ceilings: The ceiling consists of 1/2-inch drywall covered with approximately 12 inches of blown fiberglass insulation. A cross-section is shown in Figure 2-21. The roof trusses displace a minor amount of this insulation.

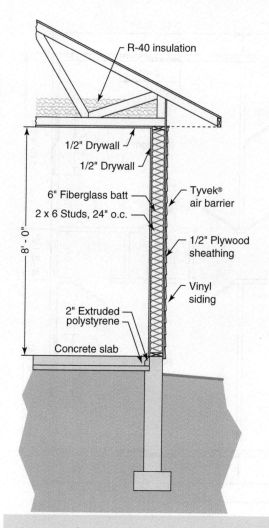

Figure 2-20 | Wall cross-section for example house.

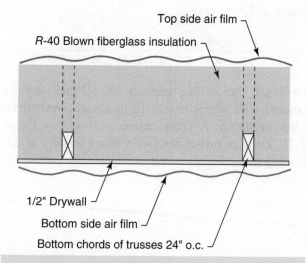

Figure 2-21 | Cross-section of ceiling in example house.

The R-values of the materials are as follows:

Material	R-value between framing	R-value at framing
Bottom air film	0.61	0.61
1/2-inch drywall	0.45	0.45
Insulation	40	28.3
Framing	0	3.5
Top air film	0.61	0.61
Total	41.67	33.47

Assuming 10% of ceiling is solid framing, the effective total R-value of the ceiling is also determined using Equation 2.4:

$$R_{\text{effective}} = \frac{(R_i)(R_f)}{p(R_i - R_f) + R_f} = \frac{(41.67)(33.47)}{0.10(41.67 - 33.47) + 33.47} = 40.7$$

This effective R-value is very close to the R-value between framing. The wooden truss chords have minimal effect on the effective total R-value.

Room-by-Room Calculations

Using information from the floor plan and the total effective R-values of the various surfaces, the design heating load of each room can be calculated. This is done by breaking out each room from the building and determining all necessary information, such as areas of walls, and windows. This information will be entered into tables to keep it organized.

To expedite the process, many of the dimensions are shown on the floor plan of the individual rooms. If such were not the case, the dimensions would have to be estimated using an architectural scale along with the printed floor plan. If the floor plan were available as a CAD file, dimensions could also be determined using the dimensioning tool of the CAD system without having to print the drawing. As a matter of convenience, the room dimensions are taken to the centerline of the common walls.

When working with architectural dimensions, it is best to convert inches and fractions of inches into decimal feet since many dimensions will be added and multiplied together. Example 2.8 illustrates this.

Example 2.8

Convert the dimension 10 ft 4 1/4 in to decimal feet.

Solution:

$$10 \text{ ft } 4\tfrac{1}{4} \text{ in} = 10 \text{ ft} + \left(\frac{4.25}{12}\right) \text{ft} = 10 \text{ ft} + 0.35 \text{ ft} = 10.35 \text{ ft}$$

Discussion:

The conversion from feet and inches into decimal feet has been performed before the calculation of areas and lengths in Figures 2-22 through 2-26.

Total Building Heating Load

The total building heating load is obtained by summing the heating loads of each room.

Living/dining room	4,384 Btu/h
Bedroom 1	3,102 Btu/h
Bedroom 2	3,102 Btu/h
Kitchen	2,278 Btu/h
Bathroom	688 Btu/h
Mechanical room/entry	2,533 Btu/h
Total	**16,087 Btu/h**

A number of simplifying assumptions were used in this example.

- The rooms were divided at the centerline of the common partitions. This ensures that all the exterior wall area is assigned to individual rooms and not ignored as being between the rooms.

- Since the two bedrooms are essentially identical in size and construction, it was only necessary to find the heat load of one, then double it when determining the total building load.

- The exterior dimensions of the rooms were used to determine areas, volumes, and so forth. This makes the calculations somewhat conservative because the exterior wall area of a room is slightly greater than the interior wall area.

- The areas and volumes of the closets between the bedrooms were equally divided up between the two bedrooms. Since a separate heat emitter would not be supplied for each closet, this part of the load is simply assigned to the associated bedrooms.

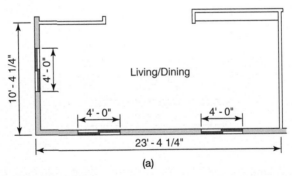

(a)

Gross exterior wall area = (10.35 ft + 23.35 ft)(8 ft) = (33.708 ft)(8 ft) = 269.7 ft²
Window area = 3 (4 ft × 4 ft) = 48 ft²
Exterior door area = 0 ft²
Net exterior wall area = Gross exterior wall area − Window and door area = 221.7 ft²
Ceiling area = (10.35 ft)(23.35 ft) = 241.7 ft²
Exposed slab edge length = (10.35 ft + 23.35 ft) = 33.7 linear ft
Room volume = (10.35 ft)(23.35 ft)(8.0 ft) = 1933.4 ft³

Room name		1 Exposed walls	2 Windows	3 Exposed doors	4 Exposed ceilings	5 Exposed floor	6 Exposed slab edge	7 Room volume	8 Air changes	9 Infiltration loss	10 Total room load
Living & dining	1	A = 221.7	A = 48	A = 0	A = 241.7	A = 0	L = 33.7	V = 1933	N = 0.5		
	2	R = 18.3	R = 3	R = n/a	R = 40.7	R = n/a	Re = 11				
	3	$\tfrac{A}{R}(\Delta T) =$ 909	$\tfrac{A}{R}(\Delta T) =$ 1200	$\tfrac{A}{R}(\Delta T) =$ 0	$\tfrac{A}{R}(\Delta T) =$ 445	$\tfrac{A}{R}(\Delta T) =$ 0	$Q_{slab} =$ 525			$Q_i =$ 1305	4384

(b)

Figure 2-22 | (a) Floor plan for living/dining room of example house. (b) Table with values for living/dining room of example house.

2.7 Computer-Aided Heating Load Calculations

After gathering the data and performing all the calculations needed for a building heating load estimate, it becomes apparent that a significant amount of information goes into finding the results. If the designer then decides to go back and change one or more of the numbers, more time is required to unravel the calculations back to the point of the change and calculate the new result. Even for people used to handling this amount of information, these calculations are tedious and always subject to errors. The time required to perform accurate heating load estimates is arguably the chief reason why some "heating professionals" do not do them.

Like many routine design procedures, heating load calculations are now commonly done using computers, and sometimes even on smartphones. The advantages of using load estimating software are many:

- Rapid (almost instant) calculations to quickly study the effect of changes
- Automated referencing of R-values, and weather data, in some programs
- Significantly less chance of error due to number handling
- Ability to print professional reports for customer presentations
- Ability to store project files for possible use in similar future projects
- Ability to import area and R-value information directly from CAD-based architectural drawings

There are many software packages currently available for estimating the heating loads for residential and light commercial buildings. They vary in cost from free to

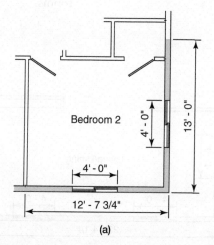

(a)

Gross exterior wall area = (12.65 ft + 13.0 ft)(8 ft) = (25.65 ft)(8 ft) = 205.2 ft²
Window area = 2 (4 ft × 4 ft) = 32 sq. ft
Exterior door area = 0 ft²
Net exterior wall area = Gross exterior wall area − Window and door area = 173.2 ft²
Ceiling area = (12.65 ft)(13.0 ft) = 164.5 sq. ft
Exposed slab edge length = (12.65 ft + 13.0 ft) = 25.65 linear ft
Room volume = (12.65 ft)(13.0 ft)(8.0 ft) = 1315.6 ft³

Room name		1 Exposed walls	2 Windows	3 Exposed doors	4 Exposed ceilings	5 Exposed floor	6 Exposed slab edge	7 Room volume	8 Air changes	9 Infiltration loss	10 Total room load
Bedroom 2	1	A = 173.2	A = 32	A = 0	A = 164.5	A = 0	L = 25.7	V = 1316	N = 0.5		
	2	R = 18.3	R = 3	R = n/a	R = 40.7	R = n/a	Re = 11				
	3	$\frac{A}{R}(\Delta T)=$ 710	$\frac{A}{R}(\Delta T)=$ 800	$\frac{A}{R}(\Delta T)=$ 0	$\frac{A}{R}(\Delta T)=$ 303	$\frac{A}{R}(\Delta T)=$ 0	$Q_{slab}=$ 401			$Q_i =$ 888	3102

(b)

Figure 2-23 (a) Floor plan for Bedroom 2 of example house. (b) Table with values for Bedroom 2 of example house.

over $500. The higher-cost programs usually offer more features such as material reference files, automated areas determination from architectural drawings, and customized output reports. The lower-cost programs offer no frills output, but can nonetheless still yield accurate results.

Figure 2-27a shows the main screen of a program named *Building Heat Load Estimator*, which is part of the more comprehensive software tool *Hydronics Design Studio 2.0*, which was co-developed by the author. This software allows the user to build a list of rooms, each of which is defined based on its thermal envelope. The total heat loss of each room as well as the overall building is continually displayed as the user interacts with the software.

Building Heat Load Estimator allows the user to define the thermal envelope surfaces with each room, using pull-down menus for materials. It also provides a point-and-click selection of common wall, ceiling, floor, window and door assemblies. An example of how a framed wall is specified is shown in Figure 2-27b.

Building Heat Load Estimator also assists the user in determining surface areas and room volumes, as shown in Figure 2-27c.

A free demo version of this software can be downloaded at **www.hydronicpros.com**.

2.8 Estimating Annual Heating Energy Use

Although heating system designers need to know the design heat loss of each room in a building before sizing a heating system, such information is often meaningless to building owners. Their interest usually lies in what it will cost to heat their building. Such estimates can be made once the building's design heating load and the seasonal efficiency of the heat source are established.

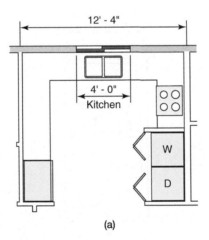

(a)

Gross exterior wall area = (12.65 ft + 13.0 ft)(8 ft) = (25.65 ft)(8 ft) = 205.2 ft²
Window area = 2 (4 ft × 4 ft) = 32 sq. ft
Exterior door area = 0 ft²
Net exterior wall area = Gross exterior wall area − Window and door area = 173.2 ft²
Ceiling area = (12.65 ft)(13.0 ft) = 164.5 sq. ft
Exposed slab edge length = (12.65 ft + 13.0 ft) = 25.65 linear ft
Room volume = (12.65 ft)(13.0 ft)(8.0 ft) = 1315.6 ft³

Room name		1 Exposed walls	2 Windows	3 Exposed doors	4 Exposed ceilings	5 Exposed floor	6 Exposed slab edge	7 Room volume	8 Air changes	9 Infiltration loss	10 Total room load
Kitchen	1	$A=82.6$	$A=16$	$A=0$	$A=135.6$	$A=0$	$L=12.3$	$V=1085$	$N=0.5$		↓
	2	$R=18.3$	$R=3$	$R=\text{n/a}$	$R=40.7$	$R=\text{n/a}$	$Re=11$				
	3	$\frac{A}{R}(\Delta T)=$ 339	$\frac{A}{R}(\Delta T)=$ 400	$\frac{A}{R}(\Delta T)=$ 0	$\frac{A}{R}(\Delta T)=$ 250	$\frac{A}{R}(\Delta T)=$ 0	$Q_{slab}=$ 401			$Q_i=$ 888	2278

(b)

Figure 2-24 | (a) Floor plan for kitchen of example house. (b) Table with values for kitchen of example house.

Degree Day Method

The seasonal energy used for space heating depends upon the climate in which the building is located. A relatively simple method for factoring local weather conditions into estimates of heating energy use is the concept of heating degree days. *The number of heating degree days that accumulate in a 24-hour period is the difference between 65 °F and the average outdoor air temperature during that period* (see Equation 2.9). The average temperature is determined by averaging the high and low temperature for the 24-hour period.

Equation 2.9:

$$DD_{daily} = (65 - T_{ave})$$

where,

DD_{daily} = daily degree days accumulated in a 24-hour period

T_{ave} = average of the high and low outdoor temperature for that 24-hour period

The total heating degree days for a month, or an entire year, is found by adding up the *daily* heating degree days over the desired period. Many heating reference books contain tables of monthly and annual heating degree days for many major cities in the United States and Canada. A sample listing of annual degree days is given in Figure 2-28.

Heating degree day data has been used to estimate fuel usage for several decades. Such data is often recorded by fuel suppliers, utility companies, and local weather stations. Degree day statistics are also frequently listed in the weather section of newspapers as well as on Web-based weather reports.

When the heating degree day method was first developed, fuel was inexpensive and buildings were poorly insulated. Building heat loss was substantially greater than for comparably sized buildings using modern construction techniques.

Estimating seasonal energy consumption using heating degree days assumes that buildings need heat input from their heating systems whenever the outside

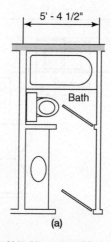

(a)

Gross exterior wall area = (5.38 ft)(8 ft) = 43.0 ft²
Window area = 0 ft²
Exterior door area = 0 ft²
Net exterior wall area = Gross exterior wall area − Window and door area = 43.0 ft²
Ceiling area = (5.38 ft)(11.0 ft) = 59.2 ft²
Exposed slab edge length = 5.38 linear ft
Room volume = (5.38 ft)(11.0 ft)(8.0 ft) = 473 ft³

Room name		1 Exposed walls	2 Windows	3 Exposed doors	4 Exposed ceilings	5 Exposed floor	6 Exposed slab edge	7 Room volume	8 Air changes	9 Infiltration loss	10 Total room load
Bathroom	1	A = 43	A = 0	A = 0	A = 59.2	A = 0	L = 5.38	V = 473	N = 0.5		↓
	2	R = 18.3	R = 3	R = n/a	R = 40.7	R = n/a	Re = 11				
	3	$\frac{A}{R}(\Delta T) =$ 176	$\frac{A}{R}(\Delta T) =$ 0	$\frac{A}{R}(\Delta T) =$ 0	$\frac{A}{R}(\Delta T) =$ 109	$\frac{A}{R}(\Delta T) =$ 0	$Q_{slab} =$ 84			$Q_i =$ 319	688

(b)

Figure 2-25 | (a) Floor plan for bathroom of example house. (b) Table with values for bathroom of example house.

temperature drops below 65 °F. While this may be true for many poorly insulated buildings, it is often incorrect for better-insulated buildings. Many energy efficient homes can maintain comfortable interior temperatures, without heat input from their space heating systems, even when outdoor temperatures drop into the 40 °F range. This is a result of greater internal heat gains from appliances, lights, people, and sunlight combined with much lower rates of heat loss due to better insulation and reduced air leakage. Internal heat gains eliminate the need for a corresponding amount of heat from the building's heating system. Passive solar buildings are a good example. On sunny days, some of these buildings require no heat input from their "auxiliary" heating systems, even when outside temperatures are below 0 °F!

Another complicating factor is that some heating degree days occur in months when most heating systems are turned off. This is especially true in northern climates where some heating degree days are recorded in June, July, and even August. Obviously, if the heating system is turned off, there will be no fuel consumption associated with these degree days. Still, any degree days that occur in these summer months are included in the annual total.

To compensate for the effect of internal heat gains, and warm-weather degree days, ASHRAE devised a correction multiplier known as **Cd factor**. This factor is plotted against total heating degree days in Figure 2-29 and will be used to estimate seasonal heating energy use. The Cd value given by this graph is representative of average conditions over a large number of buildings. The specific Cd value that applies to a given building might be higher or lower than the value shown by the graph. For example, a passive solar building in a sunny climate such as Denver is likely to have a lower Cd value than given in Figure 2-29 because it may get a major part of its heating energy from the sun. On the other hand, a building with small windows, in a shaded location, or one with very little internal heat generation may have a larger value of Cd compared to the "typical" value listed

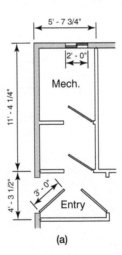

(a)

Gross exterior wall area = (4.29 ft + 11.35 ft + 5.65 ft)(8 ft) = (21.28 ft)(8 ft) = 170.2 ft²
Window area = (2 ft × 4 ft) = 8 ft²
Exterior door area = (6.66 ft × 3 ft) = 20 ft²
Net exterior wall area = Gross exterior wall area + Window and door area = 142.2 ft²
Ceiling area = (4.29 ft + 11.35 ft)(5.65 ft) = (15.64 ft × 5.65 ft) = 88.4 ft²
Exposed slab edge length = 21.28 linear ft
Room volume = (15.64 ft)(5.65 ft)(8.0 ft) = 707 ft³

Room name		1 Exposed walls	2 Windows	3 Exposed doors	4 Exposed ceilings	5 Exposed floor	6 Exposed slab edge	7 Room volume	8 Air changes	9 Infiltration loss	10 Total room load ↓
Mech. room & entry	1	A = 142.2	A = 8	A = 20	A = 88.4	A = 0	L = 21.3	V = 707	N = 1.0		
	2	R = 18.3	R = 3	R = 5	R = 40.7	R = n/a	Re = 11				
	3	$\frac{A}{R}(\Delta T)=$ 583	$\frac{A}{R}(\Delta T)=$ 200	$\frac{A}{R}(\Delta T)=$ 300	$\frac{A}{R}(\Delta T)=$ 163	$\frac{A}{R}(\Delta T)=$ 0	$Q_{slab}=$ 332			$Q_i=$ 955	2533

(b)

Figure 2-26 (a) Floor plan for mechanical room and entry of example house. (b) Table with values for mechanical room and entry of example house.

44 Chapter 2 Space Heating and Domestic Water-Heating Loads

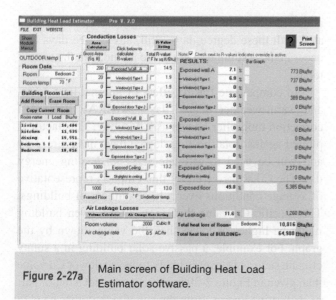

Figure 2-27a | Main screen of Building Heat Load Estimator software.

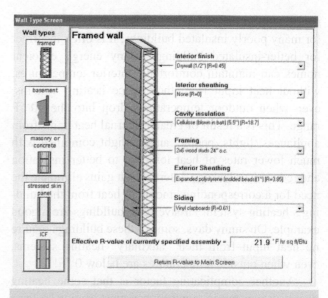

Figure 2-27b | Heat Load Pro screen for defining and specifying wall construction.

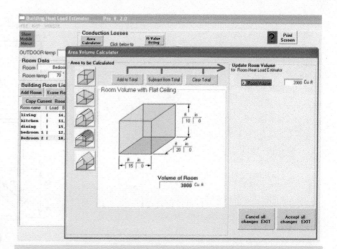

Figure 2-27c | Building Heat Load Estimator screen for assisting with room volume determination.

in Figure 2-29. This variation will directly affect the estimate of annual fuel usage.

Equation 2.10 can be used to estimate annual heating energy required by the building:

Equation 2.10:

$$E_{annual} = \frac{(Q_{design})(DD_{65})(24)(C_d)}{1,000,000(\Delta T_{design})}$$

where,

E_{annual} = estimated annual heating energy required by the building (MMBtu)

NOTE: 1 MMBtu = 1,000,000 Btu

Q_{design} = design heating load of the building (Btu/h)

DD_{65} = annual total heating degree days (base 65) at the building location (degree days)

C_d = correction factor from Figure 2-28 (unitless)

ΔT_{design} = design temperature difference at which design heating load was determined (°F)

Example 2.9

A house has a calculated design heating load of 55,000 Btu/h when its interior temperature is 68 °F, and the outdoor temperature is −10 °F. The house is located in a climate having 7,000 annual heating degree days. Estimate the annual heating energy that must be delivered to this building for a typical heating season.

Solution:

The Cd factor for a 7,000 heating degree day climate is estimated at 0.63 using Figure 2-29. This and the other given data are then substituted into Equation 2.10:

$$E_{annual} = \frac{(Q_{design})(DD_{65})(24)(C_d)}{1,000,000(\Delta T_{design})}$$

$$= \frac{(55,000)(7,000)(24)(0.63)}{1,000,000(68-(-10))} = 74.6 \text{ MMBtu}$$

2.8 Estimating Annual Heating Energy Use

CITY	°F days (base 65)
Atlanta, GA	2,990
Baltimore, MD	4,680
Boston, MA	5,630
Burlington, VT	8,030
Chicago, IL	6,640
Denver, CO	6,150
Detroit, MI	6,290
Madison, WI	7,720
Newark, NJ	4,900
New York, NY	5,219
Philadelphia, PA	5,180
Pittsburgh, PA	5,950
Portland, ME	7,570
Providence, RI	5,950
Seattle, WA	4,424
Syracuse, NY	6,720
Toronto, ONT	6,827

Figure 2-28 | Annual heating degree days for some major cities.

Discussion:

The building needs 74.6 MMBtu (or 74,600,000 Btus) delivered to it over a typical heating season. Keep in mind that *this calculation does not account for the inefficiencies of the heat source or the distribution system in producing and delivering this heat to the building.* These inefficiencies will make the *purchased energy requirement larger than the delivered energy requirement.* The lower the efficiency of the heat source, the higher the purchased energy requirement will be compared to the delivered energy requirement. These inefficiencies will be accounted for using methods covered next.

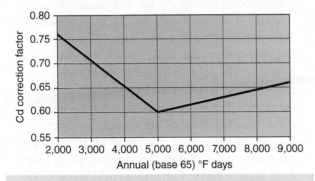

Figure 2-29 | Mean values of C_d versus annual heating degree days suggested by ASHRAE.

Estimating Annual Heating Cost

Knowing how many MMBtu a building requires over a typical heating season is not of much use to most owners. Their main concern is the cost of this energy as well as how various fuel and heat source options will affect this cost. To provide this information, the cost of available fuel options must be expressed on a "unit price" basis. The units commonly used are dollars per million Btu delivered (abbreviated $/MMBtu). The seasonal efficiency of the heat source is also factored when determining the unit price of delivered energy.

Once the unit price of the fuel options has been calculated, a simple multiplication determines the estimated seasonal heating cost in dollars.

The formulas given in Figure 2-30 can be used to convert the purchase costs of several common fuels to a unit price of dollars per million Btu delivered.

Some fuels are priced on "sliding scales." The cost per unit of fuel decreases as the monthly usage increases. This is typical of natural gas pricing from many utilities.

Electric Resistance Heat	____ cents / Kwhr x 2.93	= ____ $/MMBtu
Heat Pump	$\dfrac{\_\_\_\_ \text{cents / Kwhr x 2.93}}{\_\_\_\_ \text{average COP}}$	= ____ $/MMBtu
#2 Fuel Oil	$\dfrac{\_\_\_\_ \text{\$ / gallon x 7.14}}{\_\_\_\_ \text{AFUE (decimal)}}$	= ____ $/MMBtu
Propane	$\dfrac{\_\_\_\_ \text{\$ / gallon x 10.9}}{\_\_\_\_ \text{AFUE (decimal)}}$	= ____ $/MMBtu
Natural Gas	$\dfrac{\_\_\_\_ \text{\$ / therm x 10}}{\_\_\_\_ \text{AFUE (decimal)}}$	= ____ $/MMBtu
Firewood*	$\dfrac{\_\_\_\_ \text{\$ / face chord x 0.149}}{\_\_\_\_ \text{ave. efficiency (decimal)}}$	= ____ $/MMBtu
Wood Pellets	$\dfrac{\_\_\_\_ \text{\$ / ton x 0.06098}}{\_\_\_\_ \text{ave. efficiency (decimal)}}$	= ____ $/MMBtu

NOTES:
1. $/MMBtu = dollars per million Btu of heat delivered to building
2. Kwhr = kilowatt • hour = 3413 Btu
3. Average COP = average Coefficient Of Performance during heating season (for geothermal heat pump with low temp. distribution system generally 2.5 to 4.0)
4. AFUE = Annual Fuel Utilization Efficiency of appliance (for typical oil or gas boiler use 0.75 to 0.80)

* Assumes a 50/50 mix of maple and beech dried to 20% moisture content. Price is for 4 ft x 8 ft x 1 inch face chord split and delivered.

** Assumes 15% moisture content

Figure 2-30 | Equations for determining the unit price of various fuels based on purchase price and heat source efficiency.

Chapter 2 Space Heating and Domestic Water-Heating Loads

In such cases, use a weighted average price per unit of fuel based on usage.

In many cases, utilities assess a minimum service charge regardless of the amount of energy used. One can think of this as "meter rental." Since it has to be paid regardless of how much energy is purchased, this charge should be deducted from the bill before calculating the purchase price of the energy. However, such minimum service charges should be considered when weighing the merit of using a utility-supplied energy versus another fuel that might be available without a minimum monthly service charge. For example, if an owner decides to use natural gas supplied from a utility for space heating and does not plan to use it for other appliances, the minimum service charges associated with that fuel should be added to the estimated cost of natural gas consumed. The total cost should then be compared to that of other fuel options. Similarly, any known service fees or other fixed costs associated with a given fuel should be included in the total cost associated with that fuel when making comparison.

Example 2.10

Estimate the annual space heating cost of a building requiring 100 MMBtu/y. for the following fuel options and associated thermal conversion efficiencies.

- Electric resistance heat at $0.12 per kilo-watt-hour and 100% efficiency
- Electrically operated heat pump with seasonal average COP = 2.5
- #2 fuel oil at $2.25 per gallon and 83% annual heat source/delivery efficiency
- Propane at $2.75 per gallon and 83% annual heat source/delivery efficiency
- Natural gas at $1.25 per therm and 83% annual heat source/delivery efficiency
- Firewood at $60 per face chord and 60% annual heat source/delivery efficiency
- Wood pellets at $275 per ton and 83% annual heat source/delivery efficiency

Solution:

Substituting these numbers into the formulas of Figure 2-30 yields the results shown in Figure 2-31.

The total estimated heating cost for the building requiring 100 MMBtu/season, and based on the fuel/heat source combinations listed, are as follows:

- Electric resistance heat:
  100 MMBtu × 35.16 $/MMBtu = $3516/season
- Heat pump:
  100 MMBtu × 14.06 $/MMBtu = $1406/season

Electric Resistance Heat	12 cents / Kwhr × 2.93	= 35.16 $/MMBtu
Heat Pump	(12 cents / Kwhr × 2.93) / 2.5 average COP	= 14.06 $/MMBtu
#2 Fuel Oil	(3.50 $/gallon × 7.14) / 0.83 AFUE (decimal)	= 30.11 $/MMBtu
Propane	(2.75 $/gallon × 10.9) / 0.83 AFUE (decimal)	= 36.11 $/MMBtu
Natural Gas	(1.25 $/therm × 10) / 0.83 AFUE (decimal)	= 15.06 $/MMBtu
Firewood*	(60 $/face chord × 0.149) / 0.60 ave. efficiency (decimal)	= 14.90 $/MMBtu
Wood Pellets	(275 $/ton × 0.06098) / 0.83 ave. efficiency (decimal)	= 20.20 $/MMBtu

Figure 2-31 | Completed fuel cost worksheet

- System using #2 fuel oil:
  100 MMBtu × 19.36 $/MMBtu = $1936/season
- System using propane:
  100 MMBtu × 36.11 $/MMBtu = $3611/season
- System using natural gas:
  100 MMBtu × 15.06 $/MMBtu = $1506/season
- System using firewood:
  100 MMBtu × 14.90 $/MMBtu = $1490/season
- System using wood pellet:
  100 MMBtu × 20.20 $/MMBtu = $2020/season

Discussion:

The values used for the cost of each fuel and combustion efficiencies are not meant to skew opinions. Fuel costs vary widely, both regionally and continually with time. Sociopolitical events, and supply versus demand, can quickly change the relative rankings of commodity fuels such as oil, propane, and natural gas. The numbers used, and the results obtained from these numbers, are simply a "snapshot" based on specific assumptions. It's also worth noting that these costs are independent of the device used to convert the fuel into heat. Thus, a boiler, furnace, or small space heater operating on the same fuel, at the same assumed conversion efficiency, would produce the same estimated annual heating cost.

2.9 Estimating Domestic Water-Heating Loads

Domestic water heating is one of the largest energy consumers in a typical North American home. In northern climates, it is usually second only to the energy required for space heating. In southern climates, it is typically second behind space cooling energy use.

Many of the hydronic heating systems discussed in later chapters are configured to provide energy for both space heating and domestic water heating. The design of such systems must therefore factor in the thermal load for the latter. This section presents simple methods for estimating the total daily energy required for domestic water heating as well as the energy required during the hour of peak demand.

Daily Domestic Hot Water (DHW) Load Estimates

The daily energy requirement for heating domestic water can be estimated using Equation 2.11:

Equation 2.11:

$$E_{daily} = (G)(8.33)(T_{hot} - T_{cold})$$

where,
- E_{daily} = daily energy required for DHW production (Btu/day)
- G = volume of domestic hot water required per day (gallons)
- T_{hot} = hot water temperature supplied to the fixtures (°F)
- T_{cold} = cold water temperature supplied to the water heater (°F)

The daily usage of domestic hot water is very dependent on occupancy, living habits, type of water fixtures used, water pressure, and the time of year. The following estimates are suggested as a guideline:

- House: 10 to 20 gallons per day per person
- Office building: 2 gallons per day per person
- Small motel: 35 gallons per day per unit
- Restaurant: 2.4 gallons per average number of meals per day

Example 2.11

Determine the energy used for domestic water heating for a family of four with average usage habits. The cold water temperature averages 55 °F. The supply temperature is set for 125 °F. The assumed "average" domestic hot water usage will be 15 gallons per day per person.

Solution:

The daily domestic water-heating load is estimated using Equation 2.11:

$$E_{daily} = (4 \times 15)(8.33)(125 - 55) = 34{,}990 \text{ Btu/day}$$

The estimated annual usage would be the daily usage times 355 days of occupancy per year (assuming the family is away 10 days per year).

$$\left(34{,}990 \frac{\text{Btu}}{\text{day}}\right)\left(355 \frac{\text{occupied days}}{\text{year}}\right) = 12{,}421{,}000 \frac{\text{Btu}}{\text{year}} = 12.4 \frac{\text{MMBtu}}{\text{year}}$$

Discussion:

The cost of providing this energy can be estimated by multiplying the annual energy requirement by the cost of delivered energy in $/MMBtu, as established using methods from earlier in this section. For example, at $0.10/kwhr and 96% efficiency, electrical energy has a delivered cost of $30.52/MMBtu. The annual energy cost of providing the domestic hot water in this example would be:

$$\left(12.4 \frac{\text{MMbtu}}{\text{year}}\right)\left(30.52 \frac{\$}{\text{MMBtu}}\right) = \$378/\text{year}$$

The efficiency of 96% is typical for modern storage type electric water heater. It implies that 96% of the electrical energy used by the device leaves as heated water. The other 4% leaves as standby heat loss from the tank's jacket and attached piping.

DHW Usage Patterns

The rate at which domestic hot water is used at the fixtures within a building varies considerably with the type of occupancy, and time of day. A typical residential **domestic hot water usage profile** is shown in Figure 2-32.

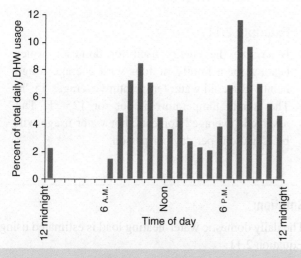

Figure 2-32 | Typical usage profile for residential domestic hot water.

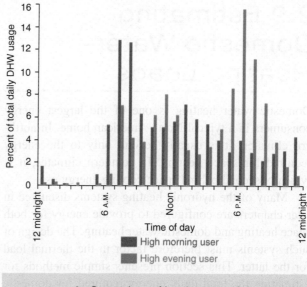

Figure 2-33 | Comparison of hourly usage of domestic hot water for a high morning user versus a high evening user.

Notice the two distinct periods of high demand: one during the wake-up period, the other during the early evening. The greatest demand in this specific profile is from 7 p.m. to 8 p.m. However, this still represents only 11.6% of the total daily DHW demand. For a family using 60 gallons of domestic hot water per day, this would represent a peak hourly usage of just under 7 gallons.

The *average* rate at which energy is required for DHW production during this peak hour (assuming cold water and hot water temperatures of 55 °F and 125 °F, respectively) can be found using Equation 2.11 modified for a single peak hour:

$$E_{peakhour} = [(0.116)(60)](8.33)(125 - 55)$$
$$= 4,058 @ 4,100 \text{ Btu/h}$$

The first factor in this equation, (0.116), represents the peak demand of 11.6% of the total daily water volume.

The occurrence of two peak demand periods, one during the wake-up period, the other following the evening meal period, is common in residential buildings. Some households tend to be "high morning users" while others are "high evening users." Figure 2-33 compares these two categories.

Peak Hourly Demand for DHW

Another important aspect of domestic water-heating loads is the **peak hourly demand**. This is simply how much domestic hot water is required, in gallons, during the hour of highest demand.

Peak hourly demand can be estimated by first identifying the hour of probable highest demand for hot water based on expected usage habits. This varies considerably from one household or building to another.

For design purposes, it can be established by discussing likely usage habits with occupants. In existing buildings, it can also be established by data logging domestic hot water flow versus time of day, over a period of several typical days.

If it is not possible to discuss or monitor domestic hot water usage habits with the occupants for whom the system is being designed, the table shown in Figure 2-34 can also be used to estimate peak hourly demand.

Figure 2-34 is based on statistics of *typical* usage as published by the U.S. Department of Energy. If the building has water-conserving showerheads and faucets, usage may be less. Likewise, if the building has high-flow fixtures, usage could be higher.

Figure 2-35 shows an example of how the table in Figure 2-34 is used.

Knowing the likely peak hourly demand assists in selecting a domestic water-heating device. All electric, gas, and oil-fired storage type water heaters sold in the United States have a **first-hour rating**. This rating, established by independent testing, is the number of gallons of domestic hot water the device can supply over one hour, starting from a normal heated standby condition. The first-hour rating includes the contribution of storage volume as well as the heating capacity of the burner or electric element. The gallons of hot water supplied as the first-hour rating are based on a 90 °F temperature rise between the incoming cold water and the leaving hot water.

Once the peak hourly demand for domestic hot water is established, common practice is to select a water-heating device with a first-hour rating that equals or slightly exceeds the peak hourly demand.

2.9 Estimating Domestic Water-Heating Loads

Use	Average gallons of hot water per use		Number of uses during peak hour		Gallons used during peak hour
Shower	12	×		=	
Bath	9	×		=	
Shaving	2	×		=	
Hands and face washing	4	×		=	
Hair shampoo	4	×		=	
Hand dishwashing	4	×		=	
Automatic dishwashing	14	×		=	
Food preparation	5	×		=	
Automatic clothes washer	32	×		=	
			Total Peak Hour Demand	=	

Figure 2-34 | DOE table for estimating peak hour demand for domestic hot water.

	Average gallons of hot water per use		Number of uses during peak hour		Gallons used during peak hour
Shower	12	×	3	=	36
Bath	9	×	0	=	0
Shaving	2	×	1	=	2
Hands and face washing	4	×	0	=	0
Hair shampoo	4	×	1	=	4
Hand dishwashing	4	×	1	=	4
Automatic dishwashing	14	×	0	=	0
Food preparation	5	×	0	=	0
Automatic clothes washer	32	×	0	=	0
			Total Peak Hour Demand	=	46

Figure 2-35 | Example of how the table in Figure 2-32 is used to estimate peak hourly demand.

Summary

This chapter has presented straightforward methods for estimating the design space heating load and domestic water-heating load of residential and light commercial buildings. Load estimating is an essential part of overall system design. It provides the foundation for all subsequent equipment selection.

Key Terms

97.5 percent design dry bulb temperature
air change method
air infiltration heat loss
blower door
building heat load estimator
building design heating load
Cd factor
conduction heat loss
conduction
degree days
design heating load
domestic hot water usage profile
effective total R-value
EnergyStar-certified
exposed surface
finite element analysis
first-hour rating

50 Chapter 2 Space Heating and Domestic Water-Heating Loads

Hydronics Design Studio 2.0
infiltration heat loss
inside air film
isotherms
outside air film
PassivHaus standard
peak hourly demand
R-value
room design heating load
smoke pencil
thermal conductivity
thermal envelope
thermal load
thermal resistance
total R-value
unit U-value

Questions and Exercises

1. One wall of a room measures 14 feet long and 8 feet high. It contains a window 5 feet wide and 3.5 feet high. The wall has an effective total R-value of 15.5. Find the rate of heat flow through the wall when the inside air temperature is 68 °F and the outside temperature is 5 °F.

2. Using the data in Appendix B, determine the total effective R-value of the wall shown in Figure 2-36. Assume that 15% of the wall area is wood framing.

3. The rate of heat flow through a wall is 500 Btu/h when the inside temperature is 70 °F and the outside temperature is 10 °F. What is the rate of heat flow through the same wall when the outside temperature drops to −10 °F and the inside temperature remains the same?

4. The perimeter of the basement of a house measures 120 feet. The masonry wall is 8 feet tall with 1 foot exposed above grade. The wall is insulated with 2 inches of extruded polystyrene insulation. The basement air temperature is maintained at 65 °F. Determine the rate of heat loss through the basement walls when the outside air temperature is 10 °F. Assume the R-value of the masonry wall is 2.0.

5. Determine the rate of heat loss through the slab-on-grade floor shown in Figure 2-37. The exposed edges are insulated with 1.5 inches of extruded polystyrene. The outside air temperature is 0 °F and the slab is maintained at an average temperature of 90 °F by embedded tubing.

6. Determine the design heat loss of the room shown in Figure 2-38. Assume the exterior walls are constructed as shown in Figure 2-36. Assume the outdoor design temperature is −10 °F and the desired indoor temperature is 70 °F.

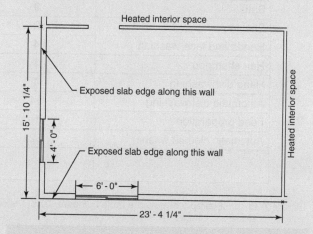

Figure 2-37 | Floor slab for Exercise 5.

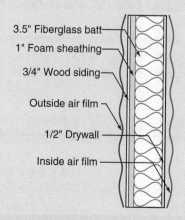

Figure 2-36 | Wall cross-section for Exercise 2.

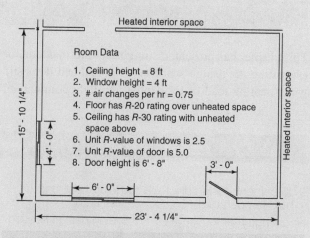

Figure 2-38 | Room floor plan for Exercise 6.

7. Using the construction information given for the example house in Section 2.6, determine the design heat loss of the entire house shown in Figure 2-39. Assume all floors are slab on grade. Estimate any dimensions you need by proportioning it to the dimensions shown.

8. Estimate the cost of domestic water heating for a typical family of five assuming cold water enters the system at 45 °F and is supplied to the fixtures at 120 °F. Assume the water is heated by an electric water heater in an area where electricity costs $0.12/kwhr.

9. What is the difference in cost to heat 60 gallons of water from 50 to 70 °F compared to heating the same amount of water from 100 to 120 °F?

10. It is determined that a family uses domestic hot water as shown in Figure 2-33. Assuming a total daily domestic hot water usage of 80 gallons, a cold water temperature of 50 °F, and a delivery temperature of 120 °F, how much energy (in Btus) is required to meet their peak hourly demand for domestic hot water?

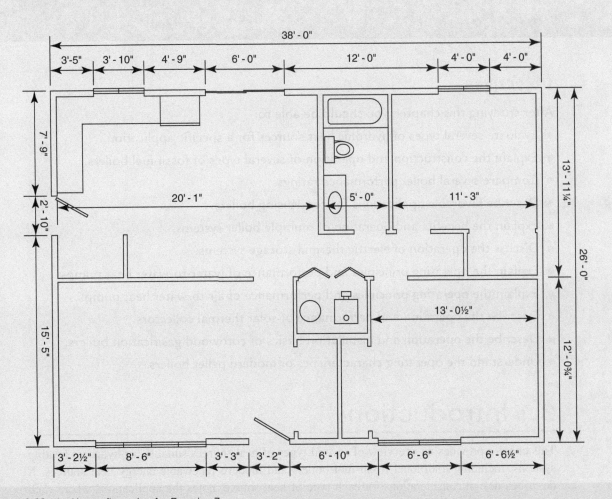

Figure 2-39 | House floor plan for Exercise 7.

Chapter 3

Hydronic Heat Sources

Objectives

After studying this chapter, you should be able to:

- Evaluate several types of hydronic heat sources for a specific application.
- Explain the construction and operation of several types of fossil-fuel boilers.
- Compare several boiler performance ratings.
- Describe favorable applications for condensing boilers.
- Explain the benefits and operation of multiple boiler systems.
- Discuss the operation of electric thermal storage systems.
- Explain the operating principles and performance of water-to-water heat pumps.
- Explain the operating principles and performance of air-to-water heat pumps.
- Describe the operation and performance of solar thermal collectors.
- Describe the operation and installation basics of cordwood gasification boilers.
- Understand the operating characteristics of modern pellet boilers.

3.1 Introduction

This chapter provides an overview of several types of heat sources suitable for hydronic heating systems, including traditional fossil-fuel boilers and modern renewable energy heat sources. It describes design considerations for each type of heat source, notes the applications where each has certain technical or economic advantages, and shows basic application schematics. Related topics such as exhaust handling, combustion air supply, and methods of piping are also discussed. Several performance ratings for heating output and efficiency are also presented.

This chapter is not meant to be an exhaustive reference on each type of heat source. The details necessary for proper installation of a specific device should be obtained directly from the manufacturer. Sometimes, the mechanical equipment codes or other ordinances in effect within a municipality will influence both equipment selection and installation. For this reason, always check with local code officials for any specific limitations affecting a proposed hydronic heat source.

Finally, although the heat source is certainly a major piece of equipment, it is still only part of the system. Even the best heat source, when misapplied, will yield disappointing performance or fail prematurely. As you read about the different heat sources, pay particular attention to temperature and flow rate limitations that ultimately affect how each type of heat source interfaces with the balance of the system.

3.2 Classification of Hydronic Heat Sources

Within this chapter, hydronic heat sources are classified as follows:

- Conventional gas- and oil-fired boilers
- Condensing gas-fired boilers
- Electric resistance boilers
- Electric thermal storage (ETS) equipment
- Water-to-water heat pumps
- Air-to-water heat pumps
- Solar thermal collectors
- Cordwood gasification boilers
- Pellet-fueled boilers

Of these, conventional gas- and oil-fired boilers are currently used in the majority of residential and light commercial hydronic systems in North America. This is due to several factors, including:

- the relatively low unit price and availability of natural gas in population centers;
- established markets for other fossil fuels such as propane and fuel oil in rural locations;
- sales, installation, and service professionals who are experienced with fossil fuel boilers;
- greater selection of boiler options compared to those available for other hydronic heat sources.

However, technical improvement, government and utility incentive programs, and political and cultural trends over the past two decades have led to increased use of renewable energy heat sources in hydronic heating and cooling systems. These trends are likely to continue. The foreseeable future for hydronics technology involves application of both traditional and renewable heat sources. In some cases, both traditional and renewable heat sources will be used in the same system. In other cases, the technical, environmental, and financial advantages of a specific heat source make it a compelling choice.

In many areas, the unit cost of heat produced by passing electricity through resistance heating elements is significantly higher than the unit cost of heat produced from fossil fuels. This has limited the use of electric boilers in areas where less expensive fuels are readily available. However, buildings serviced by electric utilities with low kilowatt-hour rates, or **time-of-use rates**, can be good candidates for electrically powered heat sources. Electric boilers are also well suited to homes with very low heating loads, where the life-cycle cost associated with their use is comparable to that of other fuel options.

Electrically driven **heat pumps**, especially those tied to ground heat sources are rapidly gaining market share and are well suited to several types of low- and medium-temperature hydronic systems.

Although relatively new to North America, air-to-water heat pumps offer competitive benefits in comparison to geothermal heat pumps and traditional fossil fuel boilers. They are likely to gain market share in North America as efforts to "decarbonize" energy supplies encourage increased use of renewably sourced electricity.

Solar thermal collectors have been used as hydronic heat sources for several decades. However, significant drops in the price of solar photovoltaic electric systems over recent years have negatively impacted the use of solar thermal collectors, especially when applied for space heating. Still, they remain a viable option for hydronic heating systems in regional markets that have a need for space heating and high solar energy availability.

Just as solar energy is a regionally appropriate renewable resource, wood is in abundant supply in many traditional hydronic heating markets, especially the Northeastern United States, and many areas of Canada. Wood is also considered to be a carbon-neutral fuel. Burning it releases about the same amount of carbon dioxide into the atmosphere as would be released if the trees from which it came died and rotted in the forest.

The wood-fired boilers discussed in this chapter include those that burn dried cordwood or wood pellets. Modern cordwood gasification boilers can extract almost twice the amount of heat from a given quantity of wood compared to older wood-burning boilers. They also operate with much lower levels of particulate emissions. State-of-the-art wood pellet boilers can operate unattended for weeks and thus provide automation approaching that of fossil fuel or electrically powered heat sources. Hydronics technology is ideally suited to control and distribution of the heat produced from cordwood gasification and wood pellet boilers.

3.3 Gas- and Oil-Fired Boiler Designs

Most boilers that operate on natural gas or fuel oil can be classified according to their physical construction and heat exchanger material. This section gives a brief description of all major types of gas- and oil-fired boilers.

Cast-Iron Sectional Boilers

Cast-iron **sectional boilers** have been widely used in residential and light commercial buildings in North America for many decades. In this type of boiler, water is heated in cast-iron chambers called **sections**. The sections are bolted together to form a boiler block. The hot gases generated in the combustion chamber near the bottom of the boiler rise through cavities between the sealed sections and transfer heat to the water within. The greater the number of sections, the greater the rate of heat output from the boiler.

Figure 3-1 shows a typical sectional **wet-base boiler**.

Many cast-iron sectional boilers intended for residential and light commercial use are sold as **packaged boilers**. This implies that components such as the circulator, burner, and controls are selected and mounted onto the boiler by the manufacturer. These components are usually selected assuming that the boiler will be used in a "typical" residential system. In some cases, one or more of these components may not be the best choice for a specific application. For example, the small circulator supplied as part of a packaged boiler may not produce the flow rate and differential pressure required by a specific distribution system. For this reason, some manufacturers also sell a "bare-bones" version of the boiler without such components. Figure 3-2 shows a typical packaged gas-fired boiler.

Most cast-iron sectional boilers are limited to use in closed-loop hydronic systems. Properly designed, these systems rid themselves of dissolved air after a few days of operation. Following this, there is very little further oxidation of the cast-iron surfaces. However, improperly designed or installed systems that allow a continued presence of dissolved air in the system's

Figure 3-1 | Example of a (a) wet-base boiler section (b) boiler block assembly. *Courtesy of Weil-McLain Corporation.*

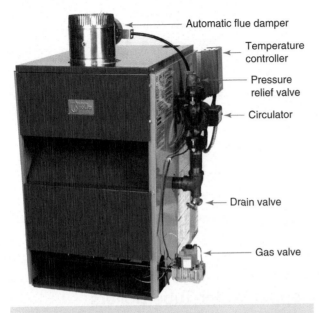

Figure 3-2 | Example of a packaged cast-iron sectional boiler. *Courtesy of ECR International.*

water will corrode cast iron. This can result from undetected (or uncorrected) water leaks, chronic opening of pressure-relief valves, misplaced air vents, and other abnormalities.

Conventional cast-iron boilers can also be severely damaged by operating at low water temperatures that allow **sustained flue gas condensation** to form on the combustion side of the sections. Details to prevent this are discussed later in this chapter, as well as throughout the remainder of the text.

When properly applied, cast-iron boilers can last for decades. Service lifetimes of 30 years or more are common.

The amount of metal in cast-iron sectional boilers makes them relatively heavy. Even a small cast-iron boiler can weigh over 300 lb. Such boilers typically contain 10 to 15 gallons of water. The combination of metal and water weight gives cast-iron boilers the ability to absorb a significant amount of heat. Such boilers are said to have a high **thermal mass**.

High thermal mass can be both a desirable and an undesirable characteristic. If the boiler is significantly oversized and experiences long off-times between firing cycles, much of the heat stored in the boiler's thermal mass can be carried up the flue by air currents passing through the combustion chamber. This can significantly lower the seasonal efficiency of the boiler. To reduce **off-cycle heat loss**, modern gas-fired high-mass boilers are equipped with flue dampers that close whenever the boiler is not firing. Some oil-fired high-mass boilers use burners equipped with air shutters that automatically close during the off-cycle for the same purpose.

A well-insulated boiler housing, or jacket as it is commonly called, also reduces standby heat loss from the boiler.

High thermal mass boilers can also have a favorable effect in systems that are divided into several small, independently controlled zones. The energy stored in their thermal mass helps prevent short cycling of the boiler's burner during **partial load conditions**. This reduces wear on the burner and improves the boiler's seasonal efficiency. The desired effect is for a high thermal mass boiler to act as a well-insulated thermal storage device during the off-cycle.

Some manufacturers design the cast-iron sections for vertical passage of flue gases. The sections shown in Figure 3-1 are an example of this geometry. To clean the flue passages in these boilers, the sheet metal top panel, flue connection, and flue gas collector assembly must be removed before a cleaning brush can be maneuvered between the sections.

Another type of cast-iron boiler design uses horizontal flue gas passages between the sections with a hinged door to access both the combustion chamber and the flue gas passages. Flue gases usually make two or more passes across the sections before exiting the boiler, allowing more heat to be transferred to the water. Such designs also make it simple to clean the combustion chamber and flue gas passages. An example of a cast-iron sectional boiler with horizontal exhaust gas flow is shown in Figure 3-3. The door of the combustion chamber is open. A technician is using a brush to clean the horizontal flue gas passages.

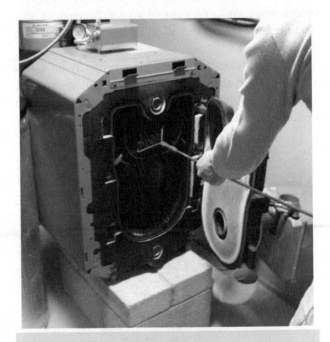

Figure 3-3 | Cast-iron sectional boiler with horizontal flue passages and a hinged combustion chamber door for simplified cleaning. *Courtesy of John Siegenthaler*

Steel Fire-Tube Boilers

Steel has been used in boiler construction for decades. In a steel **fire-tube boiler**, water surrounds a group of steel tubes through which the hot combustion gases pass. Spiral-shaped baffles known as **turbulators** are inserted into the fire-tubes to increase heat transfer by inducing turbulence and slowing the passage of the exhaust gases. The fire-tubes are welded to steel bulkheads at each end to form the overall heat exchanger assembly.

In some boilers, the tubes are oriented vertically. Figure 3-4 shows an example of a vertical fire-tube boiler. Notice the cutaway turbulators within the fire-tubes. In this design, the fire-tube assembly sits on top of the combustion chamber.

Other boiler designs are built around horizontal fire-tubes. These boilers route flue gases through multiple tubes before they reach the flue pipe connection. This allows additional heat to be extracted from the flue gases. In general, multiple-pass horizontal fire-tube boilers will have slightly higher efficiencies compared to single-pass vertical fire-tube boilers. This is due to the greater contact area between the boiler water and fire-tubes. Horizontal fire-tube designs also have the advantage of easier tube cleaning through a removable access panel at the front of the boiler. Figure 3-5 shows a cross-sectional view of a three-pass, horizontal fire-tube boiler.

Steel, like cast iron, is also subject to corrosion from system water containing dissolved air. Because of this, steel fire-tube boilers should only be used in closed-loop systems.

Although steel fire-tube boilers generally have less metal weight than cast-iron boilers of similar capacity, they often hold more water. This results in a boiler with a relatively high thermal mass.

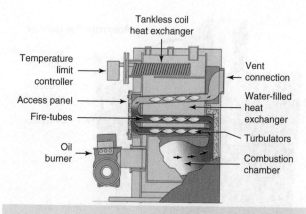

Figure 3-5 | Example of a three-pass horizontal fire-tube boiler. *Courtesy of Columbia Boiler Company.*

Copper Water-Tube Boilers

Some boilers use water-filled copper tubes as their heat exchanger. Design variations include both vertical and horizontal tube arrangements. The copper tubes are usually manufactured with fins that greatly increase the area exposed to the flue gases for improved heat transfer. The high thermal conductivity of copper compared to steel or cast iron allows significantly less surface area to yield the same rate of heat transfer. An example of a copper-tube boiler is shown in Figure 3-6a. The finned copper-tube heat exchanger used within this boiler is shown in Figure 3-6b.

The low metal and water content of copper-tube boilers provides minimal thermal mass, allowing for very fast warm-up following a cold start. However, low thermal mass provides little storage effect when the load is significantly less than the boiler's heat output. To prevent excessive thermal stress under such conditions, *all copper-tube boilers must have flow through them while being fired.* This requirement must be incorporated into system piping and control design. For example, a copper-tube boiler may require a "dedicated" circulator as shown in Figure 3-7 to provide sufficient flow through the boiler.

In applications where a copper-tube boiler is connected to an extensively zoned distribution system, an insulated "buffer tank" is often included to prevent short burner cycles under partial load operation. Such applications are discussed in later chapters.

Many copper-tube boilers use cast-iron tube headers. The internal surfaces of these headers are coated with a heat-fused glass-like material that prevents the water from contacting the cast iron. This combination of materials allows some copper-tube boilers to operate in open-loop systems in which dissolved oxygen is present in the water. Two examples of such applications would be direct heating of domestic water and swimming pools.

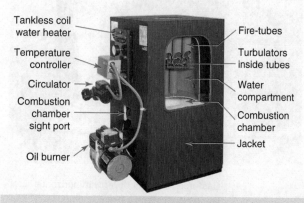

Figure 3-4 | Cutaway view of a vertical fire-tube steel boiler. *Courtesy of Columbia Boiler Company.*

3.3 Gas- and Oil-Fired Boiler Designs

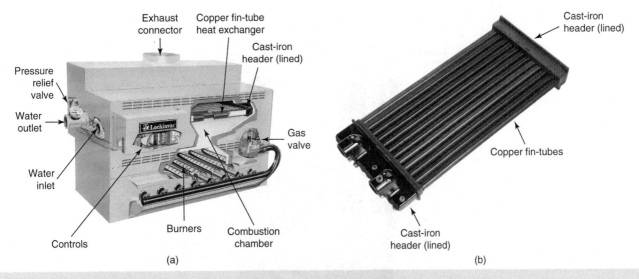

Figure 3-6 (a) Cutaway view of a copper water-tube boiler. (b) Finned copper-tube heat exchanger assembly. *Courtesy of Lochinvar Corporation.*

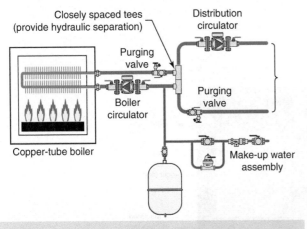

Figure 3-7 Copper-tube boiler hydraulically separated from distribution system using closely spaced tees.

Combined Boiler/Domestic Hot Water Appliances

A large percentage of residential hydronic systems supply both space heating and domestic water heating. Although several methods have been used to provide domestic hot water from a space heating boiler, one common approach uses a small storage tank that is heated by the same combustion system that also supplies space heating. Several manufacturers now offer products that use a single heat generator to provide space heating and domestic hot water. This approach saves room compared to the use of a boiler for space heating and a separate device for water heating. It also reduces installation costs because what would otherwise be field-assembled components are already assembled by the manufacturer. Figures 3-8a and 3-8b show examples of such products, which are sometimes referred to as "**combi-boilers**."

Additional Boiler Terminology

Further terminology is used to describe the specifics of how a boiler is designed and constructed.

One type of boiler design surrounds the combustion chamber with a water-filled heat exchanger. This arrangement is called a **wet-base boiler**, and it is a common configuration for oil-fired cast-iron boilers, and has the advantage of exposing more heat exchanger surface to the combustion chamber compared to other designs. The cast-iron sections shown in Figure 3-1 would be used to create a wet-base boiler.

Another common configuration is called a **dry-base boiler**. Such boilers suspend the water-filled heat exchanger above the combustion chamber. Gas-fired cast-iron boilers and oil-fired vertical fire-tube boilers are typically designed as dry-base boilers. Hot exhaust gases are drawn upward between boiler sections, or through fire-tubes, and enter a flue gas collector at the top of the boiler. Room air is always available at the burner(s). The use of an automatic flue damper to limit off-cycle draft losses is especially important for this type of boiler.

A typical cast-iron section for a dry-base boiler is shown in Figure 3-9a. Notice the similarity of this section and the upper portion of the wet-base section shown in Figure 3-1. A cutaway of the boiler assembled from these dry-base sections is shown in Figure 3-9b.

58 Chapter 3 Hydronic Heat Sources

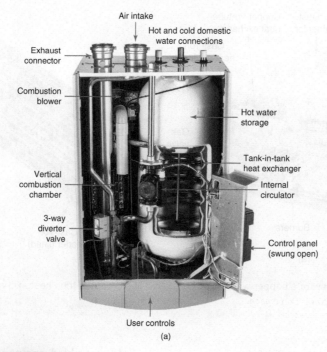

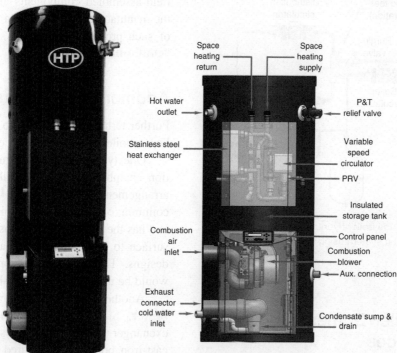

Figure 3-8 | (a) Wall-hung combination space and domestic water heating device. *Courtesy of Triangle Tube.* (b) Combination space and water heating device. *Courtesy of HTP, Inc.*

A newer approach in the gas-fired boiler industry is the wall-hung boiler. It was originally developed to conserve space in buildings such as apartments or condominiums, and to eliminate the need for a traditional chimney. Over the last decade, wall-hung boilers have grown into a major segment of the residential and light commercial boiler market. They are usually mounted on the interior side of exterior walls. Exhaust gases and combustion air travel through a coaxial sleeve, or two separate supply and vent pipes that penetrate the wall. Examples of compact wall-hung boilers are shown in Figure 3-10.

(a) (b)

Figure 3-9 | (a) A cast-iron section from a dry-base boiler. *Courtesy of Burnham Corporation.* (b) Construction of a dry-base gas-fired boiler. *Courtesy of ECR International.*

Despite their smaller size, wall-hung boilers are available with heating outputs equal to those of floor-mounted boilers intended for residential and light commercial use. Some can even provide instantaneous domestic water heating as well as space heating.

3.4 Conventional Versus Condensing Boilers

When a hydrocarbon fuel such as natural gas, fuel oil, or propane is burned, water is formed as one of the by-products. In the case of natural gas, the fundamental chemical reaction is as follows:

$$CH_4 + 2O_2 \rightarrow CO_2 + 2H_2O$$

Because of the high temperature in the combustion chamber, the water forms as a superheated vapor. As this vapor moves through the boiler's heat exchanger, it is cooled along with the other products of combustion. If sufficient heat is removed from the vapor, it may reach its **dewpoint** temperature, at which point some of the vapor condenses to a liquid. Other chemical compounds in the exhaust stream will also be present in this condensate, which makes it highly corrosive to any carbon steel or cast-iron surfaces it contacts. This flue gas condensation can also cause rapid deterioration of galvanized steel vent connector piping and masonry chimneys.

Many boilers currently used in North America were designed with the *intent* that flue gases would *not* condense within the boiler. Instead, the flue gas and the heat it contains would be carried out of the boiler and through the remainder of the exhaust system before condensing. Boilers designed with this intended mode of operation are called "**conventional**" **boilers**. Care must be taken to ensure that conventional boilers do not operate in a sustained flue gas **condensing mode**.

Figure 3-11 shows the relationship between boiler inlet temperature and boiler efficiency. Notice the steep

60 Chapter 3 Hydronic Heat Sources

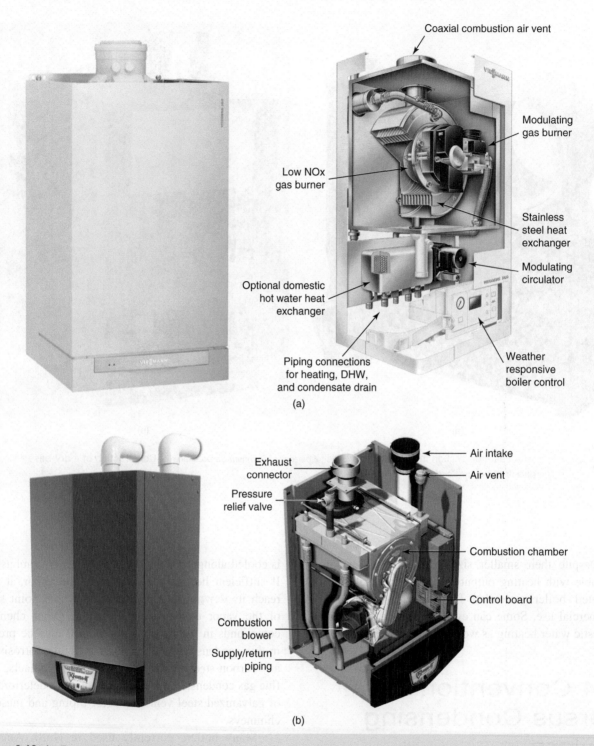

Figure 3-10 | Examples of compact wall-hung boilers. *(a) Courtesy of Viessmann Manufacturing. (b) Courtesy of Lochinvar.*

change in the slope of the curve at a temperature of approximately 130°F. This is a typical dewpoint temperature of water vapor for a boiler burning natural gas. At higher inlet temperatures, the boiler operates in a **noncondensing mode**. The water vapor and other flue gasses pass through the boiler's heat exchanger without condensing. This again, is the *intended* operating mode for a conventional boiler.

As boiler inlet temperature drops below the dewpoint, some of the water vapor in the flue gas stream condenses on the boiler's heat exchanger. The lower the inlet temperature, the cooler the heat exchanger surface, and the greater the amount of condensate formed. The exact dewpoint temperature depends on the fuel as well as the carbon dioxide content of the exhaust gases. Figure 3-12 shows how the dewpoint varies for gas- and oil-fired boilers.

3.4 Conventional Versus Condensing Boilers

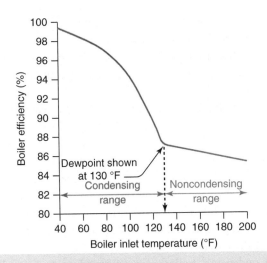

Figure 3-11 | Efficiency of a boiler in both condensing and noncondensing modes.

Figure 3-13 | Severely corroded boiler heat exchanger and flue gas collector caused by sustained flue gas condensation. *Courtesy of Dave Stroman.*

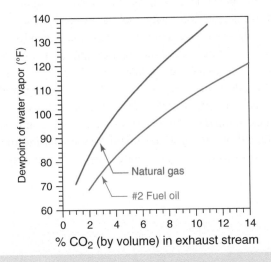

Figure 3-12 | Dewpoint of natural gas and #2 fuel oil based on percent CO_2 in exhaust gases.

In the moments following a "cold start," small droplets of the condensate form on the heat exchanger of every boiler. If the boiler temperature quickly climbs above dewpoint of the water vapor formed during combustion, this condensate evaporates, and is essentially of no concern. However, if the system the boiler is connected to prevents the boiler's inlet temperature from climbing above the dewpoint temperature, *sustained* flue gas condensation will occur. The latter can quickly and severely damage boilers not intended to operate in a sustained condensing mode as seen in Figure 3-13.

Sustained flue gas condensation can also damage steel vent connector piping connecting the boiler to the chimney. Figure 3-14 shows a galvanized steel vent connector pipe that has been corroded to the point of failure after only a few months of service.

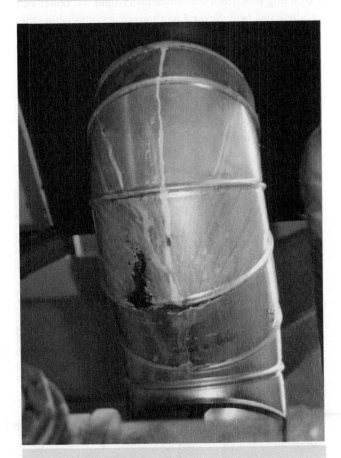

Figure 3-14 | Severely corroded galvanized steel vent connector pipe due to sustained flue gas condensation. Failure occurred after one year of service. *Courtesy of John Siegenthaler.*

A good analogy to flue gas condensation within a boiler is seen in the exhaust from a car on a cool day. When the car is first started, the temperature of the exhaust pipe and muffler is low enough to cause some of the water vapor produced in the engine to condense by the time it reaches the tailpipe. Liquid water can often be seen dripping from the tailpipe when the car is first started. After the tailpipe warms above the dewpoint temperature, the water vapor no longer condenses, and the dripping stops.

Those who repeatedly drive their car short distances in cold climates often find their tailpipes corrode very quickly. These tailpipes are succumbing to the same acid-on-steel corrosion reactions experienced in conventional boilers operating with sustained flue gas condensation.

The key to avoiding sustained flue gas condensation is maintaining the boiler inlet temperature above the dewpoint temperature of the exhaust gases. This requires a control that senses the boiler's inlet temperature, and, when necessary, prevents the distribution system from extracting heat from the water faster than the boiler can produce this heat. Several methods of accomplishing this are discussed in detail in subsequent chapters.

Condensing Boilers

Following the energy crisis of the early 1970s, higher fuel costs motivated boiler manufacturers to research methods for improving boiler efficiency. The objective was to attain the maximum heat output for each unit of fuel consumed. Analysis of combustion processes reveals that significant amounts of heat can be recaptured from the exhaust stream if the water vapor it contains can be condensed within the boiler.

Under typical operating conditions, each therm (e.g., 100,000 Btu) of natural gas produces approximately 1.15 gallons of water vapor during combustion. If this vapor could be condensed within the boiler, approximately 10,100 Btu of heat would be liberated from the exhaust gases for each therm of gas consumed. This represents about 10% of the original energy content of the natural gas, which, in a conventional boiler, is lost as part of the exhaust stream.

Condensing boilers are designed with large heat exchanger surfaces capable of extracting more heat from the exhaust gases compared to the heat exchangers in conventional boilers. With suitable inlet temperature, these heat exchangers can easily cool the exhaust stream below the dewpoint temperature, and thus cause condensation to occur. Figure 3-15 shows an example of a condensing boiler that uses a coiled stainless steel heat exchanger.

In addition to stainless steel, condensing boilers are also manufactured with aluminum and cast-iron heat exchangers. The design of a cast-iron condensing boiler allows for gradual metal loss over time due to corrosion from flue gas condensation. A counter-flow

(a)

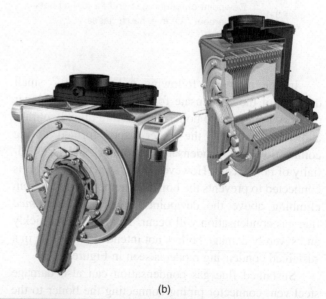
(b)

Figure 3-15 (a) Example of a condensing boiler. Notice the white condensate drain pipe in lower front corner. (b) One type of stainless steel heat exchanger assembly used in a condensing boiler. *Courtesy of Giannoni S.p.A.*

3.4 Conventional Versus Condensing Boilers

Figure 3-16 | Example of a cast-iron boiler with an external stainless steel secondary heat exchanger within which flue gas condensation can occur. *Courtesy of Weil-McLain.*

those of boilers using aluminum and stainless steel heat exchangers.

Still other boiler designs use two separate heat exchangers. Combustion takes place within the **primary heat exchanger**, where temperatures are such that flue gases will not condense when the boiler is operated within its specified application range. The primary heat exchanger can therefore be constructed of steel, copper, or cast iron. After giving up some heat at higher temperatures, exhaust gases flow from the primary heat exchanger to a **secondary heat exchanger**, which is typically made of high grade stainless steel. Additional heat is extracted from flue gases as they move through the secondary heat exchanger, and condensation forms. The secondary heat exchanger is designed to withstand the corrosive nature of the condensate. It is also equipped with a drain to remove the condensate from the boiler. Figure 3-16 shows an example of a cast-iron boiler with an externally mounted stainless steel secondary heat exchanger.

The amount of condensate formed within the boiler is highly dependent on the return water temperature of the hydronic distribution system the boiler supplies. The lower the return water temperature, the greater the amount of condensate, and the higher the boiler's efficiency as shown in Figure 3-11.

There are also condensing boilers that provide space heating and on-demand domestic hot water. An illustration of one such boiler is shown in Figure 3-17.

heat exchanger design with downward flow of exhaust gases and air purging at the end of each operating cycling minimizes metal loss rates. Testing has shown such boilers can provide design lives comparable to

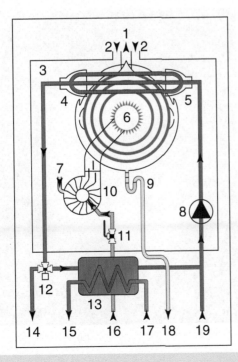

1-Flue exhaust
2-Air intake
3-Sealing chamber
4-Heating supply manifold
5-Heating return manifold
6-Environmentally frendly, premix bumer
7-Air intake
8-Circulator pump
9-Condensate siphon
10-Fan
11-Pneumatic gas valve
12-Three-way valve
13-DHW heat exchanger
14-Heating supply
15-Domestic hot water
16-Gas inlet
17-Domestic cold water
18-Condensate discharge
19-Heating return

Figure 3-17 | Internal components in a condensing capable boiler that supplies space heating and domestic hot water. *Courtesy of H.B. Smith.*

A flow detecting switch turns on the boiler whenever domestic water is drawn through the internal stainless steel heat exchanger at or above some minimum flow rate. An internal motorized diverter valve routes heated water from the boiler's main heat exchanger through the domestic water heat exchanger. The low thermal mass of the domestic water heat exchanger allows heated water to be available within seconds after the burner fires.

Oil-fired boilers designed to operate in condensing mode do exist, but are very uncommon in North America. The hydrogen content of fuel oil is lower than that of natural gas. This results in less water vapor being formed during combustion and thus less energy recovery potential. Soot formation under low-temperature operation is also more problematic due to oil's higher carbon content.

Modulating/Condensing Boilers

It's possible to build a boiler with a fixed firing rate that continually condenses its flue gases when operated at low inlet water temperatures. Such a boiler could be used to supply a relatively constant low-temperature load such as domestic water heating. However, a fixed heat output rate is not well matched to the space heating load of most buildings, which vary constantly due to changes in outdoor temperature or internal heat gain.

A more adaptable approach is to vary the heat output rate of the boiler in response to the heating needs of the building. This is called **modulation**, and is accomplished by reducing the fuel and air flow rates to the burner. Modulating a boiler is similar to changing the throttle setting on an internal combustion engine.

Nearly all currently available condensing boilers have the ability to modulate. Because of this they are often referred to as simply **"mod/con" boilers**.

Mod/con boilers use a **premix combustion system**. A variable speed blower controls the airflow rate entering a sealed combustion chamber. Natural gas or propane is metered into this air stream in proportion to the airflow rate. The slightly pressurized mixture of gas and air is forced into a burner head and ignited. Burner heads are typically made of stainless steel or porous ceramic materials. The flame pattern they create remains close to the burner's surface as seen in Figure 3-18.

The "ideal" modulating boiler would vary its heat output from 100% down to 0%. Nearly all mod/con boilers can modulate down to 20% of full-rated output. Some residential scale mod/con boilers can modulate as low as 10% of rated output. Larger mod/con boilers with multiple combustion systems can modulate as low as 2.5% of rated capacity. When loads drop below this minimum modulation rate, the burner must cycle on and off to avoid oversupplying the load.

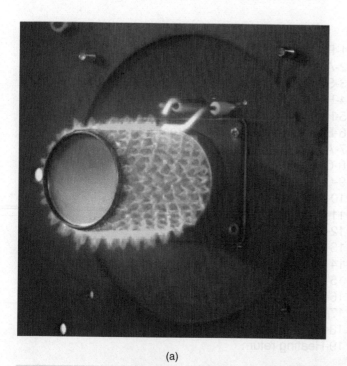
(a)

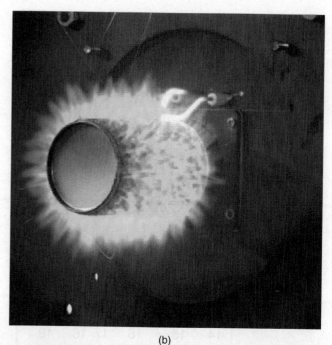
(b)

Figure 3-18 | Surface combustion on a stainless steel burner head within a mod/con boiler (a) at low input rate and (b) at high input rate. *Courtesy Heat Transfer Products.*

The ability of a boiler to reduce its heat output rate is sometimes expressed as a **turndown ratio**. The turndown ratio is the reciprocal of the lowest possible decimal percentage of full-heat output rate the boiler can maintain. For example, if a boiler can maintain stable operation down to 20% of its maximum heat output rate, it would have a turndown ratio of 5:1 (e.g., 5 being the reciprocal of 20% expressed as a decimal number). A boiler capable of modulating to 10% of its maximum heat production rate would have a turndown ratio of 10:1. The greater a boiler's turndown ratio, the better it can adjust its heat output to match what could be a constantly changing load.

Figure 3-19 shows how the heat output of a modulating boiler with a turndown ratio of 5:1 matches its heat output to a hypothetical load profile.

The upper line that borders the shaded area represents the rate at which heat is needed by the load over a period of several hours. The load begins at zero, steadily increases to a maximum value, and then reduces in various slopes and plateaus over time, eventually returning to zero.

The shaded areas represent the heat output from the modulating boiler. The pulses near the beginning and end of the load profile result from the boiler turning on at about 30% output, then quickly reducing output to the lower modulation limit of 20%, and finally turning off for a short time. Such operation is characteristic of current modulating burner technology.

The large red-shaded area shows that the boiler can accurately track the load whenever that load is 20% or more of the maximum load. This operating condition is ideal.

Mod/con boilers track heating load by monitoring water temperature. When a fixed **target water temperature** is to be maintained, the boiler reduces heat output if that temperature begins to climb and vice versa.

Most mod/con boilers can also vary the target water temperature they are attempting to maintain based on changes in outdoor temperature. The lower the outdoor temperature, the higher the target water temperature and vice versa. This technique is called **outdoor reset control**. It will be discussed in detail in later chapters.

Condensate Neutralization

The condensate produced in a condensing boiler is acidic. It typically has a pH in the range of 3.8 to 5.3, which is comparable to that of acid rain. Previous discussions have focused on the corrosive effects of this condensate on boiler heat exchangers, vent piping, and masonry chimneys. However, condensate can also degrade materials such as cast-iron drain piping and cast iron or concrete sewers. Some areas of Europe have regulations requiring condensate to be chemically neutralized prior to being discharged into a public sewer. As the use of condensing heating appliances increases in North America, there will likely be more requirements of this sort.

Fortunately, it is relatively easy to neutralize condensate by routing it through a neutralizer after it exits the boiler's condensate drain. Three examples of such neutralizers are shown in Figure 3-20. Two are commercially available, and the other, shown in Figure 3-20c, is a simple canister made of PVC (polyvinyl chloride) piping components.

The acidic condensate enters the neutralizer and slowly migrates through limestone chips or hydrolite granules, which are alkaline in nature, and react to neutralize the condensate. Eventually, the neutralized condensate flows out of the neutralizer and onward to the drain. The level of limestone chips or hydrolite granules in the neutralizer is checked annually and refilled as necessary.

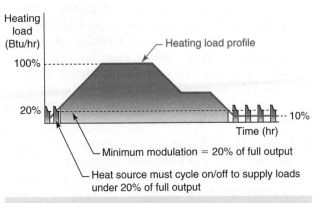

Figure 3-19 | Heat output rate of mod/con boiler relative to a hypothetical load profile.

Conventional Versus Condensing: Which Should You Use?

There is no universal answer to this question. In each situation, the answer depends on the type of hydronic distribution system the boiler will supply, the installation cost difference, and energy savings potential that would eventually displace any installed cost difference. The following issues should be weighed when considering a condensing boiler.

- Condensing boilers are typically more expensive than conventional boilers of equal capacity. Their economic viability is contingent upon recovering

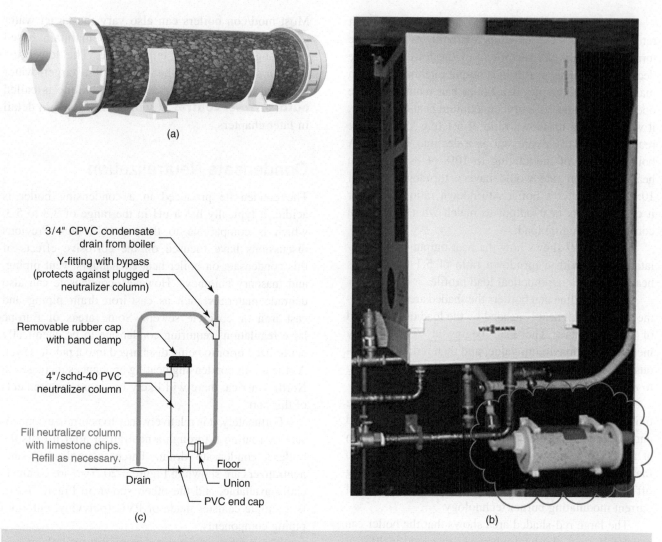

Figure 3-20 (a) A commercially available condensate neutralizer. The neutralization pellets are visible through the wall of the canister. *Courtesy of Axiom Industries Ltd.* (b) Neutralizer installed under a wall-hung condensing boiler. *Courtesy of Viessmann.* (c) An alternate design for a neutralizer fabricated from standard plumbing components.

this higher initial cost in fuel savings within a reasonable time. For this to happen, they must be applied in situations where their ability to achieve high (≥90%) efficiency can be realized over most, if not all, operating hours. Low-temperature systems such as slab-type radiant floor heating, swimming pool heating, and snowmelting consistently provide condensing boilers with low inlet water temperatures, allowing them to achieve high efficiency.

- Hydronic systems designed around higher water temperatures (150°F or above at design load conditions) may not provide inlet water temperatures low enough for a condensing boiler to actually condense its flue gases during much of its operating time. If used in such a system, a condensing boiler will operate at efficiencies very close to those of a less expensive conventional boiler, and thus may not be economically justified.

- If a condensing boiler is being considered, the installation location must have provisions for unattended condensate drainage. In a typical residential system, several gallons of condensate can be formed each day during cold weather, and must be properly disposed of. A condensate pump may be required depending on the elevation of the boiler relative to the drainage piping. Never assume that condensate will simply evaporate or run down through cracks or small holes in a concrete floor slab. The acidic condensate will eventually deteriorate any concrete with which it comes in contact.

- Condensing boilers cannot be vented into existing masonry flues. The moist/acidic exhaust products can quickly deteriorate clay flues and mortar joints. Masonry chimneys can literally be toppled by extensive, yet undetected, deterioration from moist corrosive flue gases. Furthermore, the low-temperature exhaust stream may not create sufficient draft, even with the use of power venting. This could create leakage of toxic exhaust gases into the building.

- Only materials *specifically listed for venting combustion products should be used.* Two examples of such products are listed polypropylene venting systems and listed AL29-4C stainless steel venting systems. It is the author's opinion that PVC pipe and fittings, which are manufactured for use in plumbing systems, should *never* be used for venting combustion products, even when allowed by a specific boiler manufacturer. PVC pipe is not *listed* for such purposes, and has demonstrated inadequacies when exposed to high flue gases at temperatures. A relatively new standard (ANSI/UL 1738) sets specific testing requirements for venting systems used with condensing boilers. This standard will likely be referenced by mechanical codes as well as within installation manuals for condensing boilers.
- The existing flue of a dormant chimney (e.g., one not used to vent any other device) can be used as a **chase** through which the venting and supply air piping of a condensing boiler can be routed.

3.5 Domestic Hot Water Tanks as Hydronic Heat Sources

Direct-fired tank-type domestic water heaters have been used as heat sources for hydronic space heating. Their use is generally limited to small heating loads such as apartments or residential additions.

Water heaters supplying hydronic space heating loads can be connected as either a "dedicated space heating" device, or as a "dual-use" device to supply both space heating and domestic hot water. In the latter category, the space heating distribution system may be isolated from the domestic water, or it may circulate the same potable water that is eventually supplied to fixtures. These options are shown in Figures 3-21a–c.

The use of domestic water heaters as hydronic heat sources has created long-standing controversy between manufacturers and plumbing code officials. One concern is possible contamination of potable water from materials used in the space heating portion of the system. Another concern is possible biological growth in portions of the system where water might lie stagnant for several months during warm weather. Other debated issues include:

- Domestic water heaters are not rated by the ASME (American Society of Mechanical Engineers) as space heating devices. Although they do undergo pressure testing during manufacturing, they are not held to the same standards as boilers.

- When installed so that potable water circulates through space heating circuits, all components in the space heating portion of the system must be compatible with potable water, which contains dissolved oxygen. Ferrous components such as cast-iron circulators, flow-check valves, and unlined expansion tanks would quickly corrode, producing discoloration of domestic water and potential health hazards. Use of bronze or stainless steel circulators and valves to prevent severe oxidation add significant cost to the system. Lead-based solders such as 50/50 tin lead, as well as valves or other components containing lead, which are commonly used for the non-potable water piping in hydronic heating systems, cannot be used in systems conveying potable water.

- Domestic water containing dissolved minerals or other contaminants may precipitate out of solution to foul piping, heat transfer surfaces, circulators, and valves. Such fouling may be difficult or impossible to remove. Aggressive groundwater may corrode components that would not corrode when used in closed-loop systems. The use of anticorrosion water treatments is ruled out since they would contaminate water used for domestic purposes.

- Domestic water heating tanks have less heat exchanger surface area compared to a modern boiler of equivalent capacity. This reduces their operating efficiency, especially if operated at higher temperatures (140°F or above). Water heaters also store heated water in direct contact with their flue passage and are seldom equipped with flue dampers. This increases off-cycle heat loss compared to that of a boiler operating at the same temperature.

- The **heating capacity** of residential water heaters is often substantially lower than that of boilers. Residential gas-fired water heaters are generally capable of producing 30,000 to 50,000 Btu/h. Oil-fired water heaters have slightly higher capacities depending on their burner nozzle. Although these heat outputs may be adequate for small buildings, they are often inadequate when there are simultaneous demands for both space heating and domestic water heating. For example, during the wake-up period in a typical household, the heating system may have to bring the building temperature up from nighttime setback, while several occupants are also taking showers. In such cases, it is often necessary to temporarily interrupt the space heating load while the tank attempts to supply the domestic water load. When the tank's temperature recovers, the space heating load is turned back on.

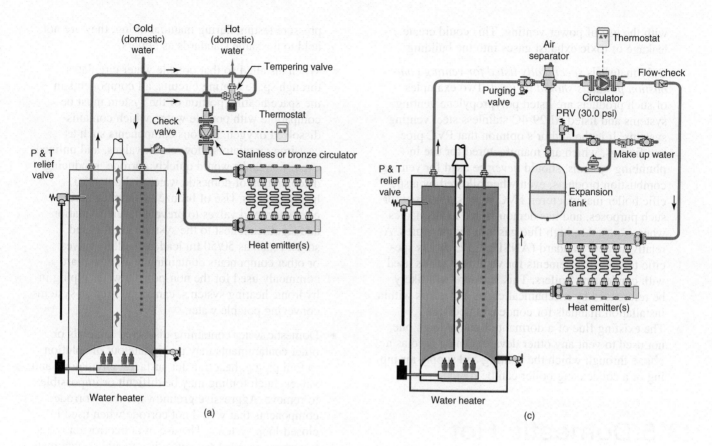

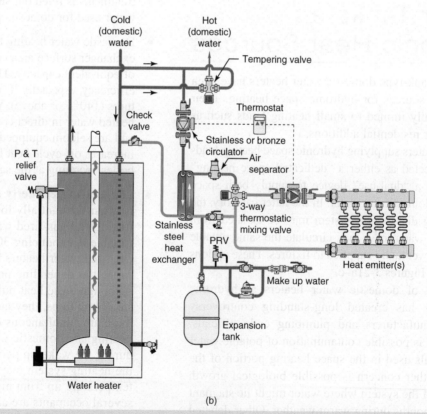

Figure 3-21: (a) Domestic water heater connected as a "dual-use" appliance for domestic water heating and space heating. Note: Potable water is circulated through the spacing heating distribution system. (b) Domestic water heater supplies space heating through a heat exchanger. (c) Water heater connected as a dedicated heat source.

- Experience has shown that tank-type water heaters will not last as long as a properly installed boiler. This is especially true in hydronic systems that operate at low water temperature. Such applications can cause flue gas condensation to remain on the tank's steel tube heat exchanger significantly longer than when the water heater is used for its intended purpose. Such condensation will rapidly corrode the heat exchanger and eventually cause it to fail. Although their purchase cost is significantly lower, reduced service life can quickly reclaim the initial savings.

- In light of these issues, it is the author's opinion that domestic water heaters should only be used as hydronic heat sources when:

A. They are set up as *dedicated* hydronic heat sources, and as such do not supply domestic hot water (as shown in Figure 3-21c). Such setup must include all safety controls that would be present on a boiler.

or

B. The space heating portion of the system is isolated from the domestic water using a suitable heat exchanger (as shown in Figure 3-21b).

Local plumbing and mechanical codes should always be checked for exclusions or special installation requirements when a domestic water heater is being considered as a hydronic heat source.

3.6 Power Venting Exhaust Systems

The traditional means of removing exhaust gases from a combustion-type heat source is to vent them up a chimney. While this is still a common approach, there are alternatives that, when properly applied, can help meet unique installation constraints that may preclude chimney venting.

One such alternative is known as **power venting**. It can be used with both gas- and oil-fired boilers and is an absolute necessity with condensing or mod/con boilers due to the low temperature of their exhaust gases.

Boilers equipped with power venting use a small blower to force exhaust gases out through an exhaust pipe. On some boilers, the exhaust blower is built into the boiler as shown in Figure 3-22. In other cases, the power venting unit is purchased as a separate assembly and installed externally to the boiler as shown in Figure 3-23.

In some installations, the vent pipe exits the building through a sidewall rather than the roof. This reduces

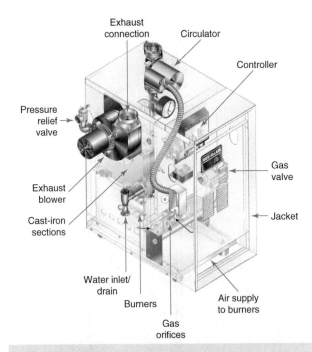

Figure 3-22 | Boiler with built-in exhaust blower for side-wall venting. *Courtesy of Weil-McLain.*

costs and simplifies installation, but must be done in a way that is aesthetically acceptable, and in a manner that will not cause frost or ice to form on adjacent surfaces, especially walking surfaces.

All power venting systems use a **draft-proving switch** to verify a proper negative pressure in the exhaust system before allowing the burner to fire. In the rare (but possible) event that the exhaust system becomes restricted or the exhaust blower fails to operate, the draft-proving switch will not allow the burner to fire. This makes power venting systems safer than a conventional chimney that cannot automatically stop the burner if draft is lost. Power venting systems are also less likely to be affected by winds or other negative pressure conditions induced within the building.

When the power venting fan is located in the boiler, the vent pipe from the boiler to the outside is under positive pressure while the boiler is firing. It must be sealed to prevent flue gases from escaping at joints and seams. Only vent pipe approved for positive pressure service should be used. An example of AL29-4C vent pipe designed for positive pressure service is shown in Figure 3-24.

When the power venting fan is at the end of the vent pipe, the latter is under negative pressure and generally does not require sealing. As always, be sure to use the manufacturer's recommended materials and installation methods.

Figure 3-23 Power venting system for an oil-fired boiler. (a) Interior vent, blower assembly, barometric damper. *Courtesy of John Siegenthaler*. (b) Outdoor vent terminal. Combustion gases exit front, cooling air induced through side slots. *Courtesy of John Siegenthaler*.

Figure 3-24 Positive pressure joining system used with AL29-4C stainless steel venting pipe. *Courtesy of Protech Systems*.

Direct Vent/Sealed Combustion Systems

A variation of power venting is known as **direct vent/sealed combustion**. This approach typically uses a coaxial air intake/exhaust assembly through which outside air is drawn for combustion and exhaust gases are routed outside. A small blower provides the pressure differential to bring air in and force exhaust products outside.

In a direct vent/sealed combustion system, no inside air is used for combustion. Figure 3-25 shows the concept when used in a small wall-hung boiler.

Notice the annular passage surrounding the center flue pipe through which outside air is drawn. The incoming air provides a cool metal surface near the building's wall.

Oil-fired boilers can also be configured for sealed combustion. The blower in the oil burner provides the necessary pressure differential to draw in combustion air and vent away exhaust gases. Because of the limited airflow available from the burner, oil-fired sealed combustion boilers generally need to be located close to the point where the vent pipe penetrates the wall.

Sealed combustion boilers are especially desirable in buildings that have low air leakage. In such cases, air infiltration alone may not provide sufficient air for proper combustion. A sealed combustion system greatly reduces the possibility that other exhaust fans in the building could cause back drafting of exhaust products from the boiler.

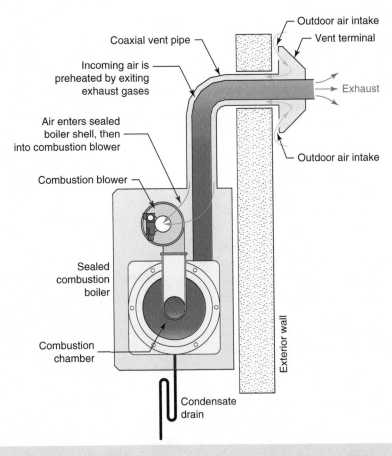

Figure 3-25 | A coaxial air intake/exhaust system used with a wall-hung sealed combustion boiler.

Sealed combustion systems also save energy since outside air rather than heated inside air is used for combustion. In coaxial air intake/exhaust systems, the incoming air is preheated by heat recovered from the exhaust gases before reaching the combustion chamber. This improves combustion efficiency.

3.7 Combustion Air Requirements

Any fuel-burning boiler requires a proper supply of air for safe and efficient combustion, exhaust gas dilution, and boiler room ventilation. Provisions to ensure that such requirements will be met are specified in the latest edition of the **National Fuel Gas Code** (NFPA No. 54/ANSI Z223.1) from the National Fire Protection Association (NFPA). The provisions of this code are widely accepted in most state and local building codes.

Under the NFPA code, the requirements for combustion and ventilation air depend on whether the boiler is located in a **confined space** or **unconfined space**. As defined in the code, unconfined space must have a minimum of 50 ft^3 volume per 1,000 Btu/h of gas input rating of all gas-fired appliances in that space. Any space not meeting this requirement is considered confined space.

Boilers located in confined space may draw air from adjacent rooms or from outside. When air will be drawn from adjacent rooms, the total volume of the boiler room and the adjacent rooms must meet the above definition for unconfined space. The boiler room must have two openings connecting it with the adjacent space. The minimum **free area** of each opening must be 1 inch^2 per 1,000 Btu/h of gas input rating of all gas-fired equipment within the boiler room, and in no case less than 100 inch^2. The top of the upper opening must be within 12 inches of the boiler room ceiling. The bottom of the other opening must be within 12 inches of the floor.

When air is drawn from outside, the size of the required openings varies depending on the type of fuel used and the orientation of supply air ducts. The NFPA code requirements are as follows: For gas-fired equipment, using vertical supply air ducting, or direct through-the-wall openings: Provide two openings between the boiler room and the outside, each having a minimum free area of 1 inch^2 per 4,000 Btu/h of gas input rating of all gas-fired equipment within the boiler room.

The 2018 version of the NFPA 54 code also allows a *single* opening between the mechanical room and outside space. This opening must have a minimum free area of 1 inch^2 per 3,000 Btu/h of gas input rating of all gas-fired equipment in the space, and not less than sum of cross-sectional areas of all vent connectors in the space. The single opening must begin within 12 inches of the mechanical room ceiling.

For gas-fired equipment, using horizontal supply air ducting: Provide two openings between the boiler room and the outside, each having a minimum free area of 1 inch^2 per 2,000 Btu/h of gas input rating of all gas-fired equipment within the boiler room.

For oil-fired equipment with either horizontal or vertical supply air ducting: Provide two openings between the boiler room and the outside, each having a minimum free area of 1 inch^2 per 5,000 Btu/h of all oil-fired equipment in the boiler room.

In all these cases, one opening, or end of duct, must be within 12 inches of the boiler room ceiling, and the other within 12 inches of the floor. These openings are intended to prevent the boiler room from overheating. They allow natural convection air currents to be established to cool the boiler room. They also allow any gases or fumes that might be present near the ceiling or floor of the mechanical room to escape.

When ducts are used to bring outside air into the boiler room, their cross-sectional area must equal the free area of the openings to which they connect. The minimum dimension of any rectangular supply air duct allowed by the NFPA code is 3 inches. The NFPA code also specifies that any screening covering these openings cannot be finer than a 1/4-inch mesh. The intent is to minimize the chance for the opening to be plugged by grass clippings or other debris.

All the above vent area requirements refer to "free area." This is the unobstructed area of the opening after accounting for the effects of any screens or louvers covering the opening. Louver manufacturers usually list the free area for each type of louver they offer. The following table lists typical free area-to-gross area ratios for different types of opening covers. These data should be used for comparison purposes and not as a substitute for specific data provided by manufacturers.

Type of opening cover:	F/G ratio:
• 1/4-inch screen	0.8
• Metal louvers	0.6
• Wooden louvers	0.2

Equation 3.1 can be used to determine the required area of the opening or duct.

Equation 3.1:

$$A_{opening} = \frac{A_{free}}{F/G \text{ ratio}}$$

where,

$A_{opening}$ = required cross-sectional area of each opening or duct (in^2);

A_{free} = free area required by the NFPA code for a given installation (in^2);

F/G ratio = ratio of free area to gross area of the louver or screen as specified by its manufacturer.

Example 3.1

The calculated free area required for venting a boiler room is 50 inch^2. The exterior of the opening will be covered by metal louvers having a free-to-gross area ratio of 0.6. What is the required gross area of each opening?

Solution:

Substituting these numbers in Equation 3.1 and solving yields:

$$A_{opening} = \frac{50}{0.6} = 83.3 \text{ in}^2$$

Discussion:

The free areas required by the code are minimums. Larger areas are also acceptable. The shape of the hole is not specified other than having a minimum dimension of 3 inches in the case of a rectangular opening.

Boilers using sealed combustion also have requirements for ventilation air to cool the space in which they are located. When the sealed combustion boiler is the only combustion appliance in the space, the requirement is to provide two openings, one within 12 inches of the ceiling, the other within 12 inches of the floor. The minimum free area of each opening must be 1 inch^2 for each 8,000 Btu/h of gas input rating of the sealed combustion boiler. This requirement essentially allows the area of the required openings to be one half that of the openings for nonsealed combustion boilers. The justification is

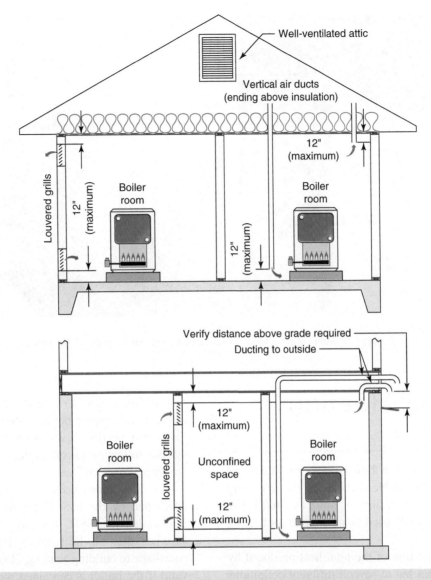

Figure 3-26 | Several possible arrangements for combustion/ventilation air supply to a boiler room.

that airflow in such situations is only for ventilation (e.g., cooling) of the boiler room, and not for combustion.

Figure 3-26 shows some possible arrangements for providing combustion/ventilation air. Keep in mind that code requirements can change with time and jurisdiction. Always check with local code officials about any requirements in addition to, or more restrictive than, what is included here.

It is also possible to install motorized dampers on combustion air openings through exterior wall. The actuator on the damper opens the damper blades when the burner is firing and closes them when the burner is off. This greatly reduces cold air entry and heat loss from the boiler room during off-cycles. Figure 3-27 shows an example of two motorized dampers, one on the upper combustion air opening and one on the lower combustion air opening.

When motorized dampers are installed on combustion air openings, mechanical codes require them to be equipped with **proving switches**. These switches close only when the damper is fully open. When two motorized dampers are used, which is common to accommodate upper and lower combustion air openings, the proving switches must be wired in series. The series-connected proving switches must interface with the boiler operating controls so that the boiler can only fire when both proving switches are closed. This safety provision prevents the boiler from firing in the event of a failed, jammed, or frozen combustion air damper.

Figure 3-27 | Example of motorized dampers on combustion air openings in exterior wall of a boiler room. *Courtesy of John Siegenthaler.*

3.8 Boiler Heating Capacity

The heating capacity of a boiler refers to its rate of *useful* heat production. In North America, this is usually expressed in Btu/h or thousands of Btu/h (MBtu/h). The word *output* is frequently used as a synonym for heating capacity.

Exactly what portion of the total heat produced by a boiler is useful has led to several different definitions of heating capacity. Each definition contains underlying assumptions about the boiler and the system in which it is used.

The performance information listed in the boiler manufacturer's literature often refers to one or more of the IBR ratings. IBR stands for the Institute of Boiler and Radiator Manufacturers. This organization was the forerunner of what is today known as the Hydronics Institute division of the Air-Conditioning Heating & Refrigeration Institute (AHRI). The two IBR ratings for boilers capacity are **IBR Gross Output** and **IBR Net Output**.

The IBR Gross Output Rating is the rate at which the boiler transfers heat to the system's water under steady-state operation. Implicit to the definition is the assumption that all heat released from the **boiler jacket** is wasted (e.g., does not offset a portion of the building's heating load).

The IBR Net Output is obtained from the IBR Gross Output Rating by multiplying by 0.85. The underlying assumption is that 15% of the boiler's gross output is released from piping between the boiler and the intended load, or is needed to warm up the thermal mass of the system following an off-cycle. The latter is called the "pickup allowance."

Another index used to describe boiler output was developed by the U.S. Department of Energy (DOE). It is called the **DOE Heating Capacity**. This rating assumes that all heat losses from the boiler jacket contribute to building heating. This assumption is usually valid if the boiler is contained within the heated space. DOE heating capacities are only provided for boilers up to 300,000 Btu/h of fuel input rating.

Because it includes jacket heat losses, the DOE Heating Capacity of any boiler will always be higher than its IBR Gross Output Rating. One or more of these capacity ratings are usually listed on the boiler's data plate and product literature.

Figure 3-28 shows a typical table of performance information for an oil-fired boiler furnished by its manufacturer. This table includes, among other data, the IBR Net Output Rating, DOE Heating Capacity, and fuel oil input rate.

The appropriate capacity rating for use in boiler selection depends on where the boiler and distribution piping are located. If both are located in heated space, the DOE heating capacity is appropriate since losses from the boiler jacket and distribution piping will offset part of the building's heating load. If the boiler is located in unheated space, but the distribution piping is mostly within heated space, the IBR Gross Output Rating can be used.

Boiler model number (1)	I = B = R Oil burner input (2)		D.O.E. Heating capacity MBH	I = B = R Net ratings (3)		Natural draft chimney size	Nozzle furnished 140 psig	A.F.U.E. Rating
	G.P.H.	MBH		Water MBH	Square feet			
SFH-365	.65	91	79	68.7	458	8 × 8 × 15	.60 80°B	86.0
SFH-3100	1.00	140	117	101.7	678	8 × 8 × 15	.85 80°B	81.0
SFH-4100	1.00	140	120	104.4	696	8 × 8 × 15	.85 80°B	86.0
SFH-3125	1.25	175	144	125.0	834	8 × 8 × 15	1.10 60°B	80.0
SFH-4125	1.25	175	149	129.6	864	8 × 8 × 15	1.10 80°B	82.5
SFH-5125	1.25	175	151	131.3	875	8 × 8 × 15	1.10 80°B	86.0
SFH-4150	1.50	210	175	152.2	1015	8 × 8 × 15	1.25 80°B	81.0
SFH-6150	1.50	210	181	157.0	1049	8 × 8 × 15	1.25 80°B	86.0
SFH-5175	1.75	245	206	179.1	1194	8 × 8 × 15	1.50 80°B	81.5
SFH-5200	2.00	280	231	200.9	1339	8 × 8 × 15	1.75 80°B	81.0

Figure 3-28   Typical listing of boiler performance information provided by manufacturer. *Courtesy of ECR International.*

If the boiler and distribution piping are both located in unheated space, such as a cold basement, crawl space, or garage, the IBR Net Output Rating should be used.

The deduction for piping heat loss assumed in the IBR Net Output Rating is a fixed 15% of the IBR Gross Output Rating. As such, it may not accurately represent piping heat loss in specific installations. For example, some state energy codes now mandate that piping carrying hot water through unheated space in new homes must be insulated to specified standards. Such insulation would likely reduce the piping's heat loss below the 15% of boiler output assumed by the IBR Net Output Rating. Using the IBR Net Output Rating to select a boiler in such situations would result in needless oversizing of the boiler. It would be better to make a separate calculation of piping heat loss, and then assess its effect on required boiler capacity. Such calculations are presented in Chapter 8, Heat Emitters.

3.9 Efficiency of Gas- and Oil-Fired Boilers

The word "efficiency" is often used in describing boiler performance. In its simplest mathematical form, efficiency is the ratio of a desired output quantity, divided by the necessary input quantity to produce that output.

Equation 3.2:

$$\text{Efficiency} = \frac{\text{Desired output quantity}}{\text{Necessary input quantity}}$$

When using Equation 3.2, both input and output quantities must be measured in the same units. The resulting efficiency will then be a unitless decimal number between 0 and 1, corresponding to percentages of 0 to 100. The higher the efficiency of a boiler, the more heat is extracted from a given amount of fuel. All other factors being equal, the higher the efficiency, the lower the boiler's operating cost.

Over the years, several specific definitions of boiler efficiency have been developed. The terms **steady-state efficiency, combustion efficiency, cycle efficiency,** and **Annual Fuel Utilization Efficiency** (AFUE) all refer to the ratio of an output quantity divided by an input quantity. However, all are based on different reference conditions. Using the wrong definition of efficiency can lead to erroneous performance predictions. Each specific definition of boiler efficiency will be discussed so the appropriate values can be used for their intended purposes.

Steady-State Efficiency

A boiler operates at steady-state efficiency only when it is being fired continuously under nonvarying conditions. All inputs such as fuel composition, air-to-fuel ratio, air temperature, entering water temperature, and so forth must remain constant. Under these conditions, the steady-state boiler efficiency can be calculated using Equation 3.3:

Equation 3.3:

$$\text{Steady-stage efficiency} = \frac{\text{Heat output rate of boiler (Btu/h)}}{\text{Energy input rate (Btu/h)}}$$

Example 3.2

A boiler runs continually under steady-state conditions for 1 h. During this time, it consumes 0.55 gallon of #2 fuel oil and delivers 65,000 Btu of heat. Determine the boiler's steady-state efficiency. Note: 1 gallon of #2 fuel oil = 140,000 Btu **chemical energy content**.

Solution:

The fuel input rate of 0.55 gallons per hour (gph) is converted into an equivalent energy input rate in Btu/h by multiplying by 140,000 Btu/gallon. The ratio of output to input is then calculated.

$$\text{Steady-state efficiency} = \frac{(65{,}000 \text{ Btu/h})}{(0.55 \text{ gal/h})(140{,}000 \text{ Btu/gal})}$$

$$= 0.844 \text{ or } 84.4\%$$

In reality, steady-state conditions seldom exist for long in either residential or commercial systems. Theoretically, they exist only while the load is equal to, or exceeds, the heating capacity of the boiler. For space heating loads, this only occurs for a small percentage of the heating season, if at all, particularly with boilers that are oversized. Just because a boiler operates at high steady-state efficiency under ideal laboratory conditions does not mean this efficiency will be maintained after it is installed. Erroneous predictions will arise if steady-state efficiency values are used in estimating seasonal performance and subsequent operating costs. This would be like estimating the average gas mileage of a car using only the highway miles per gallon rating.

The steady-state efficiency of the boiler can be estimated by dividing its DOE heating capacity by its fuel input rate.

$$\text{Steady-state efficiency} = \frac{\text{DOE heating capacity (Btu/h)}}{\text{Fuel input rate (Btu/h)}}$$

Boiler manufacturers often list the fuel input rate and DOE heating capacity in their technical literature. For gas-fired boilers, the fuel input rate is listed in Btu/h or thousands of Btu/h (MBtu/h). For oil-fired boilers, the fuel input rate is listed in gallons per hour. For #2 fuel oil, the fuel input rate in Btu/h is obtained by multiplying the gallons per hour firing rate by 140,000 Btu/gallon.

Combustion Efficiency

The combustion efficiency of a boiler is determined by *measuring* the temperature and carbon dioxide (CO_2) content of the exhaust gases under steady-state operating conditions. These readings are then used to look up the combustion efficiency on a chart based on the fuel being burned. The higher the CO_2 content and the lower the exhaust temperature, the greater the combustion efficiency. Modern electronic combustion analyzers can also measure the temperature and CO_2 content of exhaust gases and quickly calculate combustion efficiency.

Although attained through different means, combustion efficiency is essentially a field-measured value for steady-state efficiency. Because it is measured under continuous firing conditions, combustion efficiency should not be used to predict seasonal fuel usage.

Cycle Efficiency

Boilers experience heat losses when firing as well as during each off-cycle. These heat losses waste part of the heat produced by the fuel. Another efficiency measurement called cycle efficiency is used to account for these losses. Cycle efficiency is defined as the ratio of the total heat output divided by the chemical energy value of the fuel consumed, measured over an extended period. Mathematically, this is expressed as Equation 3.4:

Equation 3.4:

$$\text{Cycle efficiency} = \frac{\text{Total heat output of boiler over a time period (Btu)}}{\text{Energy content of fuel consumed over same time period (Btu)}}$$

Cycle efficiency is always less than steady-state efficiency. It is highly dependent on the percentage of time the burner is firing. This percentage is called the **run fraction** of the boiler and can be calculated using Equation 3.5:

Equation 3.5:

$$\text{Run fraction} = \frac{\text{Burner on time}}{\text{Total elapsed time}}$$

Example 3.3

At a given load condition, the burner of a boiler fires for 5 minutes, then remains off for 20 minutes before restarting. Determine its run fraction under these conditions.

Solution:

$$\text{Run fraction} = \frac{5 \text{ min}}{5 \text{ min} + 20 \text{ min}} = 0.2 \text{ or } 20\%$$

Equation 3.5 is useful when the actual firing time is known.

In other cases, it is necessary to predict what the run fraction would be under a specific heating load condition. This can be done using Equation 3.6.

Equation 3.6:

$$\text{Run fraction} = \frac{\text{Heating requirement of building (Btu/h)}}{\text{Steady-state heat output of boiler (Btu/h)}}$$

Example 3.4
Find the run fraction of a boiler with an output of 85,000 Btu/h while supplying a building heating load of 40,000 Btu/h.

Solution:

$$\text{Run fraction} = \frac{40{,}000 \text{ Btu/h}}{85{,}000 \text{ Btu/h}} = 0.47 \text{ or } 47\%$$

Figure 3-29 shows how the cycle efficiency of a conventional fixed firing rate boiler, operated in its intended noncondensing mode, decreases as its run fraction decreases. Notice that the decrease in efficiency is relatively small for run fractions from 100% down to approximately 30%. However, below 30%, run fraction cycle efficiency drops rapidly. This implies inefficient use of fuel under low run fraction conditions.

Low cycle efficiency can easily occur on days with mild outside temperature. It can also occur on cold days if the building has significant internal heat gain from people, equipment, and sunlight. Boilers with heating capacities significantly greater than their building's design heating load will operate at needlessly low cycle efficiency even during the coldest weather. The owner pays the penalty for this oversizing, both in higher initial cost, and, more importantly, greater fuel use over the life of the system.

Example 3.5
A building has a design heating load of 100,000 Btu/h when the outside temperature is 0 °F, and the inside temperature is maintained at 70 °F. Assume that boiler selected for the system has a heating capacity of 150,000 Btu/h. Using the graph shown in Figure 3-28, estimate the cycle efficiency when the outside temperature is 40 °F and internal heat gains total 20,000 Btu/h.

Solution:
The heat required of the boiler is the building's current heating load minus the current internal heat gains. This can be calculated as:

$$\text{Heating load} = 100{,}000\left(\frac{70-40}{70-0}\right) - 20{,}000 = 22{,}860 \text{ Btu/h}.$$

The run fraction required of the boiler can be calculated as:

$$\text{Run fraction} = \frac{22{,}860 \text{ Btu/h}}{150{,}000 \text{ Btu/h}} = 0.152 \text{ or } 15.2\%$$

Discussion:
From the graph in Figure 3-29, the corresponding cycle efficiency is about 72%. A boiler with a steady-state efficiency of 84.5% would operate at only 72% efficiency under the stated conditions. Keep in mind that spring and fall weather, combined with internal heat gains, can result in hundreds of hours of partial load operation during each heating season.

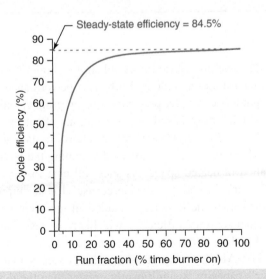

Figure 3-29 Cycle efficiency of a boiler versus run fraction. *Data source: Brookhaven National Laboratory, BNL 60816 & 80-38-HI.*

The following factors should be considered to keep a boiler's cycle efficiency as high as possible:

- Do not needlessly oversize the boiler. The author suggests a maximum oversizing factor of 10% of the (properly calculated) design heating load. Minimum oversizing allows the boiler to run for longer periods, and thus attain higher efficiency. An accurate heating load estimate is an absolute prerequisite to proper boiler selection.

- In general, boilers with low thermal mass tend to have higher cycle efficiencies due to lesser amounts of heat remaining in the boiler following burner shutdown.
- If a high thermal mass boiler is used, it should be very well insulated by its manufacturer, and it should be equipped with an automatic device that prevents airflow through the boiler during the off-cycle. A tight sealing automatic flue damper is especially important on conventional dry-base gas-fired boilers.
- A control strategy known as **heat purging** may be incorporated into systems with high-mass boilers. Flow through the boiler is maintained for several minutes after the burner has shut down to dissipate residual heat into the building, domestic water heater, or another load that can accept the heat. Some boilers are equipped with built-in controls for heat purging. In other cases, a heat purging system can be assembled from standard control components.
- The use of outdoor reset control will lower boiler water temperature as outside temperature increases, thus reducing its residual heat content and improving efficiency.

Annual Fuel Utilization Efficiency

In 1978, the U.S. Department of Energy established a test procedure for predicting the seasonal energy usage of boilers and furnaces. This procedure is incorporated into the current version of a standard known as ANSI/ASHRAE 103-2017: *Method of Testing for Annual Fuel Utilization Efficiency of Residential Central Furnaces and Boilers.*

This standard applies to all gas- and oil-fired boilers sold in the United States having fuel input rates less than 300,000 Btu/h. Its intent is to provide a uniform basis of comparison for boilers used in residential and light commercial systems.

The standard defines a performance rating called Annual Fuel Utilization Efficiency (AFUE). The AFUE rating is now the commonly used basis for expressing seasonal boiler efficiency within the United States. The Federal Trade Commission requires manufacturers to label applicable boilers with their AFUE rating. Minimum values for AFUE are often cited as part of the heating equipment performance standards in state energy codes.

AFUE values are expressed as a percent similar to steady-state efficiency. However, unlike steady-state efficiency, or cycle efficiency, AFUE values are intended to account for the many effects associated with part-load operation over an entire heating season. As such, the AFUE value is intended for use in approximating seasonal fuel usage. Equation 3.7 can be used for such an estimate:

Equation 3.7:

$$\text{Estimated seasonal fuel usage} = \frac{\text{Seasonal heating requirement}}{\text{Decimal value of AFUE}}$$

Example 3.6

Assume that the calculated seasonal heating requirement of a given home is 60,000,000 Btu (also stated as 60 MMBtu). The home is equipped with a gas-fired boiler having an AFUE rating of 82%. The estimated seasonal energy usage of the home is:

$$\text{Estimated seasonal fuel usage} = \frac{60 \text{ MMBtu}}{0.82}$$
$$= 73.2 \text{ MMBtu}$$

Discussion:

The higher the AFUE, the lower the seasonal fuel usage will be for a given seasonal heating requirement.

The test procedure for establishing a boiler's AFUE uses measured on-cycle and off-cycle performance data as input to a computer program. This program calculates the AFUE rating based on assumptions about exhaust gas composition and boiler sizing relative to its load.

AFUE values are established assuming the boiler is installed in heated space. As such, all heat losses from the boiler's jacket are considered useful heat output to that space. If the boiler is installed in unheated space, its actual seasonal efficiency will likely be less than its AFUE value.

The AFUE rating standard also assumes that the boiler's capacity is significantly greater than the design heating load of the building. Boilers sized closer to design load should yield seasonal efficiencies somewhat higher than those predicted by their AFUE rating.

3.9 Efficiency of Gas- and Oil-Fired Boilers

Boiler capacity / Design heating load	Seasonal boiler efficiency (%)
1.0	68
1.5	58
2.0	50
2.5	45
3.0	39

Figure 3-30 | Decreasing seasonal efficiency as boiler capacity is increased relative to the design heating load. Data based on a boiler with cycle efficiency described by Figure 3-28, located in a building in Syracuse, New York.

Figure 3-30 shows the theoretical effect of boiler oversizing on seasonal efficiency. The numbers are based on computer simulations of a fixed firing rate conventional boiler having an efficiency curve as shown in Figure 3-29. Notice the large drop in seasonal efficiency as the boiler size is increased significantly above the design heating load. These results obviously contradict the popular axiom that "bigger is always better."

The seasonal efficiency attained by any boiler is always influenced by the system and building in which it is installed. Listed AFUE ratings are based on a fixed set of assumptions for these system effects. As such, the AFUE rating will not necessarily predict true seasonal efficiency for all systems. Because of this, AFUE values should only be used for boiler comparison purposes.

Efficiency of Mod/Con Boilers

The instantaneous thermal efficiency of a mod/con boiler depends upon its inlet water temperature and firing rate. As previously discussed for fixed firing rate boilers, thermal efficiency increases as inlet water temperature decreases. A marked improvement is typically seen when the combustion side surface temperature of the boiler's heat exchanger drop below the dewpoint temperature of the exhaust gases.

Thermal efficiency also improves as the firing rate of a mod/con boiler decreases. As the burner decreases heat production, the ratio of heat transfer rate per unit of heat exchanger area increases. This causes the surface temperature on the combustion side of the heat exchanger to drop, further enhancing condensation. Figure 3-31 is representative of this effect. The upper curve represents the thermal efficiency of the mod/con boiler when operated at 25% of its full firing rate. Efficiency increases as inlet water temperature decreases. The lower curve represents the same effect, but with the boiler operating at maximum heat output. Operation under the latter condition keeps the combustion side of the heat exchanger surfaces at higher temperatures and somewhat reduces the rate of condensation. The area between these curves is the **performance envelope** of the boiler. Depending on operating conditions, the boiler's instantaneous thermal efficiency could be anywhere within this region.

Because its water temperature is reduced as the heating load decreases, the cycle efficiency of a mod/con boiler operating based on outdoor reset control increases as the load decreases. Representative curves comparing the time-averaged thermal efficiency of a current-generation mod/con boiler to the cycle efficiency of fixed firing rate conventional boiler are shown in Figure 3-32. The exact shape of this curve depends on the operating water temperatures in the system in which the mod/con boiler is used.

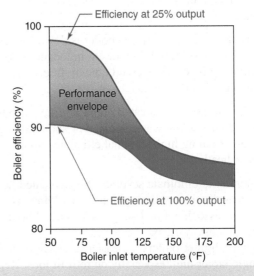

Figure 3-31 | Representative thermal efficiency of a mod/con boiler as a function of heat output rate and inlet water temperature.

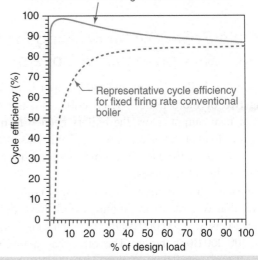

Figure 3-32 | Cycle efficiency comparison of a mod/con boiler to a fixed firing rate boiler.

3.10 Multiple Boiler Systems

When specifying a boiler for a larger home or commercial building, the designer has two options: (1) select a large boiler with sufficient capacity to meet the maximum heating load, or (2) select two or more smaller boilers that together can meet this load. The second option, known as a **multiple boiler system**, has a number of advantages including:

- Higher seasonal efficiency compared to that of a single large boiler.
- Improved ability to "track" variations in the heat load.
- Partial heat delivery if one boiler is down for servicing. This is especially important in locations where extreme temperatures could quickly freeze up nonoperational systems.
- Ability to provide elevated output during periods of high domestic hot water demand or snowmelting, while retaining high seasonal efficiency under lower space heating loads.
- Potential to eliminate several other dedicated heat sources distributed throughout the building, as well as their associated fuel supplies, electrical hookups, and exhaust systems.
- Smaller/lighter boilers are easier to install than a single larger unit, especially on retrofit jobs. Repair parts are often more readily available for smaller boilers.

Multiple boiler systems can be configured using fixed firing rate conventional boilers, mod/con boilers, or a combination of either. Each configuration will be discussed.

Multiple Boiler Systems Using Fixed Firing Rate Conventional Boilers

In a multiple boiler system, the total output is divided among two or more boilers. Each boiler constitutes a **stage** of heat output. Only the boilers needed to meet the heating load at any given time are fired. This allows the boilers that are fired to operate at higher run fractions and thus higher efficiencies.

Consider the situation illustrated in Figure 3-33. Assume that two identical buildings each have design heating loads of 300,000 Btu/h. One building has a single 300,000 Btu/h boiler. The other has a multiple boiler system consisting of three 100,000 Btu/h fixed firing rate boilers operated as stages.

At design load, the single large boiler operates continuously, and thus attains steady-state efficiency. Likewise, all three of the smaller boilers operate continuously and at steady-state efficiency.

Consider what happens as the building's load is reduced to 200,000 Btu/h. The single large boiler operates approximately 40 min/h, or 66% run fraction. The efficiency of the boiler drops to 83% due to off-cycle heat loss. Meanwhile, in the multiple boiler system, only two of the boilers are operating, but they operate continuously, at steady-state conditions, and thus still achieve steady-state efficiency of 85%.

The difference becomes more pronounced as the heating load is further reduced. At a load of 50,000 Btu/h, the single large boiler operates only about 10 minutes each hour, with a run fraction of 17%. Its cycle efficiency under this condition is 72%. In the multiple boiler system, one of the three boilers operates for 30 minutes each hour, for a run fraction of 50%. Its cycle efficiency is 83%. The other two boilers are off.

Similar scenarios will consistently prove the following: The lower the heating load, the higher the efficiency of a multiple boiler system compared to that of a single large boiler.

The number of boiler stages making up the multiple boiler system also affects seasonal efficiency. Up to a point, the greater the number of boilers, the greater the seasonal efficiency of the multiple boiler system. The more stages that are available, the better the boiler system can "track" the heating load. This is illustrated in Figure 3-34.

However, studies have shown that seasonal efficiency begins to drop off in systems using more than eight boilers. The reason is that as the number of boilers in the system increase, so does the ratio of jacket surface area to total heat output. This concept is illustrated in Figure 3-35. Higher heat losses from this extra jacket area more than offsets the cycle efficiency gains achieved through additional stages.

The red-shaded areas in Figure 3-35 show the added jacket surface area each time the boiler system is further divided from the situation shown above it. These images assume that heat output capacity per unit of boiler volume remains equal.

Modern multiple boiler systems are controlled by microprocessor-based staging controls that perform several functions including:

- Determining the proper supply water temperature for the system at any given time.
- Turning boilers on and off as needed to maintain the proper system supply temperature.
- Rotating the boiler firing sequence to accumulate approximately the same running hours on each.

3.10 Multiple Boiler Systems

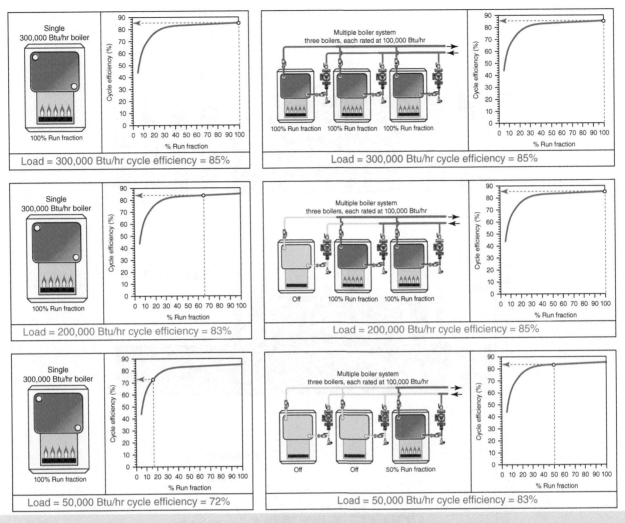

Figure 3-33 | Comparison of a single large boiler with a multiple boiler system at three different loads.

Such controls play an important role in maintaining the high efficiency of a multiple boiler system and are discussed in Chapter 9, Control Strategies, Components, and Systems.

The piping of a multiple boiler system plays a critical role in achieving high seasonal efficiency. *Boilers should be piped so that heated water is not circulated through unfired boilers.* Doing otherwise allows the unfired boilers to dissipate heat from the warm water flowing through them. Some of this heat is carried out the venting system by off-cycle air currents. Additional heat is lost by conduction through boiler jackets.

There are a number of ways to prevent hot water from circulating through unfired boilers in a multiple boiler system. One method that fulfills this requirement is shown in Figure 3-36.

Each boiler is piped in parallel with the others as shown in figure 3-39. All boilers connect to a common set of headers that in turn connect to the main system piping using closely spaced tees to provide hydraulic separation between the boiler circulators and the system circulator. Each boiler's circulator operates only while the boiler is fired, and for a short heat-purging period after the burner shuts off. Thus, there is no flow of heated water through unfired boilers. A check valve in each boiler circuit prevents flow reversal through inactive boilers that would otherwise occur when one or more other boilers are operating. Each boiler can be isolated and even removed from the system, if necessary, without affecting operation of the other boilers.

Another piping method that provides the same benefits is shown in Figure 3-37. Again, each boiler is equipped with its own circulator and check valve. Each boiler connects to common supply and return headers that are sized for very low pressure drop. These headers connect to a **hydraulic separator**, which separates the pressure differentials created by the boiler circulators from that created by the system circulator. The hydraulic separator also provides air and dirt separation for the system. Its use will be discussed in more detail in later chapters.

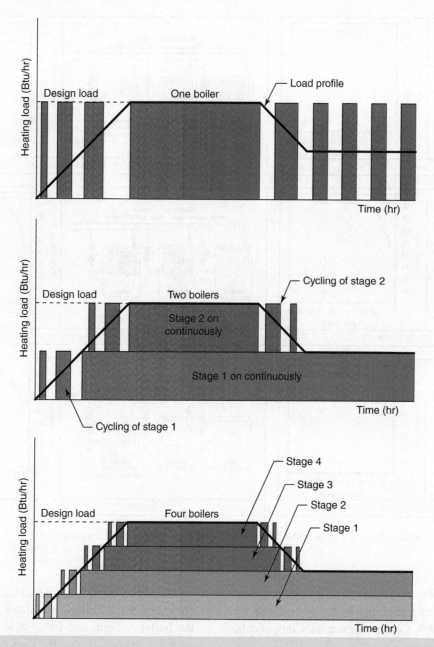

Figure 3-34 | As more stages are used, the heat output of the multiple boiler system can better "track" the load.

Another means of piping multiple boilers is shown in Figure 3-38. All the boilers are again piped in parallel. The piping leading into each boiler is equipped with a motorized ball valve, which opens only when its associated boiler is firing. A variable speed, pressure-regulated circulator provides flow through any active boiler. This circulator automatically increases or decreased flow rate depending on the number of active boilers. A hydraulic separator uncouples the pressure differential created by the boiler circulator from that created by the system circulator. No check valves are needed. This configuration has the advantage of reduced electrical energy consumption in comparison to the use of multiple small circulators for the piping configurations shown in Figures 3-36 and 3-37.

High/Low-Fire Boilers

Several manufacturers also offer boilers with two-stage heat output control built into a single enclosure. Such boilers are often called "**high/low fired**." In some cases, the gas valve on the boiler has two-stage capability

3.10 Multiple Boiler Systems

Multiple Boiler Systems Using Mod/Con Boilers

Two or more mod/con boilers can also be combined into a multiple boiler system. Such a system can stage individual boilers and modulate the heat output of each boiler in the system. This results in a **system turndown ratio** that is much higher than the turndown ratio of a given boiler within the system.

Consider the multiple mod/con boiler system shown in Figure 3-40. It consists of four boilers, each with a maximum heat output rating of 150,000 Btu/h and each with a 5:1 turndown ratio.

The maximum heat output of this system occurs when all four boilers are operating at full capacity. This yields a total heat output of 600,000 Btu/h.

The minimum (steady) heat output rate occurs when three boilers are off, and the fourth boiler is operating at its minimum stable heat output rate, which in this case is 30,000 Btu/h (e.g., 1/5th of 150,000 Btu/h).

The system turndown ratio is thus:

$$\text{TDR}_{\text{system}} = \frac{Q_{\max}}{Q_{\min}} = \frac{600,000 \text{ Btu/h}}{30,000 \text{Btu/h}} = 20 \text{ (or 20:1)}$$

This high turndown allows the boiler system to match the load whenever that load is greater than 1/20th of 600,000 Btu/h. Such precise load tracking is not possible with on/off-type boiler and represents a significant advantage of a multiple mod/con boiler system.

There a several ways multiple mod/con boiler systems can be controlled. These methods involve a combination of staging and modulation. Modern controllers can usually be set up to provide any of the methods shown in Figure 3-41.

Series staging/modulation allows a given boiler to modulate all the way to full output before staging on the next boiler.

Parallel staging/modulation maintains each boiler in the system at the same modulation rate. The group of boilers modulates "as a whole" starting from the lowest stable firing rate of the boilers and progressing to the full firing rate of each boiler. This method does not allow the system turndown ratio to be greater than the turndown ratio of an individual boiler.

Hybrid staging/modulation controls the boilers based on a trade-off between maximizing the combustion heat exchanger area per unit of heat output (to encourage condensation) and not firing the next boiler in response to a small increase in load.

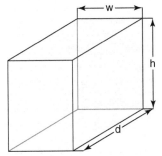

1 boilers @ 400,000 Btu/hr

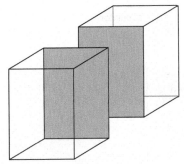

2 boilers @ 200,000 Btu/hr each

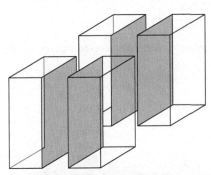

4 boilers @ 100,000 Btu/hr each

Figure 3-35 — Total jacket surface area of boilers increases as the number of boilers increases. Darker shaded areas show increase in jacket surface area from the configuration above.

(e.g., it is capable of allowing two rates of gas flow to the burners depending on the control signal it receives). In other boilers, two separate gas valves and associated burners are installed in the same enclosure.

High/low-fire boilers require less physical space than two separate boilers having the same total capacity. They typically have a single vent connection and are easier to install compared to multiple single-stage boilers.

When used in a multiple boiler system, the stages of high/low-fire boilers are typically operated in paired sequences. Modern staging controls can be configured for this operating logic.

84 Chapter 3 Hydronic Heat Sources

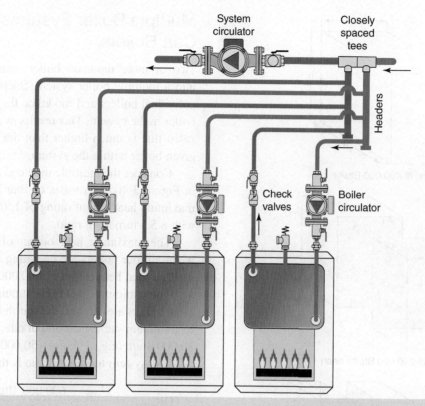

Figure 3-36 | Multiple boilers connected to a common set of headers, which connect to distribution system using closely spaced tees.

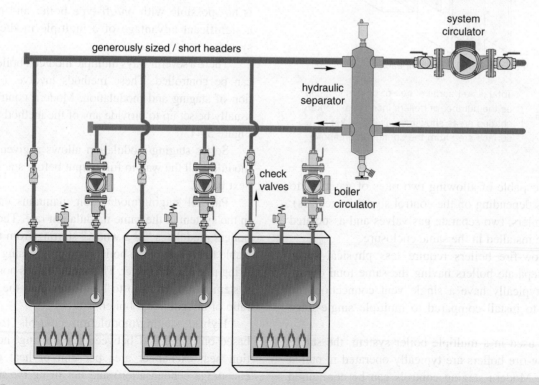

Figure 3-37 | Connecting multiple boilers using individual boiler circulators and a hydraulic separator.

3.10 Multiple Boiler Systems   85

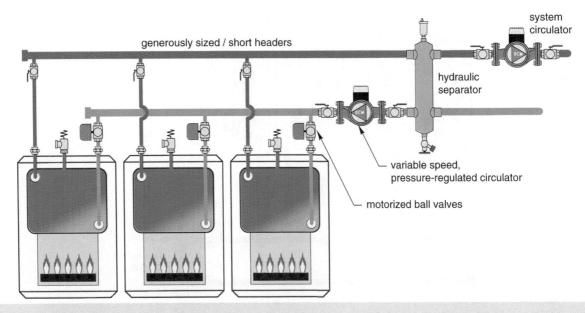

Figure 3-38 | Multiple boilers connected to a common set of headers. A motorized ball valve on each boiler opens when that boiler is firing. Flow is provided by a variable-speed pressure-regulated circulator.

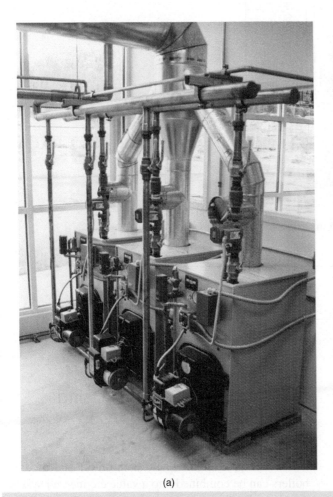
(a)

(b)

Figure 3-39 | a) Three oil-fired boilers operated as a multiple boiler system. (b) Two, induced draft gas-fired boilers operated as a multiple boiler system. Individual circulators and flow-checks used on each boiler. *Courtesy of John Siegenthaler.*

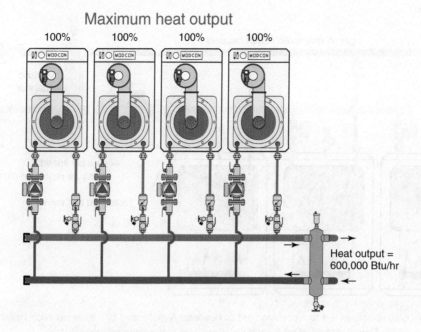

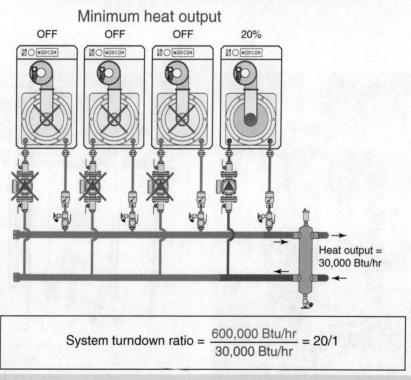

Figure 3-40 | The system turndown ratio of a multiple boiler system.

In addition to staging and modulation, modern boiler controllers periodically rotate the firing sequence on boilers in an attempt to equalize total operating hours. In theory, this allows all boilers to be serviced under approximately the same condition.

An example of a multiple mod/con boiler system consisting of three boilers and a hydraulic separator is shown in Figure 3-42.

Multiple Boiler Systems Using Conventional and Mod/Con Boilers

There are applications where both on/off and mod/con boilers can be combined into a value engineered system. An example of such a system is shown in Figure 3-43.

3.10 Multiple Boiler Systems

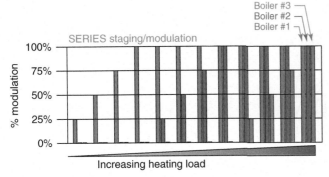

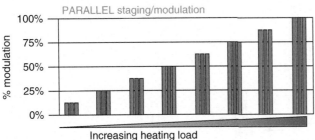

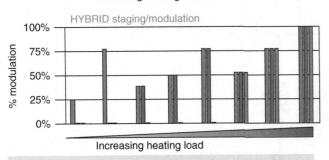

Figure 3-41 — Three possible staging/modulation control schemes for a 3-boiler system using mod/con boilers.

The supply water temperature delivered by this boiler system is based on outdoor temperature. At low load conditions, the water temperature returning from the system is very low, perhaps only a few degrees above room temperature. The mod/con boiler supplies all the required heat at these low water temperatures, which allows it to operate in high-efficiency condensing mode. Because of this advantage, it is always the first boiler to be turned on and the last boiler to be turned off by the controller. As such, this is called a "**fixed lead**" boiler.

As outdoor temperature drops, the heating load increases, and so does the required supply water temperature. This also increases the inlet water temperature to the mod/con boiler. Eventually, the inlet temperature is high enough that flue gases will no longer condense. At this point, the mod/con boiler has very little efficiency advantage over a less expensive conventional boiler.

Any additional heat output beyond that of the mod/con boiler now comes from the two conventional boilers, which operate at efficiencies comparable to that of the mod/con boiler at increased inlet temperatures. The two conventional boilers operate as stages 2 and 3 as required by the load. Their firing order is rotated by the controller to equalize total run hours on each boiler.

This approach exploits operating conditions that allow a mod/con boiler to operate at high efficiency, while at the same time reducing installation cost by not applying more expensive mod/con boilers in situations where they cannot operate in condensing mode.

Where Multiple Boiler Systems Make Sense

Multiple boiler systems are well suited to homes that have several bathrooms, particularly those having high-flow fixtures. During a high demand for domestic water heating, all stages of a multiple boiler system can go into operation at maximum heat output to provide the fastest possible recovery. Other loads in the system may be temporarily turned off during this time. Such "priority" control strategies are discussed in later chapters. An example of a multiple mod/con boiler system supplying a high-capacity indirect water heater is shown in Figure 3-44.

Because of their ability to retain high seasonal efficiency under widely varying load conditions, multiple boiler systems are also excellent for multi-load systems involving snowmelting or pool heating. The author suggests that a multiple boiler system be considered whenever the total system load exceeds 200,000 Btu/h.

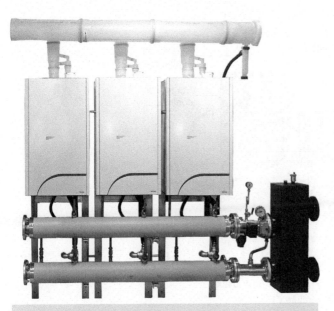

Figure 3-42 — Example of a multiple mod/con boiler system with common vent and hydraulic separator (at right). *Courtesy of SIME.*

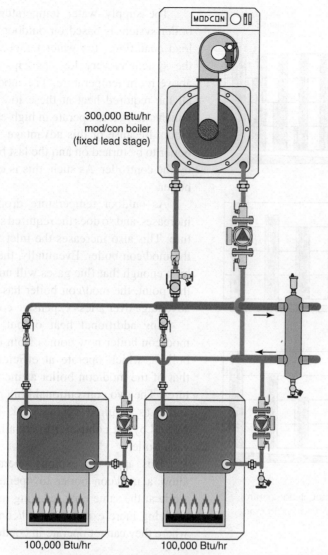
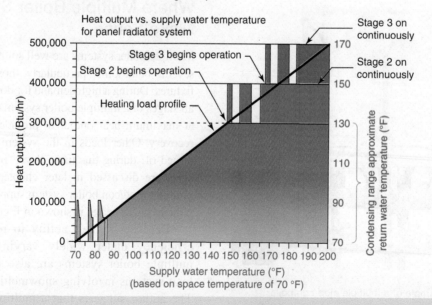

Figure 3-43 | Use of conventional and mod/con boilers in a multiple boiler system.

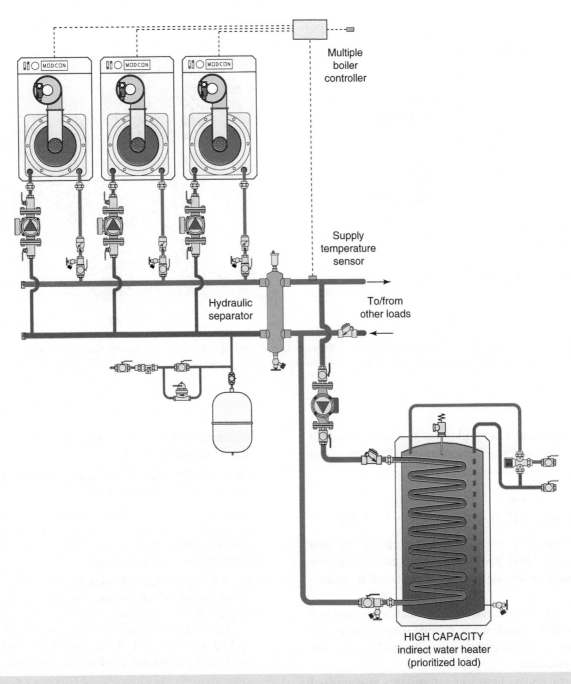

Figure 3-44 | A multiple mod/con boiler system supplying a high-capacity indirect water heater.

3.11 Electric Boilers

Electric boilers are available as compact wall-hung units. They typically contain two or more immersion-type resistive heating elements mounted in a common enclosure through which water flows. An example of a compact electric boiler is shown in Figure 3-45.

When an electrical current passes through a resistive heating element, the rate of heat output can be calculated using Equation 3.8:

Equation 3.8:

$$P = iv$$

where,
 P = power output (watts);
 i = current (amps);
 v = voltage (volts).

Figure 3-45 A compact wall-hung electric boiler. (a) External appearance (b) internal circuits supply four independently controlled heating elements. *Courtesy of Argo Division of ECR International.*

The following factors allow one to convert the output rating of electrical heating devices (usually expressed in watts or kilowatts) into Btu/h:

$$1 \text{ watt} = 1 \text{ W} = 3.413 \text{ Btu/h}$$
$$1 \text{ kilowatt} = 1000 \text{ watts} = 1 \text{ kW} = 3413 \text{ Btu/h}$$

The second conversion factor is simply the first multiplied by 1,000. These factors are used so commonly in a heating system design that they should be memorized.

With the proper controls, each heating element can be operated independently to provide staged heat output. Only those stages needed to satisfy the load at any given time are operated.

Electric boilers have some advantages as compared to combustion-type boilers. These include the following:

- Since combustion is not involved, no air supply is required for combustion or draft, and no exhaust system is necessary.
- Since there are no flue gases, there is no concern about flue gas condensation.
- On-site fuel storage is not required.
- Periodic maintenance is minimal since there is no soot to remove or fuel filters to replace.
- If used with an outdoor reset control, the water temperature in an electric boiler can be varied by controlling the on-time of the elements. This would eliminate the need for a mixing device in variable-temperature systems.
- Electric boilers can take advantage of time-of-use electrical rates where available.
- Since nearly all buildings have an electric service, the choice of an electric boiler might eliminate the need for a natural gas service and its associated monthly basic service charge.
- Electric boilers are available with much lower heat output rates than combustion-type boilers. As such, they can be well matched to buildings with very low heating loads. This approach eliminates the short-cycling and low seasonal efficiency associated with significantly oversized combustion-type boilers.

The economic viability of an electric boiler depends on the cost of electricity versus competing fuels. The economics are obviously more favorable in areas where electric utility rates are low. In commercial or industrial buildings where utilities usually impose a "**demand charge**" as well as an energy charge, the economic viability of an electric boiler is diminished because its use would significantly increase the building's demand.

If an electric boiler is being considered, be sure the building's electric service entrance is adequate to handle the load. This is particularly true in older buildings. A 200-amp/240-VAC service entrance is generally considered minimum for average-size residential applications.

3.12 Electric Thermal Storage Systems

To comprehend the concept of ETS (Electric Thermal Storage), it is important to understand the situation faced by many electrical utilities. Chief among them is the challenge of meeting the growing demands for electricity without spending hundreds of millions of dollars building new generating facilities. As the North American demand for electrical energy increases, all utilities must plan to ensure that they can supply the extra energy, especially during critical peak-demand periods. A fundamental goal is to reduce peak demand, while simultaneously increasing demand during traditional low-demand periods.

One way to do this is to offer **time-of-use rates**. In their simplest form, time-of-use rates offer customers lower cost electricity during specified "off-peak" periods. Such periods have a set beginning and ending time. In some areas, rates may vary several times during the day, as well as seasonally, during weekends, and even on holidays. The electric meter automatically records usage in each pricing category.

Because space heating is almost always the largest electrical load in all-electric buildings, it represents a major opportunity for savings if the necessary energy can be purchased from the utility during off-peak periods. This concept requires that the heat produced during these periods be stored for subsequent use during on-peak periods. Since water has excellent heat storage properties, hydronic systems using electric heating elements combined with thermal storage tanks are ideal for such applications.

Classification of ETS Systems

ETS systems are classified according to their ability to meet heating loads using energy purchased and stored during the previous off-peak period. The two common types of ETS systems are called **full storage ETS systems** and **partial storage ETS systems**.

In **full storage systems**, all energy used during the on-peak period is purchased and stored during the previous off-peak cycle. The storage tank must hold sufficient heat to supply the building's load through the entire on-peak period, which may be as long as 18 hours. Storage tank volumes of several hundred to more than 1,000 gallons are common in such applications.

In most full storage systems, the output of the electrical elements, which are immersed in the thermal storage tank, or that of an electric boiler, must be three to four times greater than the design heating load of the building. This is because these heat sources have only a few off-peak hours in which to transfer enough heat to supply the building for a 24-hour period.

In **partial storage systems**, only a portion of the on-peak heating energy is purchased and stored during the off-peak period. Once this energy is depleted, the heating elements operate as necessary to maintain comfort. Operating the heating elements during the on-peak period uses electricity at the more expensive rate, and thus this should be minimized. However, because it stores only a portion of heat required for the on-peak heat period, a partial storage system is usually significantly less expensive than a full storage system.

Figure 3-46 shows a comparison of hourly electrical demand for a house using conventional electric resistance heating versus a full storage ETS system.

Notice the high electrical demand of the ETS system during the off-peak hours when electrical rates are low. This is when the water in a hydronic-based ETS system is being heated. Ideally, sufficient heat for the entire day is acquired during this time. At 7:00 a.m., the heating elements or the electric boiler are shut off. During the midday hours, a small amount of electrical energy is needed for components such as circulators and, in some cases, blowers. This demand is very low compared to that of the conventional electric heating elements or electric boiler.

The economic merits of partial storage versus full storage systems are very dependent on the project. An economic analysis that considers the trade-off between initial cost of the equipment and the locally available

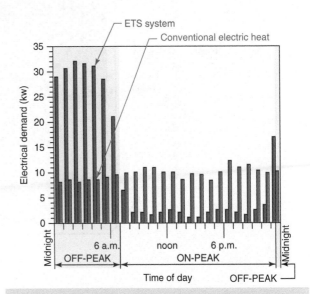

Figure 3-46 — Electric demand of full storage ETS system versus conventional electric heating. Data based on a design day in upstate New York.

on-peak/off-peak rate differential is essential in making an informed decision. The system with the lowest energy cost may not necessarily be the best choice. Instead, one should look for the lowest total life cycle cost over the estimated system life.

ETS System Layout

ETS systems do not use exotic technology. They are composed mostly of standard hydronic components and designed using fundamental concepts covered in this text. A basic schematic for an ETS system using an electric boiler as the heat source is shown in Figure 3-47. The three-way motorized mixing valve prevents potentially high-temperature water from flowing directly to a low-temperature distribution system.

In some systems, the thermal storage tank may also be equipped with a heat exchanger for domestic water heating. Other heat sources such as solar thermal collectors, solid-fuel boilers, or hydronic heat pumps might also be incorporated as alternative heat sources supplying the thermal storage tank depending on time of year and characteristics of the loads served by the system.

Thermal Storage Tanks for ETS Systems

In both types of ETS systems, the essential component is a properly sized, well-insulated thermal storage tank. Several types of tanks are suitable for such applications, including pressure-rated steel tanks and unpressurized polymer tanks. The choice must consider factors such as volume, temperature limit, corrosion, life expectancy, type of insulation, cost, and ease of installation.

Pressure-rated steel tanks are a good choice for closed-loop systems. Although their cost per gallon is usually higher than unpressurized tanks, they allow more flexibility in the remainder of the system design. Most mechanical codes classify such tanks with volumes over 119 gallons, as pressure vessels that require ASME certification. Whenever possible, tanks made to standard sizes should be considered because they will be less expensive than custom-built tanks.

Unpressurized tanks are available from several sources. Such tanks combine a structural insulated shell with a waterproof internal liner. The shell is usually assembled on site from components that fasten together.

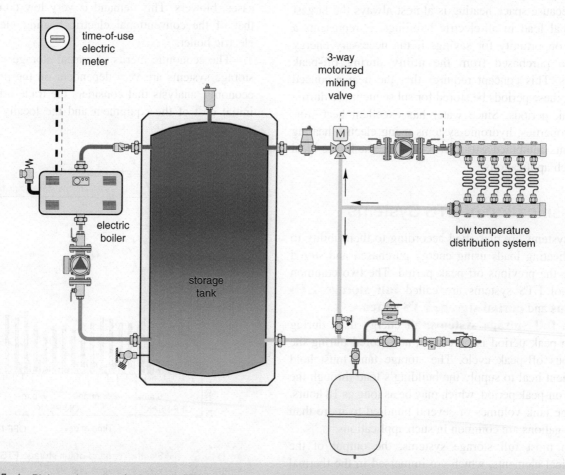

Figure 3-47 | Piping schematic of a hydronic ETS system coupled to a low-temperature floor heating distribution system.

These components will fit through standard passage doors or down a basement stairs, and thus allow a relatively large tank to be assembled at location that would not accommodate a factory-built pressure-rated tank. After the structural shell and insulation system are assembled, a waterproof EPDM (Ethylene Propylene Diene Monomer) or polypropylene liner is placed within it. When the tank is filled, static water pressure holds the liner tightly against the shell. Piping is connected to the tank using specialized **bulkhead fittings** that penetrate the tank wall and liner. Although these fittings are water resistant, it is best to keep all connections above the highest water level in the tank.

Unpressurized tanks must have a small pressure equalization hole or tube at the top of the tank. There will be some evaporation of water from the tank over time. The owner will need to periodically monitor water level and add water as required to keep the tank at its proper water level.

Heat is transferred into and out of the tank through heat exchangers. In some cases, the heat exchanger will be a coil of copper or stainless steel tubing supported by the tank shell. In other systems, an external brazed plate heat exchanger will be used.

An example of a site-assembled unpressurized tank and coiled copper heat exchanger is shown in Figure 3-48.

Tanks made of high-density polyethylene or polypropylene are also available in both cylindrical and rectangular shapes. If the tank is rectangular, its sidewalls must be structurally supported. Failure to do so will allow excessive bulging of the tank walls. Concentrated loading will also cause the polymer to "creep" over time, permanently deforming the tank.

All thermal storage tanks must be heavily insulated to limit heat loss. In the absence of codes or other mandated requirements, the author suggests that all thermal storage tanks have a complete insulation shell with a minimum thermal resistance of $R = 24$ °F·h·ft^2/Btu. Failure to properly insulate thermal storage tanks will result in excessive and uncontrolled heat loss and poor overall performance. Sprayed polyurethane foam insulation is commonly used for tank insulation. An example of a pressure-rated tank with 4 inches of rigid sprayed foam insulation is shown in Figure 3-49. Most building codes require interior applications of spray polyurethane foam to be covered with an **intumescent coating** or other jacketing that provides a barrier to limit potential fire spread.

3.13 Hydronic Heat Pump Fundamentals

A **heat pump** is a device for converting low-temperature heat into higher temperature heat. The low-temperature heat is obtained from some material called the "**source**," and then concentrated and released into another material called the "**sink**."

When used to heat buildings, heat pumps can gather low temperature heat from sources such as outdoor air, ground water, lakes or ponds, or tubing buried within the earth.

Heat pumps that extract low temperature heat from outside air are relatively common in North America. They are appropriately called **air-source heat pumps**.

(a) (b)

Figure 3-48 | Example of unpressurized tank (a) insulated structural shell; (b) suspended copper-tube heat exchanger. *Courtesy of American Solartechnics.*

Figure 3-49 Example of a pressure-rated thermal storage tank covered with spray polyurethane insulation to achieve a thermal resistance of R-24 °F·h·ft²/Btu.
Courtesy of John Siegenthaler.

Most heat pumps of this type are configured to deliver higher temperature heat through a forced-air distribution system within the building. This leads to the more specific classification of **air-to-air heat** pump. It is also possible to combine an air source heat pump with hydronic heat delivery. Such devices are called **air-to-water heat pumps**.

Heat pumps that extract low temperature heat from lakes, ponds, wells, or tubing buried in the earth use water, or a mixture of water and antifreeze, to convey heat from those sources to the heat pump. They are thus classified as **water source heat pumps**. Those that deliver heat through a forced-air systems are more specifically called **water-to-air heat pumps**. Those that deliver their heat using a hydronic distribution system are known as **water-to-water heat pumps**.

In this text, the term "hydronic heat pump" always refers to a heat pump that *delivers* heat to a flowing stream of water.

Heat Pump Refrigeration Systems

The heat pumps discussed in this text all operate using the **vapor compression refrigeration cycle**. During this cycle, a chemical compound called the **refrigerant** circulates through a closed piping loop passing through all major components of the heat pump. These major components within the heat pump are named based on how they affect the refrigerant passing through them. They are as follows:

- Evaporator
- Compressor
- Condenser
- Thermal expansion valve (TXV)

The arrangement of these components to form a complete refrigeration circuit is shown in Figure 3-50.

To describe how this cycle works, a quantity of refrigerant will be followed through the complete cycle.

The cycle begins at station (1) with cold liquid refrigerant in the **evaporator**. At this location, the refrigerant is at a lower temperature than the source media (e.g., air or water) passing across the evaporator. Because of this temperature difference, heat moves from the higher temperature source media into the lower temperature refrigerant. As the refrigerant absorbs heat, it changes from a liquid to a vapor (e.g., it evaporates). Large quantities of low temperature heat are absorbed as the refrigerant changes from a liquid to a vapor. The vaporized refrigerant continues to absorb heat until it is slightly warmer than the temperature at which it evaporates. The additional heat required to raise the refrigerant's temperature above its **saturation temperature** (e.g., the temperature where it vaporizes) is called **superheat**, and it also comes from the source media.

The vaporized and slightly superheated refrigerant vapor leaves the evaporator and flows into the **compressor** at station (2). Here, a reciprocating piston, or an orbiting scroll, both driven by an electric motor, compresses the vaporized refrigerant. This significantly increases both the pressure and the temperature of the refrigerant vapor. The electrical energy used to operate the compressor is also converted to heat and added to the refrigerant.

The temperature of the refrigerant vapor leaving the compressor is usually in the range of 120 to 170 °F, depending on the operating conditions. The hot refrigerant vapor passes into the **condenser** at station (3). In air-to-water heat pumps, and water-to-water heat pumps, the condenser is a refrigerant-to-water heat exchanger.

The hot refrigerant vapor gives up heat to the cooler stream of water flowing through the condenser, which carries the heat away from the heat pump, either directly to a load, or in some systems, to a thermal storage tank.

As it gives up heat within the condenser, the refrigerant condenses from a high pressure, high temperature vapor into a high pressure, but cooler liquid.

3.13 Hydronic Heat Pump Fundamentals

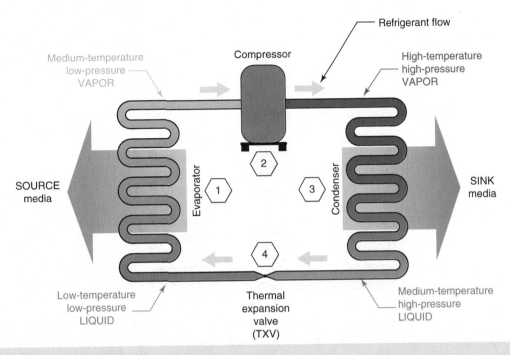

Figure 3-50 | Basic refrigeration components used in a heat pump.

The high pressure liquid refrigerant then flows through the **thermal expansion valve** at station (4), where its pressure is greatly reduced. The drop in pressure causes a corresponding drop in temperature, restoring the refrigerant to the same condition at which this description of the cycle began. The refrigerant is now ready to repeat the cycle.

The refrigeration cycle remains in continuous operation whenever the compressor is running. This cycle is not unique to heat pumps. It is used in refrigerators, freezers, room air conditioners, dehumidifiers, water coolers, soda vending machines, and other heat-moving machines. The average person is certainly familiar with these devices, but often takes for granted how they operate.

Figure 3-51 shows the energy flows created by the refrigeration cycle.

The arrow labeled (Q1) represents the low temperature heat absorbed by the evaporator. Arrow (Q2) represents the electrical energy used to operate the compressor. Arrow (Q3) represents the heat output from the heat pump's condenser.

The first law of thermodynamics dictates that, under steady-state operating conditions, the total energy input rate to the heat pump must equal the total energy output rate. Thus, the sum of the low temperature heat absorption rate (Q1) into the refrigerant at the evaporator, plus the rate of electrical energy input to the compressor (Q2), must equal the total rate of energy dissipation from the heat pump. The vast majority of this heat output will be at the condenser (Q3). A very small amount of heat output (Q4) occurs from the heat losses of components within the heat pump. The latter is typically ignored during routine performance testing. Thus, if any two of the three quantities (Q1, Q2, or Q3) are determined through measurements, the remaining quantity can be determined using the simplified relationship that Q1 + Q2 = Q3.

Reversible Heat Pumps

As previously discussed, all heat pumps move heat from a lower temperature material to another material at higher temperature. The **nonreversible heat pump** described thus far in the chapter can be used in a heating only application or a cooling only application.

In a heating only application, the heat pump's evaporator always gathers low temperature heat from some source where heat is freely available and abundant. The condenser always delivers higher temperature heat to a load. One example is a heat pump that is only used to heat a building. Another would be a heat pump that only delivers heat to produce domestic hot water.

In a cooling only application, the heat pump's evaporator always absorbs heat from a media that is intended to be cooled. The condenser always dissipates the unwanted heat to some media that can absorb and dissipate it (i.e., outside air, ground water, or soil). Examples would include a common air conditioner that can only remove heat from a building. Another would be

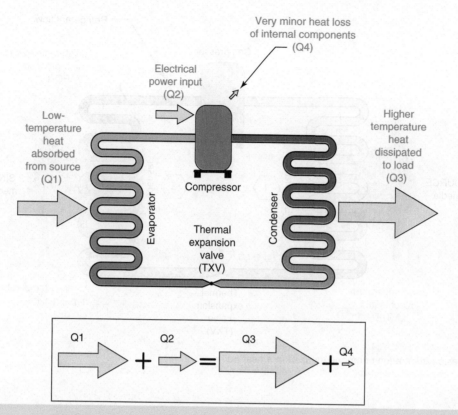

Figure 3-51 | Energy flows created by the refrigeration cycle within a heat pump.

a water-to-water heat pump that is used to chill water as part of an ice-making process.

Although there are some applications where nonreversible heat pumps can be used, one of the most unique benefits of modern heat pumps is that the refrigerant flow can be reversed to immediately convert the heat pump from a heating device to a cooling device. Such heat pumps are said to be reversible. A **reversible heat pump** that heats a building in cold weather can also cool that building during warm weather.

Reversible heat pumps contain an electrically operated device called a **reversing valve**. An example of such a valve is shown in Figure 3-52.

The type of reversing valves used in the heat pumps discussed in this text contain a slide mechanism that is moved by refrigerant pressure. The direction of movement is controlled by a small solenoid valve called a pilot solenoid valve. When the heat pump is in heating mode, the magnetic coil of pilot valve is not energized. This allows the high pressure refrigerant leaving the compressor to position the slide within the reversing valve so hot refrigerant gas from the compressor goes to the condenser, as shown in Figure 3-53a. In this mode, the condenser is the heat exchanger that transfers heat to a flowing stream off water, which carries that heat to the load.

Figure 3-52 | Example of a pilot operated reversing valve. *Courtesy of Ranco.*

When the heat pump needs to operate in cooling mode, the pilot valve is energized by a 24 VAC signal. This allows the refrigerant pressure to immediately move the slide within the reversing valve to the opposite end of its chamber. Hot gas leaving the compressor is now routed to the heat pump's other heat exchanger (e.g., what was the evaporator now becomes the condenser.) This is illustrated in Figure 3-53b.

3.13 Hydronic Heat Pump Fundamentals

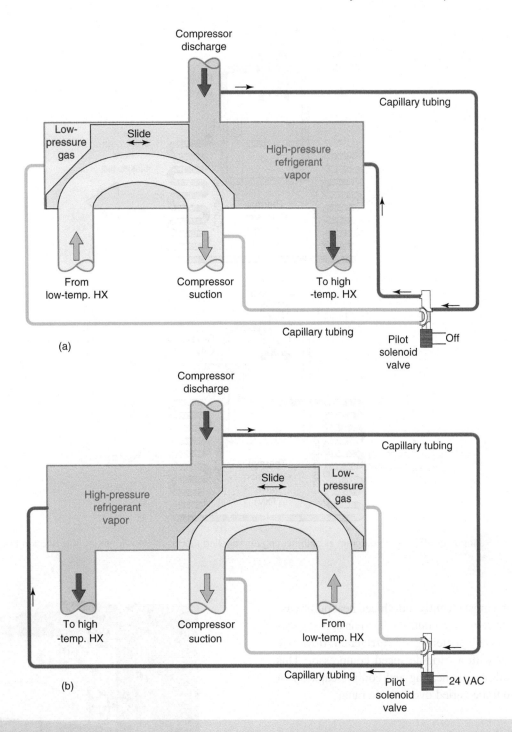

Figure 3-53 | (a) In heating mode, the pilot valve is not energized, and the slide is held to the left side of its chamber by refrigerant pressure. (b) In cooling mode, the pilot valve is powered on. Refrigerant pressure forces the slide to the right side of it chamber.

Figure 3-54 shows a heat pump refrigeration system that includes a reversing valve. Notice that the refrigerant heat exchanger on the left and right side of the refrigeration diagram change their function between serving as the evaporator or condenser, depending on the operating mode. It is customary to reference the heating mode function when describing these heat exchangers as either the evaporator or the condenser.

Some reversible heat pumps use two thermal expansion valves, in combination with two check valves. One thermal expansion valve functions in the heating mode while the other functions during the cooling mode. Other heat pumps also use a single "bi-directional" thermal expansion valve. For simplicity, the heat pump refrigeration piping diagrams shown in this text assume a single, bidirectional thermal expansion valve.

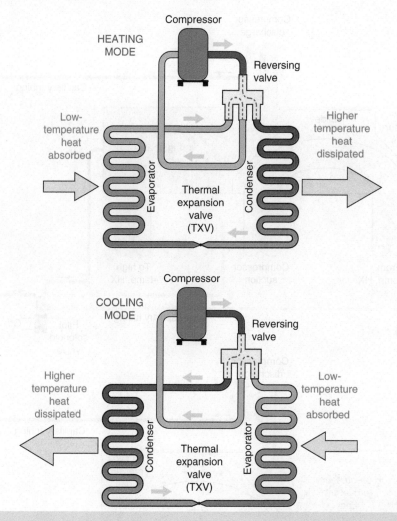

Figure 3-54 | Heat pump refrigeration diagram in both heating and cooling modes. Note refrigerant flow through the reversing valve.

Some of the most recently introduced heat pumps are also using electronically controlled expansion devices that allow precisely metering of refrigerant flow rate in combination with a variable speed compressor. This combination allows the heating and cooling capacity of the heat pump to be varied over a wide range.

3.14 Water-to-Water Heat Pumps

Currently, the most widely used hydronic heat pump is configured to extract heat from a water source, and as such is called a water-to-water heat pump. The evaporator is a water-to-refrigerant heat exchanger, and so is the condenser. Figure 3-55 shows an example of a water-to-water heat pump with the front access panel removed. Figure 3-56 shows the basic internal configuration of a water-to-water heat pump.

Figure 3-55 | Example of a water-to-water heat pump. *Courtesy of ClimateMaster Corporation.*

3.14 Water-to-Water Heat Pumps

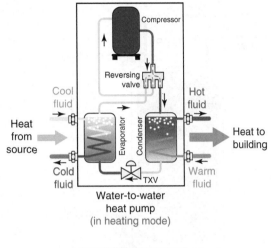

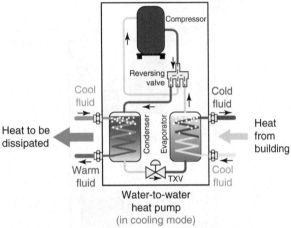

Figure 3-56 | Basic components and refrigerant flow in a water-to-water heat pump operating in either heating or cooling mode.

Most water-to-water heat pumps used for residential and light commercial building applications are on/off devices. Some are equipped with variable speed compressors, and thus modulating heating and cooling output rates are possible.

Heat pumps with fixed speed compressors usually require a buffer tank when connected to a zoned hydronic heating or cooling system. Heat pumps with variable speed compressors may or may not require a buffer tank, depending on how the distribution system is configured.

Desuperheaters

Many water-to-water heat pumps can be purchased with an optional **desuperheater**. This is a small refrigerant-to-water heat exchanger, which is installed in factory between the discharge port of the compressor and the inlet of the reversing valve as shown in Figure 3-57.

A desuperheater absorbs heat from the hot refrigerant gas leaving the compressor and transfers that heat to a stream of domestic water. This allows the heat pump to contribute to both space heating and domestic water heating. Desuperheaters are designed to extract the superheat of the refrigerant. This cools the refrigerant vapor, but not to the point where it begins to condense into a liquid. This process is called desuperheating. When the heat pump is operating in heating mode, heat transferred to domestic water at the desuperheater is not available for space heating. However, desuperheaters are designed so that only a fraction (usually 5 to 10%) of the total thermal energy in the hot refrigerant gas leaving the compressor is transferred to domestic hot water. When the heat pump operates in cooling mode, all heat transferred to domestic water is "free" heat. This is true because that heat would otherwise be dissipated to a heat sink such an earth loop, or another fluid stream that is cooling the heat pump's condenser.

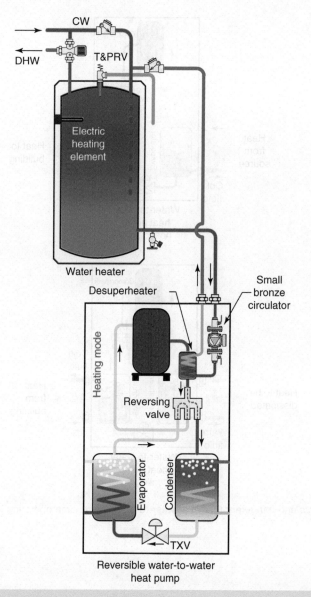

Figure 3-57 | Placement of a desuperheater heat exchanger within a reversible water-to-water heat pump, along with piping to a domestic hot water tank.

Heat pumps equipped with desuperheaters are also usually supplied with a small stainless steel circulator that creates flow between the hot water storage tank and the desuperheater. This circulator is usually wired so that it operates whenever the compressor is on. However, in some heat pumps, the desuperheater circulator can also be turned on and off depending on the operating mode. Some heat pumps also limit the temperature to which the desuperheater can heat domestic water. An upper temperature limit of 130 °F leaving the desuperheater is common.

Figure 3-57 shows water from the lower portion of water heater flowing into the desuperheater. Water leaving the desuperheater flows back to the dip tube in the cold water connection of the water heater. This arrangement allows the desuperheater to transfer heat to the cooler water in the tank, and thus, maintains maximum heat transfer.

Heat Sources for Water-to-Water Heat Pumps

Low-temperature source water for water-to-water heat pumps used for space heating and other applications can come from a variety of sources, including wells, lakes, or piping loops buried in the earth. In some applications, heat might even be extracted from a process water or wastewater stream.

Wells, ponds, or lakes are called **open-loop sources** since, at some location, the water they contain is exposed to the atmosphere. Water from such sources can be pumped through the evaporator of a heat pump where heat can be extracted from it. The chilled water can then be returned to its source where it will absorb heat and return to its original temperature.

A piping loop buried in the earth with its ends connected to the heat pump is an example of a **closed-loop source**. Fluid circulates through the buried piping loop whenever the heat pump is operating.

Open-Loop Sources

Most water-source heat pumps require 2.0 to 3.0 gallons per minute (gpm) of water flow through their evaporators, per **ton** (12,000 Btu/h) of heating capacity. During cold weather, a typical 4-ton residential system may require several thousand gallons of water to pass through the heat pump's evaporator every day. While this volume of water is readily supplied from a lake or large pond, it can place a tremendous load on a well. If that well cannot keep up with the demand, the heat pump will eventually shut itself off due to insufficient evaporator flow. The building will then be without heat. It is essential that any water source being considered to supply a water-source heat pump can consistently provide the necessary quantity of water. In the case of a well, always have a certified driller verify the well's sustained recovery rate before committing to its use in supplying a heat pump.

Since the water must be returned to the environment without contamination, it cannot be chemically treated before entering the heat pump. It is imperative that the water quality is suitable to be pumped through the heat pump's evaporator on a long-term basis. Groundwater with high concentrations of calcium, magnesium, iron, hydrogen sulfide, silts, or other materials is *not* suitable for such applications. Always have the proposed water source analyzed by a competent water testing agency, and the results approved as acceptable by the heat pump manufacturer, before committing to its use.

The temperature of the source water supplied to the heat pump is also an important consideration. Most water-to-water heat pumps have minimum inlet source water temperatures specified by their manufacturer. Modern water-to-water heat pumps can operate at conditions that will freeze water within their evaporator. When water is used as the source fluid, which is always the case for open loop systems, the heat pump must be configured to prevent freezing the evaporator. Many heat pumps have internal temperature controllers than monitor the inlet water temperature and can be set to turn off the refrigeration system if freezing conditions are imminent.

Closed-Loop Sources

In situations where open-loop water sources are not available or suitable for use, a closed-loop **earth heat exchanger** can supply low-temperature heat to a water-source heat pump. Such a heat exchanger usually consists of several hundred feet of buried polyethylene pipe through which water, or a mixture of water and nontoxic antifreeze, circulates. As this fluid passes through the heat pump's evaporator, heat is extracted from it. In the process, the temperature of the fluid drops below that of the soil surrounding the earth heat exchanger. The chilled fluid then makes another pass through the earth heat exchanger where it reabsorbs heat from the warmer surrounding soil. This process is illustrated in Figure 3-58.

Once filled and purged of air, no fluid should enter or leave the earth heat exchanger. This closed-loop approach eliminates the need for a well or other source of water at the site. It also eliminates the inevitable scaling or corrosion problems associated with pumping poor quality water from open-loop sources through the heat pump.

The type of pipe used for an earth heat exchanger is of critical importance to its longevity. Currently, high-density polyethylene pipe is the most widely used piping for earth heat exchangers. This pipe is available in a range of sizes and standard coil lengths up to 1,000 feet. It can flex to accommodate the stresses imposed by variations in soil temperature and moisture. Crosslinked polyethylene tubing (PEX) is also used for earth heat exchangers.

All joints in a polyethylene earth heat exchanger must be made using heat fusion. In this process, the two pipe ends are heated to over 500°F using specialized equipment, then fused together to form a joint stronger than the pipe itself. There can be no compromise in the materials or methods used to fabricate earth heat exchangers. When properly installed, they will last for many decades.

Figure 3-59 shows three cross-sections of typical horizontal earth heat exchanger configuration. The four-pipe square arrangement requires approximately 180 feet of trench (and 720 feet of pipe) per ton of heat pump capacity, based on a minimum fluid supply temperature of 30°F, when installed in heavy/wet soils in upstate New York. For installations in similar climates but with damp/sandy soils, the earth heat exchanger length increases to 260 feet of trench (or 1,040 feet of pipe) per ton of capacity. The additional piping is required due to the less favorable heat transfer characteristic of sandy soil. These loop lengths were calculated for a specific location (upstate New York), with

102 Chapter 3 Hydronic Heat Sources

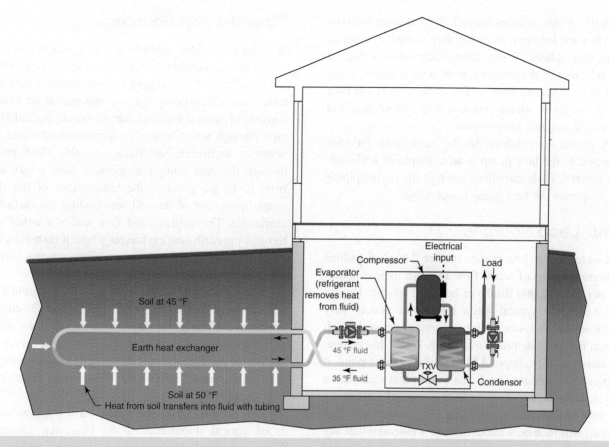

Figure 3-58 | Heat from 50 °F soil transfers into cooler fluid passing through earth loop heat exchanger.

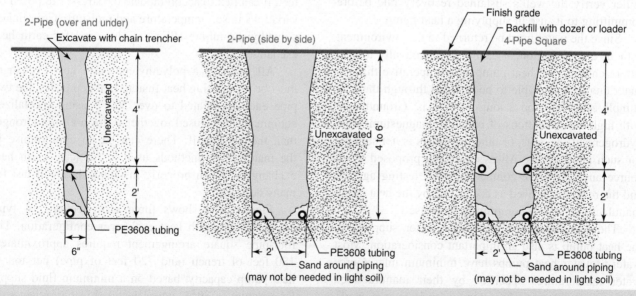

Figure 3-59 | Cross-section of some horizontal earth heat exchanger options.

specific minimum supply temperature to the heat pump (30 °F), and a specific piping geometry (four-pipe square at 5-feet average depth). The length of the earth heat exchanger will vary for other locations, soil types, piping layouts, and operating conditions. Many manufacturers of ground source heat pumps offer software to assist in properly sizing earth heat exchangers for specific installation requirements.

Another option is a vertical earth loop heat exchanger. It is constructed by boring one or more holes in the earth.

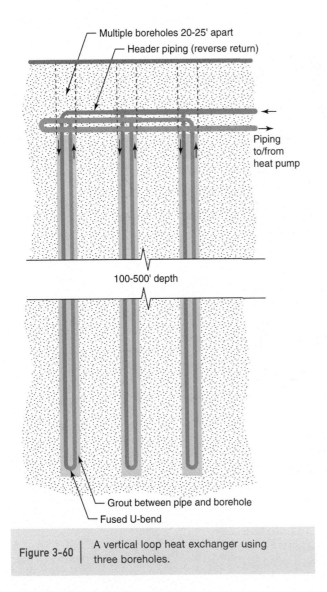

Figure 3-60 | A vertical loop heat exchanger using three boreholes.

These holes vary from 4 to 6 inches in diameter and can be up to 500-feet deep.

A piping assembly with a tight "U-bend" at its end is inserted into each **borehole**. The void between the piping and borehole is then filled with a "**grout**" consisting of bentonite clay and sand. The grout ensures good thermal conduction between the surrounding earth and the piping. When multiple vertical earth loops are installed, their respective piping assemblies are connected to common headers and the header piping extended into the building and to the heat pump. Multiple boreholes need a minimum separation of 20 to 25 feet depending on soil characteristics. Vertical earth heat exchangers are often used when the available land area or excavation conditions preclude used of horizontal trenches. Vertical earth loops can be installed under lawns, parking lots, and in some cases even under buildings. Figure 3-60 illustrates the concept of a vertical earth heat exchanger.

Complete information on designing and installing earth heat exchangers can be found in the references listed at the end of this chapter.

Performance of Water-to-Water Heat Pumps

The efficiency rating of a heat pump is called its **coefficient of performance (COP)**. COP is defined by Equation 3.9.

Equation 3.9:

$$\text{COP} = \frac{\text{Heat output of heat pump (Btu/h)}}{\text{Electrical input to run heat pump (Btu/h)}}$$

The higher the COP of a heat pump, the more heat it delivers for a given electrical energy input. A heat pump with a COP of 3.0 yields three times more heat output than the electrical energy required to operate it. This additional heat is not created by the heat pump, but instead is absorbed from the source water using the refrigeration cycle previously discussed.

Water-to-water heat pumps have COPs ranging from less than 2.0 to upwards of 5.0, depending on the temperatures of the source water as well as the load water. Because a heat pump can deliver significantly more heat than an electric boiler using the same amount of electricity, its operating cost will be significantly lower.

Figure 3-61 depicts the energy flows in a water-to-water heat pump connected to a low-temperature water source such as an earth heat exchanger. Notice the relative widths of the arrows representing the two energy flows into the heat pump and the single arrow representing energy flow out. The wider the arrow, the greater the rate of energy flow.

Both the coefficient of performance and the heating capacity of a water-to-water heat pump depend on the temperature of the entering source water and entering load water. Manufacturers rate the performance of their products at specific temperatures and water flow rates.

Figure 3-62 shows how the heating capacity of a nominal 3-ton water-to-water heat pump varies with the temperatures of the source water and the entering load water temperature (ELWT).

Notice how the heating capacity of the heat pump drops rapidly with decreasing source water temperature. The temperature of fluid from a horizontal earth loop will decrease during the heating season. Thus, the heating capacity of the heat pump in February will be lower than its heating capacity in October. This drop in heating capacity may require some type of auxiliary heat source

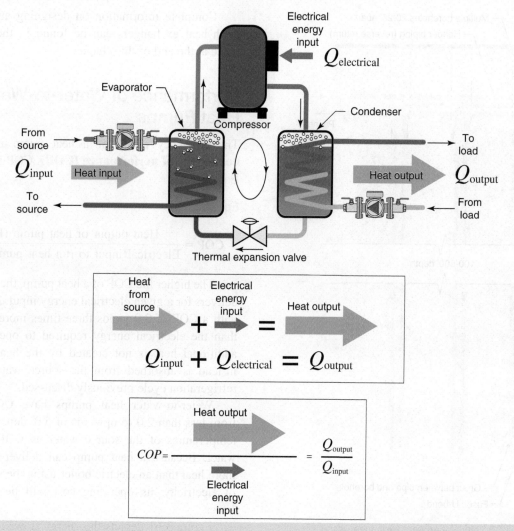

Figure 3-61 | Energy flows within a water-to-water heat pump.

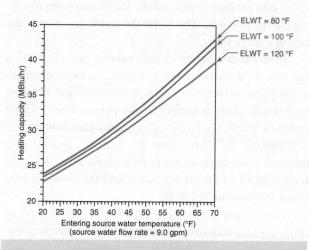

Figure 3-62 | Variations in heating capacity of a hydronic heat pump with varying source water and entering load water temperatures (ELWT).

to be incorporated into the system. The temperature of source water from a vertical earth loop will also decrease slightly over the course of a heating season.

It can also be seen that the heat pump's heating capacity decreases as the temperature of the entering load water increases. This implies that the lower the entering load water temperature is, the greater the heating capacities of the heat pump. For space heating applications, low-temperature heat emitters, such as radiant floor, wall, or ceiling panels, are good options for use with hydronic heat pumps.

Like heating capacity, the COP of a water-to-water heat pump is also affected by the entering source water temperature and entering load water temperature. The higher the source water temperature and the lower the entering load water temperature, the higher the heat pump's COP. Representative variations in COP due

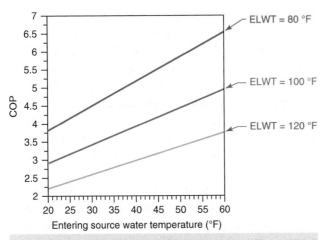

Figure 3-63 Variations in COP of a hydronic heat pump with varying source water and entering load water temperatures (ELWT).

to changes in source and load water temperatures are shown in Figure 3-63.

The COP data presented for water-to-water heat pumps typically does not include an allowance for the electrical energy need to operate circulators or pumps that drive flow through either the evaporator or condenser. This electrical usage could be substantial in the case of an open loop system where water is being lifted many feet from that static water level in a well. Likewise, the electrical power required by one or more circulators to move water through an extensive closed earth loop heat exchanger in a residential system can be several hundred watts. Since the owner pays for this electricity, it is reasonable to define a **net COP** that includes the electrical energy supplied to the heat pump and that supplied to the pump or circulator(s) associated with the operation of an open- or a closed-loop heat source.

The electrical power required to operate the *condenser* circulator may or may not be included in this calculation. An argument *for* including this power in a net COP calculation is that it is present whenever the heat pump is operating. An argument *against* including this energy is that many other hydronic heat sources such as most mod/con boilers, pellet boilers, or air-to-water heat pumps also require a dedicated circulator to move heat from the heat source to the balance of the system, and thus the electrical power supplied by the condenser circulator could be considered as approximately equivalent to that required for other hydronic heat sources.

Equation 3.10 can be used to calculate the net COP of a water-to-water heat pump including the electrical usage of the pump or circulator providing flow through the evaporator, but excluding the power used by the condenser circulator.

Equation 3.10:

$$COP_{13256} = \frac{Q}{(P_i + P_c)3.413}$$

where,
COP_{13256} = COP of heat pump as defined by ANSI 13256-2 standard (unitless)
Q = measured heat output from heat pump condenser (Btu/h)
P_i = measured electrical power to operate heat pump (watts)
P_c = assumed power demand of evaporator circulator (watts)

Example 3.7

The measured power demand of a water-to-water heat pump is 3,000 W. Its heat output is 40,000 Btu/h. The electrical power supplied to the earth loop circulator is 350 W. Calculate the COP of the heat pump only, and compare it to the net COP of the heat pump and earth loop circulator.

Solution: The COP of the heat pump only is:

$$COP = \frac{Q}{(P_i)3.413} = \frac{40,000}{(3,000)3.413} = 3.91$$

The net COP of the heat pump and earth loop circulator is calculated using Equation 3.10:

$$COP = \frac{Q}{(P_i + P_c)3.413} = \frac{40,000}{(3,000 + 350)3.413} = 3.50$$

Discussion:

The net COP is about 10% lower than the COP of the heat pump only. "Net" COP is a more accurate indicator of the merit of the heat pump since it includes an electrical power input that must be present whenever the heat pump operates. Those who design geothermal water-to-water heat pump systems should try to minimize the electrical power required by the pump or circulator(s) that create flow through the heat pump's evaporator. Use of high-efficiency circulators with electronically commutated motors is one way to do so. These circulators will be discussed in Chapter 7.

3.15 Air-to-Water Heat Pumps

Air-to-water heat pumps have an outside unit, similar to that of an air-to-air heat pump or central air conditioning system. However, the heat generated while operating in the heating mode is delivered to a hydronic distribution system within the building. When operating in the cooling mode air-to-water heat pumps deliver a stream of chilled water to an interior hydronic distribution system.

Figure 3-64 shows a modern air-to-water heat pump. The unit is mounted several feet above the ground due to heavy snow fall potential. The insulated pipes behind the unit connects it to the interior portion of the system.

The fans seen at the front of the heat pump pull outside air across the air-to-refrigerant heat exchanger on the back side.

The unit shown in Figure 3-64 is an example of a **monobloc** air-to-water heat pump. It contains all the components necessary for the refrigeration cycle within the outdoor unit. It may also contain devices such as a circulator, expansion tank, and system controller within the outdoor unit. Monobloc heat pumps are leak tested during manufacturing and factory-charged with refrigerant.

Systems designed using monobloc air-to-water heat pumps vary depending on the severity of the winter climate. In relatively mild climates, where outdoor temperature only occasionally drops below freezing, it is generally acceptable to install the heat pump with water in the piping circuit between the outdoor unit and indoor distribution system. Most monobloc air-to-water heat pumps have controls that automatically turn on the circulator when near freezing conditions are detected. Some can also turn on an auxiliary electric heating element. These components maintain the water filled portion of the heat pump above freezing, even when there is no load on the system.

In cold climates, or situations where a prolonged power outage could occur during sub-freezing weather, and no backup generator capable of running the heat pump is available, monobloc air-to-water heat pump should be installed as part of an antifreeze-protected hydronic circuit. One approach is to fill the entire system with a suitable concentration of antifreeze. Another is to isolate the heat pump from the distribution system using a generously sized heat exchanger. Figure 3-65 illustrates the latter approach.

To minimize the thermal penalty associated with the heat exchanger, it should be generously sized for a maximum **approach temperature** of 5 °F. Chapter 14 describes methods for sizing heat exchangers.

Figure 3-66 shows the fundamental refrigeration system components in a monobloc air-to-water.

The depictions in Figure 3-66 show a circulator within the heat pump. Some manufacturers provide this circulator, while others do not. In cases where it is provided, designers need to verify that the circulator is adequate for the intended flow rate. Chapter 7 provided methods for making such determinations.

Split System Air-to-Water Heat Pumps

Air-to-water heat pumps are also available as "**split systems**." The refrigerant-to-water heat exchanger, as well as the circulator, expansion tank, and any other components that contain water, are located within an *indoor* unit. The compressor, and other refrigeration components, are typically housed in the outdoor unit. Two copper refrigerant tubes, as well as electrical wiring, connect the indoor and outdoor units as shown in figure 3-68. Figure 3-67 shows the outdoor unit, the indoor unit, and the insulated refrigerant lines connecting them.

Figure 3-68 illustrates how the outdoor and indoor units of a split system air-to-water heat pump are connected.

One advantage of a split system is the absence of water in the outdoor portions of the system. This eliminates any need to freeze protect the outdoor components. However, split systems require the installation of refrigerant

Figure 3-64 — Example of a monobloc air-to-water heat pump. The unit is mounted above potential snow accumulation at installation site. *Courtesy of John Siegenthaler.*

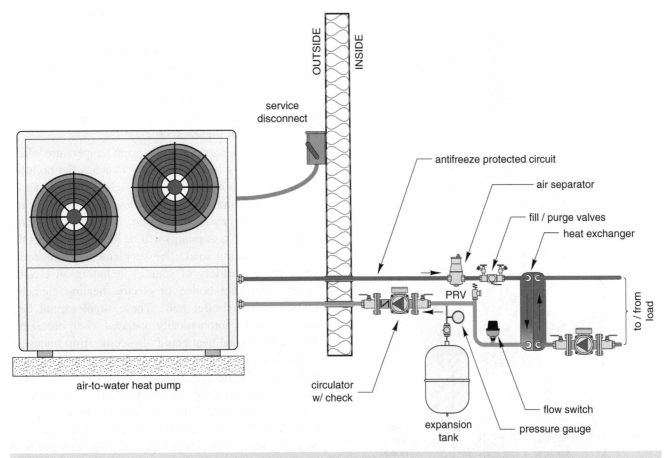

Figure 3-65 | A monobloc air-to-water heat pump isolated from the remainder of the system using a heat exchanger. The circuit between the heat pump and heat exchanger operates with an antifreeze solution.

piping, which typically necessitates the skills and tools of a refrigeration technician.

Performance of Air-to-Water Heat Pumps

As was true for water-to-water heat pumps, the heating capacity, cooling capacity, and coefficient of performance (COP) are dependent on the temperature of the source media and the sink media. For an air-to-water heat pump operating in heating mode, the source media is outdoor air. The sink media is the water entering the heat pump's condenser. The closer these two temperatures are to each other, the higher the heating capacity and COP, as illustrated in Figure 3-69.

Figure 3-70 shows a representative relationship between the heating capacity of a modern air-to-water heat pump as a function of both outdoor air temperature and the water temperature leaving the heat pump's condenser. The heating capacity is very dependent on outdoor temperature. As outdoor temperature decreases so does heating capacity. It's also apparent that heating capacity decreases as the temperature of water leaving the heat pump's condenser increases.

Figure 3-71 shows how the coefficient of performance (COP) for the same heat pump represented in Figure 3-70 varies with outdoor temperature and leaving water temperature. As with heating capacity, COP decreases as the outdoor temperature drops. There's also a very significant decrease in COP as the water temperature leaving the heat pump increases.

These performance graphs infer the importance of matching air-to-water heat pumps with low temperature hydronic distribution systems. The lower the water temperature at which the hydronic system can maintain building comfort, the better the performance of the air-to-water heat pump.

Sizing Considerations for Air-to-Water Heat Pumps

Since heating capacity is very dependent on outdoor temperature, an air-to-water heat pump cannot be selected in the same manner as an on/off boiler, which

108 Chapter 3 Hydronic Heat Sources

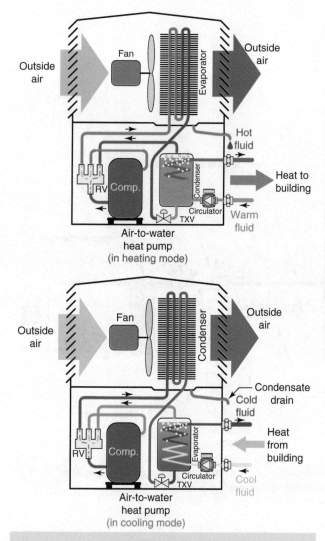

Figure 3-66 | Major components in a monobloc air-to-water heat pump.

has a fixed heat capacity. There will be times when the heating capacity of an air-to-water heat pump exceeds the building heating load. There will also be times when its heating capacity is less than the building heating load. These two possibilities imply there will also be some outdoor temperature at which the heat pump's capacity exactly matches building heating load. These conditions are illustrated in Figure 3-72.

In Figure 3-72, the balance point temperature where the heat pump's heating capacity matches the building load happens to occur an outdoor temperature of about 16 °F. At outdoor temperatures less than 16 °F, some form of supplemental heating (e.g., heat input above that supplied by the heat pump) will be required to meet the building's heating load. The supplemental heat could come from a variety of sources including an electric boiler, fossil-fuel boiler, or electric heating elements immersed in a buffer tank. These supplemental heat sources can be automatically operated when necessary. The supplemental heat could also come from manually controlled devices such as a wood stove, pellet stove, and individual electric space heaters. In most cases, occupants prefer to have fully automatic heating regardless of outdoor temperature. Figure 3-73 shows the concept for how an electric boiler could be installed in parallel with an air-to-water heat pump to provide supplemental heat input when necessary.

Notice that both heat sources can be fully isolated from the system if ever necessary. Both heat sources are also equipped with pressure relief valves. The air-to-water heat pump has an internal circulator to provide flow between it and the buffer tank. The electric boiler is shown with an external circulator. A spring-loaded check

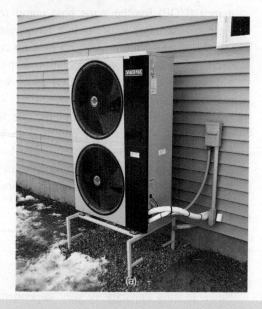
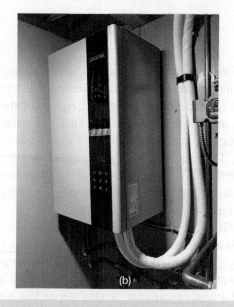

Figure 3-67 | (a) Outdoor unit of a split-system air-to-water heat pump *Courtesy of John Siegenthaler,* (b) indoor unit. The white insulation is covering refrigerant tubing that connects the outdoor unit to the indoor unit. *Courtesy of John Siegenthaler.*

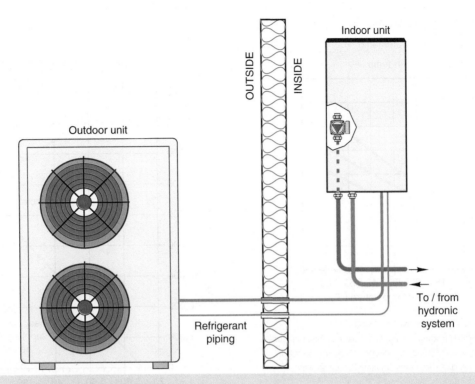

Figure 3-68 | Refrigerant tubing and electrical wiring connect the indoor and outdoor units of a split system air-to-water heat pump.

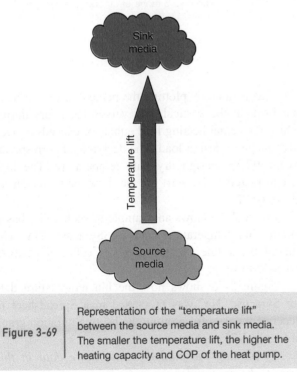

Figure 3-69 | Representation of the "temperature lift" between the source media and sink media. The smaller the temperature lift, the higher the heating capacity and COP of the heat pump.

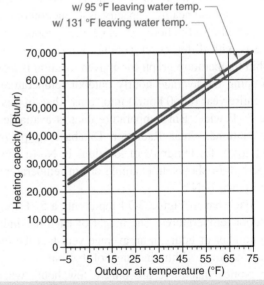

Figure 3-70 | Relationship between heat capacity and outdoor air temperature for a specific air-to-water heat pump, based on two different leaving water temperatures.

valve is installed in each heat source pipe assembly. These valves prevent flow reversal through an inactive heat source when the other heat source is operating. They also prevent reverse thermosiphon flow from the buffer tank back through either heat source. If not prevented, thermosiphon flow will dissipate heat from the tank through the piping passing through the heat sources.

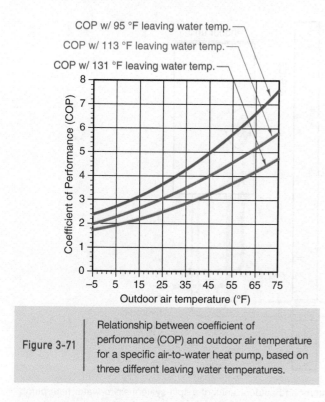

Figure 3-71 | Relationship between coefficient of performance (COP) and outdoor air temperature for a specific air-to-water heat pump, based on three different leaving water temperatures.

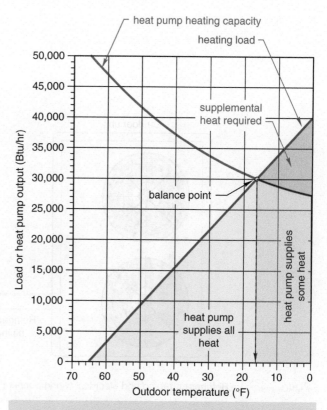

Figure 3-72 | Heating capacity of a representative air-to-water heat pump versus space-heating load over a range of outdoor air temperatures.

Bin Temperature Analysis

When selecting an air-to-water heat pump, it's important to compare the heating capacity as a function of outdoor temperature to the hours at which the outdoor temperature is with different temperature ranges called "bins." A bin temperature graph for a given location is made by grouping the average hourly outdoor temperature over an entire year (8760 hours) into "bins" that, in this case, are 5 °F wide. Bin temperature data is available from the National Solar Radiation Database (https://nsrdb.nrel.gov/) for hundreds of locations in North America. Figure 3-74 shows an example of bin temperature data for Syracuse, NY.

Each bar in Figure 3-74 represents a 5 °F range of outdoor temperatures. The height of each bar indicates the number of hours in an average year that the outside temperature is within the range of that bin.

Notice that there are only a few hours when the outdoor temperature at this upstate NY location is below 0 °F. However, there are many hours when the average outdoor temperature is in the range of 30 to 35 °F, or 40 to 45 °F. This graph reveals that the vast majority of an average heating season occurs at outdoor temperature well above the 97.5% design dry bulb temperature, which in this location is 2.0 °F.

Since heating load is usually modeled as directly proportional to the difference between inside and outside temperature, it's possible to relate the heating load to the outdoor temperature bins.

This is done by plotting the percent of design heating load on the vertical axis versus the hours during which the actual heating load equals or exceeds a given percentage of design load. The design load corresponds to the 97.5% design dry bulb temperature. The load is considered to be zero for any outdoor temperatures above 65 °F.

Figure 3-75 shows an example of such a plot based on the bin temperature data for Syracuse, NY. The curved black line on this graph is called a **heating duration curve**.

Figure 3-76 shows how the bin temperature data for Syracuse, NY, was processed in a Spreadsheet to generate the data for plotting the heating duration graph in Figure 3-75.

By plotting the outdoor temperature data in this format, the area under the curve becomes proportional to the total seasonal space heating energy required.

It's also possible to make a heating duration graph by plotting Btu/h load on the vertical axis (rather than percent of design heating load) versus the number of hours during which the load equals or exceeds the value on the vertical axis.

3.15 Air-to-Water Heat Pumps

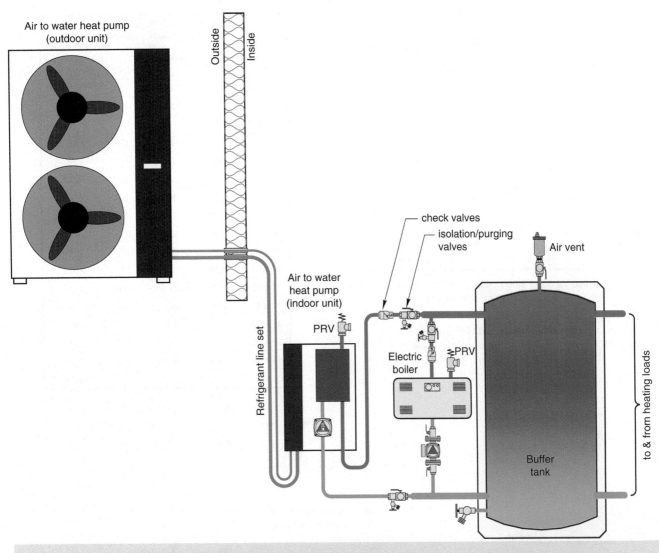

Figure 3-73 | An electric boiler installed in parallel with a split system air-to-water heat pump. The boiler provides supplemental heat input to the buffer tank and ultimately to the load.

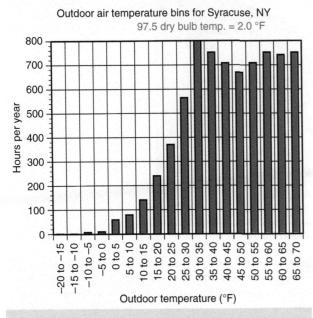

Figure 3-74 | Bin temperature data for Syracuse, NY.

A heating duration graph can be used to estimate how many hours of the heating season (e.g., whenever outdoor temperatures are < 65 °F) the space heating load equals or exceeds a given percentage of design load. For example, in Figure 3-75, the space heating load, on an average climatological year in Syracuse, NY, equals or exceeds 40% of design load for approximately 3,400 hours per year.

A heating duration graph can also be used to estimate what percentage of the total space heating energy could be supplied by a heat source having a capacity less than design load. This is illustrated in Figure 3-77.

The black curve is the heating duration curve for a specific location (based on average hourly outdoor temperature). The green line represents the heating capacity of a specific air-to-water heat pump. The left end of the green line (e.g., point A) is the heating capacity of the heat pump at design outdoor temperature.

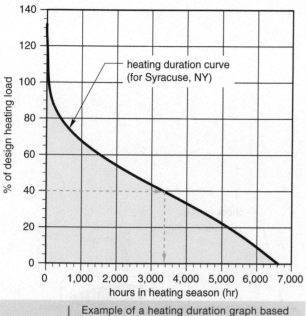

Figure 3-75 | Example of a heating duration graph based on hourly weather data for Syracuse, NY. The yellow area under the curve is proportional to the total seasonal space-heating energy required.

Bin range (°F)	Hours in bin (hours)	Average outdoor temp of bin (°F)	% of design load	Hours during which load ≥ this % of design load
-20 ≤ -15	1	-17.5	132	1
-15 ≤ -10	1	-12.5	124	2
-10 ≤ -5	7	-7.5	116	9
-5 ≤ 0	10	-2.5	108	19
0 ≤ 5	59	2.5	100	78
5 ≤ 10	79	7.5	92	157
10 ≤ 15	141	12.5	84	298
15 ≤ 20	241	17.5	76	539
20 ≤ 25	370	22.5	68	909
25 ≤ 30	564	27.5	60	1473
30 ≤ 35	800	32.5	52	2273
35 ≤ 40	752	37.5	44	3025
40 ≤ 45	708	42.5	36	3733
45 ≤ 50	669	47.5	28	4402
50 ≤ 55	709	52.5	20	5111
55 ≤ 60	753	57.5	12.5	5864
60 ≤ 65	742	62.5	0	6606

Figure 3-76 | Converting bin temperature data for Syracuse, NY, in a spreadsheet prior to plotting a heating duration curve.

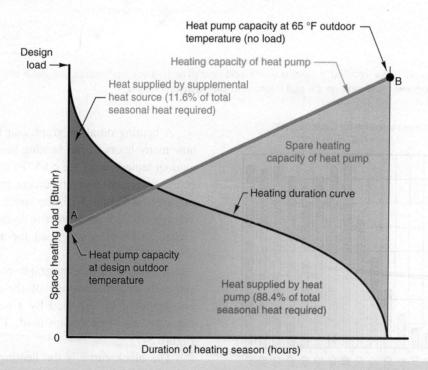

Figure 3-77 | The green line represents the heating capacity of the heat pump, from point A at design outdoor temperature, to point B at no load. The blue area under the heating duration curve represents the space heating energy supplied by the heat pump. The orange area represents the space heating energy supplied by the supplemental heat source. The percentages shown are specific to this graph.

The right end of the green line (e.g., point B) is the heat pump's capacity at a "no load" condition, which is usually assumed as 65 °F outdoor temperature. The blue and orange shaded areas on the graph represent percentages of the total seasonal heating energy that are supplied by the heat pump (shown in blue), and some type of supplemental heater (shown in orange). For this particular case, the air-to-water heat pump supplied 88.4% of the total area under the black curve (e.g., 88.4% of the total seasonal space heating energy requirement), and the supplemental heat source supplied the remaining 11.6%.

If an air-to-water heat pump with a *higher* heating capacity was used, the green line would shift upward, and the percentage of the total space heating energy supplied by the heat pump would increase. If a heat pump with a *lower* heating capacity was used, the green line would shift down and the percentage of the total energy supplied by the heat pump would decrease. These effects are shown in Figure 3-78.

The graphs in Figures 3-77 and 3-78 are based on the following two assumptions:

1. The heating energy required is based solely on outdoor temperature with no offsets for factors such as internal heat gain from sunlight, interior lighting, occupants, or equipment that may be present. As such, the total seasonal heating energy requirement is likely to be conservatively *overestimated*, especially in situations where there are significant internal heat gains.
2. The heat pump is assumed to be capable of operating at outdoor air temperatures as low as the outdoor design temperature. This may or may not be true depending on the specific heat pump and the climate where it is installed. All modern air-to-air heat pumps can operate at outdoor temperatures down to at least 15 °F. Some are rated down to outdoor temperatures of −13 °F, and a few can operate at outdoor temperatures as low as −22 °F.

If the heat pump is incapable of operating at outdoor temperatures as low as the design temperature, the graph shown in Figure 3-77 can be modified as shown in Figure 3-79.

The orange shaded area at the left of the graph shows that the supplemental heat source must assume the entire load when the outdoor temperature is lower than the minimum operating temperature of the heat pump. The colder the climate, and the higher the minimum air temperature at which the heat pump can operate, the higher the percentage of total heating load shifted to the supplemental heat source.

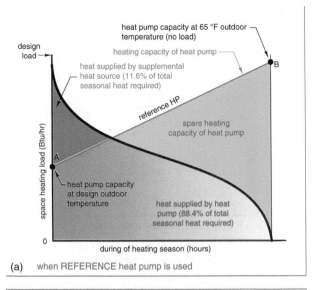
(a) when REFERENCE heat pump is used

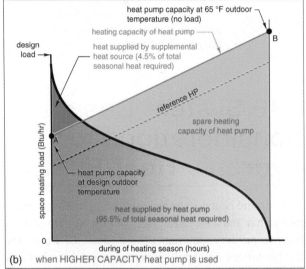
(b) when HIGHER CAPACITY heat pump is used

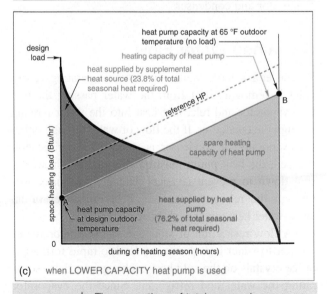
(c) when LOWER CAPACITY heat pump is used

Figure 3-78 | The proportions of total seasonal heat supplied by the heat pump and supplemental heat source vary depending on the rated heating capacity of the heat pump in comparison to the heating duration curve.

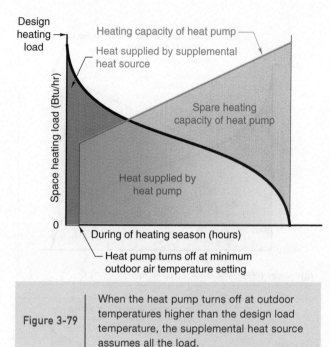

Figure 3-79 When the heat pump turns off at outdoor temperatures higher than the design load temperature, the supplemental heat source assumes all the load.

3.16 System Design Considerations for Hydronic Heat Pumps

All hydronic heat pumps have unique operating requirements that must be addressed during system design. The two primary requirements are sustained water flow and limitations on supply water temperature to both the evaporator and condenser.

Sustained Water Flow

Whenever a water-to-water heat pump is operating, refrigerant is removing heat from the water passing through the evaporator and releasing heat into the water passing through the condenser. If the flow of water through either the evaporator or condenser is stopped or significantly reduced, the heat pump must quickly and automatically shut down to prevent physical damage. Adequate flow through the refrigerant-to-water heat exchanger of an air-to-water heat pump is also necessary.

A flow restriction or stoppage in the evaporator of a water-to-water heat pump can lead to rapid formation of ice crystals on the cold heat exchanger surfaces separating the water and refrigerant. Since the evaporator contains only a small amount of water, the ice build-up can quickly choke off flow. A "hard freeze" can occur in less than one minute of operation without water flow. A number of such hard freezes will eventually rupture the copper tubing in the evaporator and result in a costly repair. Because of this possibility, manufacturers install a sensor that monitors the water temperature entering the evaporator. This sensor connects to the heat pump's internal controller, which can turn off the compressor if freezing conditions are imminent.

Maintaining water flow through the condenser whenever the heat pump is operating is equally important. A flow restriction in this stream can cause the refrigerant head pressure to increase rapidly. Eventually, this high pressure will trip a pressure safety switch that shuts down the compressor.

Air-to-water heat pumps have a similar flow requirement for their refrigerant to water heat exchangers. Inadequate flow rates will eventually cause the heat pump to turn off.

Automatic shut downs of either a water-to-water heat pump or an air-to-water heat pump typically require the electrical power to the unit to be turned off to reset the fault situation. While this is prudent and acceptable for an occasional abnormal condition, it must be avoided on a frequent basis due to improper system design.

Heat pump manufacturers usually require specific minimum flow rates for both the evaporator and condenser. In the absence of such requirements, the system should provide 2.0 to 3.0 gpm of water flow per ton (12,000 Btu/h) of heat transfer ability in both the evaporator and condenser.

Designers must be careful not to install devices such as zone valves or three-way mixing valves in series with the heat pump condenser. If such valves are present, they may restrict or totally stop water flow through the condenser causing an automatic shutdown of the heat pump's compressor.

One method for combining a hydronic heat pump with traditional valve-based zoning is to install a **buffer tank** between the heat pump and the distribution system as shown in Figure 3-80.

The buffer tank acts as a thermal reservoir between the heat pump and the distribution system. Heat is added to the tank by the heat pump and removed by the distribution system at different rates. The heat pump is simply turned on and off as necessary to maintain the buffer tank temperature within a preset temperature range.

When the building requires heat, water from the buffer tank can be circulated through the distribution system regardless of whether the heat pump is on or off. The buffer tank must be well insulated since it contains heated water. Its size depends on the capacity of the heat pump, the temperature differential at which the heat pump will be turned on and off, and the desired minimum running time of the heat pump once it is turned on.

3.16 System Design Considerations for Hydronic Heat Pumps

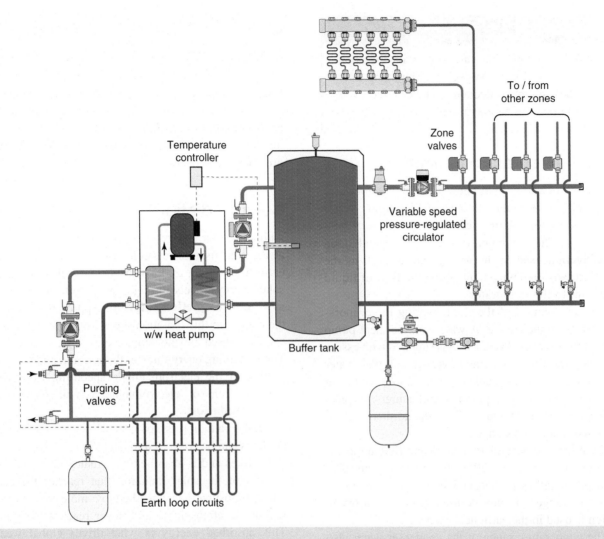

Figure 3-80 | Use of a buffer tank between heat pump and zoned distribution system.

Equation 3.10 can be used to determine the minimum volume of the buffer tank for a given set of operating conditions.

Equation 3.11:

$$V = \frac{t \times (Q_{hp} - Q_{load})}{\Delta T \times 500}$$

where,

V = minimum volume of buffer tank (gal);
t = desired minimum on-time of heat pump cycle (min);
Q_{hp} = output of heat pump while operating (Btu/h);
Q_{load} = heat flow to load when heat pump is on, if any (Btu/h);
ΔT = operating temperature differential of tank (°F).

Example 3.8

A hydronic heat pump has a rated heat output of 30,000 Btu/h. Determine the necessary buffer tank size to allow the heat pump to run for a minimum of 10 minutes while heating the buffer tank from 95 to 120 °F. Assume that the load at the time is zero.

Solution:

Inserting the data into Equation 3.10 yields:

$$V = \frac{t \times (Q_{hp} - Q_{load})}{\Delta T \times 500} = \frac{10 \times (30{,}000 - 0)}{(120 - 95) \times 500} = 24 \text{ gallons}$$

> **Discussion:**
> A tank larger than that calculated with Equation 3.11 will allow longer on-cycles (minimizing wear on the heat pump components) and/or a narrower range of operating temperature.

Water Temperature Limitations

All water-to-water heat pumps and air-to-water heat pumps have ranges of acceptable supply water temperature for their refrigerant-to-water heat exchangers. These temperature ranges depend upon the refrigerant used in the heat pump, as well as the type of expansion device controlling the flow of liquid refrigerant to the evaporator.

The temperature of the fluid entering the evaporator depends on the source. A water-to-water heat pump supplied from a drilled water well is supplied with essentially constant water temperature year-round. Well water temperature data are available for specific locations from most water source heat pump manufacturers. Typical values are between 45 and 60 °F for the northern half of the United States and Canada.

Earth heat exchangers typically yield fluid temperatures ranging from 25 to 70 °F depending on the location, depth, length, pipe size, and time of year. The lower end of this range can only be used when an antifreeze solution is used in the earth heat exchanger.

The temperature range of the condenser water also depends on the refrigerant used in the heat pump. Most hydronic heat pumps using R-410A refrigerant have a practical upper temperature limit in the range of 125 to 130 °F. There are some water-to-water heat pumps capable of supplying water at temperatures up to 140 °F. However, the higher the load water temperature, the lower the heating capacity and COP.

Entering load water temperature is mostly determined by the heat emitters used in the system. Slab-type radiant floor systems with low resistance coverings are well suited for use with hydronic heat pumps because they operate with supply water temperatures of 85 to 110 °F. By comparison, legacy finned-tube baseboard distribution systems typically require higher water temperatures at design load conditions. In some systems these temperatures are higher than what the heat pump can supply, and require the system to switch to an auxiliary heat source such as a boiler or electric resistance heating elements.

Methods for determining the required supply water temperature of various hydronic heating distribution systems are discussed in later chapters.

3.17 Solar Thermal Systems

Solar thermal systems can be configured to provide domestic water heating, space heating, and in some cases both. In this chapter, we will examine solar collectors and one type of **solar thermal combi-system** that provides space heating and domestic hot water.

Solar Radiation

Although solar energy is produced by nuclear reactions at the sun, its transmission through space and use on earth has nothing to do with nuclear radiation. Instead, solar energy travels through space and arrives at earth as **electromagnetic radiation**. Approximately half of the energy in this radiation lies within wavelengths that can be sensed by the human eye (e.g., visible light). The remaining energy lies in the infrared and ultraviolet portion of the electromagnetic spectrum. Collectively, this spectrum of electromagnetic radiation is called "**solar radiation**."

Solar radiation is not heat. However, the energy it contains can be converted to heat the instant solar radiation strikes and is absorbed by a material. A significant amount of the solar radiation that reaches the earth's outer atmosphere is absorbed by molecules of gases within the atmosphere, and never reaches the ground. This absorbed energy is what drives global weather systems.

Only the solar radiation that passes through the earth's atmosphere is useful for heating buildings or domestic water. All solar thermal systems use a **collector array** consisting of one or more **solar collectors** to intercept and capture solar energy. The two most common types of solar collectors are called **flat plate collectors** and **evacuated tube collectors**.

Flat Plate Solar Collectors

An example of a typical flat plate solar collector is shown in Figure 3-81.

The heart of a flat plate collector is its **absorber plate**. It is assembled from thin strips of copper that are metallurgically bonded to copper tubes. Several of these fin-tube assemblies are placed side by side and brazed to copper header tubes at both ends as shown in Figure 3-82.

The upper surface of the absorb plate is coated with an electroplated **selective surface**. This coating absorbs approximately 95% of the solar radiation striking it

3.17 Solar Thermal Systems

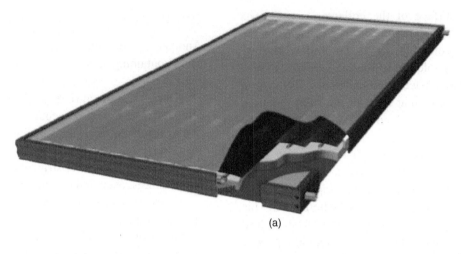

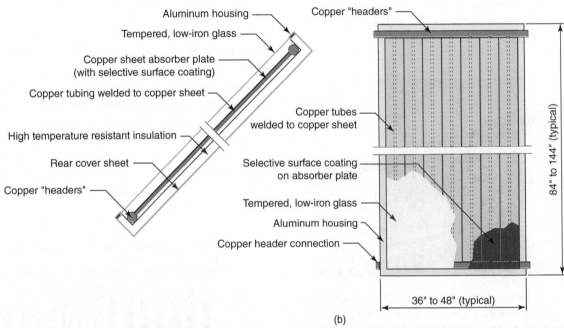

Figure 3-81 | (a) Appearance of a typical flat plate collector. (b) Cross-section of flat plate collector. *Courtesy of Caleffi North America.*

Figure 3-82 | Small section of absorber plate showing copper fin tubes with selective surface coating and brazed to copper header. *Courtesy of Solar Skies, Inc.*

and releases only about 10% of the absorbed energy as infrared radiation. The instant solar radiation strikes the selective surface, most of it is converted to heat. Most absorber plates have low thermal mass. This allows the plate's temperature to increase quickly when solar radiation strikes it. The heat generated at the plate's surface is conducted laterally across the copper fins, transferred to the copper tubes, and absorbed by the fluid within the tubes. If the fluid is flowing, it carries a high percentage of the absorbed heat out of the collector.

The absorber plate is surrounded by an insulated enclosure made of aluminum. This enclosure provides structure and a durable, weather-resistant enclosure for the absorber plate. Insulation is installed between the absorber plate and the back and sides of the enclosure to reduce heat loss.

The upper surface of the collector is called its glazing. Modern flat plate solar collectors use tempered glass with low iron oxide content for their glazing. Such glass is very tolerant of thermal stresses as well as snow loads and impacts from hailstones. Its low iron oxide content ensures that most of the solar radiation ($\geq 90\%$) that strikes the glazing passes through to the absorber plate.

A collector array consisting of five flat plate solar collectors is shown in Figure 3-83. In this system, the collectors are attached to a roof that is sloped at an angle of the local latitude plus 15° up from the horizontal. This is ideal for collectors used in solar combi-systems. Close examination of the photo also reveals sloping piping at the bottom of the collector array. This is an essential detail for **drain-back systems** and will be discussed later in this section.

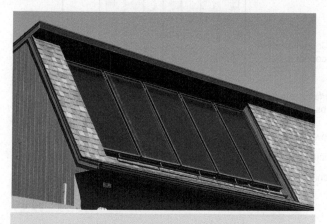

Figure 3-83 | An array of five flat plate collectors mounted at a slope of latitude plus 15°. This array is part of a drain-back solar combi-system.
Courtesy of John Siegenthaler

Evacuated Tube Solar Collectors

Another type of collector consists of several glass tubes, each containing a narrow absorber strip as shown in Figure 3-84.

During manufacturing, nearly all air in these tubes is removed. The resulting vacuum greatly reduces convective heat loss between the absorber strip and glass tube. In effect, the evacuated tube acts as a "Thermos® bottle" surrounding the absorber strip, allowing it to attain high temperatures, even under cold ambient air conditions.

Most current-generation evacuated tubes have a specialized fluid sealed within a concentric copper tubing assembly that is bonded to the absorber strip. When heated, this fluid changes phase from liquid to vapor and rises toward the top of the tube. It passes into a small copper "condenser" cylinder that fits tightly into a header assembly at the top of the collector. Heat transfers from the hot vapor through the walls of the copper condenser and into fluid circulating through the header. As heat is released, the fluid in the tubes condenses and flows back to the bottom of the tube ready to repeat the cycle.

Several evacuated tubes are mounted into a common header/frame assembly on site to create a collector assembly. Several collector assemblies can be combined to form a larger collector array as shown in Figure 3-85.

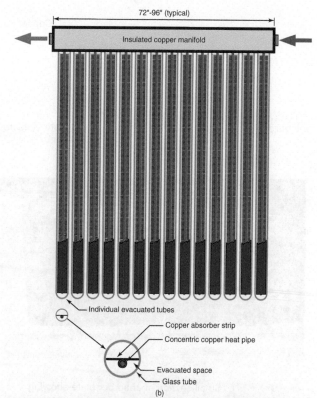

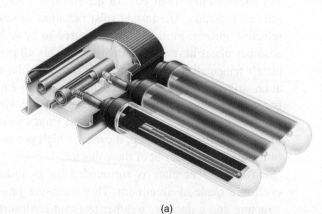

(a)

Figure 3-84 | (a) Individual evacuated tubes attached to a U-tube header. *Courtesy of Viessmann*, (b) Individual evacuated tubes attached to a flow-through header.

3.17 Solar Thermal Systems

Figure 3-85 | An array of evacuated tube collectors.
Courtesy of John Siegenthaler

Solar Collector Performance

Flat plate and evacuated tube solar collectors have different performance characteristics. Each lends itself to particular applications. When designing solar thermal systems, it is important to have a method of representing the thermal performance characteristics of both types of collectors over a wide range of operating conditions.

One method of expressing the performance of a solar collector is a numerical value for thermal efficiency. The *instantaneous* thermal efficiency is defined as the ratio of the heat output from the collector divided by the rate solar radiation strikes the panel. Equation 3.12 can be used to calculate the instantaneous thermal efficiency of a solar collector based on measured operating characteristics.

Equation 3.12:

$$n = \frac{(8.01\, Dc)f\, (T_{out} - T_{in})}{I(A_{gross})}$$

where,
- n = instantaneous thermal efficiency (decimal percent);
- D = density of fluid passing through collector (lb/ft^3);
- c = specific heat of fluid passing through collector (Btu/lb/°F);
- f = fluid flow rate through collector (gpm);
- T_{out} = temperature of fluid leaving collector (°F);
- T_{in} = temperature of fluid entering collector (°F);
- I = intensity of solar radiation striking collector (Btu/h/ft^2);
- A_{gross} = gross area of collector (based on overall height and width of collector enclosure (ft^2);
- 8.01 = a constant required by the units used.

Example 3.9

A flat plate solar collector with enclosure dimensions of 4 by 10 feet operates in bright sunlight (I = 300 Btu/h/ft^2). Water enters the collector at 100 °F and leaves at 108 °F. The water flows through the collector at 1.5 gpm. Determine the collector's instantaneous thermal efficiency under these conditions.

Solution:

The gross collector area is the overall height of the enclosure multiplied by its overall width, in this case 40 ft^2. To evaluate Equation 3.11, the density and specific heat of water at the average collector temperature of (108 + 100)/2 = 104 °F are needed. These can be found in Chapter 4 to be D = 61.8 lb/ft^3 and C = 1.0 Btu/lb/°F. Substituting these into Equation 3.11 yields:

$$n = \frac{(8.01 Dc)f\, (T_{out} - T_{in})}{I(A_{gross})} = \frac{(8.01 \times 61.8 \times 1.0)\, 1.5\, (108 - 100)}{300(40)}$$

$$= 0.495 = 49.5\%$$

Discussion:

Under these operating conditions, the collector is capturing about half (49.5%) of the solar energy striking its upward face. The remaining incident solar energy is accounted for as reflection and transmission loss of the glazing, as well as heat losses through all sides of the enclosure.

The instantaneous thermal efficiency of a collector changes whenever its fluid inlet temperature, the ambient air temperature, or the intensity of solar radiation striking it varies. To show this change over a range of operating conditions, instantaneous thermal efficiency can be shown graphically as in Figure 3-86. Data used to make this graph is obtained by physical testing in which Equation 3.12 is applied over a wide range of operating conditions.

The collector's instantaneous thermal efficiency is plotted on the vertical axis as a function of a grouping of terms that are collectively called the "**inlet fluid parameter**" (p).

Equation 3.13:

$$p = \frac{(T_i - T_a)}{I}$$

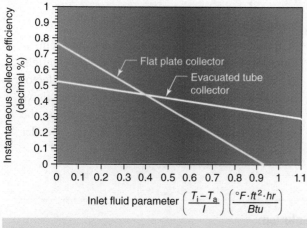

Figure 3-86 Graphical representation of instantaneous thermal efficiencies of flat plate and evacuated tube collectors.

where,

T_i = inlet fluid temperature to the collector (°F);
T_a = ambient air temperature surrounding the collector (°F);
I = solar radiation intensity striking the collector (Btu/h/ft²).

The higher the value of the inlet fluid parameter, the more severe the conditions under which the collector operates and the lower its instantaneous thermal efficiency.

Example 3.10

Assume that water at 150 °F is supplied to the flat plate and evacuated tube collectors having the efficiency lines given in Figure 3-86. During this time, the outdoor air temperature is 25 °F, and the intensity of solar radiation striking the collector is 250 Btu/h/ft². Determine the instantaneous efficiency of both collectors under these operating conditions.

Solution:

The inlet fluid parameter corresponding to these operating conditions must be determined before using the graph in Figure 3-86. Substituting in the stated operating conditions yields:

$$P = \frac{(T_i - T_a)}{I} = \frac{(150 - 25)}{250} = 0.5$$

Locating 0.5 on the horizontal axis of Figure 3-86, and then projecting up to the efficiency lines and over to the vertical axis shows the instantaneous efficiencies to be 41% for the evacuated tube collector and 35% for the flat plate collector.

Discussion:

Under these operating conditions, the evacuated tube collector is gathering a higher percentage of the solar radiation striking it compared to that gathered by the flat plate collector. However, never assume that these efficiency values hold constant for any length of time. The solar radiation intensity can vary in seconds as a cloud partially shades the collector. Likewise, the ambient air temperature and convective heat loss from the collector based on wind speed or direction can change quickly. All such changes affect the inlet fluid parameter, and thus will change the instantaneous efficiencies of both collectors.

The inlet fluid temperature also has a strong influence on instantaneous thermal efficiency. For example, consider the same two collectors operating under the same outside conditions, but with an inlet fluid temperature of 100 °F rather than 150 °F. The inlet fluid parameter is now:

$$p = \frac{(T_i - T_a)}{I} = \frac{(100 - 25)}{250} = 0.3$$

Using this inlet fluid parameter with Figure 3-86 shows the instantaneous thermal efficiency of the flat plate collector to be 52%, and that of the evacuated tube collector to be 46%. Under these operating conditions, the flat plate collector outperforms the evacuated tube collector.

Flat plate collectors will often have higher efficiencies than evacuated tube collectors in systems that supply fluid to the collectors at relatively low inlet temperatures. The converse is also true. Evacuated tube collectors will typically have higher efficiencies than flat plate collectors in systems that supply the collector with fluid at high inlet fluid temperatures.

It is possible to approximate the daily heat output of a given solar collector under standardized conditions by referencing test reports developed by the Solar Rating and Certification Corporation (SRCC). An example of one such report for a specific flat plate collector is shown in Figure 3-87.

These test reports give the estimated daily heat output for a collector, in thousands of Btu per day, under three assumed solar radiation scenarios:

- Clear Day (2,000 Btu/ft²/day);
- Mildly Cloudy Day (1500 Btu/ft²/day);
- Cloudy Day (1000 Btu/ft²/day).

SOLAR COLLECTOR CERTIFICATION AND RATING SRCC OG-100	CERTIFIED SOLAR COLLECTOR SUPPLIER: Alternate Energy Technologies, 1057 N. Ellis Road, Jacksonville, FL 32254 USA MODEL: American Energy AE-32E COLLECTOR TYPE: Glazed Flat-Plate CERTIFICATION #: 100-1999-001I

COLLECTOR THERMAL PERFORMANCE RATING

Megajoules Per Panel Per Day				Thousands of Btu Per Panel Per Day			
CATEGORY ($T_i - T_a$)	CLEAR DAY 23 MJ/m².d	MILDLY CLOUDY 17 MJ/m².d	CLOUDY DAY 11 11 MJ/m².d	CATEGORY ($T_i - T_a$)	CLEAR DAY 2000 Btu/ft².d	MILDLY CLOUDY 1500 Btu/ft².d	CLOUDY DAY 1000 Btu/ft².d
A (−5°C)	43	33	23	A (−9°F)	41	31	21
B (5°C)	38	28	18	B (9°F)	36	26	17
C (20°C)	30	20	10	C (36°F)	29	19	9
D (50°C)	13	5		D (90°F)	13	5	
E (80°C)	1			E (144°F)			

A-Pool Heating (Warm Climate) B-Pool Heating (Cool Climate) C-Water Heating (Warm Climate) D-Water Heating (Cool Climate) E-Air Conditioning

Original Certification Date: February 12, 2001

COLLECTOR SPECIFICATIONS

Gross Area: 2.965 m² 31.92 ft²
Dry Weight: 50.8 kg 112 lb.
Test Pressure: 1103 kPa 160 psig

Net Aperture Area: 2.781 m² 29.94 ft²
Fluid Capacity: 4.9 l 1.3 gal.

COLLECTOR MATERIALS

Frame: Anodized Aluminum
Cover (Outer): Low Iron Tempered Glass
Cover (Inner): None
Absorber Material: Tube - Copper / Plate - Copper Fin
Absorber Coating: Moderately Selective Black Paint
Insulation (Side): Polyisocyanurate
Insulation (Back): Polyisocyanurate

PRESSURE DROP

Flow		ΔP	
ml/s	gpm	Pa	in H$_2$O

TECHNICAL INFORMATION

Efficiency Equation [NOTE: Based on gross area and (P) = $T_i - T_a$]   Y Intercept   Slope

S I Units: η = 0.638   −4.2645 (P)/I   −0.0297 (P)²/I   0.655   −6.37   W/m²·°C
I P Units: η = 0.638   −0.7515 (P)/I   −0.0029 (P)²/I   0.655   −1.123   Btu/hr·ft²·°F

Incident Angle Modifier [(S) = 1/cos θ −1, 0°≤ θ ≤ 60°]
$K_{\alpha\tau}$ = 1.0 +0.0248 (S)   −0.0861 (S)²   Model Tested: AE-21E
$K_{\alpha\tau}$ = 1.0 −0.05   (S)   (Linear Fit)   Test Fluid: Water
                                                      Test Flow Rate: 39 ml/s   0.61 gpm

REMARKS:

February, 2008
Certification must be renewed annually. For current status contact:
SOLAR RATING & CERTIFICATION CORPORATION
c/o FSEC ♦ 1679 Clearlake Road ♦ Cocoa, FL 32922 ♦ (321)638-1537♦ Fax (321) 638-1010

Figure 3-87 | Thermal performance test report for a flat plate collector from Solar Rating and Certification Corporation (SRCC). Website: http://www.solar-rating.org.

For each solar radiation scenario, a corresponding scenario representing the temperature difference between the inlet fluid temperature and ambient air is given. The latter scenarios are intended to represent different applications of the collector such as pool heating and domestic water heating, in both warm and cool climates.

Example 3.11

Use the Collector Thermal Performance Rating table in Figure 3-87 to estimate the total daily heat output of the collector used to heat domestic water on a mildly cloudy day, in a cool climate.

Solution:

Enter the report table under the column of Mildly Cloudy, and move down to the fourth row (D) representing water heating in a cool climate. The corresponding output is given as 5,000 Btu/day per collector.

Discussion:

There is no guarantee that the values listed for a given collector and corresponding operating conditions will accurately predict that collector's performance in a given system. However, because values are given for hundreds of collector makes and models, and these values are based on standardized conditions, it is possible to make relative comparisons.

It should be noted that SRCC test reports also list values for the "Y-intercept" and slope of the collector's efficiency graph. This data make it possible to plot the efficiency line of a chosen collector in the same manner as shown in Figure 3-86.

The most recent SRCC collector test reports can be downloaded for free from the following website: http://www.solar-rating.org.

The only practical way to know which collector performs best in a given system, at a given location, is through use of solar thermal system simulation software. This software accounts for the variation in the collector operating conditions over the course of an entire year in a specific location. It accounts for differences in solar radiation striking the collector based on its mounting angles relative to the horizon and a polar North/South line. References for currently available solar thermal system simulation software are given at the end of this chapter.

Solar Thermal Combi-Systems

Solar thermal combi-systems contribute energy to both domestic water heating and space heating. They can be categorized based on the method used to protect the collector array from freezing. The two most common categories are:

- Antifreeze-based systems;
- Drain-back systems.

Each approach has strengths and limitations. Detailed information for each type of combi-system can be found in reference 1 at the end of the chapter.

The author prefers drain-back systems utilizing flat plate collector. They offer simplicity, reliability, reduced maintenance requirements, and better thermal performance in comparison to systems operating with antifreeze solutions. The remaining discussion will be limited to such systems.

Drain-Back Protected combi-Systems

One of the simplest ways to protect flat plate solar thermal collectors from freezing is to allow the water inside the collectors to drain back into heated space whenever they are not operating. This approach eliminates use of antifreeze solutions. The collectors and exposed piping refill with water at the start of each collection cycle.

Drain-back combi-systems have several advantages compared to antifreeze-based combi-systems:

- They do not require antifreeze. This reduces cost and improves thermal energy transfer due to the superior thermal properties of water versus antifreeze solutions. It also eliminates concern over maintenance of antifreeze solutions.

- Since all water drains from the collectors whenever they are not operating, there is no need to include an outside heat dump to protect the collectors during stagnation. This reduces cost and allows the system to be fully protected against stagnation during a power outage.

- The same water that flows through the collectors also flows through the space heating distribution system. There is no need of a heat exchanger between the collectors and the remainder of the system. This eliminates the thermal penalty associated with a collector heat exchanger and improves system efficiency. The result will be a greater harvest of solar energy.

- Several hardware components such as a collector heat exchanger, heat exchanger circulator, expansion tank, heat dump subsystem, collector loop purging valve, and the antifreeze solution, all of which are needed in antifreeze-based systems, are not required in drain-back systems. This simplifies the system and reduces installation cost.

An example of a closed-loop drain-back combi-system is shown in Figure 3-88.

Notice the captive air volume at the top of the storage tank. This air quickly replaces the water in the collectors and exposed piping whenever the collector circulator stops. The air enters the return pipe through an air return tube connected to the top of the thermal storage tank. If properly proportioned, this captive air volume can also serve as the expansion tank for both the collector loop and the distribution system.

It is crucial that the solar collectors and all exposed piping in a drain-back system be pitched a minimum of ¼ inch/feet toward the storage tank to allow complete drainage.

It is also necessary to size the collector circulator such that it can lift water from the static level of the thermal storage tank to the top of the collector array.

The return line from the collector array must be sized for a *minimum* flow velocity of 2 feet/s. This allows water returning from the collectors to entrain air bubbles and carry them back to the storage tank. As the return pipe fills with water, a siphon is established over the top of the collector loop. This decreases the circulator power requirement during the remainder of the collection cycle. Properly sizing the return pipe also minimizes gurgling noises and increases the water flow rate through the collectors, slightly boosting their efficiency.

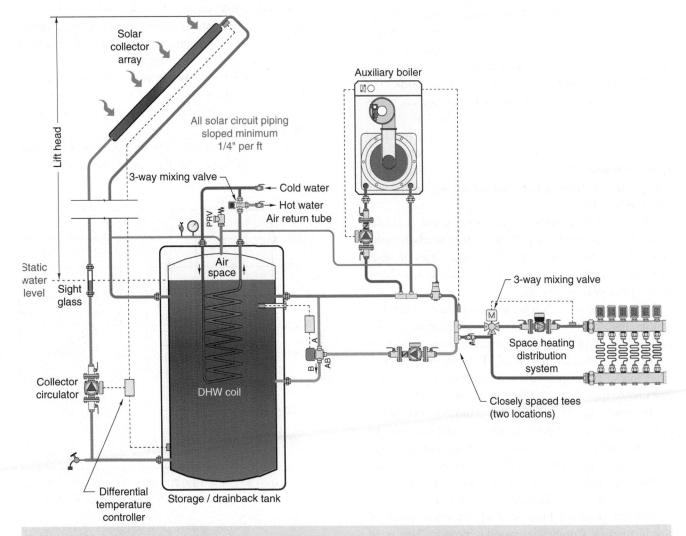

Figure 3-88 | Example of a closed-loop drain-back solar combi-system for domestic water heating and space heating.
Courtesy of ProTech Systems Div. of Simpson Dura-Vent.

3.18 Biomass Boiler Fundamentals

Solar energy can be used in many forms. The preceding section described solar thermal collectors that convert solar radiation into a heated fluid. It's also possible to convert solar radiation directly into electrical energy using solar photovoltaic modules.

One of the most omnipresent and natural solar energy conversion processes is called **photosynthesis**—the conversion of solar radiation, in combination with naturally occurring compounds such as carbon dioxide, into plant matter. That plant matter is composed of hydrocarbon molecules and water. Under the right conditions, these hydrocarbons can be converted into heat, as is possible with other hydrocarbons such natural gas, fuel oil, and propane.

There are thousands of different plant materials that can, under certain conditions, be converted to heat by combustion. Examples include grasses, seeds, nut shells, and olive pits. These are all examples of what is referred to as **biomass**. The practicality of using certain biomass materials is obviously dependent on local availability, growing requirements, and harvesting methods. For example, it's not very practical to consider heating buildings by burning olive pits in Fairbanks, Alaska.

One of the most commonly available and practical forms of biomass is wood. Its use as a heating fuel spans millennia, from crude piles of sticks burning on the ground, to fueling fully automated high-efficiency biomass boilers.

From the standpoint of chemistry, a tree is fundamentally an assembly of trillions of hydrocarbon molecules along with water and smaller quantities of other elements. *It is therefore reasonable to think of wood as a stored form of solar energy.*

Many areas of North American have abundant forests that contain vast amounts of potential wood fuel—far exceeding current demand and far more than could be practically harvested. Furthermore, responsible forestry practices allow these forests to produce sustainable quantities of wood. Well-managed forests in North American can generate approximately one ton of biomass per acre per year. Again, far more than can be used by mankind. Some of this material, such as leaves on deciduous trees, or needles on evergreen trees, is not practical as fuel. However, the heartwood and sapwood in the trunk and larger branches of trees can be the basis for several practical biomass fuels in many areas of North America. Combusting wood to produce heat is, in the author's opinion, a "regionally appropriate" method for utilizing solar energy.

Fuel Energy in Wood

Unlike other hydrocarbon fuels, wood contains some amount of water. When a live tree is harvested the moisture content of the wood can be upwards of 50%, depending on the species, location, and time of year. When harvested wood is cut, split, and stacked under cover, its moisture content will slowly drop. Cordwood that have been air-dried for at least 12 months, and kept covered to protect it from precipitation, will typically stabilize at moisture contents ranging from 12 to 20%.

The energy available from burning wood is very dependent on its moisture content. The lower the moisture content, the more energy wood can yield when combusted. Figure 3-89 shows the relationship between the energy theoretically available from wood versus its moisture content.

The **higher heating value (HHV)** plotted in Figure 3-89 is the theoretical energy available from wood assuming perfect **stoichiometric combustion** (e.g., no excess air is provided into the combustion chamber), and also assuming complete capture of the latent heat of all water vapor formed during combustion. This type of combustion can only be achieved in a laboratory.

The **lower heating value (LHV)** is based on most of the same assumptions as higher heating value, but excludes the latent heat of water vapor formed during combustion.

The HHV or LHV values do not account for the combustion efficiency of the wood burning device.

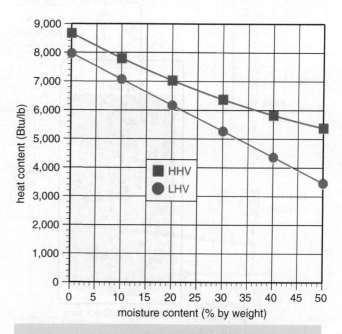

Figure 3-89 | Higher heating value (HHV) and lower heating value (LHV) of wood as a function of moisture content.

That efficiency can vary over a wide range, depending on the type of device the wood is burned in.

The fuel energy available from wood is based on *weight*, and expressed in Btu/lb. *Weight-based heating values apply equally to hardwood and softwood.* However, because most softwood species are less dense than hardwood, a greater volume of softwood is required to achieve given weight.

Wood is also a carbon-neutral fuel. Carbon is captured by trees as they grow. The amount of carbon released into the atmosphere by burning wood is the same as what would eventually be released if the tree supplying that wood died and rotted in the forest. Thus, wood burning is essentially carbon recycling. It does not release **geological carbon** into the atmosphere as does combustion of fossil fuels.

Wood-Fueled Hydronic Heating

The history of combining wood as a heating fuel with hydronic delivery systems spans many decades. A wide range of devices have been developed and used—with varying degrees of success. They range from grids of steel pipe installed in the base of fireplaces and wood stoves, to modern wood pellet boilers that operate under the control of microprocessors. The thermal efficiencies of these devices have ranged from under 40% to approximately 90%.

There has also been a wide spectrum of performance regarding particulate emissions, ranging from what most people would consider totally unacceptable, as seen in Figure 3-90a, to virtually invisible exhaust streams, as seen in Figure 3-90b.

Modern Wood-Fueled Boilers

Many people consider wood a "dirty" fuel that generates unacceptable levels of smoke. Many also view it as fuel that would only be used if "modern" fuels such as natural gas, propane, or electricity are not available. Arguably, there are centuries of experience to support this viewpoint.

However, significant progress has been made in developing methods to combust wood in ways that yield much more heat per pound of fuel, and do so at much lower levels of particulate emissions. The wood-fueled boilers discussed in this chapter employ state-of-the-art technology. The two categories of wood-fueled boilers that represent the state-of-the-art technology are cordwood gasification boilers and pellet boilers.

3.19 Cordwood Gasification Boilers

Those who have heated buildings using wood stoves, fireplaces, or standard wood-fired boilers are familiar with the routine of kindling a fire, then adding a few pieces of cordwood every two or three hours. On especially cold days, the fire is maintained at a higher rate of combustion by adding more wood, or adding it more often, to ensure sufficient heat output. Although this approach to stoking traditional wood-fired appliances has worked for centuries, it does not represent optimal conditions with respect to achieving high combustion efficiency or minimizing emissions.

Figure 3-90 (a) High levels of particulate emissions resulting from poor combustion of wood in an outdoor furnace, (b) invisible exhaust stream from a gasification type wood burner (Courtesy of Mark Odell).

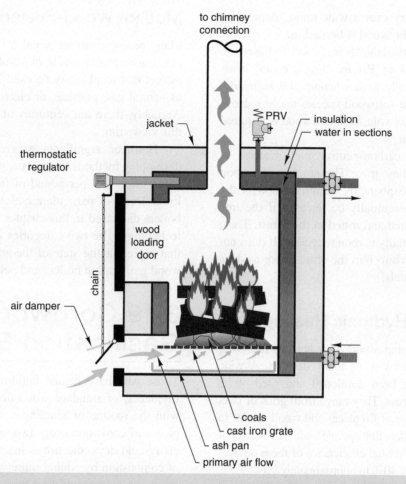

Figure 3-91 — Operation of a single stage wood-fired boiler. Air flows into boiler under a grate supporting the wood. Combustion products flow upward. Some of the pyrolytic gases released by the heated wood do not combust, but instead are carried up the chimney.

Most older wood stoves and cordwood boilers use **single stage combustion**. Figure 3-91 illustrated the processes at work in a single stage wood boiler.

After the fire is kindled, air flows upward from under a grate that supports the burning wood. Combustion occurs on the surface of the wood and to some degree within the pyrolytic gases given off by the wood as it is heated. The design of the combustion chamber, the size and moisture content of the wood, and the air damper settings of these older boilers all influence how well combustion occurs at a molecular level (e.g., where hydrocarbon compounds mix with oxygen molecules under suitable temperature and exposure time conditions).

There are several indicators that reveal how well a wood-fired stove or boiler is extracting the chemical energy from the wood it burns. One is the residue that remains in the combustion chamber at the end of a burn cycle. Heavy ash accumulation or the presence of "clinkers" (e.g., unburned but charred chunks of wood) can result from poor fuel quality (e.g., high moisture content, large amounts of bark, or both). Clinkers are also evidence of insufficient mixing of the hydrocarbon molecules in the wood with oxygen molecules under conditions that would allow more complete combustion.

Another indicator of poor combustion is creosote, a black tar-like substance that coats cooler surfaces in the combustion chamber, exhaust piping, and inside surfaces of chimneys. Creosote is formed when pyrolytic gases given off by heated wood fail to combust, and instead condense into a solid material. Creosote should also be considered as "unburnt fuel." Given the right conditions, it can be ignited. This can result in extremely dangerous chimney fires. Creosote is also wasted energy. Its presence proves that some of the potential heating energy in the wood fuel was not liberated. Figure 3-92 shows creosote coated surfaces inside a poorly performing outdoor wood furnace.

Another indicator of less than ideal combustion is heavy and persistent smoke exiting the chimney of the wood-fired device. While nearly all wood-fired devices will generate some visible smoke, especially under startup conditions, modern cordwood gasification boilers and pellet boilers will transition to relatively clean exhaust streams when properly operated under steady-state conditions.

3.19 Cordwood Gasification Boilers

Figure 3-92 | Heavy creosote accumulation on the door and within the combustion chamber of an outdoor wood furnace. *Courtesy of John Siegenthaler*

Figure 3-93 | Example of a modern cordwood gasification boiler. *Courtesy of Econoburn.*

Two-Stage Combustion

Research on wood burning has led to development of **two-stage combustion** systems. The first stage heats wood at relatively low oxygen levels and temperatures ranging from 300 to about 1500 °F. This results in gasification—the conversion of a solid fuel into a gaseous fuel. The gases produces in this reaction are called pyrolytic gases. The second stage of combustion thoroughly mixes these gases with sufficient oxygen in a high temperature chamber. This results in high efficiency combustion, much like that of other gaseous fuels. The hot gases liberated by this combustion are routed through a heat exchanger where much of the heat is transferred to water circulated through the boiler.

Modern two-stage cordwood gasification boilers can achieve thermal efficiencies in the range of 70–80%, a marked improvement over the 40–50% thermal efficiencies often present in older single stage wood burning boilers.

Figure 3-93 shows an example of a modern cordwood gasification boiler. Figure 3-94 illustrates the processes at work in such a boiler.

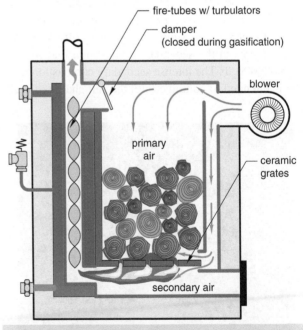

Figure 3-94 | Two-stage combustion within a cordwood gasification boiler. Wood is heated in the upper chamber to liberate pyrolytic gases. These gases are forced downward by a blower, mixed with secondary air, and combusted in the lower chamber. The hot combustion gases then pass through a steel fire-tube heat exchanger where heat is transferred to water.

The fire in cordwood gasification boilers is started by loading smaller pieces of kindling wood at the base of the boiler's upper combustion chamber. The kindling is ignited and allowed to burn for a few minutes to establish a bed of hot coals at the base of the upper chamber. Then, *unlike a wood stove*, the upper chamber is fully loaded with firewood.

During initial firing, a flue damper near the top of the upper combustion chamber is open, and the boiler operates in an updraft mode, similar to a single stage wood-fired boiler.

Once the bed of hot coals has formed and the chamber is fully loaded with cordwood, the upper damper is closed and a blower turns on to direct the pyrolytic gases downward through a slot in the ceramic plate at the base of the upper combustion chamber. This blower also forces air through holes in the side of the ceramic base plate. The hot air mixes with the hot pyrolytic gases to create a secondary combustion reaction below the slot. An example of the secondary combustion is seen in Figure 3-95.

During secondary combustion the pyrolytic gases emitted by the heated wood are thoroughly mixed with oxygen molecules. *The chemical efficiency of the secondary combustion reaction is far superior to that which occurs on or near the surface of cordwood burning in a standard wood-fired appliance, such as a fireplace or single-stage, wood-fired boiler.* Temperatures in the secondary combustion reaction can exceed 2,000 °F. The intense secondary combustion reaction is intended to continue until the full charge of cordwood is consumed. Thus, cordwood gasification boilers are designed to operate as "batch burners." *Proper operation results in the incineration of the full wood charge as hot and fast as possible.* When operated in this manner, and supplied with well-dried wood (e.g., moisture content ≤ 20), these boilers can achieve steady-state combustion efficiencies over 80%. Overall burn cycle efficiencies, which include startup, steady-state, and burnout phases, are typically in the range of 65–75%.

Need for Thermal Storage

Because of how they are designed to operate, cordwood gasification boilers often produce heat at rates far in excess of building heating loads. This is especially true when the heating distribution system is divided into multiple zones or operates under partial-load conditions.

The solution to this mismatch is not *to reduce the firing rate of the boiler, which would adversely affect both thermal efficiency and emissions. The solution is to provide thermal storage, which can absorb the excess heat production and hold it for hours if necessary, until needed by the load.*

The need for thermal storage as an integral part of a system supplied by a cordwood gasification boiler cannot be overemphasized. Experience has shown that cordwood gasification boilers operated without adequate thermal storage consistently fail to achieve the high thermal efficiency and low emissions which they are otherwise capable of.

Thermal storage is typically provided by a large, well-insulated, water-filled tank. A commonly recommended volume for thermal storage paired with a cordwood gasification boiler is approximately 130 gallons per cubic foot of volume in the boiler's primary combustion chamber. Many of the cordwood gasification boilers used in residential or light commercial systems have primary combustion chamber volumes in the range of 4 to 8 cubic feet. Thus, the recommended thermal storage tanks have corresponding volumes of approximately 500 to 1,000 gallons. Multiple smaller tanks piped in parallel are sometimes used to achieve the total required thermal storage volume.

Thermal storage tanks can be unpressurized or pressure-rated. The latter are preferred by the author when circumstances allow. Most codes require pressure-rated tanks having volumes over 119 gallons, to be ASME certified. Figure 3-96 shows an example of a large ASME rate steel storage tank.

The tank in Figure 3-96 is constructed of steel and is ASME pressure Vessel Code Section VIII certified. The exterior of the tank is coated with a primer to prevent surface oxidation.

Figure 3-95 Secondary combustion inside a cordwood gasification boiler during operation. *Courtesy of Dunkirk Metal Products, Inc.*

Figure 3-96 | A 350 gallon welded steel tank designed for thermal storage. Courtesy of Troy boiler Works.

Tanks supplied in this form must be insulated when installed to provide thermal storage for a biomass boiler. Several types of insulation products could be used. They include temperature-rated sprayed polyurethane foam, and high-density fiberglass batts. The insulation used on thermal storage tanks supplied from biomass boilers should be capable of withstanding sustained temperatures of at least 200 °F without degradation. The author recommends insulating all surfaces of thermal storage tanks to a rating of $R - 24$ °F·h·ft²/Btu.

After insulation is installed, the tank should be jacketed. Some insulation products, such a fiberglass rolls, are supplied with a reinforced foil jacketing. In other cases, fiberglass or mineral fiber insulation is jacketed with thin sheets of aluminum or plastic. Most codes require spray polyurethane foam to have a thermal barrier capable of delaying fire spread. Spray-applied intumescent coatings can provide that barrier.

Sizing Cordwood Gasification Boilers

Most boilers are sized based on the design heating load of the building they supply. *This is not the case for cordwood gasification boilers.* Instead, they are sized based on the number of batch burns the owner wants to provide on a design load day. The more batch burns the owner is willing to provide, the small the boiler. A commonly accepted and reasonable number for design day batch burns is 2. During partial load conditions, and with adequate thermal storage, there will likely be times when only one batch burn per day is required. There may even be times when a single batch burn can produce sufficient heat to supply the heating load for two or three days.

Equation 3.12 can be used to estimate the *weight* of cordwood required for each batch burn.

Equation 3.14:

$$W = \frac{[T_{\text{inside}} - (T_d + 5)](\text{UA}_b)24}{eCN}$$

where,

W = weight of firewood required for each batch burn (lb)
T_{inside} = indoor air temperature for design load conditions (°F)
T_d = outdoor design air temperature (°F)
UA_b = heat loss coefficient of building (Btu/h/°F)
24 = hours in one day
e = average efficiency of wood gasification boiler while operating (decimal %)
C = lower heating value of firewood being used (Btu/lb)
N = number of complete firing cycles per day under design load conditions

Once the weight of wood for each batch burn is calculated the required size of the boiler's primary combustion chamber can be estimated based on how densely the cordwood can be placed within it. The final step is to find a boiler with a primary combustion chamber volume close to the calculated value.

Example 3.12

Estimate the required size of the primary combustion chamber for a cordwood gasification boiler that will supply a building having a design heat loss of 50,000 Btu/h when the outdoor temperature is −3 °F and the indoor temperature is to be maintained at 70 °F. *The owner wishes to have no more than two complete firing cycles during a design day.* Assume the wood gasification boiler will be burning maple at an average moisture content of 20%, and has an average combustion efficiency of 70%, represented as 0.7.

Solution:

Several quantities need to be gathered prior to using Equation 3.12. The lower heating value of wood at 20% moisture content can be found using Figure 3-89 ($C = 6,140$ Btu/lb). The stated burn cycle efficiency of the boiler is 70% (which needs to be expressed as 0.7). The UA value for the building is the design heat loss divided by the corresponding temperature difference: UA = 50,000 (70 − (− 3)) = 685 Btu/h/°F. Putting these values into Equation 3.12 yields:

$$W = \frac{[T_{inside} - (T_d + 5)](UA_b)24}{eCN} = \frac{[70 - (-3 + 5)](685)24}{(0.7)(6140)(2)} = 130 \text{ lb}$$

The next step is to estimate the stacking density at which cordwood can be loaded into the boiler's primary combustion chamber. A suggested range of stacking density for hardwood is 15–20 lb/ft^3. For this example, a value of 17.5 lb/ft^3 will be used. The required volume of the boiler's primary chamber can now be calculated:

$$\text{volume} = \frac{130 \text{ lb}}{17.5 \frac{\text{lb}}{\text{ft}^3}} = 7.43 \text{ ft}^3$$

The final step is to check the specification sheets for possible boilers, looking for one with primary combustion chamber dimensions that yield a volume in the range of 7.4 ft^3.

Discussion:

If the owner insisted on only one batch burn during a design day, the required boiler would need a primary combustion chamber of about 15 ft^3. If the owner was willing to execute three batch burns on the same design day, the required primary combustion chamber volume would be reduced to about 5 ft^3. Keep in mind that larger boilers require larger thermal storage tanks, and usually larger chimneys, both of which can significantly increase installation cost.

Important Details for Cordwood Gasification Boilers

Cordwood gasification boilers must be protected against sustained flue gas condensation, which was described earlier in this chapter. The most common method of providing this protection is a thermally responsive three-way mixing valve. The valve measures the temperature of water entering the boiler, and adjusts itself to try to maintain that temperature above 130 °F whenever possible. It's also possible to use a **loading unit**, which combines a thermostatic mixing valve with a circulator, to provide flow between the boiler and thermal storage tank and protect it against sustained flue gas condensation.

Most boilers that operate on fossil fuels or electricity simply turn off during a power failure. This is not the case with a cordwood gasification boiler. If the boiler was burning a batch of cordwood when a power failure occurs, the combustion process keeps going, albeit at a reduced rate since the boiler's blower is off. Still, without a means of dissipating the residual heat, a cordwood gasification boiler can reach high temperatures and corresponding pressures that would cause its pressure relief valve to open. To prevent this "last resort" safety device from operating, systems using cordwood gasification boilers should be equipped with hardware and controls that can safely dissipate residual heat during a power outage. Collectively these components are called a **heat dump**.

One approach to heat dumping is allowing thermosiphon flow between the boiler and thermal storage tank. Another is to equip the system with a circulator that operates from an uninterruptible power supply, and is piped to create flow through some type of heat emitter. It's also possible to create a passive heat dump that allows hot water from the boiler to circulate through several finned-tubes mounted above the boiler. Details for constructing heat dumps are given in reference 1.

Example System

Figure 3-97 shows a piping schematic for a heating system using a cordwood gasification boiler as its primary heat source. An oil fired boiler is included as a backup heat source.

This system uses two thermal storage tanks in a "close-coupled" arrangement to meet the required storage volume. It supplies multiple space heating zones and has a subsystem for on-demand domestic water heating. During a power outage, residual heat in the boiler is passed into the thermal storage tank by thermosiphon flow through a normally open zone valve designated as (NOZV). During normal operation flow through the cordwood gasification boiler, protection against sustained flue gas condensation is provided by the loading unit. With proper design, the backup boiler can automatically cover the heating load if the cordwood

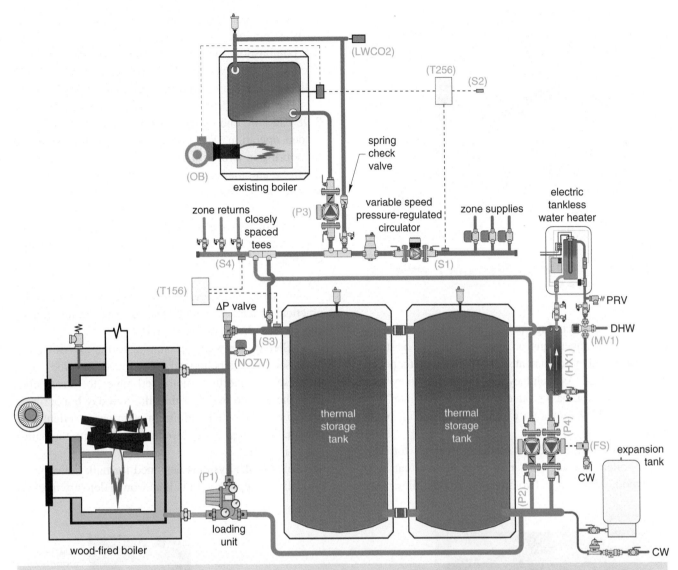

Figure 3-97 | System using a cordwood gasification boiler as its primary heat source, with an oil-fired boiler as the backup heat source.

gasification boiler is not operating and there is insufficient heat in thermal storage.

Additional and detailed information on designing hydronic systems using cordwood gasification boilers is available in reference 1.

3.20 Pellet Boilers

Although cordwood gasification boilers use advanced technology to achieve high efficiency and low emissions, they must be manually loaded and fired. In situation where this is not possible, or desired, boilers fueled by wood pellets provide an alternative. Modern pellet boilers can be operated unattended for up to several weeks, depending on the amount of fuel used. They can be configured to automatically load and ignite pellets, adjust the combustion air flow for optimal efficiency, and signal the owner when the ash drawer needs to be emptied. They are as close to a fully automatic boiler as is currently possible while using wood as a fuel.

Wood pellets are produced from wood chips and/or sawdust. The chips are shredded into fine slivers, dried, and forced through an extrusion device under very high (60,000 psi) pressure. This extreme pressure forces the natural resins within the wood cells to act as a bonding agent for the fibers. No other bonding agents are required. The extruded cylinders of compressed wood fiber, with a nominal ¼-inch diameter, randomly break into small pieces (e.g., pellets) as they exit the extrusion machine. A sample of the resulting pellets is shown in Figure 3-98.

Figure 3-98 | Wood pellets. *Courtesy of John Siegenthaler*

After manufacturing, pellets are easily handled by pneumatic and auger transport systems. They can be stored and transported by equipment similar to that used for pelletized animal feed. As such they lend themselves to existing infrastructures for mass production and transportation.

Figure 3-99 shows an example of a modern pellet boiler.

Because of the size and handling methods for fuel, pellet boilers are very different from boilers that burn cordwood. Most have two major subassemblies: One consisting of a hopper and auger system to temporarily store and then automatically feed pellets into the combustion process, and the other consisting of a combustion chamber, fire-tube heat exchanger ash collection hardware, controls, and induced draft blower.

The **day-hopper** seen on the right side of Figure 3-98 can hold enough pellets for several hours of unattended operation. The hopper can be manually loaded by pouring in two or three 40-pound bags of pellets, or it can be automatically loaded with pellets from a bulk storage device such as the reinforced fabric **bag silo** seen in Figure 3-100.

Interior storage systems for pellets use various combinations of motor-operated augers and pneumatic systems to move pellets from bulk storage to the day-hopper.

For large buildings, or where interior storage is not practical or desired, pellets can be stored in an outdoor silo such as shown in Figure 3-101.

Pellet storage silos are similar to the exterior grain bins used to store pelletized animal feed. In some cases, they use a single tube motorized auger to move pellet from the bottom outlet of the silo, to a day hopper near the boiler inside the building. In other cases, a dual tube pneumatic system is used to draw pellets into the building as needed.

Pellet boiler systems designed for bulk delivery are typically resupplied with pellets from a delivery truck as

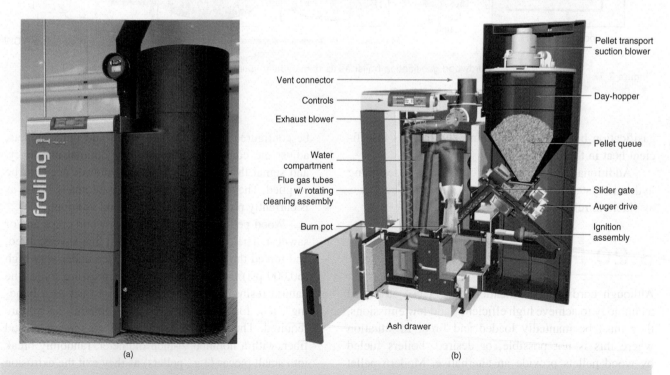

Figure 3-99 | (a) External view of modern pellet-fueled boiler with "day-hopper" for pellet queuing, (b) internal design of same boiler. *Courtesy of BioHeat USA, Inc.*

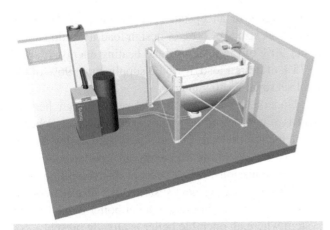

Figure 3-100 | A bag silo pellet storage system with pneumatic pellet transport to day hopper. Courtesy of Tarm Biomass.

Figure 3-101 | A pellet storage silo using underground pneumatic tubes to carry pellets into the adjacent building. Courtesy of Tarm Biomass.

shown in Figure 3-102. The truck accurately weighs the quantity of pellets delivered. The pellets are pneumatically conveyed from the truck to the bulk storage bin or silo. Pellets can be stored for months provided that the exterior storage silo is weathertight.

The Need for Thermal Storage

As is true with cordwood gasification boilers, pellet boilers attain their highest thermal efficiency and lowest emissions when operating at steady-state conditions. *One commonly cited recommendation is to designs system to achieve an average burn cycle duration of 3 hours per start up.* In some cases, burn cycles lasting upwards of 6 hours are possible, and preferable.

Achieving long duration burn cycles requires different control concepts in comparison to those used with fossil fuel or electric boilers. *A pellet boiler should* NOT *be turned on by a call for heat from one or more room thermostats. Instead, pellet boilers should be turned on and off to maintain the temperature of a thermal storage tank between some upper and lower temperature limits. This control action should be completely independent of any call from heat from room thermostats or other load devices.*

Most hydronic systems supplied by pellet boilers should include a thermal storage tank. The tank absorbs heat in excess of the load, allowing the boiler to operate at or close to steady-state efficiency for several hours. A common guideline is to provide 2 gallons of water thermal storage per 1,000 Btu/h of rated boiler heating capacity. The thermal storage tank can be unpressurized or pressurized.

One situation where water-based thermal storage may not be necessary is a pellet boiler supplying a high thermal mass single-zone floor heating system. A suggested criteria for such an application is a single-zone heated slab with a concrete volume of at least 15 cubic feet per 1,000 Btu/h of pellet boiler rating. With this situation as an exception, experience has shown that pellet boilers operated by room thermostats, or operated without thermal storage, tend to cycle frequently, and do not achieve the high efficiency and low emissions that are otherwise attainable. Figure 3-103 shows this effect based on testing involving a pellet boiler with a rated output of 85,300 Btu/h in a system with and without a 119 gallon thermal storage tank.

Pellet Boiler Sizing

When a pellet boiler is the sole heat in a system it is typically sized to the building's design heating load. However, in systems where a supplemental heat source is present, it is advantageous to size the pellet boiler for less than design heating load. Doing so allows the pellet boiler to operate at or close to steady-state conditions for more hours of the heating season.

Sizing suggestions for pellet boilers in systems having supplemental heat input range from 60 to 75% of design

Figure 3-102 | Bulk delivery of pellets. *Courtesy of Pellergy, Inc.*

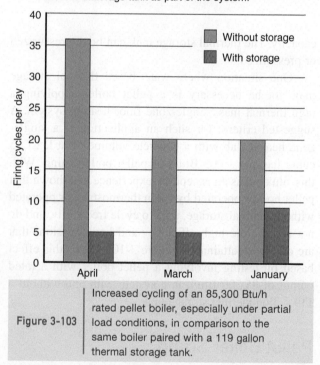

Figure 3-103 | Increased cycling of an 85,300 Btu/h rated pellet boiler, especially under partial load conditions, in comparison to the same boiler paired with a 119 gallon thermal storage tank.

heating load. These suggestions are based on analysis of bin weather data as described earlier in this chapter.

Figure 3-104 illustrates this concept using a heating duration curve.

In this graph, the pellet boiler is sized to 50% of the design heating load, yet supplies about 84% of the total space heating energy, based on average outdoor temperatures in an upstate, NY location. If the pellet boiler was sized for 60% of design load it would supply about 90% of the total seasonal heating energy. If sized to 75% of design load, it would supply about 96% of the total seasonal heating energy.

Undersizing any boiler relative to design heating load is often viewed as a "mistake." However, when a supplemental boiler is part of the system, doing so can improve the operating efficiency of a pellet boiler and lower its emissions. It is a concept that should be evaluated when planning new systems. It's also a frequent possibility when retrofitting a pellet boiler to a hydronic system where an operable boiler is already present.

Importance of Low Temperature Heat Emitters

As previously discussed, cordwood gasification and pellet boilers achieve their best performance when equipped with adequate thermal storage. That storage allows these boilers to operate for several hours at steady-state conditions. *This operating characteristic is very different from that of most fossil fuel boilers, which, under certain conditions, can turn on and off several times an hour.*

The duration of the thermal storage discharge cycle, when heat flows from thermal storage to the load, is also important. The longer the discharge cycle, the fewer burn cycles required of the boiler, and the longer the duration of those burn cycles.

Longer thermal storage discharge cycles are achieved when the heat emitters in the system can operate at low water temperatures. For example, in a system equipped with standard finned tube baseboard sized around a supply water temperature of 180 °F, it may only be possible to let the thermal storage tank to cool to 160 °F before the boiler must be re-fired. Allowing the tank to cool to a lower temperature would risk loss of builidng comfort on a cold winter day. However, if the heat emitter was a low temperature radiant panel, it might be possible to allow the same thermal storage tank to cool to 110 °F before re-firing the boiler.

Systems equipped with low temperature heat emitters are ideally matched to both cordwood gasification and pellet boilers. Methods and hardware to construct low temperature distribution systems, as well as ways of modifying existing high temperature systems for lower temperature operation are discussed in several upcoming chapters.

3.20 Pellet Boilers

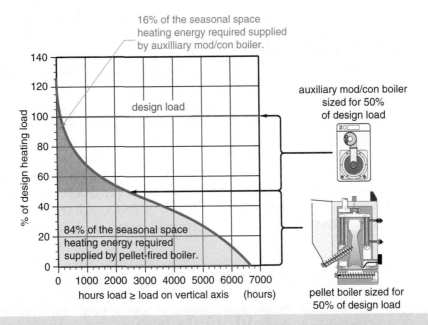

Figure 3-104 | A pellet boiler sized to 50% of design heating load can supply about 84% of the total seasonal heating energy based on the average climate in upstate NY.

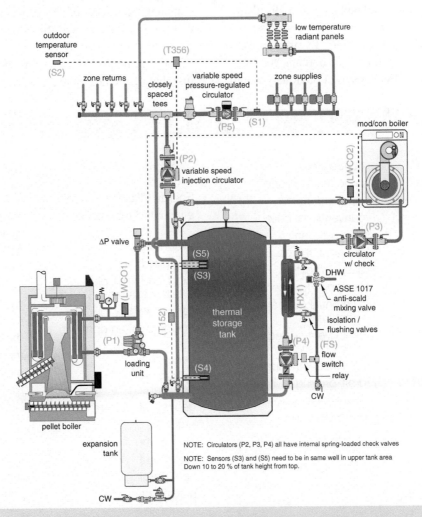

Figure 3-105 | An example system supplied by a pellet boiler as the lead heat source, with a mod/con boiler as supplemental and backup heat source. The system supplies multiple zone of low temperature heat emitters. It also supplies on-demand domestic hot water.

Example System

Figure 3-105 shows a pellet boiler system that supplies zoned space heating to low temperature heat emitters. It also supplies domestic hot water. A mod/con boiler provides supplemental and backup heating.

This system uses a loading unit to provide flow through the pellet boiler and protect it against sustained flue gas condensation. The pellet boiler is turned on when the upper tank sensor (S3) drops to some minimum temperature. The pellet boiler continues to fire until the lower tank sensor (S4) reaches some upper temperature limit. This control technique is called **temperature stacking**. It allows the thermal storage tank to be fully charged with heat each time the pellet boiler fires. Heat from the tank, or from the pellet boiler, or from both at the same time is shuttled into the distribution system using a variable speed injection circulator, which is discussed in Chapter 9. The auxiliary mod/con boiler is piped to the upper portion of the thermal storage tank. It fires, if necessary, to maintain this portion of the tank above some minimum value should the pellet boiler fail to do so. Domestic hot water is produced "on-demand" using a heat exchanger (HX1), circulator (P4) and a domestic hot water flow switch (FS1). Many of these piping and control concepts are discussed in later chapters.

Summary

This chapter has examined several types of hydronic heat sources, pointing out advantages and disadvantages of each. In some cases, the type of heat source used will be specified by the customer. In other cases, it will be dictated by the availability (or perhaps "unavailability") of certain fuels. Fuel cost will obviously weigh heavily in the decision. In all cases, the designer must consider the effects of the proposed heat source on the distribution system used. Incompatibilities do exist and are easily conceived by careless "catalog engineering." Always remember that system performance, and not just that of the heat source, is what counts.

Key Terms

absorber plate
Annual Fuel Utilization Efficiency (AFUE)
biomass
boiler jacket
borehole
buffer tank
bulkhead fitting
carbon-neutral fuel
chase
chemical energy content (of fuel)
chiller
closed-loop source
coefficient of performance (COP)
collector array
combi-systems
combustion efficiency
compressor
condenser
condensing mode
confined space
conventional boiler
cycle efficiency
demand charge
desuperheater
dewpoint (temperature)
differential temperature controller
direct vent/sealed combustion
DOE heating capacity
draft-proving switch
drain-back system
dry-base boiler
earth heat exchanger
electric thermal storage (ETS) system
electromagnetic radiation
Energy Efficiency Ratios
evacuated tube collector
evaporator
fire-tube boiler
fixed lead (multiple boiler system)
flat plate collectors
free area (of an opening)
freezestat
full storage (ETS system)
geological carbon
grout
heat dump
heat pump
heat purging
heating capacity
high/low fired boiler
higher heating value (HHV)
hydraulic separation
hydraulic separator
IBR Gross Output
IBR Net Output
inlet fluid parameter
lower heating value (LHV)
mod/con boiler
modulation
monobloc
multiple boiler system
National Fuel Gas Code
Net COP
noncondensing mode
off-cycle heat loss
open-loop source
packaged boiler
partial load condition

partial storage (ETS system)
pellets
performance envelope (of a mod/con boiler)
photosynthesis
power venting
premix combustion system
primary combustion chamber
primary heat exchanger
refrigerant

refrigeration cycle
reversible heat pump
reversing valve
run fraction
saturation temperature
sealed combustion
secondary heat exchanger
sectional boiler
sections
selective surface
sink

solar collectors
solar radiation
solid-fuel boiler
source (of heat for a heat pump)
stage
steady-state efficiency
stoichiometric combustion
storage heat exchanger
sustained flue gas condensation

system turndown ratio
thermal mass (of a boiler)
time-of-use rates
ton (of heating or cooling capacity)
turbulators
turndown ratio
unconfined space
wet-base boiler
wood-gasification

Questions and Exercises

1. Describe what is meant by the thermal mass of a boiler. What are some desirable and undesirable traits of low thermal mass boilers versus high thermal mass boilers?

2. What happens when a conventional boiler operates at an inlet water temperature lower than the dewpoint temperature of the exhaust gases?

3. Name some limitations that must be considered if a tank-type domestic water heater will be used for space heating.

4. What is the function of a draft-proving switch in a power exhaust system? Describe how it operates.

5. A boiler with a gas input rating of 100,000 Btu/h will be installed in a basement. Using the NFPA 54 requirements described in Section 3.7, Combustion Air Requirements, determine the required floor area of the basement if it is to be considered unconfined space. Assume that the height of the basement is 8 feet.

6. A boiler with a gas input rating of 100,000 Btu/h will be installed in a basement boiler room. The combustion air will be brought in through horizontal ducts. The ends of the ducts are covered with metal louvers having a free-to-gross area ratio of 0.6. Determine the size of each duct opening in square inches.

7. A boiler produces 60,000 Btu in 30 minutes. In the process, it consumes 0.5 gallons of #2 fuel oil (140,000 Btu/gallons). Determine its average efficiency during this period.

8. In 1 hour, a boiler fires for a total of 12 minutes. Use Figure 3-28 to determine the boiler's cycle efficiency for this period.

9. A house requires 75 MMBtu of heat for an entire heating season. Assume that this is provided by an oil-fired boiler with an AFUE of 75%. Estimate how many gallons of #2 fuel oil would be required over the season.

10. What is the advantage of parallel primary/secondary piping versus series primary/secondary piping when connecting boilers in a multiple boiler system?

11. An electric boiler outputs 40,000 Btu/h. It is supplied by a 240-VAC circuit. Estimate the electrical current flow to the boiler.

12. Describe the difference between parallel and hybrid modulation of a multiple mod/con boiler system.

13. Why is an air space present at the top of the thermal storage tank in Figure 3-88?

14. Why should a buffer tank always be used with a wood-gasification boiler?

15. What happens to uncombusted pyrolytic gases with a wood-fired boiler and its venting system?

16. What is the nominal buffer tank size required for a pellet boiler with a rated heat output of 100,000 Btu/h?

17. A ground source heat pump operates on 2,500 W power input, and has a corresponding heat output of 32,000 Btu/h. The ground loop circulator for this heat pump requires 280 W power input. (a) Determine the COP of this heat pump excluding power input to the circulator. (b) Determine the net COP of the heat pump and circulator combined.

18. Will the COP of an air to water heat pump increase or decreases when a heat exchanger is used between the heat pump and the buffer tank. Explain your answer.

19. Why is a heat dump necessary with a cordwood gasification boiler?

20. Why are low temperature heat emitters preferable in systems using cordwood gasification boilers or pellet boilers?

21. Estimate the required size of the primary combustion chamber for a cordwood gasification boiler that will supply a building having a design heat loss of 60,000 Btu/h when the outdoor temperature is 2 °F and the indoor temperature is to be maintained at 68 °F. *The owner wishes to have no more than two complete firing cycles during a design day.* Assume the wood gasification boiler will be burning maple at an average moisture content of 15%, and has an average combustion efficiency of 70%.

References

1. **Heating with Renewable Energy**, J. Siegenthaler, Copyright 2017 Cengage Learning, ISBN-13: 978-1-2850-7560-0

2. **Ground Source Heat Pump Residential and Light Commercial Design and Installation Guide**, Copyright 2009, International Ground Source Heat Pump Association, ISBN: 978-0-929974-07-1

3. **Solar Engineering of Thermal Processes, 3rd Edition,** J. Duffie and W. Beckman, Copyright 2006, John Wiley & Sons, ISBN: 0-471-69867-9

Chapter 4

Properties of Water

Objectives

After studying this chapter, you should be able to:

- Describe several fluid properties relevant to hydronic system design and operation.
- Explain why water is an excellent material for transporting heat.
- Demonstrate how heat and temperature are related.
- Calculate the amount of heat stored in a substance.
- Explain the difference between a Btu and a Btu/h.
- Estimate the output of a heat emitter from measurements of temperature and flow rate.
- Predict when water will boil based on pressure and temperature.
- Explain how a fluid's dynamic viscosity affects its flow rate in a hydronic system.
- Describe how air is dissolved into and released from water.
- Look up or calculate the values of several fluid properties for use in later chapters.
- Describe what incompressibility implies about flow entering and leaving a fluid-filled piping assembly.
- Understanding the process of demineralizing water within a hydronic system.

4.1 Introduction

Water is the essential material in any hydronic heating system. Its properties make possible the high efficiency and comfort associated with such systems. This chapter defines and explains several properties of water that are relevant to hydronic heating. A working knowledge of these properties is necessary to understand the behavior of hydronic systems and to make design decisions that affect this behavior, as well as for troubleshooting.

4.2 Sensible Heat Versus Latent Heat

The heat absorbed or released by a material while it remains in a single phase is called **sensible heat**. The word *sensible* means that the presence of the heat can be sensed by a temperature change in the material. As a material absorbs heat, its temperature increases. As heat is released, its temperature drops.

This process of sensible heat transfer is not the only way a material can absorb or release heat. If the material changes phase while absorbing heat, its temperature does not change. The heat absorbed in such a process is called **latent heat**. For water to change from a solid (ice) to a liquid requires the absorption of 144 Btu/lb while its temperature remains constant at 32 °F. When water changes from a liquid to a vapor, it must absorb approximately 970 Btu/lb while its temperature remains constant. The amount of energy required to vaporize water is very large. This explains water evaporating from the surface of an object (such as your skin) is a very effective means of cooling it.

4.3 Specific Heat and Heat Capacity

One of the most important properties of water relevant to its use in hydronic systems is **specific heat**. Simply put, the specific heat of any material is the number of Btus required to raise the temperature of 1 lb of that material by 1 °F. The specific heat of water is approximately 1.0 Btu/lb/°F. To raise the temperature of 1 lb of water by 1 °F, the addition of 1 Btu of heat will be required.

This relationship between temperature, material weight, and energy content also holds true when a substance is cooled. For example, to lower the temperature of 1 lb of water by 1 °F, the removal of 1 Btu of energy will be required.

The specific heat of any material varies slightly with its temperature. For water, this variation is quite small over the temperature range used in most hydronic systems. Figure 4-1 shows this variation. When designing residential or light commercial systems, this very slight variation can be safely ignored, and the specific heat of water can be considered to remain constant at 1.0 Btu/lb/°F.

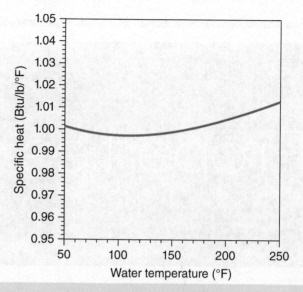

Figure 4-1 | Specific heat of water as a function of temperature.

The specific heat of water can also be calculated using Equation 4.1.

Equation 4.1:

$$c = 1.012481 - 3.079704 \times 10^{-4}(T) + 1.752657 \times 10^{-6}(T)^2 - 2.078224 \times 10^{-9}(T)^3,$$

for

$$(50 \leq T \leq 250),$$

where,
c = specific heat of water (Btu/lb/°F)
T = water temperature (°F)

The value of a material's specific heat also changes when the material changes phase (i.e., between liquid and vapor, or liquid and solid). For example, the specific heats of ice and water vapor are about 0.48 Btu/lb/°F and 0.489 Btu/lb/°F, respectively. Both values are considerably lower than that of liquid water.

The specific heat of water is high in comparison to other common materials. In fact, water has one of the highest specific heats of any known material. The table in Figure 4-2 compares the specific heats of several common materials. Also listed is the **density*** of each material, and its **heat capacity**.

The heat capacity of a material is the number of Btus required to raise the temperature of 1 ft³ of the material by 1 °F. The heat capacity of a material can be found

Material	Specific heat (Btu/lb/°F)	Density* (lb/ft³)	Heat capacity (Btu/ft³/°F)
Water	1.00	62.4	62.4
Concrete	0.21	140	29.4
Steel	0.12	489	58.7
Wood (fir)	0.65	27	17.6
Ice	0.49	57.5	28.2
Air	0.24	0.074	0.018
Gypsum	0.26	78	20.3
Sand	0.1	94.6	9.5
Alcohol	0.68	49.3	33.5

Figure 4-2 Specific heat, density, and heat capacity of some common substances.

*In this text, "density" is defined as *weight* per unit of volume, rather than *mass* per unit of volume, as used in classic physics. Its use in all subsequent equations in this text will be based on this definition, and the units of density are expressed as lb/ft³.

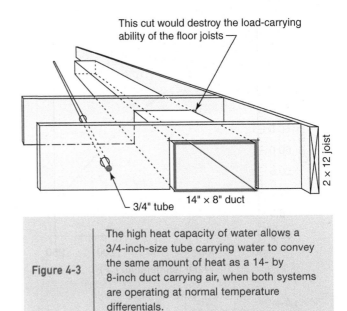

Figure 4-3 The high heat capacity of water allows a 3/4-inch-size tube carrying water to convey the same amount of heat as a 14- by 8-inch duct carrying air, when both systems are operating at normal temperature differentials.

by multiplying the specific heat of the material by its density, as shown by Equation 4.2.

Equation 4.2:

$$(\text{specific heat})(\text{density}) = \text{heat capacity},$$

$$\left(\frac{\text{Btu}}{\cancel{\text{lb}}\ °F}\right)\left(\frac{\cancel{\text{lb}}}{\text{ft}^3}\right) = \left(\frac{\text{Btu}}{°F\ \text{ft}^3}\right).$$

At first glance, the numbers in Figure 4-2 may be deceiving. For example, why is the specific heat of air greater than the specific heat of steel? The answer is that although steel has a lower specific heat, it is much denser than air (489 lb/ft³ for steel vs. about 0.074 lb/ft³ for air). Therefore, the heat capacity of steel, which reflects both its specific heat and density, is much higher than that of air.

By comparing the heat capacity of water and air, it can be shown that *a given volume of water can hold almost 3,500 times as much heat as the same volume of air for the same temperature rise.* This allows a given volume of water to transport vastly more heat than the same volume of air. Although it is not a defined scientific term, it could be said that water is almost 3,500 times more "thermally concentrated" than air.

The difference in heat capacity of water versus that of air is the single greatest advantage of using water rather than air to convey heat. It is the reason, as discussed in Chapter 1, that a 3/4-inch-size tube carrying water can convey the same amount of heat as a 14- by 8-inch duct conveying air, when both systems are operating at normal temperature differentials (see Figure 4-3).

The Fluid Properties Calculator module in the Hydronics Design Studio software can be used to find the specific heat of water as well as several water-based antifreeze solutions over a wide temperature range.

4.4 Density

In this text, the density of a material refers to the number of pounds of the material needed to fill a volume of 1 ft³. For example, it would take 62.4 lb of water at 50 °F to fill a 1-ft³ container, so its density is said to be 62.4 lb/ft³.

The classic physics definition of density is *mass* (rather than weight) per unit of volume. The mass of a given volume of a material is based on the number of molecules contained in that volume. The mass of a material, or any object, does not change regardless of whether it is located on earth, the moon, or some other location in space. However, the *weight* of a material is based on the gravitational force exerted on the material based on where it happens to be in space. Therefore, even though its mass is the same, the weight of an object is different on the moon compared to its weight on earth.

Since this text is only concerned with earth-based applications, it is convenient to simplify the definition of density to weight per unit of volume. All formulas used in this book are based on this definition.

The density of water is dependent on temperature. Like most substances, as the temperature of water increases, its density decreases. As temperature increases, each water molecule requires more space. This molecular expansion effect is extremely powerful and can easily

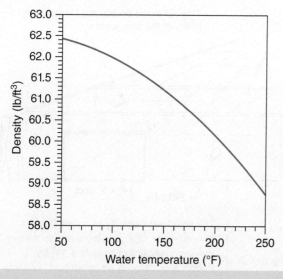

Figure 4-4 | Density of water as a function of temperature.

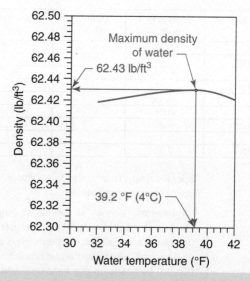

Figure 4-5 | The maximum density of water occurs at 39.2 °F.

burst metal pipes or tanks if not properly accommodated. A graph showing the relationship between temperature and the density of water is shown in Figure 4-4.

This relationship can also be represented by Equation 4.3 for temperatures between 50 and 250 °F.

Equation 4.3:

$$D = 62.56 + 3.413 \times 10^{-4}(T) - 6.255 \times 10^{-5}(T)^2,$$

for

$$(50 \leq T \leq 250),$$

where,
 D = density of water (lb/ft³)
 T = water temperature (°F)

The change in density of water or water-based antifreeze solutions directly affects the size of the expansion tank required on all closed-loop hydronic systems. Proper expansion tank sizing requires data on the density of the system fluid both when the system is filled and when it reaches its maximum operating temperature. This subject is covered in detail in Chapter 12, Expansion Tanks.

One characteristic of water that doesn't show in Figure 4-4 is that its density is maximum when its temperature is 4 °C (39.2 °F). Figure 4-5 "zooms in" on a narrow temperature range of 32 to 42 °F to reveal this characteristic. At 39.2 °F, the density of water is 62.43 lb/ft³.

Although this characteristic seldom affects design of hydronic heating systems, it does explain why an ice-covered lake, with no inlet or outlet flow disturbance, has a water temperature very close to 39.2 °F at its bottom (e.g., where the densest water settles).

The Fluid Properties Calculator module in the Hydronics Design Studio software can be used to find the density of water as well as several water-based antifreeze solutions over a wide temperature range.

4.5 Sensible Heat Quantity Equation

It is often necessary to determine the *quantity* of heat stored in a given amount of material that undergoes a specific temperature change. Equation 4.4, known as the **sensible heat quantity equation**, can be used for this purpose.

Equation 4.4:

$$h = wc(\Delta T),$$

where,
 h = quantity of heat absorbed or released from the material (Btu)
 w = weight of the material (lb)
 c = specific heat of the material (Btu/lb/°F)
 ΔT = temperature change of the material (°F)

Two conversion factors that are needed when using this equation for water are the following:

- 1 gal of water weighs approximately 8.33 lb.
- 1 ft³ of water weighs approximately 62.4 lb.

When the first of these values is factored into Equation 4.4, along with the specific heat of water = 1.0 Btu/lb/°F, the resulting equation becomes specific for use with water:

Equation 4.5:

$$h = 8.33v(\Delta T),$$

where,

h = quantity of heat absorbed or released from the water (Btu)
v = volume of water involved (gal)
ΔT = temperature change of the material (°F)

Example 4.1

A storage tank contains 500 gal of water initially at 70 °F. How much energy must be added or removed from the water to:

a. raise the water temperature to 180 °F?
b. cool the water temperature to 40 °F?
See figure 4-6

Solution:

a. Using the sensible heat quantity equation for water (Equation 4.5),

$$h = 8.33(500)(180 - 70) = 458{,}000 \text{ Btu}.$$

b. Using the same sensible heat quantity equation for water (Equation 4.5),

$$h = 8.33(500)(70 - 40) = 125{,}000 \text{ Btu}.$$

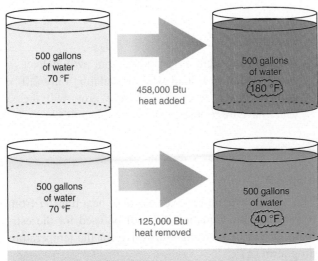

Figure 4-6 | Illustration for situation described in example 4.1. Adding and removing sensible heat causes a temperature change in water.

4.6 Sensible Heat Rate Equation

Hydronic system designers often need to know the *rate* of heat transfer to or from a stream of water flowing through a device such as a heat source or heat emitter. This can be done by introducing time into the sensible heat quantity equation. The resulting equation is called the **sensible heat rate equation**:

Equation 4.6:

$$Q = Wc(\Delta T),$$

where,

Q = rate of heat transfer into or out of the fluid stream (Btu/h)
W = rate of flow of the liquid (lb/h)
c = specific heat of the liquid (Btu/lb/°F)
ΔT = temperature change of the liquid within the device it exchanges heat with (°F)

For cold water only, Equation 4.6 can be rewritten as:

$$Q = \left[\left(\frac{8.33 \text{ lb}}{\text{gal}}\right)\left(\frac{60 \text{ min}}{\text{h}}\right)\left(\frac{1 \text{ btu}}{1 \text{ lb°F}}\right)\right]\left(f\frac{\text{gal}}{\text{min}}\right)(\Delta T °F),$$

which simplifies to

Equation 4.7:

$$Q = 500f(\Delta T),$$

where,

Q = rate of heat transfer into or out of the water stream (Btu/h)
f = flow rate of water through the device (gpm)
500 = constant rounded off from 8.33 × 60
ΔT = temperature change of the water through the device (°F)

Equation 4.7 is technically only valid for relatively cool water because the factor 500 is based on the density of water at approximately 60 °F (8.33 lb/gal). The factor 500 is easy to remember when making quick mental calculations for the rate of sensible heat transfer. However, at higher temperatures, the density of water decreases slightly. This, in turn, slightly affects its ability to transport heat.

Equation 4.8 is a more general form of Equation 4.7 that allows for variations in both the density and specific heat of the fluid. It can be used for water as well as other fluids.

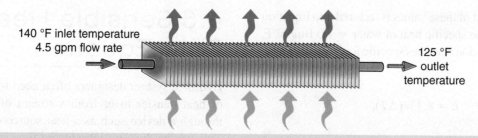

Figure 4-7 | Heat emitter for example 4.2.

Equation 4.8:

$$Q = (8.01Dc)f(\Delta T),$$

where,
- Q = rate of heat transfer into or out of the water stream (Btu/h)
- 8.01 = a constant based on the units used
- D = density of the fluid (lb/ft³)
- c = specific heat of the fluid (Btu/lb/°F)
- f = flow rate of water through the device (gpm)
- ΔT = temperature change of the water through the device (°F)

When using Equation 4.8, the density and specific heat should be based on the *average* temperature of the fluid during the process by which the fluid is gaining or losing heat.

Example 4.2

Water flows into a heat emitter at 140 °F and flows out of it at 125 °F. The flow rate is 4.5 gpm as shown in Figure 4-7. Calculate the rate of heat transfer from the water to the heat emitter using:

a. Equation 4.7
b. Equation 4.8

Solution:

Using Equation 4.7, the rate of heat transfer is:

$$Q = 500(4.5)(140 - 125) = 33{,}750 \text{ Btu/h}.$$

To use Equation 4.8, the density of water at its average temperature of 132 °F must first be estimated using Equation 4.3 or Figure 4-4.

$$D = 61.52 \text{ lb/ft}^3.$$

As previously mentioned, the variation in the specific heat of water over the temperature range used in most hydronic systems is very small. For calculation purposes, it can be assumed to remain 1.0 Btu/lb/°F.

Substituting these numbers into Equation 4.8 yields:

$$Q = (8.01Dc)f(\Delta T) = (8.01 \times 61.52 \times 1.00)$$
$$4.5(140 - 125) = 33{,}262 \text{ Btu/h}.$$

Discussion:

Equation 4.8 estimates the rate of heat transfer to be about 1.5% less than that estimated by Equation 4.7. This occurs because Equation 4.8 accounts for the slight decrease in the density of water at the elevated temperature, whereas Equation 4.7 does not. The variation between these calculations is arguably small. Equation 4.7 is generally accepted in the hydronics industry for quick estimates of heat transfer to or from a stream of water. However, Equation 4.8 will yield slightly more accurate results when the variation in density and specific heat of the fluid can be implemented into the calculation. Both equations will be used in later sections of this text.

Example 4.3

A heating system is proposed that will deliver 50,000 Btu/h to a load while circulating only 2.0 gpm of water. What temperature drop will the water have to go through to accomplish this?

Solution:

This situation again involves a *rate* of heat transfer from a water stream. Equation 4.7 will be used for the estimate. It will be rearranged to solve for the temperature drop term (ΔT).

$$\Delta T = \left(\frac{Q}{500f}\right) = \left(\frac{50{,}000}{500(2.0)}\right) = 50 \text{ °F}.$$

> **Discussion:**
>
> Although a hydronic distribution system could be designed to operate with a 50 °F temperature drop, this value is much larger than normal. Such a system would require very careful sizing of the heat emitters. This example shows the usefulness of the sensible heat rate equation to quickly evaluate the feasibility of a proposed design concept.

Proper Use of the Sensible Heat Rate Equations

Equations 4.7 and 4.8 can be used to calculate the rate of heat transfer when a *measured* flow rate and *measured* temperature change are observed. However, these equations do not "guarantee" that a specific temperature change will occur when an arbitrary value of heat transfer and flow rate are assumed.

For example, assume that the heat delivery requirement for a hydronic distribution system is 25,000 Btu/h and that a designer selects a flow rate of 2.5 gal/min. The designer then assumes that Equation 4.7 will reveal the temperature drop that *will* occur. The numbers inserted into Equation 4.7 yield the following:

$$\Delta T = \frac{Q}{500f} = \frac{25,000}{500(2.5)} = 20\,°F.$$

Although a 20 °F temperature drop is a *possibility*, under specific circumstances, *this calculated value should not be interpreted as an ensured physical result*. Instead, the calculated temperature drop should be interpreted as follows: <u>If the hydronic distribution system can release 25,000 Btu/h when supplied with water at 2.5 gpm, then the resulting temperature drop would be 20°F.</u> To see why this distinction is important, consider the following: Suppose that the temperature of the water supplied to the assumed hydronic distribution system at a flow rate of 2.5 gpm was equal to the room air temperature. Because there is no temperature difference between the water in the system and the room air, there will be no heat transfer. This would be the case regardless of flow rate. This possibility is not "covered" by Equation 4.7 or Equation 4.8. Neither is a situation where water is supplied to the hydronic distribution system at perhaps 140 °F, but the heat emitters in that system were sized to release 25,000 Btu/h when supplied with 180 °F water. When supplied with water at 140 °F and a flow rate of 2.5 gpm, the heat emitters would not be able to deliver 25,000 Btu/h, and thus, the resulting temperature drop would be significantly less than 20 °F. This physical result cannot be predicted by use of Equation 4.7 or Equation 4.8. If water was supplied to this system at 200 °F, and a flow rate of 2.5 gpm, the heat output of the heat emitters would be greater than 25,000 Btu/h, and the resulting temperature drop would exceed 20 °F. Again, this possibility cannot be predicted solely by using Equation 4.7 or Equation 4.8.

Just because various number combinations entered into Equations 4.7 and 4.8 balance the equation mathematically doesn't guarantee that a given set of physical conditions will occur. From a pure mathematical standpoint, there are infinite combinations of numbers that would balance the equation. However, many of these combinations have no counterpart in reality. For example, perhaps a designer needs to establish a rate of heat delivery of 200,000 Btu/h from a hydronic distribution system into a space maintained at 70 °F. The designer thinks this could be done by supplying water to the system at 140 °F and a flow rate of 5 gpm. Putting these numbers into Equation 4.7 yields:

$$\Delta T = \frac{Q}{500f} = \frac{200,000}{500(5)} = 80\,°F.$$

To achieve the calculated temperature drop of 80 °F, the water would have to cool from an initial temperature of 140 °F down to 60 °F. This would be impossible using any type of heat emitter in a room maintained at 70 °F. Again, this reveals the fallacy of relying on a mathematically calculated value using Equation 4.7 or Equation 4.8 to predict that some desired physical result will occur.

4.7 Vapor Pressure and Boiling Point

The **vapor pressure** of a liquid is the minimum pressure *that must be applied at the liquid's surface to prevent it from boiling*. If the pressure on the liquid's surface drops below its vapor pressure, the liquid will instantly boil. If the pressure is maintained above the vapor pressure, the liquid will remain a liquid, and no boiling will take place.

The vapor pressure of a liquid is strongly dependent on its temperature. The higher the liquid's temperature, the higher its vapor pressure. Stated another way: The greater the temperature of a liquid, the more pressure must be exerted on its surface to prevent it from boiling.

Vapor pressure is stated as an **absolute pressure** and has the units of psia. On the absolute pressure scale, zero pressure represents a complete vacuum. On earth, the weight of the atmosphere exerts a pressure on any liquid surface open to the air. At sea level, this atmospheric pressure is about 14.7 psi. Stated another way, the absolute pressure exerted by the atmosphere at sea level is approximately 14.7 psia.

Did you ever wonder why water boils at 212 °F? What is special about this number? Actually, there is nothing special about it. It just happens that at 212 °F, the vapor pressure of water is 14.7 psia (equal to atmospheric pressure at sea level). Therefore, water boils at 212 °F at elevations near sea level. If a pot of water is carried to an elevation of 5,000 feet above sea level and then heated, its vapor pressure will eventually rise until it equals the reduced atmospheric pressure at this elevation, and boiling will begin at about 202 °F.

Because water has been widely used in many heating and power production systems, its vapor pressure versus temperature characteristics have been well established for many years. These data are available in the form of steam tables. The vapor pressure of water between 50 and 250 °F can also be read from the graph in Figure 4-8.

The relationship shown in Figure 4-8 can also be generated using Equation 4.9 for water temperatures between 50 and 250 °F.

Equation 4.9:

$$P_v = 0.771 - 0.0326(T) + 5.75 \times 10^{-4}(T)^2 - 3.9 \times 10^{-6}(T)^3 + 1.59 \times 10^{-8}(T)^4,$$

for $(50 \leq T \leq 250)$,
where,

P_v = vapor pressure of water (psia)
T = temperature of the water (°F)

Closed-loop hydronic systems should be designed to ensure that the absolute pressure of the water at all the points in the system remains safely above the water's vapor pressure at all times. This prevents problems such as cavitation in circulators and valves, or "**steam flash**" in piping. The latter is a situation in which the pressure on the water drops below vapor pressure allowing the liquid water to instantly change to vapor (e.g., steam). This can cause loud banging noises in piping, and wide pressure fluctuations in the system.

4.8 Viscosity

The viscosity of a fluid is an indicator of its resistance to flow. The higher the viscosity of a fluid, the more drag it creates as it flows through pipes, fittings, valves, or any other component in a piping circuit. Higher viscosity fluids also require more pumping power to maintain a given system flow rate compared to fluids with lower viscosities.

Like all other properties discussed in this chapter, the viscosity of water depends on its temperature. The viscosity of water and water-based antifreeze solutions decreases as their temperature increases.

The relationship between the viscosity of water and its temperature is shown in Figure 4-9.

This relationship can also be described by Equation 4.10 for temperatures between 50 and 250 °F.

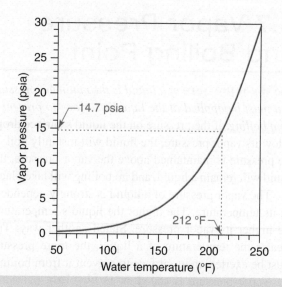

Figure 4-8 | Vapor pressure of water as a function of temperature.

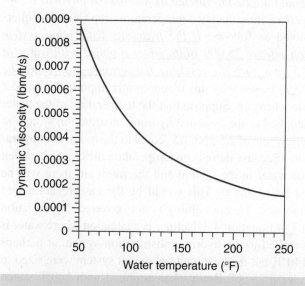

Figure 4-9 | Dynamic viscosity of water as a function of temperature.

Equation 4.10:

$$\mu = 0.001834 - 2.73 \times 10^{-5}(T) + 1.92 \times 10^{-7}(T)^2 - 6.53 \times 10^{-10}(T)^3 + 8.58 \times 10^{-13}(T)^4,$$

for ($50 \leq T \leq 250$),

where,

μ = dynamic viscosity of water (lbm/ft/s)
T = temperature of the water (°F)

Both Figure 4-9 and Equation 4.10 determine the **dynamic viscosity** of water as a function of temperature. Another type of viscosity, called kinematic viscosity, is sometimes discussed in other texts on fluid mechanics. However, it is not used in this book.

The physical units of dynamic viscosity can be confusing. They are a mathematical consequence of how dynamic viscosity is defined and do not lend themselves to a practical interpretation of what the viscosity number actually is. The units used for dynamic viscosity within this book are lbm/ft/s. These units are compatible with all equations used in this book that require viscosity data. The reader is cautioned, however, that several other units for viscosity may be used in other sources of viscosity data (such as manufacturer's literature). *These other units must first be converted to lbm/ft/s using the appropriate conversion factors, before being used in any equations in this book.*

Solutions of water and glycol-based antifreeze have significantly higher viscosities than water alone. The greater the concentration of antifreeze, the higher the viscosity of the solution. This increase in viscosity can result in significantly lower flow rates within a hydronic system, relative to those that would occur using water.

The Fluid Properties Calculator module in the Hydronics Design Studio software can be used to find the dynamic viscosity of water as well as several water-based antifreeze solutions over a wide temperature range.

4.9 Dissolved Air in Water

Water has the ability to contain air much like a sponge can contain a liquid. Molecules of the various gases that make up air, including oxygen and nitrogen, can exist in solution with water molecules. Even when water appears perfectly clear, it can contain gas molecules in solution. In other words, *just because bubbles are not visible in the water does not mean air is not present.*

When hydronic systems are first filled with water, **dissolved air** is present in the water throughout the system. If allowed to remain in the system, this air can create numerous problems, including noise, cavitation, corrosion, and improper flows. A well-designed hydronic system must be able to efficiently and automatically capture dissolved air and expel it. A proper understanding of what affects the air content of water is crucial in designing a method to get rid of it.

The amount of air that can exist in solution with water is strongly dependent on the water's temperature. As water is heated, its ability to hold air in solution rapidly decreases. However, the opposite is also true. As heated water cools, it has a propensity to absorb air. Heating water to release dissolved air is like squeezing a sponge to expel a liquid. Allowing the water to cool is like letting the sponge expand to soak up more liquid. Together these properties can be exploited to help capture and rid air from the system. The water's pressure also has a significant effect on its ability to hold dissolved air. When the pressure of the water is lowered, its ability to contain dissolved gases decreases, and vice versa.

Figure 4-10 shows the maximum amount of air that can be contained (dissolved) in water as a percent of the water's volume. The effects of both temperature and pressure on dissolved air content can be determined from this graph.

To understand this graph, first consider a single curve such as the one labeled 30.0 psi. This curve represents a constant pressure condition for the water. Note that as the water temperature increases, the curve descends. This descent indicates a decrease in the water's ability to hold air in solution. The exact change in air content can

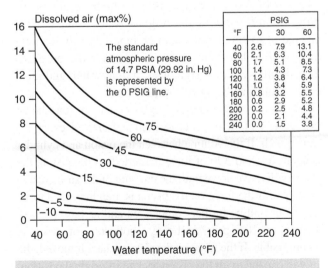

Figure 4-10 — Solubility of air in water at various temperatures and pressures. *Courtesy of Spirotherm Corporation.*

be read from the vertical axis. This trend holds true for all other constant-pressure curves.

This effect can be observed by heating a kettle of water on a stove. As the water heats up, bubbles can be seen forming at the bottom of the kettle (e.g., where the water is hottest). The bubbles are formed as gas molecules that were dissolved in the cool water are forced out of solution by increasing the temperature. The molecules join together to create the bubbles. If the kettle of water is allowed to cool down, these bubbles will disappear as the gas molecules go back into solution.

One can think of the dissolved air initially contained in a hydronic system as being "cooked" out of solution as water temperature increases. This takes place inside the heat source when it is first operated after being filled with "fresh" water containing dissolved gases.

Figure 4-10 also predicts what happens if the water temperature remains constant, say at 140 °F, and the pressure is lowered from 75.0 to 15.0 psi. The lower the pressure, the less air the water can hold in solution.

Keep in mind that Figure 4-10 indicates the *maximum possible dissolved air content* of water at a given temperature and pressure. Properly deaerated hydronic systems will quickly reduce the dissolved air content of their water to a small fraction of 1% and maintain it at this condition. Methods of accomplishing this are discussed in Chapter 13, Air & Dirt Removal & Water Quality Adjustment.

4.10 Incompressibility

When a quantity of liquid water is put under pressure, there is very little change in its volume. A pressure increase of 1.0 psi will cause a volume reduction of only 3.4 *millionths* of the original water volume. This change is so small that it can be ignored in the context of designing hydronic systems. Thus, for the purposes of hydronic system design, we can treat water, and most other liquids, as being **incompressible**.

Consider the closed-end cylinder and piston shown in Figure 4-11. It has a cross-sectional area of exactly 1 square inch. If the volume between the piston and cylinder contains air, and 5 lb of force is exerted on the piston, the piston will move down until the pressure of the air increases to 5.0 psi. If the force on the piston is then increased to 10 lb, the air will be further compressed and its pressure will increase to 10.0 psi. Air, being a collection of gases, is compressible. If the force on the piston is later removed, the air will expand to lift the piston back to its original position.

If the same cylinder is filled with liquid water, and the piston is loaded with 5 lb of force, the pressure in the

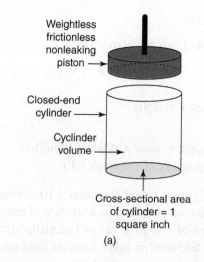

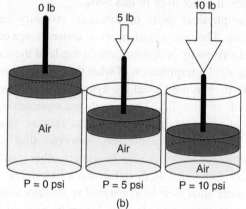

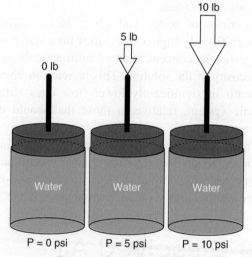

Figure 4-11 (a) A closed-end cylinder with a frictionless/weightless/nonleaking piston. (b) Increasing the force on the piston increases the pressure of air in the cylinder and decreases its volume. (c) Increasing the force on the piston increases the pressure of the liquid water, but with no perceptible change in volume.

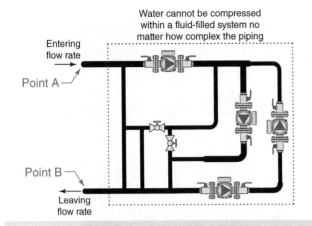

Figure 4-12 The incompressibility of water assures that the flow rate entering point A will always be the same as the flow rate leaving point B (assuming the system is fully filled with water).

water will be 5.0 psi, but the piston will not move down, because there will be no perceptible change in volume—the water is incompressible. If the force is then increased to 10 lb, the water's pressure will increase to 10.0 psi, but again there will be no perceptible change in volume.

In practical terms, this means that a given volume of liquid cannot be squeezed into a smaller volume or "stretched" into a larger volume. It also implies that *if a liquid's volumetric flow rate is known at any one point in a closed-series piping circuit, it must be the same at all other points, regardless of the pipe size. If the piping circuit contains parallel branch circuits, the flow can divide up through them, but the sum of all branch flow rates must equal the total system flow rate.*

Consider the arbitrary piping assembly shown in Figure 4-12. It contains several different branches, pipe sizes, circulators, and valves. The flow rates within the branches of such a complex maze of components may be difficult to determine and are certain to change depending on which circulators are operating and how the valves are set. However, the fact that water is incompressible guarantees that the flow rate at point (A) will always be the same as the flow rate at point (B).

The total liquid flow rate entering any portion of a fluid-filled (nonleaking) piping assembly will always equal the total flow rate exiting that assembly. This simple concept forms the basis for analyzing fluid flow in complex piping systems. It will be applied repeatedly in later chapters.

Incompressibility also implies that liquids can exert tremendous pressure in any type of filled and closed container when they expand due to heating. For example, a closed hydronic piping system completely filled with water, and not equipped with an expansion tank or pressure-relief valve, would quickly burst apart at its weakest component as the water temperature increases.

4.11 Water Quality

In addition to its thermal and hydraulic properties, the chemical properties of water can make the difference between a system that lasts for decades, and one that develops expensive corrosion issues within months of startup. It's important to establish and maintain favorable chemical properties of water and water-based solutions that circulate through the hydronic systems.

Pure water consists of only hydrogen and oxygen. It is colorless, tasteless, and odorless. Unfortunately, "pure" water does not exist in nature. Water from springs, wells, and even municipal mains contains dissolved minerals such as calcium salts and magnesium salts; sediments such as fine silica sand or organic particles; dissolved gases such as oxygen, carbon dioxide, and hydrogen sulfide. Some water sources also contain microorganisms such as bacteria or algae. There can be wide variations in these impurities from one water source to another.

In most hydronic systems, the unknown chemical characteristics of the water used to fill the system have the opportunity to interact with several materials and other chemicals already in the system. These include metals such as copper, brass, cast iron, stainless steel, and carbon steel; liquids such as glycol-based antifreeze, thread cutting oils; and greasy residue such as solder flux. This complex combination of chemicals and metals has the potential to create undesirable reactions such as scaling, oxidation, sludge formation, **galvanic corrosion**, and pitting corrosion.

Figure 4-13 shows examples of heavy scale formation on a stainless steel boiler heat exchanger. This scale greatly reduces the ability of heat to pass from the combustion gases into the water. It can also create localized hot spots that can cause warpage or blistering of the metal, and ultimately failure of the heat exchanger.

The "flakes" seen in Figure 4-13 are formed on the hottest portion of the heat exchanger due to precipitation of dissolved minerals. Expansion and contraction of the heated surfaces eventually cause the accumulated scale to flake off from these surfaces and accumulate on the baffle plate below.

Unfortunately, the chemical interactions that might take place in newly installed hydronic systems are often ignored. Many systems are just filled with whatever water is available on site. Some systems will provide

Figure 4-13 | Scaling caused by mineral deposits on a stainless steel boiler heat exchanger. Courtesy of Caleffi North America.

years of reliable service because of a fortunate set of circumstances involving the purity of the water source, the materials used in the system, the care taken when assembling the system and its subsequent operation and maintenance. Other systems experience premature component failure due to a less fortunate combination of these factors.

Internal Washing

One procedure that helps improve water quality in hydronic systems is an internal washing performed as a newly constructed system is put into service (e.g., commissioning). New systems often contain soldering flux residue, cutting oil, dirt, insects, or other residue. These materials should be cleaned out of the system to the greatest extent possible as a prerequisite to establishing and maintaining chemically stable water or a mixture of water and antifreeze. The methods and materials used for internal washing are discussed in detail in Chapter 13.

Demineralizing Water

In addition to internally cleaning the system, the mineral content of the water should be minimized. The process of doing so is called **demineralization**. As its name implies, demineralization involves removing undesirable minerals such as the dissolved calcium and magnesium compounds found in most ground water and municipal water systems. Such minerals are released from water when it is heated. As the water cools, these minerals will **precipitate** out of solution to form scale on the wetted surface of heat sources, piping, and other components. That scale can significantly reduce rates of heat transfer and fluid flow rates within the system.

In some respects, demineralization is similar to water softening. Both are **ion exchange processes**. During softening, undesirable ions in the source water, which come from dissolved salts containing calcium, magnesium, manganese, and chlorine are captured and exchanged for ions that do not precipitate out of water when it's heated.

In a typical water softening process, the source water is passed through a column containing tens of thousands of small polymer **resin beads** that are chemically synthesized to attract undesirable ions such as those mentioned above. These beads are only about 0.6 mm in diameter and highly porous. When an undesirable ion bonds to a resin bead in a water softening process, another ion, specifically a sodium ion (Na^+) is released from that bead. This ion exchange is necessary to maintain the chemical charge balance of the water.

As the source water passes through a column filled with the resin beads, it will be largely stripped of the undesirable ions that can precipitate out of solution as the water is heated, and harden into scale. The sodium ions that replace the original ions are more soluble in heated water and won't form scale.

Softened water is better than the original water from the standpoint of its ability to form scale, but it's still not pure water. The presence of sodium ions leaves softened water with relatively high electrical conductivity. This characteristic increases the possibility for galvanic corrosion between dissimilar materials in the system (such as copper and iron). Softening solves one problem (e.g., removing calcium and magnesium ions from the source water, which would otherwise cause scaling), but creates another potential problem by leaving the water in a state that's very cooperative to electrical charges. Because of this, *softened water should not be used in hydronic systems.*

The preferred ion exchange process for water used in hydronic systems is **demineralization**. Like softening, it involves passing the source water through a column filled with tiny resin beads. However, these beads are different than those used in water softeners. They are chemically structured so that when they capture an undesirable ion, they release either a **hydrogen cation**, which is chemically represented as (H^+), or a **hydroxide anion**, which are chemically represented as (OH^-). The hydrogen cation has one positive charge, and the hydroxide anion has one negative charge. When one hydrogen cation (H^+) comes in contact with one hydroxide anion (OH^-), they bond to form a neutral molecule of *pure* water (H_2O).

Figure 4-14 illustrates the ion exchange process used for demineralization.

As the resin beads absorb the undesirable ions, their capacity to absorb additional ions slowly decreases, as illustrated in Figure 4-15. Eventually, the "fully loaded" resin beads have to be discarded and replaced with new beads. The spent resin beads are not toxic. They only contain the ions that were in the source water. In most cases, they can be disposed of in recyclable trash.

The presence of ions in water determines its electrical conductivity. In the absence of all ions, water has zero electrical conductivity. *Such water is NOT desirable in hydronic systems.* Water that has been totally stripped of ions will absorb ions from surrounding surfaces. This can lead to metal erosion and other corrosion reactions. Water with zero conductivity will also prevent proper operation of low-water cutoff devices, such as those required by code in many hydronic systems.

One way to determine the extent of demineralization is based on an index called **Total Dissolved Solids (TDS)**. The fewer the ions in the water, the lower the TDS reading. TDS values are expressed in parts per million (abbreviated as PPM).

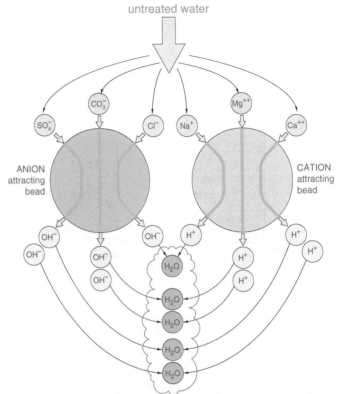

Figure 4-14 | Concept of how specially treated resin beads absorb undesirable ions in source water, and replace those ions with either a hydrogen cation (H^+), or a hydroxide anion (OH^-), which then combine to form pure water (H_2O).

Most untreated water from rivers, lakes, or wells will have a TDS value between 50 and 500 PPM. In contrast, sea water can have a TDS value as high as 40,000 PPM.

To ensure proper operation of low-water cutoff devices, the water used in hydronic systems should have a TDS reading no lower than 10 PPM. This provides sufficient mineral content to allow an electrical current to pass through the water at the probe of a low-water cutoff. The maximum suggested value of TDS for water in a hydronic system is 30 PPM. This limits electrical conductivity, and thus, discourages galvanic corrosion, and other corrosion reactions that depend on electrical currents passing through water.

The TDS value of water can also be determined using a meter such as the one shown in Figure 4-16.

Figure 4-17 shows an apparatus that can be used to demineralize water. It consists of a polymer column that is capable of being sealed and pressurized. The column has hose connection at the top and at the base, allowing water to be circulated through it.

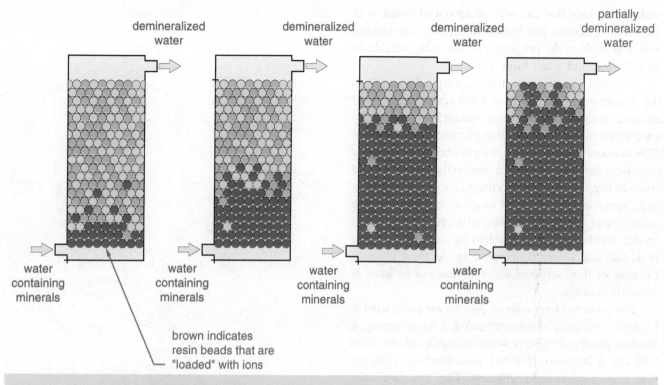

Figure 4-15 | As resin beads in a demineralization column collect undesirable minerals, their ability to provide further demineralized water decreases. Eventually, the "fully loaded" resin beads need to be replaced.

Figure 4-16 | A meter used to read the total dissolved solids (TDS) content of water. Courtesy of Caleffi North America.

Figure 4-17 | Example of a demineralizer column. Courtesy of Caleffi North America.

The column of this apparatus is filled with water-permeable bags containing thousands of the tiny chemically formulated resin beads. The bags allow water to flow through the beads so that the ion exchanger process can occur. They also allow service technicians to quickly and easily replace the beads when necessary without risk of spillage.

There are two ways to demineralize the water used in hydronic systems:

1. Fill the system with demineralized water.
2. Fill the system with source water and then demineralize it.

The first approach is good for smaller systems, where the flow rate of water through the demineralizer is sufficient to purge air from the system as the water is added. The second method is better in systems where the flow rate through the demineralizing column is insufficient to purge air from the system.

Figure 4-18 shows how the demineralizing apparatus from Figure 4-17 would be connected to a system that has already been filled with water and purged of air.

This setup requires a configuration of valves that allow water to simultaneously enter and exit the hydronic system. One such configuration uses a ball valve installed between two tees as seen in Figure 4-18. The side port of each tee connects to a ball valve fitted with an adapter to a standard hose thread. A short hose runs from the outlet valve on the system to the lower connection on the demineralization column. Another short hose runs from the upper connection on the demineralization column to the inlet valve on the system.

The inline ball valve between the system inlet and outlet valves should be closed while water is passing through the demineralizer column. The system outlet valve should be opened to allow water to flow from the system into the bottom of the demineralization column. The system's automatic make-up water system should add water to the system during this time. The water level will rise up through the resin bead column in the demineralizer and out through the loose end of the other hose. The hose connection at the inlet valve should be left slightly loose to allow air to escape as the column is filled with water. Once water reaches this connection, the hose can be fully tightened. The demineralization column and its attached hoses should now be filled with untreated water.

Turn on the system circulator(s) to create flow in the system. With the inline ball valve closed, flow will be directed through the demineralization column. Water with some degree of demineralization will be entering the system. Allow this circulation to continue for several minutes. Then draw a sample of system water from a drain valve in the system into a clean plastic or glass cup. Test the TDS level of this sample using a hand-held meter such as the one shown in Figure 4-18.

In most cases, the initial TDS level will drop rapidly as the untreated water makes its first pass through the demineralization column. The rate at which the TDS level drops will decrease with time. The ideal range for TDS is 10 to 30 PPM. Continue circulating the system water through the demineralization column, periodically sampling and testing the system water until an acceptable TDS level is achieved.

When the demineralization process is finished, the water within the demineralization column can remain in it, unless it will be subject to freezing temperatures before its next use. In the latter case, the water should be drained through the valve at the bottom of the column.

Once the demineralization process has been completed, the system should be operated with the heat source on and all air separation and venting devices active. Whenever possible, the heat source should produce an outlet temperature of 140 °F or higher. This heating and circulation should be maintained for at least one hour. Raising the temperature of the water

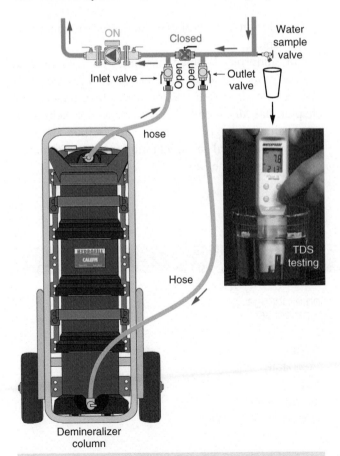

Figure 4-18 | Connecting a demineralizer column to the hydronic system using hoses and a configuration of three ball valves. *Courtesy of John Siegenthaler*

helps force dissolved gases out of solution, as discussed earlier in this chapter. These gas molecules will coalesce to form microbubbles that can be captured by a high-performance air separator. The elimination of these gases helps stabilize the pH of the demineralized water.

Stabilizing Water Treatment

A final step in helping to ensure high water quality is adding a film-forming **hydronic system stabilizer**. Several such stabilizers are available in North America. A sampling of some available products is shown in Figure 4-19.

These additives form extremely thin protective films on the inside of the piping components. They also contain chemicals to stabilize the pH of the fluid, and help protect the system against the slow but unavoidable oxygen diffusion that occurs over the system's life.

Figure 4-19 Examples of hydronic system water stabilizers. Courtesy of Caleffi North America.

Summary

The properties of water presented in this chapter describe its heat absorption, boiling point, expansion, flow resistance, and air-holding characteristics. Many of the questions that arise during system design can be answered based on a proper understanding of these fluid properties. Ignoring them can lead to serious problems such as low heat output, improper expansion tank sizing, and inadequate circulator selection.

The sensible heat quantity and rate equations are two of the most often used equations in hydronic system design. You will see them used frequently in later chapters.

In addition to understanding the fundamental thermal and hydraulic characteristics of water that play a role in system design and performance, this chapter has discussed the importance of achieving and maintaining good water quality within the system, which is vital in maintaining good performance over a long service life.

Key Terms

absolute pressure
demineralization
demineralizer column
density
dissolved air
dynamic viscosity
flow rate
galvanic corrosion
heat capacity
hydrogen cation
hydronic system stabilizer
hydroxide anion
incompressible
ion
ion exchange processes
latent heat
mass
precipitate
resin beads
sensible heat
sensible heat quantity equation
sensible heat rate equation
specific heat
steam flash
Total Dissolved Solids (TDS)
vapor pressure
viscosity

Questions and Exercises

1. How many gallons of water would be required in a storage tank to absorb 250,000 Btu while undergoing a temperature change from 120 to 180 °F?

2. How much pressure must be maintained on water to maintain it as a liquid at 250 °F? State your answer in psia.

3. If you wanted to make water boil at 100 °F, how many psi of vacuum (below atmospheric pressure) would be required? Hint: psi vacuum = 14.7 − absolute pressure in psia.

4. Determine the airflow rate (in cubic feet per minute) necessary to transport 40,000 Btu/h. based on a temperature increase from 65 to 135 °F. Base the calculation on a specific heat for air of 0.245 Btu/lb/°F and an assumed density of 0.07 lb/ft^3.

5. Assume you need to design a hydronic system that can deliver 80,000 Btu/h. What flow rate of water is required if the temperature drop of the distribution system is to be 10 °F? What flow rate is required if the temperature drop is to be 20 °F?

6. Water flows through a series-connected baseboard system as shown in Figure 4-20. The inlet temperature to the first baseboard is 165 °F. Determine the outlet temperature of the first baseboard, and both the inlet and outlet temperatures for the second and third baseboards. Assume that the heat output from the interconnecting piping is insignificant. Hint: Use the approximation that the inlet temperature for a baseboard equals the outlet temperature from the previous baseboard.

7. At 50 °F, 1,000 lb of water occupies a volume of 16.026 ft^3. If this same 1,000 lb of water is heated to 200 °F, what volume will it occupy?

8. How much heat is required to raise the temperature of 10 lb of ice from 20 to 32 °F? How much heat is then required to change the 10 lb of ice into 10 lb of liquid water at 32 °F? Finally, how much heat is needed to raise the temperature of this water from 32 to 150 °F? State all answers in Btu.

9. You are designing a thermal storage system that must supply an average load of 30,000 Btu/h for 18 consecutive hours. The minimum water temperature usable by the system is 110 °F. Determine the temperature the water must be heated at by the end of the charging period if:
   a. the tank volume is 500 gal
   b. the tank volume is 1,000 gal

10. An 18,000-gal swimming pool is to be heated from 70 to 85 °F. Assuming an oil-fired boiler with an average efficiency of 80% will be used and that all heat is retained by the pool, with no losses to the ground or air, determine the following:
    a. the number of Btu required to heat the pool
    b. the number of gallons of #2 fuel oil required by the boiler
    c. assuming a cost of $2.50 per gallon of oil, the cost of bringing the pool temperature from 70 to 85 °F

11. Recalculate the Btu used for domestic water heating in Question 13 assuming that the water is heated to only 120 °F. Estimate the annual energy savings of the reduced water temperature setting assuming the same load exists 365 days/year. Express the answer in MMBtu (1 MMBtu = 1,000,000 Btu).

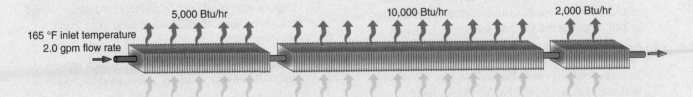

Figure 4-20 | Series-connected baseboard convectors for exercise 7.

12. Assume a hydronic system is filled with 50 gal of water at 50 °F and 30.0 psi pressure. The water contains the maximum amount of dissolved air it can hold at these conditions. The water is then heated and maintained at 200 °F and 30.0 psi pressure. Using Figure 4-10, estimate how much air will be expelled from the water.

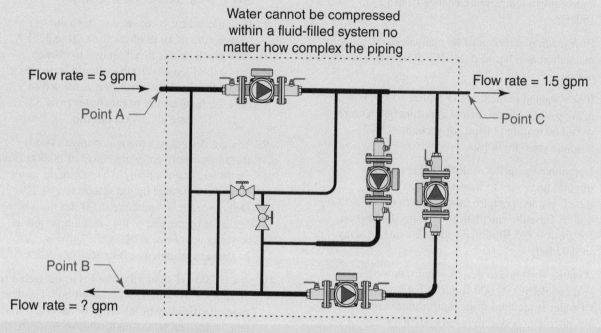

Figure 4-21 | Piping assembly for exercise 18.

13. Use the Fluid Properties Calculator module in the Hydronics Design Studio to determine the specific heat of a 50% ethylene glycol solution at 100 °F. Also find the specific heat of water at 100 °F. Assume that a flowing stream of each liquid had to deliver 40,000 Btu/h. of heat to a load with a 15 °F drop in temperature:
    a. What flow rate would be required for each fluid?
    b. Express your answers in pounds per hour.

14. Use the Fluid Properties Calculator module in the Hydronics Design Studio to determine the specific heat and density of a 30% ethylene glycol solution at 132.5 °F. Assume this fluid entered a heat emitter at 140 °F and exited the heat emitter at 125 °F, at a flow rate of 4.5 gpm. What is the rate of heat transfer from the fluid stream to the heat emitter? Express your answer in Btu/h. Compare the results to those of example 4.2. Explain the difference. Note: 1 ft^3 = 7.48 gal.

15. Examine the piping assembly shown in Figure 4-21:
    a. What is the flow rate exiting point B in the piping assembly?
    b. How would this change if only two of the four circulators were operating?
    c. Assume the flow rate exiting point B was 3.2 gpm, and the other flow rates remain as shown. What would this imply is happening somewhere with the piping assembly?

16. Explain why softened water should not be used in hydronic systems.

17. Explain why the TDS of demineralized water should not be less than 10 PPM when used in hydronic systems.

18. Describe several impurities that are often present within the piping of newly constructed hydronic systems.

Chapter 5

Pipings, Fittings, and Valves

Objectives

After studying this chapter, you should be able to:

- Describe several types of piping materials used in hydronic systems.
- Compare the temperature and pressure ratings of several piping materials.
- Describe several types of piping support devices.
- Describe the steps involved in soldering copper tube.
- Discuss the issue of oxygen diffusion through polymer pipe.
- Identify several common pipe fittings.
- Specify the exact type of pipe fitting needed for a given situation.
- Explain the use of several specialized pipe fittings.
- Calculate the expansion movement of copper pipe.
- Select the correct type of valve for a given application.
- Identify several specialized valves and explain their use in hydronic systems.
- Recognize schematic symbols for piping system components.

5.1 Introduction

This chapter presents basic information on pipings, fittings, and valves used in residential and light commercial hydronic systems. It emphasizes the proper selection and installation of these components. The intent is to acquaint the reader with the wide range of available hardware, in preparation for overall system design in later chapters.

Much of the information deals with the benchmark hydronic piping material, copper tubing. Information on common and specialized copper fittings is presented. A short primer on soft soldering is also given.

Contemporary polymer piping materials such as **Polyethylene-Raised Temperature (PE-RT)** and **Polypropylene-Random (PP-R)** tubing are also discussed. These materials have been successfully used for several years in Europe and are currently establishing significant market share in North America. The versatility they offer in certain situations is unmatched by traditional materials.

This chapter also discusses the intended application of several types of common and specialized valves used in hydronic heating systems.

Tips for professional installation of piping are also given. Small details such as level and plumb piping runs, wiped solder joints, and use of specialty fittings to reduce parts count distinguish systems that merely work from those that convey professionalism and craftsmanship.

5.2 Piping Materials

This section compares several piping materials suitable for use in hydronic heating systems. The discussion is limited to those materials that are most practical in today's market. These include copper and several types of polymer tubing.

All piping materials have strengths and weaknesses. All have conditions attached to their use that must be followed if the system is to perform as expected and last many years. No single material is ideal for all applications.

Piping materials must be selected based on considerations such as temperature and pressure ratings, availability, ease of installation, corrosion resistance, life expectancy, local code acceptance, and cost.

Designers should not feel constrained to use a single type of pipe for the entire system. Each material should be applied where it can overcome obstacles posed by other materials. For example, flexible PEX tubing may be ideal in retrofit situations where use of rigid piping materials would be difficult or impossible. On the other hand, rigid copper or straight lengths of rigid polymer tubing have obvious advantages where the piping runs must be straight or when heavier components will be supported by the tubing. A combination of flexible and rigid piping materials may provide the ideal solution in many systems.

Discussion will be limited to pipe sizes from 3/8 to 3 inches nominal diameter. Nearly all the hydronic systems within the scope of this book can be constructed of piping within this range. All analytic methods for piping design to be presented in later chapters, as well as in the Hydronics Design Studio software, also pertain to this range of pipe sizes.

Pipe Versus Tube:

The words "**pipe**" and "**tube**" are often used loosely and interchangeably. Both refer to a closed, round conduit intended for fluid (e.g., liquid or gas) conveyance.

The term "tube" generally refers to smooth, nonferrous conduit intended for fluid conveyance and having approximate diameters less than or equal to 2 inches. Examples include 1/2" copper tube, 5/8" PEX tube, and 1" CPVC tube. One exception to the above is reference to copper, which is still described as tube (rather than pipe) in nominal diameters up to 12 inches.

The term "pipe" is customarily used when referring to pipe made of ferrous materials such as steel or black iron pipe.

Copper Water Tube

Copper water tube was developed in the 1920s to provide an alternative to iron piping in a variety of uses. Its desirable features include:

- Good pressure and temperature rating for typical hydronic applications,
- Good resistance to corrosion from water-based system fluids,
- Smooth inner walls that offer low flow resistance,
- Lighter than steel or iron piping of equivalent size, and
- Ability to be joined by the well-known technique of soft soldering.

In situations where good heat transfer through the walls of the pipe is desirable, copper is unsurpassed. Examples include heat exchangers or the finned-tube element in baseboard convectors. The high thermal conductivity of copper can also be a disadvantage in situations where heat loss from piping should be minimized.

Copper tubing is available in a wide range of sizes and three common wall thicknesses. The type of copper tubing used in hydronic heating is called copper water tube. In the United States, copper water tube is manufactured according to the ASTM B88 standard. In this category, **pipe size** refers to the **nominal inside diameter** of the tube. The word "nominal" means that the measured inside diameter is approximately equal to the stated pipe size. For example, the actual inside diameter of a nominal 3/4-inch type M copper tube is 0.811 inches.

The outside diameter of copper water tubing is always 1/8 inch (0.125 inches) larger than the nominal inside diameter. For example, the outside diameter of 3/4-inch type M copper tube is 0.875 inches. This is exactly 7/8 inch or 1/8 inch larger than the nominal pipe size.

Another category of copper tubing, designated as ACR tubing, is used in refrigeration and air conditioning systems. In this category, pipe size refers to the outside diameter of the tubing. ACR tubing is not used in hydronic heating or water distribution systems. It is used for the refrigerant lines that connect the outdoor unit of an air-to-water heat pump with the indoor unit.

Copper water tube is available in three wall thicknesses designated as types K, L, and M in order of decreasing wall thickness. The outside diameters of K, L, and M tubing are identical. This allows all three types of tubing to be compatible with the same **fittings** and valves.

Because the operating pressures of residential and light commercial hydronic heating systems are relatively low, the thinnest wall copper tubing (type M) is most often used. This wall thickness provides several times the pressure rating of other common hydronic system components, as can be seen in Figure 5-1. In the absence of any building or mechanical codes that require otherwise, type M is the standard for copper tubing used in hydronic heating systems.

Two common hardness grades of copper water tube are available. Hard drawn tubing is supplied in straight lengths of 10 and 20 feet. Because of its straightness and strength, **hard drawn tubing** is the most commonly used type of copper tubing in hydronic systems. So-called **"soft temper" tubing** is annealed during manufacturing to allow it to be formed with simple bending tools. It is useful in situations where awkward angles do not allow proper tubing alignment with standard fittings. Soft temper tubing comes in flat coils having standard lengths of 60 and 100 feet. The minimum wall thickness available in soft temper copper tubing is type L.

Supporting Copper Tubing

Copper tubing must be properly supported to prevent sagging or buckling. On horizontal runs of hard temper tubing, in the absence of codes that require otherwise, the following maximum support spacing is suggested:

- 1/2- and 3/4-inch tube: 5 feet maximum support spacing;
- 1- and 1 1/4-inch tube: 6 feet maximum support spacing;
- 1 1/2- and 2-inch tube: 8 feet maximum support spacing.

These support spacings do not include any allowance for the extra weight of piping components such as circulators, expansion tanks, or other components that may be supported by the tubing. When such components are present, the piping should be supported immediately adjacent to the component.

On vertical runs, copper tubing should be supported at each floor level or a maximum of every 10 feet.

A number of different supports are available for small tubing. In residential systems, horizontal piping runs supplying fin-tube baseboard are often supported by **plastic-coated hangers** as shown in Figure 5-2. These simple devices allow the tubing to move slightly as it expands upon heating. The plastic coating reduces squeaking sounds as the tube moves through the hanger.

Nominal tube size (in)	Inside diameter (in)	Outside diameter (in)	Rated working pressure (psi)*
3/8	0.450	0.500	855/456
1/2	0.569	0.625	741/395
3/4	0.811	0.875	611/326
1	1.055	1.125	506/270
1.25	1.291	1.375	507/271
1.5	1.527	1.625	497/265
2	2.009	2.125	448/239
2.5	2.495	2.625	411/219
3	2.981	3.125	380/203

* Pressure ratings are based on service temperatures not exceeding 200 °F. The first number is for drawn (hard) tubing. The number following the forward slash is for annealed (soft temper) tubing.

Figure 5-1 | Physical data for selected sizes of type M copper water tube.

Figure 5-2 | Examples of plastic-coated metal pipe hanger. *Courtesy of John Siegenthaler.*

Some pipe hangers have pointed ends that are driven into wooden floor joists. Others have a perforated strap that can be nailed to framing.

Standoff piping supports are used to secure piping to either walls or ceilings. One type of standoff support is commonly called a "**bell hanger**" and is shown in Figure 5-3a. It is used to rigidly clamp small tubing in place with a small gap between the tubing and surface. The small gap is usually sufficient space for pipe insulation when required. The bell-shaped base is secured to the mounting surface with a single flat-head screw. Bell hangers, when fully tightened, do not allow tubing to move due to thermal expansion or contraction and should not be used to support long lengths of straight pipe.

Another type of standoff support is commonly called a **rod clamp**. It consists of a clamp that fits the outside diameter of the tubing, a base plate that fastens to the supporting surface, and a length of threaded steel rod that screws to both the clamp and base plate as shown in Figure 5-3b. The steel rod can be cut to whatever length is necessary depending on the distance between the tube and supporting structure. When the rod is oriented horizontally, the length of the rod should be limited to a few inches because of the twisting effect the weight of the piping exerts on the base plate. When the rod is vertical, it can be as long as necessary. Longer rod lengths allow for some movement of the pipe due to thermal expansion and contraction.

Copper tubing can also be supported at intermediate heights using **clevis hangers** supported by threaded steel rods attached to the ceiling structure. Clevis hangers are ideal when piping must be supported away from wall or ceiling surfaces. An example of a clevis hanger is shown in Figure 5-4. Clevis hangers also allow tubing to easily move due to thermal expansion and contraction.

In situations where several runs of piping and/or electrical conduit must be run in close proximity, a **mounting rail system** is a good option. Such systems consist of steel mounting rails to which clamps for various sizes of pipe can be secured. The rails can be mechanically fastened to walls or ceilings or hung from threaded steel rods. The pipe clamps can easily be moved along the mounting rails before being tightened. Clamps are available with rubber liners that prevent metal-to-metal contact between the clamp and the tubing. Rubber-lined clamps also help dampen vibrations imparted to the piping from components such as circulators. An example of a cushioned clamp/rail system is shown in Figure 5-5.

(a)

(b)

Figure 5-3 Examples of standoff pipe supports: (a) bell hanger; (b) rod clamp. *Courtesy of John Siegenthaler.*

Figure 5-4 Clevis hanger. Threaded steel rod connects at top. *Courtesy of Warwick Hanger.*

Figure 5-5 Example of a cushioned clamp and rail pipe mounting system. *Courtesy of Hydra-Zorb Corporation.*

Figure 5-7 Overhead piping supported by a multilevel suspended mounting rail system. *Courtesy of John Siegenthaler.*

Figure 5-6 shows the versatility of a mounting rail system in accommodating several types and sizes of pipes and/or electrical conduits. Figure 5-7 shows a multiple-level mounting rail system supporting overhead piping and electrical conduit in a mechanical room.

Soldering Copper Tubing

The common method of joining copper tubing in hydronic heating systems is **soft soldering** using a 50/50, tin/lead solder. The **working pressure rating** of the resulting joints is dependent on operating temperature and tube size and can be found in Figure 5-8. These pressure ratings are well above the relief valve settings of most residential and light commercial hydronic systems.

Figure 5-6 Example of a rail-type pipe mounting system accommodating different types and sizes of pipes and electrical conduits. *Courtesy of John Siegenthaler.*

Solder type	Service temperature	Maximum allowed pressure (psi) (1/4–1 inch tube)	Maximum allowed pressure (psi) (1.25–2 inch tube)	Maximum allowed pressure (psi) (2.5–4 inch tube)
50/50 tin/lead	100	200	175	150
	150	150	125	100
	200	100	90	75
	250	85	75	50
95/5 tin/antimony	100	500	400	300
	150	400	350	275
	200	300	250	200
	250	200	175	150

Figure 5-8 Pressure ratings versus operating temperature for soldered joints in copper tubing. *Data provided by the Copper Development Association.*

One circumstance in which 50/50 tin/lead solder cannot be used is when the tubing carries **domestic water**. *Plumbing codes no longer allow solders containing lead to be used for domestic water service.* One alternative is 95/5 tin/antimony solder. It contains no lead and has higher pressure ratings than the 50/50 solder. The melting temperature range of 95/5 solder is 452 to 464 °F. This is significantly higher and narrower than the 361 to 421 °F range of 50/50 tin/lead solder. Joints made with 95/5 require a longer heating time before the solder will flow. The narrow melting range of 95/5 also means the joint will solidify very quickly when heat is removed. In general, 95/5 solder is more difficult to work with than 50/50 tin/lead and is often used only for domestic water piping or when the pressure requirement of the joint cannot be met using 50/50 solder.

Soldering Procedure

Proper soldering results in clean, neat, and watertight joints. The attention given to such joints reflects the professional skills of the installer. Sloppy joints, even if watertight, indicate poor craftsmanship. With a little practice, making good soldered joints becomes second nature.

The following procedure describes the steps in making a good soldered joint:

STEP 1. Be sure the tube is cut square. Use a wheel cutter or chop saw with a blade suitable for cutting copper tube. Avoid the use of a hacksaw whenever possible. See Figure 5-9.

STEP 2. To remove any burrs, ream the end of the tube with a rounded file or other type of deburring tool. See Figure 5-10. Be sure to clean any chips from the tube after filing.

STEP 3. Assemble the joint and check it for proper fit and alignment. In most cases, there will be no problem with the fit. However, sometimes a fitting may be damaged or defective. If the fitting wobbles noticeably on the pipe, it should be replaced. Attempting to solder such a joint can result in a leak or poorly aligned joint.

STEP 4. Clean the fitting's socket with a properly sized fitting brush. Such brushes are available for manual use

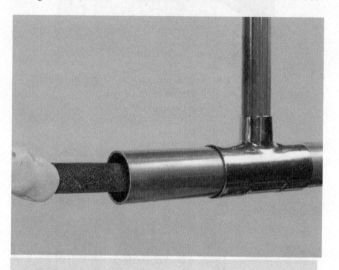

Figure 5-10 | Reaming the end of the tube with a rounded file. *Courtesy of the Copper Development Association.*

Figure 5-9 | Getting a square-cut of copper tubing using a wheel-type cutter. *Courtesy of the Copper Development Association.*

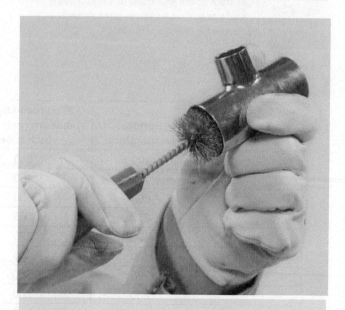

Figure 5-11 | Cleaning the fitting's socket with a steel fitting brush. *Courtesy of the Copper Development Association.*

as shown in Figure 5-11, as well as for mounting in a drill chuck. The latter saves time when many fittings need to be cleaned. Be sure to remove any pieces of the steel brush bristles remaining in the fitting after cleaning it. These small pieces of steel, if left in the joint, can create **galvanic corrosion** that eventually could cause a leak.

STEP 5. Clean the outside of the tube with abrasive cloth as shown in Figure 5-12. All oxidation, scale, paint, or dirt should be removed from the tubing surface at least 1/2 inch farther back than the edge of the fitting socket. The tube should appear bright following cleaning. Be sure to clean all the way around the perimeter of the tube.

STEP 6. Apply **paste flux** to both the fitting socket and the portion of the tube that projects inside the socket. The flux chemically cleans the surface of the copper and helps prevent oxidation of the surfaces when heat is applied. *Always use a flux brush, not your finger, to apply the flux.* Do not apply excessive amounts of flux. Only a thin film is needed. See Figures 5-13 and 5-14.

STEP 7. Assemble the tube and fitting. If possible, rotate the fitting around the tube one or two times to ensure the flux is evenly spread. It is likely there will be excess flux squeezed out of the joint. This flux serves no purpose and should be wiped off with a cloth before heating the joint as shown in Figure 5-15.

STEP 8. Apply heat to the outside of the fitting socket using the torch. Keep the blue tip of the flame just above the surface of the fitting socket as shown in Figure 5-16. Move the flame around the outside of the fitting socket to promote even heating. Heating times differ considerably with the type of torch and gas being used, as well as

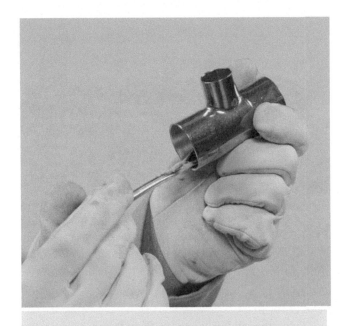

Figure 5-13 | Applying paste flux to the inside of a fitting socket. *Courtesy of the Copper Development Association.*

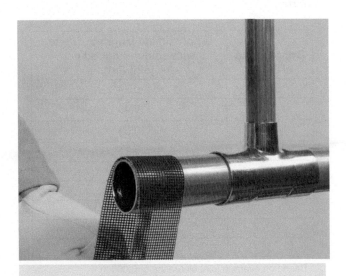

Figure 5-12 | Cleaning the outside of the tubing with abrasive cloth. *Courtesy of the Copper Development Association.*

Figure 5-14 | Applying paste flux to the outside of the tube. *Courtesy of the Copper Development Association*

pipe size, ambient temperature, and type of solder. When the flux begins to sizzle, test the joint by applying the end of the solder wire to the edge of the joint as shown in Figure 5-17. If it sticks but does not melt, continue to apply heat. If the solder immediately melts, the joint is ready to draw in the molten solder by capillary action.

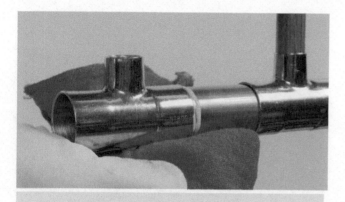

Figure 5-15 | Assemble fitting and tube, then remove excess paste flux with a cloth. *Courtesy of the Copper Development Association.*

Once the solder begins to flow, it can be continually fed into the joint. Only a small amount of solder is needed.

An experienced pipe fitter knows by sight how much solder to feed into the joint. An inexperienced person often feeds excessive solder into the joint. The excess solder can solidify into small loose particles inside the pipe. These can be carried around the system by fast-moving fluid and become lodged in valves or circulators. Figure 5-18 lists the approximate length of 1/8-inch solder wire required for joints in tube sizes from 3/8 to 3 inches. Loose-fitting joints may require more, while tight joints may require less.

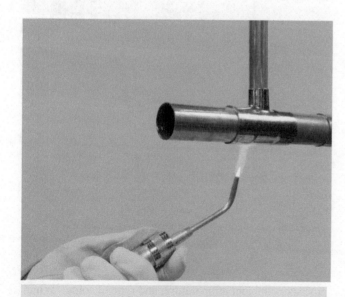

Figure 5-16 | Heating the fitting's socket during soldering. *Courtesy of the Copper Development Association.*

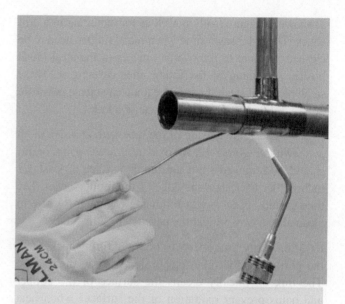

Figure 5-17 | Testing the joint for solder melting temperature. *Courtesy of the Copper Development Association.*

A properly heated joint will quickly spread the molten solder around its perimeter. When the solder forms a narrow silver ring around the visible edge of the joint, adequate solder has been applied. If the solder begins to drip off the joint, excess solder is being used. Do not feed any more solder to the joint.

STEP 9. After the solder is applied, remove the torch and carefully wipe the perimeter of the joint with a clean cloth rag to remove any excess solder or flux as shown in Figure 5-19. Be careful! All surfaces are still very hot and the solder being wiped is still molten. *Always wear gloves, long-sleeved shirts, and safety goggles when soldering.*

Copper tube size (in)	Approximate length of 1/8- in wire solder required per joint (in)*
3/8	0.24
1/2	0.40
3/4	0.84
1	1.31
1.25	1.71
1.5	2.27
2	3.65
2.5	4.94
3	6.64

* Based on average joint clearance of 0.005 in.

Figure 5-18 | Approximate length of 1/8-inch solder wire required for an average joint.

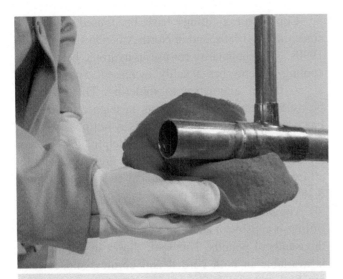

Figure 5-19 | Wiping the hot joint. *Courtesy of the Copper Development Association.*

Tube size (in)	Actual and approximate take-up distance (d) of fitting sockets (in)		Approximate thread engagement length (x) for NPTs (in)
	Actual	Approximate	
3/8	0.38	3/8	1/4
½	0.50	½	½
¾	0.75	¾	½
1	0.91	15/16	9/16
1.25	0.97	1	5/8
1.5	1.09	1 1/16	5/8
2	1.34	1 5/16	1 1/16
2.5	1.47	1 ½	7/8
3	1.66	1 11/16	1

NPT, national pipe thread.

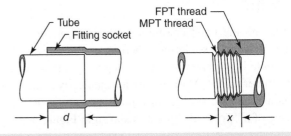

Figure 5-20 | Typical take-up allowances for copper and brass fittings.

The joint should be allowed to cool naturally before being handled or stressed. Dipping the assembly in water for rapid cooling can cause high stresses in the metal and should be avoided. Cast fittings have been known to crack under such conditions.

After the joint has cooled, one last wiping with a clean cloth moistened with a mild detergent solution will remove any remaining flux. This is an important step because flux residue left on the joint eventually causes unsightly surface oxidation. Be a professional and thoroughly clean every joint after soldering.

More Tips on Soldering

- Experience shows it is best to make up piping subassemblies on a horizontal working surface whenever possible. The solder flow is more easily controlled, and the work goes faster because it is done from a more comfortable position. After cooling, these piping subassemblies can be joined into the overall system.

- When soldering has to take place next to combustible materials, slide a flame shield pad between the joint and the combustible materials. This prevents charring wood next to the joint. *Charred wood next to soldered joints does not make a good impression on building owners. Instead, it denotes sloppy workmanship and disregard for safety.*

- Measure all tubing carefully before cutting. Allow for the take-up of the pipe into the fitting sockets. Figure 5-20 indicates the take-up allowances for copper and brass fittings in pipe sizes from 3/8 through 3 inches.

- Whenever possible, attempt to keep all piping plumb (vertical) or level (horizontal). Even a person with no plumbing background will notice the difference between plumb or level piping and pipes that simply connect point A to point B without regard to appearance. Install temporary piping supports, if necessary, during piping assembly to hold the pipe in proper alignment for soldering. Sloped piping may be required in some installations to allow for drainage. In such cases, provide a minimum slope of 1/4-inch vertical drop per foot of horizontal run.

- On fittings having both soldered and threaded connections, make up the soldered joint first. This prevents discoloration or burning of Teflon tape or joint sealing compound during soldering.

- *Always keep one end of the piping assembly open to atmosphere during soldering.* If this is not done, the air trapped inside the pipe will increase in pressure due to heating, and blow pinholes through the molten solder before it can solidify. When a piping loop is finally closed in, open a valve or other component near the final joint to allow the heated air to escape.

- Before making up the joint, put a sharp bend in the wire solder corresponding to the lengths given in Figure 5-18. This provides a visual indication of how much solder has been fed to the joint.

- Consider using MAPP gas, rather than propane, when working with 95/5 solder. MAPP gas is a stabilized mixture of methylacetylene-propadiene propane. It produces a hotter flame temperature (5,300 °F) compared to propane (2,250 °F), resulting in faster heating of joints, especially on larger components.

- Finally, remember that copper is a premium architectural material due to its luster and durability. Take a few extra minutes to clean up any excess flux or shine up oxidized tubing or components. *Your customers will notice it.*

Mechanical Joining of Copper Tubing

Although the most common method of joining copper tubing is soft soldering, there are alternatives. One example is a system in which copper fittings containing elastomer (EPDM) O-rings are mechanically compressed against the tube wall to form pressure-tight joints. A cut away of a **"press-fit" joint** is shown in Figure 5-21.

Copper press fittings have been used in Europe since the late 1980s, and in North America since the late 1990s. Their popularity for use in hydronic heating and cooling systems has steadily increased. They provide a "no flame" alternative to soldering. They can also be used to create reliable joints in copper tubing that has water trickling through it. This is not possible with soldered joints and is particularly helpful in repair or replacement situations. Press fittings are recognized by all major plumbing and mechanical codes in the United States, Canada, and Europe.

The power tool used to compress the copper press fittings is capable of exerting a crimping force of several thousands of pounds. This tool, shown in Figure 5-22, is available in both corded and cordless models and can be fitted with different size jaws to accommodate tube sizes from 1/2 to 4 inches. The tool requires about 4 seconds to uniformly compress the fitting (Figure 5-23). Once the joint is pressed (Figure 5-24), it cannot be taken apart.

Although press fittings are more expensive than standard (solder-type) fittings, the time required for joint preparation is considerably shorter. The tube ends still require reaming to remove any burrs due to cutting. However, no mechanical cleaning or fluxing is required. This reduces the installation cost of the pressed joint system compared to that of a soldered joint. No torch is required with the pressed joint that arguably is a safety advantage in many environments.

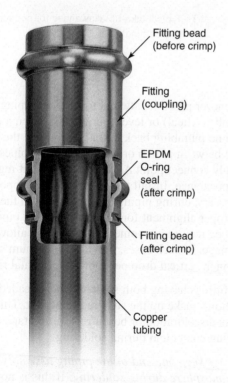

Figure 5-21 Examples of a press-fit joint in copper tubing using a mechanically compressed O-ring fitting. *Courtesy of Ridgid Tool Co.*

Figure 5-22 Tooling used for a mechanically compressed joint. The compression tool can be loaded with several different jaws depending on pipe size being joined. *Courtesy of Ridgid Tool Co.*

5.2 Piping Materials

A wide variety of fittings, valves, and other hydronic system components such as air separators and zone valves are now available for press-fit joints in hydronic systems. Some of the available fittings are shown in Figure 5-25. Press fittings are also used in water supply plumbing and gas piping systems. Some press fittings are rated to operate at pressures up to 200 psi and temperatures ranging from 0 to 250 °F. These ratings would allow press fittings to be used in most hydronic heating and cooling systems used in residential and light commercial building.

PEX Tubing

Cross-linked polyethylene tubing, also referred to as **PEX tubing**, is a product developed, refined, and extensively used in Europe over the last 70 years. In the mid-1980s, the market for PEX tubing in North America began to expand, driven by increasing interest in hydronic radiant panel heating. Worldwide, PEX tubing has proven itself a reliable alternative to metal piping in many hydronic heating applications.

Prior to cross-linking, polyethylene molecules are not bonded to each other in a way that prevents their movement. When the polyethylene is sufficiently heated, individual molecules can flow past each other. Eventually, the material "melts" and can be reformed in shape while at the elevated temperature. This is a common characteristic of many **thermoplastics** including polyethylene, polybutylene, and polypropylene, all of which can be molded into various shapes or extruded into tubing.

The cross-linking process is the essential step in transforming ordinary high-density polyethylene into a **thermoset plastic**. During cross-linking, some of the carbon/hydrogen bonds in the polyethylene molecules are broken using various chemical or electroradiative processes. The broken bonds allow adjacent polyethylene molecules to "fuse together" into a three-dimensional network. Once this occurs, the material is classified as a thermoset plastic. It can no longer be melted or reformed. Cross-linking significantly increases the temperature/pressure capabilities of ordinary high-density polyethylene tubing.

PEX tubing is capable of withstanding the repeated stresses experienced through heating and cooling while encased in a rigid material such as concrete. It is also very resistant to chemicals that can cause corrosion or scaling in metal pipe.

PEX tubing is sold in continuous coils ranging from 150 feet to more than 1,000 feet in length (depending

Figure 5-23 | A "press-fit" joint being made using a cordless compression tool. *Courtesy of Viega.*

Figure 5-24 | Completed press-fit joints using copper tube and fittings. *Courtesy of John Siegenthaler.*

Figure 5-25 | Examples of some fitting and valves available for press-fit copper piping systems. *Courtesy of Viega.*

on diameter and manufacturer). PEX is available in several diameters suitable for use in hydronic heating applications. The most commonly used nominal tube sizes range from 1/4 to 1 inch. Figure 5-26 shows samples of PEX tubing in sizes of 3/8, 1/2, 5/8, and 3/4 inches.

Figure 5-27a lists some physical data for PEX tubing that conform to the widely accepted ASTM F876 standard.

With all plastic tubing, there is a trade-off between operating temperature and allowable operating pressure.

The ASTM F876 standard for PEX tubing establishes three simultaneous temperature/pressure ratings based on operation with water. They are listed in Figure 5-27b.

The availability of long continuous coils, combined with the ability to bend around moderate curves without kinking, make PEX tubing ideal for use in radiant panel heating systems. It also allows the tubing to run through confined or concealed spaces in buildings were working with rigid copper tubing would be difficult or impossible.

Specialized fittings are available from the tubing manufacturers to transition from PEX tubing to standard metal pipe fittings, both threaded and soldered. Examples of such fittings are shown in Figure 5-28. These fittings allow PEX tubing to be connected to heat emitters, or other system components, as well as used interchangeably with metal pipe.

With reasonable care, PEX tubing is very resistant to kinking during installation. If a kink does occur, it can be easily repaired in the field. The process requires the kinked area to be heated to approximately 275 °F using a hot air gun. At this temperature, the crystalline structure of the polyethylene changes to an amorphous state, relieving the stresses created by the kink. The material returns to its normal shape. After cooling, there is virtually no indication that a kink ever existed. Because of this behavior, PEX tubing is said to have a **"shape memory."**

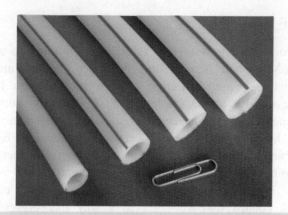

Figure 5-26 | PEX tubing in sizes of 3/8, 1/2, 5/8, and 3/4 inches. *Courtesy of John Siegenthaler.*

Nominal tube size (in)	Average outside diameter (in)	Minimum wall thickness (in)	Minimum burst pressure at 180 °F temperature (psi)
1/4	0.375	0.070	390
3/8	0.500	0.070	275
1/2	0.625	0.070	215
5/8	0.750	0.083	210
3/4	0.875	0.097	210
1	1.125	0.125	210
1.25	1.375	0.153	210
1.5	1.625	0.181	210
2	2.125	0.236	210

(a)

Temperature (°F)	Pressure rating for water (psi)
73.4	160
180	100
200	80

(b)

Figure 5-27 | (a) Physical data for several sizes of PEX tubing conforming to the ASTM F876 standard. (b) Temperature and coincident pressure ratings for PEX tubing conforming to the ASTM F876 standard.

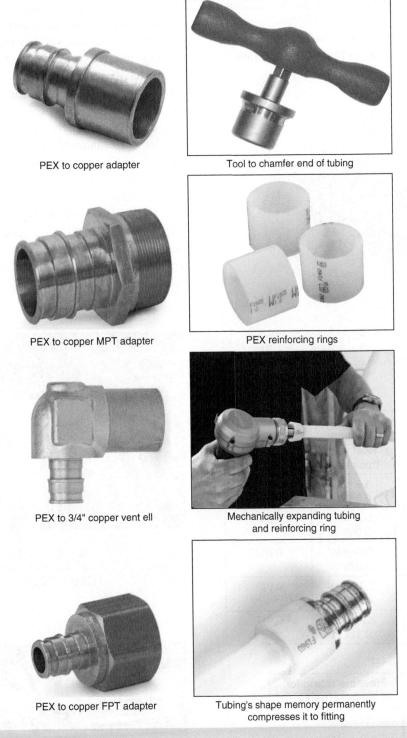

Figure 5-28 | Examples of fittings and "cold expansion" joining procedures based on ASTM-F1960 standard. *Courtesy of Uponor.*

Composite PEX-AL-PEX Tubing

Another type of tubing well suited for hydronic heating systems is called composite **PEX-AL-PEX tubing**. It consists of three concentric layers bonded together with special adhesives. The inner and outer layers are PEX. The middle layer is longitudinally welded aluminum. A close-up of the cross-section of a PEX-AL-PEX tube is shown in Figure 5-29.

The PEX-AL-PEX tubing commonly used in hydronic heating systems conforms to the standard ASTM F1281. Figure 5-30a and 5-30b lists some physical properties of PEX-AL-PEX tubing conforming to this standard.

PEX-AL-PEX tubing has temperature/pressure ratings slightly higher than ASTM F876 PEX tubing.

This is largely due to the structural characteristics of the aluminum core. The aluminum core also significantly reduces the expansion movement of PEX-AL-PEX tubing relative to other (all-polymer) tubes. This characteristic makes PEX-AL-PEX well suited for use in radiant panel construction where aluminum heat transfer plates are used as seen in Figure 5-31.

The aluminum core in PEX-AL-PEX tubing also allows it to maintain the shape it is bent to. This is appreciated in situations where the tubing needs to be routed in both curved and straight paths as shown in Figure 5-32.

Unlike PEX tubing, PEX-AL-PEX tubing cannot be repaired by heating the tube until the PEX reaches its amorphous state. Instead, most kinks can be removed by mechanically reforming the tube with a truing tool.

Figure 5-29 — Close-up of cross-section of PEX-AL-PEX tubing. Aluminum core is seen between inner and outer PEX layers. *Courtesy of John Siegenthaler.*

Nominal tube size (in)	Minimum outside diameter	Minimum total wall thickness	Minimum aluminum layer thickness (in)
3/8	12.00 mm/ 0.472 in	1.6 mm/ 0.063 in	0.007
1/2	16.00 mm/ 0.630 in	1.65 mm/ 0.065 in	0.007
5/8	20.00 mm/ 0.787 in	1.90 mm/ 0.075 in	0.010
3/4	25.00 mm/ 0.984 in	2.25 mm/ 0.089 in	0.010
1	32.00 mm/ 1.26 in	2.90 mm/ 0.114 in	0.011

(a)

Temperature (°F)	Pressure rating for water (psi)
73.4	200
140	160
180	125
210	115

(b)

Figure 5-30 — (a) Physical data for several sizes of PEX-AL-PEX tubing conforming to the ASTM F1281 standard. (b) Temperature and coincident pressure ratings for PEX-AL-PEX tubing conforming to the ASTM F1281 standard.

Figure 5-31 — Use of PEX-AL-PEX tubing in combination with aluminum heat transfer plates for radiant floor heating. *Courtesy Harvey Youker.*

Figure 5-32 — PEX-AL-PEX tubing (in this case with a black outer layer) maintains the shape to which it is bent. *Courtesy of John Siegenthaler.*

5.2 Piping Materials 171

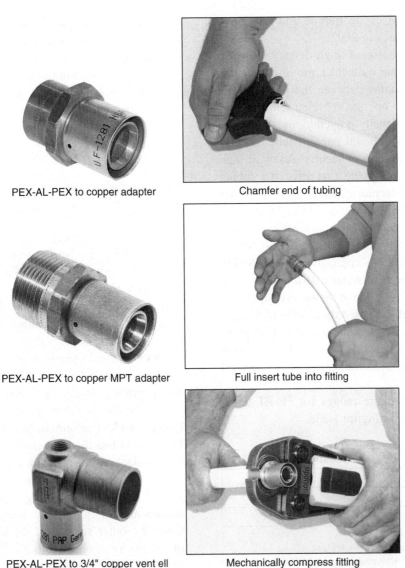

Figure 5-33 | Example of fittings and joining procedure used for PEX-AL-PEX tubing. *Courtesy of Uponor.*

A wide variety of fittings are available for connecting PEX-AL-PEX tubing to itself as well as copper tubing and other components such as valves and circulator flanges. Figure 5-33 shows some examples of these fittings, as well as the steps used to join them to PEX-AL-PEX tubing.

Polyethylene-Raised Temperature (PE-RT) Tubing

One of the newer polymer tubing products on the North American market is PE-RT, which stands for **Polyethylene-Raised Temperature**. PE-RT tubing has a 35-year history of use in Europe and was introduced in the United States in 2003.

PE-RT tubing consists of high-density polyethylene that has undergone molecular modification that creates sufficient tie chains between the polyethylene molecules to allow it to operate at higher temperatures than unmodified high-density polyethylene. A production process known as bimodal polyethylene technology is used to combine long molecular chains of high molecular weight polyethylene that give the tubing strength, with short molecular chains of low molecular weight polyethylene that provide good flexibility.

PE-RT tubing is available in nominal pipe sizes ranging from 1/4" to 6". It is manufactured to ASTM F2623 for hydronic system applications, and ASTM F2769 standards for potable water distribution use. In North America, PE-RT is made to the same copper tube sizes as PEX, and with a **dimension ratio** of 9 (e.g., outside tube diameter divided by wall thickness = 9). PE-RT has also been used for the inner and outer layers of a composite tubing in combination with a central layer of aluminum.

The pressure/temperature ratings for PE-RT under ASTM F2623 are given in Figure 5-33a.

Temperature (°F)	Pressure rating (psi)
73.5	160
180	100
200	80

Figure 5-33a | Pressure and temperature ratings for PE-RT tubing based on ASTM F2623.

These ratings make PE-RT suitable for a wide range of plumbing and hydronic system applications.

Because it is *not cross-linked*, PE-RT tubing can be heated to a melting point and reformed. This characteristic allows for joining of tubing and fittings using socket fusion. It also allows scraps to be recycled. PE-RT also has greater flexibility than PEX tubing, allowing for tighter bends.

Polypropylene Random Copolymer (PP-R) tubing

Another piping option of European origin that is now available in North America is **Polypropylene Random Copolymer (PP-R) tubing**. Available in a broad variety of sizes from 3/8 to 24 inches, PP-R tubing features a glass fiber-enhanced polypropylene core layer that, in combination with the polypropylene inner and outer layers, limits thermal expansion and allows sustained operating temperatures up to 160 °F with a corresponding pressure of 70.0 psi and temporary operating temperatures up to 195 °F for 60 days.

In North America, pressure-rated polypropylene tubing should conform to the ASTM F2389 standard. Figure 5-34 lists the physical dimensions associated with this standard in sizes up to nominal 4-inch inside diameter.

PP-R tubing is joined to polypropylene fittings by **socket fusion**. The mating surfaces of the tube and fittings are simultaneously heated in a special fixture. Once the outer surface of the tube and inner surface of the fitting are heated to 500 °F, they are separated, the heating device is removed, and the semi-molten

Nominal size*	Outside diameter (mm)**	Wall thickness (mm)**	Internal diameter (mm)**	Volume (gal/ft)	Support spacing (ft)
1/2" (SDR7.4)	20	2.8	14.4	0.017	2
3/4" (SDR7.4)	25	3.5	18	0.026	2.5
1" (SDR11)	32	2.9	26.2	0.043	3
1.25" (SDR11)	40	3.7	32.6	0.067	3.3
1.5" (SDR11)	50	4.6	40.8	0.105	3.9
2" (SDR11)	63	5.8	51.4	0.167	4.6
2.5" (SDR11)	75	6.8	61.4	0.237	4.9
3" (SDR11)	90	8.2	73.6	0.343	5.2
3.5" (SDR11)	110	10	90	0.512	5.9
4" (SDR11)	125	11.4	102.2	0.661	6.6

* SDR = standard diameter ratio = outside diameter divided by wall thickness.
** To convert (mm) dimensions to inches divide by 25.4.

Figure 5-34 | Physical properties of PP-R tubing conforming to ASTM F2389.

surfaces are immediately pushed together and held in position as they cool. The result is an extremely strong and permanent bond between tube and fitting. This sequence is shown in Figure 5-35.

Another advantage of polypropylene composite tubing is the ability to create tappings using a process called **saddle fusion**. The process begins by using a special tool to cut a hole in the side of the tube. This tool removes all cut material to prevent it from falling into the pipe. Then the outer surface of the tube and the mating surface of saddle fitting are simultaneously heated using a special tool. When the two surfaces

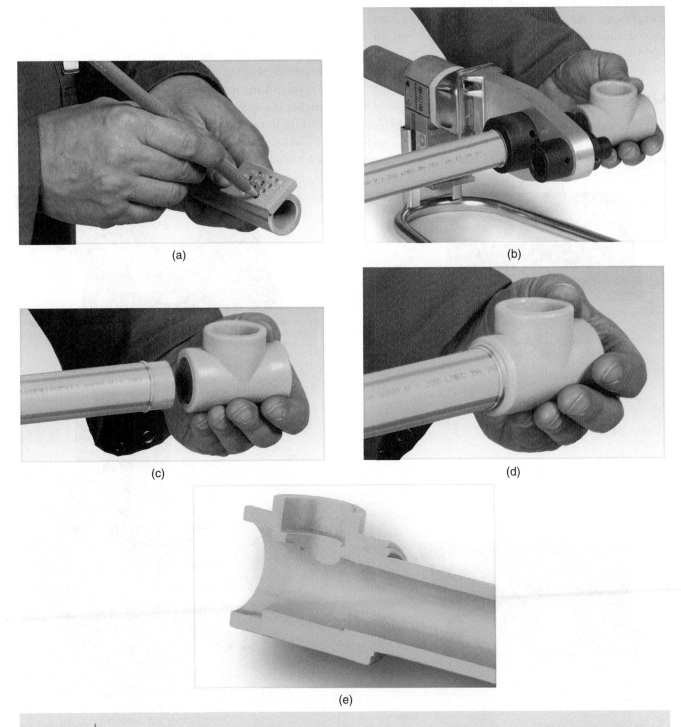

Figure 5-35 | Polypropylene PP-R pipe and fitting joined using socket fusion. (a) Tubing is marked for proper fitting insertion depth. (b) Tube and fitting surfaces heated to 500 °F. (c) Tube and fitting ready to be joined. (d) Tube and fitting pushed straight together—no twisting. (e) Cross-section of completed joint. *Courtesy of Aquatherm.*

are sufficiently heated, the heating tool is removed, and the saddle fitting is immediately pressed against the tube. The molten surfaces fuse together to form a strong and permanent joint. The sequence is shown in Figure 5-36.

A wide variety of fittings and valves are now available for use with PP-R. These include adapters for standard **national pipe thread (NPT)** threaded components. An example of a PP-R piping used in combination with metal piping components in a hydronic heating system is shown in Figure 5-37.

Oxygen Diffusion

The molecular structure of tubes made of polyethylene, polybutylene, polypropylene, or rubber compounds is such that oxygen molecules can diffuse through their tube walls and be absorbed by the fluid within. The driving force of **oxygen diffusion** is a lower concentration of dissolved oxygen molecules within the water inside the tube, compared to the concentration outside the tube. The concept is illustrated in Figure 5-38.

The pressure difference between the inside and outside of the tube has very little effect on oxygen diffusion. Diffusion can still occur when the pressure inside the tube is higher than the pressure outside the tube. Embedding the tube in concrete or other solid materials also has very little effect on oxygen diffusion. Oxygen diffusion is also temperature dependent. The higher the operating temperature, the faster oxygen can diffuse through the tube wall.

Any time oxygen enters a hydronic system containing iron or steel components, there is the

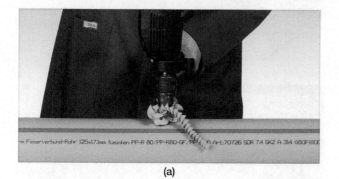
(a)

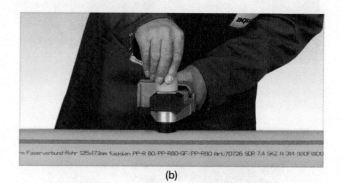
(b)

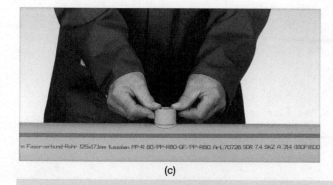
(c)

(d)

Figure 5-36 Saddle fusion of PP-R piping. (a) Hole drilled in pipe. (b) Outer surface of the tube and mating surface of the saddle fitting being heated to 500 °F. (c) Fitting and pipe are joined. (d) Cutaway of completed saddle fusion joint. *Courtesy of Aquatherm.*

potential for corrosion. It is therefore important to minimize or eliminate its entry in systems containing iron or steel components. Most manufacturers of polymer or rubber-based tubing intended for hydronic heating applications incorporate an oxygen diffusion barrier into their pipe.

The **oxygen diffusion barrier** commonly used on PEX and PE-RT tubing is a compound called **EVOH (ethylene vinyl alcohol)**. It is applied to the outside of the tube or laminated between layers of PEX or PE-RT as the tube is extruded. EVOH acts as a retarder for oxygen diffusion, reducing the rate of diffusion to a level deemed acceptable for long-term service.

Metals such as aluminum are extremely good oxygen barriers. On PEX-AL-PEX tubing, the aluminum core layer acts as the oxygen diffusion barrier. Hence, there is no such thing as "nonbarrier" PEX-AL-PEX tube. Some rubber tubings used in hydronic heating applications include aluminum foil as the oxygen diffusion barrier.

Figure 5-37 | A completed hydronic piping assembly using PP-R tubing and fittings. *Courtesy of David Morrell.*

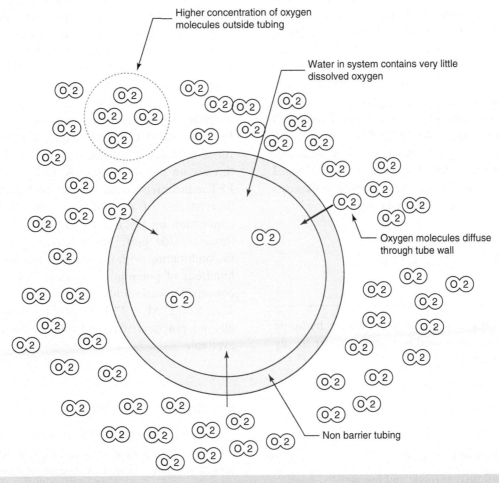

Figure 5-38 | Concept of oxygen diffusion through wall of polymer tubing. Note: The higher concentration of oxygen molecules outside of tube.

The standard recognized worldwide for oxygen diffusion protection in hydronic heating systems is DIN 4726. This standard allows a maximum oxygen diffusion rate of 0.1 mg/L of tubing water content per day. Nearly every manufacturer currently selling polymer tubing for hydronic heating applications offers "barrier" tubing that meets the DIN 4726 oxygen diffusion criteria.

Many manufacturers of PEX also offer "**nonbarrier**" **tubing**. Although intended primarily for use in domestic water plumbing systems, nonbarrier tubing has been successfully used in hydronic systems. Under the DIN 4726 standard, steel or cast-iron components such as boilers, circulators, and flow-check valves can be used if isolated from the nonbarrier tubing with a heat exchanger. Another alternative is the use of oxygen-scavenging chemicals in the system. These provisions, while possible, are not commonly used in residential and light commercial systems. Instead, tubing with an oxygen diffusion barrier has become the standard approach when iron or steel components will be used in the system and is recommended by the author.

There are several means, other than diffusion, by which oxygen can enter a hydronic system. These include fresh water flowing into the system to compensate for minor leaks, misplaced air vents, and improper sizing or placement of expansion tanks. These other forms of oxygen entry can be as much or more of a corrosion threat than oxygen diffusion through nonbarrier tubing. In other words, *the use of tubing with an oxygen diffusion barrier does not guarantee that corrosion will not occur*. However, it is reasonable to assume, with all other factors being equal, that tubing with an oxygen barrier does provide additional protection against corrosion. The slight (if any) additional cost of this protection is minor in comparison to the costs associated with repairing corrosion damage.

5.3 Common Pipe Fittings

Many of the fittings used to join pipe in hydronic systems are the same as used in domestic water supply systems. They include:

- **Couplings**: (standard and reducing);
- **Elbows**: (90° and 45°);
- **Street elbows**: (90° and 45°);
- **Tees**: (standard and reducing);
- **Threaded adapters**: (male and female);
- **Unions**.

The cross-sections of most of these common pipe fittings are shown in Figure 5-39.

Fittings for copper tubing are available in either wrought copper or cast brass. Both are suitable for use in hydronic systems. During soldering, wrought copper fittings heat up faster than cast brass fittings. They also have smoother internal surfaces. Of the two, wrought fittings are more typical in the smaller pipe sizes used in residential and light commercial systems.

Fittings for copper tubing are manufactured with solder-type socket ends, threaded ends, or a combination of the two. Standard sockets for solder-type joints create a gap of between 0.002 and 0.005 inch between the outside of the tube and the inside of the fitting. This gap allows solder to flow between the tube and fitting by capillary action.

Several abbreviations are used to describe the type of connection(s) used on pipe fittings.
The most abbreviations are as follows:

"C" implies a connection to a copper or copper alloy fitting suitable for a soldered or brazed joint

"F" or "FPT" implies female (internal) pipe threads

"M" or "MPT" implies male (external) pipe threads

"FTG" implies a connection that fits inside the socket of another nonthreaded fitting. An example would be a fitting that slides into the "C" socket of any copper fitting

These abbreviations are often combined to describe the connection type at each port of a fitting. For example: A C × FPT adapter would have a socket for a soldered joint on one side and female pipe threads on the other side. A 1"C × 3/4" C × 1/2" FPT reducer tee would have a 1-inch soldered connection on one of the end ports, a 3/4-inch soldered connection on the other end, and a 1/2-inch female threaded side port. These basic abbreviations, used in combination with the name of the fitting, allow for hundreds of potential variations. However, not every conceivable variation is manufactured. For example, a 4" C × 1/2" M × 1" F reducer tee would have virtually no practical use, and thus is not commercially available.

Specifying Fittings

Because there are thousands of fittings available for use with many types of tubing and pipe, it is necessary to have a standard method of describing them. This helps ensure accurate communication between manufacturers, wholesalers, designers, specification writers, and installers.

5.3 Common Pipe Fittings

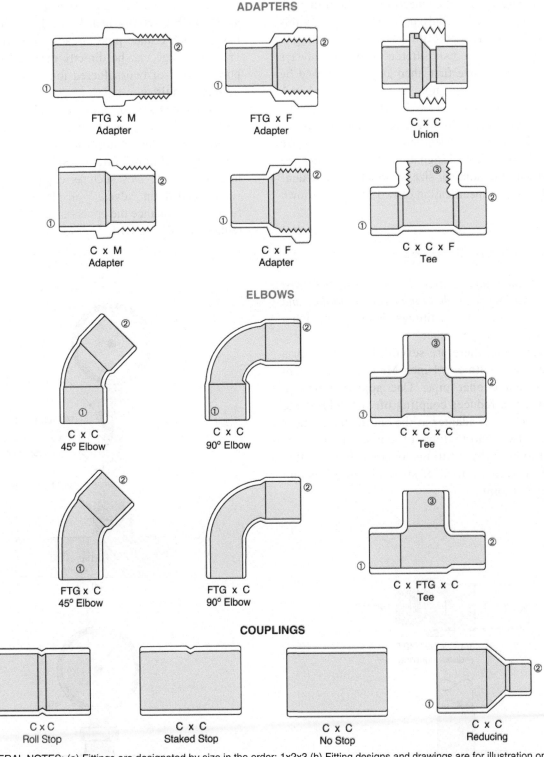

Figure 5-39 | Cross-sections of several common fittings used with copper tubing. *Courtesy of the Copper Development Association.*

The following minimum information should always be given when describing a fitting:

- Material the fitting is made of;
- Type of fitting (i.e., elbow, tee, union);
- Pipe size(s) needed on the fitting;
- A designation for the type of connections (i.e., solder − type = C, male pipe threads = M, female pipe threads = F, insert to a smooth fitting socket = FTG).

In the case where all connections on the fitting are the same pipe size, it is only necessary to specify this size once. When a two-port fitting such as a coupling or bushing needs to have two different pipe sizes, always specify the *larger* size first, then an "×," followed by the smaller size. For example, specify a 1 × 3/4 inch reducer coupling, not a 3/4 × 1 inch coupling. In the case of a tee, specify the larger of the two end ports, then the size of the other end port, and finally the size of the side port. The designations C, M, F, and are included with the nominal size information. An example of the proper specification of two fittings is shown in Figure 5-40.

Tips on Using Fittings

A well-planned piping assembly will use the least number of fittings possible to minimize both labor and materials. Minimal use of fittings also gives the job a more professional appearance.

For example, there are several ways to attach a pressure gauge with a 1/4-inch MPT threaded stem to a 3/4-inch copper pipe. One approach uses a standard tee, a reducer coupling off the side of the tee, and a C × F adapter at the end of the reducer coupling. Two short stubs of copper tubing are required to hold these fittings together. In total, this assembly requires a total of six soldered joints and one threaded joint.

A better approach is to use a special tee designated as 3/4 inch C × 3/4 inch C × 1/4 inch F. This tee has 1/4-inch female pipe threads on its side port. The pressure gauge can be directly threaded into the side port. A total of two soldered joints and one threaded joint are required.

A comparison of these assemblies is shown in Figure 5-41. The slightly higher cost of the special fitting is more than compensated for by the savings in labor and additional fittings. Similar situations occur in mounting devices such as expansion tanks, relief valves, and drain valves. Well-planned use of special fittings can enhance the appearance of the job and often reduce installation cost.

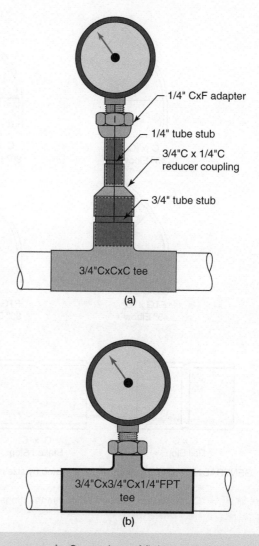

Figure 5-41 Comparison of fitting options for mounting a pressure gauge to a pipe. (a) Less efficient use of fittings. (b) Good use of a specialty fitting.

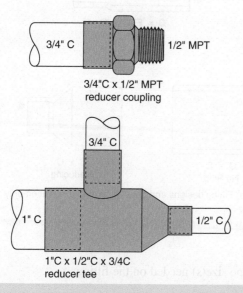

Figure 5-40 Example of proper nomenclature for two reducer fittings.

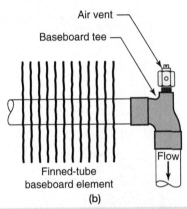

| Figure 5-42 | (a) A baseboard tee. (b) Typical location for venting air from a baseboard convector. *Courtesy of John Siegenthaler.* |

5.4 Specialized Fittings for Hydronic Systems

Several specialized metal fittings have been developed for hydronic heating systems. They include baseboard tees, diverter tees, and dielectric unions. Each is worthy of individual discussion.

Baseboard Tees

A **baseboard tee** (sometimes also called a "vent ell") is commonly used to mount an air vent at high points in a piping circuit. The fitting closely resembles a standard 90-degree elbow, but has an additional 1/8-inch FPT port as shown in Figure 5-42a.

Baseboard tees are available in pipe sizes from 1/2 to 1 1/4 inch in both wrought copper and brass, the latter being more common. The fitting's name is derived from its frequent use to mount air vents on the outlet of finned-tube baseboard convectors (Figure 5-42b). These fittings can also be used to mount a threaded temperature sensor, or a 1/8-inch brass drainage plug at a low point in the piping circuit.

Diverter Tees

A **diverter tee** is used to create flow through a branch piping path when mounted as shown in Figure 5-43.

When mounted at the beginning of the branch circuit, a diverter tee forces a portion of the entering fluid into the branch. When mounted at the end of the branch circuit, it "pulls" fluid through the branch.

The outside of a diverter tee appears similar to that of a reducing tee, as shown in Figure 5-44. However, inside there is either a cone-shaped orifice or a curved scoop. These internal details create partial obstructions that generate the pressure differentials needed to move flow through the branch piping.

Within the heating trade, diverter tees are often called "**monoflo tees**®" after the trademark brand from the Bell and Gossett Company. Another name sometimes used is **venturi fitting**. This is derived from the pressure-reducing effect of the tee when installed on the return side of the branch piping path.

A single diverter tee is often used when the branch piping path has low flow resistance. In such cases, the diverter tee is best placed at the end of the branch piping path as shown in Figure 5-45a. If the branch piping path is long or contains components that create higher flow resistance, two diverter tees can be used to create a higher pressure differential as shown in Figure 5-45b. In either case, it is crucial that diverter tees be installed in the proper direction within the piping system. This direction is indicated by an arrow on the side of the diverter tee and/or a red ring around one end port of the diverter tee. In the latter case, the fitting should always be installed with the red ring facing toward the common piping between the tees. The proper orientation of the fitting can also be determined by noting the direction of the cone-shaped orifices in Figures 5-43 and 5-45.

Figure 5-45 also shows the option of installing a valve within the branch circuit. If a zone valve is installed, flow through the branch circuit can be

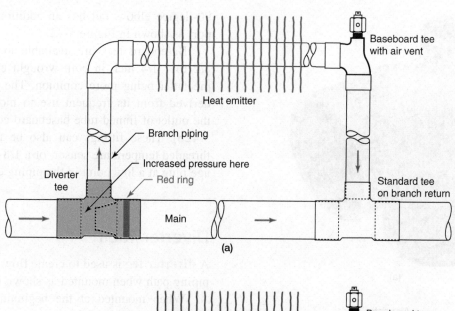
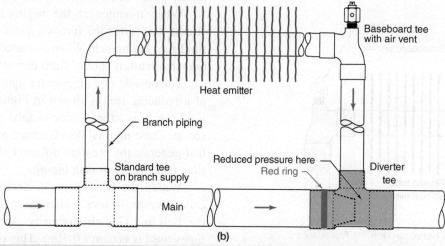

Figure 5-43 | A diverter tee used to create flow in a branch piping circuit. (a) Diverter tee mounted at the supply of the branch piping path. (b) Diverter tee mounted at the return of the branch piping path.

Figure 5-44 | Diverter tee. End view shows cone-shaped orifice. *Courtesy of Legend Hydronics.*

allowed or prevented based on the electrical signal that operates the zone valve. If a flow-regulating valve is used, flow through the branch circuit can be modulated based on room temperature. Such arrangements can be used to regulate the heat output of heat emitters on a room-by-room basis. These details are discussed under the topic of **one-pipe systems** in later chapters.

Dielectric Unions

In many hydronic systems, iron or steel components such as boilers and circulators are connected using copper tubing. If the copper and steel components are

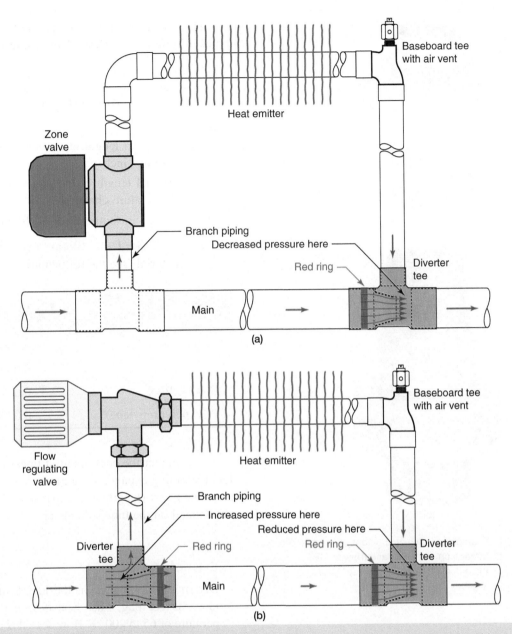

Figure 5-45 Piping assemblies using (a) single diverter tee on return side of branch and (b) two diverter tees. Valves that control flow in the branch piping are also shown.

in direct contact, and if the system's fluid is slightly conductive, a process called galvanic corrosion will cause the steel and iron components to corrode. One way to prevent this is to install an electrically insulating material called a dielectric between the two dissimilar metals. A **dielectric union** is the most commonly used fitting for this purpose.

At a glance, dielectric unions appear similar to a standard union. Closer inspection, however, shows one side of the union has steel pipe threads, while the other side has a brass socket for soldering to copper tube or a brass FPT connection. The steel and brass parts of the union are separated from each other by a nonconducting O-ring and sleeve. Besides preventing galvanic corrosion, dielectric unions also provide for simple disassembly of a piping system, the same as with standard unions. Two available types of dielectric unions are shown in Figure 5-46.

(a)

(b)

Figure 5-46 | Examples of dielectric unions (a) with FPT threads on steel side and (b) with MPT threads on steel side. *Courtesy of Watts Regulator Co.*

5.5 Thermal Expansion of Piping

All materials expand when heated and contract when cooled. In many hydronic systems, piping undergoes wide temperature swings as the system operates. This causes significant changes in the length of the piping due to **thermal expansion**. The greater the length of the pipe, and the greater the temperature change, the greater the change in length. If the piping is rigidly mounted, this expansion can cause annoying popping or squeaking sounds each time the pipe heats up or cools down. In extreme cases, the piping can even buckle due to high stresses that develop when the pipe is prevented from expanding.

The expansion or contraction movement of various types of tubing can be calculated using Equation 5.1:

Equation 5.1:

$$\Delta L = b(L)(\Delta T),$$

where,
ΔL = change in length due to heating or cooling (inch)
L = original length of the pipe before the temperature change (inch)
ΔT = change in temperature of the pipe (°F)
b = **coefficient of linear expansion** of tube material (see the following)

Tubing material	Coefficient of linear expansion (b) (in/in/°F)
Copper	0.0000094
PEX	0.000094
PEX-AL-PEX	0.000013
PP-R (polypropylene)	0.0000197

The smaller the coefficient of linear expansion, the less the tubing expands for a given change in temperature. The linear (lengthwise) expansion of the tube does not depend on the tube's diameter.

Example 5.1

Determine the change in the length of a copper tube, 50 feet long, when heated from a room temperature of 65 to 200 °F. How does this compare to the change in the length of a PEX, PEX-AL-PEX, and PP-R tube having the same length and undergoing the same temperature change?

Solution:

The change in the length of the copper tube is:

$$\Delta L = \left(0.0000094 \frac{\text{in}}{\text{in °F}}\right) \times 50 \text{ ft} \times \left(\frac{12 \text{ in}}{1 \text{ ft}}\right) \times (200 \text{ °F} - 65 \text{ °F}) = 0.76 \text{ in}$$

The change in the length for PEX tube is:

$$\Delta L = \left(0.000094 \frac{\text{in}}{\text{in °F}}\right) \times 50 \text{ ft} \times \left(\frac{12 \text{ in}}{1 \text{ ft}}\right) \times (200 \text{ °F} - 65 \text{ °F}) = 7.6 \text{ in}$$

The change in the length for PEX-AL-PEX tube is:

$$\Delta L = \left(0.000013 \, \frac{\text{in}}{\text{in °F}}\right) \times 50 \text{ ft} \times \left(\frac{12 \text{ in}}{1 \text{ ft}}\right)$$
$$\times (200 \text{ °F} - 65 \text{ °F}) = 1.05 \text{ in}$$

The change in the length for PP-R tube is:

$$\Delta L = \left(0.0000197 \, \frac{\text{in}}{\text{in °F}}\right) \times 50 \text{ ft} \times \left(\frac{12 \text{ in}}{1 \text{ ft}}\right)$$
$$\times (200 \text{ °F} - 65 \text{ °F}) = 1.60 \text{ in}$$

Discussion:

The only difference in these calculations is the value of the coefficient of linear expansion (b) for the different tubes. Notice that PEX tubing expands about 10 times as much as the copper tube for the same length and temperature change. PEX-AL-PEX tubing expands about 38% more than copper, but significantly less than PEX. The expansion of the PEX-AL-PEX tubing is largely controlled by the aluminum core layer. The expansion of the PP-R tubing is slightly more than that of the PEX-AL-PEX tubing. Because the values of b are very small, be careful to use the correct number of zeros when entering the number into calculations.

Regardless of the tubing used, expansion movement must be accommodated during installation, especially on long straight runs. In many small hydronic systems, the piping is not rigidly supported or contains sufficient elbows and tees to absorb the expansion movement without creating excessive stress in the pipe. Such situations only require piping supports that allow the tubing to expand and contract freely to prevent expansion noises.

When long straight runs of rigid tubing are required, it may be necessary to install an **expansion loop** or an **expansion offset** to absorb the movement. Both piping configurations are capable of absorbing limited amounts of movement without subjecting the pipe to excessive stress. The table and illustration in Figure 5-47 show some expansion compensating options for copper tubing.

Another alternative is to install an **expansion compensator** capable of absorbing the movement. A variety of such expansion compensator fittings are available. Those based on flexible tubing or reinforced hose require little if any maintenance. An example of a U-shaped compensator is shown in Figure 5-48. Similarly shaped compensators are available in other materials such as steel and PEX. Their ability to absorb expansion movement allows them to be much smaller than the expansion loops formed with piping as shown in Figure 5-47.

For any expansion compensator to work, the piping must be properly supported using a combination of fixed point and sleeve or roller supports. **Fixed point supports** are those that rigidly grab the pipe and do not

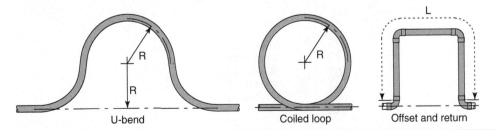

Expected expansion (inches)	R/L R = radius of U-bend or coiled loop (inches), L = total length of offset & return (inches)								
	Copper tube size								
	3/8"	1/2"	3/4"	1"	1.25"	1.5"	2"	2.5"	3"
0.5	7/44	8/50	9/59	11/67	12/74	13/80	15/91	16/102	18/111
1.0	10/63	11/70	13/83	15/94	17/104	18/113	21/129	23/144	25/157
1.5	12/77	14/86	16/101	18/115	20/127	22/138	25/158	28/176	30/191
2.0	14/89	16/99	19/117	21/133	23/147	25/160	29/183	32/203	35/222
2.5	16/99	18/111	21/131	24/149	26/165	29/179	33/205	36/227	40/248
3.0	17/109	19/122	23/143	26/163	29/180	31/196	36/224	40/249	43/272
3.5	19/117	21/131	25/155	28/176	31/195	34/212	39/242	43/269	47/293
4.0	20/126	22/140	26/166	30/188	33/208	36/226	41/259	46/288	50/314

Figure 5-47 | Dimensional requirements for expansion loops and expansion offsets for copper tubing.

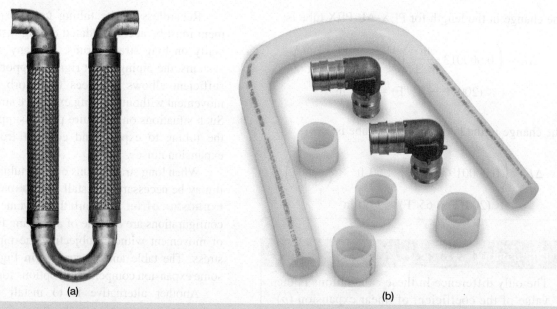

Figure 5-48 | (a) U-type expansion compensator for copper tube. *Courtesy of Metraflex Corp.* (b) A two-inch PEX expansion compensator kit. *Courtesy of Uponor.*

allow movement between the support and pipe. **Sleeve or roller supports** support the weight of the pipe, but do not inhibit expansion movement. Figure 5-49 shows an example of how the expansion compensator must absorb all expansion movement between fixed point supports.

The ability of an expansion compensator to absorb movement depends on the stress the piping is allowed to develop. Different materials and different pipe sizes affect this stress value. Manufacturers of expansion compensators list the amount of movement each compensator can accommodate.

Designers should keep in mind that cooling piping below the temperature at which it is installed will cause the piping to contract. Equation 5.1 can be used to determine the extent of this contraction based on the piping material and how much its temperature

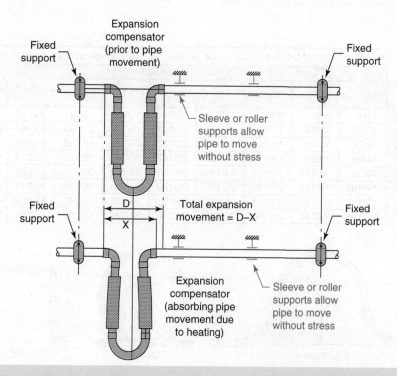

Figure 5-49 | Expansion movement absorbed by expansion compensator. Piping does not move at fixed support points.

drops below that at which it was installed. **Thermal contraction** will occur in piping serving a snowmelting system or in a chilled water cooling system. It could also occur in a heating system application if the system were not operating during cold weather. Expansion compensators should also be selected so they are able to accommodate pipe contraction due to cooling.

5.6 Common Valves

There are many types of valves used in hydronic systems. Some are common designs used in all categories of piping including hydronic, steam, plumbing, and pneumatics. Others are designed for a very specific function in a hydronic heating system. The proper use of valves can make the difference between an efficient, quiet, and easily serviced hydronic system and one that wastes energy, creates objectionable noise, or even poses a major safety threat. It is vital that designers and installers understand the proper selection and placement of several types of valves.

Most of the common valve types are designed for one of the following duties:

- **Component isolation**
- **Flow regulation**

Component isolation refers to the use of valves to close off the piping connected to a device that may have to be removed or opened for servicing. Examples include circulators, boilers, heat exchangers, and strainers. By placing valves on either side of such components, only minimal amounts of system fluid need to be drained or spilled during servicing. In many cases, other portions of the piping system can remain in operation during such servicing.

Flow regulation requires a valve to set and maintain a given flow rate within a piping system, or portion thereof. An example would be adjusting the flow rates in parallel piping branches. Valves used for flow regulation are specifically designed to remove mechanical energy from the fluid. This causes the fluid's pressure to drop as it passes through the valve. The greater the pressure drop, the slower the flow rate through the valve.

Gate Valves

Gate valves are designed specifically for component isolation. As such, *they should always be fully open or fully closed.*

The external appearance and internal design of a typical bronze gate valve are shown in Figure 5-50. When the **handwheel** is turned clockwise, the wedge

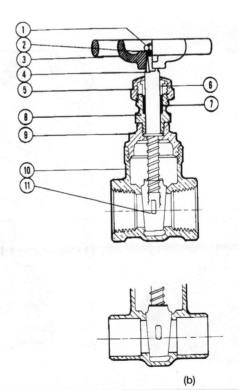

Callouts:
1. Handwheel Nut
2. Identification Plate
3. Handwheel
4. Stem
5. Packing Nut
6. Packing Gland
7. Packing
8. Stuffing Box
9. Bonnet
10. Body
11. Wedge

(a) (b)

Figure 5-50 | A typical bronze gate valve: (a) external appearance and (b) internal construction. *Photo and illustration courtesy of NIBCO, Inc.*

moves downward from a chamber in the upper body until it totally closes off the fluid passage. Near the end of its travel, the wedge seats tightly against the lower body forming a pressure-tight seal.

In its fully open position, the valve's wedge is completely retracted into the body. In this position, the valve creates very little interference with the fluid stream moving through it. This results in minimal loss of pumping energy during normal system operation. This is very desirable considering that a typical isolation valve remains open most of its service life.

Gate valves should never be used to regulate flow. If the wedge is set to a partially open position, vibration and chattering are likely to occur. Eventually, these conditions can erode the machined metal surfaces, possibly preventing a drip-tight seal when the valve is closed.

It may be necessary to occasionally tighten the **packing gland** of the valve to prevent minor leakage around the stem. A slight "snugging" of the **packing nut** is usually sufficient. Never overtighten the packing nut of any valve.

Globe Valves

Globe valves are specifically designed for flow regulation. Although their external appearance closely resembles that of a gate valve, there are major internal differences.

The fluid's path through a globe valve contains several abrupt changes in direction, as can be seen in Figure 5-51. The fluid enters the lower valve chamber, flows upward through the gap between the orifice and disc, and then exits sideways from the upper chamber. The gap between the disc and the orifice determines the flow resistance, and associated pressure drop, created by the valve. Always install globe valves so the fluid flows into the lower body chamber and upward toward the disc. Reverse flow through a globe valve can cause unstable flow regulation and noise. All globe valves have an arrow on their body that indicates the proper flow direction through the valve.

Globe valves should never be used for component isolation. The reason is that a fully open globe valve still creates much higher flow resistance compared to a fully open gate valve. Therefore, during the tens of thousands of potential operating hours when component isolation is *not* needed, the globe valve unnecessarily wastes pumping energy. Using a globe valve for component isolation is like driving a car with the brakes partially applied at all times. It may be possible, but it is certainly not efficient.

Angle Valves

An **angle valve** is similar to a globe valve, but with its outlet port rotated 90 degrees relative to its inlet port.

(a)

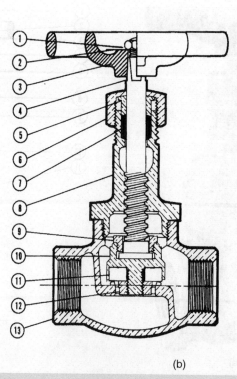
(b)

Callouts:
1. Handwheel Nut
2. Identification Plate
3. Handwheel
4. Stem
5. Packing Gland
6. Packing Nut
7. Packing
8. Bonnet
9. Disc Holder Nut
10. Disc Holder
11. Seat Disc
12. Disc Nut
13. Body

Figure 5-51 | A typical bronze globe valve: (a) external appearance and (b) internal construction. *Photo and illustration courtesy of NIBCO, Inc.*

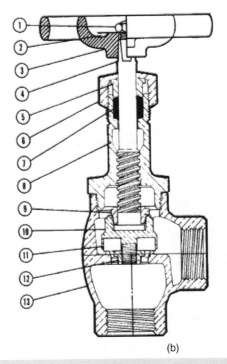

Callouts:
1. Handwheel nut
2. Identification plat
3. Handwheel
4. Stem
5. Packing gland
6. Packing nut
7. Packing
8. Bonnet
9. Disc holder nut
10. Disc holder
11. Seat disc
12. Seat disc nut
13. Body

Figure 5-52 | A typical angle valve: (a) external appearance and (b) internal construction. *Photo and illustration courtesy of NIBCO, Inc.*

This configuration is useful in place of a standard globe valve combined with an elbow. Angle valves are frequently used as flow regulation valves on the inlet of heat emitters such as panel radiators or standing cast-iron radiators. Because the path through an angle valve is not as convoluted as that through a globe valve, it creates slightly less flow resistance, and therefore, less pressure drop. However, angle valves should still not be used solely for component isolation. As with a globe valve, be sure to install angle valves so fluid flows toward the bottom of the disc. An example of an angle valve is shown in Figure 5-52.

A variation on the angle valve is known as a **boiler drain**. This valve allows a standard garden hose to be connected to its outlet port. Boiler drain valves can be placed wherever fluid might need to be drained from the system. An example of a boiler drain is shown in Figure 5-53.

Ball Valves

A **ball valve** uses a machined spherical plug, known as the ball, as its flow control element. The ball has a large hole through its center. As the ball is rotated through 90 degrees of arc, the hole moves from being parallel with the valve ports (its fully open position) to being perpendicular to the ports (its fully closed position). Ball valves are equipped with lever-type handles rather than handwheels. The position of the lever relative to the centerline of the valve indicates the orientation of the hole through the ball. An example of a ball valve is shown in Figure 5-54.

When the size of the hole in the ball is approximately the same as the pipe size, the valve is called a **full-port ball valve**. If the size of the hole is smaller than the pipe size, the valve is called a **standard-port ball valve**.

Full-port ball valves are ideal for component isolation. In their fully open position, they create very little flow interference or pressure drop. Full-port ball valves

Figure 5-53 | Example of a boiler drain valve. *Courtesy of Legend Hydronics.*

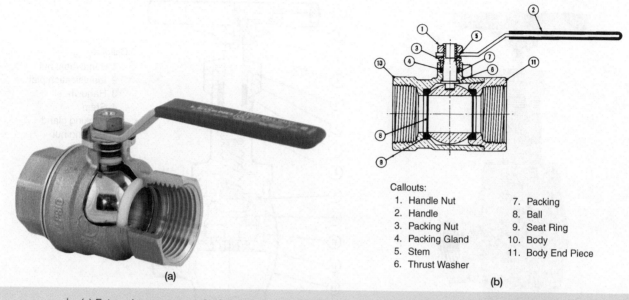

Callouts:
1. Handle Nut
2. Handle
3. Packing Nut
4. Packing Gland
5. Stem
6. Thrust Washer
7. Packing
8. Ball
9. Seat Ring
10. Body
11. Body End Piece

Figure 5-54 | (a) External appearance of a full-port ball valve. *Courtesy of Legend Hydronics.* (b) Internal construction of a standard-port ball valve. *Courtesy of NIBCO, Inc., Elkhart, IN.*

have largely replaced gate valves as the preferred valve for component isolation in residential and light commercial hydronic systems.

It is generally agreed that ball valves can be used for minor amounts of flow regulation. However, the valve should never be left in an almost closed position while operating. Doing so can cause high flow velocities through the valve that can eventually erode internal surfaces. Such operating conditions can also cause "crackling" sounds caused by flow cavitation within the valve.

Check Valves

In many hydronic systems it is necessary to prevent fluid from flowing backward through a pipe. A **check valve**, in one of its many forms, is the common solution.

Check valves differ in how they prevent reverse flow. The **swing-check valve**, shown in Figure 5-55, contains a disc hinged along its upper edge. When fluid moves through the valves in the allowed direction (as indicated by an arrow on the side of the valve's body),

Callouts:
1. Bonnet
2. Body
3. Hinge Pin
4. Disc Hanger
5. Hanger Nut
6. Seat Disc

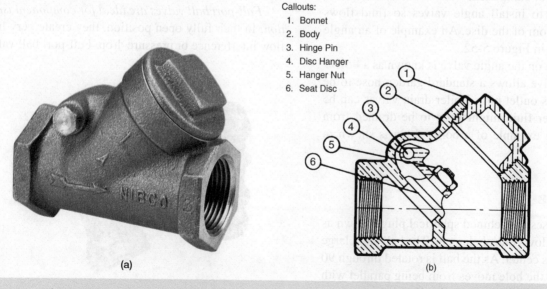

Figure 5-55 | A swing-check valve: (a) external appearance and (b) internal construction. Notice flow direction arrow on body. *Photo and illustration courtesy of NIBCO, Inc.*

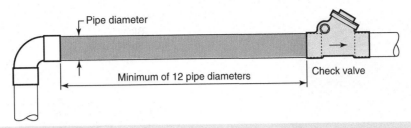

Figure 5-56 | Install at least 12 pipe diameters of straight pipe upstream of a check valve to prevent rattling sounds due to turbulence.

the disc swings up into a chamber out of the direct path of the fluid. The moving fluid stream holds it there. When the fluid stops, or attempts to reverse itself, the disc swings down due to its own weight and seals across the opening of the valve. The greater the back pressure, the tighter the seal.

For proper operation, *swing-check valves must be installed in horizontal piping with the bonnet of the valve in an upright position*. Installation in other orientations can cause erratic operation creating the potential for dangerous water hammer effects. It is also important to *install swing-check valves with a minimum of 12 pipe diameters of straight pipe upstream of the valve* as shown in Figure 5-56. This allows turbulence created by upstream components to partially dissipate before the flow enters the valve. Failure to do so can cause the valve's disk to rattle.

A **spring-loaded check valve** is designed to eliminate the orientation restrictions associated with swing check valves. These valves rely on a small internal spring to close the valve's disc whenever fluid is not moving in the intended direction. The spring action allows the valve to be installed in any orientation. The force required to compress the spring does, however, create slightly more pressure drop compared to a swing check. As with a swing check, be sure the valve is installed with the arrow on the side of its body pointing in the desired flow direction. An example of a spring-loaded check valve is shown in Figure 5-57.

As with swing checks, be sure to install at least 12 pipe diameters of straight pipe upstream of all spring-loaded check valves to dissipate incoming turbulence.

5.7 Specialty Valves for Hydronic Applications

Several valves have very specific functions in hydronic systems. Some provide safety protection and are required by plumbing and mechanical codes. Others automatically regulate the temperature and pressure at various points in the system. All these valves will be incorporated into systems discussed in later chapters.

Feed Water Valves

A **feed water valve**, also sometimes called a pressure-reducing valve, is used to lower the pressure of water from a domestic water distribution pipe before it enters a hydronic system. This valve is necessary because most buildings have domestic water pressure higher than the relief valve settings of small hydronic systems. The feed water valve allows water to pass through whenever the pressure at its outlet side drops below its pressure setting. In this way, small amounts of water are automatically fed into the system as air is vented out, or as small amounts of fluid are lost through evaporation at valve packings, pump flange gaskets, and other locations. An example of a feed water valve is shown in Figure 5-58.

Hydronic systems without feed water valves tend to slowly lose pressure as a result of air venting, and very small water losses through valves packing and pump flange gaskets. The resulting low-pressure operation can eventually cause noise and pump cavitation. In multistory buildings, the reduced pressure can also prevent circulation in upper parts of the system. The function of the feed water valve is especially important during the first few days of system operation as dissolved air is released from the fill water and vented out of the system.

Feed water valves operate by balancing the force generated by water pressure against an internal flexible **diaphragm**, with the counterforce generated by a partially compressed spring. When water pressure within the hydronic system drops below a preset level, the spring force becomes slightly greater than the diaphragm force. This imbalance causes the stem to move slightly and allow pressurized water on the inlet side of the valve to flow into the system to restore pressure. When the set pressure is restored, the valve automatically stops feeding water.

The pressure setting of feed water valves can be varied by rotating a shaft or screw. Many feed water valves come preset for 12 psi system pressure. While this is acceptable for most residential applications, it may be too low for systems in taller buildings. The correct pressure setting of a feed water valve depends on

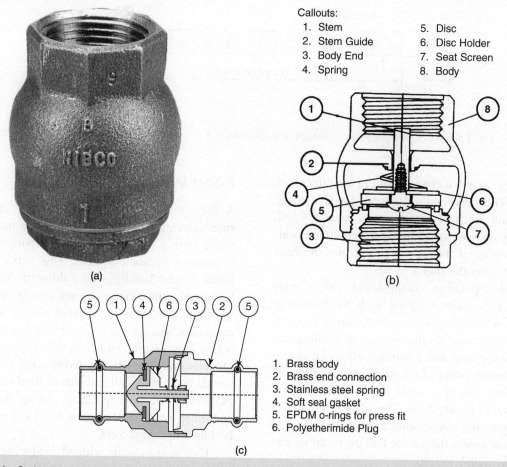

Callouts:
1. Stem
2. Stem Guide
3. Body End
4. Spring
5. Disc
6. Disc Holder
7. Seat Screen
8. Body

1. Brass body
2. Brass end connection
3. Stainless steel spring
4. Soft seal gasket
5. EPDM o-rings for press fit
6. Polyetherimide Plug

Figure 5-57 | Spring-loaded check valves: (a) external appearance and (b) internal construction. *Courtesy of NIBCO, Inc.* (c) Cross-section of a spring check using press-fit connections. *Courtesy of Bonomi USA, Inc.*

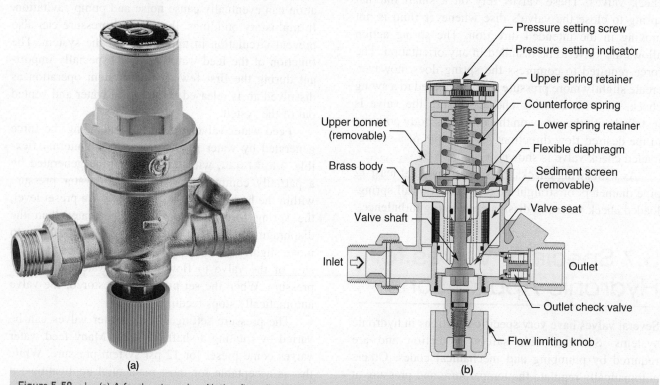

Figure 5-58 | (a) A feed water valve. Notice flow direction arrow on body. (b) Internal construction. *Courtesy of Caleffi North America.*

the height of the system, and is discussed in Chapter 12, Expansion Tanks.

Some feed water valves have a lever at the top that allows the valve to be manually opened to quickly fill the system during start-up. This lever, when lifted, temporarily disables the pressure-regulating function of the valve. Such valves require the lever to be manually returned to its normal position after the system is filled. Other feed water valves automatically stop fast-fill operation when the desired system pressure is reached.

The flow through the feed water valve when in the **fast-fill mode** is usually sufficient to **purge** small residential systems (depending on the water pressure available on the inlet side of the valve). On larger systems, it is common practice to install a full-port ball valve to provide even faster filling and purging flow. A typical piping arrangement is shown in Figure 5-59. Notice the ball valve is on the straight through piping path, while the feed water valve is installed in the branch piping. This minimizes the pressure drop through the make-up water assembly for the fastest possible purging.

Some feed water valves are equipped with removable strainers that prevent particulates in the supply water from entering the system. If the strainer becomes clogged, the valve may not supply water to the system. This is often evidenced by a slow drop in system pressure. The strainer on a feed water valve should be checked during routine system maintenance. Most feed water valves have an internal check valve to prevent reverse flow. This check valve, in combination with an isolating ball valve on the supply side of the make-up water assembly, allows the feed water valve to be isolated so the strainer can be removed, cleaned, and replaced with minimal water loss.

When installing a feed water valve, be sure the arrow on the side of the body points into the system. It is also convenient to install a pressure gauge near the outlet of the feed water valve to indicate the current system pressure. Some feed water valves are provided with integral pressure gauges.

Keep in mind that an automatic feed water valve cannot distinguish between a legitimate minor water loss versus a leak. If the latter occurs, and the make-up water system is active, water will continue to flow into the system and out the leak. This can obviously cause damage if not noticed and corrected. For this reason, some installers prefer to close the inlet water ball valve on the make-up water system after the system is fully purged and deaerated. Doing so prevents any further water entry to the system, and inevitably causes system pressure to decrease. If the pressure loss is minor, it may be of no consequence and can easily be corrected during annual inspections or even by an attentive building owner. Opinions on leaving the make-up water assembly "active" at all times vary. There is no correct answer that covers all circumstances.

Purging Valves

Hydronic heating systems, once assembled, must be purged of air and filled with fluid. Both actions are often accomplished at the same time by forcing fluid through the system at a high flow velocity. As the rapidly moving fluid progresses around the piping circuit, it pushes much of the air ahead of it—like a piston moving through a cylinder. The water stream can also **entrain** air bubbles and drag them along. Obviously, the air must exit the piping assembly as water goes in. A **purging valve** is specifically designed for this purpose.

Modern purging valves consist of two ball valve assemblies in a common body as seen in Figure 5-60a. One ball valve is "inline" with the piping, while the other serves as a side outlet port. A purging valve is typically mounted within the piping system as shown in Figure 5-60b.

During purging, the inline ball valve within the purging valve is closed, and the outlet ball valve is opened. A hose connected to the outlet port of the purging valve routes exiting flow to either a pail or drain. During purging, the fast-fill function of the feed water valve is enabled. In some systems, a separate ball valve mounted in parallel with the feed water valve is also opened. The goal in purging is to force water through the system piping as fast as possible to efficiently capture and expel air in the piping and other system components. Once the fluid stream exiting the outlet port of the purge valve is free of visible bubbles, the outlet ball valve is shut, and the fast-fill settings of the make-up water assembly are turned off. Purging

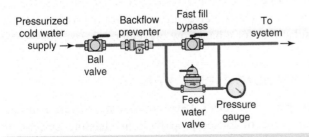

Figure 5-59 | Feed water valve installed in a typical make-up water assembly. The bypass ball valve is on the straight pipe to minimize pressure drop and maximize flow rate during fast fill or purging.

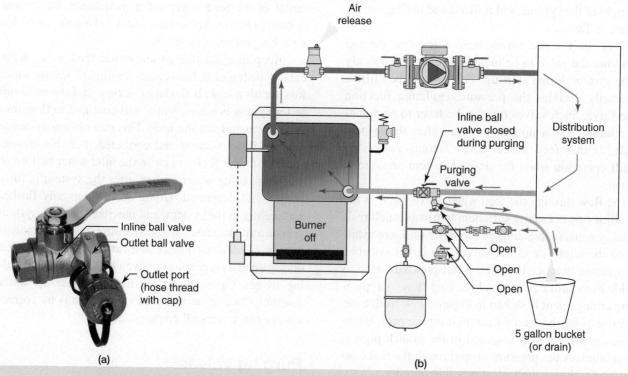

Figure 5-60 (a) A modern purging valve. *Courtesy of Webstone Corp.* (b) Use of a purging valve to fill and flush air from a hydronic system.

is discussed in more detail in Chapter 13, Air & Dirt Removal & Water Quality Adjustment.

Backflow Preventer

Consider a hydronic heating system filled with an antifreeze solution connected to a domestic water supply by way of a feed water valve. If the pressure of the domestic water system suddenly drops due to a rupture of the water main or other reason, the antifreeze solution could be pushed backward, into the domestic water piping, by the pressurized air in the system's expansion tank. This could contaminate the water in the building as well as neighboring buildings.

A **backflow preventer** eliminates this possibility. It consists of a pair of check valve assemblies in series, with an intermediate vent port that drains any backflow that migrates between the valve assemblies. An example of a small backflow preventer is shown in Figure 5-61.

When water pressure is first exerted on the inlet side of the valve, the upstream check valve and movable cartridge are pushed in the downstream direction. The end of the movable cartridge seals against a soft rubber seat to prevent leakage out the vent port. Water flow continues through the downstream check valve and into the system.

If the inlet pressure on the backflow preventer drops below the pressure in the hydronic system, springs return both check valves and the movable cartridge to their original positions. The check valves prevent any reverse flow toward the inlet. All fluid between the two check valves drains through the vent port so that no fluid remains in the valve. This prevents the possibility of potentially contaminated fluid being separated from domestic water by only a single check valve.

Most plumbing codes require backflow preventers on all hydronic heating systems that are connected to domestic water piping. Even if local codes do not require this valve, it is still good practice to use it, especially if antifreeze or other chemical treatments might be added to the system. A single check valve, or even two check valves in series, are not an acceptable substitute for a backflow preventer.

Pressure-Relief Valves

A **pressure-relief valve** is a code requirement on any type of closed-loop hydronic heating system. It functions by opening just below a preset pressure rating allowing system fluid to be safely released from the system before higher pressures can develop. The pressure-relief valve is the final means of protection in a situation where all other controls fail to limit

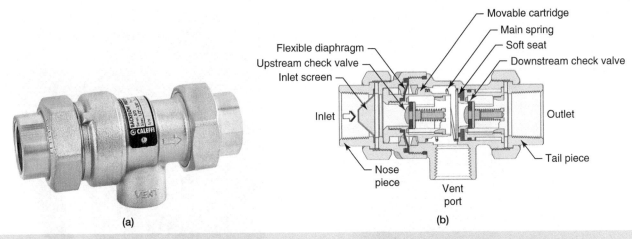

Figure 5-61 | (a) Example of a small backflow preventer. Note: Flow direction arrow and vent port marking. (b) Internal construction. *Courtesy of Caleffi North America.*

heat production. In its absence, some component in the system could potentially explode with devastating results.

Most mechanical codes require pressure-relief valves in any piping assembly that contains a heat source and is capable of being isolated by valves from the rest of the system. A pressure-relief valve is also required in any piping circuit supplied with heat through a heat exchanger.

Nearly all boilers sold in the United States are shipped with a factory-installed ASME-rated pressure-relief valve. The typical pressure rating for relief valves on small boilers is 30 psi. Valves with higher pressure ratings, as well as adjustable pressure settings are also available. They are typically used in buildings taller than three or four stories, and where the heat source is located low in the building.

The operation of a pressure-relief valve is simple. When the force exerted on the internal disc equals or exceeds the force generated by the internal spring, the disc lifts off its seat and allows fluid to pass through the valve. The spring force is calibrated for the desired opening pressure of the valve. This rated opening pressure, along with the maximum heating capacity of the equipment the valve is rated to protect, is stamped onto a permanent tag or plate attached to the valve as seen in Figure 5-62.

The lever at the top of a pressure-relief valve allows it to be manually opened. Such opening may be required to remove air when the system is filled. Manually opening the valve to verify that it is not corroded or seized is also part of annual system maintenance.

All pressure-relief valves should be installed with their shaft in a vertical position. This minimizes the chance of sediment accumulation around the valve's disc. Such sediment, if allowed to accumulate, can interfere with proper seating of the valve's disc and cause the valve to leak.

All pressure-relief valves should also have a waste pipe attached to their outlet port. This pipe routes any expelled fluid safely to a drain, or at least down near floor level. The waste pipe must be the same size as the valve's outlet port, with a minimal number of turns and no valves or other means of shutoff. The waste pipe should end 6 inches above the floor or drain to allow for unrestricted flow if necessary. Be sure to check local codes for possible additional requirements on the installation of relief valves.

Flow-Check Valves

A **flow-check valve** is a variation of the basic check valve, specifically for use in hydronic systems. Flow-check valves have a weighted internal plug that drops down over the valve's orifice. The plug's weight requires a pressure of approximately 0.3 psi to lift it from its seat. This requirement prevents hot water below the valve from rising into the distribution system due to its own buoyancy. However, the plug immediately pops open when the circulator in the circuit is turned on. Flow checks also prevent reverse flow in multizone hydronic systems.

Flow checks are available with either two or three ports. An example of the latter is shown in Figure 5-63. This design allows the valve to be installed with either a vertical or horizontal inlet pipe. The inlet port that is not used is closed off with a plug.

A small lever or screw at the top of a flow-check valve allows it to be manually opened. Doing so allows some hot water flow through the valve in case of a circulator failure. It also allows any air trapped below the valve to rise through the valve as the system is

Figure 5-62 — (a) A pressure-relief valve. *Courtesy of Watts Regulator.* (b) Always mount pressure-relief valves in vertical position. *Courtesy of Watts Regulator.*

Figure 5-63 — Example of three-port flow-check valve. The unused inlet port (left or bottom) is sealed with a plug during installation. *Courtesy of Watts Regulator.*

being filled and purged. *The lever does not allow the valve to adjust flow rate. During normal operation, the lever or screw should be in the down position.*

Flow checks are available with both cast-iron and bronze bodies. Either type can be used in closed-loop systems. Only bronze-bodied flow checks are suitable for systems where dissolved oxygen is present in the water (such as a system using nonbarrier PEX tubing).

Thermostatic Radiator Valves

Widely used in Europe, **thermostatic radiator valves** (TRVs) can provide precise room-by-room temperature control in hydronic heating systems. They are installed in the supply pipe of a heat emitter or in some cases built into the heat emitter itself. TRVs consist of two parts, the valve body and the thermostatic operator.

The nickel-plated brass valve body is available in pipe sizes from 1/2 to 1 inch, in either a **straight pattern** or **angle pattern**. In the latter, the outlet port of the valve is rotated 90 degrees relative to the inlet port. Inside the valve is a plug mounted on a spring-loaded shaft. The plug is held in its fully open position by the force of the spring. The fluid pathway through the valve is similar to that of a globe valve, thus making the valve well suited for accurate flow control.

To close the valve, the shaft must be pushed inward against the spring force. No rotation is necessary. Shaft movement can be achieved manually using a knob that threads onto the valve body, or automatically through use of a thermostatic operator. Figure 5-64 shows such a **thermostatic operator** ready to be mounted to a radiator valve.

5.7 Specialty Valves for Hydronic Applications 195

Figure 5-64 | A thermostatic operator ready to be attached to a radiator valve. *Courtesy of John Siegenthaler.*

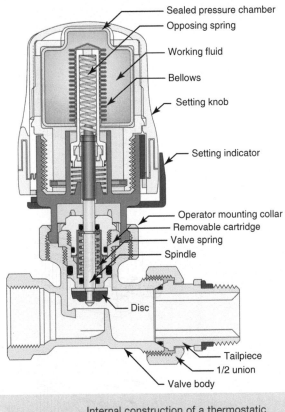

Figure 5-65 | Internal construction of a thermostatic operator attached to a straight radiator valve body. *Courtesy of Caleffi North America.*

Several types of thermostatic operators can be matched to radiator valve bodies. The most common configuration attaches the thermostatic operator directly to the valve body. A cutaway of such a combination is shown in Figure 5-65.

The thermostatic operator contains a fluid in a sealed bellows chamber. As the air temperature surrounding the operator increases, the fluid expands inside the bellows, which forces the shaft of the valve inward toward its closed position. When the valve is mounted in a pipe supplying a heat emitter, this action decreases water flow and reduces heat output. As the room air temperature decreases, the fluid within the bellows contracts, allowing the spring force to slowly reopen the valve plug and increase heat output from the heat emitter. The combination of the valve and thermostatic operator represents a fully modulating flow control element that can continually adjust flow through the heat emitter as necessary to maintain a constant (occupant determined) room temperature.

TRVs are often located close to the heat emitter. The preferred mounting position is with the valve stem in a horizontal position with the thermostatic operator facing away from the heat emitter as shown in Figure 5-66.

In no case should the valve be installed so the thermostatic operator is above the heat emitter. This will cause the valve to close prematurely and greatly limit heat output from the heat emitter. Be sure the TRV is mounted with flow passing through it in the direction indicated by the arrow on the valve body.

For versatility in different applications, most manufacturers of TRV valves offer operator assemblies with a remote setpoint dial. This allows temperature adjustments to be made at normal thermostat height above the floor. A **capillary tube** runs from the setpoint dial assembly to the operator mounted on the radiator valve. Thermostatic operators can be ordered with capillary tubes up to 12 feet long. Figure 5-67 shows an example of a radiator valve fitted with a remote setpoint adjustment. Care must be taken not to kink or break the capillary tube during installation. Severing the capillary tube renders the operator useless.

Figure 5-67a shows how a radiator valve with remote setting knob can be used to regulate flow through a towel warmer radiator and floor heating in a bathroom area.

The towel warmer and short tubing length used for floor heating create low flow resistance. This allows them to be connected in series with each other. The

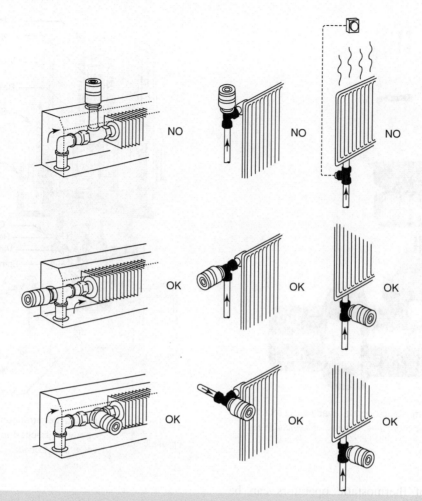

Figure 5-66 | Proper and improper mounting locations for a thermostatic operator. Do not mount the operator where it will be directly affected by convective air currents rising from the heat emitter. *Images courtesy of ISTEC, Inc., and Danfoss, Inc.*

valve is located between floor framing, and under the location of the setting knob. The capillary tube will be gently pulled from the location of the setting knob to the valve body using a string passing through the hole in the floor. The valve body should always be accessible. If a drywall ceiling is installed below it, an access panel should be provided.

TRVs can be used in a variety of piping layouts. One example is the use of such valves with fin-tube baseboard heat emitter in a **homerun distribution system** as shown in Figure 5-68.

This arrangement allows the heat output of each heat emitter to be individually controlled. The TRV on a baseboard in a bedroom might be turned down at night for comfortable sleeping, while the TRV on a bathroom baseboard is set to maintain 72 °F. The small **pressure-regulated circulator** shown in the schematic automatically adjusts its speed based on the flow demands of the system. Heated water would

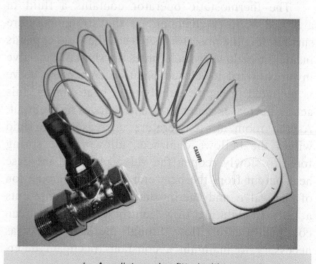

Figure 5-67 | A radiator valve fitted with a remote setpoint dial and operator assembly. Note: Capillary tubing between setpoint dial and valve operator assembly. *Courtesy of Caleffi North America.*

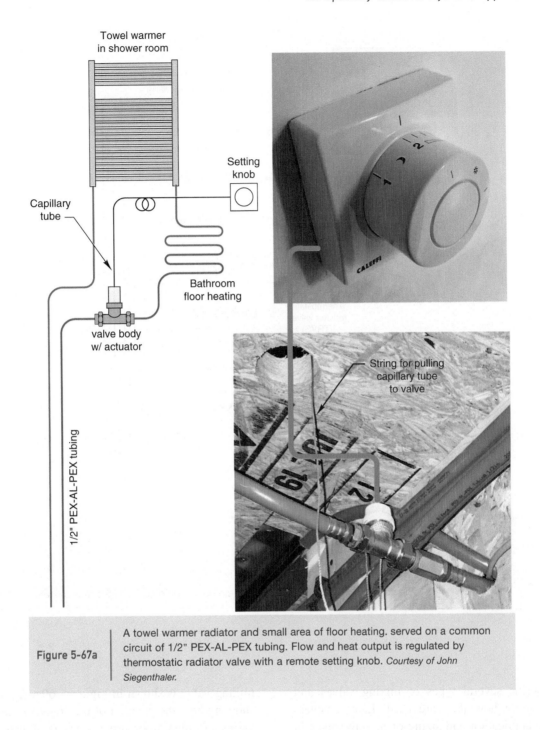

Figure 5-67a | A towel warmer radiator and small area of floor heating. served on a common circuit of 1/2" PEX-AL-PEX tubing. Flow and heat output is regulated by thermostatic radiator valve with a remote setting knob. *Courtesy of John Siegenthaler.*

always be available from the boiler (though perhaps at different temperatures depending on the outdoor temperature). This type of "homerun" distribution system is discussed in detail in Chapter 11, Distribution Piping Systems. With proper design, it offers tremendous versatility, low electrical energy use, and excellent room-by-room temperature control.

Mixing Valves

Some hydronic distribution systems require water temperatures lower than what is available from the heat source. For example, a radiant floor heating system may require water at 110 °F, while the water in a thermal storage tank supplying this load is at 150 °F. In such cases, the lower water temperature is created by blending water returning from the heat emitters with hot water from the heat source. The flow rate of each stream can be regulated by a mixing valve to attain the desired supply temperature.

There are several types of mixing valves used in hydronic heating systems. They include valves operated by electrically power **motorized actuators**, as well as valves operated by thermostatic operators. In some systems, mixing valves can be used without actuators or thermostatic operators. The sections that

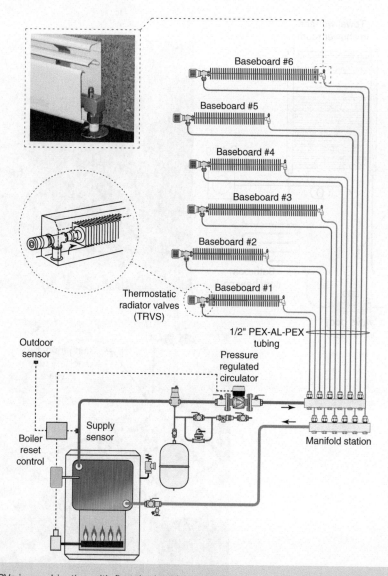

Figure 5-68 | Use of TRVs in combination with fin-tube baseboard heat emitters and homerun distribution system. *Inset images courtesy of Istec and Uponor.*

follow describe several types of mixing valves and briefly illustrate how they are used. Later chapters discuss the application and sizing of mixing valves in more detail.

Three-Way Motorized Mixing Valves

Three-way motorized mixing valves have three ports. One is the inlet port for heated fluid; another is the inlet port for cool fluid returning from the heat emitters. The third is the outlet port for the mixed fluid. The two inlet streams mix together inside the valve in proportions controlled by the position of the valve's **flow element (or spool)**. The body of a typical three-way mixing valve is shown in Figure 5-69a. An illustration showing how the position of the spool inside the valve varies the inlet floor proportions is shown in Figure 5-69b.

The position of the valve's shaft determines the amount of each fluid stream that enters the valve. The flow rate and temperature of each entering stream determines the blended temperature leaving the valve. The blended temperature will always be between the temperature of the entering hot and cold streams. Most three-port valves have an indicator scale on the valve body. The pointed end of the handle or knob indicates the setting of the valve between 0 and 10. The full rotating range of the shaft and internal spool is about 90 degrees.

5.7 Specialty Valves for Hydronic Applications 199

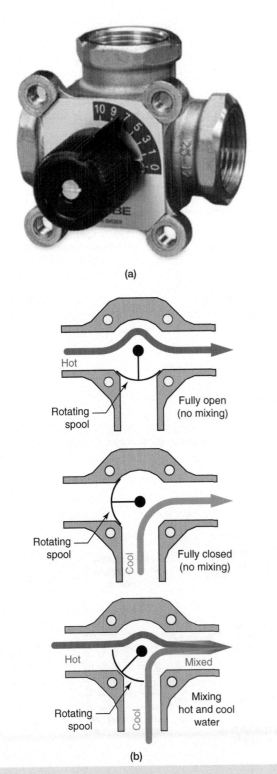

Figure 5-70 | Examples of three-way rotary mixing valves with attached controller/actuators. *Courtesy of John Siegenthaler.*

Figure 5-69 | (a) Three-way rotary mixing valve with a manual adjustment knob. *Courtesy of Paxton Corporation.* (b) Cross-section through a three-way rotary mixing valve with spool in different positions.

A three-way rotary mixing valve, such as shown in Figure 5-69a, does not adjust itself to compensate for changes in the temperature or flow rate of entering fluid streams. As such it is called a **"dumb" mixing valve**. For automatic temperature control, the valve must be equipped with a motorized actuator operated by an electronic controller. An example of three-way rotary mixing valve with a controller/actuator mounted to it is shown in Figure 5-70. The controller/actuator contains a low-voltage motor, gear train, and the necessary electronic controller to operate the valve based on inputs from temperature sensors.

A three-way motorized mixing valve can be used with a single circulator in systems where the heat source can accept cool return water from the distribution system without problems such as sustained flue gas condensation. A heated storage tank would be an example of such a heat source as seen in Figure 5-71.

As discussed in Chapter 3, Hydronic Heat Sources, conventional gas- and oil-fired boilers should not be operated with inlet temperatures that cause sustained flue gas condensation. If a three-way motorized mixing valve is used with such a boiler, it should be piped as shown in Figure 5-72 and operated by a controller that senses and reacts to boiler inlet temperature.

Using this piping and temperature sensor arrangement, the controller can limit the rate of hot water flow into the three-way valve as necessary to prevent the distribution system from extracting heat faster than the boiler can produce it. *Without a controller that senses and reacts to boiler inlet temperature, a three-way valve cannot properly protect a conventional boiler from sustained flue gas condensation.*

This system requires an additional circulator, compared to the system shown in Figure 5-71. The second circulator creates the second mixing point (e.g., mixing point #2 in Figure 5-72), at which the

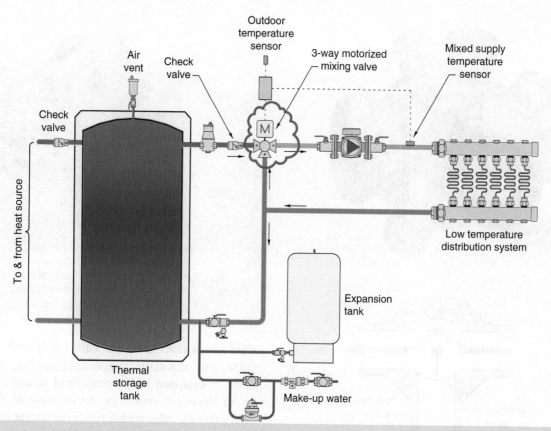

Figure 5-71 | Typical piping for a three-way motorized mixing valve when heat source is not affected by cool return water.

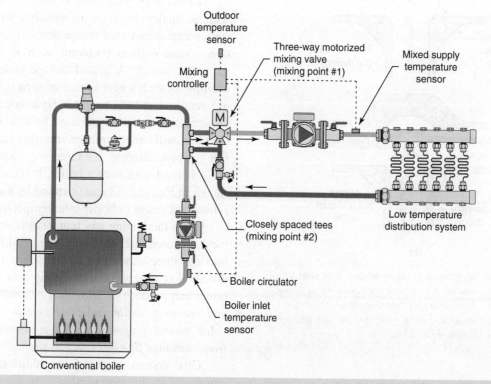

Figure 5-72 | Typical piping for a three-way motorized mixing valve when heat source is a conventional boiler requiring protection from sustained flue gas condensation.

temperature of the flow stream returning to the boiler is boosted as necessary to prevent flue gas condensation.

Four-Way Motorized Mixing Valves

Unlike three-way mixing valves, **four-way mixing valves** are specifically designed to couple low-temperature distribution systems to conventional boilers (which requires protection from sustained flue gas condensation). Figure 5-73 shows an example of a four-way rotary mixing valve, in this case with a cast-iron body.

In a four-way mixing valve, a portion of the entering hot water is blended with a portion of entering cool water to create the desired supply temperature. The remainder of the hot water is mixed with the remainder of the cool water to boost the inlet temperature to the boiler. The latter function is the main difference between three-way and four-way mixing valves.

Flow through a four-way mixing valve is controlled by a simple rotating vane. As the valve's shaft is rotated from one extreme to the other, the vane rotates through an arc of about 90 degrees.

At one end of the travel range, all water returning from the distribution system is routed back to the boiler. At the same time, all hot water entering the valve is routed to the supply side of the distribution system. No mixing takes place when the valve is set to this position. In this position, the valve is allowing the maximum possible heat transfer to the distribution system.

At the other end of the travel range, the water returning from the distribution system does not go back to the boiler. Instead, all return water is routed back to the supply side of the distribution system. No hot water enters the valve and hence no mixing takes place. In this position, there is no heat input to the distribution system. This position is useful when constant circulation through the distribution system is desired. Constant circulation is often used to help equalize surface temperatures in high thermal mass heated floor slabs.

In most cases, the valve operates at an intermediate (mixing) position between these two extremes. These valve settings are illustrated in Figure 5-73b.

Like a three-way mixing valve, four-way mixing valves do not automatically control water temperature or flow rate. For automatic control, they must be equipped with motorized actuators operated by an electronic controller. A four-way valve without such an actuator/controller is another example of a "dumb" valve that cannot adjust to variations in the temperature of the incoming fluid streams. Thus, it can neither provide a stable temperature control nor properly protect a

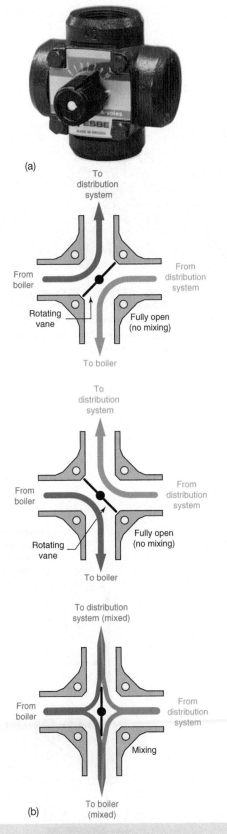

Figure 5-73 (a) Examples of a four-way mixing valve. *Courtesy of Paxton Corporation.* (b) Cross-section through a four-way rotary mixing valve with vane in different positions.

Figure 5-74 | A four-way mixing valve body with actuating motor attached. The actuating motor positions the valve's shaft based on input from an electronic controller. *Courtesy of John Siegenthaler.*

conventional boiler from flue gas condensation. Figure 5-74 shows a four-way mixing valve with an actuating motor attached to the valve body. This actuator would be operated by a separate electronic controller.

Figure 5-75 shows a system using a four-way motorized mixing valve mounted close to a boiler having low flow resistance. In this arrangement, flow through the boiler is created by the momentum exchange of water returning to the valve from the distribution system. No circulator is required between the mixing valve and the boiler. Since this system requires only one circulator, it holds a distinct advantage over a system using the same boiler and a three-way motorized mixing valve, which requires two circulators.

When the heat source and/or the piping connecting the heat source to the mixing valve has high flow resistance, a separate circulator would be required to ensure sufficient flow through the boiler. In such cases, it is simpler and less expensive to use a three-way motorized mixing valve piped as shown in Figure 5-72.

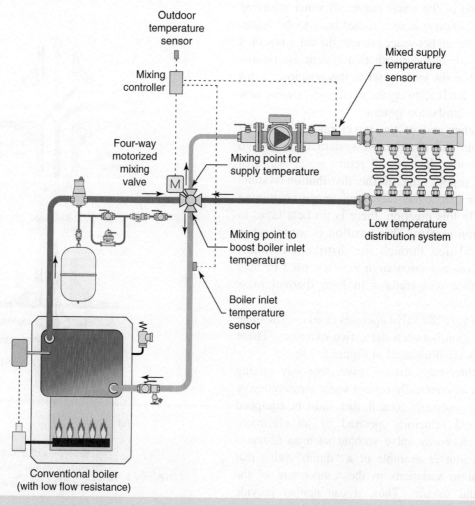

Figure 5-75 | Piping for motorized four-way mixing valve when used with low flow resistance heat source and boiler loop.

Three-Way Thermostatic Mixing Valves

Another type of mixing valve suitable for use in some hydronic systems is called a **three-way thermostatic mixing valve**. Unlike the three-way and four-way rotary mixing valves just discussed, three-way thermostatic mixing valves have internal components that automatically adjust the flow of hot and cool water to maintain a set outlet temperature. An example of such a valve is shown in Figure 5-76.

The desired mixed outlet temperature is set on the valve's knob. The two flow streams mix within the valve body and flow past the **thermostatic element**. This element does not use electricity. Instead, it creates motion based on the thermal expansion or contraction of a special wax compound sealed within it. This movement is transferred to the valve's internal spool to regulate the flow of hot and cool water entering the valve.

In some systems, the piping of a three-way thermostatic mixing valve can be identical to that of a three-way motorized mixing valve. A single three-way thermostatic valve can provide supply temperature control when the heat source can tolerate low inlet temperatures and relatively low flow rates. A storage tank supplied from solar thermal collectors, a pellet-boiler, or an electric boiler would be examples of such applications, as shown in Figure 5-77. In such applications, the three-way thermostatic valve protects the low-temperature distribution system from potentially high-temperature water in the tank.

When selecting a three-way thermostatic valve for a hydronic mixing application, it is important to consider its flow resistance. Three-way thermostatic mixing valves, especially those designed for tempering domestic hot water, often have substantially higher flow resistance than three-way rotary-type valves. *They need to be selected based on flow resistance rather than pipe size.* Later chapter will show how this is done.

Two-Way Thermostatic Valves

Although technically not classified as a "mixing" valve, **two-way thermostatic valves** can still be used to control water temperature in a hydronic distribution system through a process called **injection mixing**.

An example of a two-way thermostatically controlled valve is shown in Figure 5-78. Internally, the valve is similar to a globe valve, making it well suited for flow control. The valve body is fitted with a thermostatic operator that regulates the stem position based on temperature. The knob on the actuator sets the desired water temperature at the **sensor bulb** location.

As the temperature of the sensing bulb increases, the pressure of the working fluid inside the bulb increases. This increased pressure is passed along through a capillary tube to the flexible bellows assembly inside the actuator. The bellows respond by forcing the valve stem toward the closed position. In this way, the actuator adjusts the stem position in an attempt to maintain the preset temperature at the sensing bulb.

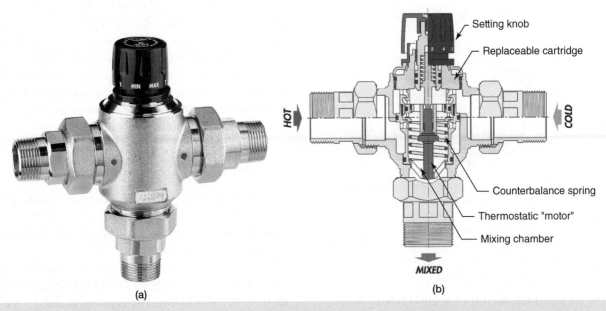

Figure 5-76 | (a) Example of a three-way thermostatic mixing valve. (b) Internal construction. *Courtesy of Caleffi North America.*

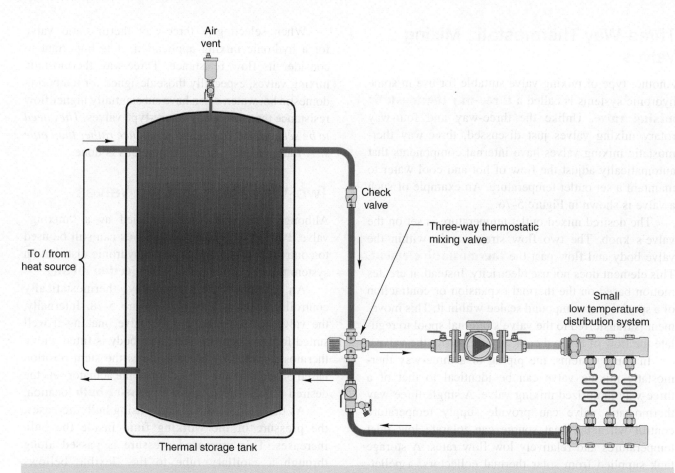

Figure 5-77 Use of a single three-way thermostatic mixing valve to protect the low temperature distribution system from excessively high water temperature. Note: The heat source is assumed to be unaffected by low-temperature water returning from the distribution system.

Figure 5-78 Example of a two-way thermostatic valve. Notice capillary tube leading to sensor bulb. The number set on dial opposite the arrowhead is the temperature the valve seeks to maintain at the sensor location. *Courtesy of Caleffi North America.*

The piping often used for injection mixing with a two-way thermostatic valve is shown in Figure 5-79. The **"flow restrictor" valve** is usually a globe valve and is partially closed to create a pressure differential between the injection risers. As the two-way thermostatic valve begins to open, this pressure differential forces cool water up the return riser and hot water down the supply riser. Mixing begins in the tee where the supply injection riser connects to the distribution system. The sensing bulb should (ideally) be located in a **sensor well** downstream of the circulator. Here it senses the final mixed temperature and provides feedback to the thermostatic operator as necessary to maintain a stable mixed temperature.

Zone Valves

Hydronic heating systems often use two or more piping circuits to supply heat to different zones of a building. One device used to allow or prevent flow through a zone circuit is called a **zone valve**.

Several types of zone valves are available. They all consist of a valve body combined with an electrically powered actuator. The actuator is the device that produces movement of the valve shaft when supplied with an electrical signal. Some actuators use small

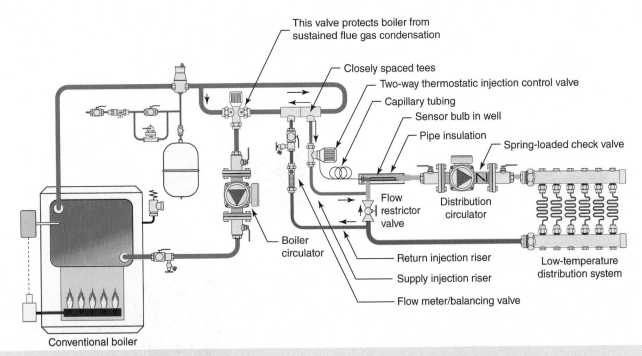

Figure 5-79 | Piping a two-way thermostatic valve for injection mixing control of the supply water temperature. The three-way thermostatic valve protects the conventional boiler against sustained flue gas condensation.

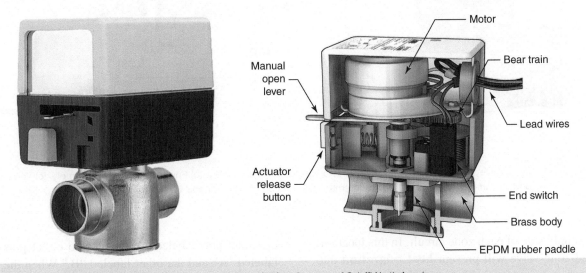

Figure 5-80 | Zone valve with detachable gear motor actuator. *Courtesy of Caleffi North America.*

electric motors and a gear train to produce a rotary motion of the valve shaft. Others use heat motors that expand and contract to produce a linear motion that moves the valve's shaft up and down. A zone valve equipped with a gear motor actuator is shown in Figure 5-80. A zone valve using a heat motor actuator is shown in Figure 5-81.

Zone valves operate in either a fully open or a fully closed position. They do not modulate to regulate the flow rate in a piping circuit. Those with gear motor actuators can move from fully closed to fully open in about 3 seconds. Those with heat motor actuators require 2 and 3 minutes from when the electrical signal is turned on, to when the valve is fully open. Since the response time of a heating system is relatively slow, the slower opening of heat motor-type zone valves does not make a noticeable difference in system performance.

A schematic for a typical four-zone system using zone valves is shown in Figure 5-82. Three of the zones are for space heating; the other supplies an indirect water heater. Notice that the zone valves are installed

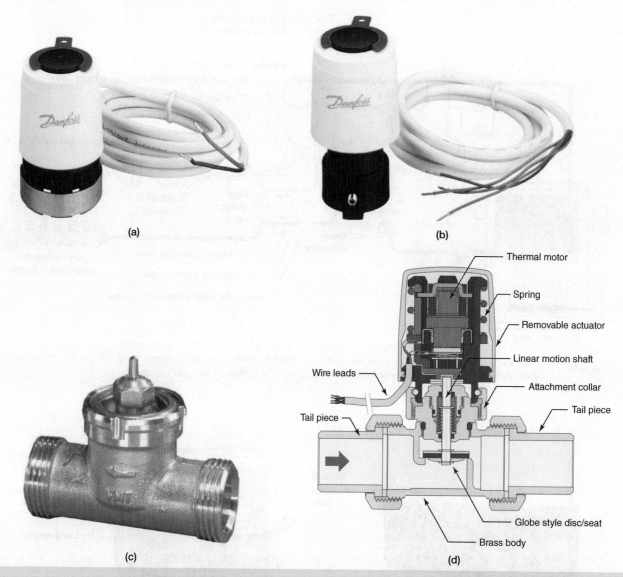

Figure 5-81 | Zone valve with heat motor actuator. (a) Two-wire actuator assembly. (b) Four-wire actuator assembly. (c) Valve body. *Courtesy of Danfoss (a, b, c,).* (d) Internal Construction of actuator and valve. *Courtesy of Caleffi North America.*

on the supply side of each zone circuit. In this location, a closed-zone valve will prevent **heat migration** due to buoyancy-driven flow of hot water into zone circuits that are supposed to be off. Zone valves mounted on the return side of the zone circuits are not as effective in stopping heat migration.

The details for designing hydronic system using zone valves are discussed in several upcoming chapters.

Three-Way Diverter Valves

There are situations in which flow must be routed through one of two possible paths depending on the operating mode of the system. A **three-way diverting valve** is designed for this purpose. Figure 5-83 shows an example of a three-way diverting valve. It has one inlet port labeled "AB" and two outlet ports, one labeled "A" and the other labeled "B." When the valve is not powered, a flow path exists from port AB to port B. When the valve is powered, the internal paddle moves and creates a flow path from port AB to port A.

A common application for a three-way diverting valve is when the hot water leaving a boiler is routed to either a space heating distribution system or an indirect water heater. This is shown in Figure 5-84.

Another example of a diverting application is shown in Figure 5-85. In this system, the two three-way diverter valve determine which set of piping mains the water-to-water heat pump delivers flow to. When the valve's actuator is deenergized, flow from the hot water return main is moved through the heat pump's condenser and then passed back to the hot water supply main.

5.7 Specialty Valves for Hydronic Applications 207

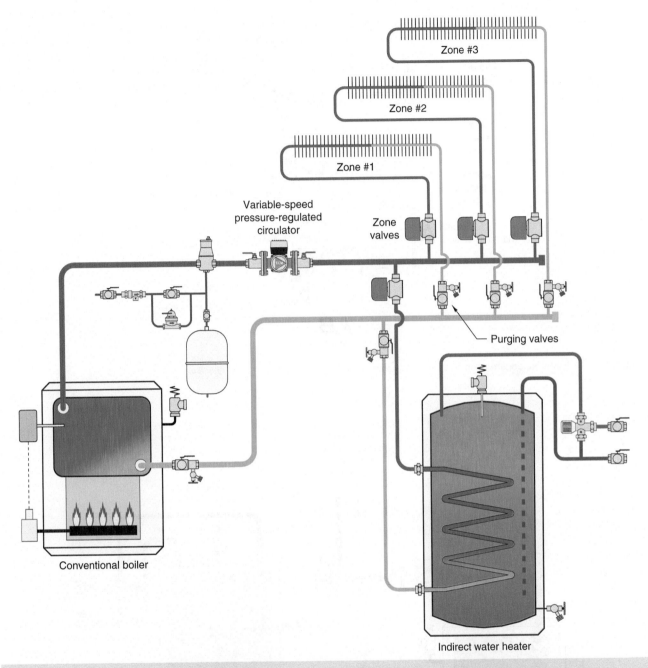

Figure 5-82 | Placement of zone valves on the supply side of zone circuits. This placement prevents heat migration from the heat source into the supply side of inactive zone circuits.

When the valve's actuator is powered on, flow is routed from the chilled water return main through the heat pump and passed back to the chilled water supply main. This diverting function allows the heat pump to be automatically switched between heating and cooling service as needed.

Three-way diverter valves were once used to zone hydronic systems in which it was necessary to maintain constant flow through the boiler, or to maintain consistent differential pressure across the system circulator. In such an application, they were typically called three-way zone valves. However, the operating cost associated with maintaining such flow under partial load conditions is now considered unacceptable given that better alternatives exist.

Most zone valves and diverting valves for residential and light commercial systems operate on 24 VAC control voltage. This is a common transformer

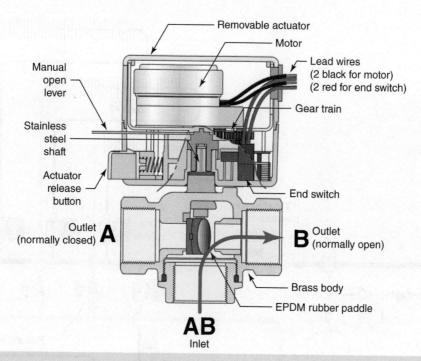

Figure 5-83 | Internal construction of a three-way diverter valve. *Courtesy Caleffi North America.*

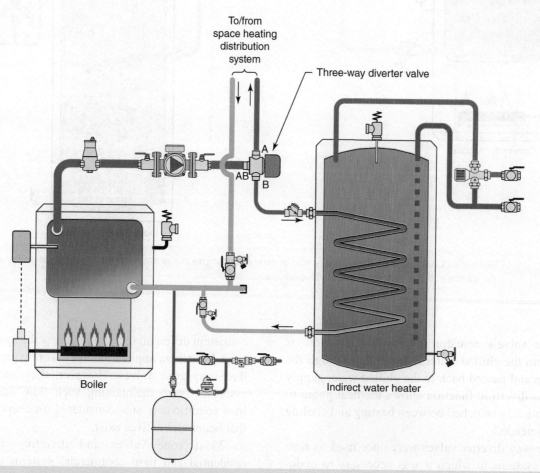

Figure 5-84 | A common application of a three-way diverting valve is to route hot water output from a boiler to either space heating or domestic water heating loads.

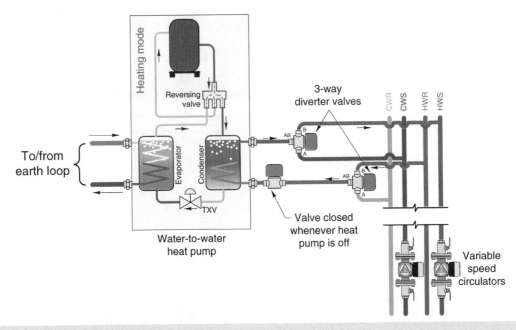

Figure 5-85 | Using two, three-way motorized diverter valves to route flow from heat pump for either heating or cooling operation.

secondary voltage used in many types of the heating and cooling systems. This allows valves to be wired with standard thermostat cable. For certain applications, it may be desirable to operate zone valves or diverting valves using standard 120 VAC or 240 VAC wiring. Some manufacturers offer actuators that are designed to operate at these voltages.

Zone valves and diverting valves are available in two-wire, three-wire, or four-wire configurations. Two wires are always required to operate the valve's actuator. The third, and in some cases fourth wire is connected to an end switch in the valve. When the valve reaches its fully open position, the end switch closes. This switch closure is used to signal other portions of the control circuit that a zone is "calling" for heat. More detailed discussion of control wiring for zone valves and diverting valves is presented in Chapter 9, Control Strategies, Components, and Systems.

When soldering a zone valve or diverting valve to copper tubing, the actuator should be removed from the valve body. The paddle that controls flow through the valve should be positioned away from the valve seats to minimize heating. Use a hot torch that can quickly heat the sockets of the valve body to minimize the potential for heat damage to other parts of the valve.

Differential Pressure Bypass Valves

Systems that use several zone valves in combination with a *fixed-speed* circulator can experience undesirable operating conditions when only one or two of the zones are on. The problem that arises is an increase in the **differential pressure** supplied by the circulator when only one or two zones are operating. The increased pressure can cause excessively high flow rates and **flow noise** in the active zone circuits.

One way to prevent this is to install a **differential pressure bypass valve (DPBV)** in systems that have more than four zones controlled by zone valves or have a circulator larger than 1/25 horsepower. This valve can be set to maintain an almost constant upper limit on the pressure differential across the distribution system. The valve functions by allowing a portion of the flow from the system's circulator to bypass the distribution system, thus limiting the pressure differential the pump can develop.

An example of a DPBV is shown in Figure 5-86a. The typical placement of a DPBV in a zoned system is shown in Figure 5-86c. The details for incorporating DPBVs into systems and properly adjusting them are discussed in Chapter 11, Distribution Piping Systems.

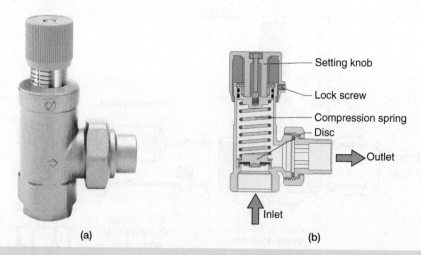

Figure 5-86 | (a) A DPBV. (b) Internal construction. *Courtesy of Caleffi North America.*

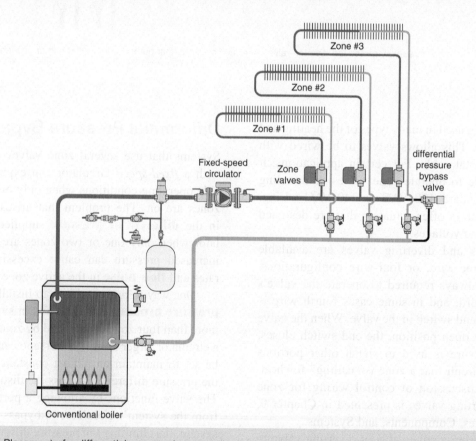

Figure 5-86c | Placement of a differential pressure bypass valve in a system using zone valves and a fixed speed circulator.

The traditional use of differential pressure bypass valves in systems using zone valve has declined as the availability of variable-speed pressure-regulated circulators steadily increases. Variable-speed pressure-regulated circulators eliminate the need for differential pressure bypass valves. They also decrease the electrical energy used by the system.

Differential pressure bypass valve can also prevent reverse flow, like a spring-loaded check valve, and provide a selectable forward opening resistance when need in a circuit. One such application is shown in figure 5-86d.

In this system, which uses a **two-pipe buffer tank**, the (1 psi) forward opening pressure of the differential pressure bypass valve forces all flow returning from the load through the thermal storage tank rather than allowing some of that flow through the nonfiring pellet boiler. This prevents inadvertent heat loss from the boiler jacket and flue. The differential pressure bypass valve also prevents **reverse thermosiphon flow** from the thermal storage

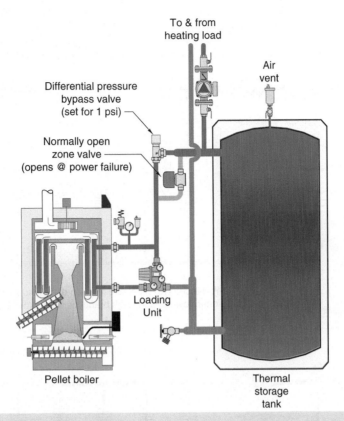

Figure 5-86d | Using a differential pressure bypass valve to prevent heat migration through a pellet boiler used with a 2-pipe buffer tank configuration.

tank through the boiler when it is off. Because of its low pressure setting (1 psi) the differentil pressure bypass valve immediately opens when the circulator in the boiler's loading unit turns on. The normally open zone valve shown in Figure 5-86d remains closed whenever electrical power is available to the system, but opens during a power failure. This allows a thermosiphon flow to develop between the boiler and tank to dissipate residual heat from the boiler.

Balancing Valves

Hydronic systems containing two or more parallel piping paths often require adjustments of the flow rates to properly balance the heating or cooling capacity of each branch. This is especially true in **two-pipe direct return systems**. An example of such a system is shown in Figure 5-88a. Without **balancing valves**, this system would produce higher flow rates in the piping paths closer to the circulator compared to those farther out in the distribution system. The proper use of balancing valves can correct for this situation as depicted in Figure 5-87b.

There are several types of balancing valves available. Three examples are shown in Figure 5-88.

Venturi-type balancing valves, which are also known as fixed orifice balancing valves, work based on a known relationship between flow rate and the pressure drop across the valve's internal venturi. A manometer is attached to the valve's pressure ports to measure the pressure drop across the valve. The flow rate through the valve is then inferred based on this measurement. The valve's knob can be adjusted to set the desired flow rate.

Some balancing valves, like the one shown in Figure 5-88c, have integral flow meters that directly read flow rate. The shaft of the balancing valve is adjusted until the desired flow rate is indicated on this flow meter.

Automatic flow balancing valves are relatively new in the North American market. They use an internal spring-loaded cartridge that maintains a specific preset flow rate over a wide range of differential pressure. This type of valve is not field adjustable; however, the internal cartridges can be changed to alter the flow rate that the valve maintains. A minimum differential pressure (typically about 2 psi) must be maintained across this type of valve to ensure that it regulates the flow rate to its specified value.

The goal of "**balancing**" a hydronic heating system is to ensure that each branch circuit delivers the proper rate of heat output. This goal is simple in concept, but more involved in practice. It will be discussed in more detail in later chapters.

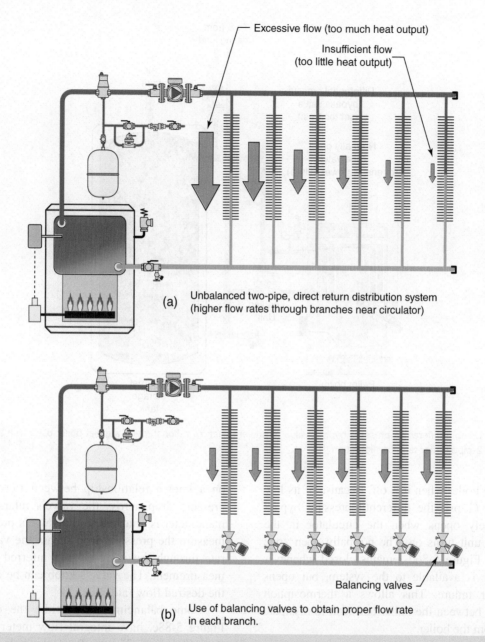

Figure 5-87 | Use of balancing valves to adjust flow rate in parallel piping branches. (a) Situation without balancing valves, and (b) situation after balancing valves are installed and properly adjusted.

Lockshield Valves

Another specialty valve widely used in hydronic systems provides the ability to isolate, flow balance, and even drain individual heat emitters. It is called a **lockshield valve** and is available in both straight and angle patterns as shown in Figure 5-89.

To adjust the flow setting of a lockshield valve, the cap is removed, and the internal plug is rotated to its closed position using an Allen key. A retaining ring that limits the travel of this plug is then adjusted to the desired setting using a blade-type screwdriver. When the valve plug is reopened, it will only move outward until it contacts the bottom edge of the retaining ring. This design allows the plug to be closed and opened repeatedly without affecting the valve's flow rate setting.

Lockshield valves can also be used to drain water from a heat emitter. This is done by first screwing a specialized drainage tool onto the valve body. The shaft of this tool engages the entire plug/retainer ring assembly within the valve and allows it to be positioned so that fluid will drain through the valve. After servicing, the heat emitter can be refilled by pumping

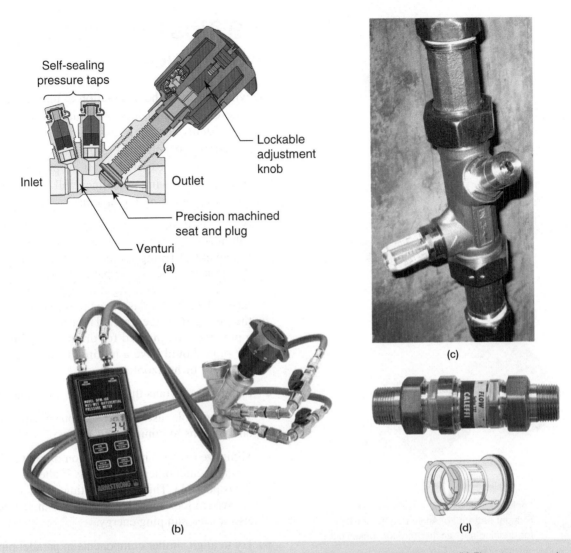

Figure 5-88 Examples of balancing valves. (a) Venturi-type balancing valve. *Courtesy of Caleffi SpA.* (b) Flow rate measuring device for venturi style balancing valve. *Courtesy of Armstrong Pumps.* (c) Balancing valve with integral flow meter. (d) Automatic flow balancing valve with preset flow rating. *Courtesy of Caleffi North America.*

fluid back in through this tool. The plug/retainer ring assembly is then screwed back into its original position in the valve body. Lockshield valves are equipped with a **half union** that allows the valve and heat emitter to be quickly separated should the radiator have to be temporarily removed from the wall.

5.8 Schematic Symbols for Piping Components

Piping schematics are the road maps of hydronic system design. They describe the relative placement of dozens (sometimes hundreds) of components that must be assembled to build a smooth operating system. Just as various geographic features are represented by small markings on a map, the components in a piping system can be represented by schematic symbols.

There is no universal standard for piping symbols to which all those in the hydronics industry subscribe. Variations are quite common among different designers and companies. Systems designers should establish a consistent set of symbols and identify those symbols using a legend on all drawings in which they are used. When this is not done, the designer's symbol for a globe valve may be interpreted as a gate valve by the installer and so forth. This can create expensive misunderstandings when the job is bid or installed.

Within this text, piping schematics use symbols based on the legend shown in Figure 5-90. Some of

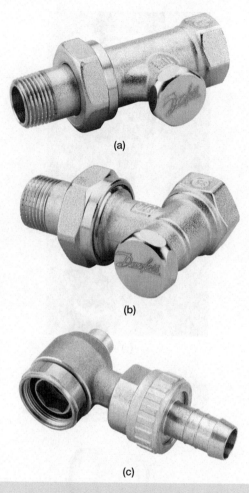

Figure 5-89 | Lockshield valves and service tool. (a) Straight pattern. (b) Angle pattern. (c) Tool for draining heat emitter through lockshield valves. *Courtesy of Danfoss.*

the components shown on this legend have not been discussed at this point, but will be in later chapters. This symbol legend is also shown as Appendix A.

The piping schematics shown in this text were all prepared using a simple computer-aided drafting program. There are many such programs available for both Windows and Macintosh operating systems. There is also "cloud based" software for drawing hydronic piping schematics, an example of which is shown in Figure 5-91.

Those who prepare piping schematics as well as other types of documentation for hydronic systems installations are highly encouraged to use such programs. They allow professional-looking drawings to be quickly prepared for design, installation, and marketing purposes.

5.9 Tips on Piping Installation

As with any trade, the installation of piping requires an attitude of craftsmanship. Well-planned and installed piping systems convey a message of professionalism. The planning of efficient piping layouts is a skill acquired through experience. It is part art and part science. The following guidelines will help in planning and installing good piping layouts.

- As a rule, piping should always be kept vertical (plumb) or horizontal (level). This ensures proper alignment of fitting joints and other components that join into the piping path. It also helps align the piping with building framing and finished surfaces. Use levels, lasers, or plumb bobs to keep the piping runs accurate. The exception is when a pipe must be pitched for drainage. In this case, the pipe should have a minimum slope of 1/4 inch/foot of horizontal run.

- Plan piping layouts that minimize joints and fittings.

  When possible, use 20-feet sections of straight pipe on long runs to minimize couplings.

- Minimize changes in the direction of piping. Do not zigzag around other components if alternative routing is possible. This adds to fitting count and makes the system look more complicated than necessary. It also wastes pumping energy.

- Try to align similar components in parallel piping branches. For example, when installing zone valves or zone circulators side by side, make sure they are aligned with each other.

- Always consider the need to remove components for servicing. Do not locate serviceable components in small, hard to get to areas, or behind other equipment.

- Always provide support adjacent to major components supported by the piping such as circulators, expansion tanks, or other relatively heavy devices.

- Plan piping assemblies that can be put together at a vise or flat working surface to minimize awkward working positions.

- Always allow for the take-up distances of threaded or soldered fittings when measuring and cutting a length of tubing. See the tables and figures in this chapter for the proper take-up allowances.

- In situations where both piping and electrical wiring are in close proximity, try to keep the piping below

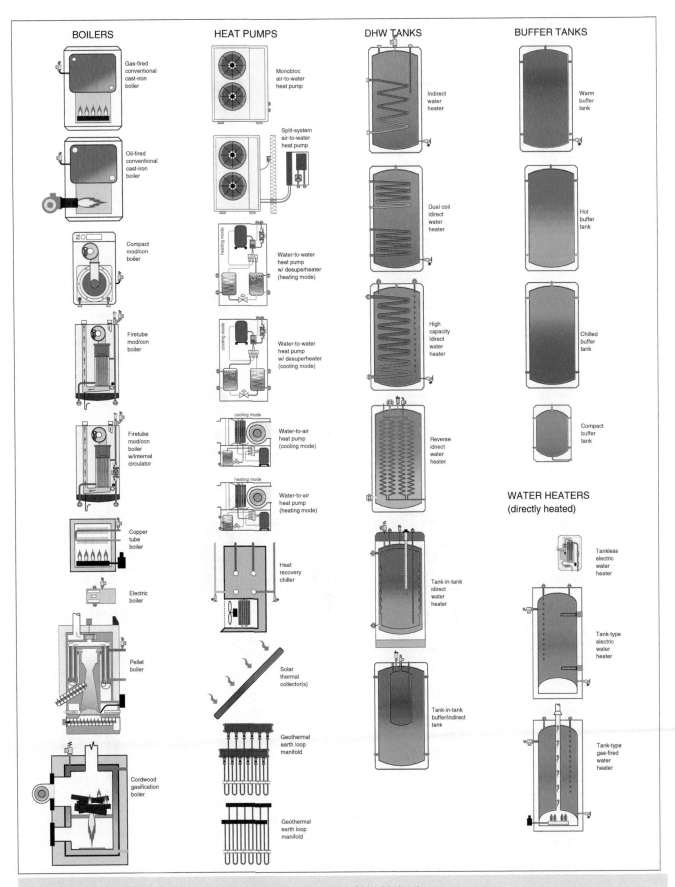

Figure 5-90a | Legend of symbols used in the piping schematics within this book.

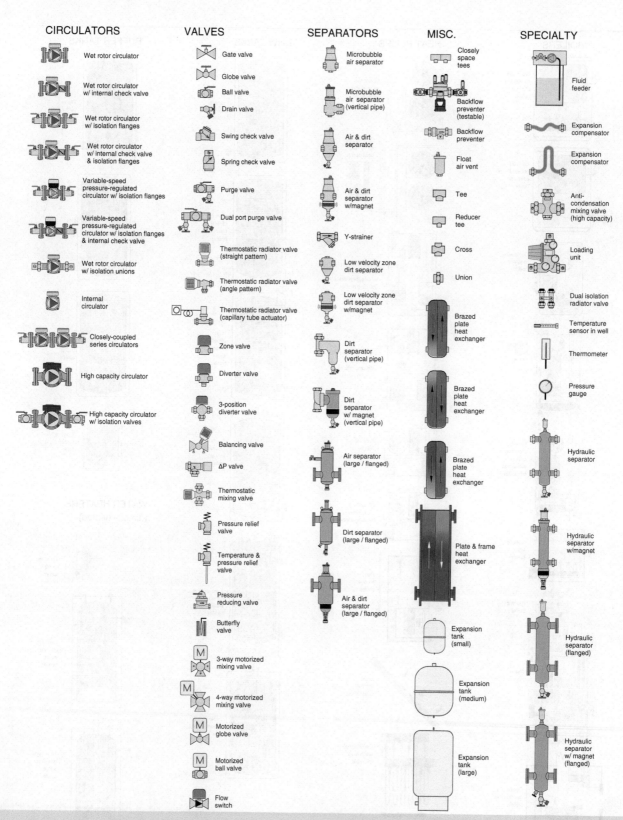

Figure 5-90b | Legend of symbols used in the piping schematics within this book.

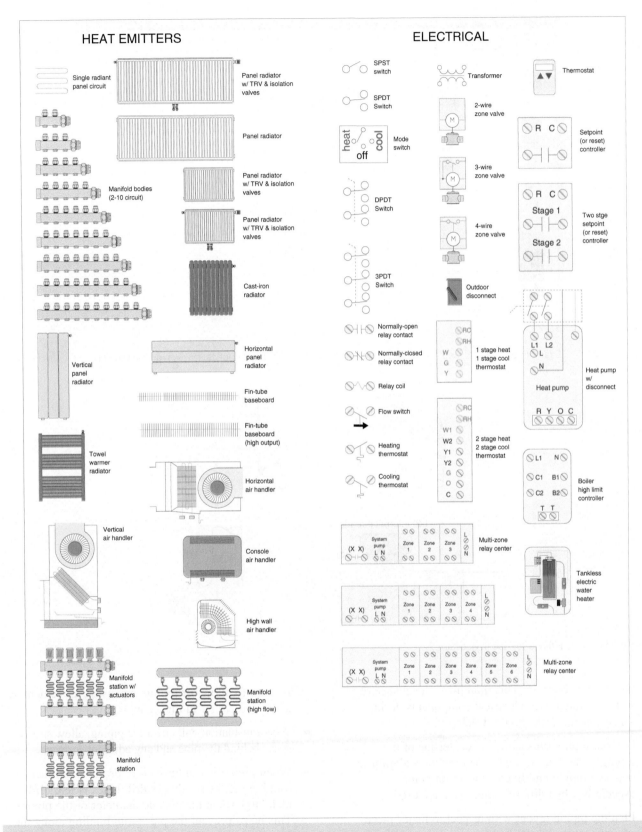

Figure 5-90c | Legend of symbols used in the piping schematics within this book.

218 Chapter 5 Pipings, Fittings, and Valves

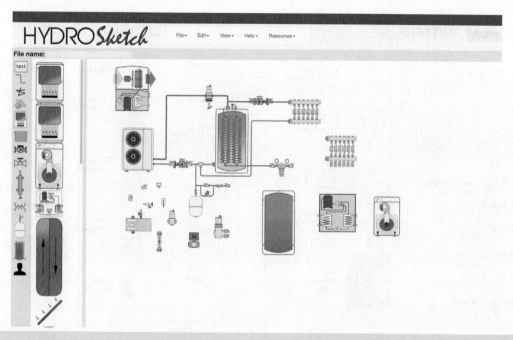

Figure 5-91 | Example of "cloud-based" software for making hydronic piping schematics. www.hydrosketch.com

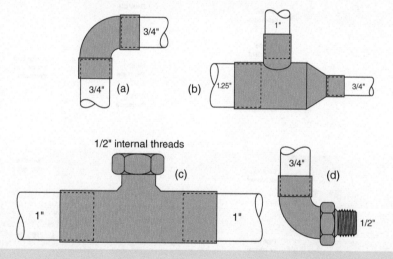

Figure 5-92 | Fittings for use in question 7.

the electrical wiring. This minimizes the chance of water dripping onto electrical components during service or if a leak should develop.

- Lay out piping runs using the centerline of the piping rather than its edges. This ensures alignment between piping and fixed connection points, especially when different pipe sizes are used.

- Plan the piping layout with respect to major pieces of equipment such as the boiler, circulator, mixing valve, and so on. The positions of these devices serve as benchmarks to which interconnecting piping can be referenced.

- Secure all piping to structural surfaces to prevent sagging and to support the weight of components supported by the piping.

- Be sure to allow for piping expansion movement as discussed earlier in this chapter.

- Where insulation will cover the piping, allow sufficient space behind the pipe and any adjacent surface.

- When piping is run through floor joists or other framing members, always drill holes at least 1/4 inch larger than the outside diameter of the piping. This helps prevent the pipe from binding along the framing as it expands and contracts. Use a chalk line or laser during layout to keep the holes precisely aligned. Plan ahead when long piping runs through many floor joists are required.

- The optimum location for a hole through a framing member such as a floor joist is halfway between the bottom edge and top edge. This location minimizes loss of strength. Never notch into the bottom of floor joists, ceiling joists, or trusses to accommodate piping. Whenever possible, keep the edge of the hole at least two inches from the edge of framing members to avoid potential punctures from nails or screws.

- Never mount circulators and adjacent piping to surfaces that can amplify vibrations, especially in proximity to living space. If possible, fasten the mounting hardware for circulators and adjacent piping to masonry or concrete walls to reduce the potential for vibration transmission.

Summary

This chapter has described both basic and specialized piping hardware used in hydronic systems. The many combinations of pipings, fittings, and valves provide almost unlimited flexibility for the creative system designer. New variations of piping fittings and valves appear regularly. Designers should make a practice of scanning trade publications and Internet sites for such new developments, many of which open new possibilities to simplify or speed installation.

Key Terms

angle pattern
angle valve
automatic flow balancing valve
backflow preventer
balancing
balancing valve
ball valve
baseboard tee
bell hanger
boiler drain
capillary tube
check valve
clevis hangers
coefficient of linear expansion
component isolation
composite tubing
copper water tube
couplings
cross-linked polyethylene
diaphragm
dielectric union
differential pressure
differential pressure bypass valve
diverter tee
domestic water
dumb mixing valve
elbows
entrain
EVOH (ethylene vinyl alcohol)
expansion compensator
expansion loop
expansion offset
fast-fill mode
feed water valve
female threads
fittings
fixed point supports
flow element (or spool)
flow noise
flow regulation
flow restrictor valve
flow-check valve
four-way mixing valve
full-port ball valve
galvanic corrosion
gate valves
globe valves
half union
handwheel
hard drawn tubing
heat migration
homerun distribution system
injection mixing
lockshield valve
male threads
monoflo tees®
motorized actuators
mounting rail system
national pipe thread (NPT)
nominal inside diameter
nonbarrier tubing
one-pipe systems
oxygen diffusion
oxygen diffusion barrier
packing gland
packing nut
paste flux
PEX tubing
PEX-AL-PEX tubing
pipe size
plastic-coated hangers
Polyethylene - Raised Temperature (PE-RT)
Polypropylene Random Copolymer (PP-R) tubing
press-fit joint
pressure-regulated circulator
pressure-relief valve
purge
purging valve
reinforced polypropylene (PP-R) tubing
reverse thermosiphon flow
rod clamp
saddle fusion
sensor bulb
sensor well
shape memory
sleeve or roller supports
socket fusion
soft soldering
soft temper tubing
spring-loaded check valve
standard-port ball valve
standoff piping supports
straight pattern
street elbows
swing-check valves
tees
thermal contraction
thermal expansion
thermoplastics
thermoset plastic
thermostatic element
thermostatic operator
thermostatic radiator valves (TRVs)
threaded adapters
three-way diverting valve
three-way motorized mixing valves
three-way thermostatic mixing valve
two-pipe buffer tank
two-pipe direct return systems
two-way thermostatic valves
unions
venturi fitting
venturi-type balancing valves
working pressure rating
zone valve

Questions and Exercises

1. Describe the differences between types K, L, and M copper tubing.

2. Describe the difference between copper water tube and ACR tubing.

3. What type of solder should be used for copper piping conveying domestic water? Why?

4. A straight copper pipe 65 feet long changes temperature from 65 to 180 °F. What is the change in length of the tube? Describe what would happen if the tube was cooled from 65 to 40 °F.

5. A hydronic system contains 50 3/4-inch and 25 1-inch soldered copper joints. Estimate the number of inches of 1/8-inch wire solder required for these joints.

6. Describe a situation where two diverter tee fittings would be preferred to a single diverter tee.

7. Write a standard description for the fittings shown in Figure 5-92.

8. What should be done prior to soldering a valve with a nonmetallic disc or washer?

9. What is the advantage of a four-port mixing valve over a three-port mixing valve when used with a conventional gas- or oil-fired boiler?

10. Describe a situation where the high thermal conductivity of copper tubing is undesirable.

11. Describe a situation where a differential pressure bypass valve installed in a multizone hydronic distribution system begins to open.

12. Describe three separate functions of a lockshield/balancing valve.

13. Why should the thermostatic actuator of a radiator valve not be mounted above the heat emitter?

14. Why should a globe valve not be used for component isolation purposes? What types of valves are better suited for this application?

15. What is an advantage of a spring-loaded check valve when compared to a swing-check valve? What is a disadvantage?

16. What is the difference between a full-port ball valve and conventional port ball valve?

17. Name two applications for a three-way diverter valve.

18. Define nominal pipe size.

19. Does the presence of an oxygen diffusion barrier on polymer tubing totally prevent oxygen entry into the system? Explain your answer.

20. What is the function of the end switch in a zone valve?

21. When can a swing-check valve offer the same protection as a backflow preventer in the make-up water line of a hydronic system?

Chapter 6

Fluid Flow in Piping

Objectives

After studying this chapter, you should be able to:
- Describe the difference between flow rate and flow velocity.
- Explain how the pressure of a liquid varies with its depth.
- Define the term head and relate it to pressure differential.
- Explain what hydraulic resistance is and how it is used.
- Use the C_v factor of a valve to calculate pressure drop.
- Calculate the hydraulic resistance of piping, fittings, and valves.
- Sketch hydraulic resistance diagrams for a given piping layout.
- Convert two or more parallel hydraulic resistances into an equivalent resistance.
- Reduce complex piping systems into simpler equivalent diagrams.
- Plot the system head loss curve of a given piping system.
- Explain several considerations that relate to pipe sizing.
- Estimate the operating cost of a given piping system.

6.1 Introduction

An understanding of how fluids behave while moving through piping is essential to designing good hydronic systems. This chapter lays the groundwork for describing fluid flow in piping. It introduces analytical methods for predicting flow rates in all parts of a piping system, including those with complex branches and many components. These methods will become a routine part of overall system design that will be used many times in later chapters.

Unfortunately, experience has shown that many of the people responsible for design and installation of hydronic systems do not understand the principles of fluid flow in piping well enough to scrutinize designs before committing time and materials to their installation. This is especially true of nonstandard projects for which there is no previous example to reference. This often leads to hesitation in trying anything beyond what is already familiar. Many existing hydronic systems could have been better matched to their buildings, installed for less money, or operated at a lower cost if their designer had had some way of evaluating how a slight change in system design would affect performance. This chapter gives designers the analytical tools needed to simulate the behavior of hydronic piping systems before committing time and money to their installation.

Many of the concepts in this chapter are drawn from classical fluid mechanics, a subject known for its share of mathematics. However, much effort has gone into distilling this theory to be concise and easy to work with design tools.

To further simplify use of this material, the Hydronics Design Studio software contains a module called **Hydronic Circuit Simulator** that can quickly evaluate many of the equations and procedures presented in this chapter. This allows the designer to experiment with the many factors involved and develop a feel for how piping systems behave under a variety of conditions.

6.2 Basic Concepts of Fluid Mechanics. What Is a Fluid?

In a technical sense, the word **fluid** can mean either a **liquid**, such as water, or a **gas**, such as air. The science of fluid mechanics includes the study of both liquids and gases. In this book, the word fluid always refers to a liquid unless otherwise stated.

As discussed in Chapter 4, most liquids, including water, are **incompressible**. This means that any given amount of liquid always occupies the same volume, with the exception of minor changes in volume due to thermal expansion. A simple example would be a long pipe completely filled with water. If another gallon of water were pushed into one end of the pipe, the exact same amount would have to flow out the other end. No additional water can be squeezed into a rigid container (such as a pipe) that is already filled.

By comparison, if the same pipe were filled with a gas such as air, it would be possible to push another gallon of air into the pipe without allowing any air out the other end. The result would be an increase in the pressure of the air within the pipe. Because of this, gases are said to be **compressible fluids**.

The fact that liquids are incompressible makes it easy to describe their motion through closed containers such as piping systems. Two basic terms associated with this motion are **flow rate** and **flow velocity**.

Flow Rate

Flow rate is a measurement of the *volume* of fluid that passes a given location in a pipe in a given time. In the North American hydronics industry, the customary units for flow rate are U.S. gallons per minute (abbreviated as **gpm**). Within this book, flow rate is represented in equations by the symbol f. The flow rate of a fluid through a hydronic system has a major impact on that system's thermal performance.

Flow Velocity

The motion of a fluid flowing through a pipe is more complex than that of a solid object moving in a straight path. The speed, or flow velocity of the fluid, is different at different points across the diameter of the pipe.

To visualize the flow velocity of a fluid, think about several small buoyancy-neutral particles that have been positioned along a line within the cross-section of a pipe and released at the same instant to be carried along with the flow. The particles carried along by the flow would move faster near the center of the pipe than near the edge. The speed of each particle could be represented by an arrow as shown in Figure 6-1a. A curve connecting the tips of all these arrows is called a **velocity profile**. The velocity profile shown in Figure 6-1a is two dimensional. However, if one imagines that this curve rotated around the centerline axis of the pipe, it forms a "nose cone" shape that represents a three-dimensional velocity profile as seen in Figure 6-1b.

The term "flow velocity" is commonly understood to mean the **average flow velocity** of the fluid as it moves through a pipe. *This average velocity would be the velocity that, if present across the entire cross-section of the pipe, would result in the same flow rate as that created by the actual velocity profile.* The concept of average flow velocity is illustrated in Figure 6-1c. *In this book, the term "flow velocity" will always mean the average flow velocity in the pipe.* The common units for flow velocity in North America are feet per second, abbreviated as either ft/s or FPS. Within equations, the average flow velocity is represented by the symbol v.

Equation 6.1 can be used to calculate the average flow velocity associated with a given flow rate in a round pipe.

Equation 6.1:

$$v = \left(\frac{0.408}{d^2}\right)f$$

where,
v = average flow velocity in the pipe (ft/s)
f = flow rate through the pipe (gpm)
d = exact inside diameter of pipe (in)

6.2 Basic Concepts of Fluid Mechanics. What Is a Fluid?

When selecting a pipe size to accommodate a given flow rate, the resulting average flow velocity should be calculated. *A pipe size should be selected that keeps the average flow velocity between 2 and 4 ft/s.*

The lower end of this velocity range is based on the ability of flowing water to move air bubbles along a vertical pipe. Average flow velocities of 2 ft/s. or higher can **entrain** air bubbles and move them in all directions including straight down. Entrainment of air bubbles is important when a system is filled and purged, as well as when air bubbles have to be captured and removed following servicing.

The upper end of this flow velocity range (4 ft/s) is based on minimizing noise generated by the flow. Average flow velocities above 4 ft/s can cause objectionable flow noise and should be avoided for any piping passing through or near living spaces. Slightly higher flow velocities in the range of 5 ft/s are acceptable from the standpoint of sound emitted from piping passing through nonliving spaces. However, flow velocities in small size copper tubing should be limited to 5 ft/s to avoid the possibility of erosion corrosion. It's also important to understand that higher flow velocities result in higher fluid drag and thus require higher input power to circulators.

Figure 6-2 tabulates the results of applying Equation 6.1 to selected sizes of type M copper, PEX, and PEX-AL-PEX tubing. Each equation can be used to calculate the flow velocity associated with a given flow rate. Also given are the flow rates corresponding to average flow velocities of 2 and 4 ft/s.

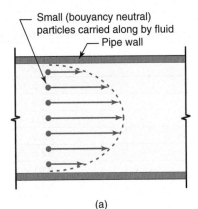
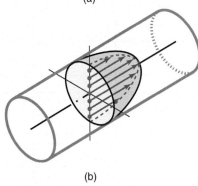
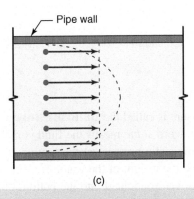

Figure 6-1 (a) Representation of actual velocity profile, (b) velocity profile in three dimensions, and (c) equivalent average velocity profile.

Example 6.1
What is the average flow velocity of water moving at 6.0 gpm through a 3/4" type M copper tube that has an inside diameter of 0.811 inch?

Solution:
Substituting these numbers into Equation 6.1 yields:

$$v = \left(\frac{0.408}{d^2}\right)f = \left(\frac{0.408}{0.811^2}\right)6 = 3.72 \text{ft/s}$$

Static Pressure of a Liquid

Consider a pipe with a cross-sectional area of 1 in², filled with water to a height of 10 feet, as shown in Figure 6-3.

Assuming the water's temperature is 60 °F, its density is about 62.4 lb/ft³. The weight of the water in the pipe could be calculated by multiplying its volume by its density.

Weight = (volume)(density)
= (cross-sectional area)(height)(density)

$$\text{Weight} = (1\text{in}^2)(10 \text{ ft})\left(62.4\frac{\text{lb}}{\text{ft}^3}\right)\left(\frac{1\text{ft}^2}{144 \text{ in}^2}\right) = 4.33 \text{ lb}$$

The water exerts a pressure on the bottom of the pipe due to its weight. This pressure is equal to the weight of the water column divided by the area of the pipe. In this case:

$$\text{Static pressure} = \frac{\text{weight of water}}{\text{area of pipe}} = \frac{4.33 \text{ lb}}{1 \text{ in}^2} = 4.33\frac{\text{lb}}{\text{in}^2} = 4.33 \text{ psi}$$

Tubing size/type	Flow velocity	Minimum flow rate (based on 2 ft/s) (gpm)	Maximum flow rate (based on 4 ft/s) (gpm)
3/8" copper	v = 2.02 f	1.0	2.0
1/2" copper	v = 1.26 f	1.6	3.2
3/4" copper	v = 0.62 f	3.2	6.5
1" copper	v = 0.367 f	5.5	10.9
1.25" copper	v = 0.245 f	8.2	16.3
1.5" copper	v = 0.175 f	11.4	22.9
2" copper	v = 0.101 f	19.8	39.6
2.5" copper	v = 0.0655 f	30.5	61.1
3" copper	v = 0.0459 f	43.6	87.1
3/8" PEX	v = 3.15 f	0.6	1.3
1/2" PEX	v = 1.73 f	1.2	2.3
5/8" PEX	v = 1.20 f	1.7	3.3
3/4" PEX	v = 0.880 f	2.3	4.6
1" PEX	v = 0.533 f	3.8	7.5
1.25" PEX	v = 0.357 f	5.6	11.2
1.5" PEX	v = 0.256 f	7.8	15.6
2" PEX	v = 0.149 f	13.4	26.8
3/8" PEX-AL-PEX	v = 3.41 f	0.6	1.2
1/2" PEX-AL-PEX	v = 1.63 f	1.2	2.5
5/8" PEX-AL-PEX	v = 1.00 f	2.0	4.0
3/4" PEX-AL-PEX	v = 0.628 f	3.2	6.4
1" PEX-AL-PEX	v = 0.383 f	5.2	10.4

*v = Average flow velocity (ft/s).
f = flow rate (gpm).

Figure 6-2 | Equations relating flow rate (f) and flow velocity (v) for type M copper, PEX, and PEX-AL-PEX tube. Flow rates corresponding to flow velocities of 2 and 4 ft/s are also given.

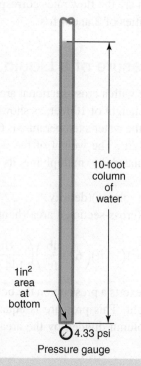

Figure 6-3 | Column of water 10 ft high in a pipe with a cross-sectional area of 1 in².

This pressure is called the **static pressure** exerted by the water. The word *static* means the fluid is not moving. In this case, the static pressure at the bottom of the pipe is due solely to the weight of the water. If a very accurate pressure gauge were mounted into the bottom of the pipe as shown in Figure 6-3, it would read exactly 4.33 psi.

Next, consider a pipe with a cross-sectional area of 1 ft² filled with water to a height of 10 feet. The weight of this column can be calculated as:

$$\text{Weight} = (1 \text{ ft}^2)(10 \text{ ft})\left(62.4 \frac{\text{lb}}{\text{ft}^3}\right)\left(\frac{1 \text{ ft}^2}{144 \text{ in}^2}\right) = 624 \text{ lb}$$

The static pressure at the bottom of the column would be:

$$\text{Static pressure} = \frac{\text{weight of water}}{\text{area of pipe}} = \frac{624 \text{ lb}}{144 \text{ in}^2} = 4.33 \frac{\text{lb}}{\text{in}^2} = 4.33 \text{ psi}$$

Notice the static pressure at the bottom of the column did not change even with a large change in the diameter of the pipe. This is explained by the fact that the larger column of water has a proportionally larger area across which to distribute its weight. Thus, the static pressure

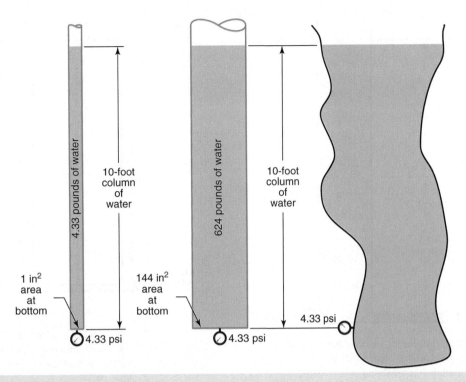

Figure 6-4 | The static pressure of a fluid at a given point depends only on depth below the surface and not on the size or shape of the container.

at a given depth below the surface of a liquid remains exactly the same regardless of the diameter of the pipe.

This same principle remains true for liquids within containers of any size and shape, as shown in Figure 6-4. Scuba divers, for example, feel a static pressure that depends only on their depth below the surface of the water, not on how wide the body of water is.

The static pressure of a liquid increases proportionally from some value at the surface to larger values at greater depths below the surface as shown in Figure 6-5. Again, the size or shape of the container makes no difference.

If the top of the system shown in Figure 6-5 were closed off, and air was pumped into the space until it attained a pressure of 5.0 psi, the static pressure of the liquid at all locations simply increases by 5.0 psi as shown in Figure 6-6.

From these observations, it is possible to develop a relationship between the depth of water within a container and the static pressure it exerts at any point below its surface.

The following approximation is useful for systems containing water:

Equation 6.2:

$$P_{static} = \frac{h}{2.31} + P_{surface},$$

where,

P_{static} = static pressure at a given depth, h, below the surface of the water (psi)

h = depth below the water's surface (ft)

2.31 = unit conversion factor for water at about 60 °F

$P_{surface}$ = any pressure applied at the surface of the water (psi)

If the water is open to the atmosphere at the top of the container, even through a pinhole-size opening, then $P_{surface} = 0$. If the container is completely closed and filled with water, $P_{surface}$ is the pressure of the water at the top of the container.

The reason Equation 6.2 is an approximation is that the density of water and, thus, the static pressure it creates within a container changes with temperature. To account for such changes and to allow for calculations with liquids other than water, Equation 6.2 can be modified as follows:

Equation 6.3:

$$P_{static} = \frac{(h)(D)}{144} + P_{surface},$$

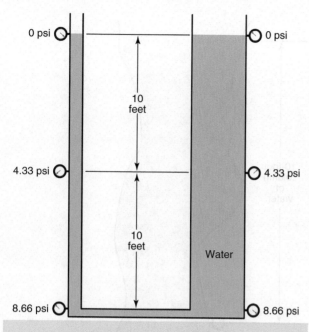

Figure 6-5 | Static pressure for various depths below the fluid surface. Notice that the width of the column, or shape of the overall container, has no effect on static pressure.

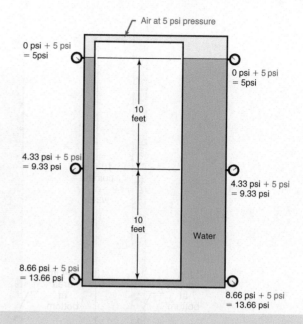

Figure 6-6 | The effect of increasing the pressure at the top of the container. All pressures below the surface increase by the same amount.

where,

P_{static} = static pressure at a given depth, h, below the surface of the water (psi)

h = depth below the water's surface (ft)

D = density of the liquid (lb/ft³)

$P_{surface}$ = any pressure applied at the surface of the water (psi)

To use Equation 6.3, the density of the liquid must be determined. The density of water over the temperature range of 50 to 250 °F can be found in Figure 4-4. The density of several other liquids commonly used in hydronic heating systems can be determined over this temperature range using the Fluid Properties Calculator module in the Hydronics Design Studio software.

Example 6.2

Find the approximate static pressure at the bottom of a hydronic system with an overall height of 65 ft, filled with cool water, and having a pressure of 10.0 psi at its uppermost point. Refer to Figure 6-7.

Solution:

Use Equation 6.2 to calculate the static pressure:

$$P_{static} = \frac{65}{2.31} + 10 = 38.1 \text{ psi}$$

Discussion:

The calculated pressure of 38.1 psi would be correct if the water temperature were about 60 °F. If, however, the same system contained water at 250 °F (having a density of 58.754 lb/ft³), Equation 6.3 would have to be used, and would yield the following:

$$P_{static} = \frac{(65)(58.75)}{144} + 10 = 36.5 \text{ psi}$$

Notice the hot water would exert slightly less pressure at the base of the system due to its lower density.

Head of a Fluid

Fluids in a hydronic system contain both thermal and mechanical energy. The thermal energy content of a fluid depends on its temperature and specific heat. We can sense the relative amount of thermal energy in a fluid by its temperature. For example, the hot water leaving a boiler contains more thermal energy than cooler water returning from the distribution system.

The mechanical energy present in a fluid is referred to as **head**. This energy depends on several factors including its pressure, density, elevation, and velocity of the fluid at some point in the system.

6.2 Basic Concepts of Fluid Mechanics. What Is a Fluid?

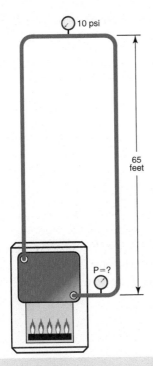

Figure 6-7 | Piping system for example 6.2.

It is helpful to divide head energy into three categories:

- **Pressure head**, which is the mechanical energy a fluid contains because of its pressure.
- **Velocity head**, which the mechanical energy a fluid contains because of its velocity.
- **Elevation head**, which is the mechanical energy the fluid contains because of its height in the system.

Total head is the sum of the pressure head, velocity head, and elevation head.

You could calculate the total head of a fluid at any point in a hydronic system using Equation 6.4:

Equation 6.4:

$$\text{Total head} = \frac{P(144)}{D} + \frac{v^2}{64.4} + Z$$

where,
- P = the pressure of the fluid (psi)
- D = the density of the fluid (lb/ft^3)
- v = the velocity of the fluid (ft/s)
- Z = the height of the point above some horizontal reference (ft)

The units for head are (ft · lb/lb). Those having studied physics will recognize the unit of **ft·lb** (pronounced as "foot pound") as a unit of energy. As such, it can be converted to any other unit of energy such as a Btu. However, engineers long ago chose to cancel the units of lb in the numerator and denominator of this ratio and express head in the sole remaining unit of feet. To make a distinction between feet as a unit of distance and feet as a unit of fluid energy, the latter can be stated as **feet of head**.

Liquids have the ability to exchange head energy back and forth between velocity head, pressure head, and elevation head. For example, when a liquid flows through a transition from a smaller to larger pipe size, its velocity decreases, while its pressure increases slightly because some of the velocity head is instantly converted into pressure head.

Example 6.3

Calculate the total head of the water at a location in the system which is 3 feet above floor level, where the static pressure of the fluid is 15.0 psi, its temperature is 140 °F, and its velocity is at 4 ft/s.

Solution:

Before using Equation 6.4, it is necessary to determine the density of water at 140 °F. This can be found using the graph in Figure 4-4. At 140 °F, water has a density of 61.4 lb/ft^3.

Substituting these numbers into Equation 6.4 yields:

$$\text{Total head} = \frac{P(144)}{D} + \frac{v^2}{64.6} + Z = \frac{15(144)}{61.4} + \frac{(4)^2}{64.4} + 3 = 38.4 \frac{\text{ft·lb}}{\text{lb}}$$

$$= 38.4 \text{ ft}$$

Discussion:

The elevation head of a fluid is always relative to some reference elevation. In this example, that reference elevation happened to be 3 ft below the point in the system where the head energy was being determined. The reference elevation can be arbitrarily chosen, but once it is selected, all calculations for head at various locations in the system should use the same reference elevation.

Relationship between Head and Pressure Change

When head energy is added to or removed from a liquid in a closed-loop piping system, there will always be an associated pressure change in the fluid. Just as a change

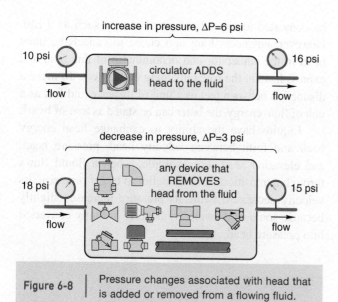

Figure 6-8 | Pressure changes associated with head that is added or removed from a flowing fluid.

in temperature is "evidence" of a gain or loss of thermal energy, *a change in pressure, that occurs without a change in velocity or elevation, is evidence of a gain or loss in head.* When head is lost, pressure decreases. When head is added, pressure increases. This concept is illustrated in Figure 6-8.

Using pressure gauges to detect changes in the head of a liquid, along a horizontal flow path of a given pipe size, is like using thermometers to detect changes in the thermal energy content of the liquid.

Equation 6.5 can be used to convert the observed change in pressure to the associated gain or loss of head energy. This equation requires the density of the liquid, which depends on its temperature.

Equation 6.5:

$$H = \frac{144(\Delta P)}{D}$$

where,
H = head added or lost from the liquid (feet of head)
ΔP = pressure change corresponding to the head added or lost (psi)
D = density of the liquid at its corresponding temperature (lb/ft³)

This equation can also be rearranged to calculate the pressure change associated with a given gain or loss in head:

Equation 6.6:

$$\Delta P = \frac{HD}{144}$$

Example 6.4

Water at 140 °F flows through a heat emitter. The observed pressure on the inlet side of the heat emitter is 20.0 psi. The observed pressure on the outlet side is 18.5 psi. What is the change in the head energy of the water as it passes through?

Solution:

The density of water at 140 °F is 61.4 lb/ft³. Substituting this and the pressure data into Equation 6.5 yields:

$$H = \frac{144(\Delta P)}{D} = \frac{144(20 - 18.5)}{61.4} = 3.52 \text{ feet of head}$$

Discussion:

Since the pressure decreases in the direction of flow, this is a head loss. If the fluid cools as it passes through the heat emitter, its density slightly increases. However, the variation in density is very small for the 10 to 30 °F temperature drop that occurs across a typical heat emitter, and thus, it has very little effect on the calculated change in head. The constraint of applying equations 6.5 and 6.6 along a horizontal flow path of a given (fixed) pipe size is to eliminate the change in pressure due to changes in velocity or elevation.

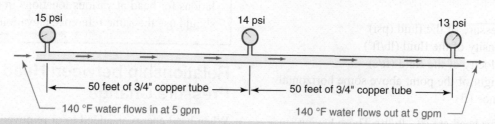

Figure 6-9 | Pressure drop in downstream direction associated with head loss created by viscous friction Note: The pipe is horizontal, and thus, the static pressure along its length is constant.

Head Loss Due to Viscous Friction

Whenever a fluid flows, an energy dissipating effect called **viscous friction** develops both within the fluid stream and along any surfaces the fluid contacts. This friction causes head energy to be converted to (or "dissipated" into) thermal energy.

In the strictest sense, no energy is lost in this process. Instead, a higher quality form of energy (head) is converted to a lower quality form of energy (heat). Before this transformation, the energy in higher quality form (e.g., head) helped move the fluid through the piping system. After being converted to heat by viscous friction, this is no longer possible. If the transformation takes place in a heated space, one could argue that the heat generated by head energy being dissipated into heat energy does help heat the space. However, in a well-designed hydronic heating system, the amount of heat produced by head loss should be a tiny fraction of the thermal energy carried along by the water.

Viscous friction develops within all fluids, but is greater for fluids with higher viscosities, or when a fluid flows over relatively rough surfaces.

The drag forces associated with viscous friction is what causes fluid molecules closer to a surface to have lower velocities than those farther away from the surface. A familiar example is the slower speed of water near the banks of a stream compared to near the center of the stream. Another is lower wind speeds near the surface of the earth. Within pipes, viscous friction gives rise to a velocity profile similar to that shown in Figure 6-1a.

Because of its incompressibility, a liquid moving at a constant flow rate through a given size pipe cannot slow down as head is dissipated due to viscous friction. Therefore, the head loss reveals itself as a drop in pressure in the downstream direction of flow as previously discussed. This concept is illustrated in Figure 6-9. Methods for calculating head loss will be detailed in Section 6.4, Hydraulic Resistance of Fittings, Valves, and Other Devices.

Fluid-Filled Piping Loops

Most hydronic heating systems consist of closed piping systems. After all piping work is complete; these circuits are completely filled with fluid. During normal operation, very little if any fluid enters or leaves the system.

Consider the fluid-filled piping loop shown in Figure 6-10a. A static pressure is present at point A due to the weight of the fluid column on the left side of the loop. It might appear that a circulator placed at point A would have to overcome this pressure to lift the water and establish flow around the loop. This, however, is not true. The reason is the weight of the fluid on the right side of the loop pushes downward at point B with exactly the same static pressure. Thus, the static pressure on both the inlet and outlet of the circulator are the same. This is true regardless of the pipe size used in the upward flowing portion of the loop versus that in the downward flowing portion. It is also true regardless of the shape, size, or complexity of the loop. Remember, static pressure is not affected by the shape of the container, only the height of the fluid column.

The fluid in the filled piping loop acts like a Ferris wheel with the same weight in each seat. The weight in the seats moving up exactly balances the weight in the seats moving down. If it were not for friction in the bearings and air resistance, this balanced Ferris wheel would continue to rotate indefinitely once started. Likewise, since the flow rate is always the same throughout a fluid-filled piping loop, the weight of the fluid moving up is always balanced by the weight of fluid moving down. If it were not for the viscous friction of the fluid, it too would continue to circulate indefinitely within a piping loop.

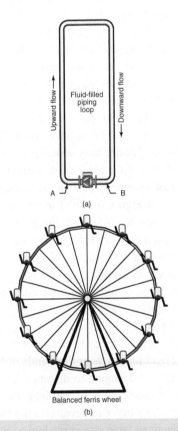

Figure 6-10 (a) A closed-loop piping system filled with fluid. The static pressure at point A equals the pressure at point B regardless of the height or shape of the loop. To maintain circulation, the circulator need only overcome the head loss due to viscous friction. (b) The fluid in the filled loop acts like a Ferris wheel with the same weight in each seat.

The following principle can thus be stated:

To maintain a constant rate of circulation in any fluid-filled piping loop, the circulator need only replace the head loss due to viscous friction. This remains true regardless of the shape, height, or pipe size(s) used in the loop.

This principle explains why a small circulator can establish and maintain flow in a filled piping loop, even if the top of the loop is several stories above the circulator and contains hundreds or even thousands of gallons of fluid. *Unfortunately, many circulators in hydronic systems are needlessly oversized because this principle is not understood.*

Flow Classifications

Fluid flow in pipes is characterized as being either **laminar** or **turbulent**. Both types of flow can enhance or detract from the performance of hydronic systems depending on where they occur.

To discuss the differences between these types of flow, it is helpful to use the concept of **streamlines**. A streamline indicates the path taken by an imaginary fluid particle as it moves along a pipe. Streamlines allow one to visualize flow as if the fluid were composed of millions of small particles, each with its own unique path.

Laminar flow is characterized by smooth, straight streamlines as the fluid moves through the pipe, as shown in Figure 6-11a. This type of flow is likely to exist when flow velocities are very low. To visualize laminar flow, think of the fluid as if it were separated into many thin layers that resemble concentric cylinders inside a pipe as shown in Figure 6-11b. The closer a given fluid layer is to the inside surface of the pipe, the slower it moves. The faster moving fluid layer tends to slide over the slower moving layers in such a way that mixing between layers is minimal.

Laminar flow creates minimal amounts of **head loss**. This is desirable because the lower head loss means less pumping power will be required to maintain a given flow rate. This is why large pipelines that carry heated or chilled water over hundreds or thousands of feet between buildings are often designed to maintain laminar flow.

Another characteristic of laminar flow is that it allows a **boundary layer** to develop along the inner pipe wall. The boundary layer is a very slow-moving layer of fluid that slides along the inner pipe wall with very little mixing effect. Such a layer creates a thermal resistance between the pipe wall and the bulk of the fluid stream, which impedes the rate of heat transfer between them. Because of this, *laminar flow is undesirable when the objective is to transfer heat between the pipe wall and fluid stream.*

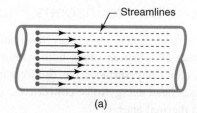
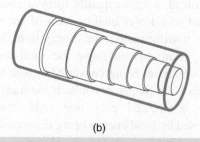

Figure 6-11 | The concept of laminar flow. (a) Velocity profile within a pipe and (b) concentric layers of fluid sliding over each other.

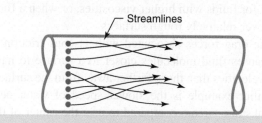

Figure 6-12 | Representation of streamlines within turbulent flow.

Turbulent flow is characterized by a vigorous mixing of the fluid as it travels down the pipe. The streamline of a single fluid particle shows it moves in an erratic path that often sweeps in toward the pipe wall and then back out into the bulk of the fluid stream as illustrated in Figure 6-12. This mixing improves heat transfer because it significantly reduces the thickness and thermal resistance of the boundary layer. Because of this, heat exchangers, heat emitters, or any other device that is intended to transfer heat between a fluid and a solid surface should operate with turbulent flow. Fortunately, the conditions present in most small- to medium-sized hydronic systems favor the development of turbulent flow.

An undesirable aspect of turbulent flow is that it creates more head loss compared to laminar flow, and therefore, requires more pumping power to maintain a given flow rate. However, the increased heat transfer capabilities of turbulent flow are usually more important than the penalty associated with the increased pumping power.

Reynold's Number

A method exists for predicting if the flow within a pipe will be laminar or turbulent. This method, derived from experimental observations is based on calculating the **Reynold's number** of the fluid. The calculated value is then compared to threshold values of 2,300 and 4,000. If the calculated Reynold's number is less than 2,300, the flow will be laminar. If the Reynold's number is 4,000 or higher, the flow will be turbulent. If the Reynold's number is between 2,300 and 4,000, the flow could be either laminar or turbulent. In this range, it is also possible for the flow to transition back and forth between laminar and turbulent.

The Reynold's number for flow in a pipe can be calculated using Equation 6.7:

Equation 6.7:

$$\mathrm{Re} = \frac{vdD}{\mu}$$

where,
v = average flow velocity of the fluid (ft/s)
d = internal diameter of pipe (ft)
D = fluid's density (lb/ft³)
μ = fluid's dynamic viscosity (lb/ft/s)

Example 6.5
Determine the Reynold's number of water at 140 °F flowing at 5.0 gpm through a 3/4-inch type M copper tube. Is this flow laminar or turbulent?

Solution:

To calculate the Reynold's number, the density and dynamic viscosity of the water must first be determined. Using the graph in Figure 4-4, the water's density is found to be 61.35 lb/ft³. Using the graph in Figure 4-9, the water's dynamic viscosity is determined as 0.00032 lb/ft/s. The density and dynamic viscosity of water can also be determined using the Fluid Properties Calculator module in the Hydronics Design studio software.

The inside diameter of a 3/4-inch type M copper tube is found in Figure 5-1 to be 0.811 inch. This must be converted to feet to match the stated units for Equation 6.7:

$$d = (0.811 \text{ in})\left(\frac{1 \text{ ft}}{12 \text{ in}}\right) = 0.06758 \text{ ft}$$

To get the average flow velocity corresponding to a flow rate of 5.0 gpm, use the appropriate equation for a 3/4-inch copper tube from Figure 6-2:

$$v = (0.62)f = (0.62)5 = 3.106 \text{ ft/s}$$

These values can now be substituted into Equation 6.7:

$$\mathrm{Re} = \frac{(3.106 \text{ ft/s})(0.06758 \text{ ft})(61.35 \text{ lb/ft}^3)}{0.00032 \text{ lb/ft/s}} = 40{,}242$$

Discussion:
This value is well above the upper threshold of 4,000, and therefore, the flow is turbulent. Note that when the quantities with the stated units are substituted into Equation 6.7, they cancel each other out. Thus, the Reynold's number is a unitless quantity. This must always be true and serves as a check that the proper units are being used in Equation 6.7.

6.3 Analyzing Fluid Flow in Smooth Pipes

This section introduces a method for predicting the head loss in hydronic systems assembled from "smooth" pipe such as copper tubing, PEX, PEX-AL-PEX, or other materials of comparable smoothness. Most residential and light commercial hydronic systems use such tubing. The smooth inside surface of such tubing creates less disturbance of the fluid stream and, therefore, reduces head loss relative to that created by steel or iron piping.

If major portions of a system will be piped with steel or iron piping, this method will somewhat underestimate head loss and should not be used without suitable corrections.

This method is based on a combination of established engineering models for fluid flow in pipes, as well as methods borrowed from electrical circuit analysis. It is based on the premise that each component in a piping assembly or circuit has an associated **hydraulic resistance** that defines its ability to dissipate head from the fluid at any given flow rate.

Similarity between Fluid Flow and Electricity

To most people, the concepts of electricity and fluid flow seem totally different. Electrical circuits involve concepts such as voltage, current, and resistance. Piping systems deal with flow rate, head, pressure, and viscous drag. What could these systems possibly have in common? The answer lies in the mathematical models that engineers have developed for describing the behavior of both types of systems.

Electrical Device Model

The classic model for describing the operating characteristics of an electrical device is **Ohm's law**. It is among the first concepts taught in a course on electrical circuits.

Mathematically, Ohm's law can be stated as follows:

Equation 6.8:

$$v_L = r_e(i)$$

where,
v_L = voltage drop across the device (volts)
r_e = electrical resistance of the device (ohms)
i = current flowing through the device (amperes)

Voltage can be thought of as the *driving force* that tries to push electrons through a device. It is measured with a voltmeter across the device as shown in Figure 6-13. Whenever a current flows through a resistance, there is an associated drop in voltage in the direction of current flow. This voltage drop can be calculated using Equation 6.8.

Current is a measurement of the rate of flow of the electrons through the device. Resistance is the ability of the device to oppose the flow of electrons.

Piping Device Model

For the following discussion, the term piping device refers to any component such as a pipe, fitting, or valve through which fluid flows. The relationship between head loss, flow rate, and hydraulic resistance for such a piping device can be described by Equation 6.9:

Equation 6.9:

$$H_L = r_h(f)^{1.75}$$

where,
H_L = head loss across a piping device due to viscous friction (feet of head)
r_h = hydraulic resistance of the piping device
f = flow rate of fluid through the piping device (gpm)
1.75 = exponent (or power) of f

The head loss can be thought of as the mechanical energy dissipated (in ft·lbs) to drive 1 lb of fluid through the device at a certain flow rate. Recall that head loss reveals itself as a pressure drop across the device. Flow rate, f, is a measurement of the rate of flow of the fluid through the device. Hydraulic resistance is the ability of the device to oppose this flow. These concepts are illustrated in Figure 6-14.

It is interesting to compare the mathematical equation for electrical devices (Ohm's law) with that for piping devices (Equation 6.9). Notice that there is a one-to-one correspondence between the physical quantities in these equations. Head loss across a piping device corresponds to voltage drop across an electrical device. Electrical current, the indicator of flow rate of electrons through an electrical device, corresponds to the flow rate through the piping device. Finally, electrical resistance corresponds to what we now call hydraulic resistance. See Figure 6-15.

Still, there is an important difference between these equations. In the electrical device equation, the voltage drop across the device is directly proportional to the current flowing through it. Thus, if the current through the device were doubled, the voltage drop across it would also double. In the case of the piping device

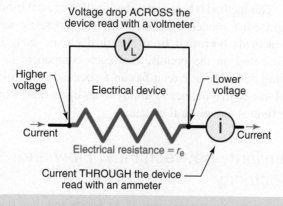

Figure 6-13 | The concept of voltage, current, and resistance for an electrical device.

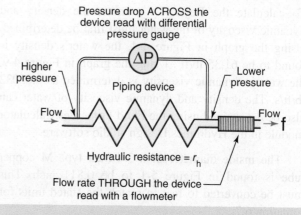

Figure 6-14 | Concept of pressure drop (associated with head loss), flow rate, and hydraulic resistance for a piping device.

$$H_L = r_h \times f^{1.75}$$
$$V_L = r_e \times i$$

H_L = Head loss (feet)
V_L = Voltage loss (volts)
r_h = Hydraulic resistance (ft/gpm$^{1.75}$)
r_e = Electrical resistance (ohms)
f = Flow rate (gpm)
i = Electrical current (amps)

Figure 6-15 Correspondence of mathematical terms between equations for piping devices and electrical devices.

model, the head loss across the device is *not* directly proportional to the flow rate through the device. This is because the flow rate is raised to the 1.75 power. If the flow rate through the piping device were doubled, the head loss across it increases by a factor of about 3.4. This difference makes the fluid device equation more mathematically complex, but still manageable. Figure 6-16 shows a comparison between representative graphs of these two equations.

Determining Hydraulic Resistance

Any component used in a hydronic piping circuit, be it a length of pipe, fitting, valve, heat emitter, or otherwise, can be represented by a hydraulic resistance. Devices that create relatively little interference with the fluid stream have low hydraulic resistances. Those that have major restrictions or twisting paths through which the fluid must flow will have higher hydraulic resistances. After the hydraulic resistances of all devices in a piping assembly are determined, they can be combined into an **equivalent resistance** similar to how electrical resistances are combined in circuit analysis.

The hydraulic resistance term, r_h, in Equation 6.9 can be expanded as shown in Equation 6.10:

Equation 6.10:
$$r_h = (\alpha c L)$$

Therefore,

Equation 6.11:
$$H_L = r_h(f)^{1.75} = (\alpha c L)(f)^{1.75}$$

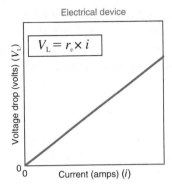

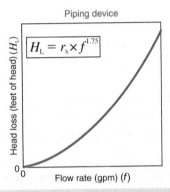

Figure 6-16 Graphical comparison between the equation describing the electrical device (Equation 6.8) and the equation describing the piping device (Equation 6.9).

where,
H_L = head loss across a device due to viscous friction (feet of head)
α = fluid properties factor based on the fluid's density and viscosity (Equation 6.12)
c = a constant based on the size of the tube hereafter called the pipe size coefficient (see table in Figure 6-18)
L = length of pipe or equivalent length of a piping device (ft)
f = flow rate of fluid through the component (gpm)

Equation 6.11 is a specialized relationship derived from the widely used Darcy-Weisbach equation. It incorporates an empirical relationship for the Moody friction factor based on turbulent flow in smooth piping. Because of this empirical factor, Equation 6.11 is only valid for turbulent flow with Reynold's numbers in the range of 4,000 to about 200,000. Fortunately, this covers most of the operating conditions found in residential and light commercial hydronic systems. At Reynold's numbers above 200,000, the equation gradually begins to underestimate head loss. At a Reynold's number of 300,000, it underestimates head loss by about 6%.

The Reynold's number for fluid flow in pipes can be found using Equation 6.7. It is prudent to verify that the Reynold's number is between 4,000 and 200,000 when using Equation 6.11 for system design.

Fluid Properties Factor (α)

The **fluid properties factor**, α (pronounced "alpha"), which is a factor in Equations 6.10 and 6.11, is defined as follows:

Equation 6.12:

$$\alpha = \left(\frac{D}{\mu}\right)^{-0.25}$$

where,

α = fluid properties factor
D = density of the fluid (lb/ft³)
μ = dynamic viscosity of the fluid (lb/ft/s)

The α value is a complete mathematical description of the fluid's physical properties for purposes of determining hydraulic resistance.

Chapter 4, Properties of Water, described how the density and dynamic viscosity of fluids vary with temperature. Because it is derived from these properties, the α value of any fluid is also dependent on temperature. Figure 6-17 is a plot of the α value for water and two concentrations of propylene glycol antifreeze over a temperature range of 50 to 250 °F. The Fluid Properties Calculator module in the Hydronics Design Studio software can also be used to quickly find density, viscosity, and the fluid properties factor (α) for several fluids commonly used in hydronic heating systems.

The c in Equations 6.10 and 6.11 is called the **pipe size coefficient**. It is a constant for a given tubing type and size. It incorporates dimensional information such as interior diameter, cross-sectional area, and appropriate unit conversion factors into a single number. A table of pipe size coefficients for several types and sizes of tubing is given in Figure 6-18.

Tube (size and type)	c value
3/8" type M copper	1.0164
1/2" type M copper	0.33352
3/4" type M copper	0.061957
1" type M copper	0.01776
1.25" type M copper	0.0068082
1.5" type M copper	0.0030667
2" type M copper	0.0008331
2.5" type M copper	0.0002977
3" type M copper	0.0001278
3/8" PEX	2.9336
1/2" PEX	0.786
5/8" PEX	0.2947
3/4" PEX	0.14203
1" PEX	0.04318
1.25" PEX	0.01668
1.5" PEX	0.007554
2" PEX	0.002104
3/8" PEX-AL-PEX	3.35418
1/2" PEX-AL-PEX	0.6162
5/8" PEX-AL-PEX	0.19506
3/4" PEX-AL-PEX	0.06379
1" PEX-AL-PEX	0.019718

Figure 6-18 Values of pipe size coefficient c for type M copper, PEX, and PEX-AL-PEX tubing for use in Equation 6.10 and Equation 6.11.

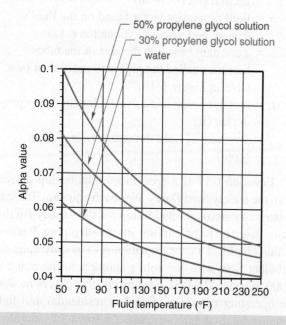

Figure 6-17 Graph of α-value of water and two solutions of propylene glycol antifreeze for temperatures of 50 to 250 °F.

Example 6.6

Using the hydraulic resistance method, determine the pressure drop created when water at 140 °F flows through 100 feet of 3/4-inch type M copper tube at a rate of 5.0 gpm. Assume that the tube is horizontal and thus that static pressure is constant along the entire tube.

Solution:

This situation requires the use of Equation 6.11. First, the value of α for water at 140 °F, as well as the pipe size coefficient (c value) for 3/4-inch copper tube, must be determined.

The value of α for water can be found using either Figure 6-17 or Equation 6.12. In this case, the latter will be used. The density of water at 140 °F is found from Figure 4-4 to be 61.35 lb/ft³. The dynamic viscosity of water at 140 °F is found from Figure 4-9 to be 0.00032 lb/ft/s. The α value can now be calculated using Equation 6.12:

$$\alpha = \left(\frac{D}{\mu}\right)^{-0.25} = \left(\frac{61.35}{0.00032}\right)^{-0.25} = (191719)^{-0.25} = 0.04779$$

The value of the pipe size coefficient c is found from the table of Figure 6-18. For 3/4-inch copper tube, $c = 0.061957$.

All data can now be substituted into Equation 6.11:

$$H_L = (\alpha c L)(f)^{1.75} = [(0.04779)(0.061957)(100)](5)^{1.75}$$
$$= 4.95 \text{ feet of head}$$

Thus far we have determined the head loss along the pipe, not the pressure drop. The final step is to convert the head loss into a corresponding pressure drop using Equation 6.6:

$$\Delta P = \frac{HD}{144} = \frac{(4.95)(61.35)}{144} = 2.11 \text{ psi}$$

Discussion:

Based on these calculations, an accurate pressure gauge at the outlet of the 100 feet of 3/4-inch copper tube should read 2.11 psi lower than the pressure at the tube's inlet.

Flow Coefficient (C_v)

Another method for predicting pressure drop due to head loss has been used within the valve industry for many years. It is based on a parameter called the **flow coefficient (C_v)** of the valve.

The C_v is defined as the flow rate of 60 °F water that will create a pressure drop of 1.0 psi through a component. For example, a valve with a C_v rating of 5.0 would require a flow rate of 5.0 gpm of 60 °F water to create a pressure drop of 1.0 psi across it. Many valve manufacturers list the C_v values of their products in their technical literature. Unless otherwise indicated, the listed C_v value for valves assumes that the valve is in a fully open position. Occasionally, C_v values will be listed for devices other than valves. The C_v value is used to describe the setting of balancing valves in the Hydronic Circuit Simulator module in the Hydronics Design Studio software.

The C_v value can be used in an equation that estimates pressure drop across a device based on the square of the flow rate of fluid through the device. The relationship is given as Equation 6.13.

Equation 6.13:

$$\Delta P = \left(\frac{D}{62.4}\right)\left(\frac{f}{C_v}\right)^2$$

where,
ΔP = pressure drop across the device (psi);
D = density of the fluid at its operating temperature (lb/ft³);
62.4 = density of water at 60 °F (lb/ft³);
f = flow rate of fluid through the device (gpm);
C_v = known C_v rating of the device.

Example 6.7

Estimate the pressure drop across a radiator valve having a C_v value of 2.8, when 140 °F water flows through at 4.0 gpm.

Solution:

The density of 140 °F water is 61.35 lb/ft³.

Using Equation 6.13, the pressure drop is:

$$\Delta P = \left(\frac{61.35}{62.4}\right)\left(\frac{4.0}{2.8}\right)^2 = 2.0 \text{ psi}$$

For smooth piping such as copper, PEX, or PEX-AL-PEX tubing, the relationship proportioning pressure drop to the square of flow rate will slightly overpredict pressure drop.

Use of Pressure Drop Charts

Many references on piping design contain charts such as the one shown in Figure 6-19. These charts show the relationship between flow rate, flow velocity, and pressure drop due to head loss for several sizes of copper tubing. They are usually based on water at a temperature of 60 °F, which is typical in domestic (cold) water distribution systems. The reference temperature of 60 °F is, however, not a typical operating condition in hydronic heating systems. The higher temperature water in such systems will have lower density and lower viscosity, resulting in less head loss, and thus smaller pressure drops.

Use of pressure drop charts based on 60 °F water will consistently *overestimate* the pressure drop in hydronic piping circuits because the water will usually be at higher temperatures. This can result in unnecessary oversizing of the system's circulator(s). *Because of this, these charts should not be used for estimating pressure drop due to head loss in hydronic heating systems.* However, they can still be used for finding the relationship between flow rate and flow velocity.

> **Discussion:**
>
> The pressure drop for the same piping using 140 °F water in example 6.6 was 2.11 psi. Comparing these results shows the pressure drop created by 60 °F water to be about 25% higher than that created by 140 °F water. This is a significant difference. The chart in Figure 6-19 indicates a pressure drop of about 2.7 psi (for water at 60 °F). This is comparable to the pressure drop found using Equation 6.11 for water at 60 °F.

Example 6.8
Use Equations 6.11 and 6.12 to recalculate the pressure drop for 100 feet of 3/4-inch copper-tubing transporting 60 °F water at 5.0 gpm. Compare the results with those of example (6.6) and with those obtained from the chart in Figure 6-19.

Solution:
Using the graph in Figure 4-4, the density of water at 60 °F is estimated to be 62.3 lb/ft³. Using the graph in Figure 4-9, the dynamic viscosity of water at 60 °F is estimated at 0.00075 lb/ft/s. The α value can be calculated using Equation 6.12:

$$\alpha = \left(\frac{D}{\mu}\right)^{-0.25} = \left(\frac{62.3}{0.00075}\right)^{-0.25} = (83067)^{-0.25} = 0.0589$$

The value of the pipe size coefficient c for 3/4-inch tubing is found in Figure 6-18: $c = 0.061957$.

Substituting these values into Equation 6.11 along with the pipe length and flow rate:

$$H_L = (\alpha c L)(f)^{1.75} = [(0.0589)(0.061957)(100)](5)^{1.75}$$
$$= 6.10 \text{ feet of head}$$

The corresponding pressure drop is found using Equation 6.6:

$$\Delta P = \frac{(6.10)(62.3)}{144} = 2.64 \text{ psi}$$

6.4 Hydraulic Resistance of Fittings, Valves, and Other Devices

Before the total hydraulic resistance of a piping circuit can be found, the individual hydraulic resistances of all fittings, valves, or other such components must be determined. One approach is to consider each fitting, valve, or other device as an **equivalent length** of copper tube of the same pipe size. *The equivalent length of a component is the amount of tubing, of the same pipe size, that would produce the same head loss (or pressure drop), as the actual component, at the same flow rate.* By replacing all components in the circuit with their equivalent length of tubing, the circuit can be treated as if it were a single tube having a length equal to the sum of the actual tube lengths, plus the total equivalent lengths of all fittings, valves, or other devices.

The piping circuit shown in Figure 6-20 contains a total of 58 feet of 3/4-inch type M copper tubing and eight 90-degree elbows. Assuming that the equivalent length of each elbow is 2 feet of 3/4-inch pipe, the entire circuit could be thought of as if it were $58 + 8(2) = 74$ feet of 3/4-inch pipe.

This method can be used for many different fittings and valves. Figure 6-21 gives the equivalent lengths of many commonly used fittings and valves.

The data in Figure 6-21 are representative of typical fittings and valves. The exact equivalent length of a specific fitting or valve will depend on its internal shape, surface roughness, and other factors. Take note that the numbers in the table are for soldered fittings. The equivalent lengths given in Figure 6-21 should be doubled for threaded fittings. The values in Figure 6-21 can be used with Equation 6.11 to determine the head loss through any type of hydronic piping circuit constructed with smooth piping.

6.4 Hydraulic Resistance of Fittings, Valves, and Other Devices 237

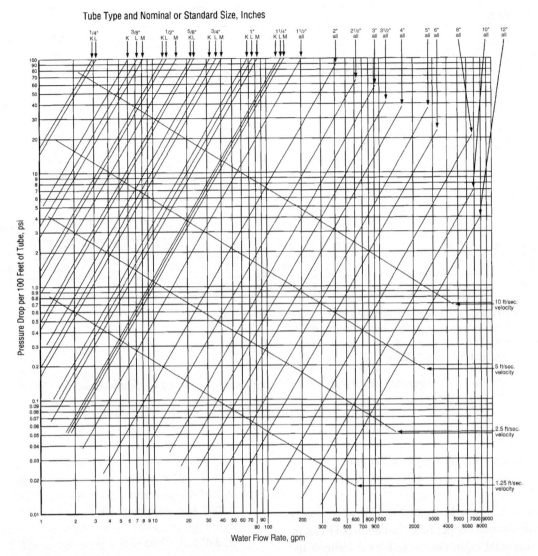

Figure 6-19 | Pressure drop chart for copper tubing carrying 60 °F water. *Courtesy of the Copper Development Association.*

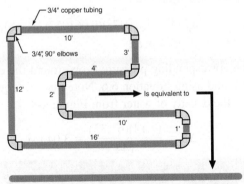

58' of tubing + 8 × (2 equivalent feet per elbow) = 74' total equivalent length

Figure 6-20 | Converting a piping circuit containing pipe and fittings to an equivalent length of straight pipe.

Example 6.9

Water at 140 °F and 6.0 gpm flows through the 3/4-inch hydronic circuit shown in Figure 6-22. Determine the head loss and associated pressure drop around the circuit.

Solution:

Start by finding the total equivalent length of the circuit by adding the equivalent lengths of the fittings and valves to the length of tubing as shown in Figure 6-23.

Chapter 6 Fluid Flow in Piping

Copper tube sizes									
Fitting or valve[1]	3/8"	1/2"	3/4"	1"	1.25"	1.5"	2"	2 1/2"	3"
90-degree elbow	0.5	1.0	2.0	2.5	3.0	4.0	5.5	7.0	9
45-degree elbow	0.35	0.5	0.75	1.0	1.2	1.5	2.0	2.5	3.5
Tee (straight run)	0.2	0.3	0.4	0.45	0.6	0.8	1.0	0.5	1.0
Tee (side port)	2.5	2.0	3.0	4.5	5.5	7.0	9.0	12.0	15
B&G Monoflo® tee[2]	n/a	n/a	70	23.5	25	23	23	n/a	n/a
Reducer coupling	0.2	0.4	0.5	0.6	0.8	1.0	1.3	1.0	1.5
Gate valve	0.35	0.2	0.25	0.3	0.4	0.5	0.7	1.0	1.5
Globe valve	8.5	15.0	20	25	36	46	56	104	130
Angle valve	1.8	3.1	4.7	5.3	7.8	9.4	12.5	23	29
Ball valve[3]	1.8	1.9	2.2	4.3	7.0	6.6	14	0.5	1.0
Swing-check valve	0.95	2.0	3.0	4.5	5.5	6.5	9.0	11	13.0
Flow-check valve[4]	n/a	n/a	83	54	74	57	177	85	98
Butterfly valve	n/a	1.1	2.0	2.7	2.0	2.7	4.5	10	15.5

1. Data for soldered fittings and valves. For threaded fittings double the listed value.
2. Derived from C_v values based on no flow through side port of tee.
3. Based on a standard-port ball valve. Full-port valves would have lower equivalent lengths.
4. Based on B&G brand "flow control" valves.

Figure 6-21 | Representative equivalent lengths of common fittings and valves (all values expressed as feet of copper tube of the same nominal size).

Notice that only tubing lengths within the flow path are included since flow does not occur in dead-end pipe branches.

The circuit can now be treated as if it were simply 104.8 feet of 3/4-inch copper tube. Equation 6.11 can be used with this total equivalent length to determine the head loss at 6.0 gpm.

The α value for water at 140 °F can be determined from either Figure 6-17 or Equation 6.12 to be 0.0478.

The value of the pipe size coefficient c for 3/4-inch copper tube is found in Figure 6-18: $c = 0.061957$.

3/4" Tubing	62 ft
6, 3/4" × 90 deg. elbows	6(2) = 12 ft
2, 3/4" side port tees	2(3) = 6 ft
1, 3/4" straight through tee	1(0.4) = 0.4 ft
1, 3/4" globe valve	1(20) = 20 ft
2, 3/4" ball valves	2(2.2) = 4.4 ft
	TOTAL = 104.8 ft

Figure 6-23 | Adding the equivalent lengths of the fittings and valves in Figure 6-22.

This data can now be substituted into Equation 6.11 to determine the head loss:

$$H_L = (\alpha c L)(f)^{1.75} = [(0.0478)(0.061957)(104.8)](6)^{1.75}$$
$$= 7.14 \text{ feet of head}$$

The corresponding pressure drop around the circuit can be found using Equation 6.6. Note that this equation requires the density of water from Figure 4-4.

$$\Delta P = \frac{(7.14)(61.35)}{144} = 3.04 \text{ psi}$$

Discussion:

Compared to example 6.6, the only additional work required in this example was looking up and adding together the equivalent lengths of the fittings and valves.

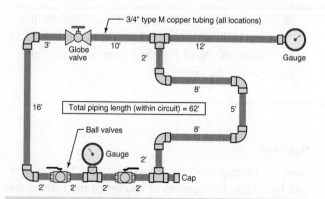

Figure 6-22 | Piping assembly for use in example 6.7.

6.5 The System Head Loss Curve

An examination of Equation 6.11 shows that for a given piping circuit, operating with a given fluid, at a constant temperature, the hydraulic resistance term (αcL) remains constant:

Equation 6.14:

$$H_L = (\alpha cL)(f)^{1.75} = (\text{number})(f)^{1.75}$$

Under these conditions, the head loss around the piping circuit depends only on flow rate. Equation 6.14 is a mathematical function that can be graphed by selecting several flow rates, calculating the head loss at each, and plotting the resulting points. Once the points are plotted, a smooth curve could be drawn through them.

Example 6.10
Using the piping circuit and data from example 6.9, plot several points representing different flow rates and the associated head loss. Draw a smooth curve through these points.

Solution:
Start by substituting the appropriate data into Equation 6.11:

$$H_L = (\alpha cL)(f)^{1.75} = [(0.0478)(0.061957)(104.8)](f)^{1.75}$$

Multiply the values of α, c, and L together to obtain a single number. This value, 0.31, is the hydraulic resistance of the piping circuit shown in Figure 6-22:

Equation 6.15:

$$H_L = (0.31)(f)^{1.75}$$

Use this equation to generate data for plotting. Simply select a few random flow rates, and substitute each into the equation to find the corresponding head loss. The values used are shown in Figure 6-24.

This data can now be plotted and a smooth curve drawn through the points as shown in Figure 6-25.

This graph is called a **system head loss curve**, or sometimes just "system curve." It represents the relationship between flow rate and head loss for a given piping circuit using a specific fluid at a given temperature. *All piping circuits have a unique system head loss curve.*

Flow rate (gpm)	Head loss (feet of head)
0.0	0
3.0	2.12
6.0	7.14
9.0	14.5
12.0	24.0
15.0	35.5

Figure 6-24 | Values of head loss calculated at several flow rates using Equation 6.15.

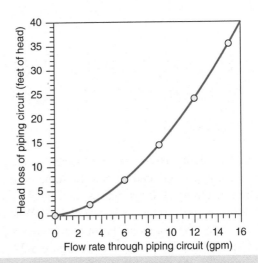

Figure 6-25 | Example of a system head loss curve plotted using data from Figure 6-24.

It could even be thought of as the analytical "fingerprint" of that circuit. Determining the system head loss curve of a piping circuit is an essential step in properly selecting a circulator for that circuit.

If additional piping, fittings, or other components were added to the piping circuit, its total equivalent length would increase, as would its hydraulic resistance. This would steepen the system head loss curve.

Figure 6-26 shows several system head loss curves plotted on the same graph. These curves were produced assuming that the total equivalent length of the circuit described in example 6.8 varied from a low of 50 feet to a high of 500 feet. Such changes in length would change the hydraulic resistance, and therefore the "steepness" of the system head loss curve. The hydraulic resistance used in Equation 6.11 to plot each curve is indicated on the graph. Note that the system head loss curve gets steeper as the hydraulic resistance of the circuit increases.

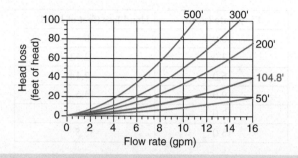

Figure 6-26 | Multiple system head loss curves for pipe circuits of different equivalent lengths. Curves get "steeper" as equivalent lengths increase. Steeper curves represent high values of hydraulic resistance.

6.6 Piping Components Represented as Series Resistors

The simplest piping assemblies are formed by connecting piping components end to end. In some cases, the collection of tubing, fittings, valves, and other components closes on itself to form a **series circuit**. When this is not the case, the components form a **series piping path**. Examples of each are shown in Figure 6-27.

To visualize the overall hydraulic resistance created by various components of a piping system, each pipe, fitting, valve, or other component can be thought of as a hydraulic resistor, and represented with a symbol like that used in electrical schematics. An example of how a series piping path consisting of several components would be represented by hydraulic resistors is shown in Figure 6-28.

As with electrical resistor diagrams, the size of a particular resistor symbol does not indicate the amount of hydraulic resistance it represents. Only numerical values can do this. The orientation of the resistor symbol or the shape of the series piping path also makes no difference. However, the sequence of hydraulic resistors forming the series piping path or circuit should always match the actual piping component layout they represent. The lines connecting hydraulic resistors are assumed to have zero hydraulic resistance.

A similar representation of piping components using hydraulic resistor symbols can be made for a series circuit, as shown in Figure 6-29.

Equivalent Resistance of Series-Connected Components

The analogy of hydraulic resistance and electrical resistance also applies to how complex resistor diagrams can be reduced to simpler equivalent diagrams. In either system, any number of series-connected resistors can be replaced by a single equivalent resistor that has a resistance equal to the mathematical sum of all the individual resistances. Equation 6.16 expresses this concept in mathematical form.

Equation 6.16:

$$r_{e(series)} = \sum_{i=1}^{n} r_i = (r_1 + r_2 + r_3 + r_4 + \cdots + r_n)$$

Where:
$r_{e(series)}$ = equivalent hydraulic resistance
n = number of hydraulic resistances in series

This concept allows hydraulic resistance diagrams such as the one in Figure 6-29 to be greatly simplified. An equivalent diagram can be drawn that combines the

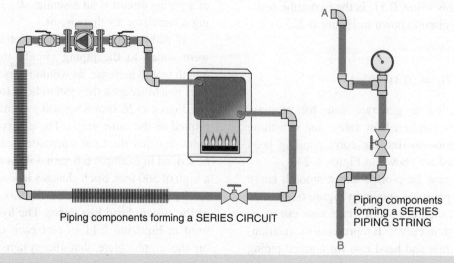

Figure 6-27 | Piping components forming (a) a series circuit and (b) a series piping path.

6.6 Piping Components Represented as Series Resistors

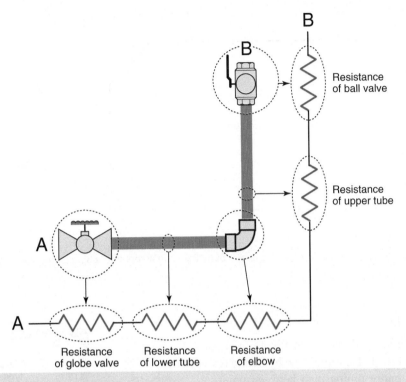

Figure 6-28 | Piping components represented as hydraulic resistors.

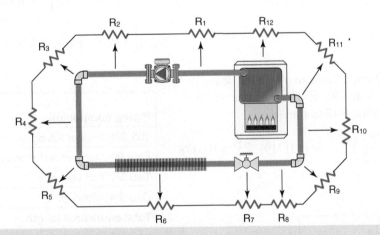

Figure 6-29 | Series circuit of piping components represented as series string of hydraulic resistors.

individual hydraulic resistance of every pipe, fitting, valve, and other component into a single equivalent hydraulic resistance as shown in Figure 6-30. The simplified diagram represents the total hydraulic resistance of all the original components, but in a much more compact form.

In electrical circuits, the resistance of each resistor is known, and the values are simply added together. In piping circuits, however, the hydraulic resistance of each component has to be calculated using Equation 6.10, before the values can be added. Although it is certainly possible to do this, it can become time consuming for complex systems.

A more efficient method can be used in the common situation where the piping circuit contains a *single size of pipe, fittings, and valves*. This method recognizes that the fluid properties factor, α, and the pipe size coefficient, c, will be the same for all components, and thus need only be determined once. The only remaining value required for each component is its equivalent length, which can be found in Figure 6-21. Finding the equivalent resistance of the circuit, therefore, becomes a matter of looking up the equivalent lengths of the fittings and valves, combining these with the total piping length, and multiplying this total equivalent length by the common values of α and c as shown in Figure 6-31.

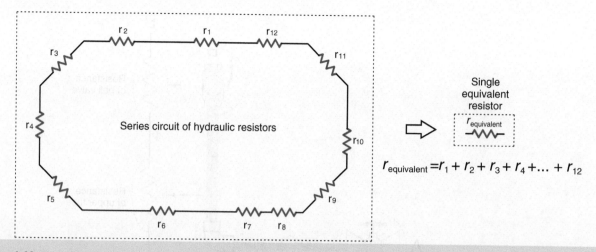

Figure 6-30 | Concept of reducing a series circuit of hydraulic resistors into a single equivalent hydraulic resistance.

Example 6.11

The series piping path shown in Figure 6-32 carries water at 140 °F. Determine the hydraulic resistance of the entire piping path between points A and B (Figure 6-33). Use this hydraulic resistance to sketch a system head loss curve.

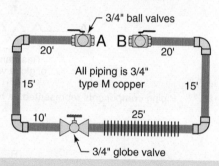

Figure 6-32 | Piping system of example 6.11.

Solution:

Figure 4-4 and Figure 4-9 can be used to obtain the density and viscosity of water at 140 °F. These values are then substituted into Equation 6.12 to find α:

$$\alpha = \left(\frac{D}{\mu}\right)^{-0.25} = \left(\frac{61.35}{0.00032}\right)^{-0.25} = (191719)^{-0.25} = 0.0478$$

Piping component	Equivalent length
105' 3/4" copper tubing =	105'
Four 90 degree x 3/4" elbows =	4(2) = 8'
Two 3/4" ball valves =	2(2.2) = 4.4'
One 3/4" globe valve =	1(20) = 20'
Total equivalent length =	**L = 137.4'**

Figure 6-33 | Adding the equivalent lengths of piping, fittings, and valves in the system shown in Figure 6-32.

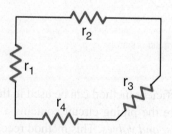

$r_{e(series)} = r_1 + r_2 + r_3 + r_4$
$r_{e(series)} = \alpha c L_{e1} + \alpha c L_{e2} + \alpha c L_{e3} + \alpha c L_{e4}$
$r_{e(series)} = \alpha c (L_{e1} + L_{e2} + L_{e3} + L_{e4})$
$r_{e(series)} = \alpha c (\text{total equivalent length})$

Figure 6-31 | Simplified method of adding hydraulic resistances *when the piping components are all the same pipe size.* In such cases, the value of α and c are the same for all components.

The pipe size coefficient (c) is found in Figure 6-18. For 3/4-inch copper tube, $c = 0.061957$.

The only remaining task is to look up the equivalent lengths of all fittings and valves and add these lengths to that of the piping. The equivalent lengths of fittings and valves are found in Figure 6-21.

The values of α, c, and L can now be substituted directly into Equation 6.10 to find the total hydraulic resistance of the piping string:

$$r = (\alpha c L) = [(0.0478)(0.061957)(137.4)] = 0.407$$

This total hydraulic resistance can now be substituted into Equation 6.11 and usetd to generate a few points for plotting the system head loss curve as shown in Figure 6-34. The resulting data are plotted and a smooth curve is drawn through the points to represent the system head loss curve as shown in Figure 6-35.

Piping Paths Containing Multiple Pipe Sizes

The piping paths shown in previous examples were relatively simple. They contained only a few fittings and valves, and all such components were of the same pipe size. A more typical hydronic piping loop can contain dozens of fittings and valves, some of which may be of different pipe sizes. The total hydraulic resistance of such circuits can still be determined using the basic principles. However, when more than one pipe size is involved, the hydraulic resistance of all piping, fittings, valves, or other components, of a given pipe size, should be determined separately, and then added together. This is because each pipe size has a different c value. These calculations can be organized as shown in Equation 6.17.

Equation 6.17:

$$r_{total} = \alpha[c_{size\ 1}(L_{pipe} + L_{components})_{size\ 1} + c_{size\ 2}(L_{pipe} + L_{components})_{size\ 2} + \ldots + c_{last\ size}(L_{pipe} + L_{components})_{last\ size}]$$

where,

r_{total} = total hydraulic resistance;
α = fluid properties factor;
C_{size} = pipe size coefficient of a given pipe size (Figure 6-18);
L_{pipe} = total length of pipe of a given pipe size (ft);
$L_{component}$ = total equivalent length of components of a given pipe size (ft).

6.7 Parallel Hydraulic Resistances

Piping systems often contain two or more piping paths connected at common points. Examples include multiple-zone distribution circuits that begin and end at the same boiler as shown in Figure 6-36, or radiant floor heating circuits that begin at a common supply manifold

$H_L = 0.407(f)^{1.75}$	
Flow rate (f), (gpm)	Head loss H_L, (feet of head)
0.0	0
5.0	6.8
10.0	22.9
15.0	46.5

Figure 6-34 | Using the equation of the system head loss curve to generate head loss versus flow rate data for the piping system of Figure 6-32.

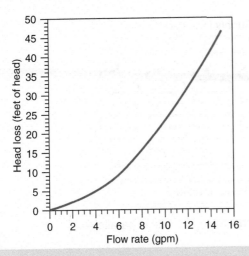

Figure 6-35 | The system head loss curve for example 6.11.

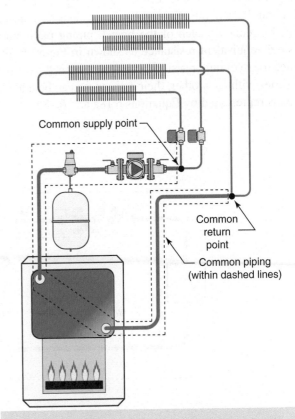

Figure 6-36 | Example of parallel piping paths formed by zone circuits.

and end at a common return manifold, as shown in Figure 6-37. The piping paths that share common points of connection are said to be **piped in parallel**. In such arrangements, the total system flow must divide up among the two or more paths at the common supply point, and then recombine at a common return point.

The analogy of electrical resistance again proves useful for analyzing how the flow divides among two or more parallel piping paths. Figure 6-38 shows a parallel piping assembly along with its associated hydraulic resistor diagram.

In the resistor diagram of Figure 6-38, the combined resistance of the two elbows in the upper piping branch is represented by a single resistor symbol, r_2. This is done as a matter of convenience because it reduces the number of resistor symbols that have to be drawn. Because this resistor now represents both elbows, its numerical value would be doubled. The three pipe segments in the upper branch are also represented by a single resistor symbol, r_3, as are the three in the lower branch, r_8. This simplifying method can be used whenever two or more identical components are configured in series. With experience, it is easy to draw simplified resistor diagrams by combining such individual resistances together.

The equivalent resistance of the entire piping assembly is a single resistor connected between points A and B. It represents the combined effect of all piping components between these points. To find this equivalent resistance, start by reducing each of the branch piping paths into a single equivalent resistance as shown in Figure 6-39. Since the hydraulic resistors within each branch path are in series with each other, their resistances can be added. This is represented by Equations 6.18a and 6.18b:

Equation 6.18a:

$$r_{e1} = \alpha c (L_1 + L_2 + L_3 + L_4)$$

Equation 6.18b:

$$r_{e2} = \alpha c (L_5 + L_6 + L_7 + L_8 + L_9)$$

Once this step is completed, reconnect the two equivalent resistances at points A and B as shown in Figure 6-40.

The arrangement will now be reduced to a single equivalent resistance as shown in Figure 6-41.

The equivalent resistance of two parallel hydraulic resistors is found using Equation 6.19.

Equation 6.19:

$$R_{equivalent_{parallel}} = \left[\left(\frac{1}{r_{e1}} \right)^{0.5714} + \left(\frac{1}{r_{e2}} \right)^{0.5714} \right]^{1.75}$$

The values of r_{e1} and r_{e2} in Equation 6.19 are the hydraulic resistances of the two parallel resistors being combined into a single equivalent resistance. This equivalent resistance then represents the entire piping assembly (between points A and B) in Figure 6-38. If this assembly were to be combined with other components to form a more complex piping system, it could still be represented by this single resistor within the resistor diagram of the overall system.

In some piping systems, there are more than two parallel piping strings. An example would be the distribution piping of a radiant floor heating system in which several piping paths originate from a common supply manifold and terminate in a common return manifold. An example is shown in Figure 6-42.

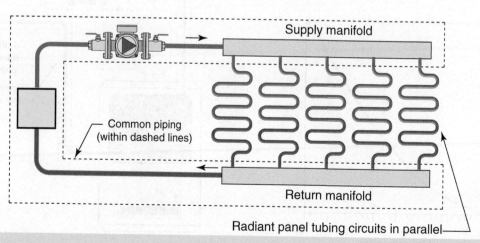

Figure 6-37 | Example of parallel piping paths formed by a manifold station.

6.7 Parallel Hydraulic Resistances

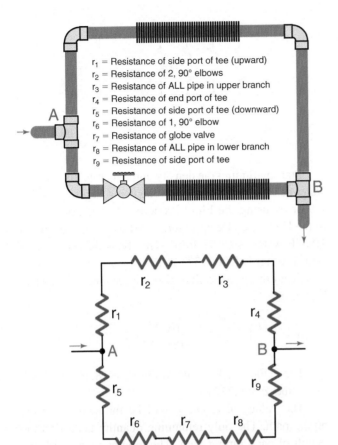

Figure 6-38 | Parallel piping assembly with associated hydraulic resistor diagram.

To find the single equivalent hydraulic resistance of this configuration, Equation 6.19 may be extended by one term as follows:

Equation 6.20:

$$R_{\text{equivalent}_{\text{parallel}}} = \left[\left(\frac{1}{r_1}\right)^{0.5714} + \left(\frac{1}{r_2}\right)^{0.5714} + \left(\frac{1}{r_3}\right)^{0.5714}\right]^{-1.75}$$

This same principle can be extended to any number of parallel resistors connected in parallel. Simply add more terms to Equation 6.19 to accommodate all parallel resistances. This can be expressed mathematically for a system with n parallel hydraulic resistances as in Equation 6.21:

Equation 6.21:

$$R_{\text{equivalent}_{\text{parallel}}} = \left[\left(\frac{1}{r_1}\right)^{0.5714} + \left(\frac{1}{r_2}\right)^{0.5714} + \left(\frac{1}{r_3}\right)^{0.5714} + \cdots + \left(\frac{1}{r_n}\right)^{0.5714}\right]^{-1.75}$$

Regardless of the number of parallel hydraulic resistances one begins with, the equivalent hydraulic resistor will always be a single resistor connected between the common points. This single resistor represents the combined effect of all original resistors.

Once the equivalent resistance of a group of parallel hydraulic resistors has been found, the flow rate through each parallel piping path can also be determined using Equation 6.22.

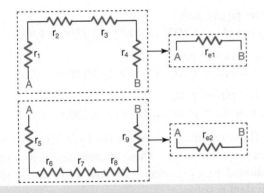

Figure 6-39 | Reduce each series group of resistors into a single equivalent hydraulic resistance.

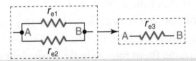

Figure 6-41 | Reducing two parallel hydraulic resistances into a single equivalent hydraulic resistance.

Figure 6-40 | A group of two parallel hydraulic resistors.

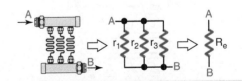

Figure 6-42 | Three parallel piping paths represented by hydraulic resistances, which are then combined into a single equivalent hydraulic resistance.

Equation 6.22:

$$f_i = f_{total}\left(\frac{R_e}{r_i}\right)^{0.5714}$$

where,
- f_i = flow rate through parallel path i (gpm)
- f_{total} = total flow rate entering the common point of the parallel paths (gpm)
- R_e = equivalent hydraulic resistance of all the parallel piping paths;
- r_i = hydraulic resistance of parallel piping path i.

Example 6.12

Using the concept of parallel hydraulic resistors, find the equivalent resistance of the radiant floor distribution system shown in Figure 6-43. All four piping paths consist of 1/2-inch PEX tubing. The average water temperature in the system is 100 °F. Assuming water enters the supply manifold at 6 gpm, determine the flow rate in each of the branch piping paths.

To simplify the example, assume that there are no valves in the manifolds, and the hydraulic resistances at the points of connection of the piping paths to the manifolds are insignificant. Also, assume that the hydraulic resistance along the manifold's length is insignificant.

Solution:

The resistor diagram shown in Figure 6-44 represents the assembly as four hydraulic resistors in parallel. After finding the values of these resistors, we can use Equation 6.21 to combine them into a single equivalent resistance.

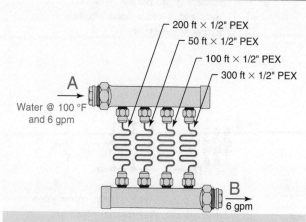

Figure 6-43 | Radiant floor distribution system for example 6.12.

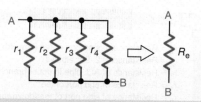

Figure 6-44 | Hydraulic resistor diagram for Figure 6-43.

Start by finding the density and dynamic viscosity of water at 100 °F. This can be done using Figures 4-4 and 4-9 or by using the Fluid Properties Calculator module in the Hydronics Design Studio software. The density of 100 °F water is 61.97 lb/ft³. The dynamic viscosity of 100 °F water is 0.0004573 lb/ft/s.

From these, the value of α can be calculated using Equation 6.12:

$$\alpha = \left(\frac{D}{\mu}\right)^{-0.25} = \left(\frac{61.97}{0.0004573}\right)^{-0.25} = 0.05212$$

The value of c is found in Figure 6-18. For 1/2-inch PEX tube, $c = 0.71213$.

The values of α and c will be the same for each piping path. The only remaining quantity needed is the length of each piping path. The lengths along with the values for α and c are now substituted into Equation 6.10:

200-feet piping path:
$$r_1 = \alpha c L_1 = (0.05212)(0.71213)(200) = 7.423,$$

50-feet piping path:
$$r_2 = \alpha c L_2 = (0.05212)(0.71213)(50) = 1.86,$$

100-feet piping path:
$$r_3 = \alpha c L_3 = (0.05212)(0.71213)(100) = 3.71,$$

300-feet piping path:
$$r_4 = \alpha c L_4 = (0.05212)(0.71213)(300) = 11.13.$$

These resulting values are the hydraulic resistances of each piping path. These resistances can now be substituted into Equation 6.21 and reduced to a single equivalent resistance:

$$R_{equivalent} = \left[\left(\frac{1}{7.423}\right)^{0.5714} + \left(\frac{1}{1.86}\right)^{0.5714} + \left(\frac{1}{3.71}\right)^{0.5714} + \left(\frac{1}{11.13}\right)^{0.5714}\right]^{-1.75}$$

$$R_{equivalent} = [(0.1347)^{0.5714} + (0.5376)^{0.5714} + (0.2695)^{0.5714} + (0.0898)^{0.5714}]^{-1.75}$$

$$R_{equivalent} = [0.318 + 0.7014 + 0.4727 + 0.2523]^{-1.75}$$

$$R_{equivalent} = [1.744]^{-1.75}$$

$$R_{equivalent} = 0.378$$

The flow rates in each piping path can now be determined by repeated use of Equation 6.22:

$$f_1 = 6\left(\frac{0.378}{7.423}\right)^{0.5714} = 1.09 \text{ gpm}$$

$$f_1 = 6\left(\frac{0.378}{1.86}\right)^{0.5714} = 2.41 \text{ gpm}$$

$$f_1 = 6\left(\frac{0.378}{3.71}\right)^{0.5714} = 1.63 \text{ gpm}$$

$$f_1 = 6\left(\frac{0.378}{11.13}\right)^{0.5714} = 0.868 \text{ gpm}$$

Discussion:

It is worth noting that the equivalent hydraulic resistance of the entire assembly (0.378) is smaller than the smallest individual hydraulic resistance (1.86). *This will always be true. It provides a way of partially checking the results of the equivalent length calculations.*

Another way of checking the results is to add the individual branch flow rates and verify that the total equals the flow rate entering the common point of the parallel paths:

$$f_{total} = f_1 + f_2 + f_3 + f_4 = 1.09 + 2.41 + 1.63 + 0.868 = 5.998 \approx 6.0 \text{ gpm}$$

Note that the very slight difference between the calculated total flow rate (5.998 gpm) and the original flow rate (6.0 gpm) is due to rounding off the calculations.

This example also shows that the larger the hydraulic resistance of a piping path, the lower its flow rate. In this example, the longest (300 feet) piping path has the lowest flow rate (0.868 gpm), while the shortest (50 feet) piping path has the highest flow rate (2.41 gpm).

6.8 Reducing Complex Piping Systems

The concept of using series and parallel resistors to represent piping assemblies is a powerful tool when properly applied. The underlying concept is to use series and parallel equivalent resistor replacements to systematically "fold" the complex piping system down to a single equivalent resistance. A system head loss curve can then be produced that will assist in choosing a circulator for the system. The single equivalent resistance can then be "unfolded" back to the individual parallel path resistances to find the flow rates in each path.

The best way to illustrate this concept is to go through an example using a typical hydronic distribution system.

Example 6.13
Use the concepts of series and parallel resistances to reduce the system shown in Figure 6-45 to a single equivalent resistance between points A and B. Find the flow rate in each branch circuit assuming the total system flow rate through the circulator is 10.0 gpm.

Solution:

Start by sketching a resistor diagram of the system between points A and B. Remember that the hydraulic resistance of two or more identical components can be lumped together and represented by a single resistor symbol. The hydraulic resistor diagram shown in Figure 6-46 uses one resistor symbol to represent the combined resistance of all the tubing segments within a given branch and another to represent the total resistance of all elbows in the branch. The third resistor symbol in each branch represents the resistance of the zone valve.

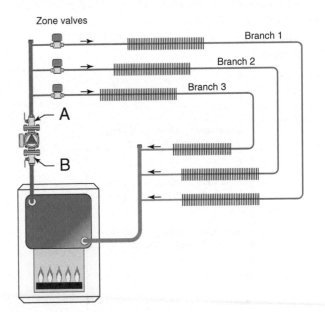

System Description:

Three zone system using zone valves and the following fittings:

Branch 1: 120 ft of 3/4" copper tube and 20, 90° elbows
Branch 2: 78 ft of 3/4" copper tube and 17, 90° elbows
Branch 3: 165 ft of 3/4" copper tube and 26, 90° elbows

Each zone valve is assumed to be equivalent to 30 ft of 3/4" copper tube.

The boiler and common header piping is assumed to be equivalent to 20 ft of 3/4" copper tube.

The system operates at an average water temperature of 160 °F.

Figure 6-45 Piping diagram to be analyzed in example 6.13.

The objective is to reduce the resistor diagram of Figure 6-46 to a single equivalent hydraulic resistance that represents the entire piping system. The first step is to calculate the hydraulic resistances of the tubing, fittings, and zone valves for each parallel piping path. Since all the tubings, fittings, and valves are of the same pipe size, this becomes a matter of totaling the equivalent length of all components within a branch and then multiplying by α and c.

Branch #1: 120 feet tubing + 20(2 feet) elbows + 30 feet zone valve equivalent length = 190 feet

Branch #2: 78 feet tubing + 17(2 feet) elbows + 30 feet zone valve equivalent length = 142 feet

Branch #3: 165 feet tubing + 26(2 feet) elbows + 30 feet zone valve equivalent length = 247 feet

The density of water at 160 °F is determined as 61.02 lb/ft³. The dynamic viscosity of water at 160 °F is determined as 0.00027 lb/ft/s. These values are then used to find the α value using Equation 6.12:

$$\alpha = \left(\frac{D}{\mu}\right)^{-0.25} = \left(\frac{61.02}{0.00027}\right)^{-0.25} = 0.04586$$

The value of c is found in Figure 6-18. For 3/4-inch smooth tube, $c = 0.061957$.

The hydraulic resistances of each branch can now be calculated:

Branch #1:
$R_1 = \alpha c L_1 = (0.04586)(0.061957)(190) = 0.5399$.

Branch #2:
$R_2 = \alpha c L_2 = (0.04586)(0.061957)(142) = 0.4035$.

Branch #3:
$R_3 = \alpha c L_3 = (0.04586)(0.061957)(247) = 0.7018$.

The resistor diagram can now be reduced as shown in Figure 6-47.

The three parallel resistors can now be reduced to a single equivalent resistance using Equation 6.21.

$$R_{equivalent} = \left[\left(\frac{1}{0.5399}\right)^{0.5714} + \left(\frac{1}{0.4035}\right)^{0.5714} + \left(\frac{1}{0.7018}\right)^{0.5714}\right]^{-1.75} = 0.07706$$

The resistor diagram can now be further reduced as shown in Figure 6-48.

The boiler is assumed to have an equivalent length equal to 20 feet of 3/4-inch copper tube. Its hydraulic resistance can therefore be calculated as:

$$r_{boiler} = (\alpha c L) = [(0.04586)(0.061957)(20)] = 0.05683$$

The hydraulic resistance of the boiler is in series with the equivalent resistance of the parallel piping paths. Therefore, the boiler's resistance will be added to the equivalent resistance of the parallel paths to get the overall equivalent resistance of the circuit as depicted in Figure 6-49.

$$r = 0.05683 + 0.07706 + 0.1339$$

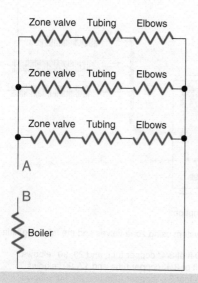

Figure 6-46 | Resistor diagram for three-zone system of example 6.13. Note that one resistor is shown to represent all the elbows in a branch, and one resistor represents all tubing segments in a branch.

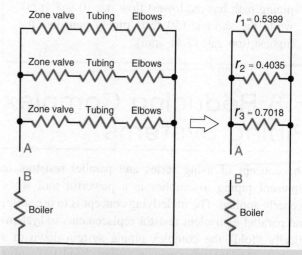

Figure 6-47 | First reduction of resistor diagram for example 6.13.

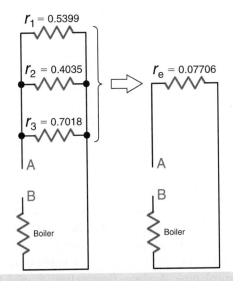

Figure 6-48 | Second reduction of resistor diagram for example 6.13.

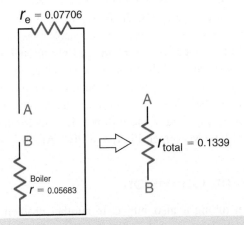

Figure 6-49 | Final simplification of resistor diagram into a single hydraulic resistor that represents the entire system between points A and B.

Assuming the circulator produces a flow rate of 10.0 gpm and all the zone valves are open, we can now find the flow rate in each parallel branch by repeated use of Equation 6.22:

$$f_1 = 10\left(\frac{0.07706}{0.5399}\right)^{0.5714} = 3.29 \text{ gpm}$$

$$f_2 = 10\left(\frac{0.07706}{0.4035}\right)^{0.5714} = 3.88 \text{ gpm}$$

$$f_3 = 10\left(\frac{0.07706}{0.7018}\right)^{0.5714} = 2.83 \text{ gpm}$$

As a check, add the individual branch flows to see if their total equals the system flow rate:

$$f_{\text{total}} = f_1 + f_2 + f_3 = 3.29 + 3.88 + 2.83 = 10.0 \text{ gpm}$$

Discussion:

A point worth noting is that the flow rate in any one circuit could be found assuming that one or both of the other circuits were closed off at the zone valves. To do this, go back to Equation 6.21 and treat the hydraulic resistance of the closed branch (or branches) as infinite. This has the same effect as simply ignoring the closed zones as parallel resistors. For example, assuming the zone valve on branch #1 was closed, one would find the equivalent resistance of parallel resistors R_2 and R_3 only and ignore the presence of R_1. The remainder of the procedure would be similar to that shown previously, knowing the branch flow rates will eventually help in properly sizing the heat emitters in each branch path.

6.9 Software-Based Circuit Analysis

The previous examples have demonstrated that determining the flow rates in systems with several parallel branches requires a considerable amount of "number crunching." Piping systems even more complex than the one represented in Figure 6-45 are often required. Although the methodology demonstrated can be used to determine flow rates in such complex systems, the necessary calculation could require hours of effort, even for those familiar with the calculation procedures. Such calculations only provide a "snapshot" of the system's performance under a given set of conditions. Any changes, such as the use of a different branch pipe size or the addition of another heat emitter, will require most of the calculations to be redone. Few people will take the time to thoroughly analyze multiple options when faced with such a task.

The only practical approach in such situations is computer analysis. To that end, the Hydronic Circuit Simulator module has been developed. A screen image from the program is shown in Figure 6-50.

The Hydronic Circuit Simulator uses a graphical user interface to configure a piping system containing up to 12 parallel branches as shown in Figure 6-51. The user can define the piping components in each branch, as well as the piping between the branches. The user can

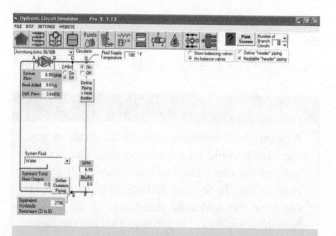

Figure 6-50 User interface screen for the Hydronic Circuit Simulator module in the Hydronics Design Studio software. *Source: Hydronicpros.*

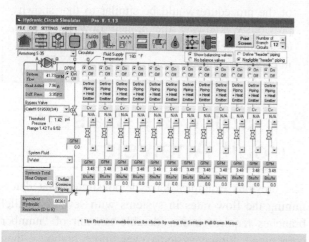

Figure 6-51 User interface screen for the Hydronic Circuit Simulator module in the Hydronics Design Studio software. *Source: Hydronicpros.*

also select from several system fluids and many available circulators.

The Hydronic Circuit Simulator uses the same analytical procedures described in this chapter to calculate the flow rates in the parallel branches. To analyze a series circuit, simply set the number of branches to one.

Each time the user changes one of the piping elements, the software recalculates the hydraulic resistance of the system and determines the flow rates in all branches based on the selected circulator. Methods for the latter are discussed in Chapter 7, Hydronic Circulators. The user can also set the hydraulic resistance of balancing valves in each of the branches to determine how the flow rates readjust. Individual branches can also be turned on and off to simulate the effect of zone valves.

6.10 Pipe Sizing Considerations

Several considerations affect the selection of a pipe size for a given application. Among these are head loss, flow velocity, potential for erosion damage, operating noise, installation, and operating cost.

Flow Velocity

When selecting a pipe size for a given flow rate, the resulting average flow velocity should be between 2 and 4 ft/s.

The lower end of this velocity range is based on the ability of flowing water to move air bubbles downward in a vertical pipe. Average flow velocities of 2 ft/s. or higher can entrain air bubbles in downward flow. The objective is to route the bubbles to an air separator where they can be collected and discharged from the system.

The upper end of this range (4 ft/s) is based on minimizing noise generated by the flow. Average flow velocities higher than 4 ft/s can cause objectionable flow noise and should be avoided for any piping passing through or near occupied space. Slightly higher flow velocities are possible in locations where flow velocity sound is not a problem.

Figure 6-2 lists the minimum and maximum flow rates corresponding to these flow velocities for common sizes of type M copper, PEX, and PEX-AL-PEX tubing.

Erosion Corrosion

Copper tubing is also subject to another problem when flow velocities exceed 5 feet per second. A condition called **erosion corrosion** can literally scrub metal off the inside wall of a tube or fittings. This effect tends to be localized at tight turns in fittings such as elbows and tees and has not been known to be a widespread problem except in extreme cases. Still, it is prudent to avoid any potential of its occurrence by sizing pipe for lower flow velocities. PEX and PEX-AL-PEX have the potential to operate at higher flow velocities without experiencing erosion corrosion, but such operating conditions are well beyond those used in hydronic systems.

Operating Cost of a Piping System

As flow through a pipe increases, the head loss created by that flow increases rapidly. Equation 6.9 can be used to show that when the flow rate through a smooth tube is doubled, the head loss increases by a factor of about 3.4. The greater the head loss, the more pumping power is required to maintain a given flow. This in turn can require a larger circulator with a greater electrical power demand and higher operating cost. Because of this, *every piping system should be thought of as having an operating cost as well as an installation cost.*

The determination of an optimal pipe size based on minimizing total owning and operating cost can be a complex process. It involves the use of specific installation costs, utility rates, circulator efficiency, estimates of hours of operation, rate of inflation of electrical rates, and more. However, since other considerations such as flow velocity often narrow down the choice to two sizes, it is prudent to estimate their operating cost and consider the results when making the final selection.

Equation 6.23 can be used to estimate the theoretical annual operating cost of any piping system transporting a fluid at a certain flow rate and head loss:

Equation 6.23:

$$E = \frac{(3 \times 10^{-6})(D)(H_L)(f)(t)(k)}{\eta_p}$$

where,
- E = annual operating cost of the piping system ($/y.)
- D = density of the fluid at the average system operating temperature (lb/ft³)
- f = flow rate through the piping circuit (gpm)
- H_L = head loss of the piping system at flow rate f (feet of head)
- t = number of hr/y during which the circulator operates (hr/y)
- k = cost of electrical energy ($/kWh)
- η_p = wire-to-water efficiency of the circulator (decimal percent)

Example 6.14

Determine the operating cost of a piping system that transports water at an average temperature of 160 °F and a flow rate of 10.0 gpm with a head loss of 9 feet, for 4,000 h/y. Assume that electrical energy costs $0.10/kWh and that the circulator has a wire-to-water efficiency of 25%.

Solution:

The density of water at 160 °F is 61.02 lb/ft³. Substituting the data into Equation 6.23:

$$E = \frac{(3 \times 10^{-6})(61.04)(9)(10)(4000)(0.10)}{0.25} = 26.37/y$$

Discussion:

Keep in mind that over the life of the system, this annual operating cost will likely add up to several hundreds of dollars.

If Equation 6.23 is combined with Equation 6.11, the head loss term, H_L, can be eliminated, and the following relationship results:

Equation 6.24:

$$E = \frac{(3 \times 10^{-6})(D)(\alpha c L)(f^{2.75})(T)(k)}{\eta_p}$$

where,
- E = annual operating cost of the piping system ($/y)
- D = density of the fluid at its typical operating temperature (lb/ft³)
- α = fluid properties factor for water at a given temperature
- c = pipe size factor for the size of copper tube used
- L = total equivalent length of a piping path (ft)
- f = flow rate through the circuit (gpm)
- t = number of h/y during which the circulator operates (h/y)
- k = cost of electrical energy ($/kWh)
- η_p = wire-to-water efficiency of the circulator (decimal percent)
- 2.75 = an exponent of flow rate

This equation can be used to calculate the annual operating cost of any piping system using a circulator to convey fluid through smooth tubing at a given average temperature.

Example 6.15

Assume a hydronic heating system has a total equivalent length of 250 feet of 3/4-inch copper tube. It conveys 140 °F water at 6.0 gpm for 4,000 h/y. Electrical energy costs $0.10/kWh, and the circulator has a wire-to-water efficiency of 25%. Determine the annual operating cost. Repeat the calculation for the same system using 1-inch copper tube.

Solution:

From the graph in Figure 4-4, the water's density is 61.35 lb/ft³. From the graph in Figure 4-9, the water's dynamic viscosity is 0.00032 lb/ft/s. The α value can be calculated using Equation 6.12:

$$\alpha = \left(\frac{D}{\mu}\right)^{-0.25} = \left(\frac{61.35}{0.00032}\right)^{-0.25} = 0.04779$$

The value of c is found in table in Figure 6-18. For 3/4-inch copper tube, $c = 0.061957$.

Substituting all data into Equation 6.24:

$$E = \frac{(3 \times 10^{-6})(61.35)(0.04779)(0.061957)(250)(6^{2.75})(4000)(0.10)}{0.25}$$

$$= \$30.08/y$$

If the same system were built using 1-inch copper tube and operated at the same 6.0 gpm flow rate, only the value of c would change in the above equation. For 1-inch tubing, $c = 0.01776$. The revised operating cost would be:

$$E = \frac{(3 \times 10^{-6})(61.35)(0.04779)(0.01776)(250)(6^{2.75})(4000)(0.10)}{0.25}$$

$$= \$8.62/y$$

In this particular case, there would be a theoretical savings of $21.46 in the first year if the 1-inch tube were used instead of the 3/4-inch tube. Such savings could quickly pay for the increased installation cost of the larger tubing. However, to obtain this energy savings, a circulator that would produce exactly the same flow rate within the larger piping would have to be used. Since only a finite selection of circulators is available, the actual energy savings will depend on the difference in wattage of the smaller circulator used to replace the larger circulator, the latter being used in combination with the smaller pipe size. The savings could be calculated using Equation 6.25.

Equation 6.25:

$$S = \frac{(w_L - w_S)(t)(k)}{1,000}$$

where,
- S = annual savings from being able to use a small circulator ($/y)
- w_L = wattage of the larger circulator (watts)
- w_S = wattage of the smaller circulator (watts)
- t = number of hours per year during which the circulator operates (hr/y)
- k = cost of electrical energy ($/kWh)

Example 6.16

By using 1-inch copper tubing rather than 3/4-inch copper tubing, a designer finds that a smaller circulator that operates at 85 watts can be used instead of a larger circulator that operates at 200 watts. Estimate the annual savings in using the smaller circulator assuming it will operate for 4000 hr/y in a location where electricity costs $0.10/kWh.

Solution:

Substituting the data in Equation 6.25:

$$S = \frac{(200 - 85)(4000)(0.10)}{1000} = 46/y$$

Discussion:

Again, it should be emphasized that this savings will accumulate year after year. The savings will also increase if the cost of electricity increases. Total savings over the life of the system could be several hundreds of dollars.

Equation 6.26 can be used to determine total savings that would occur over a period of years when the operating cost inflates at an assumed annual rate.

Equation 6.26:

$$S_{total} = S_{year1} = \left(\frac{(1+i)^n - 1}{i}\right)$$

where,
- S_{total} = total savings over a period of n years ($)
- S_{year1} = savings occurring in first year ($)
- i = assumed annual rate of inflation of energy cost (decimal percent)
- n = years over which saving is being totaled

Example 6.17

In example 6.16, the first-year savings associated with use of an 85-watt circulator versus a 200-watt circulator was $46. Assume that electrical rates increase at 4% per year. What are the total savings in operating cost over a 20-year period?

Solution:

Substituting the data in Equation 6.26:

$$S_{total} = S_{year1}\left(\frac{(1+i)^n - 1}{i}\right) = \$46\left(\frac{(1+0.04)^{20} - 1}{0.04}\right)$$

$$= \$46 \; 29.778 = \$1370$$

Discussion:

In example 6.16, the designer determined that the smaller 85-watt circulator could be used in lieu of the 200-watt circulator if the circuit was constructed 1- versus 3/4-inch tubing. Analyzing the difference in operating cost between these two options suggests that the total savings in operating cost is substantial and may far exceed the added cost of using 1-inch versus 3/4-inch size tubing for the circuit.

Selecting a Pipe Size

The following procedure incorporates the previously discussed concepts into a method for choosing an appropriate pipe size:

STEP 1. Select a tentative pipe size based on the criteria of keeping the flow rate between 2 and 4 ft/s. Use Figure 6-2 as a guide.

STEP 2. Estimate the installed cost of the system's piping, or portion thereof, using this tentative pipe size.

STEP 3. Estimate the operating cost of the system using Equation 6.24 for the tentative pipe size.

STEP 4. Estimate the installed cost of the system's piping, or portion thereof, using the next larger pipe size.

STEP 5. Estimate the operating cost of the system using Equation 6.24 for the next larger pipe.

STEP 6. If the potential savings in operating cost between steps 3 and 5 would return the higher installation cost of the large pipe in a reasonable time (suggested as 10 years or less), use the larger pipe size, and go on to step 7. If not, use the pipe size from step 1.

STEP 7. Using methods from Chapter 7, Hydronic Circulators, select two circulators, one for each pipe size being considered, that will produce the desired flow rate within the system. Assuming that a circulator that is smaller, less expensive, and that uses less power is available for use with the larger pipe size, compare the difference in installation and operating cost of the two circulators.

If the savings in installation and operating cost of the small circulator over a period of a few years exceeds the higher installation cost of the larger pipe, use the larger pipe.

Example 6.18

Assume the same pipe systems in example 6.15. The system's circulator produces 6.0 gpm and operates 4000 h/y using electrical energy purchased at \$0.10/kWh. Assume that the installed cost of 3/4-inch copper pipe costs \$0.85/ft and the installed cost of 1-inch copper pipe costs \$1.40/ft. Assume that the following circulators were selected for each pipe size.

Using 3/4-inch pipe:
   Circulator cost = \$110
   Operating wattage = 150 watts

Using 1-inch pipe:
   Circulator cost = \$65
   Operating wattage = 90 watts

Determine which pipe size should be used after considering both technical and economic factors.

Solution:

STEP 1. Check that the flow velocity in the smaller pipe is equal to or less than the 4-ft/s limit. From Figure 6-2, one finds that a 3/4-inch copper pipe can carry up to 6.4 gpm and not exceed this limit. Thus, 3/4-inch pipe is acceptable at the stated flow rate of 6.0 gpm and the potential use of the large pipe size will depend on economic considerations.

STEP 2. The installed cost of the 3/4-inch pipe will be (250 ft) (0.85/ft) = 212.50.

STEP 3. The estimated operating cost using 3/4-inch piping was calculated in example 6.15 to be \$30.08.

STEP 4. The installed cost of the 1-inch pipe will be (250 ft) (1.40/ft) = 350.00.

STEP 5. The estimated operating cost using 1-inch piping was calculated in example 6.15 to be \$8.62.

STEP 6. The theoretical savings in operating cost is: \$30.08 − \$8.62 = \$21.46/y.

The extra installation cost of the 1-inch pipe is: \$350 − \$212.50 = \$137.50.

The simple payback associated with use of the larger pipe is:

$$\frac{\$137.50}{\$21.46/\text{y}} = 6.4 \text{ years}$$

Since this is a reasonably short payback period, go on to compare costs and savings associated with the circulators.

STEP 7. The savings in purchase cost of the smaller circulator is $110 - $65 = $45.

The net increase in system cost using the larger pipe would be the higher cost of the larger pipe minus the savings due to the smaller circulator:

Net increase in system *cost* = $137.5 - $45 = $92.50.

The actual savings in annual operating cost can be calculated using Equation 6.25:

$$E = \frac{(150 - 90)(4000)(0.10)}{1000} = \$24/\text{y}$$

The time required for operating cost savings to recover the net cost increase would be:

$$\frac{\$92.50}{\$24/\text{y}} = 3.9 \text{ years}$$

Discussion:

This example shows a relatively fast return on the extra investment in the large pipe. Furthermore, any increase in electrical rates would reduce this payback period. Over a system design life of perhaps 20 years, the operating cost savings associated with the larger pipe size would return the higher initial investment several times over! The larger pipe size is well justified in this case.

Summary

This chapter has laid the foundation for describing fluid flow in piping systems. Many of the fundamentals of fluid mechanics have been condensed and presented in a way specifically suited for hydronic heating systems. Terms such as head loss, pressure drop, flow rate, flow velocity, and more have been accurately described. Such terms will be routinely used in later chapters.

The analogy between hydraulic resistance and electrical resistance represents a new approach for analyzing fluid-filled closed-loop piping systems, one that can be a powerful tool when properly applied. This approach will also be used in later chapters dealing with system design.

The reader is strongly encouraged to use the Fluid Properties Calculator module and Hydronic Circuit Simulator module in the Hydronics Design Studio software to expedite many of the calculations demonstrated in this chapter. These modules allow the reader to try many what-if scenarios related to piping system performance. This is an excellent way to gain a solid understanding of the principles in this chapter.

Key Terms

average flow velocity
boundary layer
compressible fluids
elevation head
entrain
equivalent length
equivalent resistance
erosion corrosion
feet of head
flow coefficient (C_v)
flow rate
flow velocity
fluid
fluid properties factor (α)
ft·lb (foot pound)
gas
gpm
head
head loss
hydraulic resistance
hydronic circuit simulator
incompressible
laminar flow
liquid
Ohm's law
pipe size coefficient
piped in parallel
pressure head
Reynold's number
series circuit
series piping path
static pressure
streamlines
system head loss curve
total head
turbulent flow
velocity head
velocity profile
viscous friction

Questions and Exercises

1. Find the flow velocities corresponding to the following conditions:
   a. 2.5 gpm flow rate in a 1/2-inch type M copper tube;
   b. 10.0 gpm flow rate in a 3/4-inch type M copper tube;
   c. 15.0 gpm in a tube with an inside diameter of 1.1 inches.

2. Use data from Chapter 4, Properties of Water, to determine the value of the fluid properties factor, α, for water at 120 °F. Hint: Use Equation 6.12. Compare this to the value obtained using Figure 6-17.

3. What water pressure would be required at street level to push water to the top of a 300-feet building? Express the answer in psi. If the water pressure at the top of the building had to be 50.0 psi, how would this affect the pressure at the street level?

4. Water at 150 °F moves through a pipe with an internal diameter of 1.3 inches and at a flow velocity of 3 ft/s. Determine the Reynold's number for this flow. Would the flow be laminar or turbulent under these conditions?

5. Determine the head loss when 200 °F water flows at 10.0 gpm through 400 feet of 1-inch type M copper tube. Determine the pressure drop (in psi) associated with this head loss.

6. A valve has a C_v rating of 3.5. What will be the pressure drop across the valve when 140 °F water flows through it at 6.0 gpm?

7. Find the total equivalent length of the piping path shown in Figure 6-52.

8. Find the hydraulic resistance of the piping assembly shown in exercise 7 assuming water at 120 °F will flow through it.

9. Calculate five data points representing head loss versus flow rate through the piping assembly shown in exercise 7 for 120 °F water. Plot the data and sketch a system head loss curve. Hint: use Equation 6.11 to generate the necessary head loss versus flow rate data.

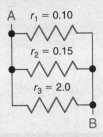

Figure 6-53 | Hydraulic resistor diagram for exercises 10 and 11.

10. Use Equation 6.20 to find the equivalent hydraulic resistance of the three parallel hydraulic resistors shown in Figure 6-53.

11. Assuming that a flow rate of 8.0 gpm enters at point A in Figure 6-53 determine the flow rate in each of the three branches. Check your results.

12. Determine the single equivalent resistance between points A and B for the hydraulic resistor diagram shown in Figure 6-54.

13. Draw a hydraulic resistor diagram for the piping assembly shown in Figure 6-55.

14. Assuming 1-inch copper tubing and fittings are used, and that the system operates with 140 °F water, determine the value of all hydraulic resistances drawn:
   a. Using series and parallel resistance concepts, reduce the resistor diagram to a single equivalent resistance.
   b. Assuming flow enters the piping assembly at 15.0 gpm, find the flow rate in each branch.
   c. Find the flow velocity in each branch.

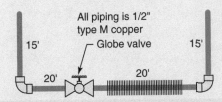

Figure 6-52 | Piping assembly for exercise 7.

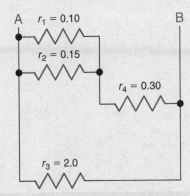

Figure 6-54 | Hydraulic resistor diagram for exercise 12.

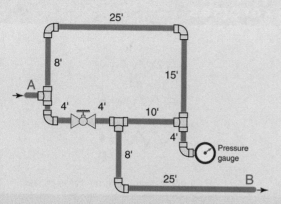

Figure 6-55 | Piping assembly for exercise 13.

15. Use the Fluid Properties Calculator program in the Hydronics Design Studio to find the answer to exercise 2. Also find the α value for the following fluids and conditions:
    a. 140 °F water;
    b. 50% propylene glycol solution at 140 °F;
    c. 30% ethylene glycol solution at 160 °F.

16. Explain the difference between head loss and pressure drop for a fluid flowing through a piping system.

Chapter 7

Hydronic Circulators

Objectives

After studying this chapter, you should be able to:

- Describe different types of circulators used in hydronic systems.
- List the main components of a centrifugal pump.
- Determine the proper placement of the circulator(s) within the system.
- Calculate the flow rate a specific circulator will produce in a given piping system.
- Estimate the flow through a circulator from the pressure differential across it.
- Work with both graphical and analytical descriptions of circulator performance.
- Predict the performance of circulators connected in series and parallel.
- Explain what cavitation is and how to predict it.
- Describe methods of avoiding cavitation.
- Select a circulator for proper performance and high efficiency.
- Describe favorable applications for pressure-regulated circulators.
- Calculated energy savings based on circulator efficiency ratings.

7.1 Introduction

Circulators create fluid motion in modern hydronic systems. Although smaller and less expensive than heat sources, circulators can prove to be just as vital to proper system performance. The wide variety of circulators available in today's market allows great flexibility in how the overall system is designed and controlled.

This chapter provides an overview of the type of circulators used in small- and medium-sized hydronic systems. It goes on to define and illustrate the concept of a pump curve. This curve is combined with the system head loss curve discussed in Chapter 6, Fluid Flow in Piping, to determine the flow rate at which the system will operate. The often-overlooked concepts of series and parallel circulators are discussed from the standpoint of customizing a circulating system to the specific needs of a piping system. The proper placement of the circulator in the system is also described. Special attention is given to the subject of cavitation and its avoidance. The use of variable-speed, pressure-regulated circulators in hydronic distribution systems is discussed. The chapter concludes with information on how to work with a new circulator efficiency rating method developed by the Hydraulic Institute.

7.2 Circulators for Hydronic Systems

Circulators come in a wide variety of designs, sizes, and performance ranges. In closed-loop, fluid-filled hydronic systems, the circulator's function is to circulate the system fluid around the piping. No lifting of the fluid is involved in such systems, as discussed in Chapter 6, Fluid Flow in Piping. This is why the term "circulator" is technically a better descriptor than "pump" relative to what the device does in a hydronic system. Still, within the hydronics industry, the terms pump and circulator are used loosely and interchangeably.

The circulators used in hydronic systems are technically classified as **centrifugal pumps**. They have a rotating **impeller** to add mechanical energy called head to the fluid. Figure 7-1 illustrates the basic construction of a centrifugal pump.

As the impeller rotates, fluid within the center opening (or eye) of the impeller rapidly accelerates through the passageways formed by the impeller vanes between the two impeller disks. The fluid's mechanical energy content is increased as it accelerates toward the outer edge of the impeller. As it leaves the impeller, the fluid impacts against the inside surface of the chamber surrounding the impeller. This chamber is called the **volute**. The fluid's former speed (or velocity head) is converted to a pressure increase (pressure head). The fluid then flows around the contoured volute and exits through the discharge port.

For this process to be continuous, the rate of fluid entering the circulator must be identical to that leaving the circulator. One might assume that a centrifugal pump continuously "sucks" fluid into the eye of its impeller. This is not true. Water entering a centrifugal pump must be *pushed* in by system pressure upstream of the inlet port. This is a very important point and is often misunderstood. If the proper conditions are not provided for fluid to be pushed into the inlet port, a very undesirable operating characteristic called **cavitation** will result. This is discussed later in the chapter.

Furthermore, hydronic circulators cannot be turned on when filled with air and expected to "suck" fluid in from some open container in which the water level is lower than the circulator's inlet. The circulator must first be filled with water and purged of most air before it can operate properly. This is called **priming**. In hydronic systems, priming is accomplished when the system is filled with water and purged of air. Details for doing this are discussed in Chapter 13, Air & Dirt Removal & Water Quality Adjustment.

Many of the circulators used in residential and light commercial systems are available with volutes made of cast iron, bronze, brass, polymer, or stainless steel. Cast-iron circulators should only be used in closed-loop systems. If the system is open to the atmosphere, at any point, it is able to absorb oxygen into system water. This oxygen can create continuous corrosion of ferrous metals. Such systems should only use circulators with bronze, brass, polymer, or stainless steel volutes since these materials are resistant to oxidation.

Centrifugal pumps can be built with differently shaped volutes while still maintaining the same internal operation. The volute's shape determines how the pump will be connected to the system's piping. Figure 7-2 shows two examples of **inline circulators**. These circulators have their inlet and discharge ports along a common centerline. An inline circulator can be placed into a piping path without need for any lateral or angular offsets between the inlet and discharge ports. The inline circulator design is by far the most common type used in residential and light commercial hydronic systems.

By modifying the shape of the volute, a design called an **end suction circulator** is created. An example of an end suction circulator is shown in Figure 7-3.

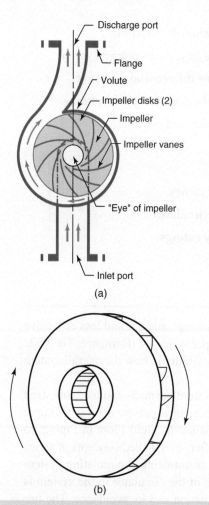

Figure 7-1 | (a) Simplified cross section of a centrifugal pump (b) and an impeller.

7.2 Circulators for Hydronic Systems

Figure 7-2 | Examples of inline circulators. (a) A wet rotor circulator. *Courtesy of Taco, Inc.* (b) Three-piece circulator. *Courtesy of Armstrong Pumps.*

Figure 7-3 | Example of an end suction circulator. *Courtesy of ITT.*

End suction circulators create a 90-degree turn in the system piping. In most cases, an offset between the centerlines of the inlet and discharge piping is also required. End suction circulators are usually floor mounted and are more common in larger hydronic systems.

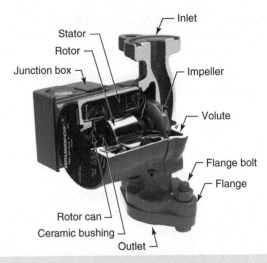

Figure 7-4 | Cutaway view of a small wet rotor circulator. *Courtesy of Grundfos Pumps Corp.*

Wet Rotor Circulators

Over the last several decades, a specialized design for small- and medium-sized circulators has been refined by a number of manufacturers specifically for use in hydronic systems. This design, known as a **wet rotor circulator**, combines the rotor, shaft, and impeller into a single assembly that is housed in a chamber filled with system fluid. An example of a wet rotor circulator is shown in Figure 7-4.

The motor of a wet rotor circulator is totally cooled and lubricated by the system's fluid. As such, it has no fan or oiling caps. The **rotor** assembly is supported on ceramic or graphite bushings in the rotor can. These bushings contain no oil, but ride on a thin film of system fluid. The rotor is surrounded by the **stator** assembly.

Some advantages of wet rotor circulators are as follows:

- There is no oiling required.
- There is no leakage of system fluid due to worn pump seals.
- Their small size makes them easy to locate and support.
- The absence of a cooling fan and external coupling makes for quiet operation.
- Several models are available with multiple-speed motors.
- They are relatively inexpensive due to fewer parts.
- They are ideal for applications where limited flow rate and head are required.
- They can be close coupled for series pumping applications.

- Many have **permanent split capacitor (PSC) motors** that can be operated over a wide speed range by suitable electronic control.
- Wet rotor design can be used with electronically commutated motors.
- The circulator will not fail due to a broken coupling between its motor shaft and impeller shaft.

The disadvantages of wet rotor circulators include the following:

- The low starting torque of the PSC motors may not be able to free a stuck impeller after a period of prolonged shutdown.
- Servicing anything not contained in the external junction box requires opening the wetted part of the circulator, resulting in some fluid loss and air entry into the system.
- They tend to have lower **wire-to-water efficiency** than circulators with air-cooled motors.

Wet rotor circulators are currently the most commonly used circulators in modern residential and light commercial hydronic systems. They are available with cast-iron volutes for use in closed hydronic systems or with bronze or stainless steel volutes for direct contact with domestic water or other open-loop applications. Most wet rotor circulators have impellers constructed of stainless steel, bronze, or polymer materials.

Three-Piece Circulators

Another common circulator design used in small- and medium-sized hydronic systems is called a **three-piece circulator**. It consists of a pump body assembly, **coupling assembly**, and motor assembly as shown in Figure 7-5.

Unlike wet rotor circulators, the motor of a three-piece circulator is totally separate from the wetted portion of the circulator. This allows the motor or coupling to be serviced or replaced without needing to open the piping system. The design of the coupling assembly between the motor and shaft varies among manufacturers. A common design employs a spring assembly that absorbs vibration or high torque between the two shafts as the motor starts. The impeller shaft penetrates the volute through shaft seals that must limit leakage of system fluid, even under high pressure. Although it is normal for these seals to experience minor fluid losses, the leakage rate of modern pump seals is so small that the fluid usually evaporates before it is seen.

The advantages of a three-piece circulator design are the following:

- Potential for longer life if bearings are properly lubricated.

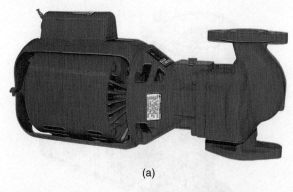

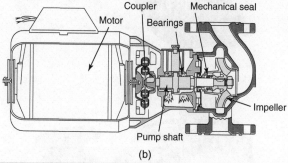

Figure 7-5 (a) Example of a three-piece circulator. (b) Internal construction. *Courtesy of ITT.*

- Easy servicing of the motor without need to open wetted portion of pump.
- Ability to produce higher starting torque to overcome a stuck impeller condition after a prolonged shutdown.
- May have higher wire-to-water efficiencies than wet rotor circulators.

The disadvantages include the following:

- Heavier construction requiring strong supports.
- Must be oiled periodically.
- More operating noise due to external motor and coupling assembly.
- Potential maintenance of mechanical seals and coupling assembly.
- Defective or worn shaft seals could allow leakage of system fluid.

Like wet rotor circulators, three-piece circulators are available in materials suitable for both open-loop and closed-loop applications.

Two-Piece Circulators

Still another circulator design combines the efficiency advantage of an external air-cooled motor, with the simplicity of a direct coupling between the motor and circulator. Such circulators are often called **two-piece circulators** because they eliminate the coupling assembly between

Figure 7-6 | Example of a two-piece circulator. *Courtesy of Armstrong Pumps.*

Figure 7-7 | Incorrectly installed circulators. The rotor shaft should always be in a horizontal orientation. *Photo courtesy of Harvey Youker.*

the motor and impeller that is used with three-piece circulators. An example of a two-piece circulator with inline inlet and outlet ports is shown in Figure 7-6. Two-piece circulators can often attain higher wire-to-water efficiencies compared to wet rotor circulators.

Circulator Mounting Considerations

Most circulators used in smaller hydronic systems are designed to be installed with their shafts in a horizontal position. This removes the thrust load on the bushings due to the weight of the rotor and impeller.

The direction of flow through the pump is usually indicated by an arrow on the side of the volute. The installer should always check that the circulator is installed in the correct flow direction. As long as the shaft is horizontal, the circulator can be mounted with the flow arrow pointing upward, downward, or horizontally. Of these, upward flow is preferred when possible because it allows the circulator to rapidly clear itself of air bubbles.

The weight of the circulator should not be supported by system piping unless the circulator is relatively light and the piping itself is well supported within a few inches of the circulator. This is especially true for three-piece circulators that create a bending effect on the pipe due to the offset weight of the motor. If the circulator is supported by system piping, the piping should not be rigidly mounted on walls that can transmit vibrations to the building structure. The use of vibration-absorbing mounting brackets is recommended. The piping supporting circulators should also not be mounted on framed partitions adjoining living or sleeping spaces where any noise transmission would be objectionable (Figure 7-7).

A traditional practice in residential installations has been to mount several small circulators that are part of a multizone system on a common header assembly directly threaded into the boiler. Threaded steel or black iron pipe in the 1.5- to 2-inch range is often used to fabricate this header. Although this pipe is strong enough to handle the weight of the circulators, the bending effect imposed on the boiler connection can be very high. For this reason, whenever several circulators are mounted on a common header, the header should be well supported to relieve this bending stress on the boiler as shown in Figure 7-8.

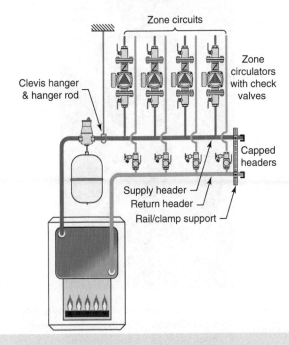

Figure 7-8 | Support options for headers supporting multiple circulators.

Connecting the Circulator to the Piping

Circulators should always be mounted so they can be removed for servicing or replacement. In North America, the most common method of connecting a circulator to piping is with bolted **flanges**. One side of the flanged joint is an integral part of the circulator's volute. The other flange is threaded onto the piping. As the flange bolts are tightened, an O-ring or gasket is compressed between the faces of the flanges to make the seal. For small- and medium-sized circulators, a two-bolt flange is common. Large circulators often use four-bolt flanges.

Circulator flanges are available in cast iron for closed system applications, as well as bronze or brass for direct contact with domestic water or other open-loop applications. A special type of flange, known as an **isolation flange**, contains a built-in ball valve. With the isolation flanges closed, the circulator may be unbolted at its flanges and removed from the system. Only a small amount of fluid within the circulator's volute is lost when the circulator is isolated and removed in this manner.

Figure 7-9 shows a circulator equipped with isolation flanges. Notice that this circulator is also installed with the electrical cable coming from the bottom of the junction box. This helps guard against the possibility of water from a nearby leak traveling along the cable and entering the circulator's junction box.

It is also possible to isolate a circulator using standard flanges combined with a gate valve or ball valve on each side of the circulator. However, the availability of modern isolation flanges such as shown in Figure 7-9 generally provides a faster and less expensive option, and as such is much more common in residential and light commercial hydronic systems.

Some small circulators available in North America as well as most small circulators sold in Europe are designed for half union connections as seen in Figure 7-10. **Half unions** are available with or without an integral ball valve for isolation.

Circulators for Chilled Water Applications

Later chapters of this text discuss chilled water cooling systems. One of the critical details in such systems is

Figure 7-9 | Circulator equipped with isolation flanges. The ball valve within the isolation flange is open when its handle is aligned with the piping. *Courtesy of John Siegenthaler.*

Figure 7-10 | Circulator with half union connections. *Courtesy of John Siegenthaler.*

Figure 7-11 | A form-fitting foam insulation shell for a small circulator. *Courtesy of John Siegenthaler.*

the avoidance of condensation on system components that convey chilled water. Any circulator in such a system needs to be rated, by its manufacturer, to operate with chilled water, down to temperatures of at least 35 °F, even lower if possible. Circulators without such ratings may develop internal condensation that eventually causes problems with their electrical circuitry.

Circulators used in chilled water cooling systems also need to be partially insulated. More specifically, the circulator's entire volute, as well as its flanges and connected piping, must be insulated. However, the circulator's motor can, which is designed to dissipate heat, should never be insulated, especially if the circulator is used to convey heated water during heating mode operation.

Some circulators can be purchased with form-fitting insulation shells. An example is shown in Figure 7-11.

The split insulation shell must be tightly fit to the volute after the circulator is installed. All seams should be sealed with a suitable caulking or tape with low vapor permeability. The objective is to prevent moisture-laden air from contacting any surfaces on the circulator or adjacent piping that could be cool enough to cause condensation.

7.3 Placement of the Circulator Within the System

The location of the circulators relative to other components in a hydronic system can make the difference between quiet, reliable operation or constant problems. One guiding rule summarizes the situation: *Always install the circulator so that its inlet is close to the connection point of the system's expansion tank.* This principle has been applied for years in commercial hydronic systems, yet is still not universally applied in smaller systems (as it should be).

To understand why this rule should be followed, one needs to consider the interaction between the circulator and expansion tank. In a closed-loop piping system, the amount of fluid, including that in the expansion tank, is fixed. It does not change regardless of whether the circulator is on or off. The upper portion of the expansion tank contains a captive volume of air at some pressure. The only way to change the pressure of this air is to either push more fluid into the tank to compress the air or to remove fluid from the tank to expand the air. This fluid

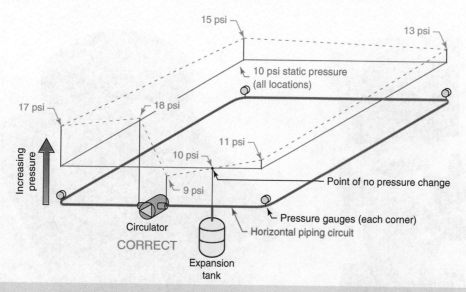

Figure 7-12 | Pressure distribution in a horizontal piping circuit. Solid (red) line is pressure distribution when circulator is off. Dashed line is pressure distribution when circulator is on. Note expansion tank is near the inlet of circulator.

would have to come from, or go to, some other location within the system. However, since the system's fluid is incompressible, and the amount of fluid in the system is fixed, this cannot happen regardless of whether the circulator is on or off. The expansion tank thus fixes the pressure of the system's fluid at its point of attachment to the piping. This is called the **point of no pressure change** within the system.

Consider a horizontal piping circuit filled with fluid and pressurized to some pressure, say 10.0 psi, as shown in Figure 7-12. When the circulator is off, the static pressure is the same (10.0 psi) throughout the piping circuit. This is indicated by the solid (red) horizontal line shown above the piping.

When the circulator is turned on, it immediately creates a pressure difference between its inlet and discharge ports. However, the expansion tank still maintains the same (10.0 psi) fluid pressure at its point of attachment to the system. The combination of the pressure difference across the circulator, the pressure drop due to head loss in the piping, and the point of no pressure change results in a new pressure distribution as shown by the dashed (green) lines in Figure 7-12.

Notice that the pressure increases in nearly all parts of the circuit when the circulator is turned on. This is desirable because it helps eject air from vents. It also reduces the chance of cavitation. The short segment of piping between the expansion tank and the inlet port of the circulator experiences a slight drop in pressure due to head loss in the piping. The numbers used for pressure in Figure 7-12 are only illustrative. The actual numbers will, of course, depend on flow rates, fluid properties, and pipe sizes.

Now consider the same system with the expansion tank incorrectly connected near the discharge port of the circulator. The dashed line in Figure 7-13 illustrates the new pressure distribution in the system when the circulator is on.

Again, the point of no pressure change always remains at the expansion tank connection. This causes the pressure in most of the system to decrease when the circulator is turned on. The pressure at the inlet port has dropped from 10.0 to 2.0 psi. This situation is not desirable because it reduces the system's ability to expel air. In some cases, it can also lead to cavitation.

To see how problems can develop, imagine the same system with a static pressurization of only 5.0 psi instead of the previous 10.0 psi. When the circulator starts, the same 9.0 psi differential will be established between its inlet and discharge ports. The pressure profile shown with dashed lines in Figure 7-13 will be shifted downward by 5.0 psi (10.0 − 5.0 = 5.0 psi). This is shown in Figure 7-14.

Notice the pressure in the piping between the upper right-hand corner of the circuit and the inlet port of the circulator is now below atmospheric pressure. If air vents were located in this portion of the circuit, the subatmospheric pressure would cause air to enter the system every time the circulator operates. The circulator is also much more likely to cavitate under these conditions.

7.3 Placement of the Circulator Within the System

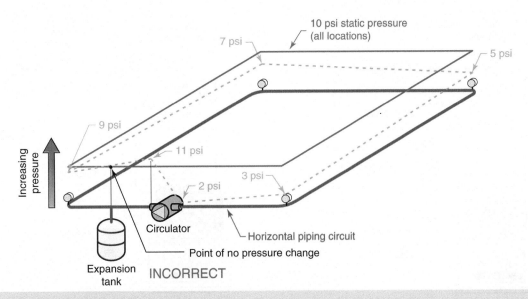

Figure 7-13 Pressure distribution in horizontal piping circuit. Solid (red) line is pressure distribution when circulator is off. Dashed (green) line is pressure distribution with circulator on. Note the expansion tank is incorrectly located near the discharge of the circulator.

Unfortunately, this latter scenario has occurred in many residential and light commercial hydronic systems. To see why, consider a piping schematic that represents many of these "traditionally piped" systems (see Figure 7-15).

In this arrangement, the circulator is pumping toward the expansion tank. Although there appears to be a fair distance between the circulator and the tank, the pressure drop due to head loss through a typical cast-iron boiler is very low. Thus, from a pressure drop standpoint, the circulator's discharge port is very close to the expansion tank. This arrangement will cause the system pressure to drop from the expansion tank connection, around the distribution system to the inlet of the circulator, whenever the circulator is operating.

One reason circulators were originally located on the return side of the boiler was to allow them to operate with cooler return water. It was believed the lower operating temperature prolonged the life of the packing, seals, and motor. This is not a concern for most currently produced wet rotor circulators that are often rated for continuous operation at fluid temperatures up to 230 °F.

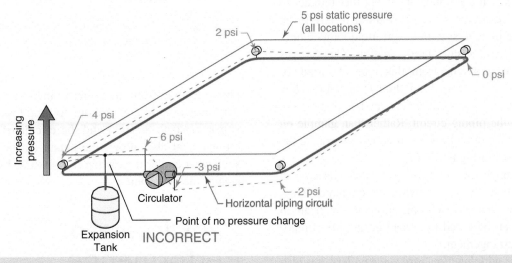

Figure 7-14 Pressure distribution in horizontal piping circuit. Solid (red) line is pressure distribution when circulator is off. Dashed (green) line is pressure distribution with circulator on. Note the expansion tank is incorrectly located near the discharge of the circulator. The pressure in some locations is subatmospheric when the circulator is on.

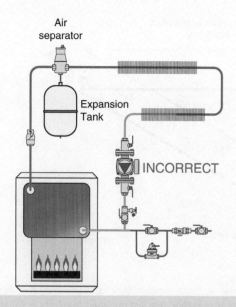

Figure 7-15 A "traditional" piping configuration in which circulator is pumping toward (rather than away from) the expansion tank.

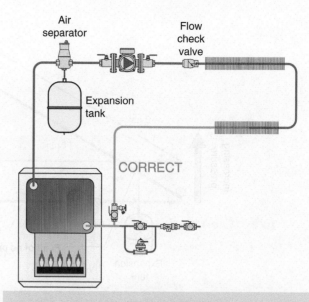

Figure 7-16 One correct arrangement of circulator relative to expansion tank. Note that the expansion tank is tapped into the circuit close to the inlet port of the circulator.

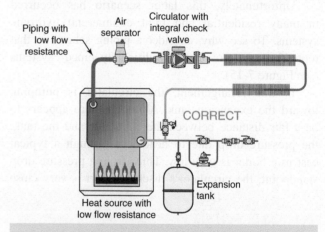

Figure 7-17 Another acceptable arrangement of circulator and expansion tank.

No modern hydronic heating system should require water heated to this temperature.

Another reason circulators were sometimes found on the inlet side of boiler was that it allowed for more compact shipment of "packaged" boilers (as seen in Figure 3-2). The installer simply assumed that this location was correct, when in reality it had nothing to do with the pressure distribution around the circuit.

Some hydronic systems that are piped as shown in Figure 7-15 have worked fine for years. Others have had problems from the first day they were put into service. Why is it that some systems work and others do not? The answer lies in a number of factors that interact to determine the exact pressure distribution in any given system. These include system height, fluid temperature, pressure drop, and system pressurization. The systems most prone to problems from this type of circulator placement are those with high fluid temperature, low static pressure, low system height, and high pressure drops around the piping circuit. Rather than gamble on whether these factors will work, it is best to arrange the piping as shown in Figure 7-16.

This arrangement will increase the pressure in nearly all parts of the system when the circulator is on. Many systems that have chronic problems with air in their piping can be cured by reconfiguring the components to this arrangement.

In systems with low flow resistance boilers, it is also possible to locate the expansion tank near the inlet to the boiler as shown in Figure 7-17. Because of the low pressure drop through the boiler and generously sized piping leading from the boiler to the circulator, there is very little drop in pressure due to head loss between the expansion tank location and the circulator's inlet. This arrangement also allows the expansion tank to remain somewhat cooler than when it is mounted on the outlet side of the heat source.

7.4 Circulator Performance

This section presents analytical methods for describing the performance of circulators. These methods are

Explanation of Circulator Head

The head produced by a circulator is a commonly misunderstood concept. Some references define it as the height to which a centrifugal pump can lift and maintain a column of water, and others define it as the pressure difference the circulator can produce between its inlet and discharge ports. Both definitions are related to the concept of head, but are both are also incomplete.

In Chapter 6, Fluid Flow in Piping, head was described as *mechanical energy* present in the fluid. For incompressible fluids such as those used in hydronic systems, the added head (e.g., added mechanical energy) reveals itself as an increase in fluid pressure between the inlet and discharge ports of the circulator. Thus, *an increase in pressure from the inlet to outlet port of an operating circulator is the "evidence" that head energy has been added to the fluid.* This concept is shown in Figure 7-18a.

The magnitude of this pressure increase depends on the flow rate through the circulator and the density of the fluid being circulated. As the flow rate through a circulator increases, the pressure rise across the circulator decreases as shown in Figure 7-18b.

In the United States, the head added to the fluid by a circulator is expressed in units of feet of head. This unit results from a simplification of the following units:

$$\text{Head} = \frac{\text{ft lb}}{\text{lb}} = \text{ft} = \frac{\text{energy}}{\text{lb}}$$

Those having studied physics will recall the unit of ft lb as a unit of energy. *Head may therefore be described as the mechanical energy added per pound of fluid passing through the circulator.*

The head added by a circulator at some given flow rate is essentially independent of the fluid being pumped. For example, a circulator producing a head of 20 feet while circulating 60°F water at 5.0 gpm will also operate at a head of 20 feet while circulating a 50%

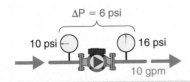

Figure 7-18a | An increase in pressure from the inlet to the outlet port of an operating circulator is the *evidence* that head has been added to the fluid.

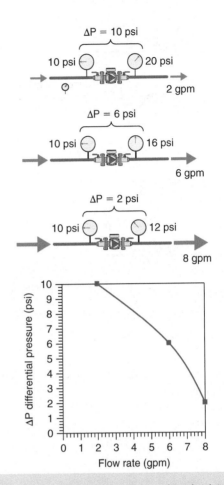

Figure 7-18b | As the flow rate through a circulator rises, the pressure difference between the inlet and outlet ports decreases.

ethylene glycol solution at 5.0 gpm. However, the pressure differential measured between its inlet and outlet ports will be slightly different because of the difference in the density of the fluids.

Converting Between Head and Differential Pressure

The head added to a fluid as it passes through an operating circulator can be determined from the differential pressure across the circulator. Figure 7-19 shows four options for equipping a circulator with gauges to read differential pressure. On some medium-sized circulators, threaded openings are often provided in the volute flanges for direct attachment of pressure gauges.

Equation 6.5 can be used to convert the pressure differential produced by a circulator to the head. This

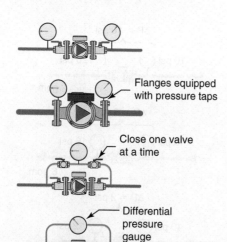

Figure 7-19 | Four options for reading the pressure differential across a circulator.

calculation requires the determination of the density of the fluid being used.

Equation 6.5:

$$H = \frac{144(\Delta P)}{D}$$

where

H = head added to fluid by the circulator (feet of head)
ΔP = pressure differential between inlet and discharge ports of circulator (psi)
D = density of the fluid (lb/ft³)

Example 7.1

Based on the pressure gauge readings, determine the head added to the fluid by the circulator in the three operating conditions shown in Figure 7-20.

Solution:

In all cases, the density of the fluid being circulated needs to be determined. The density of water at 60 and 180 °F can be found in Figure 4-4. The density of the 50% solution of ethylene glycol was found using the Fluid Properties Calculator module in the Hydronics Design Studio software.

a. 60 °F water: density = 62.31 lb/ft³
b. 180 °F water: density = 60.47 lb/ft³
c. 140 °F ethylene glycol: density = 65.39 lb/ft³

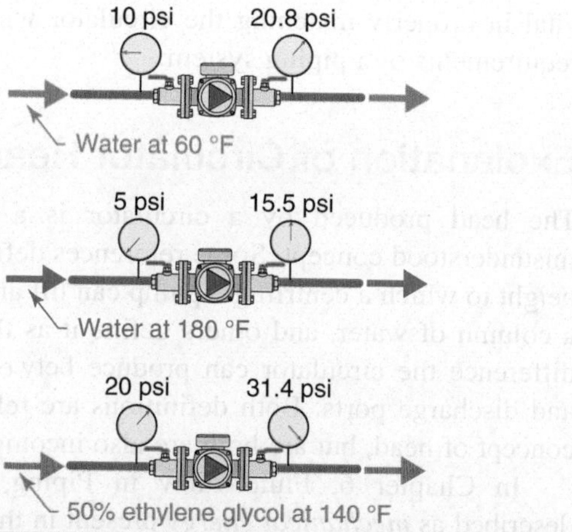

Figure 7-20 | Pressure differentials across circulators for Example 7.1.

The head produced in each situation can be calculated using Equation 6.5:

a. $H = \dfrac{(20.8-10)(144)}{62.31} = 25$ feet of head

b. $H = \dfrac{(15.5 - 5)(144)}{60.47} = 25$ feet of head

c. $H = \dfrac{(31.4 - 5)(144)}{65.39} = 25$ feet of head

Discussion:

In all three cases, the pressure rise across the circulator is different, yet the head energy imparted to each fluid is the same.

Circulator Performance Curves

A graph of a circulator's ability to add head to a fluid at various flow rates is called a **pump curve**. An example of such a graph is shown in Figure 7-21.

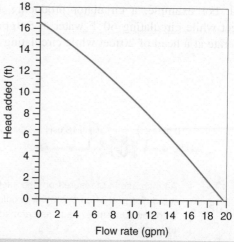

Figure 7-21 | Example of a pump curve for a small circulator.

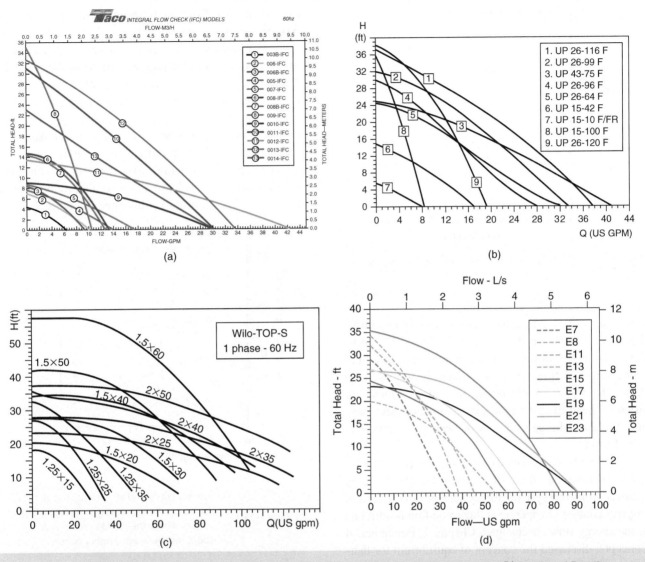

Figure 7-22 | Examples of several pump curves plotted on a common graph. (a) *Courtesy of Taco, Inc.* (b) *Courtesy of Grundfos.* (c) *Courtesy of Wilo.* (d) *Courtesy of Armstrong Pumps.*

Pump curves are developed from test data using water in the temperature range of 60 to 80 °F. For fluids with higher viscosities, such as glycol-based antifreeze solutions, there is a very small decrease in head and flow rate capacity of the circulator. However, for the fluids and temperature ranges commonly used in hydronic heating systems, this variation is so small that it can be safely ignored. Thus, *for the applications discussed in this text, pump curves may be considered to be independent of the fluid being circulated.* The reader is reminded, however, that the system head loss curve that describes head loss versus flow rate in a specific piping system is very dependent on fluid properties and temperature as discussed in Chapter 6, Fluid Flow in Piping.

Pump curves are extremely important in matching the performance of a circulator to the flow requirements of a piping system. All circulator manufacturers publish these curves for the circulator models they offer. In many cases, the pump curves for an entire series or family of circulators is plotted on the same set of axes so that performance comparisons can be made. Examples are given in Figure 7-22.

Operating Point of a Hydronic Circuit

The system head loss curve of a piping assembly was discussed in Chapter 6, Fluid Flow in Piping. Such a curve shows the head loss of the piping assembly as a function of flow rate, for a given fluid at a given temperature. It is a unique graphical description of the hydraulic characteristic of that piping assembly. An example of a system head loss curve is shown in Figure 7-23.

Notice that the pump curve in Figure 7-21 and the system head loss curve in Figure 7-23 have similar quantities on each axis. The system head loss curve shows

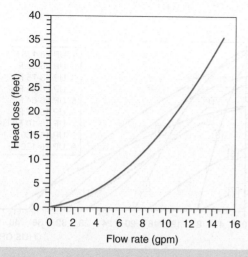

Figure 7-23 | Example of a system head loss curve for a specific piping system.

head *loss* on the vertical axis, whereas the pump curve shows head *added* on its vertical axis. The pump curve shows the ability of the circulator to add head to a fluid over a range of flow rates. The system head loss curve shows the ability of the piping system to remove, or dissipate, head from the fluid over a range of flow rates. In both cases, the gain or loss of head is expressed in feet of head.

The first law of thermodynamics states that when the rate of energy input to a system equals the rate of energy removal from the system, the system is in equilibrium and will continue to operate at those conditions until one of the energy flows is changed. Chapter 1, Fundamental Concepts, discussed the concept of equilibrium for thermal energy flows. It states that when the rate of heat input to a hydronic system exactly equals the rate of heat dissipation by the system, that system is in thermal equilibrium and temperatures will remain constant.

The concept of equilibrium also holds true for the mechanical energy (e.g., head) added or removed from the fluid in a piping system. It can be summarized as follows:

When the rate at which a circulator adds head to the fluid in a piping system equals the rate at which the piping system dissipates this head, the system is in **hydraulic equilibrium**, *and flow rate will remain constant.*

The point at which hydraulic equilibrium occurs in a given piping system, with a given circulator, can be found by plotting the system head loss curve and pump curve on the same set of axes as shown in Figure 7-24. The point where the curves cross represents hydraulic equilibrium and is called the **operating point** of the system. The system's flow rate at hydraulic equilibrium is found by drawing a vertical line from the operating point down to the horizontal axis. The head input by the circulator (or head loss by the piping system) can be found by extending

a horizontal line from the operating point to the vertical axis. An example is shown in Figure 7-24.

A performance comparison of several "candidate" circulators in a specific piping system can be made by plotting their individual pump curves on the same set of axes as that system's head loss curve. The intersection of each circulator's pump curve with the system's head loss curve indicates the operating point for that particular circulator. By projecting vertical lines from the operating points down to the horizontal axis, the designer can determine the flow rate each circulator would produce within the system. This concept is illustrated in Figure 7-25.

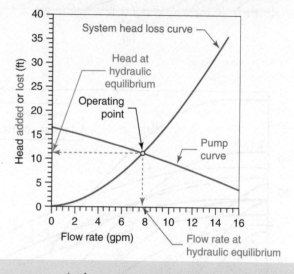

Figure 7-24 | A pump curve and system head loss curve plotted on the same axes. The operating point is where the curves cross. This is where hydraulic equilibrium occurs.

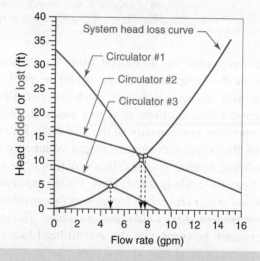

Figure 7-25 | Three pump curves plotted along with a head loss curve for a specific piping system. The vertical lines descending from the operating points indicate flow rates that would be produced by each circulator if it were used in the piping system.

Notice that even though the curves for circulators #1 and #2 are markedly different, they intersect the system curve at almost the same point. Therefore, these two circulators would yield very similar flow rates of about 7.5 and 7.8 gpm in this piping system. The flow rate produced by circulator #3, about 5.0 gpm in this case, is considerably lower.

High Head Versus Low Head Circulators

Some circulators are designed to produce relatively high heads at lower flow rates. Others produce lower but relatively stable heads over a wide range of flow rates. These characteristics are fixed by the manufacturer's design of the circulator. Factors that influence the shape of the pump curve include the diameter and width of the impeller, as well as the number and curvature of the impeller's vanes. Figure 7-26 compares the internal construction of a circulator designed for high heads at low flow rates with that of a circulator designed for lower heads over a wider range of flow rates. Notice that the impeller of the high head circulator has a relatively large diameter, but a very small separation between its disks. The low head circulator, on the other hand, has a small diameter impeller with deeper vanes, and this case no lower disk.

The pump curves for the circulators shown in Figure 7-26 are plotted in Figure 7-27. The high head circulator is said to have a "steep" pump curve. The other circulator would be described as having a relatively "flat" curve.

Circulators with **"steep" pump curves** are intended for systems having high head losses at modest flow rates. Examples would include series piping circuits containing several components with high flow resistances, earth loops for ground source heat pumps, or radiant floor heating systems using long circuits of small diameter tubing.

Circulators with **"flat" pump curves** should be used when the objective is to maintain a relatively steady pressure differential across the distribution system over a wide range of flow rates. An example of such an application is a multizone system using zone valves to control flow through individual circuits. Depending on the number of zone valves open at a given time, and the shape of the pump curve, the differential pressure across the circulator could change considerably. A circulator with a flat pump curve will minimize such changes and thus maintain relatively stable flow rates in the active zones. This is shown in Figure 7-28.

No centrifugal pump can produce a perfectly flat pump curve. However, it is possible to mimic such a perfectly flat pump curve by properly controlling the circulator's speed. This is discussed later in the chapter.

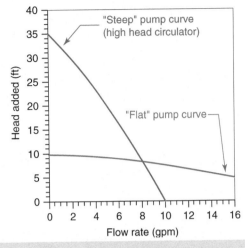

Figure 7-27 | Pump curves for circulators shown in Figure 7-26.

Figure 7-26 | Comparison between (left image) impeller design for a steep pump curve and (right image) impeller design for a flat pump curve. *Courtesy of Taco, Inc.*

Multispeed Circulators

A single-speed circulator has only one pump curve. This may be acceptable if a given circulator happens to match to the flow and head requirements of a given piping system. If not, one would need to keep looking at different model circulators until the proper match is found. Because piping systems can be created in virtually unlimited configurations, many manufacturers offer a "family" of circulators with a fairly wide range of pump curves. However, this requires wholesalers to stock an equally wide range of circulator models.

To reduce the number of circulator models needed and still cover a wide range of performance requirements, most circulator manufacturers now offer circulators

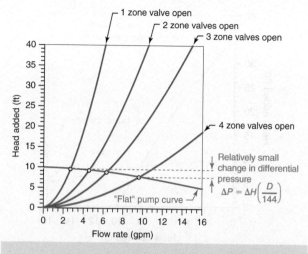

Figure 7-28 | Circulators with flat pump curves produce relatively small changes in head and differential pressure as zone circuits, controlled by zone valves, turn on and off. This is desirable.

capable of operating at three different speeds. The desired speed is selected using an external switch on the circulator's junction box. Each operating speed has a corresponding pump curve as shown in Figure 7-29.

The pump curve associated with each speed intersects the system's head loss curve at different operating points and thus produces three different flow rates. This provides the same versatility as having three separate circulators, each with a single pump curve.

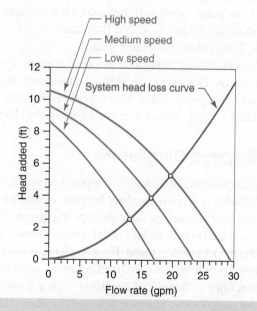

Figure 7-29 | A three-speed circulator can operate on any of its three pump curves. Each curve will produce a different flow rate within a given piping system.

Circulators in Series

Occasionally, a system requires more head than can be supplied by a single circulator. The designer has the option of finding a larger circulator that can supply the necessary head, or possibly using two smaller circulators connected in series. The second option is often the more economical choice.

When two identical circulators are connected in series, the resulting pump curve can be found by doubling the head produced by a single circulator at each flow rate.

Figure 7-30 shows the pump curve of a single circulator, as well as two such circulators connected in series. In effect, the two circulators are acting as a multistage pump with the head added by the first being doubled by the second.

Notice that the flow rate resulting from two circulators in series (about 10.5 gpm for the piping system and circulators represented in Figure 7-30) is not double that obtained with the single circulator (about 8.0 gpm). This is typical and is the result of the curvature of both the pump curves and system head loss curve. It demonstrates the fallacy of trying to double the flow rate in a system by adding a second identical circulator in series.

There are two ways of piping circulators in series with each other. The first, known as **close coupling**, is accomplished by bolting the discharge flange of the first circulator to the inlet flange of the second as shown in Figure 7-31. This is practical for small circulators. It has the advantage of requiring only one set of flanges to connect the resulting assembly to the system piping. Obviously, it is important to ensure that the flow

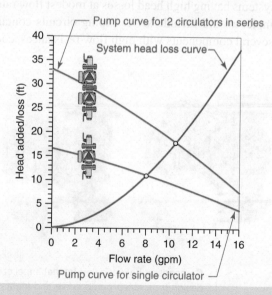

Figure 7-30 | Effect of using two identical circulators in series.

7.4 Circulator Performance

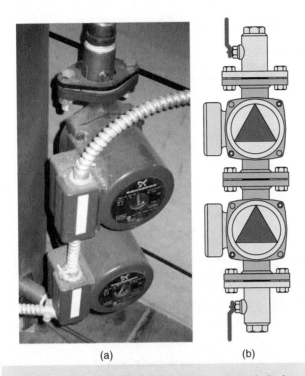

Figure 7-31 | (a) Example of two small circulators bolted flange to flange in a close-coupled series configuration. (b) Schematic representation of this assembly. *Courtesy of John Siegenthaler.*

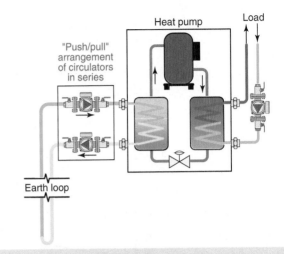

Figure 7-32 | Push/pull arrangement of circulator in series used in combination with ground source heat pump system.

direction arrow on each circulator is facing the same direction before bolting them together. The high heads and associated pressure differential resulting from such an assembly make it critical that the system's expansion tank is located near the inlet port of the first circulator. If this is not done, the likelihood of cavitation is increased.

Another way of installing circulators in series is called a **push/pull arrangement**. In this case, the two circulators are separated by several feet of piping or a major system component such as the heat exchanger coil of a heat pump. This approach distributes the pressure differential produced by each circulator more evenly throughout the system. It also can reduce the chances of cavitation compared to the previously described close-coupled arrangement. This configuration is often used for the earth heat exchanger of ground source heat pump systems where a high head is required due to the combined flow resistance of both the heat pump coil and earth heat exchanger piping (see Figure 7-32).

Circulators in Parallel

Another way of connecting two or more identical circulators is in *parallel*. The inlet ports of the circulators are connected to a common header pipe, as are the discharge ports (see Figure 7-33). This arrangement is useful when a relatively high flow rate is required at a modest head.

The pump curve for two identical circulators connected in parallel is obtained by doubling the flow rate of a single circulator at each head value. This is

Figure 7-33 | Two identical circulators connected in parallel. Electrical power to each circulator routed through service switches. *Courtesy of The Plumbing and Heating Company, Inc.*

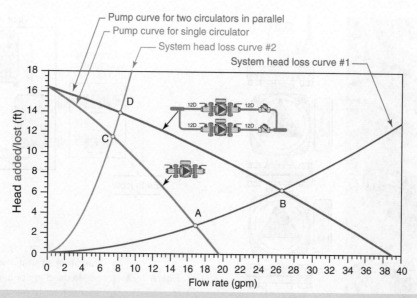

Figure 7-34 | Effect of operating two identical circulators in parallel.

illustrated in Figure 7-34. For three identical circulators in parallel, the flow rate would be tripled at each head value, and so on.

System head loss curve #1 in Figure 7-34 is for a piping system with relatively low flow resistance. Point A is the operating point for this piping system using a single circulator. Point B is the operating point using two of the same circulators in parallel. For this particular piping system and circulator combination, using two circulators in parallel increases the flow rate from about 17.0 gpm to about 26.5 gpm, a significant gain.

Points C and D in Figure 7-34 are the operating points for the same circulator and parallel circulator combination in a piping system having a higher head loss characteristic as shown by system head loss curve #2. In this system, adding a second parallel circulator results in a relatively small increase in flow from about 7.3 to 8.2 gpm, a very small gain considering twice the electrical input power would be used.

These curves demonstrate that the flow rates obtained using multiple circulators in parallel can differ considerably depending on the system head loss curve of the piping system into which they are placed. They also show that adding a second circulator in parallel will not double the system's flow rate, even for a system with low head loss characteristics.

Whenever two or more circulators are used in parallel, a check valve must be present downstream of each circulator as shown in Figure 7-35. This prevents flow reversal through a circulator that may be off because the system needs minimal flow or because of failure. Without this check valve, much of the flow produced by the active circulator would simply short-circuit backward through the inactive circulator.

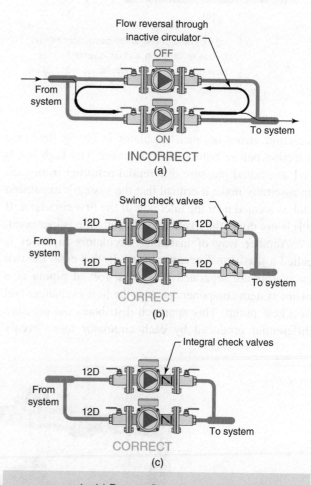

Figure 7-35

(a) Reverse flow occurs through inactive circulator. (b) Swing check valves prevent reverse flow through inactive circulator. Note that 12 diameters of straight pipe are installed between the outlet of circulators and check valves. (c) Circulators with integral check valves can be used in parallel without additional external check valves.

To minimize the possibility of chatter, install at least 12 diameters of straight pipe between the discharge ports of the circulators and the inlet of the check valves. This is indicated by the "12D" notations in Figure 7-35. Circulators with integral check valves can also be used in parallel without additional check valves.

Parallel piped circulators are also used in applications where sustained flow is critical and a backup circulator is deemed necessary. A special controller operates one circulator at a time. After a set time, the controller switches the circulator that is active to ensure relatively even wear on both circulators. If the active circulator should fail, as detected by a flow switch, this controller automatically starts the other circulator. The presence of the check valves downstream of each circulator protects against reverse flow regardless of which circulator is active.

7.5 Variable-Speed/ECM-Powered (VS/ECM) Circulators

One of the biggest advancements in hydronic heating technology over the past two decades is the development of circulators that can automatically vary their speed to adjust to differential pressure changes within the distribution system, as well as perform other specific and user-selectable functions.

These circulators use **electronically commutated motors (ECMs)**. This motor technology is based on a permanent magnet rotor in which very strong rare earth (neodymium) magnets are sealed into the rotor assembly. There are no windings in the rotor and no brushes between the rotor and stationary part of the motor. Because of the latter, ECMs are sometimes also called **brushless DC motors**. Figure 7-36 shows five examples of ECM-powered circulators suitable for most residential and light commercial applications and five ECM-powered circulators for larger systems.

Some of the circulators in Figure 7-36 have cast-iron volutes, and some have stainless steel volutes. The latter have a bright silvery appearance. All of the circulators shown can be ordered with either cast-iron or stainless steel volutes. Circulators with cast-iron volutes are commonly used in closed-loop hydronic heating or cooling systems and are less expensive than circulators with stainless steel volutes. Circulators with stainless steel volutes can be used in closed-loop systems, but, due to higher cost, are most commonly specified for open-loop hydronic systems, or systems conveying potable water, where the corrosion resistance of stainless steel is imperative.

The major components of an ECM-powered circulator are shown in Figure 7-37.

An ECM is regulated by circuitry that changes the magnetic polarity of the stator poles in a way that causes the permanent magnet rotor to spin. This effect is shown in Figure 7-38.

A microprocessor and associated solid-state switching circuitry control electrical current flow through the stator pole windings. This causes the magnetic polarity of each stator pole to change from north (N) to south (S) and vice versa, or remain off. The combined attraction and repulsion forces generated between the permanent magnet rotor and stator poles create a torque that causes the rotor to spin. The faster the stator poles change polarity, the faster the rotor spins.

ECMs are ideal when accurate speed control is needed. They are also significantly more efficient at converting electrical energy into mechanical energy compared to the PSC motors, which have been used in wet rotor circulators for several decades. ECMs also create approximately four times greater **starting torque** than PSC motors. This greatly reduces the possibility of a "stuck rotor" condition after a prolonged shutdown.

The microprocessor-based control circuitry within an ECM-powered circulator allows it to operate according to a coded **instruction set** that resides within the circulator's **EEPROM** (Electrically Erasable Programmable Read Only Memory). This instruction set is also called the circulator's **firmware**. Firmware controls the circulator's microprocessor similar to how software controls the microprocessor in a personal computer. However, the instructions contained in firmware are created and loaded into the circulator's solid-state components by its manufacturer and are typically not capable of being changed without special equipment. Another difference between software and firmware is that instructions contained in firmware are not lost during power outages. They remain ready for operation even if the circulator is unpowered for several years. The ability to regulate the operation of an ECM-powered circulator based on such firmware allows tremendous versatility in customizing the circulator for specific tasks.

By changing the impeller speed, an ECM-powered circulator can generate a wide range of pump curves as depicted in Figure 7-39a. Each time speed is decreased, the pump curve shifts down and to the left. Instead of considering such a circulator as 3-speed, 10-speed, or even 100-speed, it is helpful to think of these curves as blended together to form an **operating region** as

276 Chapter 7 Hydronic Circulators

(a) Grundfos Alpha

(b) Taco 0018

(c) Armstrong Compass H

(d) Wilo Status ECO

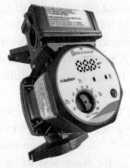
(e) Xylem Ecocirc

(f) Grundfos Magna

(g) Taco 1915

(h) Armstrong Compass R

(i) Wilo Stratos MAXO

(j) Xylem Ecocirc XL

Figure 7-36 | Examples of smaller ECM-powered variable speed circulators. (a,f) *Courtesy of Grundfos*, (b,g) *Courtesy of Taco*, (c,h) *Courtesy of Armstrong*, (d,i) *Courtesy of Wilo*, and (e,j) *Courtesy of Xylem*.

7.5 Variable-Speed/ECM-Powered (VS/ECM) Circulators

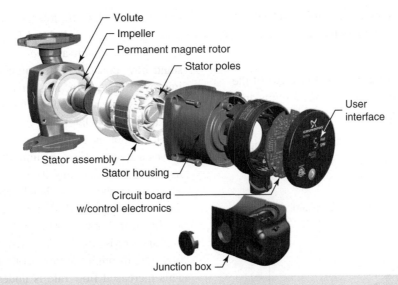

Figure 7-37 | Major components of an ECM-powered circulator. *Courtesy of Grundfos.*

depicted in Figure 7-39b. The upper and lower portions of this operating region are trimmed to avoid conditions that result in poor wire-to-water efficiency or other potential problems. The operating point for a given piping system could lie anywhere within this operating region.

Constant Differential Pressure Control

One instruction set allows an ECM-powered circulator to maintain a constant differential pressure over a wide range of flow rates. This **constant differential pressure control** mode creates the equivalent of a flat pump curve. It also eliminates the need for a differential pressure bypass valve as discussed in Chapter 5, Piping, Fittings, and Valves. When the circulator is commissioned, the installer sets the circulator for the differential pressure desired when all zone valves, or other types of valve-based zoning are open (e.g., design load conditions). The circulator then automatically varies its speed to maintain this set point.

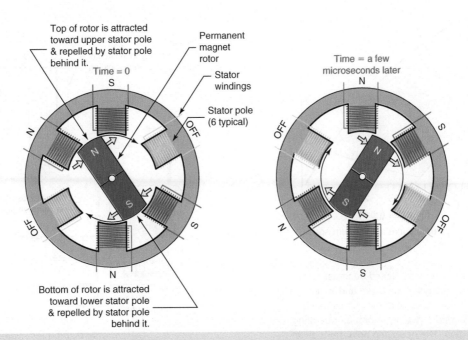

Figure 7-38 | Operating concept of an ECM. The magnetic polarity of the stator poles is regulated by microprocessor-controlled switching circuitry. This creates torque on the permanent magnet rotor causing it to spin.

Constant differential pressure control is well suited to hydronic systems that meet *both* of the following criteria:

1. Use valve-based zoning
2. Are configured so that most of the head loss of the distribution system occurs in the zone circuits and not in the common piping

The system shown in Figure 7-40 is an example. It uses zone valves on each branch circuit and has relatively short headers sized for low head loss. The cast-iron boiler also creates very little head loss. However, this system is also shown with a *fixed-speed* circulator. To understand how constant differential pressure control would improve this system, it is first necessary to understand the *undesirable* operating characteristic created by the fixed-speed circulator.

Consider an existing condition where all four zone valves are open. Next, assume that one of the zone thermostats is satisfied, and its associated zone valve closes. This causes the system head loss curve to steepen, and the operating point shifts upward along the pump curve as shown in Figure 7-41. This increases the differential pressure across the headers, which in turn *increases* the flow rate through the three zone circuits that remain on. This increased flow rate is undesirable. In some cases, increased flow rates create flow noise, cause "bleed through" flow through zone valves that are supposed to be off, or even erode valve seats or fittings within the branch circuits.

Now, consider the same system, but with a variable-speed circulator set for constant differential pressure control. When this circulator is installed, it would be set for the differential pressure desired when all four zone valves are open (e.g., design load conditions). This is the same differential pressure required from the fixed-speed circulator at design load conditions (represented by the lower operating point in Figure 7-41).

When a zone valve closes, the system head loss curve steepens and the operating point again begins

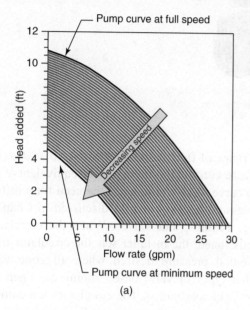

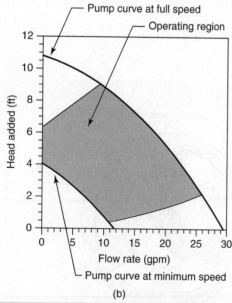

Figure 7-39 — (a) When motor speed decreases, the pump curve shifts down and to left. (b) A large group of pump curves can be blended together to form an operating region. The operating point for a given piping system could be anywhere within this operating region.

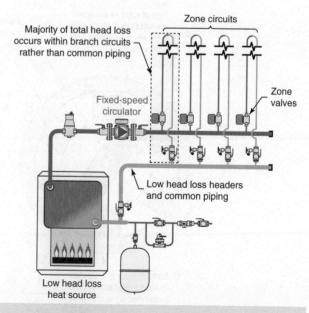

Figure 7-40 — Hydronic distribution system using valves for zoning, a fixed-speed circulator, and having low head loss common piping.

7.5 Variable-Speed/ECM-Powered (VS/ECM) Circulators

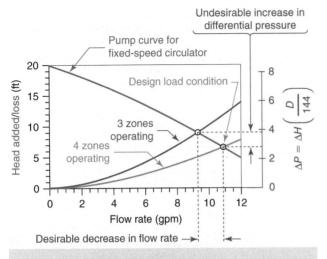

Figure 7-41 When a zone valve closes in this system, an *undesirable* increase in differential pressure across the headers occurs, along with a desirable decrease in flow rate.

circulator detects it is again operating at its original (set) differential pressure, it maintains at that speed as shown in Figure 7-42d. As other zone valves close, the process repeats itself. The net effect of this sequence is that the operating point has shifted horizontally to the left, and thus mimics what would happen if a circulator having a perfectly flat pump curve were used in the system.

When a zone valve opens, the process occurs in reverse. The initial drop in differential pressure is detected by the variable-speed circulator. It responds by increasing speed. When the original (set) differential pressure is restored, it remains at that corresponding speed.

These processes require only a few seconds and are completely regulated by the electronics and firmware within the circulator. No external pressure sensors are required. The desirable result is that the differential pressure across the headers remains almost constant regardless of the number of zones in operation at any time as shown in Figure 7-43. This allows for stable operation of each active zone.

Proportional Differential Pressure Control

Another operating mode currently built into several ECM-powered variable-speed circulators is called

shifting upward along the full-speed pump curve as seen in Figure 7-42b. However, the control circuitry within the variable-speed circulator senses this departure from the user-set differential pressure and begins reducing motor speed. This causes the pump curve to shift down and to the left as seen in Figure 7-42c. When the

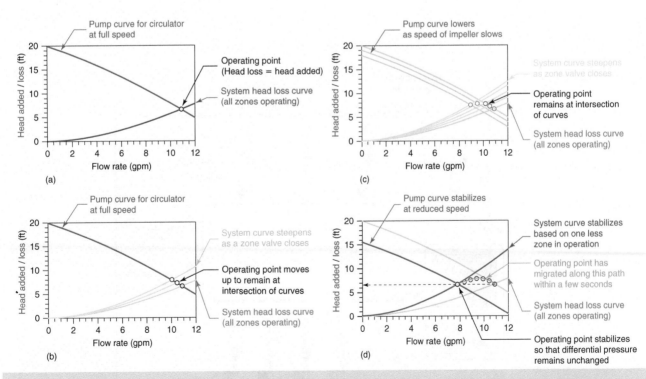

Figure 7-42 Sequence showing shifting of operating point as a zone valve closes in a system using an ECM-powered circulator set for constant differential pressure control. (a) Design load condition. (b) Operating point climbs as zone valve closes. (c) Circulator detects operating point drift and reduces speed. (d) Circulator restores original differential pressure and remains operating at that speed. Net effect is horizontal shift of operating point to the left.

proportional differential pressure control. This mode is best suited to hydronic systems where a significant portion of the total head loss occurs along the mains rather than mostly in the branch circuits. The **two-pipe reverse return piping** arrangement shown in Figure 7-44 is an example of such a system. Such systems are discussed in more detail in Chapter 11, Distribution Piping Systems.

Proportional differential pressure control also varies the circulator's speed in response to attempted changes in differential pressure detected by the circulator. However, with *proportional* differential pressure control, the operating point moves downward as it moves to the left. The target head at zero flow rate is typically one half that at design flow rate. This concept is shown in Figure 7-45. Examples of various (user set) operating point tracks (e.g., sloped lines) based on this control concept are shown in Figure 7-46.

Some ECM-based variable-speed circulators also contain an internal temperature sensor that allows them to detect a sustained drop in water temperature associated with nighttime setback. When the circulator's firmware determines that night setback is in effect, it automatically reduces flow rate.

Many ECM-based variable-speed circulators can also detect when all nonelectric zone valves in a system are closed. Under this condition, the circulator slows to a "sleep mode," where it maintains just enough

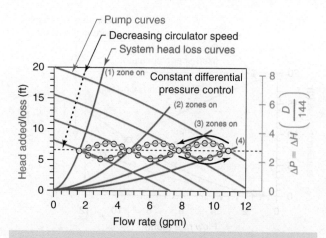

Figure 7-43 — Constant differential pressure control allows the differential pressure across the headers to remain essentially constant regardless of the number of zones operating.

differential pressure to detect when one or more zone valves reopen and then return to normal operation. During this sleep mode, a small circulator typically requires 5 to 9 watts of electrical power. In systems with electrically controlled zone valves, the circulator can be completely turned off when all zones are closed. The latter condition occurs when all end switches within the zone valves are open.

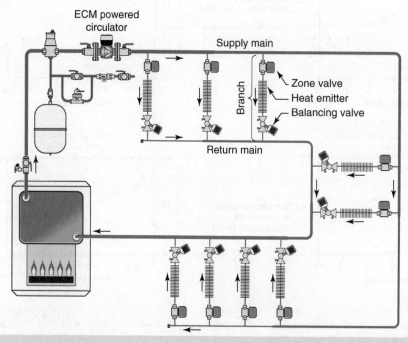

Figure 7-44 — Example of a two-pipe reverse return piping system using a variable speed circulator set for proportional differential pressure control.

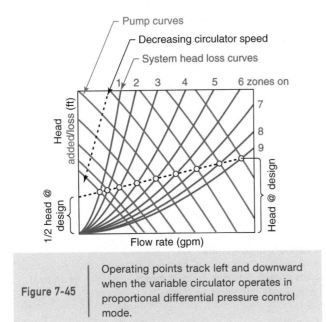

Figure 7-45 | Operating points track left and downward when the variable circulator operates in proportional differential pressure control mode.

Energy Savings Achieved Using VS/ECM Circulators

Each time the speed of a circulator decreases, so does its electrical power input. Every time the operating point shifts to the left (using constant differential pressure control) and left and downward (using proportional differential pressure control), the input wattage to the circulator's motor decreases.

Total savings depends on the amount of time the circulator operates at various speeds (and associated input wattages) during a typical heating season.

The graph in Figure 7-47 is a **flow duration curve**. It shows the number of hours in a year when the flow rate in a hydronic system using valve-based zoning is at or above a given percentage of maximum (design) flow rate. For example, the dashed lines on the graph show that the system's flow rate is at or above 50% of design flow rate about 1,080 hours/year and at or above 75% of design flow rate only 350 hours/year The reduction in system flow rate is due to a combination of partial heat load during milder weather, internal heat gains, nighttime setbacks, and setbacks in unoccupied areas. These reductions imply that the system's circulator can often operate at reduced speed (and thus reduced flow rate) while still satisfying the required heat load.

Circulator manufacturers have used flow duration curves such as that in Figure 7-47 to simulate and compare operation of various circulators over an entire heating season and estimate total electrical energy consumption. These simulations indicate that savings in *electrical energy consumption of 60% or higher are possible using*

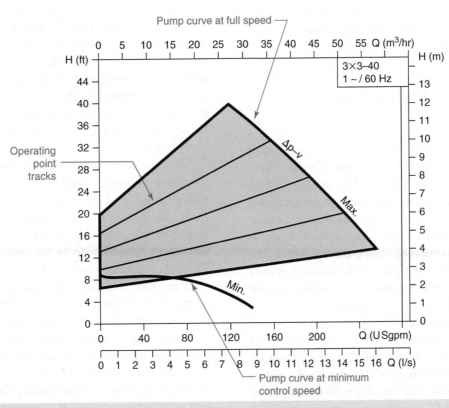

Figure 7-46 | Examples of different operating point tracks (sloped lines) that can be set using proportional differential pressure control. *Courtesy of Wilo.*

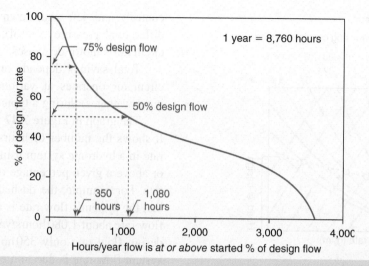

Figure 7-47 | Flow duration curve for a central European climate. Horizontal axis indicates hours in a heating season when total system flow rate is *at or above* the percentage of design flow rate shown on vertical axis.

ECM-powered variable speed circulators compared to conventional wet rotor circulators using PSC motors, and equivalent peak hydraulic performance. The implications of such savings are very significant considering that hundreds of millions of small wet rotor circulators are currently in use worldwide, with thousands more added each day. The existing circulators, most of which use PSC motors, will eventually need to be replaced. Most VS/ECM circulators are manufactured with the same flange dimensions as existing circulators and thus allow for easy replacement.

Speed Control Using 0–10 VDC Control Input

Some ECM-based circulators can also vary speed based on a 0–10 volt DC external control signal. The DC signal could come from a wide variety of controllers and be based on monitored parameters such as temperature, pressure, or flow rate. This option allows the circulator to be adapted to any control technique that can be mapped into a 0–10 volt DC analog control signal. The DC control signal does not power the circulator, it only controls speed. Power to operate the circulator's motor must still be provided as line voltage AC.

A typical configuration would keep the circulator off until the DC control signal reached 2 volts. The 0–2 volt range is often treated as a "null zone" to avoid situations where electrical interference could inadvertently change the circulator's speed. The speed would then increase in proportion to the DC input signal until reaching full speed when the control signal reaches 10 volts DC, as shown in Figure 7-48.

Magnetic Particle Protection

All of the ECM-based variable-speed circulators discussed in this section contain powerful permanent magnets within their rotor assemblies. These magnets create a strong attraction force on any ferrous (iron-containing) materials. Hydronic systems that contain cast-iron circulators, cast-iron boilers, steel expansion tanks, and steel or black iron piping also contain various amounts of iron oxides. These oxides are formed as the initial oxygen molecules in the fill water react with any iron they come in contact with. These oxides are called magnetite and appear as a brownish-gray sludge. The more ferrous metal that is exposed to the system's water and the more make-up water that passes into the system to compensate for any leaks, the higher the magnetite concentration is likely to be.

It is imperative to minimize the possibility of magnetite from becoming lodged between the permanent magnet rotor and the surrounding motor can in any ECM circulator. If such contamination occurs, the magnetite can jamb the circulator's rotor to an extent that cannot be overcome by the motor's starting torque. The only recourse would be to completely remove the rotor from the rotor can and remove all the accumulated magnetite. In some cases, the circulator will not be able to be repaired and must be replaced.

Although circulator manufactures have refined their designs to minimize this possibility, there are additional measures that can be used to further reduce the potential for magnetite to jamb ECM-based circulators. One of the best approaches is to install a magnetic dirt separator in the system, locating it upstream of the circulator whenever possible. Figure 7-48a shows an example of such a device.

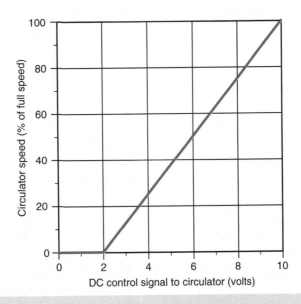

Figure 7-48 | Typical relationship between 0 and 10 VDC control input and circulator speed.

The black ring near the bottom of this device is a strong magnet that can be easily removed from the separator's brass body. As flow passes through the separator, magnetite particles are attracted to its inner brass wall that is surrounded by the magnet, and held there by magnetic force. These particles can be periodically flushed from the separator by pulling off the magnet ring and opening the drain valve at the base of the separator.

The author highly recommends that a magnetic dirt separator be installed in any hydronic system containing one or more ECM-based circulators.

Figure 7-48(a) | Example of a magnetic dirt separator. *Courtesy of Caleffi North America.*

7.6 Analytical Methods for Circulator Performance

The previous sections of this chapter have shown *graphical* methods of finding the intersection of a system head loss curve and pump curve for a circulator. In many situations, it is convenient to be able to find this operating point without having to draw the system and pump curves. This can be done by representing each curve mathematically and then solving for their intersection.

The pump curves for most circulators can be accurately represented using a second-order polynomial function such as that given by Equation 7.1:

Equation 7.1:

$$H_{\text{circulator}} = b_0 + b_1(f) + b_2(f)^2$$

where
$H_{\text{circulator}}$ = head produced by the circulator (feet of head)
f = flow rate through the circulator (gpm)
b_0, b_1, b_2 = constants that are specific for a given circulator

The numbers for b_1 or b_2 can be positive or negative.

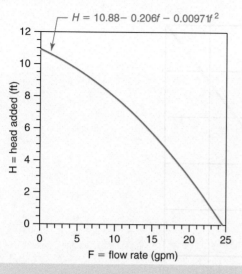

Figure 7-49 | Graph of Equation 7.2 representing a pump curve.

Equation 7.2 is an example of this equation with actual numbers for the constants b_0, b_1, and b_2:

Equation 7.2:

$$H_{circulator} = 10.88 + (-0.206)(f) + (-0.00971)(f)^2$$

When plotted, this equation produces the pump curve shown in Figure 7-49.

The values of the constants b_0, b_1, and b_2 needed for describing a particular pump curve can be determined by mathematical curve fitting. This procedure requires three data points from the pump curve, which represent three flow rates and their associated values for head added. The three flow rates constitute three "x" values, and the associated head gains are the associated "y" values. For the best accuracy, these three points should be well spread out on the pump curve.

Once these three (x, y) points are read from the pump curve, they can be entered into curve-fitting routines such as those included in graphing or spreadsheet software. Such routines will quickly generate the values of b_0, b_1, and b_2 for a given circulator. Alternatively, the values of b_0, b_1, and b_2 can be manually calculated. However, such calculations are quite time consuming. The procedure for such calculations can be found in mathematical reference books under the topic linear regression.

Analytical Method for Finding the Operating Point

Chapter 6, Fluid Flow in Piping, described how to construct the system head loss curve for a given piping system. This curve, which represents the head loss of the piping system as a function of flow rate, is represented analytically using Equation 6.9.

Equation 6.9:

$$H_L = r(f)^{1.75}$$

In this equation, r is the total equivalent hydraulic resistance of the piping system. It is the only number needed to describe the exact shape of the system head loss curve. For example, the system head loss curve used in Example 6.10 is represented by Equation 6.15:

Equation 6.15:

$$H_L = (0.31)(f)^{1.75}$$

The operating point of a piping system such as that described by Equation 6.15 and a pump curve such as that described by Equation 7.2 is the point where these two equations have the same value for $H_{circulator}$ and H_L. The manual method of finding this point involves a logical trial-and-error process called "iteration." The following steps describe this process:

STEP 1. Set up a table with one column for flow rate, f, another for the head added by the circulator, $H_{circulator}$, and a third for the head loss of the piping system, H_L.

STEP 2. Make a reasonable first guess for the system's flow rate.

STEP 3. Use Equation 7.1 with specific values for b_0, b_1, and b_2 to calculate $H_{circulator}$.

STEP 4. Use Equation 6.9 with a specific value of r to calculate H_L.

STEP 5. If $H_{circulator}$ is greater than H_L, the estimated flow rate is too low. Increase the value of flow rate f and go back to step 3.

STEP 6. If $H_{circulator}$ is less than H_L, the estimated flow rate is too high. Lower the value of flow rate f and go back to step 3.

STEP 7. Repeat this procedure until the values of $H_{circulator}$ and H_L are within 0.1 foot or less of each other. The value of flow rate, f, now approximates the actual system flow rate.

The head across the circulator is estimated by averaging the last values of $H_{circulator}$ and H_L.

Example 7.2

Find the flow rate corresponding to the pump curve described by Equation 7.2 and the system head loss curve described by Equation 6.15 using a manual iteration procedure.

Solution:

$$H_{\text{circulator}} = 10.88 + (-0.206)(f) + (-0.00971)(f)^2$$

Pump curve:

System head loss curve: $H_L = (0.31)(f)^{1.75}$

Figure 7-50 shows successive estimates of flow rate with corresponding values of circulator head and system head loss calculated from the two equations. The last estimate of 6.85 gpm showed the head added by the circulator and system head loss to be well within 0.1 foot of each other. Hence, this was the last calculation performed.

The head added by the circulator could now be estimated by averaging the final values of $H_{\text{circulator}}$ and H_L:

$$\frac{9.01 + 8.99}{2} = 9 \text{ feet of head}$$

Flow rate (ft)	$H_{\text{circulator}}$ (ft)	H_L (ft)
4.0	9.9	3.507
6.0	9.29	7.123
8.0	8.61	11.8
7.0	8.96	9.33
6.5	9.13	8.2
6.9	8.99	9.106
6.85	9.01	8.99

Close enough

Figure 7-50 | Iterative solution to find operating point in Example 7.2.

The Hydronic Circuit Simulator Module

The procedure for finding the operating point of a given piping system/circulator combination is critically important for accurately predicting system performance. Unfortunately, it is often neglected due to the time and level of mathematics involved. This often leads to needlessly oversized or undersized circulators being installed. The cost of correcting such situations can be high.

The **Hydronic Circuit Simulator module** in the **Hydronics Design Studio** software makes the process of determining the operating point of a given piping system/circulator combination fast and easy. This module contains a database of the previously discussed b_0, b_1, and b_2 constants for describing the pump curve of more than 500 current and legacy circulators. It also allows the user to quickly configure the piping system by selecting the piping, fittings, and valves in the system. Each time a change is made to any of the inputs, the module recalculates the operating point of the currently specified system and displays the resulting flow rates. The reader is encouraged to use this module as a routine design tool whenever flow rates in hydronic piping systems need to be determined. A screenshot of the Hydronic Circuit Simulator program is shown in Figure 7-51.

7.7 Circulator Efficiency

In Chapter 3, Hydronic Heat Sources, the thermal efficiency of a heat source was defined as the rate of energy output divided by the rate of energy input. A similar ratio of a desired output divided by the necessary input also applies to a circulator. In this case, the desired output is the rate at which mechanical energy (e.g., head) is imparted to the fluid. The necessary input is the rate at which mechanical energy is supplied to rotate the circulator's shaft. The circulator's **hydraulic efficiency** is the ratio of these two rates of mechanical energy transfer.

This definition of hydraulic efficiency does not include the energy losses associated with operating an electric motor, engine, or other "prime mover" to rotate the circulator's shaft and impeller. Those losses will be accounted for using a different definition called **wire-to-water efficiency**.

The hydraulic efficiency of a circulator can be calculated using Equation 7.3.

Equation 7.3:

$$\eta_{\text{circulator}} = \frac{f(\Delta P)}{1,716(\text{hp})}$$

where

$\eta_{\text{circulator}}$ = hydraulic efficiency of the circulator
f = flow rate through the circulator (gpm)
ΔP = pressure differential measured across the circulator (psi)
1,716 = unit conversion constant
hp = horsepower delivered to the circulator shaft by the motor

Although Equation 7.3 is easy to derive from basic physics, it is not easy to use because measurements of input horsepower to the circulator's shaft are difficult to obtain outside of a testing laboratory.

To provide efficiency information to designers, most circulator manufacturers plot the hydraulic efficiency of their medium- and large-sized circulators as a group of contour lines overlaid on the pump curve as shown in Figure 7-52. This information is usually not published for small circulators.

286 Chapter 7 Hydronic Circulators

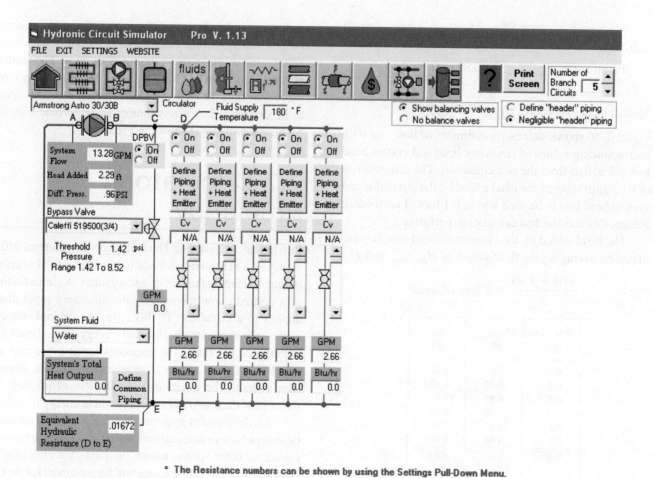

Figure 7-51 | User interface screen for Hydronic Circuit Simulator module in the Hydronics Design Studio software. *Source: Hydronicpros.*

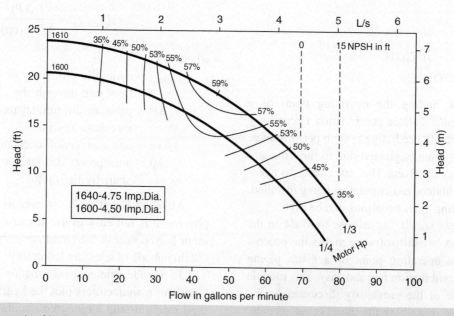

Figure 7-52 | Example of hydraulic efficiency contours plotted along with pump curves. Hydraulic efficiency information is typically available for medium and large circulators, but often not available for small circulators. *Courtesy of Taco, Inc.*

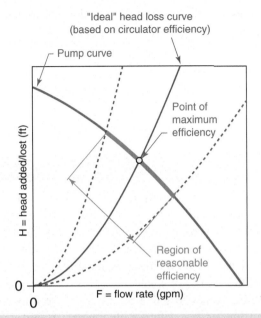

Figure 7-53 For reasonably good hydraulic efficiency, the system head loss curve should cross over the middle 1/3 of the pump curve. The "ideal" system curve (based on circulator efficiency) would always pass through the maximum efficiency point on the pump curve.

The hydraulic efficiency of most circulators is highest near the center of the pump curve as shown in Figure 7-53. For the circulator to operate at or near its maximum efficiency, the system curve must pass through this region of the pump curve. The author recommends that the operating point (e.g., the intersection between the pump curve and system curve) should fall within the middle third of the pump curve.

Wire-to-Water Efficiency of a Circulator

The hydraulic efficiency of a circulator does not factor in energy losses associated with the electric motor driving the shaft and impeller. No electric motor is 100% efficient in converting electrical energy into mechanical energy. Motor efficiencies vary considerably for different types of construction and different loading conditions. The induction motors used on larger commercial and industrial pumps have efficiencies in the range of 80% to 90% when loaded between 25% and 100% of their rated horsepower. The small PSC motors used on wet rotor circulators often have efficiencies less than 50%. A significant portion of the electrical energy input to these motors is lost before being transferred to the impeller shaft as mechanical energy. These energy losses end up as heat that is dissipated to the fluid passing through the circulator or to the surrounding air.

The combined efficiency of both the mechanical and electrical portions of a circulator can be accounted by dividing the rate of mechanical energy transfer to the fluid (e.g., head) by the rate of electrical energy input to operate the circulator. This index is called the circulator's **wire-to-water efficiency** and is especially relevant to smaller circulators that come with a specific motor. Equation 7.3 can be combined with the defining equation of motor efficiency to obtain the following relationship for wire-to-water efficiency:

Equation 7.4:

$$\eta_{\text{wire-to-water}} = \frac{0.4344\, f(\Delta P)}{w}$$

where

$\eta_{w/w}$ = wire-to-water efficiency of the circulator (decimal percent)
f = flow rate through the circulator (gpm)
ΔP = pressure differential measured across the circulator (psi)
w = input wattage required by the motor (watt)
0.4344 = unit conversion factor

The input wattage required by a circulator is not a fixed value, but depends on where the circulator operates on its curve. Therefore, to make use of Equation 7.4, one needs to know the input wattage at its operating point on the pump curve. Data for input wattage as a function of flow rate may be available from the pump's manufacturer or can be measured using a wattmeter. Figure 7-54 shows the electrical power (wattage) drawn by a small wet rotor circulator with a PSC motor over a range of flow rates.

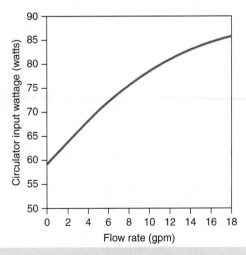

Figure 7-54 Input wattage required by a small wet rotor circulator with PSC motor.

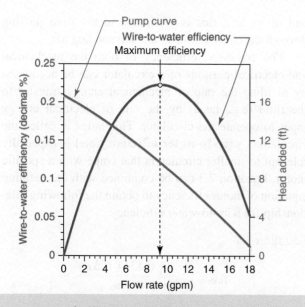

Figure 7-55 Wire-to-water efficiency and pump curve for a small wet rotor circulator. Peak wire-to-water efficiency occurs at a flow rate near the center of pump curve.

Figure 7-55 shows the pump curve of the same circulator along with its calculated wire-to-water efficiency. Notice that peak wire-to-water efficiency occurs at a flow rate that is close to the center of the pump curve. For this particular circulator, it occurs at a flow rate of about 9.3 gpm. Ideally, the intersection of the pump curve and system curve will be at or near this point.

Example 7.3

A small wet rotor circulator operates at a point where it produces a flow rate of 8.0 gpm of 140 °F water and a corresponding head of 9.7 feet. A wattmeter indicates that the circulator is drawing 85 watts of electrical input. What is the wire-to-water efficiency of the circulator?

Solution:

Equation 7.4 can be used, but since it requires the pressure difference across the circulator, not the head, Equation 6.6 is first used to convert the head to a pressure difference. The fluid density is also required and is found from Figure 4-4:

$$\Delta P = \frac{(9.7)(61.35)}{144} = 4.13 \text{ psi}$$

This pressure difference can now be substituted into Equation 7.4 along with the other data:

$$\eta_{\text{pump and motor}} = \frac{0.4344\ (8)(4.133)}{85}$$

$$= 0.169 \text{ or } 16.9\% \text{ efficient}$$

Discussion:

This value seems surprisingly low but is typical of most small, wet rotor circulators, using PSC motors. Several factors contribute to this low efficiency. One is an unavoidable decrease in impeller efficiency as its diameter is reduced. Another is that the volutes of small circulators do not contain diffuser vanes to guide the fluid off the impeller and out of the volute with minimal turbulence. Factors such as the gap between the rotor and stator, internal flow recirculation, and bushing friction also contribute to this relatively low efficiency.

Interestingly, *the full-speed wire-to-water efficiencies of small circulators using ECMs are approximately double those of small circulators using standard PSC motors.* Thus, ECM-powered circulators, even if operated as fixed-speed devices, hold significant operating cost advantages. They can provide a given flow/head capability using about half the electrical input power of a conventional wet rotor circulator with a PSC motor.

7.8 Operating Cost of a Circulator

Many people, even those in the hydronics trade, mistakenly view small circulators as having a reasonable installation cost, but an "insignificant" operating cost. In reality, the total operating cost of any circulator, over its design life, can be several times greater than its installation cost.

The annual operating cost of any device with known electrical input wattage and operating time can be calculated using Equation 7.5.

Equation 7.5:

$$C = \frac{(w)(t)(u)}{1,000}$$

where

C = annual operating cost ($)
w = input wattage (watt)
t = operating hours per year (h)
u = cost of electrical energy ($/kwhr)

Example 7.4

A small wet rotor circulator operates with a steady input power of 85 watts, for an estimated 3,500 hours/year What is its annual operating cost assuming electrical energy costs 14 cents/kwhr?

Solution:

$$C = \frac{(w)(t)(u)}{1,000} = \frac{(85)(3,500)(0.14)}{1,000} = \$41.65/year$$

Discussion:

Many people spend more on gasoline in 1 week than it costs to operate this circulator for a year. Still, one should consider the total operating cost of this circulator over its design life.

Equation 7.6 can be used to convert a first-year operating cost into a total **life cycle operating cost**, while factoring in an assumed rate of inflation on the cost of electricity.

Equation 7.6:

$$C_{total} = C_{year1}\left(\frac{(1+i)^n - 1}{i}\right)$$

where

C_{total} = total operating cost over a period of n years ($)
C_{year1} = operating cost in first year ($)
i = assumed rate of inflation of annual cost (decimal percent)
n = years over which operating cost is being totaled.

Example 7.5

What is the total life cycle operating cost of the circulator and operating conditions described in Example 7.4 over a design life of 20 years, if electrical energy costs increase at an assumed rate of 5% per year?

Solution:

$$C_{total} = C_{year1}\left(\frac{(1+i)^n - 1}{i}\right) = \$41.65\left(\frac{(1+0.05)^{20} - 1}{0.05}\right)$$
$$= \$41.65(33.066)$$
$$= \$1,377$$

Discussion:

This life cycle operating cost is probably five times or more greater than the initial installation cost of a small circulator. It demonstrates that even devices considered to have "insignificant" annual operating costs can still have very significant life cycle operating cost.

Example 7.6

Assume that the circulator in Example 7.4 could be replaced with an ECM-powered circulator that operated on 60% less electrical energy and for the same number of hours. What would be the operational cost savings over the same 20-year period?

Solution:

The first-year operating cost would be $41.65 (0.4) = $16.66.

The total 20-year operating cost would be $16.66 (33.066) = $551.

The life cycle savings would be $1,377 − 551 = $826.

This savings would likely be several times greater than the incrementally higher cost of installing the ECM-powered circulator. This is especially true in systems with valve-based zoning because the installation cost of a differential pressure bypass valve, which is required in systems using valve-based zoning and a fixed-speed circulator, would be eliminated.

Circulator Energy Rating Standard

Over the last several decades, the efficiency of boilers and other types of hydronic heat sources have been based on established industry standards. Examples include the AFUE rating for boilers and the ANSI/AHRI/ASHRAE/ISO Standard 13256-2 for geothermal heat pumps. Both of these standards were discussed in Chapter 3, Hydronic Heat Sources. However, over this same time there was no universally accepted standard for rating the overall "seasonal" efficiency of hydronic circulators. That began to change in 2016.

The **Hydraulic Institute** is a U.S.-based association formed and supported by pump manufacturers. Its mission is to create standards and assist specifiers by providing generic guidelines on selection, installation, application, and maintenance for a wide range of pumps, including circulators used in hydronic systems.

"One of the Hydraulic Institute's recent undertakings was to develop a circulator rating program that allows for energy use comparisons between circulators that are under consideration for a specific hydronic system application. This rating system is based on cooperative efforts between the Hydraulic Institute, the U.S. Department of Energy, and several manufacturers, with a common goal of developing voluntary energy performance standards for a wide range of pumps. That work has led to the development of an **Energy Rating (ER)** procedure and a publication that details that procedure. The publication is entitled "Hydraulic Institute Program Guideline for Circulator Pump Energy Rating Program HI 41.5-2018."

The ER procedure involves gathering several types of performance data for a circulator, including its **best efficiency point (BEP)**, along with electrical input power and hydraulic output power at several flow rates. These data are obtained by testing at approved labs in accordance with Hydraulic Institute standard HI 40.7. The required data are based on operating conditions that represent favorable situations, such as when the circulator is operating at its BEP, as well as less favorable conditions, such as when the circulator is operating at a small fraction of the flow rate under which it achieves maximum wire-to-water efficiency. These data are processed through several calculations that were developed to estimate the circulator's performance over a wide range of operating conditions that could be experienced in typical applications. For variable-speed circulators, these operating conditions include adjustments for different control methods such as constant differential pressure control and proportional differential pressure control.

The details of the procedure are beyond the scope of this text, but can be found in the above-cited publication.

After the data are acquired, and processed through the calculations, the results are used to create an **Energy Rating (ER) label**. Figure 7-56 shows an example of this label.

The circulator associated with the label in figure 1 is identified as having multiple control options, which are listed on the label.

The black horizontal bar in the label of figure 1 shows that the ER value for the circulator could range from a low of 91 to a high of 225, depending on the control mode under which the circulator operated. Control modes that vary the speed of the circulator in response to changes in system flow resistance or temperature generally increase the ER of the circulator. The ER for circulators that can only operate at a fixed speed will a have single ER value, and that value will always correspond to the highest fixed-speed setting.

The label gives an ER value (or range if the circulator has multiple control modes). The label also states a **Weighted Average Input Power (WAIP)** value for the circulator. *The WAIP is expressed in horsepower (not watts)* and is the result of calculations in the rating method. The higher the ER value, the lower the amount of electrical energy the circulator uses, relative to another circulator of approximately the same WAIP value.

Equation 7.6a can be used along with the information on the ER label to estimate the electrical energy saved when comparing two circulators having different ER and WAIP values.

Equation 7.6a:

$$E_s = [(ER_h)(WAIP_h) - (ER_L)(WAIP_L)](0.00746)(T)$$

where

E_s = estimated electrical energy saved by using circulator with *higher* ER rating compared to the circulator with the lower ER rating (kwhr) over time (T)

ER_h = the ER rating of the circulator having the *higher* of the two ER ratings

ER_L = the ER rating of the circulator having the *lower* of the two ER ratings

$WAIP_h$ = the WAIP value for the circulator having the *higher* ER ratings (horsepower)

$WAIP_L$ = the WAIP value for the circulator having the *lower* ER ratings (horsepower)

T = estimated operating time over which comparison is made (h)

Example 7.7

A variable-speed circulator has a "least consumptive" ER rating of 150 and a WAIP value of 0.090 horsepower. Estimate the electrical energy savings of this circulator, operated in its least consumptive mode, compared to a fixed-speed circulator with and an ER value of 70, and a WAIP value of 0.091 horsepower. Assume that both circulators would be operated in a system for 4,000 hours per year.

Solution:

Just enter the stated values in Equation 7.6a and calculate.

$$E_s = [(150)(0.90) - (70)(0.091)](0.00746)(4{,}000)$$
$$= 213 \text{ kwhr}$$

Discussion:

The estimated savings in electrical energy of 213 kwhr over the time of 4,000 hours could be easily converted to a cost-saving by multiplying the 213 kwhr by the local price of electricity in $/kwhr.

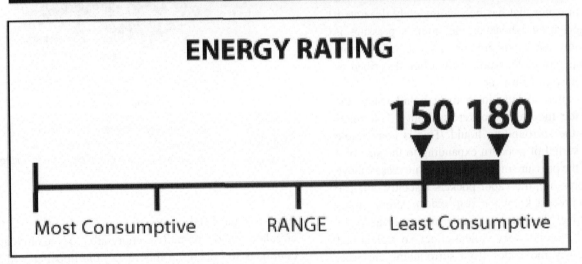

Figure 7-56 | Example of ER rating label from the Hydraulic Institute for a variable-speed circulator. *Courtesy of the Hydraulic Institute.*

Although the testing and calculations involved with establishing ER and WAIP values for a circulator are complex, the use of this information, as provided on the ER label, only requires simple arithmetic. Another calculation could be made to compare the same two circulators, but assuming that the variable-speed circulator would be operated in its "most consumptive" mode, under which its ER value is 115. This would result in an estimated electrical energy savings of 119 kwhr over the same 4,000 hour operating time.

7.9 Cavitation

Chapter 4, Properties of Water, discussed how liquids will begin boiling at specific combinations of pressure and temperature. The minimum pressure that must be maintained on a liquid to prevent it from boiling at some temperature is called its vapor pressure. The process of boiling involves the formation of vapor pockets in the liquid. These pockets look like bubbles in the liquid, but should not be confused with air bubbles. They will form even in water that has been completely deaerated, instantly appearing whenever the liquid's pressure is lowered below the vapor pressure corresponding to its current temperature. This process is often described as the water "flashing" into vapor.

When vapor pockets form from liquid water, the density inside the vapor pocket is about 1,500 times lower than the surrounding liquid. This is comparable to a single kernel of popcorn expanding to the size of a baseball. If the pressure of the liquid then increases above the vapor pressure, the vapor pockets instantly collapse inward in a process known as **implosion**. Although this sounds relatively harmless, on a microscopic level, vapor pocket implosion is a very violent effect, so violent that it can rip away molecules from surrounding surfaces, even hardened metal surfaces.

For a liquid at a given temperature, **cavitation** occurs anywhere the liquid's pressure drops below its vapor pressure. In hydronic systems, the most likely locations for cavitation are in partially closed valves or inside the circulator's impeller.

As liquid flows into the eye of a circulator's impeller, it experiences a rapid drop in pressure. If the pressure at the eye of the impeller drops below the vapor pressure, thousands of vapor pockets instantly form. The vapor pockets are carried outward through the impeller, and when the local fluid pressure rises above the vapor pressure, they implode. This usually occurs near the outer edges of the impeller. The violent implosions create the churning or crackling sounds that are characteristic of a cavitating impeller. A noticeable drop in

(a)

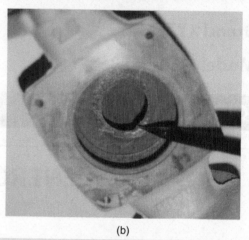
(b)

Figure 7-57 (a) Cavitation damage to a circulator's impeller. *Courtesy of ITT Bell and Gossett.* (b) Cavitation damage to circulator's volute.

flow rate and head occurs because the fluid inside the circulator is now partially compressible. If the circulator continues to operate in this mode, the impeller will be severely eroded in as little as a few weeks of operation. Obviously, cavitation must be avoided if the system is to have a long, reliable life. Figure 7-57 shows the effects of cavitation damage.

Net Positive Suction Head Available

The best way of preventing cavitation is to know what operating conditions allow it to occur, and then design the system to avoid those conditions.

Over the years, engineers have standardized a method of predicting the conditions that cause cavitation in circulator. This method requires the calculation of a quantity called the **Net Positive Suction Head Available**, or **NPSHA**. The NPSHA is a precise description of the fluid state, within a specific piping system, as it enters the circulator. It includes the effects of temperature, pressure, velocity, and the fluid's vapor pressure into a single number, which can be calculated using Equation 7.7.

Equation 7.7:

$$\text{NPSHA} = \frac{v^2}{64.4} + H_e - H_L + (p_s + 14.7 - p_v)\left(\frac{144}{D}\right)$$

where
- NPSHA = net positive suction head available at the circulator inlet (feet of head)
- v = velocity of the liquid in the pipe entering the circulator (ft/sec)
- H_e = distance to the liquid's surface above (+) or below (−) the circulator inlet (ft)
- H_L = head loss due to viscous friction in any piping components between the liquid surface and the inlet of the circulator (feet of head)
- p_s = gauge pressure at the surface of the fluid (psig)
- p_v = vapor pressure of the liquid as it enters the circulator (psi absolute)

If the pressure of the fluid entering the circulator is known, it is not necessary to know the pressure at the surface of the liquid, the distance to the liquid's surface, or the head loss due to viscous friction in the piping leading to the circulator. These effects will be accounted for in the pressure gauge reading at the circulator inlet. This allows Equation 7.7 to be simplified as follows:

Equation 7.8:

$$\text{NPSHA} = \frac{v^2}{64.4} + (p_i + 14.7 - p_v)\left(\frac{144}{D}\right)$$

where
- NPSHA = net positive suction head available at the circulator inlet (feet of head)
- v = velocity of the liquid in the pipe entering the circulator (ft/sec)
- p_i = gauge pressure of the moving fluid measured at the circulator entrance (psig)
- p_v = vapor pressure of the liquid as it enters the circulator (psi absolute)

The NPSHA indicates the difference between the total head of the fluid at the inlet port of the circulator and the head at which the fluid will boil. *NPSHA is a characteristic of the piping system in which the circulator will operate, as well as the properties of the fluid being circulated. It does not depend on the circulator being used.*

Example 7.8

Determine the NPSHA for the piping system shown in Figure 7-58.

a. Assume the water temperature is 140 °F.
b. Assume the water temperature is 200 °F.

Solution (Part A):
A number of quantities need to be determined before using Equation 7.5 to calculate the NPSHA:

The velocity of the water in the pipe entering the circulator is found using the equation for 1-inch copper tubing from Figure 6-2:

$$v = 0.367f = 0.367(8) = 2.94 \text{ ft/sec}$$

The distance to the (top) surface of the water as shown in Figure 7-58 is 15 feet. This distance is considered positive because the surface of the water is above the inlet to the circulator. Thus, $H_e = 15$ feet.

The head loss from the (top) surface of the water to the inlet of the circulator will be that created by 20 feet of 1-inch type M copper tubing carrying the 140 °F water at 8.0 gpm. This can be determined using Equation 6.11:

Equation 6.11:

$$H_L = (\alpha c L)(f)^{1.75}$$

The value of α for water at 140 °F is found from Figure 6-17: $\alpha = 0.0475$.

The value of the pipe size factor c for 1-inch type M copper tube is found in Figure 6-18: $c = 0.01776$.

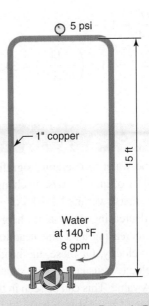

Figure 7-58 | Piping system for Example 7.7.

Substituting into Equation 6.11:

$$H_L = [\alpha c L](f)^{1.75} = [(0.0475)(0.01776)(20)](8)^{1.75}$$
$$= 0.642 \text{ feet}$$

In the case of a closed fluid-filled piping system, the pressure at the surface of the liquid would be the pressure at the top of the piping loop. Therefore, $p_s = 5$ psig.

The vapor pressure of 140 °F water can be read from Figure 4-8: $p_v = 2.9$ psia.

The density of water at 140 °F is found from Figure 4-4: $D = 61.35$ lb/ft^3.

Substituting all data into Equation 7.7 yields the following:

$$\text{NPSHA} = \frac{2.94^2}{64.4} + 15 - 0.642$$
$$+ (5 + 14.7 - 2.9)\left(\frac{144}{61.35}\right) = 53.9 \text{ feet of head}$$

Solution (Part B):

Because the water's temperature has changed, its density, vapor pressure, and head loss due to viscous friction must be recalculated.

From Figure 4-4, the density of water at 200 °F is 60.15 lb/ft^3

From Figure 4-8, the vapor pressure of 200 °F water is 11.5 psia.

From Figure 6-17, the value of α for 200 °F water is 0.043.

The revised head loss is again found using Equation 6.11 with the new value of α:

$$H_L = [(0.043)(0.01776)(20)](8)^{1.75} = 0.581 \text{ ft}$$

Substituting these changes into Equation 7.7:

$$\text{NPSHA} = \frac{2.94^2}{64.4} + 15 - 0.581$$
$$+ (5 + 14.7 - 11.5)\left(\frac{144}{60.15}\right) = 34.2 \text{ feet of head}$$

Discussion:

Increasing the water temperature significantly lowers the NPSHA of an otherwise unchanged system. Other factors that lower the NPSHA of a piping system include decreasing the system's height, lowering its pressure, increasing inlet pipe head loss, or slowing its flow velocity. Of these, the head associated with flow velocity is usually a very small part of the total NPSHA value. This can be seen in this example since the value of $v^2/64.4$ is only 0.134 feet of head.

Net Positive Suction Head Required

Every circulator has a minimum required value of net positive suction head in order to operate without cavitation. This value is called the **Net Positive Suction Head Required** or **NPSHR**. Its value depends on the design of the circulator, as well as the flow rate at which it operates. Circulator manufacturers determine values of NPSHR by lowering the NPSHA to an operating circulator until cavitation occurs. This is done for several flow rates so that a curve can be plotted.

Avoiding cavitation is simply a matter of ensuring that the NPSHA to a circulator always exceeds the NPSHR of the circulator under all possible operating conditions.

The higher the NPSHA is relative to the NPSHR, the wider the safety margin against cavitation. Common design practice is to ensure that the NPSHA to the circulator is always at least 2 feet of head higher than the NPSHR of a circulator.

Example 7.9

The circulator shown in Figure 7-59 has an NPSHR value of 3.5 feet of head. The expected operating point of the circulator is at a flow rate of 5.0 gpm with a 13-foot head gain at the circulator. The system operates at a static pressure of 3 psig. Determine if cavitation will occur.

Solution:

The pressure differential associated with a head gain of 13 feet can be calculated using Equation 6.6. Note the density of 200 °F water is 60.15 lb/ft^3:

$$\Delta P = \frac{HD}{144} = \frac{(13)(60.15)}{144} = 5.43 \text{ psi}$$

Because the expansion tank is (incorrectly) located at the outlet of the circulator, this pressure differential will *lower* the pressure of the fluid entering the circulator by 5.43 psi.

The pressure at the inlet of the circulator is the static pressure of the system (3 psig) minus the pressure drop due to the head loss around the loop (5.43 psig). The inlet pressure is thus $3 - 5.43 = -2.43$ psig. The negative sign indicates that this pressure is below atmospheric pressure. Since the pressure at the inlet of the circulator is known, Equation 7.8 can be used to calculate NPSHA. Again, several pieces of information need to be gathered before substituting into the equation.

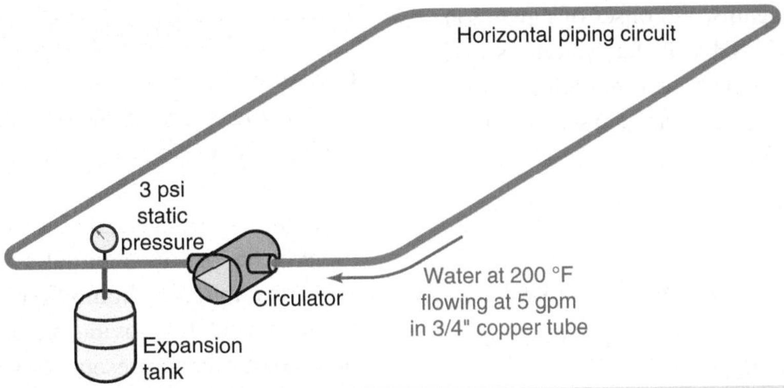

7.9 Cavitation

Figure 7-59 | Horizontal piping loop for Example 7.8.

The flow velocity corresponding to a flow rate of 5.0 gpm in a 3/4-inch copper tube is found using the appropriate equation from Figure 6-2:

$$v = 0.6211f = 0.6211(5) = 3.106 \text{ ft/sec}$$

The vapor pressure of 200 °F water is 11.5 psia.

Substituting into Equation 7.8:

$$\text{NPSHA} = \frac{3.106^2}{64.4} + (-2.43 + 14.7 - 11.5)\left(\frac{144}{60.15}\right)$$
$$= 1.99 \text{ feet of head}$$

Discussion:

Since the NPSHA of 1.99 feet is less than the NPSHR of 3.5 feet, the circulator *will* cavitate. In this example, the cavitation was caused by a combination of low static pressure and high water temperature. The latter results in high vapor pressure. The incorrectly placed expansion tank also contributed to the problem.

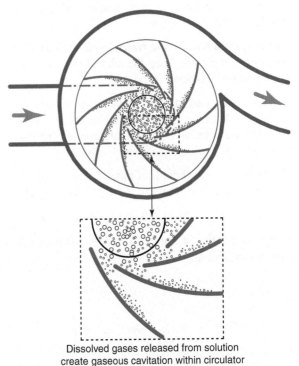

Figure 7-60 | Gaseous cavitation creates air bubbles on the low-pressure side of the impeller vanes.

Gaseous Cavitation

Another type of cavitation that can occur within circulators is called **gaseous cavitation**. It occurs when the liquid pressure at the eye of the impeller drops below the saturation pressure of any dissolved gases—such as oxygen and nitrogen—that may be contained in the water. These gases may also be present as small air bubbles entrained by the liquid. Gaseous cavitation allows bubbles to form on the lower pressure side of the impeller vanes as shown in Figure 7-60.

Gaseous cavitation is not as violent or damaging as vaporous cavitation. It often occurs intermittently when systems are first put into operation and, thus, operating with liquids containing dissolved gases. In closed-loop systems, gaseous cavitation can be eliminated by proper purging methods along with use of high-performance air separators. These are discussed in Chapter 13, Air & Dirt Removal & Water Quality Adjustment.

Guidelines for Avoiding Cavitation

A number of guidelines for avoiding cavitation can be stated based on the previous discussions. These guidelines widen the safety margin against the onset of cavitation. Most strive to make the NPSHA of the system as high as possible. The degree to which each guideline affects the NPSHA of the piping system can be assessed using Equation 7.7 or Equation 7.8.

- Keep the static pressure on the system as high as practical.
- Keep the fluid temperature as low as practical.
- Always install the expansion tank connection near the inlet side of the circulator.
- Keep the circulator low in the system to maximize static pressure at its inlet.
- Do not place any components with high flow resistance (especially flow-regulating valves) near the inlet of the circulator.
- If the system has a static water level, such as a partially filled tank, keep the inlet of the circulator as far below this level as possible.
- Be especially careful in the placement of high head circulators or close-coupled series circulator combinations because they create greater pressure differentials.
- Provide a straight length of pipe at least 12 pipe diameters long upstream of the circulator's inlet.
- Install good deaerating devices in the system.

The schematics in Figure 7-61 depict some of these guidelines in action.

Correcting Existing Cavitation

Occasionally, an installer may be called upon to correct an existing noisy circulator condition. If the circulator exhibits the classic signs of cavitation (i.e., churning or crackling sounds, vibration, and poor performance), the problem can probably be corrected by applying one or more of the previous guidelines.

The circulator should first be isolated and the impeller inspected for cavitation damage, especially if the problem has been present for some time. Look for signs of abrasion or erosion of metal from the surfaces of the impeller. If evidence of cavitation damage is present, the impeller (or rotor assembly in a wet rotor circulator) should be replaced.

Because every system is different, there is no definite order to apply these guidelines in correcting the cavitation problem. In some cases, simple actions such as increasing the system's pressure or lowering its operating temperature will solve the problem. In others, the layout may be so poor that extensive piping modifications may be the only viable option. Obviously, corrections that involve only setting changes are preferable to those requiring piping modifications. These should be attempted first.

In cases where cavitation is not severe, increasing the system's static pressure will often correct the problem. This can usually be done by increasing the pressure setting of the boiler feed water valve. Be sure the increased pressure does not cause the pressure-relief valve to open each time the fluid is heated to operating temperature.

Design "inviting" cavitation
(what's wrong)

1. Low system pressure
2. Throttling valve near circulator inlet
3. Expansion tank near circulator outlet
4. High system operating temperature
5. Lack of air separator and make up water
6. Turbulent conditions upstream of circulator

Design "discouraging" cavitation
(what's right)

1. Higher system pressure
2. Proper air purging at start up
3. Expansion tank near circulator inlet
4. Lower system operating temperature
5. Use of quality air separator
6. Straight pipe upstream of circulator

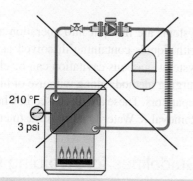
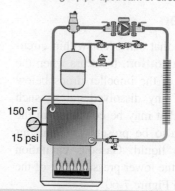

Figure 7-61 | Component arrangements representing good and bad practices for controlling cavitation.

If the system experiencing cavitation is operating at a relatively high water temperature (above 190 °F), try lowering the boiler aquastat setting by about 20 °F. In this temperature range, the vapor pressure of water decreases rapidly as its temperature is lowered. This may be enough to prevent vapor pockets (the cause of vaporous cavitation) from forming. If such an adjustment is made, be sure the system has adequate heat output at the lower operating temperature.

Always check the location of the expansion tank relative to the circulator. If the circulator is pumping toward the expansion tank, consider moving the tank near the inlet of the circulator. This may not only eliminate cavitation but also improve the ability of the system to rid itself of air.

Also look for major flow restrictions near the circulator inlet. These include throttling valves, plugged strainers, or piping assemblies made up of many closely spaced fittings. If necessary, eliminate these components from the inlet side of the circulator.

7.10 Special Purpose Circulators

Several manufacturers now offer circulators that have features needed for specialized applications. The goal is to reduce installation time by preassembling components or offering functionally equivalent components to those that would normally be individually installed at the job site. In most cases, this approach also reduces installation cost.

Temperature-Regulated Circulators

Some manufacturers offer circulators equipped with control circuitry that monitors one or more temperature sensors and varies the circulator's speed in an attempt to maintain a set point temperature or temperature differential. An example of such a variable-speed temperature-responsive circulator is shown in Figure 7-62a, b. Figure 7-62c shows a typical application where the circulator monitors the water temperature entering a conventional boiler and slows when necessary to protect the boiler against sustained flue gas condensation.

Some temperature-regulated circulators can receive input from multiple temperature sensors and can be used for injection mixing purposes. A typical piping configuration for this purpose is shown in Figure 7-63.

The circuitry and firmware within the circulator accept input from a sensor that monitors the water temperature supplied to the distribution system. It also accepts inputs from a boiler inlet temperature sensor and an outdoor temperature sensor. This information is processed by the circulator's firmware and settings. The circulator speed is regulated to maintain the boiler above conditions that would cause sustained flue gas condensation. When this priority condition is achieved and maintained, the circulator speed is adjusted to keep the supply water temperature to the distribution system as close as possible to a calculated target temperature. This type of injection mixing is described in more detail in Chapter 9, Control Strategies, Components, and Systems.

Constrained ΔT Circulators

There are also specialty circulators that are designed to vary flow rate in an attempt to maintain a fixed temperature difference between two sensor locations on the system. That temperature difference can be set at the circulator's user interface. This is called constrained ΔT control. It's appropriate in systems that have *all* the following constraints:

1. Multiple heating zones controlled with valves, or multiple secondary circulators supplied from a common primary loop.
2. Low thermal mass heat emitters.
3. Heat sources that maintain a *constant supply water temperature, at the design load value*, whenever any zone is calling for heat.

Figure 7-64 shows examples of two systems that meet these criteria.

Consider a hydronic heating system with low thermal mass heat emitters that supplies design load heat output when all zones are active and the supply water temperature remains constant at the design load value.

If the heat emitters were not oversized for the design load, all zones would, in theory, remain on until the design load condition subsided (or other factors such as internal gains or intentional thermostat setbacks began influencing the zone loads).

When design load is no longer present in one zone, and the thermostat for that zone turns off the zone valve or zone circulator, less heat would be removed from the distribution system. This change would reveal itself as an *increase in return water temperature* (assuming that the supply water temperature remains constant). The temperature difference (e.g., ΔT) between the beginning and end of the distribution system would decrease. A circulator operating based on constrained ΔT would sense this decrease and respond by reducing speed so that the design load ΔT was reestablished for the zones that remain active. This process would repeat when another zone turned off. This method of control reduces circulator energy use during partial load conditions.

When a zone turned on, *and the supply water temperature remains fixed at the design load value*, the return water temperature decreases because more heat is being removed from the distribution system. A circulator operating based on constrained ΔT would sense this increase in ΔT and respond by increasing speed to reestablish the design load ΔT.

The requirement that the distribution system have low thermal mass heat emitters implies that the temperature changes on the return side of the system would appear quickly as zones turn on and off. A high thermal mass distribution system, such as a heated concrete floor slab, could significantly delay these temperature changes because of heat being absorbed into or released from the thermal mass.

This method of control forces the active portion of the system to operate as if it is always at design load conditions. When a zone doesn't require design load heat input, the associated zone thermostat would have to cycle the zone valve or zone circulator on and off to avoid overheating the space. This, in effect, directs "pulses" of heat into each zone whenever its thermostat calls for heat. The rate of heat delivery during each pulse remains at the design load rate. The duration of each pulse is the time that the zone valve or zone circulator is on. The heat transfer rate multiplied by the time duration

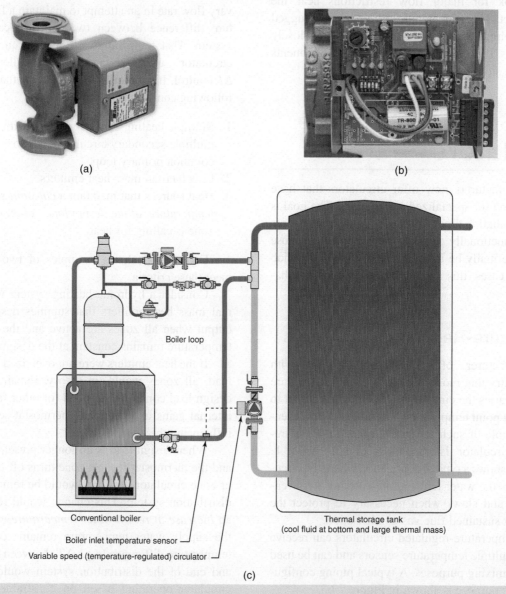

Figure 7-62   (a) Specialty circulator that monitors a temperature sensor (or two temperature sensors) and varies circulator speed to maintain a temperature set point or differential. (b) Internal circuitry and sensor connections. *Courtesy of Taco, Inc.* (c) Application where circulator monitors boiler inlet temperature to prevent sustained flue gas condensation.

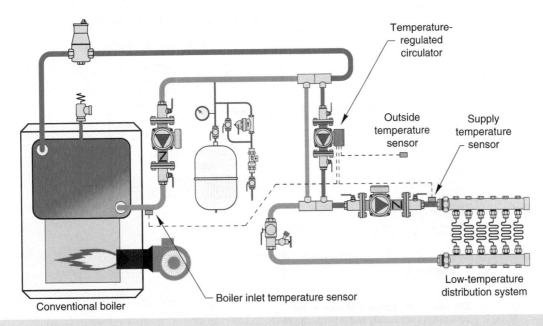

Figure 7-63 | Using a temperature-regulated circulator for variable-speed injection mixing.

of the heat input pulse determines the total heat added to the space during that time. This "pulsed" method of heat delivery has been used in millions of North American hydronic systems. It is generally acceptable if the thermostat differential is reasonable.

However, it's important to understand that not all hydronic systems meet the three previously stated constraints. Many modern systems use outdoor reset control to vary the water temperature supplied to the distribution system based on outdoor temperature. When outdoor reset control is combined with a circulator that operates based on constrained ΔT, the heat output from the distribution system decreases faster than it should based on outdoor reset control theory. This could lead to a reduction in building comfort under partial load conditions. For this reason, the author recommends that circulators using constrained ΔT control only be used in systems that meet all three of the previously stated constraints.

Circulators with Integral Flow-Checks

A standard circulator that is not running cannot prevent flow through its volutes due to pressure differentials created elsewhere in the system. Without the installation of a flow-check valve in each zone circuit (as discussed in Chapter 5, Piping, Fittings, and Valves), some of the flow driven by operating zone circulators will reverse through circulators that are off. This allows heated water to pass through zone circuits that are not in need of heat.

A flow-check valve, or spring-loaded check valve, needs to be installed in each piping circuit anytime multiple circulators are installed in parallel.

To address this common situation, some circulator manufacturers offer circulators with integral spring-loaded check valves. An example is shown in Figure 7-65.

The integral flow-check valve consists of a small spring-loaded valve housed within the outlet port of the circulator's volute. Like a stand-alone flow-check valve, its purpose is to prevent reverse flow as well as forward heat migration due to differences in fluid density. Use of such a circulator eliminates the need of a field-installed flow-check in each zone circuit. Small circulators with integral spring-loaded check valves are also used to prevent heat migration in some injection mixing systems.

The presence of an integral spring-loaded check valve reduces the head available from the circulator at any given flow rate. This is reflected in the pump curve supplied by the manufacturer as seen in Figure 7-65b.

Circulators with Bluetooth Communication

One of the newest features available on some ECM-based circulators is the ability to extract operating information such as input power, flow rate, and head using **Bluetooth communication protocol** between the circulator and a Bluetooth equipped device such as a smartphone or tablet that is within a few feet of the circulator.

300 Chapter 7 Hydronic Circulators

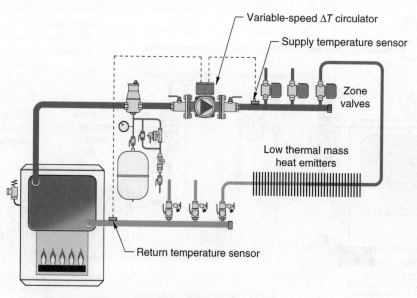

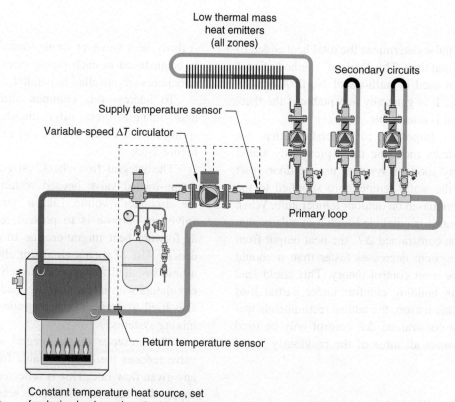

Figure 7-64 | Piping schematics appropriate for use with a constrained ΔT circulator.

7.10 Special Purpose Circulators 301

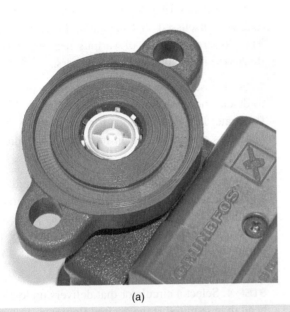
(a)

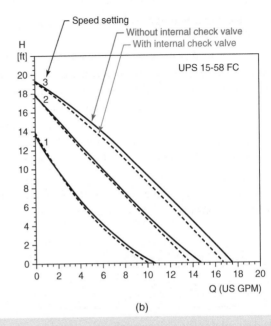
(b)

Figure 7-65 | (a) Circulator with spring-loaded check valve built into volute. (b) Variation in pump curves for a circulator used with and without removable spring-loaded check valve. *Courtesy of Grundfos.*

Figure 7-66 | A circulator communicating with a smartphone using Bluetooth protocol. *Courtesy of John Siegenthaler.*

Figure 7-66 shows an example of such a circulator that is sending continuously updated performance information to a smartphone.

In this case, the circulator has been set to operate in a proportional differential pressure mode. The circulator's current operating point is shown on the graph displayed on the smartphone, along with the control curve for its mode of operation. The numbers at the bottom of the smartphone display indicate that the circulator is currently operating at a flow rate of 1.83 gpm, a head of

9.84 feet, and input wattage of 15.98 watts, and a speed of 3,565 RPM.

Being able to receive this information directly from the circulator allows for several performance checks. For example, if the temperature difference between the supply and return of the circuit containing this circulator is known, it can be combined with the flow rate using the sensible heat rate equation discussed in Chapter 4, Properties of Water, to estimate the rate of heat delivery by that circuit. By knowing the input wattage, flow rate, and head, the wire-to-water efficiency of the circulator can also be calculated using methods discussed earlier in this chapter.

7.11 Selecting a Circulator

The ideal circulator for a given application would produce the exact flow rate and head desired while operating at maximum efficiency, and it would draw the least amount of electrical input power. It would also be competitively priced and built to withstand many years of service. These conditions, although possible, are not likely in all projects because of the following considerations:

- Only a finite number of circulators are available for any given range of flow rate and head requirements. This contrasts with the fact that an unlimited number of piping systems can be designed for which a circulator must be selected.

- Some small circulators are single-speed units that operate on a fixed pump curve. These circulators cannot be fine-tuned to the system requirements through impeller changes, motor changes, and so on, as is often possible with larger circulators. The exception being three-speed circulators where each speed represents a different pump curve.

- The choices may be further narrowed by material compatibility. For example, cast-iron circulators should never be used in open-loop hydronic applications because of corrosion.

- Purchasing channels and inventories of spare parts may limit the designer's choices to the product lines of one or two manufacturers.

The selection process must consider these limitations while also attempting to find the best match between the head and flow rate requirements of the system and circulator. A suggested procedure for finding this match is as follows:

STEP 1. Determine the desired flow rate in the system based on the thermal and flow requirements of the heat emitters. The relationship between flow rate and heat transport is described by the sensible heat rate equation and was discussed in Chapter 4, Properties of Water. The flow requirement of various heat emitters will be discussed in Chapter 8, Heat Emitters.

STEP 2. Use the system head loss curve or its equation to determine the head loss of the system at the required flow rate. Detailed methods for determining the system head loss curve were given in Chapter 6, Fluid Flow in Piping.

STEP 3. Plot a point representing the desired flow rate and associated head loss on the pump curve graph of the circulators being considered. Remember that circulators in series and parallel can be represented by a single pump curve and thus considered in situations where they offer advantages.

STEP 4. Select a circulator that delivers up to 10% more head than required at the desired flow rate. This provides a slight safety factor in case the system is installed with more pipings, fittings, or valves than what was assumed when constructing the system head loss curve.

STEP 5. Finally, when possible based on the circulators that are available, the curve of the selected circulator should (ideally) intersect the system head loss curve within the middle third of the pump curve where the circulator's wire-to-water efficiency is highest.

The following example will pull together many of the previous information and procedures.

Example 7.10

A piping system is being planned to deliver 35,000 Btu/h of heating using a design temperature drop of 20 °F. The fluid used will be water at an average temperature of 160 °F. Using the procedures in Chapter 6, Fluid Flow in Piping, the piping loop is estimated to have a total equivalent length of 325 feet of 3/4-inch type M copper tube. Determine the necessary flow rate in the system and select an appropriate circulator from the family of pump curves shown in Figure 7-67.

Solution:

The "target" flow rate for delivering 35,000 Btu/h of heat transfer using water and a temperature drop of 20 °F is determined using Equation 4.8:

$$Q = (8.01 Dc) f(\Delta T)$$

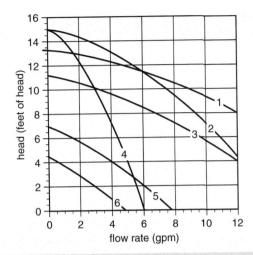

Figure 7-67 | Family of pump curves for Example 7.10.

where
- Q = rate of heat transfer into or out of the water stream (Btu/h)
- 8.01 = a constant based on the units used
- D = density of the fluid (lb/ft³)
- c = specific heat of the fluid (Btu/lb/°F)
- f = flow rate of water through the device (gpm)
- ΔT = temperature change of the water (°F)

From Figure 4-4, the density of 160 °F water is found to be 60.9 lb/ft³

The specific heat of water at 160 °F is 1.0 Btu/lb/°F.

Note: The density and specific heat values could also have been found using the Fluid Properties Calculator module in the Hydronics Design Studio software.

Equation 4.8 is now rearranged to solve for the flow rate:

$$f = \frac{Q}{8.01 \times c \times D \times \Delta T} = \frac{35{,}000}{8.01 \times 1.00 \times 60.9 \times 20}$$
$$= 3.59 \text{ gpm}$$

The head loss of the circuit can now be estimated using Equation 6.11. Doing so requires the value of the fluid properties factor (α) and the pipe size factor (c).

The value of α for water at an average temperature of 160 °F is found in Figure 6-17: $\alpha = 0.046$.

The pipe size factor for 3/4-inch copper tube is found in Figure 6-18: $c = 0.061957$.

These values are combined with the total equivalent length of the circuit and the target flow rate [using Equation 6.11] to estimate the head loss of the system:

$$H_L = [\alpha c L](f)^{1.75} = [(0.046)(0.061957)(350)]$$

$$(3.59)^{1.75} = 9.34 \text{ feet}$$

The target operating point of 3.59 gpm and 9.34 feet of head loss is now plotted on the family of pump curves, as shown in Figure 7-68. Also plotted are several other random points along the system head loss curve calculated using Equation 6.11 with different flow rates. These points allow the system head loss curve to be sketched on the same graph with the pump curves.

The pump curve for circulator 3 intersects the system curve just above the desired operating point, and thus could be used. The curve for circulator 1, although producing slightly more head than necessary, could also be used. In the latter case, the system would operate where the pump curve crosses the system curve at approximately 4 gpm. The operating points that would occur with this system head loss curve, and either pump curve do not fall within the middle third of those pump curves. This is not ideal from the standpoint of achieving the highest possible wire-to-water efficiency. However, given the circulator curves available, it is unavoidable.

The Hydronic Circuit Simulator module in the Hydronics Design Studio software could also have been used to quickly estimate the performance of several circulators in the proposed piping system. Because of its internal database of fluid properties and equivalent lengths, it would significantly reduce the time necessary to make these comparisons.

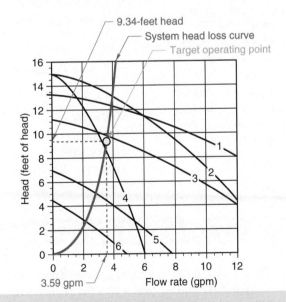

Figure 7-68 | Target operating poiint and system curve sketched on graph of multiple pump curves.

Summary

This chapter has discussed the types of circulators used in hydronic heating systems, as well as methods for describing their performance in a given piping system. The fundamental concept of determining the intersection of the pump curve and system head loss curve will be used in several later chapters as an inherent part of system design. The placement of the circulator so its inlet is near the connection point of the system's expansion tank is also a vital concept to be repeatedly applied in subsequent system design.

It's very likely that variable-speed/ECM-based circulators, which were discussed in section 7.5, will become the "new normal" for residential and light commercial hydronic systems as this edition of Modern Hydronic Heating is being used. The collective electrical energy savings afforded by these circulators is very significant, considering that millions of these circulators will soon be in used worldwide. Wet rotor circulators using traditional PSC motors will eventually be phased out of the market, although many of them will remain operational in systems for 20 years or more after their installation. Hydronic professionals should be familiar with both types of circulator motor technology to deal with both new and existing systems.

By applying the concepts presented in this chapter, the designer can select a circulator that properly supplies the flow rate and head requirements of a piping system, while operating at or near maximum efficiency and without the destructive effects of cavitation.

Key Terms

- best efficiency point (BEP)
- Bluetooth communication protocol
- brushless DC motor
- cavitation
- centrifugal pump
- close coupling
- constant differential pressure control
- constrained ΔT control
- coupling assembly
- EEPROM
- electronically commutated motor (ECM)
- end suction circulator
- Energy Rating (ER)
- Energy Rating (ER) label
- firmware
- flanges
- flat pump curves
- flow duration curve
- gaseous cavitation
- half unions
- hydraulic efficiency
- *hydraulic equilibrium*
- Hydraulic Institute
- Hydronic Circuit Simulator module
- Hydronics Design Studio
- impeller
- implosion
- injection mixing
- inline circulators
- instruction set
- isolation flange
- life cycle operating cost
- magnetic dirt separator
- magnetite
- mechanical energy (head)
- Net Positive Suction Head *Available* (NPSHA)
- Net Positive Suction Head *Required* (NPSHR)
- operating point
- operating region
- permanent split capacitor (PSC) motor
- point of no pressure change
- priming
- proportional differential pressure control
- pump curve
- push/pull arrangement
- rotor
- starting torque
- stator
- steep pump curves
- three-piece circulator
- two-piece circulator
- two-pipe reverse return piping
- volute
- Weighted Average Input Power (WAIP)
- wet rotor circulator
- wire-to-water efficiency

Questions and Exercises

1. Describe the difference between a circulator with a steep pump curve and one with a flat pump curve. Which would be better for use in a multizone system using zone valves? Why?

2. Describe the difference between the head of a circulator and the difference in pressure measured across the circulator using pressure gauges. How is this pressure difference converted to a head value?

3. A circulator with a pump curve as shown in Figure 7-69 is installed in a system operating with 140 °F water. Pressure gauges on the inlet and discharge ports have readings as shown. What is the flow rate through the circulator?

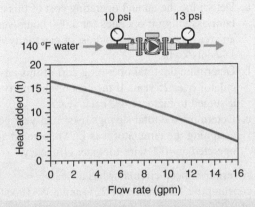

Figure 7-69 | Pressure gauge readings and pump curve for exercise 3.

4. Will using two circulators in parallel double the flow rate in a piping system? What about using two circulators in series? Justify your answer by drawing a pump curve as well as the curves for two such circulators in parallel and in series. Show the intersection of these curves with a system head loss curve sketched on the same set of axes. Estimate the flow rates at each operating point.

5. A piping system has a system head loss curve described by the following equation:

$$H_L = (0.85)(f)^{1.75}$$

where the head loss is in feet and the flow rate (f) is in gpm.
A circulator with the pump curve shown in Figure 7-70 is being considered for this system. Using iterative calculations, find the flow rate and head at the operating point of this piping system and circulator.

6. A circulator has a pump curve described by the equation:

$$H_{circulator} = 7.668 + 0.1086(f) - 0.009857(f)^2$$

Head is in feet, and flow rate f is in gpm. Assuming this circulator is used with the same piping system described in exercise 5, find the flow rate and head at the new operating point using manual iterative calculations.

7. The static pressure in the circulator of a given system is 5 psig when the circulator is off. When the circulator operates, a head of 20 feet is established across it. The fluid being circulated is 140 °F water. Determine the readings of pressure gauges tapped into its inlet and discharge ports when it is running assuming the following:
   a. The expansion tank is connected at the inlet port of the circulator.
   b. The expansion tank is connected to the discharge port of the circulator.

8. A circulator has a pump curve as shown in Figure 7-70. Sketch the resulting pump curve for the following situations:
Two of these circulators connected in series.
Two of these circulators connected in parallel.

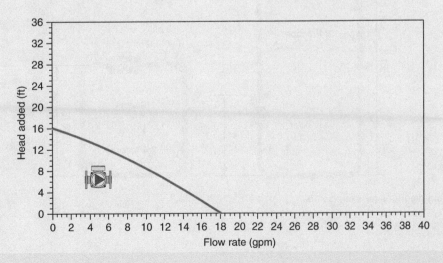

Figure 7-70 | Pump curve for exercise 5.

9. A piping system and wet rotor circulator have the system head loss and pump curves shown in Figure 7-71. The measured power consumption of the circulator while pumping water at 60 °F, and the stated conditions is 75 watts. What is the wire-to-water efficiency of the circulator?

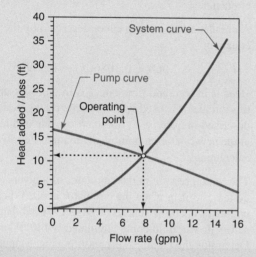

Figure 7-71 | Pump curve and system head loss curve for exercise 9.

10. Determine the NPSHA at the inlet of the circulator for the two systems shown in Figure 7-72.

11. A circulator operates in a piping circuit filled with water at 100 °F. The head developed by the circulator is 18 feet. The flow rate through the circuit is 8.0 gpm. The wire-to-water efficiency of the circulator under these conditions is 20%.
    a. Determine the annual operating cost of this circulator assuming it operates for 3,000 hours/year in an area where the current cost of electricity is 15 cents/kwhr.
    b. Determine the total operating cost of this circulator over 25 years if the cost of electricity is assumed to increase 4% each year?
    c. Determine the total savings over a 25-year period assuming the circulator was ECM powered and operated at 40% wire-to-water efficiency.

12. A variable-speed circulator has a "least consumptive" ER rating of 225, and a WAIP value of 0.080 horsepower. Estimate the electrical energy savings of this circulator, operated in its least consumptive mode, compared to a fixed speed circulator with an ER value of 70 and a WAIP value of 0.081 horsepower. Assume that both circulators would be operated in a system for 3,500 hours per year. Also calculate the cost savings of the less energy consuming circulator compared to more energy consuming circulator assuming electrical energy costs $0.18/kwhr.

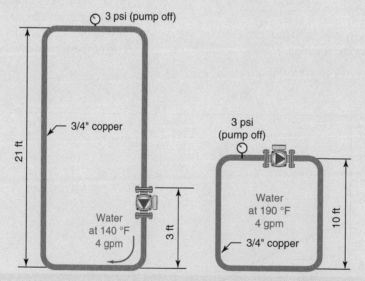

Figure 7-72 | Piping diagrams for exercise 10.

Chapter 8

Heat Emitters

Objectives

After studying this chapter, you should be able to:

- Describe several different types of heat emitters.
- Identify good applications for various types of heat emitters.
- Explain the difference between a radiator and a convector.
- Describe properly sized finned-tube baseboard convectors.
- Describe several types of fan-coils and air handlers.
- Describe different types of panel radiators.
- Describe selected fluted-channel panel radiators based on EN442 standard.
- Estimate the performance of heat emitters over a wide range of conditions.
- Estimate the rate of heat loss from copper tubing.
- Use the concept of steady thermal equilibrium to determine system operating temperature.

8.1 Introduction

Hydronic **heat emitters** release heat from the system's fluid into the space to be heated. A distinct advantage of hydronic heating is the wide variety of heat emitters available to suit almost any job requirement. They range from small **panel radiators** to systems that use the entire floor area of a building as the heat transfer surface. The latter type of system is specialized enough to be covered in a separate Chapter 10, Hydronic Radiant Panel Heating.

Hydronic system designers must select and size heat emitters for a given project, so that comfort and control are achieved in all areas of the building. Technical, architectural, and cost issues must all be considered. The technical issues involve temperature, flow rate, and heat output characteristics of the heat emitter options. Architectural issues include the appearance of the heat emitters in the building, as well as any interference they may create with furniture, door swings, and so on. Cost considerations can greatly expand or restrict heat emitter options. The designer must provide a workable, if not optimal, selection of heat emitters within a given budget. Since most heating systems are designed after the floor plan of the building has been established, the heating system designer is often faced with the challenge of making the system fit the plan. A good knowledge of the available options can lead to creative and efficient solutions. In some cases, these solutions will use two or more different types of heat emitters.

8.2 Classification of Heat Emitters

Some heat emitters directly heat the air in a room. They are properly classified as **convectors** because convective heat transfer is the primary means by which heat is released. Other types of heat emitters release the majority of their heat as **thermal radiation** and are properly referred to as **radiators**. The balance between convective heat transfer and radiative heat transfer can significantly affect the comfort achieved by various heat emitters. Each type has advantages and disadvantages that need to be understood before making a selection. Often a combination of two or more types provides an ideal match for the requirements or restraints imposed by a given project.

In this chapter, the following types of heat emitters will be discussed:

- Finned-tube baseboard convectors;
- Fan-coil convectors;
- Panel radiators;
- Other hydronic heat emitters.

These generic categories encompass most of the available types of heat emitters for residential or small commercial hydronic systems. Sometimes a specialized name is used to describe a specific heat emitter. A **kick space heater**, for example, is a small horizontal fan-coil designed to mount under a cabinet or stair tread. A **towel warmer** is a name used for certain styles of panel radiators. These specialties will be discussed within the previously listed groupings.

8.3 Finned-Tube Baseboard Convectors

Historically, the most common hydronic heat emitter used in residential and light commercial systems *in North America* is the **finned-tube baseboard convector**. It was developed during the 1940s as a flexible, modular, and quickly installed alternative to standing cast-iron radiators. Its widespread use in smaller hydronic systems is largely due to its cost advantage over other heat emitter options.

Figure 8-1 shows a cutaway view of a typical finned-tube baseboard convector. It consists of a sheet metal enclosure containing a copper tube fitted with aluminum fins. The finned tube is often called the **element** of the baseboard, since it is the component from which heat is released.

There are many styles of finned-tube baseboards available. Those intended for residential and light commercial use usually have elements consisting of either 1/2-inch or 3/4-inch copper tubing with rectangular aluminum fins about 2.25 to 2.5 inches wide and high. The enclosures are fabricated from steel sheet and vary from about 6 to 8 inches in height. Most manufacturers sell finned-tube baseboard in lengths from 2 to 10 feet long in increments of 1 foot. For long runs, combinations of these lengths are joined together.

Larger styles of finned-tube convectors are available, intended for use in larger commercial or institutional buildings. They have elements consisting of either copper tube or steel pipe in sizes from 3/4-inch up to 1.25 inches. Their fins are larger (3 to 5 inches on a side) and are available in both aluminum and steel. These larger convectors can release considerably more heat per foot of length than the smaller residential baseboards. They are also built to withstand the greater physical punishment associated with commercial or institutional usage. Because of their size and cost, they are seldom considered for use in smaller hydronic systems.

There are also contemporary variations of standard finned-tube baseboard that are intended for use in low-temperature hydronic systems using renewable energy heat sources such as heat pumps. They are discussed in a later section of this chapter.

In addition to straight lengths of finned-tube baseboard, most manufacturers offer sheet metal enclosure fittings such as end caps, inside and outside corner trim, and joint covers. These allow the baseboard to be fitted to the shape of the room when necessary. They are usually designed to snap into place over the straight length enclosure without the use of fasteners. Figure 8-2 depicts some of the available enclosure trim pieces.

Finned-tube baseboard convectors operate on the principle that heated air rises due to its lower density. This is the same effect that lifts a hot air balloon or makes smoke rise up a chimney. Air in contact with the finned-tube element is heated by natural convection. It rises through the fins and up through the slot at the top of the enclosure. Cooler air near the floor level is drawn in through a slot at the bottom of the enclosure. As long as the finned-tube element is warmer than the surrounding air, natural convection continues to "pump" air up through the enclosure. Most baseboard enclosures

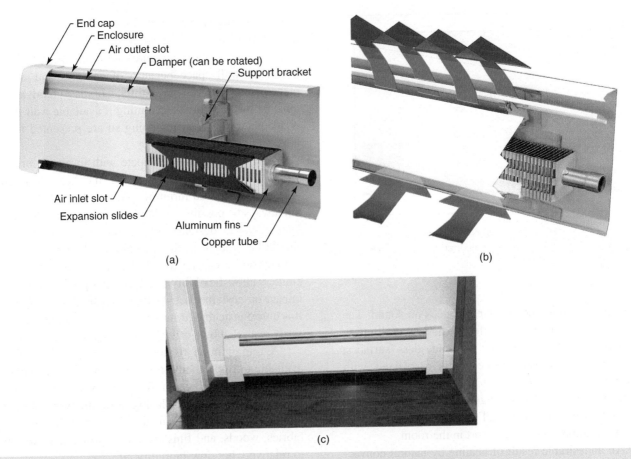

Figure 8-1
(a) Cutaway view of a typical residential finned-tube baseboard convector. *Courtesy of Mestek.*
(b) Airflow through baseboard convector due to natural convection. *Courtesy of Slant Fin Corporation.*
(c) Installed finned-tube baseboard. *Courtesy of Harvey Youker.*

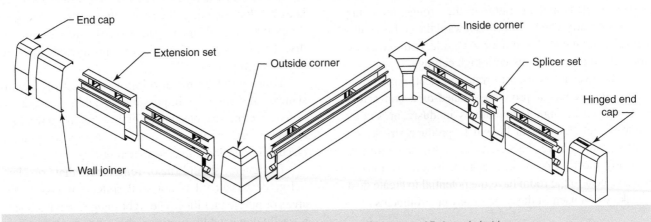

Figure 8-2 | Examples of enclosure trim parts for finned-tube baseboard. *Courtesy of Embassy Industries.*

are equipped with a hinged **damper** over their outlet slot. This damper can be manually set to limit airflow through the baseboard. When fully closed, heat output is reduced by 40% to 50%.

The upward flow of warm air can negate the effects of downward drafts from exterior walls and especially

windows. Because of this, finned-tube baseboard is usually placed along exterior walls and especially under windows. Figure 8-3 depicts the room air circulation created by finned-tube baseboard.

Heating a room by natural convection of warm air creates both desirable and undesirable effects. On the

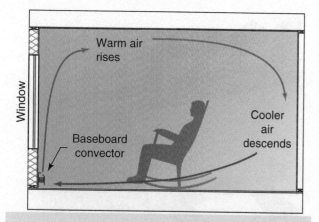

Figure 8-3 — Air circulation within a room established by natural convection currents from a finned-tube baseboard.

plus side, the slow-moving air produces no sound. The rising warm air is also an effective means of limiting condensation of water vapor on cold glass surfaces. It negates the uncomfortable drafts associated with downward movement of cool air along these surfaces. When the baseboard is arranged to blanket an exterior wall with warm air, the wall's surface temperature is increased, improving the comfort in the room.

An undesirable result of heating by natural convection is that the warmest air tends to stay near the ceiling of the room, while cooler air pools at floor level. This effect, often called **stratification**, is much more noticeable in rooms with high ceilings. It is worsened by convectors operating at high water temperatures. The higher the element temperature is, the stronger the rising air currents are. The hot air rises toward the ceiling without fully mixing with room air. As a rule, convectors are not well suited for rooms with high ceilings.

Whenever air is used to transport heat, dust movement can become a problem. Finned-tube convectors that operate at high temperatures in dusty or smoky environments have been known to produce streaks on wall surfaces above their enclosures. All heat emitters that rely mostly on air movement (e.g., convection) to transfer heat to the room have the potential to create dust streaks when used in dusty or smoky environments.

Placement Considerations

Baseboard convectors should be placed, whenever possible, along exterior walls and especially under windows. A number of other factors also enter into placement decisions. Among these are available wall space, furniture placement, and door swings.

Any room that is to be heated by baseboard convectors must have sufficient unobstructed wall space to mount the necessary length of heating element. This requirement often presents a problem, especially in smaller houses. Kitchens and bathrooms often lack the necessary wall space because of cabinets and fixture placement. However, an estimate of baseboard length must be made before determining if available wall space is adequate. Methods for doing so are presented in the next section.

Heat emitters often compete with furniture for wall space. Most homeowners believe the heating system should be fit around furniture placement. While this may not always be possible, it should be reasonably attempted by the system designer. When interference is likely, it should be brought to the attention of the owner so that a compromise can be found. The degree to which heat emitters allow flexibility of interior design can have a big impact on customer satisfaction with the overall system. It is unfortunate that furniture placement is often unnecessarily restricted by a careless placement of heat emitters.

A minimum of 6 inches of free space should be available in front of any finned-tube baseboard. Without it, heat output will be significantly lowered due to inadequate air circulation. Placing furniture very close to baseboard convectors can also lead to degradation of fabrics, woods, and finishes due to prolonged exposure to relatively hot, low-humidity air.

Door swings should also be considered before placing baseboard convectors. Baseboard convectors should not be installed within the arc of the door swing. Their thickness may prevent the door from making its normal swing back to the wall. The closer the baseboard is to the door opening, the worse the problem. Wooden doors that are left open against baseboard can also be discolored, warped, or cracked by prolonged exposure to the warm air.

Every installation can bring its own set of circumstances. It would be impossible to address all situations affecting the placement of baseboard convectors. However, a sampling of some of the most common placement considerations is shown in Figure 8-4.

Designers should also remember that residential-grade baseboard is not well suited for high-traffic areas or public facilities. The light-gauge steel enclosure is easily damaged by furniture shoved against it as seen in Figure 8-5.

Installing Baseboard Convectors

When measuring walls for baseboard convectors, it is important to account for the space required by end caps and other trim accessories. These dimensions will vary from one manufacturer to another. In some cases, trim accessories slide over the enclosure for

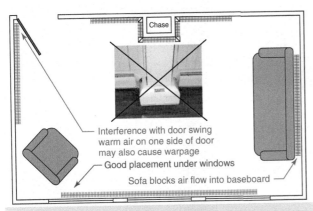

Figure 8-4 | Good and bad placement of finned-tube baseboard within a room.

Figure 8-5 | Finned-tube baseboard is easily damaged in areas with high traffic or public occupancy. *Courtesy of John Siegenthaler.*

simple adjustment of the overall assembly length. This is very helpful in situations where the enclosure runs wall to wall. In situations where stock enclosure lengths combined with sliding end caps will not allow a wall-to-wall fit, the enclosure can be carefully cut to length. The finned-tube element can be shortened as necessary by sliding some of the fins off the end of the tube and cutting the tube. To avoid sliced fingers, always wear gloves when handling the finned-tube element.

Sometimes the required length of baseboard will end 1 or 2 feet short of a corner of a room. When this happens, the enclosure can be capped with a standard end cap or the enclosure assembly without the element can be extended to the wall. This dummy enclosure may or may not contain a standard copper tube soldered to the end of the finned-tube element. Its only purpose is to continue the aesthetics of the baseboard enclosure from wall to wall.

The four basic steps of installing a baseboard convector are as follows:

1. Drill holes for the riser pipes;
2. Attach the enclosure to the wall;
3. Install the finned-tube element;
4. Install the front cover and trim accessories.

The holes for the riser pipes should be located so as to avoid drilling into floor joists, girders, wiring, or other plumbing beneath the floor. One approach is to locate a nearby object such as a plumbing stack that penetrates through the floor, measure the location of the joists or other objects to this reference object under the floor, and then transfer these measurements above the floor to locate the joists. Another method is to drill a small 1/8-inch hole and insert a rigid wire to help locate the hole from under the floor. The distance from the wire to the joists can be measured under the floor and again transferred to the top of the floor. Since floor joists are often located 16 or 24 inches on center, the position of adjacent joists can be measured easily. The location of the floor joist, or other obstructions, may require the enclosure, and/or the finned-tube element to be moved one way or the other. Always check this before fastening the enclosure to the wall.

When the centerline of one riser pipe is located, obtain the centerline of the other by measuring the assembled length of the heating element, along with any fittings or valves that are attached to it. Locate the other hole and again check that it does not fall over a floor joist or other obstruction before boring through. This method of assembling the finned-tube element, with the required fittings and valves, is simpler than trying to calculate the exact position of the holes while accounting for lengths of the finned-tube element, valves, fittings, and so forth.

The holes for the riser pipes to the finned-tube element should have a diameter about 1/2 inch greater than the nominal pipe size used. This provides a space for the tubing to expand when heated without stressing the pipe. It also allows some tolerance for errors in measurement or alignment when soldering. Because the baseboard element is installed at room temperature, it is close to its minimum length. Because of this, most of the 1/2-inch space created by the larger hole should be located on the side of the tube away from the baseboard as shown in Figure 8-6. This allows maximum room for piping expansion. An **escutcheon plate** can be installed on the riser piping to neatly cover the holes through the flooring.

The installer should be careful to mount the enclosure high enough to clear the finished flooring materials. As a rule, the bottom of the enclosure should be at least 1 inch above the subfloor. This allows sufficient space for medium-thickness carpeting and pad. The

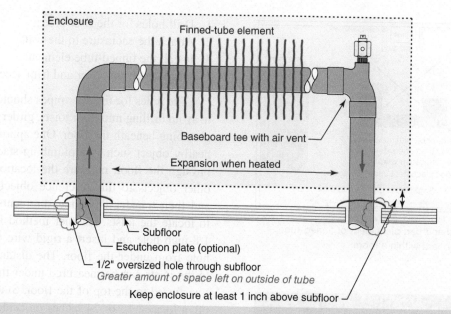

Figure 8-6 | Installation of riser piping to a finned-tube baseboard. Note: The oversized holes through subfloor and space allow for element expansion.

enclosure can be nailed or screwed to the wall. On wood-frame walls, the enclosure should be fastened at intervals not exceeding 32 inches (to every other stud in a typical wall). Allow a slight gap with the wall surface at the end of the enclosure to allow end pieces to be snapped in place.

After the enclosure is mounted, the finned-tube assembly can be placed on the cradles of the enclosure and all soldered joints made up. In some cases, the elbows and riser pipes are first soldered to the element and then this assembly is lowered into the enclosure by sliding the riser pipes down through the holes in the floor. In all cases, be sure the element rests properly on the plastic glides that allow the element to expand and contract without creating squeaking sounds.

In some installations, the supply and return piping must exit the same end of a baseboard. This approach is often used when one end of the baseboard cannot be accessed through the floor. This type of installation requires a **180-degree return bend** at the end of the finned-tube element. The return pipe then runs back over the top of the finned-tube element. A detail of this return bend fitting, with a tapping for an air vent, is shown in Figure 8-7. The return pipe also requires support within the enclosure. It is usually supported by the same brackets that support the finned-tube element.

To complete the installation, snap the front cover in place, and then install the necessary trim pieces. Be sure the damper blade on the enclosure is left in its fully open position.

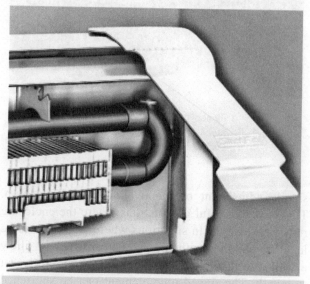

Figure 8-7 | Example of a 180-degree return bend, used when baseboard is supplied and returned from the same end. *Courtesy of Slant/Fin Corp.*

8.4 Thermal Ratings and Performance of Finned-Tube Baseboard

The heat output of finned-tube baseboard is expressed in Btu/h/ft of active element length. The **active length** of the element is the length of pipe covered with fins.

8.4 Thermal Ratings and Performance of Finned-Tube Baseboard

ELEMENT	WATER FLOW	PRESSURE DROP†	HOT WATER RATINGS Btu/hr per linear ft with 65°F entering air												
			110°F	120°F	130°F	140°F	150°F	160°F	170°F	180°F	190°F	200°F	210°F	215°F	220°F
NO. 15-75E Baseboard with 3/4" element	1 GPM	47	140*	190*	240*	190*	350	420	480	550	620	680	750	780	820
	4 GPM	525	150*	200*	250*	310*	370	440	510	580	660	720	790	820	870
NO. 15-50E Baseboard with 1/2" element	1 GPM	260	150*	200*	260*	310*	370	430	490	550	610	680	740	770	800
	4 GPM	2880	160*	210*	270*	330*	390	450	520	580	640	720	780	810	850

† Millinches per foot. *Ratings at 140 °F and lower temperatures determined by multiplying 150 °F rating by the applicable factor specified in Table E in the I=B=R Testing and Rating Standard for Baseboard radiation.

Figure 8-8 | Example of a thermal rating table for finned-tube baseboard. *Courtesy of Slant/Fin Corp.*

This is usually 3 to 4 inches less than the length of the straight enclosure, not including the end caps.

The heat output of the baseboard is dependent on the fluid temperature in the finned-tube element. Manufacturers publish tables showing the heat output of their baseboard units at several different water temperatures. An example of such a ratings table is shown in Figure 8-8.

These ratings are based on testing conducted by the Hydronics Institute using the **IBR Testing and Rating Code** for Baseboard Radiation. Output ratings are usually listed for average water temperatures ranging from 220 to 150 °F. Any output rating at water temperatures below 150 °F must be estimated using correction factors established by the testing standard, which are listed in Figure 8-9.

The correction factors multiply the output rating of the baseboard at 150 °F water temperature. They show how the heat output of finned-tube baseboard decreases rapidly at lower water temperatures. For example, the heat output of a baseboard operating with 100 °F average water temperature is only 28% of the heat output while operating at 150 °F. For this reason, finned-tube baseboard systems are seldom sized for low-temperature operation under design load conditions. However, reduced temperature operation by way of outdoor reset control can be ideal during part-load conditions and will be discussed in Chapter 9, Control Strategies, Components, and Systems.

The ratings in Figure 8-8 are also based on air entering the baseboard at 65 °F and flow rates of either 1.0 or 4.0 gpm. The heat output at 4.0 gpm is obtained by multiplying the heat output at 1.0 gpm by 1.057. The slightly improved performance is the result of increased convection between the water and the inside surface of the tube.

A footnote below the ratings table indicates that the 1.0-gpm rating should be used unless the flow rate is known to be equal to or greater than 4.0 gpm. This assumes that the flow rate through the baseboard is known. Analytical methods will soon be presented that allow baseboard performance to be adjusted for any reasonable flow rate, and thus not restricted to the values listed in rating tables for 1.0 and 4.0 gpm.

Average water temperature (°F)	Correction factor to heat output rating at 150 °F
150	1.0
140	0.84
130	0.69
120	0.55
110	0.41
100	0.28

Figure 8-9 | Correction factors for **heat output ratings** of copper tube/aluminum fin baseboard for water temperatures under 150 °F. *Based on IBR Testing and Rating Code for Baseboard Radiation.*

Another footnote below the thermal ratings table indicates that a 15% **heating effect factor** has been included in the thermal ratings. *This means that the listed values for heat output are actually 15% higher than the measured heat output established by laboratory testing.* The current rating standard for finned-tube baseboard allows manufacturers to add 15% to the tested performance provided a note indicating that this is listed along with the rating data. The origin of the heating effect factor goes back several decades. At that time, it was used to account for the higher output of baseboard convectors placed near floor level rather than at a height typical of standing cast-iron radiators. Unfortunately, it is a factor that has outlived its usefulness. Since the 15% higher heat output cannot be documented by laboratory testing, it is inaccurate to select baseboard lengths that assume its presence. *It is the author's recommendation that baseboard heat output ratings that include a 15% heating effect factor be corrected by dividing them by 1.15 before selecting baseboard lengths.* This bases selection on actual tested performance.

Analytical Model for Baseboard Heat Output

When the rating data shown in Figure 8-8 are combined with the low-temperature correction factors given in Figure 8-9, it is possible to plot the heat output of a finned-tube baseboard over a wide range of average water temperature. These data and a smooth curve drawn through it are shown in Figure 8-10.

All types of finned-tube baseboard exhibit a similar curve of heat output versus water temperature. The values on the vertical axis will vary for baseboards with different size elements, but the shape of the curve over the same range of temperature will be very similar.

A generalized way of expressing heat output for a finned-tube baseboard involves two modifications to the data shown in Figure 8-8. First, the heat output data are normalized by dividing each heat output value by the heat output at 200 °F. Second, the heat output is expressed as a function of temperature difference between the air temperature entering the baseboard and the water temperature in the baseboard. This allows heat output to be determined for situations where the baseboard is heating rooms not kept at normal comfort temperature. A factor is also included to account for heat output over a range of flow rates. This factor is based on methods used in the IBR Testing and Rating Code. The result is an empirical equation that gives the heat output of a generic finned-tube baseboard over a wide range of conditions:

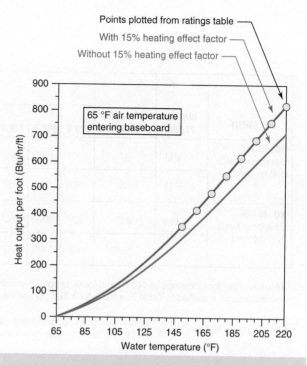

Figure 8-10 — Heat output of a finned-tube baseboard for various average water temperatures, based on ratings data from Figure 8-8, and low water temperature corrections from Figure 8-9. The upper curve includes the 15% heating effect factor. The lower curve has the heating effect factor removed.

Equation 8.1:

$$q' = B(f^{0.04})[0.00096865(T_w - T_{air})^{1.4172}]$$

where

q' = heat output in Btu/h/ft of finned-tube element length

B = heat output of the baseboard at 200 °F average water temperature, 1.0 gpm from manufacturer's literature* (Btu/h/ft)

f = flow rate of water through the baseboard (gpm)

T_w = water temperature in the baseboard (°F)

T_{air} = air temperature entering the baseboard (°F)

Note: 0.04 and 1.4172 are exponents.

The coefficients and exponents in Equation 8.1 are the result of the empirical derivation.

*If the literature indicates a 15% heating effect factor is included in the data, divide the table values by 1.15 to remove it.

Example 8.1

A baseboard has an output of 650 Btu/h/ft at an average water temperature of 200 °F, entering air temperature of 65 °F, and a flow rate of 1.0 gpm. Find the output of this baseboard while operating at an average water temperature of 135 °F, flow rate of 3.0 gpm, and having an entering air temperature of 50 °F. The heat output rating was taken from manufacturer's literature, where a 15% heating effect factor was included in the ratings data.

Solution:

The first step is to remove the heating effect factor by dividing the heat output rating of 650 Btu/h/ft by 1.15.

$$B = \frac{650}{1.15} = 565.2 \text{ Btu/h/ft}$$

This value can now be substituted along with the other data in Equation 8.1.

$$q' = 565.2(3^{0.04})[0.00096865(135 - 50)^{1.4172}]$$
$$= 565.2(1.0449)[0.5254] = 310.3 \text{ Btu/h/ft}$$

Discussion:

Although Equation 8.1 appears complex, it can easily be entered into any spreadsheet program or programmable calculator. It is a powerful analytic tool with the ability to replace dozens of lookup tables that might otherwise be used to show the baseboard's performance at nonstandard operating conditions.

Low-Temperature Finned-Tube Baseboard

Low-temperature heat sources such as condensing boilers and hydronic heat pumps are becoming increasingly common in modern hydronic systems. This has led to redesign of traditional heat emitter devices in ways that make them compatible with these low-temperature heat sources.

Figure 8-11 shows an example of one such heat emitter. This product is available in North America and provides substantially higher heat output at low average water temperatures in comparison to standard finned-tube baseboard.

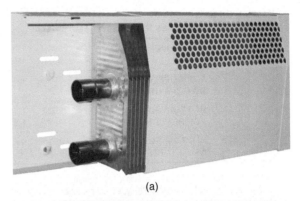

(a)

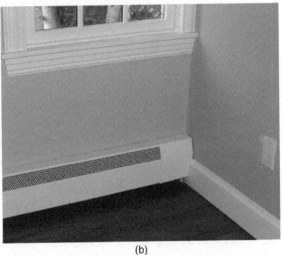

(b)

Figure 8-11 (a) Low-water temperature baseboard with large fins and dual water tubes. (b) Installed baseboard. *Courtesy of Smith's Environmental Products.*

The fins used in the low-temperature baseboard shown in Figure 8-11a have approximately three times the surface area of the fins used in standard residential finned-tube baseboard. The finned-tube element also has two ¾ inch copper tubes, compared to the single ½ inch or ¾ inch copper tube used in standard baseboard. This combination of additional fin area and tube-to-fin contact creates an element with significantly higher heat transfer potential in comparison to standard residential finned-tube baseboard.

Figure 8-12 shows the heat output table for the baseboard shown in Figure 8-11.

This rating table is based on test flow rates of 1 and 4 gpm. These are the same reference flow rates used for standard residential finned-tube baseboard. They do not imply that the baseboard *must* operate at these flows. Rather, they show the effect that increasing the flow rate through the finned-tube element slightly increases heat output. This is the result of increased convective heat transfer between the water and inner tube wall at higher flow rates.

316 Chapter 8 Heat Emitters

Heating Edge™ Hot Water Performance Ratings		Flow Rate GPM	PD in ft of H₂O	Average Water Temperature (BTU/hr/ft @AWT in °F)											
				90°F	100°F	110°F	120°F	130°F	140°F	150°F	160°F	170°F	180°F	190°F	200°F	210°F
TWO SUPPLIES	PARALLEL	1	0.0044	130	205	290	385	460	546	637	718	813	911	1009	1113	1215
		4	0.0481	155	248	345	448	550	651	755	850	950	1040	1143	1249	1352
TOP SUPPLY	BOTTOM RETURN	1	0.0088	105	169	235	305	370	423	498	570	655	745	836	924	1016
		4	0.0962	147	206	295	386	470	552	640	736	810	883	957	1034	1110
BOTTOM SUPPLY	TOP RETURN	1	0.0088	103	166	230	299	363	415	488	559	642	730	819	906	996
		4	0.0962	140	212	283	350	435	524	623	722	792	865	937	1013	1093
BOTTOM SUPPLY	NO RETURN	1	0.0044	75	127	169	208	260	311	362	408	470	524	576	629	685
		4	0.0481	85	140	203	265	334	410	472	536	599	662	723	788	850

Performance Notes: • All ratings include a 15% heating effect factor • Materials of construction include all aluminum "patented" fins at 47.3 per LF, mechanically bonded to two 3/4" (075) type L copper tubes ("Coil Block") covered by a 20 gauge perforated, painted cover all mounted to a backplate. Please see dimensional drawing for fin shape and dimensions • EAT=65°F • Pressure drop in feet of H₂O per LF.

Figure 8-12 | Heat output table of low-temperature baseboard shown in Figure 8-11. *Courtesy of Smith's Environmental Products.*

The heat output table also shows ratings for different flow arrangements through the dual tube set. Note that the highest heat output, for a given flow rate and average water temperature, occurs with parallel flow through both tubes.

A footnote below the heat output table indicates that a 15% **heating effect factor** has been included in the thermal ratings. For the same reason discussed for standard baseboards, the author recommends that the 15% heating effect factor be removed from the ratings by dividing these heat outputs by 1.15 before selecting baseboard lengths.

Figure 8-13 compares the output of the low-temperature baseboard shown in Figure 8-11 to standard residential baseboard. Both curves are based on a flow rate of 1 gpm and do not include the 15% heating effect factor. The output curve for the low-temperature baseboard is based on parallel flow through both tubes (e.g., each tube flowing in the same direction at 0.5 gpm per tube flow rate).

One can view the higher output of the low-temperature baseboard two ways:

1. A given length of the low-temperature baseboard, operating at a given condition, provides significantly higher output. Thus, a shorter length of baseboard can provide the necessary heat output for a room compared to standard residential baseboard.
2. By using the same length of Heating Edge baseboard, the average water temperature required for a given heat output drops substantially.

Point 1 is useful when the maximum heat output needs to be "squeezed" into a given wall space. For example, suppose there is only one available wall space

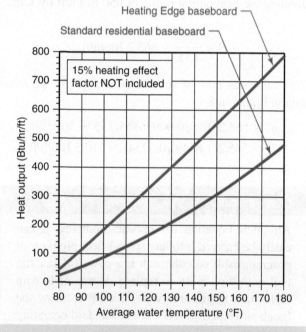

Figure 8-13 | Comparative heat output versus average water temperature for standard residential finned-tube baseboard versus the low-temperature baseboard shown in Figure 8-11. Room temperature assumed to be 70 °F.

within a room, and it measures 6 feet. At a given average water temperature, the low-temperature baseboard will yield approximately double the heat output of standard residential baseboard.

Point 2 is of particular interest in the case of renewable energy heat sources, or condensing boilers. The lower the average water temperature at which the baseboard can still provide design heating load, the higher the thermal efficiency of the heat source.

Example 8.2

A room has a design heating load of 2,250 Btu/h. At an average water temperature of 110 °F, it requires 18 feet of standard residential baseboard to meet this load. Determine the following:

a. If the average water temperature in the finned-tube element remained at 110 °F, how many feet of the low-temperature baseboard, referenced by the upper curve in Figure 8-13, and with both tubes operating in parallel, and a total flow rate of 1 gpm, would be required?

b. If 18 feet of the low-temperature baseboard was used instead of standard baseboard, with both tubes operating in parallel, and a total flow rate of 1 gpm, how much lower could the average water temperature be?

Solution:

Based on the heat output graph shown in Figure 8-13, the low-temperature baseboard at an average water temperature of 110 °F releases 250 Btu/h/ft. Thus, the required length of baseboard to meet the design heating load is:

$$L = \frac{2{,}250 \text{ Btu/h}}{250 \frac{\text{Btu}}{\text{h} \cdot \text{ft}}} = 9 \text{ ft}$$

Assuming 18 feet of wall space was available, and equipped with the low-temperature baseboard, the required heat output of each foot would be:

$$\frac{2{,}250 \text{ Btu/h}}{18 \text{ ft}} = 125 \frac{\text{Btu}}{\text{h} \cdot \text{ft}}$$

Entering the vertical axis of Figure 8-13 at 125 Btu/h/ft, draw a line to the heat output curve for low-temperature baseboard, then down to the horizontal axis to find that an *average* water temperature of approximately 92 °F will be required to produce this heat output.

Discussion:

The *supply* water temperature to the baseboard would be higher than 92 °F—how much higher depends on the flow rate. At low water temperatures, a drop of 10 to 20 °F across the heat emitter is typical under design load conditions. Thus, the supply water temperature to the low-temperature baseboard in this example would be in the range of 97 to 102 °F (e.g., half the design temperature drop added to the average water temperature). Operating the balance of the system at a supply temperature of 102 °F rather than 120 °F, at design load conditions, would significantly boost the performance of solar thermal collectors, heat pumps, or a mod/con boiler.

The heat output of the low-temperature baseboard shown in Figure 8-11, with both tubes configured for parallel flow, can be found using Equation 8.2. This equation is based on performance that excludes the 15% heating effect factor from the manufacturer's heat output table.

Equation 8.2:

$$q = 2.0063\,(f^{0.127})(T_w - T_{air})^{1.2643}$$

where

q = heat output (Btu/h/ft of finned-tube element length)

f = flow rate of water through the baseboard (gpm)

T_w = water temperature at a specific location within the baseboard element (°F)

T_{air} = air temperature entering the baseboard (°F)

Note: 0.127 and 1.2643 are exponents.

The coefficients and exponents in Equation 8.2 are the result of the curve fitting to data provided in the heat output tables.

Example 8.3

Estimate the heat output of the low-temperature baseboard of Figure 8-11 (in Btu/h/ft) at a location where the water temperature in the finned-tube element is 115 °F, and the total flow rate through the parallel-piped tubes is 3 gpm. Assume the air temperature entering the bottom of the baseboard enclosure is 67 °F.

Solution:

Substituting the stated operating conditions into Equation 8.2 yields the following:

$$q = 2.006(f^{0.127})(T_w - T_{air})^{1.2643}$$

$$= 2.006(3^{0.127})(115 - 67)^{1.2643} = 308 \frac{\text{Btu}}{\text{h} \cdot \text{ft}}$$

> **Discussion:**
>
> The calculated heat output of 308 Btu/h/ft only applies at the exact location on the finned-tube where the water temperature is 115 °F. The rate of heat output would be slightly higher upstream of this point and slightly lower downstream of this point. This continual drop in water temperature must be accounted for when determining the total heat output of the baseboard. It will be discussed next.

Heat Output from a Specific Finned-Tube Baseboard

Consider the temperature of water as it flows along a finned-tube baseboard element. Since heat is continually leaving the element, the water temperature must continually drop, as illustrated in Figure 8-14.

If the drop in water temperature was linear (e.g., at same rate along the entire element), then total heat output from the element could be accurately estimated by assuming the entire element operates at the *average* of the inlet and outlet water temperature. However, because the rate of heat transfer at every location on the element depends on the difference between the fluid temperature at that location and the surrounding air temperature, the rate of heat transfer is greater near the inlet of the element compared to near the end. This creates the curvature in the temperature profile seen in Figure 8-14. The lower the flow rate through the element, the more pronounced this curvature becomes.

When the inlet temperature and flow rate for a baseboard are known, the most accurate estimate of heat output comes from mathematically integrating the heat output functions given as Equations 8.1 and 8.2 along the length of the baseboard's element. This integration generates an equation that gives the outlet temperature of the fluid passing through the element as a function of the other operating conditions. Equation 8.3 provides the result of this integration for a *standard residential finned-tube baseboard*.

Equation 8.3:

$$T_{W_{out}} = T_{air} + [(T_{W_{in}} - T_{air})^{-0.4172} + \left(\frac{B(5.04 \times 10^{-5})}{Dc(f)^{0.96}}\right) l]^{-2.397}$$

where

$T_{W_{out}}$ = fluid temperature leaving baseboard (°F)
$T_{W_{in}}$ = fluid temperature entering baseboard (°F)
T_{air} = air temperature entering baseboard (°F)
D = fluid density (lb/ft³)
c = fluid specific heat (Btu/lb/°F)
f = fluid flow rate (gpm)
c = fluid specific heat (Btu/lb/°F)
B = heat output rating of baseboard at 200 °F water/65 °F air temp/1 gpm flow rate (Btu/h/ft)*
l = length of finned-tube element in baseboard (ft)
-0.4172, 0.96, and -2.397 are all exponents (not multipliers)

*If the manufacturer's heat output table has a footnote indicating that the heat outputs include a 15% heating effect factor, the author recommends that the manufacturer's output value at 200 °F water temperature and 1 gpm flow rate be divided by 1.15 to obtain the value of B. This removes the heating effect factor from the calculation.

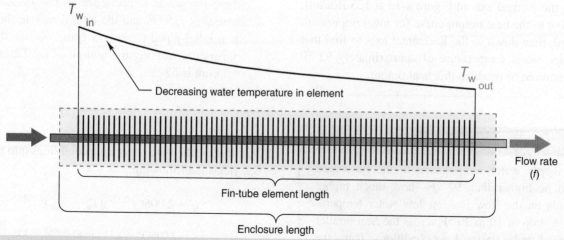

Figure 8-14 | The water's temperature is continually decreasing as the water moves along a finned-tube element within a baseboard.

Equation 8.4 provides the integration result for the low-temperature baseboard shown in Figure 8-11, and referenced to the upper curve in Figure 8-13, with both tubes operating in parallel. The heating effect factor was removed when determining this equation.

Equation 8.4:

$$T_{w_{out}} = T_{air} + [(T_{w_{in}} - T_{air})^{-2.2643}$$
$$+ \left(\frac{0.0662}{Dc(f)^{0.873}}\right)l]^{-3.7836}$$

where

$T_{w_{out}}$ = fluid temperature leaving baseboard (°F)
$T_{w_{in}}$ = fluid temperature entering baseboard (°F)
T_{air} = air temperature entering baseboard (°F)
D = fluid density (lb/ft³)
c = fluid specific heat (Btu/lb/°F)
f = fluid flow rate (gpm)
l = length of finned-tube element in baseboard (ft)
-0.2643, 0.873, and -3.7836 are all exponents (not multipliers)

Once the outlet temperature from the baseboard element is calculated, Equation 4.8 (the sensible heat rate equations) can be used to calculate the total heat output from the baseboard.

Equation 8.4 repeated:

$$Q = (8.01Dc)f(\Delta T)$$

where

Q = rate of heat transfer into or out of the water stream (Btu/h)
8.01 = a constant based on the units used
D = density of the fluid (lb/ft³)
c = specific heat of the fluid (Btu/lb/°F)
f = flow rate of water through the device (gpm)
ΔT = temperature change of the fluid as it passes through the device (°F)

Because they require the density and specific heat of the fluid passing through the element as inputs, Equations 8.3, 8.4, and 4.8 allow heat output to be determined when fluids other than water are used. In the strictest sense, the density and specific heat of the fluid flowing through the finned-tube element are both functions of temperature, and thus could be represented as temperature dependent when the integration is performed. However, for the fluids commonly used in hydronic heating systems, the variation in these properties is minor over the temperature drop range of most individual baseboards. To simplify the mathematics, these fluid properties can both be estimated at an estimated average fluid temperature within the finned-tube element.

Example 8.4

In Example 8.2, it was determined that 9 feet of low-temperature baseboard, configured for parallel flow in the dual-tube element, and operating at an average water temperature of 110 °F, and assumed water flow rate of 1 gpm, would provide the 2,250 Btu/h required for a given room at design load conditions. Use Equations 4.8 and 8.4 to assess the accuracy of this calculation.

Solution:

If 9 feet of low-temperature baseboard was releasing 2,250 Btu/h at a flow rate of 1 gpm, the temperature drop across the element could be estimated using Equation 4.8. Water at the average temperature in the baseboard (e.g., 110 °F) has a density of 61.8 lb/ft³. The specific heat of water is assumed to remain at 1.00 Btu/lb/°F. Substituting these values along with the given data into a rearranged form of Equation 4.8 yields:

$$\Delta T = \frac{Q}{(8.01Dc)f} = \frac{2,250}{(8.01 \times 61.8 \times 1.00) \times 1} = 4.55 \text{ °F}$$

If the average water temperature was 110 °F, the inlet water temperature can be estimated as follows:

$$110 + \left(\frac{4.55}{2}\right) = 112.3 \text{ °F}$$

This inlet water temperature along with the given flow rate is then entered into Equation 8.4 to calculate the outlet temperature from the finned-tube element.

$$T_{w_{out}} = T_{air} + [(T_{w_{in}} - T_{air})^{-0.2643}$$
$$+ \left(\frac{0.0662}{Dc(f)^{0.873}}\right)l]^{-3.7836}$$
$$= 65 + [(112.3 - 65)^{(-0.2643)}$$
$$+ \left(\frac{0.0662}{61.8 \times 1 \times (1)^{0.873}}\right)9]^{-3.7836}$$
$$= 107.81 \text{ °F}$$

At this point, the inlet and outlet temperatures are determined, and the flow rate is known. Thus, Equation 4.8 can be used to determine total heat output from the baseboard.

$$= (8.01Dc)f(\Delta T) = (8.01 \times 61.8 \times 1.00) \times 1$$
$$\times (112.3 - 107.81) = 2,223 \text{ Btu/h}$$

> **Discussion:**
>
> The total heat output calculated using Equations 8.4 and 4.8 (2,223 Btu/h) is very close to the 2,250 Btu/h estimated for 9 feet of Heating Edge baseboard operating at an average water temperature of 110 °F. A reasonable question then arises: Why bother with the more complex calculations associated with Equation 8.4 [or Equation 8.3 in the case of standard baseboard], when they yield results very similar to those based on average water temperature? The answer is that the more complex relationships represented in Equations 8.4 and 8.3 are able to better model situations that deviate from the assumptions used in the manufacturer's heat output table. These include operating the baseboard at flow rates other than the 1 and 4 gpm values listed in the heat output tables, as well as at different conditions for room air temperature, and use of fluids other than water (such as a water/antifreeze solution).

8.5 Sizing Finned-Tube Baseboard in Series Circuits

The most common way in which multiple finned-tube baseboards are piped is connecting them into series circuits. The outlet from baseboard #1 leads through piping that connects to the inlet of baseboard #2. The outlet of baseboard #2 leads through piping to the inlet of baseboard #3, and so on.

This section describes two methods for selecting the lengths of finned-tube baseboards *connected in a series circuit*. The first is a simple but less accurate method suitable for preliminary sizing. The second not only requires more calculations but also returns more accurate results. After discussing manual sizing procedures, a software module called **Series Baseboard Simulator** from the Hydronics Design Studio will also be discussed. The section concludes with a discussion of size methods for parallel-connected baseboard.

Sizing Method 1 (IBR Method)

Preliminary lengths of finned-tube baseboard can be selected by basing the heat output of each baseboard on the *average temperature of the fluid in the piping circuit*.

The **IBR method**, as it is often called, customarily takes this to be 10 °F lower than the outlet temperature of the heat source. The procedure is as follows:

STEP 1. Determine the average temperature of the fluid in the circuit:

Equation 8.5:

$$T_{average} = T_{heat\ source\ outlet} - 10$$

STEP 2. Look up the heat output rating of the baseboard at this temperature from the manufacturer's literature. Interpolate between the listed values if necessary. Again, it is the author's recommendation that the 15% heating effect factor be eliminated if it is included in the published ratings. Do so by dividing the listed heat output by 1.15.

Equation 8.6:

$$\text{Corrected output rating} = \frac{\text{IBR heat output rating (Btu/h/ft)}}{1.15}$$

STEP 3. Divide the design heating load of each room by the heat output per foot of element length number obtained in step 2. Round off the number to the next higher whole foot length of baseboard.

Equation 8.7:

$$\text{Baseboard length} = \frac{\text{Room design heating load (Btu/h)}}{\text{Corrected output rating (Btu/h/ft)}}$$

> **Example 8.5**
>
> A series piping circuit of finned-tube baseboard will be used to heat the rooms having the design heating loads indicated in the order they are listed. The heat output ratings of the baseboard are given in Figure 8-8. The heating effect factor should be removed from these ratings for purposes of sizing. The boiler outlet temperature is 170 °F. Use sizing method 1 to select a baseboard length for each room.
>
> Design heating load of rooms:
> 1. Living room   8,000 Btu/h
> 2. Dining room   5,000 Btu/h
> 3. Bedroom 1   3,500 Btu/h
> 4. Bedroom 2   4,500 Btu/h
> 5. Bathroom   1,500 Btu/h

Solution:

STEP 1. Using Equation 8.5, the average water temperature in the circuit is assumed to be 10 °F less than the boiler outlet temperature:

$$T_{average} = 170 - 10 = 160 \text{ °F}$$

STEP 2. The IBR heat output rating of the baseboard at 160 °F is read from Figure 8-8 as 420 Btu/h/ft. The heating effect factor, which is noted in the footnotes as being present in these ratings, is removed using Equation 8.6:

$$\text{Corrected output rating} = \frac{420 \text{ Btu/h/ft}}{1.15} = 365.2 \text{ Btu/h/ft}$$

STEP 3. The length of each baseboard is found by dividing the corrected heat output rating value by the design heat load of each room using Equation 8.7, and rounding up the result to the next whole foot length:

1. Living room   8,000/365.2 = 21.9 or 22 feet
2. Dining room   5,000/365.2 = 13.7 or 14 feet
3. Bedroom 1   3,500/365.2 = 9.6 or 10 feet
4. Bedroom 2   4,500/365.2 = 12.3 or 13 feet
5. Bathroom   1,500/365.2 = 4.1 or 5 feet

Discussion:

Although this method is simple and fast, it can lead to problems if used as the final means of sizing series-connected baseboards. Its weakness is the fact that all baseboard lengths are based on an assumed *average* water temperature within the circuit. In reality, each successive baseboard in the circuit operates at a lower temperature than the preceding baseboard. Because this method is based on average temperature, *it tends to oversize baseboards near the beginning of the circuit and undersize baseboards near the end.* This could lead to overheating some rooms and underheating others.

Another point worth noting is that the average fluid temperature is dependent on the circuit's flow rate. It may be 10 °F less than the boiler outlet temperature, or it may be 5 °F or even 15 °F less. This method does not account for this possibility. Because of these limitations, *sizing method 1 should only be used to make initial estimates of baseboard length.*

Sizing Method 2

The accuracy of sizing baseboard for series circuits can be improved by accounting for the temperature drop of the fluid as it moves from one baseboard to the next. This allows the length of each baseboard to be based on the fluid temperature at its location in the circuit, rather than an overall average circuit temperature.

This method is based on the use of water as the heat transfer fluid. It is also based on the use of standardized heat output ratings established at an entering air temperature of 65 °F. It requires a reasonable estimate of the flow rate through the baseboard. This can be made using methods from Chapter 6, Fluid Flow in Piping, and Chapter 7, Hydronic Circulators, or by using the Series Baseboard Simulator module in the Hydronics Design Studio software.

STEP 1. Using the rating data from the baseboard manufacturer, make a graph of the heat output of the baseboard versus water temperature, similar to that shown in Figure 8-15. If the ratings data contain the 15% heating effect factor (most do), remove it from each data point using Equation 8.6 before plotting the points on the graph. If water temperatures below 150 °F are expected in any of the baseboards, use the data from Figure 8-9 to calculate and plot estimated low-temperature performance data. Draw a smooth curve through all the data points.

STEP 2. Calculate the average water temperature in the first baseboard on the circuit using Equation 8.8:

Equation 8.8:

$$T_{ave\,BB} = T_{in} - \frac{\text{Room load}}{1,000f}$$

where

$T_{ave\,BB}$ = average water temperature in the baseboard (°F)

T_{in} = water temperature entering the baseboard (°F)

Room load = design heating load of the room (Btu/h)

f = flow rate of water through the baseboard (gpm)

STEP 3. Using the graph prepared in step 1, estimate the heat output of the baseboard (q') at the average water temperature found in step 2.

STEP 4. Calculate the length of the baseboard by dividing the design heating load of the room by the value of q' found in step 3:

Equation 8.9:

$$L = \frac{\text{Room load}}{q'}$$

Round off this length to the next larger whole foot. This length is designated $L_{rounded}$.

Chapter 8 Heat Emitters

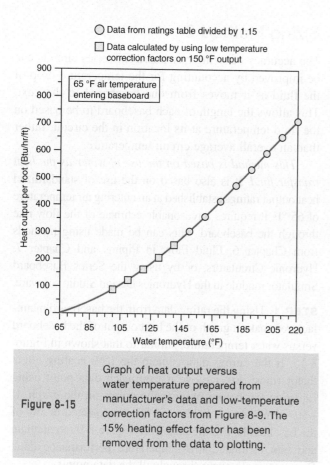

Figure 8-15 | Graph of heat output versus water temperature prepared from manufacturer's data and low-temperature correction factors from Figure 8-9. The 15% heating effect factor has been removed from the data to plotting.

STEP 6. Repeat this procedure (steps 2 through step 5) using the outlet temperature of the first baseboard as the inlet temperature for the second baseboard. Similarly, use the outlet temperature of each baseboard as the inlet temperature for the next baseboard in the series circuit. Eventually, all baseboard lengths will be determined, as will the outlet temperature of the last baseboard.

A good way to organize these calculations is to compile data as shown in Figure 8-16. This table is filled in by working across the top row to obtain the outlet temperature of the first baseboard. This value is then copied to the third column of the second row, and the process is repeated. The last value to be obtained is the outlet temperature of the last baseboard. This table can be extended, if necessary, for circuits with more than six series-connected baseboards.

Example 8.6

Using sizing method 2, repeat the baseboard selection process for the same rooms described in Example 8.2. Assume a water flow rate of 2.25 gpm. Use the graph in Figure 8-17 for heat output ratings. Note that the 15% heating effect factor has already been removed from the data shown in Figure 8-17.

Solution:

Figure 8-18 shows the results of applying the procedure for sizing method 2. Note that the outlet temperature of the first baseboard becomes the inlet temperature to the second baseboard, and so on.

STEP 5. Calculate the outlet temperature of the baseboard based on its rounded length using Equation 8.10:

Equation 8.10:

$$T_{out} = T_{in} - \frac{(L_{rounded})(q')}{500f}$$

Room name	Design heat load (Btu/hr)	Inlet temperature (°F)	Average water temperature (°F) (Equation 8.5)	Heat output of baseboard (from graph)	Calculated baseboard length (ft) (Equation 8.6)	Round baseboard length (ft)	Outlet temperature (°F) (Equation 8.7)

Figure 8-16 | Table used to organize calculations for baseboard sizing method 2.

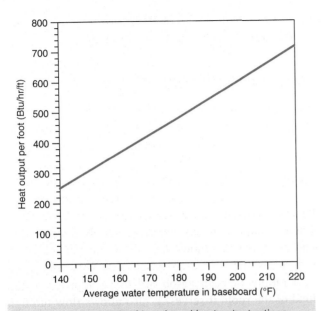

Figure 8-17 Graph of baseboard heat output ratings for Example 8.6. Note: The 15% heating effect factor has been removed from the manufacturer's data prior to plotting this graph.

Notice that the length of the first baseboard obtained using method 1 is longer than the length obtained using method 2. Conversely, method 1 calculates a shorter length for the last baseboard on the circuit compared to method 2. These results are due to the use of a single average water temperature in method 1 versus the average temperature *at each baseboard* in method 2. They reveal the tendency of method 1 to oversize baseboards at the beginning of the circuit and to undersize them near the end.

The greater accuracy of method 2 comes at the expense of more calculations. Furthermore, since the results of the first baseboard affect the input data to the second, and so on, any error in the calculations will invalidate all remaining numbers from that point on. If, for example, the flow rate through the baseboards is changed, the entire table has to be recomputed. Still, the greater accuracy obtained using method 2 justifies its use for final baseboard selection.

An ideal way to automate these calculations is to build the procedure into spreadsheet software. This allows long-series strings to be calculated error free and the results of changes in flow rates and inlet temperatures to be obtained almost instantly.

The Series Baseboard Simulator Module

The Hydronics Design Studio contains a module named Series Baseboard Simulator that executes a sizing procedure slightly more precise than even sizing method 2. It uses Equation 8.1 as well as additional algorithms and data that allow baseboard performance to be adjusted for different fluids, fluid temperatures, and a wide range of entering air temperatures.

Discussion:

It is interesting to compare the results obtained using sizing method 1 in Example 8.2, with those of sizing method 2 in this example. The best comparison can be made by looking at the unrounded lengths of baseboard calculated for each room. The unrounded lengths show the greatest differences between the calculated results and are given in Figure 8-19.

Room name	Design heat load (Btu/hr)	Inlet temperature (°F)	Average water temperature (°F) (Equation 8.5)	Heat output of baseboard (from graph)	Calculated baseboard length (ft) (Equation 8.6)	Round baseboard length (ft)	Outlet temperature (°F) (Equation 8.7)
Living	8,000	170	166.4	399.5	20.02	20	162.9
Dining	5,000	162.9	160.7	366.6	13.6	14	158.3
Bedroom 1	3,500	158.3	156.7	343.2	10.2	11	154.9
Bedroom 2	4,500	154.9	152.9	321.5	13.99	14	150.9
Bathroom	1,500	150.9	150.2	306.6	4.89	5	149.5

Figure 8-18 Table for baseboard sizing method 2 filled in with numbers for Example 8.6.

Room heat load (Btu/hr)	Baseboard lengths using method 1 (ft)	Baseboard lengths using method 2 (ft)
8,000	21.9	20.02
5,000	13.7	13.6
3,500	9.6	10.2
4,500	12.3	13.9
1,500	4.1	4.89

Figure 8-19 | Comparison of unrounded baseboard lengths obtained using baseboard sizing methods 1 and 2 for the same room heating loads. Method 2 is considered the more accurate.

The professional version of the Hydronics Design Studio can handle up to 12 baseboards in series. Each baseboard can be selected as a specific make and model from a scrollable list. The user can also select the circulator, piping, and fittings from the databases built into the module.

The Series Baseboard Simulator module determines the flow rate in the specified circuit as well as the rounded length, heat output, inlet temperature, and outlet temperature of each baseboard. The reader is encouraged to experiment with this software module to determine the effect of various system design changes on baseboard length. Figure 8-20 shows the user interface for the Series Baseboard Simulator module.

Parallel-Piped Baseboard

Although somewhat customary in older systems, finned-tube baseboards don't necessarily have to be connected into series circuits. They can also be connected in several different *parallel* piping arrangements. Such arrangements have the benefit of eliminating the sequential water temperature drop from one baseboard to the next. Instead, each baseboard receives water at approximately the same supply temperature. Parallel piping methods also allow independent flow rate adjustment for each baseboard. This allows the heat output of each baseboard to be adjusted to the load of the room it serves. It also allows flow through any baseboard to be completely turned off when no heat output is needed. Another benefit of parallel piping is that it usually lowers the circulator power required relative to series-piped systems, especially if the series circuits are long.

One of the most practical methods for connecting baseboards into parallel circuits in residential and light commercial systems is called a **homerun distribution system**. It uses small-diameter PEX or PEX-AL-PEX tubing to supply flow to and return flow from each baseboard as shown in Figure 8-21.

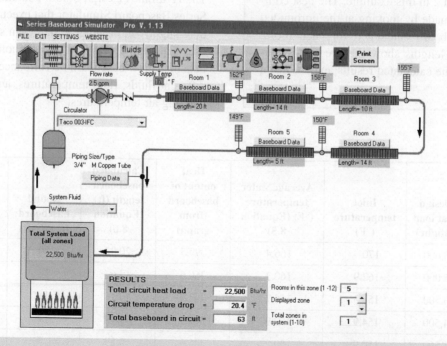

Figure 8-20 | Screenshot of the Series Baseboard Simulator module from the Hydronics Design Studio software.
Source: Hydronicpros.

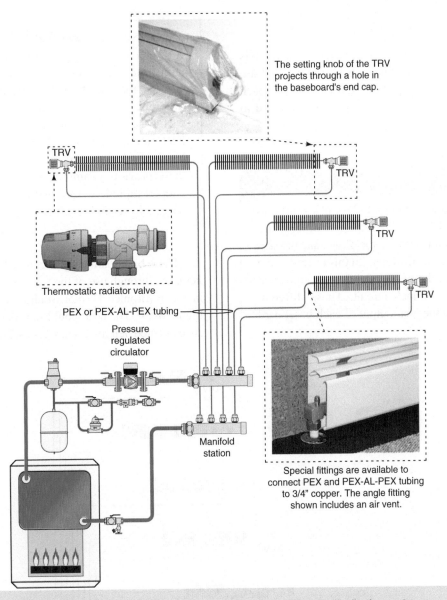

Figure 8-21 | Example of multiple baseboards connected in parallel within a homerun distribution system.

Sizing Parallel-Piped Baseboards

When baseboards are connected in parallel, each receives water at the temperature supplied to the manifold. The flow rate through each baseboard circuit needs to be determined based on the methods discussed in Chapter 6, Fluid Flow in Piping, or by using the Hydronic Circuit Simulator module in the Hydronics Design Toolkit software. The latter is much faster given the amount of parallel hydraulic resistance calculations involved.

Once the flow rate through each baseboard circuit is determined, the average water temperature within each baseboard can be calculated using Equation 8.5. The value of q', which depends on the flow rate through each baseboard, can be calculated using Equation 8.1.

The length of each baseboard is then calculated using Equation 8.6. Finally, the calculated length of each baseboard is rounded off to the next larger whole foot length.

Example 8.7

A system of five baseboards is to be piped in a homerun arrangement using 1/2-inch PEX-AL-PEX tubing in each branch circuit. The equivalent length (L_e) of each branch circuit has been determined and is listed along with the room design heat loads in Figure 8-22. The Hydronic Circuit Simulator module in the Hydronics Design Studio software was used to determine the flow rate in each branch circuit based on the stated equivalent lengths. Determine the necessary baseboard lengths for each room.

Solution:

The average water temperature in each baseboard is calculated and is listed in Figure 8-19. Also listed are the results of applying Equations 8.6 and 8.1 for each baseboard. The rounded baseboard lengths are listed in the last column.

Discussion:

Although the table in Figure 8-23 looks similar to that in Figure 8-18, there are important differences. First, in a parallel system, the flow rate in each branch is likely to be different. In this example, the differences were the result of different equivalent lengths for each branch circuit. These different flow rates affect the heat output of each baseboard and are properly accounted for by use of Equation 8.1 for each baseboard. Second, the inlet water temperature to each baseboard is the same, as seen in the fourth column of Figure 8-23. The differences in the average water temperature in each baseboard are small, as are the differences in heat output per foot of length. The final rounded baseboard lengths are slightly different from those in the previous example. This is partially due to the parallel piping, as well as the assumed equivalent lengths for each branch circuit.

These baseboard lengths could also be determined, by iteration, through use of the **Circuit Simulator Module** in the Hydronics Design Studio software, as shown in Figure 8-24. The user would start the process by entering an estimated length for each baseboard. The software

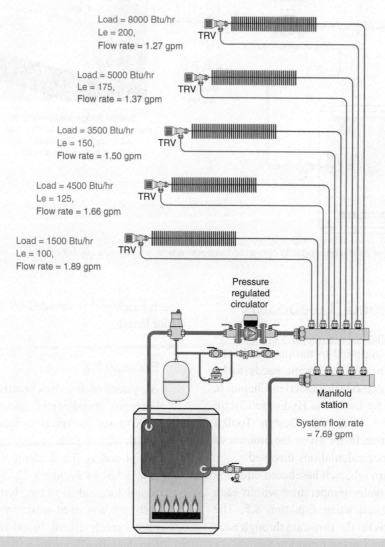

Figure 8-22 | Parallel-piped baseboards for Example 8.7.

Room name	Design heat load (Btu/hr)	Branch flow rate (gpm)	Inlet temperature (°F)	Average water temperature (°F) (Equation 8.5)	Heat output of baseboard (Equation 8.1)	Calculated baseboard length (ft) (Equation 8.6)	Round baseboard length (ft)
Living	8,000	1.27	170	163.7	365.4	21.89	22
Dining	5,000	1.37	170	166.4	381.6	13.1	14
Bedroom 1	3,500	1.5	170	167.7	390.3	8.97	9
Bedroom 2	4,500	1.66	170	167.3	389.6	11.6	12
Bathroom	1,500	1.89	170	169.2	402.5	3.73	4

Figure 8-23 | Table summarizing sizing calculations for Example 8.7. The inlet water temperature to each baseboard is 170 °F. Also note that each baseboard operates at a different flow rate based on the equivalent length of the circuit in which it is located.

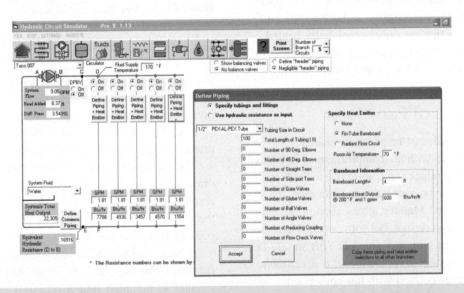

Figure 8-24 | Use of the Circuit Simulator Module in the Hydronics Design Studio to size baseboards that are piped in parallel. *Source: Hydronicpros.*

would calculate the heat output associated with each trial length. After comparing the calculated heat output of each baseboard to its associated room load, the user can adjust each baseboard length up or down until the calculated heat output is close to the design heat load of the associated room. This approach, although starting with estimated lengths for each baseboard, proceeds quickly because the thermal and hydraulic performance of the entire system is evaluated almost instantaneously as each change is made.

8.6 Hydronic Fan-Coils and Air Handlers

When rooms lack the wall space needed to mount finned-tube baseboard, or owners don't like their aesthetics, alternatives include using a hydronic **fan-coil** or **air handler**. These devices come in many different sizes, shapes, and heating capacities. They all have two common components—a finned-tube heat exchanger (often called the "**coil**") through which system fluid flows and a **fan** or **blower** that forces air to flow across the coil.

The term fan-coil applies to units that are mounted within occupied spaces. Some fan-coils use a propeller-type fan driven by a motor to create airflow across the coil. Others use a **tangential blower** to create the air movement. Room air that passes through a fan-coil discharges directly back into the room.

The term air handler applies to a unit that is mounted outside of occupied space, such as in an attic, basement, or closet. Air handlers use a squirrel cage blower to drive air across the coil and then through ducting that usually leads to several air diffusers mounted in different rooms of the building. Squirrel cage blowers create higher air pressure differentials compared to common propeller-type fans or tangential blowers. These higher pressure differentials are needed to move air through ducting.

Advantages of Fan-Coils

Unlike finned-tube baseboards, fan-coils and air handlers rely on *forced* convection to transfer heat from the coil surface to the room air. Forced convection is much more effective than natural convection in transferring heat from a surface to air. This allows the surface area of the coil in a fan-coil to be much smaller than that of a finned-tube element in a baseboard of equivalent heating capacity, and operated at the same water temperature. A fan-coil unit can often provide equivalent heating capacity using only a fraction of the wall space required by a finned-tube baseboard.

Fan-coils also have very little thermal mass and low water content. They can respond to a call for heat very quickly. This makes them well suited for spaces that need to be quickly heated from setback temperature conditions or spaces that are only occasionally heated.

Some fan-coils that are equipped with condensate drip pans can also be used for chilled water cooling.

Disadvantages of Fan-Coils

The compactness and fast-response characteristics of fan-coils do not come without compromises. First, any type of fan-coil will produce some noise due to the blower operation. Noises can be worsened by vibration, loose fitting grills, and trim. Manufacturers attempt to limit operating noise through design, but some operating noise will always be present with any fan-coil. *This should be discussed with the owners to ensure that it is acceptable.*

Another factor associated with fan-coils is dust and dust movement. Regardless of how clean a room is kept, some dust is always present in the air and will be carried into the intake of any type of fan-coil. Some fan-coils have intake filters capable of removing most of the dust before it is deposited on the coil surfaces or blown back into the room. When filters are not present, or if the unit is not periodically cleaned, dust accumulation on coil surfaces will reduce the heating capacity, especially if the unit operates at low fluid temperatures.

Fan-coils also require connection to line voltage wiring to operate the blower. Although the power requirements of the blowers in residential size fan-coils are small, the designer must ensure that a source of electricity is available at or near the proposed location of every fan-coil unit. Most fan-coils provide very little heat output during power outages, regardless of potential flow of hot water through their coils.

Wall-Mounted Fan-Coils

One type of fan-coil commonly used in residential and light commercial systems is mounted on, or recessed into, a wall or partition. Figure 8-25a shows an example of a wall-mounted fan-coil, which is also sometimes referred to as a **"console" fan-coil**.

This console fan-coil uses a tangential blower to draw air up from the bottom intake grill and force it across the finned-tube coil. A drip pan is located under the coil, allowing this unit to be used for chilled water cooling as well as heating.

Figure 8-25b shows an example of a **recessed fan-coil**. The fan and coil are located in a metal chassis that is mounted into a recessed wall cavity. Only the removable front face panel of the unit extends

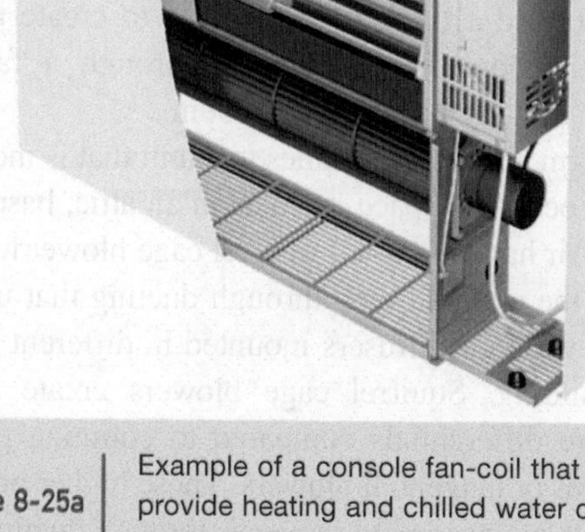

Figure 8-25a Example of a console fan-coil that can provide heating and chilled water cooling. *Courtesy of Myson.*

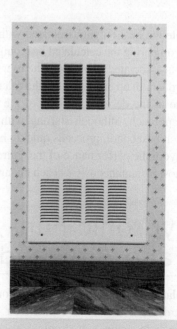

Figure 8-25b Recessed fan-coil convector. *Courtesy of Mestek, Inc.*

outside the wall surface. The fact that most of the unit is recessed into the wall gives them better protection against damage, especially in commercial buildings.

A properly sized wall cavity must be framed, and the piping and wiring routed to the cavity. The sooner this planning takes place, the easier the installation will be. It is not uncommon to find a lack of space for recessed mounting, especially in bathroom areas where plumbing and electrical systems can present interference. Recessed fan-coils are usually designed so that a base pan assembly can be mounted into the wall cavity before the walls are closed-in. The blower/coil assembly then mounts into this base pan after the wall is closed-in. Installation of recessed fan-coils in retrofit situations is usually harder than in new construction because the walls are already closed-in. *Be sure to take a careful look at retrofit jobs before committing to recessed fan-coil unit(s).*

Another type of wall-mounted fan-coil is called a **high wall cassette**, an example of which is shown in Figure 8-26.

High wall cassettes are very similar to the indoor fan-coils used for ductless heat pumps. They combine a variable-speed blower with a specially shaped coil drip pan, filter, and controls. They are typically mounted 6 to 9 inches below flat ceilings. Air is drawn into the top of the cassette and discharged through a slot at the lower front of the unit. Most high wall cassettes have motorized dampers that oscillate up and down in front of the air discharge slot. The intent of this motion is to improve mixing of the discharge air with room air. The damper closes when the fan-coil is not operating. High wall cassettes are typically turned on and off using a handheld remote. That remote can also be used to set the operating mode of the unit, blow speed, and the range of movement of the discharge air damper. The controllers build into most high wall cassettes can activate peripheral devices such as a zone valve or circulator when the unit is called to operate. Because they have drip pans, high wall cassettes can be used for chilled water cooling. The drip pan has a drainage tube that must be routed to a suitable drain. When mounted on exterior walls, the condensate drain tube is sometimes routed out the back of the fan-coil and through the wall.

Many fan-coils are equipped with controls that allow the speed of the fan or blower to be changed. In some cases, the adjustment is done using a dial or switch on the fan-coil. In other cases, the adjustment is made using a handheld remote. The ability to change the speed of the fan or blower allows some degree of control over heat output (or cooling output in chilled water systems).

Some fan-coils come with an integral thermostat that can turn the blower off once the room reaches its temperature setting. The usefulness of this thermostat depends on the other controls used in the system. For example, most fan-coil thermostats are not wired back to the remainder of the system. In such cases, the installer must provide additional control wiring to operate the heat source and circulator when the fan-coil calls for heat.

Some fan-coils also have a **low-limit aquastat** that prevents the blower from operating until the proper water temperature is present within the coil. The intent is to prevent the fan-coil from blowing unconditioned air into a room leading to potential discomfort.

Piping connections to the small fan-coils used in residential or light commercial spaces are usually made to 1/2-inch copper tube stubs or FPT connections. In some situations, the piping enters the back of the cabinet, and is thus concealed from view. If this is not possible, piping risers are routed up from the floor into the bottom of the cabinet. Tubing runs to and from fan-coils can be rigid copper, PEX, or PEX-AL-PEX tubing.

Wall-mounted fan-coils should be mounted so that airflow is not blocked by furniture. Discuss the positioning with the owners so that interference problems with furniture or other furnishings can be avoided.

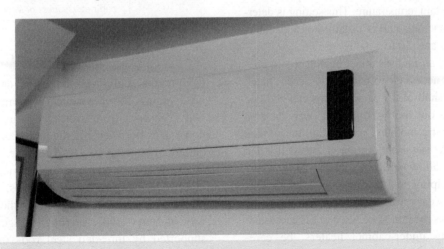

Figure 8-26 | Example of a high wall cassette fan-coil. *Courtesy of John Siegenthaler.*

The units should also not blow directly on the occupants of a room. For example, imagine trying to read a newspaper in a reclining chair that is only 2 or 3 feet away from the discharge grill of a wall-mounted fan-coil. The idea is to diffuse the air stream into the room air before it contacts the occupants.

Under-Cabinet Fan-Coils

Another type of hydronic fan-coil is designed to mount in the 4-inch "kick space" beneath a kitchen or bathroom cabinet. These units, often called **kick space heaters**, are designed for situations where wall space is not available, particularly kitchens and bathrooms. An example of a kick space heater is shown in Figure 8-27. A typical 12-inch-wide unit can yield the heat output of 8 to 10 feet of finned-tube baseboard. Wider units are available with greater heating capacities.

Air flows into the unit through the upper portion, or sides, of the front grill. It is drawn in by a tangential blower mounted at the rear of the unit. The blower reverses the airflow, sending it out through the horizontal coil in the lower part of the unit where it is heated. Some kick space heaters have air filters, while others do not. The latter are somewhat prone to dust accumulation. They should be periodically cleaned with a vacuum cleaner.

Kick space heaters are often supported by rubber grommets that reduce vibration transfer to the floor or cabinet. Connections are typically 1/2-inch copper tube stubs. An integral junction box is provided for wiring connections. Some manufacturers offer flexible reinforced hose sets that connect between the coil unit and rigid piping. Besides reducing vibration transmission, these flexible hoses allow the unit to be pulled forward, like a drawer, from its mounting position if necessary for service. Most under-cabinet heaters are equipped with a low-limit aquastat that prevents the blower from operating until the water temperature in the coil reaches a certain temperature. This setting is determined by the manufacturer and is usually not adjustable. Typical temperatures are 110 to 140 °F.

When installing an under-cabinet heater, it is sometimes convenient or necessary to provide an access hatch through the bottom shelf of the cabinet. This can be done by carefully sawing out the bottom shelf over the area where the unit will be mounted and installing wooden ledgers to support the sawn-out portion of the shelf when replaced. Be sure to allow sufficient space to fully access the side-mounted piping and wiring connections. This arrangement permits easy access for cleaning and to the air vent. It also allows the unit to be lifted out if necessary.

When installing under-cabinet heaters, be careful not to accidentally bend the exposed blades of the blower wheel. The resulting imbalance can cause noticeable vibration and associated noise.

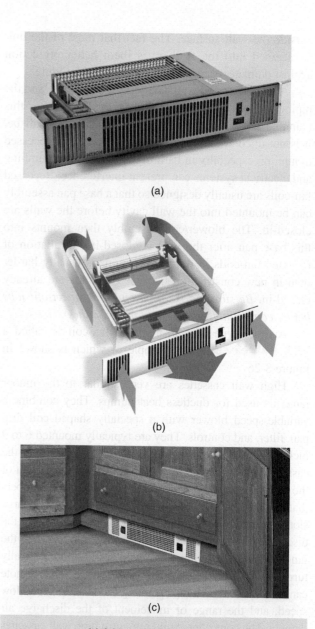

Figure 8-27 (a) A kick space heater, (b) airflow through the kick space heater, and (c) typical installation under a kitchen or bathroom cabinet. *Courtesy of Myson, Inc.*

Overhead Fan-Coils (Unit Heaters)

Commercial buildings are often designed with overhead piping for mechanical systems. One method of delivering heat from overhead piping without routing it down near floor level is the use of an overhead fan-coil commonly known as a **unit heater**.

Unit heaters are traditionally used in spaces such as service garages, machine shops, and supermarkets, where overhead mounting allows them to be out of the way of heavy traffic or rough usage. They are usually supported by threaded steel rods or perforated angle struts attached to structural members such as roof trusses or beams. Figure 8-28 shows a typical unit heater designed for angled horizontal airflow. Louvers on the

is called a **draw-through air handler**. On other air handlers, air is blown through the coil. Such units are appropriately called **blow-through air handlers**. The basic concepts for each are illustrated in Figure 8-29a. The components in a small blow-through air handler is shown in Figure 8-29b.

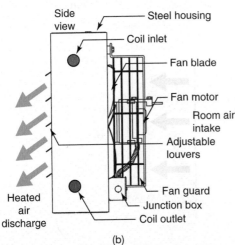

| Figure 8-28 | (a) Overhead unit heater. Note adjustable louvers at the front of the unit. (b) Side view. *Courtesy of Beacon Morris Division of Mestek, Inc.* |

front of the unit allow the discharge air stream to be directed at various angles.

Although overhead unit heaters have been used in buildings with relatively high ceilings, such installations often experience stratification in which hot air accumulates near the ceiling while cooler air pools near the floor. Good air circulation is essential to minimize such stratification.

Air Handlers

Air handlers can be used for heating or cooling. The latter requires that the air handler is equipped with a drip pan. Depending on its size, and the size of the building it serves, a single air handler combined with a ducting system could supply heating and/or cooling to an entire home or commercial building.

Some air handlers are designed so that air is pulled through the coil by the blower. This type of arrangement

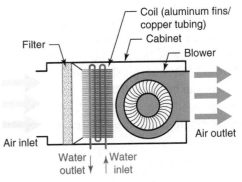
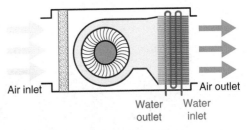
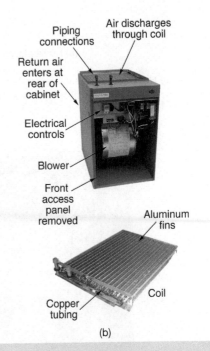

| Figure 8-29 | (a) Air handlers using "draw-through" and "blow-through" arrangement of the blower and coil. (b) Example of a small, blow-through air handler. *Courtesy of ECR International Corporation.* |

332 Chapter 8 Heat Emitters

Many modern air handlers are designed so that they can be mounted vertically or horizontally. Vertical mounting is usually preferred when the air handler is mounted within a closet, or within a small footprint area in a basement. Air passes into the bottom of the air handler, through a filter, through the coil, through the blower, and discharges at the top of the unit. Horizontal mounting is usually preferred when the air handler is mounted within an attic, or suspended from a surface such as the bottom of floor framing in a basement.

Figure 8-30 shows typical configurations for vertical and horizontal air handlers. These depictions show the air handlers equipped with "**A-coils**." An A-coil consists of two small slab coils that lean against each other at the top.

A-coils are commonly used in air handlers that can operate with either heated water or chilled water, the latter being used for cooling. A-coils are typically suspended over a drip pan that captures condensate dripping from the coil during chilled water cooling operation.

The only change required to change the unit from vertical to horizontal mounting is to rotate the coil and drain pan by 90 degrees.

Figure 8-31 shows an installed horizontal air handler. This unit is rated to provide a cooling capacity of about 36,000 Btu/h when supplied with 45 °F chilled water. It is installed within a recessed area in a fully conditioned attic space, and behind a removable access panel.

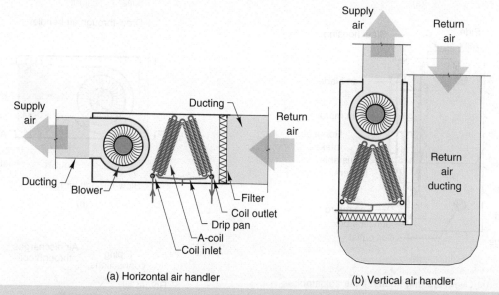

Figure 8-30 (a) Horizontal versus (b) vertical air handler mounting. A typical unit can be converted between these configurations by rotating the A-coil and drip pan. Both are shown in heating mode.

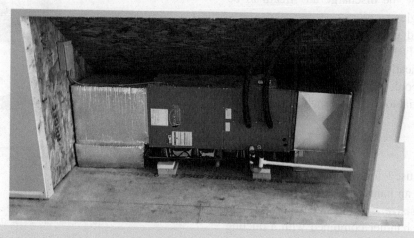

Figure 8-31 Horizontal air handler mounted over a secondary drip pan, and connected to ducting, piping, condensate drain, and electric power. *Courtesy of John Siegenthaler.*

8.6 Hydronic Fan-Coils and Air Handlers

Because it is installed over a finished ceiling, a secondary drip pan is first installed under the air handler. This pan would capture any water leaking from the unit should its internal drip pan ever fail. All captured condensate is routed to an outside drain through a sloped 3/4" PVC tube. The coil in the air handler is supplied with chilled water through 3/4" preinsulated PEX tubing, which runs seamlessly from the air handler to the basement mechanical room.

Multi-Fan Hydro-Air Distribution System

Another unique concept for creating a **hydro-air distribution system** for heating or cooling is shown, in cooling mode operation, in Figure 8-31a.

This system uses a hydronic coil, mounted in a sheet metal enclosure, to either heat or cool air as

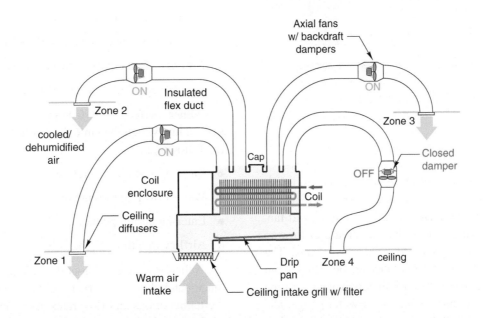

Figure 8-31a | Hydro-air system with zoned high-efficiency fans that can supply heating or chilled water cooling. *Photo courtesy of ThermAtlantic Energy Products, Inc.*

Figure 8-31b Axial flow fan connected to insulated flexible ducting supplies heating or cooling to a zone. *Courtesy of ThermAtlantic Energy Products, Inc.*

it passes through the coil. However, unlike most air handlers that rely on a single blower to create air movement, this system use a high-efficiency axial fan, mounted into a run of flex duct to create airflow to *each of several independently controlled zones*. The 6-inch-diameter axial fan operates on 66 watts input power. It can provide 301 CFM airflow when operating against a static pressure of 0.25 inch water gauge, and has an insulated shell to prevent condensation when operating in cooling mode. When a given zone calls for heating or cooling, the associated axial fan runs. When the zone is off, a spring-loaded damper within the fan closes to prevent reverse airflow. The electrical power consumed is only that of active zones, rather than that of a central blower capable of design airflow when all zones are on.

The axial fans are mounted with insulated flexible ducting attached to both their inlet and outlet, as shown in Figure 8-31b. This ducting surpasses noise allowing the system to operate with very low sound levels.

The coil enclosure is equipped with a drip pan allowing this system to provide chilled water cooling and dehumidification.

The coil enclosure can be mounted within a conditioned attic space. It can also be mounted within an insulated "dog house" enclosure in an unconditioned attic space. The insulated enclosure must be constructed to prevent water in the coil from freezing. An alternative would be to operate the coil with an antifreeze solution.

As with any chilled water fan-coil or air handler, the coil enclosure must be connected to a suitable condensate drain.

8.7 Thermal Performance of Fan-Coils

The heat output of a fan-coil or air handler is dependent on several factors. Some are fixed by the construction of the fan-coil or air handler and cannot be changed. These include the following:

- Surface area of the coil;
- Size, spacing, and thickness of the fins;
- Number of tube passes through the fins;
- Air-moving ability of the blower.

Other performance factors are dependent on the system into which the fan-coil is installed. These include the following:

- Entering water temperature;
- Water flow rate through the coil;
- Entering air temperature;
- Airflow rate through the coil.

When selecting a fan-coil or air handler, designers must match the characteristics of the available units with the temperatures and flow rates the system can provide, while also obtaining the required heating capacity. This usually involves a search of manufacturer's performance data in an attempt to find a good match among several simultaneous operating conditions. These data are often provided as tables that list the heat output at several combinations of temperature and flow rate for both the water side and air side of the unit.

An example of a thermal performance table for a wall-mounted fan-coil is shown in Figure 8-32. It lists the Btu/h heating capacity of the unit over a wide range of entering water temperatures, two different airflow rates, and two water flow rates. Each heat output value can be thought of as a "snapshot" of heating performance at specific conditions.

The heat output of the fan-coil varies continuously over a range of temperatures and flow rates. There is no guarantee that it will operate at one of the conditions listed in the table. When it is necessary to estimate the heat output of the unit at conditions other than those listed in the table, a process called **interpolation** can be used. This process is usually not as simple as averaging between the values that are above and below the desired value. Rather it sets up proportions to arrive at a more accurate estimate. The best way to illustrate interpolation is through an example.

WH SERIES Btu/hr output for Myson fan convectors—Entering air at 65°F														
	Flow Rate	Fan	Air Delivery	Entering Water Temperature (°F)										
Model	(gpm)	Speed	(cfm)	110*	120*	130*	140	150	160	170	180	190	200	210
WH 50	1.0	High	89	1,690	2,030	2,354	2,783	3,205	3,580	3,887	4,263	4,638	5,079	5,354
		Low	50	1,390	1,610	2,010	2,400	2,825	3,200	3,605	4,020	4,380	4,750	5,040
	3.0	High	89	1,996	2,456	2,866	3,309	3,717	4,094	4,470	4,913	5,323	5,664	6,142
		Low	50	1,450	1,910	2,217	2,590	2,900	3,410	3,790	4,299	4,570	4,950	5,300
WH 90	1.0	High	121	3,700	4,350	5,025	6,000	6,780	7,640	8,400	9,290	10,050	10,800	11,500
		Low	75	3,250	3,900	4,505	5,350	6,000	6,750	7,420	8,190	8,830	9,550	10,200
	3.0	High	121	3,800	4,520	5,320	6,130	7,160	7,940	8,710	9,600	10,420	11,250	12,090
		Low	75	3,350	4,020	4,800	5,580	6,380	7,050	7,730	8,500	9,200	9,980	10,790

*For Output At These Water Temperatures, Optional 107° Low Temperature Cutoff Required.

Figure 8-32 | Thermal performance data for small wall-mounted fan-coil convector. *Courtesy of Myson, Inc.*

Example 8.8

The heat output of the fan-coil represented by the data in Figure 8-32 is 3,205 Btu/h at 150 °F entering water temperature, 1.0 gpm water flow rate, and high fan speed. At 160 °F and the same flow rate, the output is listed as 3,580 Btu/h. Estimate the heat output of the fan-coil at 152 °F entering water temperature.

Solution:

Set up the known and unknown quantities as shown in Figure 8-33 and subtract the quantities connected by the brackets. Form the ratios of these differences, and equate these ratios to each other. Solve the resulting equation for the answer.

150	3205
152	x
160	3580

$$\frac{150-152}{150-160} = \frac{3205-x}{3205-3580}$$

$$\frac{-2}{-10} = \frac{3205-x}{-375}$$

$$0.2\,(-375) = 3205-x$$

$$x = 3205+75 = 3280 \text{ Btu/hr}$$

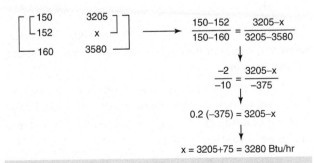

Figure 8-33 | Interpolation procedure for Example 8.8.

Discussion:

Notice that the algebraic signs + or − take care of themselves, so the answer is positive. Another way of thinking about the problem is that the temperature of 152 °F is one-fifth of the way between the numbers 150 and 160. This implies that the desired value of heat output, x, will be one-fifth of the way between 3,205 and 3,580. One-fifth of this difference is:

$$(1/5)(3{,}580 - 3{,}205) = 75$$

Adding 75 to the lower value of 3,205 gives the same answer of 3,280 Btu/h.

In some cases, **double interpolation** is required to obtain an estimate of heat output that falls between both the column and row rating points of the table. Again, an example is the easiest way to show how this is done.

Example 8.9

Make an estimate of the heat output of the fan-coil represented by the data in Figure 8-32, at an entering water temperature of 152 °F and a water flow rate of 2.5 gpm.

Solution:

Begin by interpolating between the temperatures of 150 and 160 °F two times, first for the heat output values listed at a flow rate of 1.0 gpm and again for the heat outputs listed at a flow rate of 3.0 gpm. To finish the calculation, take the results of the first two (temperature) interpolations and interpolate them between the lower flow rate of 1.0 gpm and the upper flow rate of 3.0 gpm.

The first temperature interpolation is the same as that shown in Example 8.8, with a result of 3,280 Btu/h at 152 °F.

The second interpolation is shown in Figure 8-34.

The final interpolation will be between the flow rates of 1.0 and 3.0 gpm and will use both previously calculated capacities at 152 °F entering water temperature. The final procedure is shown in Figure 8-35.

The estimated capacity of this fan-coil at 152 °F entering water temperature and 2.5 gpm water flow rate is thus 3,664 Btu/h.

$$\begin{bmatrix} 150 & 3717 \\ 152 & x \\ 160 & 4094 \end{bmatrix} \longrightarrow \frac{150-152}{150-160} = \frac{3717-x}{3717-4094}$$

$$\frac{-2}{-10} = \frac{3717-x}{-377}$$

$$-75.4 = 3717-x$$

$$x = 3717 + 75.4 = 3792.4 \text{ Btu/hr}$$

Figure 8-34 | Second interpolation procedure for Example 8.9.

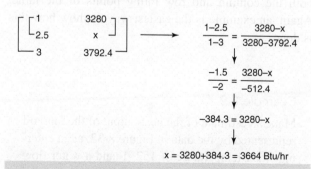

$$\begin{bmatrix} 1 & 3280 \\ 2.5 & x \\ 3 & 3792.4 \end{bmatrix} \longrightarrow \frac{1-2.5}{1-3} = \frac{3280-x}{3280-3792.4}$$

$$\frac{-1.5}{-2} = \frac{3280-x}{-512.4}$$

$$-384.3 = 3280-x$$

$$x = 3280 + 384.3 = 3664 \text{ Btu/hr}$$

Figure 8-35 | Final interpolation procedure for Example 8.9.

Discussion:

The final value should always be checked to see if it is reasonable. It should fall between the values listed in the tables both vertically and horizontally.

Another way to perform interpolation is by using a spreadsheet program. Another useful method, especially if the data will be used repeatedly, is to plot the data on a graph and draw a smooth curve through the points. The estimated performance between data points is easily read from this curve. Many manufacturers also provide design assistance software capable of quickly providing interpolated output ratings.

Some fan-coil manufacturers list the heat output of a fan-coil unit at a single reference condition consisting of a specified value for the entering temperatures and flow rates of both the water and air. Accompanying this will be one or more tables or graphs of **correction factors** that can be used to adjust the reference performance up or down for variation in temperature and flow rate of both the entering water and entering air. An example of tables listing these performance correction factors for a unit heater is shown in Figure 8-36.

General Principles of Fan-Coil Performance

The heat output of fan-coils can be estimated using several principles that can be verified by examining published ratings data. By varying one operating condition such as entering temperature or flow rate, while all others remain fixed, the designer can get a feel for which factors have the greatest effect on heat output. This can be very helpful in evaluating the feasibility of various design options.

Principle 1

The heat output of a fan-coil is approximately proportional to the temperature difference between the entering air and entering water.

If, for example, a particular fan-coil can deliver 5,000 Btu/h with a water temperature of 140 °F and room air entering at 65 °F, its output using 180 °F water can be estimated as follows:

$$\text{Estimated output at 180 °F} = \frac{(180\,°F - 65\,°F)}{(140\,°F - 65\,°F)} \times (5{,}000 \text{ Btu/h})$$

$$= 7{,}670 \text{ Btu/h}$$

8.7 Thermal Performance of Fan-Coils

Model No.	Output Btu/hr*	gpm	Final Air °F	Prssr. Drop ft/H$_2$O	Motor HP	RPM	Nominal CFM	Outlet FPM	Nom. Amps @ 115VAC
HS-108A	8,030	0.80	91	0.80	9 Watt	1,550	245	250	0.8
HS-108A	6,800	0.80	90	0.80	9 Watt	1,350	210	215	0.8
HS-118A	18,400	1.9	94	2.2	9 Watt	1,550	500	500	0.8
HS-118A	15,650	1.9	96	2.2	9 Watt	1,350	420	420	0.8
HS-125A	24,800	2.5	102	2.2	1/47	1,550	580	590	1.1
HS-125A	21,230	2.5	106	2.2	1/47	1,350	460	450	1.1
HS-136A	35,900	3.6	99	3.0	1/30	1,070	850	550	1.1
HS-136A	32,300	3.6	100	3.0	1/30	900	750	480	1.1
HS-18	13,050	1.3	95	0.005	9 Watt	1,550	395	395	0.8
HS-18	11,725	1.3	99	0.005	9 Watt	1,350	350	350	0.8
HS-24	17,400	1.8	96	0.014	9 Watt	1,550	450	450	0.8
HS-24	15,600	1.8	98	0.014	9 Watt	1,350	380	380	0.8
HS-36	26,100	2.7	103	0.09	1/47	1,550	550	550	1.1
HS-36	23,500	2.7	103	0.09	1/47	1,350	480	480	1.1
HS-48	34,800	3.5	103	0.12	1/30	1,070	750	550	1.3*
HS-48	31,300	3.5	111	0.12	1/30	900	630	460	1.3*
HS-60	43,600	4.4	105	0.17	1/30	1,070	900	650	1.3*
HS-60	39,200	4.4	112	0.17	1/30	900	700	510	1.3*

(a)

*Hot Water Conversion Factors Based on 200° Entering Water 60° Entering Air 20° Temperature Drop

Entering Air Temperature	Entering Water Temperature—20° Water Temperature Drop										
	100°	120°	140°	160°	180°	200°	220°	240°	260°	280°	300°
30°	0.518	0.666	0.814	0.963	1.12	1.26	1.408	1.555	1.702	1.85	1.997
40°	0.439	0.585	0.731	0.878	1.025	1.172	1.317	1.464	1.609	1.755	1.908
50°	0.361	0.506	0.651	0.796	0.941	1.085	1.231	1.375	1.518	1.663	1.824
60°	0.286	0.429	0.571	0.715	0.857	1.000	1.143	1.286	1.429	1.571	1.717
70°	0.212	0.353	0.494	0.636	0.777	0.918	1.06	1.201	1.342	1.483	1.63
80°	0.140	0.279	0.419	0.558	0.698	0.837	0.977	1.117	1.257	1.397	1.545
90°	0.069	0.207	0.345	0.483	0.621	0.759	0.897	1.035	1.173	1.311	1.462
100°	0	0.137	0.273	0.409	0.546	0.682	0.818	0.955	1.094	1.23	1.371

(b)

Figure 8-36 (a) Example of heat output ratings at reference conditions of 200 °F entering water and 60 °F entering air temperature and (b) correction factors for heat outputs at other entering water and air temperatures. *Courtesy of Mestek, Inc.*

A proportional relationship between any two quantities will result in a straight line that passes through or very close to the origin when the data are plotted on a graph. Figure 8-37 shows this to be the case when the heat output data from Figure 8-32 are plotted against the difference between the entering water and entering air temperatures.

This concept is useful for estimating heat output when a fan-coil is used in a room kept above or below normal comfort temperatures. For example, if a fan-coil

338 Chapter 8 Heat Emitters

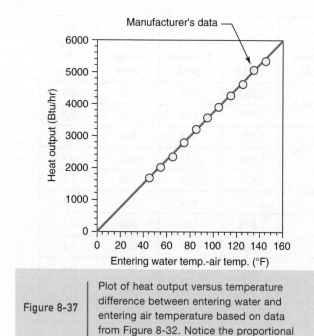

Figure 8-37 | Plot of heat output versus temperature difference between entering water and entering air temperature based on data from Figure 8-32. Notice the proportional (straight sloping line) relationship.

can deliver 5,000 Btu/h when supplied with water at 140 °F and room air at 65 °F, its output while heating a garage maintained at 50 °F could be estimated using the same proportionality method described above:

Estimated output at 50°F entering air
$$= \frac{(140\ °F - 50°F)}{(140\ °F - 65\ °F)} \times (5{,}000\ \text{Btu/h})$$
$$= 6{,}000\ \text{Btu/h}$$

Although this principle is helpful for quick performance estimates, it should not replace the use of thermal rating data supplied by the manufacturer when available.

Principle 2

Increasing the fluid flow rate through the coil will marginally increase the heating capacity of the fan-coil.

Interestingly, some heating professionals instinctively disagree with this statement. They argue that because the water moves through the coil at a faster speed, it has less time in which to release its heat. However, the time a given particle of water stays inside the fan-coil is irrelevant in a system with continuous circulation. Increased flow rate improves convective heat transfer between the fluid and the inner surface of the tubes, resulting in greater heat transfer.

Another way of justifying this principle is to consider the *average* water temperature in the coil at various flow rates. As the flow rate through the fan-coil is increased, the difference between the entering and leaving fluid temperature decreases. This implies that the *average*

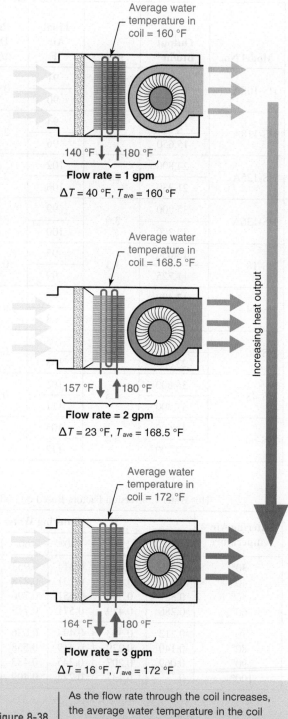

Figure 8-38 | As the flow rate through the coil increases, the average water temperature in the coil also increases. This implies that heat output always increases with increased flow rate. This is true for all hydronic heat emitters.

water temperature within the coil increases and so does its heat output. This concept, illustrated in Figure 8-38, holds true for other heat emitters including radiant panel circuits, panel radiators, and baseboard.

Those who claim that a smaller temperature difference between the ingoing and outgoing fluid implies

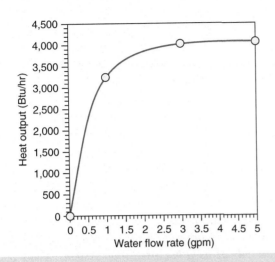

Figure 8-39 Heat output of a small fan-coil at different flow rates and constant entering water and entering air temperatures.

that less heat is being released are overlooking the fact that the rate of heat transfer depends on *both* the temperature difference and flow rate. This relationship was described by the sensible heat rate equation of Chapter 4, Properties of Water.

The rate at which heat output increases with flow rate is very "nonlinear." The rate of gain in heat output is very fast at low flow rates but decreases as the flow through the coil increases. This can be seen in Figure 8-39, which plots a manufacturer's data for heat output of a small fan-coil versus water flow rate, at constant entering water and entering air temperatures.

Notice that the fan-coil has about 50% of its maximum heat output capacity at only about 10% of its maximum flow rate. The strong curvature of this relationship means that attempting to control the heat output of a fan-coil by adjusting flow rate can be tricky. Very small valve adjustments can create large changes in heat output at low flow rates. However, the same amount of valve adjustment will create almost no change in heat output at higher flow rates. *This is also true for other types of hydronic heat emitters* and will be discussed in detail in Chapter 9, Control Strategies, Components, and Systems.

The nonlinear relationship between flow rate and heat output also implies that attempting to boost heat output from a fan-coil by operating it at an unusually high flow rate will yield very minor gains. The argument against this approach is further supported when one considers the increased head loss through the coil and higher circulator power requirements at higher flow rates.

Principle 3

Increasing the airflow rate across the coil will marginally increase the heat output of the fan-coil.

This principle is also based on forced-convection heat transfer between the exterior coil surfaces and the air stream. Faster moving air "scrubs" heat off the coil surface at a greater rate. As with water flow rates, the gain in heat output is very minor above the nominal airflow rate the unit is designed for.

Principle 4

Fan-coils with large coil surfaces and/or multiple tube passes through the coil will be capable of operating at lower entering water temperatures.

Convective heat transfer is proportional to the contact area between the surface and the fluid. If this contact area is increased, the temperature difference between the fluid and the air streams can be reduced (for a given rate of heat output). This is illustrated in Figure 8-40.

This principle can be well applied in hydronic systems that must operate with low-temperature heat sources such as solar collectors or heat pumps. In these applications, the designer may need to use fan-coils with larger coils and/or more tube passes through the fins of the coil, to drive heat from the coil at the required rate. On larger air handlers, manufacturers may offer higher performance coils with up to six rows of tubes as an option. This is usually not true on smaller residential fan-coils. For these products, the only way to increase coil surface area is to install more fan-coils.

Care must also be taken that comfort is not compromised when operating fan-coils at reduced water temperatures. It is critically important to introduce the air stream from such fan-coils into the space so that it mixes with room air before passing by occupants. Failure to do so will usually lead to complaints of "cool air" blowing from the fan-coils, even though the room is maintaining the desired setpoint temperature. As a guideline, the author recommends that fan-coil units supplying heat to occupied spaces not be operated with entering water temperatures lower than 110 °F.

Using Fan-Coils or Air Handlers for Cooling

Many large buildings use hydronic distribution systems for supplying both heated and chilled water to fan-coils or air-handlers. This can also be done in residential or

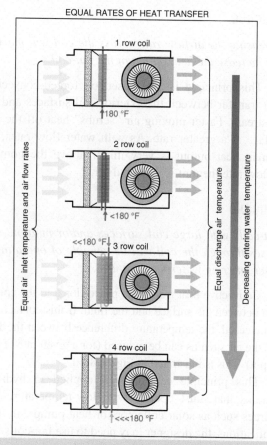

Figure 8-40 — As the size and number of tube rows in the coil increases, the entering water temperature required for a given rate of heat output decreases.

8.8 Panel Radiators

Another type of hydronic heat emitter that is gaining market share in North America is called a **panel radiator**. Having been used in Europe for many years, panel radiators are now available in hundreds of sizes, shapes, colors, and heating capacities to fit different job requirements. They can be used throughout a building or in combination with other types of heat emitters.

Most panel radiators are made of steel. Some are built of preformed steel sheets welded together at their perimeter. Others are constructed of tubular steel components. *To prevent corrosion, steel panel radiators should only be used in closed-loop systems.* Many panel radiators feature a high-quality fused powder-coat or enamel finish, in several different colors to coordinate with interior design.

Some panel radiators release a significant percentage of their heat as thermal radiation. Such panels typically have a relatively flat front and only project about 2 inches out from the wall surface to which they are mounted. Their radiant output tends to warm the objects in a room. The lower the water temperature these panels operate at, the higher the radiant heat output is as a percentage of total heat output. This is often desirable because it improves comfort and reduces room air stratification relative to units that release the majority of their heat through convection (e.g., directly heating the room air).

Other panel radiators are designed to release a high percentage of their heat output through convection. These radiators are equipped with one or more rows of fins that help dissipate heat into the surrounding air. This convective heat transfer is similar to that created of a finned-tube baseboard. These panel radiators tend to have deeper profiles that project 4 to 6 inches from the adjacent wall. This type of radiator is well suited for creating upward air currents to counteract downward drafts from large window areas. They are often operated at relatively high water temperatures that favor strong convective heat output.

Benefits of Panel Radiators

There are several benefits to using panel radiators relative to other types of heat emitters:

- Panel radiators typically require far less wall space than finned-tube baseboard sized for equivalent heat output and operating conditions. This reduces restrictions on furniture placement and usually improves aesthetics. Kitchens and bathrooms are

light commercial systems if a water-chilling device such as a hydronic heat pump is used as the heating/cooling source.

Only fan-coils or air handlers that are equipped with condensate drip pans and drains should be used for chilled water cooling applications. The drip pan acts as a catch basin for the water droplets that continually form on the coil during cooling operation. The drip pans are usually connected to a drainpipe or floor drain so that the condensate can be disposed of at the same rate it is formed. If a floor drain is not available, a condensate removal pump is necessary.

Some hydronic fan-coils are designed only for heating. They do not have condensate pans and should not be used for chilled water cooling applications. Doing so would result in condensate running out the bottom of the unit causing damage to the structure. Examples of small wall-mounted fan-coils that are equipped with drip pans, and thus suitable for chilled water cooling, are shown in Figures 8-25a and 8-26.

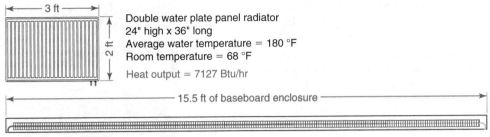

Figure 8-41 | Comparison of wall space required between a typical panel radiator and finned-tube baseboard of equivalent heat output. Both heat emitters are assumed to operate at the same conditions.

rooms where the wall space required for properly sized finned-tube baseboard is often not available. However, because panel radiators come in a wide variety of widths, heights, and thicknesses, they can often be integrated into such limited wall spaces, and still provide the necessary heat output. Figure 8-41 provides a comparison between a typical panel radiator and an equivalent length of residential-grade finned-tube baseboard when both are operated at an average water temperature of 180 °F. If the average water temperature was reduced to 110 °F, the same panel radiator would be equivalent to 18 feet of finned-tube baseboard.

- Most panel radiators contain very little water and relatively small amounts of metal. This results in low thermal mass and allows the panels to respond very quickly to variations in room air temperature or internal heat gains. The possibility of temperature overshoot in rooms with high internal heat gains from sunlight, lights, people, or heat-generating equipment gains is far less likely relative to systems that use high thermal mass heat emitters such as older cast-iron radiator or heated floor slabs. This rapid response characteristic is illustrated in Figure 8-42.

Figure 8-43 shows a sequence of infrared thermographs of a panel radiator taken over a period of approximately 4 minutes. The top image was taken 5 seconds after the panel's flow control valve was first opened. Prior to that, there was no flow through the panel, and its temperature was approximately 60 °F. The sequence of images show the surface temperatures resulting from migration of 150 °F water through the panel beginning from the valve in its upper-right corner. After 4 minutes, the panel is very close to steady-state conditions, emitting both radiant and convective heat at nearly the maximum rate possible for its current operating conditions.

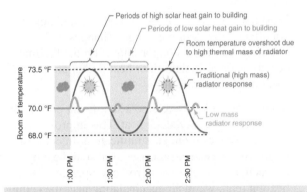

Figure 8-42 | Response time of low thermal mass panel radiators relative to high thermal mass cast-iron radiators. Note that peak temperature from high-mass radiators occurs during periods of solar gain. This is the opposite of what would be ideal.

- Panel radiators are able to operate with larger water temperature drops between their inlet and outlet, compared to some other heat emitters. In Europe, panel radiators are often sized to operate with a 20 °C (36 °F) temperature drop at design load. This is possible since occupants don't walk on wall-mounted panels, and thus wider variations in surface temperature are not as critical as they would be with floor heating. Larger temperature drops allow for lower flow rates (as discussed in Chapter 4, Properties of Water). Lower flow rates allow for smaller tubing and reduced circulator input power.

- Panel radiators can be sized to operate at relatively low water temperatures. This improves the efficiency of heat sources such as condensing boilers, solar collectors, and heat pumps. Operating panel radiators at lower water temperatures also increases the percentage of radiant versus convective heat output from the panel.

- Panel radiators are well suited to new construction and especially to remodeling. Their lightweight, easy-to-mount construction in combination with modern flexible piping materials such as PEX or PEX-AL-PEX tubing make them easy to install with minimal disruption of existing surface finishes.

- Most panel radiators are relatively durable. Their design and steel construction make them more resistant to physical damage than are most finned-tube baseboards. Most steel panel radiators also come with a high-quality powder-coat finish that provides excellent resistance to scratches or exterior corrosion. The latter is particularly important when heat emitters are located in humid spaces such as bathrooms.

- Most panel radiators are wall mounted and hence not affected by floor coverings. Changes in floor coverings can have a major impact on the thermal performance of radiant floor heating but are not an issue with panel radiators.

- Many panel radiators release a significant portion of their heat output as radiant heat. This improves comfort and reduces room air stratification. In contrast, finned-tube baseboard releases almost all heat by convection. This can create room temperature stratification (e.g., warm air accumulating near the ceiling while cool air settles at floor level), especially in room with tall ceilings. Such stratification reduces comfort and increases heat loss from the room.

Flat-Tube Panel Radiators

Some panel radiators are built as a grid of closely spaced flat steel tubes. The end of each tube is welded shut (see Figure 8-44a) to form a closed pressure-tight chamber. Holes in the rear face, near the end of each tube, connect to headers. The headers distribute flow to and from each tube as shown in Figure 8-44. The size and heating capacity of radiators constructed in this manner are determined by the length of the tubes and how many tubes are located side by side. It is a simple concept with enormous versatility.

Figure 8-45 shows an example of a **vertical panel radiator** constructed using flat tubes. Such radiators are ideal for tall but narrow wall spaces. They are especially useful in kitchens, bathrooms, or offices where horizontal wall space is often limited.

Vertical panel radiators are typically mounted with their bottom 5 to 12 inches above the floor. Most hang from rear-side mounting brackets fastened to the wall. These brackets should be fastened directly to wood-framing members or into solid masonry. In new construction with studded walls, wooden backer blocks

Figure 8-43 — Sequence of infrared thermographs of a panel radiator warming to near-steady-state condition over approximately 4 minutes. Initial panel temperature = 60°F. Entering water temperature = 160°F.

should be positioned at the proper height and location to accept the fasteners for the brackets. It is important that the heating system designer specifies the position of these framing blocks so they are installed before the wall is closed-up. Hanger locations are typically shown in manufacturer's installation instructions.

8.8 Panel Radiators

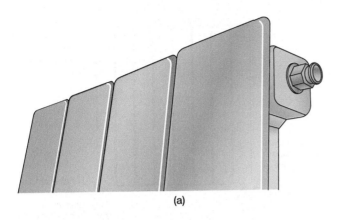

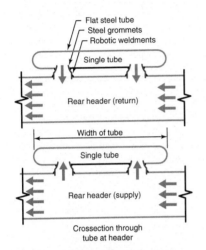

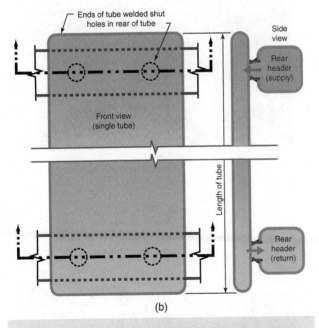

Figure 8-44 (a) Ends of flat tubes smoothly welded shut. Air vent at end of upper radiator header. *Courtesy of John Siegenthaler.* (b) Cross section through a typical flat-tube panel radiator. The length and number of tubes determine the heat output of the radiator.

Figure 8-45 Example of a vertical panel radiator. *Courtesy of Runtal North America.*

Piping connections usually consist of ½- or ¾-inch risers routed up through the floor, and connecting to threaded openings at the bottom of the radiator. This is one location where steel pipe nipples provide better protection against physical damage and are preferred over copper tubing. Their use also eliminates any visible soldered joints above the floor. Holes for the risers should be accurately located and neatly drilled about ¼ inch larger than the outside diameter of the riser pipe. This prevents noise from thermal expansion. The holes can be covered with escutcheon plates for a neat appearance.

The detail shown in Figure 8-46 is useful when flexible tubing such as PEX or PEX-AL-PEX is used to supply a panel radiator. The transition fitting to rigid steel pipe is concealed below the floor. However, the slack left in the flexible tubing allows this fitting to be pulled up through the floor if the panel ever needs to be removed for wall painting or other repair. The steel pipe nipples can be painted to match the radiator. Escutcheon plates cover the oversized holes for a simple and clean finish detail.

To provide for individual heat output control, a thermostatic radiator valve (TRV) is often mounted in the supply pipe of the panel radiator. To facilitate removal of the radiator for wall painting, it is common to mount a shutoff valve in the return pipe. A lockshield

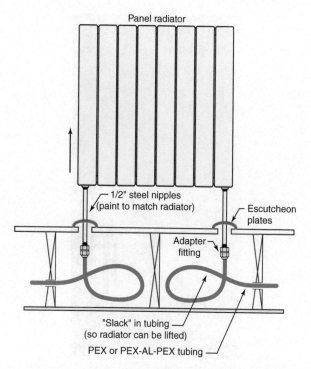

Figure 8-46 — Panel radiator installation detail using steel pipe nipples above floor and PEX or PEX-AL-PEX tubing under floor. The tubing slack allows the panel to be lifted off its mounting brackets if necessary.

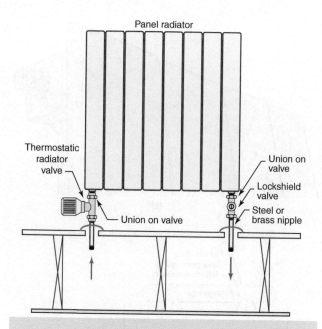

Figure 8-47 — Use of a TRV on supply riser and lockshield valve on return riser of panel radiator. These valves allow heat output control, as well as easy disconnection of the radiator if necessary.

Figure 8-48 — Examples of horizontal (flat-tube) panel radiators. *Courtesy of John Siegenthaler.*

valve with integral union, as discussed in Chapter 5, Piping, Fittings, and Valves, is a good choice for this location. The combination of a radiator valve with integral union in one riser and a drainable lockshield valve with integral union in the other allows the panel to be isolated from the system, neatly drained through a hose, and easily disconnected from the piping. It then can be lifted off its brackets for full access to the wall. A representation of these details is shown in Figure 8-47.

Horizontal (flat-tube) panel radiators are also available for wider but shorter wall spaces, such as beneath windows as shown in Figure 8-48. The mounting and riser piping details are similar to those used with vertical panel radiators.

Freestanding panel radiators are also available. They are usually supported by floor pedestals supplied by the manufacturer as shown in Figure 8-49. They are often installed at the base of glass curtain walls to control adverse drafts.

Figure 8-49 Horizontal (flat-tube) panel radiator supported by pedestal. Convection-enhancing fins are visible between "water plates." *Courtesy of Myson, Inc.*

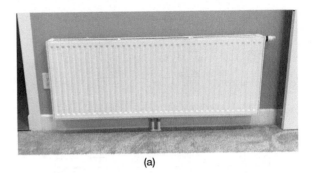

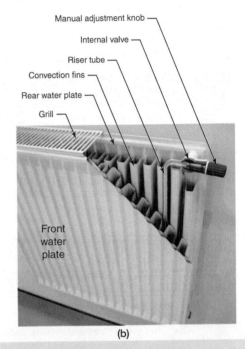

Figure 8-50 (a) Example of a fluted-channel panel radiator. (b) Cut-away showing internal construction of a fluted-panel radiator having two water plates and two rows of fins. *Courtesy of John Siegenthaler.*

Fluted-Channel Panel Radiators

Another very common style for panel radiators is the fluted-channel design such as that shown in Figure 8-50.

These radiators are made by welding two preformed steel sheets together at their perimeter to form a pressure-tight compartment. These sheets have been formed to create multiple parallel channels through which fluid can pass. These channels are called **flutes**. The forming operations also creates upper and lower manifold chambers that distribute flow to all the flutes. Several spot welds are made in the spaces between the flutes. These welds bond the front and back steel sheets together to prevent them from deforming when the radiator is pressurized. This welded assembly is called a **water plate**. It is the portion of the radiator through which water passes. Some panel radiators have a single water plate, and others have two or three water plates. Multiple water plates increase the heat output of the radiator at a given water temperature. Although a typical water plate may have several square feet of surface area, it contains a relatively small amount of water. This minimizes the thermal mass of the radiator, allowing it to respond quickly to changes in water temperature or room air temperature.

Most panel radiators also have folded steel fins welded to the back of their water plate(s). The purpose of the fins is to enhance convective heat transfer. Panel radiators with multiple water plates usually have multiple rows of fins.

The assembly of water plates and fins are joined together by a grill at the top, and side panels, to produce a clean, aesthetically pleasing finished radiator.

Figure 8-51 shows how the thickness of a fluted-steel panel radiator varies with the number of water plates and rows of fins. The thinnest radiator shown has two water plates and a single row of fins. The medium thickness radiator has two water plates and two rows of fins. The thickest radiator has three water plates and three rows of fins. The thicker the radiator (e.g., the greater the number of water plates and rows of fins), the higher the radiator's heat output per unit of frontal area.

Fluted-channel panel radiators can have a variety of piping configurations. One of the most common is called a **compact-style panel**. It has the supply and return piping connections at the bottom of the panel

346　Chapter 8　Heat Emitters

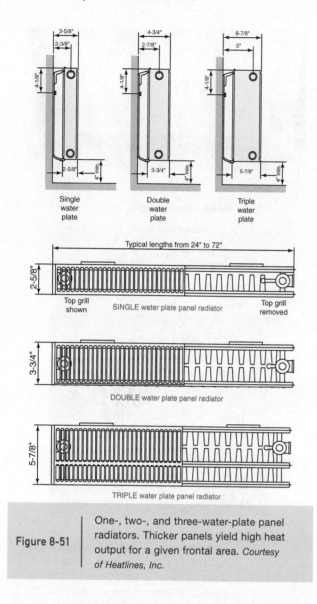

Figure 8-51 | One-, two-, and three-water-plate panel radiators. Thicker panels yield high heat output for a given frontal area. *Courtesy of Heatlines, Inc.*

(a)

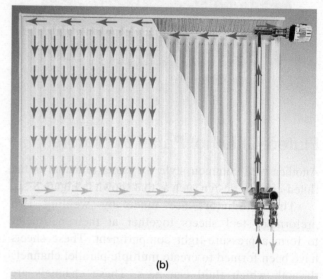

(b)

Figure 8-52 | (a) Internal construction of a fluted-panel radiator. (b) Water flow through panel radiator. *Courtesy of Heatlines, Inc.*

as seen in Figure 8-52. The supply and return piping connections are spaced 2 inches center-to-center. On some radiators, such as those shown in Figure 8-52, the piping connections are located closer to the right side of the radiator. On other radiators, such as the one shown in Figure 8-50a, the piping connections are centered on the width of the radiator.

On compact-style panel radiators, water enters at the left connection and passes up through an internal riser tube to the panel's integral flow regulating valve. If this valve is partially or fully open, flow can pass into the upper manifold, across the top of the panel, down through the flutes in the water plates, across the lower manifold, and eventually back to return connection at the bottom right of the panel. If this valve is closed, no flow can pass through the panel. *It is important not to reverse the flow direction through the panel.* Doing so can create noise within the flow regulating valve.

The location of the supply and return connections allows piping to enter from the floor or wall (as seen in Figure 8-53). It is common to install a **dual isolation valve** at the supply and return connections of the radiator. This allows it to be isolated from the remainder of the system and disconnected from the piping if necessary.

Fluted-channel panel radiators release both radiant and convective heat. Radiant heat helps warm objects in the space. It also increases the space's **mean radiant temperature**, which improves comfort. Convective heat directly warms air as it rises along the radiator. For radiators mounted under window, this rising warm air counteracts downward air current from cooler glass surfaces, which further improves comfort.

The percentage of radiant versus convective heat output from a fluted-steel panel radiator varies with the

8.8 Panel Radiators

number of water plates and rows of fins on the radiator. It also varies with the difference between the average water temperature in the radiator and the room air temperature. This difference is sometimes called "**over temperature**" or "**excess temperature**." These terms are commonly used in Europe. They can be simply interpreted as how many °F (or °C) the average water temperature is *above the room air temperature*. When the over temperature or excess temperature is zero, the heat output of the radiator is also zero. This forms the basis for developing equations that allow the radiator's heat output to be determined over a wide range of conditions (e.g., combinations of average water temperature and room air temperature).

Figure 8-54 gives the percentage of radiant to total heat output as a function of over temperature for several radiators with different water plate and fin configurations.

All the panel radiator configurations shown have higher amounts of radiant heat output when operating at relatively low differences between average water temperature and room air temperature. Radiators with more than one water plate and multiple rows of fins have lower ratios of radiant versus convective heat output. As the difference between average water temperature and room air temperature increases, so does the ratio of convective to total heat output.

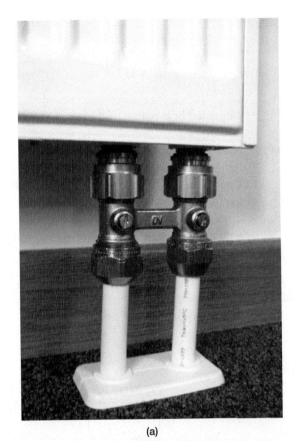

(a)

(b)

Figure 8-53 1/2-inch PEX-AL-PEX Piping entering dual lockshield valve from floor through escutcheon plate. (a) The dual lockshield valve allows the panel to be isolated from the remainder of system and removed if ever necessary. *Courtesy of John Siegenthaler.* (b) Dual lockshield angle valve allows piping to enter from wall. *Courtesy of John Siegenthaler.*

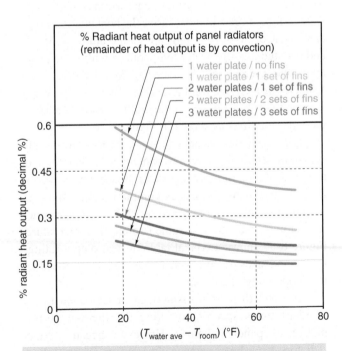

Figure 8-54 Percentage of radiant heat output to total heat output as a function of the difference between average water temperature and room air temperature for several radiator configurations.

Towel Warmer Radiators

Another variation of panel radiator is known as a **towel warmer**. These panels are designed for bathrooms or kitchens and provide a space in which to hang one or more towels. In addition to space heating, the simple appeal of a warm, dry towel after a shower or bath has proven very effective in marketing these panels. Hundreds of different styles, sizes, and colors are available. In many cases, towel warmers can be accessorized with garment knobs, mirror segments, and other amenities. When budgets allow, towel warmers are available with chrome, brass, and even 24-karat gold plating. Several types of towel warmers are shown in Figure 8-55.

Thermal Performance of Panel Radiators

Thermal ratings for panel radiators are usually given as tables that show the dimensions of the radiator, and its associated heat output at some reference average water temperature and an assumed room air temperature. When the radiator will be operated at conditions not given in the table, a correction factor is applied to determine its heat output.

For long vertical or horizontal flat-tube panels, thermal output ratings are usually expressed in Btu/h *per foot of panel length* for various panel widths and *average* water temperatures. In most cases, the heat output ratings are based on a surrounding air temperature of 65 °F, and a stated *average* water temperature within the panel (e.g., the average of the entering and leaving water temperature). Correction factors for other average water temperatures and room air temperatures usually accompany these ratings as shown in Figure 8-56.

For fluted-steel panel radiators, ratings are usually stated in Btu/h for a given panel width, height, and thickness (e.g., number of water plates and rows of fins). In North America, it is common to find the heat output of panel radiators based on 180 °F average water temperature and 68 °F assumed room air temperature. Figure 8-57 shows a typical table of heat output ratings for several sizes of fluted-steel panel radiators, based on these operating conditions.

To date, there is no American or Canadian rating standard that's specific to rating the thermal performance of panel radiators. However, because panel radiators have been extensively used in Europe, there is a European standard called **EN442**. This standard provides a basis for adjusting heat outputs over a wide range of conditions. It also involves several calculations, the first of which is given as Equation 8.11.

Equation 8.11:

$$Q_e = Q_{112}\left(\frac{\Delta T_d}{112}\right)^{1.3}$$

where
- Q_e = estimated heat output of the panel radiator (Btu/h)
- ΔT_d = temperature difference (e.g., "over temperature") determined using either Equation 8.12 or Equation 8.13 below (°F)
- Q_{112} = the output of the panel radiator when the difference between the *average* water temperature and room air temperature is 112 °F (Btu/h)
- 1.3 = an exponent (not a multiplier)

Equation 8.12:

$$\Delta T_d = \left[\left(\frac{T_{in} + T_{out}}{2}\right) - T_{air}\right]$$

Equation 8.13:

$$\Delta T_d = \frac{(T_{in} - T_{out})}{\ln\left(\dfrac{T_{in} - T_{air}}{T_{out} - T_{air}}\right)}$$

where
- ΔT_d = effective temperature difference (°F)
- T_{in} = inlet water temperature to panel (°F)
- T_{out} = outlet water temperature from panel (°F)
- T_{air} = room air temperature (°F)
- ln = natural logarithm function

Example 8.10

In Figure 8-57, a 2-water-plate radiator that's 24 inches high and 48 inches long, when operating at an over temperature of 112 °F, has a heat output of 9,500 Btu/h. Estimate its heat output assuming an inlet water temperature of 160 °F, an outlet water temperature of 140 °F, and a room air temperature of 65 °F.

Solution:

Use Equation 8.12 to calculate the value of ΔT_d:

$$\Delta T_d = \left[\left(\frac{T_{in} + T_{out}}{2}\right) - T_{air}\right]$$

$$= \left[\left(\frac{160 + 140}{2}\right) - 65\right] = 85 \text{ °F}$$

8.8 Panel Radiators 349

(a)

(b)

(c)

Figure 8-55 | Examples of towel warmer radiators. (a) Small towel warmer. (b) Tall towel warmer used as divider. *Courtesy of Vasco, Inc.* (c) Gold-plated tower warmer. *Courtesy of Myson, Inc.*

The value of Q_n for Equation 8.11 is the radiator's listed heat output at $\Delta T = 112$, which was 9,500 Btu/h.

Entering this value into Equation 8.11 yields:

$$Q_e = Q_{112}\left(\frac{\Delta T_d}{112}\right)^{1.3} = 9,500\left(\frac{85}{112}\right)^{1.3} = 6,637 \frac{\text{Btu}}{\text{hr}}$$

Discussion:

This is a relatively simple calculation. However, based on the EN442 standard, it doesn't necessarily apply all cases, as will be explained next.

As the entering water temperature drops closer to the room air temperature, or the flow rate through the panel changes, the EN442 standard introduces a modified way to calculate the difference between the average water temperature in the panel radiator and the room air temperature (e.g., the value of ΔT_d used in Equation 8.11). Use Equation 8.13 for this modified calculation.

The decision on using Equation 8.13 rather than Equation 8.12 to calculate the value of ΔT_d is based on Equation 8.14.

Equation 8.14:

$$u = \frac{(T_{out} - T_{air})}{(T_{in} - T_{air})}$$

where

T_{out} = outlet fluid temperature from panel (°F)
T_{in} = inlet fluid temperature to panel (°F)
T_{air} = room air temperature (°F)

Equation 8.14 looks at how the outlet temperature of the radiator is dropping relative to the inlet temperature. As the flow rate through the panel decreases, there would be a greater temperature drop across the radiator and thus the value of u in Equation 8.14 will decrease.

The EN442 standard uses the value of u to determine if the value of ΔT_d is calculated using Equation 8.12 or Equation 8.13. The determining criteria are as follows:

If $u < 0.7$, use formula 8.13.
If $u \geq 0.7$, use formula 8.12.

Example 8.11

Water enters the panel radiator used in Example 8.10 at 115 °F and exits at 92 °F. The air temperature in the room is 65 °F. Determine the correct ΔT_d to use in Equation 8.11.

Solution:

Start by calculating the value of u:

$$u = \frac{(T_{out} - T_{air})}{(T_{in} - T_{air})} = \frac{(92 - 65)}{(115 - 65)} = 0.54$$

Since $0.54 < 0.7$, the EN442 standard prescribes use of Equation 8.13 to calculate ΔT_d:

$$\Delta T_d = \frac{(T_{in} - T_{out})}{\ln\left(\frac{T_{in} - T_{air}}{T_{out} - T_{air}}\right)} = \frac{(115 - 92)}{\ln\left(\frac{115 - 65}{92 - 65}\right)}$$

$$= \frac{23}{\ln\left(\frac{50}{27}\right)} = \frac{23}{\ln(1.85185)} = 37.33\,°F$$

$$= 37.33\,°F$$

Now that the appropriate value of ΔT_d has been determined it can be entered into Equation 8.11 to get the estimated heat output of the radiator.

$$Q_e = Q_{112}\left(\frac{\Delta T_d}{112}\right)^{1.3} = 9,500\left(\frac{37.33}{112}\right)^{1.3} = 2,277 \frac{\text{Btu}}{\text{hr}}$$

Discussion:

This output is about one quarter of the "rated" heat output of the panel when the over temperature condition was 112 °F (e.g., average water temperature of 180 °F and room air temperature of 68 °F).

The EN442 calculation procedure can be also used "in reverse" to select a specific panel for a specific design load.

Example 8.12

Consider a room with a design load of 2,500 Btu/h when maintained at 70 °F. A panel radiator needs to be selected based on a water supply temperature of 115 °F, and while operate with a 25 °F temperature drop. Use the equations associated with the EN442 standard to select two possible radiators from the table in Figure 8-57.

Solution:

Start with Equation 8.14 to find the value of u:

$$u = \frac{(T_{out} - T_{air})}{(T_{in} - T_{air})} = \frac{(90 - 70)}{(115 - 70)} = 0.44$$

Since $u < 0.7$, use Equation 8.13 to get the value of ΔT_d:

$$\Delta T_d = \frac{(115 - 90)}{\ln\left(\frac{115 - 70}{90 - 70}\right)} = 30.83 \text{ °F}$$

Next, set up Equation 8.11 with all the known information, including the required heat output at the lower water temperature (e.g., 2,500 Btu/h):

$$2,500 = Q_{112}\left(\frac{30.83}{112}\right)^{1.3}$$

This equation can be solved for the necessary output at an over temperature ΔT_d of 112 °F, which is what the heat output ratings in Figure 8-57 are based on.

$$Q_{112} = \frac{(2,500)}{\left(\frac{30.83}{112}\right)^{1.3}} = 13,374 \frac{\text{Btu}}{\text{hr}}$$

Next, look through the table in Figure 8-57 to find a radiator with a listed output close to this value. The 3-water-plate (thickest) panel with a height of 24 inches and a length of 48 inches has an output of 13,664 Btu/h when operating at an over temperature of 112 °F. This is very close to the calculated necessary output and thus would be a suitable selection. A 3-water-plate panel that is 20 inches high and 64 inches long has a listed output of 15,829 Btu/h, which is more than adequate for this room. This radiator could also be used.

> **Discussion:**
>
> Selecting a panel radiator that has more than sufficient heat output is generally fine, assuming that the cost of the radiator is not significantly higher than that of a radiator that provide "adequate" heat output, and that the radiator can fit in the available wall space. The use of thermostatic valves on panel radiators can limit their heat output when oversized for the room's design heating load.

Fan-Enhanced Panel Radiators

Figure 8-58 shows a state-of-the-art heat emitter that combines the function of a panel radiator and convector. Inside the unit, and near its base, is a high-performance finned-tube element. Immediately above this element is a rack of several low-voltage, variable-speed "**microfans**," similar to those used in desktop computers. Each microfan only requires about 1.5 watts of electrical input power when operating at full speed. The enhanced airflow these microfans create across the finned-tube element can boost heat output by 50% during normal comfort mode, and by more than 250% during recovery from a setback condition. The microprocessor-controlled fans vary their speed as necessary based on the water temperature supplied to the radiator. Some radiators also allow room occupants to temporarily increase the fan speed to shorten the time required for room temperature to recover from a setback condition.

The significant gain in convective heat transfer allows these heat emitters to work with low supply water temperatures comparable to those required by bare heated floor slabs. The ability to operate at supply water temperatures as low as 95 °F can significantly increase the thermal efficiency of low-temperature hydronic heat sources such as solar collectors, heat pumps, and condensing boilers.

Piping for Panel Radiators

In theory, multiple panel radiators can be piped in a wide assortment of arrangements, which include both series and parallel configurations. In practice, some piping are much better suited to the flow characteristic and water temperature requirements of panel radiators.

One of the most versatile piping arrangements is called a **homerun distribution system**. It consists of a separate length of small-diameter flexible tubing, such as ½-inch PEX or PEX-AL-PEX to and from each radiator. All the supply tubes begin at a manifold, and all the return tubes end at another manifold, as shown in Figure 8-59.

Homerun distribution systems serving several panel radiators, each equipped with a thermostatic valve, are ideally suited to variable-speed pressure-regulated circulators, which were discussed in Chapter 7, Hydronic Circulators.

Notice that the manifold station shown in Figure 8-59 has a spare supply and return connection. This extra connection makes it easy to add another radiator to the system if needed in the future. Until that time the spare connection ports are simply fitted with pressure-tight caps. Multiple spare connections can also be provided depending on the anticipated extent of future system expansion.

Homerun distribution systems provide water at essentially the same temperature to each radiator, which simplifies radiator sizing. They also allow easy flow balancing or isolation if necessary. They are simple, repeatable, and adaptable to a wide variety

(a)

	MODEL	HEIGHT in	DEPTH in	Btu/h/ft Ratings @ 65°F EAT		
				215°F	180°F	140°F
PERIMETER STYLE	R-1	2.8	1.6	230	160	90
	R-2	5.7	1.6	420	300	170
	R-3	8.6	1.6	620	440	250
	R-4	11.5	1.6	820	580	330
	R-5	14.4	1.6	1,020	720	410
	R-6	17.3	1.6	1,220	860	500
WALL PANEL	R-7	20.2	1.6	1,430	1,010	580
	R-8	23.1	1.6	1,640	1,160	660
	R-9	26.0	1.6	1,850	1,300	750
	R-10	29.0	1.6	2,060	1,450	830

(b)

		EAT									
		45°F	50°F	55°F	60°F	65°F	70°F	75°F	80°F	85°F	90°F	95°F
	240°F	1.365	1.350	1.304	1.266	1.220	1.171	1.124	1.086	1.039	1	0.953
	235°F	1.343	1.305	1.267	1.219	1.171	1.124	1.086	1.038	1	0.952	0.910
	230°F	1.305	1.267	1.219	1.171	1.124	1.086	1.038	1	0.952	0.910	0.868
	225°F	1.267	1.219	1.171	1.124	1.086	1.038	1	0.952	0.910	0.868	0.826
	220°F	1.219	1.171	1.124	1.086	1.038	1	0.952	0.910	0.868	0.826	0.785
	215°F	1.171	1.124	1.086	1.038	1	0.952	0.910	0.868	0.826	0.785	0.744
	210°F	1.124	1.086	1.038	1	0.952	0.910	0.868	0.826	0.785	0.744	0.704
	205°F	1.086	1.038	1	0.952	0.910	0.868	0.826	0.785	0.744	0.704	0.664
	200°F	1.038	1	0.952	0.910	0.868	0.826	0.785	0.744	0.704	0.664	0.625
	195°F	1	0.952	0.910	0.868	0.826	0.785	0.744	0.704	0.664	0.625	0.587
	190°F	0.952	0.910	0.868	0.826	0.785	0.744	0.704	0.664	0.625	0.587	0.549
	185°F	0.910	0.868	0.826	0.785	0.744	0.704	0.664	0.625	0.587	0.549	0.511
	180°F	0.868	0.826	0.785	0.744	0.704	0.664	0.625	0.587	0.549	0.511	0.474
	175°F	0.826	0.785	0.744	0.704	0.664	0.625	0.587	0.549	0.511	0.474	0.438
AWT	170°F	0.785	0.744	0.704	0.664	0.625	0.587	0.549	0.511	0.474	0.438	0.403
	165°F	0.744	0.704	0.664	0.625	0.587	0.549	0.511	0.474	0.438	0.403	0.369
	160°F	0.704	0.664	0.625	0.587	0.549	0.511	0.474	0.438	0.403	0.369	0.334
	155°F	0.664	0.625	0.587	0.549	0.511	0.474	0.438	0.403	0.369	0.334	0.301
	150°F	0.625	0.587	0.549	0.511	0.474	0.438	0.403	0.369	0.334	0.301	0.269
	145°F	0.587	0.549	0.511	0.474	0.438	0.403	0.369	0.334	0.301	0.269	0.237
	140°F	0.549	0.511	0.474	0.438	0.403	0.369	0.334	0.301	0.269	0.237	0.207
	135°F	0.511	0.474	0.438	0.403	0.369	0.334	0.301	0.269	0.237	0.207	0.177
	130°F	0.474	0.438	0.403	0.369	0.334	0.301	0.269	0.237	0.207	0.177	0.149
	125°F	0.438	0.403	0.369	0.334	0.301	0.269	0.237	0.207	0.177	0.149	0.122
	120°F	0.403	0.369	0.334	0.301	0.269	0.237	0.207	0.177	0.149	0.122	0.096
	115°F	0.369	0.334	0.301	0.269	0.237	0.207	0.177	0.149	0.122	0.096	0.071
	110°F	0.334	0.301	0.269	0.237	0.207	0.177	0.149	0.122	0.096	0.071	0.050
	105°F	0.301	0.269	0.237	0.207	0.177	0.149	0.122	0.096	0.071	0.050	0.030
	100°F	0.269	0.237	0.207	0.177	0.149	0.122	0.096	0.071	0.050	0.030	0.011

EXAMPLE: To find the Btuh/ft Rating for an R-6 Panel at 155°F AWT and 65°F EAT, Multiply the Correction Factor (0.511) by the Btu/h/ft Rating at 215°F (1,219) e.g. (0.511) × (1,219) = 623 Btu/h/ft

Figure 8-56 (a) Thermal performance ratings of flat-tube panel radiators given as heat output per foot of panel length for several panel widths (R-1 = 1 tube, R-2 = 2 tubes, etc.). (b) Correction factors for heat output at different room air and average water temperatures. *Courtesy of Runtal North America.*

8.8 Panel Radiators 353

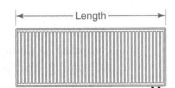

Heat output ratings (Btu/hr) at reference conditions:
Average water temperature in panel = 180 °F
Room temperature = 68 °F
Temperature drop across panel = 20 °F

1 water plate panel thickness

	16" long	24" long	36" long	48" long	64" long	72" long
24" high	1870	2817	4222	5630	7509	8447
20" high	1607	2421	3632	4842	6455	7260
16" high	1352	2032	3046	4060	5415	6091

2 water plate panel thickness

	16" long	24" long	36" long	48" long	64" long	72" long
24" high	3153	4750	7127	9500	12668	14254
20" high	2733	4123	6186	8245	10994	12368
16" high	2301	3455	5180	6907	9212	10363
10" high	1491	2247	3373	4498	5995	6745

3 water plate panel thickness

	16" long	24" long	36" long	48" long	64" long	72" long
24" high	4531	6830	10247	13664	18216	20494
20" high	3934	5937	9586	11870	15829	17807
16" high	3320	4978	7469	9957	13277	14938
10" high	2191	3304	4958	6609	8811	9913

(a)

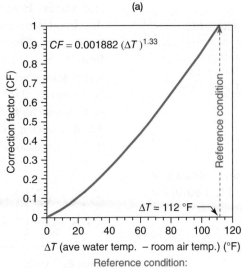

$CF = 0.001882 (\Delta T)^{1.33}$

ΔT = 112 °F

ΔT (ave water temp. − room air temp.) (°F)

Reference condition:
Ave water temperature in panel = 180 °F
Room air temperature = 68 °F

(b)

Figure 8-57 (a) Heat output ratings of fluted-steel panel radiators based on dimensions and stated operating conditions.
(b) Correction factor equation and graph for heat output of fluted-panel radiators at other operating conditions.
Courtesy Caleffi North America.

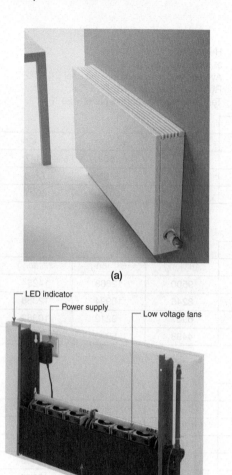

Figure 8-58 (a) Fan-enhanced panel radiator. (b) Internal construction showing low-voltage, variable-speed "microfans" that can more than double heat output relative to passive convectors of comparable design. *Images courtesy of JAGA USA.*

of heat emitters. Homerun distribution systems are discussed in more detail in Chapter 11, Distribution Piping Systems.

Reverse Return Piping

In some situations, it is desirable to use several panels of the same size, and operated as a single zone. An example is where a radiator will be placed under each of several identical windows in a large room. In these situations, it is preferable to pipe the radiators in parallel as shown in Figure 8-60. This piping arrangement uses **reverse return piping**, which encourages approximately equal flow rates and equal supply temperature to each radiator. This configuration also reduces head loss compared to piping radiators in series. Flow through the piping arrangement shown in Figure 8-60 is controlled by a single zone valve. There is generally no need of flow-balancing valves or thermostatic valves on each radiator.

Panel Radiator Piping Configurations

Another piping arrangement can be used for connecting two or three panel radiators in what appears to be a series arrangement, but in fact allows individual control of each radiator. This piping configuration requires an adjustable **1-pipe valve** to be installed on each radiator. Figure 8-61 shows how three panel radiators would be connected using these specialized valves.

The horizontal "bridge" on the 1-pipe valve contains a bypass valve, which can be adjusted to determine what percentage of the flow entering the valve passes through the radiator. When the thermostatic valve in the upper-right corner of the radiator is closed, all flow entering the 1-pipe valve passes through the bypass valve and onto the next connected radiator. Thus, each radiator on the 1-pipe circuit can be individually controlled.

The 1-pipe arrangement reduces piping lengths relative to a homerun distribution system in situations where two to three panel radiators are in proximity and need to be individually controlled by thermostatic radiator valves. However, the temperature of the water in the circuit decreases as it flows from one active radiator to the next. The farther "downstream" the radiator is from the beginning of the circuit, the lower the entering water temperature. Lower water temperatures reduce heat outputs. This is one reason that 1-pipe configuration should be limited to no more than three radiators. Another reason for this limit is to keep the head loss of the circuit reasonable.

8.9 Other Hydronic Heat Emitters

Thus far, we have examined a wide variety of heat emitters, focusing on those most widely used in modern residential and light commercial systems. There are several other hydronic heat emitters that are not as widely used. Hydronic system designers should be familiar with these less commonly used heat emitters should a suitable application present itself. This section gives a brief overview of two such heat emitters.

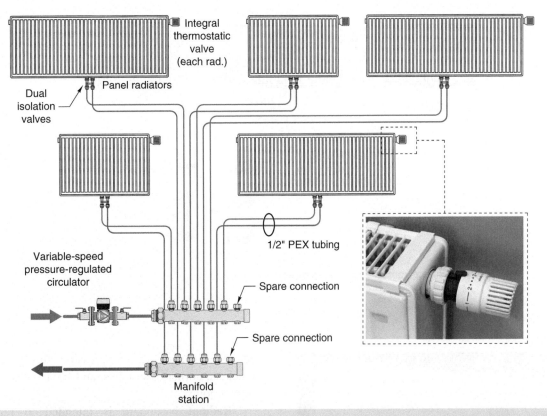

Figure 8-59 | Several panel radiators, each equipped with a thermostatic valve, and connected to a common manifold station to form a homerun distribution system. *Courtesy of John Siegenthaler.*

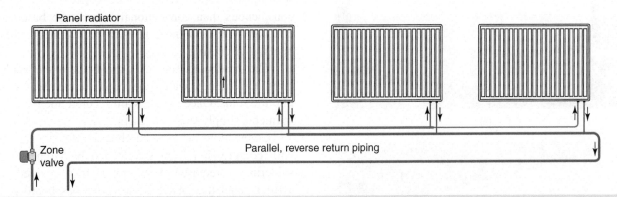

Figure 8-60 | Four identical panel radiators piped in a parallel, reverse return configuration. This reduces head loss and encourages equal flow through and equal supply temperature to all radiators.

Radiant Baseboard

One type of heat emitter that has been used, to a limited extent, in North America is called a **radiant baseboard**. The active portion of the baseboard consists of an aluminum extrusion that forms a very slim profile about 1 inch wide and 5 inches high. The rear side of the extrusion has two channels that are tightly bonded to either copper or PEX tubing. A cutaway view of a radiant baseboard is shown in Figure 8-62.

Unlike finned-tube baseboards, radiant baseboards have no fins. The majority of their heat output is by thermal radiation rather than convection. The radiant output warms the floor and objects near the floor. This enhances thermal comfort, while also reducing temperature stratification from floor to ceiling.

Radiant baseboard is typically sold in straight segments of 1 to 10 feet in increments of 6 inches. To join these segments together, as well as to accommodate corners and other details, manufacturers supply

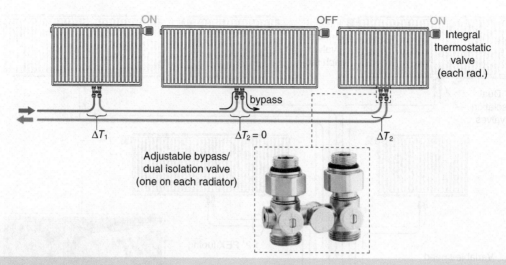

Figure 6-61 | Three panel radiators connected using 1-pipe bypass valves.

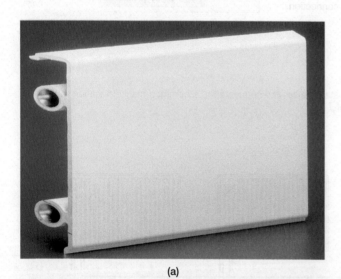

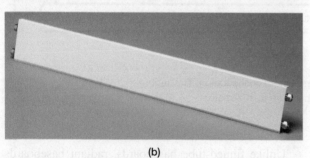

Figure 8-62

(a) Cutaway view of a radiant baseboard consisting of an aluminum extrusion with copper-lined fluid passageways. (b) Complete segment of radiant baseboard with compression fittings. *Courtesy of Radiant Technology, Inc.*

accessories such as compression fittings, mounting brackets, corner assemblies, and filler panels for covering the joints.

Radiant baseboard is intended to replace the conventional wooden baseboard at the base of the walls. *It is typically installed on all walls in the room.* The narrow profile and high-quality finish blend well with most interiors. Radiant baseboards tend to be less conspicuous than conventional finned-tube baseboards and can withstand more physical punishment without bending or denting.

The baseboard segments are held to the wall by clips. On framed walls, the clips should be fastened into the studs. Once the clips are secured, the baseboard segments simply snap into them.

Connections between baseboard segments are made with compression fittings tightened by wrenches. No soldering is required. The joint is then covered with a filler panel that matches the active baseboard panel. At corners, a short length of pre-bent PEX tubing with a tight 90-degree turn is used to connect the panels on each wall.

The double-tube design allows both supply and return connections to be made at the same end of the unit. The other end of the baseboard can be "dead-ended" at any location along the wall by installing a U-bend between the upper and lower tubes.

Radiant baseboards are typically piped using a homerun distribution system similar to that shown in Figure 8-59. Small-diameter PEX or PEX-AL-PEX tubing is used to connect each baseboard to the manifold station. Flow through individual circuits can be controlled using manifold valve actuators, which are discussed in more detail in later chapters.

Because of its smaller size and surface area, a given length of radiant baseboard has a lower heat output than the same length of finned-tube baseboard. A comparison between the heat output of a typical residential finned-tube baseboard and a 5-inch by 1-inch radiant baseboard is shown in Figure 8-63.

The lower heat output per foot of length for radiant baseboard, relative to that of finned-tube baseboard, requires longer lengths to be used to achieve a given heat output. A guideline is to allow twice the linear footage of radiant baseboard compared to that required for typical residential finned-tube baseboard.

Infloor Convectors

Occasionally, there is a need to provide convective heat output directly in front of a window or patio door that extends all the way to the floor. The convective heat output is used to "wash" the glass surface with a gentle rising layer of warm air. This discourages condensation on the inside of the glass when the outdoor temperature is very low.

Infloor convectors were developed for this specific purpose. An example of an infloor convector unit is shown in Figure 8-64.

An infloor convector combines a finned-tube element similar to that used in baseboard, with a **recessed floor box**, and upper grill. This assembly allows cool air to drop into the wall side of the floor box, reverse direction at the bottom of floor box, and pass upward through a finned-tube element. A vertical baffle within the floor box separates the short vertical column of warm air above the element from the descending column of cooler air. The buoyancy differences between the heated and cool air provide the driving force for natural convective air movement.

Although an infloor convector could be used in a residential system, such usage is rare. One of the significant restrictions is interference between the recessed floor box and floor joists. The depth of the floor box precludes "notching" the floor joists. One solution is to orient the joists in parallel with the long dimension of the floor box and carefully place them to allow the unit to be properly positioned relative to the wall. Another possibility is to use structural headers to carry the floor loading around the convector. Using an infloor convector in a concrete floor slab requires forming a cavity for the floor box and providing for the routing of supply and return piping. Again, such installations are possible, but often dismissed as overly complex or expensive in residential and light commercial buildings. Still, when glazing extends to the floor, and a pedestal-mounted panel radiator is not desired, an infloor convector provides an excellent solution for draft control.

8.10 Head Loss of Heat Emitters

When analyzing piping systems containing heat emitters, it is important to know the head loss they create. The following guidelines describe typical head loss characteristics of each of the heat emitters discussed.

Head Loss of Finned-tube Baseboard

Since the element of a finned-tube baseboard is simply a length of copper tubing with fins attached, the pressure drop will be the same as that of a standard copper tube of the same diameter and length. It can be determined using the methods given in Chapter 6, Fluid Flow in Piping.

The head loss of radiant baseboards can be determined by considering each baseboard to be a tubing circuit approximately twice as long as the baseboard extrusion. This head loss can then be combined with the head loss of the "leader" tubing connecting each radiant baseboard back to the manifold. The head loss of such circuits can also be accurately modeled using the

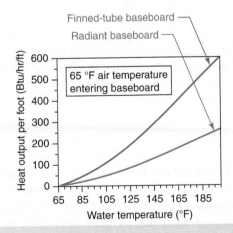

Figure 8-63 — Comparative heat output of residential-grade finned-tube baseboard and radiant baseboard. Finned-tube output does not include 15% heating effect factor.

358 Chapter 8 Heat Emitters

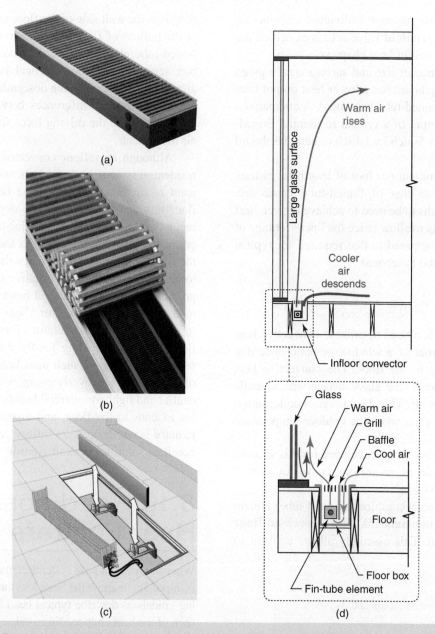

Figure 8-64 (a) Infloor convector. (b) Roll-up grill. (c) Flexible connection hoses allow element to be removed for cleaning. (d) Internal construction creates natural convection airflow across the element. Floor joist must be carefully planned to accommodate floor box. (a,b,c) *Courtesy of JAGA USA.*

methods of Chapter 6, Fluid Flow in Piping, or through use of the Hydronic Circuit Simulator module in the Hydronics Design Studio software.

Head Loss of Fan-Coils and Panel Radiators

Most manufacturers of fan-coils and panel radiators list the head loss of their products at two or three specific flow rates. If at least two points are listed, the data can be plotted and a smooth curve drawn through them for use in estimating the head loss at other flow rates. An example is shown in Figure 8-57. Keep in mind that all head loss versus flow rate curves for heat emitters will pass through zero head loss at zero flow. Also remember that head loss can be determined from pressure drop data using Equation 6.5:

$$H_{\text{loss}} = \frac{144(\Delta P)}{D}$$

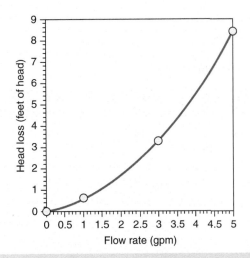

Figure 8-65 | Graph of manufacturer's data for head loss versus flow rate for small fan-coil.

8.11 Heat Loss from Copper Tubing

Although usually not intended as a heat emitter, any tube conveying fluid at a temperature greater than that of the surrounding air releases heat. If the tube runs through heated space, this heat offsets a portion of the building's heating load. If the tube runs through unheated space, the heat is lost.

Several issues related to system design can be addressed if the designer has a method for evaluating tubing heat loss. For example, the designer may want to know the heat loss and temperature drop in the tubing between the outlet of the heat source and the inlet of the first heat emitter. If the system requires long tubing runs through relatively cool spaces, significant heat loss and temperature drop can occur between heat emitters. This will obviously lower the heat output capability of the heat emitters. The designer may also want to know how much heat is released by tubing routed through heated space since this heat partially heats the building.

The heat output of a tube is affected by the material, pipe size, flow rate, fluid, and surrounding air temperature. If the tubing is insulated, its heat loss will be considerably reduced. The complexity of evaluating heat loss under all combinations of these conditions is beyond what can be presented in this book. However, the heat loss of bare copper tube can be evaluated using one of the following two methods:

Method 1

When the tube's length in feet, divided by the flow rate through it in gallons per minute, is less than 20 ft/gpm, the heat loss can be estimated using the graph in Figure 8-66.

This graph is based on extrapolation of copper tube heat loss data published in the *ASHRAE Handbook of Fundamentals*. It plots the heat loss of the tube versus the difference between the fluid inlet temperature and the air temperature surrounding the tube. When the tube is relatively short, or operates at a high flow rate, there is very little temperature drop from the inlet to the outlet. This allows the heat loss to be reasonably well estimated by using the inlet temperature when calculating the temperature difference between the fluid and air.

Method 2

When the length of the copper tube in feet divided by the flow rate in gallons per minute exceeds 20 ft/gpm, there will be a greater temperature drop along the tube from inlet to outlet. In such cases, Equation 8.15 can be used to accurately calculate the outlet temperature from the tube.

Equation 8.15:

$$T_{out} = T_{air} + \left[(T_{in} - T_{air})^{C_1} + \frac{C_2 \times L}{(8.021 cD) \times f} \right]^{1/C_1}$$

where

T_{out} = temperature of the water leaving the pipe (°F)
T_{air} = temperature of the air surround the pipe (°F)
T_{in} = temperature of the water entering the pipe (°F)
L = length of the pipe (ft)
f = flow rate of water in the pipe (gpm) ($f > 0$)
C_1 and C_2 = constants based on tube size and read from Figure 8-67
c = specific heat of the fluid (at inlet temperature) (Btu/lb/°F)
D = density of the fluid (lb/ft³)

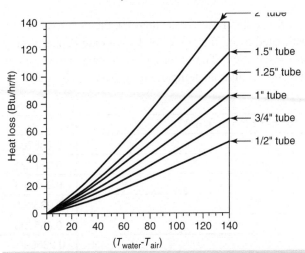

Figure 8-66 | Heat loss from bare copper tubing. Used when length in feet divided by flow rate in gallons per minute is less than 20 ft/gpm.

Be sure to copy all the significant figures shown for constants C_1 and C_2 when using them in Equation 8.15. Be sure to include the negative sign associated with each value of C_1.

Once the outlet temperature of the tube is determined, its heat loss can be calculated using Equation 8.16:

Equation 8.16:

$$Q = (8.01Dc)f(T_{in} - T_{out})$$

where

T_{out} = temperature of the water leaving the pipe (°F)
T_{in} = temperature of the water entering the pipe (°F)
f = flow rate of water in the pipe (gpm); note: $f > 0$
c = specific heat of the fluid (at inlet temperature) (Btu/lb/°F)
D = density of the fluid (lb/ft³)

Example 8.13

Water enters a 1-inch copper tube at 10.0 gpm and 160 °F. The tube is 35 feet long and is surrounded by air at 60 °F. Estimate the heat loss of the tube.

Solution:

The first step is to check the ratio of the tube length in feet divided by the flow rate in gallons per minute:

$$\frac{\text{Length}}{\text{flowrate}} = \frac{35 \text{ ft}}{10 \text{ gpm}} = 3.5 \text{ ft/gpm}$$

Since this ratio is less than 20, the heat loss can be estimated using the curve for 1-inch bare copper tubing in Figure 8-66. The entering water temperature minus the surrounding air temperature is $160 - 60 = 100$ °F.

The heat loss per foot of tube is read from the graph as approximately 57 Btu/h/ft.

The total heat loss of the 35 feet tube is thus $35 \times 57 = 1995$ Btu/h.

> **Discussion:**
>
> At the stated flow rate of 10.0 gpm, a heat loss of 1,995 Btu/h yields a temperature drop of 0.4 °F as the water flows through the tube. Therefore, the rate of heat loss at the beginning of the tube is approximately the same as that at the outlet.

Example 8.14

Water at 200 °F and 1.0 gpm enters a 105-foot long run of 3/4-inch bare copper tubing surrounded by 60 °F air. Estimate the heat loss from the tube.

Solution:

Dividing the length of the tube by the flow rate in gallons per minute yields a ratio of:

$$105 \text{ ft}/1.0 \text{ gpm} = 105 \text{ ft/gpm}$$

Since this ratio is above 20, Equation 8.15 should be used instead of the graph in Figure 8-66. The density of water at 200 °F is 60.126 lb/ft³. The specific heat is approximately 1.00 Btu/lb/°F. The values of C_1 and C_2 for 3/4-inch copper tube are found in Figure 8-67: $C_1 = -0.237721$, $C_2 = 0.03695$. Substituting these into Equation 8.15 yields:

$$T_{out} = 60 + \left[\frac{(200 - 60)^{-0.237721} + 0.03695 \times 105}{(8.01 \times 1 \times 60.126) \times 1}\right]^{1/-0.237721}$$

$$T_{out} = 60 + [0.308902 + 0.0080448] - 4.206612$$

Copper tube size (in)	C_1	C_2
1/2	−0.238285	0.02665
3/4	−0.237721	0.03695
1	−0.236284	0.04595
1.25	−0.235350	0.05475
1.5	−0.235693	0.06325
2	−0.235996	0.07985

Figure 8-67 | Table of constants (C_1 and C_2) for use in Equation 8.8.

$$T_{out} = 60 + 125.648$$
$$T_{out} = 185.648 \approx 185.7\ °F$$

The heat loss is then calculated using Equation 8.16:

$$Q = (8.01 \times 1 \times 60.126)(1)(200 - 185.648)$$
$$= 6{,}922\ \text{Btu/hr}$$

> **Discussion:**
>
> This is a significant heat loss. Enough to heat a room in a typical house. It points to the need to insulate copper tubing runs that convey hot fluids through cool surroundings.

The Pipe Heat Loss module in the Hydronics Design Studio can determine the heat loss of both bare and insulated copper-tubing operating under a wide range of operating conditions. Several selectable insulation options are included. Figure 8-68 shows a sample screen from this module.

8.12 Thermal Equilibrium

When turned on, every hydronic heating system attempts to establish operating conditions that allow the distribution system to dissipate the current rate of heat output from the heat source. If not for the intervention of temperature-limiting controls, every system would eventually stabilize at a supply water temperature where such **thermal equilibrium** exists. This supply water temperature may or may not provide the proper heat input to the building. Likewise, it may or may not be conducive to safe and efficient operation or long system life. Thermodynamics "doesn't care" about these conditions. It only "cares" about establishing and maintaining a balance between the rate of heat input and the rate of heat release.

By adjusting the size, number, or other characteristics of heat emitters in the distribution system, the designer can manipulate the steady-state supply water temperature at which the system "wants" to operate (e.g., the supply water temperature at which thermal equilibrium exists). When properly done, this allows both the heat source and distribution system to operate at conditions that are safe, efficient, comfortable, and conducive to long system life. If this tendency to operate at thermal equilibrium is disregarded, the resulting system may attempt to stabilize at a supply temperature that is either unsafe, inefficient, or that shorten the life of the heat source.

Consider the two systems shown in Figure 8-69. Each system has an identical heat source that delivers heat at a rate of 20,000 Btu/h to the circulating water and can operate over a wide range of temperature. System (a) has 31 feet of typical finned-tube baseboard for its heat emitter. System (b) contains 111 feet of the same baseboard. The temperature limit controller on each heat source is set to 200 °F, and the systems are put into continuous operation at approximately the same water flow rate.

The water leaving the heat source in system (a) climbs to a temperature of 180 °F, and stabilizes (e.g., reaches thermal equilibrium). At this temperature, the 20,000 Btu/h being added to the flowing water by the heat source is also being dissipated from the flowing water by the 31 feet of finned-tube baseboard. Note that this temperature is still 20 °F lower than the setting of the heat source's high limit controller. Even if the limit control setting was set to 220 °F, the supply water temperature would not climb higher than 180 °F. The supply water temperature will only climb to a value at which the heat output of the heat source is being dissipated by the distribution system.

When the amount of baseboard is increased to 111 feet, thermal equilibrium occurs at a supply water temperature of 120 °F. *The fact that the high limit controller is set to 200 °F is irrelevant.* The supply water temperature has no reason to climb above 120 °F since at that temperature the 111 feet of baseboard can dissipate the full 20,000 Btu/h of heat supplied by the heat source.

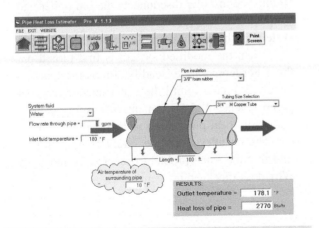

Figure 8-68 | The Pipe Heat Loss module from the Hydronics Design Studio. *Source: Hydronicpros.*

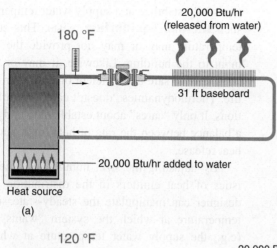
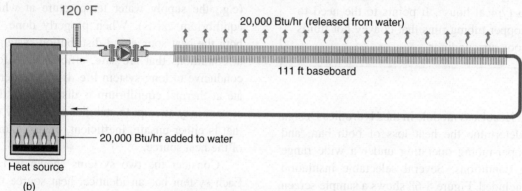

Figure 8-69 | (a) With 31 feet of baseboard as the heat emitter, thermal equilibrium is achieved at a supply water temperature of 180 °F. (b) With 111 feet of baseboard as the heat emitter, thermal equilibrium is achieved at a supply water temperature of 120 °F.

Example 8.15

A cast-iron boiler with a rated output of 60,000 Btu/h was paired with a distribution system consisting of finned-tube baseboards. The baseboards were sized to deliver the boiler's full 60,000 Btu/h output when the average water temperature in the circuit was 172 °F. The high limit control of the boiler was set to 190 °F. Each time the system was turned on, the water temperature would quickly climb until the boiler outlet temperature reached 180 °F and the water returning to the boiler settled in at 164 °F. The system was in thermal equilibrium under these conditions. The return water temperature was high enough to avoid sustained flue gas condensation within the boiler. The system operated under these conditions for 30 years.

Based on the age of the boiler, its tested efficiency of 83% and the promised efficiency of new technology, the owner decides to replace the boiler with a ground source, water-to-water heat pump. The heat pump is rated to produce 60,000 Btu/h. The installer disconnects the old boiler and connects the heat pump to the same distribution system as shown in Figure 8-70.

When the modified system is first turned on, the water temperature steadily climbs for approximately 2 minutes, until the water leaving the heat pump reaches 125 °F. An internal safety control within the heat pump then turns off the compressor. The installer resets the safety control only to find that the same problem occurs over and over. Why does this keep happening?

Solution:

The fundamental problem is that the existing distribution system is incapable of delivering 60,000 Btu/h while operating at the maximum water temperature the heat pump can provide. Although the heat pump, when operating, is adding 60,000 Btu/h to the water flowing through it, the distribution system cannot dissipate heat at this rate without receiving water at 180 °F. Thus, based on the system's inherent attempt to achieve thermal equilibrium, the water temperature continues to rise until the heat pump shuts itself off at its maximum permissible operating temperature.

Correcting this problem will require adding more heat emitter surface to the system until it is capable of delivering 60,000 Btu/h while being supplied with water at a temperature that is at least a few degrees below 125 °F, and even lower if possible.

Discussion:

Although it may seem odd to those who have read the preceding chapters, it is not uncommon for some "installers" to disconnect and remove an old heat source and simply reconnect a new heat source of the same nominal heating capacity to the existing distribution system. This example demonstrates that the temperature requirements of the distribution system must be matched to those of the heat source in order for thermal equilibrium to occur at reasonable water supply temperatures. Increasing the heating capacity of the heat pump in this example would not improve the situation. It would only shorten the time from when the heat pump was turned on to when it shuts off on its internal safety control. It is also worth noting that the flow resistance of a typical water-to-water heat pump would be significantly higher than that of the cast-iron boiler it was replacing. This could further exasperate the situation by inadvertently lowering the system's flow rate, and thus decreasing the rate of heat transfer.

Predicting Where Thermal Equilibrium Exists

It is possible to predict the supply water temperature at which thermal equilibrium will occur in a given system. Doing so relies on the fact that heat output from any hydronic distribution system, as well as a single heat emitter, is approximately proportional to the difference between the supply water temperature and the room air temperature. This can be expressed mathematically as follows:

Equation 8.17:

$$Q_{output} = c_{ds} \times (T_s - T_a)$$

Where
- Q_{output} = the heat output of the distribution system (Btu/hr)
- c_{ds} = a number that is specific to a given distribution system (Btu/hr/°F)
- T_s = temperature of fluid supplied to the distribution system (°F)
- T_a = temperature of air surrounding heat emitters (°F)

The term $(T_s - T_a)$ is called the "**driving ΔT**." It represents the temperature differential that drives heat from the water, out through the heat emitter, and into the space being heated. Anything that makes the driving ΔT larger, such as increasing the supply water temperature or decreasing room air temperature, increases the rate of heat release from the heat emitters, and vice versa.

The value of c_{ds} is determined by dividing the heat output from the distribution system at design load, by the difference between the temperature of water supplied to the distribution system at design load conditions, minus the room air temperature.

Example 8.16

A building contains a hydronic distribution system that can release 100,000 Btu/h into a 70 °F space when supplied with water at 170 °F. What is the c_{ds} value from Equation 8.17 for this distribution system.

Solution:

Rearranging Equation 8.17 and substituting in the given data yields:

$$c_{ds} = \frac{Q_{output}}{(T_s - T_a)} = \frac{100,000}{(170 - 70)} = 1,000 \text{ Btu/h/°F}$$

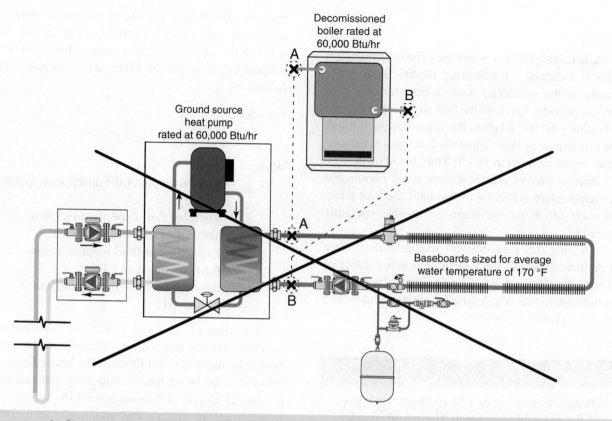

Figure 8-70 Ground source heat pump as a "cut-in" replacement for a boiler of equal heating capacity in a system using baseboard sized for an average water temperature of 170 °F. The heat pump cannot produce water at a temperature that will allow the baseboard to dissipate 60,000Btu/hr. This system cannot achieve thermal equilibrium at a supply temperature that is compatible with the heat pump.

Discussion:

A practical way to think of the c_{ds} value is the rate of heat output from the distribution system (in Btu/hr), *per degree* Fahrenheit difference between the temperature of water entering the distribution system and the room air temperature. Hence, if the supply water temperature was 130 °F and the inside air temperature was 68 °F, this distribution system would provide the following heat output to the building:

$$Q_{output} = c_{ds} \times (T_s - T_a) = 1{,}000 \times (130 - 68)$$
$$= 62{,}000 \text{ Btu/hr}$$

The relationship between supply water temperature, room air temperature, and heat output for any hydronic distribution system operating at a fixed flow rate can also be represented graphically as a **heat dissipation line** as shown in Figure 8-71, which is based on Example 8.16.

To find the water supply temperature at which this system will establish thermal equilibrium, first locate the

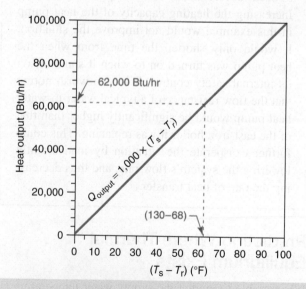

Figure 8-71 Example of a heat output line for the distribution system described in Example 8.16.

heat output rate of the heat source on the vertical axis. Project a horizontal line to the right until it intersects the heat dissipation line for the distribution system, then downward to the horizontal axis. This corresponding driving ΔT at which the system achieves thermal equilibrium is read from the lower axis. Add the room air temperature to this driving ΔT to get the required supply water temperature.

Example 8.17
A heat source with a rated output of 45,000 Btu/hr is connected to the distribution system represented by the heat dissipation line in Figure 8-71. The distribution system is to supply heat to a space that is to remain at 65 °F. Determine the supply water temperature at which the system will stabilize.

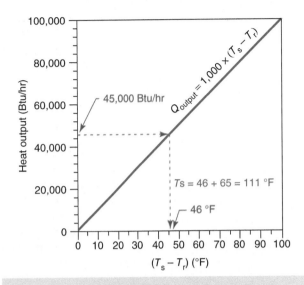

Figure 8-72 | Heat dissipation line of a given distribution system being used to find the supply water temperature required to dissipate 45,000 Btu/hr into a room maintained at 65 °F.

Solution:
The heat output rating of the heat source is located on the vertical axis. A horizontal line is drawn to the heat dissipation line for the distribution system. A vertical line is drawn from this intersection to find the corresponding driving ΔT, as shown in Figure 8-72. Finally, the interior air temperature of 65 °F is added to the driving ΔT of 46 °F to arrive at the necessary supply water temperature of 111 °F.

Discussion:
This method can also be used to find the supply water temperatures corresponding to different heating capacities of modulating boilers, or other modulating heat sources. Whenever the heat output rate of the heat source decreases, so does the supply water temperature corresponding to thermal equilibrium. Assuming that the flow rate through the system remains constant, the temperature drop (e.g., the "ΔT") of the system also decreases as the rate of heat dissipation decreases.

If the temperature-limiting control on the heat source is set *below* the temperature corresponding to thermal equilibrium, the distribution system will not get hot enough to dissipate the output of the heat source. The temperature of the water leaving the heat source will climb as the system operates, eventually reaching the temperature setting of the limit control. At that point, the heat source (burner, compressor, etc.) is turned off. The water temperature leaving the heat source then begins to decrease as heat continues to be dissipated by the circulating distribution system. Eventually, the temperature drops to the point where the heat source is turned back on, and the cycle repeats. This is a very common operating mode in many systems during partial load conditions. It can even occur under design load conditions in systems with an oversized heat source.

If the temperature-limiting control on the heat source is set *above* the temperature corresponding to thermal equilibrium, the water leaving the heat source will never achieve that temperature setting unless the load is reduced or turned off.

The slope of the heat dissipation line can be altered by changing the total surface area of heat emitters in the distribution system. In the case of panel radiators, this is done by increasing the size of radiators (height, width, or thickness), increasing the number of radiators in the system, or both. The concept is shown in Figure 8-73.

This principle holds true for any type of hydronic heat emitter. Large total surface areas allow for lower supply water temperatures. The latter increases the thermal efficiency of any heat source.

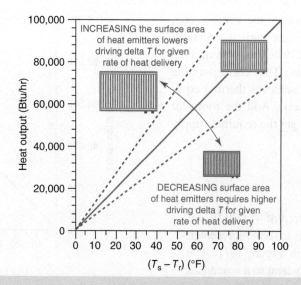

Figure 8-73 | Increasing the size and/or number of heat emitters in a system increases the total heat dissipating surface area of the distribution system. This increases the slope of the heat dissipation line, and vice versa.

Summary

This chapter has discussed several types of hydronic heat emitters suitable for use in residential and light commercial systems. They have ranged from units having mostly convective heat output such as finned-tube baseboards and fan-coils to those with a significant percentage of radiant heat output, such as panel radiators and radiant baseboards.

No one type of heat emitter is ideal for all applications. Attempts to use only one type for all kinds of jobs will inevitably lead to compromises and possible dissatisfaction. One of the benefits of hydronic heating is the ability to "mix and match" different types of heat emitters within the same overall system.

The designer needs to be aware of the range of products available and use them as resources to match the system to the specific needs of the building and the budget. Knowing the strengths and weaknesses of each type of unit can help narrow the selection during preliminary design.

Chapter 10, Hydronic Radiant Panel Heating, will present yet another spectrum of possible heat emitter selections. These can be used individually or in combination with the heat emitters discussed in this chapter.

Key Terms

1-pipe valve (for panel radiators)
180-degree return bend
A-coils
active length
air handler
blow-through fan-coils
blower
circuit simulator module
coil
compact-style panel
console fan-coil
convectors
correction factor
damper
double interpolation
draw-through fan-coil
drip pans
driving ΔT
dual isolation valve
dual lockshield valve
element
EN442 standard
escutcheon plate
fan
fan-coil
finned-tube baseboard convector
fluted-channel panel radiators
flutes
heat dissipation line
heat emitter

heat output ratings	IBR testing and rating code	panel radiator	stratification
heating effect factor		radiant baseboard	tangential blower
high wall cassette	infloor convectors	radiator	thermal equilibrium
homerun distribution system	interpolation	recessed fan-coils	thermal radiation
	kick space heater	recessed floor box	towel warmer
hydro-air distribution system	low-limit aquastat	reverse return piping	unit heater
	microfans	Series Baseboard Simulator	vertical panel radiator
IBR method	over temperature / excess temperature		water plate

Questions and Exercises

1. A finned-tube baseboard convector has the heat output characteristics given in Figure 8-10 (lower curve). Estimate the heat output of a 1-foot segment of this baseboard operating at an average water temperature of 135 °F and 3.0 gpm in a room with 60 °F air near floor level. Use Equation 8.1.

2. The following rooms are to be heated by a series string of finned-tube baseboards:

	Room design	Heating load
1.	Master bedroom	4,500 Btu/h
2.	Bedroom	3,000 Btu/h
3.	Den	8,000 Btu/h
4.	Living room	11,000 Btu/h
5.	Bathroom	1,500 Btu/h

   The series baseboards will be in the same order as the listing of the rooms. Use sizing method 1, along with the following data, to select a length for each baseboard.

   Baseboard thermal ratings given by lower curve in Figure 8-10:

   Water flow rate = 1.0 gpm

   Water temperature entering first baseboard = 180 °F

   Air temperature at floor level = 65 °F

3. Repeat exercise 2 using the more accurate sizing method 2. Compare the results with those obtained in exercise 2.

4. The output of a small fan-coil is rated at 4,500 Btu/h at an entering water temperature of 180 °F and flow rate of 1.0 gpm, in a room with 65 °F air. Estimate the heat output of the fan-coil at the following conditions:
   a. 160 °F entering water, 65 °F entering air, 1.0 gpm
   b. 140 °F entering water, 55 °F entering air, 1.0 gpm.

5. A fan-coil has the thermal performance ratings given in Figure 8-32. Interpolate between the data to estimate the heat output of the unit at 167 °F entering water temperature, 65 °F entering air temperature, and 3.0 gpm water flow rate.

6. Estimate the output of the fan-coil described in exercise 5 using principle 1 discussed in Section 8.7. How does this compare to the answer obtained in exercise 5?

7. Water enters a 1-inch bare copper tube at 9.0 gpm and 180 °F. The tube is 100 feet long. Determine the outlet temperature and rate of heat loss under the following conditions:
   a. 70 °F surrounding air temperature
   b. 35 °F surrounding air temperature.

8. Water at 200 °F enters a 3/4-inch bare copper tube at 3.0 gpm. The tube is 120 feet long and passes through a space where the air temperature is 70 °F. Estimate the heat loss from the tube using Equation 8.8.

9. Use the Series Baseboard Simulator module in the Hydronics Design Studio software to find the required baseboard lengths for the room loads described in exercise 2.

10. Repeat exercise 9 assuming the order of the room loads is reversed. Are there any changes in baseboard length? If so, explain why.

11. A distribution system has a c_{ds} value of 1,500 Btu/h/°F. Determine the supply water temperature needed to for this system to dissipate the full output of an 80,000-Btu/h boiler. Assume that the interior air temperature is to be maintained at 70 °F.

12. Assume that the boiler described in exercise 11 is a modulating boiler with a 5:1 turndown ratio. Determine the supply water temperature to the system when this boiler is operating at 50% and 25% of rated capacity. Explain why the supply water temperatures are different when the boiler operates at partial capacity.

Chapter 9

Control Strategies, Components, and Systems

Objectives

After studying this chapter, you should be able to:
- Describe the basic elements of a closed control loop.
- Explain the different control outputs and control algorithms used in hydronic heating systems.
- Describe the differences between controlling heat output using variable flow versus variable water temperature.
- Describe several basic control components such as switches and relays.
- Understand the operation of basic electromechanical temperature controls.
- Use ladder diagrams to lay out control systems.
- Properly connect both three- and four-wire zone valves.
- Describe the benefits of outdoor reset control and different ways to implement it.
- Explain the operation of several types of mixing assemblies.
- Describe the strengths and limitations of different methods of injection mixing.
- Calculate the injection flow rates needed for a given distribution system.
- Apply injection mixing in systems operating with wide ΔT heat sources.
- Explain the concept of communicating control systems.
- Describe the advantages of a Web-enabled networked control system.

9.1 Introduction

Controls are the brain of a hydronic heating system. They determine exactly when and for how long devices such as circulators, burners, compressors, and mixing valves will operate. The comfort, efficiency, and longevity of the system are as dependent on the controls as they are on any other component or subsystem.

A well-designed and properly adjusted control system can provide sophisticated operating logic to optimize comfort and energy efficiency. It will operate unpretentiously behind the scenes, require minimal attention, and help assure a long system life. Conversely, a carelessly designed control system can be a nightmare. The use of high-quality heat sources, circulators, or other components will never compensate for an improperly designed or a poorly adjusted control system.

Controls for hydronic heating represent technology that spans several decades. They range from time-proven devices such as the bimetal room thermostat to microprocessor-based controllers that have the intelligence to optimize system operation for minimum fuel use and maximum comfort. Both high-tech and low-tech control devices have their place in modern hydronic systems. Complex control systems are not necessarily "better" than simple control systems. Quite the opposite, the simpler the controls can be to accomplish a given objective, the better.

With experience, system designers can learn to integrate a wide range of control hardware for optimal performance, maximum reliability, and reasonable cost.

The operation of many hydronic heating and cooling systems is based upon the status of various electrical switches at any given time. Some of these switches are manually set, others are automatically operated by temperature, pressure, or other sensed conditions. This chapter discusses many of the basic switch-type controls used in hydronic heating and shows how they interact within an overall system. The fundamental operation of these devices also applies to hydronic cooling systems.

Electronic controllers, especially those using **microprocessors**, are now used in nearly all areas of heating and cooling technology. Hydronic system designers currently have a broad range of such controllers available to them. Many functions that previously required human intervention can now be performed automatically by such controllers. These include automatic adjustments of system water temperature based on outdoor temperature, automatic shutdown of system circulators during warm weather, even periodic exercising of components such as mixing valves and circulators to prevent seizing during nonoperational periods. The accuracy of these controllers also helps improve both comfort and energy efficiency. This chapter will acquaint you with several electronic controllers used in modern hydronic systems.

Many hydronic systems use one or more **mixing devices** to control water temperature in various parts of the system. Some of these devices such as three-way thermostatic valves and four-way mixing valves were discussed in Chapter 5, Pipings, Fittings, and Valves. This chapter will go into further detail and discuss the interaction between the piping components and the electrical/electronic controllers that operate them.

The proper documentation of control systems is essential if they are to be expediently serviced in the future. A universally accepted approach for such documentation is the **ladder diagram**. It will be introduced in fundamental terms and then expanded to demonstrate its use in more sophisticated systems.

This chapter also discusses how control technology for hydronic heating is likely to progress going forward. As good as present-generation controllers are compared to those available only 15 to 20 years ago, there are emerging technologies that will further improve control intelligence and connectivity. Many of the control devices shown and discussed in this chapter did not exist when the first edition of this book was written in 1993. In all likelihood, some of these state-of-the-art devices will be eclipsed by even newer and more advanced devices within a few years after this edition is released. This is not to imply that current-generation controls are not good at what they do. It is simply to point out that control technology is constantly evolving. Hydronic system designers should stay informed of ongoing developments in control technology.

Finally, it must be emphasized that all HVAC professionals who expect to work with modern hydronic systems must be able to specify, install, configure, and service a wide range of control hardware. Doing so is what separates true hydronic heating and cooling professionals from those who only work with the mechanical aspects of piping systems. Those who expect to design or install modern hydronic systems, but at the same time "fear" the controls used in those systems, are like those who aspire to be automotive technicians, but never intend to deal with the electrical aspects of modern vehicles. This is simply not possible in today's marketplace. To help alleviate such concerns, this chapter discusses the fundamentals of electrical circuits in the context of controls for hydronic heating and cooling systems.

9.2 Closed-Loop Control System Fundamentals

Hydronic heating systems, like all other HVAC systems, rely on fundamental control concepts to stabilize their operation and optimize their performance. It is important that hydronic heating professionals understand these basic concepts before examining the actual hardware used to implement them. This section discusses those concepts and sets the stage for more detailed discussions of specific control hardware in later sections of the chapter.

Figure 9-1 depicts the elements of a simple **closed-loop control system** for hydronic heating. Some of the boxes represent information, while others represent physical devices.

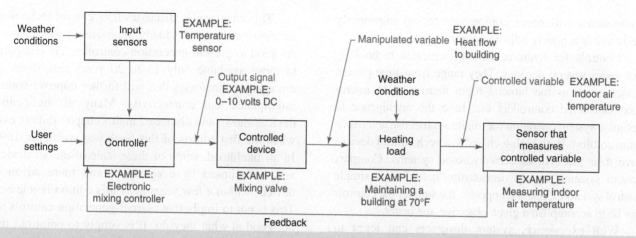

Figure 9-1 | Block diagram of a feedback control system for space-heating applications. Examples represent typical devices or measurements used for hydronic space heating.

The block in the upper left represents information that is sent to the control system by one or more sensors. In current-generation hydronic systems, this information is usually outdoor air temperature. However, a more sophisticated control system may expand this information to include measurements of wind speed, relative humidity, and solar radiation intensity. This information may be sent to the controller as an electrical signal such as a variable resistance, voltage, or current. It might also be sent as a **digitally encoded signal**.

Additional information comes into the controlled process as user **settings**. Examples include desired temperatures, pressures, flow rates, or switch status. The first three of these are examples of **analog inputs**. They all represent physical parameters that can be measured, and which vary over a continuous range of numerical values. For example, the outdoor temperature may be represented by a number anywhere within the range of −50 to 120 °F, or a flow rate may be a value anywhere between 0 and 100 gpm. The last input—switch status—is an example of a **digital input**, which can only have one of two states, represented as (on or off, 0 or 1, or open contact versus closed contact).

The block in the lower left represents a physical device called the **controller**. It accepts information from the input **sensors**. It also accepts and stores the user settings. The controller also receives information called **feedback**, sent from another sensor that measures the result of the controlled process. This result is represented by a **controlled variable**. For current-generation hydronic heating systems, the controlled variable is usually the temperature of water at some location in the system, or of conditioned (e.g., heated or cooled) space.

The controller uses this information to compare the measured value of the controlled variable with the **target value**. The latter is the desired "ideal" status of the controlled variable. Any deviation between the target value and the measured value is called **error**. In this context, the word "error" does not imply that a malfunction or failure has occurred. It simply means that there is a difference between the target value and measured value of the controlled variable.

The controller uses a stored set of instructions called a control-**processing algorithm** to generate an output signal based on the error (or lack thereof). This output may be a simple switch contact closure, a precise DC voltage, or a variable frequency AC waveform.

The output signal is passed to a **controlled device** that responds by changing the **manipulated variable** (which in a heating system is usually heat flow). The change in the manipulated variable causes a change in the process being controlled (in this case, the heat input to the building). This causes a corresponding change in the controlled variable (in this case, the inside air temperature). The change in the controlled variable is sensed by the sensing element, which provides feedback to the controller. This overall process is continuous as long as the system is operating.

Closed-loop control systems can be tuned for very stable operation. The feedback inherent to their design makes them strive to eliminate any error between the target and measured value of the controlled variable.

Control-Processing Algorithms

Most of the closed-loop control systems used for hydronic heating continually seek to eliminate any error that exists between the target value and measured value of the controlled variable. The exact procedure by which a controller uses the error information along with sensor inputs and user settings to generate an output signal is called the processing algorithm. The three common types of processing algorithms used in modern HVAC control systems are **proportional (P)**, **proportional-integral (PI)**, and **proportional-integral-derivative (PID)**.

Proportional (P) Processing

In pure proportional processing, the magnitude of the output signal depends only on the error between the target and measure values of the controlled variable. The greater the error, the stronger the output signal becomes in an attempt to eliminate this error.

This concept is illustrated in Figure 9-2. The water level in the tank is the controlled variable. *The objective of the control process is to maintain the water level at the indicated target level.* Water can flow out of the tank at different rates depending on the setting of the drain valve. Water is added to the tank by the controlled valve at the top, which is linked to a float that monitors water level. If water leaves the tank faster than the inlet valve admits it, the level drops, causing the float to open the control valve to admit water at a higher rate. The farther the float drops, the faster water enters the tank. Similarly, if the water level starts to rise, the float also rises to reduce the water inlet rate. Whenever the water inlet rate equals the water outlet rate, the water level holds constant.

Proportional controllers have what is known as a **proportional band**. This is the range of the controlled variable over which the controlled device ranges from zero to full output. In the control situation depicted in Figure 9-2, the proportional range is between the upper level, where the inlet valve is completely closed, and the lower level, where the inlet valve is fully open. If conditions are such that the water level drifts above or below these levels, the inlet valve cannot make any further compensation in its attempt to correct this error. Under such conditions, the controlled device is said to be operating outside of its proportional band. In the case of proportional temperature controllers used in heating systems, the proportional band is typically 2 to 4 °F.

Any controller that uses purely proportional processing can only maintain the exact target value of the controlled variable under one calibration condition. For example, the water level in the tank will remain at the desired target level only when the outlet flow equals the inlet flow at which the float/valve assembly is calibrated. If the outlet flow becomes slightly higher than this calibration rate, the float eventually drops to enable a higher inlet flow rate. If the higher flow rate through the tank is within the proportional range of the float/inlet valve, a balance will be reestablished, *but at a level somewhat lower than the target level.* Likewise, if the outlet flow becomes slightly less than the calibrated inlet flow, the float rises to reduce the inlet flow, and a new level will be established slightly above the target level. Both situations result in the water level being slightly different from the target value. This effect is called **offset** and is inherent to all purely proportional controllers. The narrower the proportional band of the controller, the smaller the offset will be. However, proportional bands that are too narrow can lead to erratic control action.

Most mechanical thermostats are pure proportional controllers. They are calibrated by the manufacturer to maintain a target room temperature at some assumed percentage of heat source on time. When the heating load increases, the room temperature must droop slightly for the thermostat to keep the heat source running a higher percentage of the time.

Pure proportional control can be represented mathematically by Equation 9.1.

Equation 9.1:

$$O = c_p(e) + M$$

where
- O = output signal value;
- e = error between the target and measured value of the controlled variable;
- c_p = proportional gain constant;
- M = a number representing the output value when the error is zero.

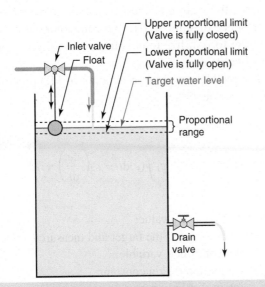

Figure 9-2 | Water level in tank is controlled by float-operated valve. Rate of water input is proportional to the rate of water drainage.

The proportional gain constant is a number that determines how aggressively the output signal increases or decreases for a unit change in error. Controllers that operate with high proportional gain produce strong changes in the output signal with relatively small changes in the error, and vice versa. Although this may seem desirable to use high proportional gain from the standpoint of control accuracy, doing so can create situations where the controller cannot "settle" to stable output condition. Instead, the controlled variable continually swings above and below the target value. This undesirable affect is called **hunting**. To avoid such situations, most controllers used in residential and light commercial hydronic systems operate with proportional gains determined by their manufacturer. These factory set values for the proportional gain are typically not adjustable by the installer.

The number M in Equation 9.1 represents the value of the control output when the error is zero. In some control systems, this number must have a nonzero value to avoid stability problems when the error is very small. For example, in some circumstances, the electrical cable between the controller and controlled device could act as an antenna. The electrical or magnetic fields generated by nearby transformers or motors could induce a slight voltage in this cable. This could cause a controlled device that interprets zero volts to mean zero error to react in an undesirable manner (e.g., to assume the controller is sending an output signal that requires a change in the controlled variable). To avoid such situations, it is common to set a minimum output signal below which the controlled device will not respond. Two common examples of such output signals are **2 to 10 VDC**, and **4 to 20 mA**.

Although pure proportional processing does have limitations, it is the dominant mathematical component underlying nearly all temperature controllers used in HVAC systems.

Proportional-Integral (PI) Processing

Fortunately, the shortcomings of purely proportional processing can be overcome by adding further mathematical sophistication to the processing algorithm. Proportional-integral processing determines the output signal based on both the amount of error, as well as *how long the error has existed*. The latter component of the processing algorithm allows the controller to eliminate the steady-state offset inherent with proportional-only control. Because of the mathematics involved, PI processing is only practical in electronic controls. It is now commonly used in many types of electronic temperature controllers including room thermostats.

The mathematical representation of proportional-integral processing is given by Equation 9.2.

Equation 9.2:

$$O = c_p(e) + c_i \int (e) dt + M$$

where
- O = output signal value;
- e = error between the target and measured value of the controlled variable;
- c_p = proportional gain constant;
- c_i = integral gain constant;
- M = a number representing the output value when the error is zero.
- t = time

Like the proportional gain constant, the integral gain constant is a number that determines how quickly the output signal changes over time as the controller seeks to eliminate error. This value is also typically factory set and not adjustable by the installer.

Proportional-Integral-Derivative (PID) Processing

Proportional-integral-derivative processing (abbreviated as PID) adds the ability to monitor the *rate of change of error* to the response provided by a PI controller. This allows the controller to predict what the error is likely to be a short time into the future and adjust the present output accordingly.

PID control is most useful in control systems that must respond to rapid changes in the inputs. Most HVAC systems do not require high-speed response. For this reason, the derivative contribution of the processing algorithm, as determined by the value of the derivative gain constant, is usually minimal. Still, the presence of some derivative component allows the controller to stabilize to the target value slightly faster than PI control.

The mathematical representation of PID processing is given as Equation 9.3.

Equation 9.3:

$$O = c_p(e) + c_i \int (e) dt + c_d \left(\frac{de}{dt} \right) + M$$

where
- O = output signal value;
- e = error between the target and measured value of the controlled variable;
- c_p = proportional gain constant;
- c_i = integral gain constant;
- c_d = derivative gain constant;
- M = a number representing the output value when the error is zero.
- t = time

9.2 Closed-Loop Control System Fundamentals

Like the gain constants used for both proportional and integral processing, the value of the derivative gain constant is usually factor-set and not adjustable by the installer.

The output signal from a PID controller at any instant is the mathematical sum of the contributions from the proportional, integral, and derivative components. The value of any of these contributions can be positive or negative at any given instant depending on the error, duration of the error, and rate of change of the error.

Figure 9-3 compares the response of P, PI, and PID processing to the same initial error between measured and target temperatures. The goal of all three processing algorithms is to eliminate the error between the measured and target temperatures. Notice how adding the integral function to the processing algorithm eliminates the offset. It is also evident that the system using full PID control stabilizes to the target value slightly faster than does the system using PI control.

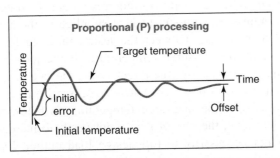
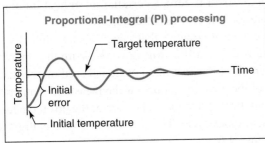
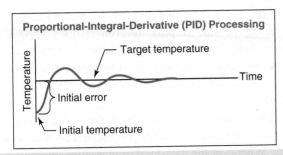

Figure 9-3 | Response of a system with an initial temperature error to Proportional (P), Proportional-Integral (PI), and Proportional-Integral-Derivative (PID) processing algorithms.

Controller Output Types

A controller can interact in several ways with a controlled device. In some cases, the output action is a simple opening or closing of an electrical contact. In other cases, the output action may be a constantly changing analog voltage or current signal from the controller to the controlled device. This section reviews the controller output types commonly used in hydronic heating control systems.

On/Off Output

In **on/off control**, the controlled device can only have one of two possible control states at any time. If the controlled device is a valve, it can be either fully open or fully closed, but never at some partially open condition. If the controlled device is a heat source, it must be fully on or completely off. On/off output signals are usually generated by the closing and opening of electrical contacts within the controller.

When used to control heating equipment, on/off controllers send heat to the load in pulses rather than as a continuous process. In such applications, all on/off controllers must operate with a **differential**. This is the change in the controlled variable between where the controller output is turned on to where it is turned off. In heating systems, the controlled variable is usually the temperature at some location within the system or the building it serves. On some controllers, the differential is below the setpoint temperature, as shown in Figure 9-4a. In other controllers, the differential is centered on the setpoint, as shown in Figure 9-4b.

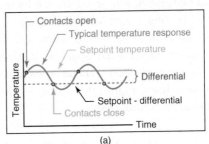
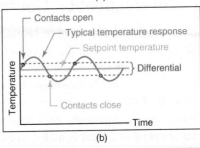

Figure 9-4 | On/off temperature controllers operate with a differential that is either (a) entirely below the setpoint value, or (b) centered on the setpoint value.

The smaller the differential, the smaller the deviation in temperature from the target value. However, if the differential is too small, the controller operates the controlled device in short but frequent cycles. This undesirable effect is called **short cycling**. In most cases, short cycling increases the wear on the equipment being controlled. This is especially true of devices such as oil burners, gas valve/ignition components, switching devices, and compressors used in heat pumps.

If the differential is too wide, there can be large deviations between the target value and measured value of the controlled variable (usually temperature). In some situations, this is acceptable, whereas in others, it is not. For example, the large thermal mass of a heated slab-on-grade floor can usually accept variations in water temperature of several degrees Fahrenheit without creating corresponding swings in room air temperature. However, a room thermostat with a differential of 4 °F or higher, for example, would cause wide swings in room air temperature that would almost certainly lead to complaints.

The term **control differential** refers to the variation in temperature at which the controller changes the on/off status of the output. On some controllers, this is a factory set value and cannot be changed by the user. On other controllers, the control differential can be adjusted over a wide range.

The term **operating differential** refers to the *actual variation in the controlled variable as the system operates*. It depends on the combined effect of the controller and the system it is controlling. In heating systems, the operating differential is usually greater than the control differential because of a time lag caused by the thermal mass of the heating system and the building. The term **overshoot** describes a situation where the controlled variable exceeds its intended upper limit. **Undershoot** occurs when the controlled variable drops lower than its intended lower limit. Figure 9-5 illustrates these terms for a temperature control system.

Pulse Width Modulation Control

When simple on/off control is used in heating systems, energy input begins only when the measured temperature drops a certain amount below the desired target value. After the heat source is turned on, the often-cool thermal mass of the heat emitters causes a further drop in temperature until the latter is warm enough to begin raising the room temperature. Heat is added to the room as a result of the room air temperature repeatedly dropping below the desired target value. Over time, the average room temperature becomes slightly less than the desired value. This effect is called **droop**, and is inherent to nonelectronic on/off temperature controls.

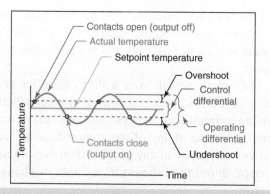

Figure 9-5 The control differential is the desired difference in the temperature between when controller output turns on and off. Operating differential is the actual temperature range experienced as the system is controlled. The operating differential is wider than the control differential due to overshoot and undershoot.

Ideally, the heating system would supply sufficient heat to keep the indoor temperature precisely at the target value. Doing so requires a control strategy that adjusts the amount of heat supplied to the room based on the error between the desired target temperature and the measured temperature. If the measured temperature is above the target value, the rate of heat input needs to decrease, but not necessarily to zero. Likewise, if the measured temperature is below the target value, the rate of heat input needs to increase, but not necessarily to full design load output. Even when the error is zero, a certain rate of heat input must be maintained to prevent droop.

A control strategy that accomplishes this is called **pulse width modulation (PWM)**. With this approach, the length of the heat input cycle is based on the magnitude of the error (e.g., how far the measured temperature is from the target value). The greater the error, the longer the heat input cycle. The concept of PWM is illustrated in Figure 9-6.

The controller generates a mathematical representation of the sloping reference lines shown in Figure 9-6. Each triangle formed by the combination of an up- and down-sloping line represents a preset **cycle time**, with typical values ranging from 10 to 15 minutes.

The controller detects when the sensed value of temperature equals the value represented by the sloping lines. Whenever the measured temperature curve crosses an up-sloping line, the control output (usually an electrical contact) closes to turn on a device such as a heat source, circulator, or zone valve. When the measured temperature crosses a down-sloping line, the output is turned off.

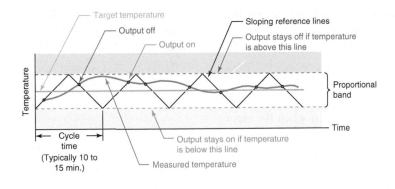

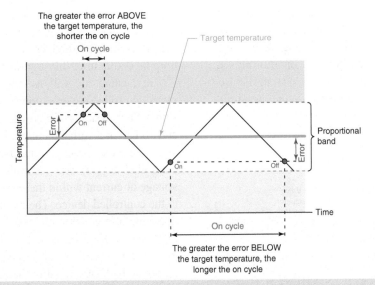

Figure 9-6 | An on/off output controlled using PWM. Note that the duration of the on versus off time is dependent on the error between the actual temperature and the target temperature.

If the measured value drifts into the red-shaded area above the proportional band, the output remains off continuously. Similarly, if the output drifts into the green-shaded area below the proportional band, the output remains on continuously. Even when the error is zero, the controller keeps the output on for 50% of the cycle length. *This is the essential difference between PWM and on/off control that allows the latter to better compensate for continuous building heat loss.*

When used to control heat output, PWM adds energy to the process in proportion to the error present. Most controllers that produce a PWM output also use integral (I) and in some cases derivative (D) processing to quickly stabilize the system and eliminate droop. When the error is small, the controller gently "nudges" the controlled device in an attempt to eliminate the error without overshooting the desired temperature. When the error is large, the controller operates the controlled device more aggressively to reduce the error quickly and eliminate any residual offset. PWM control is particularly effective when controlling air temperature in rooms heated by high mass radiant panel systems.

Floating Control

Another means of providing regulated heat input to a hydronic distribution system when the error between the target and measured value of the controlled temperature is at or close to zero is called **floating control**. This control output was developed to operate motorized valves and dampers that need to be powered open as well as powered closed. *In hydronic heating systems, floating control is commonly used to drive three-way and four-way motorized mixing valves.* Figure 9-7 depicts a floating control system in a hydronic mixing application.

One electrical contact within the controller closes to drive the valve's actuating motor in a clockwise direction. The other electrical contact closes to drive the actuator motor counterclockwise. Only one contact can be closed at any time. The actuating motor driving the valve shaft turns very slowly. Some actuating motors can take up to 3 minutes to rotate the valve shaft over its full rotational range of 90°. This slow operation is desirable since it allows the sensor ample time to provide feedback to the controller for stable operation.

Notice the floating zone in Figure 9-7. When the measured temperature is within this zone, the valve's actuating motor does not run. The error between the target temperature and actual temperature is small enough to not warrant any correcting action. The ability to hold the valve at a partially open condition is what allows some heat input to the load even when the error is at or close to zero.

As the measured temperature drifts above or below the floating zone, the controller responds with a corrective action that is proportional to the error. The greater the error, the longer the output signal that opens or closes the valve remains on. If the error drifts above or below the PWM control zone shown in Figure 9-7, the actuating motor remains on in either the open or closed mode. This usually only occurs while the system is starting up, and a large error is present between the measured and target temperatures.

Floating control is sometimes called **tristate control** because the output signal can be in any of three operating states at any time (e.g., opening, closing, or off). It is also sometimes called **three-wire control** because three wires are required between the controller and the actuating motor of the controlled device.

Modulating DC Outputs

On/off control, PWM control (as depicted in Figure 9-6), and floating control all use the opening and closing of electrical contacts to send signals between the controller and controlled device. This type of output is not suitable for all controlled devices. For example, none of these outputs could smoothly regulate the speed of a motor.

In heating as well as other industrial applications, many controlled devices require a continuous analog input signal. Such systems often use a variable voltage between 2 and 10 DC or a variable current between 4 and 20 mA DC as a continuous analog signal between the controller and controlled device. The controller generates a continuous voltage or current within these ranges and sends that signal to the controlled device. The controlled device responds by operating at a position or speed that is proportional to the input signal. For example, a control signal of 2 VDC fed to a motor speed controller means that the motor should be off. A 10-VDC signal to the same controller means the motor should be operating at full speed. A signal of 6 VDC, which is 50% of the overall range of 2 to 10 V, means that the motor should be running at 50% of full speed.

The reason these control signals do not begin at zero voltage or current is to prevent electrical interference or "noise" from affecting the controlled device. Wires that run in proximity to other electrical equipment can experience induced voltages and currents due to electrical or magnetic fields. In most situations, raising the starting threshold to 2 VDC, or 4 mA, prevents such interference. However, in some very noisy electrical environments, twisted pair wiring or shielded cable may be necessary between the controller and the controlled device. Consult the control manufacturer's recommendations regarding control wiring in such situations.

Both modulating valves and variable-speed circulators are available that can accept 2- to 10-VDC and 4- to 20-mA inputs. These inputs are for control only and do not supply the electrical power to drive the device. For example, the typical wiring of a 2- to 10-VDC modulating valve is shown in Figure 9-8. Notice that 24 VAC power must also be supplied to the valve to power the motor and the actuator's circuitry. Likewise, a variable-speed pump that is controlled by a 2- to 10-VDC or 4- to 20-mA signal requires line voltage (120 VAC) to supply operating power.

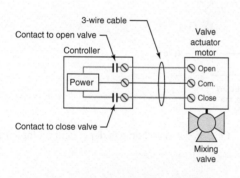

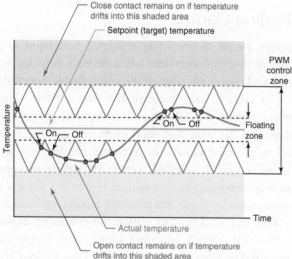

Figure 9-7. Operation of a three-wire floating control system. Possible control states include opening valve, closing valve, or holding valve in present position. The duration of the on versus off time of either contact is dependent on the error between the actual temperature and the target (setpoint) temperature.

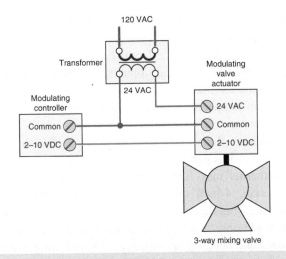

Figure 9-8 | Typical wiring for a valve actuator controlled by a 2- to 10-VDC modulating control signal.

9.3 Controlling the Output of Heat Sources

There are several methods for controlling the heat output of hydronic heat sources. The method used depends on the type of heat source, as well as the total heat output rate.

This section discusses the following common methods for controlling heat production:

- On/off control of a **single-stage heat source**
- **Multistage heat production**
- **Modulating heat production**

On/Off Control of a Single-Stage Heat Source

Many residential and smaller commercial hydronic systems use a single heat source that has only two operating modes: on or off. Whenever the heat source is on, it produces a fixed rate of heat output (see Chapter 3, Hydronic Heat Sources, for a discussion of the various heat output ratings). When the heat source is off, it produces no heat output. Such heat sources require a simple on/off control signal from their controller. This is provided by the closing and opening of an electrical contact.

The total amount of heat added to the system over time is regulated by periodically turning the heat source on and off. To illustrate this, consider the hypothetical building heating load profile shown in Figure 9-9.

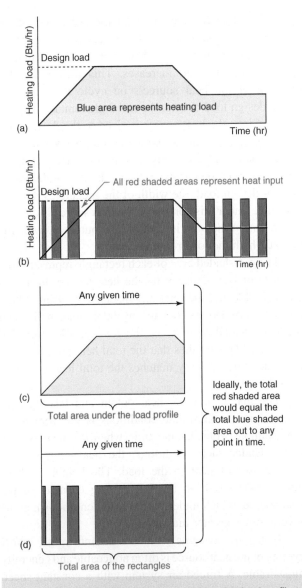

Figure 9-9 | (a) A hypothetical building heating load profile. (b) Heat pulses produced by an on/off heat source. (c) Blue-shaded area represents total building load over a given period. (d) Ideally, the total area of the red-shaded rectangles over the same period equals the blue-shaded area under the load profile.

The load begins at zero, rises at a constant rate to its maximum design value, and then decreases at a steady rate back to half its design value. The time over which this occurs can be assumed to be a few hours.

The red-shaded rectangles represent heat outputs from an on/off heat source that provides a fixed rate of heat output while operating. *The rated heat output of the heat source is assumed equal to the maximum (design) heating load of the building.* The heat source is operated by an on/off controller. In this case, that controller monitors water temperature leaving the heat source.

Notice that the height of all the red rectangles is the same. This indicates that the heat source operates at a fixed heat output rate. The width of the rectangles increases as the load increases. This means that the duration of the heat source's on cycle is increasing. When design load occurs, the rectangle remains uninterrupted. Since the heat source is sized to the design load of the building, it must operate continuously whenever design load conditions exist. As the load decreases, the rectangles become narrower. When the load stabilizes at 50% of design load, the width of the rectangles remains constant. Under this condition, the heat source is on 50% of the time. When on, its rate of heat output is twice the rate at which the load requires heat.

The red-shaded area of each rectangle represents the *quantity* of heat delivery by the heat source during its on cycle. Ideally, the total red-shaded area, from where the time axis begins, out to any later time, will equal the blue-shaded area under the load profile over that same time. This implies that the total heat production of the heat source exactly matches the total load over that period.

The match between heat output and heating load at low- and medium-load conditions is far from ideal. For example, consider the time when the load is very small. Under such conditions, the heat source sends short pulses of heat to the load. The height of these pulses corresponds to the full heating capacity of the heat source, which under partial load conditions may be several times greater than the load.

In heating systems with low thermal mass, on/off cycling of the heat source is often noticeable and generally undesirable. A forced-air system with a constant-speed blower and fixed firing rate is a good example. Under partial load conditions, the furnace produces its full rate of heat output whenever it operates. Since the air in the building has very little thermal mass for absorbing this heat, room temperature increases quickly. When the room thermostat reaches its setpoint temperature, its contacts open and the furnace is turned off. An operating cycle could be as little as 2 or 3 minutes under low-load conditions. During this time, the room temperature could increase by several degrees Fahrenheit. The occupants feel this rapid change in air temperature and usually do not like it. When the furnace is shut off, drafts from windows and doors quickly reestablish themselves and comfort rapidly declines. Eventually, the process repeats itself. The system injects repeated pulses of hot air into the space as its (rather crude) way of attempting to match heat input to heat loss.

Many hydronic systems, especially those having high thermal mass, have an advantage in this respect. The greater thermal mass of the heating distribution system allows it to temporarily absorb some of the surplus heat output generated by the heat source and spread out its delivery over time. This effect is further enhanced when the distribution circulator operates continuously. The greater the thermal mass of the system, the more stable the room temperatures will remain during partial load conditions.

Radiant floor heating using a 4- to 6-inch-thick concrete slab is perhaps the ultimate hydronic system as far as thermal mass is concerned. For example, a 2,000-ft^2 floor slab, 4-inch thick, contains over 93,000 lb of concrete. This thermal mass could absorb about 20,000 Btu while only increasing its temperature 1 °F. The air in the same building (assuming an 8-foot ceiling height) could absorb only about 290 Btu with a temperature rise of 1 °F. The thermal mass of the floor slab is almost 70 times greater than that of the interior air. This allows the slab to accept pulses of heat during partial load conditions while delivering a relatively steady heat output to the building. The thermal mass of the system dampens the temperature response of the system. A comparison of temperature swings resulting from pulsed heat input to systems with high and low thermal mass is shown in Figure 9-10.

Multistage Heat Production

Another way of attempting to match the output of the heat source to the building load is with **multistage heat production**. This approach divides the total heating output into a number of increments, or **stages**. As the load increases, these stages of heat output are turned on sequentially. When all stages are on at the same time, the heat transfer rate to the load is at its maximum value. As the load decreases, stages of heat output are sequentially turned off. The number of stages used in a system can vary from two to eight or more. Larger commercial/industrial systems tend to use more stages than residential systems. The greater the number of stages, the better the match between heat output and load.

In some systems, each stage of heat production is a separate heat source with a fixed rate of heat output. Each of these heat sources is typically turned on or off by an electrical contact within the controller. Some boilers also have two independently controlled burner assemblies that supply heat to the same heat exchanger. They are known as **HI/LO fire boilers**, and are, in effect, a two-stage heat source contained in a single cabinet. Piping configurations for multistage boiler systems were discussed in Chapter 3, Hydronic Heat Sources.

Figure 9-11 compares the heat output of a single-stage heat source to that of a two-stage and four-stage heat source. In each case, the heat source supplies heat to

9.3 Controlling the Output of Heat Sources 379

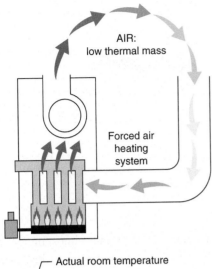

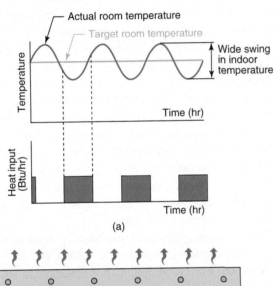

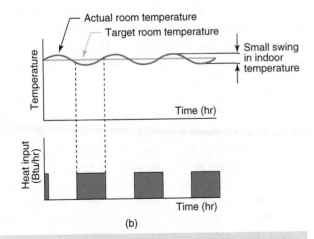

Figure 9-10 Room temperature fluctuations associated with on/off heat sources in heating systems with (a) having low thermal mass and (b) high thermal mass. Note the ability of the high thermal mass system to reduce swings in indoor temperature.

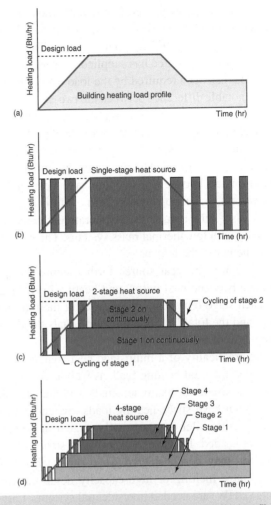

Figure 9-11 (a) Hypothetical building heating load profile and (b) heat output from a single-stage heat source. (c) Heat output from a two-stage heat source. (d) Heat output from a four-stage heat source. In each case, the design heating load is divided equally among the stages.

the same hypothetical load profile shown in Figure 9-9. The maximum heat output of each multiple heat source system is assumed to equal the maximum heating load.

During low-load conditions, the first stage of the two-stage heat source operates intermittently. The width of the rectangles becomes wider as the load increases toward 50% of design load. As the load increases above 50% of design load, the first-stage output remains on continuously, while the second stage operates intermittently to make up for the difference between the load and the output of the first stage. Maximum heat output occurs when both stages remain on continuously. This example assumes that the total heat output of both stages equals the design heating load. Any oversizing would necessitate some off-time for the second stage. As the load decreases, the second stage resumes intermittent operation. Since the load profile finally stabilizes at half of design load, the first stage remains on.

As with single-stage on/off heat input, the total area of the shaded rectangles should equal the area under the load profile up to any arbitrary point in time. This would imply that the heat source has supplied the same amount of heat that has been required by the load over that time.

The notable difference between the two-stage system and the single-stage system is the degree to which the red-shaded rectangles extend above the load profile, and the amount of "white space" between the rectangles and beneath the load profile. The two-stage system provides a closer match between heat output and heating load compared to a single-stage system. Multistage heat sources also help reduce the temperature fluctuations associated with low thermal mass systems. The more the stages, the better the match.

The four-stage heat source further improves upon the match between heat output and heating load. In this case, each stage represents one-quarter of the total heat output, and the total heat output again equals the design heating load of the building.

The practicality of a multistage heat source system depends on the total heating load, as well as the type of heat source used. For example, small- and medium-size houses often have space-heating loads low enough that a single on/off boiler or other single-stage heat source is all that is needed. However, two or more stages of heat production may be considered when other large loads such as snow-melting or high-capacity domestic water heating (DHW) are needed in residential systems. Larger homes and commercial buildings may have space-heating loads large enough to justify the use of multistage heat sources such as the multiple boiler systems discussed in Chapter 3, Hydronic Heat Sources. Systems using multiple air-to-water heat pumps, water-to-water heat pumps, or multiple electric boilers are also possible.

Modulating (Variable Output Rate) Heat Production

In an ideal space-heating system, the heat output from the heat source would always precisely match the rate of heat loss from the building. The heat output from the heat source could take on any value from zero to the full design load of the building. As the load changed due to changes in outdoor conditions or internal heat gain, the heat source would instantly readjust its output to match the load. Figure 9-12 depicts this *ideal* match between heat production and heating load. The area under the load profile is fully shaded. Unlike with single- or multistage heat production, there are neither "white spaces" under the load profile nor any instances when heat production exceeds heating load.

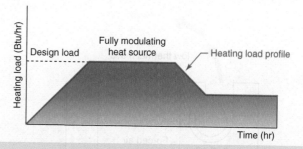

Figure 9-12 | "Ideal" match between heat output of a fully modulating heat source and hypothetical load profile.

Although this ideal condition is almost impossible to achieve, especially with combustion-type heat sources, several types of heat sources can approximate it.

Currently, there are many types of gas-fired **modulating/condensing boilers** available in North America suitable for use in residential and light commercial hydronic systems. Many of these can reduce their heat output rate (e.g., modulate) down to 10 to 20% of their full rated output. Water-to-water and air-to-water heat pumps with variable-speed compressors and thus variable heat outputs are also appearing in greater numbers on the North American market as this edition is being written. The use of modulating heat sources will undoubtedly increase in the future.

Most current-generation modulating boilers use proprietary internal controllers to adjust the rate at which air and gas are supplied to the burner. In some cases, the rate of heat output is varied based on the difference in temperature the boiler controller detects between the supply and return side of the distribution system. In other cases, it is based on the ability of the boiler to maintain a prescribed temperature on the supply side of the distribution system. If the sensed supply temperature starts to drop, the firing rate is increased and vice versa.

Because most current-generation modulating boilers cannot maintain stable heat outputs lower than 10 to 20% of rated capacity, they must cycle from their lowest firing rate to fully off under very low-load conditions. Upon restart, some modulating boilers go to a higher firing rate for a short time then quickly reduce their firing rate based on algorithms used by their internal controllers, as well as feedback from system temperature sensors.

Figure 9-13 shows how a modulating/condensing boiler with a minimum steady firing rate of 20% of its rated heating capacity, and with a maximum heating capacity equal to the design heating load, would track a hypothetical load profile. Notice that the match is

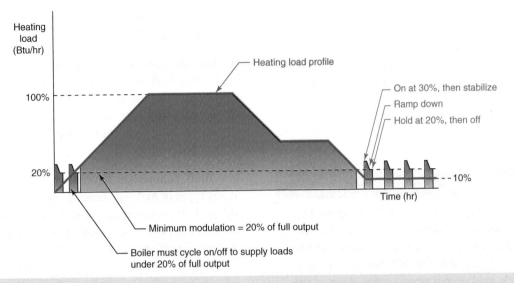

Figure 9-13 | Heat output of a modern modulating/condensing boiler with 5:1 turndown ratio as it attempts to track a hypothetical load profile. The boiler cycles whenever the load is less than 20% of its full heat output rate.

theoretically ideal provided the load is above 20% of design load. At loads under 20% of design load, the boiler must cycle on and off. At very low-load conditions, this cycling can become excessive. This short-cycling condition, which creates premature wear on heat source components, can be reduced or eliminated by:

1. Not oversizing the boiler relative to design load;
2. Using a boiler with high thermal mass and good insulation;
3. Using a **buffer tank** in combination with a low mass boiler and a highly zoned distribution system.

The sizing and placement of buffer tanks is discussed in detail in Chapter 14, Auxiliary Loads and Specialized Topics.

Combining Staging and Modulation

It has been shown that using a staged group of on/off heat sources improves the match between heat output and the load profile. This is also true of modulating heat sources. The next logical step to further improve the match between heat output and load is to combine staging and modulating. This topic, which was discussed in Chapter 3, Hydronic Heat Sources, represents the state-of-the-art in controlling heat output to any hydronic distribution system.

The ability of a **modulating heat source** to vary its heat output is described by its turndown ratio, which is defined by Equation 9.4.

Equation 9.4:

$$\text{TDR} = \frac{Q_{max}}{Q_{min}}$$

where
TDR = turndown ratio of the heat source;
Q_{max} = maximum heat output of heat source (Btu/hr);
Q_{min} = minimum stable heat output of heat source (Btu/hr).

For example, a boiler capable of delivering heat output in the range of 80,000 Btu/hr down to 16,000 Btu/hr would have a turndown ratio of 80,000/16,000 = 5:1.

When two or more identical modulating heat sources are combined into a single **heat plant**, the **system turndown ratio** of that heat plant is given by Equation 9.5.

Equation 9.5:

$$\text{TDR}_s = N(\text{TDR}_i)$$

where
TDR_s = turndown ratio of the overall heat source system;
N = number of individual heat sources;
TDR_i = turndown ratio of an individual heat source.

Example 9.1

Determine the system turndown ratio for a multiple modulating/condensing boiler system in which each of four identical boilers has a maximum heat output of 150,000 Btu/hr and a minimum stable heat output rate of 30,000 Btu/hr.

Solution:

Each boiler has a turndown ratio of 150,000/30,000 = 5, and there are four such boilers. Thus, the system turndown ratio is easily calculated using Equation 9.5.

$$TDR_s = N(TDR_i) = 4(5) = 20$$

Discussion:

An equally valid way of comprehending system turndown ratio is as follows: The maximum heat output occurs when all four boilers operate at full capacity and is equal to 600,000 Btu/hr. The minimum stable heat output occurs when three of the four boilers are off, and the remaining boiler operates at its minimum stable firing rate of 30,000 Btu/hr. The ratio of these two conditions is the system turndown ratio:

$$TDR_s = \frac{Q_{\text{max system}}}{Q_{\text{min system}}} = \frac{600,000 \text{ Btu/hr}}{30,000 \text{ Btu/hr}} = 20$$

A system turndown ratio of 20 implies that the minimum stable heat output of the system is 1/20th of its maximum heat output. Stated another way, the multiple boiler system can reduce its heat output to 5% of maximum and still maintain stable operation. This allows the boiler system to match the heating load over a very wide range of operating conditions. When the load drops to less than this minimum stable output rate, the one boiler that is operating must cycle on and off to some degree.

Staging and modulating are essential control techniques in modern hydronic systems. Their use will be demonstrated in several systems shown in later chapters.

9.4 Controlling Heat Output from Heat Emitters

Comfort is directly affected by the heat output of the system's heat emitters. The primary goal of the control system is to maintain comfort by properly regulating the heat output of the system's heat emitters. This section describes the fundamental approaches and lays the groundwork for several types of control techniques and hardware that are discussed in later sections.

There are two fundamental methods of controlling the heat output of hydronic heat emitters:

1. *Vary the water temperature supplied to the heat emitter while maintaining a constant flow rate through the heat emitter.*
2. *Vary the flow rate through the heat emitter while maintaining a constant supply water temperature to the heat emitter.*

Both approaches have been successfully used in many types of hydronic heating applications over several decades. It is important for system designers to understand the differences, as well as the strengths and weaknesses of each approach.

Variable Water Temperature Control

Chapter 8, Heat Emitters, discussed how the heat output of a heat emitter increases in approximate proportion to the difference between supply water temperature and room air temperature. This can be represented mathematically as follows:

Equation 9.5a:

$$Q_o = k(T_s - T_r)$$

where
- Q_o = rate of heat output from heat emitter (Btu/hr);
- k = a constant dependent on the heat emitters used;
- T_s = fluid temperature supplied to the heat emitter (°F);
- T_r = air temperature of room where heat emitter is located (°F).

This equation can be rearranged as follows:

Equation 9.5b:

$$Q_o = k(T_s) - [k(T_r)]$$

For a given heat emitter and room air temperature, the term in the square brackets is a constant. A representative graph of this equation where $k = 100$ and $T_r = 70°F$ is shown in Figure 9-14.

Heat output from the heat emitter occurs whenever the supply water temperature exceeds the room air temperature. For the heat emitter represented in Figure 9-14, an increase of 1 °F in supply water temperature produces an increase of 100 Btu/hr in heat output, regardless of the starting temperature. Likewise, a decrease of 1 °F in supply water temperature reduces heat output by 100 Btu/hr. The slope of the heat output versus supply water temperature line changes depending on the type of heat emitter used, but the **linear relationship** will remain.

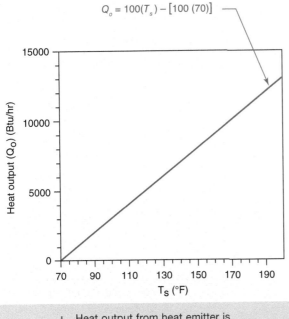

Figure 9-14 | Heat output from heat emitter is approximately proportional to difference between supply water temperature and room air temperature. A room temperature of 70 °F is assumed in this graph.

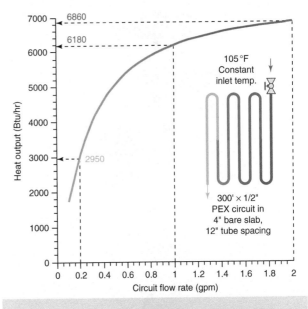

Figure 9-15 | Heat output of a floor heating circuit supplied with water at 105 °F as a function of flow rate. The circuit is a 300 feet length of 1/2-inch PEX tubing embedded in a 4-inch bare concrete slab at 12-inch spacing. The room temperature is 70 °F.

This type of linear relationship between the controlled variable (heat output) and the manipulated variable (supply water temperature) is preferable from a control standpoint. This relationship is relatively easy to mimic when designing both electromechanical and electronic controls. This approach also eliminates some complications that arise when the system flow has to vary to regulate heat output.

Variable Flow Rate Control

The graph in Figure 9-15 shows how the output of a typical floor heating circuit varies as a function of the water flow rate through it, assuming the supply water temperature is maintained at 105 °F.

This circuit's maximum heat output of 6,860 Btu/hr occurs at the maximum flow rate shown (2.0 gpm). Decreasing the circuit's flow rate by 50% to 1.0 gpm decreases its heat output to 6,180 Btu/hr, a drop of only about 10%. Reducing the flow rate to 10% of the maximum value (0.2 gpm) still allows the circuit to release 2,950 Btu/hr, about 43% of its maximum output. These numbers demonstrate that heat transfer decreases slowly at higher flow rates, but much faster at lower flow rates. This characteristic is typical of all hydronic heat emitters as discussed in Chapter 8, Heat Emitters. The relationship between the controlled variable (heat output) and the manipulated variable (flow rate) is very **nonlinear**. This characteristic makes accurate control more difficult, especially under low-load conditions. To provide accurate control under low-load conditions, the controlled device must be able to maintain stable flow at very low flow rates. Such a device is said to have good **rangeability**.

Valves with a special **equal percentage characteristic** have been developed that allow flow to develop very slowly as the valve is first opened and progress faster as the valve continues to open farther. In valves where the stem moves up and down, the equal percentage characteristic is created by the shape of the valve plug. In some valves, the plug has a specially tapered "logarithmic" shape that allows the gap between the plug and seat to form very slowly as the valve's plug begins to lift away from the seat, then faster as the stem lifts higher. In other valves, the equal percentage characteristic is created by machining a tapered slot into a cylindrical valve plug. This technique is also used with rotary valves. In the case of a ball valve, an insert with a specially tapered slot is fixed into the orifice of the ball. An example of such a **characterized ball valve** with a special equal percentage insert is shown in Figure 9-16.

The flow rate through an equal percentage valve as a function of its stem position is shown in Figure 9-17. This curve assumes that a constant differential pressure is maintained across the valve as it opens.

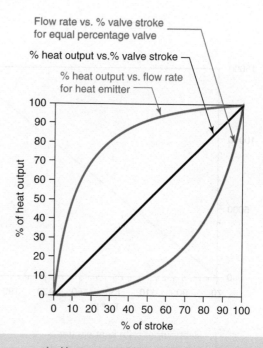

Figure 9-16 (a) A ball valve fitted with flow insert to create an equal percentage flow characteristic. (b) Close-up of equal percentage flow insert with tapered slot orifice. *Courtesy of Belimo (USA), Inc.*

Figure 9-18 Heat output of heat emitter is approximately proportional to percentage of valve stroke when flow through heat emitter is controlled by an equal percentage valve operated at a constant differential pressure. This is a desirable operating characteristic. *Note*: Supply water temperature to heat emitter is assumed constant.

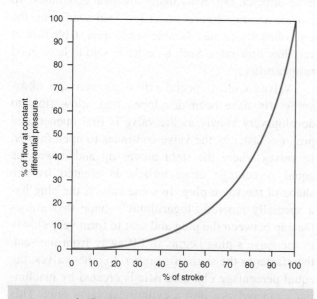

Figure 9-17 Percentage of flow versus percentage of stroke for a valve with an equal percentage characteristic operated at a constant differential pressure.

The very slow increase in flow rate when the valve first begins to open compensates for the rapid rise in heat output from the heat emitter at low flow rates. Likewise, the rapid increase in flow at higher flow rates compensates for the relatively slow rise in heat output at high flow rates. The linear net effect is desirable from a control standpoint. It is especially important under low-load conditions where a quick opening valve (i.e., a standard ball valve) would make accurate and stable heat output control very difficult. In general, *valves with an equal percentage characteristic should be used in any situation where heat output is controlled by varying the flow rate through a heat emitter.*

9.5 Outdoor Reset Control

An ideal heating distribution system would continually adjust its rate of heat delivery to match the heat loss of a building. This would allow the inside air temperature to remain perfectly stable, regardless of the load on the system.

When a valve with an equal percentage characteristic is used to control flow to a hydronic heat emitter, the resulting relationship between stem position and heat output is close to proportional, as shown in Figure 9-18.

One means of accomplishing this is to make small but continuous changes to the water temperature supplied to the heat emitters based on the outdoor temperature. As the outdoor temperature drops, and the heating load of the building increases, the water temperature supplied to the heat emitters is increased to create higher heat output. Likewise, as outdoor temperature increases, and the heating load decreases, the water temperature supplied to the heat emitters also decreases as does their heat output. This concept is called **outdoor reset control**.

When properly executed, outdoor reset control is like "cruise control" for the heating system. It allows just the right amount of heat to be released from the heat emitters to match the current heating load.

There are several benefits associated with outdoor reset control including:

- *Stable indoor temperature.* Outdoor reset control reduces fluctuation of indoor temperature. When the reset control is properly adjusted, the water temperature supplied to the heat emitters is just high enough to satisfy the current heating load. The rate of heat delivery is always maintained very close to the rate of building heat loss. This yields very stable indoor temperature compared to less sophisticated hydronic systems that deliver water to the heat emitters as if it were always the coldest day of winter. In the latter case, the flow of heated water to the heat emitters must be turned on and off to prevent overheating under partial load conditions. This often creates an easily detected and undesirable sensation that the heat delivery system is on versus off. With properly adjusted outdoor reset control, building occupants should have no sensation that heat delivery is on or off, just a sensation of continuous comfort.

- *Near-continuous* circulation. Because outdoor reset control supplies water just hot enough to meet the current heating load, the distribution circulator remains on most of the time. Flow through the distribution system stops only to prevent overheating when internal heat gains from sunlight, equipment, and interior lighting are present.

  Another benefit of near-continuous circulation is the ability to redistribute energy within a heated concrete slab. This is desirable in situations where portions of the slab experience high heat loss. One example is a floor area just inside an overhead door, such as shown in Figure 9-19. If the flow through floor heating circuits was turned off for several hours during cold weather, water in the tubing just inside the door could freeze. Continuous circulation can significantly reduce the potential for freezing.

- *Reduced expansion noise.* The combination of near-continuous circulation and very gradual changes in water temperature minimizes expansion noises from the distribution piping and heat emitters. This is especially important when PEX tubing is used with metal heat transfer plates in radiant panel heating systems. During a typical heating season, the piping and heat emitters will experience thermal expansion movement similar to that in systems not using outdoor reset control. However, when outdoor reset

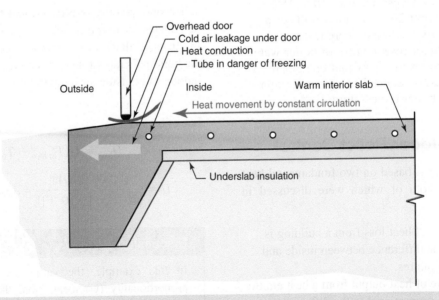

Figure 9-19 | Continuous circulation in floor circuits moves heat from the interior portion of slab to cool perimeter areas. This helps prevent freezing if the heat source is off.

control is used, the *expansion movement takes place over days, even weeks, compared to what might only be seconds in systems that simply turn the flow of hot water on and off*. Piping expansion noise is much more noticeable in systems where rapid changes in water temperature occur.

- *Reduced thermal shock.* Outdoor reset control reduces the possibility of thermal shock to either the heat source or distribution system. Hot boilers are less likely to receive slugs of cold water from zones that have been off for several hours. Wood floors over heated subfloors are less prone to thermal stress when undergoing gradual temperature changes associated with outdoor reset control.

- *Indoor temperature limiting.* When water is supplied to the heat emitters at design temperature regardless of the load, some occupants, especially those who don't pay for the heating energy they use, might choose to set the thermostat to a high temperature and open windows and doors to control overheating. Although this sounds like a foolish way to control comfort, it's often done in rental properties. However, if supply water temperature is regulated by outdoor reset control, it's just warm enough to meet the heating load with the windows and doors closed. Wasteful use of energy is discouraged.

- *Reduced energy consumption.* Outdoor reset control has demonstrated its ability to reduce fuel consumption in hydronic heating systems. The savings are a combination of reduced heat loss from boilers, reduced heat loss from distribution piping, and, in the case of condensing boilers, increased time in condensing mode operation. Exact savings will vary from one project to another. Conservative estimates of 10 to 15% are often cited. Figure 9-20 shows the results of a study in which several homes were tested with and without outdoor reset control of boiler water temperature. The results show that outdoor reset control significantly reduced energy usage relative to constant boiler temperature.

Theory of Outdoor Reset Control

Outdoor reset control is based on two fundamental heat transfer principles, both of which were discussed in previous chapters.

Principle 1: The rate of heat loss from a building is proportional to the difference between inside and outside air temperatures.

Principle 2: The rate of heat output from a heat emitter is approximately proportional to the difference between the supply water temperature and the inside air temperature.

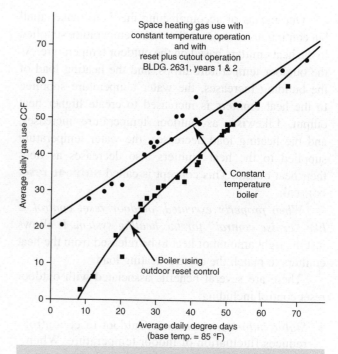

Figure 9-20 Fuel usage for hydronic heating systems with and without boiler reset control. *Source: Home Energy Magazine.*

Example 9.2

A building loses heat at a rate of 80,000 Btu/hr when the indoor air temperature is 70 °F and the outdoor design temperature is −10 °F. Express the building's heat loss according to principle 1.

Solution:

When one quantity is proportional to another, they can be equated by introducing a constant of proportionality. In this case, that constant can be determined by dividing the building's rate of heat loss by the difference between indoor and outside air temperature.

Equation 9.6:

$$Q_{loss} = U(T_{indoor} - T_{outside})$$

$$U = \frac{80,000 \text{ Btu/hr}}{(70 \text{ °F} - [-10 \text{ °F}])} = 1,000 \text{ Btu/hr/°F}$$

Discussion:

In this example, the value of the constant of proportionality (U) means that the building loses 1,000 Btu/hr for each degree Fahrenheit the outside temperature is below the indoor temperature. The design heat loss of a building divided by the difference between indoor and outside air temperature at design conditions is called the building's heat loss coefficient. Each building has a unique heat loss coefficient based on its size and construction.

Example 9.3

A hydronic heating distribution system can release heat at a rate of 80,000 Btu/hr when the supply temperature is 110 °F and the indoor air temperature is 70 °F. Express the distribution system's heat output according to principle 2.

Solution:

This situation is almost identical to that of Example 9.2. In this case, the constant of proportionality (k) is determined by dividing the heat output of the distribution system by the difference between supply water temperature and indoor air temperature.

Equation 9.7:

$$Q_{output} = k(T_{supply} - T_{indoor})$$

where

$$k = \frac{80,000 \text{ Btu/hr}}{(110\,°\text{F} - 70\,°\text{F})} = 2,000 \text{ Btu/hr/°F}$$

Discussion:

The value of k indicates that this distribution system can release 2,000 Btu/hr of heat for each degree Fahrenheit the supply water temperature is above the indoor air temperature.

When Equations 9.6 and 9.7 are combined, and assuming a balance between building heat loss and heat output from the hydronic distribution system (e.g., $Q_{loss} = Q_{output}$), they yield the following relationship:

Equation 9.8:

$$T_{supply} = T_{indoor} + \left(\frac{U}{k}\right)(T_{indoor} - T_{outdoor})$$

For the situations given in Examples 9.2 and 9.3, this becomes:

Equation 9.9:

$$T_{supply} = 70 + \left(\frac{1,000}{2,000}\right)(70 - T_{outdoor})$$

Which can be further simplified to:

Equation 9.10:

$$T_{supply} = 70 + (0.5)(70 - T_{outdoor})$$

A graph of Equation 9.10 is shown in Figure 9-21.

The line on this graph is called a **reset line**. It shows what the supply water temperature to the heat distribution system must be to maintain the desired 70 °F indoor temperature for any outside temperature between 70 and −10 °F.

The lower left end of the reset line represents a situation where no heat is needed by the building. For example, if water at 70 °F is supplied to the heat emitters in a space that is already at 70 °F air temperature, there will be no heat transfer between the heat emitters and the room air.

The upper right end of the reset line represents design load conditions. This is where the heat output from the distribution system must equal the design heat loss of the building.

Reset Ratio

The slope of a reset line is called the **reset ratio** and can be calculated as the change in supply water temperature divided by the change in outdoor temperature between

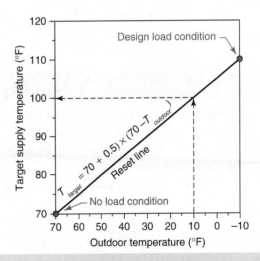

Figure 9-21: Supply water temperature required for heat output of a distribution system described in Example 9.3 to match building load described in Example 9.2. The sloping line is called reset line.

any two points on the reset line. The endpoints of the reset line are typically used to make this calculation.

Equation 9.11:

$$RR = \frac{\Delta T_{supply\ water}}{\Delta T_{outdoor}}$$

where

RR = reset ratio;
$\Delta T_{supply\ water}$ = change in supply water temperature between design load and no-load condition (°F)
$\Delta T_{outdoor}$ = change in outdoor temperature between design load and no-load condition (°F)

Example 9.4

The building described in Example 9.2 has a design heat loss of 80,000 Btu/hr when the indoor temperature is 70 °F and the outdoor temperature is −10 °F. The heat distribution system described in Example 9.3 can release 80,000 Btu/hr when the supply water temperature is 110 °F and the inside air temperature is 70 °F. Determine the necessary reset ratio for this system.

Solution:

The change in water temperature between the no-heat input and design load conditions is $110 - 70 = 40°F$. The corresponding change in outdoor temperature is $70 - (-10) = 80 °F$. Thus, the necessary reset ratio when this heat distribution system is used in this building is:

$$RR = \frac{\Delta T_{supply\ water}}{\Delta T_{outdoor}} = \frac{40}{80} = 0.5$$

Discussion:

Notice that the reset ratio is exactly the same as the mathematical slope of the reset line shown in Figure 9-21. This allows Equation 9.5 to be written in a slightly more general form as:

Equation 9.12:

$$T_{supply} = T_{indoor} + RR(T_{indoor} - T_{outdoor})$$

where

T_{supply} = desired supply water temperature to the distribution system (°F)
T_{indoor} = desired indoor air temperature (°F);
$T_{outdoor}$ = current outdoor air temperature (°F);
RR = reset ratio calculated using Equation 9.11.

Here are several other facts regarding the reset ratio:

- The greater the value of the reset ratio, the steeper the reset line.

- The reset ratio can be interpreted as the necessary increase in supply water temperature per degree drop in outside temperature.

- The reset ratio is a unitless quantity and does not change when calculated using different temperature units (provided the same temperature units are used in the numerator and denominator).

- The value of the reset ratio depends on both the heat loss characteristics of the building and the heat output capability of the hydronic distribution system in that building.

- Once the reset ratio for a particular building/distribution system combination has been calculated, it must be entered into the reset controller. This will be discussed in later sections of this chapter.

Every hydronic distribution system, in combination with the building it's installed in, yields a unique reset line. A building equipped with slab-type floor heating will have a different reset line compared to the same building using fin-tube baseboard convectors. This is due to the difference in supply water temperatures for these two types of heat emitters. A sampling of some "typical" reset lines for different types of heat emitters is shown in Figure 9-22. In general, distribution systems designed around high water supply temperatures require higher reset ratios. Systems designed around lower water temperatures require low reset ratios.

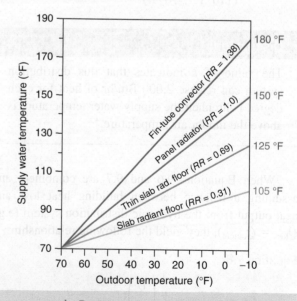

Figure 9-22 Representative reset lines for different heat emitters with different reset ratios (*RR*). Higher reset ratios result in higher supply water temperatures at a given outdoor temperature.

Compensating for Nonproportional Heat Losses

Theoretically, a heating system in which the supply water temperature is regulated according to outside reset control theory could maintain a set interior temperature while only measuring outside temperature. This can be verified by examining Equation 9.12. Once the desired inside temperature and reset ratio are set, the only remaining variable is outside temperature. This relationship holds true as long as the building's heat loss is strictly proportional to the difference between inside and outside temperature (principle 1 discussed earlier).

Unfortunately, there are many situations that partially invalidate principle 1. The effect of wind speed on building heat loss is one. Heat loss due to air infiltration increases at higher wind speeds. An outdoor reset controller that only measures outdoor air temperature cannot sense this condition and therefore cannot compensate for it.

Another effect that partially invalidates principle 1 is the presence of **internal heat gains** from sunlight, people, and appliances or equipment in the building. For example, solar heat gains through large south-facing windows may at times provide all the heat the building needs, even when it is cold outside. If the outdoor reset control only "looks at" outside temperature, it has no way of knowing that such gains have already provided the necessary heat input to the building. The controller simply goes about providing water to the system at the supply temperature it calculates based on the current outdoor temperature. The result can be significant overheating.

Internal heat gains allow a building to maintain comfortable indoor temperatures when the outdoor temperature equals or exceeds a value known as the **balance point temperature**. When the outdoor temperature is at the balance temperature, the rate of internal heat gain exactly balances the rate of building heat loss. The indoor temperature remains at the desired comfort level without any heat input from the heating system.

The balance point temperature of a building can be calculated using Equation 9.13:

Equation 9.13:

$$T_{balance} = T_{indoor} - \left(\frac{Q_{gain}}{U_{building}}\right)$$

where

$T_{balance}$ = balance point temperature (°F);
T_{indoor} = desired inside air comfort temperature (°F);
Q_{gain} = the rate of internal heat gain (Btu/hr);
$U_{building}$ = building heat loss coefficient (Btu/hr/°F).

Example 9.4

Assume that a house has a design heat loss of 40,000 Btu/hr when the outside temperature is −10 °F. and the desired inside temperature is 70 °F. Also assume that the total heat output of lights, people, and appliances in the home is 10,000 Btu/hr at some given time. What is the balance point temperature of the house under these conditions?

Solution:

The building heat loss coefficient ($U_{building}$) is calculated as follows:

$$U_{building} = \frac{40{,}000 \text{ Btu/hr}}{(70\,°F - [-10\,°F])} = 500 \frac{\text{Btu}}{\text{hr}\,°F}$$

Substituting these numbers in Equation 9.13 yields:

$$T_{balance} = T_{indoor} - \left(\frac{Q_{gain}}{U_{building}}\right) = 70 - \left(\frac{10{,}000}{500}\right) = 50\,°F$$

Discussion:

Under these conditions, the building needs no heat input from its heating system until the outside temperature drops below 50 °F. All the heat needed to maintain a normal comfort temperature of 70 °F comes from the 10,000 Btu/hr internal heat gains.

If the rate of internal heat gain drops to zero, the balance point temperature equals the desired inside temperature. The higher the rate of internal heat gain, and the smaller the building heat loss coefficient, the lower the building's balance point temperature becomes.

To compensate for internal heat gains, the reset line must be moved downward and to the right on the graph without changing its slope. This displacement of the reset line is called **parallel shifting**. The starting point of the reset line moves along the **starting point line** as shown in Figure 9-23.

The starting point of the reset line can be adjusted based on the balance point temperature of the building using Equation 9.14.

Equation 9.14:

$$T_{start} = \frac{T_{inside} + RR(T_{balance})}{1 + RR}$$

where

T_{start} = starting point temperature (along starting point line) (°F)
T_{inside} = desired inside air comfort temperature (°F);

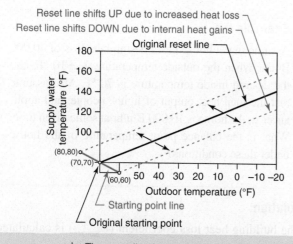

Figure 9-23 | The reset line is required to parallel shift down and right to compensate for internal heat gain. The reset line is required to shift up and left to compensate for increased load due to factors other than drop in outdoor temperature (i.e., increased wind speed). *Note:* Starting point of reset line always moves along the starting point line during parallel shifting.

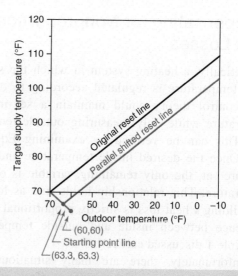

Figure 9-24 | A reset line that has been parallel shifted to a starting point of 63.3 °F to compensate for a consistent internal heat gain.

RR = reset ratio (determined from Equation 9.11);
$T_{balance}$ = balance point temperature (°F).

Example 9.5

The house used in Example 9.4 has a balance point temperature of 50 °F, a desired inside air temperature of 70 °F, and a reset ratio of 0.50. Determine the adjusted starting point temperature of the reset line, accounting for the internal heat gains.

Solution:

Substitute the numbers into Equation 9.14:

$$T_{start} = \frac{T_{inside} + RR(T_{balance})}{1 + RR} = \frac{70 + 0.5(50)}{1 + 0.5} = 63.3 \text{ °F}$$

Discussion:

The reset line with the adjusted starting point temperature of 63.3 °F is shown in Figure 9-24. Notice that the reset ratio (e.g., slope) has not changed. The reset line has simply been parallel shifted to a new starting point to compensate for the assumed internal heat gains.

The reset line can also be parallel shifted upward and to the left when the heating load becomes higher than expected. For example, suppose a strong wind develops that increases the infiltration heat loss of the building (even though the outside air temperature does not change). If the increase in heating load can be determined, it could simply be treated as a negative internal heat gain in Equation 9.13. Once a revised balance point temperature was calculated, it could be used in Equation 9.14 to calculate a revised starting point for the reset line.

Many modern outdoor reset controllers can be equipped with an indoor temperature sensor. When so configured, they *automatically* parallel shift the reset line as required to compensate for both internal heat gains and higher than normal heat loss. *This configuration is highly recommended for most residential and light commercial buildings, which have the potential for significant internal heat gain.*

If the outdoor reset controller cannot be equipped with an indoor temperature sensor, the starting point of the reset line, once calculated, can be manually set using dials or menu-driven inputs on the reset controller. However, because internal heat gains often change, this approach is not as effective as the constant and automatic feedback from an interior temperature sensor. Systems without interior temperature sensors should use an indoor thermostat wired such that it can override the output of the reset controller and prevent further heat input when internal gains push interior temperatures above the desired indoor setpoint.

Implementing Outdoor Reset Control

There are three ways of implementing outdoor reset controls in hydronic systems.

- **Heat source reset** (for on/off heat source)

- **Heat source reset** (for modulating heat source)
- **Mixing reset**

These techniques can be used individually or together.

Heat Source Reset (for On/Off Heat Sources)

The temperature of water supplied to a distribution system can be controlled by turning a heat source such as a boiler or a heat pump on and off. An outdoor reset controller can be configured to adjust the target supply water temperature up and down as the outdoor temperature varies. These controllers have an unpowered electrical contact that simply opens or closes to operate the heat source.

Figure 9-25 illustrates the control logic used by a reset controller that controls an on/off heat source. The sloping blue line represents the "**target temperature**" (e.g., the ideal supply water temperature for the distribution system based on the outdoor temperature). For example, if the outdoor temperature is 10 °F, the blue line in Figure 9-25 indicates a target temperature of 105 °F (as indicated by the blue dot).

When the outdoor reset controller is powered on, it measures outdoor temperature, and then uses this measurement along with its settings for reset ratio and parallel offset to calculate the target supply water temperature. It then compares this calculated temperature to the water temperature currently being supplied to the distribution system.

If the measured supply water temperature is equal or close to the target temperature, no control action is taken. However, if there is sufficient deviation (e.g., "error") between these temperatures, the outdoor reset controller takes action.

The settings represented in Figure 9-25 would result in the following control actions when the outdoor temperature is 10 °F:

- If the temperature at the supply sensor is above 100 °F (e.g., 105 less one half the control differential of 10 °F), the contacts in the outdoor reset controller remain open, and the heat source remains off.

- If the temperature at the supply sensor is below 100 °F, the contacts in the outdoor reset controller close to turn on the heat source. Once turned on, the heat source remains on until the supply sensor reaches a temperature of 110 °F (105 plus one half the control differential) or higher.

The control differential between the temperatures at which the heat source is turned on and off discourages short cycling. The value of the control differential can be adjusted on most outdoor reset controllers. Smaller control differentials reduce the variation in supply water temperature both above and below the target temperature. However, if the control differential is too small, the heat source will cycle on and off excessively. This is not good for the life of components such as boiler ignition systems or heat pump compressors.

Some outdoor reset controllers can be set to vary the control differential based on outdoor temperature. A graph illustrating this effect is shown in Figure 9-26.

As the outdoor temperature rises, the control differential is increased. Because the heating load decreases with increasing outdoor temperature, the duration of the heat source's on cycle also decreases. Widening the

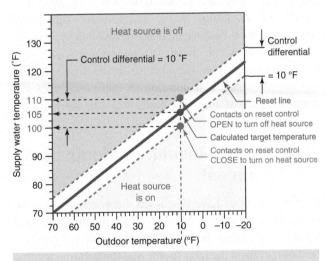

Figure 9-25 Operating logic used by an outdoor reset controller operating an on/off heat source.

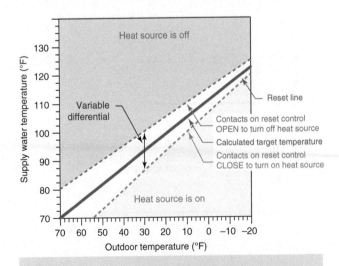

Figure 9-26 Outdoor reset controller operating with a variable differential. The differential increases as outdoor temperature increases. This reduces "short cycling" of the heat source under low-load conditions.

control differential helps protect the heat source against excessively short cycles under such conditions.

The process of measuring outdoor temperature, calculating target temperature, comparing the measured supply water temperature to the target temperature, and generating a control action based on the difference in these temperatures takes place continuously as long as the reset controller is powered on.

Minimum Supply Temperature Setting

Heat source reset control is somewhat limited when used with a conventional gas-fired, oil-fired, or solid fuel (biomass) boilers. If the water returning to the boiler is allowed to drop too low, flue gases will condense within the boiler or its venting system as discussed in Chapter 3, Hydronic Heat Sources. To prevent this, most heat source reset controllers have an adjustable minimum supply temperature. With this feature, the controller does not allow the boiler to operate below a specified minimum temperature regardless of outdoor temperature. Such an operating strategy is called **partial reset control** and is illustrated in Figure 9-27.

Notice how the lower portion of the sloping line is not accessed due to the minimum supply temperature setting.

Although partial reset control protects a conventional boiler from operating with sustained flue gas condensation, it also prevents the distribution system from operating at theoretically ideal lower water temperatures during low-load conditions. Because the water temperature supplied to the distribution system is higher than necessary during these times, some means of preventing building overheating is required. In most systems, this is done by turning off a circulator or closing a zone valve whenever the indoor thermostat senses that the room is at or above setpoint temperature.

Most heat source reset controllers also have a maximum supply temperature setting. This is used to prevent the heat source from supplying excessively hot water to the distribution system *regardless of the outdoor temperature*.

The implementation of a heat source reset control is shown in Figure 9-28. Notice that the controller has sensors that measure outdoor and supply water temperature.

In most systems, an outdoor reset controller does not replace the system's high-limit controller. The latter remains in place as a *safety device* that can stop the heat source should the outdoor reset controller ever malfunction. However, during normal operation the outdoor reset controller "preempts" operation of the high-limit controller, shutting off the heat source at lower water temperatures. One should view the outdoor reset controller as the "operating controller" while the high-limit controller acts as a "safety controller." *It is important to adjust the temperature setting of the high-limit controller high enough that it will not interfere with normal operation of the outdoor reset controller.*

Heat source reset control is largely responsible for the energy savings mentioned as one of the benefits of outdoor reset control. These saving are mostly the result of lower heat losses from boilers and lower operating temperatures for heat pumps.

On/off boilers sold in the United States after September 2012 are required to have an automatic means of adjusting water temperature downward when the space-heating load decreases and vice versa. Outdoor reset is one means of making this adjustment. Controllers that infer the heating demand by monitoring the frequency and duration of thermostat cycles is another means of adjusting boiler water temperature.

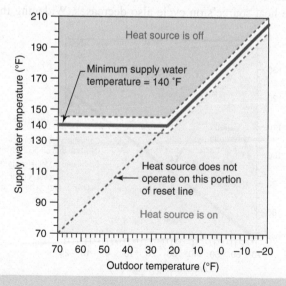

Figure 9-27 An outdoor reset controller with a minimum supply temperature setting of 140 °F and a 10 °F fixed differential. An outdoor reset controller operating in this manner is providing *partial* outdoor reset control.

Heat Source Reset (For Modulating Heat Sources)

Previous sections have discussed the concept of modulating heat sources. The goal of modulating the heat source is to obtain the best match possible between the heat output from the heat source and the building heating load. This is the same objective as previously described for outdoor reset control. *For building heating purposes,*

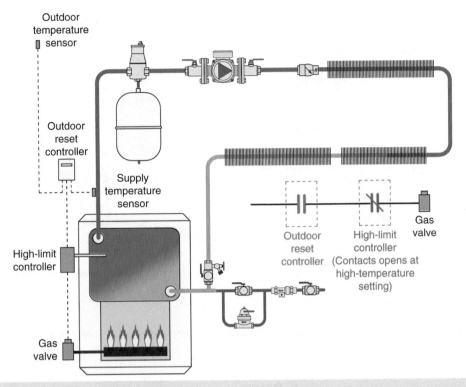

Figure 9-28 Typical installation of an outdoor reset controller for an on/off heat source. The outdoor reset controller operates the heat source as necessary to keep the boiler temperature close to the target temperature.

outdoor reset control is the logic that continually regulates the heat output of a modulating heat source.

Most modulating heat sources have their own internal outdoor reset controllers. Like other outdoor reset controllers, these internal controllers continually measure outdoor temperature, calculate the target supply water temperature, and compare it to the measured supply water temperature. Deviations between these temperatures cause the reset controller to regulate the speed of the combustion air blower on a modulating boiler or compressor speed on a modulating heat pump.

Placement of Supply Water Temperature Sensor

When heat sources are connected to a distribution system in which flow varies due to zoning there will be times when the flow rate through the heat source is different than the flow rate in the distribution system. This causes mixing that will make the temperature of water flowing onward to the distribution system different from that leaving the heat source. If the distribution system flow rate is higher than the heat source flow rate, the water temperature flowing onward in the distribution system will be lower than the water temperature leaving the heat source, and vice versa. Because of this, the supply water temperature upon which outdoor reset control is based should be measured *downstream* of the point when the heat source piping interfaces to the distribution system. This ensures that the "final" blended temperature of the water is being sensed, regardless of the differences between heat source flow rate and distribution system flow rate. Figure 9-28a illustrates this concept for different piping configurations.

Some boilers have internal controllers that can measure the water temperature supplied to the distribution system and adjust boiler firing accordingly (e.g., based on outdoor reset control). Heat sources that do not have this capability can be equipped with external controllers to provide this functionality.

Mixing Reset Control

Another method of implementing outdoor reset control is by using a mixing device between the heat source and the distribution system. Not surprisingly, this approach is called **mixing reset control**. Several mixing devices are suitable for this purpose and are discussed in detail in Section 9.7. Figure 9-29 shows the concept of a mixing assembly that acts as a "bridge" between the boiler loop and the distribution system.

394 Chapter 9 Control Strategies, Components, and Systems

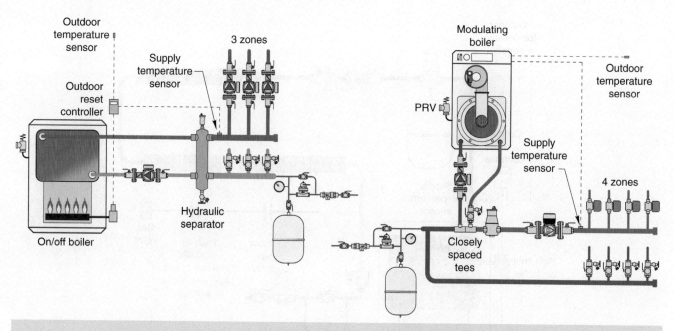

Figure 9-28a | Proper placement of supply temperature sensor in systems where the heat source is hydraulically separated from the distribution system.

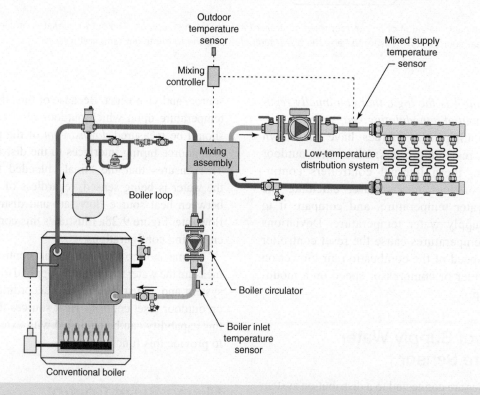

Figure 9-29 | The mixing assembly serves as a bridge between the boiler loop and distribution system. All heat reaching the distribution system must pass through the mixing assembly. Specific hardware configurations that constitute a mixing assembly will be discussed in Section 9.7.

Like a heat source reset controller, a mixing reset controller continually monitors the outdoor temperature as well as the water temperature supplied to the distribution system. It uses the outdoor temperature measurement along with its settings to calculate the target (e.g., ideal) supply water temperature for the distribution system. It then compares the target value to the actual supply water temperature. If necessary, it takes action to correct any unacceptable deviation (e.g., "error") between these temperatures.

That action is different from the relay contact closure used by reset controllers for on/off heat sources. A mixing reset controller generates one of several output signals discussed in Section 9.2. These include PWM, floating control, and 2- to 10-VDC and 4- to 20-mA signals. Some mixing controllers also generate a variable frequency AC signal that can directly control the speed of a circulator using a permanent split capacitor (PSC) motor.

Many mixing reset controllers can also protect conventional boilers from low inlet temperature conditions that would otherwise cause sustained flue gas condensation. *To provide this protection, mixing reset controllers must be equipped with a sensor to measure the boiler temperature (usually the boiler inlet temperature).* When the water temperature entering the boiler approaches the minimum temperature setting, the controller automatically reduces the rate of hot water flow from the boiler loop into the mixing assembly. This prevents the distribution systems from dissipating heat faster than the boiler can produce it.

Mixing reset control allows the water temperature supplied to the distribution system to be reduced all the way down to the room air temperature. This is called **full reset control** because the controller can operate the mixing device to produce a water temperature anywhere along the reset line. Full reset is possible when a conventional boiler is used as the heat source *if the mixing reset controller provides the previously discussed boiler protection.*

Combining Heat Source Reset and Mixing Reset

It is possible to use heat source reset and mixing reset in the same system. The supply temperature from the heat source decreases as the outside temperature increases, as does the water temperature supplied to the distribution system through the mixing assembly. Such systems increase the seasonal efficiency of the heat source as well as provide the ideal supply water temperature to the distribution system for optimal comfort. Some manufacturers offer controllers capable of handling both functions simultaneously.

Figure 9-30 shows a system that supplies heat to panel radiators as well as low-temperature floor heating.

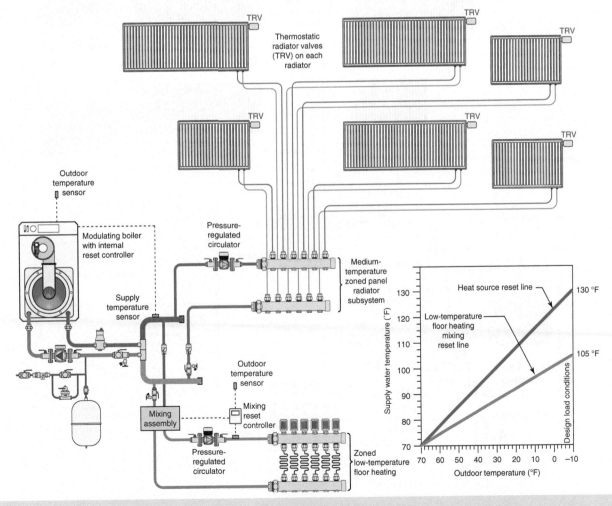

Figure 9-30 | Combined use of boiler reset control and mixing reset control. The modulating boiler operates based on the upper reset line. The mixing reset controller operates based on the lower reset line.

The system is supplied by a modulating/condensing boiler with its own internal reset controller. That controller is set for a reset line appropriate to the water temperature requirements of the panel radiators. The boiler senses the supply water temperature downstream of the closely spaced tee and modulates its firing rate as needed to maintain this temperature close to the target supply water temperature represent by the upper reset line in the graph.

A mixing assembly operated by a separate mixing reset controller lowers this water temperature to a value appropriate for the low-temperature floor heating circuits. This controller operates according to the lower reset line in the graph.

Both subsystems benefit from operating with outdoor reset control.

Figure 9-31 shows a similar system using a conventional boiler as the heat source. Because the floor heating portion of the system operates at low water temperatures, this boiler must be protected against sustained flue gas condensation. The mixing reset controller operating the mixing device must continually measure the boiler's inlet temperature and prevent it from dropping below a preset minimum value. This is accomplished by setting a minimum supply temperature condition for the heat source reset curve as seen on the reset graph.

Proportional Reset

The system shown in Figure 9-30 is an example of a modern multitemperature system. The water temperature supplied to the higher temperature portion of the distribution system was fully reset by the internal reset controller in the modulating/condensing boiler. A separate mixing reset controller and associated mixing device was used to provide full reset of the water temperature supplied to the lower temperature portion of the distribution system.

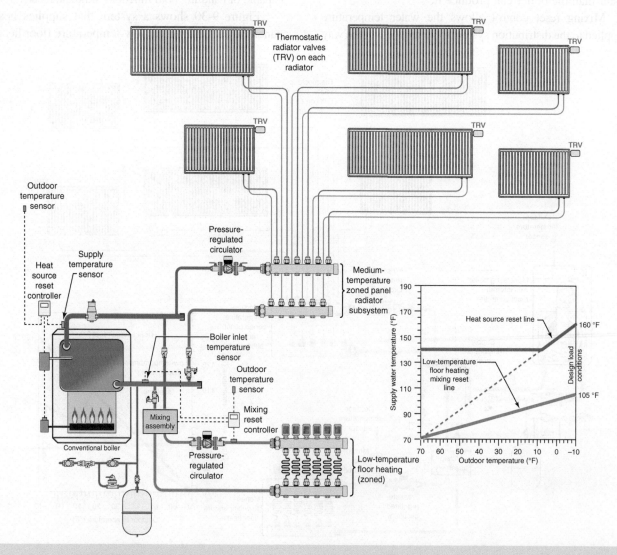

Figure 9-31 | Combined use of boiler reset control and mixing reset control. The conventional boiler operates on the upper reset line. The mixing reset controller operates based on the lower reset line.

While this configuration is fine, there are circumstances in which the low-temperature portion of the system can operate with full reset of the supply water temperature, *without the need of a separate mixing reset controller.*

When the higher of the two supply temperatures is reset, and the lower temperature is created using a manually set (nonthermostatic, nonmotorized) three-way mixing valve supplied by the higher water temperature, the lower supply temperature will also be reset. This concept is called **proportional reset** and is illustrated in Figure 9-32.

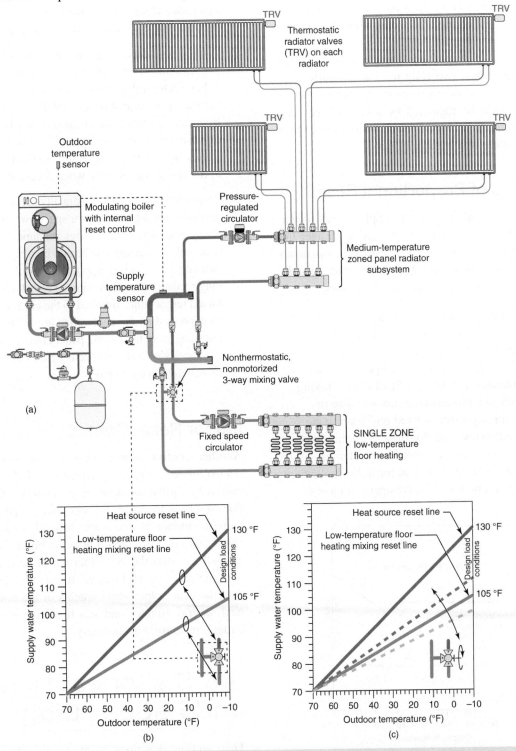

Figure 9-32 | (a) Modulating/condensing boiler supplies a zoned higher temperature distribution system, as well as a *single-zone* low-temperature distribution system. (b) Boiler operates based on the upper reset line. A manually set (nonthermostatic, nonmotorized) mixing valve creates the lower (proportional) reset line. (c) The slope of the lower reset line is adjusted by turning the stem of the three-way mixing valve.

If the higher of the two water temperatures is only partially reset due to minimum boiler temperature requirements, the lower reset line will also be partially reset. Both lines will transition from sloping to flat at the same outdoor temperature. At outside temperatures above this value, flow through the high-temperature and low-temperature subsystems must be cycled on and off to prevent overheating. An example illustrating this concept is shown in Figure 9-33.

The concept of proportional reset can be extended to produce more than two fully reset or partially reset supply temperatures. In every case, the highest water temperature would be regulated by a heat source reset controller and the lower water temperatures created by a manually set three-way mixing valves.

Limitations of Proportional Reset

Proportional reset, when properly applied, can reduce the cost of the control hardware needed to deliver the benefits of reset control in multiple water temperature systems. *However, there are limitations that must be respected when applying it.* They are as follows:

1. Since any fixed position mixing valve only provides fixed flow proportions of the incoming water streams, any variation in the temperature or flow rate of either stream causes a change in the mixed water temperature. Unlike an "intelligent" mixing device, such as a thermostatic mixing valve or motorized mixing valve, a fixed position mixing valve cannot compensate for these changing inlet conditions. For example, consider a system in which zone valves are installed for individual circuit control. Each time a zone valve opens or closes, the flow and head loss of the distribution system changes. This causes a corresponding change in the flow proportions at the manually set mixing device and a change in the mixed outlet temperature. Under certain conditions, there could be wide variations in the mixed temperature. For this reason, *proportion reset should only be considered for portions of a distribution system that are operated as single zone.*

2. A fixed position mixing valve cannot protect a conventional boiler from low-temperature operation. When such a valve supplies a low-temperature/high thermal mass radiant panel, and the system is supplied by a conventional boiler, it should only be used downstream of an **"intelligent" mixing device** that senses boiler inlet temperature and limits heat transfer to the distribution system when necessary.

3. Proportional reset with a manually set mixing valve is generally not a good idea for spaces where frequent changes in thermostat settings are anticipated, especially if that space is heated by a high mass radiant panel. Such an application would experience long recovery time following a setback because the mixed temperature supplied to the circuits would be depressed by the cooler than normal temperature returning from the radiant panel circuits. An intelligent mixing device could compensate for this. A manually set mixing device cannot.

Indoor Reset Control

Another control strategy that is increasingly used to control boiler water temperature is called **indoor reset control**. Unlike outdoor reset control, this strategy does *not* base the target boiler outlet temperature on the current outdoor temperature. Instead, it uses an algorithm that collects data on how the boiler has operated, as well as what the zone thermostat activity has been over the past 5 to 90 minutes. It uses this data to anticipate the heat output requirements of the boiler over the next several minutes and sets the firing rate and target outlet temperature accordingly.

9.6 Switches, Relays, and Ladder Diagrams

This section describes many of the basic electrical switching devices that are commonly used to build an overall control system. Virtually every heating system, hydronic or otherwise, uses one or more of these switching devices within its control system. A working

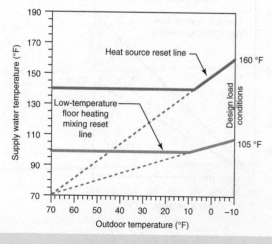

Figure 9-33 — If the heat source reset line is partially reset, the proportional reset line will also be partially reset beginning at the same outdoor temperature.

knowledge of these devices is essential in designing, installing, or troubleshooting hydronic systems. The section also presents ladder diagrams as a fundamental tool for documenting control system wiring.

Switches

The operating modes of most heating and cooling systems are determined by a specific arrangement of electrical switch contacts. Such contacts must be either open or closed at any given time. An open contact prevents an electrical signal from passing a given point in the circuit. A closed contact allows the signal to pass through. These contacts may be part of a manually operated switch or an electrically operated switch called a **relay** or **contactor**. When the settings of the various contacts allow a complete circuit to form, a current will flow, and some predetermined control action will take place.

Poles and Throws

Switches and relays are classified according to their **poles** and **throws**. *The number of poles is the number of independent and simultaneous electrical paths through the switch.* Most of the switches used in heating control systems have one, two, or three poles. They are often designated as single pole (SP), double pole (DP), or triple pole (3P). Switches with more than three poles are also available.

The number of throws is the number of position settings where a current can pass through the switch. Most switches and relays used in heating systems have either one or two throws and are called single-throw (ST) or double-throw (DT) switches.

Figure 9-34 shows a typical double-pole/double-throw (DPDT) toggle switch. Note that the printing on the side of the switch indicates that it is rated for a maximum of 10 amps of current at 250 volts AC (VAC) and 15 amps of current at 125 VAC. When selecting a switch

| Figure 9-35 | Example of a SPDT co switch. *Courtesy of John Siegenthaler.* |

for a given application, be sure its current and voltage ratings equal or exceed the current and voltage at which it will operate.

Figure 9-35 shows a single-pole/double-throw (SPDT) switch with a center off setting (SPDT c/o). The center off setting allows the middle switch contact to remain open to either of the end contacts. A typical application for this switch allows a control system to operate in either heating, cooling, or off. The latter occurs when the switch is in the center off position with its handle straight out.

The switches in Figures 9-34 and 9-35 have screw terminal connections. Similar switches are available with quick-connect terminals. Switches of this type are also available with a wide range of current ratings and voltage ratings to suit different applications. Designers need to specify switches that meet or exceed the maximum electrical current within the circuit the switch will be used in. In some cases, there will be two current ratings—one for **resistive loads** and another for **inductive loads**. Typically, the current rating for an inductive load such as a motor or transformer is lower (often half) the current rating for a resistive load such as an incandescent light or heating element.

Switches are also rated for a maximum voltage. This is the maximum allowable voltage across the any pair of contacts within the switch when the circuit is open.

Relays

Relays are simply electrically operated switches. They can be operated from a remote location by a low power electrical signal. They consist of two basic subassemblies: the **coil** and the **contacts**. Other parts include a spring, pendulum, and terminals. These components and the operation of a relay are illustrated in Figure 9-36.

The spring holds the contacts in their "normal" position. When the proper voltage is applied to the coil, it generates a magnetic force that pulls the common contact attached to the pendulum to the other position,

| Figure 9-34 | Example of a DPDT toggle switch. *Courtesy of John Siegenthaler.* |

400 Chapter 9 Control Strategies, Components, and Systems

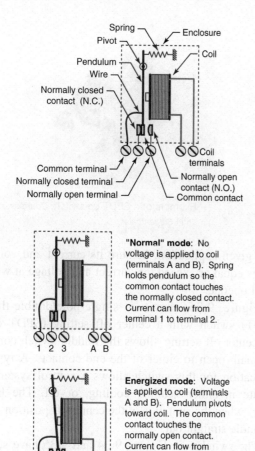

Figure 9-36 | Internal components and operation of a typical relay.

Contact designation	Switch contacts	Relay contacts
Single pole Single throw SPST	Pole 1 —o/ o—	Pole 1 —o||o—
Double pole Single throw DPST	Pole 1 —o/ o— Pole 2 —o/ o—	Pole 1 —o||o— Pole 1 —o||o—
Triple pole Single throw 3PST	Pole 1 —o/ o— Pole 2 —o/ o— Pole 3 —o/ o—	Pole 1 —o||o— Pole 2 —o||o— Pole 3 —o||o—
Single pole Double throw SPDT	Pole 1	Pole 1
Double pole Double throw DPDT	Pole 1 Pole 2	Pole 1 Pole 2
Triple pole Double throw 3PDT	Pole 1 Pole 2 Pole 3	Pole 1 Pole 2 Pole 3

- —\/\/\/— Relay coil
- —(R1)— Relay coil
- —o\|\|o— Relay contact, normally open (N.O.) (closes when coil is energized)
- —o\|\\o— Relay contact, normally closed (N.C.) (opens when coil is energized)

Figure 9-37 | Schematic symbols used to represent switches and relays having different numbers of poles and throws.

slightly extending the spring in the process. In most relays, this action takes only a few milliseconds. A click can be heard as the contacts move to their other position. When voltage is removed from the coil, the spring pulls the contacts back to their deenergized (normal) position.

Relay contacts are designated as **normally open**, or **normally closed**. In this context, the word *"normally" means when the coil of the relay is not energized.* A normally open contact will not allow an electrical signal to pass through while the coil of the relay is off. A normally closed contact will allow the signal to pass when the coil is off. Like switches, relays are also classified according to their poles and throws.

The coil is rated to operate at a specific voltage. In heating systems, the most common coil voltage is 24 VAC. Relays with coil voltages of 12, 120, and 240 VAC are also available when needed. Relays with coils designed to operate on DC voltages are also available, although less frequently used in heating control systems.

Figure 9-37 shows the schematic symbols for both switches and relays. On the DP and 3P switches, a dashed line indicates a nonconducting mechanical coupling between the metal blades of the switch. This coupling ensures that all contacts open or close at the same instant. Note the similarity between the schematic symbols for switches and relays having the same number of poles and throws.

General Purpose Relays

Some control systems require one or more relays to be installed to perform various switching or isolation functions. The type of relay often used is called a **general purpose relay**. They are built as plug-in modules with clear plastic enclosures. Their external terminals are designed to plug into a **relay socket**. These sockets have either screw terminals or quick-connect tabs for connecting to external wires.

Relay sockets allow relays to be removed or replaced without having to disturb the wiring connections. Each terminal on the socket is numbered the same as the connecting pin on the relay that connects to that terminal. This numbering is very important because it allows the sockets to be wired into the control system without the relay being present. Relay manufacturers provide wiring diagrams that indicate which terminal numbers correspond to the normally closed, normally open, common, and coil terminals. In some cases, the wiring diagram is printed onto the relay enclosure.

Figure 9-38 shows a general purpose triple-pole/double-throw (3PDT) relay along with the socket it

(a)

(b)

(c)

Figure 9-38 (a) 3PDT plug-in relay in plastic enclosure. (b) Relay socket for a 3PDT plug-in relay. Note numbers and letters designating terminals. (c) Relay mounted on socket. *Courtesy of John Siegenthaler.*

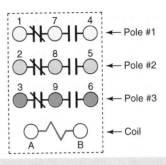

Figure 9-39 Schematic representation of relay associating contacts with terminal numbers. Coil connections are identified as A and B.

mounts into and its wiring diagram. Notice how the terminals are numbered on the bottom of the relay enclosure as well as the relay socket.

Figure 9-39 shows the wiring schematic for the relay shown in Figure 9-38. It identifies the normally open and normally closed contacts, their corresponding terminal numbers, and the coil terminals.

Many relay sockets are designed to mount to a flat surface, or to be snapped into an aluminum **DIN rail**. The latter mounting method allows relay sockets to be added or removed quickly and without fasteners. The DIN rail is a standard modular mounting system used in many types of control systems. Many small control devices are designed to be mounted to DIN rails. By mounting a generous length of DIN rail in a control cabinet, relays and other components can be easily added, removed, or moved to a different position. Figure 9-40 shows two different relay sockets mounted to a common DIN rail.

Time Delay Relays

It is sometimes necessary to incorporate a time delay between two control events. One example is keeping a

Figure 9-40 Different size relay sockets mounted to a common aluminum DIN rail. *Courtesy of John Siegenthaler.*

boiler circulator operating for a few minutes after the boiler has stopped firing to purge out the residual heat. Another is allowing a DHW subsystem to be the only load operating for a designated time once it initiates a call for heat. These and other control functions can be accomplished using **time delay relays**.

Time delay relays are available with the following operating modes:

- Delay-on-make
- Delay-on-break
- Interval timing
- Repeat cycle

These operating modes are described using timing charts in Figure 9-41. Of the four operating modes shown, the delay-on-make and delay-on-break functions are the most commonly used in hydronic heating applications.

The **delay-on-make** mode prevents the contacts from moving until a user-specified time delay period has elapsed from when the input signal is turned on. If input signal is interrupted before the delay period has elapsed, the relay automatically resets the timing circuit to begin from zero the next time the input signal is energized.

One common use of a delay-on-make function is overriding priority control after a specified time has elapsed. For example, assume a designer wants DHW to operate as the **priority load** for a maximum of 30 minutes. All other loads are to be temporarily turned off during this time. However, if the domestic water load is still operating at the end of the 30-minute period, the other loads need to be turned back on. This is desirable to prevent freezing water in remote parts of the space-heating distribution system should the DHW load fail in the on position. The wiring necessary for this action is discussed in the next section.

The **delay-on-break** mode is the opposite of the delay-on-make mode. The contacts are held in their energized position until a user-specified time has elapsed. The input voltage to the relay is maintained at all times. When an external switch closure is detected across the control switch terminals of the relay, the contacts move to their energized position and remain there. When the external control switch opens, the time delay period begins. After the time delay period has expired, the relay contacts snap back to their "normal" position. If the control switch closes before the time delay period has elapsed, the time delay period is automatically reset to zero.

Time delay relays are available in several configurations of poles and throws. One of the most common is a DPDT contact configuration with a 120-VAC input signal.

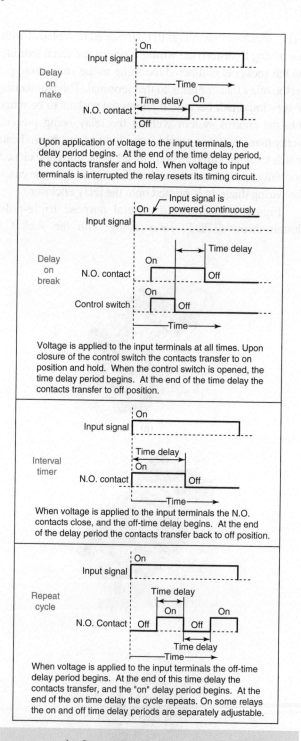

Figure 9-41 Operating modes for time delay relays. Some time delay can only provide one of these functions, others can be set to provide any of these functions.

Solid-state electronics have made it possible to combine several time delay relay functions into a single **multifunction time delay device**. An example is shown in Figure 9-42.

Notice the selector switches for both operating mode and time delay range on the multifunction time delay device. These allow a single device to be configured for a wide range of functions and time delay periods.

9.6 Switches, Relays, and Ladder Diagrams

Figure 9-42 | Example of a multifunction time delay device.
Courtesy of John Siegenthaler.

Contacts wired in parallel represent an **"OR" decision**. If any one or more of the contacts are closed, the electrical signal crosses the group of switches to operate the driven device.

When switch contacts, relay contacts, and contacts that are part of specialized hydronic controllers are physically wired together in a given manner, they create **hard-wired logic**. Such logic determines exactly how the control system functions in each of its operating modes. Some operating modes are planned occurrences while others may be fail-safe modes in the event a given component does not respond properly.

Hard-Wired Logic

One of the ways of creating operating logic for a control system is by connecting switch and/or relay contacts in series, parallel, or combinations of series and parallel. Figure 9-43 shows basic series and parallel arrangements.

Contacts wired in series represent an **"AND" decision**. For an electrical signal to reach a driven device, all series-connected contacts must be closed simultaneously.

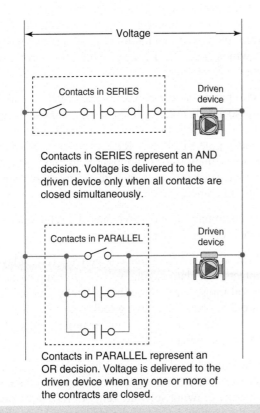

Figure 9-43 | Operating logic based on wiring of switches or relay contacts. Contacts wired in series provide "AND" logic. Contacts wired in parallel provide "OR" logic.

Ladder Diagrams

It is often necessary to combine several control components to build an overall control system. The way these components are connected to each other as well as to driven devices such as circulators, zone valves, and boilers determines how the system operates.

A ladder diagram is a standard method for developing and documenting the electrical interconnections necessary to build a control system. The finished diagram can then be used for installation and troubleshooting. Such a diagram should always be part of the documentation of a hydronic heating system.

Ladder diagrams have two basic sections, the line voltage section at the top of the ladder and the low-voltage section at the bottom of the ladder. A transformer separates the two sections. The primary side of the transformer is connected to line voltage. The secondary side of the transformer powers the low-voltage section. In North America, most heating and cooling secondary systems operate with a secondary voltage of 24 VAC.

An example of a simple ladder diagram is shown in Figure 9-44. The vertical lines can be thought of as sides of an imaginary ladder. Any horizontal line connected between the two sides is called a **rung** of the ladder. A rung connected across the line voltage section is exposed to 120 VAC. A rung connected across the low-voltage section is typically exposed to 24 VAC. The overall ladder diagram is constructed by adding the rungs necessary to create the desired operating modes of the system. The vertical length of the ladder can be extended as necessary to accommodate all required rungs.

Relays are often used to operate line voltage devices such as circulators, oil burners, or blowers based on the action of low-voltage components such as thermostats. Ladder diagrams are an ideal way to document how this is done. When a circuit path is completed through a relay coil in the low-voltage section of the ladder, one or more contacts of that relay may be used to operate devices in the line voltage section.

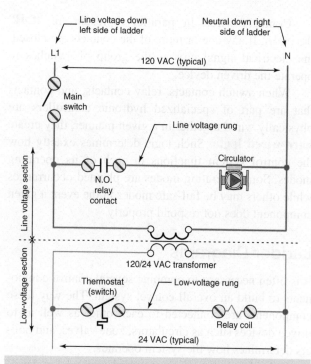

Figure 9-44 Example of a simple ladder diagram to operate a line voltage circulator using a low-voltage control circuit.

Consider a situation in which a line voltage circulator is to be operated by a low-voltage switch, such as a common thermostat. Since the circulator needs line voltage to run, it is connected across the line voltage section of the ladder diagram as shown in Figure 9-44. A normally open relay contact is wired in series with the circulator motor. When this contact is open, the motor is off. To close this contact, the coil of the relay must be energized. This requires a completed circuit path across the low-voltage section of the ladder. By wiring the relay coil in series with the thermostat, the coil is energized when the thermostat contacts are closed and deenergized when they are open. The overall operating sequence is as follows: The thermostat contacts close. Low voltage is then applied across the relay coil. The energized coil pulls the normally open contacts in the line voltage rung together. Line voltage is applied across the circulator motor to operate it.

Although this example is relatively simple, it illustrates the basic use of both the line voltage and low-voltage portions of the ladder diagram.

More complex ladder diagrams are developed by placing schematic symbols for additional components into the diagram. Some of these components might be simple switches or relays, others might be special purpose controllers that operate according to how the manufacturer designed them as well as how the adjustable parameters are set. When shown in a ladder diagram, the latter are called **embedded controllers** since they are part of a ladder diagram that documents the overall control system. An example of a ladder diagram with several rungs and an embedded controller is shown in Figure 9-45.

Drawing Ladder Diagrams

Consistency is the key to drawing ladder diagrams that will be useful to many people. The following guidelines are recommended.

- Draw the ladder diagram with line voltage section at the top and the low-voltage section at the bottom.

- Specify the supply voltage and circuit **ampacity** required for the control system represented by the ladder diagram. Show this at the top of the line voltage section.

- Always show a main switch capable of completely isolating the ladder diagram from its power source when necessary.

- Use wider lines for line voltage conductors, and thinner lines for low-voltage conductors to make the drawing easier to follow.

- When a component symbol appears in the system's piping schematic, and in the ladder diagram, be sure it has identical designations in both drawings (i.e., P1, T2, etc.). This allows for easy cross-referencing.

- Use separate abbreviations for each relay coil and its associated contacts. Use a designation such as R1–2 to identify pole #2 of relay R1. R2–1 would identify pole #1 of relay R2, and so forth. Without such designations, it is impossible to know which contacts and coils are part of the same relay, especially in complex diagrams.

- It is customary to show switches and relay contacts in their off, or deenergized, positions.

- When several of the same components (i.e., relay contacts, circulators, zone valves, etc.) appear on several rungs, they should be vertically aligned. This not only makes the drawing look better but also makes it easier to understand because of consistent symbol placement.

- Although ladder diagrams can be drawn by hand, there are many advantages to creating them using computer drawing software (e.g., CAD). For example, the vertical height of the ladder can be easily extended or shortened as needed to accommodate the necessary rungs. Component symbols and groups of symbols can be easily duplicated and moved to speed the design/drawing process.

9.6 Switches, Relays, and Ladder Diagrams

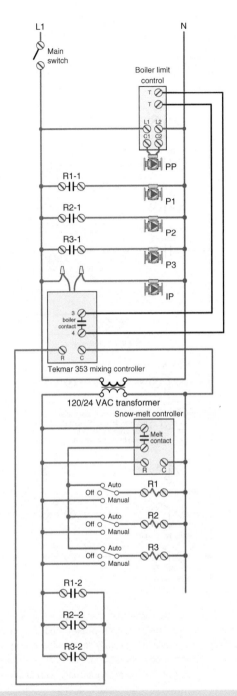

Figure 9-45 | Example of a ladder diagram with three embedded controllers.

- When the ladder diagram is completed, create a **description of operation** for the system represented by the ladder diagram. Describe each operating mode independently and make frequent reference to the component designations in the ladder diagram. This description greatly enhances the ability to understand what the system is supposed to do in each operating mode, as well as the specific components involved in each operating mode. An example of a complete description of operation in given in Section 9.10

Priority Control Strategies

Sometimes the total heating load connected to a hydronic system exceeds the output of the heat source. One way to manage this situation without increasing the size of the heat source is by prioritizing one or more of the loads. **Priority control** is a planned strategy for shedding (e.g., temporarily turning off) one or more loads when a specific condition is detected in the system. This condition might be a specified water temperature at some location, the operation of a specific load, or the occurrence of two or more simultaneous conditions within the system.

A common priority load in many residential hydronic systems is the operation of an indirect domestic water heater. A basic ladder diagram showing how the DHW load is prioritized is shown in Figure 9-46.

When the contacts of the water heater thermostat close, relay coil (R1) is supplied with 24 VAC. The normally open relay contact (R1–1), located in the line voltage section of the ladder diagram, closes to supply line voltage to the circulator that supplies the domestic water heater. At the same time, a normally closed relay contact (R1–2) opens to break low voltage (24 VAC) to the space-heating thermostats shown in the lower portion of the ladder diagram. Since power to all the space-heating thermostats has now been interrupted, these loads remain off until the priority relay (R1) is turned off. The full output of the heat source can now be directed to the prioritized load (e.g., DHW). When the priority load is satisfied, relay coil (R1) is turned off, and contact (R1–2) closes to reconnect the space-heating thermostats to 24 VAC power.

Although priority control is a useful concept, it can also lead to unexpected problems if not designed for failsafe operation. For example, imagine what might happen to a system with prioritized DHW if the prioritized load could never be met. This could happen if the tank thermostat failed in the "on" position, or if a temperature sensor was accidentally pulled out of its well in the tank. The control system represented in Figure 9-46 would respond by preventing any of the space-heating circulators from operating. In a cold climate, a couple of days of unattended operation could lead to frozen pipes.

One way to prevent this "priority lockout" is to design the control circuit with **priority override**. A delay-on-make time delay can be used as shown in Figure 9-47.

The system goes into the priority mode any time the DHW load calls for heat. Notice the line voltage time delay relay wired in parallel with the DHW circulator. This delay-on-make relay begins its timing cycle each time the DHW load initiates operation. If the DHW

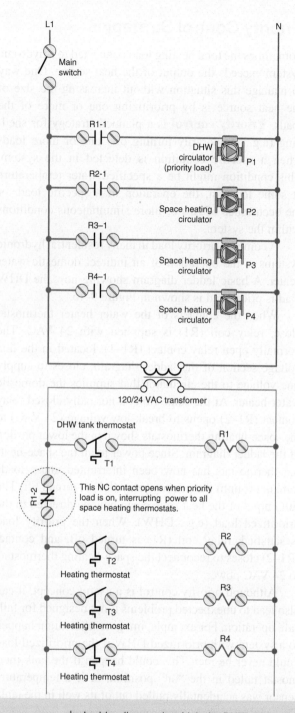

Figure 9-46 Ladder diagram in which the DHW mode is prioritized. All space-heating loads are temporarily turned off whenever the DHW load is operating.

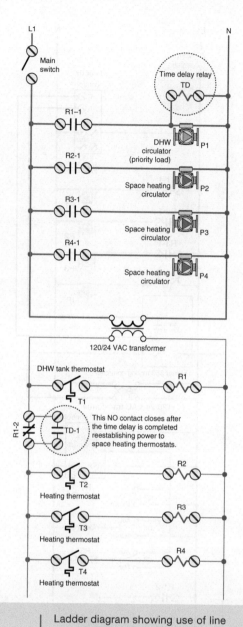

Figure 9-47 Ladder diagram showing use of line voltage (delay-on-make) time delay relay to override priority mode after a specified time has elapsed, allowing space heating to operate simultaneously with DHW.

load is still active when the time delay period has elapsed, the contacts marked TD-1 close to reestablish power to the space-heating thermostats. This allows the space-heating circuits to operate at the same time as the DHW load. When the DHW load is finally met, the time delay relay is deenergized along with the DHW circulator. The priority override process is set to repeat itself, if necessary, each time the DHW load operates.

9.7 Basic Hydronic System Control Hardware

This section discusses several of the basic control hardware devices used in modern hydronic systems. Some of these devices are classified as **electromechanical controls** and have been available for several decades. They use components such as springs, bimetal strips, magnets, and fluid-filled temperature-sensing bulbs to open and close electrical contacts based on sensed quantities such as temperature and pressure. There are literally hundreds of electromechanical controls available for specific purposes and applications in hydronic heating.

Other devices are classified as electronic controls. They use components such as solid-state temperature sensors and microprocessors to perform both simple and complex control tasks. Electronic controls represent the cutting edge of control technology. Every year, new controls with increased functionality and "intelligence" appear on the market. The ability of electronic controls, especially those using microprocessors, to execute complex and sophisticated control algorithms processing cannot be matched by electromechanical controls. Electronic controls are also capable of greater accuracy than is possible with electromechanical controls.

Keep in mind that both electromechanical and electronic controls have a place in modern hydronic systems. Each type of control has strengths and weaknesses. Fortunately, both types can often be combined in a manner that maximizes their individual strengths and provides excellent overall performance, reliability, and value.

Single-Stage Electromechanical Room Thermostats

No heating controller is more familiar by sight and name than a room thermostat. For many homeowners, it is the only control they know of, or want to know of, in the entire system. If the house is too cold, they turn it up. If it is too warm, they turn it down. The room thermostat is, however, only one of several temperature control devices, even in simple hydronic systems.

Within the context of a heating system, the function of a room thermostat is to monitor the temperature of the room in which it is located, and when necessary, close a set of electrical contacts to signal that the room is below the desired comfort temperature and therefore needs heat input. In this state, the thermostat is said to be "calling" for heat.

In essence, a single-stage heating-only thermostat is a temperature-operated switch. It can have only one of two possible states at any given time (open contacts or closed contacts). The fact that the thermostat contacts are closed does not affect the *rate* of heat input to the room. Many building occupants do not understand this. The common misconception that setting the thermostat to a high temperature will quickly increase room temperature is proof.

Most electromechanical room thermostats contain a **bimetal element** consisting of two strips of metal bonded together. Each strip is made of a different metal having a different coefficient of thermal expansion.

When heated or cooled, this bimetal element bends or rotates due to these different coefficients of expansion. The motion of the bimetal element can be used to pull contacts together or push them apart.

In a heating thermostat, the contacts are arranged so that when the bimetal element is sufficiently cooled by the surrounding air, the contacts snap together. This contact closure can be used to start the heat source, open a valve, or initiate some other control action. In a cooling thermostat, just the opposite takes place. When the bimetal element is sufficiently heated by the surrounding air, the contacts snap together, completing a circuit to start the cooling device.

Figure 9-48 shows the external appearance and internal construction of a simple heating-only bimetal thermostat. This particular model is designed to operate within a 24 VAC circuit. 24 VAC is supplied to the "R" terminal on the rear of the thermostat. This voltage passes along a wire to one side of the contacts. The other side of the contact is an adjustable screw that connects to the "W" terminal on the rear. As the temperature surrounding the thermostat decreases, the two parts of the contact get closer. At some point, the magnet exerts sufficient force across the contact gap to snap the silver-colored contact on the left to the copper-colored screw on the right. 24 VAC can now pass from the "R" terminal to the "W" terminal, and onward in the control circuit. The temperature difference (e.g., operating differential) of the thermostat can be adjusted by turning the small copper-colored screw.

The schematic symbols used to designate room thermostats for both heating and cooling applications are shown in Figure 9-49.

One bimetal room thermostat that has been used for several decades is shown in Figure 9-50. The bimetal element connects to a small glass tube containing the ends of two flexible wire leads and a small quantity of mercury is attached to the outer portion of this element. As the temperature changes, the glass tube is slowly rotated by the movement of the bimetal element. When the tube reaches a certain angle, the mercury it contains flows to the end of the tube containing the wire contacts. Because mercury is a good electrical conductor, it completes the circuit between the wire leads, and thus through the thermostat. As the room warms, the glass tube slowly rotates in the other direction. The mercury eventually flows to the other end of the tube, and the circuit is broken. This type of thermostat must be precisely leveled when mounted on the wall to ensure proper operation of the mercury switch.

Mercury is a toxic substance, and is now considered a hazardous material if ever released from a physically damaged thermostat. Because of this, thermostats

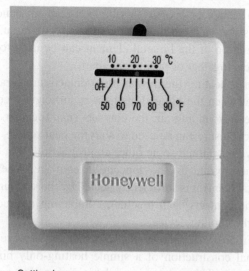

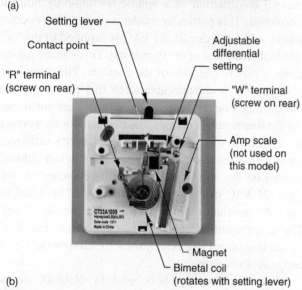

Figure 9-50 Example of an electromechanical room thermostat with an internal mercury switch. The setting appears on the upper dial. The current room temperature is indicated on the lower dial. *Courtesy of John Siegenthaler.*

Figure 9-48 (a) External appearance of a simple bimetal thermostat. (b) Internal components showing contacts open. *Courtesy of John Siegenthaler.*

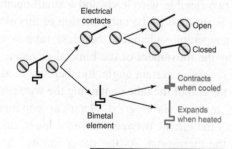

Figure 9-49 Schematic symbols used to represent heating and cooling thermostats.

containing mercury switches are no longer sold. When thermostats containing mercury are replaced, the old thermostat must be returned to a wholesaler for proper disposal.

An ideal room thermostat would operate the heating system to maintain the room precisely at a desired setpoint temperature. If the thermostat detected even the slightest drop below the setpoint, it would bring on the heating system. When the temperature of the room increased the slightest amount above the setpoint, it would stop further heat input. Unfortunately, this would result in extremely short and frequent operating cycles of the heating system. Although this is possible with electric resistance heating elements, it would cause extreme wear on boilers and heat pumps. To prevent such short cycling, room thermostats must operate with a differential as discussed in Section 9.2. The differential is the difference in temperature between where the contacts close and where they open.

Some thermostats have adjustable differentials, and others are fixed by the manufacturer. The narrower the differential, the greater the number of on/off cycles the heating system will experience during a given period of time. Narrow differentials are beneficial for comfort control if they can be obtained without creating abnormally short operating cycles of the heat source. Thermostats are available with differentials as low as 1 °F. Such thermostats are often used with electric resistance heating equipment that is not adversely affected when turned on and off many times an hour. Thermostats that directly operate a combustion-type heat source or heat pumps typically use differentials of 2 to 4 °F to reduce the number of on/off cycles and thus reduce wear on the equipment.

When a simple on/off thermostat is used, heat input begins only when the room temperature drops a certain amount below the desired setting. The thermal mass of the heating system causes a further drop in temperature while the heat source and heat emitters warm enough to begin raising the room's temperature. This causes an undershoot in room temperature as illustrated in Figure 9-51.

If the thermostat contacts remain closed until the room reaches the desired temperature, an effect known as overshoot occurs. Residual heat stored in the thermal mass of the heat emitters continues to flow into the room causing the air temperature to climb above the desired setting.

To limit undershoot and overshoot, most electromechanical thermostats use an internal **heat anticipator**. It consists of a small adjustable resistor that is usually wired in series with the thermostat contacts and thus has current flowing through it whenever the contacts are closed. The heat given off by the resistor slightly increases the temperature sensed by the bimetal element in the thermostat enclosure. This "fools" the thermostat into opening its contacts before room temperature reaches the setpoint. The residual heat in the heat emitters then causes the air temperature to drift slightly upward with minimal, if any, overshoot.

The lower the room temperature relative to the setpoint, the longer it takes the anticipator to heat the bimetal element and the longer the on cycles become. The anticipator allows the system to increase the amount of heat delivered as the difference between setpoint and actual room temperature increases. However, the longer the on cycles of the thermostat, the more heat the anticipator releases into the thermostat enclosure. This causes an undesirable effect called droop in which the room temperature drifts downward as the heating load increases. *Standard electromechanical thermostats cannot compensate for droop.*

For proper operation, thermostats with series anticipators need to be calibrated for the electrical current that flows through them while operating. This

Figure 9-52 | Heat anticipator adjustment on a low-voltage electromechanical room thermostat. Arrow is set to the amperage of the circuit. *Courtesy of John Siegenthaler.*

current depends on the circuit the thermostat is wired into and can be easily measured by connecting an ammeter in series with the thermostat circuit. Once the operating current is read, a small lever inside the thermostat is adjusted to the indicated amperage. Figure 9-52 shows the heat anticipator adjustment on a room thermostat.

Room thermostats should be mounted on interior walls away from sources of localized heat such as lights, cooking equipment, heat emitters, or direct sunlight. A mounting height of about 5 feet above the floor is typical. Choosing a thermostat location should not be considered a trivial matter. Careless placement can significantly detract from the performance of an otherwise well-installed system.

Room thermostats can be further classified as low-voltage or line voltage devices. The majority of those used in hydronic heating are designed for low-voltage operation. They typically switch a circuit operating at 24 VAC. This control voltage is a standard throughout North America. Low-voltage thermostats qualify as class 2 devices under the national electrical code and can be wired with relatively small diameter conductors in the range of 24- to 18-wire gauge.

Line voltage room thermostats are designed for the nominal 120 or 240 VAC line voltage. They are capable of switching line voltage devices such as heating elements, circulators, or fans. All line voltage thermostats have an associated ampacity rating, which is the maximum current the contacts can safely handle. Often two current ratings are given. The greater one is for resistive loads such as heating elements, the lower one for inductive loads such as AC induction motors. When selecting a line voltage thermostat, always check that these ratings meet or exceed the total amperage of all devices that draw their current through the thermostat.

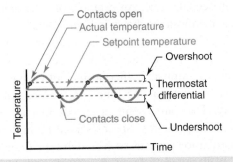

Figure 9-51 | Overshoot and undershoot of temperature as thermostat cycles heat on and off.

Two-Stage Electromechanical Room Thermostats

Occasionally, a hydronic system requires two levels of heat input. The first level provides a certain rate of heat input to the space. If the load is such that this input can maintain the setpoint temperature, the other level of heat input is not needed and therefore not called for. However, if the first level of heat input cannot maintain the room's setpoint temperature, the second level of heat input is called for in an attempt to bring the room temperature back to the desired setpoint. These levels of heat input are called stages. A thermostat capable of providing two or more levels of heat input is called a **multistage thermostat**.

In multistage heating systems, stage 1 always operates before stage 2. For example, in many heat pump systems, stage 1 of a multistage thermostat operates the compressor that drives the refrigeration cycle discussed in Chapter 3, Hydronic Heat Sources. If the room temperature continues to drop when stage 1 is operating, stage 2 of the thermostat operates electric resistance heating elements, or some other type of supplemental heat source, to provide additional heat input. Because heat from electric resistance elements is significantly more expensive than heat provided by the heat pump's refrigeration cycle, it will be used only when necessary to prevent loss of comfort.

Another common application for a two-stage thermostat is providing supplemental heat input to a room in which primary heat input is provided by a radiant floor panel. The supplemental heat may come from a baseboard, panel radiator, or other types of heat emitter. However, the supplemental heat emitter will only operate if the radiant floor panel cannot maintain the room's setpoint temperature.

Two-stage room thermostats have two sets of contacts, one for each stage. The first-stage contacts always close before the second-stage contacts. The heating device with operating priority is controlled by the first-stage contacts. The supplemental heating device is controlled by the second-stage contacts.

Two-stage thermostats usually have a fixed temperature differential of at least 1 °F between stages. This is called the **interstage differential**. If, for example, this interstage differential was 1 °F and the first-stage contacts closed at a room temperature of 69 °F, the second-stage contacts would close at 68 °F. As the room warms up, the second-stage contacts open first. If the room temperature continued to rise, the first-stage contacts will eventually open.

All multistage electromechanical thermostats create droop as more stages operate. This results from the fact that room temperature must drop below the desired setpoint before the thermostat can signal for additional heat input. *The greater the number of stages, the greater the droop.*

Electronic Thermostats

One of the first heating controls to enjoy the benefits of electronics was the room thermostat. Although they have been available for over 50 years, the most up-to-date electronic thermostats offer features and benefits that far exceed early models.

Most electronic thermostats, as well as other types of electronic temperature controls, use a **thermistor** as their temperature-sensing element. This small solid-state device resembles a glass bead with a diameter of about 0.1 inch and two wire leads. The type of thermistor commonly used in temperature controllers is called a **negative temperature coefficient (NTC) thermistor**. The electrical resistance of an NTC thermistor decreases as its temperature increases. This relationship between temperature and resistance is very predictable and repeatable. For some thermistors, a temperature change of 1 °F can result in a change in resistance of over 100 ohms, which is easily detected by the thermostat's circuitry. The small size of a thermistor bead also allows it to respond very quickly to temperature changes.

Modern electronic thermostats can be classified as either programmable or nonprogrammable. Most **programmable thermostats** can accept and store several different settings and the associated times during which these settings are to be in effect. Some programmable thermostats allow up to four different schedules per 24-hour day. Some even allow for different settings on weekends versus weekdays. Almost all programmable thermostats allow for a temporary override if the temperature setting is changed. They then revert to programmed settings when the next schedule period begins. Schedules and settings are typically entered using buttons along with an LCD display. Programmable thermostats can reduce heating costs by reducing temperature settings during unoccupied periods and/or at sleeping time. An example of a programmable electronic thermostat is shown in Figure 9-53.

Although programmable thermostats offer many features, they are not always necessary. Less expensive, nonprogrammable thermostats offer comparable accuracy, but without the time/temperature scheduling capability. They are useful in spaces that are kept at consistent temperatures. Spaces served by high thermal mass distribution systems such as slab-type floor heating are not well suited to frequent or wide setback schedules, and thus good candidates for nonprogrammable thermostats. Figure 9-54 shows a modern, electronic,

9.7 Basic Hydronic System Control Hardware

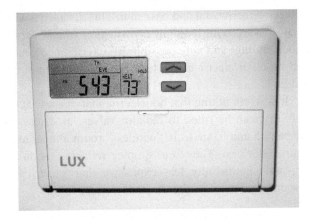

Figure 9-53 | Example of a programmable electronic thermostat that allows for multiple time/temperature schedules each day and different schedules on weekends versus weekdays. *Courtesy of John Siegenthaler.*

single-stage, nonprogrammable, heating-only thermostat suitable for use in hydronic heating applications.

Some electronic thermostats draw their very low-operating power from the 24-VAC transformer in the controlled device. This allows for both power supply and on/off control through a single pair of wires.

Other models operate on batteries that require periodic replacement.

Once an electronic thermostat is turned on, it opens and closes its relay contacts in an attempt to control the system as desired. The word *attempt* is important because no thermostat can force the inside air temperature to follow the programmed schedule. For example, imagine a situation in which a thermostat is programmed to begin a 10 °F temperature setback at 11 p.m. and resume normal comfort temperature at 5:30 a.m. the following morning. If the building has very low heat loss, and/or if the heating system has high thermal mass, the inside temperature may not drop more than 2 or 3 °F during the setback period, even with no heat input.

Wireless Thermostats

For decades, every electromechanical or electronic thermostat in a building required wiring back to the mechanical room. Although this wiring is typically low-voltage, light-gauge (24–18 AWG) cable, the time involved in routing this cable through a building, especially in a retrofit situations, can be considerable.

"True" wireless thermostats are now available that do not require wiring between the thermostat and remainder of the system. Unfortunately many thermostats that use Wi-Fi communication with devices such as a smartphone and tablet are referred to as "wireless" even though they require wires for power and control actions. This text will refer to the latter as Wi-Fi thermostats, rather than "wireless" thermostats.

True wireless thermostats are powered by batteries and generate radio frequency (RF) communication between themselves and a **base unit**. The base unit receives the RF signal from one or more thermostats and generates relay contact closures that are used to turn

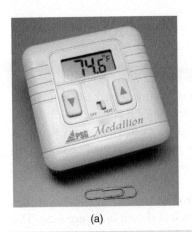
(a)

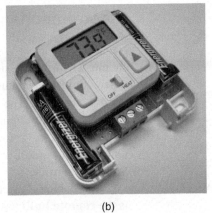
(b)

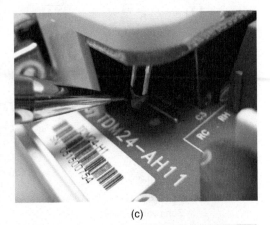
(c)

Figure 9-54 | (a) Example of a modern electronic thermostat. *Courtesy of John Siegenthaler* (b) Cover removed to show batteries and wiring terminals. (c) Close-up of thermistor temperature sensor.

on devices such as zone valves or zone circulators. The latter may require the use of external relays if the base unit is not rated to operate at line voltage.

During installation, each wireless thermostat in a multizone system must be "**paired**" with the base unit to determine which thermostat operates a given relay contact on the base unit.

Most current-generation base units can also be linked to an Internet router. Doing so allows authorized users to access the settings on each thermostat using an app on a smartphone or tablet. Figure 9-55 shows the concept of a four-zone system using true wireless thermostat to operate zone valves.

Although wireless thermostat systems do simplify installation in terms of wire routing, they bring other issues into consideration. Many are configured for multistage heating and cooling applications such as would be used when the thermostat is directly connected to an air-to-air heat pump. Some of this capability is not required in hydronic applications where the thermostat's function is limited to turning a zone valve or zone circulator on and off. Although a wireless thermostat system capable of multistage heating and cooling operation could still be used in a single-stage heating-only system, this unnecessary functionality adds cost.

Wireless thermostats also require batteries to generate radio frequency communication with the base unit. Although battery life will vary, some wireless temperature control systems list average battery life as 2 years. Some wireless thermostats are supplied with rechargeable lithium-ion batteries.

There's also limits on distance between the wireless thermostats and base unit. One manufacture lists 30 meters (about 98 feet) as the maximum separation between a thermostat and base unit. This distance is also likely to vary with building construction (metal versus wood framing) and with possible RF interference generated by other electrical or electronic devices within the building.

It is worth noting that nonelectric thermostatic actuators that can be fitted to radiator valves, as discussed Chapters 5 and 8, are truly "wireless" room temperature control devices. They are simple, reliable, modulating, and significantly less expensive than electrically powered true wireless thermostat systems.

Wi-Fi-Enabled Thermostats

There are now many offerings for thermostats that communicate with wireless routers using Wi-Fi protocol. Although many of these thermostats are referred to a "wireless," that term is literally not correct. Most Wi-Fi-enabled thermostats require wiring between the thermostat and a base unit. Thus, their wiring installation requirements are similar to other wired thermostats.

The advantage of Wi-Fi-enabled thermostats is the ability for authorized persons to access and change settings such as temperatures or setback schedules from a smartphone or tablet located anywhere that device can access the Internet. Many Wi-Fi-enabled thermostats can also report temperature trends or perceived errors, such as a room temperature that is more than a designated number of degrees below its current setpoint. The latter could be caused by some other malfunction in the system. Wi-Fi thermostats that can detect such issues usually have the ability to generate an alert e-mail or text message to a service provider. This functionality is especially beneficial for buildings that may be unoccupied for prolonged periods and subject to damage if the heating system fails to operate in cold weather.

Electromechanical Setpoint Controllers

In most hydronic systems, there are several locations where fluid temperatures are measured for safety and control purposes. The opening or closing of electrical contacts is often required when the measured temperature is above or below a predetermined temperature. One type of controller used for this purpose is called a setpoint controller.

One common electromechanical setpoint controller is the **remote bulb aquastat** shown in Figure 9-56. The temperature at some point in the system is sensed by a cylindrical copper bulb, which is attached to the controller by a **capillary tube**. This bulb contains a fluid that

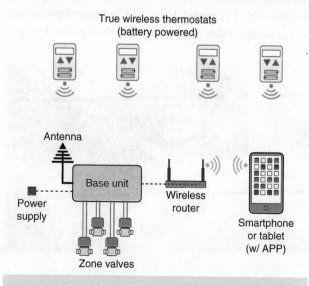

Figure 9-55 | Concept of a four-zone control system using wireless thermostats.

9.7 Basic Hydronic System Control Hardware

Figure 9-56 — A remote bulb aquastat. Note the capillary tube that connects the control's body and sensing bulb. *Courtesy of White Rogers.*

capillary tubing should be neatly coiled near the control and fastened to something solid using zip ties to prevent vibration. Care should be used in routing the capillary tube to prevent kinking or other damage. If the capillary tube is severed, the controller must be replaced.

Most aquastats can switch several amps of current at line voltage. Current and voltage ratings are typically listed on the cover plate. Aquastats can also be used in low-voltage systems.

A variation on the remote bulb aquastat is known as a **strap-on aquastat**. The difference is the location of the sensor bulb and length of the capillary tube. With a strap-on aquastat, the sensor bulb is located directly behind the metal enclosure. This allows the entire device to be strapped to a pipe using a worm-drive hose clamp.

In applications where the temperature of a surface is to be measured, the sensing bulb must be tightly secured against that surface using a clamp or tube strap. It should then be covered with insulation to minimize the effect of the surrounding air temperature.

The most accurate method of sensing a fluid temperature is to immerse the sensing bulb directly in the fluid. Some controls are available with sensing bulbs having MPT threads. These bulbs can be screwed directly into a fitting such as a tee. If the control has to be replaced, some spillage of system fluid is inevitable during removal and replacement.

Another option is to mount the sensing bulb into a **sensor well**. Such wells usually consist of a segment of copper tube with an internal diameter slightly larger than the outside diameter of the sensing bulb, closed at one end and open at the other. The open end is equipped with male pipe threads so that it can be screwed into a tapping on a tank or a tee. The closed end prevents fluid from entering the well. The sensing bulb should not be loosely inserted into the well. Instead, it should be coated with the heat-conducting paste before being inserted. This paste improves heat conduction between the bulb and the inside surface of the well, resulting in a faster response and improved accuracy. An example of a sensing well is shown in Figure 9-57.

Most remote bulb aquastats and strap-on aquastats are supplied with a SPDT switch. This allows them to make or break a circuit when the desired setpoint temperature is reached. In heating applications, aquastats are usually configured to open a set of electrical contacts to interrupt heat flow to the system when the measured temperature climbs to the setpoint value. The contacts close when the sensing bulb temperature drops through a differential below the setpoint. In most aquastats, this differential is adjustable.

increases in pressure when heated. The pressure exerted by the fluid's vapor pushes against a **bellows** assembly inside the aquastat, causing it to expand. This movement opens or closes the electrical contacts with the aquastat. The basic concept used in remote bulb aquastats is shown in Figure 9-56a.

Aquastats are available with different length capillary tubes, some up to 12 feet long, allowing the control body and electrical wiring to be located several feet from the location of the sensing bulb. The length of this capillary tube cannot be changed in the field because the working fluid would be lost. Any unused length of

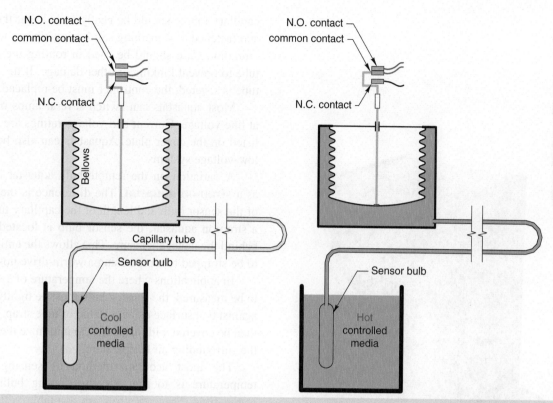

Figure 9-56a | Concept of how a remote bulb aquastat opens and closes electrical contacts in response to temperature changes at its sensor bulb.

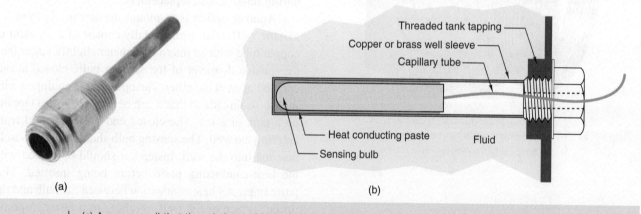

Figure 9-57 | (a) A sensor well that threads into a 3/4-inch FPT connection. *Courtesy of John Siegenthaler.* (b) Sensing bulb pushed into sensor well and surrounded with heat conducting paste.

Electronic (Single-Stage) Setpoint Controllers

An electronic setpoint control provides on/off contact closure output based on the temperature of its sensor and its programmed settings. Like its electromechanical counterpart, output is limited to an open or closed contact. However, several additional features make electronic setpoint controllers more versatile. These include wider ranges of adjustment for setpoint and differential, the ability to operate the relay on a rise or drop in temperature, and a digital readout for displaying the sensor temperature. Some electronic setpoint controls can be configured to work in either °F or °C, or to execute a programmable time delay function. Applications for these controllers include operating a heat source, circulator, diverting valve, or almost any type of electrically driven device at some setpoint temperature. An example of an electronic, single-stage setpoint control is shown in Figure 9-58.

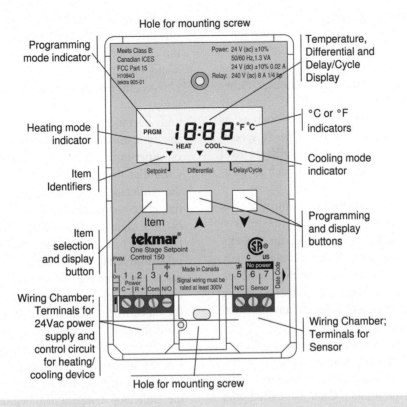

Figure 9-58 | Example of an electronic single-stage setpoint control. *Courtesy of tekmar Controls.*

Electronic setpoint controls also have some disadvantages relative to electromechanical controls. Because of the physical size of the relays used, most cannot switch electrical currents as high as those handled by electromechanical controls. In some cases, the relay contacts are only rated to operate with control voltages up to 24 VAC. If a line voltage device such as a circulator or blower is to be controlled, an additional relay will be required between the control and the device.

The solid-state components used in electronic controls, especially microprocessors, are also more susceptible to damage from voltage spikes. Although manufacturers have improved the ability of these controls to survive moderate voltage spikes, accidental connection of line voltage to low-voltage terminals or a high-voltage surge from a nearby lightning strike can still cause damage.

Electronic (Two-Stage) Setpoint Controls

Multistage control has already been discussed. The output of an electronic two-stage setpoint control consists of sequential operation of two switches or relay contacts. The first-stage contacts will always be the first to close and the last to open. The second-stage contacts close when and if necessary to deliver more heating capacity. A two-stage electronic setpoint control allows great flexibility in setting the cut-in temperature of each stage, the differential of each stage, and the interstage differential. Some two-stage controllers also allow for a user-set interstage time delay. This is the minimum time that must elapse between when the first-stage contacts close and when the second-stage contacts could close. Interstage time delays are typically used to allow time for the first-stage heat source to stabilize its operation, and "have its chance" at satisfying the control requirement, before requiring the heat source associated with the second-stage contact to start. This helps avoid short cycling the second-stage heat source.

The primary application for two-stage setpoint controls is to regulate two stages of heating (or cooling) to meet changing load conditions. *The heat sources operated by each stage do not have to be the same type or capacity.* The flexible setup allowed by electronic two-stage controls allows the operation of each stage to be matched to the operating characteristics of each controlled device.

Differential Temperature Controllers

There are situations in which a control action needs to be based on the *difference* between two temperatures, rather than the value of either temperature. This is especially true in systems using renewable energy heat sources, as well as systems that perform heat recovery. A special type of controller, appropriately named a **differential temperature controller**, is available for such applications.

A basic differential temperature controller has two temperature sensors. One is called the **source temperature sensor**, and the other is called the **storage temperature sensor**. The word "source" refers to any potential source of heat. Examples would include a solar collector, a wood-fired boiler, or a tank containing heated water. The word "storage" refers to any potential *destination* for the heat from the source.

Figure 9-59 shows an example of a modern differential temperature controller along with its wiring diagram.

All differential temperature controllers have two temperature differentials. One is called the **on-differential**, and the other the **off-differential**. The word "differential" refers to the difference in temperature between the source temperature sensor, and the storage temperature sensor.

A differential temperature controller closes its normally open contact whenever the temperature at the heat source sensor is greater than or equal to the temperature at the storage sensor, *plus the on-differential*. This logic can be represented as follows:

If $T_{source} \geq T_{storage} + \Delta T_{on}$, then relay contact closes.

Once the relay contact has closed, the controller continues to monitor the temperature at both sensors. If the temperature at the source sensor drop to, or below, the storage temperature *plus the off-differential*, the relay contact opens. This logic can be represented as follow:

If $T_{source} \leq T_{storage} + \Delta T_{off}$, then relay contact closes.

The on-differential and off-differential of most modern differential temperature controllers are adjustable over a wide range.

One of the most common applications for a differential temperature controller is operating the circulator that creates flow between an array of solar collectors and a thermal storage tank. Figure 9-60 shows a typical example.

In this application, the source sensor is mounted to the upper portion of the collector's absorber plate, or strapped to the outlet connection of the collector. The storage sensor is mounted within a **sensor well** that screws into the lower portion of the storage tank.

Figure 9-61 illustrates how the on-differential and off-differential relate to typical temperature changes at the collector and thermal storage tank during a typical sunny day.

Assuming clear sky conditions, the collector temperature sensor steadily warms during the morning, and eventually reaches a temperature that equals the temperature of the storage tank sensor plus the on-differential. At that point, the differential temperature controller turns on the collector circulator. Notice how the collector temperature dips sharply shortly after the collector circulator is turned on. This is caused by the rapid cooling effect of fluid moving through the collector. The setting for the on-differential and off-differential should be such that this dip doesn't cause the circulator to turn off. If several such "short cycles" are observed during a clear morning start up, the on-differential should be slightly decreased, and the off-differential slightly decreased.

Figure 9-61 shows a steady increase in thermal storage temperature during the mid-day period. This gain tends to level out as the afternoon continues, and the collector temperature begins to drop. When the collector temperature drops to within the off-differential of the differential temperature controller, the collector circulator is turned off. Collector temperature will typically rise slightly immediately after this occurs, since there is no fluid flow extracting heat from the absorber plate. Sometimes, this temporary temperature rise may cause the controller to restart the collector circulator. However, the "restart" cycle is likely to be very brief, and will contribute very little if any further energy transfer to storage. Collector temperature will drop rapidly in late afternoon, and the collector circulator will remain off until the next time the sun sufficiently heats the collector.

Another application in which differential temperature control is very useful is a system that combines a thermal storage tank heated by a renewable energy source with an auxiliary boiler. These systems present the following control objectives:

1. Supply heat to the load from the thermal storage tank whenever the temperature in the upper portion of the tank is high enough to do so.

2. Maintain building comfort by operating the auxiliary heat source whenever the temperature of the thermal storage tank is insufficient to meet the load.

3. Prevent heat generated by the auxiliary heat source from inadvertently entering the thermal storage tank.

9.7 Basic Hydronic System Control Hardware

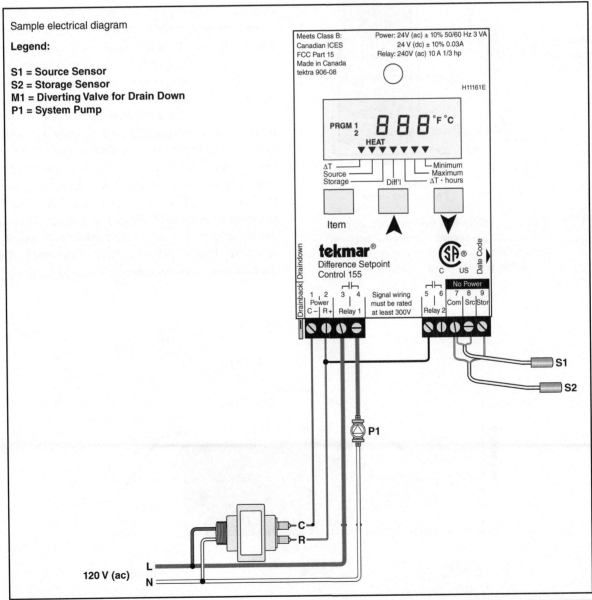

Figure 9-59 (a) Example of a differential temperature controller. (b) Typical wiring for this controller when operating a circulator.
Courtesy of tekmar Controls, Inc.

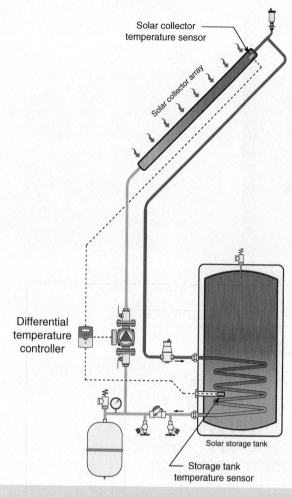

Figure 9-60 | Use of a differential temperature controller to operate a circulator in a circuit between a solar collector array and an internal heat exchanger in a thermal storage tank.

These control objectives can all be met by combining the functionality of a differential temperature controller with that of an outdoor rest controller. Figure 9-62 shows the piping concept, sensor placements, and controller wiring for this combination of controllers.

The differential temperature controller and the outdoor reset controller are powered on whenever the load is active. The differential temperature controller compares the temperature of sensor (S3) located on the upper tank header, to the temperature at sensor (S4) located on the return side of the heating distribution system. Whenever the temperature at sensor (S3) is higher than the temperature at sensor (S4) by some set value (e.g., the on-differential), circulator (P2) is allowed to operate. This transfers heat from thermal storage into the distribution system.

The temperature at sensor (S3) may eventually decrease due to depletion of heat from thermal storage, or lack of heat input from the renewable energy heat source. When the differential temperature controller "sees" that the temperature at sensor (S3) has dropped to within the temperature at sensor (S4), *plus the off-differential*, circulator (P2) is turned off.

These control actions allow the thermal storage tank, or the renewable energy heat source when it is operating, to contribute heat to the distribution system. They also prevent heat that has been added to the distribution system by the auxiliary heat source from being inadvertently transferred into thermal storage. The latter action is important because to allows the storage tank temperature to drop as low as possible while still contributing heat to the load. Circulator (P2) is only turned off when the tank can no longer contribute heat to the load.

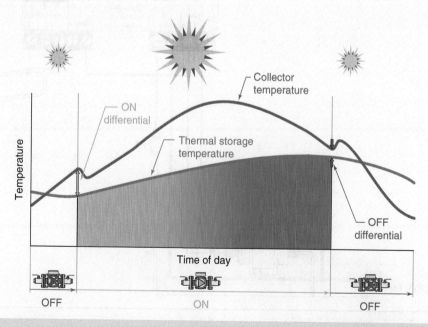

Figure 9-61 | Typical relationship between collector and storage temperature, as well as on-differential and off-differential of a differential temperature controller during a clear sky day.

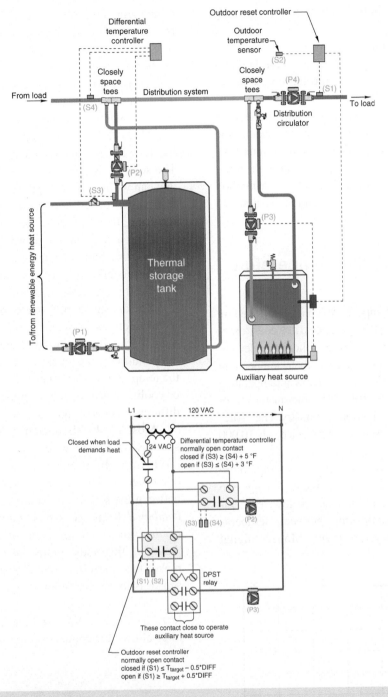

Figure 9-62 | Combined use of a differential temperature controller and outdoor reset controller to regulate heat transfer from a thermal storage tank and auxiliary heat source into a space-heating distribution system.

The controls actions of the differential temperature controller in this application can be represented by **"if/then" statements**. Assuming that the differential temperature controller (DTC) is set for an on-differential of 5 °F, and an off-differential of 3 °F, the following if/then statements apply.

If $(S3) \geq (S4) + 5°F$, then relay contact in DTC is closed.

If $(S3) \leq (S4) + 3°F$, then relay contact in DTC is open.

The off-differential of 3 °F allows the tank temperature to drop very close to the temperature of water returning from the distribution system before being taken offline. This off-differential also compensates for slight differences in sensor accuracy or variations in sensed temperature based on differences in how the two temperature sensors are mounted. To minimize the latter, the two temperatures sensors (S3) and (S4) should be mounted in the same manner (e.g., both sensors mounted in the same type of sensor wells, or both strapped to piping using identical methods).

The outdoor reset control allows the supply water temperature to drop to the lowest value where building comfort can be maintained before turning on the auxiliary heat source.

The two sets of closely spaced tees that connect the thermal storage tank and auxiliary heat sources to the distribution system allow the distribution system flow to pass through, without inducing flow through either heat source when it is not operating. This is a method for providing **hydraulic separation** and is discussed in detail in later chapters.

Sensors for Electronic Temperature Controllers

Most of the electronic temperature controllers used in hydronic systems use a **negative temperature coefficient (NTC) thermistor sensor**. These solid-state devices decrease their electrical resistance as their temperature increases, and vice versa. Figure 9-63 show a temperature versus resistance curve for a "10K" NTC thermistor sensor that's commonly used in the hydronics industry. The "10K" designation indicates the sensor's resistance of 10,000 ohms at 25 °C (77 °F).

The relationship between the sensor's temperature and its electrical resistance is not linear. Changes in resistance are much more pronounced at low temperatures. Even so, this relationship between temperature and resistance is *very repeatable*. Modern digital temperature controllers can easily translate the electrical resistance of the sensor *circuit* into an inferred temperature value based on a known and repeatable curve such as shown in Figure 9-63.

Thermistor temperature sensors contain a very small solid-state device called a **thermistor bead**, which have very fine wires attached, as seen in Figure 9-64.

These fine wires don't lend themselves to the handling and placements required in HVAC applications. The solution is to encapsulate the bead in a metal shell that can withstand more physical stress and connect it to lead wires rugged enough for installers. During manufacturing, the lead wires are soldered to the thermistor bead (or a tiny circuit board holding the bead). This assembly is then inserted into the metal shell, and a potting material is poured in to permanently seal it in place and protect it against moisture.

Sensor shells are made of a highly conductive metal such as copper, brass, or stainless steel. The high conductivity minimizes the difference between the temperature of the surface the sensor is mounted to and the temperature at the thermistor bead. For most heating or cooling system control applications, this temperature difference is very small and of no concern.

Most of the thermistor sensors used in hydronic systems have lead wires that are not long enough to reach from the installed location of the sensor to the controller. A cable must be used to complete the circuit between the sensor and controller The sensor and cable *combined* form the sensor circuit, and *the resistance of the sensor circuit is what the controller "feels."* Anything that adds significant resistance to this circuit makes the controller "think" that the temperature at the sensor is lower than it actually is.

Designers should verify the minimum wire size (e.g., the *highest* AWG #) that the controller manufacturer allows for connecting the sensor to the controller.

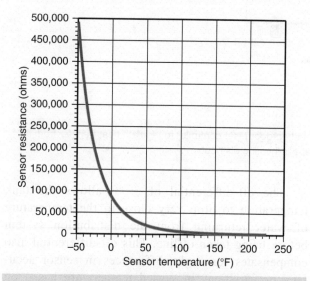

Figure 9-63 | Relationship between electrical resistance and temperature for a 10K negative temperature coefficient thermistor sensor.

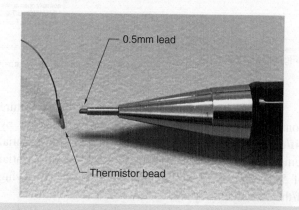

Figure 9-64 | A very small thermistor bead with attached wires. *Courtesy of John Siegenthaler*

Also verify the maximum allowed sensor cable length. A suggested minimum wire size is 18 AWG copper wire for 10K thermistor sensors. That wiring has a resistance of 0.0064 ohms per foot. If a thermistor sensor was located 250 feet from the controller, it would require about 500 feet of sensor wire. The total resistance of that wire would be about 3.2 ohms. This would add to the resistance of the sensor itself. For a 10K thermistor at 77 °F, the cable would only represent 0.03% of the circuit resistance. It would have no significant effect on the operation of a hydronic system.

Smaller diameter wires (with higher AWG #s) have higher resistances. For example, the resistance of 24 AWG copper wire is about 4 times more than 18 AWG copper. It's also more fragile.

An improper electrical bond between the sensor lead wires and the sensor cable can add significant electrical resistance to the circuit. One situation where this becomes a problem is where sensor lead wires are connected to cables using standard wire nuts, and then exposed to moisture. The moisture can collect within the wire nut and cause corrosion at the junction. Over time, this increases the resistance of the sensor circuit, and "fools" the controller into thinking the sensor temperature is lower than it actually is.

The best techniques to avoid this problem are to (1) make connections between sensor leads and cable at locations that are not exposed to moisture, and (2) make the connection using **gel-fired compression connectors**, such as those used for exterior telephone wiring connections. An example of such a connector is shown in Figure 9-65.

The unstripped ends of both wires to be bonded are inserted into the connector. The "button" on connector is then pressed down using a pliers or special compression tool. This causes a small strip of metal within the connector to pierce through the insulation on each wire, creating an electrical bond between them. The gel inside the connector surrounds this bond to provide a waterproof seal.

Sensor circuits can also be adversely affected by strong electromagnetic fields. Such fields can induce currents in sensor wiring. This can cause the controller to inaccurately interpret the temperature input. To reduce the chances of such interference, *never run sensor cable next to or in the same conduit as AC wiring*. If sensors circuit must be installed near devices such as large motors, fluorescent light ballasts, or large transformers, use cabling with **twisted pair wires**, or **shielded cable**. The latter has a metal foil layer between the exterior insulation and internal conductors. Only one end of this metal foil should be fastened to an electrical ground. The other end of the foil should be electrically isolated, and not be fastened to anything.

Temperature Sensor Mounting

Sensors are the "eyes" of temperature controllers. Those controllers can only react to what they "see." Careful sensor placement and mounting is critical in achieving accurate system control.

Some thermistor sensors have a "saddle" groove in their capsule. This allows the sensor capsule to be strapped to the rounded surface of a metal pipe suing a high-temperature-rated nylon zip tie. The mounted sensor should then be covered with a sleeve of insulation to minimize the influence of surrounding air temperature on the sensor. These details are shown in Figure 9-66.

Thermistor sensors can also be mounted into wells that have an inside diameter not more the 0.08 inches larger than the diameter of the sensor capsule. Figure 9-67 shows an example of a well (Honeywell 12137B) that is suitable to house a thermistor sensor with a capsule diameter of 0.375 inches.

These wells are available with MPT threads in 1/2-inch and 3/4-inch sizes. They are well suited to situations where the temperature of water within a tank needs to be measured. The well is threaded into a matching size tapping on the tank. If the tank tapping where the well is to be located is larger than 3/4-inch FPT, a bushing can be used to reduce to the size of the well.

Prior to inserting the sensor, a small amount of thermal paste should be injected into the end of the well using a plastic syringe with a long tip or extension tube. The sensor body should also get coated with thermal paste. The objective is to completely fill the gap between the sensor body and inside surface of the well with thermal paste. Doing so improves conductivity, and reduces the time lag between changes in water temperature, and when that change is detected by the sensor.

Figure 9-65 | Examples of gel-filled compassion connectors for low-voltage wiring. *Courtesy of John Siegenthaler.*

It's also possible to make a sensor well using copper tubing closed at the end with a soldered cap. The length of the well is whatever you need. This is especially helpful when a sensor needs to be placed at a specific height within a tank that has a tapped connection on its upper surface. Figure 9-68 shows how a well can be constructed that is suitable for a sensor having a diameter of 0.375 inches.

In this case, the sensor tube is made of 3/8-inch type M copper tube, which has an inside diameter of 0.450 inches. A thermistor sensor with a diameter of 0.375 inch fits into this tubing, leaving about 0.04-inch gap between the sensor capsule and the inside of the tube. This gap should be filled with thermal paste. The 3/8-inch copper tube can be sealed to a brass socket of the 3/4-inch dielectric union using a 3/4-inch ftg × 3/8-inch C reducer coupling. Use of the dielectric union prevents galvanic corrosion between the steel tank wall and the copper tube. It also allows the well to be easily removed from the tank if necessary.

It's important to provide strain relief for the wires connecting the sensor leads to the cable. A temperature-rated zip tie can be used to hold the sensor cable to the exterior portion of the well, or other nearby solid surface.

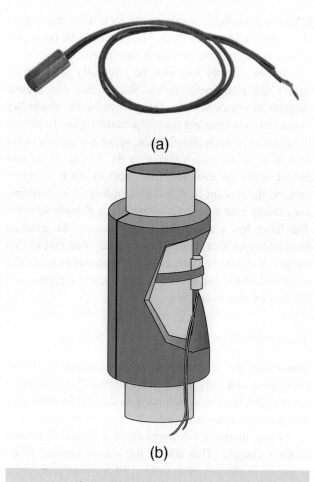

Figure 9-66 — (a) 10K thermistor sensor with concave groove for mounting to surface of a pipe. *Courtesy of tekmar Controls, Inc.* (b) Mounting of the 10K thermistor sensor using a temperature-rate zip tie and surrounding insulation sleeve. *Courtesy of tekmar Control Systems.*

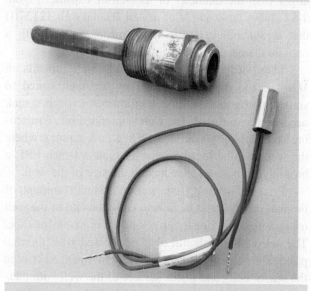

Figure 9-67 — Honeywell sensor well and 10K thermistor sensor that fits this well. *Courtesy of John Siegenthaler.*

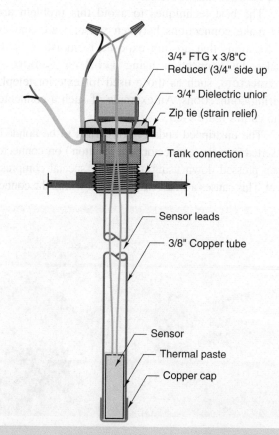

Figure 9-68 — Components used to build a sensor well suitable for attachment to a thermal storage tank with FPT connection.

It's also good practice to tag every temperature sensor in the system. Use the same designations that appear on the piping and electrical schematics for the system.

Zone Valves

Zone valves are commonly used to allow or prevent flow through two or more independently controlled zone piping circuits. Several types of zone valves are available. The most common types were discussed in Chapter 5, Pipings, Fittings, and Valves. They all consist of a valve body combined with an **actuator**. The actuator is the device that produces movement of the valve's shaft when an electrical voltage is applied.

Some actuators use small electric motors combined with gears to produce a rotary motion of the valve shaft. These actuators can fully open or close the valve in about 3 seconds. This is fast enough for rapid heat delivery, yet slow enough to prevent **water hammer** as the valve closes. Other zone valves use **heat motor actuators** that produce a linear motion when heated by an internal resistor. This type of actuator can take 2 to 3 minutes from when the operating voltage is applied to when the valve is fully open.

Although zone valves are available with several different actuator voltages, the vast majority of zone valves used in hydronic heating applications use low-voltage (24 VAC) actuators. In many systems, this allows the zone valves to be powered by the same transformer that supplies control voltage to the thermostats and perhaps other low-voltage control hardware. Figure 9-69 shows the basic low-voltage circuit used to control a zone valve using a thermostat.

The electrical power required to operate a zone valve is expressed in **VAs (volt-amps)** and is usually listed on the zone valve or its data sheet. *Care must be taken that the VA rating of the transformer supplying several zone valves is slightly greater than the total VAs drawn when all zone valves supplied by that transformer could be operating at the same time.* If this is not done, the transformer will be overloaded and eventually burn out. A rule of thumb is to add the VA rating of all zone valves and other devices to be powered, and then select a transformer that can supply at least 10 VAs more than that total.

One common approach is to source the 24-VAC power required by the zone valves from the transformer in the heat source. Before doing this, be sure that the transformer can provide the total VAs required by the low-voltage devices in the heat source as well as the VAs required by the zone valves. Manufacturers usually list the maximum **external VA rating** the heat source transformer can safely supply.

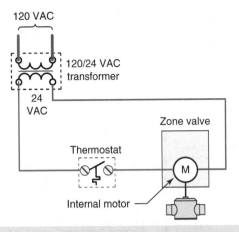

Figure 9-69 | Basic 24-VAC control circuit for a zone valve operated by a thermostat.

Many zone valve actuators are equipped with internal switches that can be used to signal other components in the system that heat is being called for by a given zone. These switches are called **end switches** because their contacts close when the actuator reaches the end of its travel and the valve is fully open. End switch contacts open as the valve starts to close.

The most common types of zone valve actuators have two, three, or four terminals or wires. Two-wire zone valves have no end switch. The difference between three-wire and four-wire zone valves is in how the end switch is wired. In a three-wire zone valve, one contact of the switch is wired in parallel with the zone valve motor. In a four-wire zone valve, the end switch is isolated (not connected to any other internal wiring). This difference determines how the zone valve interfaces with the rest of the control system. A simplified representation of two-wire, three-wire, and four-wire zone valves is shown in Figure 9-70.

When a three-wire zone valve is used, the signal passed through the end switch is supplied by the same voltage source that operates the actuator motor. This

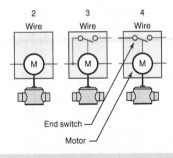

Figure 9-70 | Simplified representation of two-wire, three-wire, and four-wire zone valves. Note differences in how the end switches are wired.

voltage is provided by a 24-VAC transformer. A schematic showing the wiring of three-wire zone valves and a typical high-limit/switching relay control for an oil-fired boiler is shown in Figure 9-71.

The operating sequence of this control system is as follows: When any of the four room thermostats call for heat, the motor of the associated zone valve is energized, opening the valve. When the motor reaches the end of its travel (e.g., the valve is fully open), the end switch closes. This completes a circuit across the "T T" terminals of the high-limit/switching relay controller. This controller then starts the circulator and burner and operates as described in Section 9.7.

When a zone thermostat is satisfied, it opens the circuit to its associated zone valve. The zone valve closes, and its end switch opens. If this is the last zone valve to close, the circuit across the "T T" terminals is opened turning off the burner and circulator. If other zone valves remain open, the burner and circulator remain enabled. Burner and circulator operation can be initiated by any of the zone valves and will continue as long as any of the zone thermostats are calling for heat.

If the external VA rating of the transformer in the high-limit/switching relay controller allows, additional zone valves could be added by simply duplicating the wiring used for any of the zone valves shown. If the external VA rating of the heat source's transformer is not high enough to supply the total VAs required by all zone valves, an external transformer and isolation relay must be used. This relieves the controller's transformer from supplying the VA requirement of the zone valves. The typical wiring for this approach in combination with three-wire zone valves and a gas-fired boiler is shown in Figure 9-72. The operating sequence is similar to that just described with the exception that the high-limit/switching relay control is activated by the contact of the isolation relay, rather than the end switch of a zone valve. This isolation relay is a separate component not supplied with the high-limit/switching relay control.

When four-wire zone valves are used, the isolation relay can be eliminated. In basic systems, the isolated end switches of each zone valve are wired together in parallel and connected across the "T T" terminals of the boiler high-limit/switching relay controller as shown in Figure 9-73. This allows the circulator and boiler to operate when any one or more of the zone valves is on. In more complex systems, the isolated end switches may be used to signal a heat demand to other devices such as mixing controllers or outdoor reset controllers.

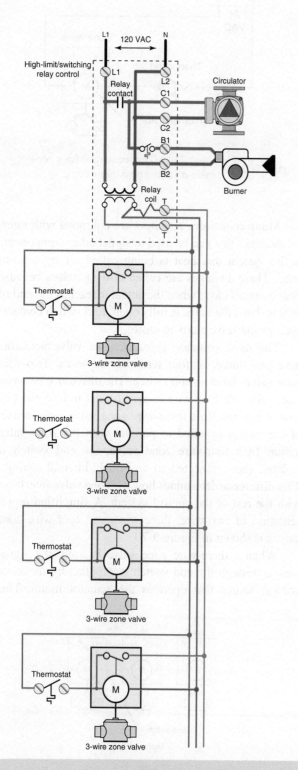

Figure 9-71 Typical wiring of three-wire zone valves in combination with a boiler high-limit/switching relay controller.

Multizone Relay Centers

A **multizone relay center** is designed to organize the wiring and control operation of several independent zone circulators or zone valves. An example of a typical

9.7 Basic Hydronic System Control Hardware 425

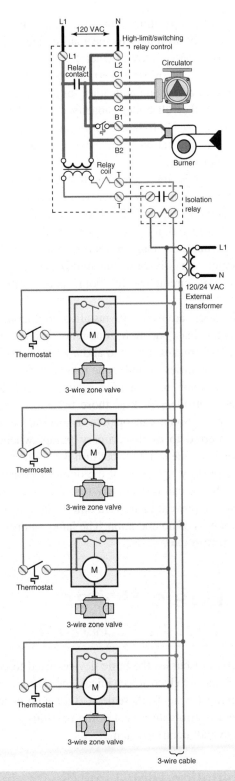

Figure 9-72 Three-wire zone valves powered by an external transformer. Isolation relay is powered up when any one or more of the zone valves are on. It provides heat demand signal to the boiler high-limit/switching relay controller.

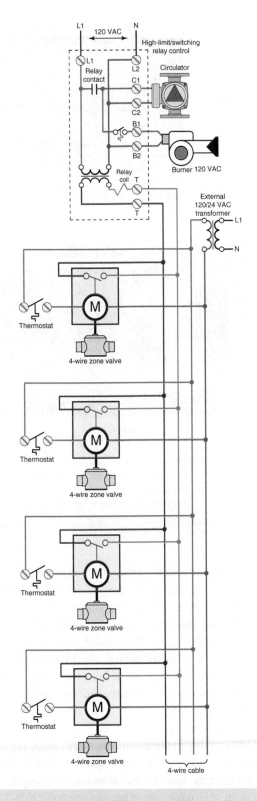

Figure 9-73 Typical wiring of four-wire zone valves in combination with a boiler high-limit/switching relay controller.

multizone relay center for a system using zone circulators is shown in Figure 9-74a. A similar controller for a system using zone valves is shown in Figure 9-74b.

Multizone relay centers are available with three to six zones. In many cases, two or more multizone relay centers can also be "daisy chained" together if needed to accommodate additional zones.

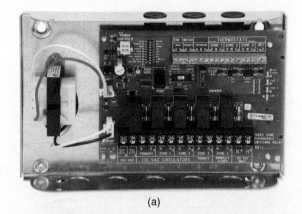

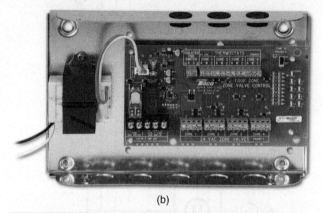

Figure 9-74 (a) Internal view of a multizone relay center designed to operate up to four-zone circulators. (b) Multizone relay center designed to operate up to four zone valves. *Courtesy of Taco.*

A multizone relay center has terminals for each room thermostat along the top of its printed circuit board. The 24-VAC power for the thermostat circuits is generated by the transformer within the relay center. When a particular zone thermostat calls for heat, an associated zone relay is turned on. In systems using zone circulators, this relay provides a line voltage output to drive the associated zone circulator. In systems using zone valves, a 24-VAC output is turned on to operate the associated zone valve.

Another relay in the multizone relay center provides an isolated contact closure that can be used to signal other equipment that a heat demand is present. The contacts in this relay are often designated "X X" and close whenever any of the zones are on. In most systems, this contact is used to turn on the heat source anytime one or more of the zone thermostats call for heat. It can also be wired to turn on a mixing control, outdoor reset control, or some other components in the system.

One of the zones in most multizone relay centers can be configured as a priority zone. When so configured, all other zones are temporarily turned off whenever the priority zone is operating. This allows the full output of the boiler to be available to the load connected to the priority zone. Most current-generation multizone relay centers also have a priority override feature.

The most common application for a priority zone is DHW. Since the full boiler capacity can be directed to DHW, hot water is produced very quickly. This is very desirable in applications that have large demands for domestic hot water.

A schematic representation that shows the basic wiring of a six-zone relay center with DHW as a priority load is shown in Figure 9-75. This schematic is a "bare bones" representation and does not include some of the options available in many current-generation multizone relay centers. In this case, the multizone relay center controls zone circulators for both space heating and DHW.

It is often advisable to install a multizone relay center with at least one more zone relay than is currently required. This makes it easy to add a zone in the future.

The capabilities of multizone relay centers continue to expand. Some now include provisions for activating other controllers such as those used for mixing or outdoor reset. Some can operate a specific circulator, usually referred to as the primary pump, whenever a space-heating zone is active, but turn it off when there is a call for priority operation of the domestic water-heating zone. Some can periodically turn on circulators for short times to prevent them from otherwise ceasing after several months of no operation. The latter control action is called **pump exercising**.

9.8 Basic Boiler Control Hardware

This section discusses the basic devices used to control boilers. Some are electromechanical, and others are electronic. In most systems, these devices are combined with the system controls discussed in the previous section to form an overall control system.

Combination High-Limit/Switching Relay Control

Because many simple residential hydronic systems use a single boiler, single circulator, and single room thermostat, they have virtually identical control requirements. The similarity of these systems allows manufacturers to design a specialized control that can coordinate the operation of all these components. Manufacturers of packaged boilers often equip their boilers with such a

9.8 Basic Boiler Control Hardware   427

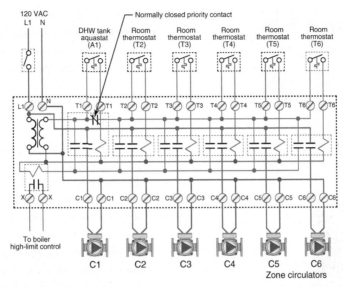

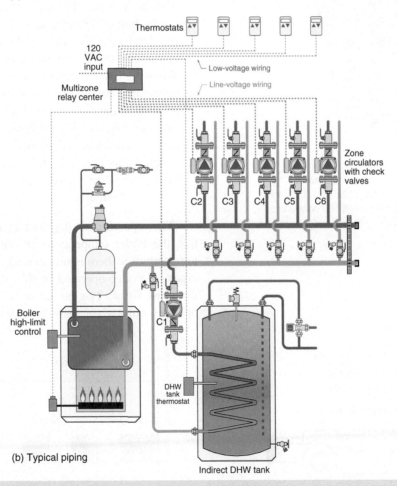

Figure 9-75 | (a) Schematic representation of a six-zone multizone relay center for zone circulators. (b) Typical piping associated with this type of control system.

control, and usually prewire it to both the burner and circulator. An example of a **high-limit/switching relay controller** for an oil-fired boiler is shown in Figure 9-76. Its circuitry is shown in Figure 9-77.

This controller incorporates several components such as a transformer, a relay, and an aquastat into a small case that mounts directly onto the boiler. These components are shown inside the dashed lines in

428 Chapter 9 Control Strategies, Components, and Systems

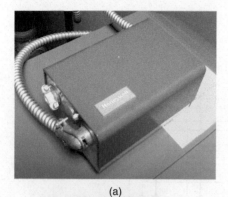

(a)

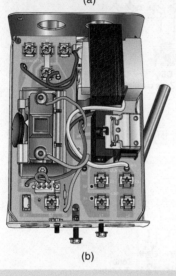

(b)

Figure 9-76 | Boiler high-limit/switching relay controller. (a) Mounted on top of boiler with sensor inserted into well under the controller. (b) Internal view, with sensor seen in rear. *Courtesy of Honeywell.*

Figure 9-77. The temperature-sensing bulb projects from the rear of the case and into a well in the boiler block.

A typical residential system with this control has the following operating sequence:

1. Whenever the system's master switch is closed, line voltage is supplied to the control.
2. When the room thermostat contacts close, a low-voltage (24 VAC) circuit energizes the coil of the internal relay.
3. The relay contacts close to supply line voltage to the system circulator. If the boiler is cool, the internal aquastat contacts are also closed, and the burner operates. If the boiler temperature is at or close to the high-limit setting of the control, the aquastat contacts are open and the burner remains off.
4. *If and when* the boiler temperature climbs to the high-limit setting, the aquastat contacts open to turn off the burner. The circulator, however, continues to run. Such operation is typical when the system load is less than the heating capacity of the boiler.
5. When the boiler temperature decreases to a certain differential below the high-limit setting (typically about 8 °F), and the demand for heat continues (as evidenced by the closed thermostat contact across the T T terminals), the burner is turned back on.
6. When the room thermostat contacts open, the circuit through the relay coil is interrupted, and both the burner and circulator are turned off.

This operation is depicted in Figure 9-78. Notice that the boiler temperature eventually drops to room temperature if there is no demand for heat over several hours. A boiler operated in this manner is said to be **demand fired**. The burner of a demand-fired boiler operates only when there is a demand for heat from

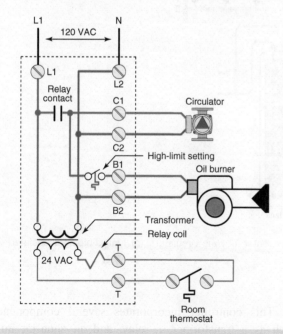

Figure 9-77 | Internal wiring and external connections for typical high-limit/switching relay controller.

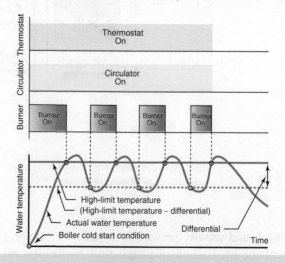

Figure 9-78 | Operation of a combination high-limit/switching relay control over a period of time.

the system. The boiler does not maintain a minimum temperature at all times. Most modern boilers are operated in this manner.

Triple Action Controller

For many years, it was common to produce domestic hot water using a **tankless coil** mounted in the same boiler that provided space heating. Cold water was heated as it passed through this copper coil on its way to a faucet. Because hot water could be needed at any time, boilers with tankless coils had to maintain a minimum water temperature at all times.

Due to constant standby heat loss from the boiler, this method of domestic hot water heating is considered inefficient by today's standards. However, many systems that use tankless coils are still operating, and individuals who might service such systems should be familiar with their operation.

Because this type of system was once common, several manufacturers developed controllers specifically for it. This type of control is called a **triple action controller**. An example of such a controller is shown in Figure 9-79. Its associated wiring is shown in Figure 9-80.

The triple action controller mounts directly to the boiler. It has a single temperature probe that inserts into a well in the boiler block. The control has two dials for temperature adjustments; one sets the boiler's high-limit temperature and the other sets the boiler's low-limit temperature. The latter is the minimum temperature the boiler will maintain regardless of load.

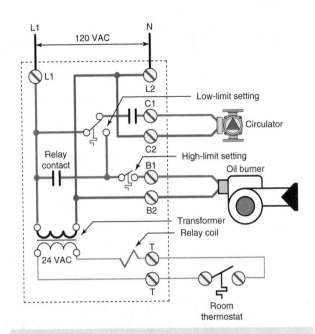

Figure 9-80 | Internal wiring of a triple action controller.

The triple action controller fires the burner if the room thermostat calls for heat, or if the boiler temperature falls below the low-limit setting. If the load on the boiler exceeds the boiler's heating capacity, boiler temperature may drop below the low-limit setting. When this occurs, the controller temporarily stops the space-heating circulator. The premise is that by shedding the space-heating load, the boiler temperature will increase, and thus be able to maintain sufficient flow of domestic hot water. The space-heating circulator restarts when the boiler temperature recovers its low-limit setting plus differential. A sequence showing control operation is given in Figure 9-81.

Manual Reset High-Limit Control

Many mechanical codes, particularly those applicable to public buildings, require a redundant temperature-limiting controller on all boilers. This controller must be capable of interrupting boiler heat production should the primary operating controller fail in the on position. Most codes also require that the redundant high-limit controller cannot automatically close its contacts once they have opened. The controller designed for this purpose is called a **manual reset high-limit (MRHL)**. An example of such a controller is shown in Figure 9-82.

The manual reset high-limit control should be mounted with its probe sensing the highest water temperature in the system. In some cases, the temperature sensor can be mounted directly into the boiler block. In other cases, the sensor mounts into a tee near the

Figure 9-79 | Example of a triple action controller. Note individual settings for high-limit and low-limit temperatures. *Courtesy of Honeywell.*

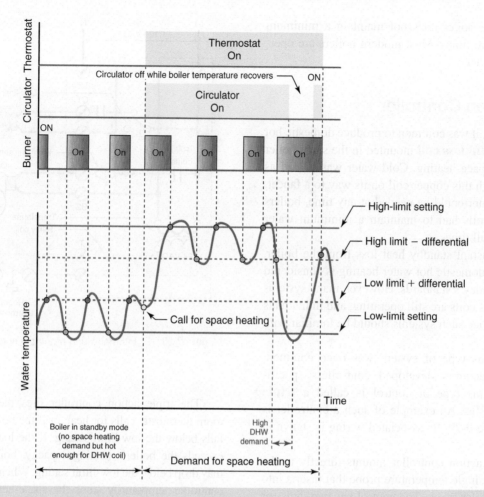

Figure 9-81 | Operation of triple action controller over a period of time.

outlet of the boiler as shown in Figure 9-82a. Typically, the MRHL is set approximately 20 °F above the primary high-limit control setting to prevent nuisance tripping due to boiler temperature overshoot. However, most mechanical codes that require this device also stipulate its maximum temperature setting, which should never be exceeded.

Low Water Cutoff Control

Another safety device required on boilers by many codes is a low water cutoff (LWCO). This device uses the conductivity of water to complete a circuit. If this circuit is broken due to the water level dropping below the probe, the normally closed contacts inside the LWCO open to interrupt the burner circuit. An example of a LWCO control is shown in Figure 9-83.

LWCOs are available for both line voltage and low-voltage (24 VAC) circuits. Line voltage controls are commonly used with oil burner circuits, while low-voltage controls are commonly used with low-voltage gas valve

(a)

(b)

Figure 9-82 | MRHL controller. (a) Mounted to sensor well on boiler outlet piping. (b) Internal construction. *Courtesy of Honeywell.*

9.8 Basic Boiler Control Hardware

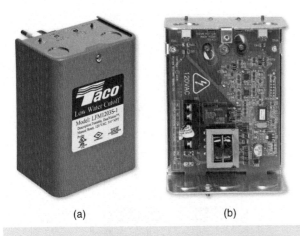

Figure 9-83 — Example of a low water cut off controller. (a) External view with probe visible in back. (b) Internal construction. *Courtesy of Taco.*

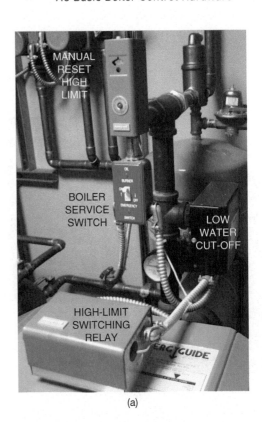

circuits. All LWCOs require constant power to operate their internal circuitry. Most also include time delay logic to prevent a nuisance trip due to an air bubble passing over the probe. Always verify any specific requirements for LWCO devices with local mechanical codes.

Figure 9-84 shows the piping near the outlet of a boiler. The boiler's high-limit/switching relay controller can be seen atop the boiler. Also visible is the MRHL controller near the top of the image and the LWCO (black box at right of image). The boiler's service switch with red faceplate is also visible. These operating and safety devices are wired in series as shown in Figure 9-84b. This configuration allows any of the controllers to stop the burner if an abnormal condition occurs.

Heat Source Reset Controllers

The concept of heat source reset control was discussed in detail in Section 9.4. From a hardware standpoint, a modern heat source reset controller is an electronic device with an on/off relay contact that turns the heat source on and off. This on/off action attempts to maintain the outlet temperature of the heat source close to a target temperature that varies according to outdoor temperature. The colder it is outside, the higher the heat source outlet target temperature. Although they are often called "boiler" reset controllers because they are most frequently used to control boilers, they can also be used with other types of heat sources such as air-to-water or water-to-water heat pumps.

Examples of modern heat source reset controllers are shown in Figure 9-85.

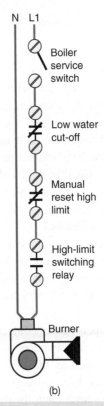

Figure 9-84 — (a) Piping near boiler outlet showing boiler's high-limit/switching relay switch relay, MRHL safety controller, LWCO safety controller, and boiler service switch. *Courtesy of John Siegenthaler.* (b) Series wiring of contact in these devices.

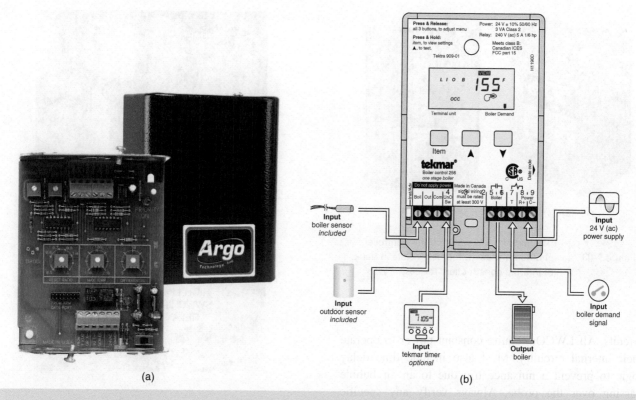

Figure 9-85 Examples of heat source reset controllers (a) using analog settings. *Courtesy of Argo*, (b) using menu-driven LCD settings. *Courtesy of tekmar.*

Because a heat source reset controller generates an on/off output, it must operate with a differential. On some controllers, the installer sets the differential, and it remains fixed at that setting regardless of outdoor temperature. An example of such operation is shown in Figure 9-25. Some heat sources reset controllers can operate with a variable differential that widens as the outdoor temperature increases as shown in Figure 9-26. The latter helps prevent short cycling the heat source under partial load conditions.

A heat source reset controller is an *operating device*, not a safety device. It does not replace the need for a high-limit controller, or for other safety controls such as an MRHL or LWCO that may be required by local mechanical codes.

The normally open output contact of a heat source reset control is often wired in series with other operating and safety controls such as the high-limit controls. In this configuration, the heat source reset control simply "preempts" the other limit controls during normal operation. Its contact opens the circuit that keeps the heat source operating at a temperature that is lower than the setting of the limit controls. If, for some reason, the heat source reset controller should fail to open its contacts, the other high-limit controls will eventually turn off the heat source. When used in this way, it is important that the settings of any other limit controls be sufficiently high to avoid interfering with the operation of the heat source reset control.

In most systems, the heat source reset controller is electrically bypassed during DHW so that the heat source can supply high-temperature water to the heat exchanger of an indirect water heater. In this mode, the boiler's high-limit controller is what limits water temperature. When the DHW cycle is finished, water temperature control is passed back to the heat source reset controller. Some heat source reset controllers have the necessary bypass logic for the DHW mode built in.

When used with a conventional boiler, the boiler reset control must be set for a minimum supply water temperature high enough to prevent sustained flue gas condensation within the boiler during partial load conditions. This type of operation is called partial reset control and is illustrated in Figure 9-27. Most modern boiler reset controls have an adjustable minimum boiler temperature that is set using a dial or through menu-driven programming. The controller will not target supply temperatures below this setting regardless of the outside temperature. The minimum operating temperature varies with the type of boiler and the type of fuel. For most gas-fired and oil-fired boilers, the minimum supply temperature should not be set below 140 °F unless otherwise allowed by the boiler manufacturer.

Multiple Heat Source Controllers (For On/Off Heat Sources)

The benefits of a multiple boiler system have been discussed earlier in this chapter, as well as in Chapter 3, Hydronic Heat Sources. The hardware necessary for controlling a multiple boiler system will now be covered. As with other controllers discussed in this chapter, the word "boiler" is frequently used to describe the controller since they are most often used with boilers. However, nearly all of these controllers can be used to operate any type of on/off heat source, and in some cases multiple *modulating* heat sources. In some systems, they can also control a combination of different heat sources such as boilers, heat pumps, or even a circulator supplied from a thermal storage tank.

The state-of-the-art in multiple heat source controllers is a microprocessor-based device that performs several functions to optimize the operation of two to eight heat sources depending on the heating load. Figure 9-86 shows an example of a multiple boiler controller that can operate up to four on/off boilers.

While operating, multiple heat source controllers attempt to keep the water temperature at its supply sensor as close to the target temperature value as possible. The target temperature depends on the type of **heat demand** the controller is receiving at the time. This demand is usually a low-voltage signal sent to the controller when a circuit is completed through another device such as a thermostat. This voltage is sent to a specific set of terminals on the multiple boiler controller. It can be thought of as a "wake up" call, signaling that some load in the system needs heat from or more of the heat sources.

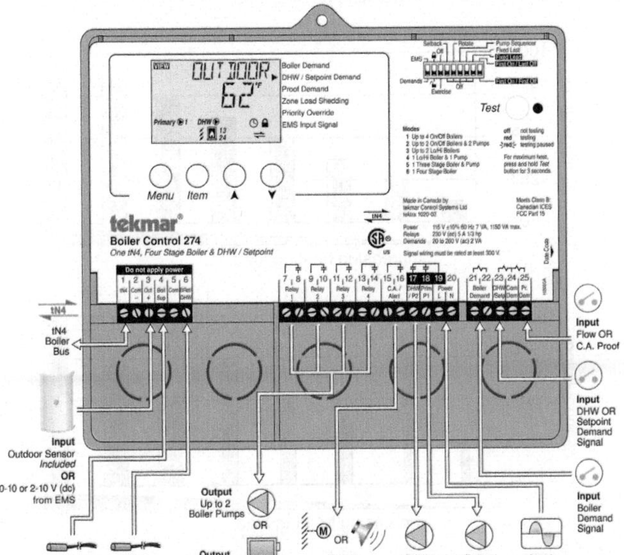

One type of heat demand is known as a **setpoint demand**. When it occurs, the controller targets a *fixed supply water temperature* that has been previously programmed into the controller by the installer. The controller does not adjust this target temperature based on outdoor temperature. Setpoint demands are commonly invoked when the load is supplied through a heat exchanger. To achieve high heat transfer rates, some heat exchangers are sized around relatively high supply water temperatures in the range of 180 to 200 °F. One example of such a load is an indirect water heater. Another is a heat exchanger used for supplying heat to a snow-melting system or a swimming pool.

The other type of demand is a **space heat demand**, which is invoked when space heating is required. Under this demand, the target temperature is calculated based on outdoor temperature and the reset ratio set on the controller. In this mode, the multiple heat source controller operates as an outdoor reset controller.

When either type of demand is present, the controller monitors the error between the calculated target temperature and the temperature measured at its supply sensor. This error, combined with PID processing, determines the operating sequence of the heat sources. After the first heat source is turned on, the controller monitors the duration of the error as well as the rate of change of the error to determine if additional heat sources need to be turned on. If the error is small and being quickly reduced by operation of the first heat source, additional heat sources will not be turned on. If the error is large, or slow to change, one or more additional heat sources will be turned on relatively quickly.

Most multiple heat source controllers can also be configured to rotate the order in which the heat sources are turned on. This is called **heat source rotation**. The goal is to allocate approximately the same total running time to each heat source so that they can be serviced and ultimately replaced at the same time.

In some systems, a specific heat source needs to be turned on first. This heat source is called the **fixed lead** heat source. In some systems, it may be a condensing boiler that operates at very high efficiency when the target supply temperature is low. In other systems, it may be a boiler that needs to fire first to establish the proper draft in the chimney and breaching system. Whatever the case, a switch or program setting on the controller allows a given heat source to serve as the fixed lead. If three or more heat sources are used, the others can still have their operating order rotated to help equalize operating time.

Some multiple heat source controllers can also be configured to operate two or more multistage heat

Figure 9-86 | Examples of a multiple boiler controller capable of operating up to four boilers. *Courtesy of tekmar Controls.*

sources. For example, a pair of boilers each having a two-stage burner represents four stages of heat input. However, the firing sequence is different from the sequence used with four identical single-stage boilers. When bringing on additional stages, the controller is typically configured to fire the HI stage of the first boiler before turning on the LO stage of the second boiler. Most modern multiple boiler controllers have this logic built in, and only require the installer to set a switch or program menu item to properly configure the controller.

Figure 9-87 shows the concepts of equal run time rotation, fixed lead, and multiple HI/LO boiler sequencing.

Multiple Heat Source Controllers (For Modulating Heat Sources)

It is also possible to stage multiple *modulating* heat source as discussed in Chapter 3. Doing so requires an analog output to each boiler rather than just relay contact closure. The most common analog output sent by the heat source controller to each heat source is 2- to 10-VDC or 4- to 20-mA signals. The span of these signals represents the range of heat output by the heat source. For example, a 2-VDC signal maintains the heat source at its minimum stable heat output rate, while a 10-VDC "tells" the heat source to operate at full output. In some cases, the lower end of this analog signal needs to be adjusted slightly to correspond to the minimum signal threshold of the heat source's internal controller.

In addition to equal run time rotation, there are other operating options available for staged/modulating heat sources. These options deal with the sequence in which modulation and staging are used. They are called **series staging/modulation**, **parallel staging/modulation**, and **hybrid staging/modulation**. All three are compared in Figure 9-88.

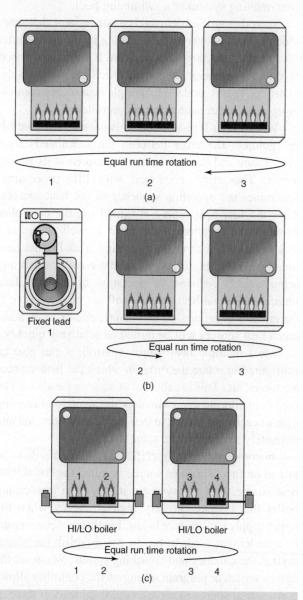

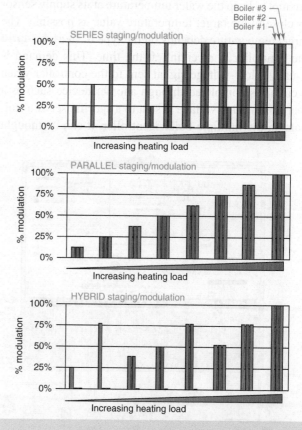

Figure 9-87 (a) Equal run time rotation operating sequence. (b) Fixed lead combined with equal run time rotation of boilers 2 and 3. (c) Multiple HI/LO boiler sequencing.

Figure 9-88 Comparison of how modulation and staging are combined with multiple modulating heat sources are used. (a) Series staging/modulation. (b) Parallel staging/modulation. (c) Hybrid staging/modulation.

Series staging/modulation allows each heat source to modulate all the way to full output before turning on the next heat source.

Parallel staging/modulation maintains each heat source in the system at the same modulation rate. The group of heat sources modulates "as a whole," starting from the lowest stable heat output rate of the heat sources and progressing to the full heat output rate. This method does not allow the system turndown ratio to be greater than the turndown ratio of an individual heat source.

Hybrid staging/modulation is typically used with modulating/condensing boilers. It controls the boilers based on a tradeoff between maximizing the combustion heat exchanger area per unit of heat output (to encourage condensation) and delaying the firing of the next boiler in response to a small increase in load.

In addition to modulating boilers, the 2–10 VDC output of a multiple heat source controller can be used to operate a variable-speed circulator. Figure 9-89 shows an example in which the fixed lead stage of the multiple heat source controller operates a variable-speed injection pump to transfer heat from a thermal storage tank into a distribution system that serves multiple zones of floor heating. The second- and third-stage output of the control each operate modulating/condensing boilers, both of which provide supplemental heat input to the distribution system when the heat output from the thermal storage tank is insufficient to meet the load. The variable-speed injection pump is a fixed lead stage, and is only allowed to operate when the temperature at the top of the thermal storage tank is a preset differential above the temperature of water returning from the distribution system. The latter constraint was discussed earlier in this chapter in the context of differential temperature controllers. The firing order of the two boilers can still be varied based on the previously discussed boiler rotation concept.

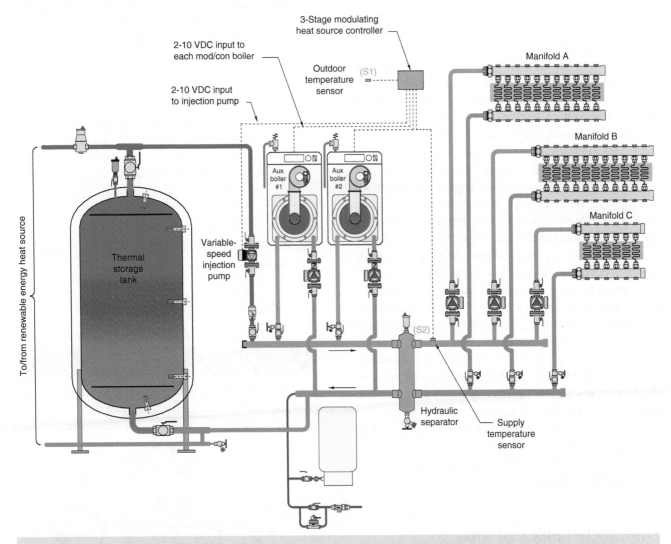

Figure 9-89 | A variable-speed injection pump provides a fixed first-stage modulating heat input from a thermal storage tank to the zoned low-temperature distribution system. Two modulating boilers provide second- and third-stage heat inputs.

9.9 Mixing Strategies and Hardware

Earlier in this chapter, the subject of regulating heat output by varying the water temperature supplied to the heat emitters was discussed. The relationship between supply water temperature and heat output is approximately proportional, making this approach preferred from a control standpoint.

This section discusses many of the mixing methods and hardware currently used in hydronic heating. Some approaches use nonelectric thermostat devices, and others used sophisticated microprocessor controllers. All have strengths and limitation as well as applications for which they are particularly appropriate.

Purposes of Mixing

One purpose of mixing is to control heat output by varying the temperature of the water supplied to the heat emitters. This is done by blending high-temperature water from the heat source with cooler water returning from the distribution system. The flow rate and temperature of these two streams determine the blended or "mixed" temperature supplied to the distribution system.

When the system is supplied by a conventional boiler, the mixing subsystem must also protect that boiler from operating at conditions that create sustained flue gas condensation. This phenomenon was previously discussed in Chapter 3, Hydronic Heat Sources. It occurs when the combustion side of the boiler's heat exchanger falls below the dewpoint temperature of the exhaust gases. This causes water vapor and other chemical compounds in the exhaust stream to condense into liquid on the surfaces of the boiler's heat exchanger. Sustained flue gas condensation often leads to rapid scaling and corrosion of the heat exchanger as well as the boiler's venting system. When properly applied, mixing can prevent this condition.

Mixing is also used to protect a low-temperature distribution system from potentially high-temperature water that may be present in an intermittent or partially controlled heat source. Examples of such heat sources include the storage tank for a solar thermal system, a solid-fuel boiler, or the storage tank of an electric thermal storage system.

These requirements will be further discussed and illustrated throughout this section.

Thermodynamics of Mixing

If the flow rate and temperature of two fluid streams that are mixed together are known, the mixed temperature can be calculated using Equation 9.15:

Equation 9.15:

$$T_{mix} = \frac{(T_{hot} \times f_{hot}) + (T_{cool} \times f_{hot})}{f_{hot} + f_{cool}}$$

where

T_{mix} = mixed temperature (°F);
T_{hot} = temperature of the hotter fluid (°F);
f_{hot} = flow rate of the hotter fluid (gpm);
T_{cool} = temperature of the cooler fluid (°F);
f_{cool} = flow rate of the cooler fluid (gpm).

The equation is based on the principles of **conservation of mass** and **conservation of energy**. In the case of mixing, conservation of mass means that the total mass of fluid entering the mixing device must equal the total mass of fluid leaving the mixing device. Similarly, conservation of energy means that the total thermal energy entering the mixing device must equal the total thermal energy leaving the mixing device.

These principles hold true regardless of the type of fluid used, as long as both streams are the same type of fluid. It also applies regardless of the device in which the two fluid streams come together.

Example 9.6

Two streams of water enter a tee as shown in Figure 9-90. The hot stream enters at 2.0 gpm and 170 °F. The cooler stream enters at 4.0 gpm and 120 °F. Determine the mixed temperature leaving the tee.

Solution:

Substituting the temperatures and flow rates given into Equation 9.15 yields the following.

$$T_{mix} = \frac{(120 \times 4) + (170 \times 2)}{4 + 2} = 136.7 \,°F$$

Discussion:

The mixed temperature always falls between the higher and lower entering temperatures.

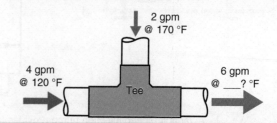

Figure 9-90 | Streams of water mixing in a tee for Example 9.6.

The physical principles underlying Equation 9.15 allow it be extended to handle more than two entering fluid streams. The general form of the equation is as follows:

Equation 9.16:

$$T_{mix} = \frac{(T_1 \times f_1) + (T_1 \times f_1) + \cdots + (T_n \times f_n)}{f_1 + f_2 + \cdots + f_n}$$

where

$T_1, T_2, \ldots, T_n$ = temperature of entering fluid streams (°F)

$f_1, f_2, \ldots, f_n$ = flow rates of entering streams (gpm)

Example 9.7

Figure 9-91 shows a manifold station serving four parallel circuits. All circuits receive water at 105 °F from the supply manifold. Because each circuit has a different length, each will have a different flow rate, and different return temperature. Use Equation 9.16 to determine the temperature of water leaving the lower manifold.

Solution:

Since the flow rates and return temperatures of each circuit have been determined, the blended outlet temperature is a relatively easy calculation.

$$T_{mix} = \frac{(T_1 \times f_1) + (T_2 \times f_2) + (T_3 \times f_3) + (T_4 \times f_4)}{f_1 + f_2 + f_3 + f_4}$$

$$T_{mix} = \frac{(99.9 \times 1.4) + (94.7 \times 1.1) + (103.9 \times 2.4) + (102.1 \times 1.7)}{1.4 + 1.1 + 2.4 + 1.7} = 101°F$$

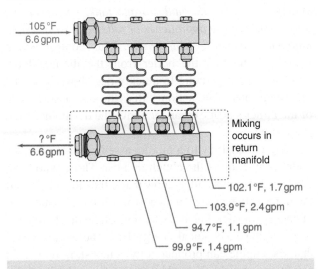

Figure 9-91 | Manifold station for Example 9.7.

Discussion:

Equation 9.15 was simply expanded by two additional terms in the numerator and two terms in the denominator to handle the four-stream situation.

Concept of a Mixing Assembly

A *mixing assembly* is a collection of hardware and control logic that collectively regulates the rate of heat transfer from the heat source to the distribution system. The mixing assembly can be composed of many different hardware options. These options will be discussed in later sections. However, all hardware options for the mixing assembly must accomplish the same two objectives:

- Regulate supply water temperature to the distribution system.
- Maintain the inlet temperature to a conventional boiler (if used in the system) high enough to prevent sustained flue gas condensation.

A piping schematic that depicts the concept of a mixing assembly is shown in Figure 9-92.

The gray box in Figure 9-92 can be thought of as a "container" for the various hardware components that make up a specific mixing assembly. The mixing assembly can also be viewed as a "bridge" between where heat is produced and where it is released. All heat that eventually reaches the heat emitters must pass through the mixing assembly. The piping schematics that follow will show how various components are assembled to build different mixing assemblies.

Boiler Protection

In systems using conventional boilers, the mixing device must protect the boiler from operating at temperatures low enough to cause sustained flue gas condensation. This is called **boiler protection**. Failure to provide it can lead to corrosion and scale formation that drastically shortens the life of the boiler and its venting system.

Unfortunately, the need to protect conventional boilers from such conditions has too often been viewed as secondary in importance to providing the desired supply temperature. Such disregard for unchecked flue gas condensation has led to serious problems in many systems.

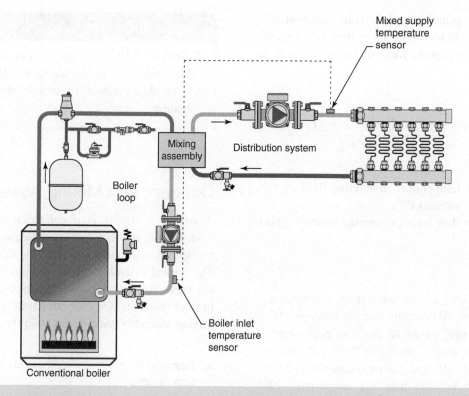

Figure 9-92 A mixing assembly serves as a "bridge" between the higher temperature boiler loop and low-temperature distribution system. The mixing assembly also acts as a "metering device" to regulate the rate of heat transfer to the distribution system.

The exact boiler inlet water temperature where flue gas condensation problems develop depends on the design of the boiler as well as its operating conditions, as discussed in Chapter 3, Hydronic Heat Sources. It is generally recognized that maintaining return temperatures of 130 °F for gas-fired boilers and 150 °F for oil-fired boilers eliminates the damaging effects of flue gas condensation. However, these temperatures may not apply to all conventional boilers, or to variations in fuel chemistry, such as the sulfur content in fuel oil. *Manufacturers should be consulted for the recommended minimum operating temperatures of their boilers.*

To properly protect a conventional boiler, the mixing assembly must determine if the distribution system is dissipating heat faster than the heat source is producing heat. If this condition is not prevented (e.g., heat dissipation exceeds heat production), the fluid temperature in the system will drop until thermal equilibrium is reestablished. If this requires the boiler to operate well below the dewpoint temperature of its flue gases, so be it! Thermal equilibrium "doesn't care" if the boiler is condensing; it only cares about balancing heat dissipation with heat production.

All mixing assemblies intended to protect a conventional boiler from sustained flue gas condensation must monitor the boiler temperature. Most monitor boiler inlet temperature. Some monitor boiler outlet temperature and use it along with other information to estimate the boiler inlet temperature. Without this temperature monitoring, the mixing assembly is "blind" to possible low-temperature operation that causes flue gas condensation.

Boiler protection is accomplished by reducing the rate of heat flow from the boiler into the distribution system, or into a thermal storage tank, whenever the boiler temperature approaches or falls below a specified minimum operating temperature. This control action partially "unloads" the boiler from the thermal load, allowing its temperature to quickly rise above condensing conditions. This action "lifts" the combustion side of the boiler's heat exchanger above the dewpoint of the exhaust gases.

A good analogy to a mixing assembly that provides boiler protection is the clutch in a car. Those who have driven a car with a clutch know that if the clutch is let out too quickly the engine "lugs down." From the standpoint of energy transfer, this is a situation where the drive train is extracting mechanical energy from the engine faster than the engine can produce it by combusting fuel. This condition is quickly sensed by an experienced driver who pushes the clutch pedal in slightly to partially unload the engine and quickly reestablish a balance between the rates of energy production and energy dissipation. A properly controlled clutch allows the full energy output of the engine to be transferred to the drive train but does not allow the drive train to pull energy away from the engine faster than it is being produced.

Similarly, a properly controlled mixing assembly acts as an intelligent "**thermal clutch**" between the heat source and the distribution system. If heat flows through the mixing assembly and into the distribution system faster than the boiler can produce it, the boiler temperature immediately drops. When the mixing controller senses this (by way of the boiler temperature sensor), it immediately responds by reducing the rate of hot water flow into the mixing assembly, which reduces the rate of heat transfer to the distribution system. *A properly controlled mixing assembly allows the full output of the boiler to reach the distribution system but does not allow the distribution system to pull heat away from the boiler faster than it is being produced.*

*To control supply water temperature and provide boiler protection, a mixing assembly must create two **mixing points**.* One mixing point regulates supply water temperature; the other boosts boiler inlet temperature.

These details, or the consequences of not having them, are illustrated in the systems of Figures 9-93 through 9-95. These systems all use a three-way mixing valve to control the temperature of the water going to the distribution system. The manifold station in each system serves six floor-heating circuits, each being 300 feet of 1/2-inch PEX tubing embedded at 12-inch spacing in a 4-inch-thick bare concrete slab. All of these systems have a conventional boiler with a rated output of 50,000 Btu/hr and a high-limit setting of 180 °F. In each case, the controller operating the mixing valve is set with the *intention* of maintaining a supply temperature of 125 °F to the manifold station.

The system in Figure 9-93 does *NOT* measure boiler inlet temperature. The water temperatures shown are those required based on thermal equilibrium, with the boiler in continuous operation. In this case, the distribution system requires water at 111.4 °F to dissipate 50,000 Btu/hr.

Because the mixing valve controller is trying to achieve a supply water temperature of 125 °F, and only has water at 111.4 °F available from the boiler, the hot port of the mixing valve is fully open, and the cool port completely shut. Hence, the mixing valve is simply "passing through" the entering water stream directly to the manifold station. No mixing occurs.

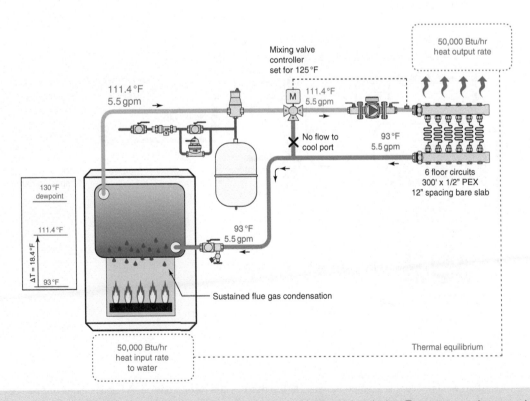

Figure 9-93 | A system where boiler inlet temperature is NOT sensed by the mixing device. Temperatures shown are based on thermal equilibrium. The boiler is operating with continuous flue gas condensation.

All water exiting the return manifold flows directly into the boiler at 93 °F. This temperature is low enough to cause sustained flue gas condensation. This is true even though the boiler is in continuous operation at full capacity. *This situation is unacceptable for any type of conventional boiler. It demonstrates that a mixing system that does not monitor and react to boiler inlet temperature cannot protect the boiler.*

The system in Figure 9-94 adds a bypass circulator between the boiler and mixing valve. The three-way mixing valve is correctly coupled to the boiler bypass piping using a pair of closely spaced tees to provide hydraulic separation. The intended purpose of the bypass circulator is to mix high-temperature water leaving the boiler into the cooler water flow returning from the distribution system and thus boost boiler inlet temperature.

The temperatures shown are those necessary for thermal equilibrium between this specific manifold station and boiler.

Although the bypass circulator increases the flow rate through the boiler, it cannot adjust the rate of heat transfer to the distribution system relative to the rate of heat production in the boiler. To dissipate the 50,000 Btu/hr added to the water by the boiler, the manifold station still requires 111.4 °F water at 5.5 gpm flow rate and returns the same flow rate at 93 °F. The hot port of the mixing valve is again fully open and passing all entering hot water directly to the manifold station. No mixing occurs in the mixing valve.

The bypass circulator and boiler loop create a second mixing point within the lower of the two closely spaced tees. This mixing slightly boosts boiler inlet temperature but not high enough to prevent sustained flue gas condensation within the boiler.

If a larger bypass circulator were used, flow rate through the boiler would increase even more, as would boiler inlet temperature. The temperature rise across the boiler would decrease in response to the higher flow rate. Logic dictates that the boiler's inlet temperature could never reach its outlet temperature, regardless of how fast flow passes through it.

Sustained flue gas condensation will again occur within the boiler. *This demonstrates that a boiler bypass circulator by itself is insufficient to protect the boiler. The missing element is a mixing system that senses and reacts to boiler inlet temperature.*

The system shown in Figure 9-95 uses a mixing valve controller that *measures boiler inlet temperature*. The operating logic within the controller gives priority to maintaining the boiler inlet temperature above a

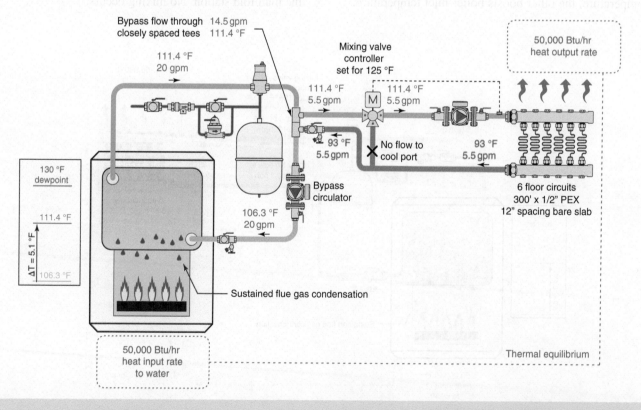

Figure 9-94 | A system where boiler inlet temperature is NOT sensed by the mixing device. A bypass circulator has been added to help boost boiler inlet temperature. The temperatures shown are based on thermal equilibrium. The boiler is still operating with continuous flue gas condensation.

9.9 Mixing Strategies and Hardware

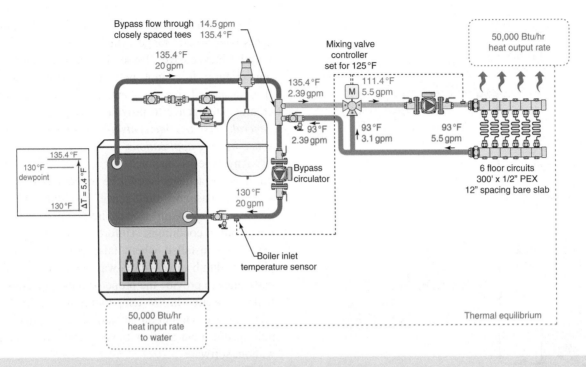

Figure 9-95 | A system where boiler inlet temperature IS sensed by the mixing device. Boiler inlet temperature is now maintained above 130 °F to prevent sustained flue gas condensation. *Note*: Supply temperature to distribution system remains at 111.4 °F because of thermal equilibrium, even though mixing controller is set for 125 °F supply temperature.

user-selected minimum value, which in this example will be 130 °F.

Notice that the water temperature entering the boiler is now 130 °F. Given the same 20.0-gpm flow rate through the boiler, the temperature rise across the boiler will be 5.4 °F, and its outlet temperature will be 135.4 °F. This high-water temperature now enters the hot port of the mixing valve.

Because hot water is now available to the mixing valve at 135.4 °F, one might assume that the mixing valve controller would adjust the valve for a mixed supply temperature of 125 °F (e.g., the setting of the mixing valve's controller). However, if the water temperature supplied to the floor circuits were to increase, so would heat output from the floor. Under such conditions, the floor circuits would release heat faster than the boiler can add heat to the water. The resulting thermodynamic imbalance would force the water temperature entering the boiler to drop. The boiler inlet temperature sensor would immediately detect this and the mixing controller would reduce the rate at which hot water enters the hot port of the mixing valve until the boiler inlet temperature increased back to 130 °F.

The net effect is that the manifold continues to be supplied with water at 111.4 °F, the manifold station releases heat at 50,000 Btu/hr, and boiler inlet temperature continues to stabilize at the controller's set minimum value of 130 °F. Thermal equilibrium between the boiler and distribution system is achieved, and boiler inlet temperature is high enough to avoid flue gas condensation.

The key to the successful strategy shown in Figure 9-95 is a mixing assembly that senses and reacts to boiler inlet temperature. This detail was lacking in the systems shown in Figures 9-93 and 9-94.

Mixing Using Manually Set Valves

A simple mixing assembly can be made using manually set valves. Three options are shown in Figure 9-96. The first uses a pair of two-way valves. The second is a single three-way mixing valve without any actuator. The third is a single four-way mixing valve without any actuator.

In all three cases, the valves are manually set to a given stem position, and remain in that position as the system operates. Assuming that the flow rate in the system is not altered through other means, these mixing assemblies all create fixed flow rates for both the hot and cool streams entering the mixing point.

If the temperature and flow rates of the hot and cool streams remained constant, so would the mixed temperature leaving the mixing point. *However, such conditions*

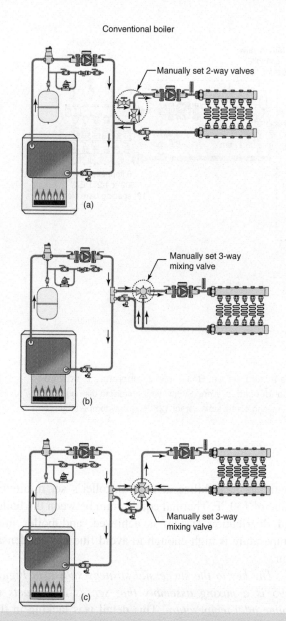

Figure 9-96 Three (nonrecommended) options for manually set mixing devices: (a) a pair of two-way valves, (b) three-way rotary valve, (c) four-way rotary valve. None of these options can respond to changes in water temperature or flow rate. None can provide stable supply temperature or boiler protection.

are very hard to sustain in real systems. The temperature of the hotter fluid varies as the heat source is cycled on and off. The temperature and flow rate of the cooler fluid returning from the distribution system also varies depending on the operating conditions at any given time. Manually set valves cannot sense these changes in temperature and therefore cannot compensate for them. The result is that the mixed temperature varies whenever the temperature or flow rate of either entering stream changes. This causes the supply temperature as well as the return temperature to the heat source to vary, in some cases over a wide range. Although this type of mixing may be inexpensive, it does not provide stable supply temperature. It cannot protect a conventional boiler from sustained flue gas condensation. It also lacks the control intelligence to provide mixing reset control. *In general, such arrangements are of little use in hydronic space-heating applications.*

Any type of manually set mixing valve that does not have an actuator should be thought of as a "**dumb valve**." It has no "intelligence" to adjust the mixing process to maintain the desired supply and return temperature conditions. Unfortunately, many systems have been installed where false confidence has been placed in dumb valves (three-way and four-way valves in particular) to provide the proper supply temperature and boiler protection in low-temperature heating systems. A mixing valve without an actuator and a controller is no more useful than a car without a driver or a computer without software.

The one application where a dumb mixing valve can be used, under specific limitations, is for proportional reset as described in Section 9.5.

Mixing with Three-Way Thermostatic Valves

Chapter 5, Pipings, Fittings, and Valves, discussed three-way thermostatic mixing valves. These valves have an internal thermostatic element that senses the temperature of the fluid leaving their mixed outlet port. The valve automatically adjusts its internal mechanism to try to maintain the outlet temperature close to the target temperature set on the valve's knob.

A single three-way thermostatic valve can be used to provide supply temperature control when the heat source can tolerate low inlet temperatures, and relatively low flow rates. A common example would be when the mixing valve is drawing water from a thermal storage tank, as shown in Figure 9-97a.

However, conventional boilers do not meet these prerequisites. When a conventional boiler is used as the heat source, a second three-way thermostatic valve should be used to monitor and boost the boiler inlet temperature as necessary to prevent sustained flue gas condensation. Figure 9-97b shows how two three-way thermostatic valves could be piped in such an application. Notice that a second circulator is now required to provide proper mixing within the thermostatic valve protecting the boiler. This valve and circulator also help maintain flow through the boiler. The latter is especially important when low mass boilers are used.

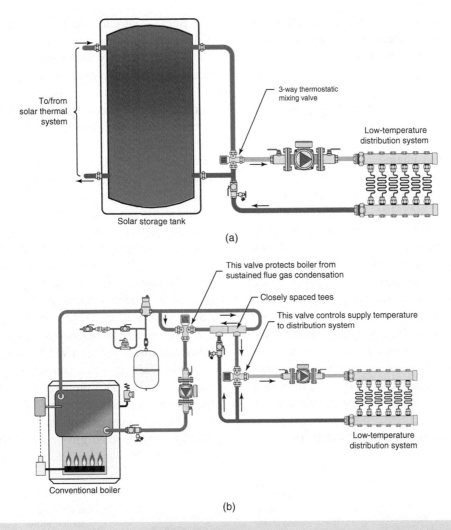

Figure 9-97 Piping for three-way thermostatic mixing valves: (a) when heat source is not damaged by flue gas condensation, or low flow rates; and (b) when heat source is a conventional boiler requiring protection from sustained flue gas condensation.

Although the use of two three-way thermostatic mixing valves is a possible approach, *it is not one recommended by the author*. There are several other mixing methods to be discussed that provide equal or better performance at lower cost, and with simpler installation requirements.

Designers are cautioned that most three-way thermostatic mixing valves currently on the North American market are intended to be used as anti-scald mixing valves on domestic hot water systems, rather than for mixing applications in hydronic heating systems. These valves have relatively low flow coefficients (e.g., low C_v values). As such, they create high head loss at relatively modest flow rates. While this characteristic is acceptable in their intended domestic hot water application, it has led to the creation of hydronic systems with grossly inadequate flow.

Some installers mistakenly "judge" the ability of a three-way mixing valve to accommodate flow based on pipe size rather than C_v value. For example, since a 1-inch size copper tube can accommodate a flow rate of about 10 gpm, they assume that a three-way thermostatic mixing valve with 1-inch FPT connections can also operate at 10 gpm. *This is a mistake*. Although the valve may indeed have 1-inch size connections, its C_v value might only be 3.0. Equation 6.13 can be used to show that forcing a 10-gpm flow rate through a valve with a C_v of 3.0 would require a differential pressure of 11.1 psi, with a corresponding head loss (assuming 100 °F water) of 25.8 feet of head. This would require a relatively large circulator with an estimated power input in excess of 240 watts. It would also likely cause excessive flow noise and possible erosion of the valve's internal components. In short, it would be a very poor choice of hardware.

In general, mixing valves should be selected with C_v values that are approximately equal to the maximum flow rate through the valve. This limits the pressure drop through the valve to about 1 psi.

Mixing with a Three-Way Motorized Valve

When a three-way rotary mixing valve is combined with a motorized actuator and controller, the resulting mixing assembly can perform several functions depending on the capabilities of the controller. Most modern mixing valve controllers provide very accurate supply temperature control. This temperature can be a fixed setpoint temperature or a temperature that is reset based on outdoor temperature. If the controller is equipped with a boiler temperature sensor, it can also protect a conventional boiler from flue gas condensation when piped as shown in Figure 9-98. Finally, some mixing valve controllers automatically detect when the system has been inactive for a few days and respond by "exercising" certain system components. The distribution circulator is turned on for a few seconds, and the mixing valve is powered fully open, then closed. This reduces the possibility of seizing during long periods of inactivity.

Some controllers used with three-way motorized valves are built into the actuating motor that attaches to the valve's body as shown in Figure 9-99. Others connect to the actuator motor using a three-wire cable.

Most mixing valve controllers generate a floating control signal (discussed in Section 9.2). This output operates a 24-VAC motor and gear train assembly in the actuator. The actuator shaft turns very slowly, taking from 2 to 3 minutes to rotate a valve across its full 90 degree range of travel. Because hydronic systems do not require wide temperature changes over short periods of time, it is not necessary for the valve shaft to rotate quickly. Slower rotation helps stabilize the system by allowing adequate time for feedback from the supply sensor located downstream of the mixing valve. The actuator motors are equipped with end switches that prevent them from stalling at either end of their travel range. Some also have CAM-operated auxiliary switches that can be used to activate a heat source or to signal other equipment as the valve begins to open.

Other mixing valve controllers generate a continuous 2- to 10-VDC modulating output and must be matched with actuators designed for this type of signal.

Although less common than either floating or 2- to 10-VDC output, some mixing valve controllers generate

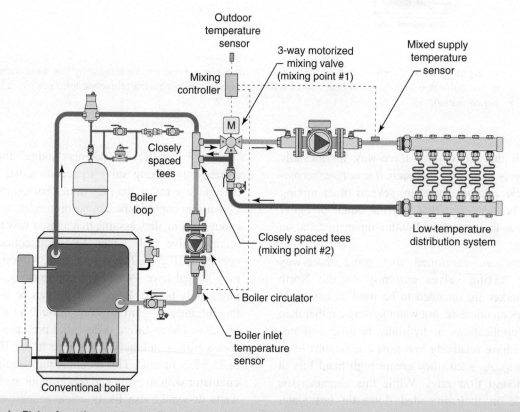

Figure 9-98 | Piping for a three-way motorized mixing valve used in combination with a conventional boiler. Note sensor measuring boiler inlet temperature. This sensor allows the controller to operate the valve so that sustained flue gas condensation is prevented.

9.9 Mixing Strategies and Hardware

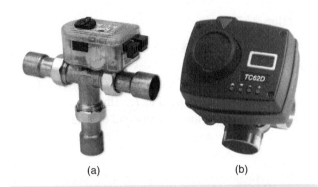

Figure 9-99 Examples of three-way valve bodies with mixing reset controller/actuating motors attached. (a) *Courtesy of Taco.* (b) *Courtesy of Paxton Corporation.*

a PWM output signal. Such controllers send low-voltage electrical pulses to a heat motor type actuator that moves the valve stem. The pulse width is continually adjusted to vary the total energy sent to the heat motor. This in turn determines where the heat motor positions the valve stem. When the pulse width is decreased or stopped, the heat built up in the actuator dissipates, and the stem begins to slowly rotate toward the closed position. The housing of the heat motor can get quite warm when the valve is near its fully open position. Full rotation of the valve can take 5 minutes or more, depending on ambient temperature.

Mixing with a Four-Way Motorized Valve

Four-way motorized mixing valves were also discussed in Chapter 5, Pipings, Fittings, and Valves. They are specifically designed to couple low-temperature distribution systems to conventional boilers (which require protection from sustained flue gas condensation). They are well suited to situations in which a conventional boiler with low flow resistance supplies a low-temperature distribution system. In such cases, they can be piped as shown in Figure 9-100.

When a four-way motorized mixing valve can be mounted close to a conventional boiler with low flow resistance, it is not necessary to install a circulator between the boiler and the mixing valve. This eliminates one of the

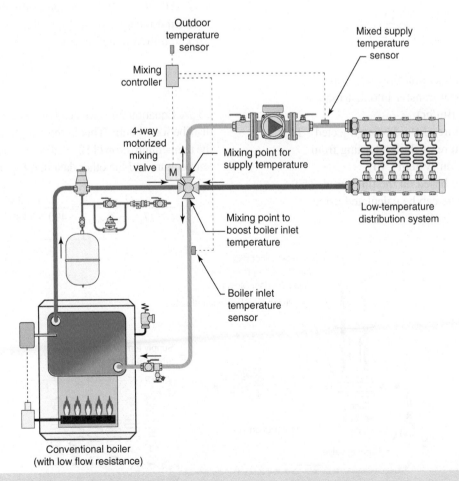

Figure 9-100 Piping for a four-way motorized mixing valve mounted close to a conventional boiler with low flow resistance. Under these conditions, no circulator is required between the boiler and the mixing valve.

circulators required when a three-way motorized valve is used.

The controllers and actuators that operate four-way motorized mixing valves are often identical to those previously described for three-way motorized mixing valves.

Injection Mixing

The term injection mixing applies to any mixing assembly in which hot water is pushed (e.g., injected) into a circulating distribution system as shown in Figure 9-101.

Since the distribution system is completely filled, cooler fluid must exit at the same flow rate hot water is injected. The greater the rate of hot water injection, the warmer the mixed supply temperature becomes and the greater the heat output of the distribution system. This simple concept applies to all forms of injection mixing regardless of the hardware used to implement it.

The injection flow rate needed to establish a given rate of heat transfer into the distribution system can be determined using Equation 9.17.

Equation 9.17:

$$f_i = \frac{Q}{k \times (T_{hot} - T_{return})}$$

where

f_i = required injection flow rate (gpm);
Q = rate of heat transfer into distribution system (Btu/hr)
T_{hot} = temperature of fluid being injected (°F);
T_{return} = temperature of fluid returning from distribution system (°F)
k = a number based on the fluid used. The exact value can be calculated using Equation 9.18.

Equation 9.18:

$$k = 8.01 \times c \times D$$

where

c = specific heat of the fluid at T_{ave} (Btu/lb/°F);
D = density of the fluid at T_{ave} (lb/ft³);
T_{ave} = average of the entering and exiting temperatures in the injection risers at design load conditions (°F)

For typical applications, T_{ave} can be assumed to be 140 °F, and the following values of k can be used:

- For water, $k = 490$.
- For a 30% solution of either propylene glycol or ethylene glycol, $k = 479$.
- For a 50% solution of either propylene glycol or ethylene glycol, $k = 450$.

Example 9.8

A radiant floor heating system requires a heat input of 120,000 Btu/hr at design load conditions. Water at 180 °F is available for injection. The supply temperature of the system at design load conditions is 110 °F, and the temperature drop across the distribution system at design load is 20 °F. What is the required injection flow rate?

Solution:

To use Equation 9.17, we need the return temperature of the distribution system. This is the supply temperature minus the temperature drop (110 − 20 = 90 °F). Substituting this temperature and the other data into Equation 9.17 yields:

$$f_i = \frac{Q}{k \times (T_{hot} - T_{return})} = \frac{120,000}{490 \times (180 - 90)} = 2.7 \text{ gpm}$$

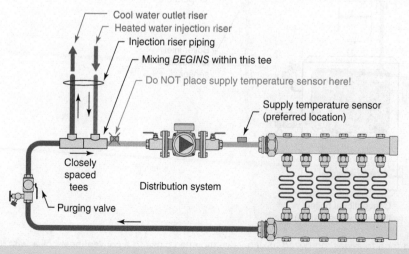

Figure 9-101 | Concept of injection mixing. Flow rate of entering hot water always equals flow rate of exiting cool water.

Discussion:

Although the rate of heat delivery is relatively high, the required injection flow rate (2.7 gpm) is quite low. This flow could pass through a 1/2-inch tube. This is a direct result of the high temperature difference between the supply and return injection risers ($T_{hot} - T_{return}$). The larger the temperature difference between the injection risers, the smaller the required injection flow rate becomes for a given rate of heat transfer. As you will soon see, this can have a profound impact on the size of the mixing hardware.

Injection Mixing Using a Two-Way Thermostatic Valve

One way to control the injection flow rate is with a two-way thermostatic valve. These valves were discussed in Chapter 5, Pipings, Fittings, and Valves. The valve is mounted into the supply injection riser as shown in Figure 9-102.

The hot water injection flow rate is controlled by the two-way thermostatic valve (V1), an example of which is shown in Figure 5-79. The knob of this valve is set to the desired supply temperature under design load conditions. The sensing bulb measures the temperature supplied to the distribution system. If the supply temperature is too low, the valve opens farther to allow more hot water into the mix point. If the supply temperature is too high, the valve begins to close allowing less hot water into the mix point.

The greater the deviation between the setpoint and the measured temperature, the more the valve stems moves. Most thermostatic valves have a proportional range of 4 to 6 °F. Half of this range is above the setpoint, and half is below the setpoint. If the supply temperature is more than 2 to 3 °F below setpoint, the valve is fully open. If the supply temperature is more than 2 to 3 °F above setpoint, the valve is fully closed.

In a typical low-temperature floor-heating system supplied from a conventional boiler operating at temperatures in the range of 160 to 180 °F, the flow rate through the injection valve (V1) is only about 15 to 25% of the flow rate in the distribution system. This allows a relatively small valve to regulate a large rate of heat transfer. However, if the heat source operates at lower supply temperatures, the injection flow rate will be significantly higher. Equation 9.17 can be used to determine the required injection flow rate.

The flow restrictor valve (V2) creates the pressure differential necessary to force flow through the injection risers as the injection valve begins to open. The greater the flow resistance created by this valve, the higher the injection flow rate for a

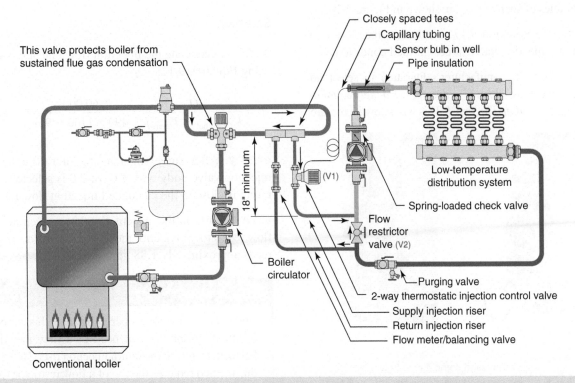

Figure 9-102 Piping for injection mixing using a two-way thermostatic valve. The three-way thermostatic valve is required to boost boiler inlet temperature to avoid sustained flue gas condensation.

given stem position of the injection valve (V1). If the flow restrictor valve creates too little flow resistance, there may not be sufficient pressure differential to move the required injection flow through the injection valve at design load. If the flow restrictor valve creates too much resistance, the necessary injection flow will be achieved when the injection valve is only partially open. This is undesirable because it forces the injection valve to operate using a small percentage of its stem travel under low-load conditions. A valve operating under such conditions is said to have low **rangeability** as discussed earlier in this chapter. These statements imply that there is an optimal setting for the flow resistance valve.

For stable and accurate control, the injection valve should be fully open when the maximum injection flow rate occurs under design load conditions. This requires the valve to be properly sized. It also requires the flow restrictor valve to be properly set.

A common mistake is to select an injection valve based on the size of the injection riser piping. This almost always results in an oversized valve. Instead, the injection valve should be selected based on its **flow coefficient (C_v value)**. The C_v value of a valve was discussed in Chapter 6, Fluid Flow in Piping. It is the flow rate (in gpm) of 60 °F water that creates a pressure drop of 1.0 psi across the valve.

The following procedure can be used to determine the necessary C_v of the injection valve as well as the C_v of the bypass valve. It assumes the use of a flow meter/balancing valve in the return injection riser as shown in Figure 9-102. Two examples of such valves are shown in Figure 9-103.

STEP 1. Use Equation 9.17 to calculate the flow rate through the injection valve at design load conditions.

STEP 2. Consult manufacturer's literature to select an injection valve body *with a C_v value approximately equal to the flow rate calculated in Step 1*. This will allow a pressure drop of approximately 1.0 psi to develop across the injection valve when it is fully open. Most manufacturers offer two-way control valves in different pipe sizes and several C_v values for each of these pipe sizes. The pipe size is just a matter of convenience. It is the C_v value that determines if the valve will function properly in this application.

STEP 3. Turn on the system circulators, then fully open the flow restrictor valve and the injection valve. Observe the flow rate indicated on the flow meter/balancing valve. If the flow rate is less than that calculated in Step 1, slowly begin closing the flow restrictor valve until the flow rate indicated on the flow meter matches that calculated in Step 1. If the indicated flow rate is already higher than that calculated in Step 1, slowly close the balancing plug on the flow meter/balancing valve until the flow rate matches that calculated in Step 1. In this case, leave the flow restrictor valve fully open.

Example 9.9

A two-way thermostatic valve will be used as the injection control device for a system supplying 60,000 Btu/hr to a radiant floor distribution system. Hot water is available from the primary loop at 160 °F. The supply temperature to the floor circuits at design load is 110 °F, with a return temperature of 95 °F. Select an appropriate valve.

Solution:

STEP 1. Calculate the injection flow rate at design load using Equation 9.17:

$$f_i = \frac{Q}{k \times (T_{\text{hot}} - T_{\text{return}})} = \frac{60{,}000}{490 \times (160 - 95)} = 1.88 \text{ gpm}$$

STEP 2. After surveying manufacturer's literature, an injection valve body with a C_v of 2.0 is selected. The C_v rating is close to the calculated injection flow rate.

STEP 3. Turn on the system circulators, then adjust the flow restrictor valve until the flow meter/balancing valve reads approximately 1.88 gpm.

Discussion:

There is a way to properly balance this type of injection system without need of a flow metering/balancing valve in the return injection riser. However, this method requires the use of a flow restrictor valve with a calibrated relationship between C_v and stem

(a)

(b)

Figure 9-103 Examples of combination flow meter/balancing valves. (a) *Courtesy of Caleffi North America.* (b) *Valve by Istec.*

> position. Although such valves are available, they are not common. The author recommends the procedure described in Steps 1 through 3 as the easiest and most reliable way to properly balance the injection mixing system.

For the most accurate control and fastest response to changing temperatures, the valve's sensor bulb should be in direct contact with the mixed fluid. Some thermostatic valves are supplied with hardware that allows the sensing bulb to be inserted into a tee. A special fitting seals against the capillary tube to prevent leakage. Another possibility is to install a sensor well into a tee, and then slide the sensor bulb into the well. This option is shown in Figure 9-102. The sensor bulb should fit snugly inside the well. It should also be coated with heat-conducting paste when inserted. If neither of these options can be used, the remaining choice is to strap the sensor bulb to the outside of the pipe downstream of the mixing point. The sensor bulb should be tightly clamped to the pipe with hose clamps, and then completely wrapped in insulation.

When the distribution system cools off, the thermostatic valve will completely open because it is set for a supply temperature significantly higher than room temperature. To prevent heat migration into the distribution system, the following details are recommended and shown in Figure 9-102:

1. Install the injection risers so that there is a drop of at least 18 inches from where the risers connect to the boiler loop and where they connect to the distribution system. This creates a **thermal trap** that discourages hot water from migrating downward through the injection risers.
2. Install a spring-loaded check valve in the distribution system downstream of the circulator, or use a circulator with an integral check valve. The forward opening resistance of the spring-check discourages heat migration into the distribution system.

Injection mixing using a two-way thermostatic valve does have some limitations. It does not provide mixing reset control nor does it allow for constant circulation in the distribution system. When this mixing technique is used in a system with a conventional boiler, a three-way thermostatic mixing valve, as shown in Figure 9-102, should be installed to protect the boiler against sustained flue gas condensation. Several of these limitations can be eliminated when a motorized two-way valve is used as the injection control device.

Injection Mixing Using a Two-Way Motorized Valve

A piping arrangement similar to that used for injection mixing with a two-way thermostatic valve can be used for a two-way *motorized* valve. These valves are typically operated by electronic controllers that generate a floating output, 2- to 10-VDC output, or a PWM output.

An example of a two-way motorized valve that can be driven by a 2- to 10-VDC signal is shown in Figure 9-104. The small cylindrical component at the top is the 24-VAC motor, which sits on top of the gear train. Wiring terminals can be seen on circuit board. See Figure 9-8 for typical wiring.

The piping and temperature sensor placement for injection mixing with a two-way motorized valve is shown in Figure 9-105.

A significant advantage of a motorized valve over a thermostatic valve is that the motorized valve can be operated by an intelligent controller. This allows the mixing assembly to provide several benefits not available with a thermostatic injection valve.

- Because the controller monitors boiler inlet temperature, it can protect a conventional boiler against low-temperature operation by reducing the injection flow through the valve when necessary.
- Most valve controllers designed for injection mixing can also provide either setpoint or outdoor reset control of the supply water temperature. The latter is desirable in many systems, especially those serving radiant panels.
- Since the controller can completely shut the injection valve when no heat input is required, the distribution system can operate with constant circulation. With the valve completely shut, there is no heat migration into the distribution system.

Figure 9-104 — Example of a two-way motorized (modulating) valve that can be used for injection mixing applications. *Courtesy of John Siegenthaler.*

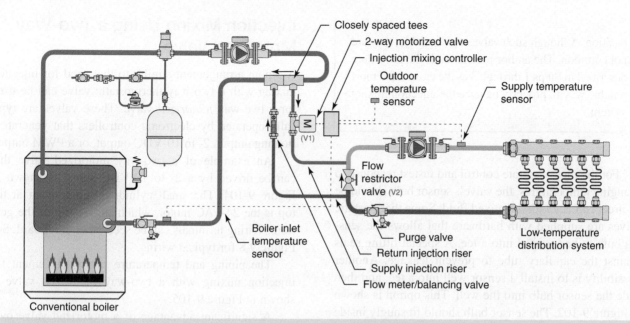

Figure 9-105 Piping for injection mixing with two-way motorized valve. Note that the controller measures boiler inlet temperature, and when necessary, can restrict hot water flow through the two-way motorized valve to prevent sustained flue gas condensation in a conventional boiler.

- Modern electronic temperature control is slightly more accurate than thermostatic control.

The sizing and selection procedure for a motorized two-way valve for injection mixing is identical to that just discussed for a two-way thermostatic valve, so is the setting of the bypass valve and combination flow meter/balancing valve.

Injection Mixing Using a Variable-Speed Pump

One of the most commonly used methods of injection mixing uses a variable-speed pump to regulate the injection flow rate. When the pump runs at low speeds, the injection flow rate is also low, as is the rate of heat transfer into the distribution system. As the speed of the injection pump increases, so does the injection flow rate and the rate of heat transfer. By regulating the speed of the injection pump, heat input to the distribution system can be varied from zero to full design load.

Most existing variable-speed injection systems use a standard line voltage circulator with a **permanent split capacitor (PSC)** motor as the injection pump. The controller modifies the electrical signal sent to the pump in a way that causes it to run at various speeds. These controllers use a microprocessor-controlled solid-state switching device known as a **triac** to vary the frequency and/or wave shape supplied to the pump.

Electronic controllers for variable-speed injection mixing using circulator with PSC motors have been in use for over 30 years. *However, these controller are not compatible with newer circulators using electronically commutated motors (ECMs). If a circulator with an ECM motor is to be used as an injection pump a controller with a 2–10 VDC, 4–20 mA, or PWM output, that matches the speed control input capability of the injection pump must be used.*

Most of the injection mixing controllers designed for use with PSC type circulator provide several features including the ability to:

- Provide full outdoor reset control of the supply temperature.
- Protect a conventional boiler against low-temperature operation by reducing injection flow when necessary.
- Coordinate the operation of the injection pump, system circulator, and heat source.
- Exercise the injection pump and system circulator during inactive periods.
- Allow the option of continuous circulation in the distribution system.

Some controllers can also coordinate their operation with other system controls such as those used for zoning or operating multiple boiler systems.

An example of variable-speed injection controller for use with a PSC-type circulator is shown in Figure 9-106.

9.9 Mixing Strategies and Hardware   451

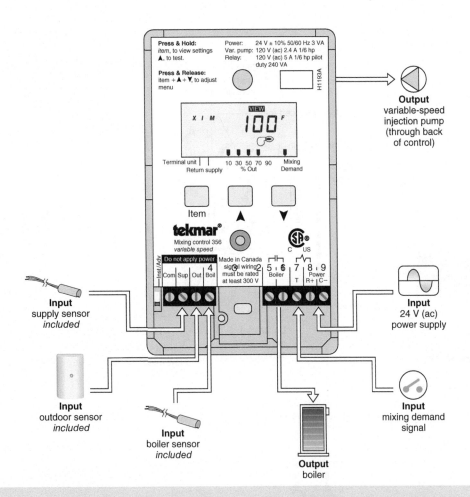

Figure 9-106 | Example of a controller designed to operate a variable-speed injection pump. *Courtesy of tekmar.*

The piping details and temperature sensor placement for a system using direct injection mixing with a variable-speed pump are shown in Figure 9-107. These piping details and sensor placements would also apply to a system using an ECM-type injection pump. Only the injection mixing controller would be different.

When the injection pump is operating, hot water in the boiler loop is drawn into the supply injection riser at point A. After passing through the variable-speed injection pump, the hot water is pushed into the tee at point B where it mixes with some of the cooler water from the return side of the distribution system. The remainder of the cooler return water leaves the tee at point C and travels up the return injection riser. *The exiting flow rate of cooler water always equals the entering flow rate of hot water.* When it flows into the tee at point D, it mixes with the bypass flow of hot water. This tee, in combination with the boiler loop circulator, provides a second mixing point that boosts the water temperature entering the boiler.

The injection pump will work equally well if placed in the return injection riser so that it pumps cool water from the distribution system back into the primary loop.

The details for this arrangement are shown in the inset of Figure 9-107. This configuration allows the injection circulator to operate at a lower temperature.

Two piping details are critically important to proper operation of this mixing assembly. Both are shown in Figure 9-107.

Detail 1: *The injection risers must drop at least 18 inches between where they connect to the boiler loop and where they connect to the distribution system.* This forms a thermal trap that discourages hot water from migrating downward into the distribution system when the injection pump is off. Without this trap, the slight pressure drop between the tees from point A to point D and from point C to point B can cause hot water migration into the distribution system when it is not needed. A standard circulator will not block this flow. A flow-check valve should not be installed in either injection riser to prevent this flow because it can cause erratic pump operation (surging) at low speeds. However, experience has shown that a circulator with a lightly loaded integral spring

452 Chapter 9 Control Strategies, Components, and Systems

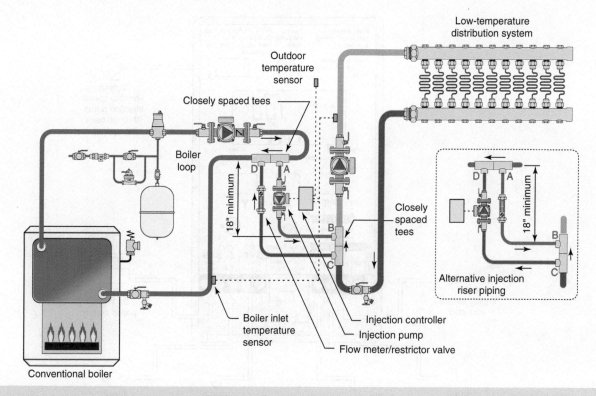

Figure 9-107 | Piping for injection mixing using a variable-speed pump.

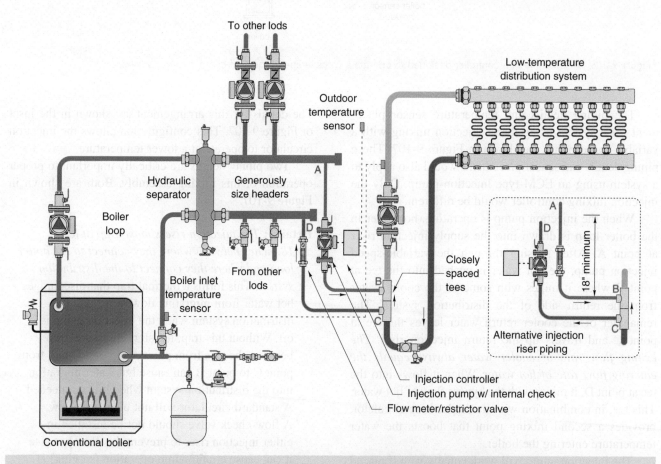

Figure 9-107a | Piping a variable-speed injection pump along with other load circulators in a system using a hydraulic separator. Note the generously sized headers on the load side of the hydraulic separator.

check valve can be used as the injection pump. The injection riser piping can change direction between the upper and lower sets of closely spaced tees provided a net drop of at least 18 inches is achieved.

Detail 2: *The tees connecting the injection risers to the boiler loop should be as close as possible. Likewise, the tees connecting the injection risers to the distribution system should be as close together as possible.* These sets of closely spaced tees create hydraulic separation between the injection pump and the other circulators in the system. This minimizes the tendency for uncontrolled hot water migration into the distribution system when the injection pump is off.

The injection mixing system shown in Figure 9-107 shows closely space tees connecting the injection risers to the boiler loop. These tees provide hydraulic separation between the boiler loop circulator and the variable-speed injection pump.

It's also possible to install a variable-speed injection pump in parallel with other fixed speed (or variable speed) circulators using common supply and return headers that connect to a hydraulic separator, or buffer tank, as shown in Figure 9-107a. With this type of configuration it's important to keep the head loss along the supply and return headers as low as possible to avoid interaction between the variable-speed injection pump and other circulators connected to the headers. Size the supply and return headers for a maximum flow velocity of 2 feet per second based on the total flow rate present when all circulators connected to the headers are operating at maximum speed.

The injection flow rate required at design load is again calculated using Equation 9.17. This flow rate does not depend on the injection control device (e.g., two-way valve versus pump). It is solely determined by the thermodynamics of the situation.

In many residential systems using conventional boilers operating at relatively high temperatures (160–180 °F), the injection flow rate is only 1 to 2.5 gpm at design load. The head loss created when this low flow rate passes through 3 to 6 feet of injection riser piping is very small. The low flow rate and low head loss would suggest that a very small circulator should be selected for the injection pump. Unfortunately, circulators with such limited capacity are not readily available at present. Instead, the majority of smaller systems using variable-speed injection mixing use a standard 1/25 horsepower wet rotor circulator with a PSC motor for the injection pump. These circulators are readily available and relatively inexpensive. However, their pumping capacity is far beyond the low flow and head loss requirements of the injection riser piping. This can lead to a situation where the injection pump only operates at a small percentage of its full speed, even under design load conditions. Like an improperly sized control valve, such a pump is said to have low rangeability and cannot provide the precise control the system is capable of delivering.

This situation can be seen by examining the pump curve of the injection pump and the system curve for the injection riser piping as shown in Figure 9-108.

If the assumed injection flow rate at design load conditions is 2.0 gpm, the pump speed must be drastically reduced to make the pump curve intersect the system curve at 2.0 gpm. Since this represents design load conditions, the pump will never need to operate at a higher speed. This wastes a very large portion of the pump speed adjustment range and compromises the ability to accurately control the supply water temperature.

One solution to this situation is to install a flow meter/restrictor valve in one of the injection risers as shown in Figure 9-107. Its purpose is to add head loss to the injection riser piping so the injection pump is forced to operate at full speed under design load conditions. As this valve is throttled, the head loss curve of the injection riser piping gets progressively steeper as shown in Figure 9-109. *The ideal valve setting causes the head loss curve of the injection riser piping to intersect the pump curve (when the pump is at full speed) at the desired injection flow rate.* This restores

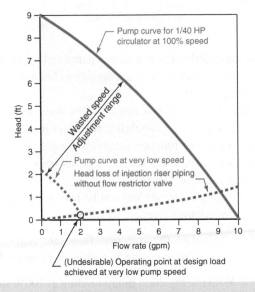

Figure 9-108 | The head loss curve of the injection risers is very shallow, and the required injection flow rate is achieved at very low pump speed, even for a small (1/40 hp) injection pump. This is undesirable because it wastes most of the injection pump's speed adjustment range. *Note:* Design injection flow rate assumed to be 2.0 gpm

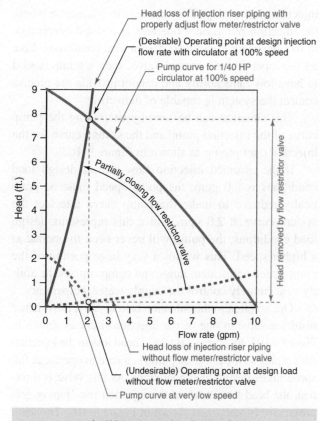

Figure 9-109 With a properly adjusted flow meter/restrictor valve, the head loss curve of the injection riser piping is much steeper. This allows the injection pump to operate at full speed under design load conditions. *Note:* Design injection flow rate assumed to be 2.0 gpm

the full rangeability of the injection pump and improves the ability of the injection mixing assembly to fine-tune supply water temperature.

Although the flow restrictor valve does not have to have a flow metering capability, the latter makes setting the flow restrictor function relatively simple. The valves shown in Figure 9-103 can be used as combined flow metering/restrictor valves.

The following procedure can be used to set the flow metering/restrictor valve.

STEP 1. Calculate the injection flow rate required at design load conditions using Equations 9.17 and 9.18.

STEP 2. With the injection pump running at 100% speed, begin closing the flow meter/restrictor valve until the indicated flow rate equals that calculated in Step 1. The balancing valve is now properly set.

The following procedure can be used if the flow restrictor valve does *not* have a flow meter, but does have a calibrated relationship between its stem position and C_v value.

STEP 1. Calculate the injection flow rate required at design load conditions using Equations 9.17 and 9.18.

STEP 2. Determine the head of the injection pump (operating at full speed) at the injection flow rate calculated in Step 1. This head is designated Hp.

STEP 3. Calculate the required C_v setting of the flow restrictor valve using Equation 9.19.

Equation 9.19:

$$C_{v_{\text{fr}}} = \sqrt{\frac{2.39 \times f_i^2}{H_p - r \times f_i^{1.75}}}$$

where

$C_{v_{\text{fr}}}$ = required C_v setting of the balancing valve;
f_i = injection flow rate at design load (gpm);
H_p = head of injection pump when operating at injection flow rate (feet);
r = a constant based on the pipe size of the injection risers*;
r = 0.058 for 3/4-inch injection risers;
r = 0.25 for 1/2-inch injection riser pipe.

*This constant assumes that the total equivalent length of the injection risers (excluding the balancing valve) is approximately 8 feet.

STEP 4. Set the balancing valve to the $C_{v\text{fr}}$ calculated in Step 3. The injection mixing assembly is now properly balanced.

Example 9.10

A variable-speed injection mixing system is planned for a radiant heating system with a design load of 100,000 Btu/hr. Hot water is available for injection at 170 °F. At design load conditions, the supply temperature of the radiant panel system is 130 °F with a 20 °F temperature drop. The injection risers will be short lengths of 1/2-inch copper tubing. The injection pump has the curve shown in Figure 9-110. Determine the required injection flow rate and the proper $C_{v\text{fr}}$ setting of the flow restrictor valve.

Solution:

The return temperature of the distribution system will be 130 − 20 = 110°F.

STEP 1. The required injection flow rate at design load is calculated using Equations 9.17 and 9.18.

$$f_i = \frac{Q}{k \times (T_{\text{hot}} - T_{\text{return}})} = \frac{100{,}000}{490 \times (170 - 110)} = 3.4 \text{ gpm}$$

9.9 Mixing Strategies and Hardware

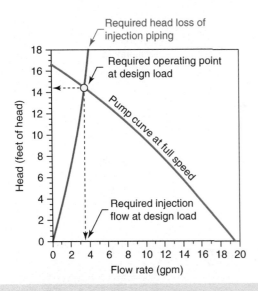

Figure 9-110 | Pump curve of injection pump in Example 9.10.

STEP 2. From the pump curve in Figure 9-108, the head loss of the injection pump while operating at 3.4 gpm is approximately 14.4 feet.

STEP 3. The C_v of the balancing valve is now calculated using Equation 9.19:

$$C_{v_{bv}} = \sqrt{\frac{2.39 \times f_i^2}{H_p - r \times f_i^{1.75}}} = \sqrt{\frac{2.39 \times (3.4)^2}{14.4 - 0.25 \times (3.4)^{1.75}}} = 1.5$$

Discussion:

When the flow restrictor valve is set to the required C_v of 1.5, the system curve of the injection riser piping intersects the pump curve at 3.4 gpm as shown in Figure 9-110. The injection mixing assembly will provide 3.4 gpm of injection flow when the injection pump is operating at full speed.

Injection Mixing in Systems with Wide ΔT Heat Sources

Systems that use conventional boilers to *consistently* supply high-temperature water for injection into low-temperature distribution systems have been discussed. Equation 9.17 has been used to demonstrate that in such systems the required injection flow rates is relatively low compared to the flow rate in the low-temperature distribution system.

There are applications for variable-speed injection mixing where the heat source may initially be at relatively high temperatures, but may not remain at those temperatures as the system operates over several hours. A good example is transferring heat from a thermal storage tank heated by a biomass boiler, solar thermal collectors, or off-peak electric resistance heating elements. At times, a tank connected to any of these heat sources may be filled with very hot water. In some cases, water temperatures could be close to 200 °F. However, as the thermal storage tank discharges heat to the load over several hours, its temperature drops significantly. This will significantly increase the required injection flow rate assuming that the design load heat transfer rate into the distribution system needs to be maintained.

Figure 9-111 shows an example of a system where a thermal storage tank supplies a low-temperature distribution system using a variable-speed injection pump as the means of controlling supply water temperature to the load.

Equation 9.17 can still be used to calculate the required injection flow rate. However, this calculation must be made at the *minimum water temperature* at which the thermal storage tank is responsible for providing the design load heat transfer rate.

Example 9.11

A thermal storage tank supplied from a pellet-fired boiler is intended to supply a design heating load of 100,000 Btu/hr. This distribution system requires 130 °F supply water temperature at design load, and operates with a 20 °F temperature drop under those conditions. The minimum useable temperature from thermal storage, based on these constraints, is 130 °F. The distribution system flow rate is 10 gpm. Determine the required injection flow rate.

Solution:

Use Equation 9.17 at design load conditions with the value of T_{hot} equal to the minimum temperature available from the thermal storage tank.

$$f_i = \frac{Q}{k \times (T_{hot} - t_{return})} = \frac{100,000}{495 \times (130 - 110)} = 10.1 \text{ gpm}$$

Discussion:

Notice that the injection flow rate essentially equals the flow rate of the distribution system. This will always be true when the temperature of the hot water being injected equals the required supply water temperature to the distribution system. No mixing can occur under this condition.

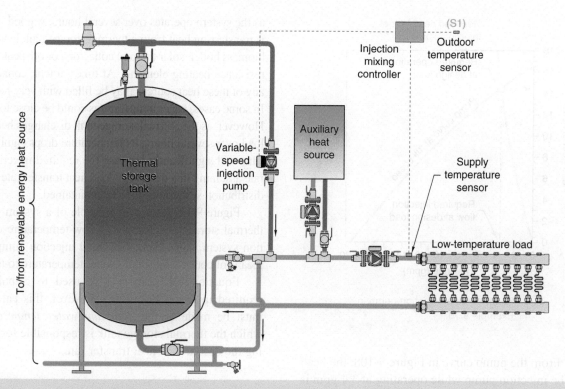

Figure 9-111 | A variable-speed injection pump provides heat input from a thermal storage tank to a distribution system. An auxiliary heat source provides additional heat input when necessary.

The full injection flow would pass to the supply side of the distribution system and the full return flow from the distribution system would pass up the return injection riser. This is the limiting condition for injection mixing, and it is fine. However, the higher injection flow rate will likely require a larger injection pump relative to the previously discussed systems supplied by boilers that consistently provide high-temperature water. When the thermal storage tank is at an elevated temperature, the injection pump would be operating at a much lower flow rate, which could be calculated using Equation 9.17. Assuming persistence of design load conditions, the injection flow rate would steadily increase as the tank temperature decreased. Under partial load conditions, the injection flow rate would decrease. The 495 value used in the denominator of Equation 9.17 is based on the density and specific heat of water at 120 °F, which is the average temperature of the distribution system at design load conditions. This is slightly different from the value of 490 used in earlier examples of Equation 9.17, which was based on an average water temperature of 140 °F.

Some Final Points on Injection Mixing

- *One of the strengths of injection mixing is that only the hot water needed for the mixing process passes through the mixing control device.* In a system with consistently high heat-source temperatures, this allows relatively small control hardware to manage significant rates of heat transfer. It also reduces head losses since the full flow of the distribution system does not have to pass through the mixing control device. Reduced head loss translates into reduced electrical power input to circulators. The life cycle cost savings can be substantial in larger systems.

- *The size of the component used in an injection mixing system is directly tied to the temperature difference between the heat source and the return side of the distribution system.* The greater this temperature difference, the lower the required injection flow and the smaller the injection hardware. Injection is well suited to systems where a conventional boiler serves as the heat source for a low-temperature distribution system. It is also a very cost-effective alternative to a three-way or four-way motorized mixing valve in larger systems.

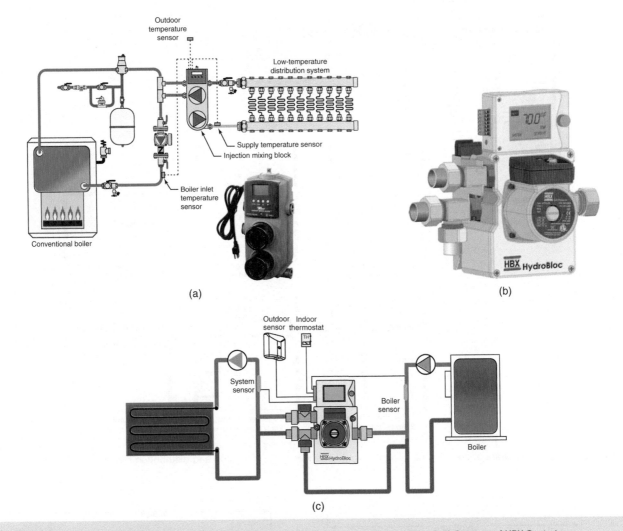

Figure 9-112 | Examples of preassembled injection mixing modules. (a) *Courtesy of Taco.* (b,c) *Courtesy of HBX Controls.*

- The piping schematics shown for injection mixing show purging valves in the distribution system. These are essential because the closely spaced tees that connect the injection risers to the primary loop do not permit good purging flow in the distribution system.

- Be sure to include the thermal trap detail wherever it has been shown on the injection piping schematics. If omitted, heat migration into the distribution system could become a problem.

Summary of Mixing Assemblies

This chapter has discussed several hardware options for mixing. Figure 9-113 shows how most of these options fit within the general concept of a mixing assembly. When properly implemented, each option controls the temperature of water supplied to the distribution system and protects a conventional boiler from sustained flue gas condensation.

9.10 Control System Design Principles

Entire books have been written describing methods for creating the control logic required for various heating and cooling system applications. This text cannot go into such detail. However, by consistently applying a few basic principles, the necessary control logic can usually be created. The principles are as follows:

Principle 1: When a device is to operate only when two or more conditions are met (as evidenced by closed switch or relay contacts), these contacts should be wired in *series*.

For example, the driven device shown in Figure 9-114a will be energized, and remain energized, only if all three series-connected contacts are closed at the same time.

458 Chapter 9 Control Strategies, Components, and Systems

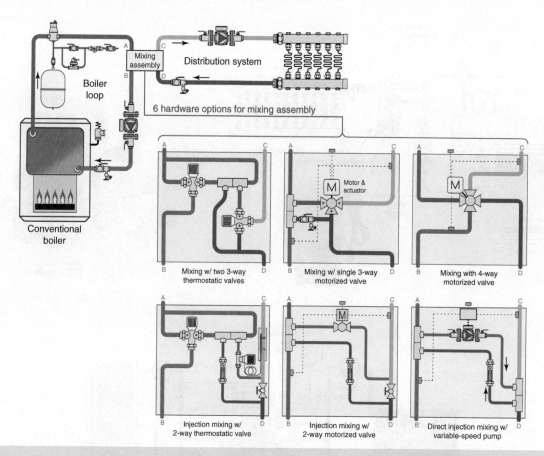

Figure 9-113 | Six hardware options for the mixing assembly between the boiler loop and distribution system.

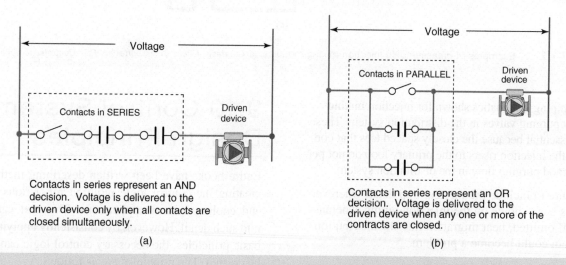

Figure 9-114 | (a) Contacts wired in series. All contacts must be closed simultaneously to operate driven device. (b) Contacts wired in parallel. The driven device is energized when any one or more of the contacts are closed.

Principle 2: When a device is to operate when *any one or more* of several possible conditions are met (as evidenced by closed switch or relay contacts), the contacts should be wired in *parallel*.

For example, the driven device shown in Figure 9-114b would be energized, and stay energized, whenever any one or more of the three contacts are closed.

Principle 3: Whenever possible, the majority of any hardwired control logic that must be created for a given system should be done using low-voltage components rather than line voltage components. This approach is less costly because the low voltage and low currents allow smaller components with lower amperage ratings to be used.

Principle 4: Whenever possible, attempt to build control systems around readily available devices that can be installed, maintained, and replaced if necessary, by any competent heating technician.

Principle 5: Never assume that the transformer in the heat source control is capable of powering several other external devices. Always check for the external VA rating. If none is stated, use a separate transformer to power the additional control devices. Control transformers with 24-VAC secondary voltage are available with VA ratings from about 20 through 100 VA. If the system may be expanded in the future, allow at least 20 VA of reserve capacity in the transformer.

Principle 6: Always document the final control system using a ladder diagram. Be sure to include a symbol legend with the diagram and identify any nonstandard devices. Provide a written description of operation for the system that describes the sequence of each operating mode (i.e., space heating, DHW, pool heating, etc.). Lack of good documentation for the control system, especially if it is customized, can lead to costly problems during installation and over the service life of the system.

9.11 Example of a Modern Control System

This section gives an example of a control system that manages the operation of a multiload, multitemperature hydronic heating system. This system is meant to show a variety of heat emitters and loads. It uses relatively basic control components such as a multizone relay center, boiler high-limit controller, and an injection mixing controller.

System Description

The example system provides two types of hydronic space heating as well as domestic hot water. Three independent zones of baseboard are controlled using on/off zone circulators. A single zone of radiant floor heating is supplied with a reduced water temperature by a variable-speed injection mixing system. DHW is provided by an indirect water heater that is configured as a priority load. When the domestic water heating load is operating, all space-heating zones are turned off.

A piping schematic of the example system is shown in Figure 9-115. Notice that all the circulators have

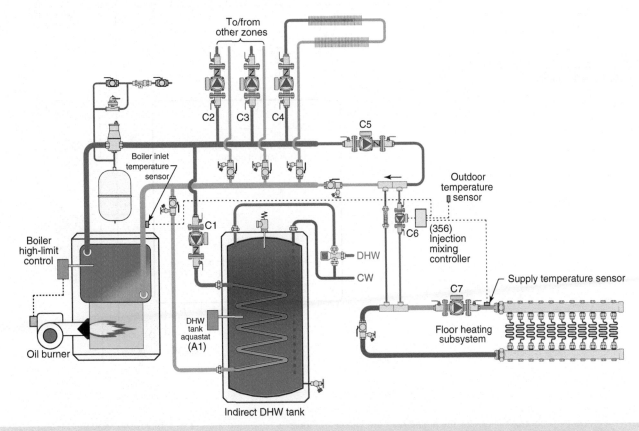

Figure 9-115 | Piping schematic for an example system supplying three fin-tube baseboard zones, low-temperature radiant floor circuits through variable-speed injection mixing, and an indirect water heater operated as a priority load.

designations such as C1, C2, and so on. This allows them to be cross-referenced between the piping and wiring schematics, as well as referenced in the description of operation.

The electrical schematic for the example system is shown in Figure 9-116.

Some of the required control logic is handled by the multizone relay center. All space-heating thermostats as well as the DHW tank aquastat are wired to this relay center. All circulators are supplied with line voltage from the outputs of the relay center. The relay center has been configured for priority control of the DHW load. Whenever the DHW load is active, all other line voltage outputs are turned off. A normally open relay contact inside the relay center connects to its (X X) terminals. This contact closes when any of the zone thermostats

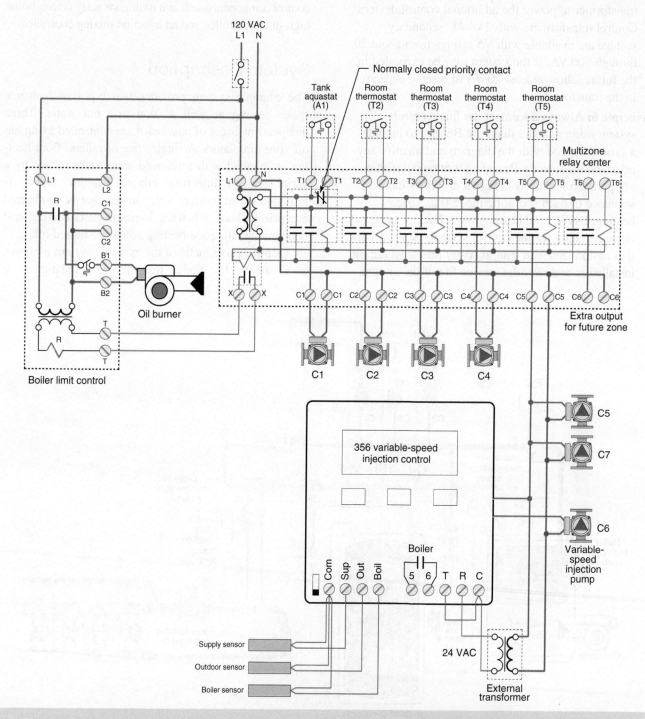

Figure 9-116 A control system for the piping diagram shown in Figure 9-105. Note the use of a multizone relay center and variable-speed injection controller. Also note the cross-referencing of the electrical devices between the piping and control wiring diagrams.

or the DHW tank aquastat calls for heat. The contact completes the thermostat circuit in the boiler's high-limit control allowing the boiler to fire. The wiring shown within the multiload relay center is representative only and will vary depending on the specific product used. It is shown so the complete circuit can be traced out when following the sequence of operation.

Description of Operation

The following text describes each operating mode of the system using references to the major components identified in the piping and wiring schematics. The description is sequential beginning with the call for heat. As you read the description of operation, refer to the piping and wiring diagrams to find the devices and circuit paths used in each operating mode.

1. DHW mode

Upon a call for heat from tank aquastat (A1), the zone 1 (priority) relay in the multizone relay center is turned on. This relay provides a line voltage output to the circulator (C1). A normally open relay contact between (X X) terminals in the multiload relay center closes to complete a circuit across the (T T) terminals in the boiler limit control, enabling the boiler to fire.

The boiler temperature is regulated by its high-limit setting and differential. Hot water from the boiler is pumped through the heat exchanger within the indirect DHW tank where it gives up heat and then returns to the boiler. The normally closed priority contact in the multiload relay center temporarily turns off all space-heating loads during the DHW heating cycle. The multiload relay center is assumed to be equipped with built-in priority override that can reestablish space heating if the DHW load is not satisfied within 30 minutes.

2. Baseboard heating mode

When any of the thermostats in the baseboard zones (T2, T3, and T4) call for heat, the associated relays in the multiload relay center are turned on. Line voltage is sent to the circulators (C2, C3, and C4), which serve the zones calling for heat. A normally open relay contact between (X X) terminals in the relay center closes to complete a circuit across the (T T) terminals in the boiler limit control enabling the boiler to fire. The boiler temperature is regulated by its high-limit setting and differential. This mode of space heating is temporarily interrupted while domestic water is being heated.

3. Floor heating mode

Upon a call for heating from thermostat (T5), the associated zone relay in the multiload relay center is turned on. It supplies line voltage to circulator (C5) and the distribution circulator (C7). It also supplies line voltage to a 30-VA 120/24-VAC control transformer. This transformer supplies 24 VAC to the 356 variable-speed injection mixing controller. When powered up, the controller boots its internal software and begins monitoring the three sensors. If the boiler temperature sensor indicates a low temperature, the injection pump will operate at a very low speed. As the boiler temperature increases, the injection mixing controller ramps up the speed of the injection pump (C6) as necessary to achieve the current target temperature at the supply sensor. The injection mixing controller calculates this target temperature using outdoor reset control. A normally open relay contact between (X X) terminals in the multiload relay center closes to complete a circuit across the (T T) terminals in the boiler limit control enabling the boiler to fire. The boiler temperature is regulated by its high-limit setting and differential. This mode of space heating is temporarily interrupted when DHW is operating.

This system is a relatively simple example and uses basic control devices and operating logic. The same system could be modified in numerous ways to add or remove functionality. For example, a boiler reset controller could be added. This controller might be capable of recognizing when the DHW load is active and respond by allowing the boiler to climb to its high-limit setting during this mode of operation. If the reset controller did not have this functionality, an external relay could be used to bypass the reset control upon a call for DHW.

Another possibility would be to configure the radiant floor distribution circulator (C7) for constant circulation provided the outdoor temperature is below the desired room temperature. Again, this could be accomplished in different ways depending on the specific injection mixing controller selected.

Still another possibility would be to add another circuit of baseboard heating that would serve as supplemental heat for the radiant floor zone. The radiant floor subsystem would be operated by the first stage of a two-stage thermostat. The second stage of that thermostat would operate the supplemental heating circuit.

The functionality of the multiload relay center could also be provided by hardwiring several relays together as shown earlier in this chapter. However, this would likely increase installation cost compared to the approach shown. The capabilities of the multiload relay center are well used in this system.

It should also be noted that some of the control functions built into the controls were not used. For example, the circulator output on the boiler limit control was not connected. There is no circulator that needs to operate when any load calls for heat. Likewise, the sixth zone of the multiload relay center was not needed. It was present because the multizone relay center was supplied with six zones. If another zone of heat was added in the future, this spare zone connection would prove handy.

9.12 Communicating Control Systems

Hydronic heating technology is constantly improving, especially in the area of controls. New products appear each year that add functionality, reduce cost, boost efficiency, or otherwise improve the benefits offered by modern hydronic heating. Rapid progress in microprocessor-based control technology has had a major impact on the hydronic heating industry over the past three decades. Further advancements in this area are virtually guaranteed. One area where such changes are expected to be fast paced is communicating control systems.

Stand Alone Versus Communicating Controllers

Many of the controllers discussed thus far were designed as **stand-alone devices** that perform a specific but limited task. Examples include temperature setpoint controllers, differential temperature controllers, boiler reset controllers, variable-speed injection mixing controllers, or multiple boiler controllers. *When several such devices are combined in one system, there is often a redundancy of hardware.* For example, a boiler reset controller, injection mixing controller, and multiple boiler controller all need to measure outside air temperature as they operate. This requires a separate temperature sensor for each stand-alone controller, even though each measures the same temperature. Although this is certainly possible, it obviously is more expensive than a system in which a single temperature sensor could be referenced by each controller that needs this information. Similar arguments can be made for other control circuitry such as boiler temperature sensors, digital displays, power supplies, and even multiple microprocessors.

Several stand-alone controllers operating simultaneously in a system may also interfere with each other, if not properly adjusted. For example, if the boiler high-limit control is set too low, it may turn off the boiler at a temperature lower than what the reset control has targeted. Likewise, if the boiler temperature is regulated by the reset control during DHW, the recovery rate of the tank may become very low during warm weather.

Communicating control systems reduce component redundancy and interference by allowing all controllers to interact with each other. In most cases, these devices are all part of a **networked control system**. Each controller is wired to a common two-wire cable called a **communications bus**. Each device attached to the communications bus has its own **digital address**, and thus can send and receive information to other devices on the network. All devices that need specific information such as boiler temperature, outdoor temperature, or differential pressure would get that information from a single-sensing device monitoring that temperature or pressure and attached to the same communications bus.

Networked controls also allows the heating system to communicate with other building systems such as those used for cooling, ventilation, lighting, and security.

Some networked control systems use industry standard **open protocols** such as **LONworks** or **BACNET**, as the "language" by which communication takes place. These protocols establish a standard method of communication between devices and can be implemented by any manufacturer. Other systems use **proprietary protocols** that limit the types of devices that can work on the networked system—usually to those produced by a single manufacturer.

A **single line drawing** showing the topology of a networked control system is shown in Figure 9-117.

Some devices shown in Figure 9-117 connect directly to the communication bus. Examples include room thermostats, digital temperature sensors/transmitters, and the boiler controller. These devices contain all the necessary electronics and firmware to communicate over the bus. Other devices, such as circulators, zone valves, and mixing valves, are connected to the bus through an **interface module**, which contain the electronics and firmware necessary to communicate with the bus, *and* produce an output signal compatible with the device being controlled. For example, one interface module may produce a 2- to 10-VDC analog output signal to drive a modulating mixing valve. Another interface module may function as a simple switch to turn a circulator on and off.

The physical location of a controller on the communications bus is irrelevant, provided the manufacturer's specifications for the maximum length and type of cable are followed. Each device attached to the communications bus has a unique digital address that allows other devices on the network to "find" and communicate with it.

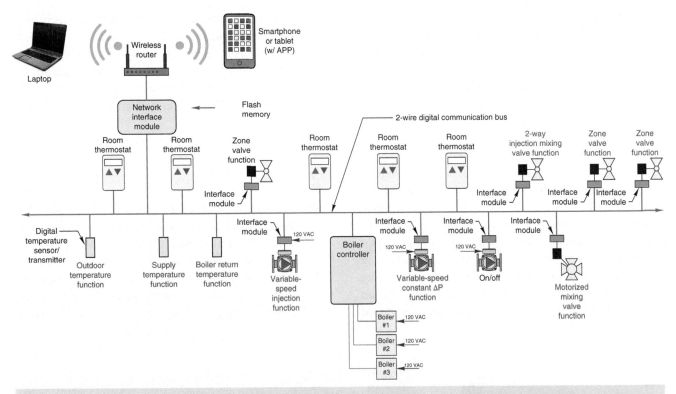

Figure 9-117 | Single line drawing of a networked control system using a two-wire communications bus. This approach reduces redundant components. It also allows each device to be monitored by a single on-site display.

One device, called the network interface module, provides a communication interface to a wireless Internet router. Software that allows the user to communicate with any device on the network can be loaded to a laptop computer, smartphone, or tablet. These device(s) communicate wirelessly with the router, which passes instructions and data to the network interface module and ultimately to any device connected to the control system network. The visual user interface is provided by the laptop, tablet, or smartphone. This eliminates the need for a visual interface on the control network, and allows access from any location with Internet service.

Networked control systems with Internet connectivity can be configured to monitor a wide range of conditions within the overall system. If one or more of those monitored conditions, such as a temperature, varies outside of a preset range, the system can automatically generate an alert and e-mail or text notification of that condition to one or more service companies. With proper authorization, a service technician can access the control system's Web page and examine a log of how the system has been operating over the last few hours, days, or even weeks. They could also make any necessary adjustments from wherever Internet access is available (worldwide).

Adaptive Control

Even sophisticated control algorithms such as PID processing work mostly with real-time inputs in present-generation controllers. A mixing reset control, for example, bases the calculated target supply water temperature on the outdoor temperature it senses at the time. It may add heat to a building at 7 a.m. while remaining oblivious to the likelihood that solar heat gains will overheat that building during the next 2 hours. This scenario is likely to repeat itself many times because the controller is not capable of learning how to anticipate and compensate for such situations.

Future control systems will use more advanced algorithms, as well as Internet access to learn when such conditions are likely, and automatically modify their actions to compensate. This is type of **fuzzy logic** is currently deployed on a limited basis in some microprocessor-based controllers.

User-Definable I/O

Most of the controllers presented earlier in this chapter have input/output (I/O) terminals that must interface with specific devices such as a line voltage circulator, a 10K thermistor, or a floating control actuator on a mixing valve. In some cases, the number of I/O terminals limits the functionality the controller can provide or the type of system in which it can be used. Some applications do not use all the capabilities offered by a given controller. These unused features add cost to the system without improving its performance. Still other systems may need capabilities that a preprogrammed controller cannot provide.

Software-based control systems can now drive generic input/output ports that can be configured to the exact requirements of the system. For example, in one system, a given pair of terminals on a device may be configured (through software) to connect to a 10K thermistor temperature sensor. On the next system, that sensor may not be needed. The same terminals may instead be configured to drive a 2- to 10-VDC modulating control valve, or perhaps turn the coil of a relay on and off. This ability allows a single mass-produced controller to be configured to perform exactly as needed in thousands of different applications. The "multiple control box" approach used in past and some current-generation control systems will likely be supplanted by software-driven universal controllers. Functionality will increase, cost will drop, and performance will continue to improve. Examples of such software-configurable control system are shown in Figure 9-118.

Zone Coordination

Most room thermostats are stand-alone devices. Any one thermostat in a system can call for heat completely independently of the other thermostats. At times, this can lead to situations in which the heat source, mixing device, or circulators are short cycled. For example, when only one zone thermostat is active, and it eventually stops calling for heat, the boiler is immediately

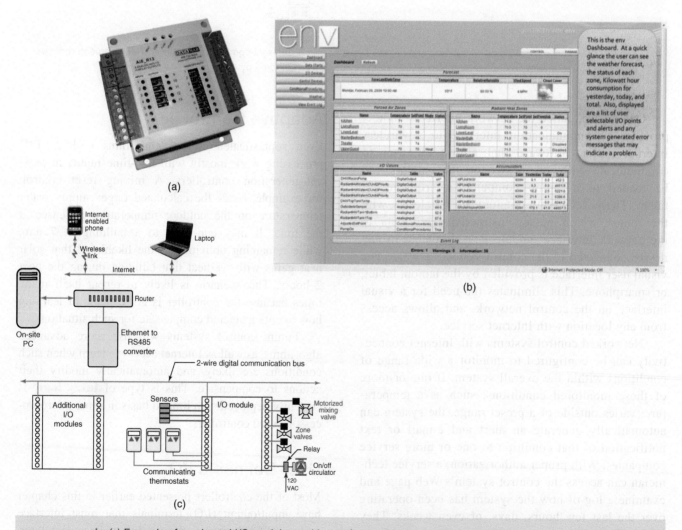

Figure 9-118 | (a) Example of a universal I/O module used in a software-driven control system. (b) User interface screen for the control system. *Courtesy of Climate Control Automation.* (c) Typical network configuration.

turned off. This would occur even if another thermostat was within a few seconds of turning on another zone, and thus requiring heat input from that boiler.

This undesirable operation can be largely eliminated when the zones are controlled by **communicating thermostats** and **coordinated zone management**. The latter type of thermostats, and the control base they connect to, can learn the operating characteristics of each zone in the system, and use this information to operate the various zones in a manner that helps level the heating load over time. The result is less cycling of the boiler and other system equipment.

A comparison between uncoordinated and coordinated zone operation is shown in Figure 9-119. For simplicity, each zone is assumed to have the same heating capacity. The colored bars represent times when each of the four zones is operating. The total demand for heat is represented by the dark-shaded area at the bottom.

Notice that the uncoordinated zone demands shown create relatively wide fluctuations in total heat demand, and that short periods of no heat demand are present. In systems using stand-alone thermostats, the heat source would be turned off during these short periods of no heat demand.

With coordinated zone management, the periods of zone heat demand are purposely made to overlap each other. This reduces variations in total system demand, which prevents boiler short cycling and improves system efficiency. Over a sufficiently long period, the total heat delivered to each zone is very close to that delivered using uncoordinated zone control.

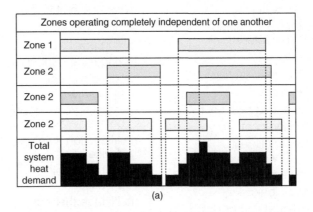

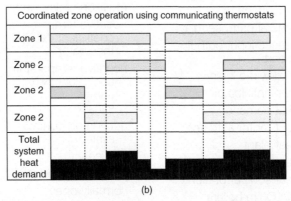

Figure 9-119 | (a) Zone operation and total heat demand for uncoordinated zoning. (b) Zone operation and total heat demand with coordinated zoning.

Summary

Without accurate and reliable control, the remainder of any hydronic heating system is virtually worthless. This chapter has provided an *introduction* and *overview* of the most commonly used concepts and hardware currently used to control modern hydronic heating systems. The hardware discussed has ranged from the capillary tube devices first developed in the 1930s to the latest Web-enabled networked control systems.

Those planning to design, install, and service modern hydronic systems need to understand a wide spectrum of currently available control hardware, as well as be familiar with legacy hardware dating back several decades. Modern control concepts such as PID control, software-definable input/output, and Web-enabled control are the building blocks of future control systems and worthy of additional study.

Key Terms

2 to 10 VDC
4 to 20 mA
actuator
ampacity
analog inputs
AND decision
BACNET
balance point temperature
base unit
bellows
bimetal element
boiler protection
buffer tank
building heat loss coefficient
capillary tube
characterized ball valve
closed-loop control system
coil
communicating thermostats
communications bus
conservation of energy
conservation of mass
contactor
contacts
control differential
controlled device
controlled variable
controller
coordinated zone management
cycle time
delay-on-break
delay-on-make
demand fired
description of operation
differential
digital address
digital input
digitally encoded signal
DIN rail

direct injection mixing
droop
dumb valve
electromechanical controls
embedded controllers
end switches
equal percentage characteristic
error
external VA rating
feedback
fixed lead heat source
floating control
flow coefficient (C_v value)
full reset control
fuzzy logic
gel-fired compression connectors
general purpose relay
generic input/output ports
hard-wired logic
heat anticipator
heat demand
heat motor actuator
heat plant
heat source reset
heat source rotation
HI/LO fire boilers
High-limit/switching relay controller
hunting
hybrid staging/ modulation
if/then statements
indoor reset control
inductive loads
intelligent mixing device
interface module
internal heat gains
interstage differential
ladder diagram
lead/lag control

linear relationship
LONworks
manipulated variable
manual reset high-limit (MRHL)
microprocessor
mixing assembly
mixing device
mixing point
mixing reset control
modulating heat production
modulating/condensing boilers
multifunction time delay device
multistage heat production
multistage thermostat
multizone relay center
negative temperature coefficient (NTC) thermistor
negative temperature coefficient (NTC) thermistor sensor
networked control system
nonlinear
normally closed (N.C.)
normally open
off-differential
offset
on-differential
on/off control
open protocols
operating differential
OR decision
outdoor reset control
overshoot
paired
parallel shifting
parallel staging/ modulation
partial reset control

permanent split capacitor (PSC) motor
poles
priority control
priority load
priority override
processing algorithm
programmable thermostats
proportional band
proportional processing (P)
proportional reset
proportional-integral processing (PI)
proportional-integral-derivative processing (PID)
proprietary protocols
pulse width modulation (PWM)
pump exercising
rangeability
relay
relay socket
relays
remote bulb aquastat
reset line
reset ratio
resistive loads
reverse injection mixing
rung
sensor
sensor well
series staging/ modulation
setpoint demand
settings
shielded cable
short cycling
single-line drawing
single-stage heat source
space heat demand

stages
stagnation pressure
stand-alone devices
starting point line
strap-on aquastat
system turndown ratio

tankless coil
target value
thermal clutch
thermal trap
thermistor
three-wire control

throws
time delay relays
triac
triple action controller
tristate control
turn-off differential

turn-on differential
twisted pair wires
undershoot
VA (volt-amps)
water hammer
Web-enabled

Questions and Exercises

1. What is the advantage of proportion-integral (PI) control over proportional only (P) control?

2. A heating system is designed to provide 180 °F water to the distribution system when the outside temperature is −15 °F. The building requires no heating when the outside temperature is 65 °F. What would be the proper reset ratio for this system?

3. Explain the difference between floating control and 2- to 10-VDC modulating control. Which can be used to operate mixing valves?

4. A building has a design heat loss of 55,000 Btu/hr when the outside temperature is −5 °F and the indoor temperature is 70 °F. Determine the balance point temperature of the building when internal heat gains are 11,000 Btu/hr.

5. Explain why a valve used to regulate heat output by varying the flow rate through a heat emitter should have an equal percentage characteristic.

6. Why do electromechanical thermostats experience droop? How do electronic thermostats compensate for droop?

7. Which is more important when selecting a two-way injection valve—pipe size or C_v value? Explain your answer.

8. What must be true of any mixing assembly capable of protecting a conventional boiler from sustained flue gas condensation?

9. What is meant by the term *priority override*? What type of device is commonly used to enable priority override in a control system?

10. What is the difference between a setpoint demand and a space-heating demand to a multiple boiler controller? What is a common application for a setpoint demand?

11. The term *error* refers to the difference between the _____ temperature, and the _____ temperature in a temperature control system.

12. Describe the difference between the reset line for a hydronic system using high-temperature finned-tube baseboard versus one using a low-temperature heated floor slab.

13. What is the difference between partial reset control and full reset control? Which is used for outdoor reset of a conventional boiler?

14. Explain the difference between the number of poles and the number of throws on a switch. Sketch a schematic symbol for each of the following switches:
    a. a DPST switch,
    b. an SPDT switch.

15. Explain the effect the thermal mass of a hydronic system has on the operation of the control system and the temperature swings inside the building. Describe a situation where a system with a high thermal mass is desirable. Describe another situation where it is undesirable.

16. Explain the operating principle of electromechanical setpoint controller using a fluid-filled sensor bulb.

17. Complete the ladder diagram shown in Figure 9-120 such that the following control action is achieved.

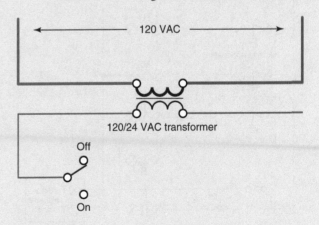

Figure 9-120 | Ladder diagram for exercise 17.

When switch S1 is set in the off position, a red 24-VAC indicator light is on. When the switch is set to the on position, a green 24-VAC indicator light is on, and a line voltage circulator is running. Label all components you sketch in the diagram.

18. A control system having the ladder diagram shown in Figure 9-121 is proposed to operate a three-zone hydronic system using a separate circulator for each zone. Identify any electrical errors in this diagram. Describe what would happen if the identified error were present when the system was turned on, or what is unsafe about the error. Sketch out how you would modify the schematic for proper and safe operation of the three-zone system.

19. What is the function of a thermal trap in an injection mixing system?

20. What is an advantage of using a motorized two-way valve and controller rather than a two-way thermostatic valve for an injection mixing application?

21. What is the difference between direct and reverse injection mixing? Which is better for smaller systems and why?

22. An injection mixing system is being designed to transfer 125,000 Btu/hr into a distribution system using water at 185 °F, both under design load conditions. The distribution system has a supply temperature of 105 °F and a temperature drop of 17 °F under design load conditions. Determine the necessary injection flow rate assuming that:
    a. The system uses direct injection mixing with a variable-speed pump,
    b. The system uses direct injection mixing with a two-way modulating valve,
    c. The system uses reverse injection mixing with a variable-speed pump.

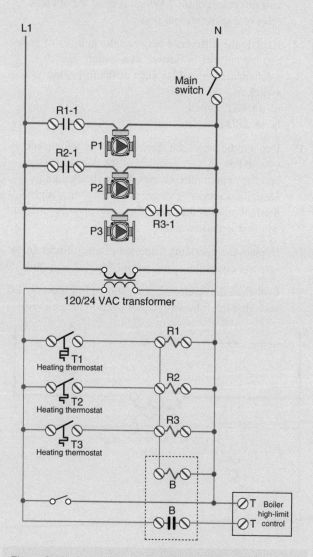

Figure 9-121 | Ladder diagram for exercise 18.

Chapter 10

Hydronic Radiant Panel Heating

Objectives

After studying this chapter, you should be able to:

- Explain what radiant heating is.
- Explain how radiant heating affects thermal comfort.
- Describe several types of hydronic radiant panels.
- Compare radiant panel heating to other methods of heating.
- Discuss the benefits of radiant panel heating.
- Identify good opportunities for radiant floor heating.
- Identify good opportunities for radiant wall and ceiling heating.
- Understand how tubing circuits are placed in radiant panels.
- Estimate the thermal performance of several types of radiant panels.
- Identify situations where supplemental heating is needed.
- Avoid common errors in the design and installation of hydronic radiant panel systems.
- Describe methods of using different types of radiant panels in the same system.

10.1 Introduction

Over the last 30 years, radiant panel heating has been one of the fastest growing sectors of the North American hydronics market. The superior comfort and fuel efficiency delivered by hydronic radiant panel heating is being discovered and sought after by increasing numbers of consumers for use in all types of buildings. Many mechanical contractors have been initiated into the broader field of hydronics heating, specifically by the need to install hydronic radiant panel systems.

This chapter describes the general theory of radiant heating as well as the specific details of using hydronics to deliver heat to various types of radiant panels. It will also show how piping and control techniques from previous chapters can be integrated in radiant panel systems.

10.2 What Is Radiant Heating?

Radiant heating *is the process of transferring thermal energy from one object to another by* **thermal radiation**. Chapter 1, Fundamental Concepts, briefly described thermal radiation as one of the three modes of heat transfer, along with conduction and convection. Whenever two surfaces are "within sight of each other," and are at different temperatures, thermal radiation travels from the warmer surface to the cooler surface. The rate of energy flow between these surfaces depends on the difference in their temperatures, the distance between the surfaces, the angle between the surfaces, and an optical property of the surfaces known as emissivity.

Thermal radiation includes both the visible and **infrared** portion of the **electromagnetic spectrum**. The wavelengths in these portions of the spectrum are such that any radiation that strikes an opaque surface does not penetrate beyond its surface (unlike X-rays and gamma rays that can penetrate beyond the surfaces of opaque objects).

All thermal radiation emitted by surfaces at temperatures lower than approximately 970°F will be in the infrared portion of the electromagnetic spectrum and as such cannot be seen by the human eye. All radiant panels discussed in this chapter operate well below this temperature, and thus only emit infrared thermal radiation.

Other than the fact that it cannot be seen, infrared thermal radiation behaves similar to visible light. It travels away from the emitting surface at the speed of light (about 186,000 miles per second) and can be partially reflected by some surfaces. *Like visible light, infrared thermal radiation travels equally well in any direction. This allows radiant energy to be delivered into occupied spaces from heated walls and ceilings as well as from heated floors.*

Like other forms of electromagnetic radiation, thermal radiation does not require a media such as a solid, liquid, or gas to move energy from one location to another. This is evidenced by the fact that sunlight passes through approximately 93 million miles of empty space before entering the earth's atmosphere.

Within a room, infrared thermal radiation emitted by a heated surface passes through the air with virtually no absorption of energy. Instead, the radiation is absorbed by the objects in the room. Most unpolished/nonmetallic surfaces absorb the majority of any thermal radiation that strikes them. The small remaining portion is reflected often to another surface where further absorption takes place. Thus, very little thermal radiation is reflected out of a room heated by a radiant panel.

It is therefore accurate to state that thermal radiation (more commonly referred to as "**radiant heat**") warms the *objects* in a room rather than directly heating the air. This difference is largely what separates radiant heating from convective heating. If the absorbed thermal radiation raises the surface temperature of an object above the room air temperature, some heat will be convected to the room air. Some heat may also be conducted deeper into the object if its interior temperature is lower than its surface temperature.

Thermal radiation is constantly emitted from our skin and clothing surfaces to any cooler surfaces around us. A large portion of the heat generated through metabolism is released from our bodies by infrared thermal radiation. Evidence of this can be seen in the **thermographic image** shown Figure 10-1.

This thermographic image was taken by a camera that detects infrared radiation rather than visible light. The colors shown are based on the temperature scale seen at the right side of the image. This scale is automatically generated based on the range of temperatures present within the image area.

Notice the relatively warm surfaces of exposed facial skin relative to body surfaces covered by clothing. Both skin and clothing surfaces are transferring thermal radiation to surrounding cooler surfaces. The body's ability to reject heat through thermal radiation greatly affects thermal comfort and will be discussed in more detail in Section 10.6.

Finally, it is important to understand that thermal radiation is entirely different from **nuclear radiation**. The latter is a type of particle radiation emitted by radioactive materials such as uranium and plutonium, and certainly not suited for direct heating of buildings. Unfortunately, many people associate the word radiation with nuclear radiation, and thus feel that any kind of

Figure 10-1 — Thermal radiation is constantly emitted by skin and clothing surfaces to any surround surfaces. The colors correspond to the temperature scale at right of image. *Courtesy of John Siegenthaler.*

radiation is inherently dangerous. This is simply not the case. *The thermal radiation emitted by low-temperature radiant panels is in no way unhealthy or harmful.*

10.3 What Is a Hydronic Radiant Panel?

A **hydronic radiant panel** *is any object warmed by passing heated water through tubing embedded in or attached to it, and which releases at least 50% of that heat to its surroundings as thermal radiation, and has a controlled surface temperature under 300°F.* The heated water is simply the material used to deliver heat to a hydronic radiant panel. If heating cable was embedded in or attached to the same object, one could refer to it as an **electric radiant panel**.

Once heat has been transferred to the materials that make up the radiant panel, its shape, orientation, surface temperature, surface properties, and surroundings determine its heat output.

The best-known type of hydronic radiant panel is a heated floor. In North America, heated floors account for over 90% of all hydronic radiant panel installations. However, *radiant panel heating is not limited to floors.* There are several established methods of incorporating hydronic tubing into walls and ceilings. In some situations, heated walls or ceilings hold an advantage over heated floors. Hydronic heating professionals should become familiar with the strengths and limitations of all such panels and use them where they are most appropriate.

Most hydronic radiant panels operate at water temperatures no higher than 200°F. Many operate with supply water temperatures in the range of 80 to 120°F, with surface temperatures in the range of 72 to 110°F. Radiant panels that operate at low supply water temperatures are well suited for use with contemporary heat sources such as modulation/condensing boilers, geothermal and air-to-water heat pumps, and thermal storage tanks associated with solar thermal collectors and biomass boilers.

10.4 A Brief History of Radiant Panel Heating

The practice of maintaining comfort by warming one or more room surfaces within occupied spaces dates back thousands of years. Archeological research shows evidence that heated stone floors were used in Asia in the 5,000-year BC time period Heated stone floors and walls were also used by the Greeks and Romans around 500 BC. The Roman **hypocausts** were structures that directed the flow of combustion gases from wood fires under raised stone floors, and up through hollow wall cavities, as seen in Figure 10-2. Variations on the hypocaust approach were used in Iraq, Algeria, Turkey, and Afghanistan in the 700 AD period. Although it is unlikely that such systems achieved the even temperature distribution or accurate control possible with modern technology, the intricacy of their construction testifies that they delivered comfort unattainable by simpler means.

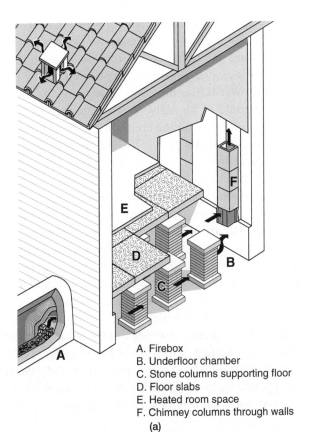

A. Firebox
B. Underfloor chamber
C. Stone columns supporting floor
D. Floor slabs
E. Heated room space
F. Chimney columns through walls
(a)

(b)

Figure 10-2 Ancient Roman hypocaust. (a) Original construction. *Universal History Archive/Getty Images.* (b) Remnants of a hypocaust. *DEA/G. DAGLI ORTI/Getty Images.*

The arrival of forced-circulation hydronic heating in the United States during the 1930s provided the needed link between producing heat in a boiler and efficiently transporting this heat to a radiant panel. During this time, steel and wrought iron pipe was threaded or welded together to form distribution circuits that were often cast into concrete floor slabs. Some of those systems are still in use today. Figure 10-3 shows an example of such a system being installed.

Rapid growth in the use of copper tubing during the 1940s further expanded the possibilities for hydronic radiant panel heating. Thousands of systems using copper tubing embedded in concrete floors as well as plaster walls and ceilings were constructed during the post–World War II housing boom. Many are still in use today. Example of an early radiant ceiling and floor panels using copper tubing are shown in Figure 10-4.

During the 1960s and 1970s, the use of hydronic radiant panel heating in America steadily fell victim to market forces that began substituting easier and less expensive heating options in place of proven comfort. As labor costs increased, and central cooling delivered by forced-air systems became readily available, hydronic radiant panel heating systems declined in popularity. The publicized failures of some hydronic systems due to corrosion or metal fatigue also discouraged further use.

Fortunately, at about the same time, new thermoplastic materials were under development in Europe that would eventually bring about a resurgence of interest in hydronic radiant panel heating. Cross-linked polyethylene (**PEX**) tubing that had been used in Europe since the early 1970s first appeared on the U.S. market during the early 1980s. PEX and other polymer-based tubing materials revolu-

Figure 10-4 | Examples of early North American hydronic radiant panels installed during the 1940s. (a) Copper tubing installed in a ceiling that will be covered by plaster. *Courtesy of John Siegenthaler.* (b) Copper tubing to be embedded in concrete floor. *Courtesy of John Siegenthaler.*

Figure 10-3 | Wrought iron piping being installed for floor heating. *Source: Byers Wrought Iron for Radiant Heating.*

tionized the installation of hydronic radiant panel heating, providing fast installation and a long, reliable service. The availability of PEX tubing was the spark that rekindled interest in hydronic radiant panel heating. That spark has now grown into a major flame. Over the last 40 years, many *billions* of feet of polymer tubing have been installed in hydronic radiant panel heating systems worldwide.

Today, hydronic radiant panel heating is being used in all types of buildings, including houses, schools, offices, aircraft hangers, fire stations, vehicle garages, and more. When combined with modern control technology, hydronic radiant panel heating represents what many consider the ultimate form of comfort heating.

10.5 Benefits of Radiant Panel Heating

A properly designed and installed hydronic radiant panel heating system offers many benefits. Some are inherent to radiant heating. Others are inherent to the use of hydronics to deliver heat to the panels. Still others are simply the elimination of undesirable side effects associated with other methods of heating.

These benefits include:

- A system that delivers *unsurpassed comfort*. Most people who have had an opportunity to compare the comfort offered by a properly installed radiant panel heating system will testify that it delivers comfort superior to other methods of heating. The underlying reasons are discussed in the next section.

- A system that is *inconspicuous*. Few people enjoy looking at portions of a heating system that, out of necessity, are located within an occupied space. Forced-air registers and even hydronic baseboard convectors can sometimes interfere with furniture placement. In contrast, many types of hydronic radiant panels are part of the building, and completely out of sight within floors, walls, or ceilings.

- A system that is *extremely durable*. Because radiant panels are literally built into the structure, they are usually well protected from physical damage. This is especially true for heated concrete floor slabs, which can withstand punishing traffic in commercial or industrial environments without damage to the embedded tubing.

- A system with virtually *no operating noise*. A properly designed and installed radiant panel system operates in virtual silence. It does not interfere with the serenity of the space it heats. Modern heat source and circulators operate with minimal noise and are usually located in a mechanical room that can be acoustically isolated from living spaces. There are no noise-generating fans or blowers located in the occupied spaces.

- A system that *minimizes drafts and dust movement*. A common complaint about forced-air heating systems is that they create drafts and redistribute dust, pollen, pathogens, odors, or other airborne pollutants within a building. Radiant panel heating creates very gentle (imperceptible) air motion within individual rooms rather than the building as a whole.

- A system that is *compatible with low-temperature heat sources*. Many types of radiant panels can operate at relatively low water temperatures, even at design load conditions. This allows low-temperature heat sources such as heat pumps, solar collectors, and condensing boilers to supply heat while operating at relatively high efficiency.

- A system that *quickly dries floors*. Radiant floor heating can quickly dry floors in areas such as bathrooms, entry foyers, and hallways, and therefore reduce the chances of slips and falls. Rapidly dried entry floors also reduce carpet soiling in others areas of the building. In buildings such as the highway garage shown in Figure 10-5, dry floors provide better conditions for maintenance, especially when work needs to be performed under vehicles.

- A system with *thermal storage*. Some hydronic radiant panels, such as a heated concrete floor slab have high **thermal mass**. Such panels can store large quantities of heat, allowing them to deliver a "surge" of heat in situations where large doors are open and cold outside air pours in. Normal comfort is restored very quickly after the doors are closed. This characteristic is particularly desirable in garages or aircraft hangers. The ability of a floor slab to store heat also protects the building against freezing for perhaps 2 or 3 days if the heat source is not operating.

- A system capable of *fast response*. Not all hydronic radiant panels have high thermal mass. Some are specifically designed for low thermal mass, allowing them to quickly "turn on" and "turn off." Radiant walls and ceilings are usually of low thermal mass construction, and well suited for situations where comfort needs to be quickly established after a prolonged temperature setback period. They can also respond quickly when significant internal heat gains occur from sunlight, occupants, or other sources.

- A system that is *easily zoned*. Hydronic radiant panel heating, like other types of hydronic heating can be easily configured for **room-by-room zoning**. Sleeping areas can be maintained cool while bathrooms are maintained warm. Heat input

Figure 10-5 | The ability to rapidly dry floors is very beneficial in buildings such as this highway garage. *Courtesy of John Siegenthaler.*

can be quickly interrupted to areas experiencing solar heat gains. Rooms that are not frequently used can be maintained at reduced temperatures to reduce energy consumption.

- A system that *reduces energy usage*. There are many documented examples of fuel savings associated with hydronic radiant panel heating. The underlying reasons include the ability to provide comfort at low interior temperatures, lower operating water temperatures, reduced air temperature stratification, reduced air leakage, and the ability to provide room-by-room zoning. A conservative estimate of fuel savings associated with the use of radiant floor heating relative to convective heating is 15%. Some buildings have demonstrated savings of over 50%.

10.6 Physiology of Radiant Panel Heating

The most sought after benefit of radiant panel heating is superior thermal comfort. Such comfort results from the interaction of several factors, not the least of which is heat exchange between the human body and a radiant panel. This section focuses on this interaction.

Our bodies constantly produce heat through a process called **metabolism**. For an adult at rest, the rate of heat production is normally in the range of 350 to 400 Btu/h. To remain comfortable, this heat must be dissipated at the same rate it is generated. *If heat is released faster than it is generated, a person feels uncomfortably cool. Likewise, if heat cannot be released as fast as it is generated, the person feels uncomfortably warm.* Under some conditions the rate of heat released from one portion of the body, the feet for example, can become disproportionately high. This will also adversely affect comfort. Maintaining comfort is thus a matter of providing environmental conditions that allow the body to release heat at the same rate it is produced. This balance also needs to be maintained without conscious effort on the part of the occupant.

Heat is released from the body through four processes:

- Evaporation of moisture
- Convection of heat to surrounding air
- Conduction of heat to objects in contact with the body
- Radiant heat transfer to cooler surfaces surround the body

Evaporation of water from the body accounts for approximately 25% of the heat released from the body during light activity under typical indoor conditions. As the indoor humidity increases, this process becomes less effective as it does during periods of hard physical work or vigorous exercise. At other times, evaporation of moisture from the skin can be extremely effective and noticeable. For example, standing in a breeze with wet skin or clothing can produce very noticeable, sometimes even excessive, cooling.

Our bodies also lose heat by convection to cooler surrounding air. About 30% of the body's heat production is lost through convection under typical interior conditions.

The speed at which air moves across our skin and clothing surfaces greatly affects convective heat loss. The faster the air moves past the body, the more effective the process becomes. This explains why a room fan that simply increases the air velocity past the body without cooling the air can improve comfort on a hot summer day. The increase in convective heat loss with increasing air speed also explains the "**windchill**" **effect** experienced on cool days.

Heat loss by conduction is usually a small percentage of the body's total heat loss. Examples include heat flowing downward from feet placed on a cool floor or heat loss to a cold car seat on a winter morning. In most interior situations, conduction heat loss is usually less than 5% of total heat loss.

Heat loss by thermal radiation is the predominant method by which the body releases heat. Under typical indoor conditions, thermal radiation from skin and clothing surfaces to surrounding surfaces accounts for about 50% of total heat loss. The rate of radiant heat transfer from the skin and clothing is very dependent on the surface temperatures in the room. Cool surfaces such as windows can serve as very effective **heat sinks** for thermal radiation. They can produce discomfort even when room occupants are surrounded by air temperatures that would otherwise be considered comfortable.

Because a large portion of the body's heat loss is by thermal radiation, comfort is greatly influenced by the temperature of the surfaces in a room. Increasing the temperature of one or more of the larger surfaces can significantly reduce the rate of heat release by thermal radiation from the body. During the heating season, when the objective is to limit the rate of heat loss from the body, moderately warmed room surfaces can greatly improve comfort.

Room Air Temperature Profiles

The priority of the body is to maintain the temperature of its critical organs in the central torso. When placed in an environment that allows heat to be released from its surface faster than it is being produced internally, the body responds by reducing blood flow to its extremities (i.e., hands and feet).

The feet in particular are most affected because they are furthest removed from the central body. To make matters worse, the feet are usually in contact with cooler surfaces that tend to draw heat away by conduction. They also have a greater surface area per unit of mass than do other parts of the body. To put it in mechanical terms, the feet are like large radiators fastened to cold walls at the far end of a hydronic distribution system. The heat they receive from the bloodstream is easily dissipated to cooler surroundings.

The head, on the other hand, usually has a good supply of heat-carrying blood. It is also insulated, to varying degrees, by hair. This combination allows the head to be comfortable at air temperatures several degrees cooler than those required near the feet. Most people feel comfortable and more alert when air temperatures at head level are in the low- to mid-60 °F range.

Figure 10-6 shows the variation of air temperature from floor to ceiling needed for ideal comfort at a relaxed activity level. Such a graph is called a **room air temperature profile**. Notice that the temperature near the floor needs to be slightly higher than average. This reduces the rate of heat loss from the lower extremities.

The degree to which a heating system provides thermal comfort is in part determined by the room air temperature profile it creates within a space. Profiles similar to that shown in Figure 10-6 are desirable. Unfortunately, many common methods of adding heat to a room result in significant deviations from this ideal.

Figure 10-7 compares typical room air temperature profiles created by three common methods of heat

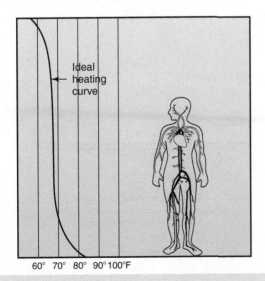

Figure 10-6 | Ideal air temperature distribution in a heated room. *Courtesy of Uponor.*

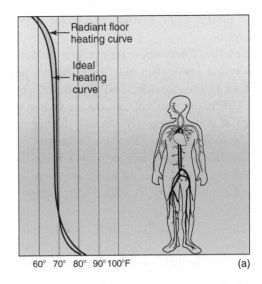

(a)

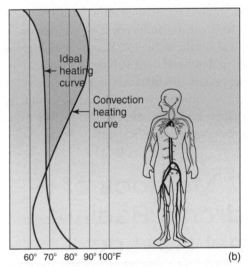

(b)

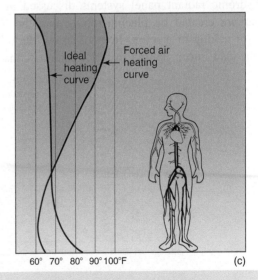

(c)

Figure 10-7 | Comparison of temperature profiles of three types of heating delivery systems. (a) Radiant floor heating profile, (b) baseboard convector profile, (c) forced-air profile. *Courtesy of Uponor.*

delivery: radiant floor heating, baseboard convectors, and forced air.

The profiles for baseboard convectors and forced-air systems reveal lower than ideal temperatures near the floor, as well as higher than ideal temperatures near the ceiling. This is the result of introducing heated air into the space at temperatures significantly higher than the desired room temperature. The "overheated air" quickly rises toward the ceiling, while the higher density cool air collects at floor level, exactly where it is least desired. This undesirable effect is called **air temperature stratification**.

In comparison, radiant floor heating creates a temperature profile very close to the ideal. Since room air is not "overheated," undesirable stratification does not occur.

The room air temperature profiles shown in Figure 10-7 can be influenced by air motion, ceiling height, and insulation levels. Rooms with high rates of air flow and well-designed ducting systems will have less temperature stratification due to vigorous mixing of room air. Rooms with tall ceilings, high register temperatures, and poor air circulation will experience greater temperature stratification.

10.7 Methods of Hydronic Radiant Panel Heating

All hydronic radiant panel systems discussed in this chapter are created by placing tubing into the floor, walls, or ceiling of a room. In some cases, the tubing is embedded into a poured material such as concrete. In other cases, it is fastened to the surface of a building material. In still other cases, the tubing is suspended in a hollow air cavity such as the space between floor joists.

When warm water is circulated through the tubing, heat is released from it and diffused outward through the materials making up the radiant panel. Most of the heat is directed into the room by the proper selection and placement of insulation. A small portion is released from the rear side of the panel. Properly designed systems allow the heat to **diffuse** (i.e., spread out) between adjacent tubes so that surface temperature variations are small. These concepts are illustrated in Figure 10-8. The rate of heat flow from the radiant panel depends on its construction, its operating water temperature, and the thermal resistance of any finish material placed over it.

Classification of Hydronic Radiant Panels

Several methods have been developed for incorporating tubing into floors, walls, and ceilings. Within this text, hydronic radiant panel installation methods are classified as follows:

1. Radiant floor panels
   a. Slab-on-grade
   b. Gypsum thin slab
   c. Concrete thin slab
   d. Above floor tube and plate
   e. Below floor tube and plate
   f. Suspended tube
   g. Plateless staple-up
   h. Prefabricated subfloor/underlayment panels
2. Radiant wall panels (tube and plate)
3. Radiant ceiling panels (tube and plate)

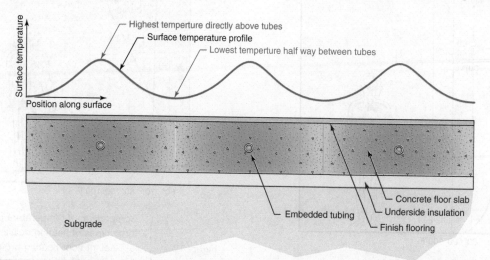

Figure 10-8 | Surface temperature profile above a heated concrete slab.

Each installation method offers technical or economic advantages in certain situations. Each method also has limitations that must be respected by designers and installers. Hydronic heating professionals should learn how to deploy each type of radiant panel in circumstances where it offers advantages. Installation methods for each of these radiant panels are detailed in upcoming sections of this chapter.

10.8 Slab-on-Grade Radiant Floors

The most economical hydronic floor heating systems are those where a concrete slab is already planned as the floor structure. In such situations, the extra cost of making the floor into a radiant panel is limited to the tubing, **underslab insulation**, and the labor to place these materials. The cross-section of a typical concrete slab-on-grade heated floor is shown in Figure 10-9. A cutaway view is also shown in Figure 10-10.

Slab-on-grade radiant panels usually operate at low water temperatures, especially "bare" slabs that are not covered with a finish flooring material. The concrete provides a good "thermal wick" to diffuse heat outward and upward through the slab. A computer simulation of such heat diffusion is shown in Figure 10-11.

The curved lines in the cross-section are called **isotherms**. Each isotherm is created by connecting points that are at the same temperature. Heat flow paths at all

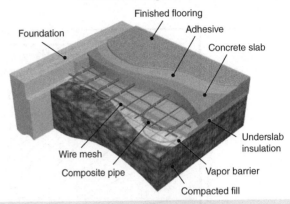

Figure 10-10 | Cutaway view of a typical slab-on-grade radiant floor panel. *Courtesy of IPEX, Inc.*

locations are perpendicular to the isotherms, as shown by the red arrows, and always in a direction of decreasing temperature. Isotherms can be thought of as "waves" of heat moving away from the tubing. Notice how heat moves laterally away from the tubing, then upward toward the surface of the floor. Some heat also moves downward and is transferred to the soil beneath the insulation.

Effect of Tube Size

The size of the tubing used in a heated slab has a relatively small effect on the slab's heat output. Larger tubes have more surface area through which heat can pass to the concrete and thus slightly increase heat output for a given water temperature. However, overall heat transfer from the floor is determined by the thermal properties and placement of *all* materials within and on the floor, rather than just the outer surface area of the embedded tubing. Figure 10-12 shows the predicted thermal performance variations for several different tube diameters in a specific concrete slab installation.

Tube sizes are primarily based on head loss considerations and circuit lengths. Larger tubing sizes allows for longer circuits without excessive head loss. Circuit lengths will be discussed later in this chapter.

Tube Depth Within a Slab

Questions often arise about the effect that tubing depth within a concrete slab has on thermal performance. Some argue, often without evidence, that tubing depth "makes no difference" on the subsequent heating performance of the slab. If this is true, it would certainly simplify installation to just leave the tubing at the bottom of the slab and cover it with concrete. However, once that concrete is poured, any opportunity for changing the tubing depth is gone, and the resulting performance

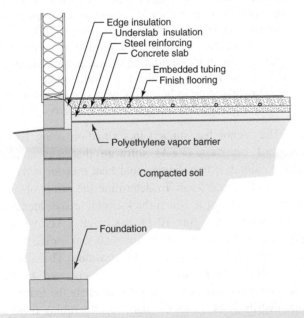

Figure 10-9 | Cross-section of a typical slab-on-grade radiant floor.

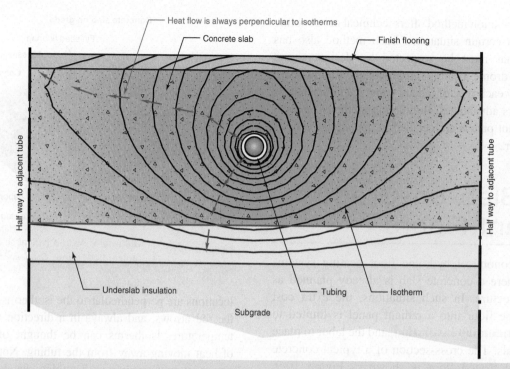

Figure 10-11 | Temperature isotherms and heat flow from warm tubing embedded in a concrete slab-on-grade floor.

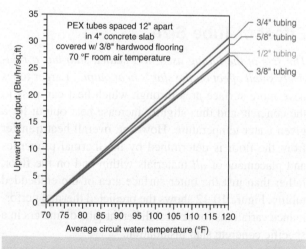

Figure 10-12 | Effect of tubing diameter on heat output from a slab-on-grade floor.

is fixed for the life of the building. The irreversible and long-term nature of this situation justifies further study to determine how much if any effect tubing depth has on slab performance. The results of such a study could then be used to make informed decisions about tube placement within the depth of the slab.

There are several ways tubing depth should theoretically effect the performance of a heated slab.

- The deeper the tubing, the greater the thermal resistance between it and the floor surface. The higher this resistance, the higher the required water temperature to maintain a given rate of heat transfer.

- The closer the tubing is to the bottom of the slab, the greater the underside heat losses should be. This is true with or without underslab insulation. Obviously, the losses are greater in the latter case.

- When the tubing ends up near the bottom of the slab more of the slab's thermal mass is above the level where heat is being added. This lengthens the time it takes to warm the floor surface to normal temperatures following a call for heat. It also lengthens the cool down time after heat input is interrupted by system controls. A warm slab often holds sufficient heat to maintain comfort for several hours after further heat input by circulation of warm water through the embedded tubing ends. The result can be significant overheating in buildings with significant internal heat gains from sunlight or other sources.

The author has studied this situation using **finite element analysis (FEA)** software that was configured to simulate two-dimensional heat transfer across a section of heated floor. To determine the effect of tube depth, a model of a 4-inch-thick concrete slab identical to that shown in Figure 10-11 was created. The vertical position of the tubing within the slab was varied while all other conditions were held constant. The shaded areas shown in Figure 10-13 were generated by the finite element simulations and approximate the temperature distributions within the slab with the tubing fixed at three different depths.

The FEA models were also run for 4-inch-thick bare concrete slabs. The surface temperature profiles for each

10.8 Slab-on-Grade Radiant Floors

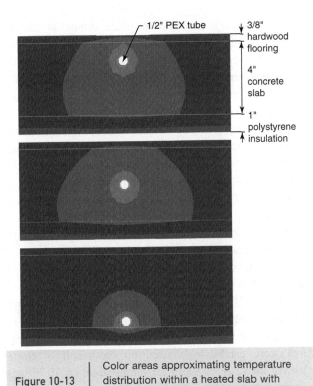

Figure 10-13 | Color areas approximating temperature distribution within a heated slab with tubing fixed at three different depths.

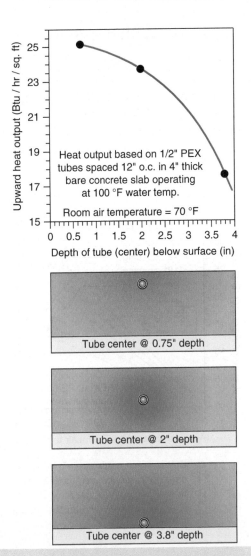

Figure 10-14 | Effect of tubing depth on upward heat output from a slab-on-grade floor.

Upward heat output (Btu/h/ft²)	Average circuit water temperature with tubing at ½ slab depth (°F)	Average circuit water temperature with tubing at bottom of slab (°F)
15	95	102
30	120	134

Figure 10-15 | Effect of tubing depth on the required average circuit water temperature for two rates of upward heat flow.

case were used to estimate the upward heat output from the floor. The results are shown in Figure 10-14.

Figure 10-14 clearly shows that *upward heat output from the floor decreases as the tubing is placed deeper within the slab*. From the standpoint of heat transfer only, the tubing should be placed near the top of the slab. However, this is impractical, especially in slabs with sawn control joints.

Considering that the tubing in most on-grade concrete slabs is fastened to the steel wire reinforcing and that this reinforcing should be placed between one-third and one-half of the slab depth from the top for structural reasons, it is reasonable to accept the thermal performance associated with tubing at approximately one-half slab thickness.

Experience has shown that heated concrete floor slabs still function when tubing is placed at or near the bottom of the slab. However, higher water temperatures are necessary to provide a given rate of upward heat flow. Figure 10-15 gives the average circuit water temperatures necessary for two rates of upward heat delivery in a 4-inch-thick bare concrete slab with tubing at mid-slab depth and at the bottom of the slab.

Higher circuit water temperature lowers the efficiency of contemporary heat sources such as geothermal heat pumps, air-to-water heat pumps, solar collectors, and condensing boilers. It also increases the rate of downward heat loss to the soil. The time required to warm the surface of the slab after a setback period also increases when tubing is located deeper in the slab.

In summary, tubing depth within a slab *does* effect thermal performance, in some cases significantly. This analysis suggests that provisions for maintaining the tubing at approximately 1/2 the slab's depth are justified and prudent.

Tube Spacing Within a Slab

The closer tubing is spaced within a concrete slab, the greater its heat output rate, all other conditions being the same. This is also true of all other radiant panel constructions discussed in this chapter. Figure 10-16 shows the relationship between tube spacing and upward heat output for a 4-inch-thick bare concrete slab with good underside insulation.

Although it's possible to place tubing at almost any spacing, the type of panel construction often favors certain spacings. For example, concrete slab-on-grade floors are often reinforced with 6×6 inch welded wire fabric. It's convenient to space tubing at multiples of this 6-inch grid (e.g., 6-, 12-, 18-, or 24-inch spacing).

Tube spacing is also based on acceptable variations in floor surface temperature. In residential applications where **"barefoot friendly" floors** are desired, tube spacing should not exceed 12 inches. Closer spacings such as 9 or 6 inches are used in areas needing higher heat output.

Wider variations in floor surface temperature are generally acceptable in industrial buildings such as garages or manufacturing facilities. Tube spacings of 18 and 24 inches can sometimes be used.

Figure 10-17 shows floor surface temperature profiles for tubing spaced 12, 18, and 24 inches apart in a 6-inch-thick bare concrete slab. These surface temperatures were established based on the result of finite element analysis. A room air temperature of 70 °F was assumed during the simulations. The water temperature in all tubes is 100 °F. Soil temperature under the slab insulation is assumed constant at 65 °F.

The wider the tube spacing, the greater the variation in floor surface temperature between the "peak" surface temperature just above the tubing, and the "valley" temperature halfway between adjacent tubes. The **average floor surface temperature** also decreases with wider tube spacing, as does the upward heat output from the floor.

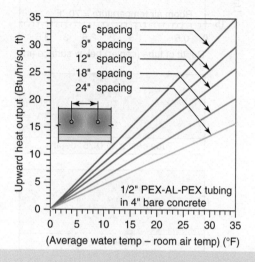

Figure 10-16 | Upward heat output from a 4-inch-thick bare concrete slab versus difference between average water temperature in tube and room air temperature, for different tube spacings.

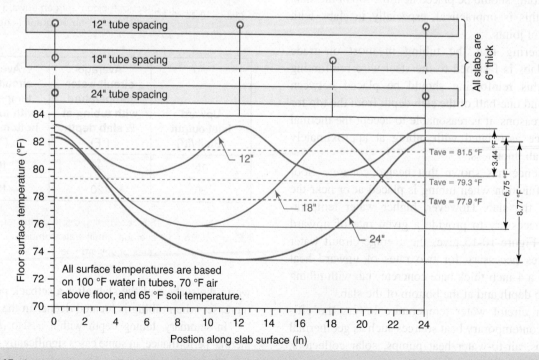

Figure 10-17 | Variation in floor surface temperature for 6-inch bare concrete slabs with different tubing spacings.

Tube spacing (in)	Average circuit water temperature (°F) required for upward heat flow of 15 Btu/h/ft²	Average circuit water temperature (°F) required for upward heat flow of 30 Btu/h/ft²
12	88.5	108.9
18	94	119.6
24	101.3	134.9

Figure 10-18 Average circuit water temperatures required to maintain an upward heat output of either 15 or 30 Btu/h/ft² in a 6-inch-thick bare concrete slab.

Figure 10-18 lists the average circuit water temperatures required to produce a fixed upward heat transfer rates of 15 and 30 Btu/h/ft² in a room being maintained at 68 °F. Wider tube spacings require higher average water temperature to achieve a given heat output. *Keep in mind that the supply temperature to floor heating circuits is typically 5 to 10 °F higher than the average water temperature.*

The "trade-off" with wider tube spacing is reduced installation cost versus operating the radiant panel at higher water temperatures. When a conventional boiler is used, the increased water temperatures will have relatively little effect on seasonal fuel efficiency. However, higher water temperatures could have a significant *negative* effect on the efficiency of heat sources such as condensing boilers, geothermal heat pumps, air-to-water heat pumps, biomass boiler systems, and solar thermal collectors.

Slab-on-Grade Installation Procedure

The installation of heating tubing in a slab-on-grade floor is an integral part of the construction sequence of a building. It must be preplanned and closely coordinated with the other trades involved in the construction.

Part of this preplanning is to develop a **tubing layout drawing**, which is described later in this chapter. Without an accurate plan showing the location and length of all piping circuits, as well as the location of the manifold stations, even an experienced installer can spend hours attempting installation by trial and error. Even then, the results are likely to be less than ideal.

The installation procedure to be described assumes that a layout plan has been prepared, and thus the installer's main function is to place the tubing according to this plan. The procedure also describes installation of a vapor barrier and underslab insulation. On some projects, these materials may be installed by other contractors. However, the heating contractor should always verify they have been properly installed before installing tubing.

All underslab plumbing or electrical conduit should be in place before beginning tubing installation. All underslab utility trenches must be filled and properly compacted. The final grade under the slab should be accurately leveled, allowing for the thickness of the slab and the underslab insulation. This is especially critical near the edges of the slab where thicker insulation is used. Any loose rocks should be raked off to provide a smooth stable surface.

Begin by placing a **vapor barrier** over the **subgrade**. Polyethylene sheeting is often used as this vapor barrier. Its purpose is to minimize water vapor migration from the soil into the slab during warm weather. Such migration, if unchecked, can discolor wood flooring placed over the slab. In areas with high radon potential, a special radon/moisture barrier may also need to be installed.

Underslab insulation is then installed over the vapor barrier. Although several options are available, **extruded polystyrene** foam is a commonly used material with a proven track record in such applications. It is available in several thicknesses and compressive stress ratings. It does not absorb moisture or outgas compounds that reduce its *R*-value over time. Designers should verify that local codes allow placement of foam insulation below concrete slabs, especially in areas where termites are present.

The insulation should be neatly placed with all tongue and groove joints interlocked. If the site is windy, place sheets of welded wire fabric or wooden planks over the foam sheets as they are placed to avoid wind uplift.

The recommended *R*-value of underslab insulation varies with geographic location. In the absence of specific codes that require higher values, the Radiant Professionals Alliance recommends that the *minimum R*-value for underslab insulation be based on Equation 10.1.

Equation 10.1:

$$R_{min} = 0.125(T_i - T_o)$$

where,

R_{min} = minimum *R*-value of underslab insulation (°F·h·ft²/Btu);

T_i = inside air temperature to be maintained at design load conditions (°F);

T_o = outside air temperature at design load conditions (°F).

Example 10.1

Determine the minimum *R*-value of underslab insulation in a climate where the outside design temperature is −5 °F, and for a space that is to be maintained at 70 °F.

Solution:

Substituting numbers into Equation 10.1 yields:

$$R_{min} = 0.125(T_i - T_o) = 0.125(70 - (-5))$$
$$= 9.4 \; (°F \cdot h \cdot ft^2/Btu)$$

Discussion:

The availability of insulation in various materials and thicknesses will in part determine the *R*-value installed. In this case, the calculated *minimum R*-value is close to that of 2-inch extruded polystyrene insulation board ($R = 10.8$). Designers should also consider the life cycle cost of thickness underslab insulation options relative to projected future fuel costs. Higher *R*-values may be justified. Always keep in mind that the underslab insulation installed will likely remain in place for the life of the building, and that "retrofitting" for higher underslab *R*-values would be extremely difficult and costly.

Underslab insulation should *not* be installed under floor areas that support high structural loads from columns or bearing walls. Such columns and **bearing walls**, if present, are generally several feet away from the building perimeter, and thus omitting a relatively small area of underslab insulation will have minimal effect. Figure 10-19 shows an area where underslab insulation has been omitted at a structure pad that will eventually support a column, as well as under a bearing wall. The black polyethylene vapor barrier remains intact across these areas.

Figure 10-20 shows 2-inch extruded polystyrene insulation with a tapered upper edge installed on the inside of a stem wall foundation associated with a slab on grade floor. The tapered edge allows concrete to be placed close to the foundation wall. With this detail, the edge of the slab will eventually be covered with framing and/or finish materials.

Most concrete slab-on-grade floors are reinforced with **welded wire fabric (WWF)** or steel rebar. When welded wire fabric is used, it should be placed over the insulation, with adjacent sheets lapped a minimum of 6 inches at all edges and wire tied together.

The next step is to use the tubing layout drawing to locate the **manifold stations** within the building. These are the locations where tubing penetrates up through the slab surface and eventually connects to a manifold station.

Manifold stations are often located where a wall cavity will eventually be built around them. It is crucial to carefully measure and place the manifold station at the precise location where such walls will eventually be built. A temporary manifold station support is constructed by driving two 3-foot pieces of steel rebar through the insulation and into the soil. A small plywood panel can then be fastened to these steel rods using conduit straps. The manifold station can then be screwed to the plywood panel as shown in Figure 10-21. After the slab is poured and the manifold station is supported within the wall cavity, the rebar can be cut off flush with the slab surface using an angle grinder, and removed. *Be sure the manifold station faces a direction where it can be accessed when the surrounding wall is completed.*

After the manifold stations are placed, use the tubing layout drawing to mark portions of each tube circuit on the foam insulation using marking paint intended for downward spraying. Corners, return bends, and offsets should be marked as shown in Figure 10-22. This greatly speeds installation and reduces the chance of mistakes as the tubing is placed. Be sure to also mark flow direction

Figure 10-19 | Underslab insulation is omitted under load bearing columns and walls. *Courtesy of John Siegenthaler.*

Figure 10-20 | Extruded polystyrene edge insulation installed along an exposed foundation wall. *Courtesy of John Siegenthaler.*

10.8 Slab-on-Grade Radiant Floors

the coil from the larger number at the other, the length can be quickly determined.

To prevent twisting, the tubing must be unrolled from the coil. *Never pull tubing off the side of a coil.* Most installers make use of an **uncoiler**, such as the one shown in Figure 10-23. This device allows the coil to freely spin as tubing is pulled from it. The uncoiler should be placed several feet away from where the tubing is being laid down. Several feet of tubing should be pulled off the coil to allow ample slack.

Connect the free end of the tubing coil to the supply manifold. Pull several feet of tubing off the coil and proceed along the supply side of the circuit, fastening the tubing along the way. Bends in PEX or PEX-AL-PEX tubing in sizes up to 3/4 inch can be made by hand, taking care not to kink the tubing in the process. Bends in larger diameter tubing or especially tight bends may require a tube bender. Be sure to verify the minimum bend radius allowed with the tubing manufacturer. PEX and PEX-AL-PEX tubing always bends easier when warm.

Figure 10-21 — Temporary manifold during tubing installation. This support must be accurately placed to ensure it places manifold within partition, or on the finish surface of an eventual wall. *Courtesy of John Siegenthaler.*

arrows to help ensure that the ends of each circuit are connected to the proper manifold.

The tubing circuits can now be placed one at a time. Look up the length of the circuit being created on the tubing layout drawing. Select a coil of tubing long enough to create the circuit *without splicing*. Most tubing sold for radiant panel applications comes with sequential length numbers printed every 3 feet along the tube. By subtracting the smaller number at one end of

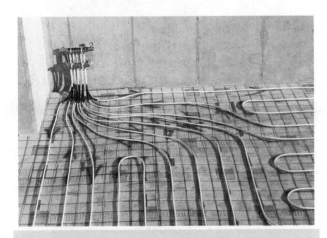

Figure 10-22 — Spray paint is used to mark portions of each tubing circuit as well as flow direction on underside insulation. *Courtesy of Harvey Youker.*

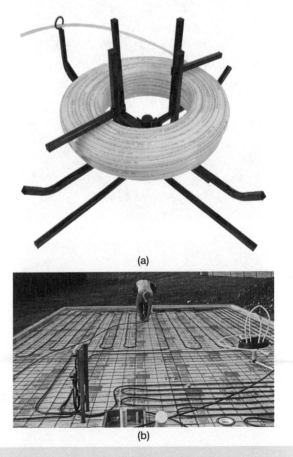

Figure 10-23 — (a) A modern uncoiler. *Courtesy of Uponor.* (b) The uncoiler, seen at right, allows tubing to be pulled off the coil without twisting. Maintain ample slack in tubing to allow efficient placement. *Courtesy of John Siegenthaler.*

Tubing is usually fastened to the welded wire fabric or steel reinforcing using wire twist ties, nylon pull ties, or some type of plastic clip. The manufacturer may require that a specific type of fastener be used to retain the warranty on the tubing. Typical spacing for ties or clips is 24 to 30 inches on straight runs and at two or three locations on each return bend. Figure 10-24 shows tubing being secured to welded wire fabric using wire bag ties.

As tubing is placed, watch for any sharp ends on steel reinforcing. Either bend the reinforcing out of the way or slightly offset the tubing to avoid chafing its surface.

Some tubing suppliers require the installation of **bend supports** where the tubing bends from horizontal to vertical under the manifold stations. Some installation instructions also call for the tubing to be sleeved where it penetrates through the slab surface. Verify the recommended details with the tubing manufacturer. Tubing that has been sleeved where it will penetrate the slab surface can be seen in Figure 10-25.

Control Joints

Most concrete slabs require **control joints**. The purpose of control joints is to intentionally weaken the slab at specific locations so shrinkage cracking occurs along the control joint rather than randomly across the slab. Control joints are usually made using a special saw the day after the slab is poured. The depth of the cut is often specified to be 20% of the slab's thickness.

The location of control joints should be marked on the underslab insulation using marking paint before fastening the tubing in place as seen in Figure 10-27.

Figure 10-26 shows how tubing should be detailed where it passes beneath a sawn control joint. The tubing is covered with a thin plastic sleeve that is centered on the eventual location of the control joint. This sleeve prevents the concrete from bonding directly to the tubing at this location. This minimizes stress on the tubing during any subsequent minor movement of the slab. Sleeving material is available from most tubing manufacturers for this purpose. The thin-wall sleeving can be slit with a utility knife and wrapped around the tubing after the circuit has been fastened in place.

The tubing and welded wire fabric should *not* be lifted where it passes beneath control joints. This ensures sufficient space between the tubing and the bottom of the saw cut.

(a)

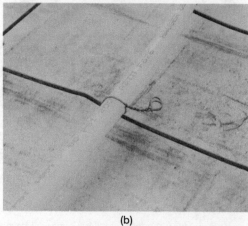

(b)

Figure 10-24 (a) Securing PEX-AL-PEX tubing to welded wire reinforcing using wire ties and "J-hook" twisting tool. *Courtesy of John Siegenthaler.* (b) Close-up of completed wire tie with loop pressed to side. *Courtesy of John Siegenthaler.*

Figure 10-25 Tubing with protective sleeving installed where it will penetrate slab surface. *Courtesy of Deven Youker.*

10.8 Slab-on-Grade Radiant Floors

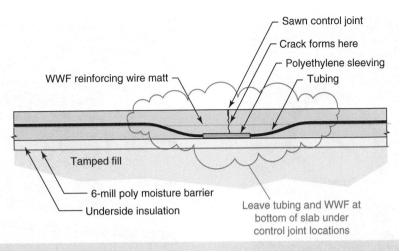

Figure 10-26 | Detail where tubing passes under sawn control joints.

Figure 10-27 | Thin wall polyethylene sleeving over tubing where it will pass under a sawn control joint. Location of the control joint is marked by the line painted on underslab insulation. *Courtesy of John Siegenthaler.*

Figure 10-28 | Hold tubing at least 8 inches away from closet flanges to prevent softening of wax ring. *Courtesy of John Siegenthaler.*

Another important installation detail is to hold tubing at least 8 inches away from any toilet drainage piping as seen in Figure 10-28. This minimizes heating of the wax ring that is eventually installed on the **closet flange**.

Pressure Testing

Although leaks due to "defective tubing" are extremely unlikely, it is possible that careless equipment handling on the job site could damage the tubing before it is embedded in concrete. To ensure that no leaks are literally cast into concrete, *all circuits should be pressure tested prior concrete placement.*

Pressure testing is best done using compressed air, which is not subject to freezing during cold weather construction. A typical pressure-testing specification requires that all tubing circuits to be pressurized to and maintained at 60.0 to 75.0 psi for at least 24 hours.

A convenient approach is to install a pressure gauge on either the supply or return manifold at each manifold station, along with a Schrader valve for adding air. Plug all remaining manifold connections and vents, then pressurize the assembly using an air compressor. An example of a temporary pressure test setup is shown in Figure 10-29.

Once the assembly has reached its test pressure, spray a soapy water solution on all manifold connections to check for any air leaks.

Figure 10-29 | Temporary installation of pressure gauge and Schrader valve for air pressure testing circuits prior to concrete placement. *Courtesy of John Siegenthaler.*

Figure 10-30 | One of several photos used to document tubing placement for future reference. Documentation is especially important where nonuniform tube routing is used. *Courtesy of Harvey Youker.*

The air test pressure reading can change to some degree with differences in air temperature or sunlight on the tubing. If the pressure drops significantly within a few hours, it is very likely the leak is at one of the manifold connections. Retighten and recheck with soap solution until the system is airtight. If the system will be left unattended for some time, the air pressure should be reduced to about 30 psi for safety. The tubing circuits should remain under slight (10–20 psi) pressure while the slab is poured. If a tube were accidentally punctured during the pour, the compressed air would give an immediate indication of the leak. The leak could then be repaired using a coupler specifically designed for that purpose.

Photo Documentation

After all circuits have been placed, several photographs of the tubing installation should be taken. Photos should be taken around each manifold station, as well as in any areas where tubing placement is tight or routed around objects such as the foundation pad in Figure 10-30. The use of digital cameras or cell phone cameras allows such photos to be quickly gathered and stored. The building owner should have access to all tubing installation photos for future reference.

Placement of Concrete

There is no standard procedure for placement of concrete slabs. From the standpoint of the tubing, the placement procedure used should minimize heavy traffic over the tubing and ensure that the tubing ends up at approximately mid-depth in the slab (other than where it passes beneath control joints).

Wheelbarrow traffic over tubing that is under pressure is generally not a problem; however, care should be taken not to pinch the tubing under the nose bar of the wheelbarrow as it is dumped. So-called "power buggies" are too heavy to be driven over the tubing and insulation. Likewise, concrete trucks should never be driven over these materials.

On small slabs, the concrete can often be placed directly from the chute of the delivery truck as seen in Figure 10-31. For larger slabs, the use of a concrete pump truck equipped with an **aerial boom** is an ideal

Figure 10-31 | Placement of concrete over tubing directly from truck chute. *Courtesy of John Siegenthaler.*

way to efficiently place the concrete with minimal heavy traffic over the tubing as seen in Figure 10-32.

As the concrete is placed, lift the welded wire reinforcing and attached tubing to approximately mid-depth in the slab (other than at control joints). Ideally, one or more workers should be assigned solely to this task. The mesh is pulled up using a **lift hook** as shown in Figure 10-33. After the coarse (stone) **aggregate** in the concrete flows under the mesh, it tends to support it quite well.

After the concrete is placed, it is finished in the usual manner. Special care should be taken not to nick the tubing with trowels where it penetrates the slab under the manifold stations. Careless operation of tools such as power trowels can sever tubing circuits and lead to costly repairs.

The installation procedures just described call for good coordination of trades at the job site. It is important for the concrete placement crew to understand details such as proper tubing depth and control joint detailing before beginning concrete placement.

10.9 Concrete Thin-Slab Radiant Floors

There are many buildings where radiant floor heating is desirable, but where slab-on-grade construction is not possible. One of the most common is a wood-framed floor deck in a residential or light commercial building.

One alternative is to install a **concrete thin-slab** radiant panel over the **subfloor**. A cross-sectional drawing of this approach is shown in Figure 10-34. A cutaway view is shown in Figure 10-35.

Thin-slab systems use the same type of PEX or PEX-AL-PEX tubing as slab-on-grade systems. The tubing is fastened directly to the wood subfloor and then covered with a special concrete mix. Underside insulation is usually installed between the floor framing under the subfloor.

As with slab-on-grade systems, the concrete thin slab provides an effective "thermal wick," allowing heat to diffuse laterally outward away from the tubing. The slab also provides moderate thermal mass to stabilize heat delivery.

Because the slab is thinner than a typical slab-on-grade floor, lateral heat diffusion is slightly less efficient. This is compensated for by operating the panel at a slightly higher average water temperature. The thermal performance of concrete thin slabs is discussed in Section 10.18.

Concrete thin-slab systems bring a number of issues into the planning process, not the least of which is the added weight of the slab on the floor deck. A typical 1.5-inch-thick concrete thin slab adds about 18.0 pounds per square foot to the **dead loading** of a floor. The floor-framing system must be capable of supporting this added load while remaining within code-mandated stress and deflection limits. Although significant, this extra weight can usually be accommodated by adjusting the spacing, depth, or width of floor framing. A competent structural designer or engineer should assess the necessary framing to support the added load. An often-cited **maximum deflection criteria** for such floors is 1/600th of the floor's clear span under full live loading.

| Figure 10-32 | Ariel boom and pump used to efficiently place concrete for a large slab-on-grade heated floor. *Courtesy of John Siegenthaler.* |

| Figure 10-33 | Mason using a lifting hook to pull welded wire reinforcing and attached tubing to approximately mid slab height. Note that hook is lifting reinforcing wire rather than directly lifting tubing. *Courtesy of John Siegenthaler.* |

488 Chapter 10 Hydronic Radiant Panel Heating

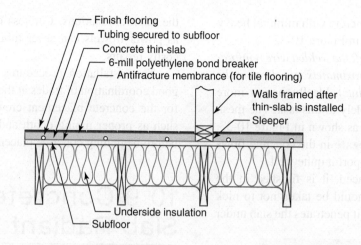

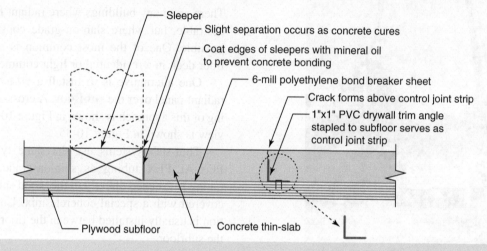

Figure 10-34 | Cross-section of a concrete thin-slab radiant floor panel.

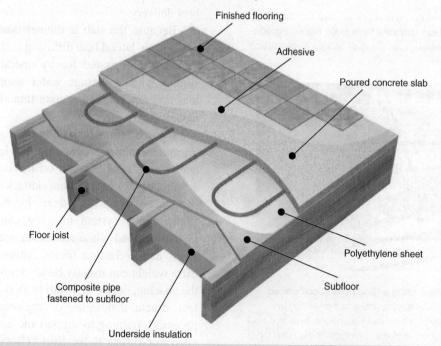

Figure 10-35 | Cutaway view of a typical concrete thin-slab radiant floor panel. *Courtesy of IPEX, Inc.*

Another consideration is the added height of the thin slab. In most installations, thin slabs add 1.5 to 2 inches the height of the floor deck. This affects the height of window and door **rough openings**, stair riser heights, and rough-in heights for closet flanges. Such adjustments are easily made if the thin slab is planned for when the building is designed. However, these adjustments can be more difficult or even impossible if the decision to use a thin-slab system is made after the building is framed.

Installation Procedure for a Concrete Thin Slab (the Youker System)

The procedure to be described for creating a concrete thin slab was developed by hydronic heating professional and longtime friend, Harvey Youker. It is called the Youker system, and provides an efficient approach to installing a concrete thin slab in new construction. As with slab-on-grade systems, some portions of the procedure to be described may be delegated to other trades.

STEP 1: Begin by sweeping the floor deck clean. Use chalk lines to mark the location of all walls and partitions. Mark the locations of any drainage pipes that will penetrate through the thin slab. Use the tubing layout plan along with a lumber crayon or chalk line to mark key areas of each circuit path such as return bends or offsets directly on the subfloor.

STEP 2: Cover the area where the thin slab will be installed with translucent 6-mil polyethylene sheeting. This serves as a **bond breaker layer** between the subfloor and the concrete. It allows for slight differential movement between the subfloor and the slab without creating high tensile stresses that can crack the concrete. Overlap adjacent polyethylene sheets about one foot, and tape the seams to minimize water absorption into the subfloor before and during the pour. *Never use asphalt saturated roofing felt or other materials that may give off odors when heated for the bond breaker layer.*

STEP 3: Install wooden **sleepers** at all locations of interior partitions or exterior walls. The sleepers should match the thickness of the exterior walls or interior partitions (usually 2 × 4 or 2 × 6) as seen in Figure 10-36.

Do not install sleepers across interior door openings. Align the sleepers to the previously snapped chalk lines, which should be visible through the polyethylene sheeting. The sleepers will serve as **screed** guides when the thin slab is poured. Eventually the exterior walls and partitions will be nailed down directly over these sleepers.

Also install 2 × 4 dams, which are 1.5-inch thick, across areas where no concrete will be placed. These

Figure 10-36 Install 2 × 4 and 2 × 6 sleepers over polyethylene bond breaker sheet at all wall and partition locations. In this case tubing circuit locations are marked on the bond breaker sheet. *Courtesy of Harvey Youker.*

include the areas under base cabinets, as well as around stairwell openings, vertical chases, and under shower stalls.

Closet flanges for toilets are easiest to install prior to pouring the thin slab. Be sure the flange is supported with its top flush with the top of the slab. Cover the flange with duct tape to prevent concrete from spilling into it during the pour. If the closet flange is not installed, block off an area large enough to accommodate the piping with a piece of 1.5-inch foam insulation secured to the subfloor. The foam can be easily removed from the slab when the closet flange is installed.

STEP 4: Temporarily support the manifold station at the location shown on the tubing layout plan. A small piece of plywood supported on short blocks works well as a temporary support. Alternatively, the manifold station may be installed *under* the floor deck.

Roll out each tubing circuit in a manner similar to that described for slab-on-grade systems other than fastening. Use a pneumatic stapler equipped with a **depth stop attachment** to fasten the tubing to the subfloor. Staplers with depth stop attachments are usually available from tubing suppliers. They place staples with their crown just touching the top of the tubing. *Never attempt to fasten tubing in place with a stapler not specifically configured for that purpose.* Slide the stapler along the tubing, placing a staple every 24 to 30 inches on straight runs and on each side of return bends. Be sure the stapled down tubing lies flat against the subfloor as seen in Figure 10-37.

If the manifold station is located *under* the floor deck, drill shallow angle holes through the subfloor that allow the tubing to gently bend downward through the

Figure 10-37 — Tubing being stapled in place over plywood floor deck that has been previously covered with polyethylene film. Stapler must be fitted with proper depth stop attachment to control staple depth. *Courtesy of Harvey Youker.*

subfloor without kinking. After tubing that will pass through such holes has been installed, use putty to fill the gaps between the tubing and subfloor to prevent leakage of concrete. This detail is shown in Figure 10-38.

STEP 5: Install control joint strips. A 1 × 1 × 1/16-inch-thick PVC drywall-trim molding works well for this purpose. These thin plastic strips can be easy to cut with a pair of snips and stapled to the subfloor. Their purpose is to prevent bonding in the lower two-thirds of the slab, and thus induce controlled cracking that divides the slab into a "mosaic" of smaller pieces. These smaller pieces are better able to adjust to slight movements in the floor deck without developing random cracks. Figure 10-39 shows a control joint strip fastened in place across a doorway opening. A narrow straight crack will form in the slab directly above the vertical edge of this strip.

Place control joint strips so the overall slab area is broken into pieces having maximum dimensions of 12 feet. Control joint strips should also be placed across doorways, narrow points in the slab, or any other location where a crack is likely to form. *A control joint should also run outward from any outside corner formed by adjoining walls, because these locations create higher stress in the concrete as it cures.*

STEP 6: Coat the edges of the wooden sleepers with **mineral oil** to prevent concrete from bonding to them. This minimizes tensile stress development as the concrete cures, which helps reduce cracking. Never use motor oil, hydraulic oil, or other lubricants as a substitute for mineral oil.

A floor deck with all tubing circuits and control joints in place, and awaiting concrete, is shown in Figure 10-40.

Figure 10-38 — Where tubing passes under floor, drill shallow angled holes through subflooring. After tubing is in place, use putty to seal any gaps between tubing and subfloor. *Courtesy of Harvey Youker.*

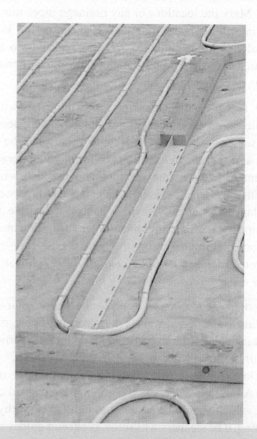

Figure 10-39 — PVC control joint strip fastened in place across a doorway opening. A narrow straight crack will form in the slab directly above the vertical edge of this strip. *Courtesy of Harvey Youker.*

10.9 Concrete Thin-Slab Radiant Floors

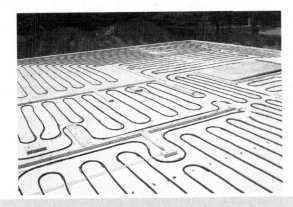

Figure 10-40 — Floor deck with all tubing circuits and control joint strips in place. Note that control joint strip near bottom of photo has been segmented for placement perpendicular to tubing. *Courtesy of Harvey Youker.*

Figure 10-42 — "Pea stone" concrete being placed over tubing. Note plywood scrap beneath nose of wheelbarrow to protect tubing. *Courtesy of John Siegenthaler.*

After all tubing circuits are stapled in place and connected to the manifold, run the same air pressure test described for slab-on-grade installations. Verify that no leaks are present.

The floor is now ready for concrete. The mix proportions suggested for a concrete thin slab are given in Figure 10-41. This mix can be supplied by most batch plants. The small #1A "pea stone" aggregate along with the **superplasticizer** and **water-reducing agent** yield good flow characteristics and allow the concrete to fully encase the tubing. This is accomplished without adding access water that can cause shrinkage cracks, loss of strength, and dusty finished surfaces. The **Fibermesh®** additive provides tensile reinforcement without need for reinforcing steel.

The concrete can be placed using wheelbarrows as shown in Figure 10-42, or in some cases, directly from the chute of the concrete truck. For second floor applications, it can be pumped or transported by conveyor.

The concrete is screeded flat using the sleepers as depth guides as shown in Figure 10-43. The slab is then floated and finished in the usual manner. A very smooth finish should be specified in areas that will receive vinyl flooring. Areas that will be covered by ceramic tile or carpet do not require an ultra-smooth finish. After troweling, the slab should be properly cured by maintaining moist conditions at its upper surface.

Wall framing can begin the following day. Walls can be nailed directly to the previously placed sleepers. Holes for wiring and plumbing can be drilled through the sleepers without need of drilling through concrete. Narrow shrinkage cracks will form between the sleepers and the slab edge, as well as over the control joint strips

Mix design for 1 cubic yard of 3,000 psi thin slab concrete	
Type 1 Portland cement	517 lb
Concrete sand	1,639 lb
#1A (1/4-inch maximum) crushed stone	1,485 lb
Air entrainment agent	4.14 oz
Hycol (water-reducing agent)	15.5 oz
Fibermesh	1.5 lb
Superplasticizer (WRDA-19)	51.7 oz
Water	About 20 gal

Figure 10-41 | Mix proportions for concrete thin slab.

Figure 10-43 — Concrete is screeded flat using sleepers as depth guides. *Courtesy of John Siegenthaler.*

within 2 or 3 days. This is desirable in that it minimizes random shrinkage cracks in the slab. An example of a **controlled shrinkage crack** across a doorway is seen in Figure 10-44.

The final step is to install **underside insulation** between the floor framing to minimize downward heat loss. The amount of insulation needed depends on the temperature of the space beneath the floor, and the R-value of any finish floor materials above the thin slab. A conservative rule of thumb is to provide a minimum of 10 times the finish flooring R-value as underside insulation. *This insulation is essential for proper performance, and should never be compromised.*

The following *minimum* R-values are suggested:

- Floors over heated space: R-11.
- Floors over partially heated basements: R-19.
- Floors over vented crawlspaces or other exposure to ambient conditions: R-40.

The insulation can be any suitable product that is relatively easy to install, will stay in place, and will retain its R-value. Fiberglass batts are a good choice. The batts should be neatly tucked up against the underside of the subfloor. If installed in floor framing above a basement, be sure loose fibers from the surface of the batts cannot drift into the space below. If that space is partially heated, **kraft paper faced batts** can be used with the facing at the bottom to contain any loose fibers. However, if the space below is not heated, the facing on the bottom of the batt could trap condensed moisture. This situation must be avoided. In such cases, use a vapor permeable layer such as polyolefin housewrap under the batts. Poly-wrapped fiberglass batts supported by wire stays are another option. With the exception of unheated crawlspaces, wiring and plumbing should be installed under the insulation. This minimizes warming of electrical cables and prevents cold water from being unintentionally heated on its way to faucets.

Avoiding Lightweight Concrete

There are several types of **lightweight concrete** currently available. They are primarily used to increase the fire resistance rating and decrease sound transmission through floors. Most formulations use special lightweight aggregates such as expanded shale, vermiculite, and polystyrene beads in place of the usual crushed stone. Depending on the materials and mix proportions used, some lightweight concretes achieve densities as low as 25 lb/ft^3. This is approximately one-sixth the density of standard concrete.

Unfortunately, *the thermal resistance of most lightweight concrete is substantially higher than that of standard concrete.* This is undesirable from the standpoint of diffusing heat away from embedded tubing in a thin-slab radiant panel. The R-value of lightweight concrete of various densities is shown in Figure 10-45.

The greater the R-value of the slab material, the poorer its ability to diffuse heat away from the tubing and disperse it across the slab. This is evident in Figure 10-46, which shows the results of finite element analysis of thin slabs constructed with lightweight concrete of various densities.

Figure 10-44 | A (desirable) controlled shrinkage crack forms directly above all control joints strips within 2 or 3 days after placing concrete. *Courtesy of John Siegenthaler.*

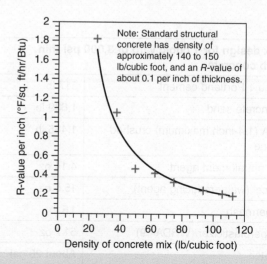

Figure 10-45 | Thermal resistance (R-value) versus density of lightweight concrete.

The lower the concrete's density, the lower the floor surface temperatures for a given water temperature in the embedded tubing. This is shown in Figure 10-47, which shows the surface temperature profile of five different densities of concrete.

The lower the surface temperature profile, the less heat emitted from the slab. In this case, the slab constructed of 25 lb/ft^3 concrete has an upward heat output of only 21.5% that of the slab constructed of 150 lb/ft^3 concrete. This is a very significant loss in performance.

Higher slab resistance also increases the temperature gradient across the floor surface, from a maximum temperature directly above the tubes to a minimum temperature halfway between adjacent tubes. This undesirable effect is called **striping**. In extreme cases, striping can damage flooring materials such as vinyl tile or carpet. Poor **heat diffusion** due to low thermal conductivity of the slab material also reduces overall heat output.

Because of its reduced ability to diffuse heat, lightweight concrete with a density less than 110 lb/ft^3 is generally not suitable for radiant panels.

10.10 Poured Gypsum Thin-Slab Radiant Floors

Poured gypsum underlayments have been used for many years to level uneven floor surfaces, as well as to decrease sound transmission and increase the fire resistance of floor assemblies. They are now widely used for thin-slab radiant floor heating.

Poured gypsum underlayment consists of a mixture of gypsum cement, sand, water, and additives to improve flexibility. These materials are mixed on-site and installed by a trained crew. A cross-section of a gypsum thin slab is shown in Figure 10-48. A cutaway of the same system is shown in Figure 10-49.

The primary advantages of gypsum-based underlayments are the speed and ease of installation relative to concrete. A trained crew of three can install about 2,500 ft^2 of gypsum-based thin slab in one day. Interior work can usually resume the next day. Gypsum underlayments are also less prone to cracking than is a concrete thin slab. Their ability to decrease air infiltration by sealing small cracks along the base of walls and partitions is also an advantage.

The final compressive strength of the gypsum underlayments used in floor heating applications is typically 2,500 to 3,000 psi, depending on the mix proportions. This is adequate to support foot and light equipment traffic during construction. Care must be taken, however, not to gouge the completed floor with heavy or sharp objects.

When dry, a 1.5-inch-thick gypsum slab adds about 14.4 lb/ft^2 to the dead loading of the floor. This is slightly less than the 18 lb/ft^2 dead loading of a 1.5-inch concrete thin slab. Still, *the added load must be taken into account when the floor framing that will support it is specified. The added height of the thin slab must also be accounted for during building design.*

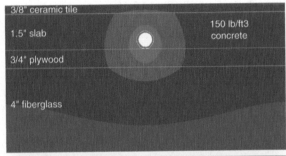
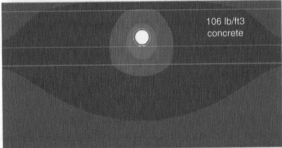
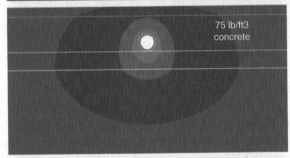
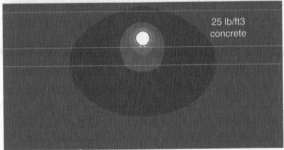

Figure 10-46 | Heat diffusion through thin slab assembly as determined by finite element analysis. Each slab consists of different density concrete as indicated. All other materials and operating conditions are identical.

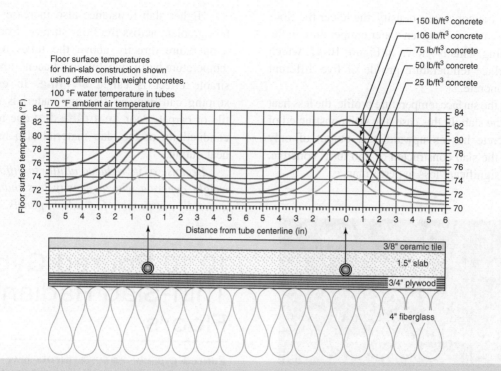

Figure 10-47 | Surface temperature profile simulation for thin-slabs made of lightweight concretes of various densities. All temperature profiles assume 100°F water in tubing. Higher surface temperature profiles result in greater heat output.

Gypsum underlayments are not designed to function as the permanent wearing surface. They must be covered by some type of finish flooring. After full curing, they can still be damaged by prolonged contact with water. A minor but persistent leak from a plumbing fixture or manifold connection can cause the gypsum slab to soften similar to how gypsum wallboard reacts to prolonged contact with water.

Gypsum underlayments also have slightly lower thermal conductivity relative to standard concrete. This results in slightly less heat output under the same operating conditions.

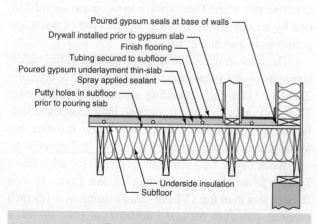

Figure 10-48 | Cross-section of a thin-slab radiant floor panel made with poured gypsum underlayment.

Installation Procedure

In contrast to concrete thin slabs, gypsum thin slabs are usually installed near the end of the construction sequence, after interior drywall is in place. Because gypsum underlayments are not waterproof, they should never be poured in situations where they will be exposed to precipitation.

The installation begins by stapling the tubing circuits directly to the subfloor and pressure testing as previously described with concrete thin slabs. Polyethylene bond breaker sheets are not used with gypsum thin slabs. A pneumatic stapler with depth stop attachment must be used to ensure that staples do not damage the tubing. Figure 10-50 shows a properly installed pneumatically driven staple that just touches the top to the tubing. These staples should be positioned every 24 to 30 inches along the length of the tubing, with two or three staples at each return bend.

When applied, poured gypsum underlayments have a fluid consistency similar to a pancake batter. This allows excellent encasement of the tubing. It also allows the material to flow into (and through) very small gaps. It is important to seal off any holes or cracks in the subfloor that exceed 1/8 inch in size. Putty works well as a filler material for small holes and gaps.

The underlayment installer begins by spraying the subfloor with a combination sealer and bonding agent as shown in Figure 10-51. This material limits water

10.10 Poured Gypsum Thin-Slab Radiant Floors

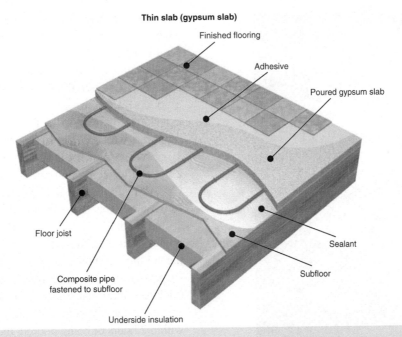

Figure 10-49 | Cutaway view of a typical thin-slab radiant floor panel made with poured gypsum underlayment. *Courtesy of IPEX, Inc.*

absorption and improves the bond between the subfloor and slab.

The gypsum underlayment is prepared in batches in an outside mixer, then pumped into the building through a hose as shown in Figure 10-52.

To help ensure a level finished surface when installed over tubing, it's best to install poured gypsum underlayment in two layers, referred to as "**lifts**." To place the

Figure 10-50 | A properly installed pneumatically driven staple that just touches the top of the tubing. *Courtesy of John Siegenthaler.*

Figure 10-51 | Combination sealer/bonding agent sprayed on subflooring prior to pouring thin slab. *Courtesy of Andrew Wormer.*

Figure 10-52 | Gypsum-based underlayment is mixed outside and pumped in through a hose. *Courtesy of Andrew Wormer.*

first lift, the installer moves back and forth across the room as the material flows from the hose as shown in Figure 10-53a. The materials consistency allows it to flow around the tubing, and into any small gaps. This characteristic also allows for self-leveling and provides a good seal against air leakage along the lower edge of exterior walls. A wooden float is used to keep the material even with the top of the tubing as seen in Figure 10-53b. The first lift is usually firm enough to walk on 2 hours after being poured.

As it cures, the first lift undergoes slight vertical settlement. The top edge of the tubing is just visible at this point. A second lift of the same material is then poured to a minimum depth of 3/4-inch over the top of the tubing as shown in Figure 10-54. A wooden float suspended on two small pins is used to ensure uniform thickness and a flat finished surface.

Although gypsum slabs can usually be walked on within 2 hours of being poured, they can take several days to fully cure. Curing time depends on the temperature and humidity at the job site. If the heating system is operational, a slight amount of heat can be applied to the slab as soon as it sets up to expedite curing. The water temperature in the tubing during curing should never exceed 120°F.

(a)

(b)

Figure 10-53 | (a) Installer places first lift of gypsum-based underlayment. Note its thickness is approximately the same as the tube's diameter. (b) Installer lightly floats surface to ensure uniform thickness. *Courtesy of Andrew Wormer.*

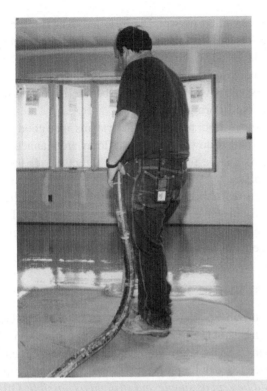

Figure 10-54 — Installer places second lift of gypsum-based underlayment over partially hardened first lift. *Courtesy of Andrew Wormer.*

Gypsum underlayment contains considerably more water than does an equal volume of concrete. Approximately 0.16 gallon of water must eventually evaporate from each square foot of floor covered with a 1.5-inch-thick gypsum slab. Therefore, a 2,000-ft² thin slab pour requires the evaporation of about 320 gallons of water. *It is essential to ventilate the space where the slab is poured and to maintain the walls and ceilings warm enough to prevent condensation as this water evaporates.* This is especially true if the slab is poured during cold weather. Failure to properly ventilate the space can lead to severe moisture condensation on cool surfaces such as windows, doors, or any uninsulated walls and ceilings.

The moisture content of a gypsum slab should always be checked before installing finish flooring. This can be done by taping down all edges of a 2 ft × 2 ft piece of clear polyethylene sheeting, and verifying that moisture droplets do not accumulate on the underside during the following 48 hours.

Manufacturers often specify that gypsum slabs be sealed with specific products to protect them against moisture from spills or mop water that may penetrate the finish floor. The sealant also binds any fine particles at the surface, which prevents them from working up through carpeting over time. Be sure to follow any sealing recommendations before installing the finish flooring.

The final step is to install insulation between the floor framing to minimize downward heat loss. The amount of insulation needed depends on the temperature of the space beneath the floor and the R-value of the finish floor. A conservative rule of thumb is to provide a minimum of 10 times the finish flooring R-value as underside insulation. This insulation is essential for proper performance and should never be compromised.

The following *minimum* R-values are suggested:

- Floors over heated space: R-11.
- Floors over partially heated basements: R-19.
- Floors over vented crawlspaces or other exposure to ambient conditions: R-30.

The insulation can be any suitable product that is relatively easy to install, will stay in place, and will retain its R-value. Fiberglass batts are a good choice. The batts should be neatly tucked up against the underside of the subfloor. If installed in floor framing above a basement, be sure loose fibers from the surface of the batts cannot drift into the space below. If that space is partially heated, kraft paper faced batts can be used with the facing at the bottom to contain any loose fibers. However, if the space below is not heated, the facing on the bottom of the batt could trap condensed moisture. This situation must be avoided. In such cases, use a vapor permeable layer such as Tyvek® housewrap under the batts. Poly-wrapped fiberglass batts supported by wire stays are another option. With the exception of unheated crawlspaces, wiring and plumbing should be installed beneath the insulation. This minimizes warming of electrical cables and prevents cold water from being unintentionally heated on its way to the faucets.

10.11 Above-Floor Tube and Plate Systems

In some buildings, slab or thin-slab-type radiant panels are not possible due to one or more of the following reasons:

- The floor framing cannot support the additional dead loading of a slab.
- The wall framing is already completed and was not adjusted for the added thickness of the slab.
- The finish flooring requires extensive nail penetrations of the subfloor.
- The slab materials are not available or cannot be easily placed.

One alternative is called a tube and plate system. Instead of concrete or gypsum, this system uses preformed aluminum **heat transfer plates** to laterally conduct heat away from the tubing.

Systems that locate the heat transfer plate above the subfloor are appropriately called **above-floor tube and plate systems**. If the heat transfer plates are installed beneath the subfloor, the system is called a **below-floor tube and plate system**. The latter type of system is discussed in the next section.

A cross-section of an above-floor tube and plate system is shown in Figure 10-55. A cutaway view is shown in Figure 10-56. A topside view is shown in Figure 10-57.

Although there are slight differences from one manufacturer to another, most heat transfer plates are formed from 0.020- to 0.025-inch-thick aluminum sheets. A typical plate is 5 or 6 inches wide and 24 inches long. The groove up the center of the plate wraps around the lower portion of a 3/8 or 1/2-inch PEX or PEX-AL-PEX tube. Most plates are formed so the tube snaps tightly into this groove. Good contact between the tube wall and plate is essential for good heat transfer.

The "wings" of the plates conduct heat laterally away from the tube. This part of the plate is typically supported by plywood sleepers that have been secured to the subfloor using construction adhesive and mechanical fasteners as shown in Figures 10-55 and 10-56.

The material installed over the plates depends on the type of finish floor planned. If the floor will be covered with vinyl sheet, ceramic tile, carpet, or other materials requiring a smooth **substrate**, a layer of 1/4 or 3/8-inch plywood is installed over the tube and plates before installing the finish flooring materials. This layer adds some resistance to upper heat flow. It also increases the height of the floor slightly; however, it is essential when a smooth substrate is needed under the finish flooring.

Above-floor tube and plate systems are a good choice where nailed-down wood finish floors are planned. Such flooring can be fastened in place directly over the tube and plates system. This approach minimizes upward thermal resistance between the plane where the heat is laterally dispersed and the top of the finish floor. Lower resistance decreases the necessary water temperature and helps improve overall operating efficiency.

Installation Procedure

As with any radiant panel installation, a circuit layout drawing should be prepared before installation. The drawing should show the location of tubing circuits so the sleepers can be properly located. *The circuit routing should be kept as simple and straight as possible.* Offsets and other complex curves will have to be made using a router, and they do not permit efficient placement of heat transfer plates. An example of a tubing layout drawing for an extensive above-floor tube and plate installation is shown in Figure 10-58a. A small portion of this drawing showing the tubing within a single room is shown in Figure 10-58b.

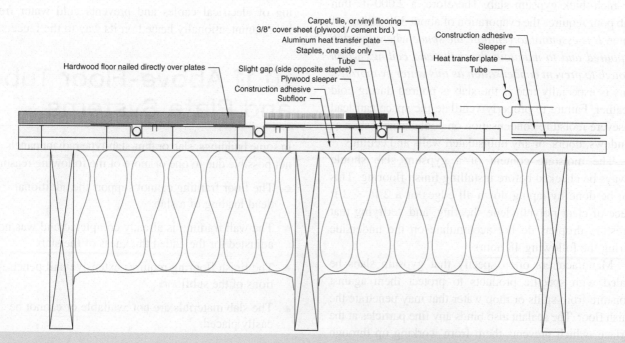

Figure 10-55 | Cross-section of an above-floor tube and plate radiant panel showing different finish floor options.

10.11 Above-Floor Tube and Plate Systems

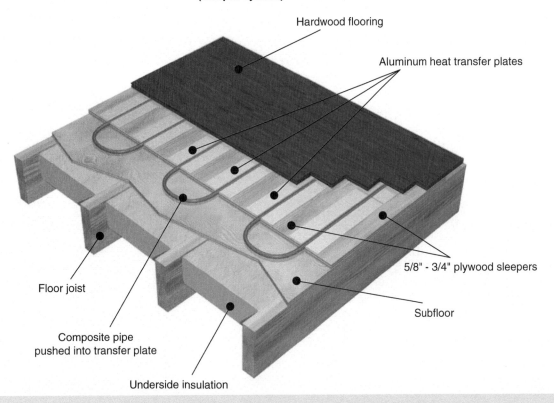

Figure 10-56 | Cutaway view of a typical above-floor tube and plate radiant floor panel. *Courtesy of IPEX, Inc.*

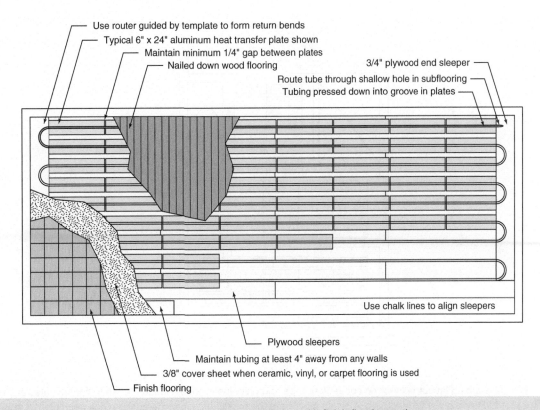

Figure 10-57 | Top view of an above-floor tube and plate installation with finish flooring options.

500 Chapter 10 Hydronic Radiant Panel Heating

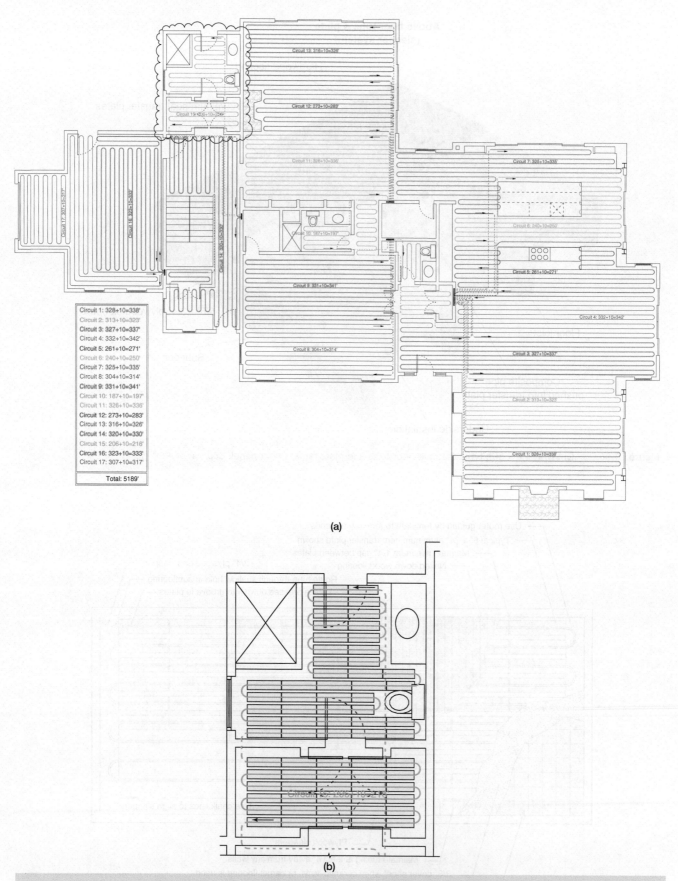

Figure 10-58 (a) A tubing layout plan for an above-floor tube and plate installation. Notice the table that identifies each tubing circuit and gives its length. (b) A single room from the overall drawing where tubing has been placed at 8-inch spacing. 6 × 24-inch heat transfer plates are also shown. Dashed green lines indicated tubing routed under floor.

If nailed-down wood flooring will be installed, the sleepers should be placed perpendicular to the direction of the flooring. Determine the preferred direction of the flooring, and plan sleeper installation accordingly. If there is no strong preference on flooring direction, plan the tubing layout to allow the warmest portion of the circuit to run parallel to exterior walls.

Sleepers are typically cut from 3/4-inch plywood or 3/4-inch oriented strand board (OSB). In most applications that use 1/2-inch PEX or PEX-AL-PEX tubing, the sleepers are cut 3/4 inch narrower than the tube spacing. For example, if the tubing will be placed 8 inches on center, the sleepers are sawn to 7.25 inches wide. This creates a 3/4-inch-wide by 3/4-inch-deep channel between adjacent sleepers that accommodates the groove in the plate. On jobs where a large area will be covered with sleepers, the installer may choose to have the sleepers fabricated ahead of time using a panel saw.

Vacuum the floor deck and snap chalk lines on the floor to mark the edges of the sleepers. Install 3/4-inch plywood or OSB end boards where return bends will be required. The channels to accommodate tubing at return bends will eventually be cut with a router.

When installing the sleepers, be sure to apply construction adhesive on the bottom side and place them carefully along the chalk lines. Immediately fasten them in place with a pneumatic nailer or screw gun. The construction adhesive helps prevent squeaks in the floor due to subsequent shrinkage. Figure 10-59a shows 7.25-inch-wide sleepers installed so that tubing can be placed at 8-inch spacings. The ends of the sleepers, rather than the end board, have been cut to a semicircular shape to accommodate the return bends. This alternative to routering the end board is fine but does require the aluminum plates to begin farther out from the wall.

When channels for return bends will be cut in the end board the use of a 2 horsepower (or larger) plunge router fitted with a 3/4-inch cylindrical bit is recommended. Specialized jigs are available to speed this procedure as shown in Figure 10-59b.

In many installations, the tubing circuit has to eventually pass back across the room. It is easiest to route the tubing down through a shallow hole in the subfloor and then back to the required location through the floor framing. This eliminates a situation where the tubing runs parallel to nailed-down wood flooring and, as such, would be more likely to be punctured when the floor is installed.

After all wood fabrication and installation is complete, the floor should again be vacuumed before placing the heat transfer plates. Place each plate into the channel between adjacent sleepers, then slide the plate tight to one side of the channel. The slight gap on

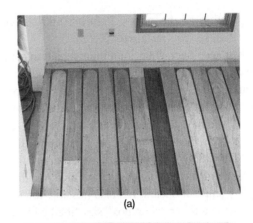

(a)

(b)

Figure 10-59 (a) 3/4-inch plywood sleepers have been glued and screwed to floor deck. Note edge strips to keep tubing a few inches away from wall. *Courtesy of Harvey Youker.* (b) Return bends can also be cut after sleepers are installed using a router mounted to a specialized jig. *Courtesy of Warwick Hanger.*

the opposite side of the plate allows the formed groove in the plate to open slightly as the tube is pressed in. Staple the wing on the tight side of the plate to the sleeper *with two or three light-gauge staples. Never staple both wings of a plate down since this prevents the groove from expanding ass the tube in pressed in place.* Adjacent plates should be at least 1/4 inch apart to allow for expansion. Some designs are based on wider end-to-end separation of the plates. Figure 10-60 shows heat transfer plates being installed.

When necessary, sheet aluminum heat transfer plates can be cut using a "score and snap" method. Use a utility knife and speed square to score across the bottom side of the plate and carefully around the convex side of the groove. Place the score line directly over a rigid supporting edge with the groove side up, then snap the plate with a sharp blow. The plate will break along the score line, producing a clean, straight cut without the distortion caused by snips or shears. It's also possible to cut plates cleanly using a miter "chop box" saw fitted with an abrasive cut off wheel.

The methods used for placing tubing in the plates vary among manufacturers. Some recommend a thin bead of silicone caulk to be placed along the bottom of the grooves before pressing in the tubing. The silicone spreads out as the tubing is pressed in place to provide a thin elastic layer that helps minimize expansion sounds as the tubing heats and cools. It is more commonly specified when PEX tubing is used. If silicone caulk is used in the plate grooves, it is essential to press the tubing into position quickly before the caulk begins to cure.

PEX-AL-PEX tubing has a lower coefficient of expansion than PEX tubing and does not require silicone caulk in the plate groove. *The author highly recommends the use of PEX-AL-PEX tubing with all metal tube and plate systems.*

Be sure to uncoil sufficient tubing to reach the manifold stations before beginning to press the tubing into the plates. Align the tubing with the grooves and simply walk along the tube to press it in place using foot pressure. If necessary, a wooden block and mallet can also be used to gently tap the tube into the grooves. Figure 10-61 shows PEX-AL-PEX after being pressed into the plate grooves. It also shows shallow angle holes used to route tubing through the subfloor. Again note that only one side of each plate has been stapled to the sleepers.

The manifolds can be located under or between the floor framing. They can also be located in vertical stud cavities above the floor. In the latter case, it is often convenient to route the tubing down through the bottom plate and then back up through shallow holes in the subfloor.

After all tubing is placed and connected, it should be air pressure tested as previously described. Maintain the tubing under pressure as the cover sheet and finish flooring is installed.

Traditional (nailed-down) hardwood flooring should be as dry as possible before installation. A moisture level of 4 to 8% is suggested. Once the wood is nailed in place, the floor heating circuits should be operated for at least 10 days before sanding and finishing the floor. This helps drive residual moisture from the flooring and sleepers. It also helps the wood to adjust to its environment before finish sanding and finishing.

If the finish flooring will be ceramic tile, stone, vinyl, carpet, or glue-down hardwood, it is necessary to install a 3/8-inch **cover sheet** over the tubing and plates. The cover sheet may be plywood, or in the case

Figure 10-60 Aluminum heat transfer plates being stapled to sleepers *on one side only*. Two or three light gauge staples are sufficient. A minimum end gap of 1/4 inch should be left between adjacent plates. *Courtesy of Harvey Youker.*

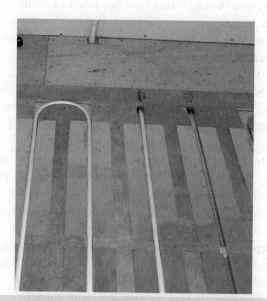

Figure 10-61 PEX-AL-PEX tubing after being pressed into plate grooves. Shallow angle holes where tubing passes through subfloor are also visible. *Courtesy of Harvey Youker.*

of ceramic tile or stone, it may be cement backer board. Figure 10-62 shows an example of a thin plywood cover sheet being installed over tubing and plates.

It is imperative not to damage the tubing during installation of the cover sheet. Lines should be drawn on each sheet to indicate both straight tubing runs and return bends. The staples used to secure the cover sheet should be held at least 2 inches away from the tubing centerlines.

When nailed-down hardwood flooring is installed, all nails should be driven into the exposed portion of the sleepers and not through the heat transfer plates (see Figure 10-63). This minimizes the potential for expansion sounds. It also keeps nails away from the tubing. Be especially careful when nailing in the area of return bends.

Figure 10-63 | Hardwood flooring being nailed to the exposed potion of plywood sleepers. *Courtesy of Harvey Youker.*

(a)

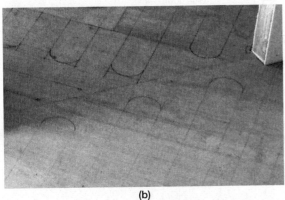

(b)

Figure 10-62 | (a) Plywood cover sheet being installed over tubing and plates to provide smooth substrate for finishing flooring. This system used double tube plates.
(b) Lines should be drawn to indicate tubing location prior to stapling cover sheet. *Courtesy of Harvey Youker.*

Do not install tubing and plates in areas that will be covered by cabinets, islands, or other fixed objects. Such objects block most of the heat output. Food stored in cabinets over heated floors can also spoil faster than normal.

An above-floor tube and plate system only adds 1 to 2 lb/ft^2 to the floor's dead loading. This is usually insignificant. However, it also adds 3/4 to 1 1/8 inches to the floor height. This should be accounted for when planning stair risers, windows and door rough openings, and so forth.

The final step is to install insulation between the floor framing to minimize downward heat loss. The amount of insulation needed depends on the temperature of the space beneath the floor and the *R*-value of the finish floor. A conservative rule of thumb is to provide a minimum of 10 times the finish flooring *R*-value as underside insulation. This insulation is essential for proper performance and should never be compromised.

The following minimum *R*-values are suggested:

- Floors over heated space: *R*-11.
- Floors over partially heated basements: *R*-19.
- Floors over vented crawlspaces or other exposure to ambient conditions: *R*-30.

The insulation can be any suitable product that is relatively easy to install, will stay in place, and will retain its *R*-value. Fiberglass batts are a good choice. The batts should be neatly tucked up against the underside of the subfloor. If installed in floor framing above a basement, be sure loose fibers from the surface of the batts cannot drift into the space below. If that space is partially heated, kraft paper faced batts can be used with the facing at the bottom to contain any loose fibers. However, if the space below is not heated, the facing on the bottom of the batt could trap condensed moisture. This situation must be avoided. In such cases, use a vapor-permeable layer such as Tyvek® housewrap under the batts. Poly-wrapped fiberglass batts supported by wire stays are another option. With the exception of unheated crawl-spaces, wiring and plumbing should be installed beneath the floor insulation. This minimizes warming of electrical cables and prevents cold water from being unintentionally heated on its way to the faucets.

Tube and plate systems, especially those covered by solid sawn wood flooring, should be equipped with an outdoor reset control. As discussed in Chapter 9, Control Strategies, Components, and Systems, outdoor reset control provides gradual changes in water temperature as outdoor conditions vary. This minimizes any potential for expansion noises from the plates. It also minimizes rapid temperature changes in the wood flooring that can induce stress.

10.12 Below-Floor Tube and Plate Systems

It is also possible to install heat transfer plates and tubing *below* a subfloor. In most cases, the tubing is cradled by the heat transfer plates, which are stapled to the bottom of the subfloor. A cross-sectional view of a system using aluminum heat transfer plates is shown in Figure 10-64. A cutaway view is shown in Figure 10-65.

The feasibility of a below-floor tube and plate system depends on several considerations including:

- Are the spaces between the floor joists readily accessible and free from other mechanical, electrical, or structural components?
- Can the tubing be easily pulled though the floor framing or girders without need for extensive drilling?
- Are there, or will there be, nails piercing through the underside of the subfloor?
- What finish floor materials are (or will be) installed?
- Are there any materials in an existing floor (such as asphalt-saturated felt) that could outgas if heated?

If the spaces between the floor framing are not relatively free of other obstacles such as wiring, plumbing, or structural bracing, installation of tubing circuits will be slow, difficult, or in some cases nearly impossible. *It is crucial to inspect the underside of the floor deck for such obstacles before committing to this type of system.*

The structural design of some buildings can also result in numerous girders crisscrossing the floor deck. Drilling holes for tube routing can be slow. It can also affect the structural integrity of the girders. Buildings with complex floor framing are often not well suited to below-floor tube and plate systems.

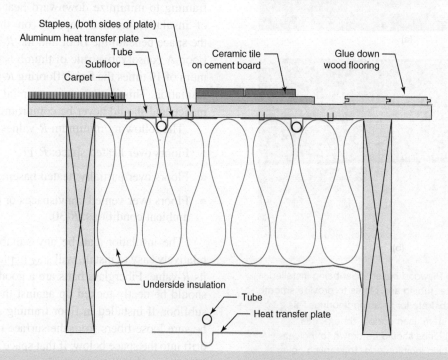

Figure 10-64 | Cross-section of a below-floor tube and plate radiant panel with different finish flooring options.

10.12 Below-Floor Tube and Plate Systems

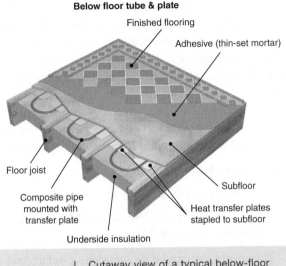

Figure 10-65 | Cutaway view of a typical below-floor tube and plate radiant floor panel. *Courtesy of IPEX, Inc.*

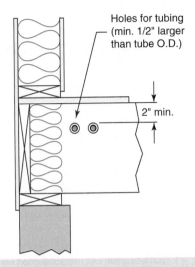

Figure 10-66 | Location of holes in floor joists when pulling tubing for a below-floor tube and plate installation.

The presence of nails puncturing through the underside of the subfloor is also a problem. If the nails are already in place, the installer may have to cut or grind them flush with the floor—a very time-consuming process. If the tubing and plates are installed first and nails are then driven through the subfloor, multiple tube punctures are virtually certain.

In retrofit applications, it is essential to determine what materials are currently in place in the floor deck before committing to a below-deck tube and plate system. Although not visible from above or below the floor deck, a layer of asphalt-saturated felt may be installed under hardwood flooring. If heated, it could outgas compounds that create persistent and objectionable odors in the house. Multiple thick layers of wooden flooring are also common in older homes, and they present considerable upward resistance to heat flow.

Installation Procedure

The installation of an underfloor tube and plate system begins by drilling two holes side by side through the floor joists to allow the tubing to be pulled to each of the joist cavities. These holes should be at least 1/2-inch larger than the outside diameter of the tubing (Figure 10-66). They should be kept in a straight line by snapping a chalk line along the joists or using an alignment laser. The holes should be far enough beneath the subfloor to avoid any nails that might be driven in from above. *Never notch the underside of floor framing to accommodate the tubing since this can severely weaken the framing.*

The preferred method of pulling tubing into each joist cavity varies among installers and manufacturers. One approach is to thread the tubing through a line of holes in the joists until it reaches the farthest joist cavity as shown in Figure 10-67. The end of the circuit is then routed back through the other set of holes until it reaches the manifold location. Tubing is pulled from the uncoiler and then routed up and back along the first joist cavity. The tubing and plates are pushed up tight against the subfloor and stapled in place. Be sure each plate lies tight against the subfloor, and install at least eight staggered staples on *each side* of a typical 2-foot-long plate. A light-gauge pneumatic or electrically powered stapler is well suited to this task. Allow at least 1/4-inch spacing between ends of adjacent plates. Any excess tube is fed forward toward the next joist cavity. Notice the gentle return bends where the tubing moves from one joist cavity to another.

The author recommends PEX-AL-PEX tubing (rather than PEX tubing) for all underfloor tube and plate systems. The coefficient of expansion of PEX-AL-PEX tubing is very close to that of the aluminum heat transfer plates. This relationship minimizes the possibility of expansion sounds as the tubing and plates change temperature.

Be sure the tubing is aligned with the groove in the plates to prevent chaffing. Also, be sure the tubing is free to expand and contract where it passes through holes in the framing. Tubing bends that are tightly bound in place can cause noises as the tubing expands and contracts.

Air pressure test the system after all circuits are in place. If a circuit was punctured by a nail, it can be spliced using a coupling supplied by the tube manufacturer.

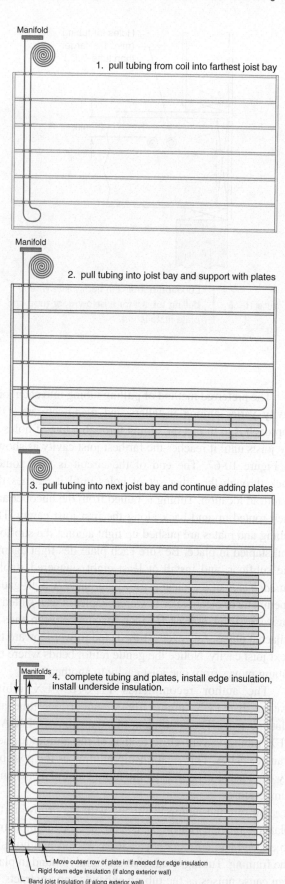

Figure 10-67 | Tubing installation sequence for a below floor tube and plate radiant panel.

The final step is to install insulation between the floor framing to minimize downward heat loss. The amount of insulation needed depends on the temperature of the space beneath the floor and the R-value of the finish floor. A conservative rule of thumb is to provide a minimum of 10 times the finish flooring R-value as underside insulation. This insulation is essential for proper performance and should never be compromised.

The following minimum R-values are suggested:

- Floors over heated space: R-11.
- Floors over partially heated basements: R-19.
- Floors over vented crawlspaces or other exposure to ambient conditions: R-30.

The insulation can be any suitable product that is relatively easy to install, will stay in place, and will retain its R-value. Fiberglass batts are a good choice. The batts should be neatly tucked up against the underside of the subfloor. If installed in floor framing above a basement, be sure loose fibers from the surface of the batts cannot drift into the space below. If that space is partially heated, kraft paper faced batts can be used with the facing at the bottom to contain any loose fibers. However, if the space below is not heated, the facing on the bottom of the batt could trap condensed moisture. This situation must be avoided. In such cases, use a vapor permeable layer such as Tyvek® housewrap under the batts. Poly-wrapped fiberglass batts supported by wire stays are another option. With the exception of unheated crawlspaces, wiring and plumbing should be installed beneath the floor insulation. This minimizes warming of electrical cables and prevents cold water from being unintentionally heated on its way to the faucet.

Extruded Underfloor Plates

Below-floor tube and plate systems can also be built using **extruded aluminum heat transfer plates**. The extruded plates are generally sold in 4- to 8-foot lengths and are screwed into the underside of the subfloor.

Some plates require the tube to be inserted from the top side. Others allow the tubing to be installed after the plates are fastened in place. In the latter case, the extruded channel grips the tubing after it has been tapped in with a mallet. Examples of both types of extruded plates are shown in Figure 10-68.

Figure 10-69 shows extruded aluminum plates mounted under a subfloor. Notice the gentle bends in the PEX tubing, which is pressed into the plates after they have been mounted. Underside insulation has yet to be installed.

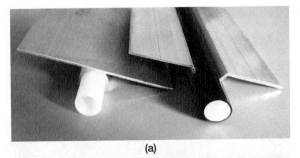
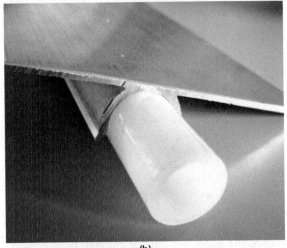
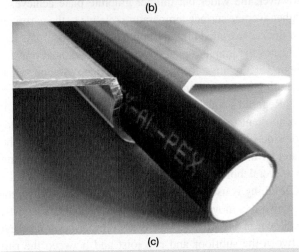

| Figure 10-68 | (a) Examples of 4-inch-wide extruded aluminum heat transfer plates. The plate on the left allows tubing to be pushed into position after the plate has been mounted. The plate on the right wraps under the tube. *Courtesy of John Siegenthaler.* (b) Close up of tube channel opening to bottom. *Courtesy of John Siegenthaler.* (c) Close up of tube channel opening to top. *Courtesy of John Siegenthaler.* |

Graphite Heat Transfer Plates

One of the newest materials used for heat dispersion in radiant panels is **graphite**. Long used in for semiconductor cooling in devices such as flat screen TVs, laptop computers, and cell phones, graphite is a natu-

| Figure 10-69 | Extruded aluminum heat transfer plates with PEX tubing pushed into grooves. This installation is awaiting underside insulation. *Plates by Radiant Engineering.* |

rally occurring, nontoxic, nonflammable mineral. It is purified and fabricated into flexible sheets. These sheets have excellent thermal conductivity in the plane perpendicular to the thickness of the material.

An example of a graphite-based heat transfer plate for radiant panel applications is shown in Figure 10-70. The formed polyethylene channel near the center of the plate enhances contact between the graphite sheet and tubing.

Graphite heat transfer plates are currently sold in 6-inch-wide by 4-foot lengths. The graphite sheet is 0.030-inch thick, and easily cut with a pair of utility scissors or a utility knife as seen in Figure 10-71. The edges of the graphite sheet are not sharp, and thus not capable of chaffing tubing.

A typical below-floor tube and plate installation using graphite heat transfer plates is shown in Figure 10-72. Notice that staple placement secures both the graphite sheet and the polyethylene channel to the subfloor. Secure fastening of the channel helps ensure good

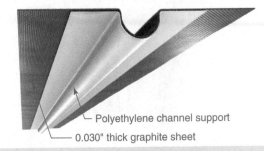

| Figure 10-70 | A graphite heat transfer plate with polyethylene form-enhancement channel. *Courtesy of Watts Radiant.* |

508 Chapter 10 Hydronic Radiant Panel Heating

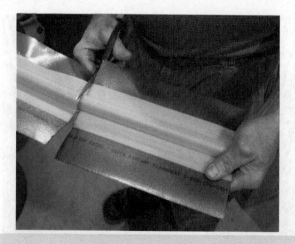

Figure 10-71 | Graphite heat transfer plates can be cut with a utility scissors or knife. *Courtesy of Watts Radiant.*

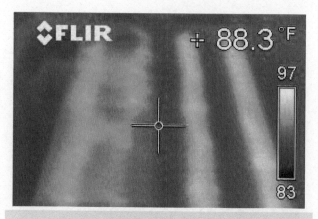

Figure 10-73 | Side by side thermal imaging of graphite heat transfer plates (left) versus extruded aluminum plates (right). The tubing and plates are mounted under 3/4-inch plywood. The water temperature in the tubing, and tube spacing are the same on both sides of the image. The graphite plates deliver a higher average surface temperature. *Courtesy of Watts Radiant.*

contact between the graphite and the tubing, which is vital to good performance.

Figure 10-73 shows an infrared thermograph taken during side by side testing of below floor tube and plate systems mounted under 3/4-inch-thick plywood. The tube and plate combination used on the left side of this image is the graphite heat transfer plate installation shown in Figure 10-72, with tubing spaced 8 inches apart. The system on the right uses 4-inch-wide extruded aluminum plates with the same tube spacing and water temperature.

The thermograph shows that both the graphite and aluminum have good lateral heat dispersion characteristics.

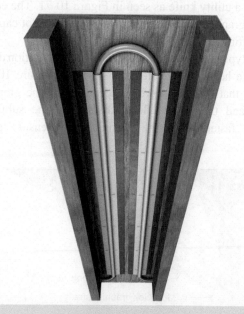

Figure 10-72 | Typical installation of graphite heat transfer plates in an underfloor tube and plate system. *Courtesy of Watts Radiant.*

However, the wider but thinner graphite plate achieves a higher average surface temperature relative to the narrower but thicker extruded aluminum plates. Analysis of this test indicates the graphite plate system increased heat transfer about 15% relative to the aluminum plate system, while operating under the same conditions.

Regardless of the type of heat transfer plates used, it is important to keep the tubing return bends on the interior side of band joist insulation. This helps protect the water in the tubing from freezing if it is not circulating for several hours during cold weather. When PEX-AL-PEX tubing is used, the return bend can be carefully bent downward to move it inward and away from the rim joists, as shown in Figure 10-74. It is also important to air seal between the subfloor and rim joist and between the rim joist and sill to minimize any cold air leakage into the heated joist cavity. Spray polyurethane foam is a good option for both insulating and air sealing the rim joist area.

10.13 Extended Surface Suspended Tube Systems

Another approach to underfloor tube placement is to suspend the tubing within the air cavity between floor joists. Slotted aluminum plates are then fastened to the tubing to enhance convective heat transfer. A cross-section of this concept is shown in Figure 10-75.

10.13 Extended Surface Suspended Tube Systems

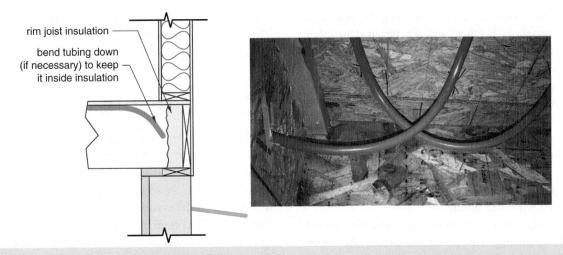

Figure 10-74 | Bend tubing down, if necessary, to ensure that it stays within rim joist insulation. *Courtesy of John Siegenthaler.*

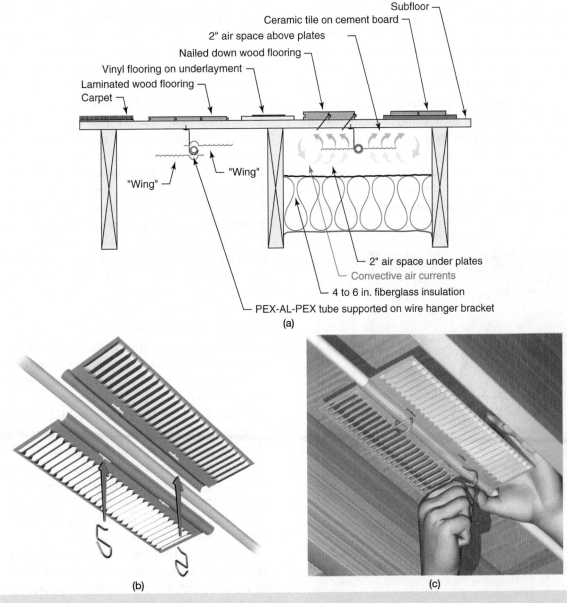

Figure 10-75 | (a) Cross-section of a typical extended surface suspended tube installation. (b, c) Twist lock attachment system for aluminum convector plates. *Courtesy of Ultra-Fin.*

Each plate consists of two matching "wings" that partially overlap each other, and are fastened together with either pop rivets or preformed steel wires. Once fastened, the plates are tightly clamped to the tubing for good heat transfer.

An example of an installation (minus the underside insulation) is shown in Figure 10-76.

When heated water passes through the tubing, the plates conduct heat away from the tubing and create convective air circulation within the joist cavities. The convective heat transfer is significantly higher than that achieved with bare tubing.

The supply water temperature is in part determined by the plate spacing. The closer the plates, the lower the supply water temperature for a given rate of heat transfer. Closer tube spacing enhances the efficiency of low-temperature hydronic heat sources such as condensing boilers and heat pumps.

The elimination of air leakage into and out of the closed joist spaces is also crucial to the proper performance of any suspended tube system. The **band joist** area at the perimeter of the floor should be tightly sealed against leakage and well insulated. Spray foam insulation is the preferred material for such sealing. Penetrations of the joist spaces by plumbing pipes, ducts, conduits, or similar hardware should also be well sealed.

Suspended tube systems can be used in situations where nail points through the subfloor associated with hardwood flooring are either present or will occur during construction. Since the tubing is located approximately 2 inches below the subfloor, it is unlikely to be damaged by normal floor nailing procedures.

10.14 Plateless Staple-Up Systems

Another approach to hydronic radiant floor heating that has been used, in some cases with limited success, is called a **plateless staple-up system**. Tubing is stapled directly to the bottom of the subfloor without heat transfer plates. A cross-sectional view of this system is shown in Figure 10-77a. A cutaway view is shown in Figure 10-77b.

In the author's opinion, plateless staple-up systems should be carefully scrutinized, and used only in situations where limited upward heat output of 10 Btu/h/ft² or less is sufficient at design load conditions, and when the heat source can operate efficiently at water temperatures of at least 150°F. The latter constraint would rule out use of plateless staple-up systems with modulating/condensing boilers and most renewable energy heat sources.

Because heat transfer plates are not present to disperse heat laterally away from the tubing, the flooring materials directly above the tubing can reach significantly higher temperatures than they would in a plated or suspended tube system. This can lead to high **temperature gradients** across the floor, especially during transient conditions when hot water begins flowing through tubing stapled to a cool floor deck. These gradients create high thermal stresses in the flooring materials, which can result in cracked flooring, warped flooring, or loss of bond between the subfloor and flooring. High temperature gradients can also discolor certain flooring materials directly above the tubing. Limited lateral heat diffusion also limits the overall heat output from the floor relative to the other systems discussed.

Figure 10-78 shows comparative finite element thermal simulations of two below-floor systems. The simulation models both have 1/2-inch PEX tubing spaced 8 inches apart. The underside insulation is 4 inches of fiberglass pushed up against the subfloor. The flooring is 3/8-inch ceramic tile installed above a 3/4-inch plywood subfloor. One simulation includes 6-inch-wide aluminum heat transfer plates; the other is plateless staple-up. Other than the plates, or lack thereof, the two floor constructions are identical.

The surface temperature profile for both systems is plotted for water temperatures of 110 and 140°F. Notice the relatively flat "plateaus" in the temperature profile for the plated system, versus the steep slopes in the profile for the plateless system. The latter indicates a rapid drop in floor surface temperature as one moves 2 or 3 inches

Figure 10-76 — Aluminum convection-enhancing plates fastened to suspended PEX-AL-PEX tubing in joist cavities. Relatively close spacing allows reduced supply water temperatures. Underside insulation, (not yet installed), must allow a minimum 2-inch air space under plates for proper convection.
Courtesy of John Siegenthaler.

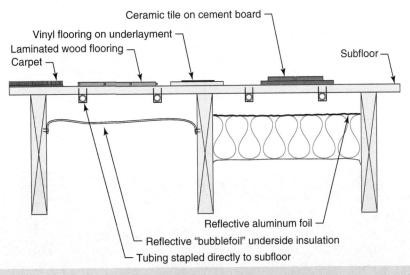

Figure 10-77a | Cross-section of a plateless staple-up radiant floor panel, with underside insulation options.

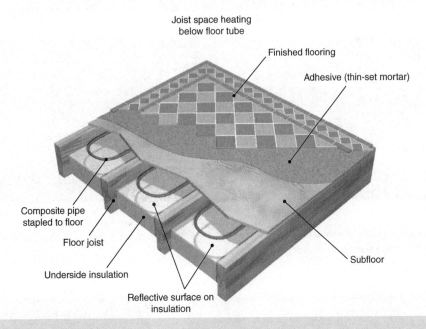

Figure 10-77b | Cutaway view of a typical plateless staple-up radiant floor panel. *Courtesy of IPEX, Inc.*

away from the tube. This undesirable effect is known as "**striping**." It is the direct result of poor heat dispersion away from the tubing. As water temperatures are increased, striping becomes more pronounced. This strong temperature gradient induces stress on the finish flooring materials, and it ultimately may cause premature failure of those materials.

The heat output of each system is proportional to the area under the temperature profile curves. *For the conditions simulated, the plated system yields approximately three times more heat output than the plateless system when operated at the same water temperature.*

10.15 Prefab Subfloor/Underlayment Panels

The use of an aluminum layer to disperse heat away from a tube has also been applied in prefabricated flooring panels. In some cases, these panels serve as the structural subfloor as well as the substrate for the tubing. In other cases, they are fastened to the existing subfloor to form a grooved serpentine path ready to accept tubing.

The product shown in Figure 10-79 is a structural panel that installs directly to floor framing. It

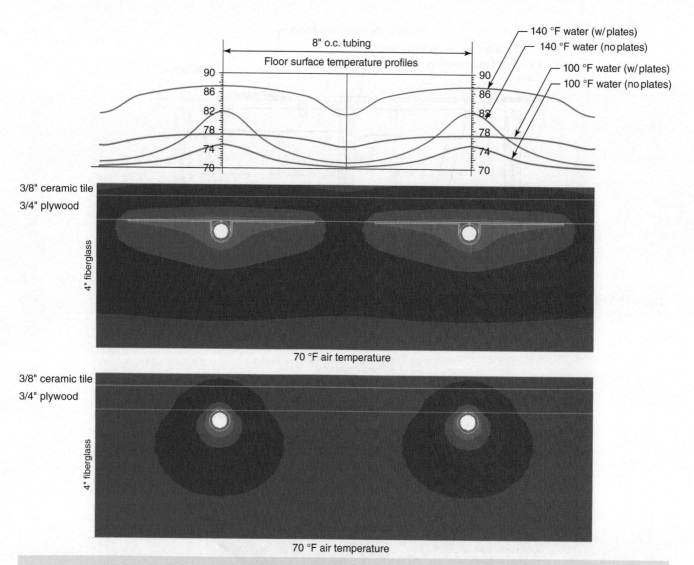

Figure 10-78 | Thermal simulation of below floor tubing installation with and without 6-inch-wide aluminum heat transfer plates on tubing. Surface temperature profiles shown at top.

Figure 10-79 | (a) Prefabricated 4 × 8-foot panel with tubing grooves and aluminum upper layer. (b) Panel being installed on floor framing to serve as both structural floor deck and heat diffusion system. Note temporary placement of alignment pins to ensure tube channel alignment. *Courtesy of Warmboard, Inc.*

is approximately 1-inch thick, and supplied in 4 × 8-foot sheets with various types of circuit routing grooves. A 0.025-inch-thick aluminum layer formed to the groove patterns is bonded to the top of the panel. When the tubing is pressed into the grooves, the result is a low mass radiant floor panel with performance comparable to the previously discussed above-floor tube and plate system. The installation time required for this type of product is significantly less than the previously discussed above-floor tube and plate systems.

Suppliers of such panels provide panel layout and circuit drawings as shown in Figure 10-80. These drawing identify the type and placement of each panel, as well as the eventual tubing circuit paths required. They greatly speed installation.

Another approach bonds plywood strips to a flat layer of aluminum as shown in Figure 10-81. These panels are placed over an existing structural subfloor to form serpentine circuit patterns. Panels are available with both straight grooves and semicircular return bends. Some panels use 5/16-inch PEX tubing, and only add 1/2-inch tubing to the height of the floor. They are particularly well-suited to retrofit applications where minimum gain in floor height is desirable.

After the panels are fastened to the floor, a bead of silicone caulk is laid along the aluminum sheet at the bottom of the groove. The tubing is then uncoiled and driven into the groove with a mallet. The silicone forms an elastic bond between the outer surface of the tube and the aluminum sheet.

Still another type of panel bonds a preformed aluminum sheet to a dense polystyrene base as seen in Figure 10-82.

The panels come in various sizes. They are designed to support residential and light commercial foot traffic. The panels can be placed directly over existing subfloors. Preformed return bend channels at the ends of certain panel types allow for fast installation. Once the panel is fastened in place, tubing is uncoiled and "walked" into place using light foot pressure. The grooves in the panel are designed to hold the tubing once it is pressed into place.

After the tubing is placed and pressure tested these panels should be covered with 3/8-inch plywood or cement board to provide a smooth substrate for glued-down hardwood, ceramic tile, stone, carpet, or vinyl flooring.

The combination of an aluminum skin and foam base gives these panels relatively low thermal mass. This allows them to respond quickly to rapidly changing heating loads. These panels can also be used to construct radiant walls and ceilings, which are covered later in this chapter.

10.16 Radiant Wall Panels

Because radiant heat travels equally well in any direction, it can be delivered from walls and ceilings as well as floors. In some cases, heated walls or ceilings may be possible in circumstances that would rule out floor heating, an example being a room where the owner wants to cover the floor with thick pad and carpet.

There are several advantages to the radiant wall panel system described in this section. They include:

- *Fast response.* Most radiant walls have very low thermal mass. This is especially true about the wall system described later in this section. Low mass radiant panels can respond quickly to changes in load caused by internal heat gain. They can also quickly reestablish comfort in a room that has been maintained at a setback temperature for several hours.

- *Higher heat output.* The surface temperature of a radiant wall can be higher than that of a radiant floor without producing discomfort. A typical radiant floor panel is usually not operated above 85 °F average surface temperature to avoid complaints of hot feet. A radiant wall is not subject to these limitations. When operated at an average surface temperature of 110 °F, the radiant wall can deliver close to twice the heat output per square foot of a radiant floor.

- *Easy to retrofit.* Radiant walls are relatively easy to retrofit into existing rooms. The system described later in this section only adds 0.75 to 1.25 inches to the thickness of an existing wall. It is also very low in weight and does not present the structural concerns that must be addressed with floor heating systems.

- *Not affected by changing floor coverings.* Over the life of the building, walls are far less likely than floors to be covered with finish materials or objects having high thermal resistance. However, it is still prudent to consider possible furniture placement with respect to a proposed radiant wall. Placing upholstered furniture, an entertainment center, or other large objects against a heated wall could block a substantial portion of its heat output.

The radiant wall system described in this section is shown in cross-section in Figure 10-83.

Installation Procedure

The system to be described requires a flat, smooth surface as a **substrate**. In new wood-frame construction, it can be a layer of 1/2-inch plywood or 7/16-inch oriented

514 Chapter 10 Hydronic Radiant Panel Heating

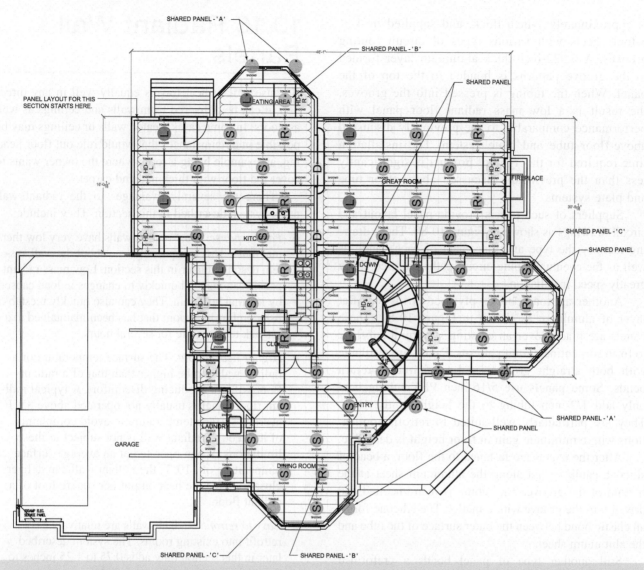

Figure 10-80 | Example of a panel placement drawing. Colored dots and letters identify the type of panel to be installed. *Courtesy of Warmboard, Inc.*

Figure 10-81 | Installation of prefabricated radiant underlayment panels. *Courtesy of Foley Mechanical, Inc.*

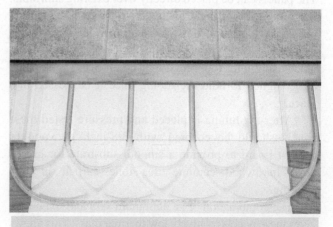

Figure 10-82 | Preformed panel with aluminum upper surface and polystyrene base. *Courtesy of Roth Industries.*

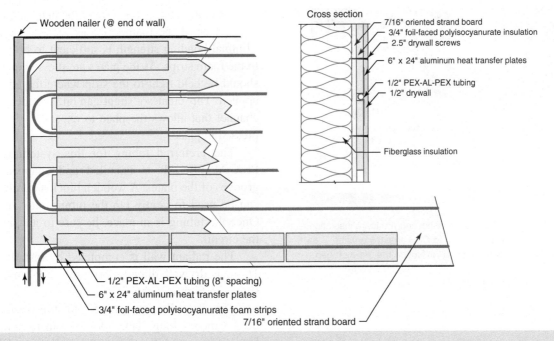

Figure 10-83 | Radiant wall constructed using aluminum heat transfer plates.

strand board (OSB) fastened directly to the studs. In retrofit applications, the existing wall surface can likely serve as the substrate. This substrate should cover the entire area that will become a radiant wall panel.

To "steer" heat output into the room, the wall behind the radiant panel must be well insulated, especially if it is an exterior wall. In the latter case, the R-value installed behind the heat transfer plates should be approximately 50% higher than a normal exterior wall. This keeps the outward heat loss approximately the same as a normal (unheated) exterior wall. The foam insulation strips used as part of this system add about R-5 to the backside R-value of the wall. If the radiant wall is built on an interior wall, the stud cavity should be insulated to R-11.

Once the substrate layer is established, snap horizontal chalk lines on it beginning at least 4 inches above the floor, and then every 8 inches vertically. These lines mark the top edges of foam insulation strips that will be bonded to the substrate. These strips will support the aluminum heat transfer plates. The distance from the floor to the lowest line should consider the height of the baseboard molding that is typically installed at the base of walls. The first line needs to be high enough above the floor to avoid any nails used to secure the baseboard.

The foam insulation strips are made by ripping 3/4-inch-thick foil-faced polyisocyanurate insulation board into 7.25-inch-wide strips with a table saw.

The strips are then bonded to the substrate using solvent-based **contact adhesive**. A light coat of adhesive applied with a paint roller is sufficient. Be sure the strips are accurately aligned with the chalk lines before pressing them to the wall; 3/4-inch-wide wood spacer blocks set along the top edge of each successive row also help keep the foam strips aligned. Once the two surfaces coated with contact adhesive touch, a strong bond is instantly created. Further movement of the strip is not possible. No mechanical fasteners are needed to hold the strips in place.

The installed foam strips are shown in Figure 10-84. The 3/4-inch-wide gap between adjacent strips will accommodate the tubing and grooved portion of the heat transfer plates.

Besides adding backside R-value, these strips have very low thermal mass. This allows the radiant wall to warm up quickly following a cold start. The aluminum facing on the foam strips also enhances lateral heat diffusion away from the tubing, and provides a good bonding surface for adhesives. The polyisocyanurate foam is thermally stable to temperatures of 250 °F. This is far higher than any temperature it would experience in this application.

To accommodate wall receptacles, 1 1/4-inch-deep plastic junction boxes can be screwed directly to the substrate. The boxes should be centered between adjacent rows of tubing. The front of the box will finish flush with the 1/2-inch drywall. Electrical wiring can be routed out the back of the boxes through the substrate. Do not install heat transfer plates within

Figure 10-84 | Foil-faced polyisocyanurate foam insulation strips bonded to OSB sheathing using contact adhesive. Notice the 3/4-inch gap between strips to accommodate tubing. *Courtesy of John Siegenthaler.*

4 inches of junction boxes to minimize warming the receptacles and wiring. Mounting the junction boxes horizontally helps keep them a few inches away from the tubing.

At the ends of the wall, the foam strips are cut short to allow space for the return bends. A vertical 3/4 × 2-inch wood strip should be installed at each end of the wall to provide for solid fastening of the drywall.

The next step is to bond the aluminum heat transfer plates to the foam strips. A light coat of contact adhesive is applied to the back of *one side* of the plate and the matching surface of the foam strips as seen in Figure 10-85. Minimal bonding strength is sufficient to hold the plates to the foam strips. As the plates are pressed in place, they should be held tight to the strip where the adhesive was applied. This creates a slight gap on the other side of the channel that allows the plate to expand as the tubing is pressed in place.

The preferred tubing for this system is 1/2-inch PEX-AL-PEX. It is uncoiled and snapped into the grooves of the plates. A wooden block and rubber mallet can be used to gently tap the tubing into the grooves. Once the tubing is in place, be sure to air pressure test the circuit.

The radiant wall is completed by installing 3/8- or 1/2-inch drywall directly over the tube and plates as shown in Figure 10-86.

The drywall screws used with this system should be 2.5 inches long. This allows them to penetrate the foam strips and fully engage the underlying OSB layer. Notice that the screws installed in Figure 10-86 are located half way between adjacent tubes. This keeps them safely away from the tubing, and allows them to avoid penetrating the 6-inch-wide aluminum heat transfer plates. The latter detail helps minimize any potential for expansion sounds as the wall heats up and cools down.

The screws should be placed 8 inches apart vertically (centered between tubes), and either 8 or 12 inches apart horizontally. The latter spacing is determined by the wall

Figure 10-85 | Contact adhesive being applied to one side of foil-faced foam strips using a narrow paint roller. *Courtesy of John Siegenthaler.*

Figure 10-86 | Drywall being installed over the tube and plate assembly to complete the radiant wall. Note the plates held back near the junction box. *Courtesy of John Siegenthaler.*

framing. If the studs are 16 inches apart, use 8-inch screw spacing such that every other vertical column of screws anchors into the studs. If the studs are 24 inches apart using 12-inch horizontal screw spacing to again allow every other vertical screw line to fall directly over a stud. This screw pattern holds the drywall tightly against the aluminum plates for good conductive heat transfer. Once the drywall is fastened in place the wall can be taped and finished as usual.

This radiant wall construction is very scalable. Panels of various sizes can be built in strategic locations. Examples include the leg space below a breakfast bar, the walls surrounding a dressing area, the wall area below the windowsills, or a part height wall that encloses a stairwell. When installed in a high-moisture environment such as a pool enclosure or walk-in shower, the wall should be covered with cement board and ceramic tile rather than drywall.

Thermal Performance of the Radiant Wall

The heat transfer through the radiant wall system just described was modeled using finite element analysis. Figure 10-87 shows the isotherms in the system under steady-state operation.

Analysis of the finite element simulation indicates the thermal output of this radiant wall construction can be approximated using Equation 10.2.

Equation 10.2:

$$Q_{room} = 0.8(T_{w(ave)} - T_r)$$

where,
- Q_{room} = heat transfer rate to room side of wall (Btu/h/ft^2);
- $T_{w(ave)}$ = average water temperature in the wall circuit (°F);
- T_r = room air temperature (°F).

Example 10.2

Assume that the construction method discussed in this section is used to build a radiant wall panel that is 30-feet long and 4-feet high. The wall is supplied with water at 120°F and operates with a temperature drop of 8°F. The room the wall is releasing heat to is at 70°F. What is the wall's total heat output to the room?

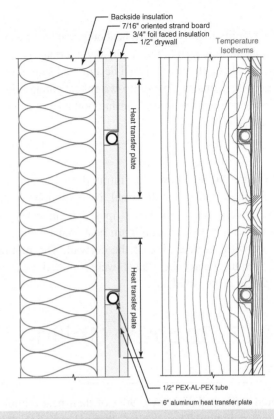

Figure 10-87 Temperature isotherms within a low-mass tube and plate radiant wall panel.

Solution:

The average water temperature associated with these operating conditions is the supply temperature minus half the circuit's temperature drop:

$$T_{w(ave)} = T_{supply} - \frac{\Delta T}{2} = 120 - \frac{8}{2} - 116°F$$

The total area of the wall is 30 × 4 = 120 ft^2.

The room side heat output from the wall (per square foot of area) is estimated using Equation 10.2.

$$Q_{room} = 0.8(116 - 70) = 36.8 \text{ Btu/h/ft}^2$$

The total heat output is the heat output per square foot of area multiplied by the total wall area:

$$Q_{total} = \left(36.8 \frac{\text{Btu}}{\text{h} \cdot \text{ft}^2}\right) 120 \text{ ft}^2 = 4{,}416 \text{ Btu/h}$$

Discussion:

This is a substantial heat output rate per unit of area given the relative low supply water temperature.

> The ability to provide good heat output at low water temperatures, combined with low thermal mass, makes this radiant wall panel design well suited for use with geothermal water-to-water heat pumps, air-to-water heat pumps, condensing boilers, and solar thermal collectors.

Figure 10-88 shows an infrared thermograph of the finished radiant wall panel in operation. Notice the relatively consistent red stripes that indicate good heat dispersion by the aluminum heat transfer plates in contact with the rear side of the drywall. The minimal heat output where the plates were omitted near the receptacle is also evident.

10.17 Radiant Ceiling Panels

Heated ceilings deliver over 90% of their heat output as thermal radiation. They "shine" thermal radiation down into the room much as a light fixture shines visible light downward. They offer many of the same benefits as radiant walls, including:

- *Low thermal mass.* Most radiant ceiling panels have low thermal mass allowing them to quickly warm up following a cold start. This characteristic makes radiant ceilings a good choice in rooms where quick recovery from setback conditions is desirable.

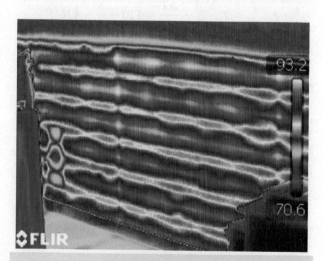

Figure 10-88 Infrared thermograph of the radiant wall panel shown in this section during operation. The temperatures represented by the colors are shown on the scale at the right side of the thermograph. *Courtesy of John Siegenthaler.*

Equally important is the fact that low thermal mass allows heat output to be quickly suspended when necessary. This can help limit overheating in situations where significant solar heat gain can suddenly occur.

- *Higher heat output.* Because occupants are not in contact with them, radiant ceilings can be operated at higher surface temperatures than radiant floors. This allows greater heat output per square foot of ceiling. For example, a ceiling operating at an average surface temperature of 102 °F releases approximately 55 Btu/h/ft^2 into a room maintained at 68 °F. This is almost 60% more heat output than a radiant floor with a mean surface temperature limit of 85 °F. *Higher heat output per square foot often allows a room to be properly heated without need of covering the entire surface of the ceiling.* This is especially true for the higher temperature ceiling panels discussed later in this section.

- *Not affected by changing floor coverings.* The ceiling is arguably the least likely surface of a room to ever be covered. Thus, the performance of a heated ceiling is very unlikely to be compromised by future changes in floor covering or furniture.

- *Warms objects in the room.* The radiant energy emitted from a heated ceiling is absorbed by the surfaces in the room below. This includes unobstructed floor area as well as the surfaces of objects in the room. The upward facing surfaces tend to absorb the majority of the radiant energy; the top of beds, tables, and furniture are slightly warmer than the room air temperature. The surface temperature of floors below a radiant ceiling is often 3 to 4 °F above what it would be in a room heated by convection.

- *Easy to retrofit.* Radiant ceilings are usually easier to retrofit into existing rooms than are radiant floors. They add very little weight to the structure, and result in minimal loss of headroom.

Hydronic radiant ceiling heating has been used for several decades in the United States. Early versions used copper tubing covered with steel lathe and plaster. An example of an early system was shown in Figure 10-4. Although very labor intensive to construct by modern standards, these radiant panels provided good performance. Some are still in use today.

One modern version of a hydronic radiant ceiling panel uses aluminum heat transfer plates is shown in Figure 10-89. Other than its orientation, this panel is identical to the radiant wall panel described in the previous section. The installation procedure is also essentially the same.

Figure 10-90 shows this radiant ceiling design during installation. The foil-faced foam strip and aluminum heat

10.17 Radiant Ceiling Panels 519

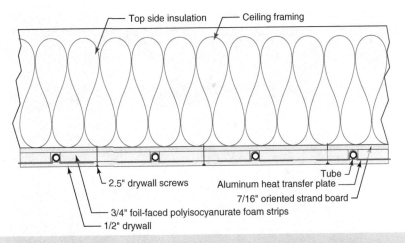

Figure 10-89 | Cross-section of radiant ceiling assembly using aluminum heat transfer plates.

transfer plates are held in place by contact adhesive. The 1/2-inch PEX-AL-PEX tubing is held in place by the formed groove in the plates.

Figure 10-91 shows the completed ceiling. Visually it is indistinguishable from a standard drywall ceiling. However, an infrared thermograph of the same ceiling area shown in Figure 10-92 verifies that the lower surface of the drywall is quickly warming. This thermal image was taken a few minutes after warm water began flowing through ceiling circuit, from left to right. Notice that the left side of the image shows more uniform warming of the areas below the heat transfer plates.

It is not necessary to locate manifold stations for radiant ceiling circuits above those circuits. The tubing circuits can be routed to interior partitions, turned downward, and connected a manifold station within a wall cavity, or under even under the floor framing. Figure 10-93 shows an installation where several radiant ceiling circuit are routed down through a wall cavity and eventually connect to a manifold station mounted under the floor. Proper purging of the circuits when the system is filled will entrain the air in the piping and flush it from the system. A high-efficiency microbubble air separator helps ensure that these circuits will not collect air at high points.

Figure 10-90 | Installation of a low-mass tube and plate radiant ceiling. *Courtesy of John Siegenthaler.*

Figure 10-91 | Finished appearance of radiant ceiling panel. *Courtesy of John Siegenthaler.*

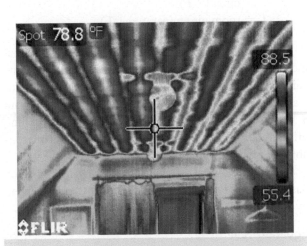

Figure 10-92 | Infrared thermograph of radiant ceiling panel as it warms up from left to right. *Courtesy of John Siegenthaler.*

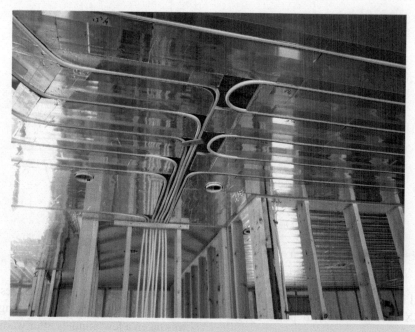

Figure 10-93 | Routing radiant ceiling tubing circuits to a partition and then downward to a manifold station in basement. *Courtesy of John Siegenthaler.*

Thermal Performance of a Heated Ceiling

The heat output of the radiant ceiling panel shown in Figure 10-89 can be estimated using Equation 10.3:

Equation 10.3

$$Q_{room} = 0.71(T_{w(ave)} - T_r)$$

where,

Q_{room} = heat transfer rate to room side of ceiling (Btu/h/ft^2);
$T_{w(ave)}$ = average water temperature in the wall circuit (°F);
T_r = room air temperature (°F).

A quick comparison with Equation 10.2 reveals that the only difference between the two equations is the value of the coefficient (e.g., 0.71 for ceilings vs. 0.8 for walls). This is due to lower convective heat transfer from a heated ceiling versus heated wall. Still, the ceiling offers relatively high output per square foot at reasonable supply water temperatures.

The heat output of a heated ceiling can also be based on its **average surface temperature** as related by Equation 10.4:

Equation 10.4:

$$Q_{room} = 1.6(T_{s(ave)} - T_r)$$

where,

Q_{room} = heat output from ceiling (Btu/h/ft^2);
$T_{S(ave)}$ = average surface temperature of the ceiling (°F);
T_R = air temperature in the room (°F).

Example 10.3

What is the required average surface temperature of a heated ceiling that needs to deliver 45 Btu/h/ft^2 to a 68 °F room below?

Solution:

Equation 10.4 can be rearranged to solve for the required average surface temperature:

$$T_{S(ave)} = T_R + \frac{Q_{room}}{1.6}$$

The given data can then be substituted in and the result calculated:

$$T_{S(ave)} = T_R + \frac{Q_{room}}{1.6} = 68 + \frac{45}{1.6} = 96.1 \,°F$$

> **Discussion:**
>
> The surface temperature at a given location on the ceiling may be somewhat higher or lower than this calculated mean surface temperature. Surface temperature variation depends on the spacing of the tubing as well as the heat dispersion characteristic of the ceiling panel.

The maximum surface temperature of a wall or ceiling constructed of drywall should not exceed 120 °F to avoid thermal degradation of the gypsum. Some references also suggest that the maximum average surface temperature of a heated ceiling should not exceed 100 °F when the ceiling height is 8 feet or 110 °F when ceiling height is 9 to 12 feet. These limitations are based on thermal comfort rather than material limitations.

The relatively high heat output ability of radiant ceiling panels often allow for situations where it is not necessary to heat the entire ceiling area of a room. A heated perimeter band 4- to 8-foot-wide may provide all the heat output necessary. This reduces the amount of tubing and heat transfer plates needed and therefore lowers cost.

An example of a partially heated ceiling using the construction shown in Figure 10-89 is shown in Figure 10-94. In this project, the radiant ceiling panel extends out approximately 4 feet from the wall. It provides high radiant output near a wall with large glass area. Architectural edge treatment gives the appearance of a coffered ceiling.

When a radiant ceiling panel is constructed on the bottom of an exposed ceiling, upward heat loss must be minimized. Excessive top side heat losses can raise attic temperatures high enough to melt snow on the roof, which often leads to large ice accumulations at the eaves. The upward loss is driven by the temperature difference between the heated layer of the ceiling (the horizontal plane of the heat transfer plates) and the air temperature above the insulation. This differential is usually significantly higher than that with a nonheated ceiling. *A conservative recommendation is to increase the R-valve above a heated ceiling approximately 50% above the R-value planned for a nonheated ceiling.* If the heated ceiling is below heated space, a minimum of *R*-19 insulation should be installed above it.

10.18 Tube Placement Considerations (Floor Panels)

An important part of planning a hydronic radiant floor heating system is deciding how to place individual tubing circuits. The eventual goal is to generate a tubing layout plan that can be easily followed during installation, and serve as permanent documentation of the installation. Experienced installers can attest that such a plan saves hours of frustration and prevents costly installation errors at the jobsite.

The tubing layout for a given system must simultaneously account for many factors including:

- Dimensions of the spaces;
- Design heating loads of the spaces;
- Type of floor panel being used;
- Ease of tubing installation;
- Any objects that interfere with tubing placement;
- Position of the manifold stations;
- Position of any control joints;
- How the building will be zoned for two or more temperatures;
- Type(s) of floor coverings to be used;
- Length of the piping circuits;
- Horizontal spacing of the tubing;
- Position of the exterior walls; and
- Possibility of fasteners being driven into the floor.

With all these considerations, any one designer would probably produce a tubing layout plan that differs somewhat from that of another designer. *Several different layouts for the same building could be all acceptable*, although there might be some variations in comfort, hydraulic efficiency, length of tubing required, and the ease or cost of installation.

Tubing layouts are developed by applying several general principles to the extent allowed by a specific

Figure 10-94 | Radiant ceiling panel using tube and plate construction extends approximately 4 feet from exterior wall. *Courtesy of Chip Agnew.*

building. Often these principles involve thermal and hydraulic performance considerations that are detailed in later sections of this chapter. In some cases, the principles "compete" against each other for inclusion in the final design. Tradeoffs are often required to accommodate the needs of each building.

- *Principle 1: Install the tubing so the warmest fluid is first routed along the exterior wall(s) of the room.* This allows the highest heat output to occur where the losses from the room are greatest. This is analogous to placing baseboard convectors at the base of exterior walls. The higher floor surface temperatures help compensate for the cooler surfaces of windows and exterior walls. They also create convective currents that counteract the downward draft present along exterior walls.

Figure 10-95 shows three different routing paths for the tubing in a room having one, two, or three exposed walls. These patterns are called **serpentine patterns** because of how the tubing zigzags across the floor area. Notice in all three cases, the warmest water is first routed along the exposed walls. The circuit then works its way back toward the interior area of the room.

These circuit routing patterns work well for rooms in which a single circuit can cover the entire floor area. Larger rooms are often laid out with one or more perimeter circuits combined with one or more interior circuits. The perimeter circuit(s) usually consist(s) of closely spaced tubing covering a band extending 3 to 4 feet in from exterior walls. The interior circuit typically uses wider tube spacing over the remaining floor area. An example of such a layout is shown in Figure 10-96.

Another circuit routing pattern called a **counterflow** is shown in Figure 10-97. Unlike the three patterns shown in Figure 10-95, this tubing pattern is not designed to bias heat output toward exterior walls. Instead, *it is intended to provide the same average heat output over the entire floor area.* This pattern is useful for circuits not installed next to exterior surfaces, especially in larger buildings. In most cases, it is used for interior circuits in combination with serpentine perimeter circuits.

- *Principle 2: Whenever possible, arrange the tubing to maximize straight runs and minimize return bends and corners.* This speeds installation and reduces the possibility of kinking the tubing. The narrower a room is compared to its length, the more important this becomes. Figure 10-98 illustrates this principle for a room twice as long as it is wide.

- *Principle 3: Limit individual circuits to the lengths shown in Figure 10-99:*

These lengths limit the head losses of the circuits to reasonable values that can be handled by small circulators. For designs in which circulator power is to be minimized, these lengths can be shortened by 25%.

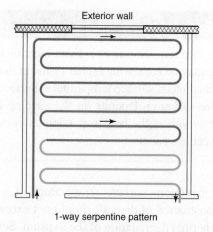

1-way serpentine pattern

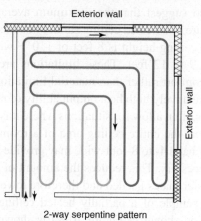

2-way serpentine pattern

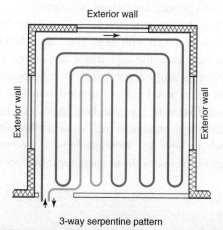

3-way serpentine pattern

Figure 10-95 Serpentine tubing layout patterns. Note that warmest portion of circuit runs adjacent to exterior walls.

The circuit length required to cover a certain floor area can be approximated by dividing the floor area (ft^2) by the spacing of the tubing (feet). The results of these calculations are shown for several tube spacings (in inches) in Figure 10-100.

- *Principle 4: Design circuits so room-by-room flow control is possible.* This is especially important for rooms that will be maintained at different-

10.18 Tube Placement Considerations (Floor Panels)

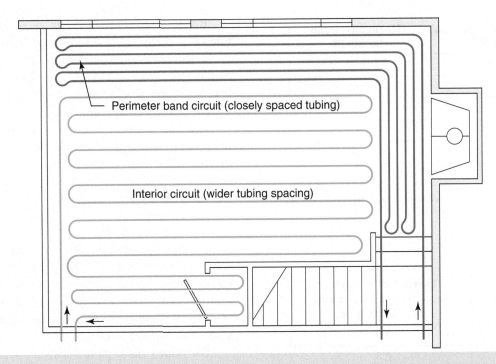

Figure 10-96 | Combination of perimeter circuit using closely spaced tubing with an interior circuit using wider tube spacing.

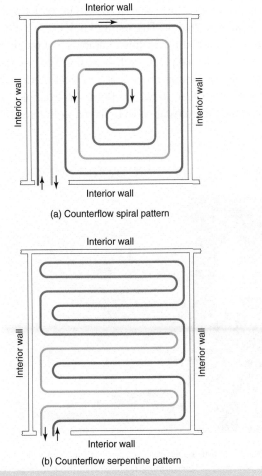

Figure 10-97 | Example of a counterflow tubing layout pattern.

temperatures. For example, it usually makes sense to use separate piping circuits for a bathroom and an adjacent bedroom. Many occupants will prefer the bathroom somewhat warmer for showers and baths, and the bedroom slightly cooler for sleeping.

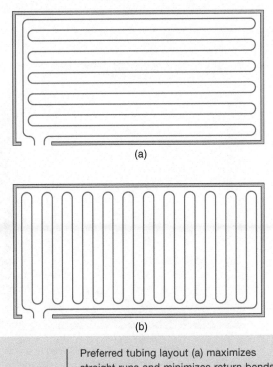

Figure 10-98 | Preferred tubing layout (a) maximizes straight runs and minimizes return bends and corners. (b) Less desirable but still possible tubing layout.

Tubing size (nominal ID) (in)	Maximum circuit length (ft)
1/4	125
3/8	250
1/2	300
5/8	450
3/4	600

Figure 10-99 Maximum recommended circuit lengths for radiant panel circuits based on tube size.

Tube spacing (in)	Linear feet of tubing required per square foot of floor area (ft)
4	3
6	2
8	1.5
9	1.33
12	1.0
18	0.67
24	0.5

Figure 10-100 Linear footage of tubing required for various tube spacings.

Room-by-room circuit layout allows room temperature to be controlled in several ways. First, by adjusting the balancing valves on the manifold, the flow rate through each circuit can be altered. This allows some control over heat output and therefore room temperature as discussed in Chapter 9, Control Strategies, Components, and Systems.

Second, if frequent changes in individual room temperature are desired, a manifold valve actuator can be attached to each manifold valve. This device opens its associated valve when supplied with a 24-VAC signal through a room thermostat. When this signal is not present, the actuator closes the manifold valve and prevents flow through that circuit.

It is also possible to use nonelectric thermostatic valves to control individual radiant panel circuits. Such valves are an option to electric thermostats and valve actuators. They require no wiring or batteries. They vary the flow rate through the radiant panel circuit to maintain a set room comfort level.

For heated slabs, these valves are typically located in small boxes mounted into recessed wall cavities. The tubing circuit serving each room passes through the associated valve body within the box, typically on the return (lower temperature) side of the circuit on its way back to a manifold station. The setting knob extends through the face plate, and has settings numbers 1 through 5, with 3 representing a typical comfort setting. The box is mounted so that the setting knob is at a typical thermostat height of 4.5 to 5 feet above the floor. An example of a wall-mounted thermostat valve is shown in Figure 10-100a.

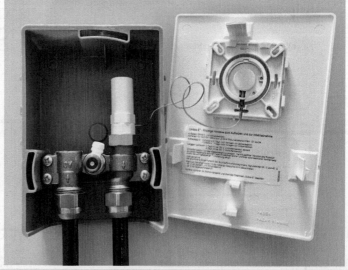

Figure 10-100a A thermostatic valve enclosed in a wall-mounted box. A radiant panel tubing circuit is routed through this valve on its way back to the manifold station. This hardware provides temperature regulation in the space served by the radiant panel circuit. *Courtesy of John Siegenthaler.*

For radiant circuits installed in framed floors the thermostatic valve may be mounted in a recessed wall box, as shown in Figure 10-100a, or the valve may be mounted in an *accessible location* under the floor. When the valve is located below the floor, the thermostatic operator on the valve connects to a remote setting dial using a stainless steel capillary tube. The setting dial is mounted to a wall at a nominal 4.5 to 5 foot height as seen in Figure 10-100b.

Designers should remember that individual room thermostats, or thermostatically regulated valves, are not needed simply because some rooms need to be maintained at different temperatures. This can usually be handled by circuit flow rate adjustments. Individual room thermostats or thermostatically regulated valves make sense when *frequent changes* in individual room temperatures are desirable.

The principle of room-by-room circuit layout is often the dominant concept that determines how a building is divided into individual circuits. Figure 10-101 shows an example of room-by-room circuit layout (principle 4) for a house. These circuits use closely spaced tubing in areas near exterior walls with large windows (principle 1) and are organized into three manifold stations (principal 7). Showing circuits in different colors on the circuit layout drawing aides in quickly tracing their routing through the building.

- *Principle 5: For slab-on-grade applications minimize locations where the tubing has to pass beneath* **control joints**. In most buildings, some tubing will have to cross beneath sawn control joints. When it does, it should be held low in the slab to prevent any possible contact with the saw as discussed in Section 10.8. It should also be protected by a plastic pipe sleeve to relieve stresses that could result if the slab flexes slightly at the control joint.

- *Principle 6: For slab-on-grade systems, tubing placement should correspond with the grid formed by welded wire reinforcing whenever possible.* This allows for efficient placement and fastening of tubing. Standard welded wire reinforcing uses a 6×6-inch grid size. Tube spacings of 6, 12, and 18 inches work well with this reinforcing. 9-inch tube spacing is also possible when necessary. In this case, some tubing will need to be secured half way between adjacent steel reinforcing wires in the 6×6 inch welded wire fabric.

- *Principle 7: Install manifold stations in central locations to minimize tubing runs between the manifold station and the room(s) to be heated.* Think of the manifold station as the hub of a wheel, with the supply and return pipes of individual circuits as the spokes. By keeping the manifold station centrally located to a group of rooms, long, closely spaced

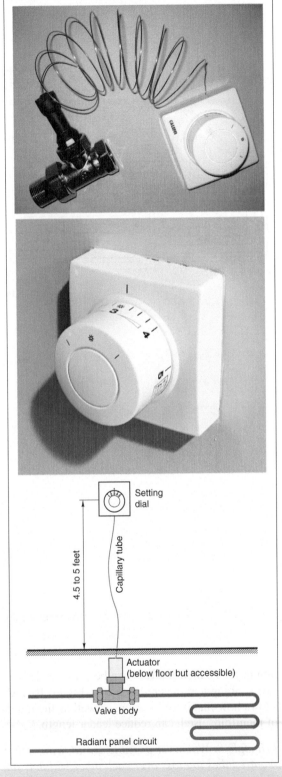

Figure 10-100b

A nonelectric thermostatic valve connected to a wall-mounted setting dial by a capillary tube. The valve is installed below floor level in an accessible location. The radiant panel circuit passes through the valve on its way back to the manifold station. This hardware provides temperature regulation in the space served by the radiant panel circuit.
Courtesy of John Siegenthaler.

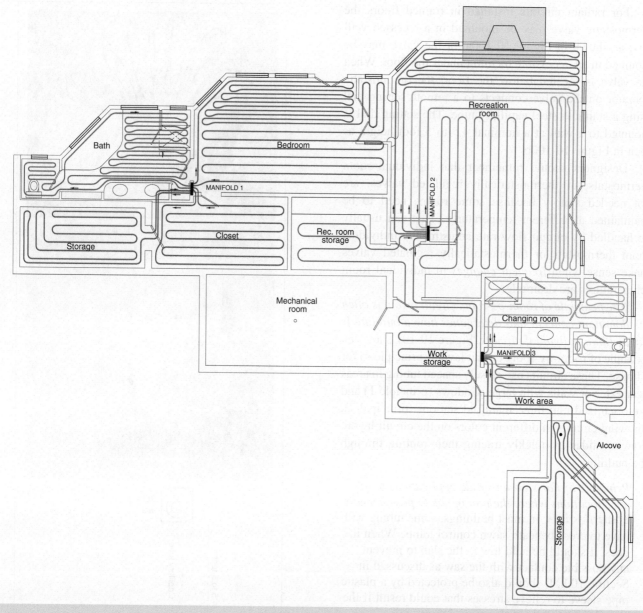

Figure 10-101 | Example of room-by-room radiant floor circuit layout.

tubing runs called **leaders** can be avoided. This saves tubing and thus reduces cost. Figure 10-102 shows how manifold location, as well as the number of manifolds used, can reduce leader length.

- *Principle 8: Whenever possible, avoid layouts that require tubes to cross over each other within a slab.* This is an absolute must for thin-slab systems. If unavoidable, crossovers can be made when the tubing is installed in a 4-inch or thicker concrete slab.

- *Principle 9: Avoid routing tubing where fasteners are likely to penetrate the floor.* This reduces the chance of the tubing being punctured after it is installed. Fastener penetrations are likely beneath interior partitions, under doors with thresholds, or where equipment will be secured to the floor. There is no way to guarantee that a fastener will not penetrate a tube at some point in the life of the building. An accurate layout plan showing the location of the tubing, along with avoiding high probability areas is the best defense.

One solution that has worked on numerous slab-on-grade projects is to coordinate with the framing contractor so that interior partitions over tubed areas are bonded to the slab using construction adhesive rather than mechanical fasteners. This allows simpler layouts and faster installation because the tubing can pass repeatedly beneath interior partitions. It is vitally important that all trades involved in the project are informed about not penetrating the slab at certain locations.

10.18 Tube Placement Considerations (Floor Panels)

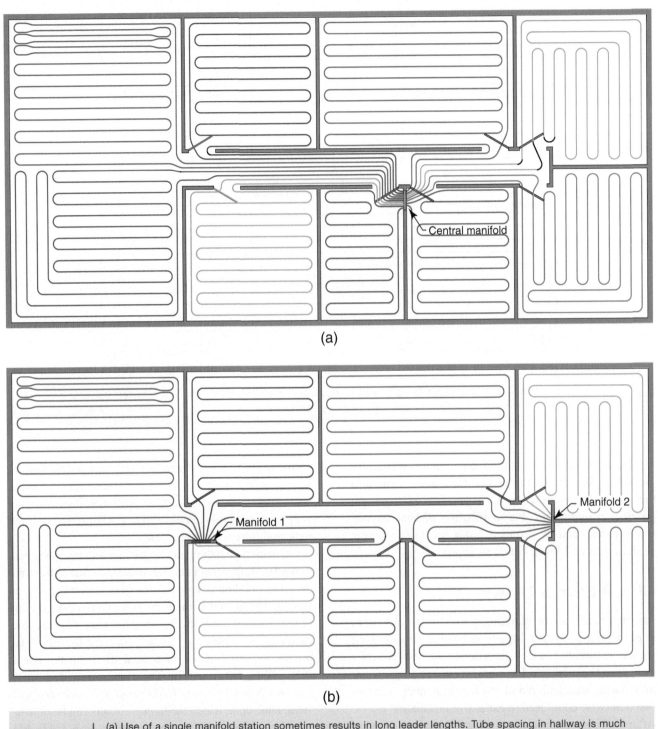

Figure 10-102   (a) Use of a single manifold station sometimes results in long leader lengths. Tube spacing in hallway is much tighter than needed for heating load. (b) Use of two manifold stations reduces leader lengths and total tubing required.

Figure 10-103 is a comparison of two different tubing arrangements for a group of rooms with a single exposed wall. In Figure 10-103a, the tubing has been arranged so only one passage beneath the interior partition separating rooms is required. Assuming the location of the crossing is avoided, fasteners could be driven as required to anchor the partition to the slab. However, *because the circuit passes from one room to the next in a series, the heat output of the floor in the first room will be the greatest, while that in the last room will be the least.* Since the rooms may have nearly identical loads, this could overheat the first room and underheat the last room. This piping arrangement also requires more bends and corners and thus will take more time to install.

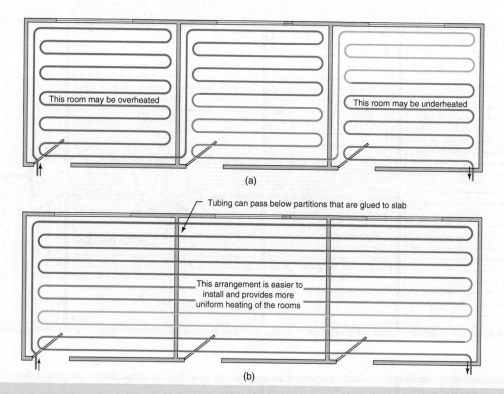

Figure 10-103 | Comparison of tubing layouts. (a) Pattern that minimizes routing under partitions to minimize potential puncture from fasteners, (b) preferred pattern allowed by securing partitions to slab with construction adhesive.

Figure 10-103b shows the tubing circuits repeatedly crossing under the partitions that have been secured to the slab with construction adhesive. This arrangement provides a more equitable distribution of heat to each room. It also can be installed faster because of longer straight runs and fewer bends and corners.

Circuit Layout Drawings

An important part of planning a radiant panel system is producing a **circuit layout drawing**. This drawing shows the routing and flow direction of all tubing circuits as well as the manifold locations. It can save significant time during installation and resolves a number of issues that otherwise could lead to expensive installation errors. After the project is complete, this circuit layout drawing also provides permanent documentation of the installation. This can help avoid damage in situations where the building undergoes future renovations. *A tubing layout drawing should be prepared for every radiant panel installation.*

This section describes circuit layout drawings for radiant floor heating systems; however, the same techniques can usually be applied to radiant wall and ceilings. It also assumes the drawings will be produced using **computer-aided drafting (CAD)** software. Several companies now offer relatively inexpensive CAD software for Windows® and MacOS® computers that are very capable of producing these drawings. Although hand-drawn circuit layouts are still possible, they are very time consuming compared to CAD-produced drawings, and thus seldom practical.

Circuit layout drawings require an accurate floor plan of the building. Such plans are usually available from the owner or building designer. If the floor plan was prepared using CAD software, it may be possible to obtain a copy of the drawing file from the designer. Such files are now routinely distributed as e-mail attachments. They are usually available in one of several file formats such as PDF, DWG, DXF, or EPS. These and other file formats can be opened by many CAD programs, and used as a "template" upon which the tubing circuits can be drawn.

If the floor plan is not available as a CAD file, radiant heating designers need to recreate a simplified version of the floor plan using their own CAD system. This drawing must be to scale and accurately place all walls, partitions, cabinets, and doorways. It should also show plumbing fixtures such as baths, shower stalls, and toilets. It is usually not necessary to draw in windows or other architectural features that do not directly affect circuit layout. The most common scale for such drawings is 1/4 inch = 1 feet.

Before the circuits can be drawn, it is necessary to determine the required tube spacing for each circuit.

This can be done using methods presented later in this chapter, or through use of manufacturer-supplied radiant heating sizing software.

Start the layout drawing by deciding where the manifold station(s) will be located (see principle 7). The manifold locations serve to "anchor" the floor circuits. Their location should minimize the distance to the various room circuits, as well as provide easy access when needed. Place a black rectangle on the plan where each manifold station will be located. The scaled length of the rectangle should accurately represent the length of the manifold.

It is preferable to draw each tubing circuit with a unique color. This allows the circuit to be easily seen on the monitor, as well as in color printouts. In many CAD systems, it is also preferable to draw each circuit on a different **drawing layer**. This makes it easy to isolate each circuit when measuring its length. Assign a unique name to each circuit and its corresponding drawing layer for quick reference.

Drawing circuits involve simultaneous application of the principles discussed earlier. The designer tries to apply the principles "on the fly." As the circuit takes shape, the designer may see a problem developing such as the circuit being too long. In this case, the CAD software allows for quick and easy modifications. If the designer is not sure of the change, most CAD systems allow them to copy and paste the existing drawing elements to a new drawing window before making any changes. If the change does not work out, the new drawing window can simply be discarded.

This ability makes it possible for designers to try different circuit routings in minimal time, and then decide which layout is best.

Begin each circuit by creating a new drawing layer and assigning the name of the circuit as the name of that layer. The first line representing a tube is usually drawn parallel to the exterior wall(s) of the room. To avoid carpet tack strips or future holes drilled for cables and so forth, keep the line at least 4 inches (to scale) away from the wall.

Proceed by laying out multiple parallel lines at the desired tube spacing distance. These lines can later be trimmed and connected using a fillet tool or semicircle tool in the CAD program. For the best accuracy, avoid square bends or sharp vertices in the tubing pattern. Use a radius drawing tool to make all such changes in direction.

Different CAD systems offer different drawing tools that will affect the manner in which the circuit is drawn. With experience, the designer will learn to optimize the use of these drawing tools to quickly generate accurate drawings.

An example of a drawing sequence in which a grid of parallel lines is systematically trimmed and filleted into a final circuit path is shown in Figure 10-104. An exploded view of the completed circuit is shown in Figure 10-105. The latter illustration shows how the overall circuit can be viewed as simply a collection of lines and arcs. The total length of these lines and arcs is the total circuit length (on the floor plane).

When the circuit has been completed, its length can usually be measured by the CAD software. The exact procedure varies from one CAD system to another. Some require the collection of line segments and arcs that make up the circuit to be joined into a "**polyline**." Once this is done, the length is instantly calculated by the software. Because each circuit, and only that circuit, is drawn on a separate drawing layer, the circuit's length is simply the sum of the lengths of all line segments and arc on that layer.

Keep in mind that the CAD software only measures the length of the tubing drawn on the floor plane. When installed, slightly more tubing will be required for the risers to the manifold station. *Experience has shown that adding 10 feet to the measured length (on the floor plane) will be sufficient for the riser to a typical manifold station located 2 feet above the floor.* Part of this added length also represents a safety factor in case some portions of the circuit are not installed in the exact position shown on the drawing.

If the length of a completed circuit is too long based on the limits suggested in principle 3, the circuit routing must be modified. This may require dividing a single long circuit into two shorter circuits.

One can approximate the circuit length required to cover a certain floor area by dividing the floor area (ft^2) by the spacing of the tubing (feet). The results of these calculations are shown for several tube spacings in Figure 10-100. The square footage of a given floor area can be quickly determined using the CAD system to draw a rectangle or polygon around the area, and then reading the area as one of the "attributes" of that shape.

Once all circuits have been drawn onto the floor plan template, the designer should construct a table on the drawing, listing the names and lengths of all circuits. If all circuits will use the same size and type of tubing, a note at the bottom of the table should specify this information. If different tubing sizes are used, be sure to note the size of each circuit along with its length.

Figures 10-106, 10-107, and 10-108 are examples of tubing layout plans created for three different types of radiant floor panel installations: slab-on-grade, thin-slab, and above-floor tube and plate systems. In each case, a list indicating the circuit name and estimated length is given. Also shown are flow direction arrows for each circuit. The latter helps ensure the circuits are properly connected to the supply and return manifolds.

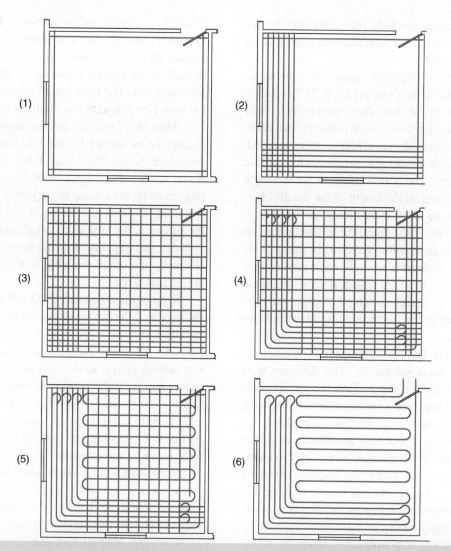

Figure 10-104 | One approach to drawing a radiant floor circuit using a CAD program. Return bends are shown in opposing color to emphasize they are created as separate objects that are eventually joined to line segments.

Readers are encouraged to study these layouts, and look for details such as control joint crossings, omission of tubing under base cabinets, clearance between tubing and closet flanges, closer tube spacing near cooler surfaces, flow direction, and manifold placement.

Specialized software is now available that can automatically generate tubing circuits within floor plans. An example of such software is shown in Figure 10-109. Look for continued advances in such software that will further simplify the preparation of tubing layout drawings.

Finally, radiant heating professionals should recognize that tubing layout drawings represent a good marketing tool. They demonstrate to the owner that the designer is taking the time to "customize" the layout to the exact needs of the building. They also demonstrate that the system is being accurately documented for future reference. Some radiant heating professionals even frame the circuit layout drawing and mount it in the mechanical room along with the company contact information. Professional workmanship along with a well-documented design often leads to referrals.

10.19 Radiant Floor Panel Circuit Sizing Procedure

Because hydronic radiant floor panels are an integral part of a building rather than a manufactured heat emitter assembly, their thermal performance is affected by many factors including:

- Construction of the panel (e.g., slab-on-grade, thin-slab, and below floor tube and plate system.);
- Type of floor covering used (for floor panels);

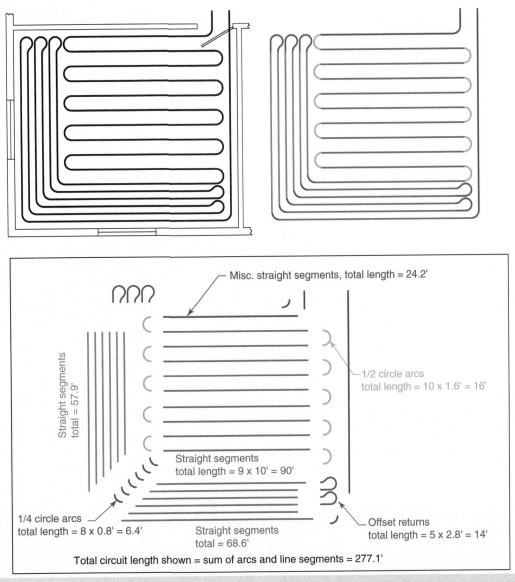

Figure 10-105 | How a radiant panel circuit can be thought of for length measuring purposes. Deconstruct the circuit into a collection of line segments and arcs. Determine the length of all lines and arcs, then add them up.

- Type and amount of the back side insulation;
- Size, spacing, and depth of the tubing;
- Water temperature and flow rate in the panel circuits; and
- Surface and air temperatures in the room being heated.

The simultaneous interaction of all these factors is difficult (sometimes impossible) to theoretically model. Another complication is that every site-built radiant floor panel is constructed somewhat differently. For example, if the tubing depth varies in two otherwise identical slab-on-grade systems, so does their performance. The possibilities for building hybrid radiant panels using different materials, dimensions, and installation techniques are virtually unlimited. Thus, a general performance model for every imaginable type of hydronic radiant panel would be very complex and of little use in routine system design.

This section takes a more practical approach. After describing some fundamental concepts, it presents a step-by-step method for sizing the radiant floor panel circuits for a given building. The selection of available panels includes slab-on-grade, thin-slab, and tube and plate floor systems. The method allows the designer to determine factors such as tube spacing and supply water temperature for each panel circuit based on the heating output it must deliver. It also identifies situations where panel output will not be sufficient to handle the load, and it shows how to determine the amount of supplemental heating needed.

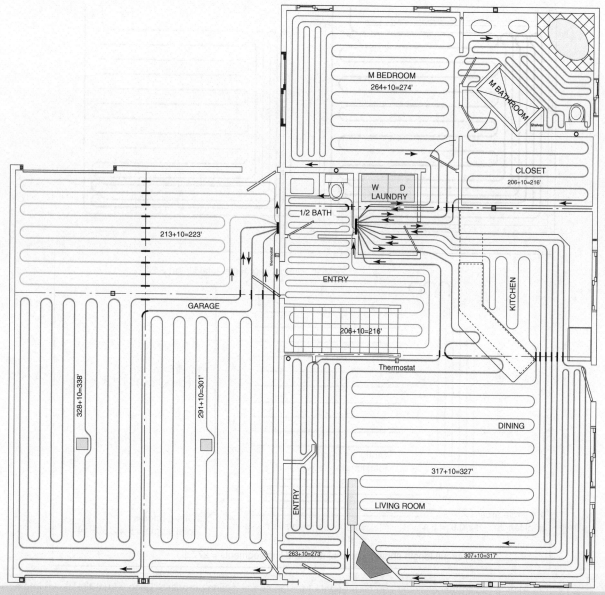

Figure 10-106 | Tubing layout for slab on grade house.

The procedure is presented in a step-by-step manner to demonstrate a logical approach to sizing. Although the required calculations are relatively simple, they can be time consuming for larger projects. As with other design procedures discussed in this text, these procedures and equations could be implemented using software.

Concept of Heat Flux

The useful heat output of a radiant panel is expressed as the rate of heat flow into the room per unit of area of radiant floor panel surface. This is called the **room side heat flux**. In the United States, the common units for room side heat flux are Btu/h/ft^2.

The required room side heat flux at design load conditions can be calculated using Equation 10.5:

Equation 10.5:

$$q' = \frac{Q_R}{A_p}$$

where,

q' = room side heat flux under design load conditions (Btu/h/ft^2);
Q_R = design heat load of room (excluding underside losses of heated floor) (Btu/h);
A_p = area of radiant panel (ft^2).

10.19 Radiant Floor Panel Circuit Sizing Procedure

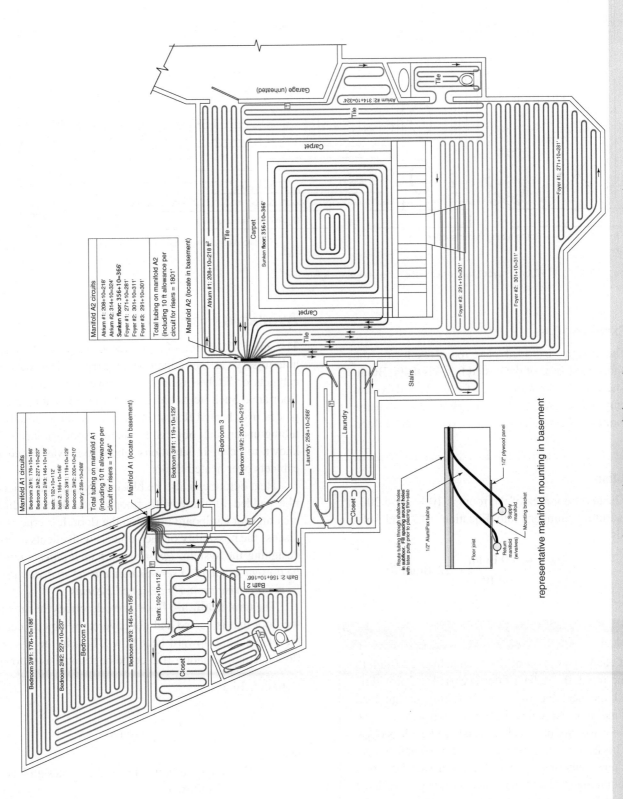

Figure 10-107 | Tubing layout plan for a thin-slab system using poured gypsum underlayment.

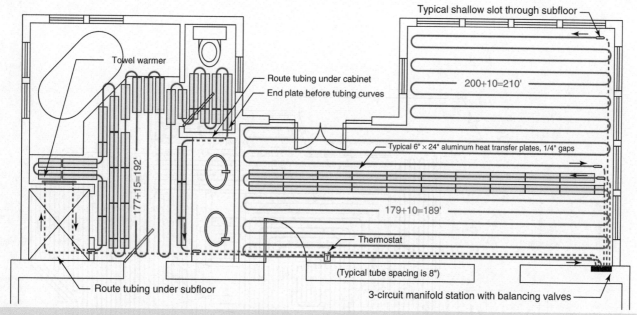

Figure 10-108 | Tubing layout for above floor tube and plate system.

Example 10.4

The design heat load of a room (excluding the underside losses of the heated floor) is 4,400 Btu/h. The floor in the room is 10-feet wide and 20-feet long and contains tubing throughout. Determine the room side heat flux requirement of the radiant floor.

Solution:

The heated floor area of the room is $10 \times 20 = 200$ ft^2. Substituting numbers into Equation 10.2 and solving yields:

$$q' = \frac{Q_R}{A_p} = \frac{4{,}400 \text{ Btu/h}}{200 \text{ ft}^2} = 22 \text{ Btu/h/ft}^2$$

Discussion:

If each square foot of floor can deliver 22 Btu/h to the room, the radiant floor is capable of providing all the heat the room needs during design load conditions. Note that the underside heat loss of the floor was *not* included in the design heat loss (Q_R). This heat loss will be accounted for in another step. Unless otherwise stated, the room side heat flux requirement of a radiant panel is always calculated based on design load conditions. During milder weather, the rate of heat output of the radiant panel decreases (as it would for any heat emitter). This also reduces the room side heat flux.

Available Floor Area

The entire floor area of some rooms may not be available to serve as the radiant panel. Fixed objects such as cabinets or kitchen islands may cover a portion of the floor area. Such items block most upward heat output if placed over heated floors. There is no point in installing tubing under such objects. Premature spoilage has been known to occur in situations where food was stored in cabinets over heated floors.

Heat output can also be blocked by movable objects like padded furniture with floor-length skirting, thick area rugs, or cardboard boxes on the floor. It is crucial to discuss the possibility of such objects with clients before designing the system. *Ignoring that such objects might be present can lead to serious deficiencies in heat output from otherwise properly designed systems.*

The term **available floor area** describes the area of the floor that is heated and not covered by objects capable of blocking upward heat flow. Thus, in the case of floor heating, Equation 10.5 could be restated as follows:

Equation 10.6:

$$q' = \frac{Q_R}{A_\text{available}}$$

where,

q' = required upward heat flux at design load (Btu/h/ft^2);

Q_R = design heat loss of room (excluding underside losses of floor) (Btu/h);

$A_\text{available}$ = floor area available for heat output (ft^2).

10.19 Radiant Floor Panel Circuit Sizing Procedure   535

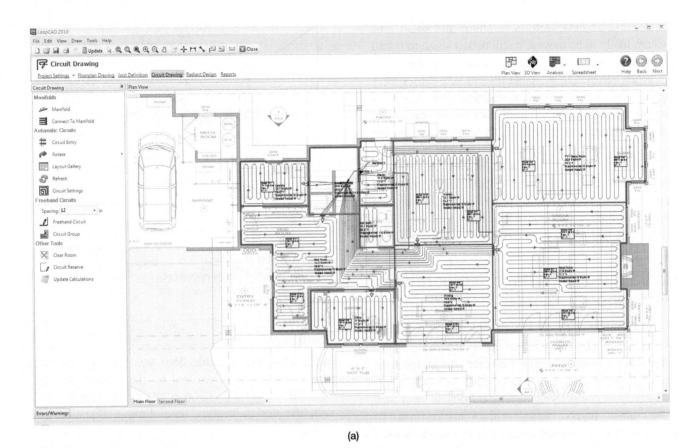

(a)

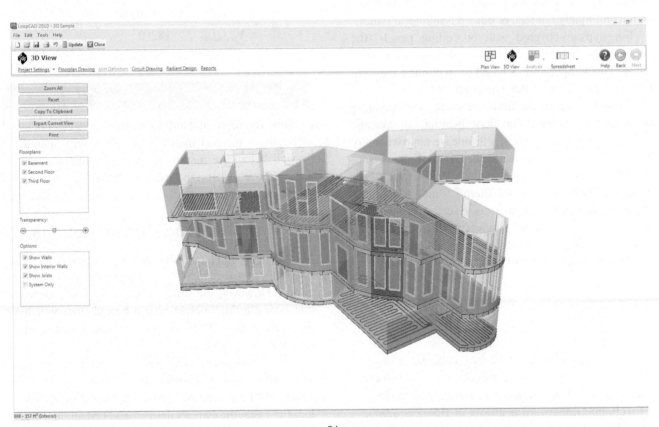

(b)

Figure 10-109 Software for automating circuit layout drawings. Circuits are viewable in both (a) 2D and (b) 3D. *Courtesy of Avenir Software, Inc.*

Because objects such as furniture and small area rugs are likely to be moved, it still makes sense to install tubing under floor areas where they are to be placed.

When calculating available floor area, a conservative approach is to subtract the estimated area covered by thick area rugs, padded furniture with floor-length skirting, or commercial inventory stored directly on the floor from the room's gross floor area. Since this decreases the available floor area, the required upward heat flux increases. In rooms where the value exceeds 35 Btu/h/ft^2, the floor panel may not be capable of providing the full heating load of the room. This is due to limitations in floor surface temperature that will now be discussed.

Radiant Panel Surface Temperature Limitations

The room side heat flux delivered by a radiant panel can be limited by physiological or material considerations. In the case of floor heating, the average surface temperature of the floor should not exceed 85 °F in rooms *where prolonged foot contact with the floor is likely*. In areas such as bathrooms, foyers, entries, or the outer perimeter of rooms, the average surface temperature of the floor should not exceed 92 °F. These temperatures are based on avoiding uncomfortably hot feet, or other physiological issues.

For gypsum-covered wall or ceiling panels, the surface temperature of the panel should not exceed 120 °F. This temperature is based on avoiding long-term thermal degradation of the gypsum board.

Surface temperature limits produce corresponding limits in heat output. For floor heating, an average surface temperature of 85 °F will release approximately 35 Btu/h/ft^2 into a room maintained at 68 °F. Increasing the average surface temperature to 92 °F, raises heat output to approximately 48 Btu/h/ft^2 into a room at 68 °F. A radiant wall with an average surface temperature of 120 °F releases about 93 Btu/h/ft^2 into a room at 68 °F. A radiant ceiling with an average surface temperature of 120 °F releases about 83 Btu/h/ft^2 in a 68 °F room.

Example 10.5
The 14 × 16-foot room shown in Figure 10-110 contains a built-in 2 × 12-foot cabinet. A sofa with floor-length skirting also covers a 3 × 6-foot area of the floor. The room's design heat loss (excluding downward heat loss of floor) is 4,000 Btu/h. Assuming that the room will be entirely heated by the floor, what is the room side heat flux requirement?

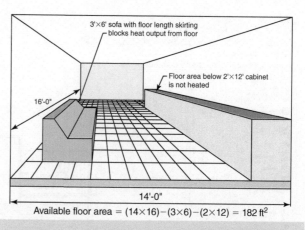

Figure 10-110   Available floor area is floor area that is heated and not blocked by furniture or built-in cabinets.

Solution:
In this case, the available floor area is the overall floor area minus the area covered by the cabinet and sofa: 14 × 16 − 2 × 12 − 3 × 6 = 182 ft^2. Combining this with the design heating load through Equation 10.6 yields:

$$q' = \frac{Q_R}{A_{available}} = \frac{4{,}000 \text{ Btu/h}}{182 \text{ ft}^2} = 22 \text{ Btu/h/ft}^2$$

Discussion:
This required upward heat flux is well under 35 Btu/h/ft^2, and therefore the floor will operate well below the maximum surface temperature of 85 °F, even at design load conditions.

Flooring Covering Considerations

Most people understand that draping a thick wool blanket over a cast-iron radiator would greatly reduce its heat output. However, the same people may not understand that laying a thick carpet over a heated floor will have the same effect. They simply are not accustomed to thinking of the floor as the "radiator."

Floor coverings can have a major impact on the thermal performance of a radiant floor heating system. This effect must be accounted for when the system is designed. In some cases, the owner's desire to use certain floor coverings may even eliminate radiant floor heating as an option.

From the standpoint of heating, the ideal floor covering is no covering. Bare concrete slab floors are ideal situations for floor heating. Painted, stained, or patterned concrete surfaces all perform essentially the same as a bare slab.

When floor coverings are present, they affect heat output in two ways. First, floor coverings with low R-values allow greater upward heat flow from a given radiant floor operating at a given average water temperature. Viewed another way, the greater the R-values of the floor covering(s), the higher the average water temperature in the tubing circuits must be to create the desired rate of upward heat flow.

Second, the R-value of floor coverings affects the variation in surface temperature between the tubes. Higher R-value coverings will reduce the variation between the highest floor surface temperature directly above the tube and the lowest surface temperature halfway between the tubes. This effect is shown in Figure 10-111.

The areas under the surface temperature profile curves in Figure 10-111 are proportional to heat output. Notice that the area under the curve for the ceramic tile floor is considerably greater than the area under the curve for the carpet-covered floor. Although the higher R-value carpeted floor covering creates less variation in surface temperature between the tubes, it also significantly reduces the heat output from the floor.

A table listing some R-values of floor coverings (and covering systems) is given in Figure 10-112.

Floor covering description	Total R-value of floor covering, or floor covering group (°F h · ft²/Btu)
Bare slab	0
Asphalt or rubber tile	0.05
1/8" sheet vinyl	0.20
3/8" ceramic tile (thin set)	0.30
1/4" carpet (nylon) (glue down)	0.95
1/4" carpet over 1/4" rubber pad	1.44
3/4" marble	0.42
3/8" hardwood	0.52
1/2" hardwood	0.70
3/4" hardwood	1.05

Figure 10-112 | Thermal resistance of various floor coverings.

Thermal Model of a Radiant Panel Circuit

A **thermal model** is a mathematical relationship that can be used to predict the thermal performance of a device. They are used for both design and simulation of that device. A thermal model was developed for fin-tube baseboard as Equation 8.1.

This section presents a generalized thermal model of a radiant panel circuit. The mathematical form of the model allows it to be "calibrated" to any type of hydronic radiant panel circuit including those used in floors, walls, or ceilings. Such a model is helpful in predicting the performance of a single radiant panel circuit. It is also useful as part of a larger analytical procedure used for balancing several simultaneously operating radiant panel circuits within a specific building.

This thermal model determines the heat output of a radiant panel circuit by tracking the temperature of the fluid moving through the circuit, and the circuit's flow rate. If the temperature of the fluid entering and leaving the circuit is known, along with the flow rate, the total heat dissipation of the circuit can be calculated using Equation 4.8.

To derive this model, it is helpful to think of the circuit as being "stretched out" into a straight line as shown in Figure 10-113.

This derivation also assumes that heat transfer across the centerline between adjacent tubes is negligible as depicted in Figure 10-114.

In the strictest sense, this can only be true if the water temperature in adjacent tubes was exactly the same and the panel construction was perfectly uniform. However,

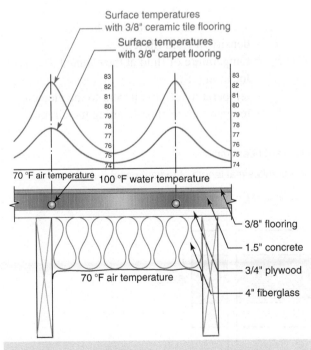

Figure 10-111 | Surface temperature profiles for thin-slab radiant panel covered with 3/8-inch ceramic tile or 3/8-inch carpet. More heat is released at the same water temperature when the flooring has lower R-value.

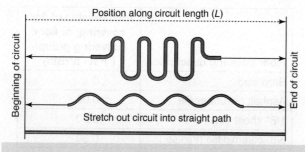

Figure 10-113 | Concept of a radiant panel circuit "stretched out" into a straight line.

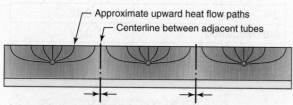

Assumption: no heat transfer across centerline between adjacent tubes (close approximation in serpentine circuits)

Figure 10-114 | The mathematical model assumes no heat transfer across the centerline between adjacent tubes within the radiant panel. This is a reasonable approximation for serpentine circuit patterns.

for consistently built serpentine circuit patterns, this is a reasonable assumption that allows an otherwise very complex three-dimensional heat transfer process to be modeled with relatively simple mathematics.

Figure 10-115 shows a very short element along the length of the circuit. Under steady-state conditions, the energy that leaves through the upper and lower surfaces of short element must be evidenced by an associated drop in fluid temperature. This is represented by the differential equation given as Equation 10.7.

Equation 10.7:

$$f \times (8.01 \times c \times D) \times (-dT_w) = a \times (T_w - T_{air}) \times (dL)$$

where,
- f = flow rate (gpm);
- c = specific heat of fluid (Btu/lb/°F);
- D = density of fluid (lb/ft³);
- T_w = temperature of fluid at the element location (°F);
- T_{air} = temperature of room air adjacent to the element;
- L = distance from beginning of circuit (feet);
- a = a number that depends on the specific panel construction;
- 8.01 = constant based on the units used.

The solution to the differential Equation 10.7 is given as Equation 10.8.

Equation 10.8:

$$T_w = T_{air} + (T_{in} - T_{air}) \times e^{-\left(\frac{a \times L}{f \times (8.01 \times D \times c)}\right)}$$

where,
- f = flow rate (gpm);
- c = specific heat of fluid (Btu/lb/°F);
- D = density of fluid (lb/ft³);
- T_w = temperature of fluid at the element location (°F);
- T_{air} = temperature of air adjacent to the element;
- T_{in} = temperature of fluid entering the circuit;

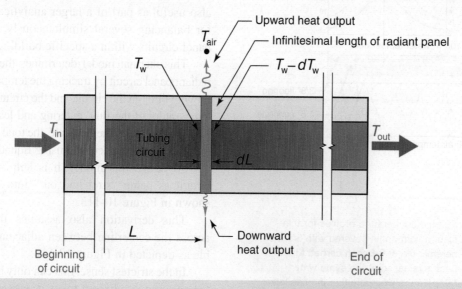

Figure 10-115 | Energy balance on differential element of length along circuit path.

L = distance from beginning of circuit (feet);
a = a number dependent on the specific panel construction;
8.01 = constant based on the units used;
e = 2.71828 (base of natural logarithm system).

The value of "a" for several radiant floor panel configurations is given in the next subsection.

Example 10.6

Assume a radiant panel consists of a 4-inch-thick bare concrete slab. The underslab insulation allows downward heat output to be 10% of upward heat output. The 1/2-inch PEX tubing is located at 12-inch spacing, and approximately mid-depth of the slab. For this specific construction, the value of "a" in Equation 10.8 can be assumed to be 0.97. Further assume that water flows into the circuit at 1.2 gpm and at an entering temperature of 110 °F. The air temperature above the panel is 70 °F. Determine the outlet temperature and total heat output of this circuit, assuming its length is 300 feet.

Solution:

All the data given are substituted into Equation 10.7 and reduced.

$$T_w = 70 + (110 - 70) \times e^{-\left(\frac{0.97 \times 300}{1.2 \times (8.01 \times 61.9 \times 1)}\right)}$$

$$T_w = 70 + (40) \times e^{-(0.489)} = 94.5 \text{ °F}$$

Once the outlet temperature is determined, the total heat output from the circuit can be calculated using Equation 4.8:

$$Q = (8.01Dc)f(\Delta T) = (8.01)(61.9)(1))1.2(110 - 94.5)$$
$$= 9,222 \text{ Btu/h}$$

Discussion:

Although this example calculated the outlet temperature and total heat output for a specific circuit length (300 feet), Equation 10.8 can, in theory, be used to determine the temperature at *any* position along a circuit of *any* length. For example, it could be used to determine the temperature every 25 feet along a circuit that was 1,000 feet long and constructed with 3/8-inch PEX tubing. Keep in mind, however, that such a circuit would be very impractical due to high head loss.

Figure 10-116 shows a plot of the temperature at any position along the length of the circuit used for this example. The plot was generated by evaluating Equation 10.8 at several length positions.

Notice the slight curvature in the graph. This happens because heat is being dissipated at a slightly higher rate near the beginning of the circuit compared to near the end.

The Effect of Flow Rate on Temperature Drop and Heat Output

Chapters 8 and 9 both discussed that heat output from a hydronic heat emitter is not a linear function of flow rate. Heat output increases rapidly at low flow rates, and more slowly at high flow rates. This also holds true for radiant panel circuits. Figure 10-117 shows the effect for the circuit analyzed in Example 10.6 at three different flow rates and a constant inlet water temperature of 110 °F. In each case, the circuit's outlet temperature was calculated using Equation 10.8, and then its total heat output was calculated using Equation 4.8.

The relationship between total heat output and flow rate supports the fact that flow rate reductions can sometimes be made without substantial changes in heat output.

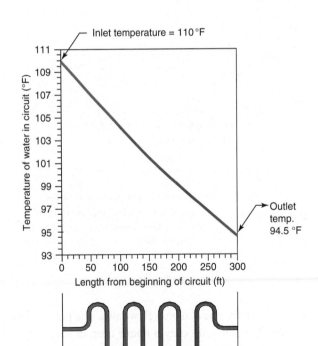

Figure 10-116 — Temperature at various positions along the length of a 300-foot circuit of 1/2-inch PEX tubing spaced at 12 inches within a 4-inch bare concrete slab. Water is supplied to the circuit at 110 °F and at a flow rate of 1.2 gpm.

540 Chapter 10 Hydronic Radiant Panel Heating

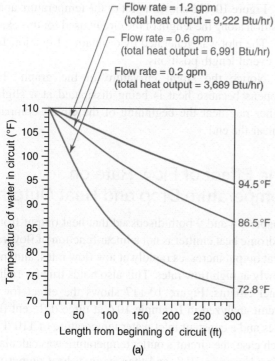

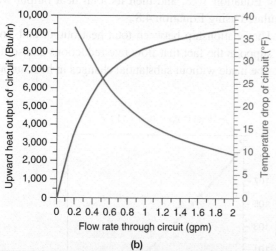

Figure 10-117 (a) Effect of flow rate on outlet temperature for radiant panel circuit described in Example 10.6. (b) Effect of flow rate on upward heat output and circuit temperature drop for the radiant panel circuit described in Example 10.6.

This is especially true at higher flow rates. The potential of reducing flow rate with only a minor drop in heat output allows for reduced circulator power input and reduced operating cost. Designers are cautioned, however, that reduced flow rates create higher circuit temperature drops and consequently higher variations in floor surface temperature. *In cases where floor heating is expected to deliver "**barefoot friendly floors**" circuit temperature drops should be limited to 15 °F at design load conditions.*

In other applications where floors are not subject to this scrutiny, a circuit temperature drop of 20 to 25 °F under design load is generally acceptable and will likely reduce circulator size and life cycle operating cost.

Step-by-Step Circuit Sizing Procedure (Floor Heating)

The following procedure can be used to determine the tube size, tube spacing, and supply water temperature for several types of radiant floor constructions. The circuits are designed on a room-by-room basis. Examples of the calculations are given in each step.

The calculations in this procedure can be kept organized using a worksheet. That sheet will contain many columns. Examples of portions of such a worksheet are shown at various stages in the procedure. The full worksheet will be presented at the end of the procedure.

STEP 1: Determine the design heating load of each room.

The first step in sizing any heating system is to determine the design heating load of each room in the building. Methods and data for doing so are available in Chapter 2, Heating Load Estimates.

Note: In this procedure, the design heating load of the room should not include the heat loss from the underside of the heated floor. This loss will be calculated separately in a later step. The symbol Q_R will be used to designate the design heat loss of a room excluding downward heat loss from the heated floor.

STEP 2: Determine the available floor area of each room.

Available floor area is floor area that is heated and is not blocked from releasing heat into the room. It does not include floor area covered by built-in cabinets, furniture with floor-length skirting, thick area rugs, or boxes stored on the floor. The symbol $A_{available}$ will be used to designate the available floor area of each room.

STEP 3: Determine the required upward heat flux for each room.

This is done using Equation 10.6:

$$q' = \frac{Q_R}{A_{available}}$$

Where:

q' = required upward heat flux under design load conditions (Btu/h/ft^2);

Q_R = design heat loss of room (excluding underside losses) (Btu/h);

$A_{available}$ = floor area available for heat output (ft^2).

STEP 4: Establish a maximum allowable floor surface temperature for each room.

A maximum average floor surface temperature of 85 °F is recommended for rooms with prolonged foot contact. In bathrooms, hallways, and entry foyers, this temperature may be increased as high as 92 °F.

STEP 5: Determine the *average* floor surface temperature needed to deliver the required upward heat flux to each room.

This temperature can be calculated using Equation 10.8a.

Equation 10.8a:

$$T_s = T_{room} + 0.5q'$$

where,
- T_s = required average floor surface temperature (°F);
- T_{room} = desired room air temperature (°F);
- q' = required upward heat flux for the room (Btu/h/ft²).

Example 10.7

A room heated by a radiant floor is to be maintained at 67 °F at design load conditions. The required upward heat flux has been calculated as 29 Btu/h/ft². What average floor surface temperature is required to provide this heat input?

Solution:

Substituting the numbers into Equation 10.8a and solving yields:

$$T_s = T_{room} + 0.5q' = 67 + 0.5(29) = 81.5\ °F$$

Discussion:

This temperature is the average floor surface temperature. Areas of the floor closer to the embedded tubing will have slightly higher surface temperature. Likewise, areas several inches away from a tube will have slightly lower surface temperatures.

A sample of what the first seven columns of a worksheet to organize the preceding calculations might look like, along with some sample data, is shown in Figure 10-118.

STEP 6: Determine if the room requires supplemental heat.

If the required floor surface temperature in Step 5 is greater than the maximum allowable surface temperature from Step 4, then supplemental heat is required for that room. Identify all such rooms.

The amount of supplemental heat required for a given room can be calculated using Equation 10.9:

Equation 10.9:

$$Q_{sup} = Q_R - A_{available}\left(\frac{T_{floor\ max} - T_{room}}{0.5}\right)$$

where,
- Q_{sup} = required supplemental heat input at design load (Btu/h);
- Q_R = design heat loss of room (excluding downward loss from floor) (Btu/h);
- $A_{available}$ = available heated floor area of room (ft²);
- $T_{floor\ max}$ = maximum allowable floor surface temperature (°F);
- T_{room} = desired room air temperature (°F).

Room name	Design heat loss (Btu/h)	Available floor area (ft²)	Room temperature (°F)	Required upward heat flux (Btu/h/ft²)	Maximum allowable floor surface temperature (°F)	Required floor surface temperature (°F)
Room 1	4,000	200	70	20	85	80
Room 2	8,000	350	70	22.9	85	81.5
Room 3	5,000	120	70	41.7	85	90.9
Room 4	3,000	120	70	25	85	82.5

Figure 10-118 | Using a table to keep calculations organized.

Example 10.8

Room 3 in Figure 10-118 will require a floor surface temperature of 90.9 °F to deliver the design load heat input from the floor. Assuming prolonged foot contact with the floor is likely, the maximum allowable surface temperature is 85 °F. Therefore, supplemental heating is needed. The amount needed is calculated using Equation 10.9 assuming that the desired room temperature is 70 °F.

$$Q_{sup} = Q_R - A_{available}\left(\frac{T_{floor\ max} - T_{room}}{0.5}\right)$$

$$= 5{,}000 - 120\left(\frac{85 - 70}{0.5}\right) = 1{,}400\ \text{Btu/h}$$

Discussion:

In this room, the radiant floor panel provides 3,600 Btu/h of the 5,000 Btu/h total design load. The remainder is provided by supplemental heat. The latter could be delivered through another hydronic heat emitter supplied by the same heat source. It might also be supplied from a totally separate heating device.

For rooms requiring supplemental heat, the maximum allowed upward heat flux from the floor should be recalculated using Equation 10.9a:

Equation 10.9a:

$$q'_{limit} = \frac{(Q_R - Q_{sup})}{A_{available}}$$

where,

q'_{limit} = maximum allowed upward heat flux (Btu/h/ft^2);

Q_{sup} = required supplemental heat input to room at design load (Btu/h);

Q_R = design heat loss of room (Btu/h);

$A_{available}$ = available heated floor area of room (ft^2).

Note: The value of q'_{limit} rather than q' should be used in all subsequent calculations for the room (Steps 7 and on) that reference q'.

Example 10.9

Room 3 in Figure 10-118 required 1,400 Btu/h of supplemental heat. What is the limiting value of upward heat flux?

Solution:

Substituting values into Equation 10.9a and solving yields:

$$q'_{limit} = \frac{(Q_R - Q_{sup})}{A_{available}} = \frac{(5{,}000 - 1{,}400)}{120}$$
$$= 30\ \text{Btu/h/ft}^2$$

Discussion:

The limiting heat flux of 30 Btu/h/ft^2 is based on not exceeding an average floor surface temperature of 85 °F. Again, use q'_{limit} rather than q' in all subsequent calculations for the room that reference q'.

STEP 7: Determine the tube spacing based on the required upward heat flux requirement of each room.

Suggested tube spacings are listed in the table in Figure 10-119. These spacings are conservative; they limit variations in floor surface temperature to acceptable ranges for residential "barefoot friendly" floors. Spacing for industrial floors, where greater variations of floor surface temperature are acceptable, could be 3 inches wider than the spacings listed in Figure 10-119. In some industrial applications, it is even possible to use tube spacing of 24 inches.

Required upward heat flux (Btu/h/ft^2)	Spacing if 1/2" tubing is used (in)	Spacing if 5/8" tubing is used (in)	Spacing if 3/4" tubing is used (in)
<10	15	18	18
10–20	12	12	15
20–30	9	9	12
30–40+	6	6	9

Figure 10-119 | Suggested tube spacings for radiant floor panel circuits.

When the system is supplied by a renewable energy heat source, such as a heat pump, the author recommends that the maximum tube spacing be limited to 12 inches. This helps reduce the required water temperature for a given heat output, and improves the thermal performance of the renewable energy heat source.

STEP 8: Estimate the total length of tubing for each room.

The length of tubing required on the floor in each room can be estimated using Equation 10.10.

Equation 10.10:

$$L = 12\frac{(A_{available})}{S}$$

where,

$A_{available}$ = available heated floor area of room (ft^2);
S = tube spacing (in).

Example 10.10

A room has an available floor area of 260 ft^2. The tube spacing selected is 9 inches. Estimate the amount of tubing needed in the room.

Solution:

Substituting the numbers into Equation 10.10 and solving yields:

$$L = \frac{12(A_{available})}{S} = \frac{12(260)}{9} = 347 \text{ ft}$$

Discussion:

Equation 10.10 proportions tubing to floor area assuming straight runs of tubing "wall-to-wall." It does not precisely account for the length of return bends or risers to wall-mounted manifold stations. The next step will add length to the circuit to account for leaders from the tubing circuit on the floor of the room to the manifold station.

STEP 9: Estimate the leader length required for the circuit to reach the manifold station.

The **leader length** of a circuit is the additional tubing needed to connect the floor circuit to a manifold station as shown in Figure 10-120.

Estimating the leader length requires the designer to know where the manifold station the circuit connects to

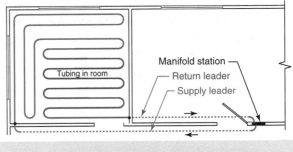

Figure 10-120 | "Leaders" are tubing lengths between the room that the circuit heats and the manifold connection (shown as dashed lines).

is located. Knowing this location, the designer estimates the additional tubing needed and adds it to the length of tubing for the room. Note: Additional circuit leaders may be needed based on the results of Step 10.

STEP 10: Determine if multiple tube circuits are needed in the room.

In many rooms, a single tube circuit will suffice. However, in larger rooms with close tube spacing, the length of a single circuit (including its leader length from Step 9) may be excessive. The length of a given circuit, including its leaders, should be limited to the values given in Figure 10-99.

If the total circuit length, including leaders, determined in Steps 8 and 9 exceeds these maximum circuit length values, divide the total tubing length by the maximum circuit length value and round up to the next whole number. This determines the number of tube circuits needed in the room.

Divide the total "in-room" tubing length by the number of circuits, then add an estimated leader length to each circuit. This will be the estimated length of the circuits for that room. In the unlikely but possible event that these circuit lengths still exceed the maximum circuit length allowed for a given tube size, divide the total in-room tubing into one additional circuit and add estimated leader lengths. Be sure to record all circuit lengths, including estimated leader lengths. These lengths will be needed when estimating the head loss of the circuits.

Example 10.11

A room needs 280 feet of 1/2-inch tubing. The estimated leader length (supply + return) is 80 feet. Can this work as a single circuit?

Solution:

The total length of a single circuit would be the 280 feet of in-room tubing, plus the 80 feet estimated leader length, or 360 feet. This is longer than the maximum recommended circuit length of 300 feet for 1/2-inch tubing. Therefore, the in-room tubing should be divided into two equal circuits of 280/2 = 140 feet. Assume the estimated leader length for each is 75 feet. Now the total length of each circuit is 140 + 75 = 215 feet, which is acceptable.

Example 10.12

A room requires 850 feet of 1/2-inch tubing. How many tubing circuits are necessary, and what is the approximate length of each of these circuits? Assume the leader length of each circuit is 60 feet.

Solution:

The recommended maximum circuit length for 1/2-inch tubing is 300 feet. Dividing this into 850 total feet yields 850/300 = 2.8. Rounding this up yields three circuits. Dividing 850 by 3 yields circuit lengths of 283 feet. Now assume that the leader length for each circuit is 60 feet. The total estimated length of each circuit is now 283 + 60 = 343 feet, which is too long for a 1/2-inch tube. To create equal but shorter circuits, divide the in-room length by 4 rather than 3; that is, 850/4 = 213 feet. Assuming that a nominal 60-foot leader length for each circuit yields a total estimated circuit length of 213 + 60 = 273 feet, which is acceptable.

Discussion:

This example demonstrates that long leader lengths can add significant tubing to the system that is "unproductive" since it is not necessarily installed into the radiant panel. Long leaders increase both material cost and circuit head loss, both of which are undesirable. *A good designer always attempts to minimize leader lengths.* In some cases, this requires the use of additional manifold stations. The designer should also attempt to keep the length of circuits installed in the same room reasonably close to each other.

STEP 11: Calculate the average water temperature in the tube circuit(s) based on floor covering and tube spacing.

Use Equation 10.11 along with the data given in the table of Figure 10-121 to determine the average water temperature in the tube circuits.

Equation 10.11:

$$\overline{T}_w = T_{room} + \frac{(q')}{k}$$

Slab-on-grade floors	(k) For 6" tube spacing	(k) For 9" tube spacing	(k) For 12" tube spacing	(k) For 18" tube spacing	(k) For 24" tube spacing
$R_f = 0$	1.13	1	0.882	0.69	0.563
$R_f = 0.5$	0.74	0.67	0.62	0.52	0.436
$R_f = 1.0$	0.53	0.5	0.47	0.41	0.355
$R_f = 1.5$	0.42	0.40	0.38	0.34	0.3
$R_f = 2.0$	0.34	0.33	0.32	0.29	0.263

Figure 10-121a | The k-values for slab-on-grade floor circuits.

Thin-slab floors	(k) For 6" tube spacing	(k) For 9" tube spacing	(k) For 12" tube spacing
$R_f = 0$	1.2	0.97	0.787
$R_f = 0.5$	0.74	0.66	0.57
$R_f = 1.0$	0.54	0.49	0.44
$R_f = 1.5$	0.43	0.40	0.36
$R_f = 2.0$	0.35	0.33	0.31

Figure 10-121b | The k-values for thin-slab floor circuits.

Above-floor tube and plate floors*	(k) For 8" tube spacing	(k) For 12" tube spacing
$R_f = 0.5$	0.58	0.55
$R_f = 1.0$	0.48	0.44
$R_f = 1.5$	0.41	0.36
$R_f = 2.0$	0.33	0.30

* All values assume 6-inch-wide aluminum plates installed at indicated tube spacing. The value of R_f must include the R-value of any cover sheet over plates, plus the R-value of any finish flooring materials.

Figure 10-121c | The k-values for above-floor tube and plate floor circuits.

Below-floor tube and plate**	(k) For 8" tube spacing	(k) For 12" tube spacing
$R_f = 0$	0.52	0.36
$R_f = 0.5$	0.42	0.32
$R_f = 1.0$	0.34	0.24
$R_f = 1.5$	0.29	0.21
$R_f = 2.0$	0.26	0.19

** All values assume 6-inch-wide aluminum plates installed below 3/4-inch plywood subfloor at indicated tube spacing. R_f is the total R-value of floor coverings installed on top of 3/4-inch plywood.

Figure 10-121d | The k-values for below-floor tube and plate circuits.

where,
$\overline{T}_w$ = average water temperature required in tubing at design load (°F);
T_{room} = desired room air temperature (°F);
q' = required upward heat flux (Btu/h/ft²);
k = constant based on tube spacing and R-value of floor covering from the table in Figure 10-121.

The data used to estimate the average circuit water temperature assumes the presence of good underside insulation, specifically that the downward heat loss is limited to no more than 10% of the upward heat output. This will typically be the case when the insulation recommendations described in previous sections of this chapter are followed.

It is also possible to use the k values in the tables of Figure 10-121 to determine the "a" values for Equation 10.8. The following relationship holds when the downward heat loss from the floor is 10% of the upward heat output.

Equation 10.12:

$$a = 1.10k\left(\frac{s}{12}\right)$$

where,
a = value needed in exponent of Equation 10.7 (Btu/h/ft/°F);
k = value from Figure 10-122 (Btu/h/ft²/°F);
s = tube spacing (in);
1.10 = coefficient that assumes downward heat loss is 10% of upward heat output.

If the downward heat loss is greater or less than 10%, change the coefficient (1.10) in Equation 10.12 to a corresponding value. For example, if the downward heat loss was known to be 5% of upward heat output, change the coefficient from 1.10 to 1.05. The lowest possible value for the coefficient would be 1.00, and this would only hold true if there was no downward heat loss. In the author's opinion, the downward heat loss from a heat floor should not exceed 10% of the upward heat output.

Example 10.13

A slab-on-grade floor panel with tubing spaced at 12-inch centers must create a required upward heat flux of 25 Btu/h/ft². The floor coverings have a total R-value of 0.75. The room is to be maintained at 70°F. Determine the average water temperature for the circuits:

Solution:

The finish floor R-value of 0.75 is not listed in the table for slab-on-grade systems. Therefore one must interpolate between the value of k for $R = 0.5$ and $R = 1.0$. In this case, the interpolation is a simple average: $(0.62 + 0.47)/2 = 0.545$. Substituting into Equation 10.11 yields:

$$\overline{T}_w = T_{room} + \frac{(q')}{k} = 70 + \frac{(25)}{0.545} = 116°F$$

Discussion:

The temperature calculated is the average water temperature required in the circuit. This will eventually be converted to a supply temperature based on the selected temperature drop of the circuits. Using Equation 10.11 and the data in

Figure 10-121, it is possible to estimate the average circuit water temperature for many heated floor systems having floor covering resistances between 0 and 2.0. Interpolate when necessary for various floor covering resistances and tube spacings.

R-value of edge insulation (°F h·ft²/Btu)	Value of m
0	0.83
5	0.39
7.5	0.30
10	0.23

Figure 10-122 | The m-values for different R-values of slab edge insulation.

It is emphasized that heat output from floors having floor covering resistances over 2.0 will be very marginal. The term "floor warming" is perhaps more appropriate than floor heating.

STEP 12: Estimate the downward heat loss from the floor panel.

The method used depends on whether the floor is slab-on-grade or built into a suspended floor deck (i.e., thin-slab or tube and plate systems).

For slab-on-grade floors use Equation 10.13:

Equation 10.13:

$$Q_d = mL[q'(R_f + 0.5) + T_{room} - T_{outdoor}]$$

where,

- Q_d = downward and edge heat loss of slab (Btu/h);
- m = a value based on the edge insulation used (see Figure 10-122);
- L = length of exposed slab edge (feet);
- q' = required upward heat flux at design load (Btu/h/ft²);
- R_f = R-value of floor covering(s) (°F h·ft²/Btu);
- T_{room} = room temperature at design load (°F);
- $T_{outdoor}$ = outdoor design temperature (°F).

Example 10.14

Determine the downward and edge loss from a heated slab on grade with 20 feet of exposed length, R-7.5 edge, and underside insulation. The slab releases 30 Btu/h/ft² into a room maintained at 70°F at design load. The floor covering has an R-value of 0.75. The outdoor design temperature is 0°F.

Solution:

Looking up the value of m in Figure 10-122 and substituting, along with the other data, into Equation 10.13 yields:

$$Q_d = mL[q'(R_f + 0.5) + T_{room} - T_{outdoor}]$$
$$= 0.30(20)[30(0.75 + 0.5) + 70 - 0]$$
$$= 645 \text{ Btu/h}$$

Discussion:

Equation 10.13 is a special adaptation of Equation 2.7, and it accounts for both edge and downward heat loss from a heated slab.

For radiant panels built into suspended floor decks (i.e., for thin-slab and tube and plate systems) use Equation 10.14 to estimate downward heat loss.

Equation 10.14:

$$Q_d = \frac{A_f}{R_d}[q'(R_f + 0.5) + T_{room} - T_{under}]$$

where,

- Q_d = downward heat loss (Btu/h);
- A_f = heated floor area (ft²);
- R_d = total R-value of insulation installed beneath the radiant panel (°F h ft²/Btu);
- q' = required upward heat flux at design load (Btu/h/ft²);
- R_f = R-value of floor covering(s) (°F h·ft²/Btu);
- T_{room} = room temperature at design load (°F);
- T_{under} = air temperature under the heated floor (°F).

Example 10.15

A 200-ft² thin-slab system delivers an upward heat flux of 25 Btu/h/ft² to a room maintained at 70°F. The floor covering has a resistance of R = 0.5. The underside of the floor deck is insulated to R-19. The temperature below the floor deck is 60°F. Estimate the downward heat loss from the floor.

Solution:

Substituting the data into Equation 10.14 and calculating yields:

$$Q_d = \frac{A_f}{R_d}[q'(R_f + 0.5) + T_{room} - T_{under}]$$

$$= \frac{200}{19}[25(0.5 + 0.5) + 70 - 60] = 368 \text{ Btu/h}$$

Discussion:

Equation 10.14 is a special adaptation of the conduction heat loss equation from Chapter 2, Heating Load Estimates.

STEP 13: Determine the required flow rate through each circuit.

This step requires the designer to select a temperature drop for the circuits. For residential and light commercial systems where "barefoot friendly" floors are expected, the temperature drop should be between 10 and 15 °F. For larger commercial or industrial systems, it can be from 20 to 25 °F. The higher the temperature drop, the lower the circuit flow rate; however, excessively high temperature drops can create noticeable variations in floor surface temperature from one side of the room to the other.

Once the temperature drop is selected, calculate the flow rate of each circuit using Equation 10.15.

Equation 10.15:

$$f = \frac{(Q_R + Q_d)}{(N)(b)(\Delta T)}$$

where,
- f = circuit flow rate (gpm);
- Q_R = design heat loss of room excluding downward loss from Step 1 (Btu/h);
- Q_d = downward heat loss of radiant panel from Step 12 (Btu/h);
- N = number of circuits in the room;
- ΔT = selected temperature drop of circuits (°F);
- b = fluid capacitance factor (for water b = 495, for 30% glycol b = 479, for 50% glycol b = 450).

Example 10.16

A large room has a design heat loss, excluding downward loss, of 20,000 Btu/h. The downward heat loss from the edge and bottom of the floor slab is estimated at 1,800 Btu/h, using methods from Step 12. Because of the size of the room, it is divided into four circuits. The designer has selected a circuit temperature drop of 20 °F. The system operates with water in the tubing. Estimate the flow rate in each circuit.

Solution:

Substituting the data into Equation 10.15 yields:

$$f = \frac{(Q_R + Q_d)}{(N)(b)(\Delta T)} = \frac{(20,000 + 1,800)}{(4)(495)(20)}$$

$$= 0.55 \text{ gpm/circuit}$$

Discussion:

Equation 10.15 is a special adaptation of the sensible heat rate equation discussed in Chapter 4, Properties of Water. The calculated flow rate provides for the upward heat output of the floor, as well as the rear side heat loss, without exceeding the specified circuit temperature drop.

STEP 14: Estimate the supply water temperature for each circuit.

This is found by adding *one-half* the circuit temperature drop to the average circuit water temperature determined in Step 11. It is expressed as Equation 10.16.

Equation 10.16:

$$T_{supply} = T_w + \frac{\Delta T}{2}$$

For example, if a circuit has an average water temperature of 105 °F and a temperature drop of 20 °F was selected, the supply temperature would be 105 + (20/2) = 115 °F.

STEP 15: Calculate the average of the supply temperatures of all circuits.

Add up all the supply temperatures of all circuits and divide by the number of circuits.

Check to see if all the supply temperatures fall within ±5 °F of the average. If this is the case, then all circuits can usually be supplied by a single water temperature. The suggested value of this supply temperature is the average supply temperature plus 2.5 °F.

Individual circuits that have supply temperature requirements slightly lower than the system supply temperature may require a slight reduction in flow rate to reduce output. Circuits with supply temperature requirements no more than 2.5 °F above the system supply temperature are generally able to provide sufficient output due to safety factors inherent in the design procedure.

The designer should identify any circuits with supply temperatures more than 5 °F above or below the average circuit supply temperature. Those circuits with higher supply temperature requirements can often be modified by returning to Step 11 and using a closer tube spacing to reduce the supply temperature requirement. Likewise, circuits with supply temperatures more than 5 °F below the average may be candidates for wider tube spacing.

In buildings with similar types of radiant panel construction throughout, it is preferable to adjust tube spacing, when necessary, so all circuits fall within the ±5 °F range of the average supply temperature. This allows the system to be designed for a single supply temperature, which lowers installation cost and reduces complexity.

When the building uses two or more types of radiant panels with widely different supply temperature requirements, it may not be possible to configure circuits for a single supply water temperature. Instead, the system can be designed to supply two or more temperatures. Piping and control techniques for this are discussed in the next section.

Figure 10-123 is a worksheet that can be used to organize the calculations from this procedure for a project with up to 20 rooms. A computer spreadsheet could also be developed based on this worksheet along with the equations and data given in this section.

Software-Assisted Design

Almost every company that currently sells tubing for radiant panel systems now offers software that can assist in the sizing of circuits. In most cases, the software is specific to the hardware offered by that company and may use calculation algorithms that are somewhat different from those given in this section.

The Hydronics Design Studio software contains a module called the **Hydronic Circuit Simulator**. This module is capable of simulating the interrelated thermal and hydraulic performance of up to 12 parallel radiant panel circuits. A screen shot of this module configured with 10 radiant panel circuits is shown in Figure 10-124.

The user defines the length, tube type and size, construction, and finish floor resistance of each circuit. The user also selects the circulator, common piping, fluid, and supply temperature used for the system.

The software models the hydraulic performance of the system using the methods given in Chapter 6, Fluid Flow in Piping, to determine the flow rate in each circuit. It then uses the thermal performance models given earlier in this chapter to determine the corresponding heat output of each circuit.

The software also includes the ability to turn any circuit on or off to simulate the use of zone valves or manifold valve actuators. A differential pressure bypass valve can be selected and set, as can balancing valves on any of the radiant panel circuits.

When any setting, selection, or numerical input is changed, the software repeats all calculations to determine the new point of thermal and hydraulic equilibrium.

10.20 System Piping and Temperature Control Options

There are several options for connecting hydronic heat sources to radiant panel distribution systems. The method used depends on the water temperature requirement of the radiant panel(s) as well as any other heat emitters used in the system. It also depends on the operating temperature and flow requirements of the heat source.

In most cases, the system piping method and control technique used are mutually dependent. One cannot be used without the other. Some of the mixing assemblies and control techniques discussed in Chapter 9, Control Strategies, Components, and Systems, will be applied in the schematics shown in this section. These schematics show the general concept but not necessarily every individual component required by the system or local codes. The reader is encouraged to cross reference previous chapters on piping and controls to fill in the details needed to complete the system. More methods for incorporating radiant panels along with other heat emitters and loads will be shown in Chapter 11, Distribution Piping Systems.

Manifold Stations

A common, subassembly in radiant panel systems is the **manifold station**. It consists of one **supply manifold**, one **return manifold**, and in some cases, several accessories such as **circuit-balancing valves**, **manifold valve actuators**, flow meters, purging valves, air vents,

room name	design heat loss (Btu/h)	available floor area (sq. ft)	room air temp. (°F)	required upward heat flux (Btu/h/ sq. ft)	max. allowable floor surface temperature (°F)	required floor surface temperature (°F)	supplemental heat? (Y/N)	amount of supplemental heat needed (Btu/h)	maximum allowable upward heat flux (Btu/h/ sq. ft)	tube spacing selected (in)	length of tubing in room (ft)	leader length for circuit (ft)	total circuit length (ft)	number of circuits in room	average circuit water temp. (°F)	downward heat loss of panel (Btu/h)	circuit flow rate (gpm)	circuit supply temp. (°F)

Figure 10-123 I Suggested spreadsheet table for organizing radiant sizing calculations.

550 Chapter 10 Hydronic Radiant Panel Heating

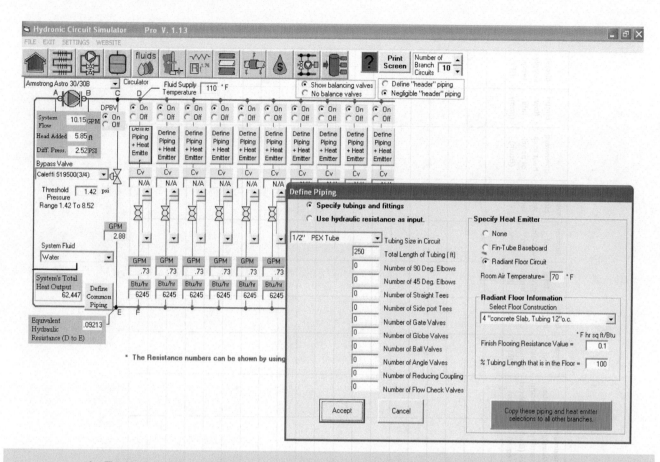

Figure 10-124 | The Hydronics Circuit Simulator module in the Hydronics Design Studio can simulate the thermal and hydraulic performance up to 12 parallel connected radiant panel circuits. *Source: Hydronicpros.*

isolation valves, and thermometers. Several of these accessories will be described later in this section.

In most systems, all radiant panel circuits begin at a supply manifold and end at a return manifold. Each circuit is therefore in parallel with the other circuits attached to the same manifold.

The number of piping circuits supplied by a manifold station can range from 1 to 20 or more depending on the total flow rate required by the circuits and the diameter of the manifold. In residential systems, manifolds with 2 to 12 circuits are common. In commercial or industrial buildings with large open spaces, larger manifold stations supplying 10 to 25 circuits may be used.

The supply and return manifolds are usually supported by a common set of brackets. This assembly can be mounted in several locations. In residential and light commercial buildings with slab floors, it is often mounted in the stud cavity of an interior wall and covered by an access panel. In basements, garages, or other utility spaces where exposed hardware is acceptable, manifold stations are often mounted directly to a wall surface. They can also be mounted horizontally on the underside of floor joists and accessed from the basement. Figure 10-125 shows several examples of manifold stations.

One of the simplest and lowest cost manifold stations uses **valveless manifolds** such as shown in Figure 10-126. Valveless manifolds are also available in brass, as well as polymer materials. *Valveless manifolds are ideal when all circuit attached to the manifold station have lengths that are within ±10% of each other, and will operate as a single zone.*

When circuit lengths are significantly different, or when circuits supplied from the same manifold station will operate as separate zones, a **valved manifold station** should be used. Valved manifolds allow individual flow adjustment of each circuit. Many also allow manifold valve actuators to be attached. These allow on/off flow control of each circuit using a low voltage (24 VAC) signal from a control system. Figure 10-127 shows examples of a valved manifold station constructed primarily of brass, and configured for five circuits. This type of manifold body is available with 2 to 12 circuit connections.

Manifold Accessories

Manifold stations can be "accessorized" to suit the needs of individual systems. The accessories available depend on manufacturer and model. Figure 10-128 shows the

10.20 System Piping and Temperature Control Options

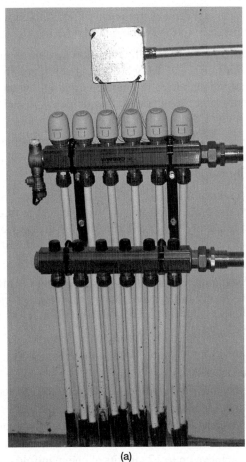

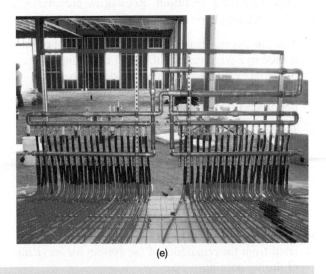

Figure 10-125 (a) Manifold station mounted to wall surface above slab. *Courtesy of Harvey Youker.* (b) Manifold station mounted in a stud cavity. *Courtesy of Dominic Kovacevic.* (c) Manifold station mounted in wall cavity box with access panel. *Courtesy of Caleffi North America.* (d) Manifold mounted horizontally under floor framing. *Courtesy of John Siegenthaler.* (e) Large commercial manifold stations with reverse return piping. *Courtesy of Northeast Radiant, Inc.*

Figure 10-126 | Example of a copper valveless manifold. *Courtesy of Uponor.*

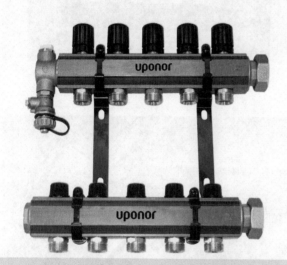

Figure 10-127 | Example of a brass manifold station with individual balancing valves for each of its five circuits. *Courtesy of Uponor.*

possibilities for hardware that can be attached or used in conjunction with one valved brass manifold.

The following manifold accessories are available from several suppliers:

- *Pipe Adapters.* These allow the manifold to be connected to a wide variety of piping in various sizes. Some adapters also provide a 90-degree bend to allow piping to fit into limited space within a cabinet or stud cavity.
- *Balancing valves.* These allow the flow rate of each circuit to be adjusted. They are essential when the manifold station serves circuits of widely different lengths. Proper balancing allows the heat output of each circuit to be adjusted to the needs of the space it serves.
- *Isolation valves.* These allow the manifold to be isolated from the remainder of the system for servicing or during purging.
- *Purging valves.* These allow the manifold station to be purged independently of the rest of the system. This may or may not be necessary depending on the remainder of the system piping.

- *Circuit-return temperature thermometers.* Indicate return temperature of each individual circuit.
- *Inlet-temperature thermometer.* Used to read water temperature to the supply manifold.
- *Flow meters.* These allow the flow rate through each circuit to be read. May be used in combination with supply temperature thermometer and individual circuit-return temperature thermometers to estimate each circuit's rate of heat delivery.
- *Air vents.* Either float type or manually operated venting mechanisms to assist with air removal at each manifold.
- *Differential pressure bypass.* This component is used in combination with valve actuators to maintain relatively stable differential pressure between supply and return manifold as flow through individual circuits starts and stops.
- *Valve actuators.* These devices screw onto manifold valve heads and provide linear movement to open and close manifold valves in response to a low voltage (24 VAC) signal.

Figure 10-129 shows a manifold station with several of these accessories mounted.

Some manifold stations are assembled on the job site using modular components. One approach offers segmented manifolds with two, three, and four connections each. These are joined together in various combinations to achieve up to 12 circuits at one manifold station.

Another approach uses individual circuit segments joined together end-to-end to make up the required number of circuits. O-rings are used to seal between each section as well as the end caps. This type of manifold is commonly made of high temperature compatible polymer materials.

Manifold Valve Actuators

Many manifold systems also allow electric valve actuators to be screwed onto the valves of the return manifold. The manifold shown in Figure 10-129 is an example. Manifold valve actuators allow a low voltage (24 VAC) thermostat circuit to open and close individual manifold valves. When screwed onto the manifold, they force the spring-loaded valves to their closed position and hold them there. When a given actuator is powered on, it retracts its shaft and allows the manifold valve to open. Valve actuators make it possible to independently control flow in each circuit served by the manifold station. This is particularly useful when each room needs to be controlled as a separate zone. Two or more valve actuators can also be powered through a common circuit for simultaneous operation. Figure 10-130 shows

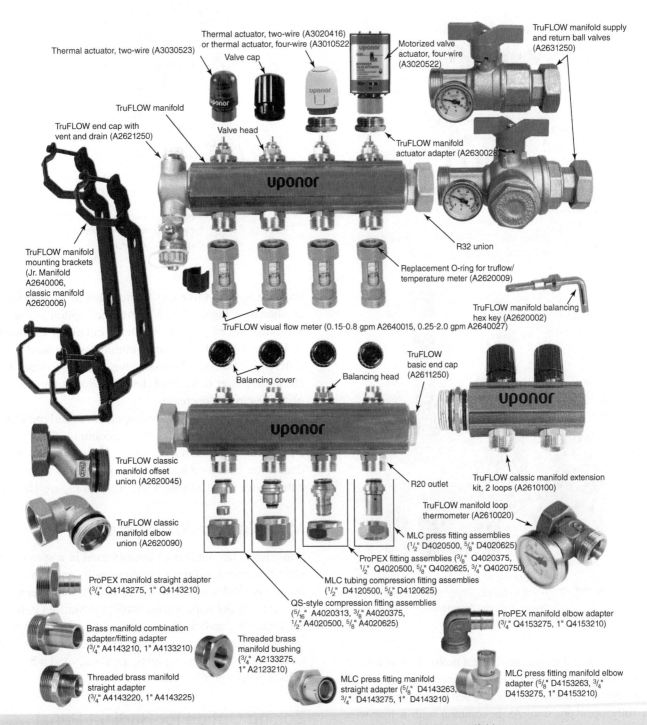

Figure 10-128 | Sampling of the accessories available for use with a modern valved brass manifold. *Courtesy of Uponor.*

a valve actuator ready to be attached to a valved return manifold.

Most manifold valve actuators use a **heat motor** to create the linear shaft movement. Heat motors are wax-filled cylinders with low wattage electric heating elements. When power is applied, the wax within the heat motor warms and expands. This expansion produces the shaft movement. Heat motor type valve actuators typically take 2 to 3 minutes from when power is first applied to when they are fully open. The internal construction of a heat motor type manifold valve actuator is shown in Figure 10-131.

Manifold valve actuators are available in both two-wire and four-wire configurations. Two wires are

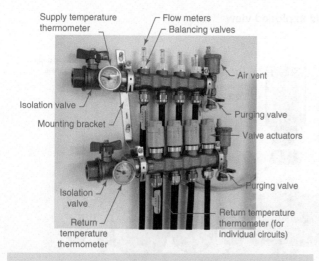

Figure 10-129 | Brass manifold with several accessories mounted. *Courtesy of John Siegenthaler.*

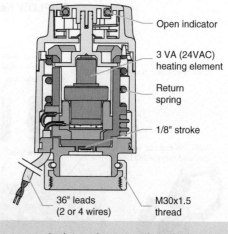

Figure 10-131 | Internal construction of a heat motor type manifold valve actuator. *Courtesy of Caleffi North America.*

Figure 10-130 | Example of a 24-VAC valve actuator ready to be mounted to a manifold valve. *Courtesy of John Siegenthaler.*

always required to operate the heat motor. The additional two wires in four-wire actuators connect to an isolated **end switch**. This switch closes when the actuator reaches it fully open position. As is common with a four-wire zone valve, the end switch closure is used to signal other equipment in the system that the zone circuit is open, and ready for flow of heated water. Figure 10-132 shows typical wiring for a four-wire manifold valve actuator.

When several independently controlled manifold valve actuators are used on one or more manifolds, some form of differential pressure control is necessary to prevent wide variations in differential pressure as the zone circuits turn on and off. If a fixed speed circulator is used to supply flow to the manifold, a differential pressure bypass valve can be installed at the manifold as shown in Figure 10-133a. Another option is use of a variable speed pressure regulated circulator as shown in Figure 10-133b. The latter eliminates the need for a differential pressure bypass valve. It also reduces the electrical power demand of the circulator as the number of active circuits decreases.

Although it is possible to operate several manifold valve actuators in parallel from a common thermostat, a practical limit is three actuators. If the zone has more than three circuits a zone valve can be used in combination with valve actuators as shown in Figure 10-134.

Multiple Manifold Systems

Sometimes, the number of circuits used in a system, or their placement within a building, will require two or more manifold stations. When all circuits can operate at the same supply temperature, multiple manifold stations are piped in parallel. *Never pipe multiple manifolds in a series*. The large pressure and temperature drops that would result from series piping are far beyond what can deliver proper performance.

Zoning Option for Radiant Panel Distribution Systems

Many of the zoning techniques, discussed in previous chapters, can be used with radiant panel distribution systems. Common approaches include zoning with circulators, zone valves, and electric valve actuators. Zoning can also be done using nonelectric thermostatic valves. Examples of zoned radiant panel systems will be shown in later sections of this chapter as well as in Chapter 11, Distribution Piping Systems.

10.20 System Piping and Temperature Control Options

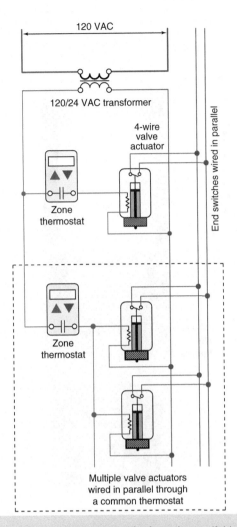

Figure 10-132 | Typical wiring for four-wire manifold valve actuators. Multiple actuators can be wired in parallel for simultaneous operation.

Constant Circulation

Slab-on-grade floor heating systems are well suited for buildings such as vehicle service garages or fire stations. These buildings typically have one or more large overhead doors. The slab areas just inside these doors experience higher rates of heat loss compared to other areas along the perimeter of the building, especially if there is no insulation between the edge of the slab and the outside pavement. One technique that is effective in preventing these floor areas from becoming "cold spots" is to maintain constant circulation through all floor circuits, even when no heat is being added to the slab from the system's heat source. Constant circulation helps in redistributing heat from the interior "core" areas of the building to the floor areas adjacent to the overhead doors. It is also a way to guard against freezing in these areas of higher heat loss, which might otherwise occur in systems operating with water if circulation were to stop for several hours during extreme cold weather. The latter could occur if the thermostat controlling the circulator(s) were to be configured for a deep temperature setback lasting several hours.

One way constant circulation can be accomplished is through use of a **three-way diverter valve** as shown in Figure 10-135.

When heat input is needed, the diverter valve is powered on, typically a 24-VAC signal from a thermostat or other zone controller. In the "on" mode heated water passes into the "A" port of the valve and out through the "AB" port. No flow passes through the "B" port.

When no further heat input is required, the diverter valve is turned "off." Flow now passes into the "B" port and out through the "AB" port. No flow passes through the "A" port, and thus no heat is added to the floor circuits.

Using this arrangement, the circulator can operate continuously throughout the heating season, while heat input is regulated by a thermostat or other zone controller.

Constant circulation can also be achieved in systems using two-way, three-way, and four-way mixing valves, or in systems using a variable speed injection pump. Piping arrangements for all of these hardware configurations were shown in Figure 9-102. The controls used on systems with any of these hardware configurations must be able to stop the flow of heated water between the heat source and floor circuits, without stopping flow through the latter, whenever the thermostat is not calling for heat in the zone.

Multi-Temperature Radiant Panel Systems

When two or more types of radiant panel construction are used in the same building, it is often necessary to supply two or more water temperatures simultaneously. An example would be a house with a heated basement slab, and below-floor tube and plate radiant panels on the first and second floors.

Multi-temperature radiant panel systems can be constructed several ways, depending on the heat source used, the operating temperatures of the radiant panels, and the capabilities of the mixing devices.

If one radiant panel subsystem can operate at the water temperature leaving the heat source, that subsystem does not require a mixing device. The other portion(s) of the system that require(s) lower water temperature would be supplied through one or more mixing devices. One example of such a system is shown in Figure 10-136.

In this system, the modulating/condensing boiler is configured to supply water at a temperature based on the upper outdoor reset line. The boiler measures the water temperature downstream of the closely spaced tees,

556 Chapter 10 Hydronic Radiant Panel Heating

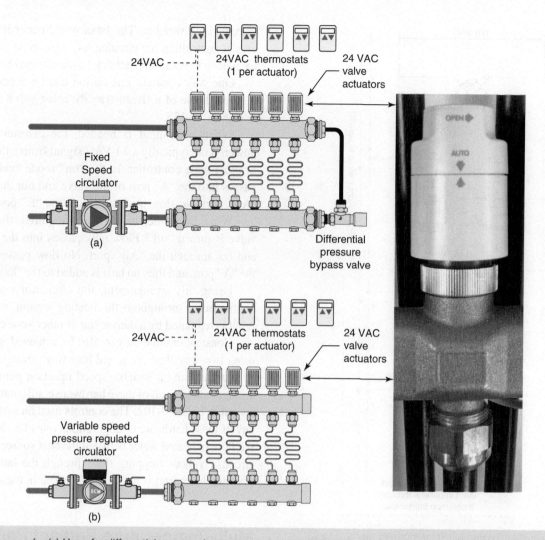

Figure 10-133 | (a) Use of a differential pressure bypass valve in combination with a fixed speed circulator. (b) Use of a variable speed pressure regulated circulator. *Courtesy of John Siegenthaler.*

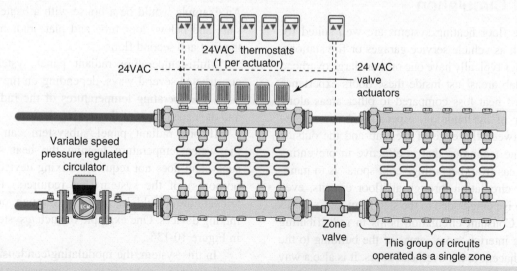

Figure 10-134 | A zone valve used to operate a manifold segment with more than three circuits in a single zone. Manifold valve actuators are used to operate other independent circuits.

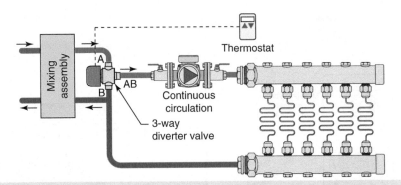

Figure 10-135 | Use of a three-way diverter valve to regulate heat input to a radiant panel distribution system operating with constant circulation.

which provide hydraulic separation between the boiler circulator and the distribution circulators. The firing rate of the boiler is automatically adjusted to maintain the supply water temperature close to the target value determined by the upper reset line and the current outdoor temperature.

The two lower temperature manifold stations are independently zoned using circulators with built-in check valves. They are both supplied through a single three-way motorized mixing valve, which operates based on the lower outdoor reset line. This mixing valve is powered on when either of the lower temperature manifold stations are active.

Figure 10-137 shows a similar, "four-temperature" system configuration using a conventional boiler, and valve-based zoning for two higher temperature and two lower temperature manifold stations. This system also supplies panel radiators directly from the boiler (e.g., without a mixing device). Separate mixing assemblies are used to provide the necessary supply water temperature to each group of manifold stations. An indirect water heater is the fourth subassembly supplied by the boiler.

Although the schematic in Figure 10-137 appears complex, it is more easily understood when viewed as fours independent "**hydronic subassemblies**," each operating at a different supply water temperature.

Each subassembly is connected directly across the low flow resistance headers. These headers are sized for a maximum flow velocity of 2 ft/s. to minimize any hydraulic interaction between subassemblies.

The two mixing assemblies could be any of those described in Chapter 9 that are capable of providing outdoor reset control. Each mixing assembly monitors supply water temperature, boiler inlet water temperature, and outdoor temperature. When necessary each mixing assembly can independently reduce heat transfer to its associated distribution system to prevent sustained flue gas condensation in the boiler.

Variable speed pressure-regulated circulators are used for proper differential pressure control within each subassembly since zoning is done using valves.

This system also shows that it is possible to combine electrically operated zone valves and manifold valve actuators, along with nonelectric thermostatic radiator valves.

The panel radiator subassembly is supplied directly from the headers, and thus operates at the supply water temperature supplied from the boiler. The boiler is controlled based on partial outdoor reset. Its reset controller is set to a minimum supply temperature of 140°F to prevent sustained flue gas condensation.

When the aquastat of the domestic water heater calls for heat, the boiler water temperature is raised to 170°F regardless of outdoor temperature. If the domestic water heater is a priority load, all other loads are temporarily turned off while it is active. If it is not a priority load, the panel radiators will receive water at 170°F during this time. This is usually not a problem because the domestic water heating cycle is relatively short. However, if the designer wanted to prevent this, another mixing assembly could be added to supply the panel radiator subassembly. The two radiant panel subassemblies would be unaffected because their mixing assemblies would compensate for the high boiler water temperature.

The approach of adding "subassemblies" to a common heat plant with low flow resistance header piping allows designers unlimited possibilities for building multi-load/multi-temperature systems that combine radiant panel heating along with other types of heat emitters and heat sources. Figure 10-138 illustrates this concept by "exploding" the system of Figure 10-137 into four subassemblies.

Additional examples of how radiant panel subassemblies can be incorporated into multi-zone/multi-temperature systems will be presented in Chapter 11, Distribution Piping Systems.

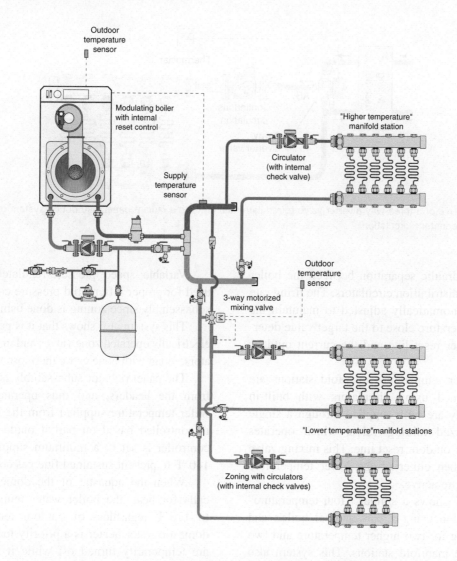
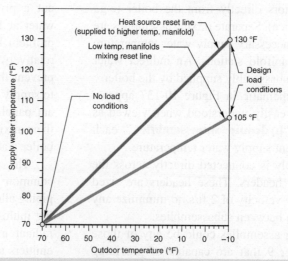

Figure 10-136 | Higher temperature manifold is supplied without a mixing device. The two lower temperature manifolds are supplied through a mixing device (three-way motorized mixing valve), and zoned using circulators.

10.20 System Piping and Temperature Control Options 559

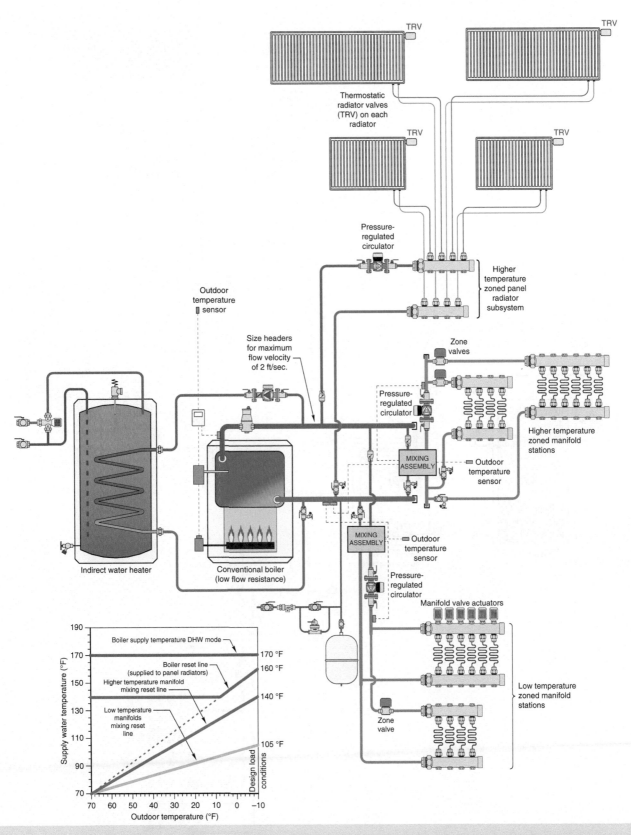

Figure 10-137 — A multi-temperature system supplying zoned manifold stations at two different mixed supply temperatures, and a panel radiator subsystem supplied directly from the boiler. An indirect water heater is also supplied directly from the boiler.

560 Chapter 10 Hydronic Radiant Panel Heating

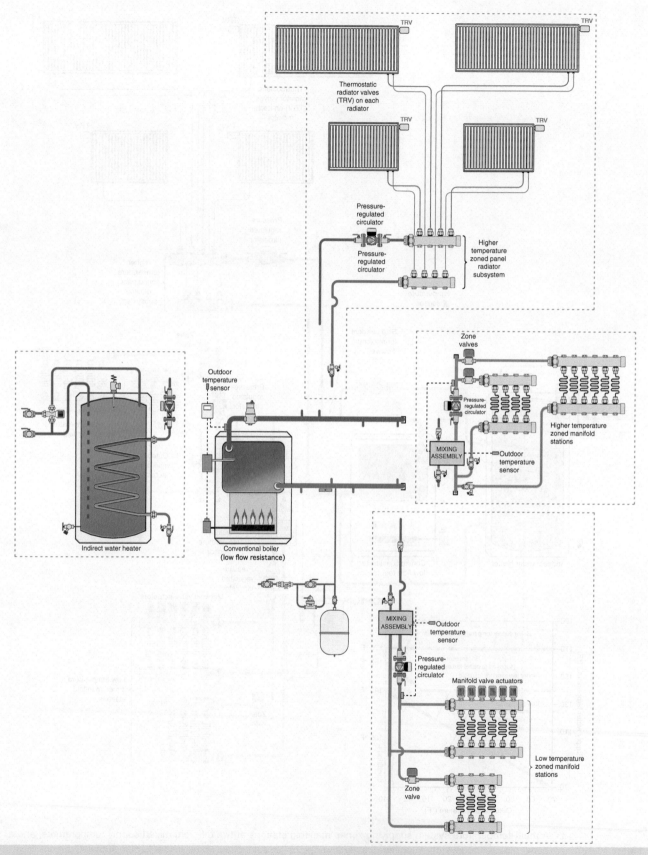

Figure 10-138 | Four temperature system "exploded" into independent subassemblies.

Summary

Because hydronic radiant panels are an integral part of a building, they require more planning relative to other types of hydronic systems. The time committed to this planning is time well spent. It will return itself many times over.

A final point is that hydronic radiant panel heating systems tends to be "forgiving." Minor differences between calculated loads and estimated heat outputs from radiant panel tend to be insignificant in most projects. This statement is not meant to endorse careless design or installation, but rather to calm concerns over obtaining a precise match between calculated heat output and heating load in every room. Remember, the methods used to calculate both the heating load and radiant panel output are subject to inaccuracies. Neither can totally account for all the nuances of every building in which these systems are used. These calculations do however serve as a guide to judge the reasonableness of a proposed design and to check for large or unanticipated performance problems.

The technology of hydronic radiant panel heating is constantly evolving. New products and design methods appear each year. The reader is encouraged to contact manufacturers of flooring heating equipment to stay abreast of the latest materials and design methods.

Key Terms

- above-floor tube and plate system
- aerial boom
- air temperature stratification
- aggregate
- available floor area
- average floor surface temperature
- band joist
- barefoot friendly floors
- bearing wall
- below-floor tube and plate system
- bend supports
- bond breaker layer
- circuit balancing valve
- circuit layout drawing
- closet flange
- concrete thin-slab
- contact adhesive
- control joint
- controlled shrinkage crack
- counterflow serpentine
- counterflow spiral
- cover sheet
- dead loading
- depth stop attachment
- diffuse
- drawing layer
- electric radiant panel
- electromagnetic spectrum
- extruded aluminum heat transfer plates
- extruded polystyrene
- fibermesh®
- finite element analysis
- graphite
- heat diffusion
- heat motor
- heat sink
- heat transfer plate
- Hydronic Circuit Simulator
- hydronic radiant panel
- hydronic subassemblies
- infrared radiation
- isotherms
- kraft paper faced batt
- leader length
- leaders
- lifts
- lift hook
- lightweight concrete
- manifold station
- manifold valve actuators
- maximum deflection criteria
- mineral oil
- nuclear radiation
- polyline
- poured gypsum-underlayment
- radiant heat
- return manifold
- room air temperature profile
- room side heat flux
- rough openings
- serpentine patterns
- sleepers
- striping
- subfloor
- subgrade
- substrate
- superplasticizer
- supply manifold
- suspended tube system
- thermal model
- thermal radiation
- thermographic image
- three-way diverter valve
- tubing layout drawing
- uncoiler
- underside insulation
- underslab insulation
- valved manifold station
- valveless manifold
- vapor barrier
- water-reducing agent
- welded wire fabric (WWF)

Questions and Exercises

1. Why is it desirable to have warmer surfaces and air near the lower extremities of the body?

2. What is an advantage of radiant floor heating compared to finned-tube baseboard convectors in rooms with high ceilings?

3. Describe at least three considerations when locating a manifold station within a building.

4. Why is insulation needed beneath a radiant floor panel installed in a wood-frame floor, even when the space beneath the floor is maintained at normal comfort temperatures?

5. Describe the function of control joints in thin-slab installations.

6. What are some of the architectural considerations (things the building designer needs to know about) associated with using a thin-slab floor heating system?

7. Why are gypsum-based underlayments installed in two lifts rather than a single pour?

8. Why should floor piping circuits route the warmest water near the exterior wall?

9. A kitchen measures 10×12 feet with an island measuring 3×5 feet. The kitchen has a design heating load of 4,000 Btu/h when it is 68 °F inside and 0 °F outside. What is the required average upward heat flux from a floor heating system at design load conditions?

10. Describe what happens to the supply water temperature requirement of a floor heating system when a bare concrete slab floor is covered with 3/8-inch ceramic tile. How does this compare to the supply water temperature requirement if the floor is covered with 1/4-inch carpet?

11. The kitchen described in exercise 9 has a slab-on-grade floor with one 10-foot wall and one 12-foot wall exposed to the outside. The edges of the slab are insulated with 2 inches of extruded polystyrene insulation. The slab is expected to operate at 105 °F. Estimate the downward and edgewise losses from the slab.

Chapter 11

Distribution Piping Systems

Objectives

After studying this chapter, you should be able to:

- Assess several considerations when planning a zoned hydronic system.
- Discuss the pros and cons of zoning with building owners.
- Explain the concept of hydraulic separation.
- Implement hydraulic separation in several ways.
- Identify the type of distribution piping used in existing systems.
- Explain the iterative processes involved in designing a distribution system.
- Assess the suitability of a particular distribution system topology for a given application.
- Use previously discussed analytical tools to design a distribution system.
- Explain sizing and selection of differential pressure bypass valves.
- Understand the fundamentals of primary/secondary piping.
- Select appropriate hardware for various types of distribution systems.
- Identify proper applications for variable-speed pressure-regulated circulators.
- Determine the distribution efficiency of various heat delivery systems.
- Convey the inherent advantage hydronic distribution systems have based on high distribution efficiency.
- Understand how complex distribution systems can be divided into simpler and standardized piping topologies.

11.1 Introduction

The purpose of a hydronic distribution system is to transport heated water to each heat emitter so that it can properly heat its assigned area. The decision of what type of distribution system to use must consider factors such as the following:

- Number of heating zones in the building
- Method of controlling zone heat output

- Water temperature requirements of the heat emitters
- Head loss characteristics of the heat emitters
- Installation cost of the piping system
- Operating cost of various options
- Size and operating cost of the system's circulators and
- Ability to expand the system in the future

This chapter discusses all the standardized piping approaches applicable to residential and light commercial systems. These approaches show how components such as piping, valves, circulators, heating sources are arranged relative to each other. These arrangements are sometimes called **topologies**. The strengths and weaknesses of each topology are examined. Design procedures are discussed and examples are given.

This chapter ties together much of the material from earlier chapters into true system design. It is important that the reader be familiar with concepts such as series and parallel piping arrangements, head loss, hydraulic resistance, pumps curves, thermal performance of heat emitters, controls, and other material presented in earlier chapters.

A poorly designed distribution system can ruin the performance of any hydronic system. All too often, people, including heating professionals, focus their attention on hydronic heat source selection and simply assume that heat source will perform as advertised regardless of the distribution system it is connected to. They soon discover that the performance of any heat source can be crippled by an inadequate distribution system. *Sadly, the vast majority of complaints about insufficient or poorly proportioned heat delivery are traceable to poorly designed or installed distribution systems, rather than improperly performing heat sources.*

Another of the author's observations over 40 years of practice is that some in the heating trade seem determined to "reinvent the wheel" each time a new hydronic distribution system is being planned. This is not only unnecessary but also very likely to create problems. This is not to suggest that each new distribution system cannot contain two or more piping topologies that are shown in this chapter. Rather, it is to discourage indiscriminate "morphing" of established piping arrangements without a clear understanding of underlying principles such as hydraulic separation, differential pressure control, head loss through valves, prevention of heat migration through inactive portions of the system, and control of supply water temperature.

The best hydronic system designers know these underlying principles well. They also know when each of the distribution systems discussed in this chapter is appropriate and apply them where they have advantages.

11.2 Zoning Considerations

A **zone** *is any area of a building for which indoor air temperature is controlled by a single thermostat (or other type of temperature controller).* A zone may be as small as a single room, or it may be as large as an entire building.

Multizone systems have the potential to provide two important benefits not available in single-zone systems, namely:

1. The ability to meet individual comfort requirements through independent control of room temperature.
2. The ability to conserve energy by reducing the temperature of unoccupied zones and thus the rate of heat loss from them.

Multizone hydronic heating systems are easy to create, much easier than their forced-air counterparts. Several hardware options exist for implementing zoning in different types of distribution systems. Most have been mentioned in previous chapters and will be shown in several of the piping schematics in this chapter.

Before selecting a distribution piping system, it is important to evaluate how zoning can best be implemented in the building the piping system will serve. The objective is to maximize the benefits of zoning and at the same time keeping the system affordable, energy efficient, reliable, and as simple as possible.

In some cases, hydronic zoning has been "oversold." For example, people sometimes think that the more zones a system has, the better it is designed. This is not necessarily true. Just because it is *possible* to create a separate zone for every room does not mean it is the best choice. In some cases, the extra piping and control hardware necessary for **room-by-room zoning** adds complexity and cost without returning tangible benefits. Dividing a system into an excessive number of zones may even create operational problems such as heat source short-cycling. It can also add to system maintenance requirements, again without returning any benefit to the owner.

System designers should consider the following when deciding how to best implement zoning.

Effect of Thermal Mass

Occupants sometimes assume that zoned systems are somehow able to lower the temperature of a room (or group of rooms) down to some setback temperature as soon as the thermostat is turned down. Likewise, they might assume that any given room will quickly warm back to normal occupied temperatures as soon as the thermostat is turned up.

No occupied space can drop in temperature by several degrees over 1 to 2 minutes simply by stopping heat input. Similarly, no (practical) heating system can instantly raise the temperature of an occupied space by several degrees. The time required to cool down or warm up a zone is significantly affected by the thermal mass of the heat emitter(s) and the building itself. The greater the thermal mass of the heating system and building components, the slower the decrease in indoor temperature when heat input to the zone is turned off. This is illustrated in Figure 11-1.

The equation in Figure 11-1 is used to approximate room air temperature versus time, once heat input is turned off. The rate at which this temperature decreases is determined by the total thermal mass within the room (h) and the room's overall heat loss coefficient (UA). The lower the ratio of heat loss divided by thermal mass, the slower the room's air temperature will drop.

A very well-insulated building with high thermal mass might only experience a temperature drop of 3 to 5 °F over a cold midwinter nighttime lasting 8 to 10 hours, even with its heating system completely off. If one were to reduce the thermostat setting in such a building from 70 to 60 °F at bedtime, the inside air temperature would not decrease to the reduced setpoint before morning. The energy savings potential of such a situation is very limited. If the thermostat were turned down 10 °F or even 20 °F, the results would be the same.

The thermal mass of the heat emitter also affects the recovery period when the setback ends. Low thermal mass heat emitters such as finned-tube convectors can usually restore a room that has been set back 10 °F to normal comfort in perhaps 15 to 30 min. High thermal mass systems such as a heated concrete slab-on-grade floor can take several hours to bring a space back to normal comfort temperature following a prolonged setback period. This is illustrated in Figure 11-2, which shows the simulated heating of both medium-mass and high-mass radiant floor panels, starting at an initial temperature of 60 °F, and having a heat input rate of 60,000 Btu/h.

It is important not to use heat emitters having significantly different thermal mass in the same zone when that zone will operate with frequent changes in

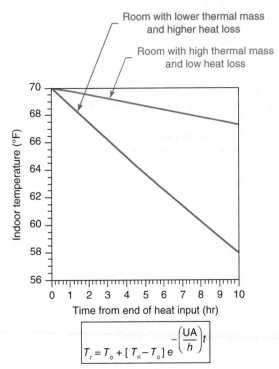

$$T_r = T_o + [T_{ri} - T_o] e^{-\left(\frac{UA}{h}\right)t}$$

where
T_r = temperature of room (°F)
T_o = outdoor temperature
T_{ri} = initial temperature of room when heat input stops (°F)
e = 2.718281828
UA = heat loss coefficient of room (Btu/hr/°F)
h = heat capacitance of room's thermal mass (Btu/°F)
t = elapsed time without heat input (hr)

Figure 11-1 | Temperature drop within two rooms of different thermal mass and heat loss.

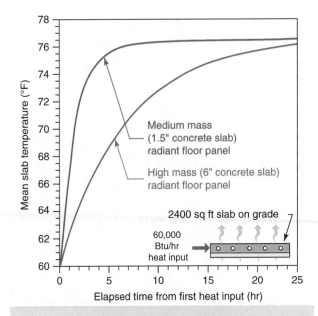

Figure 11-2 | Slab temperature versus time for 2400 ft.² slabs of different thicknesses while warming from an initial temperature of rate of heat input 60 °F with constant 60,000 Btu/h input.

setpoint temperature. Because their response times during both cool down and warm up are so different, neither type of heat emitter is likely to produce acceptable comfort during the transition periods between stable setpoint temperatures.

It is also important to realize that the energy conservation potential of a high thermal mass-zoned system is very dependent on the duration of any reduced thermostat setting. If a zone can be maintained at a reduced temperature for several days, the energy conservation potential can be significant. However, *attempting to create significant temperature changes on a daily basis in zones with high thermal mass heat emitters is futile.* The energy savings will be very small and the potential for temperature overshoot and undershoot will be significant. This characteristic should be explained to occupants, who often do not understand how high-mass systems respond to changes in the thermostat settings.

Interzone Heat Transfer

Another factor affecting zoning performance is **interzone heat transfer**. This refers to heat transfer through interior partitions as the building attempts to equalize temperature differences from one room to another. If, for example, a particular room is kept at 65 °F, while an adjacent room is kept at 75 °F, heat will flow from the warmer to the cooler room through the wall separating them as shown in Figure 11-3. This partially defeats attempts to maintain temperature differences by zoning. The greater the thermal resistance of the exterior envelope of the building, the harder it is to maintain significant temperature differences between rooms separated by uninsulated interior partitions.

Zoned areas in which it is desired to maintain temperature differences of 10 °F or more should have insulated partitions separating them from other zones. Doors separating such areas must remain closed when these temperature differences are to be maintained.

Occupant Management of Zoned Systems

Another consideration that helps determine the cost-effectiveness of zoning is the willingness of the occupants to regulate the system. It might seem obvious that an owner who is willing to pay more for an extensively zoned system would also be willing to regulate it. However, experience shows this is not always true, especially when nonprogrammable thermostats are used. Today's fast-paced lifestyles mean people often forget to regulate their heating systems, especially when their schedules vary from day to day.

In many hydronic systems, different rooms within a single zone (e.g., controlled by a single thermostat) can be maintained at different temperatures. This requires that the heat emitters in those rooms be piped to allow independent flow adjustments. This approach is usually

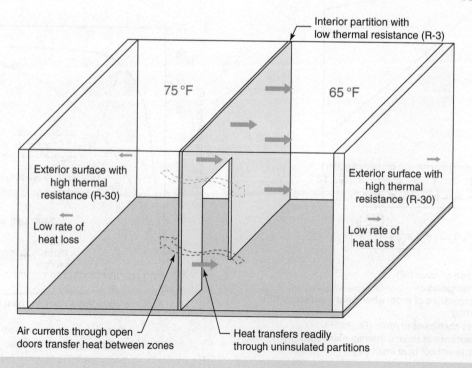

Figure 11-3 | Interzone heat transfer. Heat flows through uninsulated partition and open doors that separate one zone from another. *Courtesy of Caleffi North America.*

adequate when the goal of the occupants is to maintain several rooms at different (but stable) temperatures.

Room-by-room zoning is justified in situations where several rooms are to have both different and *frequently changing* air temperatures. These distinctions should be carefully discussed with the eventual owners before committing several hundred dollars for extensive zoning controls that may seldom be used.

Effect of Solar Heat Gain

Rooms with large southerly windows may at times be totally heated by solar heat gains. Rooms in the same building that do not receive solar gain may still require heat input. *Well-planned zoning separates rooms with likely solar heat gain from those without such gain.* It also accounts for the fact that the rooms receiving solar heat gain can change over the course of the day as the sun moves across the sky. This type of zoning is crucially important in buildings designed to take significant advantage of solar heat gain.

It is also important to use heat emitters with low thermal mass in areas subject to large and highly variable heat gains from the sun or other internal heat sources. The low-mass heat emitters can quickly start and stop heat input as necessary. This is especially important in buildings that have very low rates of heat loss.

Desired Sleeping Comfort

Many people prefer to sleep in cool bedrooms. Good zoning design allows for this possibility. It is common to zone a residential hydronic system so one or more bedrooms are controlled as a single zone. It is also common to create one zone for the master bedroom, with a separate zone for the master bathroom. This allows the bathroom to remain comfortable even when the bedroom is cool.

Effect of Activity Level

Exercise rooms are also good candidates for separate zoning, as are rooms that are frequently unoccupied (guest rooms, hobby rooms, etc.). Rooms where occupants have moderate-to-high activity levels should be zoned separately from rooms with sedentary occupants.

Zoning of Unoccupied Areas

It seldom makes sense to maintain normal human thermal comfort requirements in spaces that are infrequently occupied. Basements, if heated, often fall into this category and are normally set up as a separate zone. Garages, if heated, are usually maintained at air temperatures of 45 °F to perhaps 60 °F and again are usually set up as a separate zone. In cold climates, the garage zone is usually designed to operate with antifreeze solutions. This allows the garage zone to be completely turned off, if desired, without the risk of freezing.

Zoning of Transitory Areas

Entry vestibules with exterior doors and doors into fully heated space are good candidates for separate zoning. In some cases, the goal is to maintain the vestibules at higher temperatures to help buffer interior spaces from cold air gusts when the exterior door is opened. When floor heating is used in such areas, the floor surface temperature is often maintained several degrees above that of other interior areas to help dry tracked-in water and snow.

Effect of Heat Gains from Lighting/Equipment

It is common for rooms with several electrical appliances to require less heat input than adjacent rooms without such equipment. Rooms that contain heat-producing equipment such as computers, vending machines, intense lighting, or cooking facilities are usually good candidates for separate zoning.

Effect of Exterior Wall Exposure

Rooms with minimal exterior exposure will have small heating loads relative to their floor area. When such rooms are controlled as separate zones, overheating can be minimized. Rooms that have no surfaces exposed to the outside typically do not need heat emitters.

When properly planned, zoning can significantly enhance owner satisfaction while also improving the energy efficiency of the system.

11.3 System Equilibrium

All hydronic systems, regardless of how they are designed or installed, attempt to simultaneously stabilize at two unique operating conditions called **thermal equilibrium** and **hydraulic equilibrium**.

Thermal equilibrium occurs when the active portion of the distribution system is dissipating heat to the building at exactly the same rate as the heat source is adding heat to the circulating fluid. Thermal equilibrium was introduced in Chapter 1, and discussed in more detail in Chapter 8.

Hydraulic equilibrium occurs when the head energy added to the fluid as it passes through the circulator(s) is exactly the same as the head energy dissipated by all the components the fluid is flowing through. Hydraulic equilibrium was introduced and discussed in Chapter 7, Hydronic Circulators.

A system operating under thermal and hydraulic equilibrium may or may not provide the proper heating of the building. In animated terms, the system "does not care" if proper heating of the building is occurring; it only "cares" about adjusting itself to and maintaining itself at these equilibrium conditions. Likewise, the conditions under which thermal and hydraulic equilibrium are established may or may not be conducive to safe operation or a long and reliable system life. Again, the system "doesn't care" about these conditions. It only cares about establishing and maintaining thermal and hydraulic equilibrium.

The task of the designer is to assemble components into a system so that thermal and hydraulic equilibrium both occur at operating temperatures and flow rates that allow the building to be properly heated. This needs to be accomplished within an allowable budget, with the highest possible fuel efficiency, and so the resulting system will have a long life with minimal maintenance.

11.4 The Concept of Iterative Design

Situations where the selected equipment depends on the overall performance of the system, yet the system performance depends on the selected equipment, occur frequently in many areas of engineering. The solution to such apparent paradoxes lies in a technique known as **iteration** in which the designer repeatedly analyzes both the hydraulic and thermal performance of a system using "candidate" hardware options and operating conditions, until a combination is found that yields acceptable heat distribution.

For hydronic systems the process of iterative design typically goes as follows:

STEP 1. Make a tentative selection of heat emitters for the rooms to be heated based on their design heat loads and on an estimated average water temperature of the distribution system. A common practice is to assume the average water temperature in the system will be about 5 to 10 °F lower than the supply temperature from the heat source.

STEP 2. The system flow rate for water can be estimated using Equation 11.1:

Equation 11.1:

$$f_s = \frac{q}{500(\Delta T)}$$

Where:
f_s = initial estimated system flow rate (gpm)
q = total design heating load served by the system (Btu/h)
ΔT = intended temperature drop of the system (°F)

Equation 11.1 is a rearranged form of the sensible heat rate equation introduced as Equation 4.7.

When a fluid other than water is used, this equation can be modified as follows:

Equation 11.2:

$$f_s = \frac{q}{8.01 D c (\Delta T)}$$

Where:
f_s = initial estimated system flow rate (gpm)
q = total design heating load served by the system (Btu/h)
ΔT = intended temperature drop of the system (°F)
D = the fluid's density at the average system temperature (lb/ft^3)
c = the fluid's specific heat at the average system temperature (Btu/lb/°F).

If the heat emitters are connected in series, the system flow rate will pass through each heat emitter. If the heat emitters are connected in parallel, the flow rate through each heat emitter can be proportioned based on load using Equation 11.3:

Equation 11.3:

$$f_i = f_s \left(\frac{Q_i}{Q_t} \right)$$

Where:
f_i = estimated flow rate through heat emitter i (gpm)
f_s = initial estimated system flow rate (gpm)
Q_i = design heat output of heat emitter i (Btu/h)
Q_t = total design load of the system (Btu/h).

STEP 3. Select a piping distribution system to connect the heat emitters with the heat source. Several options are shown in later sections of this chapter.

Tubing size/type	Minimum flow rate (based on 2 ft/s) (gpm)	Maximum flow rate (based on 4 ft/s) (gpm)
3/8" copper	1.0	2.0
1/2" copper	1.6	3.2
3/4" copper	3.2	6.5
1" copper	5.5	10.9
1.25" copper	8.2	16.3
1.5" copper	11.4	22.9
2" copper	19.8	39.6
2.5" copper	30.5	61.1
3" copper	43.6	87.1
3/8" PEX	0.6	1.3
1/2" PEX	1.2	2.3
5/8" PEX	1.7	3.3
3/4" PEX	2.3	4.6
1" PEX	3.8	7.5
1.25" PEX	5.6	11.2
1.5" PEX	7.8	15.6
2" PEX	13.4	26.8
3/8" PEX-AL-PEX	0.6	1.2
1/2" PEX-AL-PEX	1.2	2.5
5/8" PEX-AL-PEX	2	4.0
3/4" PEX-AL-PEX	3.2	6.4
1" PEX-AL-PEX	5.2	10.4

Figure 11-4 | Minimum and maximum flow rates for different types and sizes of tubing, based on flow velocities between 2 and 4 ft/s.

STEP 4. Select tube sizes based on establishing a flow velocity between 2 and 4 ft/s. Figure 11-4 gives the flow rates that correspond to these flow velocities for various types and sizes of tubing. For other tube types and sizes, use Equation 6.1 to establish the relationship between flow rate and flow velocity, then select a tube size to keep the flow velocity between 2 and 4 ft/s.

STEP 5. Using the methods of Chapter 6, Fluid Flow in Piping, the piping system can now be described by a hydraulic resistance diagram. After reducing this diagram to a single equivalent hydraulic resistance, the system curve can be sketched.

STEP 6. The pump curves of one or more **"candidate" circulators** can now be matched against the system curve. Recall from Chapter 7, Hydronic Circulators, that the intersection of the system curve and the pump curve yields the operating flow rate for the combination. It is likely that the system flow rate produced by any given circulator will not exactly match the flow rate estimated in step 2. This is normal and to be expected as part of the design process.

STEP 7. If all heat emitters are in a series circuit with the circulator, this step can be skipped. If some or all of the heat emitters are located in parallel piping branches, the methods of Chapter 6, Fluid Flow in Piping, can be used to find the flow rate in each branch. Steps 5 through 7 can also be handled using the **Hydronic Circuit Simulator module** in the **Hydronics Design Studio** software.

STEP 8. The thermal performance of each heat emitter can now be verified at the actual flow rate and entering water temperature present at their location in the system. If the heat output of a given heat emitter is too low, a number of options exist. These include selecting a higher output heat emitter, increasing system operating temperature, or modifying the piping to obtain a higher flow rate. If the output of a heat emitter is more than 10% above design load, a lower capacity heat emitter can be used, or in the case of a parallel system, a balancing valve can be used to reduce the flow rate through it.

STEP 9. Depending on what, if any, changes are made to the system, the designer may need to go back to step 3 and repeat the process of determining the new flow rates in the modified system. This is iteration. In most cases, only one or two iterations are needed to find a design that can provide the necessary heat output with the selected heat emitters, piping system, and circulator.

11.5 Single Series Circuits

The simplest distribution system is a single piping circuit that progresses from the heat source, through each heat emitter, and back to the heat source. System operation is often controlled by a single regulating device such as a room thermostat. This type of distribution system is called a **single series circuit**.

Single series circuits are appropriate for small buildings in which all rooms experience similar changes in heating load. An example of a single series circuit of finned-tube baseboard convectors is shown in Figure 11-5.

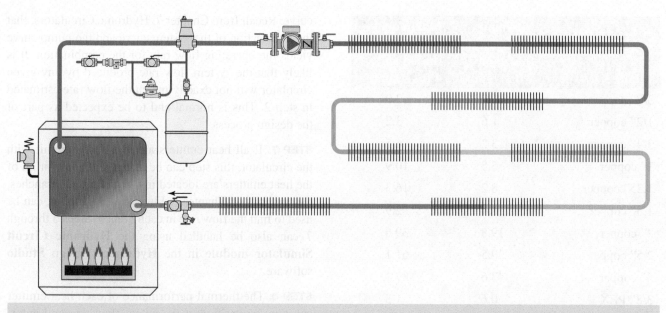

Figure 11-5 | Example of a single series circuit distribution system.

Single circuit systems can also be designed around a combination of heat emitters including panel radiators and fan-coils. When different types of heat emitters are used, they should have comparable thermal mass.

Because heat input to the entire building is regulated based on the temperature at one thermostat location in the building, overheating or underheating of rooms other than where the air temperature is sensed is possible. It is crucial that all heat emitters be sized for the load of their respective rooms and to the water temperature at their location within the piping circuit. The latter condition is often overlooked when heat emitters are sized based on a single average water temperature. Chapter 8, Heat Emitters, demonstrated the differences in sizing finned-tube baseboard units based on a single average water temperature versus the temperature of the water at the location of each baseboard. Using the single average water temperature method, the baseboards near the beginning (hottest) part of the circuit tend to be oversized and thus put out too much heat. Those near the end of the circuit tend to be undersized and thus put out too little heat. These tendencies will also be true for other types of heat emitters if they are sized based on the average water temperature. *To prevent this, the designer must keep track of the fluid temperature as it cools from one heat emitter to the next.* Begin at the location of the first heat emitter and progress around the circuit treating the outlet temperature of each heat emitter as the inlet temperature to the next.

If a piping segment between heat emitters is longer than 40 feet, or is otherwise expected to have a significant heat loss, the designer should calculate the temperature drop along the pipe and correct the inlet temperature to the next heat emitter accordingly. Methods for estimating pipe heat loss were presented in Chapter 8, Heat Emitters.

The **Series Baseboard Simulator module** in the Hydronics Design Toolkit software can be used to quickly determine the required lengths of finned-tube baseboard in a series circuit, or portion thereof.

The limiting factors in designing single circuit systems are usually temperature drop and flow resistance. In the United States, series circuits have traditionally been designed around a temperature drop of approximately 20 °F under design load conditions. However, *there is nothing sacrosanct about this number.* Systems can be designed to operate properly with both smaller and larger temperature drops.

In Europe, hydronic distribution systems are routinely operated with temperature drops of 30 to 40 °F under design load conditions. The advantage of having a higher temperature drop is that the flow rate can be relatively low. This in turn allows small-diameter piping and a smaller less power-consuming circulator to be used. The disadvantage of a high-temperature drop system is that heat emitters near the end of the series circuit often need to be significantly larger to meet a given load with the lower temperature water. There is also the need of maintaining the return temperature to a conventional boiler high enough to prevent sustained flue gas condensation.

Distribution systems that operate with low temperatures drops usually have relatively high flow rates that may require increased pipe sizes to keep the flow velocity under 4 ft/s. These systems may also require larger circulators. However, because the heat emitters will operate at higher average temperatures, they may be slightly smaller compared to those in systems operating at higher temperature drops.

It is impossible to say what the optimum temperature drop is for a single series circuit. The answer depends on the cost of piping, circulators, heat emitters, electricity, and several other factors. The designer may choose to explore different possibilities based on various combinations of supply water temperature and circuit flow rate using the procedures and resources described just ahead.

Series circuits that contain one or more heat emitters with high flow resistance characteristics must be carefully evaluated for potential problems. *A heat emitter with high flow resistance can greatly restrict flow rate through the entire series circuit, limiting its total heat output.*

Another potential problem is excessive flow velocity through the small tubes or valves in the restrictive heat emitter. This can create flow noise and eventual erosion corrosion. A high head circulator may be required. High head differentials across a circulator encourage cavitation, especially in high-temperature systems. Circulators capable of generating heads in excess of 30 feet should be carefully scrutinized for cavitation potential. In many cases, a different type of distribution system will prove more appropriate for use with heat emitters having high flow resistance characteristics.

Another drawback of a single-zone series circuit is that control of heat output to individual rooms is very limited. In the case of finned-tube baseboards, heat output can be reduced up to about 50% by closing the dampers on the enclosure. When fan-coils are used, the blower speed can be reduced to slightly limit heat output. Both these methods, however, require manual adjustments to the heat emitters in response to changing load conditions. Many occupants either do not know these adjustments can be made or soon tire of making them. The resulting overheating and underheating is begrudgingly tolerated to the detriment of future referrals for the installer.

Design of Single Series Circuits

The iterative design procedure discussed in Section 11.4 can be adapted to the design of single series circuits. The following is a brief summary of this procedure, specifically tailored to series circuits.

STEP 1. Determine the design heating load of each room served by the series circuit.

STEP 2. Select a tentative supply temperature and temperature drop for the circuit based on the heat emitters used. For example, for a baseboard circuit supplied by a conventional boiler, the supply temperature is often selected in the range of 170 to 190 °F, along with an assumed temperature drop of 20 °F.

STEP 3. Use the average water temperature of the system to select *tentative* sizes of heat emitters for each room. The average water temperature is the supply temperature minus half the assumed temperature drop.

STEP 4. Once the tentative sizes of the heat emitters are established, they can be located within the floor plan of the space being heated. In most cases, the preferred locations will be under windows on exterior walls. However, every building presents different circumstances. It may be necessary to locate some heat emitters in less traditional locations to accommodate furniture or other requirements of the owners.

STEP 5. Once the heat emitters are located, sketch a pipe routing path that accommodates both the building construction and heat emitters. Think about where it will be possible (or not possible) to route the piping under floors, through partitions, and so on, to access each heat emitter. Once the pipe routing has been determined, estimate the length of the series circuit as well as the number of elbows and other fittings or valves it will likely contain. Do not worry about getting the exact length or number of fittings; a reasonable estimate is fine. Be sure to include an allowance for the piping, fittings, and valves near the heat source.

STEP 6. Knowing the total heating load served by the circuit and the estimated temperature drop, use Equation 11.1 or 11.2 to calculate the **target flow rate**.

STEP 7. Using the flow rate from step 6, select a tube size for the circuit using the data in Figure 11-4 or Equation 6.1.

STEP 8. Use the methods from Chapter 6, Fluid Flow in Piping, to construct a system head loss curve for the series circuit. The Hydronic Circuit Simulator module in the Hydronics Design Studio software can also be used for this step.

STEP 9. Select a circulator such that the intersection of the system head loss curve and pump curve (e.g., the operating point) falls close to the target flow rate determined in step 6. An operating point having a flow rate within ±10% of the target flow rate is generally acceptable.

STEP 10. Knowing the system flow rate and supply temperature, verify the heat output and sizing of each heat emitter on the circuit. If finned-tube baseboard is used for all heat emitters, the Series Baseboard Simulator module in the Hydronics Design Studio software can be used. If other types of heat emitters are present, their performance and temperature drop will have to be evaluated individually.

It may be necessary to change the size of one or more heat emitters based on the results of this step. If such is the case, the designer may have to iterate the calculations starting with step 3. The designer may also choose to explore options such as reduced supply water temperature or a larger circuit temperature drop before establishing the final hardware selection and operating conditions.

11.6 Single Circuit/ Multizone (One-Pipe) Systems

A variation on the single series circuit allows for one or more of the heat emitters connected to a single piping circuit to be independently controlled. This piping arrangement involves the use of **diverter tees** as discussed in Chapter 5, Piping, Fittings, and Valves. These tees are used to divert a portion of the water flowing in the main piping circuit through a branch piping path that includes the heat emitter. An example of the resulting single circuit/ multizone (one-pipe) system is shown in Figure 11-6.

The diverter tees can be used individually or in pairs. When two diverter tees are used, they set up a greater pressure difference across the branch circuit. This in turn creates a higher flow rate through the branch piping path. Single diverter tee arrangements are often sufficient for low-resistance heat emitters such as a few feet of finned-tube baseboard or a panel radiator. Double diverter tees may be needed on heat emitters with higher flow resistances such as fan-coils. Double diverter tees are also suggested to overcome buoyancy forces when the heat emitter is located several feet below the distribution circuit. When two diverter tees are used, there should be at least one foot of tube between the tees to allow turbulence created by the upstream tee to partially dissipate before the flow enters the downstream tee.

Several types of valves can be used to control the flow rate through the branch piping path containing the heat emitter. By regulating the flow, the heat output from the heat emitter can be varied or turned completely off.

The branch valve could be a simple manually adjusted globe or angle valve adjacent to or even built into the heat emitter. If automatic control of the heat output is desired, a nonelectric thermostatic radiator valve can be used as shown in Figure 11-6. Such valves were discussed in Chapter 5, Piping, Fittings, and Valves. Electrically operated zone valves are another possibility for the branch valves.

Systems employing diverter tees are often called **one-pipe systems**. They offer zoning flexibility at relatively low cost. Any one or more of the heat emitters on the circuit can be selected to have independent

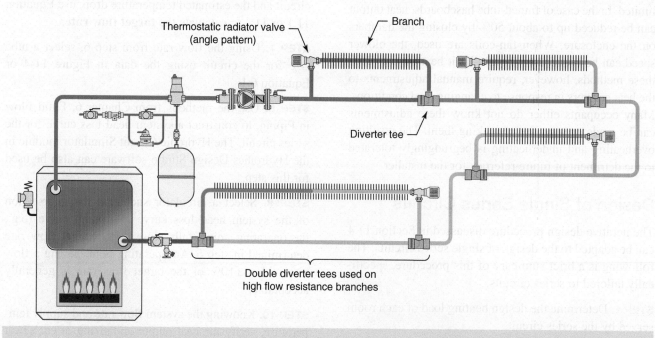

Figure 11-6 | Use of one or two diverter tees to induce flow in branch circuits. Heat output of each heat emitter is regulated by a thermostatic radiator valve.

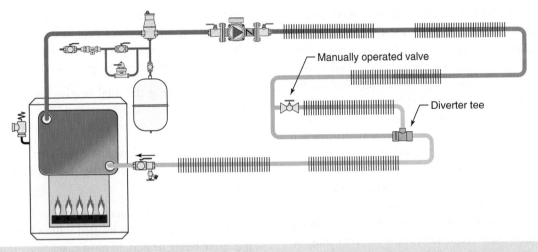

| Figure 11-7 | Use of diverter tee and a manually adjusted valve to limit heat output of one heat emitter in an otherwise series distribution circuit. |

temperature control simply by piping it into the circuit using diverter tee(s) and some type of flow regulating valve.

Figure 11-7 shows a series circuit system in which one heat emitter is connected using a diverter tee arrangement while the other baseboards are piped in series. This is a convenient way to limit the heat output of one heat emitter without significantly affecting the remainder of the heat emitters. This arrangement could be used, for example, in a guest bedroom that only needs to be heated to normal comfort temperatures a few days each year.

Diverter tees can also be used to supply a piping subassembly such as the one shown in Figure 11-8. In this case, the subassembly is a group of four panel radiators, all located in the same heated space. The radiators are connected in a parallel reverse-return arrangement to ensure the same supply temperature and reduce head loss. Parallel reverse-return systems will be discussed more in Section 11.11.

When nonelectric thermostatic radiator valves are used for individual room temperature control, the designer must provide controls to operate the heat source and distribution circulator whenever the building might need heat. One approach is to equip the system with an outdoor reset control that automatically operates the heat source and distribution circulator when the outdoor temperature falls below a preset value. When such a control system is used, the thermostatic valves on the heat emitters serve as temperature-limiting devices for their respective rooms.

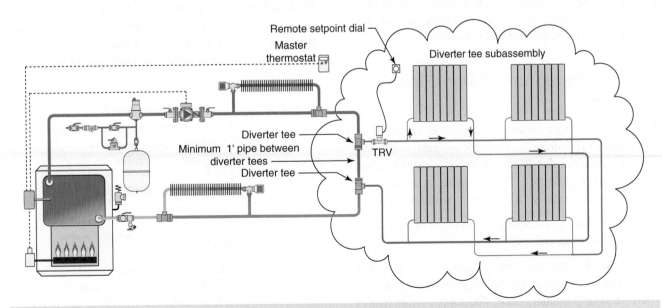

| Figure 11-8 | Use of diverter tees to route flow through a subassembly of several heat emitters. |

Another possibility is to use an electric **master thermostat** that serves as a temperature-limiting "switch" for the circulator and boiler. As long as the master thermostat is calling for heat, the boiler is enabled to fire, and the circulator is operating. If, for any reason, the master thermostat is satisfied, the boiler and circulator are off, and none of the branch circuits can deliver heat. This strategy is also helpful in that the entire distribution system can be put into a setback mode by lowering the setting on the master thermostat.

Another benefit of one-pipe systems with branch control valves is that they allow some heat emitters to be intentionally oversized without losing the ability to control individual room temperatures. This may be desirable when an infrequently occupied room needs to be quickly warmed following a setback. Intentional oversizing also allows a space to be heated higher than the normal comfort temperature if desired. An example would be when someone wants to quickly warm a bathroom to perhaps 75 °F while taking a bath, and then return the room to a normal 68 to 70 °F setting. Keep in mind, however, that there must be flow of heated water in the main distribution circuit for individual heat output control of any heat emitter located in a branch that connects to the main circuit using diverter tees.

Whenever diverter tees and control valves are used to reduce or totally stop water flow to a heat emitter, it is crucial that the heat emitter and the piping leading to it cannot freeze during cold weather. Some thermostatic radiator valves have a freeze-proof minimum setting that attempts to maintain the air temperature in the space above freezing, even when the setting knob appears fully closed. If a manual control valve is used, it should be left slightly open during very cold weather. If there is a chance, the valve may be unintentionally closed or it may be a good idea to remove the handle.

Design Procedure for One-Pipe Systems

The iterative design concept discussed in Section 11.4 also applies to the design of one-pipe systems. Additional steps are required due to the more complex piping and parallel circuit branches. The following procedure can be performed manually, or incorporated into software.

STEP 1. Determine the design heating load of each room to be served by the one-pipe circuit.

STEP 2. Select a tentative supply temperature and overall circuit temperature drop for the circuit based on the heat source and heat emitters used. The supply temperatures and circuit temperature drops for one-pipe systems are often similar to those used for series circuits.

STEP 3. Use the *average* water temperature of the system to select *tentative* sizes of heat emitters for each room. The average water temperature is the supply temperature minus half the assumed temperature drop.

STEP 4. Knowing the total heating load served by the circuit, and the estimated temperature drop, use Equation 11.1 or 11.2 to calculate the target system flow rate.

STEP 5. Using the flow rate from step 4, select a tube size for the circuit using the data in Figure 11-4, or Equation 6.1.

STEP 6. Once the tentative sizes of the heat emitters are established, they can be located within the floor plan of the space being heated. In most cases, the preferred locations will be under windows on exterior walls. However, it may be necessary to locate some heat emitters in less traditional locations to accommodate furniture or other requirements of the owners.

STEP 7. After the heat emitters are located, sketch a pipe routing path that accommodates both the building construction and heat emitters. Think about where it will be possible (and not possible) to route the piping under floors, through partitions, and so on, to access each heat emitter.

Also sketch *each* diverter tee subassembly showing the heat emitter, estimated size and length of the main piping and branch piping, fittings, valves, and the number of diverter tees used to connect it to the main. An example is shown in Figure 11-9. This information is needed to determine the hydraulic resistance of each subassembly.

STEP 8. Use the methods described in Chapter 6, Fluid Flow in Piping, to determine the hydraulic resistance of the components in each subassembly. Do this for both the branch circuit and the main piping between the tees. Include the resistance of the diverter tees in the overall

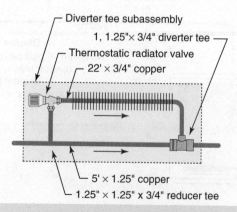

Figure 11-9 | Identifying the piping elements in a diverter tee subassembly.

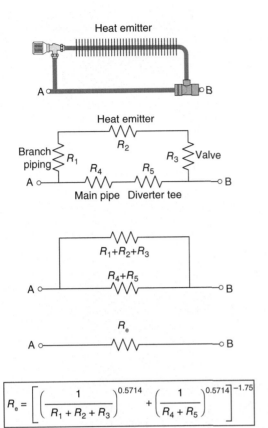

Figure 11-10 | Reducing a diverter tee subassembly into a single equivalent hydraulic resistance.

resistance of the main pipe. This hydraulic resistor diagrams associated with this are shown in Figure 11-10.

STEP 9. Repeat step 8 for each of the diverter tee subassemblies in the distribution circuit, reducing each subassembly to a single equivalent hydraulic resistance.

STEP 10. Add the equivalent hydraulic resistances of all diverter tee subassemblies along with the hydraulic resistance of the other piping and fittings in the circuit. Use this information to generate a system head loss curve. This concept is shown in Figure 11-11.

STEP 11. Select a circulator so that the intersection of the system head loss curve and pump curve (e.g., the operating point) falls close to the target flow rate determined in step 4. An operating point having a flow rate within ±10% of the target flow rate is generally acceptable.

STEP 12. Use the system flow rate and the parallel hydraulic resistances of each subassembly to calculate the flow rate through each heat emitter.

STEP 13. Use the inlet temperature and flow rate to the first heat emitter to calculate its heat output.

STEP 14. Calculate the temperature leaving the diverter tee subassembly using Equation 11.4. Typically, this temperature is used as the supply temperature to the next heat emitter. However, if there is a long length of tubing leading to the next heat emitter, it may be necessary to calculate the temperature drop along this tube. Subtract this temperature drop from the outlet temperature of the upstream subassembly to determine the inlet temperature to the next subassembly. Heat loss from base copper tubing can be determined using methods presented in Chapter 8, Heat Emitters.

Equation 11.4:

$$T_{out} = T_{in} - \frac{Q_i}{500 \times f_s}$$

Where:

T_{out} = temperature leaving the subassembly (°F)
T_{in} = temperature of water entering the subassembly (°F)
Q_i = heat removed by the heat emitter (Btu/h)
f_s = total flow rate entering the subassembly (gpm)
500 = a constant for water*

* For fluids other than water, replace the number 500 with $(8.01 \times D \times c)$

Where:

D = density of the fluid at average circuit temperature (lb/ft³)
c = specific heat of fluid at average system operating temperature (Btu/lb/°F)
8.01 = a constant based on units

STEP 15. Repeat steps 10 through 12 for each diverter tee subassembly, in sequence, until reaching the end of the circuit.

STEP 16. After determining the heat output of each heat emitter, it may be necessary to make changes to the hardware or operating conditions, and then reevaluate the results by iterating the procedure starting with step 2. The designer may also choose to explore options such as reduced supply water temperature, or a larger circuit temperature drop before establishing the final hardware selection and operating conditions.

11.7 Multicirculator Systems and Hydraulic Separation

All distribution systems discussed thus far in this chapter have used a single circulator. While this is fine for certain applications, single circulator systems are

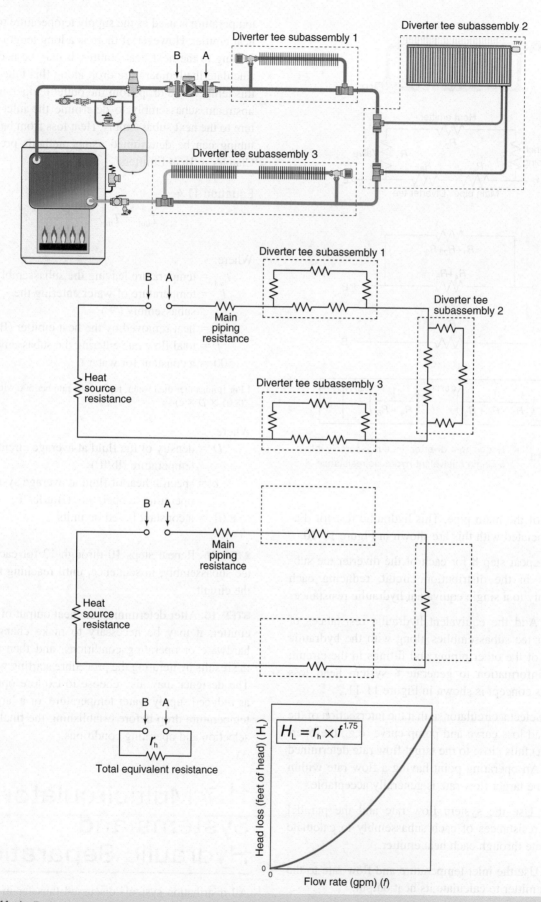

Figure 11-11 | Reducing a one-pipe distribution system into a single equivalent hydraulic resistance.

somewhat limited, especially in modern multiload hydronic systems. Systems with multiple circulators greatly expand the range of possible system designs. They also bring about issues that are not present in single circulator systems. Issues must be properly addressed and detailed in order for multiple circulators to operate correctly within the system.

Hydraulic Separation

When two or more circulators operate simultaneously in the same system, they each attempt to establish differential pressures based on their pump curve. *Ideally, each circulator in a system will establish a differential pressure and flow rate that is unaffected by the presence of another operating circulator within the system.* This very desirable condition is called **hydraulic separation** and can be achieved in several ways.

Conversely, the lack of hydraulic separation can create very *undesirable* operating conditions in which circulators interfere with each other. The resulting flows and rates of heat transport within the system can be greatly affected by such interference, often to the detriment of proper heat delivery and comfort.

The degree to which two operating circulators interact with each other depends on the head loss of the piping path they have in common. This piping path is called the **common piping** since it is shared by both circuits. If the common piping has low flow resistance, it is incapable of creating any significant head loss.

Figure 11-12 shows a very simple system with two circuits, each with its own circulator. The two circuits

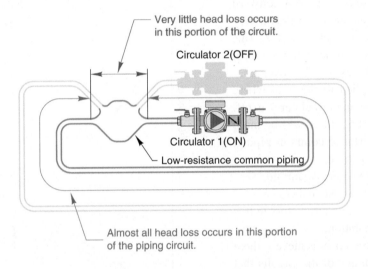

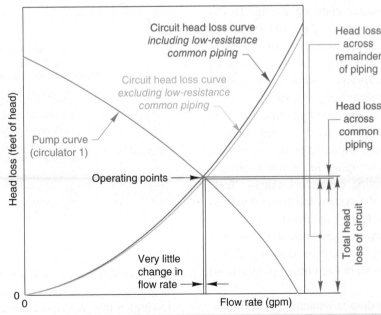

Figure 11-12 | Low-resistance common piping can be thought of as a "bulge" in the piping that is incapable of creating any significant head loss. The vast majority of the circulator's head must therefore be dissipated in the remainder of the circuit.

"overlap" at a device that has very low flow resistance. The graph in the lower portion of Figure 11-12 shows two head loss curves for the operating circuit. One includes the head loss of the common piping, the other does not. Notice that there is very little difference in the two head loss curves, and thus very little change in flow rate when the flow resistance of the common piping is included in the system curve.

Next, consider the situation when one circulator is initially operating, and thus a stable flow rate is established in its associated circuit; then, the circulator in the other circuit is turned on. This situation is shown in Figure 11-13.

When the second circulator is turned on, the flow rate through the common piping increases. However, because the flow resistance of the common piping is very low, there is very little increase in head loss across it. This very slight increase in head loss causes a very small leftward shift in the operating point for circuit 1 and thus a very small decrease in the flow rate through circuit 1. This very small drop in the flow rate will not affect the heat delivery capability of circuit 1 to any practical extent.

A similar logic can be applied to the case where circulator 2 is initially operating and circulator 1 is off. When circulator 1 is then turned on, there will be more flow through the low flow resistance common piping, resulting in slight shift in the operating point of circuit 2 and a very slight drop in its flow rate. Again, this very small drop in the flow rate is of no practical concern.

Thus, it can be stated that the two circulators in this system have good hydraulic separation.

When good hydraulic separation is achieved, there is very little interaction between any of the circuits that share the low flow resistance common piping. *This is true regardless of the pump curves of the simultaneously operating circulators. Thus, it is possible to combine large circulators with small circulators, and/or fixed-speed circulators with variable-speed circulators provided the circuits they are in are hydraulically separated.*

When two or more circuits are hydraulically separated, the vast majority of the head created by each circulator will be dissipated within its respective circuit, rather than in the common piping. In animated terms— the simultaneously operating circulators cannot "feel" each other's presence in the system and thus operate as if they were each in an independent circuit.

For all practical purposes, one can think of (and design) circuits that are hydraulically separated as if they were completely separate from each other as illustrated in Figure 11-14.

Having stated that hydraulic separation is desirable, it is still worthwhile to consider a situation in which hydraulic separation is NOT present, and observe the consequences.

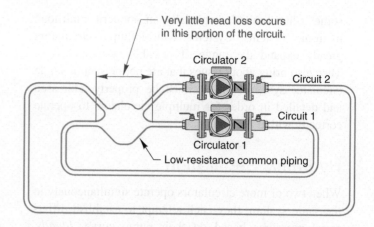

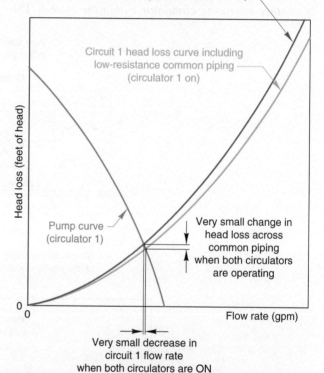

Figure 11-13 When both circuits operate simultaneously, there is a very slight increase in head loss across the low flow resistance common piping. This results in a very slight decrease in flow rate through circuit 1. Very adequate hydraulic separation has been achieved between these circuits.

Consider the system shown in Figure 11-15. The larger circulator is sized to move sufficient flow through the higher flow resistance circuit including the high flow resistance heat source. When operating, the flow created by the large circulator creates a pressure drop of 5.0 psi between the supply and return headers.

11.7 Multicirculator Systems and Hydraulic Separation

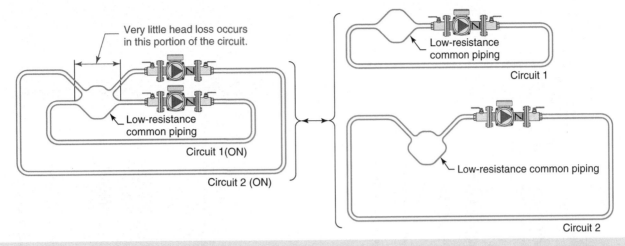

Figure 11-14 | For practical purposes, circuits that are hydraulically separated from each other can be visualized, and designed, as if they were totally independent of each other.

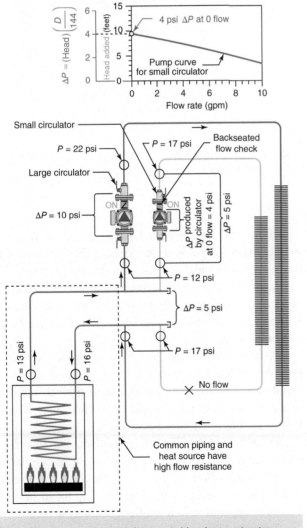

Figure 11-15 | Two circuits, one with a larger circulator and the other with a smaller circulator are joined by high flow resistance common piping path. The lack of hydraulic separation causes flow in the circuit with the smaller circulator to stop whenever the larger circulator operates.

When the smaller circulator is turned on, it can only produce 4.0 psi pressure differential between its inlet and outlet ports—even at zero flow. Because the reverse pressure differential across the smaller circulator, caused by the larger circulator is 5 psi, the check valve in the smaller circulator is held shut. Although the smaller circulator will continue to run, it is said to be "deadheaded" since there is no flow through it. Under such a condition, it will dissipate its input wattage as heat. This heat will be absorbed by the water in the circulator's volute as well as dissipated by the circulator's body. Small wet rotor circulators can typically withstand this condition for a while, although it is not a condition that should be allowed by proper design. The solution is to create hydraulic separation between all circulators that could be operating simultaneously.

Creating Hydraulic Separation

Any device or combination of devices that create very low flow resistance can provide hydraulic separation between two or more circulators. Examples of such devices include the following:

- Closely spaced tees
- A tank (which also serves other purposes in the system)
- A hydraulic separator
- A low flow resistance heat source

Examples of such devices creating hydraulic separation between various circulators are shown in Figures 11-16 through 11-18.

The common piping in Figure 11-16 consists of the closely spaced tees and "generously sized" headers. Together, these details create hydraulic separation between all three circulators. *The headers should be sized so that*

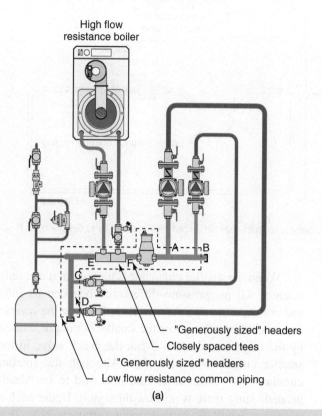
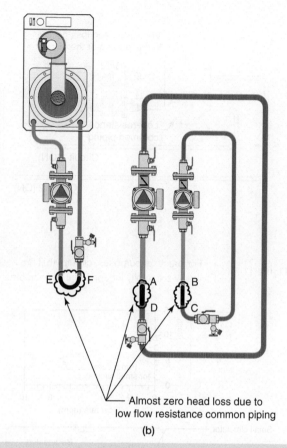

Figure 11-16 | Hydraulic separation accomplished using closely spaced tees and generously sized headers. (a) Low flow resistance common piping. (b) Concept of visualizing system as three independent circuits for design purposes.

the maximum flow velocity at design load is not more than 2 ft/s. This slow flow velocity creates very little head loss.

The closely spaced tees allow the boiler circulator to only "see" the flow resistance of the boiler and piping between the boiler and closely spaced tees. The boiler circulator does not assist in moving flow through the distribution circuits. Likewise, the two distribution circulators are only responsible for moving flow through their respective circuits and do not assist in moving flow through the boiler.

Figure 11-16b shows how the overall system in Figure 11-16a can be visualized as three completely independent circuits for the purpose of establishing flow rates, pipe sizes, head loss, and circulator selection. This concept of **circuit separation** is a powerful technique that is possible when all circuits are hydraulically separated. It allows what often appear to be complex piping systems to be quickly simplified for analysis.

Figure 11-17 shows a buffer tank and generously sized headers serving as the low flow resistance common piping that provides hydraulic separation between the boiler circulator and each of the distribution circulators. This demonstrates that hydraulic separation can sometimes be accomplished as an ancillary function to the main purpose of the device. Notice also that fixed-speed and variable-speed circulators, all of different sizes, can be combined onto the same generously sized header system. Interaction between these circulators will be very minimal because of the low flow resistance common piping.

A heat source with low flow resistance can also provide hydraulic separation between two or more simultaneously operating circulators. A typical sectional cast-iron boiler is an example of such a heat source. Figure 11-17a illustrates this concept.

From the standpoint of hydraulics, the low flow resistance heat source can be thought of as similar to a buffer tank. Both create very low flow resistance, and either can serve as the common piping to provide hydraulic separation in a multicirculator system.

This combination of a sectional cast-iron boiler with multiple-zone circulators has been used successfully for decades. Hydraulic separation was present, but not necessarily "recognized" as a desirable condition. When the North American hydronics industry began to substitute boilers with compact heat exchangers,

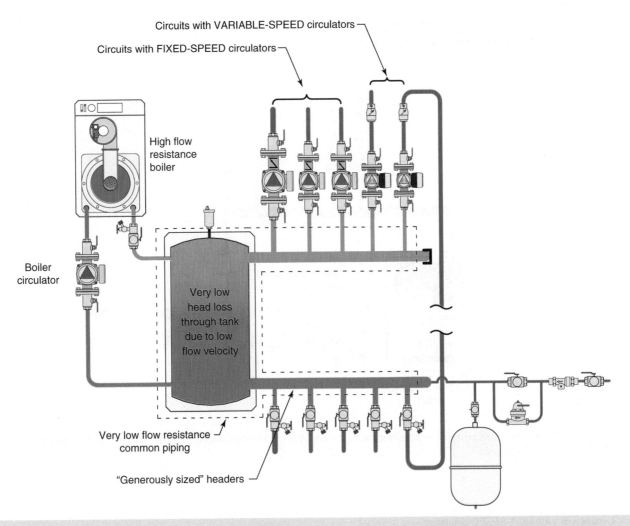

Figure 11-17 | Buffer tank and generously sized headers provide hydraulic separation between all circulators, both fixed and variable speed.

and thus much higher flow resistance, into the same multicirculator systems, the previously unrecognized benefit of hydraulic separation was lost, and many installations suffered from erratic flow problems that had not be present in earlier systems. Fortunately, the benefit of hydraulic separation was "rediscovered" and is now widely applied in modern systems.

Still another method of providing hydraulic separation is using a device appropriately called a "**hydraulic separator**." This type of device is relatively new to the North American hydronics market, but has been used in Europe for many years. Figure 11-18 shows a hydraulic separator installed in a multicirculator system. Figure 11-19 shows its external and internal construction.

Hydraulic separators, which are sometimes called **low loss headers**, create a zone of low flow velocity within their vertical body. The diameter of the body is typically three times the diameter of the connected piping. This causes the vertical flow velocity in the body to be approximately 1/9th that of the connecting piping. Such low velocity creates very little head loss and very little dynamic pressure drop between the upper and lower connections. Thus, a hydraulic separator also provides hydraulic separation in a manner similar to a buffer tank, only smaller.

The reduced flow velocity within a hydraulic separator allows it to perform two additional functions. First, air bubbles can rise upward within the vertical body and be captured in the upper chamber. When sufficient air collects at the top of the unit, the float-type air vent allows it to be ejected from the system. Thus, a hydraulic separator can replace the need for a high-performance air separator.

Second, the reduced flow velocity allows dirt particle to drop to the bottom of the vertical body. A valve at the bottom can be periodically opened to flush out the accumulated dirt. Thus, the hydraulic separator serves as a dirt removal device. Some modern hydraulic separators also have removable rare-earth magnets that assist in gathering iron oxide particles within the flow stream.

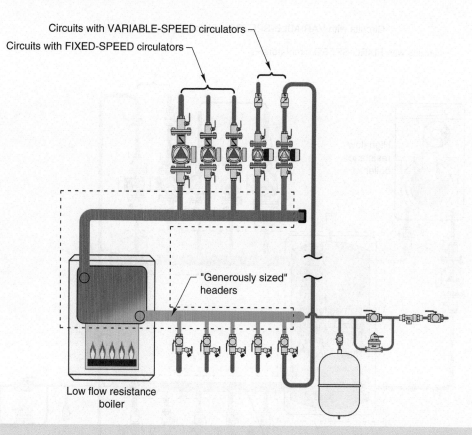

Figure 11-17a | Hydraulic separation of multiple circulators is provided by a low flow resistance boiler in combination with generously sized headers.

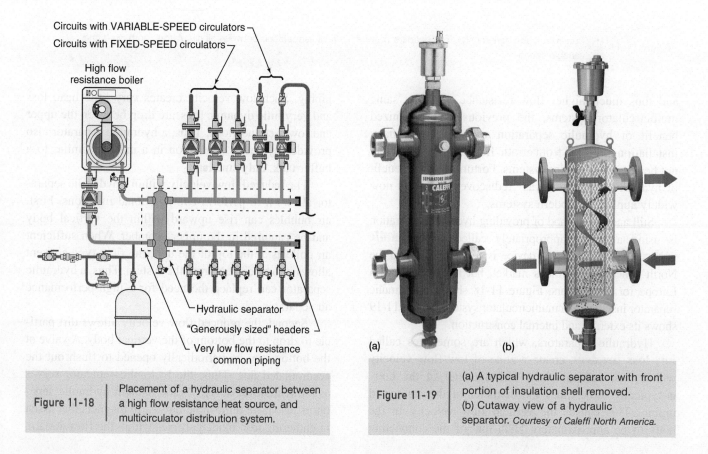

Figure 11-18 | Placement of a hydraulic separator between a high flow resistance heat source, and multicirculator distribution system.

Figure 11-19 | (a) A typical hydraulic separator with front portion of insulation shell removed. (b) Cutaway view of a hydraulic separator. *Courtesy of Caleffi North America.*

The author strongly recommends the use of magnetic particle separation in any system containing circulators with electronically commutated motors.

Some hydraulic separators contain an internal **coalescing media** that enhances their ability for air and dirt removal. Coalescing media are discussed in more detail in Chapter 13, Air Removal, Filling, and Purging.

Hydraulic separators are available in several pipe sizes from 1 inch to over 12 inches.

Given the surface area of their bodies, hydraulic separators should always be insulated to minimize heat loss to their surroundings. This is especially true of larger hydraulic separators, which may have more surface area than a modestly sized radiator, and would otherwise needlessly overheat the mechanical room.

Mixing Within Hydraulic Separating Devices

All devices used for hydraulic separation can have different flow rates and different entering temperatures between their primary side, which is connected to the heat source, and their secondary side, which connects to the distribution system. This implies that mixing often takes place within the hydraulic separating device. Such mixing will affect the fluid temperature supplied to the distribution system.

There are three possible mixing scenarios that apply to all hydraulic separating devices:

1. The flow rates are the same on both sides of the hydraulic separator.
2. The secondary-side flow rate is higher than the primary-side flow rate.
3. The secondary-side flow rate is lower than the primary-side flow rate.

The first law of thermodynamics, applied to a hydraulic separating device operating at steady-state conditions (e.g., stable flow rates and entering fluid temperatures), can be stated as follows: The rate of energy carried into the hydraulic separating device must equal the rate of energy that is carried out of the device. Figures 11-20 through 11-22 show that these three scenarios apply to a hydraulic separator and give the equations that can be used to calculate the outlet temperature of the mixed fluid stream.

When the primary- and secondary-side flow rates are equal, and the hydraulic separating device is piped as shown in Figure 11-20, there is no significant mixing between the higher and lower temperature fluids. The entering hot stream passes through the upper portion of the hydraulic separator and on to the distribution

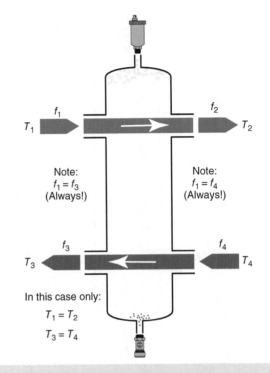

Figure 11-20 | Primary flow rate equals secondary flow rate. No mixing takes place within the hydraulic separator. *Courtesy of Caleffi North America.*

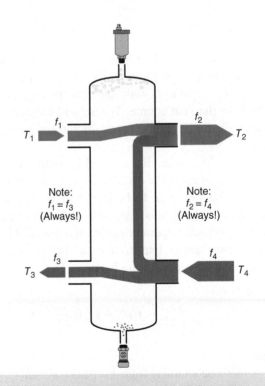

Figure 11-21 | When the secondary side flow rates are higher than the primary side flow rate, mixing occurs within the hydraulic separator that lowers the outlet temperature supplied to the distribution system. *Courtesy of Caleffi North America.*

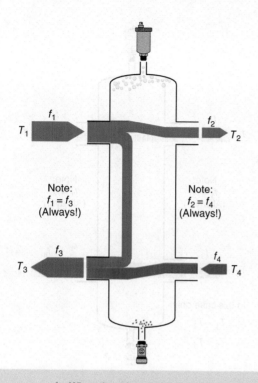

Figure 11-22 When the primary side flow rate is greater than the secondary side flow rate, mixing occurs with the hydraulic separator that increases the temperature of the fluid returning to the heat source.

system. Likewise, the flow returning from the distribution system passes through the lower portion of the hydraulic separator, with minimal mixing, on its way back to the heat source. The air and dirt removal functions of the hydraulic separator remain active at all times.

A more like scenario, especially when multiple circulators are used in the distribution system, is for the secondary-side flow rate to be higher than the primary-side flow rate. In this case, there will be mixing within the hydraulic separator as shown in Figure 11-21. The resulting supply temperature to the distribution system can be calculated using Equation 11.5.

Equation 11.5:

$$T_2 = \left(\frac{(f_4 - f_1)T_4 + (f_1)T_1}{f_4} \right)$$

Where:
f_4 = flow rate returning from distribution system (gpm)
f_1 = flow rate entering from boiler(s) (gpm)
T_4 = temperature of fluid returning from distribution system (°F)
T_1 = temperature of fluid entering from boiler (°F)

Example 11.1

At design load, the flow rate through a distribution system connected to the secondary side of a hydraulic separator is 25.0 gpm. Water returns from the distribution system at 120 °F and enters port 4 of the hydraulic separator. At the same time, the boiler flow rate is 10.0 gpm and the water temperature supplied to port 1 is 160 °F. What is the mixed water temperature supplied to the distribution system? Also, what is the water temperature returning to the boiler?

Solution:

The mixed water temperature supplied to the distribution system is found using Equation 11.5:

$$T_2 = \left(\frac{(f_4 - f_1)T_4 + (f_1)T_1}{f_4} \right)$$

$$= \left(\frac{(25 - 10)120 + (10)160}{25} \right) = 136 \,°F$$

Since there is no mixing in the lower portion of the hydraulic separator (see Figure 11-21), the water temperature returning to the heat source is the same as that returning from the distribution system, 120 °F.

Discussion:

In this case, the mixing resulting from the relatively high flow rate in the distribution system compared to that through the boiler causes a substantial difference between the water temperature at the boiler outlet and that supplied to the distribution system. Any changes in flow rate on the secondary side due to circulators turning on and off, or changing speed, will affect the mixing proportions and hence change the water temperature supplied to the distribution system.

If the water temperature supplied to the distribution system is being controlled by either a set-point device or outdoor reset controller, the supply temperature sensor for that controller should be on the outlet side of the hydraulic separator to ensure that the final mixed water temperature to the distribution system is being measured.

The third possible scenario is when the primary-side flow rate is higher than the secondary-side flow rate. This could occur when, for example, a multiple boiler system is supplying a zoned distribution system, and the load of that distribution system is suddenly reduced. In this scenario, the mixing point changes from the top to the bottom of the hydraulic separator as shown in Figure 11-22.

As with the previous scenario, the first law of thermodynamics governs the mixed outlet temperature, which can be calculated using Equation 11.6.

Equation 11.6:

$$T_3 = \left(\frac{[f_1 - f_2]T_1 + [f_4]T_4}{f_1}\right)$$

Where:
T_3 = temperature of fluid returned to the boiler(s) (°F)
f_1 = flow rate entering from boiler(s) (gpm)
f_2, f_4 = flow rates of the distribution system (gpm)
T_1 = temperature of fluid entering from boiler(s) (°F)
T_4 = temperature of fluid returning from distribution system (°F)

Example 11.2

Assume that the boiler supply temperature is 170 °F and the boiler flow rate into port 1 of the hydraulic separator is 15.0 gpm. Water returns from the distribution system and enters port 4 of the hydraulic separator at 100 °F and 10.0 gpm flow rate. What is the water temperature returned to the boiler?

Solution:

Substituting these operating conditions into Equation 11.6 yields:

$$T_3 = \left(\frac{[f_1 - f_2]T_1 + [f_4]T_4}{f_1}\right)$$

$$= \left(\frac{(15 - 10)170 + (10)100}{15}\right) = 123.3 \,°F$$

Discussion:

Notice that the boiler inlet temperature is about 23 °F higher than the return temperature of the distribution system. This again is due to mixing within the hydraulic separator.

If the system uses a conventional boiler, one might consider the boost in boiler return temperature beneficial because it moves the boiler operating condition away from potential flue gas condensation. However, this temperature boost can quickly diminish if flow through the distribution system increases, or if the return temperature of the distribution system drops. Use of a hydraulic separator alone does not prevent flue gas condensation under all circumstances. The only way to ensure such protection is to install automatic mixing devices on the load circuits that monitor boiler return temperature and reduce hot water flow into those mixing devices when necessary to prevent the boiler from dropping below a predetermined minimum return temperature. The details for proper boiler protection were discussed in Chapter 9, Control Strategies, Components, and Systems.

Hydraulic Separation—Summary

Designers should think of hydraulic separation as a physical concept that can be accomplished to various degrees rather than an all-or-nothing condition. "Perfect" hydraulic separation is rarely present in practical hydronic systems. In most systems, using one of the previously described methods of hydraulic separation, there will still be slight interaction between simultaneously operating circulators. However, the methods discussed in this section, when properly applied, can all keep such interactions to an insignificant level where they will not create detrimental operating conditions.

The ability to prevent detrimental interaction between simultaneously operating circulators is critically important, especially in modern multiload/multitemperature systems. The hardware options used to accomplish hydraulic separation can be seen in many of the schematics within this chapter and throughout the book.

11.8 Multizone Systems—Using Circulators

Multiple-zone systems using multiple circulators have been used in both residential and light commercial buildings for many decades. They consist of two or more parallel piping circuits, each with its own circulator and check valve, connected to a common heat source. Each

586 Chapter 11 Distribution Piping Systems

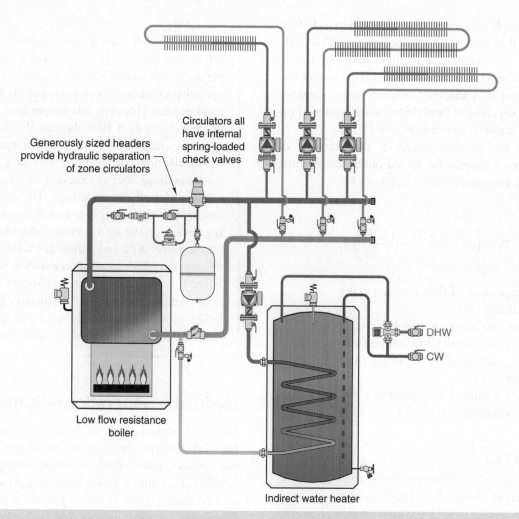

Figure 11-23 | Schematic of a four-zone system using zone circulators.

zone circuit can have several heat emitters. They can be connected in series, or in one of several other piping arrangements described in this chapter. A schematic representation of a four-zone system is shown in Figure 11-23. In this case, three of the zones are used for space heating, and the remaining zone is used for domestic water heating.

Notice the zone supplying the domestic water heater is connected closest to the boiler. While this is not absolutely necessary, it does reduce heat loss from the outer portions of the common piping when only domestic water heating is required, during the summer for example. Also notice the check valve installed just upstream of where the return piping from the indirect water heater joins the return header. This minimizes heat migration along the return header during warm weather when only the domestic water heating load is likely to be active.

Hydraulic separation is provided by the combination of a low flow resistance boiler (such as one built of cast-iron sections) and generously sized headers (sized for a maximum flow velocity of 2 ft/s under design load flow).

Importance of Check Valves

It is crucial that a check valve is used on every zone circuit in this type of system. The check valves prevent water from flowing backward through inactive circuits when other circuits are operating. Without these valves, warm return water flowing backward through inactive zone circuits will cause heat to be released in areas where it is not needed.

Reverse flow through an inactive zone circuit is shown in Figure 11-24.

The type of check valve used is also important. It should either be a **spring-loaded check valve** as shown in Figure 5-57 or a flow-check valve as shown in Figure 5-63. Either of these valves can prevent both reverse flow and undesirable **thermosiphoning** of

11.8 Multizone Systems—Using Circulators

Systems that use circulators for zoning have the following benefits:

- The head loss and flow rate requirements of a given zone circuit tend to be less than for single circuit systems. This allows the use of smaller piping and circulators. Small wet rotor circulators with permanent split capacity (PSC) motors that draw between 50 and 90 watts of electrical power, or circulators with ECMs that draw between 10 and 50 watts are generally adequate for most properly designed residential zone circuits.

- In a multiple-circulator system, the failure of one circulator only stops heat flow to one of the zone circuits. In a single circulator system, the entire building could be without heat.

- Zone circuits supplied from a common header all operate at the same supply water temperature. This keeps the average water temperature of each circuit higher than if all heat emitters were served by a single series circuit. The higher average water temperature may allow the size of the heat emitters to be reduced.

- If a heat source and common piping with low flow resistance are used, hydraulic separation is achieved, and flow rates within each zone will remain fairly stable regardless of what zones are operating. This allows for better balancing and consistent heat output from each zone circuit.

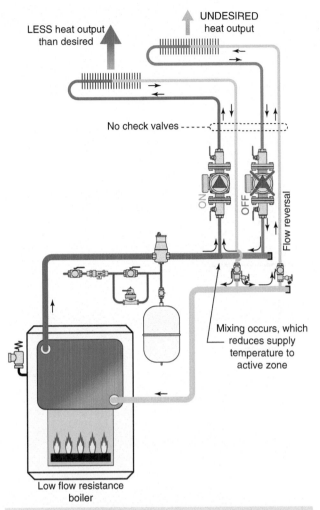

Figure 11-24 | Reverse flow through inactive zone circuit when check-valves are not used.

hot water through inactive zone circuits. The latter occurs because the higher temperature water in the heat source has a lower density than the cooler water in the distribution piping. Without a valve that imposes a small resistance to forward flow, hot water will migrate upward into inactive zone circuits, causing heating when it is not needed. This effect is especially undesirable during the summer when the boiler operates only for domestic water heating, and heat leaking from the heat emitters adds to the building's cooling load.

Another option is the use of circulators with a built-in spring-loaded check valves such as that shown in Figure 7-65. These circulators eliminate the need for a separate check valve, and are ideal for multi-zone applications. If possible, orient the circulator for horizontal flow or downward flow to minimize the possibility of air being trapped between the internal check valve and the impeller. Also be sure that each zone circuit is fully purged to minimize the possibility of air entrapment.

Design of Multizone Systems Using Zone Circulators

The design of a multizone system using zone circulators is similar to the procedure used for a single circuit system. The difference is in how the flow resistance of the common piping is treated.

Because the pressure drop across the common piping is a function of flow rate, it will change depending on how many zone circuits are operating. When the flow resistance of the common piping is low, as is desired, the pressure drop across it will change very little as various zone circuits turn on and off. *This is hydraulic separation at work.* The result will be very small changes in flow rate within the active zone circuits regardless of how many zone circuits are operating.

A conservative design approach is to include the flow resistance of the common piping, assuming all zones are operating, into the overall resistance of each zone circuit as it is analyzed. This approximates the situation that is likely to occur under design load conditions. This concept is illustrated in Figure 11-25.

588 Chapter 11 Distribution Piping Systems

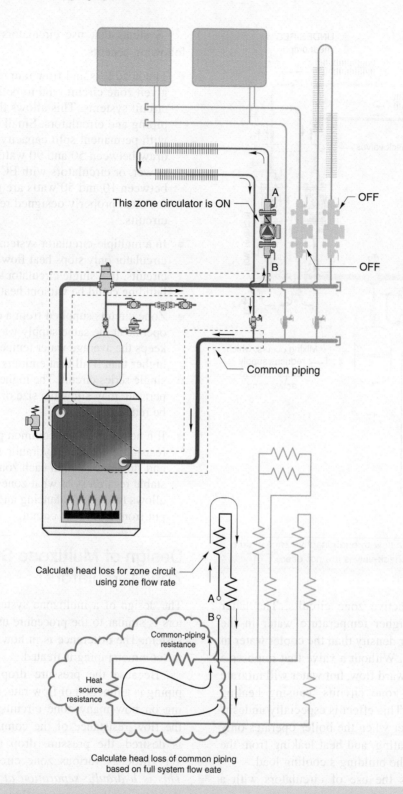

| Figure 11-25 | Include head loss of the common piping segment at full system flow rate along with the head loss of an individual zone circuit at that zone's flow rate. |

Equation 11.7 can be used to modify the hydraulic resistance of the common piping segment to represent full flow conditions:

Equation 11.7:

$$r_{\text{modified}} = r_{\text{common}} \left(\frac{f_{\text{total}}}{f_{\text{zone}}}\right)^2$$

Where:
- r_{modified} = hydraulic resistance of the common piping segment at full system flow
- r_{common} = hydraulic resistance of the common piping segment as normally calculated
- f_{total} = estimated system flow rate through the common piping when all zones are on (gpm)
- f_{zone} = estimated flow rate in the zone circuit being designed (gpm)

Example 11.3

Using manual calculations, estimate the flow rate in the zone circuit shown in Figure 11-26 when all zones are operating. The heat source is a boiler with a hydraulic resistance of 0.3. The flow from all zones passes through this boiler. The tentative circulator selection is a Grundfos UP15-42. The zones are intended to have a temperature drop of 15 °F. The design heating load supplied by the boiler is 70,000 Btu/h. The design load supplied by the zone is 25,000 Btu/h. The water temperature supplied from the boiler is 160 °F.

Solution:

It is necessary to assemble information from Chapter 6, Fluid Flow in Piping, to develop the hydraulic resistance diagram of the zone circuit.

The average water temperature in the zone will be about:

$$T_{\text{ave}} = 160 - \frac{15}{2} = 152.5 \text{ °F}$$

The α value of water at this temperature is estimated from Figure 6-17 as 0.047.

The c-value for the 3/4-inch copper tube from Figure 6-18 is 0.061957.

The c-value for the 1.25-inch copper tube from Figure 6-18 is 0.0068082.

The total equivalent length of the zone piping and fittings is $150 + 30(2) + 83 = 293$ feet.

The total equivalent length of the common piping is $16 + 4(.6) + 2(3) + 2(5.5) + 30 = 65.4$ feet.

The hydraulic resistances of the zone circuit and common piping can each be calculated using Equation 6.10: $r = (\alpha c L)$.

For the zone circuit: $r_1 = \alpha c L = (0.047)(0.061957)(293) = 0.853$

For the common piping: $r_2 = \alpha c L = (0.047)(0.0068082)(65.4) = 0.0209$

A hydraulic resistor diagram showing these resistances is given in Figure 11-27.

The total system flow rate can be estimated using Equation 11.1:

$$f_{\text{system}} = \frac{70,000}{500(15)} = 9.33 \text{ gpm}$$

The target flow rate needed by the individual zone circuit is found using Equation 11.3:

$$f_{\text{zone}} = 9.33\left(\frac{25,000}{70,000}\right) = 3.33 \text{ gpm}$$

The hydraulic resistance of the boiler and common piping need to be added together and then modified using Equation 11.7:

$$r_{(\text{boiler} + \text{commonpiping})} = 0.3 + 0.0209 = 0.3209$$

$$r_{\text{modified}} = (r_{\text{boiler} + \text{commonpiping}})\left(\frac{f_{\text{total}}}{f_{\text{zone}}}\right)^2$$

$$= 0.3209\left(\frac{9.33}{3.33}\right)^2 = 2.52$$

Notice that the modified hydraulic resistance of the common piping is considerably greater than the original calculated value. This accounts for the additional head loss when the total flow of all zones passes through the common piping segment. It also indicates marginal hydraulic separation.

The total hydraulic resistance of the zone circuit is found by adding the modified resistance of the boiler plus common piping to that of the zone piping:

$$r_{\text{zone}} = r_{\text{modified}} + r_1 = 2.52 + 0.853 = 3.37$$

The revised system head loss curve and pump curve for the Grundfos UP15-42 pump are shown in Figure 11-29.

The system curve can now be represented as $H_L = 3.37(f)^{1.75}$.

This system head loss curve is plotted in Figure 11-28a along with the pump curve for the Grundfos UP15-42 circulator to obtain the approximate flow rate in the zone circuit when all zones are operating.

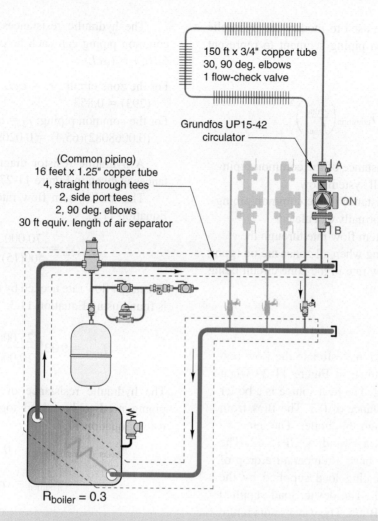

Figure 11-26 | Piping schematic for Example 11.3.

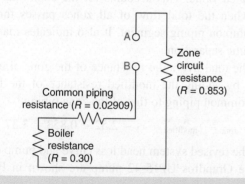

Figure 11-27 | Hydraulic resistance diagram for example 11.1.

Discussion:

Notice the predicted flow rate in the zone (2.36 gpm) is significantly less than the target flow rate for the zone circuit (3.33 gpm) based on obtaining a 15 °F drop. Therefore, a larger circulator should be considered. The pump curve for a Grundfos UP26-99 circulator is shown in Figure 11-28b. Notice that it intersects the system head loss curve at about 3.6 gpm. This is slightly higher than the target flow rate and would be acceptable.

However, if the designer desired to stay with a smaller circulator, they could go on to evaluate the use of larger tubing in the zone to reduce head loss. The equivalent length of the 3/4-inch flow check (83 feet) is especially high. A design temperature drop of 20 or 25 °F would also reduce the necessary flow rate in the zone and thus reduce head loss. The Hydronic Circuit Simulator module in the Hydronics Design Studio software could quickly evaluate these options.

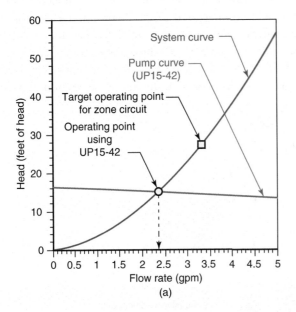

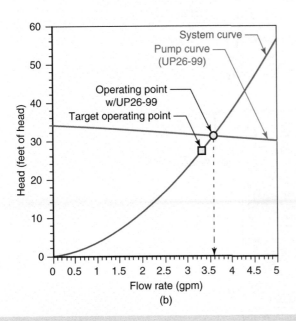

Figure 11-28 System curve and pump curve for Example 11.1.

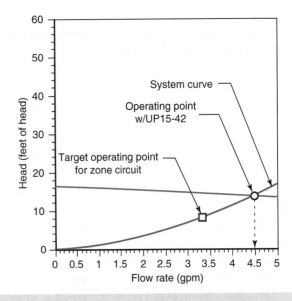

Figure 11-29 System curve and pump curve for Example 11.1 using with lower hydraulic resistance.

Example 11.4

Repeat Example 11.3 assuming a cast-iron sectional boiler with a hydraulic resistance of 0.002 is used instead of the higher flow resistance boiler.

Solution:

The total hydraulic resistance of the boiler and common piping is now:

$$r_{(\text{boiler}+\text{common piping})} = 0.002 + 0.0209 = 0.0229$$

This resistance is modified for full flow conditions using Equation 11.7:

$$r_{\text{modified}} = r_{(\text{boiler}+\text{common piping})} \left(\frac{f_{\text{total}}}{f_{\text{zone}}}\right)^2$$

$$= 0.0229 \left(\frac{9.33}{3.33}\right)^2 = 0.179$$

The total hydraulic resistance of the zone circuit is found by adding the corrected resistance of the boiler plus common piping to that of the zone piping:

$$r_{\text{zone}} = r_{\text{modified}} + r_1 = 0.179 + 0.853 = 1.033$$

Discussion:

The Grundfos UP15-42 circulator now yields a flow rate of about 4.5 gpm compared to only 2.36 gpm when used with the high flow resistance boiler of Example 11.1. This significant increase is due solely to the lower hydraulic resistance of the cast-iron sectional boiler. It demonstrates the need to carefully scrutinize multizone/multicirculator systems using heat sources having high hydraulic resistance.

The higher flow rate will result in a higher-than-expected average circuit temperature, which slightly increases heat output. The 4.5 gpm flow rate through the 3/4-inch tubing will not exceed a flow velocity of 4 ft/s, and thus not generate excessive flow noise. In summary, the higher flow rate would improve the zone's thermal performance without creating undesirable operating conditions. The designer could choose to stay with the tentative circulator selection and accept the added performance.

The other option would be to search for a smaller, lower wattage circulator that would yield a flow rate close to the "target" flow rate of 3.33 gpm initially calculated. The latter option, if achievable, would lower the operating cost of the system over its service life. Finally, using the boiler with low hydraulic resistance would improve the hydraulic separation of the zone circuits.

In summary, multizone systems using circulators provide one way to deliver the benefits of zoning. To minimize interaction between circulators designers should always "generously" size the common piping for minimal flow resistance. Doing so helps ensure good hydraulic separation between all circulators, and thus ensures stable zone flow rates regardless of which zones are operating.

If a heat source with high flow resistance is used, it should be interfaced to the distribution system using one of the previously described devices for hydraulic separation. Figure 11-30 shows how this is done using a pair of closely spaced tees. It is also possible to use a buffer tank to provide hydraulic separation between

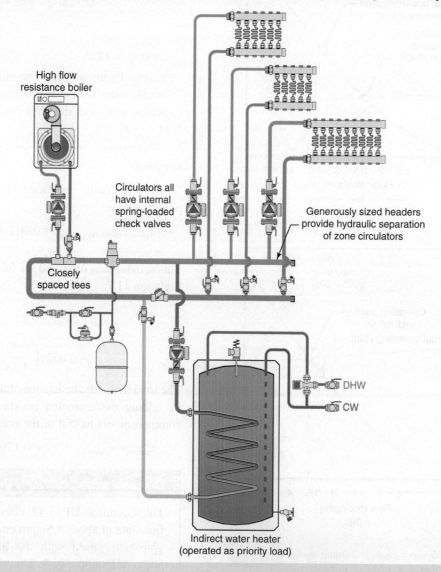

Figure 11-30 | Connecting a high flow resistance heat source using generously sized header piping and closely spaced tees to achieve hydraulic separation.

a heat source with high flow resistance and the headers associated with the zone circulators. The buffer tank also provides thermal mass to prevent the heat source from short cycling.

This system configuration is relatively common. The indirect water heater is operated as a **priority zone**. When the domestic water heating mode is active, all zone circulators supplying space heating are temporarily turned off, and the boiler supplies higher temperature water. When the domestic water heating mode is off, the boiler operates based on outdoor reset control at setting appropriate for the type of heat emitters used.

11.9 Multizone Systems Using Zone Valves

Another common approach to hydronic zoning uses electrically operated zone valves and a single fixed-speed circulator. A piping schematic showing one way to configure such a system is shown in Figure 11-31.

When the room thermostat in a given zone calls for heat, its associated zone valve begins to open. When the valve reaches its fully open position, an internal end switch closes to signal the circulator and heat source to operate. When the thermostat in the zone is satisfied, power to the zone valve is shut off and the zone valve closes. The wiring of various types of zone valves was discussed in Chapter 9, Control Strategies, Components, and Systems.

Since each zone valve remains closed until a call for heat from a room thermostat opens it, there is no need for check valves as with systems using zone circulators.

In the system of Figure 11-31, one of the zones supplies heat to the indirect water heater. If the indirect water heater will be controlled as a priority zone, all space heating zones will be temporarily turned off while the domestic water heating mode is active.

The configuration shown in Figure 11-31 uses a fixed-speed circulator, which has been sized to handle the full design load flow rate assuming all space heating zones are operating. If the indirect water heater is *not*

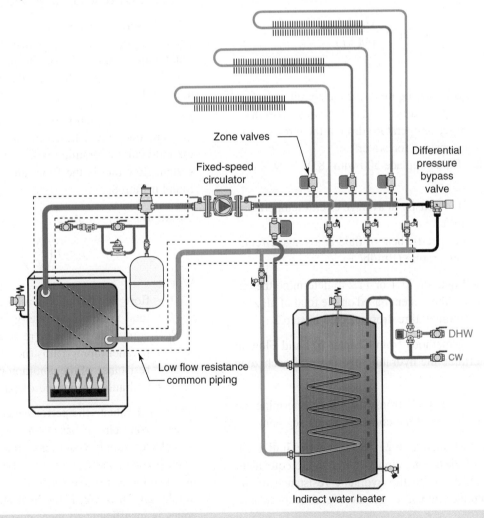

Figure 11-31 | Example of a four-zone system using zone valves.

operated as a priority load, this circulator must provide the required flow of *all* zone circuits (e.g., space heating and domestic water heating).

The flow rate and differential pressure produced by the circulator depend on the number of zones open at any given time. This characteristic must be dealt with using a means of **differential pressure control**, which is discussed later in this section.

Design Procedure for Multiplezone Valve System

STEP 1. Determine the design heating load of each zone circuit.

STEP 2. Select a target temperature drop for the zone circuits under design load conditions.

STEP 3. Determine the target flow rate through each zone circuit using Equation 11.8.

Equation 11.8:

$$f_{circuit} = \frac{Q_{circuit}}{500(\Delta T)}$$

Where:

$f_{circuit}$ = flow rate required in zone circuit (gpm)
$Q_{circuit}$ = design heating load of the circuit (Btu/h)
ΔT = target temperature drop of the circuit (°F)
500 = a constant for water*

*For other fluids, replace 500 with $(8.01 \times D \times c)$.

Where:

D = fluid density at average circuit temperature (lb/ft³)
c = the specific heat of fluid at average circuit temperature (Btu/lb/°F)

STEP 4. Use Equation 11.1 or 11.2 to determine the target flow rate for the system based on the total of all zone loads and the temperature drop selected in step 2.

STEP 5. Using methods from Chapter 6, Fluid Flow in Piping, determine the hydraulic resistance of each zone circuit.

STEP 6. Using methods from Chapter 6, determine the hydraulic resistance of the common piping segment.

STEP 7. Use Equation 6.21 to reduce the individual resistances of each zone circuit into a single equivalent resistance. Add this to the hydraulic resistance of the common piping segment to get the overall resistance of the system.

Equation 6.21 repeated:

$$R_{equivalent_{parallel}} = \left[\left(\frac{1}{r_1}\right)^{0.5714} + \left(\frac{1}{r_2}\right)^{0.5714} + \left(\frac{1}{r_3}\right)^{0.5714} + \ldots + \left(\frac{1}{r_n}\right)^{0.5714}\right]^{-1.75}$$

Where:

$R_{equivalent_{parallel}}$ = equivalent resistance of n parallel hydraulic resistors
$r_1, r_2, r_3, \ldots, r_n$ = hydraulic resistances of each zone circuit

STEP 8. Use Equation 11.9 to plot the system head loss curve.

Equation 11.9:

$$H_L = r_s(f_s)^{1.75}$$

Where:

H_L = head loss of system (feet of head)
r_s = overall hydraulic resistance of system
f_s = total flow rate in system (gpm)

STEP 9. Select a candidate circulator and plot its pump curve over the system resistance curve. If the operating point falls within ±10 of the target system flow rate calculated in step 3, the circulator would typically be acceptable.

STEP 10. Use Equation 6.22 to determine the flow rate in each zone circuit based on the system flow rate and the individual hydraulic resistances of each circuit. The system flow rate is the flow rate at the operating point found in step 8.

Equation 6.22 repeated:

$$f_i = f_s\left(\frac{R_e}{r_i}\right)^{0.5714}$$

Where:

f_i = flow rate through parallel path i (gpm)
f_s = total system flow rate (gpm)
R_e = equivalent hydraulic resistance of all the parallel piping paths (not including the resistance of the common piping)
r_i = hydraulic resistance of parallel piping path i

STEP 11. Evaluate the performance of the heat emitters in each zone circuit based on the zone flow rates and the selected supply water temperature. If the heat output of each zone equals or exceeds the design load requirements of the respective zones, there is no need to change the design. However, if the heat output of one or more zones is insufficient to meet design load requirements,

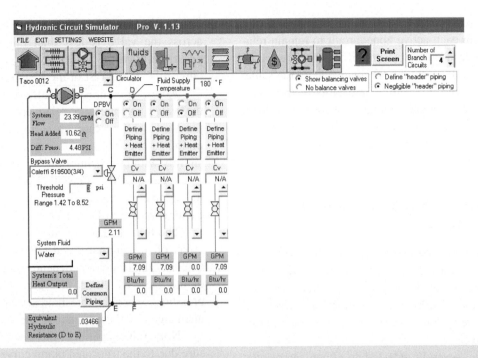

Figure 11-32 Screenshot of the Hydronic Circuit Simulator program module from the Hydronics Design Studio. This screen shows the software configured for a four-zone system with balancing valves and differential pressure bypass valves. *Source: Hydronicpros.*

changes need to be made. These may include increasing the supply water temperature, reducing the hydraulic resistance of the zone, or using a different circulator or a combination of these changes.

The Hydronic Circuit Simulator module in the Hydronics Design Studio software can simulate the operation of several parallel zone circuits. A screenshot of the module configured for a four-zone system is shown in Figure 11-32. The user can designate the number of zone circuits, the piping components in each zone and the common piping segment, the circulator, the fluid, and the supply temperature at which the system operates. The program module instantly finds the hydraulic equilibrium point of the user-specified system and reports all branch flow rates. It can also report the heat output from each branch based on selections of finned-tube baseboard or radiant floor panel heat emitters.

Use a Circulator with a Flat Pump Curve

When zoning with a fixed-speed circulator and *any* type of valve (e.g., electric or thermostatic) to regulate flow in each zone circuit, it is important that the circulator has a relatively "flat" pump curve. The flatter the curve, the more stable the flow rates in each zone circuit as the other circuits cycle on and off. This concept is illustrated in Figure 11-33 for a system having three parallel zone circuits.

When a circulator with the flat curve is used, there is minimal change in the differential pressure across the

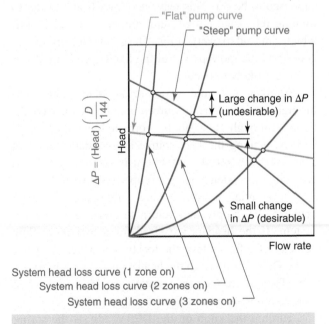

Figure 11-33 The differential pressure between the supply and return header changes based on the number of zone circuits operating. Using a circulator with a "flat" pump curve reduces the change in differential pressure.

circulator as zone circuits turn on and off. This allows the flow rate in each active zone circuit to remain relatively stable because the pressure difference between its starting point and ending point remains about the same. If a circulator with a "steep" curve is used, there is a substantial drop in the differential pressure as additional zones turn on. This, along with other undesirable characteristics, can be corrected by using a differential pressure bypass valve.

Applying Differential Pressure Bypass Valves

Differential pressure bypass valves (DPBV) were discussed in Chapter 5, Piping, Fittings, and Valves. Their purpose is to limit the differential pressure imposed on the distribution system by a *fixed-speed* circulator as the system flow rate decreases due to closing zone valves.

There are several undesirable effects associated with not limiting the differential pressure on a system when a small percentage of the total zone circuits are operating. One is an increase in flow rate in the active zone circuits that can lead to velocity noise. Besides the obvious acoustical annoyance, high flow velocities can cause erosion of valve seats as well as other piping components.

Another repercussion of unregulated differential pressure is partial opening of zone valves that are supposed to be off. This leakage allows heated water to migrate into zone circuits that are supposed to be off, which can lead to overheating. The greater the number of zones and the more powerful the fixed-speed circulator, it is more likely to occur.

Another potential problem is inadequate flow through low-mass heat sources under low-load conditions. Although this can be corrected in other ways, such as making the heat source a secondary circuit in a **primary/secondary system**, it is less likely to need correction in systems using a properly set DPBV.

Differential pressure bypass valves are connected between the supply and return headers of the zoned system. If the headers are sized for a maximum flow velocity of 2 ft/s or less, the location at which the ports of the DPBV connect to the headers is essentially irrelevant. The ends of the headers often provide a convenient installation location as shown in Figure 11-31.

Turning the knob on the DPBV adjusts the spring tension holding the plug against the seat. The pressure where the plug just starts to lift away from the seat is called the **threshold differential pressure**. Most DPBVs have a calibrated scale for setting the threshold differential pressure. When the pressure differential across the valve is less than this threshold, no flow passes through the value. In this situation, the valve is "invisible" to the rest of the system.

Above the threshold differential pressure setting the valve maintains a relatively constant differential pressure as flow through it increases. Still, like any piping component, head losses through a DPBV increase slightly with increasing flow.

When a DPBV is installed as shown in Figure 11-31, the pump curve can be thought of as being partially truncated as shown in Figure 11-34.

The truncated pump curve represents the combined effect of the circulator and DPBV acting together. The extent of the truncation depends on the threshold setting of the valve as shown in Figure 11-35. At differential pressures below the threshold setting, the pump curve is unchanged by the presence of the valve.

Notice the very slight increases in differential pressure due to the truncated pump curve when the system's operating point is above the threshold setting of the DPBV. Compare this to the changes in differential pressure that would occur if no DPBV were present.

The following procedure can be used to determine the threshold pressure setting of the DPBV as well as the maximum flow rate through the DPBV if all zones are closed simultaneously. This setting procedure allows the DPBV to remain "invisible" to the system at design load conditions (when all zone circuits are assumed active), yet come online quickly as the zone's valves begin to close.

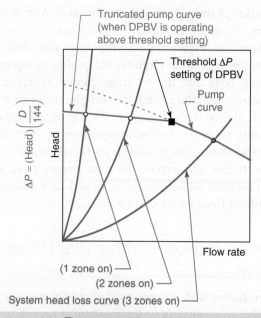

Figure 11-34 | The upper portion of the pump curve can be thought of as being partially truncated when the differential pressure bypass valve becomes active.

11.9 Multizone Systems Using Zone Valves

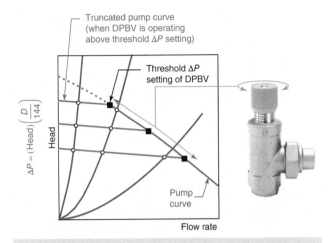

Figure 11-35 | Changing the threshold setting of the DPBV changes how the pump curve is truncated.

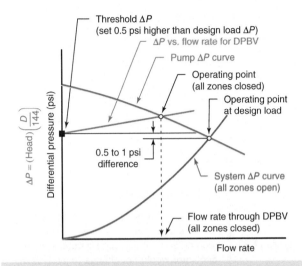

Figure 11-36 | Determining the flow rate through a DPBV when all zone circuits are closed.

STEP 1. Determine the pressure differential across the circulator with all zones on. Do this by finding the intersection of the pump curve and system curve, and then convert the head to differential pressure using Equation 6.6 (repeated):

Equation 6.6 repeated:

$$\Delta P = \frac{HD}{144}$$

Where:
ΔP = pressure differential (psi)
H = head at pump curve/system curve intersection (feet)
D = density of fluid being circulated (lb/ft^3)

STEP 2. Set the threshold differential pressure of the valve to 0.5 psi higher than the value determined in step 1. This allows the DPBV to remain fully closed at design load conditions, so all circulator flow is directed into the distribution system for maximum heat transfer.

Owing to their European origin, some DPBV valves are calibrated in bars or KPa **(Kilo-Pascals)** rather than psi. Use the following conversion factor to convert to psi.

1.0 psi = 6.895 KPa = 0.068 bar

STEP 3. Determine the flow rate through the DPBV when all zones are off. Do this by first plotting the pressure drop versus flow rate relationship of the DPBV, beginning at the pressure threshold setting on the vertical axis (see the green line in Figure 11-36). Next, plot the *differential pressure* curve of the circulator versus flow rate. This is not the same as the pump curve of the circulator, which plots *head* added by the circulator against flow, but can easily be based on the pump curve by converting head to differential pressure using Equation 6.6. This is shown as the red curve in Figure 11-36. The intersection of the green line and red curve in Figure 11-36 occurs where the differential pressure produced by the circulator, with all zones closed, matches the differential pressure drop across the DPBV. The flow rate at this condition is read by dropping a line to the horizontal flow axis.

STEP 4. Select a DPBV from manufacturer's literature that is rated to carry the flow rate determined in step 3.

As a practical matter, most residential and small commercial systems can work with either 3/4- or 1-inch size DPBVs. Larger systems may require 1.25-inch size valves, which tends to be the largest pipe size for which these valves are available in North America.

This sizing/setting procedure is conservative in that it allows the DPBV to begin opening at a relatively high percentage of system load. If the designer's objective is solely to prevent flow noise, the valve's threshold pressure could likely be set higher. In some systems, the DPBV is set while the circulator is operating by slowly opening the valve from its fully closed position under a minimal loading condition, until excessive flow noise goes away. Such a procedure is acceptable in smaller systems, but does not provide the range of almost constant differential pressure operation that is provided by the previously described approach.

Zone Valve Systems Using Variable-Speed Pressure-Regulated Circulators

An "ideal" circulator for a system using valve-based zoning and low flow resistance common piping would have a perfectly flat pump curve and thus maintain a

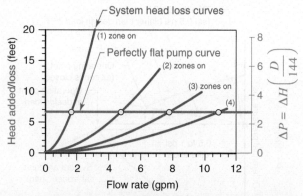

Figure 11-37 A perfectly flat pump curve would be ideal for a system using zone valves. The head (and differential pressure) produced by the circulator would not change regardless of which zones are operating.

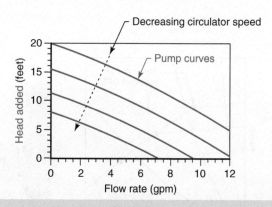

Figure 11-38 When circulator motor speed is reduced, pump curves shift left and downward.

constant differential pressure regardless of how much flow is passing through it. This concept is shown in Figure 11-37. Such a circulator would allow any given zone circuit to turn on or off without affecting flow in the other zone circuits.

Although it is possible to build a circulator with *relatively* flat pump curves, the curve of a centrifugal pump will always have some drop in head (and hence differential pressure) as flow rate increases.

By varying the speed of the circulator, it is possible to shift the pump curve as shown in Figure 11-38. When the circulator speed is decreased, the pump curve shifts down and to the left. When speed is increased, the curve shifts up and to the right.

At certain speeds, the pump curves of the variable-speed circulator cross over the system head loss curves so that the head (and differential pressure) at the operating points remains constant. Examples of these pumps curves and the corresponding system head loss curves are shown in Figure 11-39. The "net effect" is a horizontal tracking of the operating points, which is the same effect that would be produced by a circulator with a perfectly flat pump curve.

In the past, the "intelligence" necessary to keep the pump operating at the right speed in such an application was provided by an external **differential pressure controller**. Such a device monitors the pressure difference across the piping mains, compares it to a "target" differential pressure, and outputs a control signal based on the error that exists between the measured pressure and target differential pressure. This control signal was sent to a **variable-frequency drive**, which in turn regulated frequency and voltage of the AC waveform sent to the motor to control its speed. This type of variable-speed control system was developed for larger commercial/industrial systems. Although this approach remains in use in such systems, its cost and complexity typically precludes its use in residential and light commercial hydronic systems.

Fortunately, the recent introduction of small wet rotor circulators with **electronically commutated motors** (ECMs) has made it both practical and economically attractive to implement constant differential pressure control in smaller hydronic systems with valve-based zoning. Such circulators were discussed in Section 7.5. *In the author's opinion, the availability of such circulators one of the biggest technological achievement in residential and light commercial hydronics technology to occur over the last 40 years.*

The ECMs and improved hydrodynamic design of current generation **variable-speed pressure-regulated circulators** provide wire-to-water

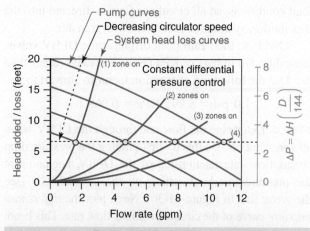

Figure 11-39 By properly varying the speed of the circulator, the intersection of the pump curve and current system head loss curve can be maintained at a constant differential pressure.

efficiencies that are approximately double those of standard wet rotor circulators with PSC motors. This, as well as the ability to reduce speed and input wattage under partial load conditions, typically allows annual electrical energy savings in excess of 60% when applied in hydronic distribution systems using valve-based zoning.

The use of a variable-speed pressure-regulated circulator also eliminates the need for a differential pressure bypass valve. Differential pressure is now regulated by varying the speed of the circulator rather than by throttling away the excess head energy from a fixed-speed circulator using a differential pressure bypass valve.

A variable-speed pressure-regulated circulator can be used in any system with valve-based zoning (e.g., systems using electric zone valves, thermostatic radiator valves, manifold valve actuators, or a combination of this hardware).

Readers will find several examples of pressure-regulated circulators in the schematics of this chapter and throughout this text.

11.10 Parallel Direct-Return Systems

Parallel distribution systems, also sometimes called **two-pipe systems**, are constructed so that every heat emitter receives water from a common **supply main** and returns it to a common **return main**. Two schematics for a **parallel direct-return system** are shown in Figure 11-40. The system in Figure 11-40a uses a fixed-speed circulator and operates as a single zone. The system in Figure 11-40b uses a variable-speed pressure-regulated circulator in combination with zone valves to provide four independently controlled zones.

Assuming that heat loss from the supply main is minimal, a parallel direct-return system supplies water at approximately the same temperature to each **branch**. In this respect, it is similar to the header-type zone valve systems discussed in Section 11.8. However, there are some distinct differences between these two types of distribution systems.

One is the length of mains piping between the beginning and ends of the parallel branches. In a header-type zone valve system, each circuit usually begins from a common supply header and ends at a common return header. The length of header piping between the starting points and ending points of zone circuits is very short, usually not more than a few feet, and has insignificant head loss. However, in a parallel direct-return system, the distance between adjacent branches can be many feet. These longer piping segments can generate significant head loss that must be accounted for when the system is designed.

The words "direct-return" indicate that the first branch supplied by the hot water main is also the closest to the circulator on the return side of the system (see Figure 11-40).

The length of the piping path from the discharge port of the circulator, through the branch, and back to the inlet side of the circulator significantly affects the flow rate through that branch. In a situation where all heat emitters have identical hydraulic resistances, the farther the branch is from the circulator, the lower its flow rate as shown in Figure 11-41. The lower the flow rate through a given heat emitter, the lower its heat output. *This situation must be compensated for to avoid serious heat distribution problems.*

By installing **balancing valves** in each branch, the tendency of a direct-return system to allow more flow through the branches closer to the circulator and less flow in the more remote branches can be corrected as shown in Figure 11-42.

By partially closing the balancing valve on a given branch, the hydraulic resistance of that flow path can be increased to compensate for its distance away from the circulator. If the hydraulic resistance of all branch paths is about the same, the balancing valves closer to the circulator are usually closed more than those farther out along the piping mains.

The amount a given balancing valve must be closed depends on the hydraulic resistance of the heat emitter and the other piping component in the branch, as well as the position of the branch in the system. For example, the valve on a high flow resistance heat emitter near the circulator may not need to be closed as much as the valve on a low hydraulic resistance branch farther out in the distribution system.

The subject of hydraulic balancing is discussed in more detail in Chapter 14, Auxiliary Loads and Specialized Topics.

The pipe size of the supply and return mains at any point in the system should be determined by the flow rate present at that point. As the supply main passes by each branch, a portion of its flow is routed through that branch. This reduces the flow that continues farther out in the system. Similarly, flow in the return main increases as it collects additional flow from each branch it passes on its way back toward the circulator. Proper design keeps the flow velocity in all locations under 4 ft/s to avoid flow noise.

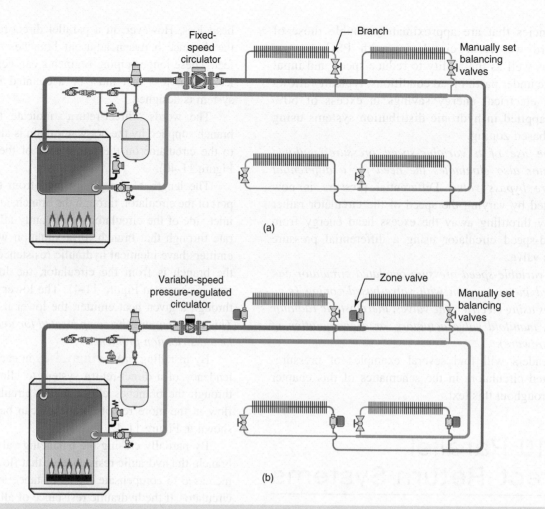

Figure 11-40 Piping schematics of parallel direct-return distribution systems. (a) Using fixed-speed circulator and manual balancing valves. (b) Using variable-speed pressure-regulated circulator, manual balancing valves, and zone valves.

Because each branch receives water at approximately the same temperature, and since each branch only serves a single heat emitter, all heat emitters can be selected based on the same supply water temperature. This makes parallel direct-return systems well suited for low-temperature applications where temperature drops often must be held to a minimum. It also eliminates the need to correct for temperature drops from one heat emitter to the next as is necessary in series circuit and one-pipe distribution systems.

Another advantage of parallel systems is that each heat emitter can be separately controlled. The flow through any heat emitter can be modulated or turned on and off by devices such as thermostatic radiator valves or zone valves as shown in Figure 11-40b.

As the flow through individual heat emitters is reduced or stopped, the system flow rate decreases, and the head across a fixed-speed circulator increases. This in turn increases the flow rate in the branches that remain open. If only one or two branches are open, and a constant-speed circulator is used, flow noise could develop. To avoid this, a differential pressure bypass valve (DPBV) can be installed and set to prevent the circulator from operating at high pressure differentials. The DPBV can be selected and sized as discussed in Section 11.9.

Variable-speed pressure-regulated circulators are also well suited for use in parallel direct-return systems. Such circulators would typically be configured for proportional differential pressure operation as discussed in Chapter 7, Hydronic Circulators. Proportional differential pressure control is illustrated in Figure 11-43. This operating mode was specifically developed to account for the greater head losses associated with longer piping mains in a parallel direct-return system, versus the insignificant head losses in a header type system with good hydraulic separation. *The pressure-regulated circulator should be configured to provide the maximum required*

11.10 Parallel Direct-Return Systems

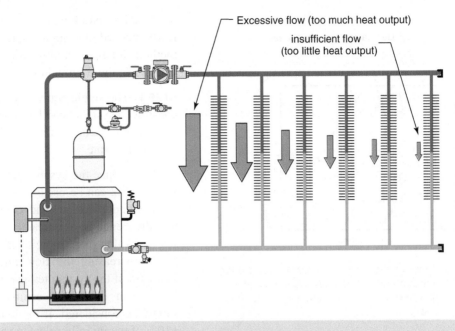

Figure 11-41 | In an "unbalanced" parallel direct-return system, where each branch has the same hydraulic resistance, the farther a given branch circuit is from the circulator, the lower its flow rate.

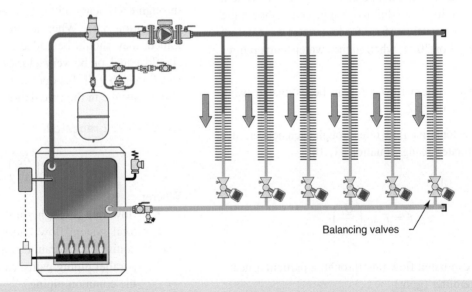

Figure 11-42 | The undesirable situation of Figure 11-41 can be corrected by installing and properly adjusting balancing valves within each branch. *Note:* This scenario assumes the goal is to have equal flow rates through each branch.

head and, by association, differential pressure, under design load conditions assuming full flow through all branches.

Manual Design Procedure for Parallel Direct-Return Systems

Parallel direct-return systems can be analyzed using the concepts of series and parallel hydraulic resistances as discussed in Chapter 6, Fluid Flow in Piping. The following procedure outlines a typical approach. As the reader will see, this can be a long procedure.

STEP 1. Heat emitters should be selected based on providing design heat output at a nominal operating flow rate and supply water temperature.

STEP 2. A tentative piping layout should be sketched for the particular heat emitters used and their placement in the building. The lengths of the piping segments and the number/type of fittings required within the branches and main piping should be estimated.

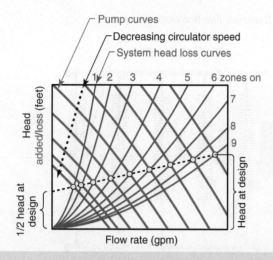

Figure 11-43 | Pressure-regulated circulators configured for proportional differential pressure control are ideal for parallel direct-return systems using valve type zoning.

STEP 3. Select a target temperature drop for the system under design load conditions. Suggested temperature drops range from 10 to 15 °F for low water temperature systems to 25 or 30 °F when higher water temperatures are used.

STEP 4. The system flow rate can be estimated using Equation 11.1 or 11.2.

STEP 5. The tentative flow rate through each heat emitter can be estimated using Equation 11.10:

Equation 11.10:

$$f_i = f_{initial}\left(\frac{Q_i}{Q_t}\right)$$

Where:
- f_i = estimated flow rate through a particular heat emitter (gpm)
- $f_{initial}$ = first estimate of system flow rate from step 4 (gpm)
- Q_i = design heat output of the particular heat emitter (Btu/h)
- Q_t = total design heating load served by the system (Btu/h)

STEP 6. Pipe sizes can now be selected based on keeping the flow velocity in all pipes less than or equal to 4 ft/s. Use the formulas in Chapter 6, Fluid Flow in Piping, to verify flow velocity versus flow rate for various types and sizes of tubing.

STEP 7. The pipe sizes and their estimated lengths can now be combined with the sketch of the piping layout to construct a hydraulic resistance diagram.

STEP 8. The hydraulic resistor diagram can be systematically reduced to a single equivalent resistance using the methods for series and parallel hydraulic resistors from Chapter 6, Fluid Flow in Piping.

STEP 9. The single equivalent resistance can be used to construct a system head loss curve. This can be overlaid with the pump curve of a candidate circulator to find the system flow rate.

STEP 10. Once the system flow rate is found, Equation 11.1 or 11.2 can be used to verify the temperature drop present under design load conditions. If the temperature difference is significantly *greater* than the initially selected value (e.g., more than 10%), a circulator capable of increasing the flow rate is worth considering. Likewise, if the system temperature drop is significantly *lower* than the initial estimate, a smaller circulator should be considered. If the temperature drop is within ±10% of the initially selected value, the circulator being considered is probably suitable.

STEP 11. Knowing the system flow rate, the flow rate through each heat emitter can be found by repeated use of Equation 6.22. When all these flow rates are determined, they should be added up as a check. Their total should equal or be very close to the system flow rate found in step 9. If this does not check out, there is probably a mistake in the calculations.

Equation 6.22 repeated:

$$f_i = f_s\left(\frac{R_e}{r_i}\right)^{0.5714}$$

Where:
- f_i = flow rate through parallel path i (gpm)
- f_s = total system flow rate (gpm)
- R_e = equivalent hydraulic resistance of all the parallel piping paths, excluding resistance of the common piping
- r_i = hydraulic resistance of parallel piping path i

STEP 12. The flow rates through the individual parallel piping paths and heat emitters should be checked for excessive flow velocity (over 4 ft/s). These flow rates should also be checked against the tentatively estimated values. If a given flow rate is higher than its initial estimate, it can probably be reduced using a balancing valve. If it is too low, a number of corrective options are available. These include using a larger branch piping size or a different heat emitter with a lower flow resistance.

STEP 13. After checking the flow rate and expected thermal performance of each heat emitter, and making any changes to the pipe sizes or heat emitters, the

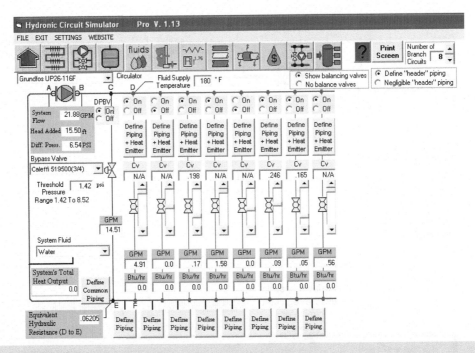

Figure 11-44 | Hydronic Circuit Simulator module configured to analyze an eight-branch parallel direct-return system including balancing valves. *Source: Hydronicpros.*

hydraulic resistor diagram can be modified to reflect these changes. The modified resistor diagram can then be reanalyzed beginning with step 8. This process can be repeated until the resulting system performs reasonably close to expectations. With practice, this can usually be accomplished within two iterations of the calculations.

The Hydronic Circuit Simulator module in the Hydronics Design Studio software can also be used to simulate the performance of a parallel direct-return circuit. The user can define piping components in the branches as well as the mains piping segments between branches. The software performs the previously described calculations whenever one or more of the inputs are changed. Several variations of a proposed system can be analyzed in a few minutes. An example of the software configured to analyze an eight-branch parallel direct-return system with balancing valves is shown in Figure 11-44.

11.11 Parallel Reverse-Return Systems

Another method for laying out a parallel system is known as **parallel reverse-return piping**. Figure 11-45 shows a modification of the direct-return system of Figure 11-40, making it a reverse-return system.

Notice that the branch closest to the circulator along the supply main is now the *farthest* from the circulator along the return main. Likewise, the farthest branch on the supply is the closest on the return. This arrangement *helps* correct the inherent flow balancing problems of the parallel direct-return systems, especially when the heat emitters all have similar hydraulic resistances. *However, a reverse-return system is not guaranteed to be "self-balancing."*

Parallel reverse-return systems usually require different pipe sizes in various parts of the system. The supply main gets progressively smaller as one moves away from the circulator. The return main gets progressively larger as one moves toward the circulator. The commonly accepted approach is to size the supply main and return main piping so that the head loss per unit of length is reasonably consistent—typically in the range of 4 feet of head loss per 100 feet of pipe. *When this practice is followed, and when all branches have the same or very similar hydraulic resistance, the reverse-return system will be close to "self-balancing."* However, most systems do *not* meet these conditions, and thus are not necessarily self-balancing. For this reason, the author still recommends the installation of balancing valves in each branch.

The ideal arrangement for a parallel reverse-return distribution system is to route both the supply and return mains around the perimeter of the area to be heated as shown in Figure 11-46. If this is not done, there may be a need for considerable extra piping to achieve the reverse-return layout as shown in Figure 11-47. This extra piping, all sized to carry the full design load flow

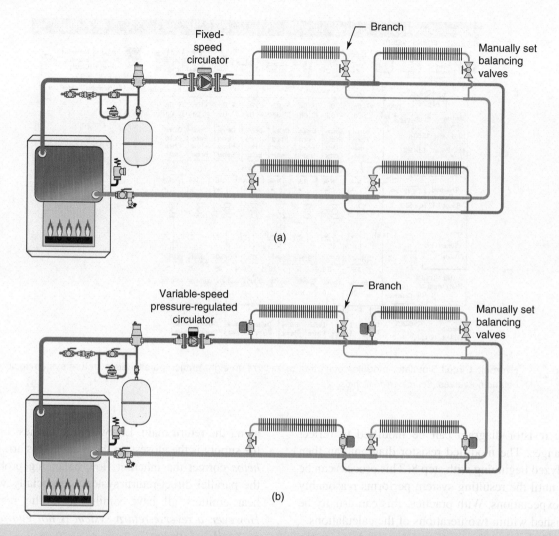

Figure 11-45 Piping schematics of parallel reverse-return distribution systems. (a) Using fixed-speed circulator and manual balancing valves. (b) Using variable-speed pressure-regulated circulator, manual balancing valves, and zone valves.

rate, could add substantial cost to the system. It would also increase extraneous heat loss from the system.

Reverse return piping is well applied when two or more identical devices need to be piped to achieve approximately equal portions of a total flow rate. When two identical devices are used, the total flow would ideally divide so that 50% of the total passes through each device. When three identical devices are present, the total flow would (ideally) divide so that one third of the total flow would pass through each device.

Figure 11-48 shows four identical solar collectors piped into an array using reverse return. The internal headers at the top and bottom of the absorber plate in each collector are joined to form the "supply main" and "return main" of the array. Because the pipe size is the same all along these headers, but the flow rates are different, the head loss per unit of length of header is *not* the same. This means the flow rates through the collectors will not be identical. The outer two collectors will operate at slightly higher flow rates than the two inner collectors. However, the difference in flow rates will be small and have very little effect on overall performance. Thus, it is acceptable to group up to eight collectors in this manner without balancing valves.

Most of the design concepts discussed for parallel direct-return systems are also applicable to reverse-return systems. Parallel reverse-return systems are a good choice for low-temperature distribution systems. Variable-speed pressure-regulated circulators are also well suited to parallel reverse-return systems where valve-based zoning is used. If a fixed-speed circulator is selected, a differential pressure bypass valve should be installed to reduce the potential for flow noise when only one or two zones are active. This valve would also prevent **dead-heading** the circulator if all zone circuits were closed simultaneously.

11.11 Parallel Reverse-Return Systems

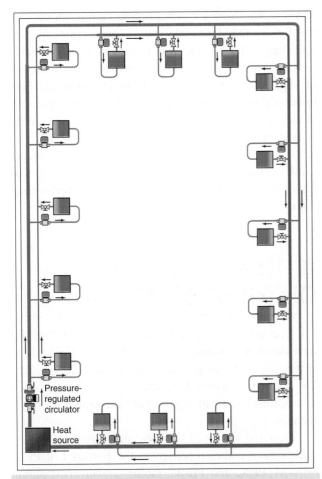

Figure 11-46 | Efficient use of parallel reverse-return piping around the perimeter of a building.

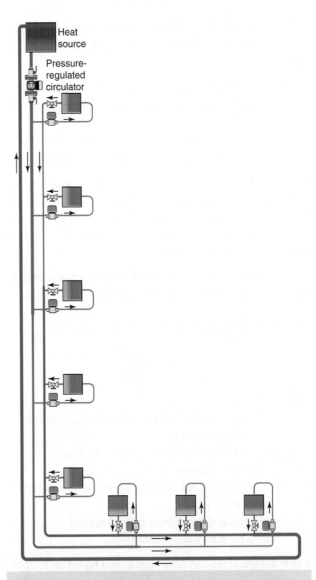

Figure 11-47 | Inefficient (but possible) use of parallel direct-return piping in a "dead-end" layout.

The standard approach of setting up a hydraulic resistance diagram of the system is not very useful for reverse-return systems. If the number of parallel branches exceeds two, the resulting hydraulic resistor diagram cannot be reduced to a single equivalent resistance using the methods of Chapter 6, Fluid Flow in Piping. The flow rates can still be solved using sophisticated mathematical iteration, but the analytical methods involved in solving multiple nonlinear equations required are beyond the scope of this book.

Manual Design Procedure for Parallel Reverse-Return Systems

Since the equivalent resistance methods of Chapter 6, Fluid Flow in Piping, are usually not applicable to reverse-return systems, a somewhat simplified method will be presented. This procedure will result in a conservative (e.g., slightly oversized) circulator being selected. The first four steps of this method are identical to those used for direct-return systems.

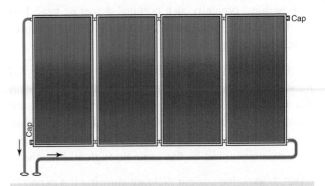

Figure 11-48 | Parallel reverse-return piping of four identical solar collectors. No balancing valves are required. Although there will be slight variations in flow rate through the collectors, balancing valves are not necessary.

STEP 1. Heat emitters should be selected based on providing design heat output at a nominal operating flow rate.

STEP 2. A tentative piping layout should be sketched for the heat emitters selected and their placement in the building.

The lengths of the piping segments and the number and type of fittings required should be estimated.

STEP 3. Select a tentative temperature drop for the system under design load conditions. A practical value for parallel systems is 10 to 30 °F. The system flow rate can then be estimated using Equation 11.1 (for water) or Equation 11.2 (for other fluids).

STEP 4. The tentative flow rate through each heat emitter can be estimated using Equation 11.10:

Equation 11.10 repeated:

$$f_i = f_{system}\left(\frac{Q_i}{Q_s}\right)$$

Where:
 f_i = estimated flow rate through a given heat emitter (gpm)
 f_{system} = estimate of system flow rate from step 3 (gpm)
 Q_i = design heat output of a given heat emitter (Btu/h)
 Q_s = total design heating load served by the system (Btu/h)

STEP 5. Pipe sizes can now be selected based on keeping the flow velocity in all pipes less than or equal to 4 ft/s. However, if all branches of the system have identical (or very similar) hydraulic resistance, the piping for the supply main and return main should instead be sized so that the head loss per unit of piping length is as constant as possible—typically at a nominal 4 feet of head loss per 100 feet of pipe. This allows the system to be closer to self-balancing.

STEP 6. *Identify the branch with the highest expected flow resistance.* This will often be the branch with the most restrictive heat emitter. If all branches are identical, and the main piping is sized for a constant head loss per unit of length as described in step 5, the flow resistance will be the same through all branches.

STEP 7. Sketch a system diagram showing the expected flow rates (as calculated in step 3) present in all piping segments. An example is illustrated in Figure 11-49.

STEP 8. Determine the head loss of the circuit path formed by the branch with the highest resistance and the associated supply and return mains.

STEP 9. The head loss obtained in step 8, along with the full system flow rate, sets a conservatively high operating point for the circulator. Circulators suitable for this system will have pump curves that pass through or slightly above this operating point. Since the flow resistance of the other branches will be lower than the branch used for design purposes, the system will likely operate at flow rates slightly higher than determined by this method. The higher flow rates in some branches can be corrected through the use of balancing valves as discussed in Chapter 14, Auxiliary Load and Specialized Topics.

11.12 Home-Run Distribution Systems

All hydronic distribution systems discussed thus far were developed around the assumption that rigid tubing or pipe would be used to convey heat from the mechanical room to various heat emitters. They have been successfully used with rigid tubing and piping for many decades.

Over the last three decades, many heating professionals have come to recognize the potential of PEX, PEX-AL-PEX, and PE-RT tubing as *universal* piping materials for residential and light commercial hydronic systems. In addition to radiant panel heating applications, the temperature/pressure rating of these polymer tubes, along with their inherent flexibility, offers new possibilities for piping traditional higher temperature heat emitters such as cast-iron radiators, finned-tube baseboards, and panel radiators.

One approach that takes advantage of the flexibility of these tubes is called a **home-run distribution system**. An example of such a system is given in Figure 11-50.

In a home-run system, a separate supply and return run of small-diameter (usually 3/8 or 1/2 inch) PEX, PE-RT, or PEX-AL-PEX tubing is routed from a manifold station to each heat emitter. The small flexible tubing can be routed through the framing cavities of the building much like an electrical cable. The ability to "fish" tubing through closed framing cavities, as seen in Figure 11-51, is a tremendous advantage over rigid tubing, especially in retrofit situations. Home-run systems are also an excellent way of using up the shorter "remnant" lengths of tubing often left over from radiant panel installations.

Besides fast and simple installation, home-run systems provide many of the same benefits as other types of parallel distribution systems:

11.12 Home-Run Distribution Systems

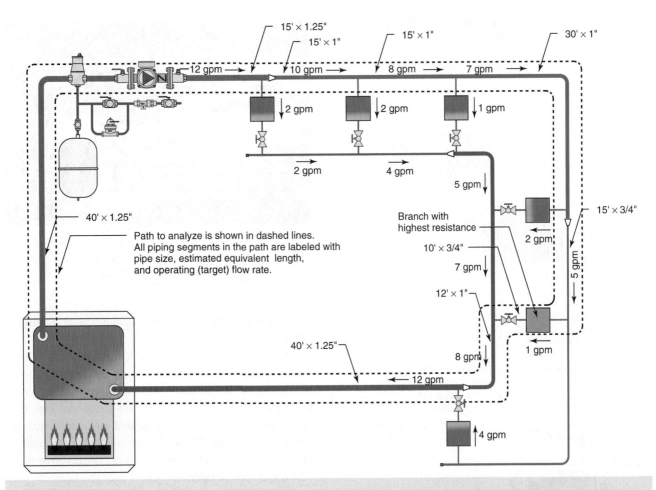

Figure 11-49 | Identifying the flow path of greatest resistance when analyzing a parallel reverse-return system.

- They allow the heat output of each room to be individually controlled through a number of options that will be discussed.
- They deliver water at approximately the same temperature to each heat emitter, thus simplifying heat emitter sizing.
- They are adaptable to almost any combination of panel radiators, fin-tube convectors, towel warmers, cast-iron radiators, or finned-tube baseboards.
- Balancing valves at the manifold station or at each heat emitter can be used to compensate for the flow resistances of different home-run circuits serving different heat emitters, allowing for system balancing.
- They often require less circulator power than do series-type distribution systems of comparable ability.

The last advantage can be demonstrated by comparing two systems of equivalent heating ability, one piped as a single series circuit and the other piped as a home-run distribution system.

The system shown in Figure 11-52 has been designed, based on information in previous chapters, to supply 10,000 Btu/h. from each of five fin-tube baseboards connected in a single series circuit. The piping circuit has a total equivalent length of 280 feet of 3/4-inch copper tubing. The supply water temperature is 180 °F and the flow rate is 5.0 gpm.

Using the methods in Chapter 6, Fluid Flow in Piping, the head loss and associated pressure drop of the circuit operating under these conditions was determined. The latter was calculated at 5.51 psi.

Using Equation 7.4, and an assumed 22% wire-to-water efficiency for the wet rotor circulator, the estimated input wattage to the circulator is 54.4 watts.

For comparison, consider the system shown in Figure 11-53. It uses a zoned home-run distribution system to supply the same 50,000 Btu/h. output using the same 180 °F supply water temperature, 20 °F circuit temperature drop, and 5.0 gpm flow rate. However, because of its parallel versus series layout, the pressure

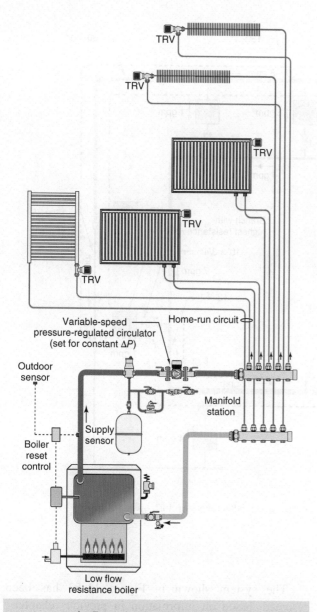

Figure 11-50 Example of a simple home-run distribution system with thermostatic radiator valves on each heat emitter and variable-speed pressure-regulated circulator set for constant ΔP operating mode.

Figure 11-51 Small-diameter PEX or PEX-AL-PEX tubing is easily "fished" through framing cavities. *Courtesy of John Siegenthaler.*

drop around this distribution system is significantly lower than that of the series circuits (1.68 psi versus 5.51 psi). This allows the required circulator input power to drop to 16.6 watts, assuming the same 22% wire-to-water circulator efficiency.

The difference in circulator input power between the series and home-run distribution system is 37.8 watts. This may seem insignificant. However, if supplied for 3000 hours per year in a location where the current cost of electrical energy is $0.12/kWhr, and is inflating by 4%/year, this additional power requirement increases operating cost by $404 over a 20-year period.

Furthermore, the home-run system could be zoned using valves (either manifold valve actuators as shown or thermostatic radiator valves at each heat emitter). This would allow the fixed-speed circulator to be replaced with a variable-speed pressure-regulated circulator with an ECM. Assuming the latter reduces electrical energy consumption by a conservative 60%. The difference in operating costs over the same 20-year period would be approximately $510. This demonstrates the inherent hydrodynamic advantage of a home-run system compared to a traditional single series circuit.

Zoning Control Options for Home-Run Distribution Systems

Home-run distribution systems allow several methods of zoning control. The approach shown in Figure 11-53 uses **electric valve actuators** to open and close manifold valves for each circuit. These actuators are controlled by 24-VAC thermostats in each room. The end switch in each actuator is used to call for boiler operation and turn on the circulator.

Another approach uses a thermostatic radiator valve on each heat emitter as shown in Figure 11-50. These nonelectric valves, which were discussed in Chapter 5, Piping, Fitting, and Valves, modulate flow through their respective heat emitters to regulate heat output. They are available in various configurations for use with finned-tube baseboards and panel radiators.

Keep in mind that thermostatic radiator valves are nonelectric and therefore cannot signal for the heat source to operate or for the circulator to run. The distribution

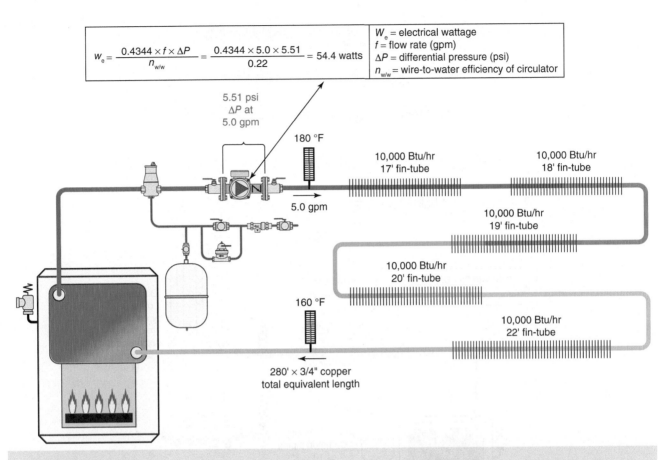

Figure 11-52 | A series piping circuit that delivers 50,000 Btu/h of heating and requires 54.4 watts of circulator input power.

circulator must operate whenever heat might be needed by any zone. This is usually accomplished by a setpoint controller, or boiler reset controller, that senses outdoor temperature and turns on the distribution circulator and heat source whenever the outdoor temperature is below a preset starting value (usually about 55 °F).

Stacked Home-Run Systems

There may be situations in which a home-run distribution system needs to be installed in a tall building such as a three-story house. If home-run circuits were routed from each heat emitter to a single manifold, such as in the basement, significant lengths of tubing would end up as "leaders" between the heat emitters on the upper floors and the manifold.

One alternative is a "stacked" home-run distribution system in which manifold stations on each floor level are supplied by a common set of vertical piping risers as shown in Figure 11-54. The designer should select a central location in the building where the riser piping can be routed from the basement mechanical room to the top floor. Plan the vertical risers so they will be reasonably close to the manifold station on each floor. This approach allows relatively short home-run circuits between each heat emitter and the manifold stations.

To ensure fast air venting in this type of system, install float-type air vents along with isolation/servicing valves at the top of each vertical riser. To make sure they work properly in what may be a relatively tall system, adjust the feed water valve to ensure a minimum of 5.0 psi static pressure at the top of the system.

Any home-run system that uses valve-based zoning must have a means of differential pressure control. This can be accomplished with the combination of a fixed-speed circulator and differential pressure bypass valve, or by using a variable-speed pressure-regulated circulator. The latter approach takes advantage of modern circulator technology and significantly lowers operating cost. If a fixed-speed circulator is used, it should have a relatively "flat" pump curve.

Designers should also remember that home-run distribution systems can also be used as a *subsystem* that

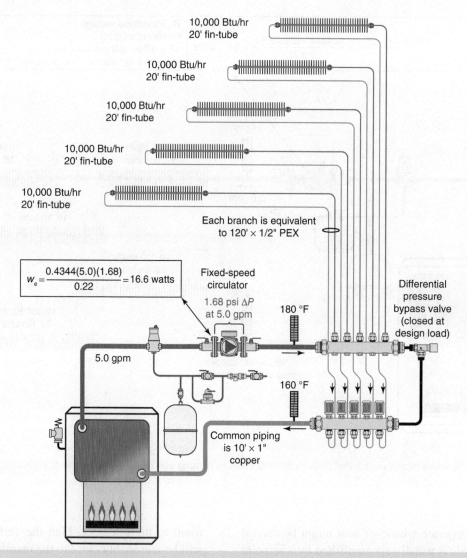

Figure 11-53 | A home-run distribution system that delivers 50,000 Btu/h of output and requires 16.6 watts of circulator input power.

is part of a larger overall system. An example is shown in Figure 11-55.

Design Procedure for Home-Run Systems

From the standpoint of piping "topology," home-run distribution systems are very similar to the zone valve systems discussed in Section 11.8. The differences are essentially in the lengths and types of tubing used. The design procedure described in Section 11.8 can thus be used to design a home-run system or subsystem. The Hydronic Circuit Simulator module in the Hydronics Design Studio software can also be used to model the flow performance of a home-run system.

11.13 Primary/ Secondary Systems

Primary/secondary (P/S) piping is *one way* to achieve hydraulic separation between simultaneously operating circulators. Its origins date back to the 1950s, when it was deployed primarily in large commercial heating systems, and chilled water cooling systems. However, the increasingly sophisticated expectations placed on residential and light commercial systems over the last three decades prompted many designers, including the author, to use P/S piping as a flexible and expandable "backbone" upon which to build modern multiload/multitemperature systems.

11.13 Primary/Secondary Systems 611

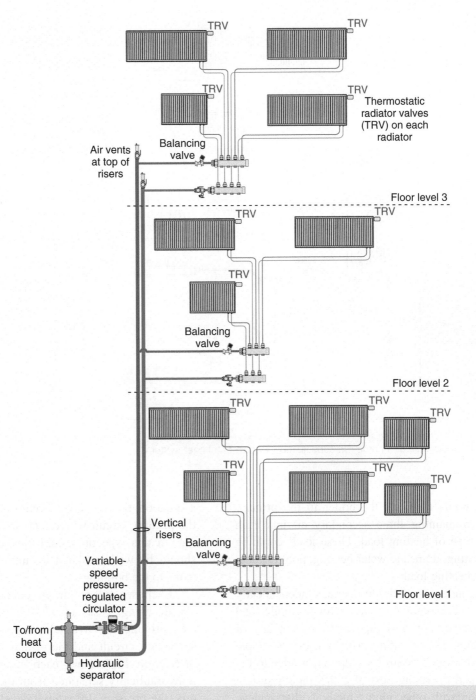

Figure 11-54 | A stacked home-run distribution system using one manifold station for each floor level of the building.

In the second edition of this text, P/S piping was treated as the *only* piping method by which hydraulic separation of simultaneously operation circulators could be achieved. Given the advances in hydronics technology since that edition was written, P/S piping remains an option, *but not an exclusive solution* for providing hydraulic separation. Several other options for achieving the same benefits provided by P/S piping were discussed earlier in this chapter.

An example of a relatively simple P/S system is shown in Figure 11-56. It consists of a **primary circuit** that connects to four **secondary circuits**. One of the

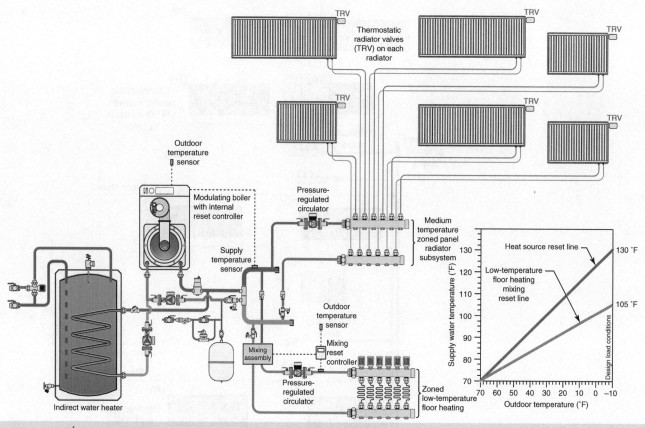

Figure 11-55 | Example of a home-run subassembly as part of a larger system.

secondary circuits connects the boiler to the primary circuit. The remaining three secondary circuits each serve some type of heating load. These loads could be for space heating, domestic water heating, pool heating, or any other heating load.

Each secondary load circuit operates independently of the others. The secondary circuit for the boiler and the primary circulator must operate whenever any of the load circuits are operating. Although this is simple to accomplish from a control standpoint, it adds to the installation and operating cost of a P/S piping system in comparison with other piping methods that achieve equivalent hydraulic separation of circulators, but without need of a primary circulator.

The primary circuit could be very short or quite long. It may consist of no more than 5 to 10 feet of pipe near the boiler. In such situations, the minimum length of the primary circuit is usually limited by the need to install the necessary piping components and the lengths of straight pipe both upstream and downstream of each set of closely spaced tees.

Another possibility is for the primary circuit to travel completely around the inside of a building. In such a situation, the secondary circuits are simply tapped into the primary circuit where they are located in the building. In this type of system, the series primary circuit should be well insulated to minimize extraneous heat gains to the building.

Each secondary circuit connects to the primary circuit using a set of **closely spaced tees**. These tees provide the hydraulic separation between each secondary circuit and the primary circuit as discussed in Section 11.7. Thus, each circulator in the system is hydraulically separated from the others. Because of this, the pressure differential developed by each circulator only drives flow through its associated circuit and does not "assist" in driving flow through any of the other circuits. Thus, from the standpoint of hydraulics, one can visualize the primary circuit, and each secondary circuit, as "stand-alone" entities, not influenced by the other surrounding circuits. This is illustrated in Figure 11-57.

The degree of hydraulic separation achieved using closely spaced tees, as with other forms of hydraulic separation, is not perfect. However, very acceptable hydraulic separation can be achieved by adhering

11.13 Primary/Secondary Systems

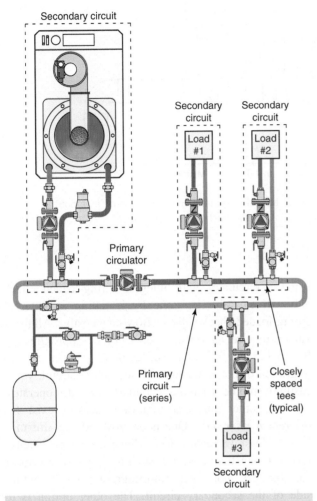

Figure 11-56 | A P/S piping systems using a series primary circuit.

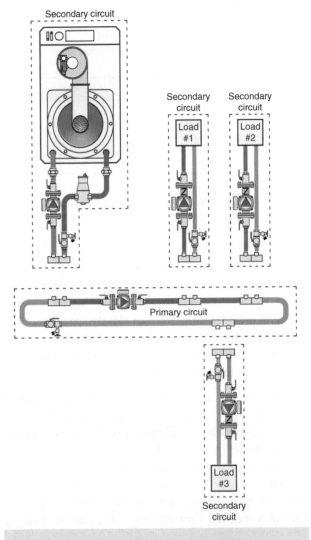

Figure 11-57 | Because of the hydraulic separation provided by the closely spaced tees, each circuit can be hydraulically analyzed as if it were a completely separate entity.

to the following guidelines, which are illustrated in Figure 11-58:

- The distance between the centers of the closely spaced tees should not exceed four times the nominal diameter of the primary circuit piping.
- Allow a minimum of eight primary pipe diameters of straight pipe upstream up the first tee.
- Allow a minimum of at least four primary pipe diameters of straight pipe downstream of the second tee.

The minimal pressure drop associated with a pair of closely spaced tees can also be accomplished using a special P/S fittings as shown in Figure 11-59a. Notice that the side ports of this fitting are very close together. They have been created using a **T-Drill® tool**. Some installers also use T-Drills® to install two closely spaced copper tubes directly into the tube wall of the primary circuit as shown in Figure 11-59b.

Preventing Thermal Migration In P/S Systems

All P/S systems require protection against the tendency for hot water to migrate into and through inactive secondary circuits (e.g., circuits in which the secondary circulator is not operating).

Two factors cause hot water to migrate into and through secondary circuits:

1. The tendency of lower density hot water to rise into the supply side of a piping circuit installed above a hot water source, while higher density cool water descends through the return side of the same circuit.

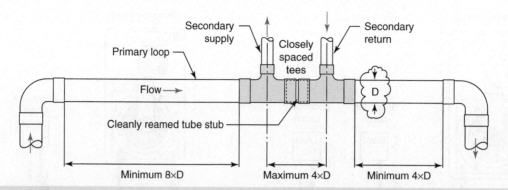

Figure 11-58 | Installation details for closely spaced tees in a P/S system. Note the required lengths of straight pipe both upstream and downstream of tees. Note: Tees may be mounted in vertical or horizontal pipes. Tees may also face in different directions.

2. The fact that the hydraulic separation provided by the closely spaced tees is not perfect allows a very slight pressure drop between closely spaced tees. This pressure drop creates a very slight tendency to induce flow in the secondary circuit whenever there is flow in the primary circuit.

The supply and return side of every secondary circuit in a P/S system must include detailing to prevent heat migration when its circulator is off. Various options for preventing heat migration into inactive secondary circuits are shown in Figure 11-60.

The supply side of each secondary circuit needs to be equipped with a valve that can prevent buoyancy-induced forward flow. This can be done by using a circulator with integral spring-loaded check valve. It can also be done by installing an external spring-loaded check valve or flow-check valve. All these valves require an opening pressure of approximately 0.3 to 0.5 psi, which is sufficient to prevent upward migration of hot water into the secondary circuit, yet is easily overcome when the secondary circulator operates.

Two other anti-heat migration options exist for the *return riser only*. One is an **underslung thermal trap** at least 18 inches deep. Since hot water "wants" to rise rather than descend, such a piping arrangement "discourages" hot water from migrating into the return side of the secondary circuit. The other option is to install a swing-check on the return side of the secondary circuit. *It is important to note that neither the*

Figure 11-59 | (a) Example of a P/S fitting. *Courtesy of John Siegenthaler.* (b) Use of a T-Drill® to create a P/S interface.

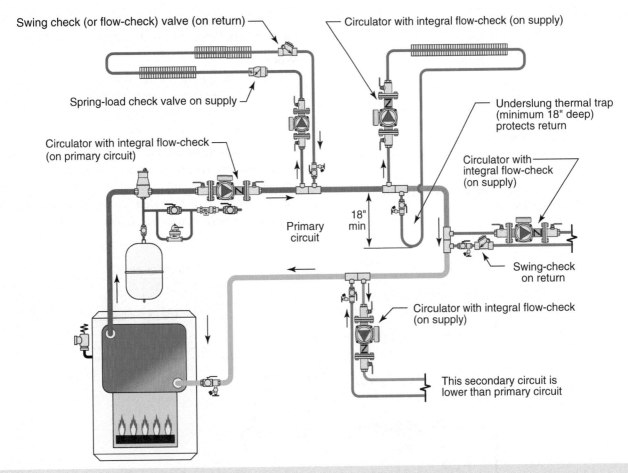

Figure 11-60 | Options for preventing thermal migration into inactive secondary circuits.

underslung thermal trap nor the swing-check valve can prevent hot water from rising into the supply side of the secondary circuit.

It is often desirable to locate a primary circuit high within the mechanical room. This allows the secondary circuits to drop down away from the primary circuit, even if they eventually must turn and go up in the building. A thermal trap is automatically created in the return side of the secondary circuit whenever it drops at least 18 inches below the primary circuit. The supply side must still be protected with either a spring-loaded check valve or flow-check valve.

Series Primary Circuits

The type of primary circuit shown in Figure 11-56 is more specifically called a *series* **primary circuit**. This word *series* comes from the fact that the secondary circuits are arranged in a sequence around the primary circuit. This arrangement causes a drop in temperature of the fluid moving through the primary circuit whenever it passes an operating secondary circuit.

This temperature drop is not necessarily a problem, provided it is recognized and accommodated. For example, a multiload system may include both high-temperature and low-temperature heat emitters. If these will be supplied by a series P/S system, the secondary circuit serving the higher temperature heat emitters would be the first supplied by the primary circuit (e.g., closest to the heat source). The lower temperature heat emitters would be supplied by a secondary circuit connected farther along the primary circuit. If more than two supply temperatures are required, the secondary circuits supplying these loads would be arranged along the primary circuit in order of decreasing supply water temperature. This technique may allow the overall temperature drop along the primary circuit to be greater than the common 20 °F often assumed. The larger the temperature drop, the lower the flow rate, and the lower the circulator input power requirement.

The drop in water temperature as the primary circuit flow passes an operating secondary circuit can be calculated using Equation 11.11.

Equation 11.11:

$$\Delta T_p = \frac{Q_s}{500 f_p}$$

Where:
- ΔT_p = temperature drop in the primary circuit as it passes an operating secondary circuit (°F)
- Q_s = rate of heat output to the secondary circuit (Btu/h)
- f_p = flow rate in the primary circuit (gpm)
- 500 = a constant for water

When a fluid other than water is used, replace the factor 500 in Equation 11.11 with $(D \times c \times 8.01)$

Where:
- D = the fluid's density at the average system temperature (lb/ft³)
- c = the fluid's specific hear that at the average system temperature (Btu/lb/°F)

To determine the inlet temperature to a given secondary circuit, subtract the temperature drop(s) created by any upstream operating secondary circuit(s) from the supply temperature provided by the heat source. This procedure is demonstrated in Example 11.5.

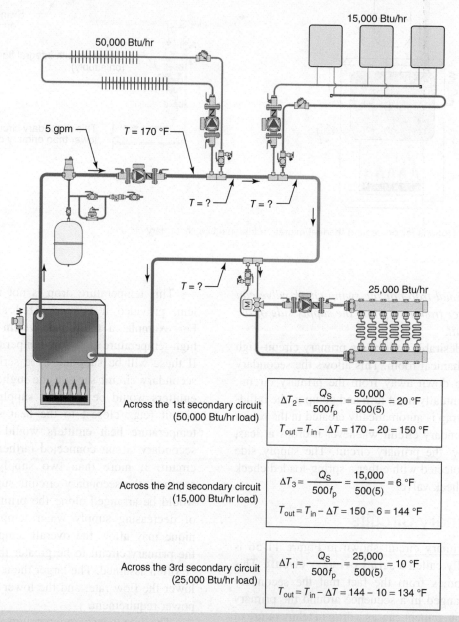

Across the 1st secondary circuit (50,000 Btu/hr load)

$$\Delta T_2 = \frac{Q_S}{500 f_p} = \frac{50,000}{500(5)} = 20 \text{ °F}$$

$T_{out} = T_{in} - \Delta T = 170 - 20 = 150 \text{ °F}$

Across the 2nd secondary circuit (15,000 Btu/hr load)

$$\Delta T_3 = \frac{Q_S}{500 f_p} = \frac{15,000}{500(5)} = 6 \text{ °F}$$

$T_{out} = T_{in} - \Delta T = 150 - 6 = 144 \text{ °F}$

Across the 3rd secondary circuit (25,000 Btu/hr load)

$$\Delta T_1 = \frac{Q_S}{500 f_p} = \frac{25,000}{500(5)} = 10 \text{ °F}$$

$T_{out} = T_{in} - \Delta T = 144 - 10 = 134 \text{ °F}$

Figure 11-61 Determining the temperature drop across each operating secondary circuit.

Example 11.5

Determine the water temperatures in the primary circuit between the secondary circuits as shown in Figure 11-61.

Solution:

Use Equation 11.11 to determine the temperature drop across each load, and then subtract the drop from the inlet temperature to get the outlet temperature.

$$\Delta T_1 = \frac{Q_s}{500 f_p} = \frac{500}{500(5)} = 20 \text{ °F}$$

$$T_{out} = T_{in} - \Delta T = 170 - 20 = 150 \text{ °F}$$

$$\Delta T_2 = \frac{Q_s}{500 f_p} = \frac{1{,}500}{500(5)} = 6 \text{ °F}$$

$$T_{out} = T_{in} - \Delta T = 150 - 6 = 144 \text{ °F}$$

$$\Delta T_3 = \frac{Q_s}{500 f_p} = \frac{2{,}500}{500(5)} = 10 \text{ °F}$$

$$T_{out} = T_{in} - \Delta T = 144 - 10 = 134 \text{ °F}$$

Discussion:

When a conventional boiler is used as the heat source, the designer should verify that the water temperature returning from the primary circuit under design load conditions is high enough to prevent sustained flue gas condensation. This is especially important when the primary circuit operates with a large temperature drop.

In some respects, a series P/S system is similar to a one-pipe diverter tee system. Both use a single circuit to transport heated water to several loads. Both can be configured to control each branch or secondary circuit as a separate zone. The difference is that a secondary circulator, rather than a diverter tee, is used to create the differential pressure that induces flow through the heat emitters in a P/S system. One advantage of using a secondary circulator is that it can produce far higher flow rates in larger secondary circuits in comparison with those attainable using diverter tees. A secondary circulator also uses electrical energy only when flow is needed in its circuit. A diverter tee, by contrast, dissipates mechanical energy (e.g., head) through its orifice even when its branch piping path is closed off.

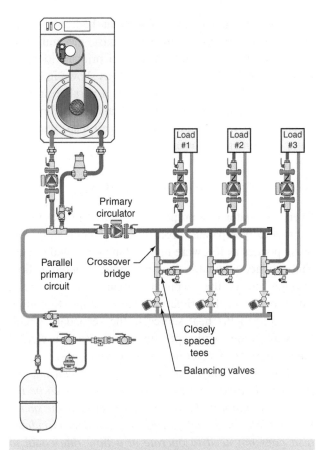

Figure 11-62 | Example of a parallel primary circuit. Each secondary circuit operates at the same supply temperature.

Parallel Primary Circuits

When several load circuits have approximately the same supply temperature requirement, and they are going to be supplied by a common primary circuit, the latter should be a **parallel primary circuit** as shown in Figure 11-62.

In a parallel P/S system, each secondary circuit is supplied from a pair of closely spaced tees that connect to a **crossover bridge**. The parallel arrangement of the crossover bridges assures the same water temperature is delivered to each secondary circuit. The closely spaced tees ensure there will be hydraulic separation between simultaneously operating circulators.

Each crossover bridge should be equipped with a balancing valve so flow rates can be proportioned to the load served by each bridge. Equation 11.3 can be used to estimate the flow needed in each crossover bridge.

Parallel P/S systems were originally developed for chilled water cooling systems in which all air handlers need to operate within a relatively narrow range of water temperature. They work equally well for heating applications, especially in systems with low supply water temperatures.

However, given the evolution in hydronics technology, and the options that now exist, the author no long recommends the use of parallel primary loops. A properly applied hydraulic separator, as shown in Figure 11-18 provides the same benefits as a parallel primary loop (e.g., hydraulic separation of all circulators, and equal supply water temperature to each load) with less complication and at lower cost.

Sizing The Primary Circulator

Every circulator in a properly configured P/S system is hydraulically separated from the other circulators, and thus operates as if it were installed in an isolated circuit. The primary circulator does not assist in moving flow through any of the secondary circuits, or vice versa. The function of the primary circuit is simply to convey the output of the heat source to the secondary circuit "pickup" points, while operating at or close to a selected temperature drop under design load conditions.

Contrary to myths that exist in the industry, the primary circulator does not necessarily have to be the largest circulator in the system. It may even be the smallest circulator in some systems! The primary circulator also does not need to operate at a flow rate equal to or greater than the total flow rate of all the secondary circuits that can operate simultaneously.

The flow rate necessary to deliver the full output of the heat source using a specific temperature drop around the primary circuit can be found using Equation 11.12:

Equation 11.12:

$$f_p = \frac{Q_{hs}}{500(\Delta T)}$$

Where:
 f_p = required flow rate in the primary circuit (gpm)
 Q_{hs} = heat output rate of heat source (Btu/h)
 ΔT = intended temperature drop of the primary circuit (°F)
 500 = a constant for water

When a fluid other than water is used, replace the factor 500 in Equation 11.12 with $(D \times c \times 8.01)$.

Where:
 D = the fluid's density at the average system temperature (lb/ft^3)
 C = the fluid's specific hear at the average system temperature (Btu/lb/°F)

Example 11.6

A primary circuit is connected to a boiler having an output of 250,000 Btu/h. The circuit is intended to operate with a temperature drop of 15 °F. What is the necessary primary circuit flow rate? How does the flow rate requirement change if the primary circuit is operated with a 30 °F temperature drop?

Solution:
Substituting values into Equation 11.12 (assuming a 15 °F temperature drop) yields:

$$f_p = \frac{Q_{hs}}{500(\Delta T)} = \frac{250,000}{500(15)} = 33.3 \text{ gpm}$$

The required flow rate in the primary circuit when the temperature drop is increased from 15 to 30 °F is:

$$f_p = \frac{Q_{hs}}{500(\Delta T)} = \frac{250,000}{500(30)} = 16.7 \text{ gpm}$$

Discussion:

Doubling the temperature drop of the primary circuit cuts the required flow rate in half. Not only does this reduce the required pipe size, it also reduces the size of the circulator. The smaller circulator will operate on lower wattage, and thus reduce the operating cost of the system. Considering that the primary circuit circulator must operate whenever any secondary circuit operates, such energy savings over the life of the system can be substantial.

Example 11.7

Assume that operating the primary circuit at the lower flow rate in Example 11.6 allows an 80-watt primary circuit circulator to be used instead of a 200-watt circulator. Also assume the primary circuit circulator operates 3000 hours per year in an area where the present cost of electricity is $0.12/kWhr, and the annual inflation rate of that cost will be 4% per year. Estimate the total savings in electrical energy cost over a period of 20 years.

Solution:

The operating cost of both circulators for the first year can be determined using Equation 11.13:

Equation 11.13:
$$O_1 = (P)(t)(c)$$

Where:
- O_1 = operating cost first year ($)
- P = operating power (kW)
- t = operating time (h)
- c = cost of electricity ($/kWh)

The first year operating cost for the 200-watt circulator is:

$$O_1 = (P)(t)(c) = (0.2\,\text{kW})\left(3000\,\frac{\text{hr}}{\text{y}}\right)\left(\frac{\$0.12}{\text{kWhr}}\right) = \$72$$

The first year operating cost for the 80-watt circulator is:

$$O_1 = (P)(t)(c) = (0.08\,\text{kW})\left(3000\,\frac{\text{hr}}{\text{y}}\right)\left(\frac{\$0.12}{\text{kWhr}}\right) = \$28.8$$

The difference in first year operating cost is $72 − $28.8 = $41.2.

Using Equation 7.6, this difference in first year operating cost can be extrapolated into an estimated total savings after 20 years:

$$\text{Saving}_{\text{total}} = C_{\text{year1}}\left(\frac{(1+i)^n - 1}{i}\right)$$

$$= \$41.20\left(\frac{(1+0.04)^{20} - 1}{0.04}\right)$$

$$= \$41.20(29.778) = \$1227$$

Discussion:

The savings over a 20-year period is several times the installed cost of either circulator. This significant cost savings is also associated with relatively small low-wattage circulators. The savings would be proportionally higher in larger systems using higher wattage circulators.

Once the primary circuit flow rate and pipe size are determined, the designer should estimate the head loss of the primary circuit based on the piping components it contains. A system head loss curve can be constructed using the methods of Chapter 6, Fluid Flow in Piping. The equivalent lengths of the closely spaced (straight through) tees should be included when constructing the system head loss curve for the primary circuit. However, do not include the resistance of any piping components from the secondary circuits.

A circulator can then be selected by overlaying pump curves for candidate circulators on the system curve to find an operating point that is reasonably close to the target flow rate. The Hydronic Circuit Simulator module in the Hydronics Design Studio software could also be used to model the performance of a proposed primary circuit design.

Finally, notice that a primary circuit circulator can be selected without reference to the secondary circuits. Likewise, each secondary circulator can be selected based solely on the flow and head loss requirements of the secondary circuit it serves. This is made possible by hydraulic separation.

A similar procedure can be used to select a circulator for a parallel primary circuit. The hydraulic resistance of the parallel crossover bridges and main pipe segments can be modeled using methods from Chapter 6, Fluid Flow in Piping, or by using the Hydronic Circuit Simulator module. Again, the flow rate and head loss requirements of the secondary circuits do not affect the selection process for the primary circuit circulator.

Indirect Water Heaters In P/S Systems

There are two basic ways to pipe an indirect water heater into a P/S system:

1. As a secondary circuit
2. As a parallel circuit to the primary circuit

If piped as a secondary circuit, the primary circulator and the circulator serving the indirect water heater must both operate whenever domestic water heating is required. A failure of either circulator will prevent domestic water heating. This also adds to the annual operating cost of the system. Furthermore, the entire primary circuit will be warmed whenever domestic water heating is required. This increases heat loss from piping. During summer months, when the indirect water heater may be the only load operating, such heat loss could add significantly to the building's cooling load.

Based on these considerations, it is best to connect indirect water heaters as a separate circuit as shown in Figure 11-63. When domestic water heating is active, it is only necessary to turn on the DHW circulator. There

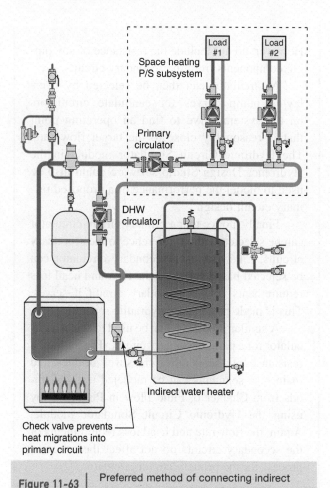

Figure 11-63 | Preferred method of connecting indirect water heaters in a series P/S system.

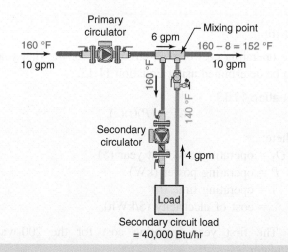

Figure 11-64 | Temperatures and flow rates for a given secondary circuit operating at a flow rate less than the primary circuit flow rate.

pressurized make-up water in the desired direction through the secondary circuit as air is blown out through the open drain port of the purging valve. When the flow exiting the purging valve is bubble free, the secondary circuit has been purged of bulk air, and the side port of the purging valve can be closed.

The primary circuit should also include valves for purging. The preferred approach is to first purge the primary circuit, and then purge each secondary circuit one at a time.

Flow Reversal In P/S Systems

One question that often arises regarding P/S systems is what happens if the flow rate in a secondary circuit is greater than the flow rate in the primary circuit?

From the standpoint of conservation of mass, the flow direction between the closely spaced tees must reverse whenever the secondary circuit flow rate is higher than the primary circuit flow rate.

Those new to P/S piping systems may view the primary circulator as having to produce enough flow to "feed" all the operating secondary circulators. Although the primary circuit does supply *heat* to each secondary circuit, it does not supply *flow* to each secondary circuit. Supplying heat to a secondary circuit is evidenced by a drop in temperature of the primary circuit flow as it passes a pair of closely spaced tees. This is entirely different from supplying flow to the secondary circuit.

To assess the effect of flow reversal through a set of closely spaced tees, we will start with a "typical" situation in which the primary circuit flow rate is higher than the secondary circuit flow rate. Figure 11-64 shows a subsystem in which the primary circuit serves a

---

is no need to operate the primary circulator during this mode. If domestic water heating is treated as a priority load, the control system would turn the primary circulator off (if it were already running) whenever the domestic water heating load was active. A check valve is installed in the primary circuit just upstream of where the return pipe from the indirect water heater tees in. This valve prevents reverse heat migration into the primary circuit when the domestic water heating mode is active.

Purging Provisions For P/S Systems

The closely spaced tees connecting a secondary circuit to the primary circuit are counterproductive when air is being purged from the system. A strong purging flow in the primary circuit will not induce a similar flow in a secondary circuit.

The solution is to install a means of independently purging each secondary circuit. One approach is to install a purging valve near the end of each secondary return riser. This detail has been shown on many of the piping schematics in this section. During purging, the inline ball of the purging valve is closed. This forces

secondary circuit having a design load of 40,000 Btu/h. Assume the primary circuit flow rate is 10.0 gpm and that the water arriving at the upstream tee is at 160 °F. Also assume that the circulator/piping design used for the secondary circuit allows a flow rate of 4.0 gpm to develop in that secondary circuit.

Since the secondary circuit load dissipates 40,000 Btu/h. from the water, the associated temperature drop in the secondary circuit can be calculated using Equation 11.14:

Equation 11.14:

$$\Delta T_s = \frac{Q_s}{500 \times f_s}$$

Where:
ΔT_s = temperature drop in secondary circuit (°F)
Q_s = heat dissipated from secondary circuit (Btu/h)
f_s = flow rate in secondary circuit (gpm)
500 = a constant for water

When a fluid other than water is used, replace the factor 500 with $(D \times c \times 8.01)$

Where:
D = the fluidchar/0x2019s density at the average system temperature (lb/ft^3)
c = the fluidchar/0x2019s specific heat at the average system temperature (Btu/lb/°F)

The temperature drop around this secondary circuit is:

$$\Delta T_s = \frac{Q_s}{500 \times f_s} = \frac{40{,}000}{500 \times 4} = 20 \text{ °F}$$

The temperature returning from the load at design conditions is $160 - 20 = 140$ °F.

The temperature drop along the primary circuit can also be determined using Equation 11.11:

$$\Delta T_p = \frac{Q_s}{500 \times f_p} = \frac{40{,}000}{500 \times 10} = 8 \text{ °F}$$

The water temperature leaving the downstream tee can be determined using Equation 9.15, which is based on conservation of thermal energy at a mixing point:

$$T_{mix} = \frac{(T_{hot} \times f_{hot}) + (T_{cool} \times f_{cool})}{f_{hot} \times f_{cool}}$$

In this equation, f_{hot} and f_{cool} are the flow rates of the two streams being mixed, and T_{hot} and T_{cool} are the temperatures of these streams as they enter the mixing point. For this situation being analyzed, the mixing point is the downstream tee. There is a bypass flow from the upstream to the downstream tee of 6.0 gpm at 160 °F. So $T_{hot} = 160$°F and $f_{hot} = 6.0$ gpm. T_{cool} and f_{cool} are the temperature and flow rate of the secondary circuit return

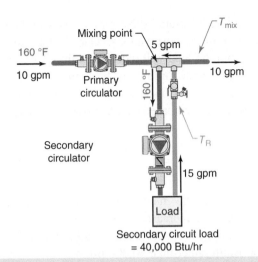

Figure 11-65 Flow reversal between closely spaced tees because the flow rate through the secondary circuit being greater than flow through the primary circuit.

($T_{cool} = 140$°F and $f_{cool} = 4.0$ gpm). Substituting these numbers into Equation 9.15 yields:

$$T_{mix} = \frac{(160 \times 6) + (140 \times 4)}{6 + 4} = 152 \text{ °F}$$

Next, consider this subassembly with one change: A larger circulator has been installed in the secondary circuit, and the secondary flow rate has increased from 4.0 to 15.0 gpm as shown in Figure 11-65.

Notice that flow between the closely spaced tees has reversed. This reversal is the only possibility given the flow rates in the surround pipes.

The flow rate in the primary circuit, for all practical purposes, does not change. This is because the hydraulic separation provided by the closely spaced tees prevents the primary circuit from "feeling" any significant change in its hydraulic resistance.

The flow reversal between the tees creates a mixing point, which is now at the tee on the left. This mixing point reduces the water temperature supplied to the secondary circuit. The extent of this reduction can again be assessed using the mixed flow equation at the upstream tee.

$$T_{mix} = \frac{(160 \times 10) + (T_R \times 5)}{10 + 5}$$

Where:
T_{mix} = supply temperature to the secondary circuit (°F)
T_R = return temperature from the secondary circuit (°F)

The return temperature from the secondary circuit is not yet known because it is dependent on heat transfer from the heat emitters in that circuit. We can, however, make two more observations. One is that the heat output from the load will increase slightly based on the increased flow rate through the heat emitters. This effect was discussed in Chapters 8 and 9. For finned-tube baseboards, the increased heat transfer can be estimated as follows:

$$\left(\frac{f_{higher}}{f_{original}}\right)^{0.04} = \left(\frac{15}{4}\right)^{0.04} = 1.054 = 5.4\%$$

Assuming the heat emitters were finned-tube baseboards, the increased heat transfer due to higher flow rate implies the secondary circuit can now dissipate 42,160 Btu/h rather than 40,000 Btu/h.

The temperature drop across the secondary circuit is now:

$$\Delta T_{secondary} = \frac{42,160}{500 \times 15} = 5.62 \,°F$$

Hence, the return temperature from the secondary circuit (T_R) is:

$$T_R = T_{mix} - 5.62$$

This can be combined with mixed flow equation and solved for the mixed supply temperature:

$$T_{mix} = \frac{(160 \times 10) + ([T_{mix} - 5.62] \times 5)}{10 + 5}$$

The result is $T_{mix} = 157.2\,°F$. Hence, we have estimated that the supply to the secondary circuit is about 2.8 °F lower due to mixing at the upstream tee.

Knowing this, we further recognize that the lower supply temperature also lowers heat output from the secondary circuit. How much? Here is an estimate, assuming that the heat emitters are finned-tube baseboard and the room air temperature is 70 °F.

$$\left(\frac{T_{mix} - 70}{160 - 70}\right)^{1.42} = \left(\frac{157.2 - 70}{160 - 70}\right)^{1.42} = 0.956$$

$$= 4.6\% \text{ reduction}$$

Interestingly, the increased flow rate (by itself) would increase heat output about 5.4%, while the decreased supply temperature (by itself) would lower heat output about 4.6%. One effect essentially offsets for the other, and any net change in heat output is very small.

Figure 11-66 shows the resulting temperatures and flow rates for the flow reversal situation.

In summary: It is possible to have flow reversal between the closely spaced tees that link a secondary circuit to the primary circuit. This results in mixing in the upstream tee that lowers the supply temperature to the secondary circuit. However, the combined effect of the lowered supply temperature, and the increased heat transfer

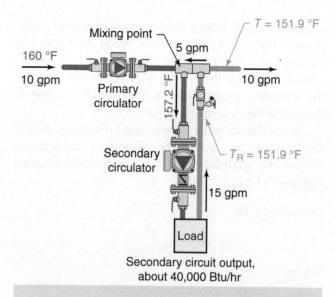

Figure 11-66 The stable flow rates and temperatures resulting from a 15.0 gpm flow rate in secondary circuit. Rate of heat delivery is essentially unchanged by operating the secondary circuit at 15.0 gpm versus the original 4.0 gpm.

from the heat emitters due to higher flow rate, is essentially a wash. The only significant difference is a needless increase in both installation and operating cost for the larger circulator needed to increase flow in the secondary circuit.

11.14 Distribution Efficiency

There is virtually no limit to the ways in which hydronic systems can be configured. However, it is reasonable to assume that some design strategies could use less electrical energy to operate the distribution system compared to other designs. This was demonstrated in Section 11.12 by comparing the circulator wattage needed by a home-run distribution system to that required by a single series circuit system of equivalent heat delivery capacity.

The concept of **distribution efficiency** provides a means by which the electrical energy consumption of a distribution system, per unit of heat delivery capacity, can be calculated and compared to that of another distribution system. Systems with higher values of distribution efficiency will use less electrical energy to convey a given rate of heat delivery to a building.

Distribution efficiency, which was first described in Chapter 1, Fundamental Concepts, is defined by Equation 11.15:

Equation 11.15:

$$e_d = \frac{Q_d}{p_d}$$

Where:

e_d = distribution efficiency (Btu/h/watts)
Q_d = total heat delivered by the distribution system (Btu/h)
P_d = electrical power required to operate the distribution system (watts)

Example 11.8

A system contains four 85-watt circulators, which are all operating at design load conditions, and collectively deliver 120,000 Btu/h of heat to the building. What is the distribution efficiency of this system?

Solution:

Substituting in the given performance information yields:

$$e_d = \frac{Q_d}{P_d} = \frac{120{,}000 \text{ Btu/hr}}{4 \times 85 \text{ W}} = 353 \frac{\text{Btu/hr}}{\text{W}}$$

Discussion:

Although the units for the results could be further manipulated to yield a unitless number, it is more intuitive to leave them as shown. This allows distribution efficiency to be interpreted as the number of Btu/h of heat delivery *per watt* of electrical input power to operate the distribution system.

The value of distribution efficiency from Example 11.8, by itself, is not very useful. To judge the performance of this distribution system, it needs to be compared to another distribution system.

Example 11.9

Consider a forced-air delivery system associated with a water-to-air ground source heat pump. The heat pump delivers 4 tons (e.g., 48,000 Btu/h) of heating capacity. The unit's blower draws 800 watts while operating. Determine this system's distribution efficiency.

Solution:

$$e_d = \frac{Q_d}{P_d} = \frac{48{,}000 \text{ Btu/hr}}{800 \text{ W}} = 60 \frac{\text{Btu/hr}}{\text{W}}$$

Discussion:

Compare the distribution efficiency of the hydronic system in Example 11.8 with that of the forced-air system of Example 11.9. The hydronic system delivers heat to the load using only about 17% of the electrical energy required by the forced-air system. Another way of comparing these results is that the hydronic distribution system delivers about 5.9 times more thermal energy, per watt of electrical input power, compared to the forced-air distribution system. These examples also demonstrate that distribution efficiency can be used to compare the heat conveyance ability of hydronic and forced-air distribution systems on a common basis.

Example 11.10

A hydronic system in a large house has 40 circulators (yes, it's an actual installation). The circulators are of different sizes, but assume an average power requirement is 90 watts per circulator. The building has two boilers, each rated at 200,000 Btu/h. Determine the distribution efficiency of this system assuming all 40 circulators are operating at design load conditions.

Solution:

$$e_d = \frac{Q_d}{P_d} = \frac{40{,}000 \text{ Btu/hr}}{40 \times 90 \text{ W}} = 11.1 \frac{\text{Btu/hr}}{\text{W}}$$

Discussion:

This demonstrates that it is possible to create hydronic distribution systems with relatively low distribution efficiency. In this case, the overuse of circulators for both zoning and P/S piping was responsible for the low value of distribution efficiency.

Example 11.11

The circulator shown in Figure 11-67 provides flow to a home-run distribution system that supplies 62,500 Btu/h to a 2,500 ft² house at design load conditions. Its ECM operates on 22 watts at this condition. What is the distribution efficiency of this system?

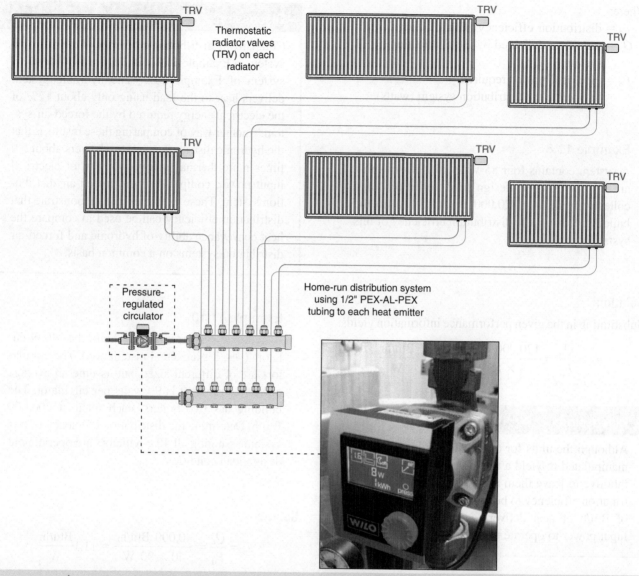

Figure 11-67 | Small variable-speed, pressure-regulated circulator with 22-watt peak power demand. *Courtesy of John Siegenthaler.*

Solution:

$$e_d = \frac{Q_d}{P_d} = \frac{62,500 \text{ Btu/hr}}{22 \text{ W}} = 2841 \frac{\text{Btu/hr}}{\text{W}}$$

Discussion:

This example demonstrates what is possible when a modern circulator is combined with a simple, yet state-of-the-art distribution system. To put it in perspective, this system delivers heat using only about 2% of the electrical energy required by the forced-air delivery system in Example 11.9, and only about 12% of the electrical energy required by the hydronic delivery system described in Example 11.8.

Designers should use the concept of distribution efficiency when comparing alternative designs for hydronic distribution systems. Those who market hydronic heating should also use the concept to demonstrate the major advantage of using properly designed hydronic delivery systems versus either forced-air delivery, or a "wall-full-of-circulators" approach to hydronic heating. Unfortunately, the operating cost of a distribution system is often neglected by those who focus solely on the cost of fuel used by the heat source. Doing so overlooks a decisive advantage of modern hydronic heating.

> Finally, one might argue that since the electrical energy consumed by a distribution equipment ultimately becomes heat within the building, it should be added to the amount of heat delivered by the hydronic heat source when calculating distribution efficiency. For example, in Example 11.9, the 800 watts used by the blower converts to 800 × 3.413 = 2730 Btu/h of additional heat input. The "modified" distribution efficiency would therefore become:
>
> $$e_d = \frac{Q_d}{P_d} = \frac{[48,000 + 2730]\text{Btu/hr}}{800 \text{ W}} = 63.4 \frac{\text{Btu/hr}}{\text{W}}$$
>
> Although the "modified" distribution efficiency is slightly higher, the change is relatively insignificant in comparison with the distribution efficiencies produced by state-of-the-art hydronic distribution systems. The author would also emphasize that the "goal" of any heating distribution system is not to *produce* heat, but rather to *deliver* heat. Systems that use higher wattage hardware in meeting this goal should not be rewarded through the use of skewed definitions that do not reflect the fundamental purpose of a distribution system.

11.15 Hybrid Distribution Systems

The diverse loads served by many modern residential and light commercial systems often dictate that the distribution system contains elements from two or more of the basic distribution systems discussed in this chapter. This section presents examples of such systems. Readers are encouraged to study the details for techniques such as hydraulic separation, buffering of zoned distribution systems, mixing methods, and different subassemblies that have been assembled to meet specific project requirements.

An Extensive System

Figure 11-68 is an example of a heating system for a large custom home. All heat is provided by a single oil-fired sectional cast-iron boiler. The basement floor slab is heated with embedded tubing. The main floor uses above-floor tube-and-plate radiant floor panels. The second floor uses a combination of above-floor tube-and-plate panels in bathroom areas and panel radiators supplied by a home-run distribution system in other areas. The system also supplies heat to garage floor circuits that operate with antifreeze. Finally, this system also supplies domestic hot water using an indirect water heater.

This system was designed for very specific customer requirements and preferences. At first glance, it may appear intimidating in its overall complexity. However, the concept of hydraulic separation, as well as its intended purposes, allows the overall schematic to be thought of as several independent subsystems as illustrated in Figure 11-69.

The combination of the low flow resistance cast-iron sectional boiler and generously sized/short header piping provides hydraulic separation of all circulators connected to the headers. One of these circulators (P1) is a relatively powerful fixed-speed circulator serving the indirect water heater. The remaining circulators are operated at variable speeds as either injection mixing pumps or pressure-regulated distribution circulators. Thus, hydraulic separation is essential in avoiding interference between circulators. It is also worth noting that hydraulic separation is provided without use of P/S piping.

Figure 11-70 shows the electrical schematic for the system of Figure 11-68. Again, divide the overall drawing into subsystems to better understand its functioning. In this case, all calls for heat come to the multizone relay center, which in turn sends 120 VAC power to the necessary circulators and injection mixing controllers. A boiler reset controller manages the boiler's outlet temperature during both space heating and domestic water heating modes.

A Simple But Sophisticated System

The system shown in Figure 11-71a uses a low-ambient air-to-water heat pump with a variable-speed compressor to supply space heating, domestic water heating, and chilled water cooling. The heat pump is a split system. The outdoor unit is connected to the indoor unit by two refrigerant lines and a communications cable. No water is present outside the building, and therefore no antifreeze is used in the system.

This system uses a **reverse indirect tank** to provide thermal mass for buffering its extensively zoned heating distribution system. This tank also enables domestic water heating. The heat pump maintains the water in this tank at an elevated temperature. The tank is supplied from its manufacturer with several internal copper or stainless steel coils. Domestic water passes through these coils and absorbs heat from the surrounding water.

The extent of domestic water heating depends on the temperature of the water in the tank shell. If the heat pump maintains this temperature between 110 and 120 °F the domestic water can be heated to a delivery

626 Chapter 11 Distribution Piping Systems

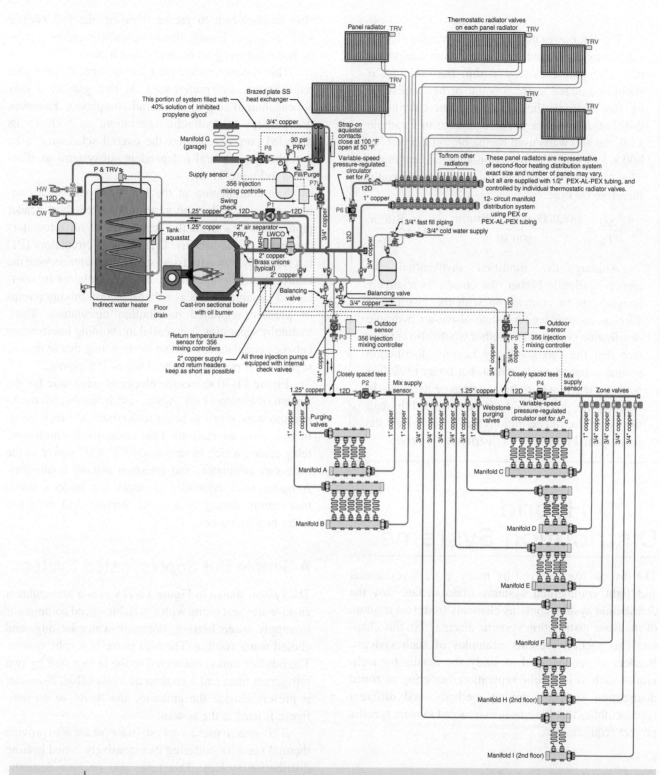

Figure 11-68 | Piping schematic for a multiload/multitemperature hydronic system.

temperature in the range of 105 °F in a single pass through the internal coils. However, if the tank temperature is based on outdoor reset control, the extent of domestic water heating diminishes as outdoor temperatures increase. Outdoor reset control of the tank temperature will significantly improve the seasonal COP of the heat pump. However, it also requires the temperature of the preheated domestic hot water to be "topped off" by another heat source. One option for providing this supplemental heat is an electric tankless water heater, piped as shown in Figure 11-71b.

This system uses several types of heat emitters including panel radiators, a towel warming radiator, and two small areas of radiant floor heating. All three panel

11.15 Hybrid Distribution Systems

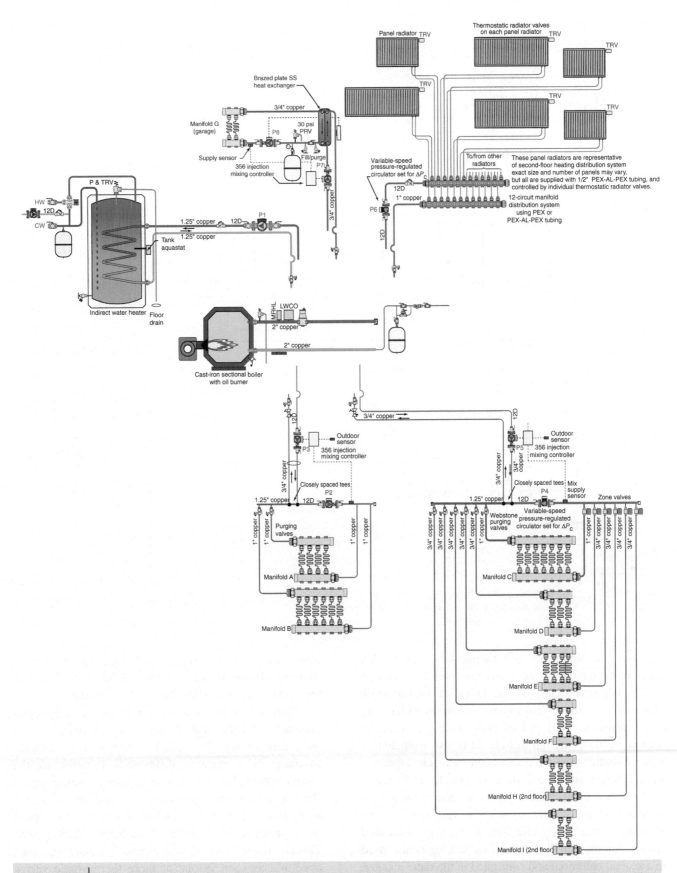

Figure 11-69 | Separating the schematic of Figure 11-68 into independent subsystems based on function and hydraulic separation.

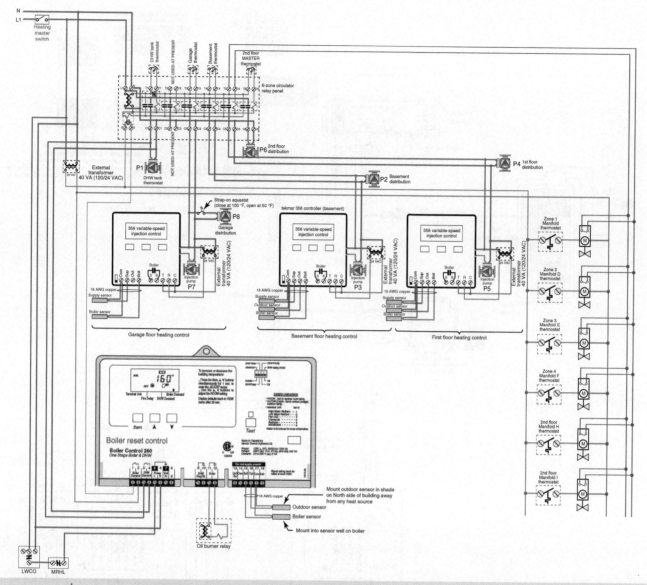

Figure 11-70 | Electrical schematic for the system shown in Figure 11-68.

radiators are equipped with thermostatic valves that sense room air temperature and regulate flow through their associated radiators to maintain a set comfort level. The towel warming radiator is piped in series with tubing that provides a small area of floor heating to a master bathroom. Flow through this circuit is also regulated by a thermostatic valve. That valve and its attached thermostatic actuator are located in an accessible cavity under the floor. A capillary tube connects the actuator to a wall-mounted setting dial. This valve also responds to room air temperature to regulate flow through its associated circuit. Another small area of floor heating is also used in a second bathroom and controlled in the same manner.

The panel radiators and the other heat delivery circuits are configured as a home-run distribution system using 1/2" PEX tubing. The manifold supplying these home-run circuits is equipped with three "spare" connections to allow other heat emitters to be easily added to the system, such as when the home is expanded.

During heating mode operation the variable-speed pressure-regulated distribution circulator (P1) remains on continuously. Its speed automatically changes to maintain a constant differential pressure as the thermostatic valves on the heat emitter circuits regulate flow. If all heat emitter circuits are closed, this circulator shifts to a "sleep" mode in which it runs at a low speed on approximately 9 watts of electrical input power. The circulator "wakes" into normal pressure-regulated operation as soon as one or more thermostatic valves allow flow to occur.

The heat pump and its associated circulator are turned on when the water temperature at sensor (STS) in

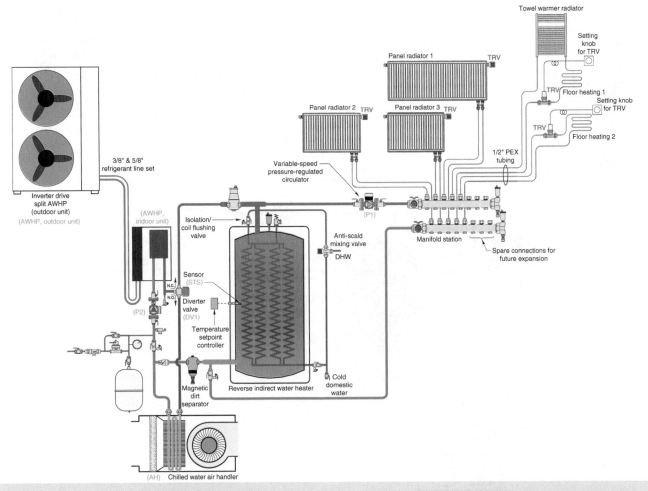

Figure 11-71a | Piping schematic for a multiload system for heating and cooling using a split-system air-to-water heat pump. Water temperature in tank is regulated by a temperature setpoint controller.

the tank drops to a predetermined minimum value (based on either setpoint or outdoor reset control).

The diverter valve (DV1) is powered on whenever the heat pump operates in heating mode. It directs flow leaving the heat pump to the header piping above the tank. If none of the heat emitters are active, all flow passes into the connection at the top of the tank. If some heat emitters are active, the flow stream divides at the upper tank header, some passing to the active heat emitter circuits and the rest into the tank. This "two-pipe" tank configuration allows the possibility of hot water passing directly from the (operating) heat pump to any active heat emitters.

During cooling mode, the diverter valve remains unpowered. Flow leaving the heat pump goes to the chilled water coil in the air handler. The heat pump's inverter drive compressor varies its speed to maintain a preset chilled water outlet temperature of 50 °F.

If the temperature of the buffer tank drops below a preset minimum value while the system is in cooling mode, the heat pump temporarily switches to heating mode to restore the tank temperature. This is necessary to ensure adequate domestic hot water during cooling mode operation. Given the warm outdoor temperatures present during the cooling season, the time required for the heat pump to restore the tank temperature is likely to be short, perhaps only 5 to 10 minutes. The lack of cooling during this short time is unlikely to be noticed.

This system uses a magnetic dirt separator to maintain the flow stream entering the heat pump as clean as possible. It also uses a high-efficiency air separator placed to receive the hottest water in the system—just downstream of the heat pump.

A spring-loaded check valve is used to prevent chilled water migration into the lower tank header. All piping and components in the chilled water are insulated and vapor sealed.

The domestic water coil assembly in the reverse indirect tank is equipped with isolation/flushing valves at its inlet and outlet connections. These valves allow the coil assembly to be isolated from the remainder of the domestic plumbing and flushed with a mild acid solution to remove any accumulated scale.

630 Chapter 11 Distribution Piping Systems

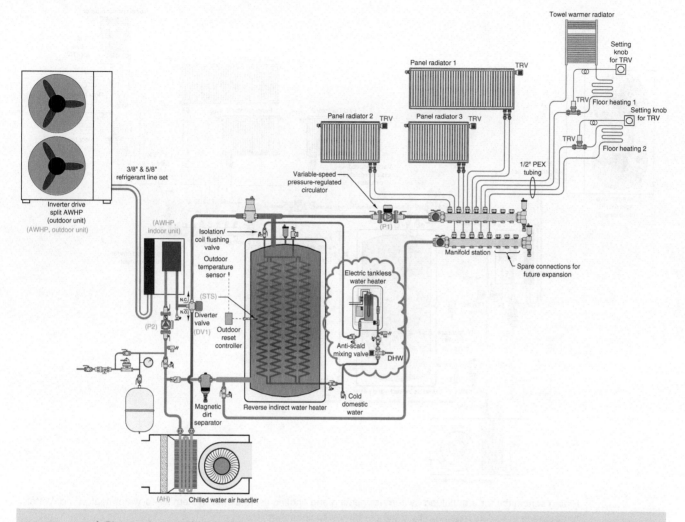

Figure 11-71b | Piping schematic for a multiload system for heating and cooling using a split-system air-to-water heat pump. Water temperature in tank is controlled based on outdoor reset control. An electric tankless water heater provides supplement heat for domestic hot water.

A thermostatic mixing valve is installed to ensure that the domestic hot water temperature supplied to the fixtures doesn't exceed 110 °F.

The control wiring for the system shown in Figure 11-71a is shown in Figure 11-72.

The electrical schematic in Figure 11-72 is structured as a ladder diagram. It uses three 3PDT relays (RH, RC, and R3) to achieve the necessary "hard wired" control logic for space heating, cooling, and domestic water heating priority during cooling operation. This type of control system documentation and the components involved were discussed in Chapter 9, Control Strategies, Components, and Systems.

Several independent circuits supply power to this schematic. A 120 VAC/15 amp circuit supplies power through a master switch (MS) for heating circulator (P1) and transformer (X1). This transformer reduces this voltage to 24 VAC to supply the cooling thermostat (TC), the temperature setpoint controller (T150), and the diverter valve (DV1). Separate 240 VAC circuits supply the air handler, the outdoor unit of the air-to-water heat pump, and the indoor unit of this heat pump.

A master switch controls power to all control circuits. If the master switch is off, the control system cannot operate the heat pump, circulators, air handler, or other electrical hardware in the system.

A heating/cooling mode switch (HCMS) determines if the system operates in heating mode, cooling mode, or remains off.

When the heating/cooling mode switch (HCMS) is set to heating, relay coil (RH) is powered on. Relay contact (RH-1) closes to supply 120 VAC to the variable-speed pressure-regulated circulator (P1). This circulator automatically adjusts speed and input power to maintain a preset differential pressure. As the thermostatic valves on the heat emitter circuits open, circulator (P1) automatically detects an "attempt" to reduce differential pressure. It responds by increasing

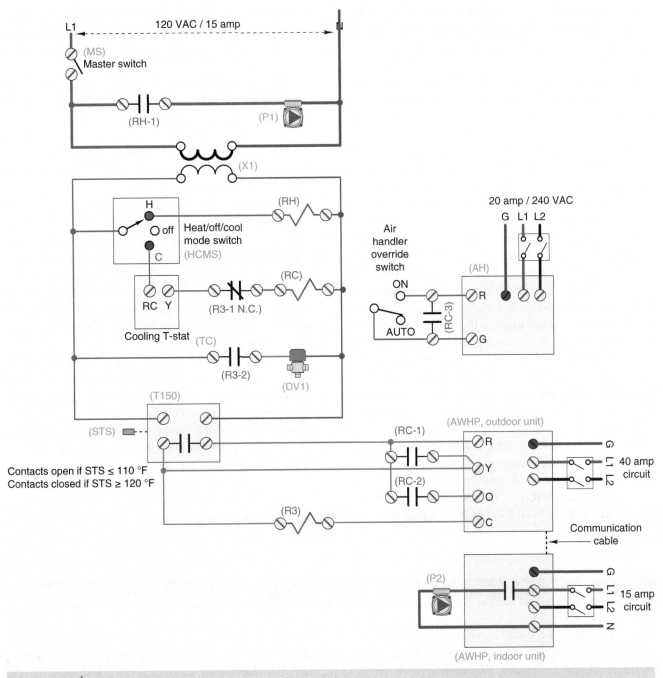

Figure 11-72 | Electrical schematic for the system shown in Figure 11-71a.

motor speed to restore the preset differential pressure, and vice versa.

When the heating/cooling mode switch (HCMS) is set to heating, 24 VAC is also sent to the temperature setpoint controller (T150). This controller monitors the temperature at sensor (STS) at the mid-height of the reverse indirect tank. When this temperature is 110 °F or lower, the contacts in the (T150) controller close. This allows 24 VAC from the heat pump's internal transformer to pass to the heat pump's (R) terminal to its (Y) terminal. The heat pump then turns on circulator (P2). After an internal flow sensor verifies adequate flow, the heat pump's compressor is also turned on. Relay coil (R3) is energized whenever the temperature setpoint controller (T150) calls for the heat pump to operate. Relay contact (R3-2) closes to pass 24 VAC from transformer (X1) to the diverter valve (DV1), which opens its normally closed port to allow flow from the heat pump to the tank header. The heat pump, circulator (P2), and diverter valve (DV1) continue to operate until the temperature at sensor (STS) in the tank reaches 120 °F, at which point they turn off. This operation maintains the temperature of the water in the reverse indirect water heater high enough to heat domestic water to a delivery temperature of 105 °F after passing through the internal coils.

When the mode switch is set to cool, 24 VAC from transformer (X1) is passed to the (RC) terminal of the cooling thermostat. When this thermostat calls for cooling, 24 VAC is passed to the thermostat's (Y) terminal. This 24 VAC signal continues through the normally closed relay contact (R3-1 N.C.) and energizes the coil of relay (RC). Relay contact (RC-1) closes to allow 24 VAC from the heat pump's internal transformer to reach the heat pump's (Y) terminal. This turns on the heat pump's circulator (P2). When flow through the heat pump is verified by an internal flow switch, the heat pump's compressor turns on. Relay contact (RC-2) passes 24 VAC from the heat pump's internal transformer to its (O) terminal. This energizes the heat pump's reversing valve for cooling mode operation. Relay contact (RC-3) closes to allow a low voltage circuit in the air handler (AH) to turn on the blower. In this mode, the heat pump continuously monitors the chilled water temperature leaving the indoor unit. The compressor speed varies as necessary to maintain this temperature close to 50 °F. This mode of operation continues until the cooling thermostat (T1) is satisfied.

Relay (R3) is used to temporarily interrupt cooling operation if the temperature setpoint controller (T150) determines that the tank requires heating to allow uninterrupted availability of domestic hot water. Relay contact (R3-1 N.C.) breaks the 24 VAC signal from the cooling thermostat to relay coil (RC) during this mode of operation. Relay contact (R3-2) closes to supply 24 VAC from transformer (X1) to energize the diverter valve (DV1).

The air handler (AH) is also wired to a manually operated switch that allows it to be turned on at times when cooling is not needed. This allows for whole house air circulation to distribute solar heat gains. The air handler (AH) continues to operate as long as this manual switch is in the ON setting, regardless of which mode the system is in.

In summary, this system is relatively simple, both in terms of piping and wiring. It provides multiple zones of heating and a single zone of cooling, both sourced from a modern variable-speed air-to-water heat pump. It takes advantage of a low-power, variable-speed pressure-regulated circulator to provide flow through a home-run distribution system. A single tank serves as the buffering mass for the zoned heating distribution system and as a means of providing domestic hot water.

A variation of the system, shown in Figure 11-71b, uses an outdoor reset controller, rather than a setpoint controller, to reduce the water temperature in the tank under partial load conditions. This variation would increases the seasonal COP of the heat pump, but it also requires a supplemental heater to boost the delivery temperature of domestic hot water. In Figure 11-71b, that supplemental heat is provided by a tankless electric water heater.

Chapter 15 will provide additional information on chilled water cooling systems.

Summary

The wide variety of distribution piping systems discussed in this chapter allows systems to be designed for the exact requirements of the buildings they serve.

Those new to hydronic design are advised to design their initial systems around one of the distribution systems topologies presented in this chapter. Then, and with care and experience, they may find it beneficial to merge the fundamental topologies of one distribution system type with those of another to create customized systems. The system shown in Figure 11-68 is an example. It combines the fundamental topology of a multiplezone system using circulators, with subsystems such as the home-run topology for the panel radiators, and multizone/zone valve topology for some of the radiant panel heat emitters. When combining these topologies, it is critical to verify that the final design includes details for preventing flow through inactive circuits, limiting thermal migration, pressure relief protection, proper expansion tank placement, differential pressure control for systems with valve-based zoning, and achieving hydraulic separation of multiple circulators.

Key Terms

- balancing valves
- bar (pressure)
- branch
- candidate circulator
- circuit separation
- closely spaced tees
- coalescing media
- common piping
- crossover bridge
- dead-heading
- differential pressure bypass valves (DPBV)
- differential pressure control
- differential pressure controller
- distribution efficiency
- diverter tees
- drainback freeze protection
- electric valve actuators
- electronically commutated motors
- home-run distribution system
- hydraulic equilibrium
- hydraulic separation
- hydraulic separator
- Hydronic Circuit Simulator
- Hydronics Design Studio
- interzone heat transfer
- iteration
- Kilo-Pascals
- low loss header
- master thermostat
- multifunction tank
- one-pipe systems
- parallel direct-return system
- parallel distribution systems
- parallel primary circuit
- parallel reverse-return piping
- piping topology
- primary circuit
- primary/secondary (P/S) piping
- primary/secondary system
- priority zone
- return main
- reverse indirect tank
- room-by-room zoning
- secondary circuits
- self-balancing
- Series Baseboard Simulator module
- series primary circuit
- single series circuit
- siphon
- spring-loaded check valve
- supply main
- T-Drill® tool
- target flow rate
- thermal equilibrium
- thermosiphoning
- threshold differential pressure
- topologies
- two-pipe system
- underslung thermal trap
- variable-frequency drive
- variable-speed pressure-regulated circulators
- zone

Questions and Exercises

1. A thermometer attached to the supply pipe of a hydronic distribution system shows the temperature of the water rising steadily even after the system has been operating for several minutes. What can you conclude about the heat output of the boiler relative to the heat transferred to the load?

2. When is hydraulic equilibrium established in a hydronic system? Are there certain types of hydronic systems where hydraulic equilibrium will not be established?

3. Will a hydronic heating system, in all cases, properly heat a building when it attains both thermal and hydraulic equilibrium? Explain your answer.

4. Discuss the feasibility of using a setback thermostat with a daily setback schedule in a well-insulated building equipped with a heated 6-inch concrete slab. When would this make sense? When would it not make sense?

5. What is the main advantage of a one-pipe distribution system relative to a single series circuit?

6. A three-zone distribution system is to supply a total load of 90,000 Btu/h with a design temperature drop of 15 °F. The design heating loads of the three zones are 25,000 Btu/h, 50,000 Btu/h, and 15,000 Btu/h. Determine the system design flow rate and the design flow rate in each of the zone circuits.

7. Describe why check valves are necessary in each zone circuit of a multicirculator zoned system.

8. What is the principal advantage of a reverse-return rather than a direct-return parallel distribution system?

9. Why are parallel distribution systems good for use with low-temperature heat emitters?

10. What is the purpose of a differential pressure bypass valve in parallel direct- or reverse-return systems? When would it typically operate?

11. What is the advantage of designing a series P/S system for a high-temperature drop in the primary circuit? What precaution must be observed if a conventional boiler is used?

12. What circumstances suggest the use of a parallel primary circuit rather than a series primary circuit?

13. Describe what happens to the flow rate through a particular zone circuit when additional zone circuits (containing zone valves) turn on. Assume the system

uses a fixed-speed circulator. What happens to the system flow rate as additional zone circuits open up? Justify your answer.

14. Describe why a fixed-speed circulator with a relatively "flat" pump curve is best suited for multizone systems using zone valves. What can happen if a circulator with a "steep" pump curve is used in such a system?

15. Why is it necessary to "correct" the hydraulic resistance of the common piping segment in a multizone/multicirculator system using Equation 11.5?

16. When the calculated flow rate through a particular parallel branch circuit is significantly lower than the initial design value, what are some things that can be done to increase it?

17. A diverter-tee arrangement operates as shown in Figure 11-73. Calculate the outlet temperature from the diverter tee.

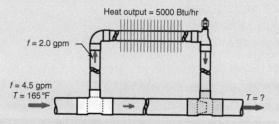

Figure 11-73 | Piping and operating conditions for exercise 17.

18. List three different hardware options for achieving hydraulic separation.

19. Describe three distinct advantages of a home-run distribution system in comparison with a single series circuit system.

20. Describe what happens when the flow rate in a secondary circuit is higher than the flow rate in the primary circuit.

21. What is the fundamental reason for placing tees as close as possible at a P/S circuit interface?

22. Why is it necessary to install lengths of straight tubing upstream and downstream of a pair of closely spaced tees in a P/S system?

23. Why is it necessary to use purge valves in each secondary circuit rather than a single purging arrangement in the primary circuit?

24. Can an underslung thermal trap protect the supply side of a secondary circuit from thermal migration?

25. Does the primary circulator have to produce a flow rate at least as great as the sum of the secondary circuit flow rates?

26. What are two ways of zoning individual heat emitters in a home-run distribution system?

27. Why is it preferable to route the piping for a parallel reverse-return system around the perimeter of a building?

28. True or false: The only way to achieve hydraulic separation is to install a hydraulic separator in the system. Explain.

29. Why are balancing valves necessary in each crossover bridge of a parallel primary circuit?

30. Why can't a swing-check valve prevent heat migration into the supply side of a secondary circuit?

31. Does a hydraulic separator create "perfect" hydraulic separation of multiple circulators? Explain.

32. When a variable-speed pressure-regulated circulator is used in a home-run distribution system, what operating mode should the circulator be set for?
    a. Constant differential pressure
    b. Proportional differential pressure
    Justify your answer.

33. A heating distribution system delivers 85,000 Btu/h at design load. Under these conditions, the distribution system is operating two circulators rated at 60 watts each, and four circulators rated at 80 watts each. What is the distribution efficiency of this system?

34. The system described in exercise 33 is modified to make it a P/S system. This adds one more circulator that requires 80 watts input power. What is the distribution efficiency of this system at design load?

35. A primary circulator operates for an estimated 2800 hours per year. When operating, it produces a flow rate of 10.5 gpm and a differential pressure of 4.0 psi. Under these conditions, the circulator's wire-to-water efficiency is 22%. Electrical energy at the site currently costs $0.13/kWhr and is expected to increase at a steady rate of 5% per year over the next 20 years.
    a. Determine the first year operating cost of this circulator.
    b. Determine the total operating cost of this circulator over the next 20 years under the assumptions given.

36. True or false: A differential pressure bypass valve should always be installed when a variable-speed pressure-regulated circulator is used in a two-pipe reverse-return distribution system.

Chapter 12

Expansion Tanks

Objectives

After studying this chapter, you should be able to:

- Explain the purpose of an expansion tank in a closed-loop hydronic system.
- Describe two different types of expansion tanks.
- Explain why standard expansion tanks become waterlogged.
- Estimate the volume of fluid in a hydronic system.
- Determine the proper location of the expansion tank within a system.
- Calculate the required expansion tank volume for a given system.
- Determine the required air pressurization of a diaphragm-type tank.
- Determine the required expansion tank volume in a low-temperature system.
- Make use of the Expansion Tank Sizer software module.

12.1 Introduction

All liquids used in hydronic heating systems expand when heated. This **thermal expansion** is an unavoidable and extremely powerful force of nature. Upon heating, each of the trillions of fluid molecules contained in the system becomes slightly larger. From a macroscopic perspective, one might think there is an increase in the amount of fluid in the system. This is not true. *The same molecules just take up more space when at a higher temperature.* The volume of the fluid has increased, but the total mass of the system's fluid has not changed.

For all practical purposes, liquids are **incompressible**. A given number of liquid molecules cannot be compacted or squeezed into a smaller volume without tremendous force. Any container completely filled with a liquid and sealed from the atmosphere will experience a rapid increase in pressure as the liquid is heated. If this pressure is allowed to build, the container will eventually burst, in some cases violently.

To prevent this from occurring, all hydronic heating systems must be equipped with a means of accommodating the volume increase of their fluid as it is heated. In systems that are open to the atmosphere, such as a non-pressurized thermal storage tank, the volume increase can be accommodated by extra space at the top of the tank. This allows the expanding fluid to "park" its extra volume as shown in Figure 12-1.

636 Chapter 12 Expansion Tanks

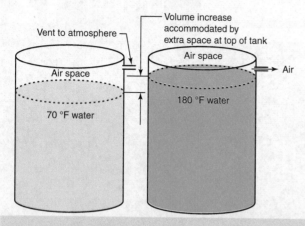

Figure 12-1 | Extra space at top of an open tank accommodates expansion of fluid.

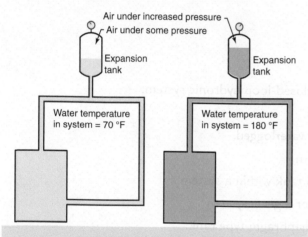

Figure 12-2 | An expansion tank provides the volume to accommodate fluid expansion in a closed-loop hydronic system.

In a more typical *closed-loop* hydronic system, the extra space is usually provided by a separate chamber called the **expansion tank**. This tank contains a volume of air that is compressed, somewhat like a spring, as the system's fluid expands against it. This concept is shown in Figure 12-2.

This chapter discusses the types of expansion tanks used for closed hydronic systems. It also gives methods for sizing, pressurizing, and locating the tank in the system. A software module called Expansion Tank Sizer in the Hydronics Design Studio software is also introduced as an aid for determining tank size and pressurization.

12.2 Standard Expansion Tanks

One type of expansion tank used in very early hydronic heating systems consisted of a simple open-top drum placed at the high point of the system. It served as a chamber into which water would rise when heated. In most systems, this tank was located in the building's attic.

This arrangement presented several issues. First, the open tank allowed fluid to escape from the system by evaporation. The lost water would have to be replaced with fresh water, which was often carried to the attic in buckets and poured into the open-top tank. The fresh water also contained dissolved oxygen, which caused corrosion of iron-containing components in the system.

Second, the height of the tank limited the pressure and thus the upper operating temperature of the system.

Finally, since a tank located in a cold attic was often some distance away from the heated parts of the system, it could freeze during cold weather. This could lead to a ruptured tank and a mess in the spaces below. Considering these limitations, such tanks are not used in modern hydronic systems.

The next advancement in expansion tank technology was the use of a closed tank located above the boiler as shown in Figure 12-3.

The air in this type of **standard expansion tank** is initially at atmospheric pressure. When the system is filled with fluid, the air is trapped inside the tank and partially compressed. The higher the system piping rises above the expansion tank, the more the entrapped air is compressed. As the water expands upon heating, additional fluid volume enters the tank, further compressing the air.

If the tank is properly sized, the air pressure in the tank should be about 5.0 psi lower than the system's pressure-relief valve rating when the system reaches its maximum operating temperature. The 5.0 psi safety

Figure 12-3 | A standard expansion tank supported from the ceiling of a mechanical room. This tank is equipped with sight glass for monitoring the water level and special anti-thermosiphon fitting at bottom. *Courtesy of John Siegenthaler.*

margin is to prevent the relief valve from leaking just below its rated opening condition. It also allows the relief valve to be located a few feet below the inlet of the expansion tank as is typical in many systems.

Standard expansion tanks were used on thousands of early hydronic systems. Many are still in use today. These tanks were often hung from the ceilings of mechanical rooms.

One inherent issue with standard expansion tanks is that the air and water they contain are in direct contact. As the water in the tank cools off, it can reabsorb some of the air back into solution. Slight thermosiphoning flow in the piping between the tank and the boiler can carry this cooler water and the dissolved air it contains back to the system. Upon reheating, the dissolved air will again come out of solution, but now in the system piping. Eventually, this air will be captured and expelled from the system through air vents. An automatic feed water valve will then admit small amounts of water to make up for the lost air. The net effect over many heat-up/cool-down cycles is that the air in the expansion tank is replaced by water. Eventually, the tank becomes **waterlogged**, which means the tank has become completely filled with water.

When waterlogging occurs, there is no longer a cushion of air for the system's water to expand against. This causes the system's pressure-relief valve to release small amounts of water each time the system heats up. The system's feed water valve replaces this water as soon as the system cools and its pressure decreases. This repeated sequence of events can move many gallons of fresh (oxygen-containing) water through a system during a single heating season. It greatly increases the chance of serious corrosion damage, especially in systems containing ferrous metals such as steel and cast iron.

Standard expansion tanks typically need to be drained and refilled with air at least once each year to prevent the problems associated with waterlogging. Special drain valves that allow air to enter the tank at the same time water is being drained are available for this purpose. These valves also isolate the tank from the rest of the system during this draining operation.

A specialized boiler fitting is often used with standard expansion tanks. This fitting, shown in Figure 12-4, allows air bubbles that accumulate at the top of the boiler to rise into the expansion tank. It contains a short dip tube that extends down below the top of the boiler sections to pick up hot water without air bubbles. This fitting helps return air that may have migrated into the system back to the expansion tank.

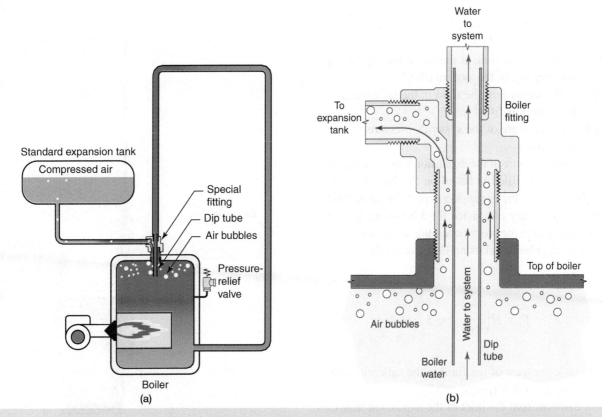

Figure 12-4 | (a) Installation of a standard expansion tank. (b) Special boiler fitting used to capture air bubbles in boiler and channels them to the standard expansion tank.

The pipe leading from the boiler fitting to a standard expansion tank should be sloped at least 1 inch in 5 ft. This allows the bubbles to rise toward the tank. This piping should also be at least 10 ft long to minimize heat migration toward the tank. Such heating can increase the pressure in the tank.

A specialized tank fitting is sometimes used to connect the riser pipe from the boiler to the tank. This fitting must be selected based on the diameter of the expansion tank. It helps minimize thermosiphoning flow between the tank water and the water in the system piping.

Sizing Standard Expansion Tanks

The size of a standard expansion tank can be calculated using Equation 12.1:

Equation 12.1:

$$V_t = \frac{V_s\left(\frac{D_c}{D_h} - 1\right)}{14.7\left(\frac{1}{P_f + 14.7} - \frac{1}{P_{RV} + 9.7}\right)}$$

Where:

V_t = minimum required tank volume (gallons)
V_s = fluid volume in the system (gallons) (see Section 12.4)
D_c = density of the fluid at its initial (filling) temperature (lb/ft^3)
D_h = density of the fluid at the maximum operating temperature of the system (lb/ft^3)
P_f = static pressure of the fluid in the tank when the system is filled (psi)*
P_{RV} = rated pressure of the system's pressure-relief valve (psi)

*The static pressure at the expansion tank connection when the system is filled is based on the height of the system piping above the expansion tank, the type of fluid used, and a customary allowance for 5.0 psi pressure at the top of the system for proper operation of the air vents. This pressure can be found using Equation 12.2:

Equation 12.2:

$$P_f = H\left(\frac{D_c}{144}\right) + 5$$

Where:

P_f = static pressure of the fluid in the tank when the system is filled (psig)
H = distance from the inlet of the expansion tank to top of the system (feet)
D_c = density of the fluid at its initial (fill) temperature (lb/ft^3)

Example 12.1

Calculate the minimum size standard expansion tank for the system shown in Figure 12-5. Water is added to the system at an initial temperature of 60°F. The maximum operating temperature of the system is 200°F. The boiler is equipped with a 30 psi rated relief valve. The estimated system volume is 35 gal.

Solution:

Before using Equation 12.1, the pressure in the tank when the system is filled must be found using Equation 12.2. This equation requires the density of the water at 60°F, which can be read from Figure 4-4, calculated using Equation 4.3, or found using the Fluids Properties Calculator module in the Hydronics Design Studio software. Since the water density at the maximum operating temperature of 200°F is eventually required, it is also determined at the same time:

D_c = 62.36 lb/ft^3 (water density at 60°F)

D_h = 60.13 lb/ft^3 (water density at 200°F)

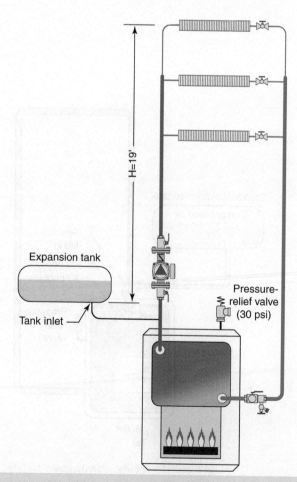

Figure 12-5 | System used in example 12.1.

The pressure at the inlet of the tank is now calculated using Equation 12.2:

$$P_f = H\left(\frac{D_c}{144}\right) + 5 = 19\left(\frac{62.36}{144}\right) + 5 = 13.23 \text{ psig}$$

This value and the remaining data are now substituted in Equation 12.1:

$$V_t = \frac{35\left(\frac{62.36}{60.13} - 1\right)}{14.7\left(\frac{1}{13.23 + 14.7} - \frac{1}{30 + 9.7}\right)}$$

$$= \frac{1.298}{0.156} = 8.3 \text{ gal}$$

Discussion:

It is interesting to consider the effect of the expansion tank's height on its volume requirement. If the tank were located closer to the top of the system, the static pressure caused by fluid in the piping above the tank would decrease. This would result in less initial compression of the air in the tank and thus more room for fluid. In this example, if the tank were moved 10 ft higher, its size could be reduced to 5.1 gal. A lower maximum fluid temperature would also decrease the tank size required.

It should be noted that Equation 12.1 determines the *minimum* volume of a standard expansion tank. It is possible to use a tank with a larger volume. As the tank's volume increases above its minimum calculated value, the increase in the tank's air pressure as the system's fluid expands is reduced. A larger tank, however, provides no tangible benefit to the performance or longevity of the system. The only reason to use a larger tank would be to use a standard production tank size rather than a custom-made tank.

Standard expansion tanks are seldom used in modern residential or light commercial systems. Their maintenance requirements, size, weight, cost, and associated fittings and valves make them less desirable than the diaphragm-type tanks discussed in the next section.

12.3 Diaphragm-Type Expansion Tanks

Many of the shortcomings of standard expansion tanks can be avoided by separating the air and water in an expansion tank. During the 1950s, a new type of expansion tank with an internal flexible diaphragm became available. On one side of the diaphragm is a captive volume of air that has been pre-pressurized by the manufacturer. On the other side is a chamber for accommodating the expanded volume of system fluid. As more fluid enters the tank, the diaphragm flexes, allowing the air volume to be compressed. An example of a small **diaphragm-type expansion tank** and a sequence illustrating the movement of its diaphragm is shown in Figure 12-6.

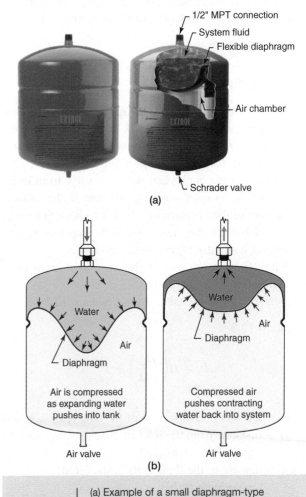

Figure 12-6
(a) Example of a small diaphragm-type expansion tank. *Courtesy of Amtrol, Inc.*
(b) Flexing of the diaphragm as expanding fluid enters, or contracting fluid exits the tank.

There are several advantages to using diaphragm-type expansion tanks rather than standard expansion tanks:

- Since the diaphragm separates the fluid from the air, the air cannot be reabsorbed by the fluid. Diaphragm-type expansion tanks do not have to be periodically drained to prevent waterlogging.
- By avoiding waterlogging, the possibility of accelerated corrosion caused by addition of fresh water to make up for relief valve losses is no longer a factor.
- The air pressure in the tank can be adjusted to match the static pressure of the system before the system is filled with water. No significant amount of water enters the tank until the system begins warming up. This results in a significantly smaller and lighter tank.
- Because the air volume is captive, the tank can (theoretically) be mounted in any orientation.
- The special boiler and tank fittings required with standard expansion tanks are no longer needed.

Sizing a Diaphragm-Type Expansion Tank

A properly sized diaphragm-type expansion tank allows the pressure-relief valve to reach a pressure of about 5.0 psi lower than its rated opening pressure when the system reaches its maximum operating temperature. The 5.0 psi safety margin prevents the relief valve from leaking just below its rated opening pressure. It also allows for the inlet of the pressure-relief valve to be slightly (up to 4 ft) below the connection point of the expansion tank, and thus at a slightly higher static pressure.

The first step in sizing a diaphragm-type expansion tank is to determine the proper **air-side pre-pressurization** of the tank, using Equation 12.3.

Equation 12.3:

$$P_a = H\left(\frac{D_c}{144}\right) + 5$$

Where:
- P_a = air-side pressurization of the tank (psi)
- H = distance from the inlet of the expansion tank to top of the system (feet)
- D_c = density of the fluid at its initial (cold) temperature (lb/ft^3)

The proper air-side pre-pressurization is equal to the static fluid pressure at the inlet of the tank, plus an additional 5.0 psi allowance at the top of the system. *The pressure on the air side of the diaphragm should be adjusted to the calculated pre-pressurization value before fluid is added to the system.* This is done by either adding or removing air through the Schrader valve on the shell of the tank. A small air compressor or bicycle tire pump can be used when air is needed. An accurate 0 to 30 psi pressure gauge should be used to check this pressure as it is being adjusted. Many smaller diaphragm-type expansion tanks are shipped with a nominal air-side pre-pressurization of 12 psi. However, the air-side pressure should always be checked before installing the tank.

Proper air-side pressure adjustment ensures that the diaphragm will be fully expanded against the shell of the tank when the system is filled with fluid, *but before it is heated*, as illustrated in Figure 12-7.

Failure to properly pre-pressurize the tank can result in the diaphragm being partially compressed by the fluid's static pressure before any heating occurs. If this occurs, the full expansion volume of the tank will not be available as the fluid heats up. An under-pressurized tank will act as if undersized and possibly allow the relief valve to open each time the system heats up. This situation must be avoided.

Once the air-side pre-pressurization is determined, Equation 12.4 can be used to find the *minimum* required volume of the expansion tank:

Equation 12.4:

$$V_t = V_s \left(\frac{D_c}{D_h} - 1\right)\left(\frac{P_{RV} + 9.7}{P_{RV} - P_a - 5}\right)$$

Where:
- V_t = minimum required tank volume (gallons) not "acceptance volume"
- V_s = fluid volume in the system (gallons) (see Section 12.4)
- D_c = density of the fluid at its initial (cold) temperature (lb/ft^3)
- D_h = density of the fluid at the maximum operating temperature of the system (lb/ft^3)
- P_a = air-side pressurization of the tank found using Equation 12.3 (psi)
- P_{RV} = rated pressure of the system's pressure-relief valve (psi)

Example 12.2

Determine the minimum size diaphragm-type expansion tank for the system described in example 12.1. Water is added to the system at an initial temperature of 60 °F. The maximum operating temperature of the system is 200 °F. The boiler is equipped with a 30 psi rated relief valve. The estimated system volume is 35 gal.

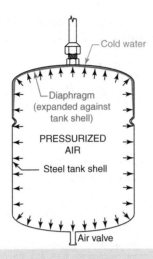

Figure 12-7 A properly pre-pressurized diaphragm-type expansion tank will have its diaphragm fully expanded against the tank's shell when the system is filled with water, but before the water is heated. This ensures that the full working volume of the tank is available to accept the expansion volume of the heated water.

Discussion:

Notice the required volume of the diaphragm-type tank is less than half that required by the standard expansion tank of example 12.1. This is due to pre-pressurization on the air side of the tank.

Some references on sizing expansion tanks refer to a tank's "**acceptance volume**." This is the volume of water the tank can accept for an assumed change in pressure. *Equation 12.3 calculates the actual volume of the tank shell, not the acceptance volume.*

Sizing Diaphragm-Type Tanks for Low-Temperature Systems

The sizing method just discussed has a very conservative assumption built in. It assumes that the *entire volume* of system fluid reaches the maximum fluid temperature simultaneously. This is never the case because the fluid cools as it passes through the heat emitters or through mixing devices.

The cooler the average temperature of the fluid, the less expansion takes place.

In small systems, operating with single supply temperatures and a typical temperature drop in the range of 20 °F, this assumption has minimal effect. Seldom will it allow a smaller volume tank to be used. However, low-temperature radiant panel heating systems often present a different situation.

In many of these systems, a large percentage of the system volume is contained in the radiant panel circuits, some of which only operate at temperatures in the range of 100 to 110 °F. This significantly reduces the total fluid expansion.

Sizing expansion tanks based on the assumption that all water in such systems reaches the upper boiler water temperature results in tanks that are needlessly oversized. Although an oversized tank will not cause operating problems, it certainly increases installation cost.

A more accurate method divides the system's fluid volume into two portions:

1. The portion that could operate up to the maximum temperature of the heat source
2. The portion that will only operate up to the maximum temperature of the distribution circuits

Solution:

As in example 12.1, the density of water at both 60 and 200 °F is required. This information can be read from Figure 4-4, calculated using Equation 4.3, or found using the fluids program in the Hydronics Design Studio.

$$D_c = 62.36 \text{ lb/ft}^3$$

$$D_h = 60.13 \text{ lb/ft}^3$$

The proper air-side pre-pressurization of the tank is calculated using Equation 12.3:

$$P_a = H\left(\frac{D_c}{144}\right) + 5 = 19\left(\frac{62.36}{144}\right) + 5 = 13.23 \text{ psig}$$

This information is now substituted into Equation 12.4:

$$V_t = 35\left(\frac{62.36}{60.13} - 1\right)\left(\frac{30 + 9.7}{30 - 13.23 - 5}\right) = 4.38 \text{ gal}$$

Equation 12.5 is a modified version of Equation 12.4 that determines the minimum expansion tank volume in such a low-temperature system.

Equation 12.5:

$$V_t = \left[V_{high}\left(\frac{D_c}{D_{high}} - 1\right) + V_{low}\left(\frac{D_c}{D_{low}} - 1\right) \right] \times \left(\frac{P_{RV} + 9.7}{P_{RV} - P_{air} - 5}\right)$$

Where:
- V_t = minimum required tank volume (gallons)
- V_{high} = volume of fluid contained in the higher temperature portion of the system (gallons)
- V_{low} = volume of fluid contained in the lower temperature portion of the system (gallons)
- D_c = density of the fluid at its initial (cold) temperature (lb/ft³)
- D_{high} = density of the fluid at the maximum operating temperature of the higher temperature portion of the system (lb/ft³)
- D_{low} = density of the fluid at the maximum operating temperature of the lower temperature portion of the system (lb/ft³)
- P_{air} = air-side pressurization of the tank found using Equation 12.3 (psig)
- P_{RV} = rated pressure of the system's pressure-relief valve (psig)

Example 12.3

Assume that a large radiant floor heating system contains 200 gal of water in the floor circuits and 15 gal of water on the boiler side of the mixing device. At design conditions, the water on the boiler side of the mixing devices reaches 180°F and that supplied to the radiant floor circuits reaches 110°F. Assume that the air-side pressurization of the tank is 13.5 psi, and the system's relief valve is rated at 30 psi. The system is initially filled with water at 60°F. Calculate the minimum required volume of a diaphragm-type expansion tank using the following:

a. Standard method (Equation 12.4)
b. Modified method (Equation 12.5)

Solution:

The densities of the water at the three temperatures are

$$D_c = 62.36 \text{ lb/ft}^3$$
$$D_{low} = 61.84 \text{ lb/ft}^3$$
$$D_{high} = 60.60 \text{ lb/ft}^3$$

Using the standard method (Equation 12.4), the minimum expansion tank volume required is

$$V_t = V_s\left(\frac{D_c}{D_h} - 1\right)\left(\frac{P_{RV} + 9.7}{P_{RV} - P_a - 5}\right)$$

$$= 215\left(\frac{62.36}{60.06} - 1\right)\left(\frac{30 + 9.7}{30 - 13.5 - 5}\right) = 21.6 \text{ gal}$$

Using the modified sizing procedure, the minimum expansion tank volume required is

$$V_t = \left[15\left(\frac{62.36}{60.06} - 1\right) + 200\left(\frac{62 + 36}{61.84} - 1\right) \right]$$

$$\times \left(\frac{30 + 9.7}{30 - 13.5 - 5}\right) = 7.3 \text{ gal}$$

Discussion:

Accounting for the fact that only 7% of the system's water volume reaches the higher temperature, the modified procedure reduces the required expansion tank volume by about 66% compared to that calculated by the standard method. Even the modified method is conservative because it assumes that the 200 gal of water reaches 110°F at the same time the other 15 gal reaches 180°F. In reality, the average system water temperature in the floor circuits is likely to be about 10°F cooler due to the drop in temperature as heat is released into the floor. The average temperature of the higher temperature water is also less than 180°F.

Compatibility of the Expansion Tank and System Fluid

It is very important that the diaphragm material used in the tank is chemically compatible with the fluid and/or any dissolved gases in the system. Incompatibilities can result in the diaphragm being slowly dissolved by the fluid. This can create sludge in the system that gums up system components.

When the diaphragm eventually ruptures, the tank quickly waterlogs, and the vicious cycle between the relief valve and feed water valve begins.

Most modern diaphragm-type tanks use a butyl rubber or **EPDM** diaphragm. Although such tanks are generally compatible with glycol-based antifreeze solutions, the installer should always verify this compatibility if it is not clearly stated in product specifications. Tanks are also available with special, high-temperature, glycol-compatible diaphragms for specific use in the collector circuit of solar thermal systems.

Another compatibility issue involves the presence of dissolved oxygen in the system water. Most expansion tanks sold for use in hydronic heating systems are intended for installation in closed-loop, oxygen-tight systems. In most such tanks, system water contacts the plain steel tank shell. If the system allows the presence of dissolved oxygen in the water, the shell will corrode and eventually fail. Dissolved oxygen could be present in any type of open-loop system, or in systems using oxygen-permeable tubing. In such cases, an expansion tank rated for use in open-loop systems should be selected. Such tanks often have a polypropylene liner that prevents contact between the system water and the steel tank shell.

Pressure and Temperature Ratings

Expansion tanks have maximum pressure and temperature ratings. This information is stamped on the label of the tank. Typical ratings are 60 psi and 240 °F. These ratings are usually adequate for residential and light commercial systems. They must be observed, however, in nontypical applications. For example, if an expansion tank with a 60 psi pressure rating was used on a domestic hot water tank with a 150 psi relief valve, the tank could rupture before the relief valve opened.

Selection, Mounting, and Service

Diaphragm expansion tanks are available in a variety of sizes and shapes. Smaller tanks are designed to be hung from their 1/2-inch MPT piping connection, and having volumes from 1 to about 14 gal, are usually adequate for standard residential and light commercial hydronic systems. Because of the potential water weight in the tank, the piping supporting these tanks should itself be well supported.

One common mounting method is to hang a diaphragm-type expansion tank from the bottom of the system's air separator. Many air separators have 1/2-inch FPT bottom tapping for this purpose. An example of this type of mounting is shown in Figure 12-8a. Notice how the system piping is routed diagonally across the corner space to create clearance for the tank.

If necessary for spatial reasons, the tank can also be located several feet away from the point where its connecting piping taps into the system. From a pressure standpoint, the horizontal distance between the tank and where the piping from it taps into the system is of no concern. However, lowering the tank inlet several feet below the connection point will add static pressure to the tank and slightly reduce its expansion absorption capability.

When the tank does not hang vertically downward from a well-supported pipe or piping component, the tank itself should be well supported by wall brackets, as shown in Figure 12-8b.

Although the captive air volume in a diaphragm-type tank would theoretically allow it to function in any orientation, there are other concerns to address whenever the tank is not hung vertically from its inlet connection. One is that air bubbles that could migrate toward the tank can be trapped on the waterside of the tank's shell if it is mounted with its piping connection facing down or horizontally. The oxygen in this air will corrode the tank's steel shell. Another concern is mechanical stress on the tank's inlet connection, especially if the tank is not properly supported. These factors support the author's recommendation that *small diaphragm-type expansion tanks should only be mounted vertically with their inlet connection at the top.*

The pressure in an expansion tank is partially affected by the temperature of the captive air volume. The warmer the shell of the tank, the warmer the captive air inside will be. Although the pressure increase due to heating is generally compensated for by conservative assumptions in the sizing equations, installation details that minimize tank heating are still desirable.

Installing several feet of piping between the tank inlet and the system tap-in point will reduce tank warming. This is especially desirable if the system operates at high water temperatures.

When larger expansion tanks are required, floor-mounted models are an option. An example of floor-mounted diaphragm-type expansion tank is shown in Figure 12-9. The Schrader valve for air pressurization is usually located at the top of floor-mounted expansion tanks. The piping connection is located inside of the support collar at the bottom of the tank.

Figure 12-8 (a) Expansion tank hung from its 1/2-inch MPT connection from the bottom of an air separator. (b) Expansion tank secured using wall brackets. *Courtesy of John Siegenthaler.*

644 Chapter 12 Expansion Tanks

Figure 12-9 — Example of a floor-mounted diaphragm-type expansion tank (in rear). Smaller diaphragm-type expansion tank that is designed to be hung from upper connection is shown in the foreground. *Courtesy of Watts, Inc.*

It is also possible to combine two or more expansion tanks in parallel. The combined volume of the two tanks should at least equal the minimum tank volume calculated with Equations 12.4 and 12.5, or the Expansion Tank Sizer software module. Parallel-mounted tanks should have their inlet connections at the same height and have equal air-side pre-pressurization. Be sure each tank is properly supported and that the air valves are accessible in case the air pressure has to be adjusted.

It is also good practice to install an isolation valve near the inlet of any expansion tank. This allows the tank to be easily isolated from the system if it ever needs replacing. A small drainage fitting or valve between the isolation valve and tank is also desirable to relieve any pressure and drain some fluid before disconnecting the tank.

Figure 12-10 shows a unique combination valve designed specifically for piping expansion tanks to smaller hydronic systems. It includes valving to isolate the tank when necessary, a shutoff valve for make-up water, a pressure gauge, and a valve for releasing water pressure from the expansion tank before removing it.

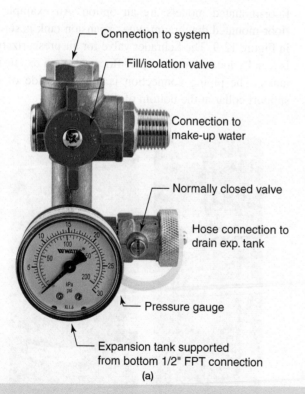
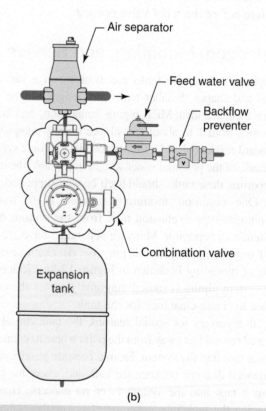

Figure 12-10 — (a) Combination valve for connecting, isolating, and draining an expansion tank, as well as a connection point for make-up water. *Courtesy of Watts, Inc.* (b) Typical installation of this valve.

Figure 12-11 shows a small spring-load check valve assembly designed to mount between a hanging expansion tank and the piping connecting it to the system. When the 1/2-inch MPT-threaded connector for the expansion tank is screwed into this fitting, the internal check valve mechanism is forced open, allowing flow and pressure equalization between the tank and the system. When the tank is unscrewed, the spring-load check valve mechanism closes to prevent all but a minor loss of system fluid.

Most diaphragm-type expansion tanks will retain their captive air volumes for decades. However, there is always the possibility that a diaphragm could leak. If this occurs, the tank will eventually lose its air and fill with system fluid.

Such failure is easy to detect during a routine service check. Simply press in on the stem of the tank's Schrader valve. If a stream of water comes out, the diaphragm has failed and the entire tank needs to be replaced. Gently tapping one's fingers on the side of a tank with an intact diaphragm will produce a hollow sound. If a "thud" sound is heard, it may indicate that the tank has filled with water. Check for water coming out of the Schrader valve to be sure.

If a new tank is needed, be sure the air-side pressure is adjusted before water is allowed to enter the tank.

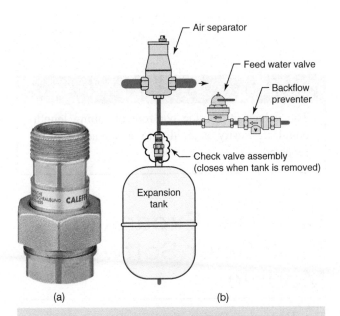

Figure 12-11 (a) Expansion tank connection fitting with internal spring-loaded check valve. *Courtesy of Caleffi North America.* (b) Check valve closes to prevent leaks when tank is removed.

12.4 Estimating System Volume

The sizing equations for both standard- and diaphragm-type expansion tanks require an estimate of system volume.

In a typical hydronic system, most of the fluid is contained in the piping and heat source. The fluid volume contained in type-M copper, PEX, and PEX-AL-PEX tubing can be calculated by estimating the total length of each tube type and size used in the system, then multiplying these lengths by the appropriate volume factors from Figure 12-12.

For other pipe sizes and materials, Equation 12.6 can be used to estimate the volume factor:

Equation 12.6:

$$V = 0.04085(d_i)^2$$

Where:
V = volume of one foot of the piping (gal/ft)
d_i = exact inside diameter of the piping (in)

The volume of the heat source can vary over a wide range. Small copper tube boilers may only contain 1 or 2 gal of fluid. Cast-iron sectional boilers may hold 10 to 15 gal. The only reliable source for these data is the manufacturer's specifications.

The volume of any type of storage tank, such as the buffer tank described in Chapter 3, Hydronic Heat Sources, must also be included. Systems that employ large pressurized storage tanks will require relatively large expansion tanks.

Most modern heat emitters contain relatively small amounts of water. In the case of either finned-tube or radiant baseboard, the volume can be estimated using the appropriate pipe volume factors in Figure 12-12. Small kick space heaters or wall-mounted fan-coils contain very little fluid, probably not more than 1/4 gal each. The fluid content of panel radiators varies with their design and size. Standing cast-iron radiators can contain significant quantities of water. Again, manufacturer's specifications are the best source for component volume data.

The volume of the diaphragm-type expansion tank itself is not entered into the system volume. This is because there is no fluid in a properly pre-pressurized diaphragm-type tank until the system is heated and the fluid begins expanding.

	Copper (Type M)		PEX		PEX-AL-PEX
	gal/ft		gal/ft		gal/ft
3/8" copper	0.008272	3/8" PEX	0.005294	3/8" PEX-AL-PEX	0.004890
1/2" copper	0.01319	1/2" PEX	0.009609	1/2" PEX-AL-PEX	0.01038
		5/8" PEX	0.01393	5/8" PEX-AL-PEX	0.01658
3/4" copper	0.02685	3/4" PEX	0.01894	3/4" PEX-AL-PEX	0.02654
1" copper	0.0454	1" PEX	0.03128	1" PEX-AL-PEX	0.04351
1.25" copper	0.06804	1.25" PEX	0.04668		
1.5" copper	0.09505	1.5" PEX	0.06516		
2" copper	0.1647	2" PEX	0.1116		
2.5" copper	0.2543				
3" copper	0.3630				

Figure 12-12 | Volume factors for tubing.

Example 12.4

Estimate the volume of the system in Figure 12-13.

Solution:

The volume of all piping is estimated using data from Figure 12-12. This is then added to the volume estimates for the heat emitters and boiler.

The total piping volume is:

$$V_{piping} = (55 + 70)(0.06804) + (45 + 40)(0.0454)$$
$$+ (35 + 35)(0.02685) + (25 + 25)(0.01319)$$
$$= 14.9 \text{ gal}$$

The total estimated volume of the branch piping and heat emitters is:

$$8(1) = 8 \text{ gal}$$

Assume that the boiler has a volume of 12 gal.

The total system volume is the sum of these volumes:

$$14.9 + 8 + 12 = 34.9 \text{ gal}$$

Discussion:

The values of volume per foot of tubing length could be easily entered into a spreadsheet for repeated use.

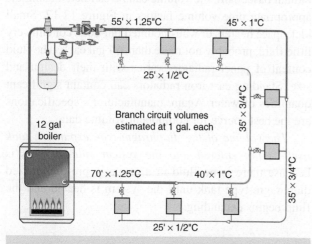

Figure 12-13 | Piping diagram of system for example 12.4.

12.5 The Expansion Tank Sizer Software Module

The Hydronics Design Studio software contains a module called the **Expansion Tank Sizer** that can be used to size diaphragm-type expansion tanks. The user interface is shown in Figure 12-14.

This module allows the user to select up to four different types/sizes of tubing and enter the associated lengths. Another input is used for entering miscellaneous volumes such as those for the boiler, heat emitters, tanks,

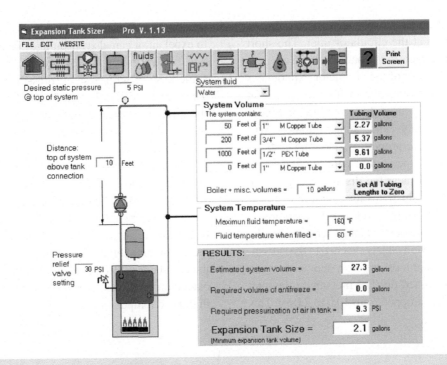

Figure 12-14 | Screenshot of the Expansion Tank Sizer module from the Hydronics Design Studio. *Source: Hydronicpros.*

and so forth. The user can vary the pressure at the top of the system, the relief valve setting, and the distance from the expansion tank connection to the top of the system. The user can also select from one of several fluids. When a fluid other than water is used, the program corrects for the density differences and estimates the volume of antifreeze required based on its concentration.

The module determines the minimum expansion tank size as well as the proper air-side pressurization. It also reports the estimated system volume, and the volume of antifreeze required (if an antifreeze solution is selected as the system fluid).

12.6 Point of No Pressure Change

The placement of the expansion tank relative to the circulator significantly affects the pressure distribution in the system when it operates. *The expansion tank should always be connected to the system near the inlet port of the circulator.*

To understand why, one needs to consider the interaction between the circulator and expansion tank. In a closed-piping system, the amount of fluid (including that in the expansion tank) is fixed. It does not change regardless of whether the circulator is on or off. The expansion tank contains a captive volume of air at some pressure. The only way to change the pressure of this air is to either push more fluid into the tank compressing the air or remove fluid from the tank expanding the air. This fluid would have to come from, or go to, some other location in the system. However, since the system's fluid is incompressible, and the amount of fluid in the system is fixed, this cannot happen regardless of whether the circulator is on or off. The expansion tank thus fixes the pressure of the system's fluid at its point of attachment to the piping. This is called the **point of no pressure change**.

Now consider a horizontal piping circuit filled with fluid and pressured to approximately 10 psi, as shown in Figure 12-15. When the circulator is off, the pressure is the same (10 psi) throughout the piping circuit. This is indicated by the solid horizontal pressure line shown above the piping.

When the circulator is turned on, it immediately creates a pressure differential between its inlet and outlet ports. However, the pressure at the point where the expansion tank is connected to the circuit remains the same. The combination of the pressure differential across the circulator, the flow resistance of the piping, and location of the point of no pressure change gives rise to the new **dynamic pressure distribution** shown by the dashed line in Figure 12-15.

Notice how the pressure increases in nearly all parts of the circuit when the circulator is operating. This is desirable because it helps eject air from vents. It also helps keep dissolved air in solution and minimizes the potential for cavitation at the circulator inlet. The short segment of piping between the expansion tank connection point and the inlet port of the circulator experiences a very slight drop in pressure due to flow resistance in the piping. The numbers used for pressure in Figure 12-15 are illustrative only. The actual numbers will depend on flow rates, fluid properties, and pipe sizes.

Next, consider the same system with the expansion tank (incorrectly) located near the discharge port of the circulator. Figure 12-16 illustrates the pressure distribution that will develop when the circulator is operating.

The point of no pressure change remains at the location where the expansion tank is attached to the system. This causes the pressure in most of the system to decrease when the circulator is turned on. The pressure at the inlet port of the circulator has dropped from 10.0 to 2.0 psi. This situation is not desirable since it reduces the system's ability to expel air. It can also encourage circulator cavitation.

To see how problems can develop, imagine the same system with a static pressurization of only 5 psi. The same 9 psi differential will be established across the circulator when it operates, and the entire pressure profile shown with dashed lines in Figure 12-16 will shift downward by 5 psi as shown in Figure 12-17.

Notice that the pressure in the piping between the upper-right corner and the circulator inlet is below atmospheric pressure when the circulator is operating. If there are air vents or slightly loose valve packings in this portion of the circuit, air will be sucked into the system every time the circulator operates. The absolute pressure at the inlet of the circulator is also lower and thus the potential for vapor cavitation is increased, especially if the system operates at high fluid temperatures.

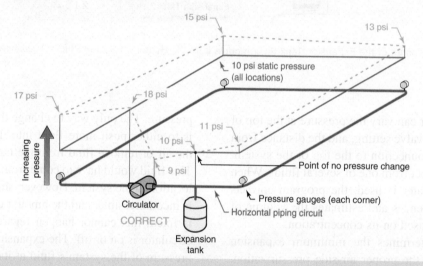

Figure 12-15 | Pressure distribution in a horizontal piping circuit with circulator on and off. Note that the expansion tank is (correctly) located near the inlet port of the circulator.

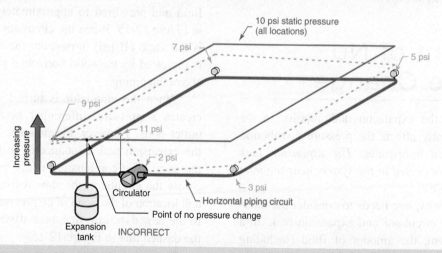

Figure 12-16 | Pressure distribution in a horizontal piping circuit with circulator on and off. Note that the expansion tank is (incorrectly) located near the discharge port of the circulator.

12.6 Point of No Pressure Change

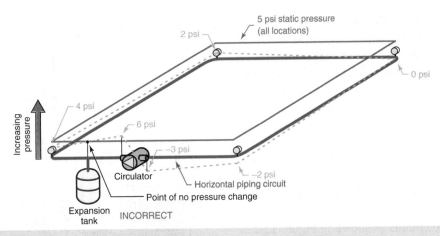

Figure 12-17 Pressure distribution in a horizontal piping circuit with low static pressure. Note that the expansion tank is (incorrectly) located near the discharge port of the circulator. The pressure in a portion of the piping system becomes subatmospheric when the circulator operates.

Many residential hydronic systems that are over 30 years old were installed with the circulator pumping *toward* rather than away from the point where the expansion tank is connected (e.g., the point of no pressure change). A piping arrangement typical of this vintage is shown in Figure 12-18.

In this arrangement, the circulator is pumping *toward* the expansion tank. The pressure drop created through a typical cast-iron or steel boiler is very low. Thus, from the standpoint of pressure change, the circulator's outlet is very close to the point of no pressure change. Whenever the circulator operates, this arrangement causes a drop in system pressure from the expansion tank connection point, around the distribution system to the inlet port of the circulator.

Some hydronic systems piped as shown in Figure 12-18 have worked fine for years. Others have had problems from the first day they were put into service. Why is it that some systems work and others don't? The answer lies in a number of factors that interact to determine the exact pressure distribution in any given system. These include system height, fluid temperature, pressure drop, and system pressurization. The systems most prone to problems are those with high fluid temperature, low static pressure, low system height, and high pressure drops around the piping circuit.

Rather than gamble on whether problems will occur, it is best to arrange the piping as shown in Figure 12-19. This arrangement causes an *increase* in pressure in nearly all parts of the system when the circulator operates.

Many systems with chronic "air problems" can be corrected by simply rearranging the piping so that the circulator pumps away from the point of no pressure change.

In systems with low flow resistance heat sources, it is also possible to locate the expansion tank near the inlet to the boiler as shown in Figure 12-20. Because of the low pressure drop through the boiler and generously sized piping leading from the boiler to the circulator, there is very little drop in pressure due to head loss between the expansion tank's connection location and the circulator's inlet. This arrangement also allows the expansion tank to remain somewhat cooler than when it is mounted on the outlet side of the heat source. *Do not use this arrangement when there will be a significant pressure drop between the point of no pressure change and the circulator inlet.* This includes most situations where a modulating/condensing boiler with a compact heat exchanger is used as the heat source. The greater the pressure drop between these points, the smaller the safety margin against vapor cavitation in the circulator.

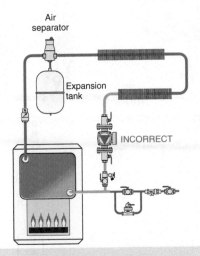

Figure 12-18 Typical piping configuration of an older residential system. Notice the incorrect placement of the expansion tank relative to the circulator.

When multiple zone circulators are used, the header they connect to should be located close to and just downstream from the point where the expansion tank is connected as shown in Figure 12-21. The header pipe should also be generously sized to minimize the pressure drop between the point of no pressure change and any of the zone circulators.

In primary/secondary systems, the expansion tank is connected to the primary circuit with the primary circulator pumping away from it. *The primary circuit is "seen" as the expansion tank by the secondary circuits.* Accordingly, all secondary circulators should "pump into" their respective circuits as shown in Figure 12-22. This causes the pressure in the secondary circuits to increase when their circulator is operating.

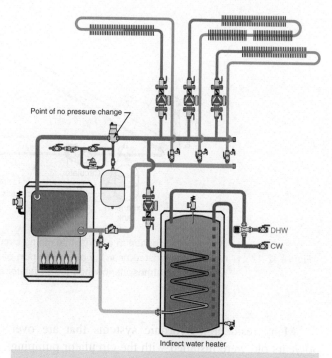

Figure 12-21 | Placement of expansion tank in a system using multiple zone circulators.

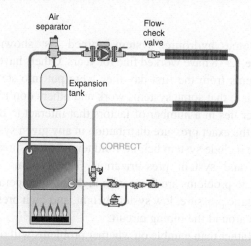

Figure 12-19 | Proper arrangement of circulator and flow-check valve relative to expansion tank. Note that the inlet port of circulator is now close to the expansion tank connection.

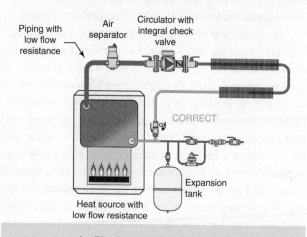

Figure 12-20 | Placement of expansion tank near inlet of heat source. Only recommended when boiler and piping upstream of circulator have low flow resistance.

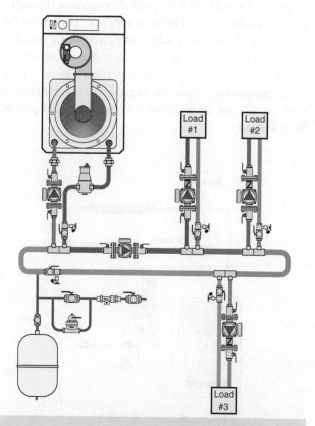

Figure 12-22 | Placement of expansion tank in primary loop of a primary/secondary system.

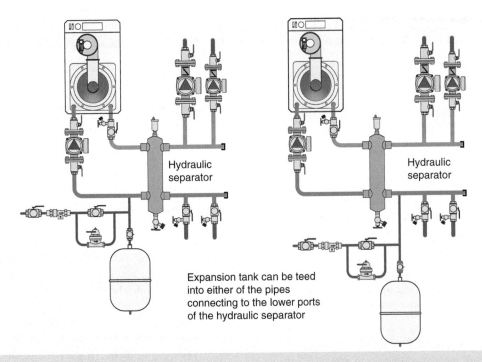

Figure 12-23 | Two equivalent connection points for an expansion tank in systems with a hydraulic separator.

In systems with a hydraulic separator, the expansion tank can be teed into either of the pipes connecting to the lower ports of the separator as shown in Figure 12-23. *It is not a good idea to connect the expansion tank to the low point of the hydraulic separator, since sediment will collect at that point, and is likely to drop down into the expansion tank.*

In many systems, the make-up water system is teed into the piping connecting the expansion tank to the system. Because this is the point of no pressure change, additional make-up water will not be pulled into the system when the circulator operates. This detail is shown on many of the piping schematics in this text.

Finally, *it is not advisable to install two or more expansion tanks at different locations within a system.* When the circulator is operating, the dynamic pressures between such points could vary in ways that create undesirable conditions such as air being drawn into the system through vents, circulator cavitation, or unnecessary opening of a pressure-relief valve. This precaution does not apply to the use of multiple expansion tanks, piped in parallel, and connected to the same point in the system.

Summary

The expansion tank is an important part of any closed-loop hydronic heating system. Its sizing, pressurization, and location within the system are easily determined, but unfortunately sometimes misunderstood or ignored. When these factors are determined by "guestimating," serious problems can develop that lead to unsatisfactory system performance and even premature failure. Those who design and/or install hydronic systems should always review piping diagrams, as well as existing installations, for the proper location of the expansion tank.

Key Terms

acceptance volume
air-side pre-pressurization
diaphragm-type expansion tank
dynamic pressure distribution
EPDM
Expansion Tank Sizer
incompressible
point of no pressure change
standard expansion tank
thermal expansion
waterlogged

Questions and Exercises

1. Explain why air slowly disappears from standard expansion tanks.

2. Can a standard expansion tank be placed lower than the boiler outlet? Justify your answer with a sketch.

3. Calculate the minimum size *standard* expansion tank for a system containing 50 gal of water where the uppermost system piping is 35 ft above the base of the boiler, and the inlet of the expansion tank is 8 ft above the base of the boiler. The water temperature when the system is filled is 50 °F. The maximum operating temperature of the system is 170 °F.

4. Calculate the minimum size diaphragm-type expansion tank for the same system described in exercise 3. Compare the results with those of exercise 3.

5. Describe what happens if the diaphragm in an expansion tank ruptures. How could you check the tank to see if it is ruptured?

6. How does heat that migrates from the system piping to the expansion tank affect the pressure of the air in the tank?

7. Why is it necessary to adjust the pressure on the air side of a diaphragm-type expansion tank on each system? What can happen if this is not done, and the pressure in the diaphragm is less than the static fluid pressure at its mounting location?

8. Why should the expansion tank be located near the inlet port of the circulator?

9. How does the use of glycol-based antifreeze affect the required size of an expansion tank? Why is this so?

10. The pressure-relief valve on a hydronic system with a standard expansion tank opens each time the system heats up. How could the expansion tank be causing this situation? What can be done to correct it?

11. An expansion tank is connected 10 ft upstream from the inlet port of a circulator as shown in Figure 12-24. The pipe between the tank connection and the inlet port of the circulator is 3/4-inch copper with 150 °F water flowing through at 8 gpm. Use methods from Chapter 6, Fluid Flow in Piping, to determine the pressure at the inlet port of the circulator assuming the water pressure at the tank connection is 15 psi.

12. In sizing expansion tanks, an allowance for 5 psi fluid pressure at the top of the system is made. Why?

13. A radiant floor heating system contains 215 gal of water in the floor circuits and 20 gal of water

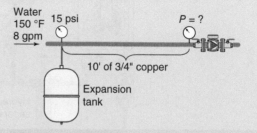

Figure 12-24 | Piping arrangement for exercise 11.

on the "hot" side of the mixing device. The boiler and higher temperature piping operate at 190 °F at design load. The floor circuit operates with a supply temperature of 120 °F at design load. The pressure-relief valve is rated at 30 psi, and there is 15 ft of system height above the inlet to the tank with 5 psi pressure at the top of the system. Determine the required expansion tank volume using Equation 12.5. Recalculate the expansion tank volume using Equation 12.4. Why are the results different?

14. Is an expansion tank required in an open-loop hydronic system? How is the expansion of the fluid accommodated in such a system?

15. What is the purpose of the boiler fitting used with standard expansion tanks?

16. Why should expansion tanks be sized so that they reach a pressure of about 5 psi below the rated pressure of the system's pressure-relief valve, as the system reaches maximum temperature?

17. Look up the exact inside diameters of 3/4- and 1-inch Schedule 40 steel pipes in a piping handbook. Calculate the volume of fluid they contain per foot.

18. Use the Expansion Tank Sizer module in the Hydronics Design Studio software to determine the size of a diaphragm-type expansion tank required for the system described in example 12.3. Resize the tank assuming that propylene glycol in concentrations of 30% and 50% is used as the system fluid. What can you conclude from this?

19. Use the Expansion Tank Sizer module to determine the minimum size of a diaphragm-type expansion tank located 40 ft below the top of a distribution system containing 100 gal of water. Assume that the system is filled with 60 °F water and has a maximum operating temperature of 200 °F.

20. Use the Fluid Properties Calculator module in the Hydronics Design Studio to determine the density of a 30% ethylene glycol solution at 50 and 170 °F and repeat exercise 4 assuming that this fluid is used in the system.

21. Explain why an expansion tank should not be connected to the bottom port of a hydraulic separator.

22. Why should circulators on secondary circuits within a primary/secondary piping system be installed so that they "pump into" their respective circuits?

23. Describe what type of problem(s) may develop if the air side of a diaphragm-type expansion tank is not properly pre-pressurized.

24. Describe what can happen to the dynamic pressure distribution in a system with two expansions tank connected at different locations within the system.

Chapter 13

Air & Dirt Removal & Water Quality Adjustment

Objectives

After studying this chapter, you should be able to:
- Discuss the different forms in which air can exist in a hydronic system.
- Diagnose some of the problems caused by air in hydronic systems.
- Explain the operation of various devices for removing air from a system.
- Describe where air venting devices should be placed in a system.
- Design hydronic systems that will not experience air problems.
- Explain the operation of a microbubble air separator.
- Explain the importance of dirt removal.
- Describe methods and hardware used for dirt removal.
- Understand the use of magnetic particle separation.
- Describe good applications for a combined air and dirt separator.
- Correct chronic air problems in existing systems.
- Describe the construction and operation of a purge cart.
- Add antifreeze to a system using a purge cart.
- Describe the operation of an automatic fluid feeder.
- Understand the importance of proper water quality in hydronic systems.
- Describe the differences by physical and chemical water quality.
- Explain the concept of and processes used for demineralizing water.

13.1 Introduction

For a hydronic heating system to deliver silent comfort, it must be free of air. If this is not the case, problems ranging from occasional "gurgling" sounds in the pipes to complete loss of heat output can occur.

Air problems are often bewildering. Just when the problem appears to be fixed, it can recur. Many owners and installers eventually give up, thinking the system's air problem simply cannot be corrected. The installer hopes the same mysterious problem will not occur on the next job.

A lack of understanding often prevents the true cause of the problem from being diagnosed, corrected, and most importantly, avoided in the future!

At some point, every hydronic system contains a mixture of water and air. This is especially true when the system is first filled and put into operation. *If properly designed, a closed-loop hydronic system should rid itself of most air within a few days of initial start-up.* The system should then maintain itself virtually air-free throughout its service life. Systems that experience chronic problems with entrapped air usually contain one or more classic design or installation errors.

This chapter provides the facts of how air gets into the system, how it behaves during system operation, and most importantly, how to get rid of it.

This chapter also describes the importance of dirt and magnetic particle separation and removal. It compares different types of dirt removal devices and shows how they are applied in typical systems.

Figure 13-1 | A cast-iron circulator volute severely damaged by oxygen-rich water. *Courtesy of Bob Rohr.*

13.2 Problems Created by Entrapped Air

There are several problems that can be traced to excessive air within hydronic heating systems. They include the following.

Flow Noise: One of the most sought-after benefits of hydronic heating is the silent conveyance of heat. Occupants should not hear flow as it travels through tubing and heat emitters. Properly deaerated water traveling through piping at velocities of 4 ft/s. or less makes very little (essentially unnoticeable) sound. However, a mixture of water and air is much more acoustically active. Sounds due to entrapped air are often most noticeable when a flow begins through piping due to disturbance of stationary air pockets. Air-filled cavities within the system act as acoustic chambers, especially if the water level in the device is below the level of the incoming water. Noise is also generated when **dissolved gases** are released from improperly **deaerated** fluid due to a reduction in pressure. In systems containing excessive dissolved gases, this is likely to occur with valves or circulators.

Accelerated Corrosion: Air contains oxygen, and oxygen in contact with ferrous metals creates **oxides** (e.g., corrosion). In systems with chronic air problems, oxygen is constantly available to ferrous metals and corrosion can proceed at several times its normal rate. Figure 13-1 shows the results of oxygen-induced corrosion of a cast-iron circulator volute.

The oxygen that causes corrosion is not the oxygen that forms water molecules (e.g., the "O" portion of the H_2O molecule). Rather, it is **free oxygen** (O_2) molecules that are either contained in air pockets or bubbles or dissolved into the fluid as discussed in Chapter 4, Properties of Water.

The following chemical reactions can occur in hydronic systems containing ferrous (iron-containing) components and free oxygen.

$$O_2 + Fe + 2H_2O \rightarrow Fe(OH)_2 + H_2$$
$$3Fe(OH)_2 \rightarrow Fe_3O_4 + H_2 + 2H_2O$$

The compound Fe_3O_4 is called **magnetite** and appears as a dark sludge within the system. If oxygen continues to be present in the system, magnetite will be converted to **hematite** (Fe_2O_3), which can cause pitting corrosion throughout the system.

Later sections of this chapter will discuss the importance of capturing magnetite and removing it from the system.

Inadequate Flow: Circulators transfer maximum head energy to **incompressible fluids** (e.g., liquids). A mixture of water and air is not an incompressible fluid. Although most circulators can still move water with some entrained air, head energy transfer is not as efficient as with fully deaerated water. This decreases flow and reduces the rate of heat conveyance by the system, in some cases to the point of complaints.

Complete Loss of Flow: A large pocket of air trapped within a system can totally stop flow. This effect is often called **air binding**. It occurs in a number of ways. One of the most common occurrences is when an air pocket displaces water within the volute of a circulator, and thus prevents the circulator from moving fluid through the system. Another possibility is when the maximum head of a circulator is insufficient to lift the water over the top of a piping circuit that has a large air pocket at its top as shown in Figure 13-1a.

Poor Heat Transfer: Air has much lower specific heat and density than water. When air displaces water away from heat transfer surfaces within boilers or heat emitters, the rate of heat transfer can be significantly reduced. Air pockets collecting in the upper portions of heat emitters can significantly lower their heat output. Air pockets trapped in boilers can lead to "hot spots" that create high thermal stress on the boiler's heat exchanger, as well as reduced heat transfer.

Gaseous Cavitation: Dissolved air can also lead to **gaseous cavitation** as discussed in Chapter 7, Hydronic Circulators. Gaseous cavitation occurs when the pressure on the fluid drops below the **saturation pressure** of the dissolved gases it contains. In circulators, this typically occurs in the eye of the impeller and/or along the trailing edges of the impeller vanes as illustrated in Figure 13-2. Although gaseous cavitation is not as damaging as vaporous cavitation, it still creates noise and reduces the flow a given circulator can produce.

Circulator Damage: Wet-rotor circulators have ceramic bushings that depend on system fluid for lubrication. Because air is much lower in density than water, it tends to accumulate near the pump shaft around the bushings. The mixture of air and water creates a foam-like solution that does not produce the same lubricating effect as a film of fluid. This is likely to cause premature bushing failure. Such failure typically requires the wet-rotor circulator to be replaced.

Circulators with internal spring-load check valves installed in vertical piping with upward flow are especially susceptible to air pockets. If a sufficient volume of air collects in the volute, the circulator may lose its ability to displace this air through the slight opening resistance of the check valve. Under such conditions, the circulator is operating without proper lubrication and will eventually fail. Complete loss of flow is symptomatic of such a condition.

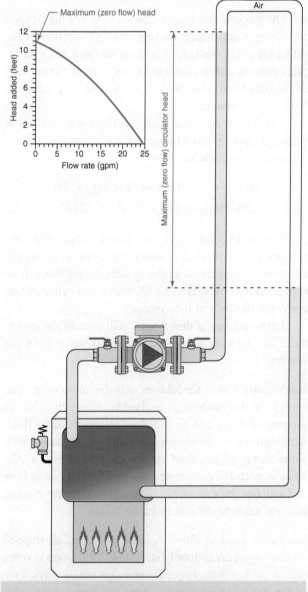

Figure 13-1a — Air binding occurs when maximum head of circulator is insufficient to displace water over top of circuit.

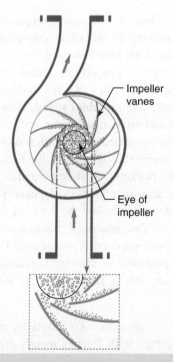

Figure 13-2 — Gaseous cavitation occurs in circulators when the pressure on the fluid drops below the saturation pressure of the dissolved gases contained in the fluid.

13.3 Types of Entrapped Air

In hydronic systems, air is found in three forms:

- Stationary air pockets,
- Entrained air bubbles, and
- Air dissolved within the fluid.

All three forms of air can exist simultaneously, especially when the system is first put into operation. Each exhibits different symptoms in the system.

Stationary Air Pockets

Since air is lighter than water, it tends to migrate toward the high points of the system. These points are not necessarily at the top of the system. **Stationary air pockets** can form at the top of heat emitters, even those located low in the building. Air pockets also tend to form in horizontal piping runs that turn downward following a horizontal run. A good example is when a pipe is raised to cross over a structural beam, and then dropped down as shown in Figure 13-3.

When a system is first filled with water, these high points are dead-ends for air movement. In some cases, this trapped air can displace several quarts of fluid that eventually must be added to the system. Even after a system is initially purged, stationary air pockets can reform as residual air bubbles, merge together, and migrate toward high points. This is especially likely in components with low flow velocities where slow-moving fluid is unable to push or drag the air along with it. Examples of such components include large heat emitters, large-diameter piping, and storage tanks.

Entrained Air Bubbles

When air exists as bubbles, a moving fluid may be able to carry them along (e.g., entrain them) through the system. **Entrained air** can be both good and bad. It is good from the standpoint of transporting air from remote parts of the system back to an air removal device; however, it is bad if the air cannot be separated from the fluid within the air removal device. How well a fluid entrains air is best judged by its ability to move bubbles downward, against their natural tendency to rise. Simply put, if the fluid moves downward faster than a bubble can rise, it will carry the bubble in its direction of flow. If air entrainment through a downward flowing pipe is desired, it is crucial that the fluid's flow velocity is greater than the bubble's rise velocity as illustrated in Figure 13-4.

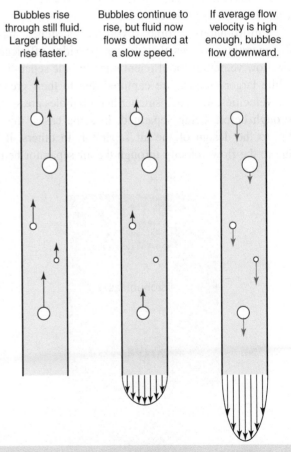

Figure 13-4 | The speed of the downward flowing fluid determines if bubbles continue to rise or are pulled along (entrained) in the direction of the flow.

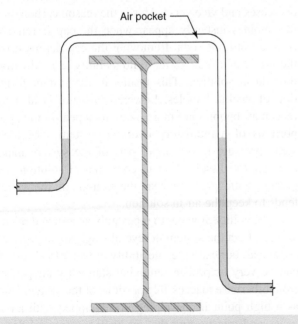

Figure 13-3 | A typical location where a stationary air pocket can form.

The **rise velocity** of a bubble through a fluid depends on the bubble's diameter and density, as well as the density and viscosity of the surrounding fluid. *The larger its diameter, the faster the bubble rises.* The higher the viscosity of the surrounding fluid, and the lower its density, the slower the bubble rises. Of these factors, the diameter of the bubble has the greatest effect on rise velocity. A bubble twice the diameter of another will rise at four times the velocity through the same fluid. The difference in bubble rise velocity can be observed in a glass filled with a carbonated drink.

One type of bubble often present in hydronic systems is called a **microbubble**. They are so small it is often difficult to see a single microbubble. Dense groups of microbubbles make otherwise clear water appear cloudy. They can often be observed, momentarily, in a glass of water filled from a faucet with an aerator device as shown in Figure 13-5.

Microbubbles can also be formed as dissolved air comes out of solution upon heating or when a component generates significant turbulence. *They have very low rise velocities and are easily entrained by moving fluids. This characteristic makes it more difficult to separate microbubbles from the fluid.*

Unfortunately, some hydronic systems have air-separating devices that do not provide sufficiently low flow velocities for efficient microbubble separation. While larger bubbles are captured due to their greater rise velocities, the much smaller microbubbles are swept through before being separated. In some cases, this is due to the design of the air separator. In others, it is due to the flow velocity through the air separator being too high. Eventually, the microbubbles will merge into larger bubbles that can be separated and ejected from the system, but this may take several days of operation.

Air Dissolved Within the Fluid

Perhaps the least understood form in which air exists in a hydronic system is as dissolved air. Molecules of the gases that make up air including oxygen and nitrogen can exist "**in solution**" with water molecules as discussed in Chapter 4, Properties of Water. These molecules cannot be seen, even under a microscope. Although water may appear perfectly clear and free of bubbles, it can still contain a significant quantity of air in solution.

The amount of air that exists in solution with water is strongly dependent on the water's temperature and pressure. The curves shown in Figure 13-6 show the maximum dissolved air content of water as a percentage of volume over a range of temperature and pressure.

Note that as the temperature of water increases, its ability to hold air in solution *decreases*. This explains why air bubbles appear on the lower surfaces of a pot of water being heated on a stove. It also explains the formation of microbubbles along the inside surfaces of a boiler's heat exchanger. As the water nearest these surfaces is heated, some of the air comes out of solution as microbubbles as depicted in Figure 13-7. The opposite is also true. *When heated water is cooled, it will absorb air back into solution.*

The pressure of the water also markedly affects its ability to contain dissolved air. When the pressure of the water is lowered, its ability to contain dissolved air decreases and vice versa. This is the reason carbon dioxide bubbles instantly appear when the cap is removed from a bottle of soda. Removing the cap depressurizes the liquid and reduces the fluid's ability to hold carbon dioxide in solution. This results in the instant formation of visible bubbles. Lowered pressure is also one reason air bubbles are more likely to appear in the upper portions of a multistory hydronic system. The lower static pressure in the upper part of the system makes it easier for dissolved air to come out of solution. The greater static pressure near the bottom of the system tends to keep the air in solution.

The ability of water to repeatedly absorb and release air can affect the system in several ways, some good and some bad. For example, the ability of water to absorb air can be very helpful in removing stationary air pockets from otherwise inaccessible portions of the system (such as a high point in the piping not equipped with an air vent), and transporting that air back to a central deaerating device. On the negative side, the ability of water to absorb air is also the primary cause of waterlogging

Figure 13-5 | Microbubbles slowly rising through a glass of water. *Courtesy of John Siegenthaler.*

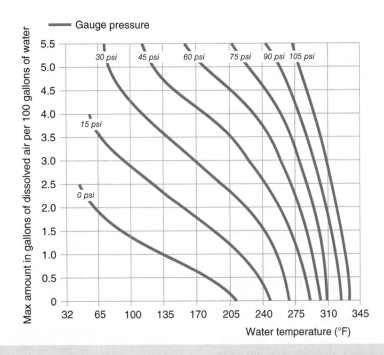

Figure 13-6 | Curves showing the maximum solubility of air in water at different temperatures and pressures.

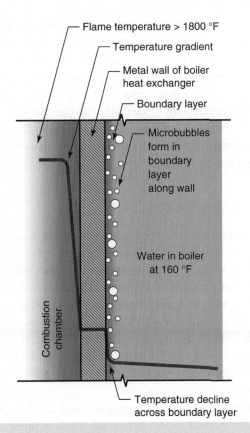

Figure 13-7 | Microbubbles form along the inside (water side) wall of the boiler's heat exchanger. Their presence decreases the rate of heat transfer from the wall to the water.

in standard expansion tanks as discussed in Chapter 12, Expansion Tanks.

It is always desirable to minimize the dissolved air content of the system's fluid. This is accomplished by establishing conditions that "encourage" air to come out of solution (e.g., high temperatures and low pressures). When such conditions exist, the resulting microbubbles must be captured and ejected from the system. Devices for doing this are discussed in the next section.

13.4 Air Removal Devices

The air removal devices used in small hydronic systems can be classified as either **high point vents** or **central air separators**.

High point vents are intended to release air from one or more high points in the system piping where it tends to accumulate in stationary pockets. Typical locations for such vents are at the top of each heat emitter, at the top of distribution risers, or wherever piping turns downward following an upward or horizontal run. High point vents are particularly useful for ejecting air immediately after the system has been filled with fluid (e.g., at start-up or following service).

A central air separator is a device intended to remove entrained air from a flowing fluid, as well as maintain the system at the lowest possible air content. It is usually mounted near the outlet of the heat source and has the entire system flow passing through it.

There are several types of devices used as high point vents and several others used as central air separators. All types relevant to residential and light commercial systems will be discussed separately.

Manual Air Vents

The simplest type of high point venting device is a **manual air vent**. These components are small valves with a metal-to-metal seat. They thread into 1/8- or 1/4-inch FPT tappings and are operated with a screwdriver, square head key, or the edge of a dime. They are sometimes referred to as "coin vents" or "**bleeders**." When their center screw is rotated, air can move up through the valve seat and exit through a small side opening. An example of a manual air vent is shown in Figure 13-8.

Manual air vents are commonly located at the top of each heat emitter. They can also be mounted into baseboard tees as shown in Figure 13-9, or any fitting with

(a)

(b)

Figure 13-8 | (a) Example of a manual air vent. *Courtesy of Bell & Gossett.* (b) A manual air vent mounted on top of a towel warmer radiator. *Courtesy of John Siegenthaler.*

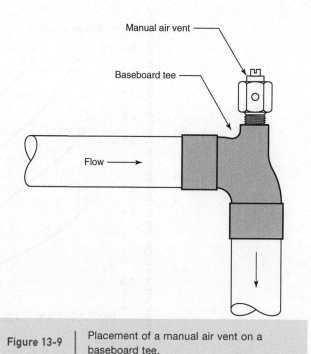

Figure 13-9 | Placement of a manual air vent on a baseboard tee.

the appropriate threaded tapping. Because of their small size, care should be taken that joint compound or Teflon tape does not cover their small inlet port.

After a manual vent is opened and any air below it is released, a small stream of water will continue to flow through the vent until it is closed. While it is operating, a person should stay near the vent with a can ready to catch any ejected fluid before it stains a carpet or floor. This can be difficult (especially for one person), if all manual vents are left open as the system is filled. It is better to move from one vent to the next, operating them in sequence.

Hygroscopic Air Vents

Another type of small high-point venting device is called a **hygroscopic air vent**. An example of such a device is shown in Figure 13-10. Figure 13-11 shows this device installed at the top of a cast-iron radiator.

Hygroscopic air vents contain several cellulose fiber discs. When that disc is dry, air can pass through it and exit the vent. When moisture reaches the disc, it expands very quickly to stop further flow from the device. The location and thickness of the fiber discs are illustrated in Figure 13-12.

Hygroscopic air vents can be used in either automatic or manual mode. When the knob is opened one turn from its fully closed position, as shown in Figure 13-13a, it operates the same as a manual air vent. Any pressurized air at the base of the vent exits through a small hole at the side of the vent's brass body.

13.4 Air Removal Devices

Figure 13-10 | Example of a hygroscopic air vent. *Courtesy of Caleffi North America.*

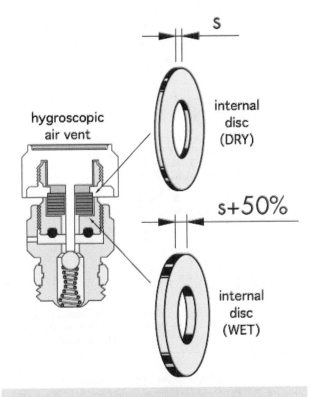

Figure 13-12 | Swelling of fiber washers within a hygroscopic air vent when exposed to water. *Courtesy of Caleffi North America.*

Figure 13-11 | A hydroscopic air vent installed at the upper portion of a cast-iron radiator. *Courtesy of Caleffi North America.*

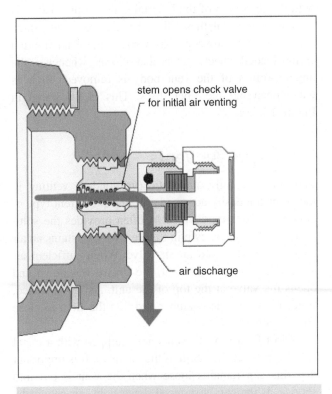

Figure 13-13a | Air released from a hygroscopic air vent when opened one turn. *Courtesy of Caleffi North America.*

When the knob is fully closed, an internal O-ring seals off the side port. However, if air is present at the vent, the fiber discs will dry and allow air to pass through them. This air is discharged under the vent's knob, as shown in Figure 13-13b. Once the air has been vented and water reaches the fiber discs, they swell very quickly to seal off any further discharge.

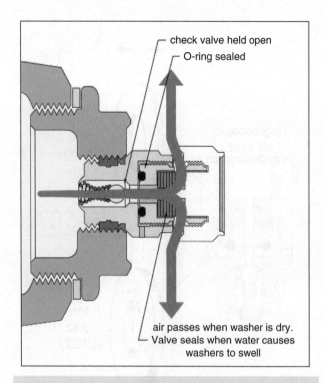

Figure 13-13b — Air released from a hygroscopic air vent when knob is closed and fiber discs are dry. *Courtesy of Caleffi North America.*

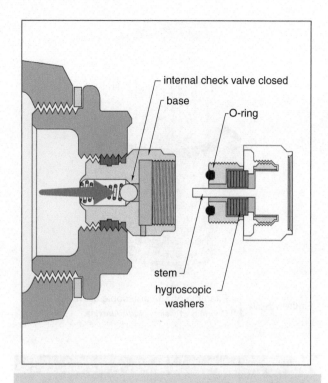

Figure 13-13c — A small spring loaded check valve within a hygroscopic air vent prevents water leakage when the outer portion of the vent is removed to replace the fiber washers. *Courtesy of Caleffi North America.*

Minerals or sediment in the system water can interfere with the operation of the internal hygroscopic disc. It is recommended that these discs be replaced every three years. Some hygroscopic air vents contain an internal spring-loaded check valve that closes whenever the upper portion of the vent body is removed, such as when changing the fiber discs. This is illustrated in Figure 13-13c.

Float-Type Air Vents

The need for fully automatic (unattended) venting in locations not easily accessed requires a different type of device. A **float-type air vent** often provides the solution. This device, shown in Figure 13-14, contains an air chamber, a float, and an air valve. When sufficient air accumulates in the chamber, the float drops down and opens the valve at the top of the unit. As air is vented, water rises into the chamber and lifts the float to close the air valve.

Most float-type air vents are equipped with a metal cap that protects the stem of the air valve. It is important that this cap remains loose when the vent is put into service. If the cap is screwed down tight, air cannot be ejected.

Most float-type vents come with MPT threads and are designed to thread into threaded fittings such as baseboard tees or reducer tees in the same manner as manual air vents. They are available in different sizes and shapes that allow mounting in both horizontal and vertical orientations. Low-profile designs are available that allow mounting within the enclosures of heat emitters such as finned-tube convectors or fan coils. Some float-type air vents are available with larger FPT piping connections that allow higher air venting rates when installed on central deaerators or at the top of large tanks.

Most automatic and float-type air vents will allow air to enter the system if the fluid pressure at their location falls below atmospheric pressure. This can result from a number of factors, most notably improper placement of the expansion tank relative to the circulator. It is a deceptive problem because it usually occurs only when the circulator is operating. The best way to prevent this from happening is to make sure there is always at least 5 psi of positive pressure at all locations where float type vents are mounted.

Air Purgers

One type of central air separating device is known as an **air purger** or **air scoop**. This cast-iron component, shown in Figure 13-15, is equipped with an internal

13.4 Air Removal Devices 663

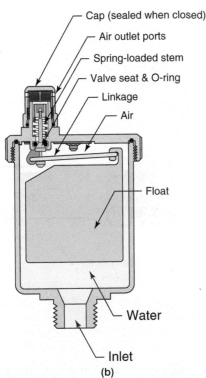

| Figure 13-14 | (a) Example of float type air vents. *Courtesy of Caleffi North America.* (b) Internal construction. *Courtesy of Caleffi North America.* |

baffle that directs bubbles into a separation chamber where they can rise into a float-type air vent screwed into the top of the chamber. The internal baffle also creates a region of lowered pressure that assists in bringing dissolved air out of solution. An arrow on the side of the air purger indicates the direction it must be installed in relative to flow.

To be reasonably effective, air purgers must operate at inlet flow velocities not exceeding 4 feet per second. If the flow velocity is higher, small air bubbles will remain entrained in the flow and be swept through the purger.

The preferred location for an air purger is near the outlet side of the heat source, about 12 pipe diameters upstream of the inlet port of the circulator. This allows the water entering the purger to be as hot as possible while also being at a relatively low pressure. The air's solubility in water is low under these conditions, and thus bubbles are more likely to form. If make-up water is added through the bottom of the purger, air bubbles that may enter with the water are more likely to rise into the upper chamber and be vented before being carried through the system.

Microbubble Air Separators

Microbubble air separators have been available in the North American hydronics market for several decades, but not as long as air purgers. Still, since they were introduced, microbubble air separators have established a strong reputation as very effective air removal devices. As flow passes through them, a region of reduced velocity and enhanced turbulence is created through an inserted component called a **coalescing media**. This media creates thousands of small reduced-pressure areas in which dissolved gas molecules can form into microbubbles, and where microbubbles can merge into larger bubbles.

Eventually, the bubbles reach a size where the adhesion forces holding them to the coalescing media are overcome by buoyancy forces. The bubbles then rise along the coalescing media toward the upper venting chamber. The coalescing media helps keep the bubbles from being swept out of the air separator by the passing flow. The upper portion of the air separator is a float type air venting device that eventually ejects the accumulated air from the system.

Figure 13-16 shows three types of microbubble air separators in sizes typical of those used in residential and light commercial systems. All three separators have brass bodies. A variety of connection options exist including soldered copper, FPT thread, press fit, and two-bolt flanges that can be directly attached to small circulator volutes.

A microbubble air separator is capable of maintaining the system fluid in an **unsaturated state of air solubility**. This means *that the water is always ready to absorb additional air from any areas of the system where it may be present*. Once absorbed, the air is transported by system flow back through the microbubble air separator where it can be separated and ejected from the system. The fluid then returns to its unsaturated state and is ready to absorb more air if available. This ability to continually seek out and collect air is very helpful for removing residual air pockets from all areas of the system, especially inaccessible areas that may not be equipped with vents.

Research has shown that microbubble air separators can reduce the dissolved air content of the system fluid

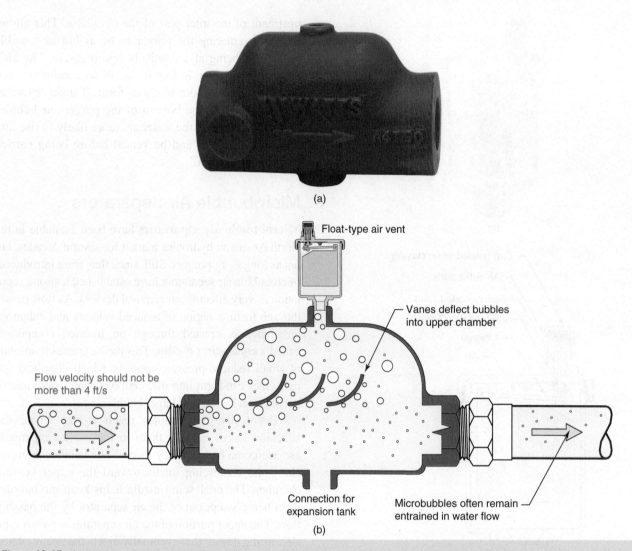

Figure 13-15 | (a) Example of a cast-iron air purger. *Courtesy of Watts.* (b) Operation of an air purger.

below 0.4%. The amount of oxygen present in this small air content is of virtually no significance as far as corrosion is concerned.

To achieve the best air separation efficiency, the inlet flow velocity to microbubble air separators should be no higher than 4 feet per second. As with air purgers, microbubble separators should be placed where the entering flow is at the highest possible temperature and lowest possible pressure. This is typically near the outlet of the heat source, as shown in many schematics throughout this text.

13.5 Correcting Chronic Air Problems

Properly designed and installed hydronic systems do not require constant manual air venting. A complaint about recurring air noises is symptomatic of one or more underlying design or installation errors. Many of these errors have little intuitive connection with the problems they create and thus are hard to recognize, especially by a novice installer. What follows is a list of the potential causes of recurring air problems, a short description of how the problem develops, and what to do to correct it. Keep in mind that any given air problem may be the result of almost any combination of these factors.

Potential Cause 1: Expansion tank located on discharge side of circulator

Description: If the expansion tank is located on the discharge side of the circulator, the pressure in certain locations in the system may drop below atmospheric pressure while the circulator is on. If a float-type air vent or hygroscopic air vent is located in this portion of the system, air will be sucked into the piping each time the circulator operates. Air can also be sucked in through loose valve packings or micro-leaks at threaded joints.

13.5 Correcting Chronic Air Problems

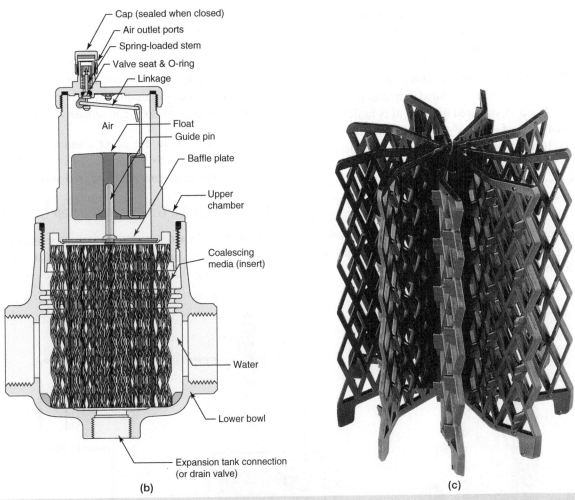

Figure 13-16 (a) Examples of microbubble separators. *Products by (left to right) Watts, Caleffi, Spirotherm.* (b) Internal construction. (c) Coalescing media insert. *Courtesy of Caleffi North America.*

Corrective Action: This problem is best solved by relocating the expansion tank to the inlet side of the circulator as discussed in Chapter 12, Expansion Tanks. This fix has proven very successful in many systems with chronic air problems. The location of the expansion tank relative to the circulator is definitely one of the first things to check on a problem system.

Potential Cause 2: Low system pressurization

Description: Low system pressurization is usually caused by not setting the make-up water system properly. If, for example, the shutoff valve on the make-up water line is left closed, the system pressure will eventually drop due to minor losses at valve packings or by air removed at vents. When the pressure at the top of the system drops below atmospheric pressure, air can be drawn in through vents, valve packings, or micro-leaks in threaded joints. This problem can be worsened by having the expansion tank on the wrong side of the circulator as described earlier.

Corrective Action: Make sure the feed water valve is adjusted so that a minimum static pressure of 5 psig is maintained at the top of the system.

Potential Cause 3: Waterlogged expansion tank

Description: If there is no air cushion in the expansion tank, or if the tank is too small, the system's pressure-relief valve can open each time the system heats up. This allows fresh water (containing dissolved air) to enter the system during each heating cycle.

Corrective Action: If the system contains a diaphragm-type expansion tank with a ruptured diaphragm, the tank must be replaced. A ruptured diaphragm is indicated by water flowing out of the tank's air valve when its stem is depressed. If the tank is suspected of being undersized, use the appropriate equations in Chapter 12, Expansion Tanks, to verify the minimum acceptable tank volume. Be sure the air side of the tank is properly pressurized. If the system contains a standard expansion tank, it may require periodic draining to relieve a **waterlogged** condition. Another option is to replace the standard tank with a diaphragm-type tank.

Potential Cause 4: Air goes in and out of solution but is not vented from system

Description: It is possible for air to come out of solution in the form of microbubbles, and then be absorbed back into solution without being captured and ejected from the system. The microbubbles that form as air comes out of solution are not easily captured by common air purgers, especially if the flow velocity through the purger is too high.

Corrective Action: A microbubble air separator should be installed to efficiently separate and remove dissolved air.

Potential Cause 5: Lack of high point vents or central deaerator

Description: High points in the system that do not have either a manual or automatic air vent can persistently collect air. If the system does not have a microbubble air separator, the air pockets can remain in place for weeks.

Corrective Action: Install either manual or automatic air vents at all high points and on all heat emitters. Be sure the caps on automatic vents are loose so that air can escape.

Potential Cause 6: Unentrained air bubbles

Description: When persistent gurgling sounds are heard in piping, especially piping with downward flow, it is likely that the flow velocity is too low to entrain the air

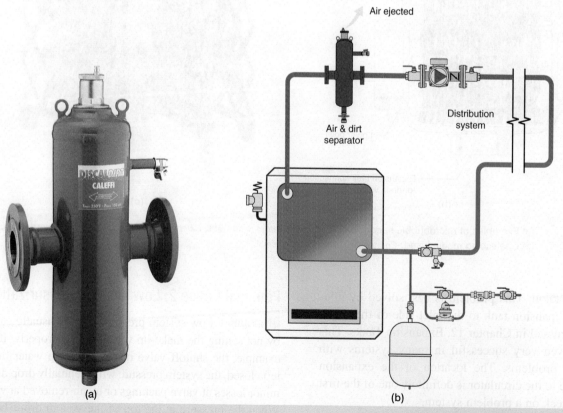

Figure 13-17 | (a) Combined air and dirt separator. *Courtesy of Caleffi North America.* (b) Typical mounting within a heating system.

and transport it to a central deaerator. The low flow velocity can be caused by a number of factors. These include but are not limited to significantly oversized piping, insufficient circulator size, or excessively long piping circuits.

Corrective Action: *All down-flowing piping should be sized to maintain a flow velocity of at least 2 feet per second to effectively entrain air bubbles.* If this is not possible, automatic high point vents must be installed at locations where air can collect.

Potential Cause 7: Air vents or leaks at the top of open-loop systems

Description: This problem is often associated with hydronic thermal storage systems that have unpressurized tanks. Consider the schematic shown in Figure 13-18. Because this is an open-loop system, the pressure in all piping above the water level in the tank will drop below atmospheric pressure when the circulator stops. Air will then try to enter the piping at any point it can (above the water level). Air vents located above the water level are the most likely entry points. However, valve packings, pump flange gaskets, and even microleaks at threaded joints can also admit air.

Corrective Action: This type of system should not have air vents or any other device that could allow air entry in any piping above the water level in the tank. All piping should be designed for a flow rate of at least 2 feet per second to effectively entrain air bubbles and bring them back to the tank where they can be vented.

Some open thermal storage tanks are not designed for piping connections below the water level. In this case, it is possible to route the piping connection through the tan wall, using a **bulkhead fitting** located just 2 to 4 inches above the highest water level in the tank, as shown in Figure 13-19.

The piping that is above the water level in the tank is called **gooseneck piping**. Once it is filled with water, that water will remain in the piping even though its pressure is slightly below atmospheric pressure. *It is critically important not to install any vents, valves, or other components that could potentially allow air into the gooseneck piping.* This portion of the piping should also be kept as short as possible to minimize head loss. A **bidirectional purging valve** should be installed in the piping between the gooseneck and circulator inlet. This valve allows the air initially in the gooseneck piping to be purged back into the tank water. It also allows air to be purged out of the remaining piping in the system. A tee should be installed in the piping, returning water to the lower portion of the tank, as shown in Figure 13-19. This tee allows the returning flow to be directed horizontally, and thus helps reduce vertical mixing currents within the tank. Such currents, if allowed to occur, tend to cause mixing that reduces beneficial temperature stratification within the tank.

Whenever a circulator is used to convey water from an open thermal storage tank, that circulator should be placed as low as possible relative to the water level in such a tank, as shown in Figures 13-18 and 13-19. This increases the static pressure at the circulator's inlet, which boosts the net positive suction head available and decreases the potential for cavitation.

Potential Cause 8: Improper purging during initial start-up

Description: Inadequate purging of piping can leave large amounts of bulk air in the system. The entrapped air pockets may even be large enough to cause air binding and complete loss of circulation.

Corrective Action: It is essential to remove the bulk air in the system by **forced-water purging** during initial startup. The next section describes methods for doing so.

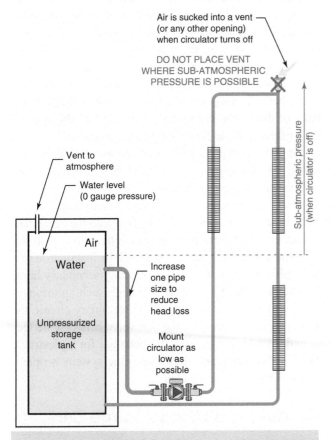

Figure 13-18 Schematic of an open-loop system showing areas of **subatmospheric pressure**. Never install vents or other devices that could allow air entry in any area where pressure could be subatmospheric.

13.6 Filling and Purging a System

Over the years, many installers have developed their own methods of filling and purging hydronic systems as part of their start-up procedure. This section will show some common approaches, as well as an approach using a tool called a purge cart.

Gravity Purging

A portion of the air inside an empty hydronic system can be expelled by opening the feed water valve to fill the system and allowing the air to rise up to and out of the high point vent(s). This approach is known as **gravity purging**. Air pockets will likely form at all high points in the system. If all these points are not equipped with a venting device, it may be difficult for the system's circulator to establish flow through the system after filling it, especially if a low head circulator is used. The air pockets may take several minutes to exit through low-capacity air vents. This method, although simple in concept, is slow in execution, especially if high point vents are not properly located. Figure 13-20 shows some potential problems when gravity purging alone is used.

Forced-Water Purging

The amount of air that can be quickly expelled from a hydronic system during filling is substantially increased when the entering fluid has sufficient velocity to entrain air bubbles and carry them to an outlet valve. The greater the water pressure available from a well, water main, or dedicated purging pump, the faster forced-water purging can push air out of the system.

One method of forced-water purging is to include a purging valve on the boiler inlet line as shown in Figure 13-21, and numerous other schematics in this text.

To fill and purge the system:

1. Close the inline ball portion of the purging valve.
2. Open the outlet port of the purging valve.
3. Connect a hose to the outlet port of the purging valve and fix the end of it into a bucket or over a floor drain.
4. Lift fast-fill lever on the feed water valve.
5. Fully open the feed-water bypass valve, if present.

Water will now be flowing into the system as fast as possible. Because the inline ball within the purging valve is closed, the entering water first fills the boiler, and then is forced out through the remainder of the distribution piping. Some of the bulk air in the system will be ejected through the air separator, and the remainder of it will be forced around the distribution system and exit through the outlet port of the purging valve.

When the water stream exiting the discharge hose is free of air bubbles for at least 30 seconds, most of the bulk air will have been purged from the system. At this point, the fast-fill lever is returned to normal position, and the feed-water bypass valve, if present, is fully closed.

In systems with two or more branch circuits, it is best to purge each branch individually. This is done by closing valves in all but one branch while purging. When flow exiting the purge valve of one branch circuit is free of air bubbles, the valve in the next branch circuit is opened and the valve in the purged branch is closed. This process is repeated until each branch has been purged. This allows the maximum possible flow rate through each branch to dislodge and entrain as much air as possible. This is especially helpful on radiant panel systems or home run distribution systems having several parallel circuits. Manifold valves can be used to open and close each circuit as needed.

After all branches have been purged individually, open all branch valves and continue purging. The reduced hydraulic resistance of the fully open distribution system maximizes purging flow and helps dislodge any remaining air pockets in larger piping and components. If the system has a differential pressure bypass valve, it should be temporarily closed during purging.

Placing the purging valve near the inlet connection of the boiler helps ensure that debris such as small solder balls or dirt in the piping are flushed out of the system, rather than into the boiler's heat exchanger.

It is very important to install purging valves on the return side of every secondary circuit in a primary/secondary system as shown in Figure 11-56. The closely spaced tees that couple each secondary circuit to the primary circuit will *not* induce sufficient purging flow into the secondary circuit, even if there is a strong flow in the primary circuit.

The author does not recommend the installation of a ball valve between the tees at a primary/secondary interface for use during purging. Although such an arrangement will allow for purging of the secondary circuit, the residual pressure drop of the ball valve between the tees is undesirable. It increases the potential for undesirable flow in the secondary circuit once the system is operational. This must be avoided.

Building and Using a Purge Cart

For $400 to $500 worth of materials, it's possible to build a device that makes short work out of filling and purging residential and light commercial hydronic systems. It also serves as a tool for filtering system water, as well as adding antifreeze. This device is called a **purge cart**. It consists of a minimum 1-horsepower swimming pool

13.6 Filling and Purging a System

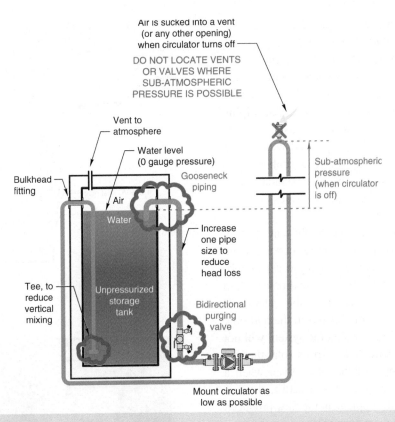

Figure 13-19 | Piping details for hydronic circuits connected to a non-pressurized thermal storage tank.

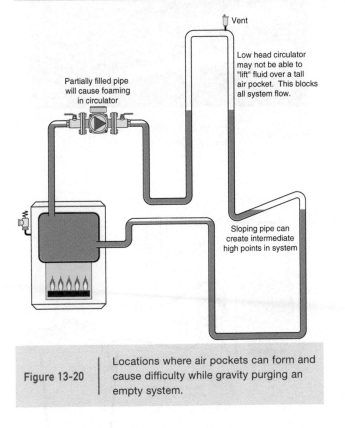

Figure 13-20 | Locations where air pockets can form and cause difficulty while gravity purging an empty system.

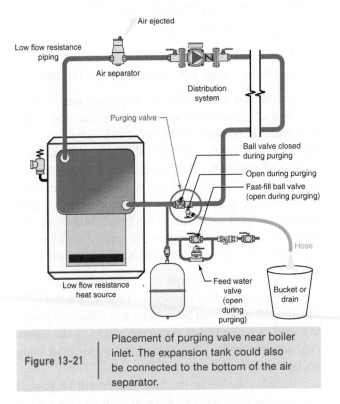

Figure 13-21 | Placement of purging valve near boiler inlet. The expansion tank could also be connected to the bottom of the air separator.

pump mounted on a 30-gallon plastic trash barrel as shown in Figure 13-22. The hoses that connect the purge cart to the system should be reinforced to accommodate a pressure of at least 40 psi.

By installing a union on the pump's intake pipe, the assembled purge cart can be easily disassembled for transport as seen in Figure 13-23. *Author's note: Technically, anything called a "cart" should probably* have wheels. However, after having used this "cart" on many projects, I've concluded they would add significant cost without adding significant convenience.

The piping arrangement that works best with a purge cart is shown in Figure 13-24. The layout is very similar to that shown for use with a purge valve. The piping connections for the purge cart hoses are usually made using 1-inch full-port ball valves equipped with 1-inch barbed PVC connectors. The reinforced rubber hoses are secured to the barbed connectors using standard hose clamps. This piping allows for high flow rates and thus effective purging of bulk air.

The system's make-up water assembly may or may not be present, or active, when using a purge cart to fill and purge. In some cases, the building's supply water plumbing may not be operating by the time the hydronic system needs to be commissioned. In other cases, the water quality at the site may not be suitable for use in the hydronic system. Still another possibility is that the system will not have a make-up water system. A drain-back solar thermal combi-system is an example of the latter. Even when a make-up water assembly is present and active, there is seldom any need to use it to supplement the flow produced by a purge cart with a 1-horsepower or larger pump.

Filling and Purging a System Using a Purge Cart

The following procedure describes how the purge cart is used to fill the system and remove entrapped air bubbles.

STEP 1. Before turning on the purge cart pump, make sure that all hoses are secured, the suction line of the pool pump is primed, and the inlet and outlet valves on the system are open. The barrel should be almost full of water, and at least one parallel branch of a multi-branch system should be open. If the system volume is more than about 25 gallons, have a hose or some 5-gallon buckets ready to add water to the barrel. Make sure that the return hose is held or otherwise secured firmly inside the barrel, or you could get a short but intense shower!

STEP 2. Turn on the purge cart pump. The fluid level in the barrel will drop rapidly. In 10 to 15 seconds, the initially full barrel will be reduced to a minimum pumping level of 4 to 6 inches above the bottom of the barrel. Add water to the barrel as needed to keep air from being sucked into the foot valve.

Initially, the air in the heat source will be pushed ahead of the water and exit through the return hose. After a few seconds of pump operation, a solid water stream will appear in the return hose. Hold the end of the return hose just below the surface of the fluid in the barrel to prevent air from being entrained down into the fluid. You

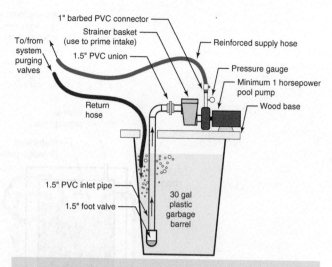

Figure 13-22 | The components of a purge cart.

Figure 13-23 | Purge cart parts broken down for transportation.

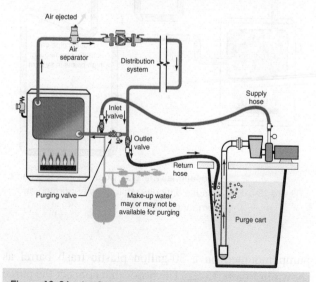

Figure 13-24 | Connecting a purge cart to the system.

may notice that some occasional air bubbles exiting the return hose as air pockets are displaced.

STEP 3. Operate the purge cart for at least 10 minutes after a solid stream of water first returns to the barrel. On multiple branch systems, open one branch at a time to achieve a high flow velocity in each branch. Be aware that zone valves may make a whining sound due to the high flow rate. If multiple zone circulators are used, purge one zone at a time by closing the isolation flanges flanking these circulators. After each zone circuit has been purged individually, open all the circuits to increase the system flow rate. Allow the purging to continue for 2 or 3 more minutes.

STEP 4. The shutdown procedure is simple. Close the outlet valve on the system to stop return flow to the barrel. In a few seconds, the purge cart pump will have added all the fluid it can to the system as it reaches its maximum (no flow) pressure differential. This will probably be around 20 psi. Now close the inlet valve, and then shut off the purge cart pump. If the system pressure is higher than desired, drain some water from the system. At this point, nearly all (non-dissolved) air bubbles should be out of the system. The remaining (dissolved) air will be removed by the microbubble air separator over several subsequent heating cycles. Be sure that the feed water valve or other automatic feed water device is set to add water to the system as the remaining air is collected and vented.

Pre-Filtering Using a Purge Cart

The purge cart can also be used to pre-filter system water to remove some sediments before that water is pumped into the system. This is accomplished by recirculating the fluid through a cartridge filter assembly as depicted in Figure 13-25. A number of different filter cartridges are available with various filtration abilities. It is best to start with a coarser filter and work down to smaller particle size limits. Some very fine sediments may still not be completely removed.

Adding Antifreeze Using a Purge Cart

The purge cart can also be used to mix and inject antifreeze into a system. Before adding antifreeze, however, it is always advisable to operate the system for a short time with water. This will confirm the presence of any leaks and the proper functioning of system components. If a leak is found, or a system component must be removed, at least antifreeze will not be lost or spilled. Glycol-based antifreezes tend to leave a slippery film on components and tools that eventually turn sticky. It is much easier to correct any piping problems without this added mess.

It is also advisable to internally clean the piping system to remove contaminant such as solder flux, residual oil

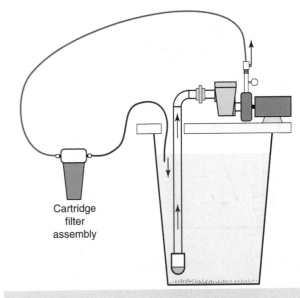

Figure 13-25 Setup for pre-filtering system water using purge cart and cartridge filter.

films created during manufacturing, or other debris. This process is described in more detail later in this chapter.

STEP 1. Calculate the system volume using the methods presented in Chapter 12, Expansion Tanks.

STEP 2. If the system uses an automatic feed water system, temporarily turn it off to prevent drained water from being automatically replaced.

STEP 3. Calculate the amount of antifreeze required using Equation 13.1:

Equation 13.1:

$$V_{\text{antifreeze}} = (\%)(V_s + V_{b\,\text{min}})$$

where,
$V_{\text{antifreeze}}$ = volume of antifreeze required (gallon)
% = desired volume % antifreeze required for a given freeze point (decimal %)
V_s = calculated volume of the system (gallon)
$V_{b\,\text{min}}$ = minimum volume of fluid required in the barrel of the purge cart to allow proper pumping (gallon)*

*The minimum volume of fluid required in the barrel of the purge cart will depend on how the foot valve is located. It should be determined by experimenting with the purge cart (using water). Reduce the water level in the barrel until the pump begins to draw air into the foot valve. Make sure that the return hose is located to minimize turbulence around the foot valve. Using a permanent marker, draw a prominent line on the side of the barrel at least 1 inch above this level. This is the minimum operating level of the barrel. Write "Minimum Operating Level" on the outside of the barrel just above this line. Carefully

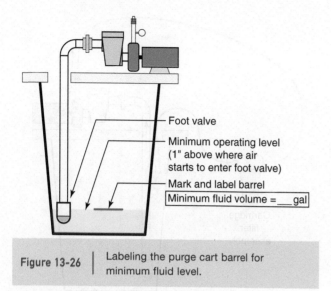

Figure 13-26 | Labeling the purge cart barrel for minimum fluid level.

measure out the water in the barrel and note its volume. Write this minimum operating volume (in gallons) on the side of the barrel as shown in Figure 13-26.

STEP 4. Drain an amount of water from the system equal to the required volume of antifreeze calculated in Step 3. This water should be removed from a low point in the system such as the boiler drain valve.

STEP 5. Attach the purge cart hoses to the system as shown in Figure 13-24. Keep both inlet and outlet valves closed.

STEP 6. Fill the barrel to its minimum operating level with water, then pour in the required volume of antifreeze calculated in Step 3. Turn on the purge cart pump. Open the inlet and outlet valves to the system.

STEP 7. Follow the same purging procedure described earlier in this section. Allow the mixture to circulate through the system for at least 10 minutes to thoroughly mix the water and antifreeze. Open and close valves as necessary to purge branch circuits one at a time and then all together to obtain maximum purging flow.

STEP 8. When no visible air is being returned from the system to the barrel, close the outlet valve on the system. Let the purge cart pressurize the system, then close the inlet valve. If the system pressure is higher than desired, drain some of the fluid back into the barrel through the return hose.

STEP 9. The system should now contain the desired concentration of antifreeze and be purged of all but dissolved air. Turn off the purge cart and disconnect the hoses. Reopen the shutoff valve on the make-up water line.

STEP 10. The volume of mixed fluid in the barrel should be very close to the minimum operating fluid level from which this procedure began. This mixed fluid should be neatly poured into clean/sealable containers and labeled. Be sure to include the exact fluid used, the percentage mixture, and the date the system was filled. This fluid can be saved for the next job that will use the same fluid and percentage mixture. It can then be used to make up the minimum operating volume of the barrel. This will reduce the amount of the same antifreeze used on the next job to just that required by the system. In such a case, Equation 13.1 should now be modified to Equation 13.2:

Equation 13.2:

$$V_{\text{antifreeze}} = \frac{(\%)(V_s)}{(1 - \%)}$$

Where:

$V_{\text{antifreeze}}$ = volume of antifreeze required (gallon)
% = desired volume % antifreeze required for a given freeze point (decimal %)
V_s = calculated volume of the system (gallon)

13.7 Make-Up Water Systems

Almost all hydronic systems experience minor water loss over the course of a heating season. These losses often go unnoticed because the water evaporates before visible drops can form. The presence of scale or discoloration near valve stem packings, pump flange gaskets, air vents, or threaded joints usually indicates minor water losses. Eventually, such losses allow the system pressure to drop. This may lead to other problems such as cavitation or air entry into the system.

To "make up" for these minor losses, it has been customary to install an automatic **make-up water assembly** between the building's cold water plumbing system and the heating system. This assembly usually consists of a pressure-reducing feed water valve, backflow preventer, and at least one isolation valve. Examples of such an assembly are shown on many piping schematics throughout this text, as well as in Figure 13-27.

Note that the feed-water (pressure-reducing) valve has been piped to bypass the fast-fill ball valve. This arrangement minimizes the pressure drop through the latter during purging, which slightly increases the available purging flow rate.

During normal operation, the make-up water assembly automatically feeds small quantities of water to the system to replace the minor water losses and maintain the pressure set on the feed water valve.

Unfortunately, a standard make-up water assembly can also be a liability. Since it reacts to *any* decrease in system pressure, it will continuously feed water if a leak develops in the system. In situations where occupants

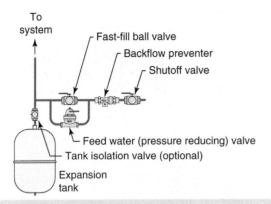

Figure 13-27 | The components in a typical make-up water assembly.

are not present, or not aware of the leak, considerable water damage to the building could occur. If the leak occurs in a system filled with antifreeze, an automatic make-up water system could eventually add enough water to dilute the solution to the point of freezing. This could lead to further equipment damage.

Opinions vary on the merits of leaving an automatic make-up water assembly "active," versus turning it off until the system pressure needs a boost. Arguments can be made both ways. The author's opinion is that an automatic make-up water assembly should be turned off once all portions of the system have been in operation for about 2 weeks. This allows sufficient time for removal of the dissolved air initially in the system. When the system is periodically serviced, the make-up water system should be activated to restore the system to normal pressure. The owner should also be advised of the need to periodically check system pressure and turn on the make-up water system when necessary to maintain pressure. This minimizes the possibility of uncontrolled water supply should a leak develop in the system.

Fluid Feeders

An alternate means of automatically adding fluid to a hydronic system is shown in Figure 13-28. This automatic **fluid feeder** uses a small positive displacement pump operated by a pressure sensing switch to automatically pump fluid into the system when the pressure drops below a set minimum value.

The amount of fluid that can be fed into the system is limited by the storage volume of the fluid feeder. The unit shown in Figure 13-28 holds approximately 6 gallons. It can be placed on the floor or supported by wall brackets. By monitoring the fluid level in the reservoir, it is possible to know how much fluid has been added to the system. If the reservoir is emptied, the fluid feeder provides an electrical contact closure that can activate an alarm and prevent the feeder pump from further operation. Most fluid feeders can also be used to premix water with glycol or other additives, or to circulate the fluid through a cartridge filter assembly before pumping it into the system.

The use of a fluid feeder eliminates the need to pipe a hydronic system to a potable water system. It also eliminates the potential to feed unlimited amounts of water into a system should a leak develop. Considering these benefits, look for the increasing use of such devices in residential and light commercial hydronic systems.

13.8 Dirt & Magnetic Particle Separation

Older hydronic systems often used cast-iron boilers, steel piping or copper tubing, low head circulators with mechanically coupled motors, and either cast-iron radiators of finned-tube baseboard. All of these components

(a)

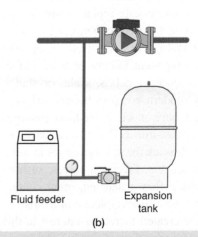
(b)

Figure 13-28 | (a) Example of an automatic fluid feeder. *Courtesy of Axiom Industries Ltd.* (b) Use of a fluid feeder to maintain pressure and volume of earth loop circuit in ground source heat pump system.

could tolerate a certain amount of dirt in the system. The presence of some dirt in the system was usually accepted as inevitable, and there were very few measures undertaken to minimize or eliminate it.

Some of the dirt could have originated from the components used to assemble the system. Many of these components are handled, transported, and stored multiple times between manufacturing and installation. During that time, they can be exposed to insects, sand, sawdust or drywall dust. Careless handling, storage, or installation often puts this debris inside a new system.

Progressive corrosion of ferrous metals such as steel or cast iron is another source of debris in hydronic system. Such corrosion is especially aggressive in systems that lack oxygen diffusion barriers on polymer pipe, or use components, such as cast-iron circulators, in open-loop applications. The latter is specifically warned against by circulator manufacturers, yet occurs due to lack of knowledge, or simply defaulting to a lower cost cast-iron circulator instead of a more expensive corrosion-resistance stainless steel circulator.

The combination of oxygen and ferrous metal creates a dark brown colored sludge known as magnetite. The sludge typically settles in areas of the system operating at low flow velocity, such as inside cast-iron boiler sections or radiators. Over time, the accumulation of magnetite can coat heat transfer surfaces reducing heat transfer rates and equipment efficiency. Magnetite can also accumulate inside circulators, especially those with electronically commutated motors. The result can be jammed rotors that require the circulator to be replaced. Figure 13-29 demonstrates the attraction force of the rare earth magnets in the rotor of an electronically commuted circulator. Figure 13-30 shows a heavy accumulation of magnetite on this type of permanent magnet rotor.

Systems with persistent leaks, or with undersize or waterlogged expansion tanks, can exacerbate oxygen-based corrosion by allowing fresh water, containing dissolved oxygen, to continuously replace water losses in what is supposed to be a closed-loop system.

As hydronics technology has advanced, so has the potential for serious component failures or poor performance due to the presence of debris, scale, or sludge in the system. Many modern compact boilers and heat pump heat exchangers have much small fluid passages compared to sectional cast-iron boilers. These passages are more easily clogged, as are the small passages in some brazed plate heat exchangers. Precision valves, such as those used for automatic flow balancing, often contain small mechanisms that are also less tolerant of debris.

These issues have created increased interest in dirt and magnetic particle separation in modern hydronic systems. Several devices have been developed to

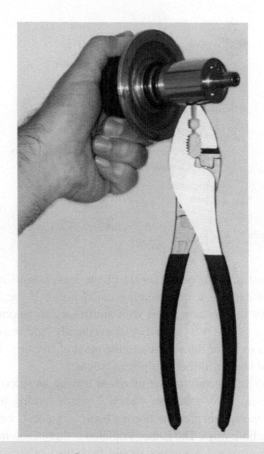

Figure 13-29 Strong attraction force created by rare earth magnets in the rotor of an electronically commutated circulator. *Courtesy of Caleffi North America.*

Figure 13-30 Heavy accumulation of magnetite on the rotor of an electronically commuted circulator. *Courtesy of Caleffi North America.*

allow dirt and magnetic particle separation to be easily integrated into new systems, as well as retrofit into older systems.

Basket Strainers

One type of dirt separation device that has been used for many decades is called a **basket strainer**, or sometimes as a **Y-strainer**. These devices force all flow moving along a pipe flow through a basket-shaped stainless steel screen. Depending on how fine the screen is, various particles that are carried along with the flow are captured inside the basket. When the accumulation of dirt is deemed unacceptable, the strainer body is temporarily isolated from the remainder of the system and opened. The screen is removed, flushed of debris, and reinstalled into the strainer. Figure 13-31 shows a cross section of a typical basket strainer.

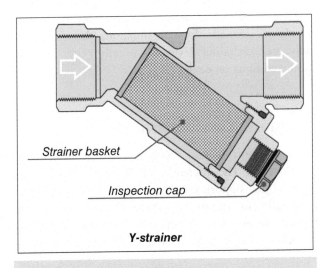

Figure 13-31 | Cross section of a basket strainer. *Courtesy of Caleffi North America.*

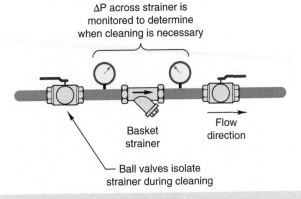

Figure 13-32 | Typical installation of a basket strainer with pressure gauges and isolation valves.

The most common way to install a basket strainer is between a pair of pressure gauges and a pair of isolating ball valves, as shown in Figure 13-32.

The two-pressure gauges are used to determine the pressure drop across the strainer due to accumulation of debris. When the pressure drop across the strainer reaches some predetermined value, it's time to clean the screen. The two-ball valves allow the basket strainer to be isolated while it is opened and cleaned.

Due to their design, basket strainers always create some non-trivial pressure drop. This is evidence of head energy, imparted to the fluid by a circulator, being dissipated. The loss of head energy always has an associated operating cost, as described in Chapters 6 and 7.

The ancillary hardware needed to properly install and service a basket strainer also adds to installation cost and time relative to other dirt separation options. Finally, most currently available basket strainers are not equipped for magnetic particle separation.

Low-Velocity-Zone Dirt Separators

A more contemporary approach to dirt separation is to create an area of low fluid velocity and allow dirt particles to simply "fall" out of the flow stream due to gravity. In this type of separator, the flow velocity is reduced, often by a factor of 9 or more. This reduces the tendency for dirt particles to remain entrained in the flow stream. A specially designed media within the **low-velocity-zone dirt separator** further impedes dirt entrainment. The dirt particles drop out of the active flow region of the separator and collect in the bowl at the bottom of the separator. When a valve at the bottom of this bowl is opened, the accumulated dirt is "blown down" (e.g., expelled) to a hose or bucket. A cut-away illustration of a low-velocity-zone dirt separator is shown in Figure 13-33.

A low-velocity-zone dirt separator has significantly less head loss and pressure drop in comparison to a basket strainer of equal pipe size, even when the latter has a clean screen. Figure 13-34 compares the pressure drop of one low velocity separator with 1-inch piping connections to that of a 1-inch pipe size basket strainer with a clean screen, and the same basket strainer with a screen that is 70% loaded with debris.

Lower head loss and pressure drop reduces the circulator power required for a given flow rate. This reduces long-term operating cost.

Because sediment accumulates below the flow stream, low-velocity-zone dirt separators can operate for relatively long periods between blowdowns. Furthermore, flow through the system does not need to stop during the blowdown procedure.

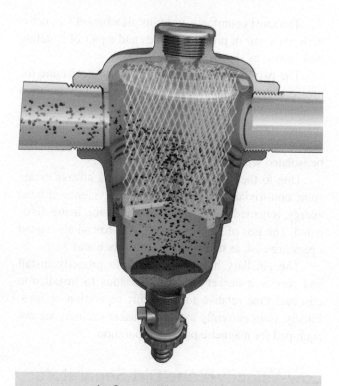

Figure 13-33 | Cut-away illustration of a low-velocity-zone dirt separator. *Courtesy of Caleffi North America.*

flow rates up to 4 feet per second. Eventually, these separators can remove particles as small as 5 micrometers (approximately 0.0002 inch). This dimension is less than 1/10th the diameter of a human hair, and significantly smaller than the particle size that can be captured by a typical Y-strainer.

Dirt separators should be installed upstream of critical components such as boilers, heat pumps, heat exchangers, and circulators. In most systems, a single dirt separator is sufficient to protect most or all of the components in a system, provided those components all share the same fluid.

Several companies now offer combined air & dirt separators. The upper portion of the component provides microbubble air separation. The lower bowl of the component provides low-velocity-zone dirt separation. While these combined-function devices reduce installation time and expense, the ideal placement of an air separating device (e.g., highest fluid temperature and low pressure), versus that of a dirt separating device (e.g., upstream of heat sources, heat exchangers, etc.) do not always coincide. In such situations, the efficacy of individual separating devices is usually better than the combined function device.

Magnetic Dirt Separators

Most hydronic systems contain cast iron or steel components. The presence of these ferrous metals creates the opportunity for iron oxides to form, especially in situations where the fluid contains higher concentrations of dissolved oxygen. Iron oxide particles, also known as magnetite, are attracted to

The ability of a low-velocity-zone dirt separator to remove particles varies with the number of passes the fluid within the system makes through the separator. Larger particles are separated first, with progressively smaller particles being captured after additional passes. Some separators of this type can remove nearly 100% of small sand particles in sizes greater than 100 micrometers (approximately 0.004 inches) when operating with

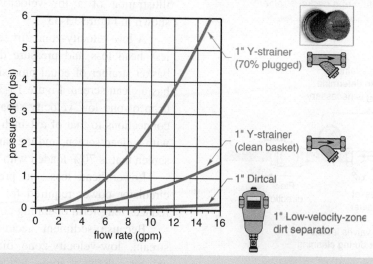

Figure 13-34 | Comparison of pressure drop versus flow rate for a 1-inch pipe size basket strainer versus a 1-inch pipe size low-velocity-zone dirt separator.

magnetic fields. The most common location of such fields is within wet-rotor circulators. Although circulator manufacturers have continually improved the ability of their products to keep magnetic particles out of the fluid-filled space between the rotor and rotor can, experience shows that the possibility for such accumulation is still present. In extreme cases, the iron oxide particles can accumulate to the point of jamming the circulator's rotor, especially after the circulator has been off for several months. This usually requires the circulator to be replaced.

High-efficiency wet-rotor circulators with electronically commutated motors create the highest attraction of magnetic particles due to the very strong rare earth magnets within their rotors. These circulators also have smaller gaps between the rotor and rotor can to improve motor efficiency. These smaller gaps present a greater risk if magnetic particles can migrate into them.

Even if the rotor of a circulator is not jammed, the presence of iron oxide in the gap between the rotor and rotor can will significantly reduce the circulator's wire-to-water efficiency.

Fortunately, methods and hardware have been developed to trap magnetic particles before they can create problems in the system. This is possible using **magnetically enhanced, low-velocity-zone dirt separators**.

The ability of a low-velocity-zone dirt separator to capture iron oxide particles is improved by installing powerful and removable rare-earth magnets on or within the separator. Iron oxide particles are attracted to these magnets, improving the capture efficiency of the separator. When the magnet is temporarily removed from the separator, the iron oxide particles, along with other debris, can be flushed out through the drain valve at the bottom of the separator bowl.

Figure 13-34 shows a magnetically enhanced low-velocity-zone dirt separator. In this product, the black band around the lower portion of the brass bowl holds the powerful magnets in place. The brass body of the separator is non-magnetic, and thus, does interfere with the magnetic attraction of the iron oxide particles.

The magnets used in this collar maintain their strength over time, allowing the separator to continue capturing iron oxide particles as they form.

Magnetically enhanced dirt separators are also available for mounting in vertical piping, and in a wide variety of pipe sizes suitable for both residential and commercial systems.

The author recommends the use of magnetically enhanced dirt separators in all systems using circulators with electronically commuted motors. An example of such a separator is shown in Figure 3-35.

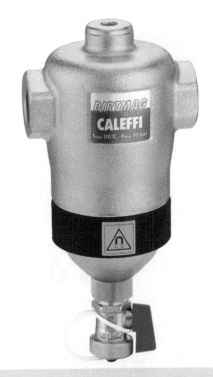

Figure 13-35 | Example of a magnetically enhanced low-velocity-zone dirt separator. *Courtesy of Caleffi North America.*

13.9 Internal Cleaning of Hydronic Systems

Another requirement in achieving high quality and reliable hydronic systems is to internally clean the piping and components to remove contaminants other than air, dirt, or iron oxide particles. These contaminants include soldering flux, cutting oil used in threading pipe, pollen, insect deposits, road salt, and other internal films that are the result of manufacturing, transportation, and storage. These contaminants, if left in the system, can react with metals or antifreeze solutions to create potential corrosion or sludges within the system.

It is relatively simple to internally clean hydronic systems during their commissioning. Specialized solutions that the author refers to as **hydronic detergents** are available that can break down the oil and grease films found in hydronic systems. One example of such a product is shown in Figure 13-36.

Internal washing of most residential and light commercial systems can be done using a clean 5-gallon plastic pail, a small submersible pump, two double-ended hoses, and a bidirectional purge valve. Figure 13-37 shows an example of the latter.

Set up these components as shown in Figure 13-38.

Figure 13-36 — Example of a hydronic detergent used to internally wash hydronic systems. *Courtesy of Fernox.*

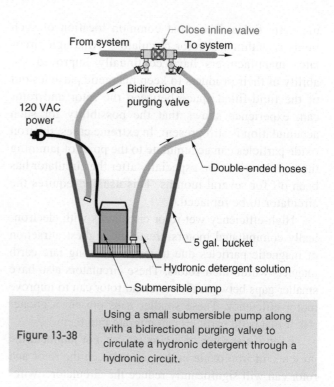

Figure 13-38 — Using a small submersible pump along with a bidirectional purging valve to circulate a hydronic detergent through a hydronic circuit.

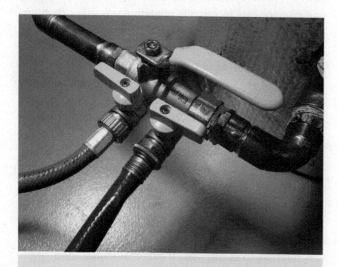

Figure 13-37 — Example of a bidirectional purging valve. The two side port straddle the inline ball valve. The inlet and outlet valves are closed and the inline ball valve is fully open.

The internal washing procedure starts by filling the system with water. Next, put just enough water in the bucket to allow the submersible pump to operate without drawing air into its inlet. Pour in enough hydronic detergent to achieve the manufacturer's recommended concentration based on the volume of water in the system. Close the inline ball of the bidirectional purging valve and open the two side ports. Hold or clamp the end of the return hose to ensure that all return flow stays in within the bucket. Turn on the submersible pump to circulate the hydronic detergent through the system. If possible, turn on the heat source and bring the system up to normal operating temperature. This enhances the cleaning ability of the hydronic detergent solution to break down oils and greases and suspends them in the circulating fluid. Maintain circulation for a minimum of one hour. Be sure that all branch circuits are open.

At the end of the wash cycle, the fluid in the system should be completely drained and disposed of based on the detergent supplier's directions. Most of the currently available hydronic detergents are biodegradable and can be disposed of in sewer drains. The piping should get a final water rinsing to remove as much of the wash solution as possible.

13.10 Water Quality in Hydronic Systems

Water is the "life-blood" of all hydronic systems. As such, it's important to maintain its physical and chemical properties such that systems operate reliably over long service lives.

Pure water contains only hydrogen and oxygen. It is colorless, tasteless and odorless. However, pure water does not exist in nature. Due to its solvent nature, all water from wells, springs or municipal distribution systems contains other materials. Examples include dissolved minerals such as calcium and magnesium, sediment such as fine silica sand or organic particles, dissolved gases such as oxygen, carbon dioxide and hydrogen sulfide, and microorganisms such as bacteria or algae. There can be wide variations in the impurities contained in water from one geographic location to another.

There are also wide variations in the materials used to construct hydronic systems. Examples include copper, steel, cast-iron, brass, bronze, stainless steel, lead, and polymers such as polyethylene and polypropylene. In addition to the chemical properties, these materials contribute to the system, components made of these materials can carry impurities such as soldering flux, cutting oil, manufacturing oil films, dirt, insects, pollen, sawdust, and drywall dust into a system as it is assembled.

Some hydronic systems also contain *intentionally added* chemicals, such as propylene glycol or ethylene glycol antifreeze solutions. These chemicals can interact with other chemicals in the water, as well as the wetted internal surfaces of solid materials.

The dissolved oxygen content of the water also plays a role in determining chemical reactions that might take place within the system. The presence of dissolved oxygen should be minimized through proper purging and ongoing capture and elimination of dissolved gases using microbubble air separators, as discussed earlier in this chapter.

Older hydronic systems, especially those converted from steam to hot water, often contain sludge formed by oxidized iron. This sludge contains very fine particles that can become lodged between moving parts in valves and circulators, which impedes or totally stops the required motion.

There are virtually unlimited combinations of initial water chemistry, contaminants within the system, added chemicals, materials of construction, and dissolved oxygen content, which can potentially set up undesirable chemical reactions within the system.

Categorizing Water Quality

The "ideal" water for use in hydronic systems would contain very little soluble substances such as ions of calcium, magnesium, or sodium. It would also be free of insoluble substances such as oils, grease, fine sand or metallic particles. It would be non-aggressive toward any of the metal, polymer or elastomeric materials in the system. It would have very low electrical conductivity and not contain microorganisms. Unfortunately, water in this "ideal" form is impossible to find from common sources such as wells or municipal water systems.

Water quality describes the extent that "real" water varies in its characteristics relative to "ideal" water. In the context of hydronic systems, water quality can be divided into two categories:

- **Physical water quality**
- **Chemical water quality**

Physical water quality is maintained by capturing and eliminating air, dirt, and magnetic particles from the system. Techniques and hardware for this have been described earlier in this chapter.

Chemical water quality is maintained by modifying or eliminating various chemical substances in water to prevent unintentional and undesirable chemical reactions that could shorten system life or create a need for excessive maintenance. *Ideally, every new hydronic system* should have its chemical water quality adjusted to optimal conditions.

Unfortunately, many North American hydronic systems have been installed with very little attention to chemical water quality. Many have been filled with whatever water is available on site. Some will provide several decades of operation because of a "fortunate" set of circumstances involving the water chemistry, the materials used in the system, the care taken when assembling the system, and its subsequent operation and maintenance. Other systems will experience premature component failure due to unanticipated chemical reactions that lead to scaling, corrosion, component replacement, and leakage.

If on-site water goes into a system equipped with good air and dirt removal details, that system will probably operate fine *initially*. However, over time, problems associated with poor *chemical* water quality can appear. That appearance may take the form of visible discoloration or odor when a sample of water is drawn from the system. It might also be detected as a low pH value indicating that the water has turned acidic. It might also appear as small "pinhole" leaks through copper tubing or other thin-wall components. The latter usually indicates widespread internal corrosion that may require extensive component replacement and repiping.

Undesirable Chemical Characteristics of Water

There are several undesirable characteristics that may be present in the water available at the installation site. One of those characteristics is **turbidity**, which is a measure of the optical clarity of water. Turbidity is caused by the scattering of light due to very fine particles—less than 1 micron in diameter—suspended within water. These particles make the water appear cloudy when a sample is viewed through a clear container. Turbidity should not be confused with water color. It is possible to have a sample of water that shows a color tint due to the specific ions it contains, but without turbidity.

Turbid water can be treated before it is added to a hydronic system by using a filtering system capable of removing sub-micron size particles. One common approach is called **step down filtration**. Two or more **spin-down filters** are piped in series with a **sub-micron cartridge filter**, as shown in Figure 13-39.

Spin-down filters are different than cartridge filters. The latter use a disposable "cartridge" filtering element. A spin-down filter is designed to allow the internal screen to be flushed, in place, using pressurized water. A ball valve downstream of the filter is closed, and the flushing valve at the base of the filter is opened. Pressurized water enters from the upstream side to flush accumulated sediment out the lower valve.

Larger particles in the range of 700 down to 15 microns are captured by one or two spin-down filters. Smaller particles that pass through the upstream filter(s), are captured by the sub-micron filter. This approach prolongs the life of the sub-micron filter cartridge.

Another undesirable water condition is the presence of microbes that can live and reproduce within copper tubing when water temperatures remain between 75 and 135 °F. These microbes feed on carbon contained in drawing lubricant residue within the tubing. Once established, these microbes will acidify the grain boundaries in the copper causing slow erosion that eventually leads to pinhole leaks.

Some microbes also feed upon iron within a hydronic system. Their presence is usually indicated by water with a reddish or brown color. Some bacteria will also produce slimes inside the system and pungent odors when the water is sampled. Cutting oils and residual solder flux can also serve as food sources for certain microbes.

The best defense against microbial corrosion is to remove their food source by internally cleaning the system as described in the previous section.

Scaling is one of the most common undesirable characteristics of water in hydronic systems. It is caused by minerals such as calcium and magnesium, which enter a hydronic system with the fill water. These minerals typically exist as chemical compounds called salts, which are dissolved in the water. Examples include: calcium carbonate ($CaCO_3$), calcium phosphate ($Ca_3(PO_4)_2$), calcium sulfate ($CaSO_4$), calcium chloride ($CaCl_2$), magnesium chloride ($MgCl_2$), and sodium chloride ($NaCl$).

The common term used to describe the presence of these salts in water is **hardness**. The greater the amount of dissolved salts, the "harder" the water.

Salts are formed when positively charged ions called **cations** combine with negatively charged ions called **anions**. The resulting chemical bond produces a compound that is electrically neutral. However, most salts dissolve in water causing the electrically neutral molecules to separate into positive and negatively charged ions, as depicted in Figure 13-40.

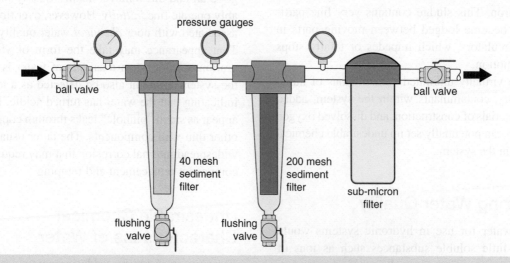

Figure 13-39 | Step down filtering system using two spin-down filters and a sub-micron cartridge filter to reduce turbidity of water before it enters a hydronic system.

13.10 Water Quality in Hydronic Systems

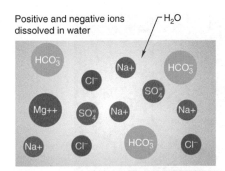

Figure 13-40 | Representation of positively charged ions (a.k.a. cations) and negatively charged ion (a.k.a. anions) dissolved in water.

The ability of water to contain these ions depends on its temperature. Some salts like sodium chloride (NaCl) are *more* soluble in warmer water. The higher the water temperature, the greater the amount of dissolved salt ions it is capable of holding **in solution** (e.g., without formation of any solid particles).

Salts formed from calcium or magnesium become *less* soluble as water temperature increases. If water containing these salts is heated above the saturation temperature, some amount of these compounds will precipitate out of solution to form solid deposits on surrounding surfaces.

The higher the temperature of the surface in contact with the water, the greater the potential for scale formation. This is why scale often forms on the bottom of tea kettles. It's also why scale forms on the wetted internal surfaces of boiler heat exchangers, as seen in Figure 13-41.

Mineral scaling can significantly reduce the heat transfer rate across heat exchangers, as well as within boilers and other hydronic heat sources, and thus it needs to be minimized.

Water Softening

The most common treatment for "hard" domestic water containing dissolve minerals is to "soften" it using an **ion exchange process**. This is what occurs in a typical residential water softener. As water flows through the softener, positively charged calcium ions (Ca^{++}), magnesium ions (Mg^{++}), and manganese ions (Mn^{++}) in the water make their way into thousands of tiny polymer resin beads contained in a vertical cylinder. Figure 13-42 shows a magnified view of these resin beads, which are porous and only about 0.6 mm in diameter.

The resin beads are made of organic polymers that have a higher attraction for calcium, magnesium and manganese ions compared to their attraction for sodium ions. The chemical reaction that occurs within the resin bead column exchanges *two* sodium ions (Na^+) for each calcium, magnesium or manganese ion captured. The net result is an *exchange* of sodium ions for calcium, magnesium and manganese ions. The sodium ions bond to other ions within the beads, creating salts such as sodium sulfate. These salts do not have the scaling potential of the ions originally in the water, but they do leave the water with relatively high electrical conductivity, which is *not* desirable within hydronic systems because it can accelerate undesirable

Figure 13-41 | Mineral scale deposits formed on the water side of a boiler heat exchanger. *Courtesy of Caleffi North America.*

Figure 13-42 | Magnified view of resin beads used for ion exchanger processes. *Courtesy of Caleffi North America.*

conditions such as galvanic corrosion. For this reason, *water softening is not recommended for water that will be used in hydronic systems.*

Demineralizing Water

The preferred process for removing *nearly all* minerals from water is known as demineralization. This process causes the undesirable cations and anions in the water to be exchanged with either *hydrogen cations (H^+)* or *hydroxide anions (OH^-)*. The latter two ions then recombine to form H_2O (e.g., pure water). In effect, the original cations and anions "disappear" from the source water and are replaced with molecules of pure water.

Although the results are quite different, the *process* of demineralizing water is similar to that of softening water. The source water containing minerals is passed through a vertical column filled with very small polymer resin beads, as illustrated in Figure 13-43.

The resin beads used for demineralization are different from those used for water softening. They are created with specific ions that are chemically bonded to porous polymer materials. These ions are "fixed" within the beads. They can react with other ions but cannot leave the beads.

Some beads are formulated to attract positively charged cations such as calcium (Ca^{++}), magnesium (Ma^{++}), and sodium (Na^+). When a cation such as (Ca^{++}) enters a bead, it is captured, and as a result, *two* hydrogen cations (H^+) are released from the bead. The two positive charges of the entering cation (Ca^{++}) are balanced by the positive charges of the two exiting cations (H^+).

A similar ion exchange occurs when a single magnesium cation (Mg^{++}) enters a resin bead and is exchanged for *two* hydrogen (H^+) cations.

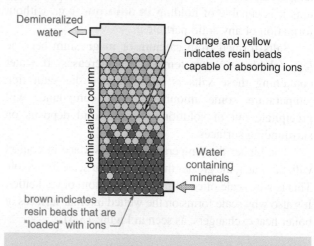

Figure 13-43 | Illustration of a demineralization column containing resin beads that absorb undesirable ions as water passes through.

When a sodium cation (Na^+) is captured within a bead, only one (H^+) cation is released.

In all cases, the total positive charges of the cations captured equal the total positive charges of the hydrogen cations released from the bead. Thus, the bead always remains electrically neutral.

Other beads are created with ions that attract negatively charged ions (e.g., anions) in the water. For example, when a chlorine anion (Cl^-) enters a bead, the resulting reaction captures it and releases a single hydroxide anion (OH^-). When a carbonate ion (CO_3^{--}) enters a bead, it is captured, and the resulting reaction releases *two* hydroxide ions (OH^-). The total negative charges of the captured anions are balanced by the total negative charges of the anions leaving the bead.

The final reaction between the released ions is the key to why demineralization, rather than softening, is used for water destined for hydronic systems. The hydrogen cations (H^+) released from the beads immediately react with the hydroxide anions (OH^-) released from other beads to form (H_2O) (pure water), as shown in Figure 13-44.

The resin beads used in demineralization eventually reach a point where they are fully loaded with undesirable ions and can no longer exchange undesirable ions for those that combine to form pure water. At that point, the resin beads either have to be regenerated or replaced. Regeneration uses processes involving strong acid and base chemicals, and is not practical for the amount of beads used to demineralize water in smaller hydronic systems. Instead, it is more cost effective to discard the fully loaded beads, and replace them with new beads. The spent beads are not toxic. They only contain the ions that were in the source water. In most cases, they can be disposed of in recyclable trash. Most of the hardware used for demineralizing water for hydronic systems package the beads in small easily handled water-permeable bags making disposal and replacement very easy.

The ion content of demineralized water can be assessed by measuring its electrical conductivity. The more ions the water contains, the higher its conductivity.

In the absence of *all* ions, water has zero electrical conductivity. Water in this state is *NOT* desirable in hydronic systems because it seeks to reabsorb ions from surrounding surfaces, which can lead to metal erosion and other corrosion reactions. Water with zero conductivity would also prevent proper operation of low water cutoff devices, such as those required by code in many hydronic systems.

A common way to express the extent to which water has been demineralized, which is based on electrical conductivity, is through an index called **Total Dissolved Solids** (abbreviated as TDS). The fewer the ions in the water, the lower the TDS reading. TDS values are expressed in parts per million (abbreviated as PPM).

To ensure proper operation of low-water cutoff devices, the water used in hydronic systems should have a TDS reading of at least 10 PPM. This provides sufficient conductivity for proper operation of a low-water cutoff, and prevents metal erosion. The maximum suggested value of TDS for water in a hydronic system is 30 PPM. This limits electrical conductivity, and thus, discourages galvanic corrosion and other corrosion reactions that depend on electrical currents passing through water.

Figure 13-44	Specifically formulated resin beads create ion exchange that strips undesirable ions out of water and replaces them with ions that recombine to form pure water.

Figure 13-45	Example of a meter to measure total dissolved solids (TDS) and pH of a liquid. *Courtesy of Caleffi North America.*

The TDS value of water can be determined using an inexpensive meter such as the one shown in Figure 13-45. This meter can also be set to read the pH value of the water.

Demineralizing Devices

One way to fill a hydronic system with demineralized water is to purchase that water from an industrial water treatment company. However, the logistics of sourcing this water, transporting it to the installation site, and minimizing waste tend to often make this option less cost-effective than demineralizing the water found at the site.

Equipment for on-site demineralizing is available from several North American suppliers. One example of a **demineralizing column** is shown in Figure 13-46.

A demineralizing column is essentially a container that can be filled with several permeable bags containing resin beads. The upper lid unscrews to allow access to these bags within the cylindrical body. Valves with hose thread connections allow hoses to connect between the column and the system as shown in Figure 13-47.

Demineralizing Procedure

The following procedure assumes that the system has been pressure tested and confirmed to be free of leaks. It also assumes that the system has been flushed of physical debris and chemically cleaned as previously described in this chapter.

Begin by connecting short double-ended hoses, such as those used to connect washing machines, between the demineralizing column and a pair of inlet/outlet burning valves as shown in Figure 13-47. Leave the hose connection at the inlet purging valve slightly loose to allow air to escape as the demineralizing column is filled with water from the system.

Close the inline ball valve between the inlet/outlet purging valves.

Open the outlet purging valve to allow water to flow from the system into the bottom of the demineralizing column. The system's automatic make-up water system should supply this flow. The water level will rise up through the resin bead column forcing air ahead of it. Once the water stream has filled the inlet hose, tighten the hose connection at the inlet purging valve.

Turn on the system circulator(s) to create flow within the system. With the inline ball valve closed, all flow will be forced through the demineralizing column.

Allow circulation through the system to continue for several minutes. Then draw a sample of system water from another drain valve in the system into a clean plastic or glass cup. Take a TDS reading of this water sample using a handheld meter such as the one shown in Figure 13-45.

In most cases, the initial TDS reading will drop rapidly as the untreated water makes its first pass through the demineralizing column. The rate at which the TDS level drops will decrease with time. Continue circulating the system water through the demineralizing column until the sample shows a TDS reading in the range of 10 to 30 PPM. At that point, the system water has been adequately demineralized. The inline ball valve can be opened. The inlet/outlet purging valves can be closed, and the hoses disconnected.

The water remaining within the demineralizing column is demineralized water. It can be left in the cylinder and used on another system, unless there's a possibility of freezing, in which case the water should be drained through the lower valve.

Figure 13-46 Example of a commercially available demineralization column. *Courtesy of John Siegenthaler.*

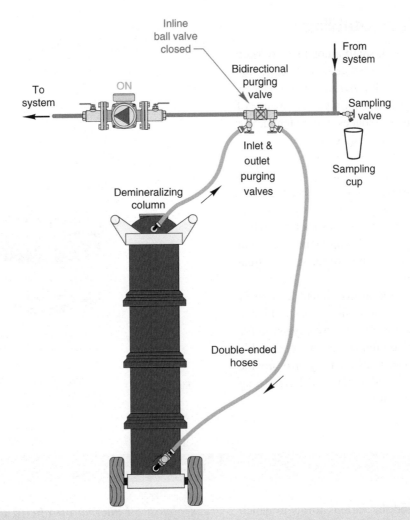

Figure 13-47 | Piping and hose connections to enable water in system to be demineralized. *Courtesy of John Siegenthaler.*

Smaller, cartridge-style demineralizers are also available for treating water admitted to the system over time through the automatic make-up water system.

Final pH Adjustment of Demineralized Water

The **pH of a substance** is a number that expresses its acidity or alkalinity. A pH of 7.0 is neutral (e.g., neither acidic or alkaline). pH values lower than 7.0 indicate acidity, while pH values greater than 7.0 indicate alkalinity.

Demineralizing water can lower its pH. Although demineralization removes most ions from water, it does not remove dissolved gases such as nitrogen, oxygen, and carbon dioxide. Demineralized water will react with carbon dioxide to form a weak carbonic acid (H_2CO_3), which lowers the water's pH.

It is possible for the pH value of demineralized water to *initially* drop into the range of 5 to 6 depending on the amount of carbonic acid formed. This pH value is lower than desired. However, the pH of demineralized water in closed-loop hydronic systems has demonstrated a tendency to rise with time. That time can be minimized by operating the system at its design heating temperature after the water has been demineralized. This helps drive dissolved carbon dioxide out of solution to form microbubbles which can then be captured by the microbubble air separator and ejected from the system.

If system components such as the boiler require a specific pH value *when the system is installed*, it is possible to increase the water's pH by adding a "pH boost" chemical. This chemical may be part of an overall water treatment formulation, such as a film-forming water stabilizer.

Film-Forming Water Stabilizers

Regardless of the water used, small amounts of oxygen can enter closed-loop hydronic systems over time due to molecular diffusion through polymer tubing, gaskets and pump seals. Good system detailing, such as use of microbubble air separators, reduces the potential for oxygen to negatively affect water quality.

The use of a **film-forming water stabilizer** product, specifically formulated for the multiple metals found in many new hydronic systems, further increases the long-term protection against oxygen-related corrosion. Figure 13-48 shows an example of such a product specifically formulated for closed-loop hydronic system applications. One gallon of this solution can treat 50 gallons of system water.

Film-forming water treatments create extremely thin molecular films on metal surfaces that minimize interaction between those metals and any oxygen in the system water. These films do not create any measurable reduction in convective heat transfer. They also provide pH buffering of demineralized water, and long-lasting stabilization of the system's water.

Figure 13-48 | Example of a film-forming water stabilizer product. *Courtesy of Rhomar Water.*

Summary

This chapter has discussed the various forms in which air can enter and exit in a hydronic heating system. A thorough understanding of how air behaves will allow many of the classic complaints associated with hydronic heating (poor heat output from certain heat emitters and gurgling sounds) to be avoided entirely. This knowledge also gives installers the ability to diagnose and correct previously installed systems exhibiting the classic symptoms of entrapped air. The bottom line is simple: *If properly designed and installed, all hydronic systems should automatically rid themselves of almost all internal air after a few days' operation. They should then maintain themselves virtually air-free throughout their service life.*

Dirt separation and magnetic particle separation have also been discussed. High quality hydronic systems use air, dirt, and magnetic particle separation to ensure long life of critical components, and sustain the initial performance of heating and cooling source equipment.

Finally, this chapter has discussed the importance of high water quality in hydronic systems. It has also described hardware and procedure for demineralizing water.

Key Terms

air binding
air purger
air scoop
anions
basket strainer
bidirectional purging valve
bleeders
bulkhead fitting
cations
central air separators
chemical water quality
coalescing media
deaerated
demineralizing column
dissolved gases
entrained air
film-forming water stabilizer
float-type air vent
fluid feeder
forced-water purging
free oxygen
gaseous cavitation
gooseneck piping
gravity purging
hardness
hematite (Fe_2O_3)
high point vents
hydronic detergents
hygroscopic air vent
in solution
incompressible fluids
ion exchange process

low-velocity-zone dirt separator
low-velocity-zone dirt separator
magnetically enhanced, low-velocity-zone dirt separators
magnetite (Fe_3O_4)
make-up water assembly
manual air vent
microbubble
microbubble air separators
oxides
pH of a substance
physical water quality
purge cart
rise velocity (of a bubble)
saturation pressure
spin-down filters
stationary air pockets
step down filtration
sub-micron cartridge filter
subatmospheric pressure
Total Dissolved Solids
turbidity
unsaturated state of air solubility
waterlogged
Y-strainer

Questions and Exercises

1. Describe why a waterlogged expansion tank can lead to recurring air problems in a hydronic system.

2. What flow velocity is necessary to entrain air bubbles with the flow in a downward flow direction?

3. Will a bubble rise faster through water or a mixture of water and antifreeze? (Hint: Consider the viscosity of each fluid.)

4. Describe why microbubbles form on the inside surfaces of boiler sections as the boiler heats up. How does this affect heat transfer from the boiler wall to the fluid?

5. Describe the difference between an air purger and a microbubble air separator.

6. Describe how a microbubble air separator is able to remove air pockets from remote areas of a system.

7. Why should a 3/4-inch purge valve not be installed in large-diameter main piping?

8. A system is filled with water at 60 °F and 15 psi. The water contains the maximum amount of dissolved air it can hold at these conditions. The water is then heated to 160 °F and maintained at 15 psi. How many gallons of air have been given off in the process?

9. What is the best location for a central deaerator in a typical residential or light commercial system? Why?

10. Describe a situation in which air is drawn into the system through a vent each time the circulator operates.

11. How would you test an installed diaphragm expansion tank to see if its diaphragm has ruptured?

12. A customer describes air noises in flow passing through the air purger. What might be preventing the air from being separated and vented?

13. Describe a situation where a purge cart would be desirable over the use of forced-water purging through a feed water valve.

14. Does a purge cart remove dissolved air from water? Why?

15. Why is it necessary to hold the end of the return hose to a purge cart below the fluid level in the barrel?

16. A hydronic system consists of a 15-gallon sectional cast-iron boiler, 250 feet of 3/4-inch copper tubing, eight heat emitters holding 0.5 gallons each, and a 2-gallon expansion tank. It is desired to protect the system from freezing to temperatures as low as 0 °F. The minimum usable volume of the purge cart used to fill the system is 4 gallons. How much propylene glycol is needed to produce a 20% solution for this system? Use the Fluid Properties Calculator module in the Hydronics Design Studio software to determine the freezing point of this mixture.

17. What is the advantage of piping the feed water (pressure-reducing) valve so that it bypasses the fast-fill ball valve in a make-up water assembly?

18. What are two advantages of using an automatic fluid feeder instead of a make-up water assembly?

19. Explain why the use of a low-velocity-zone dirt separator could lower the operating cost of a hydronic system in comparison to that of the same system using a basket strainer.

20. Explain why circulators with electronically commutated motors need protection against magnetite.

21. T/F The ideal water in a hydronic system would have a TDS value of less than 5 PPM.

22. T/F Purging a system with hot water will remove all necessary internal dirt and films.

23. Describe how it is possible for a demineralizer to create some pure H_2O.

24. Which type of water treatment should be used for water intended for hydronic systems (and why)? a. softened water, b. demineralized water, c. strained water, d. pre-boiled water

25. Explain why water with high electrical conductivity is not desirable in hydronic systems.

Chapter 14

Auxiliary Loads and Specialized Topics

Objectives

After studying this chapter, you should be able to:

- Describe how heat exchangers are used in hydronic systems.
- Explain how heat exchanger performance is determined.
- Describe multiple methods for heating domestic water in hydronic systems.
- Calculate the energy requirement associated with domestic water heating.
- Understand how to add a storage tank to a boiler with a tankless coil.
- Describe methods for intermittent garage heating.
- Explain how a hydronic system can heat a swimming pool.
- Describe installation details for snow and ice melting (SIM) systems.
- Discuss control options available for SIM systems.
- Estimate the load associated with a SIM system.
- Identify appropriate applications for buffer tanks.
- Size a buffer tank for a given application.
- Describe three piping configurations for buffer tanks.
- Explain the benefits of minitube distribution systems.
- Describe heat metering systems and their application.
- Understand the fundamentals of hydronic system balancing.
- Explain two different procedures for hydronic system balancing.

14.1 Introduction

Although space heating is the primary load served by most residential and light commercial hydronic systems, they can also provide heat to other **auxiliary loads**. Examples include domestic water heating, garage heating, pool heating, and snow melting. Through proper design, a single heat source, or multiple heat source system, can serve as the universal heat source for the multiload system. This approach reduces installation costs relative to using separate heat sources for each load. It may also reduce fuel consumption when boilers serve as the heat source as the result of increased duty cycle and thus higher efficiency.

The feasibility of supplying space heating as well as auxiliary heating loads depends on their magnitude, and when they occur. The best results are obtained when auxiliary heating loads do not occur at the same time as peak space heating loads. A good example is heating a pool during mild weather when space heating loads are small or nonexistent. If an auxiliary heating load is small in comparison with the space heating load, it might be met using the surplus capacity of the heat source, especially during milder fall and spring weather.

Another possibility is to prioritize loads. This approach, described in Chapter 9, temporarily suspends heat input to predetermined low priority loads so full boiler capacity can be directed to a high priority load. Once the high priority load is satisfied, heat output is shifted back to the lower priority loads. Such intelligent management of boiler capacity minimizes the total boiler capacity needed.

This chapter also covers specialized components that do not fit within the context of earlier chapters, but are nonetheless very useful in many modern hydronic systems. Heat exchangers are one example of such components.

Finally, this chapter discusses specialized topics and applications in modern hydronics technology. Examples include heat metering and minitube distribution systems. These topics are often a melding of the fundamentals discussed in earlier chapters. In the right situations, they offer significant benefits and cost reductions relative to "standard" approaches.

14.2 Liquid-to-Liquid Heat Exchangers

Several hydronic loads require heat to be transferred from water heated by a boiler or other heat source to another liquid operating at a lower temperature. **Liquid-to-liquid heat exchangers** allow for such heat transfer while also preventing the two liquids from mixing.

There are several types of heat exchangers used in residential and light commercial hydronic systems. Fundamentally, all the hydronic heat emitters discussed in Chapters 8 and 10 are heat exchangers. A stream of water carries heat to these heat emitters. They release the heat to room air (by convection), and to room surfaces and objects within the room (by thermal radiation). The thermal performance characteristics of these heat emitters is similar to those of liquid-to-liquid heat exchangers described in this chapter.

Heat exchanger selection involves both thermal and hydraulic performance issues. Thermal issues relate to the ability of the heat exchanger to transfer heat from one liquid to another at a specific rate, given the entering temperatures and flow rates. Hydraulic issues relate to the head loss and pressure drop associated with moving liquids through the heat exchanger at the required flow rates. Failure to address either issue can result in poor performance.

Selection must also consider chemical compatibility between the materials in the heat exchanger and the liquids passing through it. For example, some heat exchangers are partially constructed of carbon steel. If the so-called "dead" water within a closed hydronic system is circulated through such a heat exchanger, corrosion due to oxidation is seldom a problem. However, if *potable* water were circulated through the carbon steel portion of the heat exchanger, the dissolved oxygen in that water could quickly corrode the steel. Not only would this damage the heat exchanger, it would also contaminate the fresh water. To avoid this, heat exchangers designed to heat (or cool) fresh water must be made of materials such as copper, cupronickel, and stainless steel.

Pool water with high concentrations of chlorine can also be very corrosive to certain materials, including some grades of stainless steel. The heat exchangers used for heating pools often use titanium or titanium/stainless steel alloys for all surfaces that contact pool water.

Heat exchangers are often categorized as being "internal" or "external." In the context of hydronic systems, **internal heat exchangers** are typically helical coils made of a metal tubing with high thermal conductivity. These coils are typically suspended within a liquid-filled container such as a thermal storage tank. One fluid passes through the tubing, the other fluid contacts the outer surface of the tubing. An **external heat exchanger** is any heat exchanger that is not immersed in a fluid inside another component such as a tank. Unless the term "internal" is used to describe the heat exchanger, assume that it is external and is typically just referred to as a heat exchanger.

Internal Heat Exchangers

Figure 14-1 shows an example of a copper coil heat exchanger that was designed to be suspended within a nonpressurized thermal storage tank, such as used in systems supplied by a cordwood gasification boiler. This heat exchanger is made by simultaneously coiling four separate lengths of copper tubing to form the overall coil. The four separate tubes are joined to headers at each end of the coil to provide one inlet and one outlet connection. This configuration allows a relatively high

Figure 14-1 | A copper heat exchanger coil with four parallel circuits, and designed to be suspended within a thermal storage tank. *Courtesy of American Solartechnics.*

surface area to be created while keeping the head loss of the coil much lower compared to a single tube coil of the same surface area.

Figure 14-2 shows how two heat exchanger coils, one for heat input and the other for heat extraction, would be suspended within a nonpressurized thermal storage tank.

Piping from each coil passes through the tank's sidewall above the highest water level. The piping penetrations are sealed to limit evaporative water loss, but are typically *not* intended to be installed below the water level.

Note how the flow direction through each coil is opposite the direction in which tank water moves by buoyancy. Flow inside the heat input coil is from top to bottom. The water absorbing heat from this coil becomes more buoyant than the surrounding water and slowly rises. Flow inside the heat extraction coil is from bottom to top. The water adjacent to this coil is cooled as it gives up heat. Its density increases causing it to flow downward. These relative flow directions between the fluid providing heat, and the fluid absorbing heat create a condition called **counterflow heat exchange**. A fundamental concept of applying heat exchangers is that counterflow operation always yields a higher rate of heat transfer compared with operating the same heat exchanger with both fluids moving in the same direction. The rationale for this will be discussed shortly.

Flow inside an internal coil heat exchanger is typically driven by a circulator. As such, heat transfer between the fluid and the internal wall of the heat exchanger is classified as **forced convection**. Heat transfer between the outer surface of the coil and the tank water is driven solely by **natural convection**.

The rate of heat transfer due to forced convection or natural convection can be calculated using Equation 14.1.

Equation 14.1:

$$q = hA(\Delta T)$$

where,

q = rate of heat transfer by convection (Btu/hr)

h = **convection coefficient** (Btu/hr/ft² °F)

A = surface area separating the fluids exchanging heat (ft²)

ΔT = difference in temperature between the fluid providing heat and the fluid absorbing heat (°F)

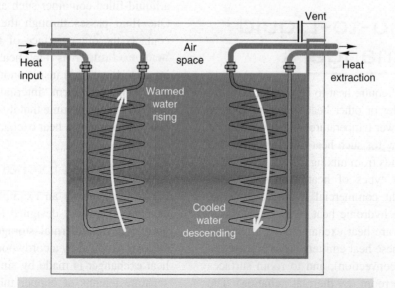

Figure 14-2 | A nonpressurized thermal storage tank with two internal coil heat exchangers, one for heat input and the other for heat extraction.

Although Equation 14.1 is mathematically simple, the process of estimating the convection coefficient (h) can be very complex. The value of (h) depends on the physical properties of the fluid (e.g., its density, dynamic viscosity, specific heat, and thermal conductivity). It also depends on the flow velocity and geometry of the surface that's accepting or providing heat to the fluid. Heat transfer textbooks provide methods for estimating the value of (h) for both forced convection and natural convection. These methods often involve empirical equations based on dimensionless numbers such as the Reynolds number discussed in Chapter 6.

Figure 14-3 gives some practical ranges for the value of the convection coefficient (h) for natural or forced convection involving air or water. The wide range of values imply that widely different rates of convective heat transfer are possible depending on the fluids and the operating conditions of the heat exchanger.

Convective heat transfer increases with increasing fluid speed. This happens because a layer of fluid called the boundary layer, which clings to surface gaining or loosing heat, gets thinner as the fluid's velocity increases. The boundary layer as well as the relative velocity of a fluid moving through a pipe is shown in Figure 14-4.

The thinner the boundary layer is, the lower the thermal resistance between the bulk of the fluid stream and the surface. Less thermal resistance allows for higher rates of heat transfer between the fluid molecules in the bulk of the stream and the tubing wall.

The relationship between convective heat transfer and fluid speed across a surface can be seen in many hydronic systems. For example, the faster water flows through a finned-tube baseboard, the higher its heat output, assuming all other conditions remain the same. This effect was described in Chapter 8. Figure 14-5 illustrates this effect for a typical residential grade finned-tube baseboard.

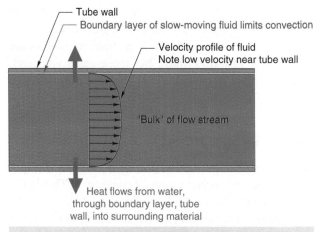

Figure 14-4 | Flow velocity profile within a pipe. Boundary layers are seen between the bulk of the flow stream and inner tube wall.

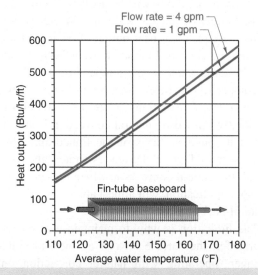

Figure 14-5 | Differences in heat output of residential finned-tube baseboard at flow rates of 1 and 4 gallons per minute.

At any given water temperature, the heat output occurring at a flow rate of 4 gallons per minute (gpm) is higher than at a flow rate of 1 gpm. This increase, although slight, is due to a thinner boundary layer between the bulk of the fluid stream and the inner tube wall of the finned-tube element.

In general, the faster one or both fluids flow through any type of heat exchanger, the higher the rate of heat transfer between those fluids. *From the standpoint of heat transfer only, there is no such thing as the water moving "too fast" through any type of heat exchanger.* However, other considerations such as pumping power, noise, or long-term erosion of materials impose practical limits on flow velocity through heat exchangers.

application	range of convection coefficient (h) (Btu/hr•ft²•°F)
natural convection involving air	1–5
forced convection involving air	3–100
convection (natural or forced) involving water	20–3,000

Figure 14-3 | Convection coefficient ranges for air and water undergoing forced and natural convection heat transfer.

Figure 8-39 also showed that changes in the rate of heat transfer based on changes in flow rate are very "nonlinear." Large gains in heat transfer rate occur at relatively low flow rates, but these gains are much smaller as flow rates increase beyond the normal range of operation.

Natural vs. Forced Convection

Natural convection is typically a much "weaker" form of heat transfer compared to forced convection. This is the result of much slower fluid motion created by buoyancy differences in the fluid versus faster fluid motion created by a circulator or blower. Slower fluid motion increases boundary layer thickness, which creates greater resistance to heat transfer.

The difference between forced convection and natural convection explains why a small wall-mounted fan coil that's only 32 inches wide and 24 inches tall can provide the equivalent heat output of 57 feet of typical residential finned-tube baseboard, when both are operating at the same water supply temperature and flow rate. This comparison is shown, to scale, in Figure 14-6.

The rate of heat transfer from the baseboard is significantly limited by natural convection heat transfer between its outer surfaces of the finned-tube element and the surrounding air. Designers should always keep in mind that natural convection, where present in a heat exchanger application, will almost always be what limits the rate of heat transfer.

This limitation is why, to maintain a given rate of heat transfer, the surface area of an internal coil heat exchanger, operating with natural convection on its outer surface, must be much larger than the surface area of other types of heat exchangers operating with forced convection in both fluid streams.

External Heat Exchangers

Liquid-to-liquid heat exchanger design has evolved over more than a century based on available materials and production methods. The three common types of liquid-to-liquid heat exchangers are:

- Shell & tube heat exchanger
- Shell & coil heat exchanger
- Flat plate heat exchanger

Figure 14-7 illustrates the basic construction of these heat exchangers.

Shell & tube, as well as shell & coil heat exchangers, though widely used for many decades in industrial applications, many involving high pressures and temperatures, are not commonly used in modern hydronic heating or cooling systems; the one exception being shell & tube heat exchangers used for heating swimming pools, which are discussed later in this chapter. The flat plate heat exchanger design now dominates the hydronics market.

Flat Plate Heat Exchangers

Flat plate heat exchangers are made by assembling several preformed metal plates into a stack, and then sealing the perimeters of those plates together. The plates are preformed to create narrow channels between them. One fluid passes from one end of the heat exchanger to the other through the odd-numbered channels (1, 3, 5, 6, etc.). The other fluid passes from one end of the heat exchanger to the other through the even-numbered channels (2, 4, 6, 8, etc.). This concept is simplistically illustrated in Figure 14-8.

Flat plate heat exchangers can be divided into two major categories:

- Brazed plate heat exchangers
- Plate & frame heat exchangers

Brazed Plate Heat Exchangers

As the name implies, the plates that make up a brazed plate heat exchanger are brazed together along their perimeter and at contact points across their inner surfaces. This metallurgically seals the flow channels and creates

FAN-COIL
output = 6,700 Btu/hr
average water temperature = 110 °F
room air temperature = 68 °F
forced convection on air side

FINNED-TUBE BASEBOARD
output = 6,700 Btu/hr
average water temperature = 110 °F
room air temperature = 68 °F
natural convection on air side

|← 57'–0" →|

Figure 14-6 A scaled comparison of a fan-coil unit measuring 24 inches high by 32 inches long, operating with forced convection on the air side of its internal heat exchanger, and releasing the same amount of heat at 57 feet of residential finned-tube baseboard operating with the same average water and entering air temperature. The baseboard operates with natural convection on the air side of its heating element.

14.2 Liquid-to-Liquid Heat Exchangers

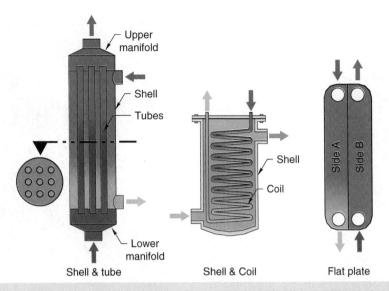

Figure 14-7 | Basic construction of three types of heat exchangers. Note that all are shown in counterflow operation.

Figure 14-8 | Concept of alternating flow channels for two fluids passing through a flat plate heat exchanger. *Courtesy of FlatPlate, Inc.*

Figure 14-9 | Construction of a brazed plate heat exchanger. The four threaded studs on front face are for attaching the heat exchanger to a bracket. *Courtesy of FlatPlate, Inc.*

two separate pressure-rated compartments within the heat exchanger. Once the plates are brazed together, the heat exchanger cannot be modified or opened. Figure 14-9 shows an exploded view of a typical brazed plate heat exchanger.

Prior to assembly, the plates are pressed to create ridges, which create the gaps necessary to form the flow channels, and increase turbulence for improved heat exchange. Most of the plates also have four circular openings, two at each end, which form internal fluid passage when the plates are pressed together. Figure 14-10 shows some typical plates made of stainless steel, and coated with a copper alloy to allow brazing that seals the patterned plates into two isolated flow channels. After the plate stack is assembled, it is pressed together and heated in an oven to approximately 2,000 °F to braze the surfaces together.

Most manufacturers of brazed plate heat exchangers have standardized plates sizes. Typical plate dimensions are 3×8 inches, 5×12 inches, and 10×20 inches. For a given plate size, heat exchange ratings are increased by adding plates to the "stack" that become the overall heat exchanger.

Brazed plate heat exchangers are commonly described by the nominal size of their plates, and the number of plates in the stack. For example, a $5 \times 12 \times 40$ brazed plate heat exchanger has 40 plates, of nominal dimension 5 inches by 12 inches. One unique plate forms the back of the heat exchanger (e.g., it has no holes through it). Another unique plate forms the front

Figure 14-10 Partial stack of preformed stainless steel plates with copper coating for brazed bonding to form internal flow channels within a brazed plate heat exchanger. *Courtesy of Alfa Laval.*

Figure 14-11 Examples of stainless steel brazed plates heat exchangers. Largest heat exchanger is 5" × 12" × 100 plates, medium size heat exchanger is 5" × 12" × 40 plates, and smallest heat exchanger is 3" × 8" × 10 plates. *Courtesy of John Siegenthaler.*

of the heat exchanger, and transitions to the four piping connections.

Figure 14-11 shows three brazed plate heat exchangers: A small 3 × 8 inch × 10 plate, a 5 × 12 inch × 30 plate, and a 5 × 12 inch × 100 plate. The largest heat exchanger has 1.25-inch MPT connections. The medium size exchanger has 1-inch MPT connections. The small heat exchanger has ¾-inch FPT connections.

Most of the brazed plate heat exchangers used in hydronic heating and cooling systems are constructed of 316 stainless steel. This allows them to operate with potable water having up to some amount of chlorine. When chloride concentrations are low, some manufacturers offer heat exchanger made of less expensive 304 stainless steel. Both of these materials are compatible with propylene glycol antifreeze solutions.

Brazed plate heat exchangers are used for many applications other than liquid-to-liquid heat exchange in hydronic systems. Examples include their use as the condenser or evaporator in refrigeration systems, engine oil cooling, beer brewing, and process heating or cooling associated with food production. Because of this broad range of application, most brazed plate heat exchangers have pressure and temperature ratings well above those required in hydronic applications.

Plate & Frame Heat Exchangers

Plate & frame heat exchangers could be described as the "big brother" of brazed plate heat exchangers. They use the same concept of a stack of preformed stainless steel plates to separate the two fluids in alternating channels. However, instead of brazing, plate & frame heat exchangers use gaskets, placed around the perimeter and some interior areas of each plate, to seal the fluid channels. The stack of plates is assembled on a frame between two thick steel pressure plates. When the required number of plates have been loaded onto the frame, several threaded steel rods are used to pull the two pressure plates together and compress the stack into a pressure-tight assembly, as shown in Figure 14-12.

Plate & frame heat exchangers are commonly used when the required rate of heat transfer is higher than what can be provided by a large-brazed plate heat exchanger.

One distinction of plate & frame heat exchangers is their ability to be disassembled if necessary for cleaning or other service. The tension rods that hold the thick pressure plates together are removed and the plate stack is carefully pried open. Individual plates or gaskets can be replaced, followed by reassembly of the plate stack, and tightening of the tension rods.

Another distinction of plate & frame heat exchangers is that the frame and tension rods are typically sized so that plates can be added to the exchanger, if ever necessary, due to expansion of the original system, or a significant change in operating conditions. The elongated frame and tension rods can be seen to the left of the plate stack in Figure 14-12.

Figure 14-12 | Assembly of multiple preformed and gasketed stainless steel plates between two pressure plates to form a plate & frame heat exchanger. *Courtesy of Alfa Laval.*

Both brazed plate and plate & frame heat exchangers have several advantages over shell & tube, or shell & coil heat exchangers. These include:

- Less material required to generate a given internal heat transfer area.
- Thinner metal walls separating the fluids, which improves conduction heat transfer.
- Compact size relative to other heat exchanger designs of equivalent performance.
- Lower fluid volume and faster response to temperature changes compared to other designs.
- Surface patterns that encourage turbulent flow and thus create higher convection coefficients relative to other designs.
- Ability to operate with very small temperature differences between the two fluids.
- Less prone to the formation of fouling films on internal surfaces compared to other designs.
- Ability to operate with a **temperature crossover**, where the temperature of the cooler fluid exits the heat exchanger at a higher temperature than that of the existing hot fluid. Temperature crossovers are not possible with some other heat exchanger designs.

These advantages are significant, and have led to the current dominance of flat plate heat exchangers in hydronic heating and cooling applications.

Installation Details for Flat Plate Heat Exchangers

To achieve optimal performance, it is essential to keep the internal surfaces of any heat exchanger as clean as possible. Any accumulation of debris or **fouling films** on the internal surfaces adds thermal resistance between the fluid supplying heat, and the fluid accepting heat. This decreases the rate of heat transfer through the heat exchanger. Deposits within the heat exchanger can also cause pitting corrosion over time.

The spaces through which fluids pass in a flat plate heat exchanger, especially those in small-brazed plate heat exchangers, are narrow. When a heat exchanger is installed in a system, especially an older system, where dirt or other contaminants could be present, a **low velocity zone dirt separator** should be installed upstream of both fluid inlets. Figure 14-13 shows suggested piping details.

These separators can prevent debris such as dirt, solder balls, metal chips, or iron oxide particles from entering and possibly lodging within the heat exchanger.

Note that both separators in Figure 14-11 are installed *upstream* of the circulators that provide flow through the heat exchanger. As such the separators protect both the heat exchanger and the circulators.

The circulators have been equipped with purging flanges on their discharge ports. These flanges, in combination with the isolation/purging valves shown, allow each side of the heat exchanger to be isolated from the attached piping and chemically cleaned by circulating a hydronic detergent through the heat exchanger. Procedures for such cleaning were described in Chapter 13. Figure 14-14 shows a typical setup where a small submersible pump is used to circulate a hydronic detergent through one side of a brazed plate heat exchanger.

The brazed plate heat exchangers commonly used in residential and light commercial hydronic systems are typically mounted above floor level in mechanical rooms. Although they are relatively light in comparison to plate & frame heat exchangers, it's important to properly support them to avoid undue stress on the piping connected to them.

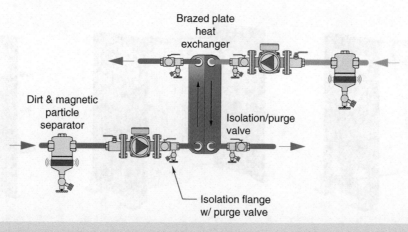

Figure 14-13 | Use of dirt/magnetic particle separators upstream of both sides of a flat plate heat exchanger. The use of an isolation/purge flange on each circulator outlets, in combination with an isolation/purge valve on the piping leaving heat exchanger, allows each side to be isolated and flushed if necessary.

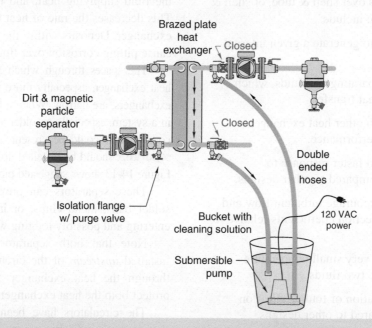

Figure 14-14 | Use of isolation/purging flanges and isolation/purging valves in combination with hoses and a small submersible pump to circulate a cleaning fluid through one side of a flat plate heat exchanger.

Heat exchangers with 5×12 inch, (or larger) plates should be supported by some type of bracket. Figure 14-15a shows an example of a fabricated bracket supporting an insulated 5×12 inch $\times 100$ plate heat exchanger. Figure 14-15b shows a 5×12 inch $\times 40$ plate heat exchanger supported on a small steel angle bracket screwed to a plywood wall. Figure 14-15c shows a brazed plate heat exchanger that was supplied with threaded mounting studs on its end plate. Those studs allow the heat exchanger to be supported by a simple metal bracket.

Small 3×8 inch brazed plate heat exchangers are light enough that they can usually be supported directly by connected piping, as shown in Figure 14-16. That piping should itself be supported within a few inches of the heat exchanger.

Plate & frame heat exchangers are quite heavy, often weighing several hundred pounds. They are typically mounted on a concrete pad that elevates them a few inches above the mechanical room floor. Expansion bolts are used to secure them to the pad, as shown in Figure 14-16a.

Figure 14-15 | Different methods of supporting brazed plate heat exchangers. *Courtesy of John Siegenthaler.*

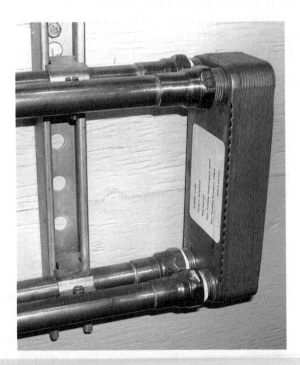

Figure 14-16 | A small 3"x 8" brazed plate heat exchanger supported by piping, with nearby support of that piping. *Courtesy of John Siegenthaler.*

Figure 14-16a | A large plate & frame heat exchanger supported by and fastened to a concrete pad. *Courtesy of John Siegenthaler.*

Thermal Performance of Heat Exchangers

The rate of heat transfer through any heat exchanger depends on several operating conditions. These include the entering temperature and flow rate of each fluid stream, and the physical properties of both fluids (e.g., their specific heat, density, viscosity, and thermal conductivity).

In all cases, the rate of heat transfer is higher when the flow through both sides of the heat exchanger is turbulent rather than laminar. Chapter 6 introduced the concept of Reynolds number as a means of determining if a flow stream is laminar or turbulent. Reynolds numbers over 4000 are generally indicative of turbulent flow. The higher the Reynolds number, the higher the degree of turbulence.

The liquid-to-liquid heat exchangers used in hydronic systems are configured so that the two fluid streams flow mostly parallel to each other. The two parallel streams could be moving in the same direction or in opposite directions. When the two streams flow in the same direction, the configuration is called **parallel flow** or sometimes as **cocurrent flow**. When the two streams flow in opposite directions, the configuration is called **counterflow**. These two flow configurations, along with representative temperature changes of both fluid streams, are shown in Figure 14-17.

The *rate* of heat transfer across any heat exchanger depends on the temperature difference between the fluid providing heat, and the fluid absorbing that heat. The greater this temperature difference is, the higher the rate of heat transfer.

From Figure 14-17, it's apparent that the temperature difference between the hot fluid and cool fluid changes considerably depending on where it is measured within the heat exchanger. For the parallel flow heat exchanger, the temperature difference at the left side, where both fluids enter, is very large. But this difference decreases rapidly as the fluids exchange heat and move toward the outlet ports. There's also a variation in the temperature difference between the fluids as they move through the counterflow heat exchanger.

To evaluate the rate of heat transfer across the entire heat exchanger, it is necessary to establish an "average" temperature difference. This average temperature difference is a value that *if* present at all locations separating the fluids within the heat exchanger would yield the same overall heat exchange rate as would occur with the highly variable temperature difference.

Heat transfer theory can be used to derive this "average" ΔT. It's called the **log mean temperature difference (LMTD)**, and can be calculated using Equation 14.2.

Equation 14.2:

$$\text{LMTD} = \frac{(\Delta T)_1 - (\Delta T)_2}{\ln\left[\frac{(\Delta T)_1}{(\Delta T)_2}\right]}$$

where,
  LMTD = log mean temperature difference (°F)
  $(\Delta T)_1$ = temperature difference between the two fluids at one end of the heat exchanger (°F)
  $(\Delta T)_2$ = temperature difference between the two fluids at the other end of the heat exchanger (°F)
  ln [ ] = the natural logarithm of the quantity in the square brackets.

Equation 14.2 applies to both parallel flow and counterflow heat exchangers. It's also important to understand that it makes no difference which end of the heat exchanger is assigned to be $(\Delta T)_1$ as long as the other end is assigned to be $(\Delta T)_2$.

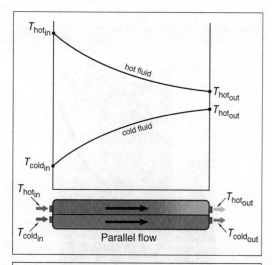

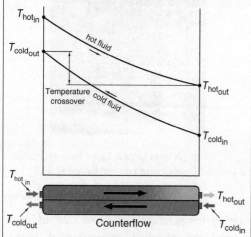

Figure 14-17 | Temperature changes in fluids as they pass through a heat exchanger in parallel flow and counterflow arrangements.

Example 14.1

Calculate the LMTD for the heat exchanger shown in Figure 14-18.

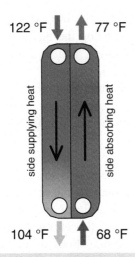

Figure 14-18 Inlet and outlet temperatures for determining LMTD of a flat plate heat exchanger.

Solution:

Let $(\Delta T)_1$ be assigned to the top end of the heat exchanger. Thus $(\Delta T)_1 = 122 - 77 = 45$ °F. This means that $(\Delta T)_2$ is at the bottom end of the heat exchanger. Thus $(\Delta T)_2 = 104 - 68 = 36$ °F. Putting these values for $(\Delta T)_1$ and $(\Delta T)_2$ into formula 4-3 yields:

$$\text{LMTD} = \frac{(\Delta T)_1 - (\Delta T)_2}{\ln\left[\frac{(\Delta T)_1}{(\Delta T)_2}\right]} = \frac{45 - 36}{\ln\left[\frac{45}{36}\right]} = \frac{9}{0.223} = 40.3 \text{ °F}$$

Reversing the ends of the heat exchanger representing $(\Delta T)_1$ and $(\Delta T)_2$ would make $(\Delta T)_1 = 36$ °F, and $(\Delta T)_2 = 45$ °F. Putting these values into Equation 14.2 yields:

$$\text{LMTD} = \frac{(\Delta T)_1 - (\Delta T)_2}{\ln\left[\frac{(\Delta T)_1}{(\Delta T)_2}\right]} = \frac{36 - 45}{\ln\left[\frac{36}{45}\right]} = \frac{-9}{-0.223} = 40.3 \text{ °F}$$

Discussion:

The algebraic signs for the quantities in equation 14.2 automatically compensate for which end of the heat exchanger is selected as $(\Delta T)_1$ and $(\Delta T)_2$. In the case where $(\Delta T)_1$ and $(\Delta T)_2$ are equal, the ratio $(\Delta T)_1/(\Delta T)_2$ would be 1, and the $\ln(1) = 0$. Dividing by 0 is undefined in any equation. In this special case, the LMTD = $(\Delta T)_1 = (\Delta T)_2$.

Heat transfer theory can be used to prove a very important concept in the application of heat exchangers. Namely, that *the LMTD of a heat exchanger configured for counterflow will always be higher than that of the same heat exchanger configured for parallel flow, and having the same entering and existing conditions for both flow streams*. This implies that *heat exchangers should always be configured for counterflow when the goal is to maximize the rate of heat transfer*. The differences in LMTD between counterflow and parallel flow configuration can also affect the required size, and thus the cost of the heat exchanger.

Overall Rate of Heat Transfer (LMTD Method)

The heat transfer rate through a heat exchanger is directly proportional to the LMTD. It is also directly proportional to the surface area that separates the two fluids. Thus, the heat transfer rate is proportional to the multiplication of internal surface area (A) times the LMTD. This can be expressed as the proportionality:

Proportionality 14.3:

$$q \propto A(\text{LMTD})$$

where,

q = rate of heat transfer across the heat exchanger (Btu/hr)
A = internal surface area separating the two fluids within the heat exchanger (ft²)
LMTD = log mean temperature difference established by the current conditions (°F)

This proportionality can be changed to an equation (e.g., with an = sign) by introducing a constant called U.

Equation 14.4:

$$q = UA(\text{LMTD})$$

The constant U is called the **overall heat transfer coefficient** of the heat exchanger. To make the units in Equation 14.4 consistent, the units of U must be Btu/hr/ft² °F. The numerical value of U is the rate of heat transfer through the heat exchanger (in Btu/hr), per square foot of internal surface area, per °F of LMTD.

The U-value of a heat exchanger involves knowing the convection coefficient for both fluid-to-surface interfaces, as well as the thermal conductivity of the wall separating the fluids. In some cases, additional thermal resistances called **fouling factors** are also included in the calculation of the U-value.

Figure 14-19 shows how the U-value for a heat exchanger where the two fluids are separated by a flat plate can be viewed as a sum of several thermal resistances.

The green lines (and curves) in Figure 14-19 represent temperature gradients. The bulk temperature of the hot fluid is represented by the dot labelled (T_1). There is a temperature drop from (T_1) to (T_2) across the convective boundary layer. The extent of this temperature drop depends on the characteristics of the boundary layer. Laminar flow would result in a relatively thick boundary layer and a larger temperature drop from (T_1) to (T_2). Highly turbulent flow would decrease the boundary layer thickness and reduce the temperature drop across it.

Another temperature drop, (T_2) to (T_3), occurs across the resistance associated with a fouling layer. If the heat exchanger is new, or otherwise free of any dirt or accumulated films, the temperature drop from (T_2) to (T_3) would be zero. In that case, the thermal resistance of the fouling film would be zero. If operating conditions allow a fouling film to develop the thermal resistance called fouling factor increases, and so does the temperature drop from (T_2) to (T_3).

Another temperature drop occurs across the metal wall of the heat exchanger that separates the two fluids. Because of the high thermal conductivity of the wall, the temperature drop from (T_3) to (T_4) will be very small, perhaps only a fraction of one degree F. In some situations, this temperature drop is so small relative the resistance of the fluid boundary layers and fouling factors that it's deemed insignificant and not factored into the calculation of the overall heat transfer coefficient (U).

Similar temperature drops occur on the right side of the metal wall, across any fouling film, and the convective boundary layer of the cool fluid. Notice that the temperature drop from (T_5) to (T_6) is smaller than that from (T_2) to (T_3). In this case, the smaller temperature drop is associated with a "thinner" boundary layer

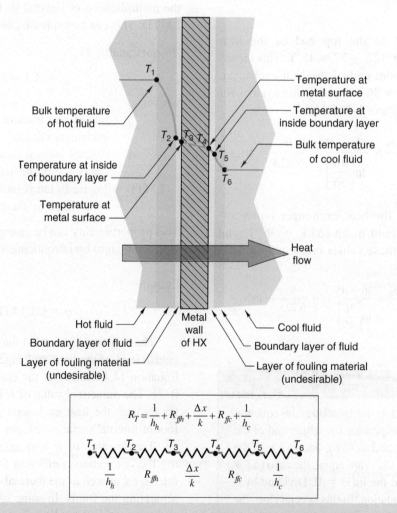

Figure 14-19 | Temperature gradient across the wall of a heat exchanger based on the thermal resistances of boundary layers, fouling films, and metal wall separating the fluids.

on the cooler side of the heat exchanger. The thinner boundary layer could be caused by one or more factors including higher flow velocity, higher Reynolds number, or a surface texture that encourages a higher degree of turbulent flow.

Each of the temperature drops occurs because of a resistance to heat flow. The diagram at the bottom of Figure 14-19 shows these five thermal resistances, with the temperatures (T_1 through T_6) present between them. The numerical value of each thermal resistance can be estimated based on several factors such as fluid properties, flow velocity, temperature, Reynolds numbers, and more. Some of these calculations can be complex and require iteration. Procedures for these calculations are given in heat transfer textbooks.

The overall rate of heat transfer depends on these five thermal resistances. Equation 14.4 can be expanded to show how these resistances influence the overall rate of heat transfer. That expansion is shown as Equation 14.4.

Equation 14.4:

$$q = UA(\text{LMTD}) = \frac{A(\text{LMTD})}{R_T} = \frac{A(\text{LMTD})}{\left[\frac{1}{h_h} + R_{ffh} + \frac{\Delta x}{k} + R_{ffc} + \frac{1}{h_c}\right]}$$

where,

q = rate of heat transfer across the heat exchanger (Btu/hr)
A = internal surface area separating the two fluids within the heat exchanger (ft²)
LMTD = log mean temperature difference established by the current conditions (°F)
h_h = convection coefficient on the hot side of the heat exchanger (Btu/hr·ft² °F)
R_{ffh} = thermal resistance of fouling factor on hot side (°F·hr·ft²/Btu)
Δx = thickness of metal wall of heat exchanger (ft)
k = thermal conductivity of metal wall of heat exchanger (Btu/hr·ft °F)
R_{ffc} = thermal resistance of fouling factor on cool side (°F·hr·ft²/Btu)
h_c = convection coefficient on the cool side of the heat exchanger (Btu/hr·ft² °F)

From Equation 14.4, it follows that the overall heat transfer coefficient (U) can be written as follows as:

Equation 14.5:

$$U = \frac{1}{\frac{1}{h_h} + R_{ffh} + \frac{\Delta x}{k} + R_{ffc} + \frac{1}{h_c}}$$

The values of U are dependent on the fluids exchanging heat, the flow-conditions, and the thermal conductivity of the material separating the two fluids. For water-to-water heat exchangers, a guideline range for values of U is 150–300 Btu/hr/ft² °F.

Numerical values for fouling factors are given in heat transfer reference books or website. The units for fouling factors are the same as for insulation R-values, or other thermal resistances. In the IP units system, those units are (°F·hr·ft²/Btu). Figure 14-20 lists some fouling factors relevant to hydronic heating and cooling applications.

These values show a range of thermal resistances associated with different water conditions. Note that the fouling factor associated with demineralized water is only 17 percent of that associated with untreated river water or untreated cooling tower water. The presence of minerals in flow streams passing through heat exchangers, especially when those fluids are at elevated temperatures, can significantly increase fouling factors and thus decrease rates of heat transfer. This again points to the importance of using high-efficiency dirt separators as well as demineralized water for optimal system performance.

Example 14.2

A brazed plate heat exchanger is used to heat domestic water. Boiler water enters the primary side of the heat exchanger at 150 °F and flow rate of 10 gpm. Cold domestic water enters the secondary side of the heat exchanger at 50 °F and 6 gpm. The plates in the heat exchanger are 0.02 inches thick and made of 316 stainless steel, which has a thermal conductivity of 29 Btu/hr/ft °F. The heat exchanger is new and therefore there are no fouling films on either side of the plates. The heat exchanger is configured for counterflow, as shown in Figure 14-21. Prior calculations by the designers indicate that the convection coefficient on the hot side of the heat exchanger is 250 Btu/hr/ft² °F, and on the cool side is 100 Btu/hr/ft² °F. Assume that the heat exchanger is operating under steady state conditions and insulated so that external heat loss is negligible. Determine the rate of energy flow through the heat exchanger and the required internal area of the plates.

Solution

The sensible heat rate equation from chapter 4 can be used to determine the rate of heat transfer into the heat

Fluid	Fouling Factor (°F·hr·ft²/Btu)
River water (untreated)	0.003
Cooling tower water (untreated)	0.003
Cooling tower water (treated)	0.00102
Demineralized water	0.00051

Figure 14-20 Sample of fouling factor thermal resistances.

exchanger. The values of density and specific heat can be based on the average water temperature on the hot side of the exchanger $(150+135)/2 = 142.5$ °F. From figure 4-4, the density of water at 142.5 °F is 61.3 lb/ft³. From figure 4-1, the specific heat of water at this temperature is 1.00 Btu/lb/°F.

The sensible heat rate equation (Equation 4.8) can be used to calculate the rate of heat transfer.

Equation 4.8 (repeated from Chapter 4):

$$q = (8.01Dc)f(\Delta T)$$

where,
 q = rate of heat transfer (Btu/hr)
 D = density of the fluid (lb/ft³)
 c = Specific heat of the fluid (Btu/lb/°F)
 f = flow rate (gpm)
 ΔT = temperature changer of the fluid (°F)

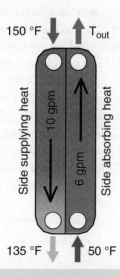

Figure 14-21 Fluid temperatures and flow rates through a flat plate heat exchanger to be analyzed using the LMTD method.

8.01 = constant to balance the units on each side of the equation

Putting the values into Equation 4.8 yields:

$$q = (8.01Dc)f(\Delta T) = (8.01 \times 61.3 \times 1.00)(10)(150-135)$$
$$= 73,650 \frac{\text{Btu}}{\text{hr}}$$

Since the heat exchanger is operating under steady state conditions, and with insignificant external heat loss, the hot side must be releasing heat at the same rate as heat is being absorbed by the cool side. This allows the sensible heat rate equation to be used again, in a rearranged form, to determine the outlet temperature on the cool side.

$$T_{out} = \frac{73,650}{(8.01 \times 62.4 \times 1.00)(6)} + 50 = 74.6 \text{ °F}$$

Since both inlet and outlet temperatures are known, the LMTD can be calculated as:

$$\text{LMTD} = \frac{(\Delta T)_1 - (\Delta T)_2}{\ln\left[\frac{(\Delta T)_1}{(\Delta T)_2}\right]} = \frac{75.4 - 85}{\ln\left[\frac{75.4}{85}\right]} = \frac{-9.6}{-0.1198} = 80.1 \text{ °F}$$

The U value for the heat exchanger can also be determined based on formula 14.5:

$$U = \frac{1}{\left[\frac{1}{h_h} + R_{ffh} + \frac{\Delta x}{k} + R_{ffc} + \frac{1}{h_c}\right]}$$

$$= \frac{1}{\left[\dfrac{1}{250 \dfrac{\text{Btu}}{\text{hr}\cdot\text{ft}^2\cdot\text{°F}}} + 0 + \dfrac{0.00167 \text{ ft}}{29 \dfrac{\text{Btu}}{\text{hr}\cdot\text{ft}\cdot\text{°F}}} + 0 + \dfrac{1}{100 \dfrac{\text{Btu}}{\text{hr}\cdot\text{ft}^2\cdot\text{°F}}}\right]}$$

$$= 71.1 \frac{\text{Btu}}{\text{hr}\cdot\text{ft}^2\cdot\text{°F}}$$

Note that the units for the thickness of the heat exchanger plates was changed from 0.02 inches to 0.00167 ft. This is necessary to maintain consistent units in all terms, yielding the units of Btu/hr/ft² °F for the result.

Equation 14.3 can now be set up and solved for the required internal surface area of the heat exchanger.

$$q = UA(\text{LMTD}) = 73,650 \frac{\text{Btu}}{\text{hr}} = \left(71.1 \frac{\text{Btu}}{\text{hr}\cdot\text{ft}^2\cdot\text{°F}}\right) A(80.1 \text{ °F})$$

$$A = \frac{73,650 \dfrac{\text{Btu}}{\text{hr}}}{\left(71.1 \dfrac{\text{Btu}}{\text{hr}\cdot\text{ft}^2\cdot\text{°F}}\right)(80.1)} = 12.9 \text{ ft}^2$$

14.2 Liquid-to-Liquid Heat Exchangers

> **Discussion:**
>
> The designer can now consult manufacturer's specifications to locate candidate brazed plate heat exchangers with internal surface areas of approximately 12.9 ft².

Figure 14-22 shows the temperature gradients from the hot fluid to the cold fluid for this example.

The temperature gradient is based on the LMTD of 80.1 °F. Approximately 28.5 percent of this overall LMTD occurs across the boundary layer of the hot fluid. Only about 0.4 percent occurs across the stainless steel wall separating the fluids. About 71.1 percent of the LMTD occurs across the boundary layer of the cool fluid. Since the example stated that there were no fouling films, no layers representing these films are shown in Figure 14-22, and there is no temperature drop associated with them. *The temperature drops are directly proportional to the thermal resistance of the boundary layers and the metal wall.* In this case, the temperature drop across the latter is extremely small. Around 71.1 percent of the overall resistance to heat transfer is caused by the boundary layer on the cool side, and thus, 71.1 of the overall temperature drop occurs across this boundary layer. This is directly attributable to the lower convection coefficient on the cool side (100 Btu/hr•ft²•°F) versus the higher convection coefficient on the hot side (250 Btu/hr•ft²•°F). This demonstrates that the rate of heat transfer across a heat exchanger operating with a high convection coefficient on one side and a low convection coefficient on the other side is significantly limited by the latter.

This example did not delve into the complexities associated with calculating the two convection coefficients, (h_h) and (h_c). Procedures for doing so are given in most heat transfer textbooks. In many cases, these values are determined by empirical equations for specific and often simplified geometry situations.

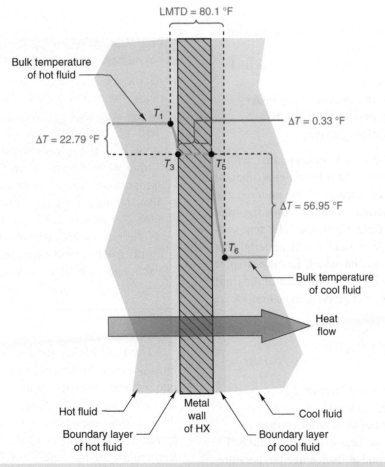

Figure 14-22 | Temperature gradient across a heat exchanger operating with no fouling films, and with significantly different convection coefficients on both sides of the metal wall.

Overall Rate of Heat Transfer (Effectiveness Method)

The LMTD method of analyzing heat exchanger performance is useful when the entering and leaving temperatures of both flow streams are known, or can be calculated based on an energy balance across the heat exchanger. However, when there is insufficient information to calculate the LMTD, the analysis must resort to iterative calculations. These involve making "guesstimates" for the unknown temperatures, calculating heat transfer rates based on these guesstimated temperatures, finding the "error" between the calculated rates of heat transfer on the two sides of the heat exchanger, correcting the guesstimated temperatures, and repeating the procedure. This is possible but very tedious. In these situations, another method of heat exchanger analysis is preferred. That method is based on the concept of "effectiveness" of a heat exchanger.

The effectiveness of a heat exchanger is defined by Equation 14.6.

Equation 14.6

$$\varepsilon = \frac{q_{actual}}{q_{max}}$$

where,

ε = effectiveness of the heat exchanger (decimal number between 0 and 1)
q_{actual} = Actual rate of heat transfer across the heat exchanger (Btu/hr)
q_{max} = Maximum possible rate of heat transfer across the heat exchanger given the operating conditions (Btu/hr)

Effectiveness is based on the concept that with a hypothetical heat exchanger having *infinite* internal area, one of the two fluid streams could undergo the maximum possible temperature change, from the inlet temperature of the cold fluid, up to the inlet temperature of the hot fluid. Which fluid could undergo this temperature change depends on which has the smaller **capacitance rate**, which is the specific heat of the fluid multiplied by its flow rate, and given as Equation 14.7.

Equation 14.7:

$$C = (8.01cD)f$$

where,

C = fluid capacitance rate (Btu/hr °F)
8.01 = a constant based on units
c = specific heat of the fluid at its average temperature in the heat exchanger (Btu/lb/°F)
D = density of the fluid at its average temperature in the heat exchanger (lb/ft³)
f = flow rate of the fluid (gpm)

Example 14.3

Consider a situation where one side of a heat exchanger operates with water at a flow rate of 5 gpm. The other side operates with 50 percent solution of propylene glycol, and a flow rate of 6 gpm. The average temperature of both fluids is 120 °F. Determine the capacitance rate of each fluid stream.

Solution:

The capacitance rate for the water side is:

$$C_{water} = \left(5\,\frac{gallon}{minute}\right)\left(1\,\frac{Btu}{lb \cdot °F}\right)\left(\frac{8.33\,lb}{gallon}\right)\left(\frac{60\,minute}{hour}\right) = 2{,}499\,\frac{Btu}{hr \cdot °F}$$

The capacitance rate for the side operating with propylene glycol is:

$$C_{50\%PG} = \left(6\,\frac{gallon}{minute}\right)\left(0.88\,\frac{Btu}{lb \cdot °F}\right)\left(\frac{8.54\,lb}{gallon}\right)\left(\frac{60\,minute}{hour}\right) = 2{,}705\,\frac{Btu}{hr \cdot °F}$$

Discussion:

Of these two capacitance rates, the one associated with the water side is smaller, and will be designated as C_{min}. The larger capacitance rate will be designated as C_{max}. It's important to note that the physical properties of the two fluids were referenced to an *estimated* average temperature in the heat exchanger. Also note that the units for capacitance rate are Btu/hr °F. One can interpret the value of capacitance rate as the rate of heat transfer (in Btu/hr) that each fluid stream could carry into or out of the heat exchanger per °F- change in temperature of that stream.

The fluid having the lower of the two capacitance rates is one of the factors limiting the maximum possible rate of heat transfer. *That rate, for a hypothetical infinitely large heat exchanger, would be the smaller of the two fluid capacitance rates (e.g., C_{min}), multiplied by the inlet temperature of the hot fluid minus the inlet temperature of the cold fluid.* This allows the rate of heat transfer across the heat exchanger to be written as Equation 14.8.

Equation 14.8:

$$q = \varepsilon(C_{min})(T_{hot\,in} - T_{cold\,in})$$

where,
- q = actual rate of heat exchange across the heat exchanger (Btu/hr)
- ε = effectiveness of the heat exchanger (unitless value between 0 and 1)
- C_{min} = the smaller of the two fluid capacitance rates (Btu/hr•°F)
- $T_{hot\,in}$ = inlet temperature of the hot fluid (°F)
- $T_{cold\,in}$ = inlet temperature of the cold fluid (°F)

Although Equation 14.8 appears relatively simple, there is additional effort involved to determine the value of effectiveness (ε).

Two additional quantities need to be defined in order to determine the value of effectiveness (ε).

One is called the **capacitance rate ratio**. This is simply the *smaller* fluid capacitance rate of the two fluid streams, divided by the *larger* fluid capacitance rate, stated as Equation 14.9.

Equation 14.9:

$$C_{ratio} = \frac{C_{min}}{C_{max}}$$

where,
- C_{ratio} = capacitance rate ratio (unitless)
- C_{min} = *smaller* of the two capacitance ratios for the two flow streams involved (Btu/hr/°F)
- C_{max} = *larger* of the two capacitance ratios for the two flow streams involved (Btu/hr/°F)

The other quantity that's needed in order to determine effectiveness is called NTU, which stands for **number of transfer units**, and is defined as formula 14.10.

Formula 14.10:

$$\text{NTU} = \frac{UA}{C_{min}}$$

where,
- NTU = number of transfer units (unitless number)
- U = overall heat transfer coefficient for the heat exchanger (Btu/hr•ft²•°F)
- A = internal area of the heat exchanger (ft²)
- C_{min} = *smaller* of the two capacitance rates of the two flow streams (Btu/hr•°F)

One can think of the NTU as the ratio of the ability of a heat exchanger to transfer heat (based on its construction, internal area, and the overall heat transfer coefficient) divided by the ability of the "least able" flow stream to convey heat into or out of the heat exchanger. For example, a heat exchanger with a large internal area, and operating under favorable convection conditions, but also operating with a low capacitance rate on one side, would have high NTU. Because of its large internal area and high convection coefficients, it would be very "effective" in transferring heat to (or from) the fluid with the lower capacitance rate, possibly even approaching the performance of a hypothetical infinite heat exchanger, which would have an effectiveness of 1.0.

Once the capacitance rate ratio (C_{ratio}), and NTU are calculated, the effectiveness for a specific type of heat exchanger can be calculated based on either formulas or graphs in heat transfer reference books.

The effectiveness of any *counterflow* heat exchanger can be calculated using equation 14.11a or 14.11b. Formula 14.11a is used in the typical case where $C_{ratio} < 1$. Formula 14.11b is used in the less typical case where $C_{ratio} = 1$ (e.g., both fluid streams have the same capacitance rate ratio.

Equation 14.11a: (use when $C_{ratio} < 1$)

$$\varepsilon = \frac{1 - e^{(-\text{NTU})(1 - C_{ratio})}}{1 - (C_{ratio})e^{(-\text{NTU})(1 - C_{ratio})}}$$

Equation 14.11b: (use when $C_{ratio} = 1$)

$$\varepsilon = \frac{\text{NTU}}{1 + \text{NTU}}$$

The exponent of (e) in Equation 14.11a is the same in the numerator and denominator, and thus only has to be calculated once.

The effectiveness of a counterflow heat exchanger as a function of the NTU and the capacity rate ratio (C_{min}/C_{max}) can also be found using Figure 14-23, which is based on Equation 14.11a.

Example 14.4

A liquid-to-liquid heat exchanger, operating in a counterflow configuration, has an internal area of 20 ft². It is operating with the two previously described flow streams of water at 5 gpm, and 50% propylene glycol at 6 gpm. The water stream enters at 150 °F, and the propylene glycol solution enters at 60 °F. An analysis of the convection coefficients and fouling factors on both sides of the heat exchanger has been used to establish the overall heat transfer coefficient (U) as 150 Btu/hr•ft²•°F. Determine the rate of heat transfer across the heat exchanger.

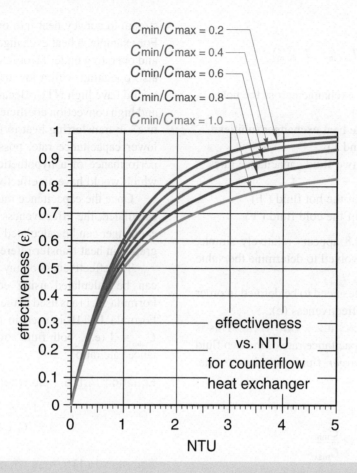

Figure 14-23 | Effectiveness of a counterflow heat exchanger as a function of the number of transfer units (NTU), and capacitance rate ratios, at which it operates.

Solution:

The two fluid capacity rates have already been determined as:

$$C_{water} = \left(5 \frac{gallon}{minute}\right)\left(1 \frac{Btu}{lb \cdot °F}\right)\left(\frac{8.33 \ lb}{gallon}\right)\left(\frac{60 \ minute}{hour}\right) = 2{,}499 \frac{Btu}{hr \cdot °F}$$

$$C_{50\%PG} = \left(6 \frac{gallon}{minute}\right)\left(0.88 \frac{Btu}{lb \cdot °F}\right)\left(\frac{8.54 \ lb}{gallon}\right)\left(\frac{60 \ minute}{hour}\right) = 2{,}705 \frac{Btu}{hr \cdot °F}$$

Of these two capacitance rates, the one for the water side is smaller. Thus, the capacitance rate ratio is:

$$C_{ratio} = \frac{C_{min}}{C_{max}} = \frac{2{,}499}{2{,}705} = 0.924$$

The number of transfer units is:

$$NTU = \frac{UA}{C_{min}} = \frac{(150)20}{2{,}499} = 1.2$$

These values can now be entered into Formula 14-11a:

$$\varepsilon = \frac{1 - e^{(-NTU)(1 - C_{ratio})}}{1 - (C_{ratio})e^{(-NTU)(1 - C_{ratio})}}$$

$$= \frac{1 - e^{(-1.2)(1 - 0.924)}}{1 - (0.924)e^{(-1.2)(1 - 0.924)}} = \frac{0.0872}{0.1565} = 0.557$$

The final step is to apply Formula 14.8:

$$q = \varepsilon(C_{min})(T_{hot \ in} - T_{cold \ in})$$

$$= 0.557(2{,}499)(150 - 60) = 125{,}280 \frac{Btu}{hr}$$

Discussion:

As with previous examples, the "behind the scenes" efforts of determining the convection coefficients needed to calculate the overall heat transfer coefficient (U) were not shown. Heat transfer reference books provide methods for determining these convection coefficients based on fluid properties and the flow characteristics of each fluid stream. Any assumed fouling factors would also be included in the calculation of (U), and thus be factored into the calculation of (NTU).

Software-Based Heat Exchanger Selection

The procedures introduced thus far; the LMTD method, and the effectiveness (ε) method, can be used for manual calculations of heat exchanger performance. Both of these approaches demonstrate that the mathematics necessary for analyzing heat exchangers can be complex. As such, they don't lend themselves to repeated "what if" scenarios when a designer is trying to select an appropriate heat exchanger.

Many manufacturers now offer software that can rapidly evaluate the necessary mathematics for sizing and selection of their heat exchangers. In some cases, this software is accessed online. In other cases, it's necessary to download and install the software.

One example of online-based sizing and selection software is FlatPlateSELECT, available at **http://flatplateselect.com/site/hx/chooseapp.aspx**. Figures 14-24a and b show some examples of this software.

Some heat exchanger sizing and selection software produce a list of several specific models that can meet the requirements based on input data. Some also show an **oversurface percent**, which is the percent by which the surface area in the suggested heat exchanger exceeds the minimum surface area required for the specified heat exchange rate. In some cases, the extra surface area is necessary to limit the pressure drop of one or both sides of the heat exchanger. Some software also provide performance data such as the overall heat transfer coefficient, flow velocity, and rate of heat transfer. Physical data on the candidate heat exchangers, including volume and dimensions are also listed.

Note: To download a PDF file version of this report, navigate to the Print/Save page.

Model: FG5X12-120(1-1/4"MPT)			
Selection ID	NUD6E5E3Q	Model size	5x13
Application	Snow melt	Nominal surface (ft²)	45.2
Load (Btu/h)	260,000	Dimensions	5.1W x 13.3H x 11.0D
Log mean temp. diff. (°F)	12.3	Plate construction	Single wall
Overall HTC (Btu/h·ft²·°F)	469	Net weight (lb)	39.9
Oversurface percent	0.6		

Design Conditions	Side A - Liquid	Side B - Liquid
Fluid type	**Water**	**Propylene glycol**
Fluid conc.		50
Fluid mass flow rate (lb/min)	289.2	249.0
Entering fluid temp. (°F)	140.0	110.0
Leaving fluid temp. (°F)	125.0	130.0
Fluid flow rate (GPM)	35.2	29.1
Fluid fouling factor (h·ft²·°F/Btu)	0.00010	0.00010
Model Parameters		
Number of channels	59	60
Velocity (ft/s)	0.57	0.46
Pressure drop (psi)	1.7	1.4
Heat transfer coef. (Btu/h·ft²·°F)	1,942	706
Internal volume (ft³)	0.177	0.180

Ratings at Varying Conditions					
Percent difference	**-15%**	**-7½%**	**0%**	**7½%**	**15%**
Pressure drop (psi) (Side A)	1.3	1.5	1.7	2.0	2.2
Pressure drop (psi) (Side B)	1.0	1.2	1.4	1.5	1.8
Load (Btu/h)	**221,000**	**240,500**	**260,000**	**279,500**	**299,000**
Fluid flow rate (GPM) (Side A)	29.9	32.5	35.2	37.8	40.4
Fluid mass flow rate (lb/min) (Side A)	245.8	267.5	289.2	310.9	332.6
Fluid flow rate (GPM) (Side B)	24.7	26.9	29.1	31.3	33.5
Fluid mass flow rate (lb/min) (Side B)	211.6	230.3	249.0	267.6	286.3
Entering fluid temp. (°F) (Side A)	140.0	140.0	140.0	140.0	140.0
Entering fluid temp. (°F) (Side B)	110.0	110.0	110.0	110.0	110.0
Leaving fluid temp. (°F) (Side A)	125.0	125.0	125.0	125.0	125.0
Leaving fluid temp. (°F) (Side B)	130.0	130.0	130.0	130.0	130.0
Oversurface percent	7.3	3.8	0.6	-2.3	-4.9

Disclaimer

This software and the generated calculations provided herein are estimates only and should be treated as such. GEA PHE Systems North America, Inc. always strives to give complete and accurate information, but cannot provide any guarantees. This software and its output are provided "as is" and any express or implied warranties, including, but not limited to, the implied warranties of merchantability and fitness for a particular purpose are disclaimed. In no event shall GEA PHE Systems North America, Inc. be liable for any direct, indirect, incidental, special, exemplary, or consequential damages (including, but not limited to, procurement of substitute goods or services; loss of use, data, or profits; or business interruption) however caused and on any theory of liability, whether in contract, strict liability, or tort (including negligence or otherwise) arising in any way out of the use of this software, even if advised of the possibility of such damage.

(b)

Figure 14-24b | Detailed thermal and hydraulic operating conditions. *Courtesy of FlatPlate, Inc.*

The ability of this software to evaluate several "what if" scenarios in a fraction of the time required for manual calculations makes it an essential design tool.

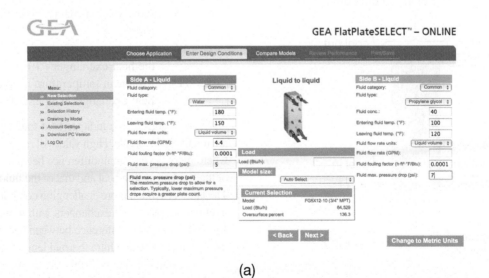

(a)

Figure 14-24a | User interface of sizing and selection software for brazed plate heat exchangers.

Heat Exchanger Thermal Operating Indices

Two indices are commonly used to describe the actual performance, or performance target, of heat exchangers. They are:

- Approach temperature difference
- Thermal length

By definition, there can be two approach temperature differences associated with any operating heat exchanger. They are illustrated in Figure 14-25.

Although it is possible for both approach temperature differences to have the same value, this is usually not the case. Therefore, when referring to approach temperature difference, it is important to state which of the two combinations of temperatures is being used. *In this textbook, unless otherwise stated, approach temperature difference #1 will be used (e.g., hot fluid inlet temperature minus cold fluid exit temperature).*

A theoretically infinite heat exchanger would have an approach temperature difference of 0. Any real heat exchanger will have an approach temperature difference greater than zero. Some brazed plate and plate & frame heat exchangers can operate at approach temperature differences as low as approximately 2 °F, although this requires a relatively large internal area, and thus a large heat exchanger. The expense of such a large heat exchanger may not be justified based solely on a very small approach temperature difference. More typical approach temperature differences for flat plate heat exchangers operating at design load conditions, range from 4 to 10 °F. A typical minimum approach temperature difference for shell & tube heat exchanger is 10 °F.

When a heat exchanger is present between a heat source and load, the approach temperature difference should be selected based on how it will affect the performance of the heat source. For example, the coefficient of performance (COP) of water-to-water heat pump, or air-to-water heat pump, is very dependent on the load water temperature (e.g., the water temperature leaving the heat pump's compressor). If the heat produced by a heat pump needs to pass through a heat exchanger, the approach temperature difference should be as low as practical and economically justifiable. A suggested *maximum* approach temperature difference for heat pump applications is 5 °F.

In a case where a conventional boiler operating at 180 °F is transferring heat to a load operating at 110 °F, the approach temperature difference can be much higher without significantly affecting the boiler's efficiency. This would allow use of a much smaller and less expensive heat exchanger.

Thermal Length of a Heat Exchanger

Another concept that helps describe the expected duty of a heat exchanger is called thermal length. It's defined by Equation 14.12.

Equation 14.12:

$$\theta = \frac{\Delta T}{\text{LMTD}}$$

where,

θ = (Greek letter theta) = thermal length (unitless)
ΔT = temperature change of fluid on one side of the heat exchanger (°F)
LMTD = log mean temperature drop at which the heat exchanger is operating (°F)

Low values of (θ) indicate heat transfer conditions that are relatively favorable. High values of (θ) represent situations where there is very small difference between the temperature of the fluid providing heat and that of the fluid absorbing heat. The latter is a much more challenging condition and generally requires flat plate heat exchangers with longer plates. Figure 14-26 illustrates the difference between heat transfer requirements having low and high thermal lengths (θ).

Heat transfer theory limits the thermal length (θ) of a single pass shell & tube heat exchanger to 1.0. Flat plate heat exchangers can operate with thermal lengths up to approximately 10.0. This makes flat plate heat exchangers

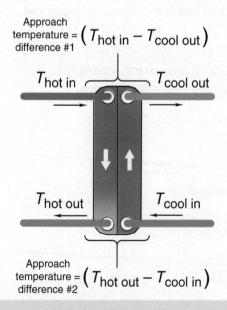

Figure 14-25 | Approach temperature difference at both ends of a flat plate heat exchanger.

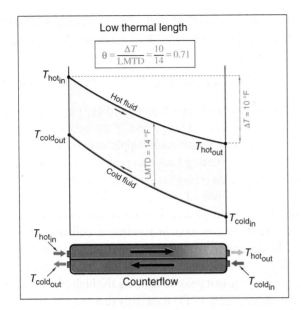

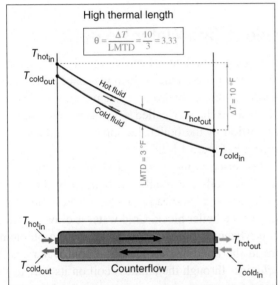

Figure 14-26 Temperature profile comparison for a counterflow heat exchanger operating at low thermal length conditions, and high thermal length conditions.

better suited to situations where the temperature differences between the fluids is very small. Manufacturers can advise on optimal plate patterns and lengths based on the intended thermal length of the heat exchanger.

14.3 Domestic Water Heating

Almost all residential and some light commercial buildings require domestic hot water (DHW). It is a common auxiliary load to be combined with hydronic space heating. This section presents a simple method for estimating daily and peak hourly DHW heating loads. It also discusses different hardware options for heating domestic water using a hydronic system. Special devices and piping configurations are also shown for high-capacity domestic water heating.

Daily DHW Load Estimates

The daily energy requirement for DHW production can be estimated using Equation 14.13:

Equation 14.13:

$$E_{\text{daily}} = (G)(8.33)(T_{\text{hot}} - T_{\text{cold}})$$

where,

E_{daily} = daily energy required for DHW production (Btu/day)

G = volume of hot water required per day (gallons)

T_{hot} = hot water temperature supplied to the fixtures (°F)

T_{cold} = cold water temperature supplied to the water heater (°F)

The daily usage of domestic hot water is heavily dependent on occupancy, living habits, type of water fixtures used, water pressure, and the time of year. The following estimates are suggested as a guideline:

- House: 10 to 20 gallons/day/person
- Office building: 2 gallons/day/person
- Small motel: 35 gallons/day/unit
- Restaurant: 2.4 gallons/average number of meals/day

Example 14.5

Determine the energy used for domestic water heating for a family of four with average usage habits. The cold water temperature averages 55 °F. The supply temperature is set for 125 °F.

Solution:

The daily domestic water heating load is estimated using Equation 14.13:

$$E_{\text{daily}} = (4 \times 15)(8.33)(125 - 55) = 34{,}990 \text{ Btu/day}$$

The estimated annual usage would be the daily usage times 355 days of occupancy/y. (Assuming the family is away 10 days/y.)

$$\left(34{,}990 \frac{\text{Btu}}{\text{day}}\right)\left(355 \frac{\text{occupied days}}{\text{y}}\right)$$
$$= 12{,}421{,}000 \frac{\text{Btu}}{\text{y}} = 12.4 \text{ MMBtu/y}$$

Discussion:

The cost of providing this energy can be estimated by multiplying the annual energy requirement by the cost of delivered energy in $/MMBtu (see Chapter 2 for determining delivered energy cost on a $/MMBtu basis). For example, at $0.10/kWh and 100 percent efficiency, electrical energy has a delivered cost of $29.30/MMBtu. The annual energy cost of providing the domestic hot water in this example would be:

$$\left(12.4 \frac{\text{MMBtu}}{\text{y}}\right)\left(29.3 \frac{\$}{\text{MMBtu}}\right) = \$363/\text{y}$$

DHW Usage Profiles

The rate at which DHW is used varies considerably with the type of occupancy. A typical residential **domestic hot water usage profile** is shown in Figure 14-27.

Note the two distinct periods of high demand: One during the wake-up period, the other during the early evening. The greatest period of usage in this profile is from 7 to 8 p.m. However, this still represents only 11.6 percent of the total daily DHW demand. For a family using 60 gallons of DHW per day, this would represent a peak usage of just under 7 gallons/hour. The *average* rate at which energy is required for DHW production during this peak hour (assuming cold water and hot water temperatures of 55 and 125 °F, respectively) can be found using Equation 14.13 modified for a single peak hour:

$$E_{\text{peak hour}} = [(0.116)(60)](8.33)(125 - 55) = 4{,}058 \text{ Btu/hr}$$

The first factor in this Equation (0.116) represents the peak hour demand of 11.6 percent of the total daily water volume. If the heat source could supply an extra 4058 Btu/hr above the space heating load, this peak DHW requirement could be met, even on continuous basis if required.

The occurrence of two peak demand periods, one during the wake-up period, the other following the evening meal period, is common in residential buildings. Some households tend to be "high morning users" while others are "high evening users." Figure 14-28 compares these two categories. Note that peak demand for the high evening user is only about 15.5% of the total daily demand.

Tankless Coil Water Heaters

One "legacy" method of providing DHW using a gas-fired or oil-fired boiler, which was once widely used in North American hydronic systems, was to insert a copper heat exchanger called a **tankless coil** into the upper portion of the boiler. Examples of tankless coils are shown in Figure 14-29.

When inserted into a special chamber in the boiler block, these coils are totally surrounded by hot boiler water. A gasketed bulkhead provides a watertight seal at the side of the boiler block. Cold water is drawn through the coil whenever hot water is drawn from a plumbing fixture in the building. The entering cold water makes a *single pass* through the tankless coil on its way to the hot water fixtures.

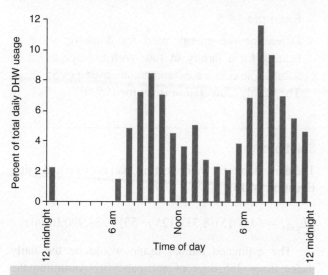

Figure 14-27 | Typical usage profile for residential domestic hot water.

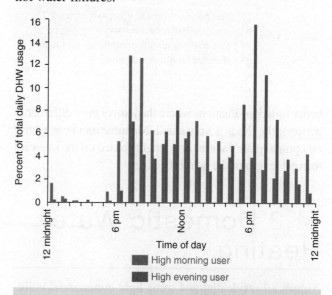

Figure 14-28 | Comparison of hourly usage of domestic hot water for a high morning user versus a high evening user.

cost of fuel, is making tankless coil water heaters economically obsolete. Boilers with such coils are still available, but are mostly used in replacement jobs where the customer chooses lower cost over longer-term energy savings. Eventually, the type of tankless coil water heating described above will disappear from the industry, either by choice or by energy code mandate.

Combining a Tankless Coil With a Storage Tank

Some of the disadvantages of traditional tankless coil water heaters can be offset by a retrofit strategy in which a storage tank is added to the system. A schematic of this concept is shown in Figure 14-30. This modification will significantly improve annual fuel efficiency and reduce temperature fluctuations at fixtures.

For this arrangement to function properly, the triple aquastat control on the boiler must be replaced with a conventional high limit control. After this modification is made, the boiler should fire only upon a call for heat from the tank's thermostat, or a demand from the space heating system. It no longer needs to maintain a constant minimum operating temperature.

Upon a call for domestic water heating from the tank's aquastat, a small stainless steel or bronze circulator moves cooler water from near the bottom of the storage tank through the tankless coil. The heated water from the coil is then returned to the top of the storage tank. The circulator continues to run until the tank aquastat is satisfied. The larger the volume of the storage tank, and the wider the differential of the tank's aquastat, the longer the boiler's operating cycle. Longer firing cycles improve boiler efficiency. When the storage tank reaches its setpoint temperature, the boiler and tank circulator are shut off. As domestic hot water is drawn from the storage tank, its temperature drops until the tank's aquastat closes to repeat the cycle.

As an option, the tank circulator can be configured to run for two or three minutes after the boiler shuts off. This can be done using a delay-on-break time delay relay as discussed in Chapter 9. The temporarily sustained circulation purges residual heat from the boiler, moving it into the storage tank before it is lost up the chimney or through the boiler jacket.

Since the tank circulator moves fresh water, it must be suitable for use in an open system. Circulators with stainless steel volutes are available for such applications. Some plumbing codes may require such circulators to be NSF (National Sanitation Foundation) listed for direct contact with potable water. *Cast-iron circulators should never be used for this duty.*

Figure 14-29 (a) Examples of tankless coil water heaters. *Courtesy of Amtrol, Inc.* (b) A tankless coil being inserted into a boiler. *Courtesy of Weil-McLain Corp.*

Since hot water may be required at any time of day, the boiler water surrounding a tankless coil must always be kept hot. A triple aquastat control, as described in Chapter 9, Control Strategies, Components, and Systems, is used to fire the boiler whenever its water temperature drops below a preset lower limit. During warm weather, when there is no demand for space heating, the boiler water must be kept at a relatively high temperature solely for supplying DHW. The standby heat losses associated with this requirement substantially lower fuel efficiency. Boiler efficiencies as low as 35 percent have been observed during warm weather operation of tankless coil systems.

Tankless coils can also cause noticeable variations in the temperature of hot water delivered to fixtures, especially if the fixture is operated for several minutes. This is caused by fluctuations in boiler temperature between firing cycles. Since the amount of water contained in the coil is very small, it has very little thermal mass to help absorb energy without a significant temperature change. This problem can be partially compensated for by installing an anti-scald rated thermostatic mixing valve on the outlet of the coil.

The inherent inefficiency of maintaining hot water in the boiler at all times, along with the increasing

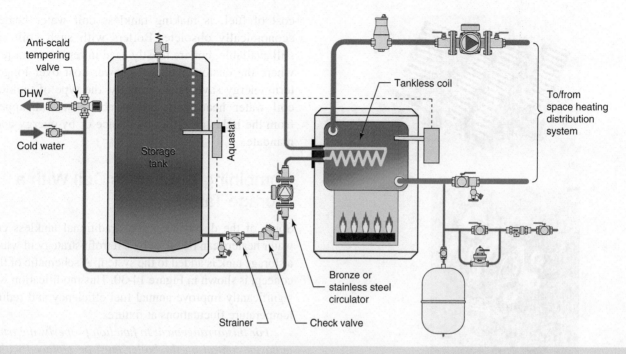

Figure 14-30 | Adding a storage tank to a conventional tankless water heating system.

The tank's aquastat can be a conventional electromechanical control or an electronic setpoint controller. The sensor bulb or thermistor should be mounted into a sensor well inserted into the tank water.

If the storage tank includes an electric heating element, it could be wired and used as a backup DHW heat source in the event the boiler is nonoperational.

Indirect Water Heaters

Indirect water heaters were first used in North America in the late 1970s as heating professionals began recognizing the limitations of tankless coil water heating.

An indirect water heater is an insulated storage tank that does not produce its own heat, as does a typical **direct-fired water heater**. Instead, it uses an internal heat exchanger to transfer heat from a flowing stream of hot boiler water into domestic water. In most (but not all) indirect water heaters, hot boiler water flows through the internal heat exchanger coil, which is surrounded by potable water.

Since it is in contact with potable water, the tank's shell is usually constructed of stainless steel, or lined with a heat-fused porcelain-like glazing. The heat exchanger could be made of stainless steel, copper, cupronickel, or carbon steel coated with the same porcelain-like glazing applied to the tank shell. One example of an indirect water heater is shown in Figure 14-31.

A stainless steel internal heat exchanger coil in the lower portion of the tank is immersed in the coolest potable water. This increases the LMTD between the outer surface of the coil and the water surrounding it, which increases the rate of heat transfer. The tank's pressure vessel is also made of stainless steel and covered with a layer of foamed-in-place insulation. The outer jacket of modern indirect water heaters is usually made of painted steel or a molded polymer.

In Figure 14-31b, the circuit supplying hot boiler water to the heat exchanger in the water heater is in parallel with the space heating circuit. Both circuits have a circulator with an internal check valve to prevent reverse flow. An additional (optional) check valve is shown upstream of the purging valve in the space heating return pipe. If present, this valve reduces heat migration into the space heating distribution system. It is particularly helpful during warm weather, when the only need for heating is for domestic hot water.

Also note that hot water from the boiler is supplied to the *upper* coil connection. This, in combination with buoyancy-induced rising water in the tank, creates counterflow heat exchange for improved heat transfer efficiency.

This piping arrangement allows the DHW load to operate as either **priority load** (e.g., the only load active when it is on), or a nonpriority load (e.g., operating simultaneously with other loads that may be on). Prioritized domestic water heating was discussed in Chapter 9, Control Strategies, Components, and Systems.

Another style of indirect water heater is shown in Figure 14-32. In this **tank-in-tank** design, boiler water passes between the outer carbon steel shell and the inner stainless steel tank. This approach creates a large surface area between the two fluids allowing for rapid heat transfer when suitable rates of heat input are available from the heat source. It also reduces the head loss associated with circulating boiler water through the tank's heat exchanger.

14.3 Domestic Water Heating

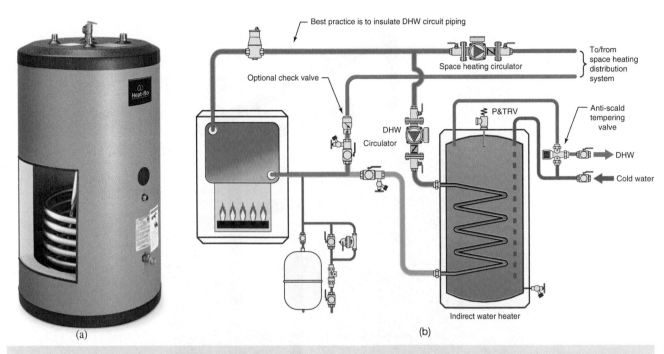

Figure 14-31 | (a) Example of an indirect water heater with internal coiled heat exchanger. *Courtesy Heat-Flo.* (b) Typical piping in a system that supplies both DHW and space heating.

Yet another type of indirect water heater is shown in Figure 14-33. This design circulates boiler water through the insulated steel shell, while potable water makes a single pass through multiple suspended copper coils. This leads to the name **"reverse" indirect water heater** because this approach reverses the locations of the two fluids exchanging heat relative to those in a standard indirect tank. The coils are connected in parallel. Each coil is joined to an upper and lower header within the tank. This allows for a relatively large heat exchange surface area without excessive pressure drop.

The water in the shell of a reverse indirect water heater is always maintained at a suitable temperature to provide domestic water. The large surface area of the internal coils allows for rapid heat transfer in domestic water when the tank is connected to a high-capacity heat source.

Indirect water heaters offer several benefits not provided by other means of water heating. These include the following:

- The ability to hold sufficient heated water in reserve to meet high flow rate loads. Just as a heated concrete slab can deliver high rates of heat output when necessary, an indirect water heater can supply high flow/short duration DHW demands without the need to operate the heat source. Because heat is stored in the tank, the rate of heat transfer to such loads can be temporarily higher than the rate at which the heat source can generate heat.

- The stored water provides thermal mass that allows the boiler to operate for longer cycles, and thus at higher efficiencies. When the indirect water heater's thermostat is satisfied, the boiler and tank circulator turn off. They can remain off, in some cases for several hours, while the hot water demands of the building are provided directly from the water stored in the tank. Standby heat loss from the boiler is substantially reduced since it does not have to remain at a constant elevated temperature, as is required with a tankless coil.

Figure 14-32 | Example of an indirect water heater with tank-in-tank design. *Courtesy Weil-McLain Corp.*

714 Chapter 14 Auxiliary Loads and Specialized Topics

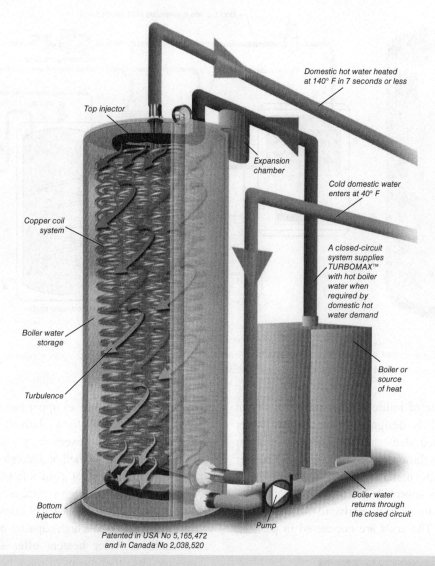

Figure 14-33 | Example of an indirect water heater with boiler water in shell and domestic water in suspended coils. *Courtesy of Thermo 2000, Inc. Patented in USA & Canada.*

- The low pressure drop and relatively large piping connections on most indirect tanks offer far less flow resistance than does a tankless coil. This allows high flow rates to fixtures without an accompanying drop in water pressure.

- In sustained flow situations, a properly selected indirect water heater, operated as a priority load, can transfer the full Btu/hr output of the system's heat source into the domestic water. In contrast, most direct-fired water heaters have sustained delivery limitations imposed by the capacity of their heating element or burner. Figure 14-34 compares the sustained DHW flow capability of several common residential water heaters. Relatively low heat input rates are the reason such tanks require large volumes to meet high demand applications. This is especially true for electric water heaters.

- A modern boiler and indirect water heater can provide DHW at higher thermal efficiencies than conventional direct-fired water heaters. Boiler heat exchangers are larger and designed with better heat transfer geometry compared to the heat exchangers in most direct-fired water heaters. Boiler heat exchangers can also be cleaned during their service life to help maintain that efficiency advantage.

- Indirect tanks are not exposed to the high-temperature gradients present in direct-fired tanks. The high temperature differences between the combustion gases and water create stresses that eventually take their toll on the heat exchanger materials in direct-fired water heaters. The higher temperature heat exchanger surfaces also allow increased precipitation of minerals dissolved in the water. Some of these precipitants bond to the surface of the heat exchanger further reducing its heat transfer efficiency. Direct-fired tanks are also subject to corrosion caused by flue gas condensation, especially when operated at reduced temperatures.

Heater type	Heat output rate (Btu/hr)	Hot water recovery rate* (gpm/gph)**
Electric (3.8 kW)	12,970	1/60
Electric (4.5 kW)	15,360	1/61
Electric (6.0 kW)	20,480	1/60
Gas-fired (30–50 gallons)	30,000–40,000	0.8–1.1/48–66

*Results are based on cold water temperature of 50 °F, and outlet temperature of 125 °F.
**gpm = gallons/minute, gph = gallons/hour.

Figure 14-34 | Typical heat input and recovery rates of residential direct-fired water heaters.

Piping and Settings for Indirect Water Heaters

Although there are different ways to pipe an indirect water heater into a system, *the goal is always to get heat from the boiler(s) into the water heater with minimal losses in between.* Whenever possible, indirect water heaters should be located close to boilers to minimize heat loss through interconnecting piping. The piping between the boiler and tank should also be insulated. Remember that hot water will likely be flowing through this piping on the hottest day of the year. The building's cooling system does not need the added load created by heat loss from long lengths of hot uninsulated piping.

Figure 14-35 shows one possible piping arrangement for a system in which a boiler supplies the indirect water heater, and several zones of space heating.

This system uses a single variable-speed circulator to supply three zones of space heating and the indirect water heater. The latter is operated as a priory load. Whenever the indirect tank calls for heat, any active zone valves in the space heating circuits are temporarily turned off. This allows the full heat output of the boiler to be available to the indirect water heater.

When this piping configuration is used, it is critical to ensure that the variable-speed circulator, in combination with the zone valve and piping associated with the indirect water heater, allows for sufficient flow between the boiler and the coil inside the indirect water heater. This flow rate should be adequate to transfer the full heat output rate of the boiler into the indirect water

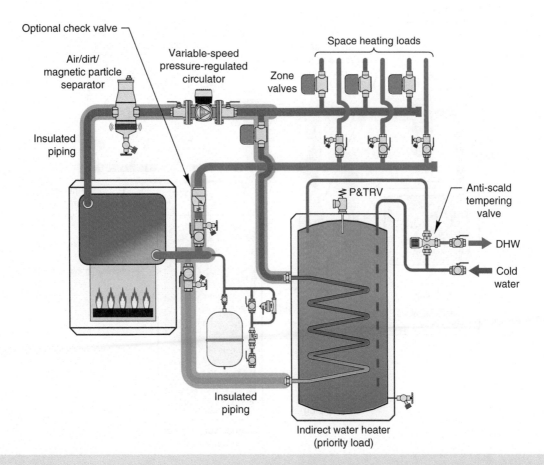

Figure 14-35 | Piping for an indirect water heater in a system with multiple zone valves and a variable-speed pressure-regulated circulator. DHW circuit can be operated as a priority load or simultaneously with space heating.

heater. Undersized piping or insufficient head from the circulator could result in a heat transfer "bottleneck" between the boiler and indirect water heater. This will limit the rate of heat transfer between the two, and if severe, may cause the boiler to operate in short cycles. These conditions must be avoided. The methods for pipe sizing discussed in Chapter 6, and for selecting circulators, discussed in Chapter 7 should be used to verify that adequate flow will be available. Check manufacturer's specification sheets for determining the necessary flow rate.

Note that the zone supplying the indirect water heater is connected *closest* to the boiler. While this is not absolutely necessary, it does reduce heat loss from the outer portions of the common piping when only domestic water heating is required, during the summer for example. Also note the check valve installed just upstream of where the return pipe from the space heating portion of the system joins the return header. This check valve minimizes heat migration along the return header during warm weather.

Another approach to the system configuration shown in Figure 14-36 is to use a separate fixed speed circulator to supply flow for the indirect water heater, in combination with a variable-speed pressure-regulated circulator for the space heating zones. The indirect water heater would again be operated as a priority load.

The author highly recommends that every water heater be equipped with an **ASSE 1017-rated anti-scald tempering valve**, *whether required by local code or not.* This valve protects occupants from potentially scalding hot water should a temperature controller malfunction or be improperly set. In residential applications, the outlet temperature setting of the anti-scald valve should not exceed 120 °F.

The water temperature maintained inside an indirect water heater can be controlled by an aquastat, an electronic setpoint controller, or by a more extensive control system that manages all loads in the system. This temperature may be determined by personal preference, or in some situations mandated by local code or other governmental requirements. If there is concern over potential Legionella bacteria in the domestic hot water, a minimum tank temperature of 140 °F is either maintained continuously or for a specified duration on a daily basis to kill the bacteria. Again, an ASSE-1017-rated anti-scald tempering valve should be

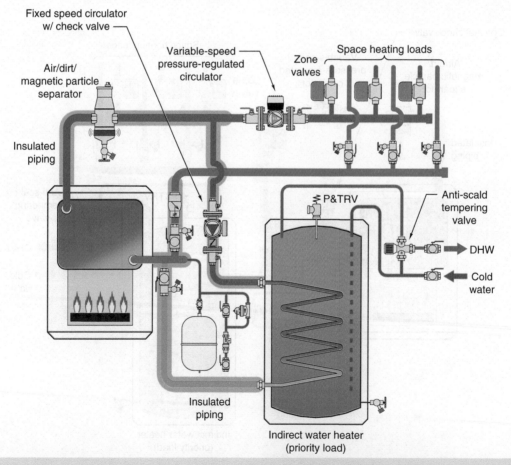

Figure 14-36 | Piping for a system with multiple zone valves and a variable-speed pressure-regulated circulator for space heating, and a fixed speed circulator to supply the coil of an indirect water heater.

installed, and the domestic water temperature supplied to the building's plumbing distribution system should never exceed 120 °F.

Several other piping schematics involving indirect water heaters are shown throughout this book. Readers are encouraged to study all these schematics looking for the details discussed in this section.

On-Demand Domestic Water Heating

Another approach to domestic water heating uses a brazed plate heat exchanger sourced from a buffer tank to create domestic hot water seconds before it is required at a fixture.

The assembly shown in Figure 14-37 can be used in residential and light commercial systems that have a heated buffer tank. That tank may be associated with renewable heat sources, such as solar thermal collectors, biomass boilers, or heat pumps. It could also be heated by a fossil fuel boiler.

When the domestic hot water drawn by the plumbing distribution system exceeds 0.7 gpm, a flow switch, rated for use with domestic water, closes its contact. This energizes the coil of a relay, which turns on a circulator to draw hot water from the top of the buffer tank through the primary side of the heat exchanger. After passing through the heat exchanger, this flow returns to the lower portion of the buffer tank. Cold domestic water flows through the secondary side of the heat exchanger in a counterflow direction, absorbing heat from the primary side.

Figure 14-38a shows an example of a flow switch suitable for use with domestic water. The switch closes its contact at flow rates of 0.7 gpm or higher, and opens its contact at flow rates of 0.4 gpm or lower. This switch is designed to thread into a 3/4" stainless steel tee, as shown in Figure 14-38b.

The high surface-area-to-volume ratio of the brazed plate heat exchanger, combined with a minimal amount of piping connecting it to the buffer tank, allows heated domestic water to be available within three to five seconds of the flow switch contact closure. This is significantly faster than the response time required by a gas-fired tankless water heater.

This approach also has the advantage of storing very little domestic hot water when it is not operational, which reduces the potential for Legionella growth.

The system shown in Figure 14-37 assumes that the water in the buffer tank is maintained hot enough to fully heat the domestic water on a single pass through the brazed plate heat exchanger. This may or may not be the case depending on the heat source, the space heating emitters, the required domestic hot water delivery

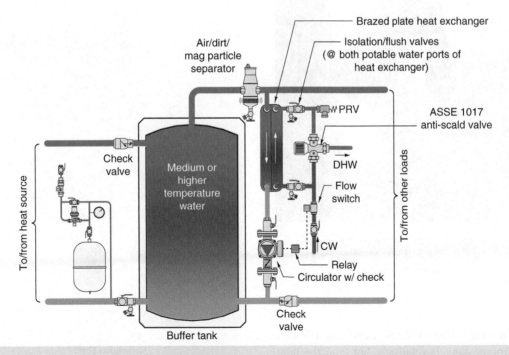

Figure 14-37 Using a stainless steel brazed plate heat exchanger to transfer heat from a buffer tank to domestic water on an "on demand" basis. This assembly assumes that the buffer tank is always slightly hotter than the required delivery temperature of domestic hot water. No auxiliary heating is used.

1017 rated anti-scald valve should always be installed to protect the plumbing distribution system from hot water delivery temperatures over 120 °F.

This application is a good example of when a heat exchanger with a high thermal length is appropriate. The domestic water needs to undergo a large temperature change, but the LMTD between this water and water from the buffer tank will be relatively small. This application favors a brazed plate heat exchanger with relatively long plate length.

The system shown in Figure 14-37 includes two isolation/flushing valves installed at both domestic water ports of the heat exchanger. These valves, which must be rated for use with domestic water, allow the heat exchanger to be isolated from the rest of the domestic water system and flushed with a cleaning solution. Figure 14-39 shows a brazed plate heat exchanger fitted with these valves on the side that will operate with domestic water.

Another detail is the use of two spring-check valves adjacent to the buffer tank. The check valve near the upper left connection prevents reverse thermosiphoning between the buffer tank and heat source. The valve in the lower right limits heat migration into the return side of the space heating circuits.

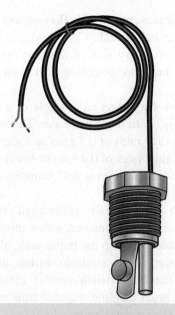

Figure 14-38a | Example of a domestic water flow switch. *Courtesy of Harwil Corp.*

Figure 14-38b | Domestic water flow switch threaded into the side port of a 3/4" stainless steel tee. *Courtesy of John Siegenthaler.*

Figure 14-39 | Stainless steel brazed plate heat exchanger fitted with two isolation/flushing valves on the side that will operate with domestic water. *Courtesy of John Siegenthaler.*

temperature, and the sizing of the heat exchanger. When the thermal storage tank is relatively hot, and the heat exchanger is sized for a close approach temperature difference, there is no need for a temperature boost after the domestic water leaves the heat exchanger. An ASSE

The brazed plate heat exchanger and its associated piping should be kept as close to the buffer tank as possible. This minimizes the amount of system water that needs to be displaced at the start of a DHW draw, and helps keep the response time to only a few seconds. The heat exchanger and its associated piping should also be insulated to reduce heat loss to the mechanical room.

Figure 14-40 shows an extension of this approach. An electric tankless water heater has been added to boost the final DHW deliver temperature when necessary.

This configuration is useful when the buffer tank operates at temperatures below the required delivery temperature for domestic hot water. The heat exchanger can still provide significant preheating, leveraging the low starting temperature of cold domestic water relative to the water temperature in the buffer tank. Combination isolation/flushing valves are also installed on the piping connected to the tankless water heater to allow for isolation and cleaning. It would also be possible to use a tank-type water heater to provide the final temperature boost.

Hydronic Solutions for High-Capacity Water Heating

One trend in custom residential construction has been increased demand for luxury bathrooms. The North American plumbing industry has done a great job of promoting such bathrooms. The essence of such promotions is surrounding oneself with lavish amounts of warm water, be it in a deep whirlpool tub, or a simulated "tropical downpour showering experience." Figure 14-41 shows a high flow shower enclosure with multiple spray heads that is typical in modern luxury bathrooms.

For decades, the ability to produce domestic hot water using a hydronic system was viewed as ancillary to space heating. The traditional thinking was as follows: Since the customer wants hydronic space heating, and therefore needs a boiler, why not spend a bit more to equip that boiler with a tankless coil or indirect water heater for domestic water heating? This reasoning was sound, and remains sound, for average houses that don't require the high flow rates of domestic hot water that some new custom homes now demand.

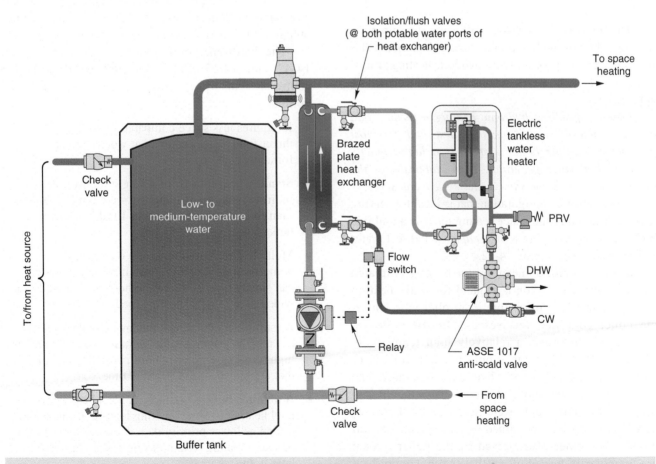

Figure 14-40 | Modification of the system shown in Figure 14-37 that uses a thermostatically controlled tankless electric water heater to provide a temperature boost to preheated domestic water leaving the heat exchanger.

Figure 14-41 | A luxury master bathroom shower can create a high demand for domestic hot water. This shower is one of several in the house. *Courtesy of John Siegenthaler.*

The increased domestic hot water consumption in larger homes with multiple luxury bathrooms has presented a new opportunity for hydronic heating professionals. It does, however, require a different marketing approach.

Instead of selling hydronic heating with the "add on" of domestic water heating, first and foremost promote a multiple boiler system as a high-capacity domestic hot water generator. Then elaborate on the other heating loads the same boiler system can serve.

One hardware configuration that is ideal for serving high demand domestic water heating loads is a multiple boiler system supplying a *high-capacity* indirect water heater as shown in Figure 14-42.

In this system, domestic water heating is treated as a priority load. Upon a demand for water heating from the tank's temperature controller, the multiple boiler controller receives a **setpoint demand**. The controller targets a relatively high temperature (typically 180 °F) at the supply temperature sensor located downstream of the hydraulic separator. The controller fires and modulates the boilers required to achieve this high supply water temperature. If necessary, all boilers may be turned on and operated at full output. Any other loads served by the boiler system are temporarily turned off so that the maximum heat production from the boiler system is transferred to the indirect water heater.

It is vitally important that the heat exchanger within the indirect water heater can release heat to the domestic water at a rate that's at least equal to the full output of the boiler system, while operating at a reasonable boiler outlet temperature (suggested as no higher than 200 °F).

Failure to use an indirect tank with sufficient coil area will limit the rate of hot water production, and cause short cycling of the boiler system.

It is also vitally important that the piping and circulator used allow the full output of the boiler system to be transferred to the heat exchanger within the indirect water heater.

It is not uncommon for high-capacity domestic water heating sub-systems to require 1.5-inch, 2-inch, or even 2.5-inch piping between the boiler and indirect water heater. Flow rates of 20 to 90 gpm may be required.

If a suitable indirect water heater cannot be sourced, a brazed plate heat exchanger in combination with an insulated storage tank provides an alternative, as shown in Figure 14-43. Again, the heat exchanger, piping, and circulators must all be sized to transfer the full heating output of the boiler system to the domestic water.

The hardware configurations shown in Figures 14-42 and 14-43 and associated multiple boiler control systems offer several benefits relative to other options:

- It requires less space compared to multiple direct-fired water heaters that would be required for equivalent capacity.

- Standby heat loss from a single high-capacity indirect tank, or insulated storage tank, is lower than that from several direct-fired hot water tanks.

- Multiple direct-fired tanks are only useful for domestic water heating. A multiple boiler system can supply heat for domestic hot water as well as a wide range of other heating loads.

- A multiple modulating/condensing boiler system has a high system turn-down ratio and can maintain high seasonal efficiency until partial load condition that are prevalent much of the time.

- A multiple boiler system is an ideal heat source for pool heating or snow melting loads that can be controlled as lower priority than domestic water heating or space heating. Examples of this are discussed later in this chapter.

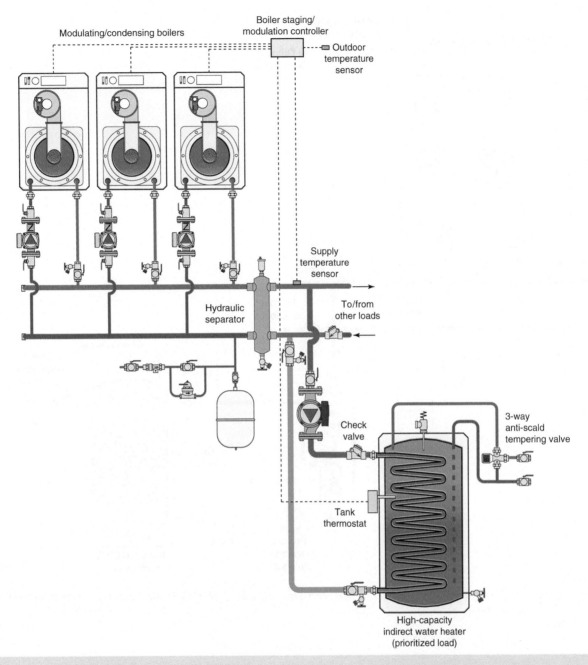

Figure 14-42 | A multiple boiler system combined with a high-capacity indirect water heater for high-capacity domestic water heating.

- This approach offers a good combination of high-capacity water heating when needed for high demands, as well as a modest, but necessary amount of storage to handle small DHW draws without always needing to operate a combustion system.

- Use of a multiple boiler system and high-capacity indirect water heater is a "starting point" from which to promote hydronics technology in less traditional hydronics markets. Once this subsystem is in place for high-capacity domestic water heating, the system can be expanded, at present or in the future, to handle a wide range of other heating loads.

- Unlike gas-fired tankless water heaters, which may require 30 seconds or more to deliver domestic hot water following a cold start, DHW is *immediately available* from the indirect tank or storage tank.

- Unlike gas-fired tankless water heater, there is no minimum DHW flow rate required to initiate water heating.

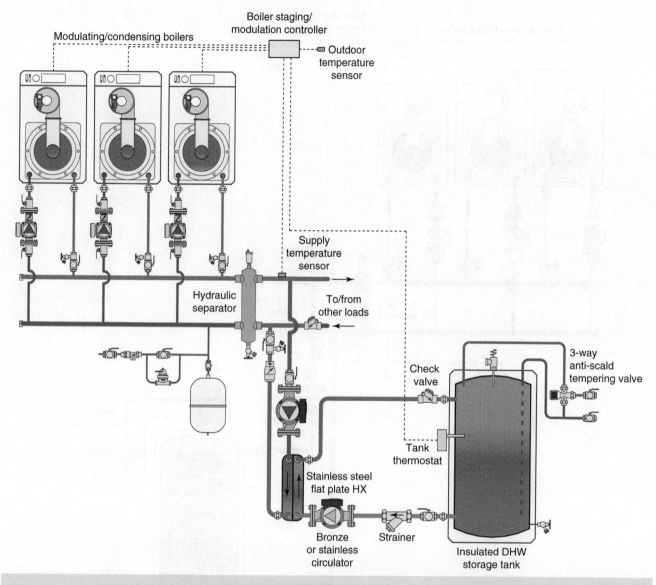

Figure 14-43 | Using a stainless steel brazed plate heat exchanger and storage tank as substitute for high-capacity indirect water heater.

Performance of Indirectly Fired Storage Water Heaters

Most manufacturers of indirect water heaters provide two types of rating information. One is called the **first hour rating**; the other is called the **continuous flow rating**.

The first hour rating is the amount of water the tank can deliver within one hour of the initial demand at a specified delivery temperature. This rating assumes the tank is at the specified delivery temperature when the demand begins. The two domestic hot water delivery temperatures that are usually referenced are 110 and 140 °F.

The first hour rating is influenced by both the volume of the tank, and its ability to transfer heat to the domestic hot water once the tank's temperature begins to drop. This rating is strongly dependent on the boiler capacity available to the tank, as well as the water temperature supplied to the heat exchanger by the boiler. In most cases, the manufacturer provides several first hour ratings along with the associated heat input rates and a specified temperature of boiler water.

The continuous flow rating is the gallons per minute, or gallons per hours, that the tank can deliver under *sustained* demands with a fixed heat input rate and boiler water temperature. This rating is an indicator of the tank's steady-state heat transfer ability rather than its storage capacity.

An example of a manufacturer's rating table showing both ratings is shown in Figure 14-44.

Some manufacturers also provide information on head loss versus flow rate through the tank's internal heat exchanger. This information is needed to properly size the circulator and piping between the heat source and tank. An example of such data is shown in Figure 14-45.

Sizing An Indirect Water Heater

The usage pattern and quantity of DHW needed is often difficult to estimate given variations in lifestyle, schedules, activities, and fixtures/appliances installed in any given building. It is common to estimate the *likely* demand of DHW during the peak hour of a typical day, and base water heater selection on this.

The data given in Figure 14-46 can be used to estimate the highest hourly demand for domestic hot water. This data reflects usage by average occupants drawing water through average fixtures. It does not represent possible usage in applications having high-demand fixtures such as multiple head showers and large whirlpool baths. Estimates of the DHW demand of such fixtures are best obtained directly from fixture manufacturers.

When selecting an indirect tank, be sure to check manufacturers' ratings for the heat transfer capability of the tank's heat exchanger, as well as the necessary inlet water temperature and flow rate required to achieve that performance.

When one or more boiler(s) serves as the heat source, the author recommends selecting an indirect

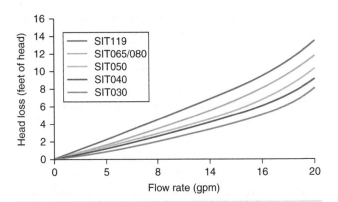

Figure 14-45 | Head loss versus flow rate for coiled heat exchanger inside a family of indirect water heaters. *Courtesy of Lochinvar Corporation.*

Usage	High flow fixture (gal./usage)	Low flow fixture (gal./usage)
Food preparation	5	3
Hand dishwashing	4	4
Automatic dishwashing	15	15
Clothes washer	32	21
Shower or bath	20	15
Face and hand washing	4	2

Figure 14-46 | Estimated usage of domestic hot water for high flow and low flow fixtures in residential applications.

Model Number	Capacity (gal.)	Heat Source Water Volume (gal.)	Stand by Loss (°F/hr)	Continuous Delivery (GPH)	1st Hour Delivery (gal.)	Min Coil Load (Btu/hr)	Flow Rate (GPM)	Friction Loss (ft. Hd.)
SIT030	27	1.1	1.2	160	183	99,000	14.0	3.9
SIT040	40	1.6	0.7	181	208	115,000	14.0	4.5
SIT050	52	1.7	0.6	209	255	133,000	14.0	5.3
SIT065	67	1.9	0.5	263	327	154,000	14.0	5.7
SIT080	82	2.1	0.5	285	358	171,000	14.0	6.1
SIT119	113	3.2	0.4	349	459	216,000	14.0	6.5

Performance data is based on IBR test results. All ratings are based on 180 °F boiler water temperature.

Figure 14-44 | Example of first hour and continuous domestic hot water ratings of an indirect water heater. *Courtesy of Lochinvar Corporation.*

water heater capable of transferring the full heat output of the boiler(s) without exceeding a boiler outlet temperature of 180 °F.

When one or more heat pump(s) serve as the system's heat source, the author recommends sizing the DHW heat exchanger, whether it be an indirect water heater, or an external brazed plate heat exchanger, to accept the full heat output from the heat pump(s) without exceeding a supply water temperature from the heat pump(s) of 125 °F.

This temperature, (125 °F), may or may not be high enough to fully heat domestic water to the desired delivery temperature. When it is not, some form of supplemental heat input, such as from a tankless or tank-type electrical water heater will be necessary.

When high DHW delivery rates are required, the domestic water heating load should be prioritized over any other loads supplied by the system. The author also recommends establishing an estimated peak demand for domestic hot water based on the fixtures used in the building, and how the owner intends to operate them. The estimate should be stated as the number of gallons per minute of domestic hot water supplied, along with a stated temperature rise. For example, a peak demand of 20 gallons per minute, heated through a 70 °F temperature rise (i.e., from 50 °F cold water temperature to a delivery temperature of 120 °F). The estimate should then be reviewed with the owner to verify that it meets their expectations. The owner should confirm, in writing, that they accept this proposed peak DHW demand as one of the design goals of the system. Doing so establishes a mutually agreeable objective and allows the thermal capacity of the equipment to be selected accordingly.

Keep in mind that the size of the boiler, or multiple boiler system, may be determined by this DHW supply requirement, rather than by the design heat loss of the building.

Example 14.6

An owner and designer agree that the domestic water heating subsystem for a 10,000 ft.² custom home with six bathrooms will be capable of delivering a continuous domestic hot water flow of 18 gpm at a supply temperature of 120 °F. The lowest cold water temperature at the site is 45 °F. Determine the heat source capacity required for this load.

Solution:

The sensible heat rate Equation 4.7 can be used for this calculation:

$$Q = 500f(\Delta T) = 500(18)(120 - 45) = 675,000 \text{ Btu/hr}$$

Discussion:

The result clearly shows that significant heat source capacity is required to meet this DHW requirement. It is unlikely that the design space heating load of even a 10,000 ft.² home would be this high. A properly sized multiple boiler system with three boilers could supply this domestic hot water load on a continuous basis. If three boilers were used, and each boiler had a turndown ratio of 10:1, the system turndown ratio would be 30:1. This would allow the boiler system to meet the peak domestic water heating load and still retain high seasonal efficiency for space heating at reduced firing rates. The controls in this system would be configured to make domestic water heating the priority load.

14.4 Intermittent Garage Heating

The option of heating a garage is often appealing to homeowners in cold and snowy climates. In some cases, the owner's desire is to maintain the garage at a relatively cool temperature sufficient to melt snow and ice from cars, or perhaps allow the garage to be used as a workshop. In other cases, the intent is to heat the garage only when necessary, and thus minimize fuel usage.

The intended usage of a garage should be considered when selecting the type of heat emitter(s). For example, radiant slab heating is ideal in situations where a relatively constant temperature will be maintained. The heated floors will significantly improve comfort, as well as quickly melt snow and ice from vehicles and dry the floor. However, the large thermal mass of such systems makes them very slow to react to sudden changes in thermostat setting, and therefore, unsuitable for situations where frequent thermostat adjustments will be made. In contrast, an overhead unit heater has very little thermal mass and can react quickly to changes in thermostat setting. It would be a better choice when the primary goal is to quickly warm the air in a garage.

Arguments can be made both for and against the use of antifreeze in garage heating applications. If the garage is to remain above freezing at all times, one might assume that antifreeze is not needed. On the other hand, if the heat is to be totally shut off during cold weather, the distribution system is likely to experience subfreezing temperatures and therefore must contain antifreeze.

The author's position is that antifreeze should be used in all hydronic garage heating applications. Unanticipated events such as power outages or equipment breakdowns can lead to frozen pipes in a matter of hours. A hard freeze can burst piping and create a very expensive repair. The extra cost associated with using antifreeze is a small price to pay to be sure the garage heating circuits will not be damaged by freezing temperatures.

If the entire hydronic system will be filled with an antifreeze solution, the garage can be heated by creating another zone circuit. This is generally a good approach in smaller systems. However, if the garage zone is relatively small in relationship to the remainder of the system, the cost of installing and maintaining antifreeze in the entire system may not be justified. In such situations, the garage zone can be supplied through a brazed plate heat exchanger. An example of such a system supplying garage floor heating is shown in Figure 14-47.

This system uses a modulating/condensing boiler as the heat source. It supplies two independent space heating zones (using zone circulators). It also supplies an indirect water heater, which is configured as a priority load. When the temperature sensor in the indirect tank signals calls for heat, the boiler operates at an elevated setpoint temperature, the DHW circulator turns on, and the circulator between the boiler and hydraulic separator turns off. This is a common control mode that is now integrated into most modulating/condensing boilers. It allows for a high rate of heat delivery to the indirect tank. During this mode of operation, the two space heating zones, and the garage zone are temporarily turned off. When the domestic water heating mode is off, and either of the two space heating zones, or the garage zone call for heat, the boiler operates based on outdoor reset control.

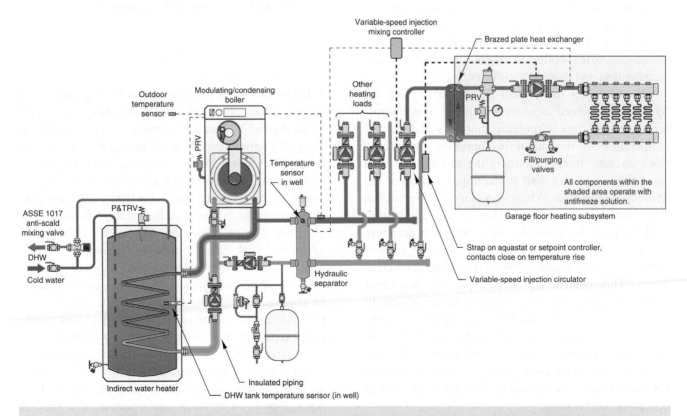

Figure 14-47 | System that supplies garage floor heating as a separate zone through a brazed plate heat exchanger. All piping components outside of heated space are filled with an antifreeze solution.

The garage slab is heated by embedded tubing. The tubing and all piping back to the brazed plate heat exchanger operates with a 30–40% solution of propylene glycol antifreeze. This allows the garage zone to be completely turned off without risk of freezing any portion of the heating system.

The heat exchanger and all piping to the right of it form an independent closed piping circuit. As such, it requires components associated with any closed loop, such as an expansion tank, air separator, pressure-relief valve, and a means for filling and purging. A dual port purging valve allows hoses to be connected to the pump in the antifreeze solution, and push air out of the circuit.

The heat exchanger is piped into the system as a parallel branch circuit. Hydraulic separation of the three circulators is provided by generously sized header piping (sized for a maximum flow velocity of 2 ft/sec under full design load flow), in combination with a hydraulic separator. The latter also hydraulically separates the boiler circulator from the other circulators, and provides air and dirt separation for the system. All circulators are equipped with internal check valves to prevent reverse flow and heat migration.

The heat exchanger is piped for counterflow to maximize heat transfer rates. In this case, the heat exchanger should be sized for a relatively close approach temperature difference (suggested as no more than 5 °F) under design load conditions. This allows the modulating/condensing boiler to operate at relatively low temperatures and thus higher efficiency.

Heat transfer to the heat exchanger is regulated by a variable-speed injection circulator. The controller operating this circulator monitors the supply temperature to the manifold supplying the garage floor circuits. It also monitors the temperature of a sensor mounted in a well in the upper portion of the hydraulic separator. When necessary, the controller slows the injection pump to prevent the temperature at the top of the hydraulic separator from dropping below a value where the two space heating circuits could no longer maintain comfort in their respective zones. This control action is especially important when the garage floor is warming from potentially very cold conditions, and thus capable of absorbing heat from the system at much higher rates. Without this control action the water temperature to the two space heating zones could be significantly depressed for several hours as the garage slab warms to normal operating temperature.

Another important detail is use of a strap-on aquastat, or electronic setpoint controller, monitoring the temperature of the piping leaving the water side of the brazed plate heat exchanger. The purpose of this controller is to verify flow of heated water through the water side of the heat exchanger *before allowing the circulator on the distribution side of the heat exchanger to operate*. This protects against the possibility that antifreeze at very low sub-freezing temperatures could flow through the heat exchanger before heat is supplied from the heat source. Without this control action, there is a possibility of freezing the water in the heat exchanger. A hard freeze could potentially rupture the heat exchanger. When the strap-on aquastat or setpoint controller detects a temperature of perhaps 80 °F or more, its contacts close to allow the distribution circulator to operate. If the water temperature on the water side of the heat exchanger drops to perhaps 40 °F, as could happen if the boiler fails, the contacts in the aquastat or setpoint controller open to stop the antifreeze circulation through the heat exchanger before freezing could occur.

The expansion tank in the anti-freeze protected portion of this system can be intentionally oversized. The extra tank volume provides a reserve of antifreeze solution under pressure, and therefore, able to enter the system to make up for minor losses such as when air is vented from this portion of the system.

14.5 Pool Heating

A boiler, or multiple boiler system, that supplies space heating and domestic hot water to a building can usually be configured to heat a swimming pool or spa. This is especially appropriate for pools heated from late spring to early fall, when space heating loads are very low or nonexistent. The subsystem needed for pool heating can usually be installed at a lower cost than a dedicated pool heater operating on the same fuel as the boiler(s).

Because pool water is often highly chlorinated, or treated with other disinfecting chemicals, a suitable heat exchanger must be used between the heat source and the pool water. That heat exchanger must be compatible with the chemicals in the pool water. In most cases, this requires that all portions of the heat exchanger that come in contact with pool water be made of titanium or a titanium/stainless steel alloy. Never assume that standard 316 or 304 stainless steel heat exchangers can be used in pool heating applications. Figure 14-48 shows an example of a shell & tube heat exchanger specifically designed for pool heating applications.

Figure 14-48 Example of a shell & tube heat exchanger design for pool heating. *Courtesy of Weil-McLain Corp.*

Figure 14-49 shows a piping schematic for a system in which a multiple mod/con boiler array provides space heating, high-capacity domestic water heating, and pool heating.

The pool heating subsystem shown in Figure 14-49 uses the pool filter pump to move pool water through the tube bundle in the heat exchanger. Bypass piping is shown along with a manually operated bypass valve. The shell side of the heat exchanger is also piped to CPVC unions and ball valves. This detailing allows the heat exchanger to be removed from service if necessary, while still maintaining the ability to circulate pool water through the filter.

The hot side of the pool heat exchanger is piped in parallel with the space heating subsystem and the indirect water heater subsystem. The piping is arranged so boiler water and pool water flow through the heat exchanger in opposite directions (e.g., counterflow) to maximize the rate of heat transfer.

A swimming pool containing thousands of gallons of water, and typically maintained at temperatures between 75 and 90 °F represents a huge, and relatively low-temperature thermal mass. As such, it has the potential to "dominate" the water temperature in the entire system. If the pool heating subsystem is operating at the same time as space heating or domestic water heating, it could depress the water temperature in the system far below those needed for proper heat delivery to the other loads.

One of the simplest ways to avoid this is to make pool heating the lowest priority load in the system. As such, it is only allowed to operate when both the domestic water heating subsystem and the space heating subsystem are inactive. Figure 14-50 shows one way to do this using simple hard-wired relay logic, as discussed in Chapter 9.

This ladder diagram makes domestic water heating the highest priority load, space heating the next highest priority load, and pool heating the lowest priority load. This control logic would interrupt pool heating if either space heating or domestic water heating was required. For pools heated from late spring to early fall, it is relatively unlikely that space heating loads would interrupt pool heating under ordinary circumstances. Thus, the most likely interruption would be for domestic water heating, and such interruptions are likely to be for relatively short periods. The very high thermal mass of the pool "cooperates" with this prioritization concept. Following an initial warm up to nominal operating temperature, heat input interruptions lasting one or two hours, if necessary, would create essentially undetectable changes in pool temperature.

If the pool heat exchanger is sized for a low approach temperature difference, the relatively low-temperature pool water allows the mod/con boilers shown in Figure 14-49 to operate at high efficiency. The author recommends that the heat exchanger be sized so that the full boiler output can be transferred through it while operating at a maximum approach temperature difference of 20 °F.

Pool Heating Loads & Heat Source Requirements

Many factors affect the heating requirement of a swimming pool. These include:

- Radiational cooling to the sky (outdoor pools)
- Heat gain from solar radiation (outdoor pools)
- Convective heat loss to the air
- Conduction losses through the sides and bottom of the pool
- Evaporation losses from the water surface
- Allotted time during which the pool needs to be raised from some unheated condition to the desired temperature.

The dominant form of heat loss from pools is evaporation. Each pound of water that evaporates from the surface removes about 960 Btus from the pool. Evaporation losses increase with increasing wind speed across the water surface. They also increase as the relative humidity of air above the pool decreases. Evaporative

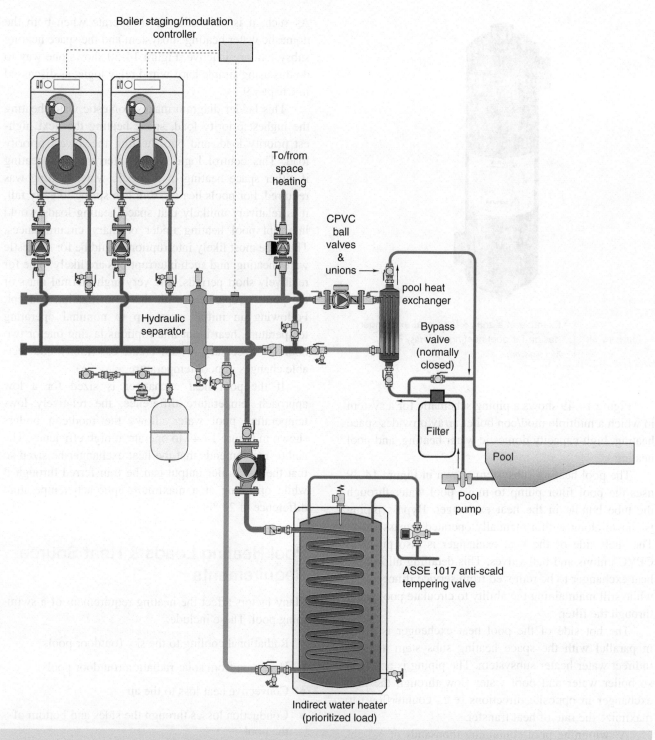

Figure 14-49 | Piping schematic for pool heating as ancillary load to space and domestic water heating. Heat for the pool is lowest priority load, and is supplied through shell & tube heat exchanger.

heat loss changes almost constantly for outdoor pools. Estimates range from 30 to well over 100 Btu/hr/ft² of pool surface area. A pool cover will significantly reduce evaporation as well as convective losses and is highly recommend for any heated pool application.

The amount of boiler capacity allotted to pool heating can also depend on owner expectations. A typical request will be the ability to raise the pool temperature from some initial unheated condition, to a desired average temperature, within a specified time. The greater the temperature rise, and the shorter the allotted time, the great the required boiler capacity. The following example demonstrates how to quantify this situation.

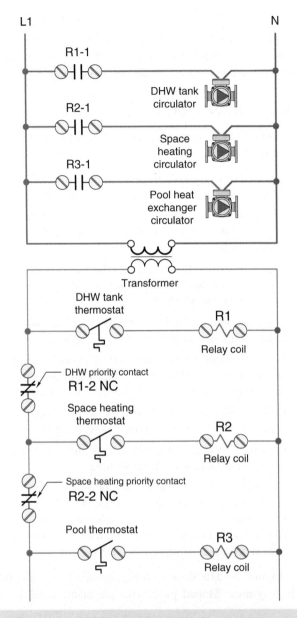

Figure 14-50 Ladder diagram showing three levels of priority control. DHW has highest priority, space heating second highest priority, and pool heating the lowest priority.

Example 14.7

A pool measures 16 feet wide, 32 feet long, and has an average depth of 4.5 feet. Assuming an unheated average water temperature of 60 °F, determine the required heat source capacity to:

a. raise the pool to 85 °F over 24 hours.
b. raise the pool to 85 °F over 48 hours.

Because the pool remains covered, and is fully insulated, assume that heat losses through the pool walls and by evaporation are insignificant.

Solution:

The sensible heat quantity equation introduced in Chapter 4, Properties of Water, can be used to calculate the total heat that must be added to the pool for the necessary 25 °F temperature rise. However, to use this equation, the volume of the pool must first be determined.

$$V_{pool} = (16 \text{ ft})(32 \text{ ft})(4.5 \text{ ft})\left(\frac{7.49 \text{ gal}}{\text{ft}^3}\right) = 17{,}257 \text{ gal}$$

The sensible heat quantity equation (Equation 4.5) can now be used:

$$h = 8.33v(\Delta T) = 8.33(17{,}257)(85 - 60 \text{ °F}) = 3{,}594{,}000 \text{ Btu}$$

If this quantity of heat must be supplied in 24 hours, assuming the heat source is operating continuously, the required heat source capacity is:

$$q = \frac{3{,}594{,}000 \text{ Btu}}{24 \text{ hr}} = 149{,}800 \frac{\text{Btu}}{\text{hr}} = 150{,}000 \frac{\text{Btu}}{\text{hr}}$$

If this amount of heat needs to be supplied over 48 hours, again assuming the heat source is operating continuously, the required heat source capacity is:

$$q = \frac{3{,}594{,}000 \text{ Btu}}{48 \text{ hr}} = 74{,}900 \frac{\text{Btu}}{\text{hr}} = 75{,}000 \frac{\text{Btu}}{\text{hr}}$$

Discussion:

The lower capacity requirement (75,000 Btu/hr) is typical of the design heating load in a 3000 to 4,000 ft² house with reasonable insulation. If the pool heating must be accomplished in half the time (e.g., 24 vs. 48 hours), it is likely that a heat source significantly larger than needed for space heating would be required. A multiple modulating/condensing boiler system would be a good choice for such an application. Its suitability is further enhanced if the home has a high domestic water heating load or a driveway snow melting system.

Pool Heat Exchanger Placement

In most climates where hydronic heating is used, outdoor nonenclosed pools are not heated in winter. In such systems, it is likely that pool water would be drained from the secondary side of the pool heat exchanger when the pool is closed for winter. If the pool heat exchanger is located in an area where freezing could occur, both sides of the heat exchanger must be protected from freezing. Although it is possible to isolate and drain the hydronic system side of the pool heat exchanger each fall, doing so requires adding fresh water to the overall system each spring. This increases the potential for corrosion.

Another option is to protect the hydronic side of the pool heat exchanger and any associated outside piping with antifreeze. This is especially viable in systems where all subsystems operate with antifreeze. When the majority of the system operates with water, it is likely that a second heat exchanger may be required.

Yet another option is to locate the pool heat exchanger within heated space and provide drainage provisions for the pool water side. *If the pool heat exchanger is located in a basement, or other space that is below the level of the pool water, it is possible that a leak in the pool side piping, or heat exchanger, could drain a large amount of the pool water into that space. The author highly recommends that any such space be equipped with a floor drain that could handle the potential volume and flow rate of such a leak.* Any electrical or electronic equipment that could be damaged from such a leak should be isolated.

14.6 Hydronic Snow Melting

Another sector of the North American hydronics market that has enjoyed growth over the last three decades is hydronic snow and ice melting systems (also known as **SIM systems**). Modern SIM systems use PEX, PE-RT, or PEX-AL-PEX tubing embedded in pavements to melt snow and ice.

SIM systems can be large stand-alone installations, or smaller subassemblies tied into hydronic space heating systems. The potential applications for SIM systems include driveways, walkways, parking areas, exterior steps, patios, wheelchair ramps, loading docks, and any other areas that must be cleared of snow and ice.

Benefits Offered by SIM Systems

Hydronic SIM systems offer several benefits over traditional methods of snow removal. These include:

- The ability to provide fully automatic and unattended snow and ice removal whenever required.
- The ability to remove snow without creating banks or piles that often lead to drifting and/or damage to landscaping.
- The elimination of sanding and the associated mess created when sand is tracked into buildings or swept from pavements in spring.
- The elimination of salting and its potential damage to landscaping and the surrounding environment.
- Less pavement damage due to frost action, salting, or physical damage from plowing, which is especially important for surfaces covered with pavers.
- Reduced likelihood of slips, falls, or vehicular accidents due to safer walking and driving surfaces.
- Improved property appearance in winter due to lack of snow banks and sand/salt residue.
- The ability to use almost any fuel or heat source to provide the energy needed to operate the SIM system. In commercial and industrial installations, it may even be possible to use "waste heat" from cooling or refrigeration processes to supply the SIM system.

Figure 14-51a shows tubing installed in a sloped driveway area. Sloped pavements are prime candidates

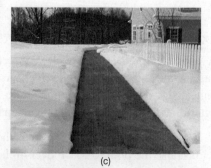

Figure 14-51 (a) Tubing for snow and ice melting being installed in a sloping driveway. *Courtesy of John Siegenthaler.* (b) Sloping driveway surface while SIM system is operating. *Courtesy of Clay Thornton.* (c) Melted walkway is free of snow and ice. *Courtesy of Gary Todd.*

for hydronic SIM systems. Figure 14-48b shows the completed pavement area while the SIM system is active. The melted walkway shown in Figure 14-48c demonstrates a smaller but strategic use of SIM systems.

SIM System Classifications

There are several approaches to designing a hydronic SIM system. They vary in the rate of heat delivery to the surface and the type of controls used to start and stop the melting process. The following SIM system classifications have been developed by the industry.

- *Class 1 SIM.* **Class 1 SIM** systems are sufficient for most residential walkway and driveway areas. The rate of heat delivery to the surface is typically 80 to 125 Btu/hr/ft^2 depending on the geographic location of the system.

    Class 1 SIM systems typically allow snow to accumulate on the surface during a heavy snowfall, especially if the melting process is not automatically started. This snow layer is beneficial because it acts as an insulator between the heated pavement surface and the outside air, and thus reduces evaporative and convective heat loss. The majority of the heat reaching the surface melts the underside of the snow layer instead of being dissipated in other ways.

- *Class 2 SIM.* **Class 2 SIM** systems are sufficient for most retail and commercial paved areas. The rate of heat delivery to the surface is generally 125 to 250 Btu/hr/ft^2 depending on geographical location. Class 2 systems are designed to prevent snow from accumulating on the surface as it falls once the pavement is up to normal operating temperature.

- *Class 3 SIM.* **Class 3 SIM** systems have the highest heat delivery rate to the pavement surface, typically 250 to 450 Btu/hr/ft^2 depending on the geographical location. Class 3 systems are used in areas that must be kept clear of snow and ice at all times, and where the surface should be dry soon after the snowfall ends. Examples include helicopter pads, toll plazas, sloped pavements in parking areas, and entrances to hospital emergency rooms.

    Figure 14-52 gives suggested heat delivery rates for all three SIM system classifications in several geographic locations.

Melted Pavement Construction Details

This section discusses construction details used to incorporate snow and ice melting in different types of paved surfaces. The designer should always check to see if local codes require variations on the methods presented.

Suggested output (Btu/hr•sq.ft.)

City	Class 1	Class 2	Class 3
Albuquerque, NM	71	82	167
Amarillo, TX	98	143	241
Boston, MA	107	231	255
Buffalo, NY	80	192	307
Burlington, VT	90	142	244
Caribou, ME	93	138	307
Cheyenne, WY	83	129	425
Chicago, IL	89	165	350
Colorado Springs, CO	63	63	293
Columbus, OH	52	72	253
Detroit, MI	69	140	255
Duluth, MN	114	206	374
Falmouth, MA	93	144	165
Great Falls, MT	112	138	372
Hartford, CT	115	254	260
Lincoln, NE	67	202	246
Memphis, TN	134	144	212
Minneapolis-St. Paul, MN	95	155	254
Mt. Home, ID	50	90	140
New York, NY	121	298	342
Ogden, UT	98	216	217
Oklahoma City, OK	66	81	350
Philadelphia, PA	97	229	263
Pittsburg, PA	89	157	275
Portland, OR	86	97	111
Rapid City, SD	86	102	447
Reno, NV	98	154	155
St. Louis, MO	122	152	198
Salina, KS	85	120	228
Sault Sainte Marie, MI	78	144	213
Seattle-Tacoma, WA	92	128	133
Spokane, WA	87	127	189
Washington, DC	117	121	144

Data source: Section 55.4 1987 *ASHRAE Systems Handbook*

Figure 14-52 Recommended rates of heat transfer for Class 1, Class 2, and Class 3 SIM systems in various geographical locations.

Drainage Provisions

It is very important that all melted pavements be detailed for proper drainage of melt water. The heat delivery rates used with Class 1 and Class 2 systems assume that most of the melted snow is drained from the surface as a liquid rather than being removed by evaporation. *If the SIM system is forced to evaporate a high percentage of the melted water, energy usage will be significantly increased.*

Improper drainage can also allow melted water to accumulate at low points on the surface, or where melted pavement adjoins nonmelted areas. When melting stops, puddle water can quickly turn to dangerous ice.

Pavements must be sloped toward drains capable of routing the melted water to a drywell, storm sewer, or other discharge means acceptable to local codes. Drainage piping should not run through the heated pavement because the cold water will absorb heat from it. Instead, drainage piping should be routed beneath the underside insulation where it is protected from freezing. When not under the melted pavement, the drainage piping should be either buried below frost level or covered with a layer of extruded polystyrene to control frost penetration.

Trench drains are often installed along the low areas in melted pavements. If the pavement slopes toward a building, the melt water must be intercepted by the trench drain before it can flow into the building. Likewise, be sure melt water running toward a street will be collected by a drain before it contacts unheated pavement. Figure 14-53 shows some examples of pavement drainage concepts.

On some projects, the drainage system will not be installed by the same company that installs the rest of the SIM system. In such cases, be sure the necessary drainage details are clearly documented, communicated, and coordinated with those who are responsible for their installation. *The lack of proper drainage can severely compromise an otherwise well-planned SIM system. Keep in mind that the same drainage provisions that handle melted water in winter must also handle peak rainfall rates year-round.*

Subsurface Conditions

When planning a SIM system, designers should always evaluate the soils under the area to be melted. Failure to address subsoil conditions can lead to unanticipated conditions that could damage both the pavement and embedded tubing.

If the local water table is within 3 feet of the surface, it has the potential to draw heat downward at a high rate as the SIM system operates. High water table situations require drainage. A properly detailed "French drain" constructed around the perimeter of the paved area is a common solution.

If bedrock is present under the melted area, it is important to slope or channel the rock surface so water percolating downward can be drained away. Otherwise, the bedrock may pond water under the melted pavement. It is also important to install at least 1-inch (nominal R-5) extruded polystyrene insulation, having a suitable compressive stress rating, between the bedrock and upper (heated) layers of the pavement system to reduce downward heat conduction.

Soils containing high amounts of clay or silt are often saturated with water in winter. When these soils freeze, the expanding ice crystals create powerful forces that can easily crack and heave pavements. If such soils are present, the base layer of the pavement system should consist of 6 to 9 inches of 1/2 to 1-inch sized crushed stone. The soil surface beneath this layer of crushed stone should be sloped so water will be drained away rather than accumulate within the stones. Be sure a complete drainage path is provided to daylight or a suitable drywell or storm sewer. Also be sure the water cannot freeze as it passes through that drainage system. This stone layer should be compacted to form a flat

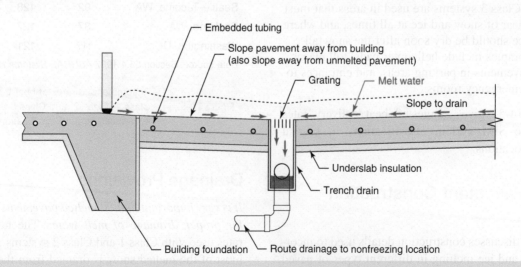

Figure 14-53 | Melted pavement slopes toward trench drain.

and stable surface for the insulation board that will be installed above it.

When pavements will be placed over areas of disturbed or otherwise unstable soil, a **geotextile fabric** should be incorporated into this base layer. This very strong nondeteriorating structural fabric helps spread high loads over larger areas to prevent eventual depressions in the pavement. Such depressions could eventually damage embedded tubing.

No snow melting system can correct for poor pavement design. Be sure to involve knowledgeable professionals in the pavement planning process.

SIM Installation in Concrete Pavement

Figure 14-54 shows the construction details used for a typical SIM system in a concrete driveway or walkway.

If the soil at the installation site has good drainage characteristics, the base layer of the pavement system generally consists of 6 to 9 inches of compacted gravel. Moisture percolating down through this base layer will continue downward into the subsoil.

In cold climates, or projects where the heated pavement will be "**idled**" at temperatures just above freezing, install a minimum 1-inch (nominal R-5) extruded polystyrene insulation over the compacted gravel base. This insulation greatly reduces downward heat loss. It also shortens the time required for the pavement to reach melting temperature following a cold start. Be sure the insulation board lies flat against the compacted gravel base at all locations so the pavement is fully supported.

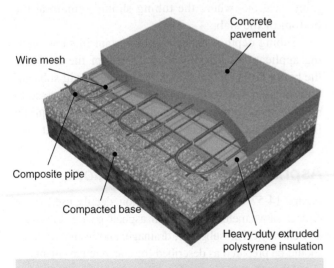

Figure 14-54 | Installation of tubing for SIM system in concrete slab pavement. *Courtesy of IPEX Corp.*

The **compressive load rating** of the extruded polystyrene insulation should be selected to match loads that may be imposed on the pavement. A 25-psi rated extruded polystyrene insulation board is the minimum rating for pavements subject to light vehicular traffic. If heavier traffic is possible, insulation with a compressive load rating of 40 to 100 psi may be required. Manufacturers of extruded polystyrene insulation can provide guidance on selecting the proper compressive load rating for a given pavement application.

Welded wire fabric (WWF), or a grid of steel rebar is now installed over the insulation. Be sure to overlap all sheets of WWF by at least 6 inches and tie them together with wire ties.

The tubing can now be secured to the steel reinforcing using wire twist ties spaced 24 to 30 inches apart. *Tube spacing should never exceed 12 inches in Class 1 systems or 9 inches in Class 2 and 3 systems.* Wider spacing can produce uneven melting patterns that may not completely clear tubing the pavement of snow before the melting cycle is complete.

Tube circuits should be planned so as not to exceed the maximum circuit lengths given in Figure 14-64.

The warmest portion of the circuit should be routed in the areas with the highest melting priority. For example, the tire track area of a driveway should have a higher melting priority than the edges of the driveway. Do not install tubing closer than 6 inches to the edge of the pavement.

Figure 14-55 is an example of a tubing layout for a driveway. As with floor heating systems, it is advisable to prepare a tubing layout drawing before installing any tubing. The CAD-based methods, discussed in Chapter 10, Hydronic Radiant Panel Heating, can be used to quickly develop such a drawing.

The tubing should be protected by polyethylene sleeving wherever it crosses a control joint location. At such locations, the tubing should be held to the bottom of the slab to ensure it will not be damaged when the control joints are sawn.

Once placed, all tubing circuits should be pressure tested with compressed air at 75 psi for at least 24 hours prior to placing the concrete.

In locations where tubing passes through a foundation wall, it should be protected by rigid schedule-40 PVC or steel pipe sleeves. The sleeves provide space around the tubing, allowing it to move without pinching, should the soil or pavement shift vertically. The cut ends of the sleeves should be smoothed to prevent chafing the tubing. The soil beneath the sleeved area should be fully compacted to minimize settling. The sleeves

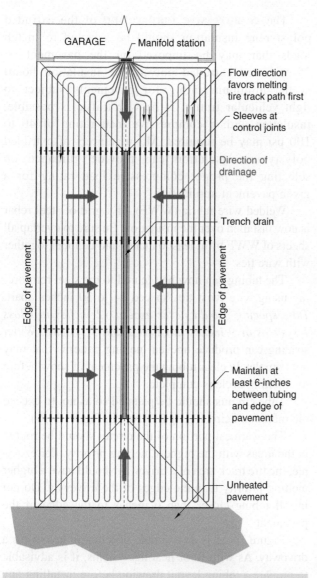

Figure 14-55 Example of tubing layout plan for melted driveway. Note flow direction in circuits and pavement slope.

Figure 14-56 Use of a compressive "donut" collar to seal a nominal 1" tube passing through a 3" hole in a concrete basement wall. A 2-inch extruded polystyrene insulation is applied to the inner surface of the wall. *Courtesy of John Siegenthaler.*

should be cast or tightly sealed into the foundation wall and pitched slightly downward on the exterior side to discourage water intrusion. The soil beneath the outside end of the sleeves should be well-drained so that any water percolating down from the surface cannot accumulate. Finally, once the tubing has been connected, the space between the inside of the sleeves and the tubing should be sealed with an elastomeric caulk to prevent water or insect entry.

It is also possible to use a **compressive "donut" collar** to seal between the outside diameter of the tubing and a hole drilled through a foundation wall. Figure 14-56 shows an example. In this case, the compressive donut collar is sealed against a 3-inch diameter hole through a poured concrete basement wall. A 2-inch thick layer of extruded polystyrene insulation has been placed against the inside surface of that wall.

Air-entrained concrete with a minimal 28-day compressive stress rating of 4,000 psi is often specified for exterior slabs. As with a slab-on-grade floor heating system, the tubing and reinforcing mesh should be lifted during the pour such that the top of the tubing is 1.5 to 2 inches below the finish surface of the pavement. The exception is under control joint locations where the tubing should remain at the bottom of the slab.

Tubing depth is especially important in snow melting applications. Leaving the majority of the circuit at the bottom of a typical 6-inch exterior slab will substantially increase the time required to warm the surface to melting temperature. It will also increase the required fluid temperature and create greater downward heat loss.

Asphalt Pavement Installations

Figure 14-57 shows the construction details used for a typical snow melting system in asphalt paved driveways or walkways. The subgrade, drainage, and insulation layer should be prepared as described for concrete pavements.

After this, a layer of sand or stone dust approximately 2 inches thick should be placed on the insulation. This material helps protect the PEX or PEX-AL-PEX tubing from the hot (250–350 °F) asphalt when it is placed. A mat of welded wire fabric (WWF) is then laid

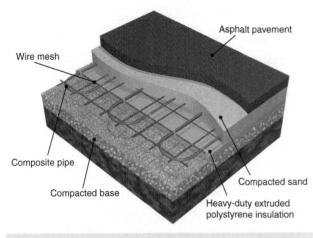

Figure 14-57 | Installation of tubing for SIM system in asphalt pavement. *Courtesy of IPEX Corp.*

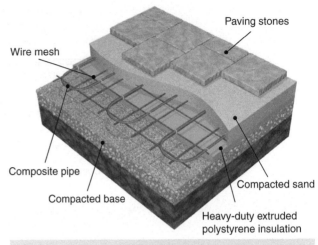

Figure 14-58 | Installation of tubing for SIM system in area covered by pavers. *Courtesy of IPEX Corp.*

out over the sand layer. The tubing circuits are tied to the WWF using wire twist ties, and all circuits are pressure tested.

After pressure testing, an additional 2-inch layer of sand or stone dust is placed over the tubing circuits. The circuits should remain under pressure as this material is placed and leveled by hand. Power loaders should not be driven directly over the tubing circuits. After the material is placed and leveled, it should be fully compacted.

Although opinions differ on necessity, some installers prefer to circulate cool water through the tubing circuits as the hot asphalt is placed to further protect them from the elevated temperatures. The final step is to place the asphalt pavement as normal.

SIM Installation for Paver-Covered Surfaces

Surfaces consisting of loosely laid pavers or tiles are easily damaged by conventional methods of snow removal, and therefore, well suited to hydronic snow melting. Figure 14-58 shows the material assembly used for such installations.

Unlike concrete or asphalt surfaces, water can seep down between pavers across the entire melted area. This water cannot be allowed to accumulate under the pavers because subsequent freezing can cause heaving. In areas with low permeability soil, the base layer below the insulation should be detailed for efficient drainage. A 6- to 9-inch-deep layer of compacted crushed stone placed over a slightly sloping subgrade is appropriate. Drainage piping that removes the accumulated water to a nonfreezing disposal point is necessary.

Extruded polystyrene insulation is impermeable to water. To allow water drainage, nominal 1/2-inch gaps should be left between adjacent sheets of insulation. Alternatively, several sheets of rigid insulation can be stacked and drilled to form a grid of 1-inch diameter holes spaced 12 inches apart. In either case, the gaps or holes should be covered with strips of water-permeable **filter fabric**. This allows water to drain into the crushed stone layer without carrying sand or stone dust with it.

Note: It is critical that foam board insulation installed below paver-covered surfaces have sufficient bearing strength (60 psi minimum compressive stress rating is suggested). It is also crucial that the foam insulation lies absolutely flat on a fully compacted subgrade. Failure to do so can result in settling of pavers subjected to heavy concentrated loads.

A layer of sand or stone dust about 2 inches thick is now placed over the insulation and compacted. A mat of WWF is then laid out over this layer. The tubing circuits are tied to the WWF and pressure tested.

Next, an additional 2-inch layer of sand or stone dust is placed over the tubing circuits. The circuits should remain under pressure during the time this material is placed, leveled by hand, and fully compacted. Power loaders should not be driven directly over the tubing circuits. The base layer is now ready for the pavers.

If heavy vehicle traffic is anticipated on the paver-covered surface, the use of a sunken concrete sub-slab is recommended. This installation technique is shown in Figure 14-59.

The sunken **concrete sub-slab** provides additional structural support and helps spread concentrated loads over a greater area of insulation. This load-spreading

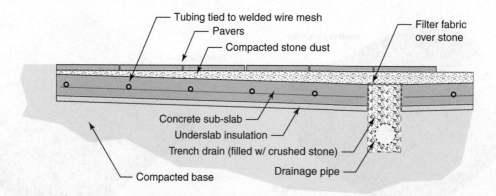

Figure 14-59 | Cross-section of SIM system in area covered by pavers, and subject to heavy loads.

effect reduces the compressive strength required of the insulation board relative to that required when no slab is present.

The sub-slab should be slightly sloped to promote drainage of the water that percolates down between the pavers. This drainage water should be collected in a trench filled with crushed stone and equipped with a drainage pipe as shown in Figure 14-59. Because the sub-slab will be covered with stone dust, it does not need to be troweled.

SIM System Design and Control Options

There are several ways to design and control SIM systems. They differ in how melting is initiated, as well as how pavement temperature is regulated before, during, and after melting. They also have different hardware requirements.

The approach used must consider the owner's expectations, the size of the area being melted, and the class of system being designed. The following is a brief discussion of the main issues and options available to address them.

Freeze Prevention Options

Some SIM installations are "dedicated" systems. Snow/ice melting is the only function of the system. It is common to fill such systems with an antifreeze solution (typically a 40–50 percent mixture of inhibited propylene glycol and water).

Figure 14-60 shows an example of a SIM system using a dedicated modulating/condensing boiler.

It is important to ensure adequate flow through the pavement circuits in a SIM system. The rates of heat transfer required, even in Class 1 SIM systems, require significantly higher circuit flow rates compared to those used with radiant panels that provide space heating. These flows require good hydraulic separation between any heat source with moderate or high hydraulic resistance and the distribution system. In the system shown in Figure 14-60, this is achieved using a hydraulic separator, which also provides air and dirt separation for the system.

The system in Figure 14-60 also includes a fluid feeder, an example of which is shown in Figure 13-24. This component monitors system pressure and automatically turns on a small positive displacement pump to add premixed antifreeze solution to the system if the pressure drops due to small fluid losses from air removal or from flushing the lower valve of the hydraulic separator. This prevents dilution of the antifreeze and maintains adequate pressure for proper system operation.

In other systems, snow melting is one of several loads served by a common boiler (or multiple boiler system). Such systems need to be designed one of two ways:

1. The entire system operates with an antifreeze solution of suitable concentration for the SIM portion of the system.

2. The SIM portion of the system, which operates with an antifreeze solution is separated from the remainder of the system, which operates with water, using a heat exchanger.

In the latter case, brazed plate heat exchangers are commonly used to separate the SIM subsystem from the remainder of the system. Figure 14-61 shows an example of this approach.

When a SIM system first starts during cold weather, the antifreeze solution returning to the heat exchanger can be very cold. To avoid the possibility of freezing water on the heat input side of heat exchanger, it's necessary to prevent the antifreeze from circulating through the other side of the heat exchanger until heat input is verified on the water side. This is

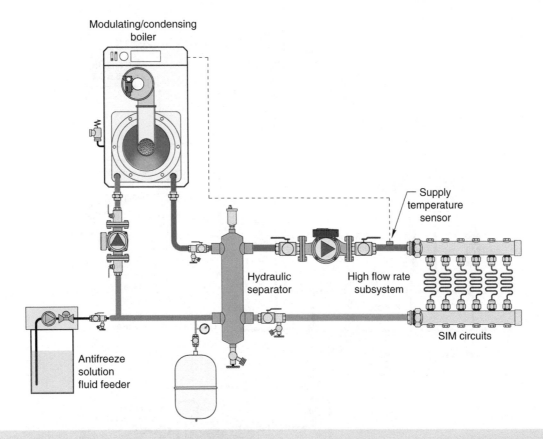

Figure 14-60 | Mod/con boiler supplying a manifold for snowmelting circuits. Entire system operates with antifreeze. A hydraulic separator decouples boiler circulator from distribution circulator. System pressure is maintained by fluid feeder.

the same concept discussed for intermittent garage heating. It can be done with a setpoint temperature controller or strap-on aquastat mounted on the outlet pipe from the water side of the heat exchanger. The contacts in setpoint controller or aquastat close at a temperature that verifies that the water side is at a temperature well above freezing. The SIM distribution circulator(s) can then operate. If the temperature of the water leaving the primary side of the heat exchanger drops to perhaps 40 °F or less, the contacts in the setpoint controller or aquastat open to turn off the SIM distribution circulators, and avert a potential freeze.

If the boiler(s) supplying a multiload system do not have sufficient capacity to operate the SIM subsystem simultaneously with the other loads, such as space heating or domestic water heating, those loads should be set up as having higher priority than other loads in the system. Methods for doing so were discussed previously for pool heating subsystems.

Boiler Options for SIM Systems

When a conventional boiler supplies heat to a SIM system, that boiler must be protected against sustained flue gas condensation. The system must use a mixing controller that measures boiler inlet temperature and reduces the rate of heat transfer to the distribution circuits when necessary to keep the boiler operating above the dewpoint of the flue gases. The mixing device can be any of those previously discussed for space heating applications as long as boiler inlet temperature protection is provided.

Modulating/condensing boilers are well suited to hydronic snow melting systems. When a condensing boiler is used as a dedicated SIM heat source, it is not necessary to install a mixing device between the boiler and the distribution system. The lower the operating temperature of the condensing boiler, the higher its efficiency.

Pavement Temperature Control

When a SIM system starts up from a very cold temperature, it may take several hours for the surface to reach a temperature where melting begins. To decrease this lag time, some SIM controllers "idle" the pavement at a temperature close to 32 °F. If the pavement is idled just above 32 °F, it will generally be free of frost and "black ice." This is an important safety advantage, especially in public areas. Idling the pavement just below freezing reduces standby heat loss,

738 Chapter 14 Auxiliary Loads and Specialized Topics

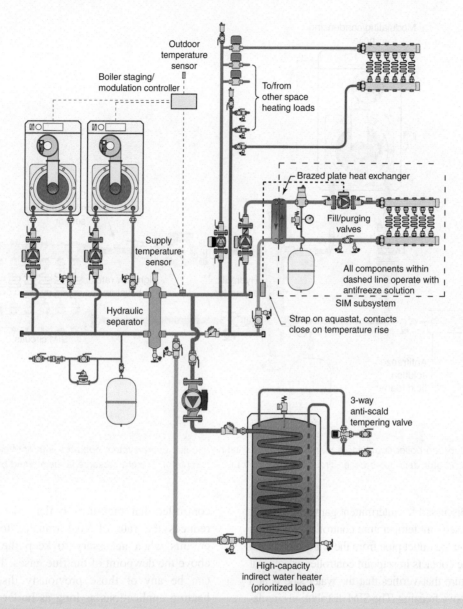

Figure 14-61 | A SIM subsystem supplied through a heat exchanger.

but still allows rapid warm-up to melting temperature when snow or freezing rain begins falling. Most SIM controllers with idling capability allow the installer to select the desired idling temperature. These controllers typically initiate the idling mode when outside air temperature drops within a few degrees of freezing. Frozen precipitation can occur at these air temperatures. Allowing the system to idle at higher outdoor air temperatures is unnecessary and wastes fuel.

To idle the pavement, the controller must sense pavement temperature. Some controllers use a small thermistor sensor located in a well that's embedded in the pavement. The well usually is made of copper tubing with a cap soldered on the end. The position of the slab temperature sensor is crucial to the proper performance of the control system. It is typically located 1 inch below the top of the pavement, and halfway between adjacent tube circuits. The open end of the well should lead to an accessible location so the sensor can be replaced if it ever fails. Other SIM controllers sense pavement temperature through the same device that detects precipitation. Always follow any specific installation instructions provided by the manufacturer for the pavement temperature sensor.

On/Off Control Options

SIM systems can be turned on and off by a simple manually operated switch. This is acceptable, provided someone pays attention to the weather and turns the system on and off as needed. If the attending person forgets to turn the system off, it could run indefinitely (or at least until running out of fuel). The cost of unnecessary operation can be high, especially on larger systems. This possibility is the principal argument against manual start/manual stop control systems.

A somewhat more sophisticated method of control uses manual start combined with automatic shutoff after a set time has elapsed. The operating time is selected by the person turning on the system based on the amount of snow and previous experience with the system. Overriding conditions such as very cold air temperatures or sustained air temperatures above freezing may also be used to turn off the system. The goal is to turn the system off as soon as melting is complete and the majority of the surface is dry. The latter is important to prevent the formation of dangerous "black ice."

SIM controllers are available that can automatically detect frozen precipitation on the slab surface and initiate melting operation. They also terminate heat input when melting is complete. Some can also be configured to idle the pavement at a specific temperature when desired.

Some SIM systems use a **snow/ice pavement sensor** that mounts into a sleeve that's cast into the pavement. The top of this sensor is flush with the top of the pavement. It detects when frozen precipitation is present and measures the pavement temperature. They must be used with specific controllers that can monitor the signals they produce. An example of a snow/ice pavement sensor is shown in Figure 14-62.

Another type of sensor used in SIM systems is called a **snow switch**. This device measures air temperature and detects the presence of precipitation. It does *not* measure pavement temperature. When the outdoor air temperature is below a preset value, typically about 38 °F, and precipitation is falling (as liquid or frozen), the device closes a set of electrical contacts to turn on the SIM system. Snow switches are mounted near, but not on the pavement being melted. They are often installed at the top of a vertical conduit attached to the building or embedded in the ground. They should be located where they will not be tampered with, damaged, or affected by drifting. An example of a snow switch is shown in Figure 14-63.

Both types of snow detectors have a small electrically heated cell at their top surface. The heat generated is just enough to melt any frozen precipitation that falls on a small area of the sensor. The presence of water on the sensor is detected by electrical conductivity.

SIM controllers that monitor pavement temperature tend to reduce fuel usage by allowing the system to maintain the pavement surface just warm enough for sufficient melting. Without pavement temperature sensing, some SIM systems may heat the pavement to a

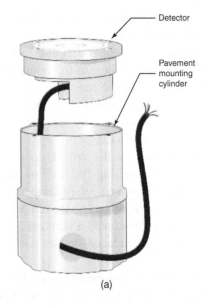

(a)

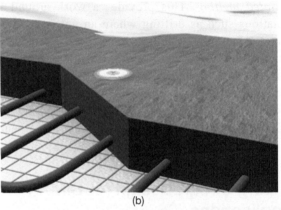

(b)

Figure 14-62 | (a) Snow/ice pavement sensor that detects precipitation and pavement temperature. (b) Detector surface is flush with finished pavement. *Courtesy tekmar Control Systems.*

Figure 14-63 | Snow switch sensor detects near freezing air temperature and the presence of precipitation. *Courtesy ETI, Inc.*

temperature higher than necessary. Although the rate of melting increases, so does unnecessary heat loss from the pavements, especially when the pavement is not covered by snow. Many SIM controllers can also "lock out" snow melting operation when outdoor temperatures are above or below preset values.

With all SIM systems, it is desirable to supply heat to the pavement for a minimum of five hours after frozen precipitation stops falling. Some of this time may be needed to melt any accumulation on the surface. The sustained heat input also helps dry the surface to prevent subsequent formation of ice.

Some SIM controllers also provide ΔT protection of melted slabs. When a melting cycle begins, the controller measures the supply and return temperature to the circuits. Heat input is regulated so the temperature drop of the circuits does not exceed a nominal 20 to 30 °F. This reduces thermally induced stresses within the cold slab.

It is the author's recommendation that all snow melting systems should have the option of manually initiated melting. This provides a work around for situations such as drifting where an automatic snow detector may not be calling for melting, even though snow is present on other portions of the pavement. A wind up timer switch that can be manually started to initiate melting, and yet turn off automatically is a good solution. It is also good to provide an indicator light, mounted in a noticeable location, that remains on while the SIM system is active.

Tubing Circuit Design for SIM Applications

Selecting the proper tube size, spacing, and flow rate is an important part of designing a SIM system. The information that follows gives suggested guidelines and analytical tools that can be used to evaluate the tradeoffs, and help optimize the system.

Flow Requirements

The flow rate required for a SIM circuit to deliver a given amount of heat to the pavement can be determined using Equation 14.14.

Equation 14.14:

$$f = \frac{q}{k(\Delta T)}$$

where,

f = required flow rate (gpm)
q = rate of heat output required (Btu/hr)
k = a constant based on the concentration of antifreeze used (see Figure 14-64).

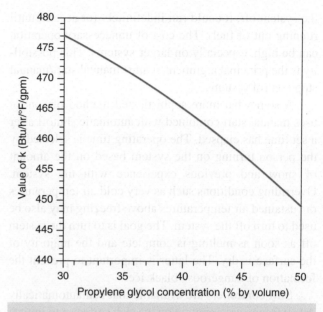

Figure 14-64 | Value of constant k (for use in Equation 14.14) for various concentrations of propylene glycol.

Example 14.8
Determine the circuit flow rate required to deliver 22,000 Btu/hr using a 40 percent solution of propylene glycol in a pavement circuit operating with a 20 °F temperature drop.

Solution:

The value of k for the 40 percent solution of propylene glycol is read from Figure 14-64 as 465. Substituting the values into Equation 14.14 yields:

$$f = \frac{22,000}{456 \times 20} = 2.36 \text{ gpm}$$

Discussion:

The rate of heat delivery required for snow melting is considerably higher than that required for a typical floor heating system. To deliver more heat without excessive temperature drop, the circuit flow rate must be increased. The use of glycol-based antifreeze solutions, which have lower specific heats than water, further increases the flow rate requirement.

To accommodate higher flow rates, most SIM systems require larger diameter tubing relative to the sizes used in radiant panels for space heating. Residential and small commercial SIM systems typically use 5/8 or 3/4-inch tubing. Larger commercial installations often use 1-inch tubing. In areas such as steps, it may not be possible to use 3/4-inch or larger tubing due to bending limitations. The common solution is to use multiple circuits of 1/2- or 5/8-inch tubing.

Figure 14-65 lists the *maximum* suggested circuit lengths for various sizes of PEX and PEX-AL-PEX tubing in SIM applications. The lengths assume that the system operates with a 50 percent propylene glycol solution, and that circuit head loss is equivalent to that of a 300-foot circuit of 1/2-inch PEX or PEX-AL-PEX operating with water. The latter is a common circuit size/length limitation used with radiant panel heating systems.

SIM Circuit Temperature Drop

Most SIM systems are designed for circuit temperature drops of 20 to 25 °F during steady state operating conditions. Temperature drops over 25 °F, when used in standard SIM systems, may result in uneven melting patterns on the pavement surfaces. The areas of the pavement above the warmest portion of the embedded circuits will melt and dry faster than those areas above the cooler portion of the circuits.

Circuit temperature drops over 25 °F can be used provided that flow through the pavement circuits is periodically reversed. This helps even out the overall energy delivery to all portions of the pavement.

There are several ways to achieve **periodic flow reversal**, including periodically cycling a four-way valve from one extreme to the other as shown in Figure 14-66.

Operating snow melting circuits with 25 to 30 °F temperature drops decreases flow requirements and reduces pumping power.

Estimating Heat Output

The output from a heated exterior pavement depends on several factors, many of which can change as the system operates. These include air temperature, wind speed, relative humidity, snow coverage, pavement drainage, tubing placement, average circuit temperature, circuit flow rate, soil temperature, and underside insulation. An exact engineering model that can account for all these factors would be very complex.

Figure 14-67 lists heat outputs and associated average circuit temperatures required to melt snow at a rate of 1/4-inch per hour. Values are given for "snow-free" surfaces characteristic of Class 2 and Class 3 systems, as well as snow-covered surfaces characteristic of Class 1 systems. *All data assumes tubing installed at 12-inch spacing in a concrete slab with approximately 2 inches of concrete coverage.*

The data in Figure 14-67 show that wind speed and air temperatures have significant effects on the heat delivery rates and circuit temperatures necessary to maintain snow-free surfaces. Much of the energy delivered to a melted but wet snow-free surface goes into evaporative and convective heat loss rather than just melting snow. Increasing wind speed and decreasing air temperature also increases the required heat delivery rates to snow-covered surfaces, but to a smaller degree because the snow layer largely blocks convective and evaporative heat loss.

It is the author's opinion that systems requiring fluid supply temperatures in excess of 150 °F are impractical in residential and light commercial systems. To reduce the circuit temperatures required, designers should use closer tube spacing.

Tube Spacing Considerations

Figure 14-68 can be used to estimate the change in heat output of slabs having tubing spacings other than 12 inches. Multiply the slab's estimated heat output using 12-inch spaced tubing by the factor found in Figure 14-64 to estimate the output for closer tube spacings.

Tubing size	1/2" tubing	5/8" tubing	3/4" tubing	1" tubing
Maximum circuit length	210'	310'	410'	530'

Figure 14-65 | Recommended maximum circuit lengths for use in SIM systems.

Example 14.9

A Class 1 SIM system using tubing spaced at 12 inches can deliver an output of 124 Btu/hr/ft² at a given circuit temperature. Estimate the output at the same average fluid temperature if the tubing is placed at 9-inch spacing.

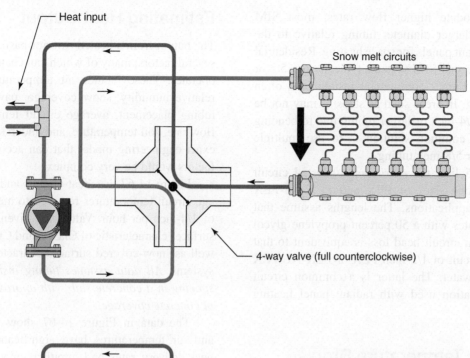

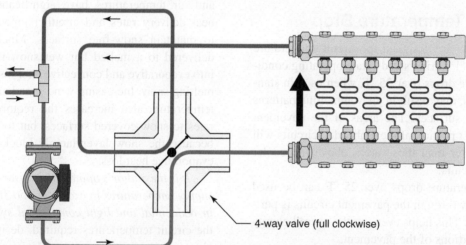

Figure 14-66 | Use of four-way motorized valve for periodic flow reversal through SIM circuits.

All data is for tubing spaced 12 inches apart in concrete slab. Also assumes 80% relative humidity		Air temperature = 0°F			Air temperature = 10°F			Air temperature = 20°F			Air temperature = 30°F		
		windspeed (mph)			windspeed (mph)			windspeed (mph)			windspeed (mph)		
		5	10	15	5	10	15	5	10	15	5	10	15
To maintain snow-free surface	Heat delivery rate (Btu/hr.sq.ft)	308	363	414	275	317	354	241	268	292	204	214	223
	Average circuit temperature (°F)	187	215	240	171	192	210	154	167	179	135	140	144
Surface covered w/ snow during melting	Heat delivery rate (Btu/hr.sq.ft)	208	208	208	202	202	202	195	195	195	189	189	189
	Average circuit temperature (°F)	137	137	137	134	134	134	131	131	131	127	127	127

Figure 14-67 | Heat delivery rates and average fluid temperatures for SIM circuits using 12-inch spacing. Data given at different air temperature and wind speeds for both snow-free and snow-covered melting surfaces. Data Source: *ASHRAE 1999 Fundamentals*.

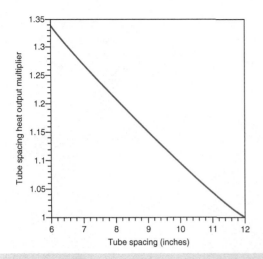

Figure 14-68 | SIM circuit heat output multipliers. Multiplier is for heat output at tube spacings less than 12 inches, based on the same average circuit temperature.

Solution:

The spacing correction factor from Figure 14-64 is 1.15. Thus, the estimated output is:

$$Q_{9'' \text{ spacing}} = (124 \text{ Btu/hr/ft}^2)1.15 = 143 \text{ Btu/hr/ft}^2$$

Discussion:

Closer tubing spacings increase the rate of heat delivery to the pavement for a given operating temperature. Viewed another way, *closer tube spacings allow a given rate of heat delivery to be achieved at lower average circuit temperature.* Reducing the average circuit temperature will improve the performance of condensing boilers and other low-temperature heat sources. It also decreases system heat loss. Closer tube spacing also increases the amount of tubing required, the amount of antifreeze needed, and the size of the expansion tank. In most systems, it also increases the number of manifold connections. The tradeoffs in cost versus performance should be evaluated.

The following tube spacings are suggested based on the class of the SIM system. *Tube spacings over 12 inches are never recommended for SIM systems because they can create uneven surface melting patterns.*

- Class 1 systems: 9 to 12 inches
- Class 2 systems: 6 to 9 inches
- Class 3 systems: 4 to 6 inches

14.7 Buffer Tanks

When low mass heat sources are combined with extensively zoned distribution systems, it is possible for the heat source to **short-cycle** when only one or two zones operate. *This happens because the rate of heat production is much greater than the rate of heat release at the heat emitters.* The low thermal mass heat source, in combination with the low water content of modern heat emitters and distribution systems, cannot absorb much of this excess heat without experiencing a rapid rise in temperature. This causes the heat source to reach its high-temperature limit very rapidly. In some systems, heat source on-cycles lasting less than one minute have been observed. When the heat source turns off, the low thermal mass components also cool rapidly as heat is released at the heat emitter(s). This leads to repeated short off-cycles.

Frequent on/off cycling can be tolerated by electric resistance heating elements, but it is very undesirable for components such as gas valves, oil burners, ignition systems, and the compressors used in heat pumps. *The life expectancy of such hardware will be significantly reduced if short-cycling is present.* Short-cycling also reduces combustion efficiency and raises the concentration of pollutants in the exhaust stream of combustion-type heat sources.

One way of achieving fewer but longer on/off cycles is by adding thermal mass to the distribution system using an insulated **buffer tank**. This thermal mass can absorb the excess heat produced when the heat source operates with minimal loading on the system.

Buffer Tank Piping

There are many possible ways to pipe a buffer tank. The "best" approach often depends on the heat source characteristics, and how the thermal mass of the buffer tank will be "engaged" with the remainder of the system. The following piping configurations will be discussed.

- 4-pipe buffer tank configuration
- 3-pipe buffer tank configuration
- 2-pipe buffer tank configuration

Figure 14-69 shows these configurations.

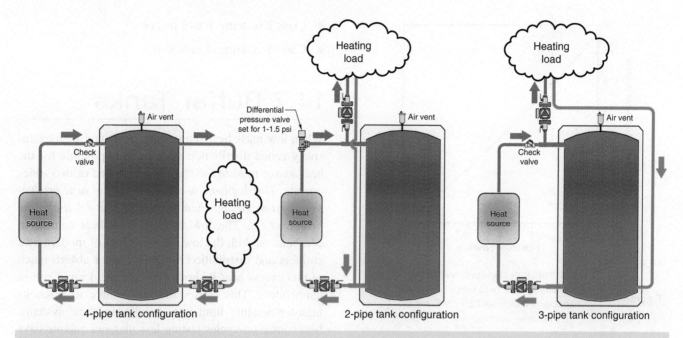

Figure 14-69 | Three piping configurations for buffer tanks.

The 4-pipe buffer tank configuration is by far the most common in North America. The heat source adds heat on one side, while the load removes heat from the other side. This piping configuration allows excellent hydraulic separation between the heat source circulator and the load circulator(s).

One characteristic of a 4-pipe configuration is that all heat from the heat source must pass through the tank on its way to the load. This isn't a problem when the buffer tank temperature is being maintained within a reasonable temperature range. However, this arrangement retards heat transfer from the heat source to the load if the tank is allowed to cool substantially.

A check valve should always be installed on the heat source side of the system to prevent reverse thermosiphoning from the heated tank back through the heat source circuit when it is off. If allowed to occur, reverse thermosiphoning can dissipate a substantial amount of heat from the tank over a period of several hours when the heat source is off.

The 2-pipe configuration places the load *between* the buffer tank and the heat source. This allows the possibility of passing heat directly from the heat source to the load when both are operating at the same time. This **direct-to-load heat transfer** is very desirable when expediting heat to the distribution system to recover a building from setback conditions.

The 2-pipe configuration also reduces the flow rate into and out of the buffer tank when the heat source and load are operating at the same time. The flow rate into and out of the buffer tank will always be the *difference* between the flow rate through the heat source and the flow rate to the load. An example of this is shown in Figure 14-70.

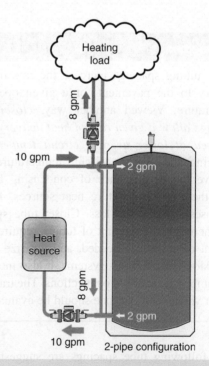

Figure 14-70 | With a 2-pipe buffer tank configuration, the flow rate into and out of the buffer tank is the difference between the heat source flow rate and the load flow rate. In this case, 10 gpm − 8 gpm = 2 gpm.

Assuming that the flow rate through the heat source is 10 gpm, and the piping to the load is operating at 8 gpm, the flow rate into and out of the tank will be 10 − 8 = 2 gpm.

Reducing the flow rate into and out of the buffer tank helps preserve beneficial temperature stratification within the tank. A well-stratified tank maintains the hottest water at the top of the tank, and the coolest water at the bottom. Although the thermodynamic justification for why this is better is beyond the scope of this text, suffice it to say that the "usefulness" of thermal energy stored in a highly stratified thermal storage tank is greater than that of the same tank, containing the same total heat, but with the water fully mixed to the same average temperature as that in the stratified tank.

One limitation of the 2-pipe configuration is that a differential pressure valve that creates a forward opening resistance of 1 to 1.5 psi, or a motorized ball valve needs to be installed in the heat source piping to prevent flow that's returning from the load from passing through the heat source when it's off. It's also necessary to keep the tees that connect to the supply and return piping to the load as close as possible to the tank to provide good hydraulic separation between the heat source circulator and the circulator delivering flow from the tank to the distribution system.

Two-pipe buffers should only be used when the heat source is turned on and off based on buffer tank temperature. The reasoning is as follows: If the heat source flow rate and load flow rate are about the same, there will be very little flow through the buffer tank. This could lead to the heat source shutting off based on satisfying the space heating thermostat, without adding or removing much heat to or from the tank. In this scenario, the tank is not "engaged" in the energy flow processes. However, when the heat source is controlled directly from tank temperature, it will continue to run even after the space heating thermostat is satisfied, storing heat that's immediately ready to go to the next zone requesting it. The controller that turns the heat source on and off based on temperature could be a setpoint controller, or an outdoor reset controller, both of which were discussed in Chapter 9.

A 3-pipe buffer tank borrows details from both the "classic" 4-pipe configuration, and the 2-pipe configuration. It provides the direct-to-load heat transfer ability of the 2-pipe configuration, while also forcing return flow through the lower portion of the tank, and thus ensuring that the tank's thermal mass is engaged. This approach is well-suited for use with hydronic heat pumps (either geothermal water-to-water, or air-to-water). It eliminates the need for the differential pressure valve required by the 2-pipe configuration, but still needs a check valve to prevent reverse thermosiphoning.

Designers should not create systems that engage buffer tanks for loads that are "self-buffering." Examples include indirect water heaters or large heated floor slabs that are controlled as a single zone. When paired with appropriate controls, both of these loads have ample thermal mass to accept heat directly from the heat source, without causing short-cycling. Placing a buffer tank between such loads and the heat source adds standby heat loss to the system. It might also delay transfer of energy from the heat source to the load because of the need to warm the buffer tank. It also adds unnecessary installation cost.

Figure 14-71 shows a system where a small buffer tank is used for the space heating system, which has several independently controlled panel radiators. The system also has an indirect water heater controlled as a priority load. Both tanks are piped separately. The small buffer tank can provide some heat to the distribution system when the system operates in priority domestic water heating mode.

In some applications, a buffer tank can also provide hydraulic separation between multiple circulators. Recognizing this possibility allows the system to be constructed without the need for a hydraulic separator or closely spaced tees to achieve equivalent hydraulic separation.

The three buffer tank piping configurations shown in Figure 14-69 can all provide hydraulic separation between the heat source circulator and load circulator. However, *in the case of the 2-pipe tank configuration, it is essential to keep the tees supplying the load as close to the tank as possible.* The header between these tees and the tank connections should be as short as possible and generously sized. *In a 3-pipe tank configuration, the tee joining the heat source supply and load circuit supply should also be kept as close to the tank as possible.*

Figure 14-72 shows an example where the buffer tank, piped in a 3-pipe configuration, provides thermal mass for a highly zoned distribution system, as well as hydraulic separation between the heat pump circulator and the variable-speed distribution circulator.

Another system concept uses a reverse indirect tank as both a buffer tank and source of domestic hot water. Figure 14-73 shows an example.

This system has a split system air-to-water heat pump as its heat source. The heat pump is turned on and off based on the temperature of a sensor in the buffer tank. This on/off logic could be based on temperature setpoint with associated differential, or it could be based on outdoor reset control. Heat pump operation is completely independent of the status of the zone valves.

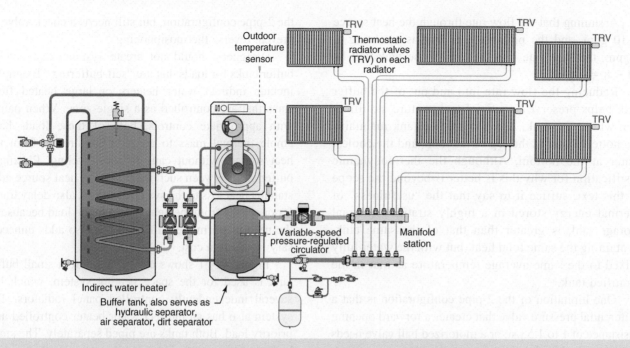

Figure 14-71 | Small tank under boiler serves as buffer tank, hydraulic separator, and air/dirt separator.

The water in the buffer tank is always maintained as some elevated temperature. The buffer tank is a "tank-in-tank" reverse indirect tank. It consists of an outer 119-gallon shell. A 40-gallon stainless steel tank is suspended in the upper portion of this shell, and completely surrounded by "system water" in the shell. Domestic cold water passes into the inner tank whenever domestic hot water is drawn from a fixture. The temperature to which the domestic water is heated within the inner tank depends on how fast it is drawn from the fixture, as well as the temperature of the system water in the shell. Under most conditions, the initial hot water drawn from the inner tank will be very close to if not equal to the temperature of the system water in the upper portion of the tank. This configuration is well suited for situations where frequent but relatively short duration draws of hot water are anticipated.

The heated domestic water leaving the inner tank passes into a thermostatically controller electric tankless water heater, which can boost it to the desired delivery temperature if necessary. The hot water leaving the tankless heater passes through an ASSE 1017 anti-scald mixing valve to ensure that the hot water delivered to the plumbing fixtures doesn't exceed 120 °F.

The system water in the shell of the tank provides the buffering mass for the zoned space heating distribution system.

This approach provides two major functions that often require separate tanks in other systems, one for domestic water heating and the other for buffering space heating loads. This reduces cost, space requirements in the mechanical room, and standby heat loss.

Sizing a Buffer Tank

The required volume of a buffer tank depends on the rates of heat input and release, as well as the allowed temperature rise of the tank from when the heat source is turned on, to when it is turned off. The greater the tank's volume, and the wider the operating temperature differential, the longer the heat source cycles will be.

Equation 14.16 can be used to calculate the volume necessary given a specified minimum heat source on-time, tank operating differential, and rate of heat transfer:

Equation 14.16:

$$V = \frac{t(q_{hs} - q_{load})}{500(\Delta T)}$$

where,

V = required volume of the buffer tank (gallons)

t = desired duration of the heat source's "on cycle" (minutes)

q_{hs} = heat output rate of the heat source (Btu/hr.)

q_{load} = rate of heat extraction from the tank (can be zero) (Btu/hr.)

ΔT = temperature rise of the tank from when the heat source is turned on to when it is turned off (°F)

14.7 Buffer Tanks

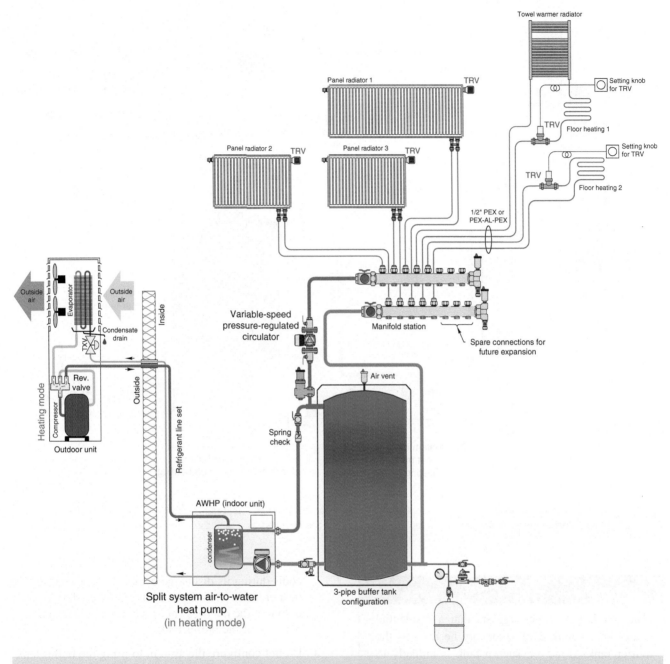

Figure 14-72 | A buffer tank piped in a 3-pipe configuration is supplied by a split system air-to-water heat pump, and serves multiple panel radiators configured into a homerun distribution system.

Example 14.10

A boiler with a heat output rating of 75,000 Btu/hr is expected to have a minimum on-cycle of 10 minutes when it operates simultaneously with the smallest load in the system—a 2,000 Btu/hr towel warmer radiator. The burner turns on when the buffer tank temperature drops to 120 °F, and off when the tank reaches 160 °F. What is the smallest buffer tank required for this situation buffer tank volume to accomplish this?

Solution:

Substituting the numbers into Equation 14.7 yields:

$$V = \frac{t(q_{hs} - q_{load})}{500(\Delta T)} = \frac{10(75,000 - 2,000)}{500(160 - 120)} = 36.5 \text{ gallons}$$

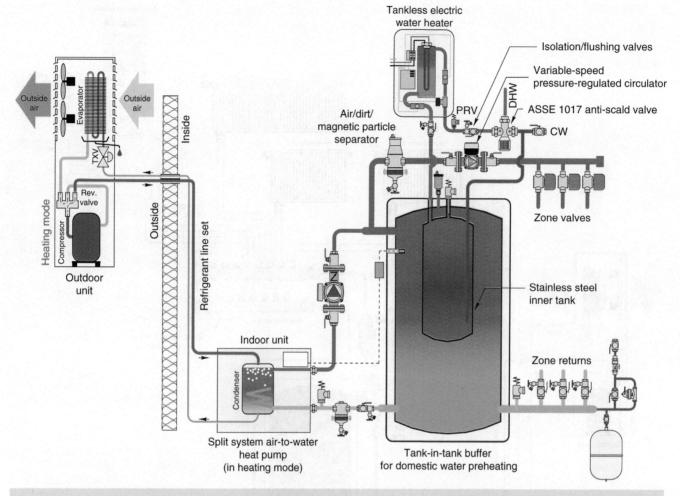

Figure 14-73 | A "tank-in-tank" buffer tank, supplied by a split-system air-to-water heat pump. Domestic water is preheated inside the inner stainless steel tank. Any final heating is supplied by a tankless electric water heater.

Discussion:

The required volume suggests that a residential water heater tank could probably be used as the buffer tank. However, such a tank is unlikely to have the necessary piping connections. Several North American suppliers now offer buffer tanks with volumes ranging from 25 to 119 gallons. These tanks are usually supplied with piping connection sizes and locations that allow them to be piped in 2-pipe, 3-pipe, or 4-pipe configurations.

Design Summary for Buffer Tanks

The following concepts should be used by designers when considering use, size, and placement of buffer tanks in hydronic systems:

- Heat emitters supplied from a buffer tank will release heat at progressively slower rates as the tank cools during its discharge cycle. To compensate, heat emitters should be sized to provide design load output at the *average* (not maximum) water temperature of the buffer tank.

- Do not configure the system to bring the buffer tank online during a call for domestic water heating. The thermal mass of the DHW tank itself will buffer the boiler.

- Do not configure the system to bring the buffer tank online when other high thermal mass loads such as pool/spa heating or snow melting are active. These loads are self-buffering.

- The wider the operating temperature differential of the buffer tank, the fewer (and longer) the heat source cycles will be. However, wide temperature differentials also create wider variations in heat emitter output. These will "average out" over time, but they will be more noticeable to occupants.

- Always install a spring check valve in the circuit between the buffer tank and heat source to prevent reverse thermosiphoning during off cycles. A circulator with an integral check valve can also be used for this purpose.

- Buffer tanks should be well insulated to minimize standby heat loss. A *minimum* jacket insulation of R-10 °F·h·ft²/Btu is suggested, with R-24 °F·h·ft²/Btu preferred. Buffer tanks should always be installed within a heated space.

- Keep in mind that buffer tanks, with appropriate piping details, can provide "double duty" as hydraulic separators between the heat source circulator(s) and the distribution circulator(s).

- Keep in mind that reverse indirect water heaters can often be used as buffer tanks for space heating loads, as well as for heating domestic water.

- Buffer tanks should have connections for an air vent at top and a provision for draining water near the bottom. The drain valve could connect directly to the tank or to piping connected to the lower portion of the tank.

- The temperature maintained in a buffer tank, as well as the temperature differential between "heat source on" and "heat source off" status can be based on either setpoint control or outdoor reset control.

14.8 Minitube Distribution Systems

Variable-speed injection mixing was extensively discussed in Chapter 9, Control Strategies, Components, and Systems. The reader will recall that a relatively low flow rate of hot injection water can establish a high rate of heat transfer into a low-temperature distribution system. Example 14.11 demonstrates this relationship.

Example 14.11

What injection flow rate of 175 °F water is required to transfer heat at 250,000 Btu/hr into a large radiant floor system with a return water temperature of 85 °F?

Solution:

Applying the sensible heat rate equation from Chapter 4, Properties of Water, yields:

$$f = \frac{q}{500(\Delta T)} = \frac{250,000}{500(175 - 85)} = 5.6 \text{ gpm}$$

Discussion:

This flow rate could be handled by a 3/4-inch tube. The large temperature drop between the supply and return injection risers is what enables this relatively low flow rate to deliver such a high rate of heat transfer.

Figure 14-74 shows a piping schematic for a **minitube distribution system** designed to take advantage of the large temperature difference available between a boiler and the relatively cool water returning from a heated slab-on-grade floor. Note the relatively small diameter tubing between the boiler loop and the manifold station. One of these "**minitubes**" carries hot water from the boiler loop to the manifold station, the other carries cool water back to the boiler loop.

Flow through the minitubes is controlled by a variable-speed injection pump. The faster the injection pump runs, the greater the proportion of hot water injected at the mix point (A) in the manifold station piping, and the warmer the supply water temperature to the distribution circuits becomes. Heat input to the manifold station can be regulated from zero (when the injection pump is off) to full design load (when the injection pump is running at full speed).

The supply and return minitubes are coupled to the boiler loop as well as the manifold station piping using pairs of closely spaced tees. These tees provide hydraulic separation between the variable-speed injection pump and the other two circulators in the system.

Multizone Minitube Systems

When planning a floor heating system for a large building, it is often necessary to locate manifold stations far from the mechanical room. These situations are well suited to multizone minitube distribution systems. Using this approach, each manifold station has its own variable-speed injection mixing system and minitube piping. Each mixing system draws hot water from a common header.

Figure 14-75 shows piping for a three-zone system sourced from a multiple boiler array. The hydraulic separator isolates the pressure dynamics of the boiler circulators from the three variable-speed injection circulators. In combination with the generously sized headers, it also isolates the pressure dynamics of each variable-speed circulator from the others.

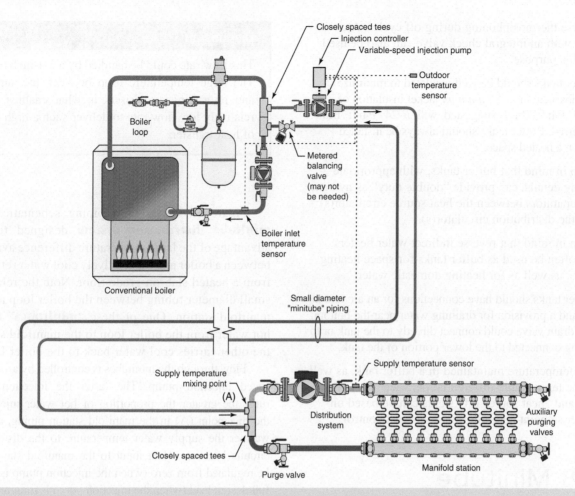

Figure 14-74 | Single zone minitube distribution system for radiant floor heating.

Benefits of Minitube Distribution Systems

Minitube distribution systems offer several benefits relative to other methods of supplying remotely located manifold stations:

- They use much smaller piping between the mechanical room and manifold station. Besides saving on installation cost, small tubing is also easier to route and support. It reduces system volume, which in turn may reduce expansion tank size, and the required volume of antifreeze.

- If desired, each manifold station in a minitube system can operate at a different supply water temperature or outdoor reset schedule. This flexibility is very useful if different areas of the building have different floor coverings or widely varying load characteristics.

- If desired, the distribution circulator at each manifold station can operate continuously during the heating season. This is of particular benefit in larger garage-type buildings. It transports stored heat from interior areas of the slab to localized "heat sinks" such as adjacent to large overhead doors, or the wet floor under a vehicle shedding snow. Should the boiler be inoperable, continuous circulation also helps prevent freezing in these high heat loss areas.

- Minitube systems improve the **rangeability** of variable-speed injection mixing systems. When variable-speed injection mixing is done using short injection risers, a throttling valve must be installed in the return riser to remove a significant portion of the pump head. This detail, which was discussed in Chapter 9, is necessary to force the circulator to operate at full speed under design load conditions, and thus utilize the full range of speed control. However, in a minitube system, the vast majority of the head loss occurs in the minitube piping. *If the head loss of the minitube circuit at design injection flow rate is approximately equal to the head added by the injection pump at full speed, it is not necessary to install a balancing valve in the minitube circuit.*

- Since the manifold station circulator is not responsible for flow between the mechanical room and

14.8 Minitube Distribution Systems 751

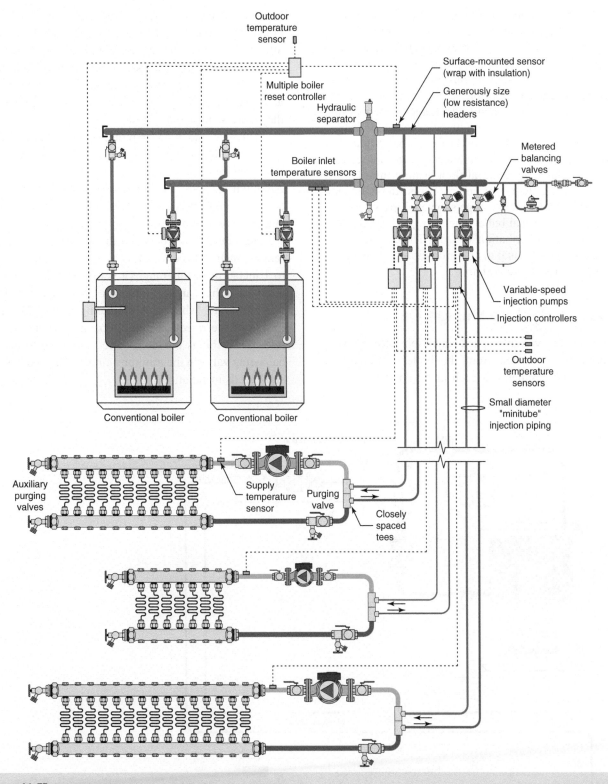

Figure 14-75 | Multizone minitube distribution system.

manifold station, it may be possible to use a smaller circulator. This reduces both initial cost and life cycle operating cost.

- The small minitube piping loses less heat to surrounding air than would larger diameter distribution piping of equivalent heat transport capacity. Still, prudence dictates that the supply minitube be insulated to minimize heat loss. The return minitube can also be insulated, although energy savings will be minimal given the small surface area and low operating temperature. Any minitubes passing through unheated space should be insulated.

- The minitube concept is highly scalable. A small system using 1/2-inch or even 3/8-inch minitube piping may be ideal for larger residential space heating and/or snow melting applications. It is also well suited for retrofit applications given the ease of routing small diameter flexible tubing from the mechanical room to distant manifolds. Larger systems using 1.5- to 2-inch minitube piping are capable of transporting several million Btu/hr from a boiler plant to a low-temperature load. Such systems are well suited to large agricultural floor heating installations such as multiple hog or poultry barns supplied from a central boiler plant. Large snow melting systems using conventional boilers are also good candidates for minitube distribution systems. The cost-saving relative to traditional distribution piping increases with the size of the system.

Minitube System Design Details

Because each variable-speed injection circulator is hydraulically separated from the boiler circulator(s), as well as the circulator at each manifold station, each minitube circuit can be designed as if it is a "stand-alone" hydronic circuit as shown in Figure 14-76. This allows each injection circulator and the piping used for each minitube circuit to be selected independent of the others. From the standpoint of hydraulics, the boiler loop and manifold station circuit can also be designed as if they were stand-alone entities.

The size and type of piping used for each pair of minitubes should be selected based on the injection flow requirements associated with its manifold station. Although any type of tubing with suitable temperature and pressure ratings could be used for the minitubes, the author recommends PEX-AL-PEX tubing. It allows fast, seamless installation between the mechanical room and remote manifold stations. It is available in sizes from 3/8 to 1 inch. Given this range, it is often possible to select a tube size that can accommodate the design load injection flow rate, while also producing a head loss close to that of the injection pump, operating at design flow rate. When this is the case, there is no need to install a balancing valve in the minitube circuit. This reduces cost and speeds installation.

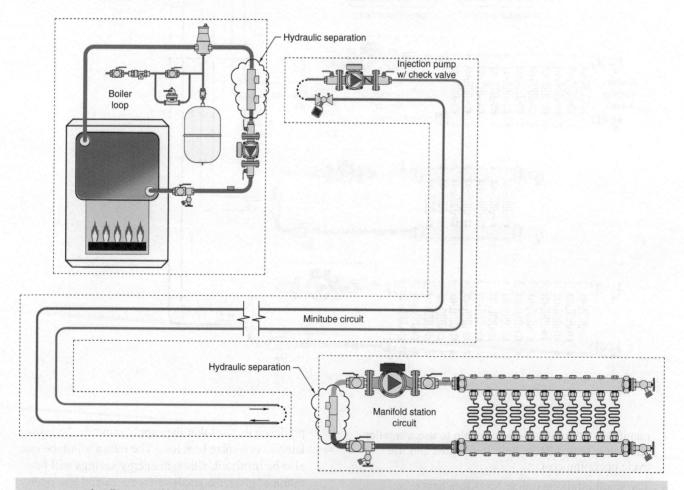

Figure 14-76 Because of hydraulic separation, each minitube circuit can be designed as if it is a stand-alone entity. Boiler loop and manifold station circuits can also be designed as stand-alone entities.

The circulators at each manifold station in a minitube system should be operated continuously during the heating season. The constant flow past each supply sensor helps stabilize temperature control. With the injection mixing control(s) properly configured for outdoor reset, the water temperature supplied to the distribution circuits is just warm enough to meet the prevailing heating load. Constant flow through the distribution circuit has other benefits as discussed in Chapters 9 and 10.

If internal heat gains are present in a given zone, and if the injection controller has indoor temperature sensing, it may totally stop the injection pump for a time. During this time, it is important to prevent hot water from slowly migrating through the supply minitube and potentially entering the mixing point in the continuous flow manifold system. The preferred solution, when using the piping shown in Figure 14-75, is to use an injection pump with an integral check valve.

When a conventional boiler is used in a minitube system, it is important to ensure that it operates at a temperature high enough to prevent sustained flue gas condensation. This is especially true when the load is a large slab with very high thermal mass. Most currently available variable-speed injection controllers have the logic necessary to monitor the boiler inlet temperature and slow the injection pump when necessary to prevent the load from extracting heat faster than the boiler is generating it. Be sure to install the boiler temperature sensor as specified by the controller manufacturer to provide this protection.

The boiler or multiple boiler system used to supply a minitube system can operate with outdoor reset control. This is in addition to use of mixing reset control logic for regulating the speed of each injection pump. Boiler reset will reduce heat loss from the boiler under partial load conditions and thus increase seasonal boiler efficiency. It will also reduce heat loss from the minitubes under partial load conditions.

The method used to fill and purge a manifold station that is part of a minitube distribution system depends on the length of the minitube circuit. In cases where the minitube circuit is relatively short, it is likely that purging flow from the mechanical room portion of the system can be sent *simultaneously* through *both* minitubes, and used to purge the distribution circuits in the same manner as described in Chapter 13.

If the minitube circuit is long, its head loss may not allow for sufficient purging flow from the mechanical room. In such cases, inlet and outlet purging valves should be installed at the manifold station as seen in Figures 11-74 through 11-76. Water can be added directly to the manifold station through a high-capacity hose or by using a purging cart.

Figure 14-77 | Example of large (500,000 Btu/hr) manifold station served by 1-inch minitube supply and return piping. *Courtesy of John Siegenthaler.*

The author and others have successfully deployed minitube systems in many buildings including municipal highway garages, churches, and large industrial facilities. The largest known installation to date is a 170,000-ft^2 manufacturing facility having 13 independently controlled minitube-supplied manifold stations. An example of one of the large manifold stations used in this facility is shown, under construction, in Figure 14-77. Notice the 1-inch copper supply and return minitubes at the top of the manifold, which at design load delivers 500,000 Btu/hr to the floor circuits.

14.9 Heat Metering

There are many buildings, including apartments, condominiums, or leased retail or office spaces, where owners or tenants pay directly to utilities for the electrical energy and natural gas they use. For electrical energy, each unit within the building is equipped with its own meter and is billed by the utility for the kilowatt·hours used. For thermal energy, the *customary* practice in North America has been to install a separate gas meter for each unit. The gas supplies a spacing heating appliance (e.g., furnace or boiler) within each unit. It may also supply a domestic water heating appliance in some or all of the units. The local utility issues another invoice for the gas used within each unit. Figure 14-78 shows a medical building where separate electrical and gas meters are used for each independently operated unit within the building.

Figure 14-78 Separate electric and natural gas meters for each office space within a multiunit building. *Courtesy of John Siegenthaler.*

The use of separate electric and gas meters for each unit in a building is a viable approach, but it also has several disadvantages:

- The design heating load of apartments, condominiums, or small offices is often significantly lower than the rated heat output of the smallest available boiler or furnace. This leads to oversized equipment, short-cycling, more frequent service needs, and decreased seasonal efficiency.
- Each owner or tenant must pay a minimum monthly service charge for a separate gas meter.
- Fuel supply piping and venting provisions are required for each unit in the building.
- Floor space is required within each unit to accommodate equipment and provide access for maintenance. This reduces otherwise usable space.
- Any noises generated by heating appliances is present within the unit.
- Each heat source must be individually maintained. This requires access to each unit, as well as the potential for odors, spills, etc., during service work.
- If there is a combustion heat source within each unit, the potential for carbon monoxide leakage increases compared to a scenario where combustion is isolated to one mechanical room separated from the occupied units.
- The seasonal energy use of several individual combustion heat sources, especially if they are oversized, is very likely to be higher than that of a central boiler system with a high turndown ratio.

A better approach is to centralize heat production and distribute the heat throughout the building using a hydronic system. Cost allocation is handled by measuring the total thermal energy transferred to each unit, and invoicing for that energy on a periodic basis. This approach is called **heat metering**. In North America, it's also referred to as **Btu metering**.

Heat metering eliminates many of the disadvantages just described. Furthermore, studies have shown seasonal energy use reductions of 20 percent or more in multi-occupant buildings using heat metering compared to situations where the cost of energy for space heating and domestic hot water is not individually allocated.

Heat Metering Theory

The total heat transferred to a load depends on the flow rate of the fluid conveying the heat, as well as the density and specific heat of the fluid, and the temperature drop the fluid undergoes as it flows through the load.

If the flow rate, inlet temperature, and exit temperature of the fluid stream were all *constant*, the *rate* of heat transfer could be calculated using the sensible heat rate equation.

Equation 4.8 (repeated from Chapter 4)

$$q = (8.01 Dc) f (\Delta T)$$

where,

q = rate of heat transfer (Btu/hr)
D = density of the fluid (lb/ft^3)
c = Specific heat of the fluid (Btu/lb/°F)
f = flow rate (gpm)
ΔT = temperature changer of the fluid (°F)
8.01 = constant to balance the units on each side of the equation (e.g., Equation 4.4)

The total *quantity* of heat transferred would be this rate multiplied by the time over which that rate exists, as given by Equation 14.17.

Equation 14.17:

$$H = q(t)$$

where,

H = total amount of heat transferred to space (Btu)
q = steady rate of heat transfer (Btu/hr)
t = time over which steady rate of heat transfer persists (h).

Unfortunately, steady flow rates and fixed temperature drops rarely occur in hydronic systems. These quantities typically vary with time as heat sources warm up and cool down, zone valves open and close, etc.

In such cases, the theoretical method of calculating the total heat delivered to a load turns to a calculus technique known as integration. If both the flow rate of the fluid (f), and the temperature drop it undergoes (ΔT), are known as functions of time (t), the total heat transferred can be calculated using Equation 14.18.

Equation 14.18:

$$H_{total} = \int_{t=0}^{t=now} [(8.01Dc)f\Delta T]dt$$

where,
H_{total} = total amount of heat transferred to load (Btu)
f = fluid flow rate as a function of time (gpm)
ΔT = temperature change of fluid as a function of time (°F)
D = density of fluid, at its average temperature in the process (lb/ft^3)
c = specific heat of the fluid at its average temperature in the process (Btu/lb/°F)
8.01 = a constant required by the units used for the other factors
t = time (h)

Figure 14-79a shows a graph that plots the rate of heat delivery as a function of time. Figure 14-79b shows

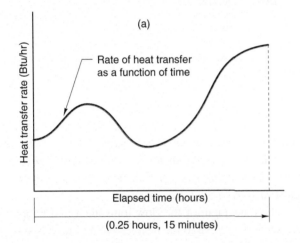

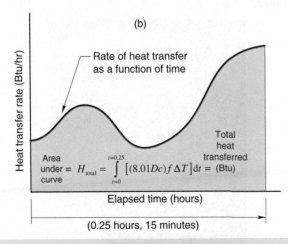

Figure 14-79
(a) Rate of heat transfer as a function of time. (b) Blue shaded area under the curve represents *total amount of heat* transferred; in this case, over a 15-minute period.

the area under this curve. That area is what equation 14.18 calculates, and it represents the *exact* total heat transferred to the load over the elapsed time shown on the graphs (e.g., 0.25 hours or 15 minutes).

Situations in which the rate of heat transfer can be precisely described by a mathematical function are very rare. Thus, the use of Equation 14.18 is not practical in most applications. However, the concept that the area under the curve represents the total amount of heat transferred is very useful. What's needed is a practical way to *approximate* this area. This can be done by dividing the area under the curve into several small rectangular areas as shown in Figure 14-80.

Each rectangle in Figure 14-80 is 1 minute "wide" along the horizontal time axis. The height of each rectangle reaches the black curve at the mid-width of the rectangle. This implies that the height of each rectangle is the *average* rate of heat transfer over that one-minute period (in Btu/hr). Multiplying the height of the rectangle by its width gives the area of the rectangle, which is an *approximation* of the heat transferred to the load over that minute. The total area of all 15 rectangles in Figure 14-80 is an approximation of the blue shaded area in Figure 14-79b. This allows Equation 14.18 to be replaced by the approximation of Equation 14.19.

Equation 14.19:

$$H_{total} = \sum_{time=0}^{time=now} (8.01Dc)f(T_{source} - T_{storage})\Delta t$$

where,
H_{total} = total heat transferred from source to storage during a specific measurement period (Btu)
8.01 = a constant based on the units used
D = density of the fluid transferring heat (at the average temperature of the process) (lb/ft^3)
c = specific heat of the fluid transferring heat (at the average temperature of the process) (Btu/lb/°F)
f = flow rate of the fluid transferring heat (gpm)
T_{supply} = higher temperature of the fluid stream (supplied from the heat source) (°F)
T_{return} = lower temperature of the fluid stream (returning to the heat source) (°F)
Δt = short time increment (hours)

As the time increment (Δt) gets smaller, and thus the "width" of each rectangular area seen in Figure 14.80 decreases, the approximation of the area under the curve

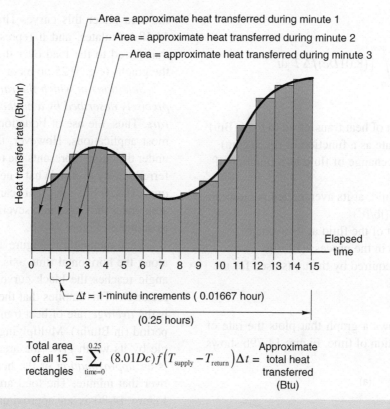

Figure 14-80 | Area under the curve is approximated by a collection of rectangular areas, each one minute "wide," and having a height approximately equal to the rate of heat transfer during that each one minute period.

gets closer and closer to the exact area represented by Equation 14.18. Modern heat meters take several "snapshots" of flow rate and ΔT every minute to comply with an accuracy standard that will be discussed shortly. This technique of approximating the heat transferred over many short time increments, and then totaling them, is the mathematical basis for how modern heat meters operate.

Heat Metering Hardware

The knowledge and hardware for measuring the quantity of heat transferred across a given boundary in a mechanical system has existed for several decades. Figure 14-81 shows a mechanical **heat meter** that has been operating in a German hotel since the 1960s. Note the units of MWh (megawatt-hours) on the face of the meter. This is a unit of heat: 1 MWh = 3,413,000 Btu.

Electronically based heat meters have been available worldwide since the 1970s. The first generation of analog electronic meters has since given way to self-contained digital electronic meters, an example of which is shown in Figure 14-82.

*All electronic heat metering systems use a **flow transducer** and two temperature sensors to measure fluid flow rate and temperature change. That data is fed to a **heat calculator unit** that mathematically processes it into the instantaneous rate of heat flow, and the total heat transferred over a given time.*

The placement of the flow transducer and temperature sensors relative to the heat source and load is shown in Figure 14-83.

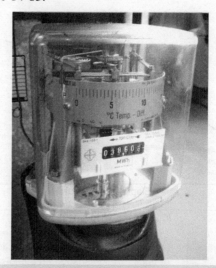

Figure 14-81 | A mechanical heat meter that has operated since the 1960s. *Courtesy of John Siegenthaler.*

14.9 Heat Metering

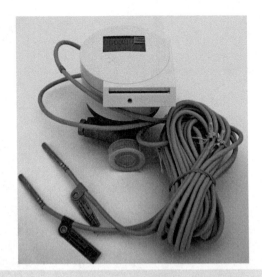

Figure 14-82 | Example of a modern, self-contained, digital electronic heat meter. *Courtesy of ISTEC Corporation.*

Modern heat meters typically use **resistance temperature detectors (RTDs)** for both temperature measurements. The electrical resistance of RTDs change with changes in temperature in a very precise and repeatable manner. The two RTDs are attached to cables of a specified length, and those cables are *factory attached* to the heat calculator unit of the heat metering system.

The heat calculator unit "feels" the combined resistance of each RTD, its cable, and any electrical resistance associated with the connection between the cable and the terminal of the heat calculator unit. Although the resistance of the cable and connection terminal are very small in comparison to that of the RTD, they are not zero. Some heat metering standards require each heat calculator unit to be calibrated to both RTD circuits to account for the combined resistance of the RTD, its cable, and the connection terminal. *For this reason, it is very important not to alter the length of either temperature sensor cable.* Doing so would void the calibration and introduce inaccuracies. If the full length of the cable(s) is not needed, neatly coil the extra length and secure the coil using zip ties, as seen in Figure 14-84.

Care should also be taken not to damage the RTDs or their cables. Because the RTDs, their cables, and the heat calculator unit are all calibrated as a system, a damaged RTD or cable could require both RTDs, their cables, and the heat calculator unit to be replaced.

Because some heat meters can be used as the basis of billing customers for thermal energy use, they must be protected against tampering. Examples of tampering would be disconnecting a sensor circuit, or temporarily removing a sensor from its mounting fixture, with the intent of reducing the amount of heat the meter records.

When the sensor cables are factory-attached to the heat calculator unit, the terminals where the RTDs cables connect are "sealed" as seen in Figure 14-85.

The sealant prevents oxidation of the terminals, which could alter the accuracy of the temperature sensing circuit. It would also provide visual evidence of any attempt to disconnect the temperature sensing cables.

The RTDs should be mounted into fittings that place their tip directly in the flow stream. This eliminates temperature differences that could otherwise occur if the RTD were loosely fit into a standard sensor well. Figure 14-86 shows an example of an RTD being inserted into this type of fitting. The small O-ring on the sensor body provides a pressure-tight seal.

Some of the RTDs used in heat metering systems have a small hole drilled in the upper (exposed) portion of their body. This is another anti-tampering provision. After the sensor is screwed in place, a fine stainless steel wire is passed through the hole and wrapped around the fitting. A lead seal is then fixed to the two ends of this wire as shown in Figure 14-87. Any attempt to unscrew the sensor would require the lead seal to be broken or the wire to be severed, and thus indicate potential tampering.

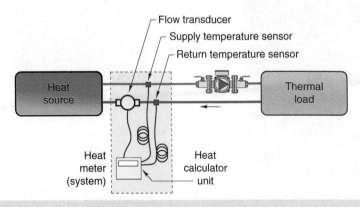

Figure 14-83 | Placement of flow transducer and temperature sensors needed for heat metering between a heat source and thermal load.

Figure 14-84 | Any extra cable leading to RTDs in a heat metering system should be neatly coiled and secured with zip ties. Never cut or otherwise modify either cable. *Courtesy of John Siegenthaler.*

Figure 14-86 | An RTD temperature sensor being inserted into a fitting that places the tip of the sensor in the center of the flow stream. *Courtesy of Caleffi North America.*

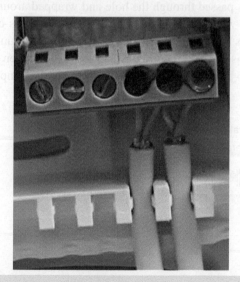

Figure 14-85 | The leads from RTDs in a heat metering system are sealed to terminal in the heat calculator unit to prevent tampering. *Courtesy of John Siegenthaler.*

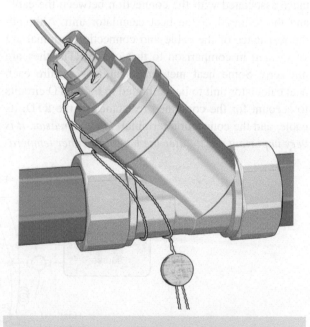

Figure 14-87 | Fine stainless steel wire and lead seal are used to prevent tampering with the RTD temperature sensors after they have been installed as part of a heat metering system. *Courtesy of Caleffi North America.*

Some heat calculator units record each time their access panel is opened, which is another indicator of potential tampering. Many also provide a means of locking the access panel, or requiring a special tool to open it.

Most of the heat calculator units used in modern heat metering systems can also be configured for the specific fluid used in the system. Most default to water as the assumed fluid, but can be reprogrammed for various concentrations of propylene glycol and ethylene glycol antifreeze. Because there are differences in the density and specific heat of these fluids, and those properties are used in calculating the rate of heat transfer, installers should always verify that the heat calculator unit is properly configured for the fluid used in the system.

ASTM E3137/E3137M-17 standard:

Since heat meters are often used to bill customers for energy use, it's essential for them to meet an accepted standard of accuracy. Without such compliance, designers, as well as those billed for energy use, would likely question or dispute the accuracy of the recorded energy use. This would certainly limit acceptance of heat metering and thus limit its potential market growth.

Efforts to develop a widely accepted American standard for heat metering were undertaken several years ago by ASTM committee E44.25. These efforts produced a detailed document entitled *Standard Specification for Heat Meter Instrumentation* (**ASTM E3137/E3137M-17**). The latest version of this standard was issued in February 2018. It covers several categories of heat metering equipment and the accuracy of that equipment.

This Standard makes references to several previous heat metering standards such as EN1434, which is extensively recognized in Europe, and OIML (*Organisation Internationale De Metrologie Legale*) R75-1 Edition 2002, which is an International Recommendation that establishes the metrological characteristic required of certain measuring instruments, as well as methods for testing their conformity.

ASTM E3137/E3137M-17 establishes maximum permissible measurement variations for the flow transducer, temperature sensors, and heat calculator unit. It classifies these elements of a heat metering system based on their accuracy at maximum and minimum flow rates and temperature differentials.

The details of this standard are too extensive to cover in this text. Interested readers can obtain the standard at: https://www.astm.org/Standards/E3137.htm

The ASTM E3137/E3137M-17 standard will likely be referenced by the majority of North American engineering firms that include heat metering specifications for the systems they design.

Applying Heat Meters

Heat meters can be deployed in several locations in many types of hydronic heating and cooling systems. They are often installed to record the heat output of renewable energy heat sources such as solar thermal collectors or hydronic heat pumps. Some heat meters can automatically record heat supplied for spacing heating, as well as the heat absorbed by a chilled water cooling system.

Figure 14-88 shows a possible application where one heat meter records the heating energy and chilled water cooling effect (e.g., heat removal) supplied by a geothermal heat pump. Three other heat meters record the heating and cooling energy supplied to each of three independent load circuits.

Readings from multiple heat meters can sometimes be combined to determine conditions such as heat loss from piping or thermal storage tanks. They can also be combined with electrical power and energy meter readings to determine equipment performance, or used for diagnosing problems in a system.

For example, the *instantaneous* COP of the heat pump in Figure 14-88, and its associated circulators, could be determined by combining the thermal power reading from the heat meter (in kBtu/hr), with the electrical power input to the heat pump and its associated circulators, in kilowatts, using the following equation.

$$COP_i = \frac{Q_0}{P_e \times 3.413}$$

where,
COP_i = instantaneous COP of heat pump + circulators
Q_o = heat output of heat pump (kBtu/hr)
P_e = electrical power input to heat pump + circulators (kW)
3.413 = conversion factor from kilowatts to kBtu/hr

The *average* COP of the heat pump and its associated circulators *over a period of time* can be determined in a similar manner. It would be calculated by dividing the total heating or cooling energy recorded by the heat meter, by the total kilowatt•hours of electrical energy supplied to the heat pump and circulators over the same period. This ratio can be expressed as follows.

$$COP_a = \frac{H}{E \times 3.413}$$

where,
COP_a = average COP of heat pump + circulators over time period
H = heat output of heat pump over time period (kBtu)

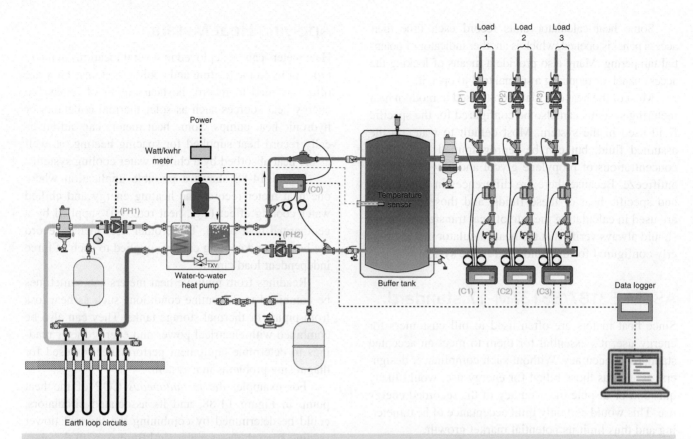

Figure 14-88 | Example of a system using four networked heat meters, and one electrical meter, to determine heat pump performance, total heat to each of three loads, and other calculated performance information.

E = electrical energy input to heat pump + circulators (kWh)

3.413 = conversion factor

The three heat meters on the load circuits could be used to invoice each "client" for the total heating energy used. Furthermore, the sum of the three load heat meters, subtracted from the total heat output from the heat pump, over a period of time, would yield the thermal losses associated with the buffer tank and adjacent piping.

The data collected by modern heat meters can also be accessed without being physically present at the meter. For example, the four heat meters shown in Figure 14-88 are networked to a common data logger by a simple 2-conductor cable. The readings from all four meters can be accessed through a computer or mobile device connected to the data logger. This eliminates the need for a person to enter each space where a heat calculator unit is located, just to read that meter. Some heat meters can also be accessed through wireless networks.

Heat Interface Units

Heat meters are commonly used in systems where a single heating plant (usually a multiple boiler system) provides heat to several independently occupied units within a building. Examples include apartment buildings, condominiums, office spaces, or retail shops. Heat from the central plant is transported throughout the building using a 2-pipe distribution system. A third pipe carrying cold domestic water is routed along with these two pipes. A compact assembly called a **heat interface unit (HIU)**, which is located in each building unit, connects to these three "mains." All heat, as well as the volume of cold domestic water passing from the 3-pipe building mains into each HIU is accurately metered. Each building unit is periodically invoiced for the thermal energy and total domestic water used.

A brazed plate heat exchanger within each HIU allows the primary water supplied from the building's central heating plant to be separated from the water used for space heating or domestic hot water within each unit. The need to separate the heating plant water from domestic water is obvious. The separation of heating plant water from the water used for space heating within each unit ensures that any leak in the space heating subsystem within a building unit can be isolated and repaired without disrupting service to other HIUs in the building.

Figure 14-89 shows an example of a modern HIU that uses two brazed plate heat exchangers—one for space heating and the other for heating domestic water.

Figure 14-89 | Example of a modern heat interface unit (HIU). *Courtesy of Caleffi North America.*

The HIU is typically mounted within a wall cavity in the building unit, close to the distribution pipes from the central plant. It has connections for the heating supply and return pipes, and a third connection for cold domestic water supply.

When space heating is required, a modulating valve within the HIU regulates the flow of hot water from the building mains through the primary side of one brazed plate heat exchanger in the HIU. This flow regulation maintains a desired water temperature leaving the secondary side of the heat exchanger, and supplying the unit's space heating distribution system.

A flow switch detects when there is a demand for domestic hot water. A second modulating valve within the HIU then begins regulating flow of hot water from the building mains through the primary side of a second brazed plate heat exchanger. This flow regulation is based on providing a desired domestic hot water temperature leaving the secondary side of the heat exchanger.

The total space heating and domestic hot water energy used is based on the flow rate and temperature drop of the water from the building mains, between where it enters and leaves the HIU. The controller within the HIU records the total thermal energy used.

The HIU shown in Figure 14-89 also includes a circulator and expansion tank for the space heating distribution system within the building unit.

Figure 14-90 shows a *conceptual* piping arrangement for an HIU that provides the functions just described. It does not show exact hardware devices or their placement.

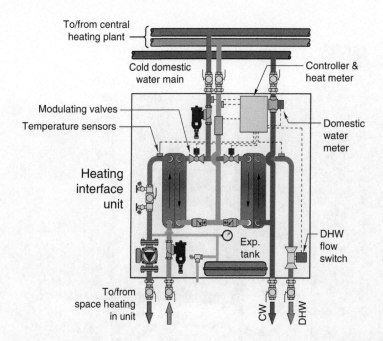

Figure 14-90 | Representative component configuration in an HIU that provides space heating and domestic hot water.

HIUs allow for individually metered thermal energy for space heating and domestic hot water without the need for combustion heat sources within each unit. This reduces cost and increases safety by eliminating the need for gas piping and venting of combustion appliances dispersed throughout a building.

Systems based on a central heating plant serving multiple HIUs can be designed around several types of heat sources. A multiple boiler system is one of the most common heating plants to date. However, heat input from renewable energy heat sources such as heat pumps, solar thermal collectors, or biomass boilers is certainly possible. Another possibility is when heat to the entire building is provided from a district heating system. The building's heating plant would be a large plate & frame heat exchanger that receives hot water from insulated piping that's typically buried under streets or sidewalks. Large capacity heat meters are used to record the total heat transferred from the district heating system to the building. Such systems are common in many European cities. Their use in North America is likely to increase.

Invoicing for thermal energy

Because heat metering is relatively new in North America, there are no widely accepted standards or legal requirements for how thermal energy can be invoiced to clients of heat metered systems. This situation will likely change as heat metering becomes more widely accepted and implemented.

In the absence of specific legal requirements that govern billing for thermal energy, the owners of buildings in which heat metering is used can establish their own agreements for how occupants will pay for the thermal energy they use, as well as cost associated with maintaining the system. The potential occupants of each building unit using heat metering would then decide if the owner's agreement for thermal energy invoicing is acceptable as part of the overall decision to purchase or rent the building space.

Any agreement covering invoicing for thermal energy should address the following:

- The current cost of fuel for the central plant that supplies hot water for heating and/or chilled water for cooling. This should include all fixed and variable fees, surcharges, tariffs, etc., associated with the fuel used by the system.

- How the cost of fuel and any associated fees may increase or decrease over time.

- Heat loss (or heat gain in cooling mode) from components on the building owner's side of the heat meters. This would include heat loss (or gain) from the central plant equipment as well as heat loss from distribution piping up to the points where heat passes through individual heat meters for each building space. The overall energy input to the system will always be more than the sum of all heat meter readings due to equipment inefficiencies and heat loss/gain from the building's distribution system.

- The current and projected maintenance costs for the system.
- How any required maintenance or service on the heating plant or distribution system will be initiated and paid for.
- How any required maintenance or service cost on the client's side of the heat meter will be initiated and paid for.
- How the administrative costs for reading energy meters and issuing invoices will be allocated.
- Any applicable regulations dealing with sales tax or potential profit on metered energy.
- How a contingency fund to cover maintenance or service on the central portion of the system is funded and managed.

These issues are only for consideration when drafting heat metering agreements. They do not imply legal advice. An attorney should be consulted to draft a specific agreement that includes relevant details for each project, and meets all legal requirements applicable to the installation.

14.10 Introduction to Balancing

Many of the hydronic systems discussed in this text contain two or more parallel branches. The flow created by a single circulator will divide up among these branches based on their hydraulic resistance. However, the percentage of total system flow that passes through a given branch may not be the flow necessary for proper heat delivery. Thus, it is often necessary to adjust valves within each branch to achieve flow rates equal or close to those necessary for proper heat delivery. This process is called "balancing" the system. This section discusses the basic concepts involved with **hydronic balancing**. It also demonstrates two of several balancing procedures that are applicable to parallel direct-return or reverse-return systems. Additional references that cover balancing in more detail are cited at the end of this section.

Why Balance?

Perhaps the most immediately noticed consequence of a poorly balanced hydronic system is the lack of comfort. Achieving and maintaining comfort is arguably the most desired and widely promoted benefits of hydronic heating. It is foolish to design and install a system that has the potential of delivering such comfort, and then limit its performance due to improper balancing.

Besides the lack of comfort, there are several other undesirable consequences associated with a poorly balanced system. They include:

- Frozen pipes due to insufficient heat delivery
- Shrinkage cracks in wood and drywall surfaces
- Condensation on windows
- Growth of mold and mildew
- Increased potential for respiratory illnesses
- Poor mental alertness
- Wasted energy due to higher heat losses
- High flow velocities in piping components creating noise and possible erosion
- Excessive energy use by circulators due to overflow conditions
- Circulators operating at low efficiency
- Circulators operating at unnecessarily high differential pressure

The Purpose of Balancing

Although most hydronic heating professionals agree that balanced systems are desirable, there are many opinions on what constitutes a balanced system. The following are some of those "opinions."

- The system is properly balanced if all branches experience the same temperature drop.
- The system is properly balanced if the ratio of the flow rate through a branch divided by the total system flow rate is the same as the ratio of the required heat output from that branch divided by the total system heat output.
- The system is properly balanced only when all branches are identically constructed (e.g., same type, size, and length of tubing, same fittings and valves, same heat emitter).
- The system is automatically balanced if constructed with a reverse-return piping layout.

Although some of these opinions have partial validity, none of them are totally correct or complete. It follows that any attempt at balancing a system is pointless without a definition and "end goal" for the process.

In this text, the working goal of balancing is as follows:

A properly balanced hydronic system is one that consistently delivers the proper rate of heat transfer to each space served by the system.

Balancing causes simultaneous changes in the hydraulic operating conditions (e.g., flow rates, head losses, pressure drops), as well as the thermal operating conditions (e.g., fluid temperatures, heat emitter outputs). These operating conditions interact as the system continually seeks both hydraulic equilibrium and thermal equilibrium, as discussed in previous chapters.

A Theoretical Approach to Balancing

Considering that there are often hundreds, sometimes thousands of piping components in some hydronic systems, and that nearly all of them have some influence on flow rates and heat transfer, it's apparent that a theoretical approach to balancing that would account for all these details would be complicated. Computer simulation of a virtual hydronic system is the only plausible approach to handling the mathematics involved in determining simultaneous thermal and hydraulic operating conditions. This is especially true given the fact that these operating conditions will change, throughout the system, whenever the setting of a single balancing valve changes.

Chapters 8, 9, and 10 all discussed the fact that changing the flow rate through any type of hydronic heat emitter affects its heat output. Figure 14-91 demonstrates this effect for two common heat emitters, a radiant floor panel circuit, and a 20-foot length of finned-tube baseboard, both of which are being supplied at a constant water temperature.

A truly accurate computer simulation of a hydronic system would use thermal/hydraulic modeling equations for each heat emitter, each balancing valve, and each piping segment in the system. It would also model circulator performance.

Based on the current settings of each balancing valve, the computer simulation would determine the flow rate through each section of piping and heat emitter. It would then use this flow information in combination with a thermal model of each heat emitter to determine heat output.

As the balancing valves within each branch of the virtual system are further adjusted, the software would generate immediate feedback on how the heat output of each heat emitter changed as a result of those adjustments. The goal would be to adjust the balancing valves in the virtual system so that the output of each heat emitter is at, or very close to, the necessary design heating load of the space it serves.

The Circuit Simulator module in the Hydronics Design Studio can simulate the thermal and hydraulic behavior of user-defined parallel direct-return systems containing up to 12 crossovers. It can also determine the effect of changing balancing valve setting, as well as the influence of a differential pressure bypass valve if present.

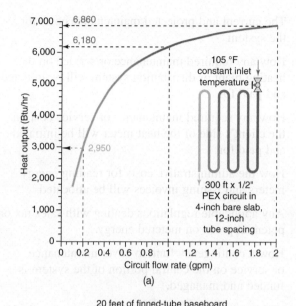

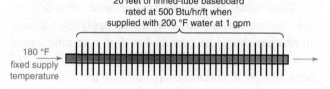

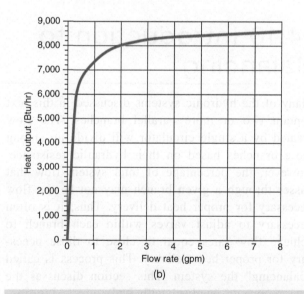

Figure 14-91 (a) Heat output versus flow rate for a radiant floor circuit supplied at a constant water temperature of 105 °F. (b) Heat output versus flow rate for a 20-foot length of finned-tube baseboard supplied at a constant water temperature of 180 °F.

Simulation software that could model *any* user-defined hydronic system represents the ultimate tool for assessing the effects of balancing valve settings. Such software does not currently exist. Thus, most *practical* balancing procedures have a more limited goal: *To adjust the balancing valves so that flow rates in all parts of the system are as close as possible to a set of calculated values*. These calculated flow rates are based on calculated heating loads and a selected fluid temperature drop under design load conditions. Adjusting the system to achieve a predetermined flow through each branch is the goal of the procedures discussed later in this section.

Although establishing a set of calculated flow rates, one for each parallel branch, does not guarantee that the desired heat transfer occurs at all heat emitters, it is an important part of ensuring that the system operates reasonably close to expectations.

Balancing Systems with Manually Set Balancing Valves

Consider the parallel direct-return distribution system shown in Figure 14-92. It consists of seven **crossovers** connected between a **supply main** and **return main**. Circulation is created by a fixed-speed circulator. For now, *assume that all crossovers have identical piping components, and ideally would all operate at the same flow rate.* The vertical piping near the circulator, and the closely spaced tees where heat is injected, can be considered to have insignificant head loss.

Assume that when this system is first turned on, all the balancing valves are fully open. Because of the head loss along the supply mains and return mains, the head energy available to each crossover is different. The **"most favored crossover"** nearest the circulator will experience the highest available head and thus operate the highest flow rate. The **"least favored crossover"** at the far right side of the system will experience the lowest available head, and thus have the lowest flow rate. This undesirable flow distribution, shown in Figure 14-93, is what balancing is meant to correct.

Figure 14-94 illustrates the decreasing head available to each crossover based on head loss along the supply and return mains. The upper sloping line represents head loss due to flow moving from left to right along the supply main. The lower sloping line represents head loss due to flow moving from right to left along the return main. The distance between upper and lower sloping lines, at the location of a given crossover, represents the head available to drive flow through that crossover.

To achieve equal flow, the balancing valve within each crossover must dissipate the difference between the head available across the mains, and the head necessary to achieve the desired flow rate through each crossover. This is depicted in Figure 14-95 for a system with identical piping components in each crossover.

In many systems, the hydraulic resistance of each crossover is different. The design flow rate requirement may also vary from one crossover to another. It is still possible to balance such systems for the desired flow rate in each crossover. *In each case, a properly adjusted balancing valve will dissipate the difference between head available across the mains, at the location of the crossover, and the head required to sustain the design flow rate through the crossover.* This is shown in Figure 14-93 for a system with three different crossovers.

Balancing Procedure

One method of balancing a system with manually set balancing valve is called the "**preset method**." It calculates the necessary C_v value of each balancing valve based on the design load flow rate required through each crossover, at that crossover's location within the system.

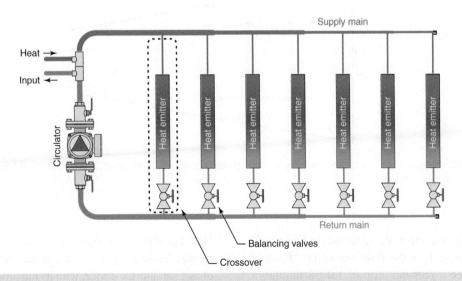

Figure 14-92 | A parallel direct-return system with seven identical crossovers.

766 Chapter 14 Auxiliary Loads and Specialized Topics

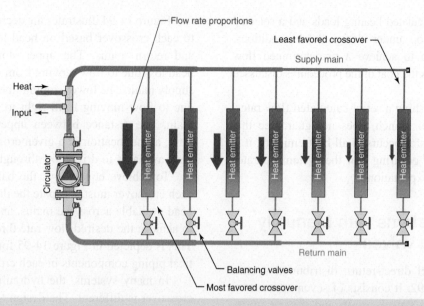

Figure 14-93 | With all balancing valves open, the flow rates through each crossover will be different. The farther a crossover is from the circulator, the lower its flow rate.

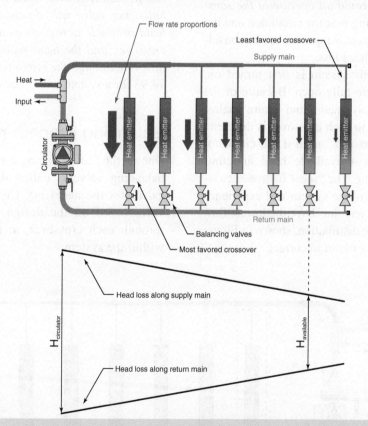

Figure 14-94 | The head available to drive flow through the different crossovers is different because of head loss along the supply and return mains.

The **Cv rating** of a valve was discussed in Chapter 6, Fluid Flow in Piping. It is the flow rate of 60 °F water required to produce a pressure drop of 1.0 psi through a *fully open* valve. As the valve's stem is moved toward the closed position, the valve's **Cv setting** (as opposed to its *Cv rating*) decreases. Thus, an adjustable valve can be thought of as a variable Cv device, with its maximum value (e.g., Cv rating) available when the valve is fully open.

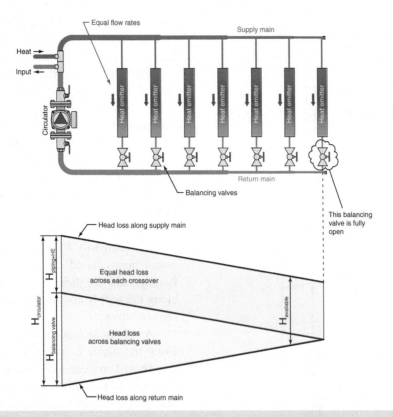

Figure 14-95 | Each balancing valve is set to dissipate the *difference* between the head available across the mains (at a given location), and the head needed to achieve the desired flow through the crossover.

Some balancing valves have a built-in scale that shows the current Cv setting based on stem position. Other valves come with a chart that shows their Cv setting based on the number of turns of the valve's stem starting from its fully closed position. Either type of valve can be used for the preset method of balancing.

Consider the typical crossover piping shown in Figure 14-94. The total head loss from the supply main to the return main consists of the head loss across the piping, fittings, heat emitter, and zone valve, plus the head loss across the balancing valve. For simplicity, the head loss of the piping, fittings, heat emitter, and zone valve (but not the balancing valve) are combined into one quantity called $H_{(piping + HE)}$. Thus, Equation 14.20a will be:

Equation 14.20a:

$$H_{mains} = H_{(piping + HE)} + H_{BV}$$

where,

H_{mains} = head available between the mains (feet of head)

$H_{(piping + HE)}$ = head loss of piping components and heat emitter in crossover (feet of head)

H_{BV} = head that must be dissipated by balancing valve in crossover (feet of head)

To achieve the desired flow rate through a crossover, its balancing valve must dissipate the difference between ΔH_{mains} and $\Delta H_{piping + HE}$. Thus, Equation 14.20b will be:

Equation 14.20b:

$$H_{BV} = H_{mains} + H_{(piping + HE)}$$

The necessary Cv setting for the balancing valve to dissipate this head can be found using Equation 14.10. Thus, Equation 14.21 will be:

Equation 14.21:

$$Cv_{BV} = \frac{1.52 \times f}{\sqrt{\Delta H_{BV}}} = \frac{1.52 \times f}{\sqrt{(\Delta H_{mains} - \Delta H_{piping + HE})}}$$

where,

Cv_{BV} = required Cv setting of the balancing valve;

f = desired flow rate through the crossover (gpm)

ΔH_{BV} = head dissipation required of balancing valve (feet of head)

To evaluate Equation 14.21, it is necessary to know the head loss across the mains at the location of the

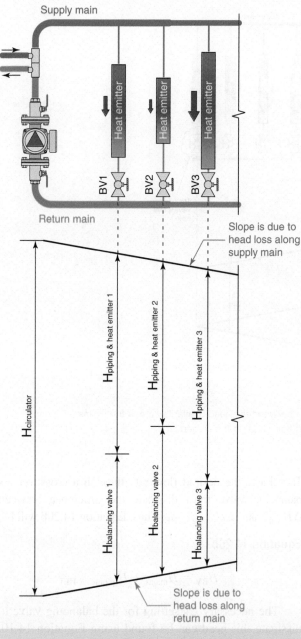

Figure 14-96 | To achieve desired flow, the balancing valve must dissipate the difference between the available head at a crossover minus the head loss of the piping and heat emitter.

crossover, as well as the head loss of the piping and heat emitter in that crossover when operating at design flow rate.

The head loss of the piping and heat emitter within the crossover can be found, at design load flow rate, using methods from Chapter 6, Fluid Flow in Piping. The total head loss along the supply main, out to the location of a crossover, and back along the return main can also be found by applying these methods to each segment of the mains.

Start by making a diagram of the system, which indicates all piping sizes, lengths and design load flow rates. Using the method of hydraulic resistance, calculate the head loss for each segment of the supply and return mains based on the design flow rates required. Finally, add the head losses of each piping segment along the supply and return mains, out to the location of the crossover.

An alternate method of *approximating* the head loss along the mains is to assume that the mains piping is sized based on a constant head loss per foot. A commonly accepted value is 4 feet *of head loss per 100* feet *of pipe*. Our further discussions will assume the mains are sized using this criteria. Based on this assumption, the head loss across the mains at some location downstream of the circulator can be estimated as shown in Figure 14-98.

The hydraulic separator in Figure 14-98 is where heat is added to the distribution system. However, the hydraulic separator has insignificant head loss. The available head at a given crossover is estimated by subtracting an assumed 4 feet of head loss per 100 feet of mains piping from the head produced by the circulator. This relationship is given as Equation 14.22:

Equation 14.22:

$$H_{available} = H_{circulator} - 2L(0.04)$$

where,

$H_{available}$ = head available at a given crossover location (feet of head)

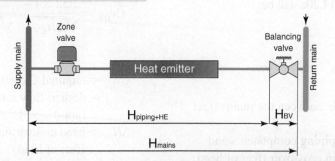

Figure 14-97 | Head loss across the mains is the sum of the head loss across the piping and heat emitter, plus the head loss across the balancing valve.

14.10 Introduction to Balancing

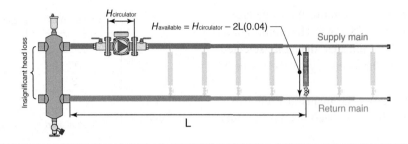

Figure 14-98 | Method for estimating the head loss along the mains assuming mains piping is sized for a head loss of 4 feet of head loss per 100 feet of pipe.

$H_{circulator}$ = circulator head at design flow rate (feet of head)
L = length of supply or return main from hydraulic separator out to the crossover (feet)
0.04 = assumed head loss of 4 feet/100 feet of main piping

Example 14.12

The system shown in Figure 14-99 requires balancing. The system is operating with water at an average temperature of 140 °F. The design load temperature drop assumed for each crossover is 20 °F. The required design heating load of each crossover is indicated. The supply and return main piping has been sized for a head loss of 4 ft/100 ft. The circulator used in the system has the pump curve shown in Figure 14-97. Determine:

a. The required design flow rate in each crossover.

b. The required Cv setting of the balancing valve in each crossover.

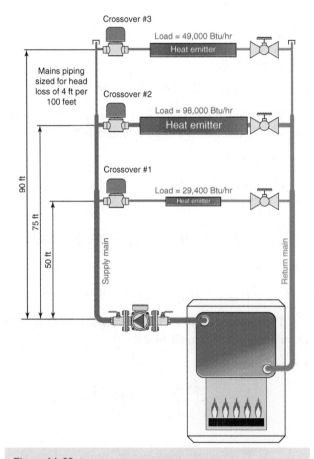

Figure 14-99 | System for Example 14.12.

Solution:

The required design load flow in each crossover is determined using the sensible heat rate Equation 4.8. The value of $(8.01Dc)$ for water at an average temperature of 140 °F is calculated as 490.

For crossover 1:

$$f_1 = \frac{Q_i}{(8.01Dc)(\Delta T_d)} = \frac{29{,}400}{(490)(20)} = 3.0 \text{ gpm}$$

For crossover 2:

$$f_2 = \frac{Q_1}{(8.01Dc)(\Delta T_d)} = \frac{98{,}000}{(490)(20)} = 10.0 \text{ gpm}$$

For crossover 3:

$$f_3 = \frac{Q_1}{(8.01Dc)(\Delta T_d)} = \frac{49{,}000}{(490)(20)} = 5.0 \text{ gpm}$$

Using these flow rates, the head loss of the piping components and heat emitter in each crossover (exclusive of the balancing valve) are determined using the head loss calculation methods from Chapter 6, Fluid Flow in Piping. Figure 14-101 lists the values for these head losses that will be assumed for this example.

The system flow rate is the total of the three crossover flow rates, in this case 18 gpm. The pump curve shown in Figure 14-100 indicates that the selected circulator adds 12 feet of head energy to the water at this flow rate.

The head available across the mains at each crossover location can now be calculated using Formula 14-22:

For crossover 1:

$$H_{mains} = H_{circulator} - 2L(0.04) = 12 - 2(50)(0.04) = 8 \text{ feet}$$

For crossover 2:

$$H_{mains} = H_{circulator} - 2L(0.04) = 12 - 2(75)(0.04) = 6 \text{ feet}$$

For crossover 3:

$$H_{mains} = H_{circulator} - 2L(0.04) = 12 - 2(90)(0.04) = 4.8 \text{ feet}$$

The required head loss across each balancing valve can now be calculated using Formula 14-20b:

For crossover 1:

$$H_{BV1} = H_{mains} - H_{(piping + HE)} = 8 - 5 = 3 \text{ feet}$$

For crossover 2:

$$H_{BV2} = H_{mains} - H_{(piping + HE)} = 6 - 2 = 4 \text{ feet}$$

For crossover 3:

$$H_{BV3} = H_{mains} - H_{(piping + HE)} = 4.8 - 3 = 1.8 \text{ feet}$$

The required Cv of each balancing valve can now be calculated using Formula 14-21:

For crossover 1:

$$Cv_{BV1} = \frac{1.52 \times f_1}{\sqrt{\Delta H_{BV1}}} = \frac{1.52 \times 3}{\sqrt{3}} = 2.63$$

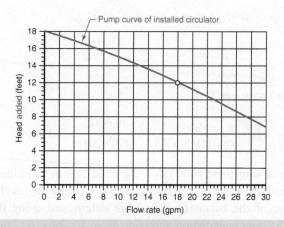

Figure 14-100 Pump curve for circulator used in Example 14.12.

crossover	$\Delta H_{piping + HE}$ (ft)
1	5
2	2
3	3

Figure 14-101 Calculated head loss of piping and heat emitter in each crossover for Example 14.12

For crossover 2:

$$Cv_{BV2} = \frac{1.52 \times f_2}{\sqrt{\Delta H_{BV2}}} = \frac{1.52 \times 10}{\sqrt{4}} = 7.6$$

For crossover 3:

$$Cv_{BV3} = \frac{1.52 \times f_3}{\sqrt{\Delta H_{BV3}}} = \frac{1.52 \times 5}{\sqrt{1.8}} = 5.66$$

The balancing valve in each crossover could now be set to its required Cv value using a scale built into the valve's stem, or using a graph/table indicating Cv versus stem position.

Discussion:

This procedure is based on an *estimate* of head loss along the mains. As such, it is not guaranteed to yield the exact design flow rates when the balancing valves are set to the calculated Cv values. If metered balancing valves are used, each crossover flow rate could be "fine-tuned" to the required value.

The Role of Differential Pressure Control in Balancing

The system analyzed in Example 14.12 includes zone valves on each crossover. Although the balancing valves were adjusted to produce the calculated flow rates in each crossover when all zones are open, these flow rates will vary when one or more of the zone valves closes.

To minimize this variation, it is advisable to use some means of differential pressure control in systems with valve-based zoning as discussed in Chapter 11, Distribution Piping Systems.

In systems using a constant speed circulator, a differential pressure bypass valve (DPBV) can be used. It should be set for a threshold differential pressure approximately 0.5 psi higher than the circulator's differential pressure when all zone valves are open. This setting allows the DPBV to remain closed under design load conditions. When one or more of the zone valves close, the differential pressure between the mains will increase slightly, and some flow will pass through the DPBV, which will stabilize the differential pressure between the supply and return mains. The flow rate in the zones that are on will remain relatively stable.

A variable-speed pressure-regulated circulator could also be used for differential pressure control. In systems where there is significant head loss along the mains piping (i.e., the system shown in Example 14.12), the pressure-regulated circulator should be set for *proportional* differential pressure control mode. In systems with insignificant head loss between crossovers (i.e., a homerun system using a manifold station), the circulator would be set for *constant* differential pressure control mode.

Balancing Using Compensated Method

Another method for balancing systems that use differential pressure type balancing valves is called the **compensated method**. It can be used in relatively simple systems such as shown in Figure 14-102, or in more complex systems such as shown in Figure 14-103.

As system piping becomes more complex, it is important to establish consistent terminology. Terms such as "**branch**," "**riser**," "**mains**," "crossover," or "**building headers**" can mean different things to different people. The terminology shown in Figure 14-104 will be used for our discussion of balancing.

The differential pressure type balancing valves used for this method of balancing become more subject to inaccuracies due to turbulence when operated at pressure drops below 3 KPa (0.435 psi). It follows that this type of balancing valves should be selected so they will operate at pressure drops higher than 3 KPa (0.435 psi). Using this criterion along with the definition of Cv leads to Equation 14.23 for determining the *maximum Cv* rating of a differential pressure-balancing valve.

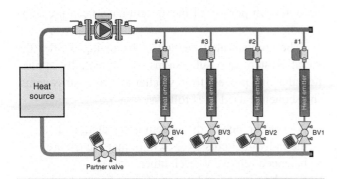

Figure 14-102 | Example of a relatively simple system to be balanced using the compensated method. Note the partner valve in the return main.

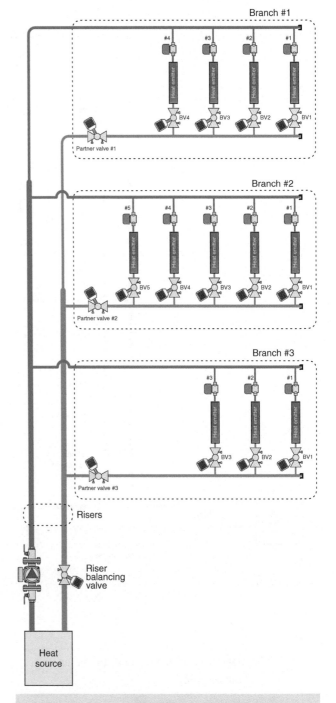

Figure 14-103 | Example of a more complex system to be balanced using the compensated method. Note the multiple partner valves.

Equation 14.23:

$$Cv_{BV(max)} = \frac{f}{\sqrt{\Delta P}} = -\frac{f}{\sqrt{.435}} = 1.52 \times f$$

where,

Cv_{BVmax} = *maximum Cv* rating of a differential pressure type balancing valve;

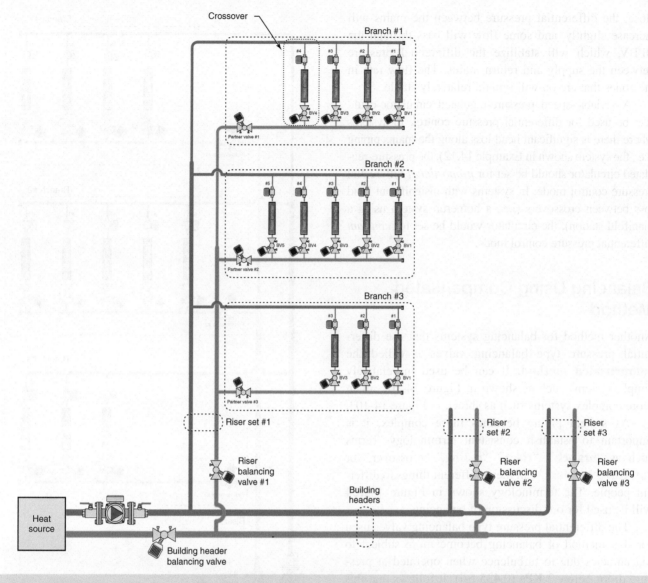

Figure 14-104 | Terminology used to describe portions of a complex piping assembly.

f = design flow rate through the balancing valve (gpm).

To further minimize the effects of turbulence, a minimum of 10 pipe diameters of straight piping should be installed upstream of every balancing valve.

The procedure to be presented for the compensated method of balancing assumes that the heat source is hydraulically separated from the distribution system. This prevents flow rate changes within the distribution system from inherently altering flow rates through the heat source. There are several ways to achieve this hydraulic separation. They include use of a hydraulic separator, closely spaced tees, or a buffer tank. Each of these methods was discussed in Chapter 11, Distribution Piping Systems.

As is true with the preset method, the compensated method of balancing assumes that the design flow rates through each portion of the system have been predetermined.

The compensated method of balancing divides the overall system into subsystems. It sequentially applies the balancing process from smaller to larger subsystems. The sequence used is as follows:

1. A given branch of the system is selected, and the individual crossovers within that branch are balanced relative to each other.
2. Each branch served by a set of risers is then balanced relative to the other branches on that set of risers.
3. If the system contains multiple sets of risers, each set of risers is then balanced relative to each other.

The compensated balancing method, when used with differential pressure balancing valves, required at least one, and preferably two differential pressure meters that can accurately display the differential pressure across any balancing valve they are attached to. An example of such a meter attached to a differential pressure type balancing valves was shown in Figure 5-89b.

In most cases, this method of balancing is best performed by two technicians working as a team. Each would have a differential pressure meter that can accurately display the differential pressure across any valve it is attached to. The two technicians will often be in different locations within the building where specific balancing valves are located, and will need to stay in communication with each other as the balancing procedure progresses.

Before beginning the balancing procedure, the system should be purged, and all balancing valves set to their fully open position. If there are multiple riser sets in the system, select the riser set closest to the circulator, and temporarily shut-off flow to the other riser sets. Since a riser must have at least two branches, select the branch farthest from the circulator. The balancing valves on all crossovers with this branch will be adjusted before moving to a different branch.

Consider the branch configuration shown in Figure 14-105. It consists of four crossovers, each serving a heat emitter, and each containing a differential pressure type balancing valve. Another balancing valve, known as the **partner valve** is located on the return pipe of the branch.

The following procedure can be used to adjust the balancing valves in this branch.

STEP 1: Identify the "least favored crossover" within the branch. If the terminal units are similar or identical, the least favored crossover will likely be at the far end of the branch. If the crossovers have significantly different piping, the least favored crossover will be the one with the highest head loss at design flow conditions. Identifying this crossover may require calculating the head loss of each crossover in the branch at its associated design flow rate. Once the least favored branch is identified, the balancing valve in this branch is called the **reference valve**.

STEP 2: Set the reference valve for a minimum pressure drop that ensures stable and accurate readings. The commonly accepted value for this minimum pressure drop is 3 KPa (0.435 psi). This minimizes the pressure drop (and head dissipation) across the balancing valve while still maintaining the valve's accuracy. If the Cv of this balancing valve was selected using Equation 14.23, the valve should be close to, or at, its fully open setting, and thus at its minimum head dissipation setting. Once this setting is made, lock the handwheel of the reference valve in this position. If the differential pressure across the valve cannot be set as low as 3 KPa (0.435 psi), leave the reference valve fully open.

STEP 3: Attach a differential pressure meter to the *reference valve* as shown in Figure 14-105.

STEP 4: Adjust the *partner valve* on the branch return pipe so that the desired design flow rate is passing through the reference valve, and record the pressure drop across the reference valve. This pressure drop is now called the **reference pressure**.

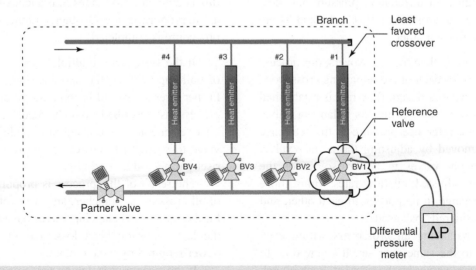

Figure 14-105 | Differential pressure meter attached to least favored crossover. The balancing valve in this crossover is called the reference valve.

STEP 5: Move to the next crossover (#2 in Figure 14-105). Adjust balancing valve BV2 so that the desired design flow rate is established through crossover #2. This will likely cause some change in the pressure drop reading across the reference valve, which implies that the flow rate through crossover #1 has changed. This *undesirable* change will be corrected in the next step.

STEP 6: Adjust the *partner valve* on the branch return pipe so that the pressure drop across the reference valve is *restored to the reference pressure*. This restores the design load flow rate across the reference valve. The flow rates in crossovers #1 and #2 are now proportionally balanced to each other. When one changes, the other will change proportionally. This is desirable.

STEP 7: Move to the next crossover (#3 in Figure 14-105). Adjust balancing valve BV3 so that the design flow rate is established through crossover #3. This will cause a change in the differential pressure read across the reference valve, which means there has been a change in flow rate.

STEP 8: Again, adjust the *partner valve* so that the reference pressure is reestablished across the reference valve. This restores the design load differential pressure and flow rate. The flow rates in crossovers #1, #2, and #3 are now proportional to each other.

STEP 9: Move to the next crossover (#4 in Figure 14-105). Adjust balancing valve BV4 so that the desired design flow rate is established through this crossover. As before, this will likely cause some change in the pressure drop across the reference valve.

STEP 10: Adjust the partner valve so that the reference pressure is reestablished across the reference valve. This restores the design load differential pressure and flow rate. The flow rates in crossovers #1, #2, #3, and #4 are now proportional to each other.

If there were more than four crossovers, this process would be repeated for each of the remaining crossovers. In each case, the desired design flow rate is established in a given crossover by adjusting its balancing valve. Any change in the reference pressure at the reference valve is then removed by adjusting the partner valve. When the balancing valves in all crossovers of the branch have been adjusted, all flow rates within those crossovers will change in proportion to each other, and the branch is considered "balanced."

Proceed by balancing all crossovers within each branch attached to the same riser set. It's now time to balance the branches serves by the riser set. Each branch connected to the riser set can now be considered as if it were a single entity as shown in Figure 14-106. None of the crossover balancing valves in any of the branches need further adjustment.

The partner valve in the farthest branch now becomes the new reference valve.

STEP 1: Adjust the stem of the new reference valve (partner valve #1 in Figure 14-103) so that it will create a minimum pressure drop of approximately 3 KPa (0.435 psi). Lock the handwheel of this valve at that setting. If the pressure drop cannot be lowered to 3 KPa, simply leave the valve in its fully open position.

STEP 2: Connect the differential pressure meter across this valve as shown in Figure 14-103.

STEP 3: Adjust the *riser partner valve* until the desired design load flow is passing through partner valve #1. Once this flow rate is established, the pressure drop across partner valve #1 becomes the new reference pressure.

STEP 4: Move to branch #2, and adjust partner valve #2 for the desired design load flow rate through branch #2. This will cause a change in the flow rate and differential pressure across partner valve 1.

STEP 5: Reestablish the reference pressure across partner valve #1 by adjusting the *riser partner* valve. Once this is done, the flow rates into branches #1 and #2 are in proper proportion.

STEP 6: Move to branch #3, and adjust partner valve #3 for the desired design load flow rate through branch #3. This will cause a change in the flow rate and differential pressure across partner valve #1.

STEP 7: Reestablish the reference pressure across partner valve #1 by adjusting the riser partner valve. Once this is done, the flow rates in branches # 1, # 2, and # 3 are in proportion to each other, and the overall balancing procedure is completed.

If the system has multiple risers connected to a set of building headers, the partner balancing valve on the farther riser set would become the reference valve. The process just described would be repeated, balancing each riser set one at a time using the building mains partner valve to reestablish the reference pressure each time a riser partner valve is adjusted.

The result of this process is proportional balancing of all crossovers, branches, and riser sets in the system. Furthermore, the balancing valves have been set to have the least parasitic head loss, and thus minimize the power required by the circulator.

If the system uses zoning valve or modulating thermostatic radiator valves on the individual crossovers, there will still be variations in flow rate between the

14.10 Introduction to Balancing   775

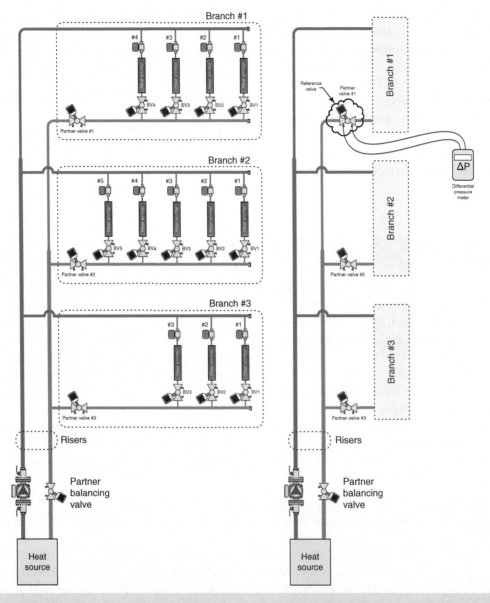

Figure 14-106 | Once each branch has been balanced, the riser needs to be balanced. Each branch can now be treated as a single entity. None of the crossover balancing valves within any branch will require further adjustment. The differential pressure meter is connected across the partner valve in the branch furthest from the circulator.

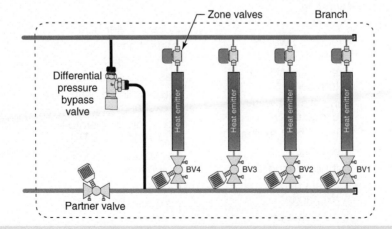

Figure 14-107 | A differential pressure bypass valve installed across each branch when zone valves are used in the crossovers.

crossovers on a given branch when one or more of those zone valves closes. This variation can be reduced by installing a differential pressure bypass valve on each branch as shown in Figure 14-107. The threshold differential pressure on the bypass valve should be set 0.5 psi higher than the differential pressure measured across the supply and return piping of the branch after the system has been fully balanced.

This section is but an introduction to the topic of hydronic balancing. Other methods are available for balancing more extensive systems, as well as system using specialty-balancing devices. Further information on hydronic balancing is available in the following references:

1. Petitjean, R. Total Hydronic Balancing (ISBN 91-630-2626-0).
2. *idronics #8, Balancing Hydronic Systems*, Caleffi North America, PDF file available from www.caleffi.us.

Summary

The ability to supply several auxiliary loads using the same heat source that provides space heating is a unique feature of hydronic heating systems. It reduces installation costs relative to using separate heat sources for each load. It also increases seasonal boiler efficiency by creating greater run fractions. Furthermore, it reduces the number of fuel supplies, exhaust systems, as well as maintenance requirements.

Wherever possible, hydronic system designers should try to maximize the use of the heat source by sharing its heat output as necessary. Prioritized load shedding can usually provide an acceptable means of managing boiler capacity in the event of excess loading.

The other unique application topics discussed in this chapter can significantly enhance the performance and appeal of hydronics technology in specific situations.

By making use of these, as well as the piping and control concepts covered in previous chapters, creative designers can develop systems to match very specific requirements. These systems can be efficient, reliable, and cost-effective over a long service life.

Key Terms

approach temperature difference
ASSE 1017-rated anti-scald tempering valve
ASTM E3137/E3137M-17
auxiliary loads
branch
brazed plate heat exchangers
Btu metering
buffer tank
building headers
capacitance rate
capacitance rate ratio
Class 1 SIM
Class 2 SIM
Class 3 SIM
cocurrent flow
compensated method
compressive "donut" collar
compressive load rating

concrete sub-slab
continuous flow rating
convection coefficient
counterflow
counterflow heat exchange
crossovers
Cv rating
Cv setting
direct-fired water heater
direct-to-load heat transfer
domestic hot water usage profile
external heat exchanger
filter fabric
first hour rating
flat plate heat exchanger
flow transducer
forced convection
fouling factors

fouling films
geotextile fabric
heat calculator unit
heat interface unit (HIU)
heat meter
heat metering
hydronic balancing
idled
indirect water heaters
internal heat exchangers
least favored crossover
liquid-to-liquid heat exchangers
log mean temperature difference (LMTD)
low velocity zone dirt separator
mains
minitube distribution system
minitubes
most favored crossover

natural convection
number of transfer units (NTU)
overall heat transfer coefficient
oversurface percent
parallel flow
partner valve
periodic flow reversal
plate & frame heat exchangers
preset method
priority load
rangeability
reference pressure
reference valve
resistance temperature detectors (RTDs)
return main
reverse indirect water heater
riser
setpoint demand

shell & coil heat exchanger
shell & tube heat exchanger
short-cycle
SIM systems
snow switch
snow/ice pavement sensor
supply main
tank-in-tank
tankless coil
temperature crossover
thermal length

Questions and Exercises

1. Describe one of the advantages of a flat plate heat exchanger relative to other types of heat exchangers.

2. Describe why all heat exchangers should be piped for counterflow.

3. Estimate the cost of domestic water heating for a typical family of five assuming cold water enters the system at 45 °F and is supplied to the fixtures at 130 °F. Assume the water is heated by an electric water heater in an area where electricity costs $0.12/kWh.

4. Describe two disadvantages of tankless coil water heaters relative to systems with storage tanks.

5. Why do indirect water heaters have the potential to produce hot water much faster than residential electric water heaters?

6. Describe "priority control" as it applies to domestic water heating in a multiload hydronic system.

7. Why is it a good idea to use antifreeze in garage heating circuits, even if the garage thermostat will be constantly kept above freezing?

8. Water enters one side of a flat plate heat exchanger at 159 °F and a flow rate of 10 gpm. It exists that side at 122 °F. Water enters the other side of the heat exchanger at 50 °F and at a flow rate of 12 gpm. Determine the rate of heat transfer across the heat exchanger (assuming no heat loss to its surrounding). Also determine the LMTD of the heat exchanger under these operating conditions.

9. Describe the operation of a boiler capable of producing 200,000 Btu/hr at 200 °F water temperature when it supplies all its output to an indirect water heater with an internal heat exchanger capable of sinking 150,000 Btu/hr when supplied with 200 °F boiler water. What is the limiting factor on DHW production in such a situation? What could be done to correct this situation?

10. What is an advantage of sensing the temperature of a snow melting slab versus sensing only air temperature and the presence of precipitation?

11. Name two benefits of "idling" a snow melting slab.

12. Why is it better to drain melt water from the surface rather than allow the system to evaporate it?

13. Why should asphalt not be placed directly on PEX or PEX-AL-PEX tubing when constructing a SIM system?

14. How do the flow rates used in melted pavement circuits compare to those used in radiant floor heating applications?

15. Why is it recommended that heat be supplied to a melted surface for at least five hours after snow stops falling?

16. When is it permissible to use circuit temperature drops in excess of 25 °F for SIM circuits? Describe the hardware details necessary for such a situation.

17. Determine the volume of a buffer tank for a system in which the heat source generates 100,000 Btu/hr and the smallest load releases heat at 2500 Btu/hr. The heat source is to have a minimum on-time of eight minutes. The buffer tank operating range is from 110 to 170 °F.

18. Describe a situation where it is beneficial to avoid routing flow through a buffer tank in a multiload system.

19. Calculate the minitube flow rate needed to transfer 125,000 Btu/hr to a manifold station that supplies 110 °F water to floor heating circuit that in turn operates with a 20 °F temperature drop. Assume water is available from the primary loop at 180 °F. What size minitube piping is needed for this situation? What flow rate and tube size would be needed to supply the manifold station if the 110 °F water was produced in the mechanical room rather than at the manifold station?

20. What are three advantages of using a single boiler or multiple boiler system to supply *all* the loads in a multiload system versus using a "dedicated" heat source for each load?

21. Describe three ways in which heat metering of apartments is preferred over installing a separate heating system in each apartment.

778 Chapter 14 Auxiliary Loads and Specialized Topics

22. What are the three physical parameters that must be measured in order to calculate the rate of heat transfer by a stream of fluid?

23. Water flows into the heating distribution system of an apartment at 140 °F and 2.5 gpm. The same water exits the system after dropping in temperature by 13 °F. Determine the rate of heat transfer from the water stream to the apartment. Assuming these conditions were maintained for 7.5 minutes, what is the total thermal energy transferred. How would the results change if a solution of 40 percent propylene glycol was operating under these same conditions?

24. What are two advantages of a networked heat meter versus a stand-alone heat meter?

25. Describe the difference between balancing using the preset method compared to balancing using the compensated method.

26. Does a balanced reverse-return system containing zone valves still need a means of differential pressure control? Justify your answer.

Chapter 15

Fundamentals of Hydronic Cooling

Objectives

After studying this chapter, you should be able to:

- Explain how hydronic cooling can be used in residential and light commercial buildings.
- Discuss the benefits and advantage of hydronic cooling.
- Understand the difference between sensible and latent cooling.
- Describe several temperature and humidity indices for air.
- Demonstrate basic skills using a psychrometric chart.
- Describe several types of chilled-water sources suitable for hydronic cooling.
- Describe different types of chilled-water terminal units.
- Discuss central versus zoned cooling systems.
- Describe the difference between 2-pipe and 4-pipe distribution systems.
- Understand the importance of insulating chilled-water piping to avoid condensation.
- Explain how radiant cooling works and what its limitations are.

15.1 Introduction

Most of the information presented in this text deals with hydronic heating. This is consistent with how hydronics technology has traditionally been used in residential and light commercial buildings. The vast majority of hydronic systems are heating only systems.

Although properly designed and installed, hydronic heating systems have many advantages, especially in regards to comfort and energy efficiency, they lack any ability to maintain comfort during hot and humid weather. This inability has undoubtedly lead to many lost sales opportunities for hydronic heating systems. Some potential clients who are initially receptive to the benefits of hydronic heating, instead opt for forced air systems because they provide both heating and cooling, albeit at reduced comfort and efficiency.

In situations where budgets can support it, completely separate systems have been installed for cooling buildings equipped with hydronic heating. These separate cooling systems are often installed by firms other than the one providing hydronic heating. When the controls or hardware for separate heating and cooling systems overlap, such as at room thermostats, issues can arise such as which firm is responsible to ensure compatibility, proper operation, and eventual service. In most cases, it is better for a single mechanical contractor to have sole responsibility for the installation and service of the overall heating and cooling system(s) in a building.

Changing Energy Landscape

From the early 1900s, through the end of the 20th century, fossil fuels such as natural gas, propane, and fuel oil were the dominant energy source for hydronic heating. Millions of boilers are currently operating on these fuels in North America and much of the world.

Like many markets that use energy to provide a benefit, the hydronics market has been and continues to be shaped by public perception, government policy, and economic competition. As the 21st century unfolds, social attitudes and government policies around the world are increasingly focused on climate change, with a prevailing emphasis on reducing **carbon emissions**.

There are widely varying opinions on how **decarbonization** should be dealt with. They range from complete dismissal of any need to act on the subject, to proposals that would radically change how an average person would eat, remain comfortable in their home or workplace, travel, or even use their leisure time.

It is not the intent of this text to endorse any specific view on how carbon reduction could be, or should be dealt with. However, this text was written to equip heating professionals to deal with concepts that allow hydronics technology to evolve to meet future market needs. To that end, *it appears highly likely that the energy used by future hydronic heating and cooling systems will be increasingly supplied through electricity and less by the burning of fossil fuels.*

As discussed in Chapter 3, Hydronic Heat Sources, this trending toward renewably sourced electrical energy will encourage the use of heat pumps, both geothermal water-to-water and air-to-water as heat sources for hydronic systems. These heat pumps come equipped with reversing valves enabling them to operate as chillers (e.g., sources of low temperature water suitable for building cooling). *Nearly all hydronic systems that have a heat pump as their heat source will have the ability to provide chilled-water cooling.* This chapter presents methods for using this ability to provide hydronic-based cooling in residential and light commercial buildings.

15.2 Benefits of Chilled-Water Cooling

There are several benefits to hydronic-based chilled-water cooling:

- Minimally invasive installation;
- Reduced electrical energy usage;
- Available sources of chilled-water;
- Zoning possibilities;
- Eliminating coil frosting;
- Variety of delivery methods;
- Low refrigerant volume;
- Possibility of using rejected heat;
- Adaptability to thermal storage.

Each of these will be discussed in further detail.

Minimally Invasive Installation

Chapter 4, Properties of Water, described how water can absorb almost 3,500 times more heat than the same volume of air. This has profound implications regarding the size of the piping required to convey chilled-water through a building, compared to the size of ducting required to move an equivalent cooling effect using air.

For example, a 3/4-inch tube carrying chilled-water at a flow rate of 6 gallons per minute through a hydronic cooling circuit that absorbs enough heat to warm the water by 15 °F is conveying 45,000 Btu/h. To absorb heat at this rate using forced air would require a duct with a cross-section area of 240 square inches (based on a typical 1,000 feet per minute air velocity in a trunk duct, and a 30 °F temperature rise in the air stream). Figure 15-1 shows two possible duct configuration juxtaposed with a 3/4-inch tube carrying chilled-water. Either of these ducts would be very difficult to conceal within the normal framing cavities of residential or light-commercial building.

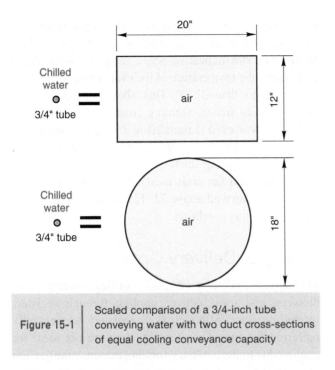

Figure 15-1 Scaled comparison of a 3/4-inch tube conveying water with two duct cross-sections of equal cooling conveyance capacity

Reduced Electrical Energy Use

Chilled-water distribution systems can use significantly less electrical power compared to an equivalent forced-air distribution system. The concept of **distribution efficiency**, first introduced in Chapter 1, Fundamental Concepts, can be used to demonstrate this superiority. Recall that distribution efficiency is defined as follows:

Equation 15.1:

$$\text{distribution efficiency} = \frac{\text{rate of thermal energy delivery (Btu/h)}}{\text{electrical power supplied to distribution system (watts)}}$$

Example 15.1

Published data for the blower in a geothermal water-to-air heat pump indicates that a 3/4-horsepower motor is required to deliver approximately 1,500 CFM airflow. The estimated electrical input power to this motor when operating at full capacity is 690 watts. The rated total cooling capacity of this heat pump is 53,000 Btu/h (based on 60 °F entering source water temperature). Compare the distribution efficiency of this forced air distribution system to that of an equivalent chilled-water distribution system.

Solution:

The heat pump's forced air delivery system has a distribution efficiency of:

$$\text{distribution efficiency} = \frac{53,000 \text{ Btu/h}}{690 \text{ watts}} = 76.8 \frac{\text{Btu/h}}{\text{watt}}$$

An equivalent chilled-water cooling circuit will have assumed equivalent length of 200 feet of 1-inch copper tubing. It will operate with a 15 °F chilled-water temperature rise (45 °F to 60 °F. It uses a standard wet rotor circulator with a PSC motor operating at an assumed 22% wire-to-water efficiency. The water flow rate required for delivering 53,000 Btu/h under these conditions can be calculated using the sensible heat rate equation from Chapter 4, Properties of Water:

$$f = \frac{q}{500(\Delta T)} = \frac{53,000}{500(15)} = 7.0 \text{ gpm}$$

The associated pressure drop of the assumed circuit operating at this flow rate is 5.2 psi. The input power supplied to the circulator under these conditions can be estimated using Equation 15.2:

Equation 15.2:

$$w = \frac{0.4344 f (\Delta P)}{e}$$

where,
w = electrical input power to circulator (watts);
f = flow rate through circulator (gpm);
ΔP = pressure increase across circulator (psi);
e = wire-to-water efficiency of circulator (decimal %).

For the assumed chilled-water distribution system, the estimated electrical input to the circulator is defined as follows:

$$w = \frac{0.4344 f (\Delta P)}{e} = \frac{0.4344(7)(5.2)}{0.22} = 71.9 \text{ watt}$$

The distribution efficiency of the chilled-water distribution system can now be calculated using Equation 15.1.

$$\text{distribution efficiency} = \frac{53,000 \text{ Btu/h}}{71.9 \text{ watts}} = 737 \frac{\text{Btu/h}}{\text{watt}}$$

> **Discussion:**
>
> Based on the calculated distribution efficiencies, the chilled-water circuit is delivering cooling using about 10% of the electrical energy of the forced air system (e.g., (76.8/737) = 0.104).
>
> There may or may not be a blower in the terminal unit used in the chilled-water system. If a blower is present, its power input must be added to that of the circulator when calculating distribution efficiency.
>
> It's important to realize that *all* electrical input to a cooling distribution system ultimately becomes a heat gain to the building. When energy efficiency is a primary design goal, it is imperative to minimize the electrical input power required to operate the cooling distribution system. Hydronic delivery options can provide significantly lower electrical energy requirements compared to thermally equivalent "all air" systems.

Available Sources of Chilled-water

A wide variety of currently available devices can be used to created chilled-water for cooling buildings. These include non-reversible **dedicated chillers**, as well as reversible heat pumps. The latter category includes **water-to-water heat pumps** and **air-to-water heat pumps**, both of which were discussed in Chapter 3, Hydronic Heat Sources. In some circumstances, it is also possible to use cold water from a lake to directly cool a building. These options will be discussed in more detail later in this chapter.

Zoning Possibilities

The same versatility that allows for easy zoning in hydronic heating systems can be applied to hydronic cooling systems. Zoned systems can be designed around electrically operated zone valves in combination with **variable-speed pressure-regulated circulators**. They can also be designed around individual zone circulators.

Designers should always verify the lowest-rated operating temperature of circulators that are being considered for use in chilled-water cooling systems. Circulators that are not compatible with a minimum water temperature of 40 °F should not be used.

Eliminating Coil Frosting

Most of the air handlers and fan coils used in traditional refrigeration-based cooling systems have **direct expansion ("DX") evaporator coils**. Liquid refrigerant flows into these coils and evaporates as it absorbs heat from the air stream passing across the coil. Under certain conditions, the temperature of the evaporating refrigerant can be lower than 32 °F. This allows moisture in the passing air to freeze, forming frost on the coil. Frost buildup is worsened if the airflow rate across the coil is lower than the rated airflow. The latter is often the result of improperly sizing ducting that restricts airflow. The air handlers and fan coils used in chilled-water cooling systems operate well above 32 °F, and thus eliminate this potential for this condition.

Variety of Delivery Options

Traditional ducted forced-air cooling systems use blowers, and most deliver cooling through registers located on floor. Although floor registers are generally preferred for forced air heating, they are not ideal for cooling due to the tendency for temperature stratification in occupied spaces (e.g., warmer air near ceiling level and cooler air near floor).

When centralized chilled-water cooling is used, the air handler is usually connected to ducting, leading to registers located high on walls or in the ceiling. This helps break up temperature stratification and improve comfort.

Other types of chilled-water terminal units, such as radiant panels, can deliver much of the cooling load with reliance on fans or blowers. This significantly reduces operating cost relative to "all-air" systems. Radiant panel cooling is discussed later in this chapter.

Low Refrigerant Volume

Hydronic cooling systems typically contain far less refrigerant compared to **direct expansion (DX)** or **variable refrigerant flow (VRF)** cooling systems. This is important for several reasons.

First, even a modest size VRF system can contain hundreds of feet of copper tubing to distribute refrigerant throughout a building. Figure 15-2 shows an example of this tubing, as well the many brazed and compression joints connecting the tubing and other components.

A leak in a VRF system could mean the loss of many pounds of refrigerant. Not only is this expensive, it is also dangerous, potentially even fatal, if it should occur in confined and occupied spaces. Chilled-water cooling systems do not circulate refrigerant throughout buildings, and thus eliminate this potential issue.

Second, the type of refrigerants used in current generation VRF systems may not be the same as those

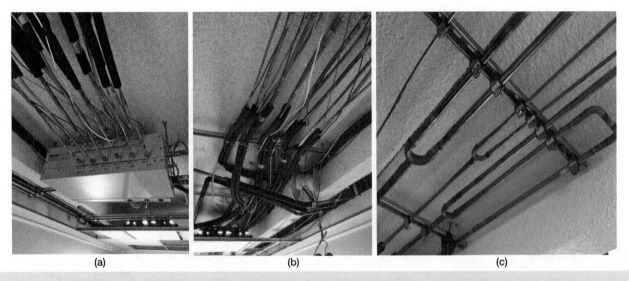

Figure 15-2 | Copper tubing used in a portion of a commercial VRF system. *Courtesy of John Siegenthaler.*

allowed by future regulations. As of this writing, the refrigerant market is rapidly evolving to keep pace with directives aimed at reducing global warming. There is no guarantee that currently installed VRF systems will be compatible with future refrigerants or the oils used with these refrigerants. Incompatibility could require a major changeout in equipment. In contrast, hydronic cooling systems remain compatible with virtually any device or method of supplying chilled-water at suitable temperatures.

Third, most VRF systems require blowers within each **terminal units** to which refrigerant is piped. Hydronic cooling is adaptable to low power demand delivery systems such as radiant panels.

Fourth, most VRF systems use **proprietary controls** that must be commissioned and serviced by trained technicians. This limits the building owner's choices on maintenance as well as parts availability and cost. Hydronic systems typically use readily available components that can be sourced from multiple suppliers.

Finally, VRF systems use more electrical power to deliver refrigerant through tubing compared to the power required to deliver chilled-water. Figure 15-3 compares the electrical power used to distribute thermal energy, *as a percentage of compressor power*, for chilled-water hydronic systems and VRF systems.

Possible Use of Rejected Heat

The heat rejected from a chiller or heat pump producing chilled-water may be of use for simultaneous heating loads within the building. One example is pre-heating (or fully heating) domestic water using the heat

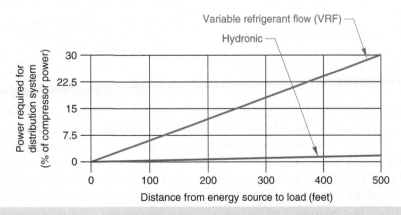

Figure 15-3 | Comparison of the electrical energy needed to convey thermal energy using a VRF system versus a hydronic system, and expressed as a percentage of compressor power.

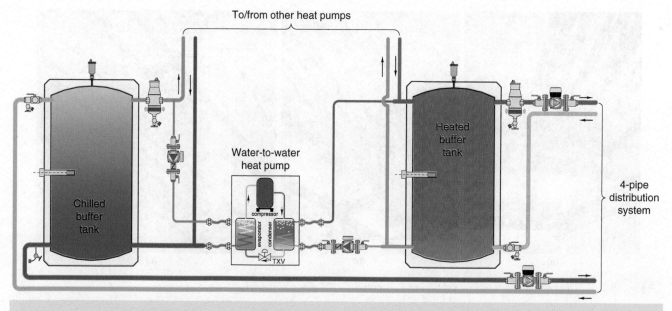

Figure 15-4 Using a water-to-water heat pump between a hot and cold buffer tank. The heated water and chilled-water are both simultaneously useful to the building's heating and cooling system.

rejected from a water-to-water heat pump supplying chilled-water for cooling. Another is using the heat rejected from a chiller or heat pump to warm a swimming pool. In buildings where there are a simultaneous heating and cooling loads, it is possible to employ a water-to-water heat pump for heat recovery as shown in Figure 15-4.

When both the chilled-water and **heat of rejection** from the heat pump are useful, the effective coefficient of performance (COP_e) of the heat pump is given by Equation 15.3.

Equation 15.3:

$$COP_e = (2 \times COP_h - 1)$$

where,
COP_e = effective "net" COP of the heat pump, considering total energy flows
COP_h = coefficient of performance of heat pump as measured or calculated.

Thus, a heat pump applied in this manner, and operating with a measured heating COP of 4.0, would have an effective COP_e of $(2 \times 4) - 1 = 7$. The useful thermal output, considering both heated and chilled-water, is 7 times greater than the electrical power required to operate the heat pump. From the standpoint of thermal performance this is one of the best possible applications for a heat pump.

Adaptability to Thermal Storage

As described in earlier chapters, water is one of the best material on earth for storing thermal energy. Some chilled-water cooling systems can be designed to exploit this ability. For example, in utility districts that offer time-of-use electric rates, it may be possible to operate a chiller or heat pump during off-peak periods—typically at night—to draw down the temperature of a well-insulated storage tank filled with water. That water could then be used to cool a building during the subsequent on-peak period. This could greatly reduce operating cost, especially in commercial buildings that are charged for both energy use and peak power demand.

15.3 Thermodynamic Concepts Associated with Cooling Systems

All cooling systems are designed to condition interior air so that human thermal comfort is achieved and maintained. That conditioning involves changing the temperature of the air and altering its moisture content. Most cooling systems are designed to establish and maintain the temperature and **relative humidity** of interior air at some set combination of conditions deemed to provide human comfort. A typical combination would be an air temperature of 75 °F and a relative humidity of 50%.

Temperature & Humidity Indices:

To appreciate how any cooling system affects the temperature and humidity of interior air, it's important to understand several specific physical quantities that are used to describe the thermodynamic condition of air. Those quantities include:

- Dry bulb temperature;
- Wet bulb temperature;
- Dewpoint temperature;
- Relative humidity;
- Absolute humidity (aka humidity ratio).

Dry Bulb Temperature

One of the simplest indexes that describes the condition of air is called **dry bulb temperature**. It's simply the temperature of air read from a thermometer after it has stabilized within an interior or exterior space. The word "bulb" makes reference to mercury bulb thermometers that were extensively used for temperature measurements prior to development of electrically based temperature measuring instruments.

Wet Bulb Temperature

Anyone who has stood in a breeze while wearing wet clothes can attest to the strong cooling effect created as the breeze evaporates water from their clothes. This same effect (e.g., cooling caused by evaporating water) can be used to determine the moisture content of air. That moisture content is related to a specific thermodynamic index called **wet bulb temperature**, which is the temperature below which the air being sampled cannot hold any additional water vapor.

The classic device for determining the wet bulb temperature of air, as well as the air's dry bulb temperature, is a **sling psychrometer**, an example of which is shown in Figure 15-5.

This instrument has two mercury bulb thermometers, one of which has a small cotton "sock" over it bulb. These thermometers are mounted to a plate that is attached to a handle. A shaft inside the handle allows the plate holding the thermometers to be spun around, similar how a cowboy would swing a lasso.

The first step in using a sling psychrometer is to saturate the small cotton sock with water. The person using the instrument holds the handle and swings the plate holding the thermometers in a circular path for about 30 seconds. This causes the water on the sock to evaporate until it reaches equilibrium with the moisture content of the surrounding air. Both thermometers are then immediately read. The one with the sock indicates the air's current wet bulb temperature. The other thermometer gives the air's dry bulb temperature. These temperatures can be used along with a **psychrometric chart** to determine several indexes related to the moisture content of the air. Psychrometric charts are described later in this chapter. Although slight psychrometers can still be used as described above, there are now electronic instruments that can measure and report dry bulb and wet bulb temperatures, as well as relative humidity, with less preparation and operating requirements.

Dewpoint Temperature

The ability of air to contain water vapor depends on the air's temperature. The warmer the air, the greater its ability to "hold" water vapor. If a given sample of air is progressively cooled, it eventually reaches a **saturation condition** at which it can no longer retain the moisture in vapor form. The temperature at which this occurs

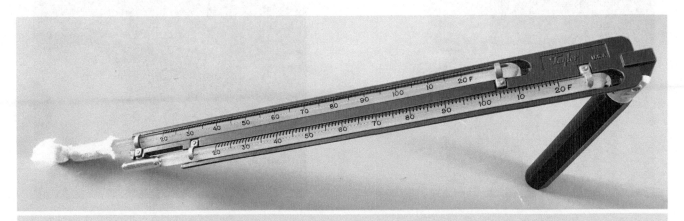

Figure 15-5 | Example of a sling psychrometer for simultaneously measuring the dry bulb and wet bulb temperature of air.
Courtesy of John Siegenthaler.

is called the **dewpoint** of the air. If the air is cooled to temperatures below the dewpoint, the water vapor it contains will continue to condense into liquid. This liquid water is called **condensate**.

The formation of **condensation** in cooling systems is very desirable when it occurs *intentionally*, such as on the surfaces of a chilled-water coil within an air handler. The **dehumidification** of air as it passes through the air handler's coil is an intended action, which improves thermal comfort. However, *unintentional* condensation on surfaces of piping or other components in a chilled-water cooling system must be avoided. For example, if air contacts a surface that cools it to (or below) its dewpoint temperature, water vapor will condense into liquid water on that surface. The circulator shown in Figure 15-6 is an example of unintentional condensation.

This circulator is part of a chilled-water cooling system. It is conveying water at a temperature that is well below the dewpoint of the surrounding air. Notice that the circulator's volute, as well as both isolation flanges are covered with liquid water droplets formed as moisture from the surrounding air condensed. The circulator's motor can, which dissipates heat due to electrical resistance and mechanical friction inside the motor, remains warm enough to stay above the dewpoint temperature, and thus remains dry. Although this condensate doesn't impair the circulator's performance, it will quickly lead to surface oxidation as seen in Figure 15-7.

As it continues to form, condensate eventually drips from oxidized surfaces and stains any materials it falls on. Repeated formation of condensation also encourage mold growth, and can eventually destroys materials such as wood and drywall. To maintain the aesthetic quality of the installation, it is imperative to insulate circulator volutes, as well as all other components that come in contact with chilled-water.

Figure 15-8 shows an example of a circulator that can be ordered with a **form-fitting insulation shell**.

Figure 15-6 | Condensation forming on the volute and flanges of an uninsulated circulator conveying chilled-water. *Courtesy of John Siegenthaler.*

Figure 15-7 | Surface oxidation of the cast iron volute of an uninsulated circulator conveying chilled-water. *Courtesy of John Siegenthaler.*

15.3 Thermodynamic Concepts Associated with Cooling Systems

(a)

(b)

Figure 15-8 | A form-fitting insulation shell for a small circulator. *Courtesy of John Siegenthaler.*

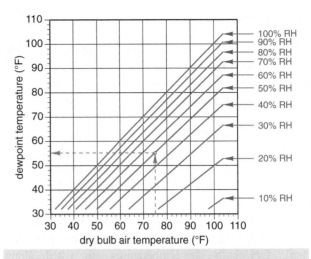

Figure 15-9 | Graph for determining the dewpoint temperature of air based on its dry bulb temperature and relative humidity.

This two-piece shell snaps together over the circulator's volute. After it is placed, all seams should be carefully sealed with silicone caulk to ensure that moisture-laden air cannot contact the volute. The adjacent isolation flanges and piping should also be insulated, and that insulation should be sealed against the circulator's insulation shell.

The dewpoint temperature of air is generally considered the same as the air's current wet bulb temperature, and can be read from a psychrometric chart. It can also be read from Figure 15-9.

Absolute Humidity

There are several indices used to represent the amount of water vapor contained in air. One that is particularly useful for cooling system design is called **absolute humidity**. It is commonly expressed as the pounds of water contained in one pound of dry air. Perfectly dry air would have an absolute humidity of 0. However, perfectly dry air is not naturally occurring. Air at 70 °F and 50% relative humidity has an absolute humidity of 0.0078 lb/lb indicating that one pound of air at this condition contains 0.0078 pounds of water, and 0.9922 pounds of the gases that make up air.

Absolute humidity can also be stated in *grains* of water per pound of dry air. There are 7,000 grains in one pound. Thus, air at 70 °F and 50% relative humidity also has an absolute humidity of approximately 55 grains per pound of dry air. The absolute humidity of air can be read from the vertical right side axis of a psychrometric chart. Absolute humidity is also sometimes called **specific humidity**.

Relative Humidity

When people discussing the weather refer to "humidity," they are usually referring to *relative* humidity. This is the ratio of the actual water mass in a given mass of air divided by the maximum mass of water the air could hold at its current temperature. For example, air at 60% relative humidity only contains 60% of the water vapor that it is capable of holding at that temperature. A relative humidity of 100% means that the air is saturated, and holding the maximum amount of water possible given the air's temperature. Air reaches 100% relative humidity at the dewpoint temperature. Relative humidity is plotted as a series of curves on a psychrometric chart.

Psychrometric Chart

Willis Carrier is considered by most historians to be "the father of air conditioning." In 1905, he developed the first psychrometric chart that quantified the properties of air. Although the psychrometric chart has been refined since its inception, modern versions still bear close resemblance to the original version developed over a century ago. Several versions of modern psychrometric charts are available online. Figure 15-10 shows one example.

Figure 15-11 shows a limited approximation of a psychrometric chart.

The previously discussed thermodynamic properties of air are all represented on a modern psychrometric chart. Dry bulb air temperatures are listed along the chart's bottom horizontal axis (shown in red in Figure 15-11). Wet bulb temperatures are listed along the curved upper left side of the chart (shown in blue in Figure 15-11). The relative humidity of air is indicated by the curves that span from lower left to upper right (shown in green in Figure 15-11). The absolute humidity (lb H_2O/lb dry air) is listed along the right side vertical axis of the chart. Lines that are perpendicular to the bottom and right-side axis allow dry bulb temperature and absolute humidity values to be projected into the field of the chart. Sloping lines are also drawn from the wet bulb temperatures indicated on the chart's upper left curved edge.

When any two of the thermodynamic properties plotted on this chart are known, the remaining properties can be determined. For example, if the dry bulb and wet bulb temperature of the air are known, the psychrometric chart can be used to find relative humidity and absolute humidity.

Some psychrometric charts also plot the density of air, the enthalpy of the air when fully saturated with moisture, and other information.

Cooling Loads

The starting point for designing any chilled-water cooling system is to establish the building's **cooling load**. Those loads involve several influencing factors, each of which can change with time. These include:

- Building occupancy;
- Solar heat gain through windows and opaque exterior surfaces of buildings;
- Operation of interior equipment (computers, office machines, production equipment, cooking, etc.);
- Lighting schedules;
- Ventilation requirements during occupied and unoccupied periods;
- Building thermal mass;
- Combined temperature and humidity (e.g., **enthalpy**) of outside air.

Some of these influencing factors do not add moisture to interior spaces. Examples include:

- Heat transfer from outside to inside the building by conduction through exposed building surfaces;
- Heat gain from sunlight through windows;
- Heat gain from interior lighting;
- Heat gain from electrical equipment operating within the conditioned space;
- A *portion* of the metabolic heat output from occupants.

These factors are grouped together into an overall term called the **sensible cooling load**. The word "sensible" implies that these factors can be "sensed" (e.g., detected), by changes in temperature of indoor air.

Other factors that influence cooling load add moisture to indoor air. Examples include:

- Respiration/perspiration from occupants;
- Cooking;
- Showering or bathing;
- Moisture emitted from interior plants;
- Laundry, or drying clothes inside the building;
- Washing interior objects or surfaces;
- Infiltration of outside air (unintentional);
- Incoming ventilation air (intentional).

These factors contribute to the building's **latent cooling load**. In this context, the word "latent" implies "hidden." These factors do not change the temperature of interior air, but can have a profound impact of human thermal comfort as well as the **total cooling load** of the building.

The building's total cooling load, at any time, is the sum of the sensible cooling load and latent cooling load.

Total cooling load is almost never equal to the sum of the *maximum* expected rate of heat gain (sensible or latent) due to each of the previous stated influencing factors.

For example, the thermal mass of buildings having large exposed interior masonry walls can absorb some of the internal heat gains, especially if that mass has been cooled prior to the occurrence of those heat gains. This causes a time delay between the occurrence of the internal gain and when that energy becomes part of the instantaneous cooling load.

The diversity of the factors that influence total cooling loads has been extensively studied. This has led to development of detailed methods for determining hourly cooling loads. The often-cited reference for methods of calculating cooling loads in commercial buildings is chapter 18 in

15.3 Thermodynamic Concepts Associated with Cooling Systems

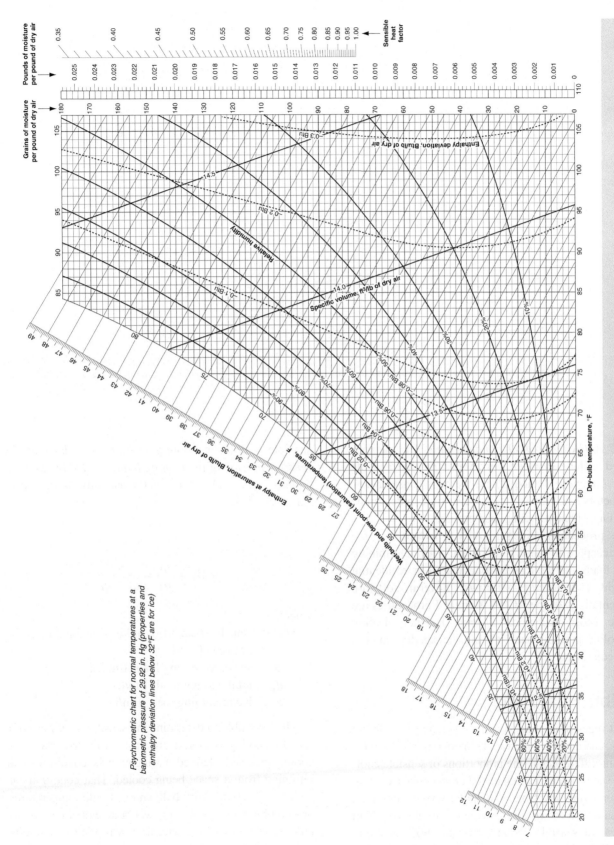

Figure 15-10 | A psychrometric chart showing several thermodynamic properties of air. Courtesy of RSESs.

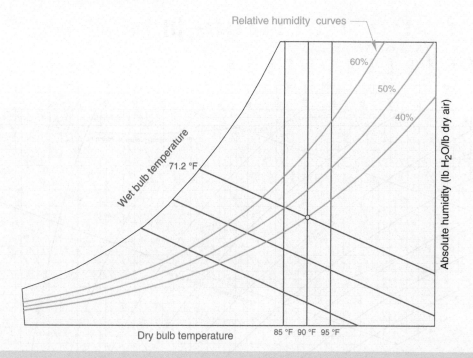

Figure 15-11 A portion of a psychrometric chart showing lines for dry bulb temperature, wet bulb temperature and curves for relative humidity.

the 2017 *ASHRAE Book of Fundamentals*. The methods presented include *Radiant Time Series* method and the *Transfer Function* Method. These methods consider how all the factors listed above interact over time to determine the building's hourly cooling load profile.

The procedures given in **ACCA Manual J** are widely accepted in the HVAC industry for determining cooling loads (as well as heating loads) in residential and light commercial buildings. Several software implementations of the ACCA Manual J procedures are available. Most mechanical systems designers dealing with cooling systems use software to expedite cooling load calculations.

Sensible Heat Ratio

Different types of cooling systems, operating in different locations, and under a wide variety of conditions, will have significant differences in the proportions of sensible cooling load versus latent cooling load. For example, a cooling system removing heat from a computer server room in an arid climate such as Santa Fe, NM, operates at a very high percentage of sensible cooling, probably 90% or more. A gym filled with exercising people on a hot day in New Orleans, LA, will have a much different cooling load, perhaps 60% sensible cooling and 40% latent cooling.

An index called **sensible heat ratio** has been developed to describe these different proportions of sensible and latent cooling load. *It is defined at the sensible portion of the cooling load, divided by the total cooling load*, and can be mathematically expressed as Equation 15.4.

Equation 15.4:

$$\text{SHR} = \frac{q_s}{q_T} = \frac{q_s}{q_s + q_L}$$

where,
SHR = sensible heat ratio (unitless decimal number between 0 and 1);
q_s = sensible cooling load (Btu/h);
q_T = total cooling load (Btu/h);
q_L = latent cooling load (Btu/h).

It is possible to determine the sensible heat ratio of a specific cooling process using a psychrometric chart. The cooling process is defined starting with the condition of air returning from a space being cooled. That condition can be a combination of dry bulb and wet bulb temperatures, or a combination of either the wet bulb temperature or dry bulb temperature along with a known relative humidity. Either of these combinations can be used to establish a point on a psychrometric chart. Another point on the chart is established based on the *desired* condition of the air *supplied* to the space being cooled. After these two points are plotted on the psychrometric chart, a line is drawn between them. That line is called the **process line**. Ftttigure 15-12 shows an example.

The sensible heat ratio is found by constructing another line, parallel to the process line, and passing through a reference point on the chart. The reference point is defined as (80 °F dry bulb temperature and 50% relative humidity). This reference point is usually identified, to some extent, on most psychrometric charts. The parallel line is extended from the reference point to the upper right portion of the chart, and passes through a scale labeled sensible heat ratio or sometimes as sensible heat factor.

Knowing the sensible heat ratio helps determine the type of cooling equipment used. Cooling scenarios with high sensible heat ratios are well suited to cooling using thinner cooling coils or radiant panels. Cooling scenarios with low SHR will require significantly more dehumidification and favor thicker, multi-row cooling coils in air handlers.

15.4 Chilled-Water Sources

Several devices are now available that can produce chilled-water at rates suitable for cooling residential and light commercial buildings. They include:

- Geothermal water-to-water heat pumps;
- Air-to-water heat pumps;
- Heat recovery chillers;
- DX condensers coupled to flat-plate heat exchangers.

Although there are distinct differences between these devices, they all operate on electricity, and they all use the basic refrigeration cycle to absorb heat from water, or mixture of water and antifreeze, and dissipate that heat to the outside environment (e.g., either outside air or the earth). The term **chiller** is often used to informally describe any of these devices.

When a building is located near a large lake, it may also be possible to use that lake as a source of chilled-water for cooling. Such applications are obviously limited by the distance between the building and the lake. They may also be limited or even prohibited by federal, state, or local laws and regulations. Any initial evaluation into the possibility of lake water cooling must start with a thorough review of such regulations and any required permitting.

Heat Pumps as Chillers

Geothermal water-to-water heat pumps, as well as air-to-water heat pumps were discussed in Chapter 3, Hydronic Heat Sources. These heat pumps use revers-

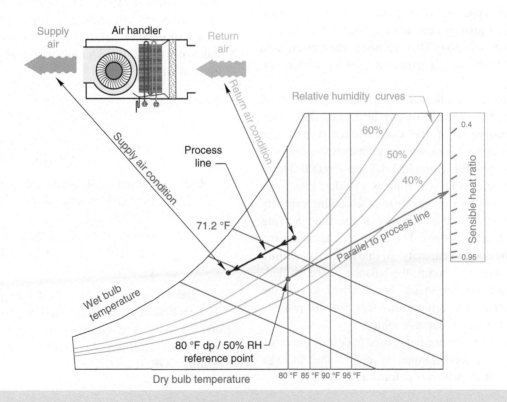

Figure 15-12 | Using a psychrometric chart to determine sensible heat ratio.

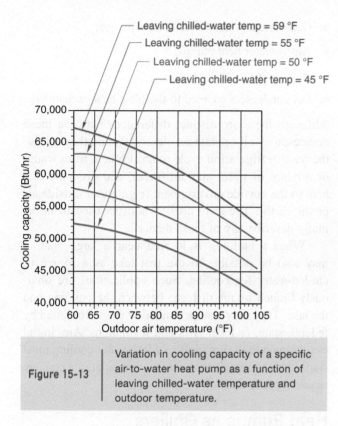

Figure 15-13 Variation in cooling capacity of a specific air-to-water heat pump as a function of leaving chilled-water temperature and outdoor temperature.

ing valves to determine which of the two refrigerant heat exchangers they contain serves as the evaporator for the refrigerant cycle. When operated in cooling mode, both type of heat pumps use evaporating refrigerant to absorb heat from a stream of water or an antifreeze solution. That stream is then circulated through a hydronic distribution system within the building.

When used as chillers it is important to understand that the cooling capacity of any heat pump decreases as the temperature of the chilled-water it produces decreases. Figure 15-13 shows representative **cooling capacity** graphs for a nominal 4.5 ton (54,000 Btu/h) air-to-water heat pump operating as a chiller.

Each of the four curves shows the cooling capacity of the air-to-water heat pump at a specified leaving chilled-water temperature. The lower that temperature, the lower the cooling capacity. To maximize the cooling capacity of the heat pump, the balance of the chilled-water cooling system should be designed to use the highest chilled-water temperature that can still provide adequate sensible and latent cooling.

Another performance index used to describe cooling performance of heat pumps is called the **Energy Efficiency Ratio**, which is defined as follows.

Equation 15.5:

$$\text{EER} = \frac{q_c}{W_e} = \frac{\text{cooling capacity (Btu/h)}}{\text{electrical input power (watts)}}$$

EER = Energy Efficiency Ratio of the heat pump (Btu/h/watt);
q_c = cooling capacity of the heat pump (Btu/h);
w_e = electrical input wattage to the heat pump (watts).

The higher the EER of a heat pump, the lower the electrical power required to provide a given rate of cooling.

Like cooling capacity, the EER of a heat pump is an *instantaneous value* for a given set of operating conditions. If any of those conditions change, so does the EER. Figure 15-14 shows the EER of the same air-to-water heat pump for which cooling capacity is plotted in Figure 15-13.

Notice that the EER follows a similar trend to cooling capacity. For an air-to-water heat pump, the EER decreases with the leaving water temperature. It also decreases as the outdoor temperature to which heat is being dissipated increases. This again points to the need to design chilled-water cooling systems for the highest chilled-water temperature that provides adequate sensible and latent cooling.

Another cooling performance index for heat pumps that is commonly used in Europe, but not as commonly used in North America is called the **cooling COP**. This performance index is calculated by converting the units of watts in the denominator of Equation 15.5 into units of Btu/h.

Equation 15.6:

$$\text{COP}_c = \frac{q_c}{w_e \times 3.413} = \frac{\text{EER}}{3.413}$$

COP_c = cooling COP of the heat pump;
EER = Energy Efficiency Ratio of the heat pump (Btu/h/watt);
q_c = cooling capacity of the heat pump (Btu/h);
w_e = electrical input wattage to the heat pump (watts).

The cooling COP of a heat pump is the ratio of its beneficial output (e.g., Btu/h of cooling capacity), divided by the necessary electrical input power converted from watts to Btu/h. As such, it can be compared to heating COP.

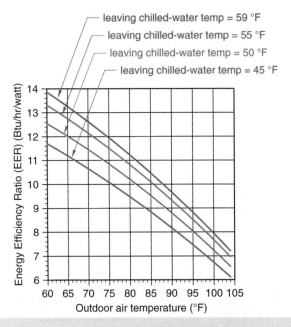

Figure 15-14 | Variation in Energy Efficiency Ratio (EER) of a specific air-to-water heat pump as a function of leaving chilled-water temperature and outdoor temperature.

Many air-to-air central cooling systems, as well as air-to-air heat pumps have a performance rating called **Seasonal Energy Efficiency Ratio (SEER)**. This performance index is established based on a standard testing procedure and processing of the test data. It is meant to give an estimate of the seasonal average value of EER over a specific simulated cooling season. It is also meant to be used as a way of comparing the cooling performance of different air conditioning systems or air-to-air heat pumps. The higher the SEER of a given unit, the lower its expected operating cost. Many energy codes now set minimum values of SEER for new equipment installations.

SEER is not used to describe the performance of water-to-water or air-to-water heat pumps. The reason is that the testing and calculation procedure used to establish SEER does not reflect the conditions that either of these heat pumps could be operating with, and how those conditions could vary over the cooling season.

Multiple Chiller Systems

When cooling loads are greater than the cooling capacity of a single heat pump, multiple heat pumps can be used. The piping concepts are the same as used for multiple heat sources.

Figure 15-15 shows how three air-to-water heat pumps could be combined into a 3-stage chilled-water plant.

Each heat pump is operated as a stage of cooling capacity. A staging controller, similar to one that would be used for a multiple boiler system, is used to turn each heat pump on and off as necessary to maintain a target chilled-water supply temperature to the distribution system. As the temperature of the chilled-water supplied to the distribution system rises, additional heat pumps are turned on, and vice versa. When a heat pump is turned on so is its associated high Cv zone valve or motorized ball valve. A variable-speed pressure-regulated circulator automatically adjusts flow depending on the number of active heat pumps. The closely spaced tees isolate the pressure dynamics of this circulator from the circulator providing flow in the distribution system. Eliminate last sentence.

Because monobloc heat pumps are used, the entire system operates with an antifreeze solution. A fluid feeder maintains an appropriate pressure in the system, adding small amounts of pre-mixed antifreeze solution to make up for pressure drops due to air venting. Similar piping configurations could be used to connect multiple water-to-water heat pumps.

Heat Recovery Chillers

Another device that is similar to an air-to-water heat pumps, but with extended capabilities is called a **heat recovery chiller**. These units have an additional refrigerant-to-water heat exchanger that can transfer some or all of the heat of rejection produced during cooling mode operation to a stream of water or antifreeze solution instead of dissipating all that heat to outside air. The heated water or antifreeze solutions can then be used for loads such as space heating, domestic water heating, pool heating, or other process heating. Any heat not transferred to this warm fluid stream is dissipated to outside air by a condenser coil and fan. Figure 15-16 shows an example of a 5-ton rated heat recovery chiller. One set of pipes connects to the chilled-water distribution system, the other set connects to the heated water distribution system.

Heat recovery chillers are ideal in applications where *simultaneous heating and cooling* are required, or possible. Potential applications include:

- Cooling a building while simultaneously heating a swimming pool;
- Cooling the core area of a building while simultaneously heating perimeter areas;

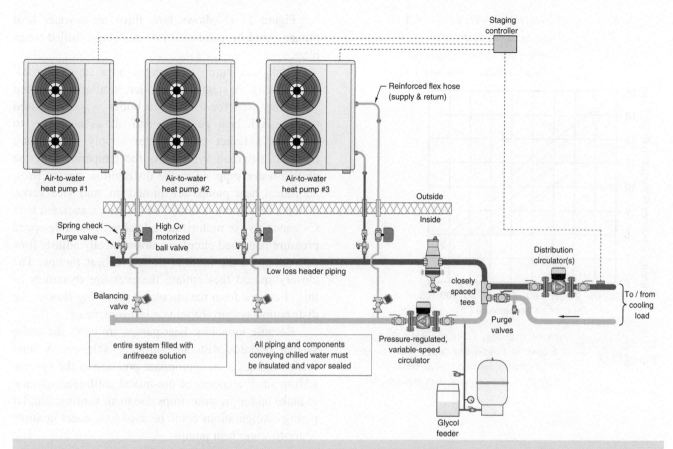

Figure 15-15 | Multiple air to water heat pumps piped for staged heating and cooling capacity. Flow through heat pump provided by a variable-speed pressure-regulated circulator.

Figure 15-16 | A heat recovery chiller for providing simultaneous heated fluid stream and chilled fluid stream. *Courtesy of Multi-Aqua, Inc.*

- Cooling a building while providing domestic water heating;
- Cooling a building while providing hot water for an industrial process, such as laundry, food preparation, greenhouse soil warming.

When both the heated stream and chilled stream are being fully utilized, the effective COP of the heat recovery chilled can be calculated using Equation 15.3.

Air-Cooled Condensers With Direct Expansion Heat Exchangers

It's also possible to produce chilled-water by combining an outdoor **condenser unit** with an external brazed plate refrigerant-to-water heat exchanger. The heat exchanger serves as the evaporator for the refrigeration cycle. The condenser would be the same as or very similar to a standard "non-reversible" condenser for central air conditioning. These condensers do not have reversing valves. Thus, systems that use them are "cooling only." Figure 15-17 shows the system concept.

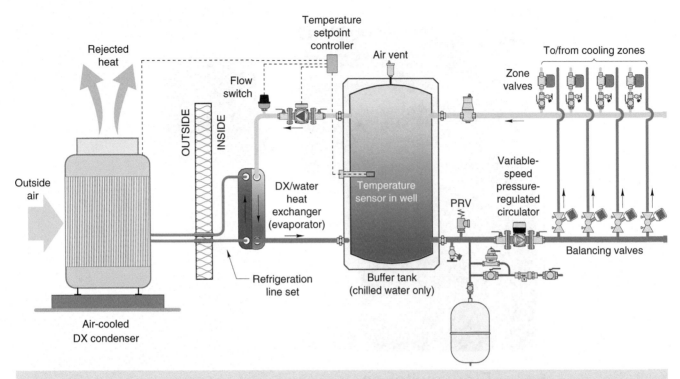

Figure 15-17 | Using an air-cooled condensing unit coupled to a brazed plate heat exchanger to provide chilled-water for a multi-zone cooling system.

The brazed plate heat exchanger must be properly sized based on the refrigerant used, the desired evaporator temperature, the **superheat** setting of the thermal expansion valve in the condenser unit, and the required chilled-water temperature and flow rate. Designers are encouraged to contact manufacturers of flat-plate heat exchanger to assist with a proper selection based on these parameters.

The buffer tank in this system is configured solely for cooling. The coolest water will tend to stratify to the lower portion of this tank. Connecting the supply header for the distribution system to the lower portion of the tank ensures that the coolest available water is sent to the load.

The air separator is installed on the *return* side of the chilled-water distribution system. The slightly higher water temperature at this location will have a slightly lowest solubility for dissolved gases, and thus be the preferred location from which to gather and eject these gases from the system.

The zone valves are also shown on the *return* side of the zone circuits. The slightly warmer water temperature at this end of the circuits will reduce the possibility of condensation on the zone valves. However, it is still imperative to properly insulate the valve bodies, but not the actuators, of these zone valves.

This system also includes a flow switch that verifies water flow through the heat exchanger prior to allowing the compressor in the condenser unit from starting. This is a safeguard against possibly of freezing the water in the heat exchanger should adequate water flow not be present. Figure 15-18 shows an example of two flow switches installed in a chilled-water cooling system supplied by two independent air-cooled condensers.

Lake Source Cooling

Anyone who has gone swimming in a lake in the northern half of the United States understands that the water temperature gets significantly cooler just a few feet below the surface, even on a hot summer day. Measurements have shown that the water temperature at depths of approximately 40 feet or more below the surface of such lakes experience very little temperature variation on an annual basis. This is illustrated in Figure 15-19.

Water attains its maximum density at a temperature of 39.2 °F. In cold climates, water at this temperature will accumulate in the lowest areas of lakes that are at least 40 feet deep. Water at this temperature is very adequate

Figure 15-18 Two flow switches installed in a chilled-water cooling system are used to verify flow through the chiller's evaporator prior to allowing compressor operation. *Courtesy of John Siegenthaler.*

for chilled-water cooling systems if it can be accessed from shore.

Lake source cooling has been successfully applied on a large scale. Examples include the chilled-water cooling system at Cornell University in Ithaca, NY, which draws water from Cayuga Lake. The inlet pipe is supported 10 feet above the lake bottom, and 250 feet below the lake surface. Several buildings in the downtown region of Toronto, Ontario, Canada are also cooled by water drawn for several hundred feet below Lake Ontario.

The feasibility of lake source cooling on a smaller scale should always start with assessment of any codes, rules, or regulations regarding the use of or access to lake water. Different requirements may apply to "navigable waters" versus lakes in which boating is not allowed. If these criteria allow further pursuit of the project, water temperature measurements should be taken at several depths below the lake surface to determine available water temperatures during months when cooling is needed. Designers should also investigate how to transition the required piping from the shore into the lake, and how to protect it against freezing in winter.

Assuming the project reaches the design phase, it can be configured as either a direct or indirect system. **Direct lake water cooling systems** draw water from the lake and routes it through a heat exchanger that's usually located within the building being cooled. **Indirect lake water cooling systems** use a special type of heat exchanger submerged in the lake.

Figure 15-20 shows one plausible concept for a direct lake water cooling system.

This system uses a shallow well pump to provide flow of lake water. That pump may be within the building being cooled or a separate building closer to the lake shore. If water will remain in the system during winter, the pump and piping must be protected against freezing. Shallow well pumps are limited in their ability to lift water above the surface of the lake. If the lift requirement is not more than 20 feet above the lake surface, and the piping to and from the lake is sized for relatively low head loss, a shallow well pump should suffice. The objective is to keep the pump as low as possible relative to the lake surface.

Lake water is drawn into a **foot valve** at the end of a high-density polyethylene (HDPE) pipe. Foot valves use a special type of check valve enclosed within a

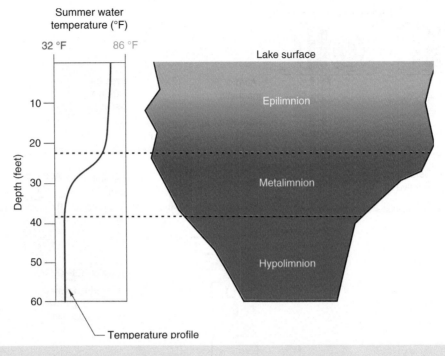

Figure 15-19 | Temperature stratification within a lake. Coolest water settles to bottom of lake.

perforated protection screen. They are used to prevent reverse flow when the pump is off. The end of the pipe with the foot valve is secured to a concrete ballast block using stainless steel hardware. This anchoring is important to prevent the HDPE pipe, which has slightly lower density than cold water, from floating. The ballast block also keeps the foot valve above silt on the lakebed. Depending on the length of pipe required, it may be necessary to use additional concrete ballast blocks to ensure the piping in the lake stays on, or close to, the lakebed. The piping must make the transition to shore at depths that prevent the water it contains from freezing in winter.

The incoming lake water passes into an **automatic backwash filter**. This device monitors the differential pressure across its inlet and outlet ports. When the pressure differential reaches a set limit, the filter initiates an automatic cleaning cycle. In this application, the filter needs a minimum water flow of 30 gpm at 30 psi pressure to provide proper flushing. This flow must come from a separate water source routed through a pilot-operated solenoid valve that opens simultaneously with the waste valve on the filter. Any accumulated sediment is flushed off the internal stainless steel strainer media and blown out through a waste pipe. A filtering criteria of 100 microns is suggested to minimize any accumulation of sediment on the internal surfaces of the heat exchanger. The frequency of filter operation depends on the characteristics of the lake. If the intake pipe is located deep within a lake that has minimal surface water flow, the intake water should remain relatively clean. However, if the intake pipe is closer to the surface in a lake with significant surface water flow, or other weather-related disturbances, the water may be turbid at times, and thus require more backwashing of the filter.

Water passes from the automatic filter to the pump. Ideally, this pump should be constructed with a stainless steel volute to provide maximum corrosion resistance. Full port ball valves should be installed to isolate the pump if maintenance is required. A priming line may also be required to add water to the intake pipe when the system is being commissioned.

From the pump, lake water passes into one side of a stainless steel flat-plate heat exchanger. On small projects, this heat exchanger will likely be a brazed plate configuration. On larger projects, a plate & frame heat exchanger may be needed. In either case, the

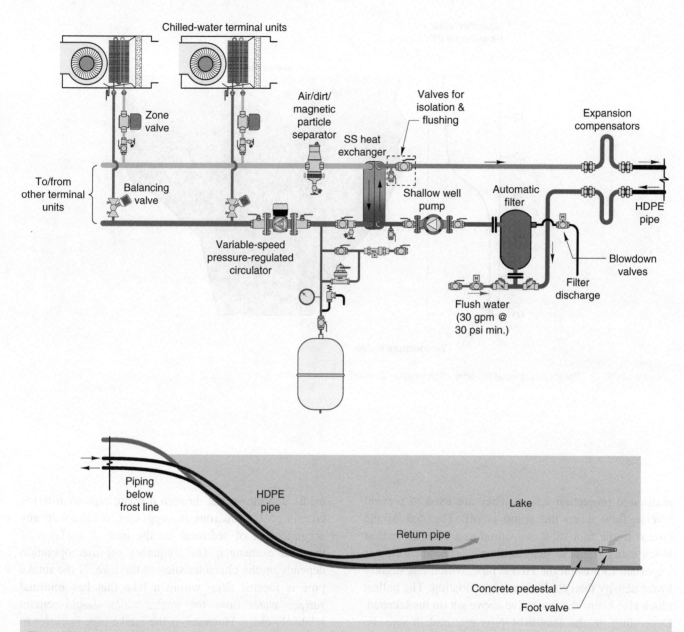

Figure 15-20 | Example of a direct lake water cooling system using an automatic backwash filter for the lake water.

heat exchanger should be sized so that the approach temperature difference between the incoming lake water and the chilled-water leaving the other side of the heat exchanger is not more than 5 °F. The lower the approach temperature difference, the lower the "thermal penalty" associated with having the heat exchanger. Lower temperature gains across the heat exchanger may allow for smaller coils in air handlers, lower flow rates, and lower circulator operating costs. The lake side of the heat exchanger is equipped with valving that allows it to be isolated, flushed, and chemically cleaned if necessary.

After passing through the heat exchanger, the lake water returns to the lake through another HDPE pipe. Like the supply pipe, the return pipe should be kept on or near the lakebed using concrete ballast blocks. It should discharge into the lake several feet away from the foot valve, and preferably in a direction that carries the water away from the foot valve.

The load side of the heat exchangers consists of a 2-pipe chilled-water distribution system. The ECM-based variable-speed circulator operates on differential pressure control, automatically changing speed in response to the opening and closing of the zone valves controlling flow through each air handler.

15.4 Chilled-Water Sources

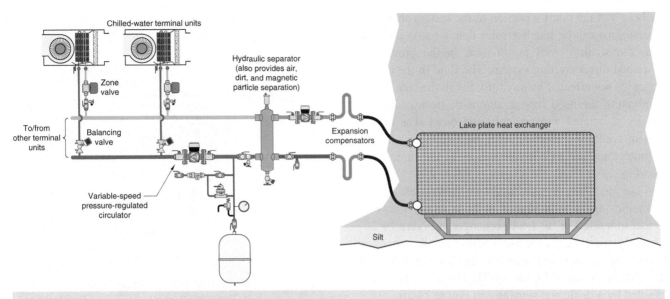

Figure 15-21 | Example of an indirect lake water cooling system using a submerged lake heat exchanger.

The electrical energy used by the lake water pump could be further reduced by using a variable-speed pump that responds to either the temperature differential across the lake side of the heat exchanger or to the temperature of the lake water leaving the heat exchanger. Either criteria would allow the flow rate of lake water through the heat exchanger to be adjusted based on the current cooling load on the other side of the heat exchanger.

Designers should also keep in mind that the lake water circuit might also be used to supply low-temperature heat to a water-source heat pump for space heating or domestic water heating.

Figure 15-21 shows the concept for an indirect lake water cooling system. This system used a special heat exchanger that's designed to be submerged in the lake.

The **lake heat exchanger** is made of multiple stainless steel plate assemblies. Each assembly has two stainless steel plates that are specially patterned to create flow channels and are welded together along their perimeter. Each plate assembly is connected to a supply and return header. The overall assembly is welded to a stainless steel base that supports the plates several inches above the surface they rest on. HDPE tubing is routed from the headers to the shore. Figure 15-22 shows a lake plate heat exchanger with attached HDPE pipe, being lowered into a lake.

The indirect lake water cooling system shown in Figure 15-21 is a completely closed system. As such,

Figure 15-22 | A lake heat exchanger with attached HDPE piping below lowered into a lake. *Courtesy of AWEB supply.*

the height of the interior portion of the system above the lake surface can be much greater than what is possible with the shallow well pump used in the previously described direct cooling system. The fluid in the closed system is not affected by any turbidity in the lake water, eliminating the need for the automatic backwash filter and its associated water supply and drain. The closed system could also operate with an antifreeze solution to protect any piping near the surface of the shore from freezing.

The distribution system shown to the left of the hydraulic separator in Figure 14-21 is nearly identical to that to the left of the heat exchanger in Figure 14-20. The only difference is the use of the hydraulic separator to provide air, dirt, and magnetic particle separation in the system shown in Figure 14-21. These functions are handled by a dedicated separator in Figure 14-20.

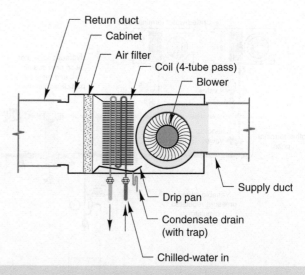

Figure 15-23 | Construction of a typical chilled-water air handler.

15.5 Chilled-Water Terminal Units

There are several devices that can absorb heat from interior air and transfer that heat to a stream of chilled-water. They are often referred to as **terminal units**. Some have been used for many decades, while others are relatively new. Some are delivered to a project ready to be fastened to a wall or ceiling, and then piped to the chilled-water distribution system. Others are integrated into the construction of the building.

Air Handlers for Chilled-Water Cooling

One of the most common terminal units is an **air handler**. These devices were discussed, in the context of hydronic heating, in Chapter 8, Heat Emitters. Further discussion of how they are used in chilled-water cooling systems is appropriate.

An air handler suitable for cooling applications consists of a chilled-water coil, a **condensate drain pan**, blower, air filter and enclosure, as illustrated in Figure 15-23.

Never use an air handler that does not have a condensate drip pan in a chilled-water cooling application.

Air handlers are available in a wide range of rated cooling capacities from about 1 ton (e.g., 12,000 Btu/h), to several hundred tons. A single small air handler with a cooling capacity of 2 to 6 tons can be used, along with a ducted distribution system, for central chilled-water cooling in average size houses. In larger houses, or commercial buildings, multiple air handlers can provide zoned cooling, and reduce the need for large and long trunk ducts.

Smaller air handlers can often be configured for vertical or horizontal mounting. Figure 15-24 shows a nominal 3-ton chilled-water air handler that has been configured for horizontal mounting. This same air handler could be quickly reconfigured for vertical mounting in other applications.

This air handler is mounted within the building's thermal envelope. As such, it is well protected against freezing, assuming the building is reasonable heated in winter. When air handlers with water coils are mounted in unconditioned attics or other unconditioned spaces, they must protected from freezing. If the coil in the air handler located in an unconditioned space operates with water, it should be isolated and drained in winter. Another option is to operate the coil with a suitable solution of antifreeze. Some designers rely on electrically power heat tracing cable to keep the water filled portions of an air handler above freezing. Although this is plausible, it can fail to provide protection during prolonged power outages, and for that reason is not recommended by the author.

The coil in this air handler is supplied by a pair of 3/4-inch **pre-insulated PEX** tubes. Pre-insulated PEX tubing is available in nominal pipe sizes ranging from 1/2 inch to 2 inches, and in coil lengths up to 100 feet. It use expedites installation and reduces the possibility of gaps between smaller pieces of insulation. Only PEX tubing with an oxygen diffusion barrier should be used for these closed loop system applications.

15.5 Chilled-Water Terminal Units

Figure 15-24 | A 3-ton rated horizontal air handler used for chilled-water cooling. 3/4-inch pre-insulated PEX tubing supplies the coil in the air handler. *Courtesy of John Siegenthaler.*

The condensate drain pan inside the air handler is connected to a 3/4-inch PVC drain pipe that routes the condensate outside or to a drain line within the building. That pipe must have a water filled **condensate P-trap** to prevent the blower from sucking air, or sewer gases, into the air stream. Figure 15-25 shows how a typical P-trap is typically installed.

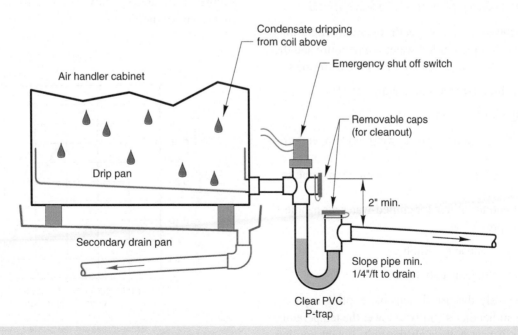

Figure 15-25 | Use of a P-trap, emergency shut off switch, and secondary drain pan for drainage of condensate from a chilled-water air handler, and protection of spaces underneath the air handler.

Several companies offer complete P-trap assemblies for air handlers. They are typically supplied with a flexible brush that can be used to clean the inside of the trap at the end of each cooling season. Some condensate P-trap assemblies are available with an **emergency shut off switch**, which detects if the water level in the P-trap rises above normal due to a clog, and is wired to interrupt operation of the chilled-water circuit supplying the air handler.

The air handler shown in Figure 15-24 is mounted above a **secondary drain pan**. This detail is also shown in Figure 15-25. The secondary drain pan is intended to capture any condensate that could leak from the air handler if the primary drain pan within the air handler failed, or if the P-trap was clogged causing the water level inside the air handler to overflow the internal drip pan. The secondary drain pan should also be piped to drain outside the building or above a floor drain within the building. In some installations, the secondary pan drain pipe can be connected the main condensate drain pipe downstream of the P-trap— *but only when that drain pipe leads to outside the building, or terminates above a floor drain. Never tee the secondary pan drain pipe into the main condensate drain pipe if the latter is connected to a building sewer or venting system.* The author recommends the use of a secondary drain pan and emergency shut off switch in the P-trap of any air handler mounted above finished interior space.

Thermal Ratings for Air Handlers

Most air handlers are rated at specific operating conditions for both the incoming chilled-water temperature and the incoming air. Those typical conditions are as follows:

- Entering chilled-water temperature: 45 °F
- Incoming air conditions

  Dry bulb temperature = 80 °F, wet bulb temperature = 67 °F

  Dry bulb temperature = 75 °F, wet bulb temperature = 63 °F

The thermal output ratings for chilled-water air handlers are usually expressed as:

- Total cooling capacity (Btu/h);
- Sensible cooling capacity (Btu/h).

As previously discussed, sensible cooling capacity refers to the air handler's ability to lower the temperature of the air as it passes through the unit. Latent cooling capacity is a measure of the air handler's ability to remove moisture from the air stream. Total cooling capacity is the sum of sensible cooling capacity and latent cooling capacity. Latent cooling capacity can be obtained by subtracting sensible cooling capacity from total cooling capacity.

Other performance measures for air handlers include:

- Head loss or pressure drop across the coil as a function of the flow rate through the coil.
- Airflow rate produced by the blower as a function of the **external static pressure** of the ducting system it supplies. This is typically stated in cubic feet per minute (CFM), versus the static pressure of the duct system. The latter is usually given in **inches of water column (w.c.)**. If the air handler has a multiple speed motor, this information is given for each speed setting.

Figure 15-26 shows an example of airflow rate produced by a nominal 3-ton air handler with a three-speed motor, versus the static pressure developed by its blower. The static pressure at which the air handler operates is determined by the design of the ducting system it supplies.

A typical airflow rate for a chilled-water air handler is 400 CFM per ton of delivered cooling capacity.

Fan Coils:

Fan coils were also discussed in Chapter 8, Heat Emitters. Those equipped with condensate drip pans are suitable for use in chilled-water cooling systems. **Console fan coils**, such as the unit shown in Figure 15-27 are typically mounted to walls and a few inches above the floor.

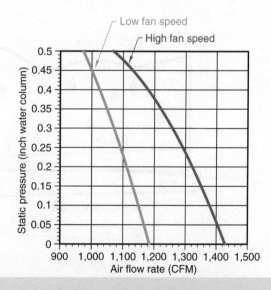

Figure 15-26 Static pressure developed by the blower in a specific air handler as a function of air flow rate and blower speed.

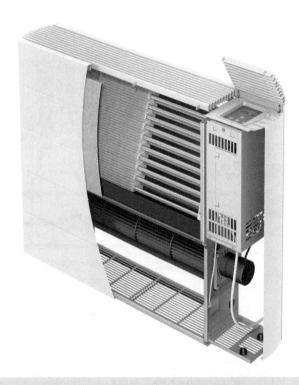

Figure 15-27 | Example of a console fan coil equipped with a drip pan, and thus capable of chilled-water cooling. *Courtesy of Myson, Inc.*

Fan coils, equipped with condensate drip pans, are also available in "**high wall**" configurations, an example of which is shown in Figure 15-28.

Most fan coil are intended to deliver cooling (or heating) to a single room. As such, they are well suited to zoned chilled-water cooling systems.

The electrical controls used in fan coils vary. Some have integral control panels that can be used to set desired room temperature, setback periods, and fan speeds. Others use handheld remotes for these settings. The type of controls available on the fan coil will determine how the balance of system is designed. For example, if the room temperature is set at the fan coil, rather than at a room thermostat, the electrical wiring in the system must ensure that the circulator is running, and that a zone valve that controls flow to the fan coil is open, whenever the unit calls for heating or cooling.

2-Pipe and 4-Pipe Air Handlers and Fan Coils

Air handlers and fan coils are also specified as being configured for 2-pipe or 4-pipe distribution systems. A **2-pipe air handler** or **2-pipe fan coil** contains a *single coil* that could be used for either heating or cooling. In heating mode, heated water flows through the coil. In cooling mode, chilled-water flows through the coil. Several 2-pipe air handlers are typically combined on a single distribution system, such as that shown in Figure 15-21.

The 2-pipe distribution systems can operate with either heated water or chilled-water. All air handlers or fan coils connected to a 2-pipe distribution system can only operate in the mode (e.g., heating or cooling) that the overall system is set for at a given time.

A "4-pipe" air handler or fan-coil contains *two separately piped coils:* one for heating and one for cooling. Each coil has a supply and return connection to the appropriate distribution mains. A 4-pipe system allows each air handler to operate in either heating or cooling, independent of the other air handlers in the

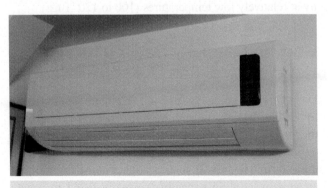

Figure 15-28 | Example of a high wall fan-coil equipped with a chilled-water coil. *Courtesy of John Siegenthaler.*

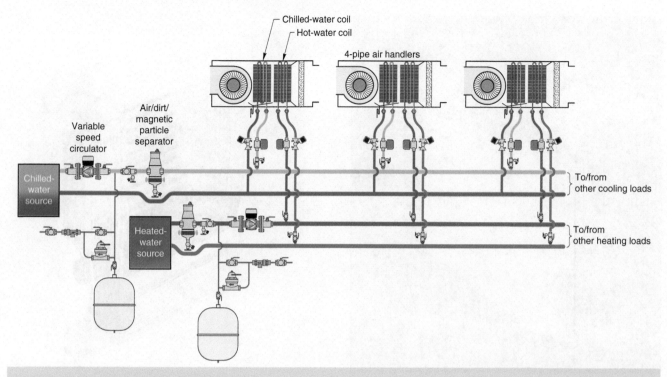

Figure 15-29. Piping for 4-pipe air handlers. Two separately piped coils in each air handler allow it to operate in either heating or cooling mode, independent of the other air handlers on the system.

system. Figure 15-29 shows an example of a distribution system using 4-pipe air handlers.

Notice that the heated water and chilled-water circuits are *completely separate*. As such, each should be equipped with make-up water systems, expansion tanks, air/dirt/hydraulic separators, and safety devices.

Another option that allows a 2-pipe air handler or fan-coil to provide either heating or cooling is shown in Figure 15-30.

This piping arrangement uses two **3-position motorized diverter valves** for each air handler. These valves have an internal ball with an "elbow-shaped" flow passage machined into it. The actuator can rotate this ball to three specific orientations, which are illustrated in Figure 15-29. These ball positions allow (or prevent) flow as follows:

- Flow allowed between the air handler coil and heating supply and return mains.
- Flow allowed between the air handler coil and cooling supply and return mains.
- No flow allowed through the air handler coil.

When an air handler is off, the ball inside the 3-position diverter valve rotates to block flow through it. This "3rd position" function eliminates the need for a two-way motorized valve at each handler to block flow when the unit is not operating.

This approach is well suited to systems designed around a relatively short 4-pipe manifold located in the mechanical room. The diverter valves, balancing valves, and purging valves shown in Figure 15-29 are located close to the 4-pipe manifold. Two pipes connect between this assembly and each air handler. This piping arrangement eliminates the need to route a 4-pipe distribution system close to every air handler or fan coil in the system. An example of this piping concept is shown in Figure 15-31.

When selecting an air handler or fan coil for this piping configuration, it should have a "deep" multi-row coil to enable good dehumidification at moderate chilled-water temperatures (50 to 55 °F), as well as good heating capacity at relatively low temperatures (100 to 120 °F).

Cooling Capacity Ratings for Air Handlers and Fan Coils

Designers planning to use air handlers or fan-coils in chilled-water cooling systems need to reference the **total cooling capacity** and **sensible cooling capacity** ratings of units under consideration. Since these terminal units deliver a cooled and dehumidified air stream, cooling capacity is usually listed in a way that allows sensible and latent cooling capacity to be determined based on the chilled-water temperature

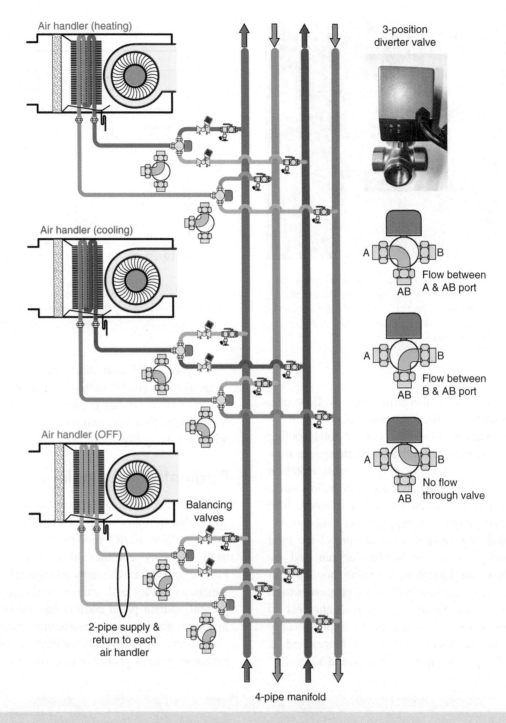

Figure 15-30 Using several a 2-pipe air handlers in combination with a 4-pipe distribution system and two 3-position diverter valves for each air handler.

and flow rate supplied, along with typical indoor air temperature, relative humidity, and flow rate conditions. Latent cooling capacity is usually not listed, but easily determined by subtracting sensible cool capacity from total cooling capacity. Figure 15-32 shows an example of cooling capacity rating for a series of fan-coil units.

The cooling capacity data shown in Figure 15-32 is for specific operating conditions. The chilled-water supply temperature is 45 °F, the chilled-water return temperature is 54 °F, and the enter air temperature is 81 °F. The performance data is also is based on a relatively humidity of 50%.

Cooling capacity increases as chilled-water temperature decreases, especially when the supplied chilled-water temperature is well below the dewpoint of air entering the terminal unit.

Figure 15-31 | Example of a 4-pipe manifold assembly in a mechanical room, supplying six 2-pipe air handlers. A pair of 3-position diverter valves is required for each air handler. *Courtesy of ThermAtlantic.*

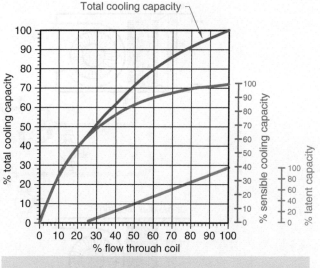

Figure 15-33 | Percent of sensible, latent, and total cooling capacity of an air handler versus percent of design flow rate passing through coil.

Cooling capacity is also influenced by flow rate. Figure 15-33 shows a typical relationship.

The non-linear relationship between total cooling capacity and flow rate shows that small changes in flow rate at low flow rates have significantly more effect on cooling capacity than equal changes at high flow rates. This characteristic is similar to that of hydronic heat emitters operated at constant supply water temperature.

If a modulating valve is used to control the total cooling capacity of an air handler or fan coil by controlling flow rate through it, that valve should use an internal flow control element with an **equal percentage characteristic**. This valve characteristic attempts to make changes in total cooling capacity approximately proportional to the valve stem travel range, and is discussed in Chapter 5, Pipings, Fittings, and Valves.

Figure 15-33 also shows that the sensible cooling capacity of the coil begins to rise at lower flow rates, while the coil's latent cooling capacity begins at approximately 25% of total flow rate, and rises in a quasi-linear manner with increasing flow.

Radiant Panel Cooling

Chilled-water can also be passed through tubing circuits embedded in floors, walls, or ceilings. This approach is called **radiant panel cooling**. It has been successfully used in many commercial buildings, especially in Europe and Asia. Although infrequently used in North American homes and smaller commercial buildings at present, radiant panel cooling has the potential to gain increased acceptance as geothermal water-to-water and air-to-water heat pumps bring chilled-water capacity to hydronic systems primarily intended for heating

2-PIPE MODELS			Model				
Parameter	Metric	Units (IP)	S2 200 / S2 200i	S2 400 / S2 400i	S2 600 / S2 600i	S2 800 / S2 800i	S2 1000 / S2 1000i
Cooling/heating	Total cooling (45/54/81°F)	btuh med (min - max)*1	2,491 (na - 3,106)	4,641 (2,560 - 7,234)	7,098 (3,924 - 9,589)	8,156 (4,505 - 11,261)	8,770 (4,812 - 12,660)
	Sensible cooling	btuh med (min - max)*1	1,877 (na - 2,491)	3,651 (2,014 - 5,869)	5,153 (2,833 - 7,200)	6,279 (3,481 - 9,247)	6,757 (3,651 - 9,896)
	Flow rate	gpm med (min - max)*1	0.6 (na - 0.7)	1.0 (0.6 - 1.6)	1.6 (0.9 - 2.1)	1.8 (1.0 - 2.4)	1.9 (1.1 - 2.8)
	Pressure drop	ft of hd med (min - max)*1	3.4 (na - 4.0)	1.4 (0.6 - 2.7)	3.3 (0.9 - 5.7)	2.9 (0.8 - 6.0)	3.7 (4.6 - 7.1)
	Heating (176/167/68°F)	btuh med (min - max)*1	6,859 (4,607 - 8,326)	12,352 (6,927 - 17,710)	17,982 (10,066 - 24,841)	21,804 (12,352 - 31,631)	23,442 (12,455 - 36,647)
	Flow rate	gpm med (min - max)*1	0.8 (0.5 - 0.9)	1.4 (0.8 - 2.0)	2.0 (1.1 - 2.8)	2.5 (1.4 - 3.6)	2.7 (1.4 - 4.2)
	Pressure drop	ft of hd med (min - max)*1	1.1 (0.4 - 2.6)	1.4 (0.7 - 2.0)	3.6 (1.2 - 6.4)	4.0 (1.3 - 7.2)	5.6 (1.8 - 12.2)

Figure 15-32 | Cooling capacity ratings for a series of console fan coils. *Courtesy of Myson, Inc.*

Figure 15-34 | Low mass radiant ceiling panel under construction. See additional details on this panel in Chapter 10, Hydronic Radiant Panel Heating. *Courtesy of John Siegenthaler.*

Radiant ceilings, in particular, are ideal for absorbing heat from the occupied space below. An example of a radiant ceiling that can provide radiant heating and radiant cooling is shown, under construction, in Figure 15-34. After all tubing has been installed and pressure tested, this ceiling will be finished with standard 1/2-inch drywall. More details on this system were presented in Chapter 10, Hydronic Radiant Panel Heating.

It is critically important that the chilled-water supplied for radiant panel cooling be at temperatures above the current dewpoint temperature of the room. This prevents water vapor in the air from condensing on the radiant panel. This also applies to components that convey chilled-water to and from those panels, such as piping above a cooled ceiling. *The chilled-water supplied to radiant cooling panels should be a minimum of 3 °F above the current dewpoint temperature of the room.*

Designers should also consider how the dewpoint temperature varies in different spaces within a building. For example, on a humid day, the dewpoint temperature within a frequently used entry vestibule is likely to be higher than that within an interior building space. This is also true for spaces used for food preparation, commercial laundry, washing, or exercising. Such areas should be zoned and use independent controls for regulating chilled-water supply temperature.

This dewpoint constraint only allows a radiant panel to absorb the majority of the sensible cooling load. Another terminal unit such as a central chilled-water air handler is needed to handle the remainder of the sensible load, and all of the latent cooling load.

One approach for keeping radiant cooling panels above the dewpoint is to use a 3-way motorized mixing valve operated by a **dewpoint controller**. Whenever the cooling system is active, this controller repeatedly samples the dry bulb temperature and relative humidity of the interior space it is responsible for. It calculates the current dewpoint temperature, and operates the mixing valve's actuator to keep the supply water temperature to the panel approximately 3 °F above that dewpoint temperature. This mixing valve would be piped the same way as if it was controlling the supply temperature for heating, as shown in Figure 15-35.

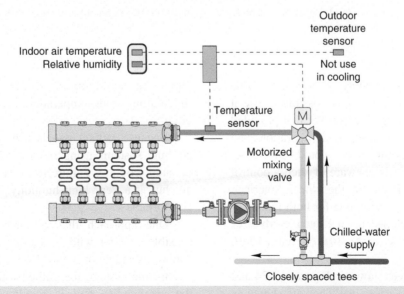

Figure 15-35 | Concept of dewpoint controller for three-way motorized mixing valve to maintain chilled-water temperature to panel about 3 °F above current room dewpoint temperature.

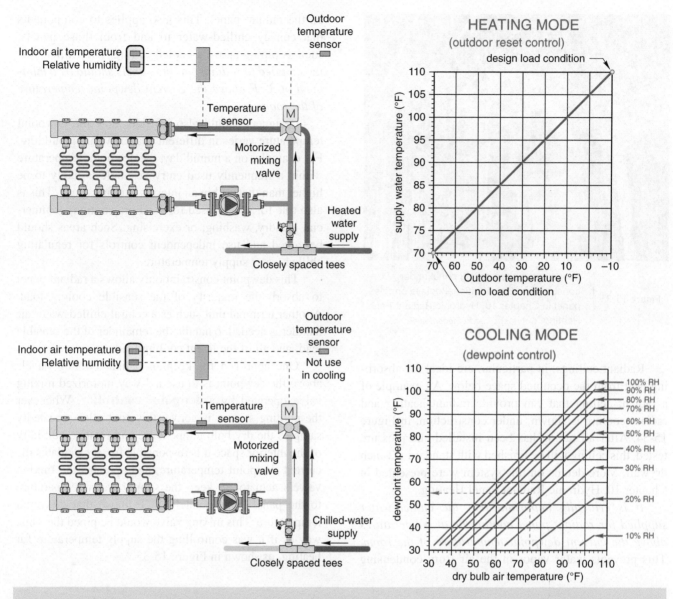

Figure 15-36 | Concept of the same piping configuration between controlled based on outdoor reset control in heating and dewpoint control in cooling.

Because this piping assembly is the same as one that would be used for heating, it is possible to use it for both heating and cooling. A controller that provided outdoor reset control for heating mode operation, and **dewpoint control** for cooling mode, would be ideal. As of this writing, the author is unaware of such a control that's commercially available on the North American market. However, the necessary code for both modes of operation could be developed for a **micro controller** such as the **Raspberry Pi®** or **Arduino®** Figure 15-36 shows the concept.

The closely spaced tees shown in Figures 15-34 and 15-35 may or may not be necessary. They are shown as a possible means of hydraulically separating the piping assembly from the remainder of the system. This would allow the possibility of continuous circulation with modulating water temperature through the radiant panel circuits.

Because radiant panels can only provide sensible cooling, systems using them need to provide a means of latent cooling. Failure to do so will result in cooler air, but at a very high humidity level that will not be comfortable.

One approach that combines radiant panels for sensible cooling with a central air handler for latent cooling is shown in Figure 15-37.

In this system, the chilled-water flow rate through the air handler coil is regulated by variable-speed circulator that responds to an input signal from a **relative humidity controller**. That controller monitors the

15.5 Chilled-Water Terminal Units 809

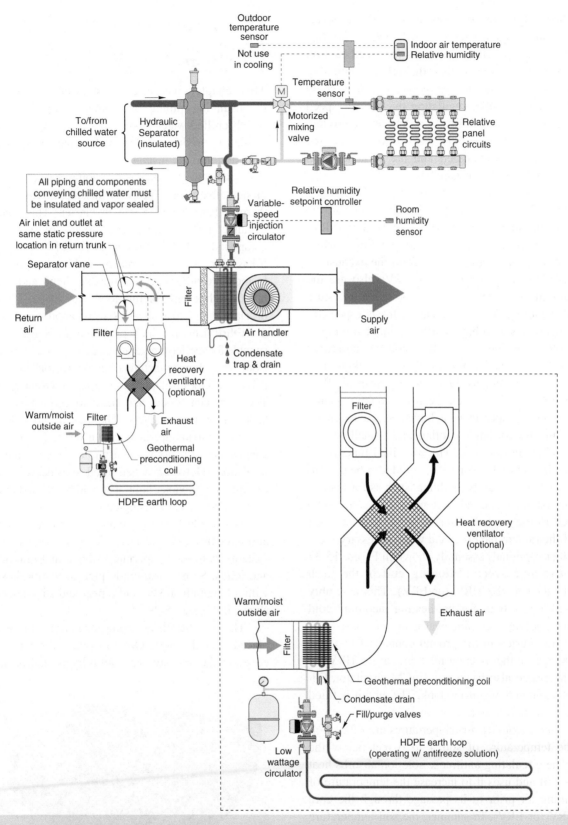

Figure 15-37 | Piping concept where a radiant panel provides majority of sensible cooling load, while an air handler provides remainder of sensible cooling an all of latent cooling load. An (optional) heat recovery ventilator is shown for providing fresh air to intake side of air handler.

relative humidity in the space being cooled and generates an output a signal such as 2–10 VDC, 4–20 mA, or PWM that the circulator interprets for speed control. There's also a possibility that the relative humidity sensor associated with the dewpoint controller could transmit to the controller regulating the variable-speed circulator. Another possibility is a "master controller" that provides all the outputs needed to operate the mixing valve, the air handler, and the variable-speed circulator supplying the air handler.

Figure 15-37 also shows how a **heat recovery ventilator (HRV)**, or **Energy Recovery Ventilator (ERV)** can be integrated into the system. Both HRVs and ERVs introduce fresh air into buildings, while recovering heat (or cooling effect) from the exhaust air stream. During the cooling season, ERVs also use the lower humidity of exhaust air to absorb some moisture from the incoming fresh air stream. This is especially helpful in climates with higher summertime humidity.

The blower in the air handler could be configured to operate continuously at a relatively low "background" air flow rate sufficient to distribute the fresh air flow from the HRV or ERV. When a call for cooling occurs, the blower could operate at a higher speed to create sufficient air flow for latent cooling, and a small portion of the sensible cooling load. The combined function of the air handler and HRV (or ERV) could also be used to circulate air throughout the building, such as for distributing solar or other internal heat gains. It could also be configured to use heated water to supplement the heat output of the radiant panels if and when necessary.

Another (optional) assembly shown in Figure 15-37 is a **geothermal preconditioning coil** in the fresh air supply duct to the HRV (or ERV). This assembly is a closed-loop circuit for pre-heating incoming cold air, or pre-cooling incoming warm air. It consists of HDPE piping buried in the ground connected to a coil heat exchanger in the fresh air inlet duct of the HRV (or ERV). The circuit also has a small ECM circulator, fill/purge valve, and an expansion tank. The circuit is filled with an antifreeze solution.

In winter, incoming air temperatures are often well below the temperature of soil 4 to 6 feet below the surface. The circulating antifreeze solution absorbs heat from this soil and uses it to increase the temperature of the incoming air prior to its passage through the core of the HRV (or ERV). In summer, the soil temperature can be well below that of incoming air. The circulating antifreeze pre-cools the incoming air.

Operation of the geothermal preconditioning assembly could be controlled based a differential temperature between the soil and incoming outside air temperature. A low power ECM circulator provides flow for this assembly at a very low operating cost.

15.6 Chilled-Water Piping Insulation

This chapter has already stressed the importance of insulating piping and other components through which chilled-water passes. It's also important that the insulation product used provide a barrier against vapor diffusion into the insulation.

Elastomeric foam is one commonly used insulation product that, when properly installed, provides both a thermal barrier and a vapor barrier. This material is manufactured by extruding a mixture of specific rubber compounds, PVC, and a non-CFC/HCFC foaming agent. It has a closed cellular structure making it highly resistant to moisture diffusion. It is flexible, light, and can easily be cut with a knife or fine tooth saw blade. The R-value of elastomeric foams varies from approximately 3.5 to 5 (°F•h•ft^2/Btu) per inch of thickness. Some grades are suitable for service temperatures up to 220 °F. It is available from several manufacturers, in many thickness, and as cylindrical pipe insulation or sheet, as shown in Figure 15-38. The latter can be cut and fitted to components such as tanks or heat exchangers. The typical stock length for elastomeric foam pipe insulation is 6 feet. For most interior applications, it is not necessary to cover elastomeric foam insulation. However, when used outside, or otherwise exposed to ultraviolet light, it should be painted or covered with a UV-resistance wrapping.

When elastomeric foam insulation is used for anti-condensation purposes in chilled-water cooling systems, it is very important to bond and seal all seams and joints. Some elastomeric pipe insulation is supplied with a longitudinal slit, and a peal and stick adhesive, as shown in Figure 15-39.

The pre-slit/self-adhering insulation can be quickly placed over the pipe. Once in place, the release strips covering the adhesive are carefully pealed back as the

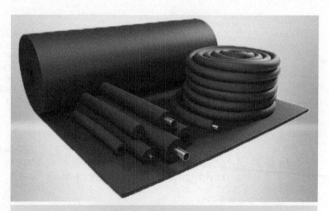

Figure 15-38 | Examples of sheet and pipe insulation products made of elastomeric foam. *Courtesy of Armacell, Inc.*

15.6 Chilled-Water Piping Insulation

using a miter saw with a fine tooth blade, or a sharp utility knife. A reciprocating electric knife designed for cutting meat also works well, and allows for straighter cuts on unsupported pieces. Patience and careful workmanship will produce professional looking and long lasting results.

Sheets of elastomeric foam insulation, as well as 1/8-inch thick self-adhering "tapes" of elastomeric foam are available. These materials can be neatly wrapped around and bonded to surfaces that don't lend themselves to standard piping insulation. For example, Figure 15-41 shows a brazed plates heat exchanger that has been fully wrapped with 1/2-inch thick elastomeric foam sheet to prevent condensation when it operates below the dewpoint of the surrounding air. The copper tubing and fittings that connect to this heat exchanger will also be insulated with 1/2-inch thick elastomeric foam insulation. That insulation will be bonded to the sheet insulation on the heat exchanger.

Figure 15-42 gives recommend thicknesses of elastomeric foam insulation for preventing surface condensation on different pipe sizes, while operating under varying chilled-fluid temperature and surround air conditions. Designers should verify that these thicknesses meet any local mechanical or energy code requirements for chilled-water piping systems.

Insulation should also be as continuous as possible where piping is supported. When elastomeric foam insulation is used, it is poor practice to simply apply the insulation and then attach a mechanical support to the insulated pipe. Doing so compresses the insulation at the supporting clamp due to the weight of the pipe and

Figure 15-39 | Example of pre-slit self-adhering pipe insulation. *Courtesy of Armacell, Inc.*

two edges are aligned and pressed together. The bond is instant and non-reversible.

All butt joints should also be bonded using a suitable rubber-based contact cement. Each end of the insulation to be butted must coated with this cement. The cement then needs about 5 minutes to "tack up" before the ends are pressed together. Both ends should be carefully aligned as they are brought together. The bond is instant upon contact and non-reversible.

The author recommends fitting pieces of elastomeric foam insulation to elbow and tee fittings using miter cuts as shown in Figure 15-40.

For 90° elbows and tees, use a 45° miter cut. For 45° elbows, use a 22.5° miter-cut. The insulation can be cut

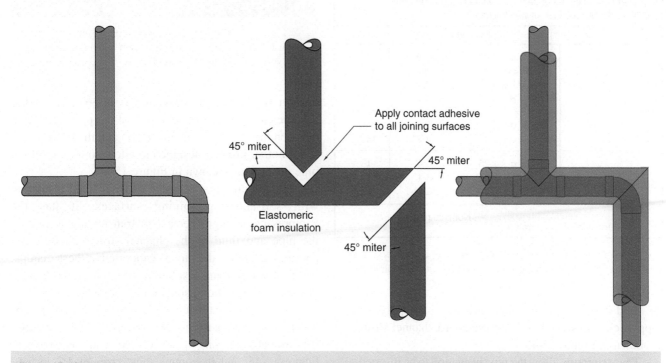

Figure 15-40 | Joining elastomeric foam pipe insulation using miter cuts at elbows and tees.

Figure 15-41 A brazed plate heat exchanger wrapped with 1/2-inch thick elastomeric foam sheet insulation. *Courtesy of John Siegenthaler.*

Figure 15-43 Polymer clips for supporting insulated piping from channel strut with minimum disruption of the insulation. *Courtesy of John Siegenthaler.*

Suggested wall thickness of elastomeric foam insulation to prevent surface condensation (R-value of approximately 7.0 °F•hr•ft²/Btu per inch of wall thickness).

	fluid temperature in pipe ➡	50 °F	35 °F
normal conditions 85 °F / 70% RH	3/8" ≤ d ≤ 1.25"	3/8"	1/2"
	1.25" < d ≤ 2"	3/8"	1/2"
	2" < d ≤ 2.5"	3/8"	1/2"
	2.5" < d ≤ 6"	1/2"	3/4"
mild conditions 80 °F / 50% RH	3/8" ≤ d ≤ 2.5"	3/8"	3/8"
	2.5" < d ≤ 6"	1/2"	1/2"
severe conditions 90 °F / 80% RH	3/8" ≤ d ≤ 1.25"	3/4"	1"
	1.25" < d ≤ 3.5"	3/4"	1"
	3.55" < d ≤ 6"	3/4"	1"

Relative humidity — Dry bulb air temperature — Nominal pipe size

Figure 15-42 Recommended thicknesses of elastomeric foam insulation to prevent condensation on piping.

the fluid it contains. One alternative is to use polymer support clips designed to fit into standard channel struts as shown in Figure 15-43.

These supports allow the pipe insulation to be inserted from both sides, with a gap of only 1/8 inch at the center support ribs. After the insulation is inserted, a bead is silicone caulk should be used to seal the perimeter seam between the outside of the insulation and the bracket.

Another product designed to allow continuity of the pipe insulation is shown in Figure 15-44.

This product is a segment of rigid foam insulation with self-adhering joining surfaces. It has the compressive strength needed to transfer the weight of the pipe and fluid to the channel strut without being compressed. Once the pipe is supported at these clamps, the elastomeric foam insulation is bonded to both sides of the rigid insulation using contact cement.

Another product that is well suited to smaller chilled-water cooling systems is **pre-insulated PEX tubing**. It is available in coils up to 100 feet long, as shown in Figure 15-45, and in nominal pipe sizes of 1/2 inch to 2 inches.

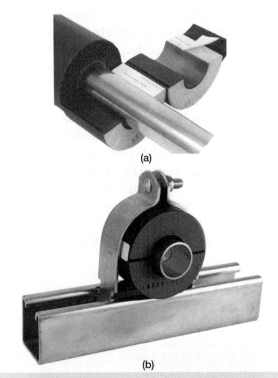

Figure 15-44 | Self-adhering segment of rigid foam insulation for supporting insulated piping with no loss of insulation continuity. *Courtesy of ZSi, Inc.*

Figure 15-45 | Example of pre-insulated PEX tubing supplied in coils. *Courtesy of Uponor.*

The thickness of the insulation on pre-insulated PEX tubing varies with the size of the tubing. Larger tubing sizes typically have thicker insulation. When used in closed-loop cooling (or heating) systems, the PEX tubing should be specified with an oxygen barrier 3/4-inch and 1-inch tube sizes are typically suitable for chilled-water piping to air handlers ranging from 2 to 6 tons (24,000 to 72,000 Btu/h). Small console air handlers with cooling capacities up to 1 ton (12,000 Btu/h) can usually be piped with 1/2 inch pre-insulated PEX. These pipe sizes should be verified using the methods presented in Chapter 6, Fluid Flow in Piping.

Figure 15-46 | Nylon straps used to suspend pre-insulated PEX tubing within joist cavities. *Courtesy of John Siegenthaler.*

Pre-insulated PEX can be routed through and along building framing just like other flexible tubing products. It speeds installation in comparison to use of rigid pipe with custom fitted elastomeric foam insulation, especially in close confines. It can be supported using nylon strapping as seen in Figure 15-46.

Pre-insulated PEX tubing can be transitioned to copper tubing or other components using a variety of fitting systems such as those using cold expansion, compression, or push-to-fit connections.

Chilled-water systems can also use fiberglass insulation. To prevent water vapor migration as well as abrasion damage, the fiberglass must be covered. One suitable covering that is usually installed on fiberglass pipe insulation is known as **Foil Scrim Kraft**. It's a reinforced wrapping containing a layer of aluminum foil as a vapor barrier. Another suitable covering system uses thin sheets and pre-shaped shells of PVC for both physical protection and as a vapor barrier. All seams in the PVC jacketing must be bonded.

15.7 Example Systems

This section describes three systems that combine chilled-water cooling with hydronic heating. These systems demonstrate concepts such as single zone cooling with multi-zone heating, as well as multi-zone cooling and heating. All three systems are based on components and piping methods described earlier in this chapter, as well as previous chapters.

System #1: Single Zone Cooling/ Multi-Zone Heating

Modern hydronics technology makes it easy to create multi-zone distribution systems. This is true for both heating and cooling. However, adding zones increases the cost and complexity of any hydronic system. This is especially true with zoned hydronic cooling because it typically requires a chilled-water terminal unit in each zone. Unlike the use of a panel radiator, or separate floor heating circuit, each terminal unit needs electrical power and piping for condensate drainage, in addition to the chilled-water supply and return piping.

Although room-by-room zoning for both heating and cooling is certainly possible, this approach is often unnecessary, especially if interior doors are left open, and the desired temperatures or relative humidity levels in different areas of the building are essentially the same.

One approach that considers the tradeoff between functionality, complexity, and cost is to combine multiple heating zones with a single cooling zone. The heating portion of the system could use any of the heat emitters and distribution piping systems described in earlier chapters. Cooling is provided by a single chilled-water air handler and an appropriate ducting system. Supply air registers can be placed high on walls or in ceilings, either of which is ideal for cooling.

Figure 15-47 shows a system that provides multi-zone heating, single zone cooling, and domestic water heating.

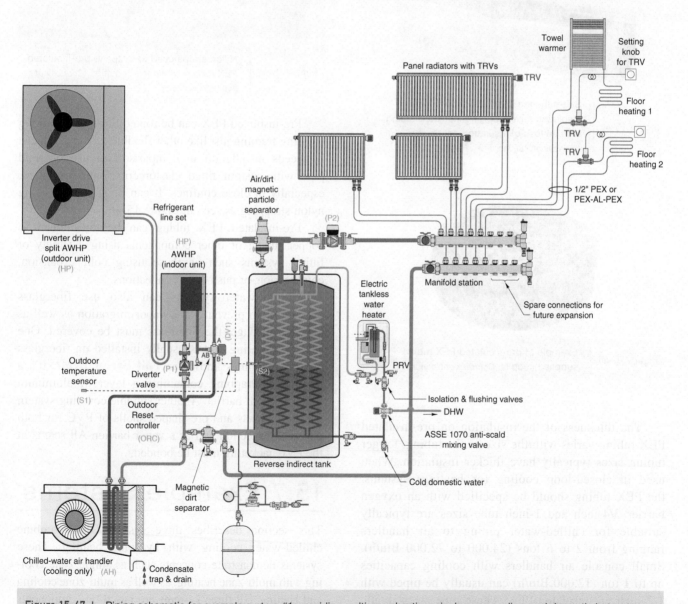

Figure 15-47 | Piping schematic for example system #1, providing multi-zone heating, single zone cooling, and domestic hot water.

This system uses a low ambient split system air-to-water heat pump as its main heat source and chiller. The heat pump has an inverter drive compressor that allows variable heating and cooling output down to approximately 40% of rated capacity. The heat pump also has internal controls that allow the user to set a desired chilled-water outlet temperature. The compressor speed is automatically varied to maintain this chilled-water outlet temperature while the heat pump operates in cooling mode. The heat pump can maintain steady operation, at reduced capacity during cooling mode, and full capacity if necessary during heating mode. This is helpful because the design cooling loads of houses in cold climates are often significantly lower than design heating loads. If the cooling capacity of the air handler is not less than the minimum cooling capacity of the heat pump, it is not necessary to use a buffer tank during cooling mode operation.

Since the heat pump is a split system, there's no need to protect the outdoor unit from freezing, and thus the system can operate with water rather than an antifreeze solution.

A motorized diverter valve with a spring-return actuator directs flow leaving the heat pump to either the heating or cooling portion of the system. It should be configured with its normally closed port, designated as "A" supplying the heating portion of the system. The normally open port, designated as "B" would supply the coil in the air handler during cooling mode. This allows the diverter valve to prevent reverse thermosiphoning when the tank is warm and the heat pump is off.

A **reverse indirect water heater** serves as both a buffer tank and domestic water pre-heater. The target water temperature in the tank is regulated by an outdoor reset controller. At design load conditions, this controller attempts to keep the tank at a target temperature of 120 °F. The target temperature decreases as the outdoor temperature increases, which increases the heat pump's COP. The minimum target temperature for the tank is set at 90 °F.

The heat pump and its associated circulator are turned on whenever the tank temperature drops to 5 °F below the current target temperature, and off when the tank reaches 5 °F above the current target temperature.

Domestic cold water passes through the copper heat exchanger coils in the tank whenever domestic hot water is drawn at a fixture. The temperature of the domestic water leaving the coils depends on the tank's current temperature. At or near design load conditions, the domestic water could leave the tank at temperatures in the range of 115 to 120 °F. Under partial load the temperature of the domestic water leaving the tank may only be 85 to 90 °F. A **tankless electric water heater** adds just enough heat to boost the domestic hot water temperature to the desired delivery temperature. The water heater is equipped with combination isolation / flushing valves and a pressure relief valve. Similar valving is provided to allow the coils inside the indirect tank to be isolated and chemically cleaned if necessary. An anti-scald mixing valve ensures that the water temperature supplied to the building's plumbing system will not exceed 120 °F.

The heating distribution system is configured for five independently controlled zones. Three zone are panel radiators equipped with integral thermostatic radiator valves. One zone is a towel warmer piped in series with a short floor heating circuit. On this zone, a thermostatic radiator valve with a capillary tube connects the valve body with a wall mounted adjustment dial. The fifth zone is also a floor heating circuit with a capillary-style thermostatic valve. All zone circuits and heat emitters have been sized to provide design load output to their respective spaces when supplied with 120 °F water. This eliminates the need for mixing valves. The five "homerun" zone circuits of 1/2-inch PEX tubing connect to a manifold station that has three spare connections to allow additional heat emitters to be easily added in the future. Flow to the manifold station is provided by a variable-speed pressure-regulated circulator operating in constant differential pressure mode.

In cooling mode, the motorized diverter valve directs the chilled-water leaving the heat pump through the coil in the air handler. The heat pump's compressor automatically adjusts speed to maintain a chilled-water outlet temperature of 50 °F.

The heat pump circulator must be selected based on the desired flow rate through the coil of the air handler. The head loss of the heat pump, the air handler's coil, and the interconnecting piping must be evaluated at this desired flow rate as part of selecting an appropriate circulator.

The distribution circulator would operate continuously during the heating season. Its speed would automatically adjust as the thermostatic radiator valves open, close, or modulate flow. If all zones are closed, the circulators shifts in "sleep" mode with an input power of 10 watts or less. The circulator's impeller is still spinning in this mode. This enables a slight no-flow reference differential pressure to exist. When one of the thermostatic valves begins to open the differential pressure across, the circulator will drop. This change provides the "wake up" call for the circulator to resume normal constant differential pressure control.

The systems controls are configured so that highest priority is given to maintaining the target temperature within the indirect tank. If the system is operating in cooling mode, and a call for tank heating occurs, the heat pump temporarily turns off for up to 3 minutes, and then restarts in heating mode. The diverter valve is energized to direct flow from the heat pump to the indirect water heater. The heat pump brings up the temperature within the water heater and then shuts off for another 3-minute period. It then reverts back to cooling mode. The short "off periods" are typically built into the heat pump's internal control logic. They prevent the potential for rapid cycling between heating and cooling mode, especially under conditions when there are large differences in refrigerant pressure within the heat pump.

Figure 15-48 shows one possible electrical control schematic for system #1. The circuitry is based on several assumptions regard the specific hardware used. For example, it assumes the heat pump is turned on in heating mode when a circuit is completed between the heat pump's R and Y terminals. For cooling operation it assumes that connections are required between the R and Y terminal, and the R and O terminals. It assumes that circulator (P1) is powered through and controlled by the heat pump. It also assumes that the air handler is powered by a 240-VAC circuit, and is turned on by a connection between its R and G terminals. Another assumption is that the diverter valve is power by 24 VAC.

Some heat pumps, air handlers, and other control devices may have requirements different from these assumptions. The control concepts and ladder diagramming described in Chapter 9, Control Strategies, Components, and Systems, can be used to alter the circuitry shown in Figure 15-48 for the specific devices used in a system

The following **description of operation** was developed for system #1 based on the piping and electrical schematics shown in Figures 15-47 and 15-48.

Description of Operation (Example System #1)

1. Power supply: Whenever the master switch (MS) is closed, 120 VAC is supplied to the control circuit. The heat pump (HP) is supplied by a 240 VAC/30 amp dedicated circuit through a disconnect switch. The air handler (AH) is supplied by a 240 VAC/15 amp dedicated circuit through a disconnect switch. The tankless electric water heater is supplied by a 240-VAC / 60 amp dedicated circuit.

2. Heating mode: Whenever the master switch (MS) is closed, 120 VAC is supplied to one pole of a DPDT mode switch (MOS). If this switch is set to "heat," 120 VAC is passed to distribution circulator (P2), which operates continuously, in constant differential pressure mode. The second pole of the mode switch (MOS), when set to heat, does nothing. Heated water is supplied to any space heating circuit with a partially open, or fully open thermostatic radiator valve. If all thermostatic valves are closed, circulator (P2) reverts to a low power "sleep" mode. If the mode switch (MOS) is set to off, space heating and cooling are disabled. However, when necessary, the system will still operate the heat pump (HP) in heating mode to maintain the water temperature within the reverse indirect water heater.

Whenever the master switch (MS) is closed, 24 VAC from transformer (X1) is also passed to the outdoor reset controller (ORC) powering it on. The (ORC) controller begins measuring the outdoor temperature at sensor (S1), and the temperature of the reverse indirect tank at sensor (S2). Based on these temperatures, and the controller's settings, the (ORC) calculates the target water temperature in the tank. When the measured water temperature at sensor (S2) drops 5 °F or more below the current target temperature, the normally open relay contact in the (ORC) closes. This passes 24 VAC from the heat pump's R terminal to its Y terminal, which turns the heat pump on in heating mode. Heat pump circulator (P1) is also turned on by the controls within the heat pump whenever the heat pump is called to operate. When the measured water temperature at sensor (S2) in the tank is 5 °F or more above the target temperature the relay contact in the (ORC) opens, turning of the heat pump and circulator (P1).

When the relay contact in the (ORC) closes, 24 VAC from the heat pump's R terminal also passes to relay coil (RH). Relay contact (RH-1) closes to pass 24 VAC to the diverter valve (DV1) which opens the flow path between the heat pump and reverse indirect tank. When the relay (RH) is off this flow path is blocked, preventing reverse thermosiphoning from the tank.

Relay contact (RH-1 NC) is used to prioritize heating mode over cooling mode, even when the system's mode switch (MOS) is set for cooling. This is necessary to ensure domestic water pre-heating within the coils of the reverse indirect tank. Whenever the (ORC) calls for heating, relay contact (RH-1 NC) opens, which turns off the heat pump if it is operating in cooling mode. The system's air handler (AH) continues to run if the cooling thermostat is calling for cooling. This allows circulation of house air even when cooling mode operation of the heat pump is temporarily disabled.

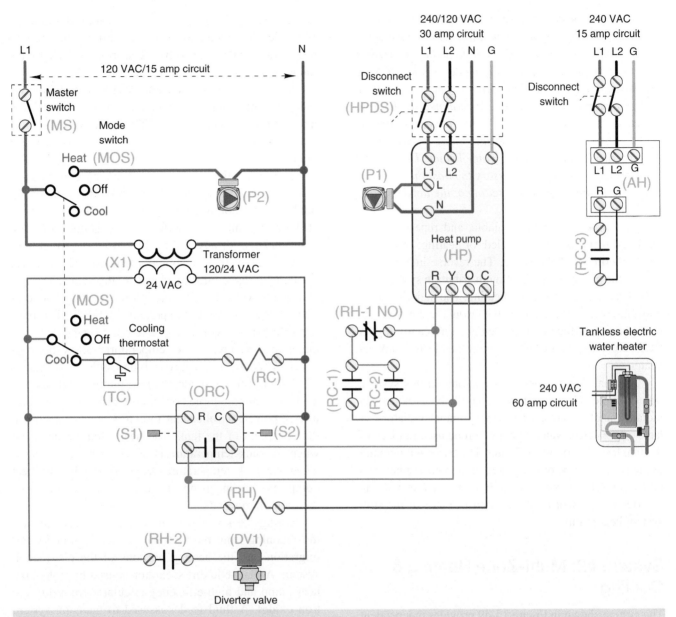

Figure 15-48 | Electric control schematic for example system #1. Other wiring variations are possible based on specific equipment selections.

The heat pump's control panel is used to set an upper limit temperature that is 8 °F higher than the highest target temperature setting of the (ORC) controller. This provides a secondary high limit temperature that would turn the heat pump off if the (ORC) failed to do so at a lower temperature.

Cooling mode: Whenever the master switch (MS) is closed, 120 VAC is supplied to the primary winding of transformer (X1). 24 VAC from the transformer's secondary winding is passed to the common terminal of one pole of the DPDT mode switch (MOS). If this switch is set to "cool," 24 VAC is passed to the cooling thermostat (TC). When this thermostat calls for cooling, 24 VAC is passed to relay coil (RC). Relay contact (RC-1) closes to connect between the heat pump's R and Y terminals, turning on the heat pump's compressor. Relay contact (RC-2) also closes connecting the heat pump's R and O terminals, which energizes the heat pump's reversing valve for cooling mode operation.

Relay contact (RC-3) also closes upon a call for cooling from the thermostat. It completes a circuit between the R and G terminals in the air handler allowing it to operate.

The water temperature leaving the heat pump in cooling mode is set at the heat pump's control panel. The compressor within the heat pump varies its speed as necessary to maintain the set chilled-water outlet temperature.

Domestic water heating: *Whenever the master switch (MS) is closed, the temperature within the reverse indirect tank is monitored and maintained by the (ORC) controller, regardless of how the mode switch (MOS) is set. Whenever domestic hot water is being used at or above some minimum flow rate determined by the flow switch in the tankless electric water heater, pre-heated water from the coils in the reverse indirect tank passes through the tankless electric water heater. That heater operates its elements as necessary to boost the domestic hot water to the temperature setting of the tankless electric water heater.*

System #1 is simple, scalable, and repeatable. No electrical thermostats are needed for heating, and only a single thermostat for cooling. The reverse indirect tank buffers the highly zoned heating distribution system, and provides domestic water pre-heating. The majority of the domestic water temperature "lift" would come from the heat pump, and thus be provided at significantly higher COPs compared to a COP of 1.0 for the electric tankless heater.

There are many possible variations on this system. For example, all space heating could be provided by radiant panel circuits, or solely by panel radiators. A tank-type electric water heater could be used in place of the tankless electric water heater. The target temperature in the tank could be based on setpoint control rather than outdoor reset control. A monobloc heat pump with an inverter compressor could be used in place of the split system heat pump.

System #2: Multi-Zone Heating & Cooling

The system shown in Figure 15-49 provides independent control of heating and cooling to two zones. The system can only operate the zones in the same mode (e.g., either heating or cooling) at any time. This constraint is generally not a problem in most houses., especially in climates where there is typically a "deadband" period of several days between the end of the heating season and the beginning of the cooling season, and a similar deadband after the cooling season ends and when the heating season begins.

The heating and cooling source in this system is a single speed monobloc air-to-water heat pump. The heat pump supplies a buffer tank piped for a 3-pipe configuration and is used in both heating and cooling modes. The entire system operates with a solution of propylene glycol antifreeze.

In heating mode, a variable-speed pressure-regulated circulator supplies two independent zones of radiant panel heating. Flow through each manifold station is controlled by a zone valve. Each manifold station also has a flow balancing valve. The zone valves to the fan-coils remain closed during heating mode.

In cooling mode, the same circulator supplies two independently controlled fan-coils. Each fan-coil circuit has a flow balancing valve. The zone valves to the heating manifold stations remain closed during cooling mode.

During heating mode, the heat pump and its associated circulator are turned on and off based on an outdoor reset controller that monitors the temperature at the mid-point of the buffer tank. This control action is independent of any call for heat from the room thermostats. The warmer the outdoor temperature, the lower the target temperature calculated by this controller. The maximum water temperature during heating mode is 110 °F. The minimum temperature is 80 °F. The reset controller operates on a 10 °F differential centered on the target temperature. These temperatures allow the heat pump to operate at a relatively high COP.

During cooling mode, the heat pump and its associated circulator are turned on and off based on a temperature setpoint controller that monitors another sensor at the mid-point of the tank. The heat pump is turned on when the tank temperature is 60 °F or higher, and off when the tank temperature drops to 45 °F. The heat pump operates independently of calls for cooling from the room thermostats.

Spring check valves are used to prevent reverse thermosiphoning from the tank, and to limit chilled-water migration into the heating portions of the distribution system. A magnetic dirt separator is used to protect the heat pump and high-efficiency circulator from dirt and iron oxides. A fluid feeder is used to maintain system pressure as dissolved air is captured and vented.

System #3: Multi-Zone/ Simultaneous Heating & Cooling

Some large homes or commercial buildings may require heating in certain areas at the same time that other areas require cooling. For example, a room in the core area of the building, with no exposed building envelope surfaces, and experiencing constant heat gain from people or equipment, may require cooling much of the year. At the same time, the perimeter of the building may require heating. These situations can be handled in several ways. One approach is based on multiple heat pumps serving two buffer tanks, one tank for heating and the other for cooling. This concept is shown in Figure 15-50.

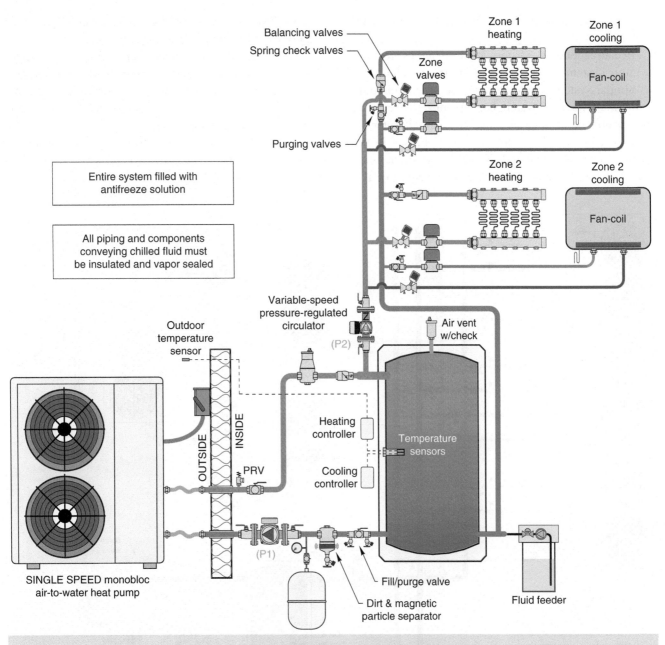

Figure 15-49 | Piping schematic for example system #2. Multiple heating and cooling zones.

This system is supplied by three independently controlled monobloc air-to-water heat pumps. Any of these heat pumps can be turned on, whenever needed, in either heating mode or cooling mode.

The heat pumps are connected to a 4-pipe extended manifold that leads back to the two buffer tanks. Each heat pump has two 3-position diverter valves that connect the heat pump to either the heating supply and return pipes, or the cooling supply and return pipes. If the heat pump is off, the diverter valves go to their "off" position, preventing flow through the heat pump.

Flow through any active heat pump and its target buffer tank is provided by a variable-speed pressure-regulated circulator. These circulators operate in constant differential pressure mode, automatically adjusting speed as necessary based on the number of active heat pumps.

Both buffer tanks provide hydraulic separation between the heat pump circulators and the variable-speed distribution circulators.

Heating is provided by zoned radiant panels. Cooling is provided by zoned fan-coils.

The controls for this system are based on maintaining each buffer tank within a suitable temperature range that allows adequate heating or cooling. At the beginning of the cooling season, and when there is no expectation of heating capability, the controls could be set to focus

820 Chapter 15 Fundamentals of Hydronic Cooling

solely on maintaining the chilled-water buffer tank temperature.

Because monobloc air-to-water heat pumps are used, the entire system operates with an antifreeze solution.

Similar systems could be designed around multiple geothermal water-to-water heat pumps, or even a combination of air-to-water and geothermal water-to-water heat pumps. The number of heat pumps in the system could be increased if additional capacity is needed. It could also be reduced for lower capacity. If a single heat pump was used the controls would have to set a priority of cooling overheating, or vice versa.

The controls could also be programmed to monitor the total operating hours accumulated on each heat pumps, and adjust the start up/shut down order so that each heat pump undergoes approximately the same number of run hours each year.

It would also be possible to incorporate a water-to-water heat pump between the two buffer tanks, as shown in Figure 15-4.

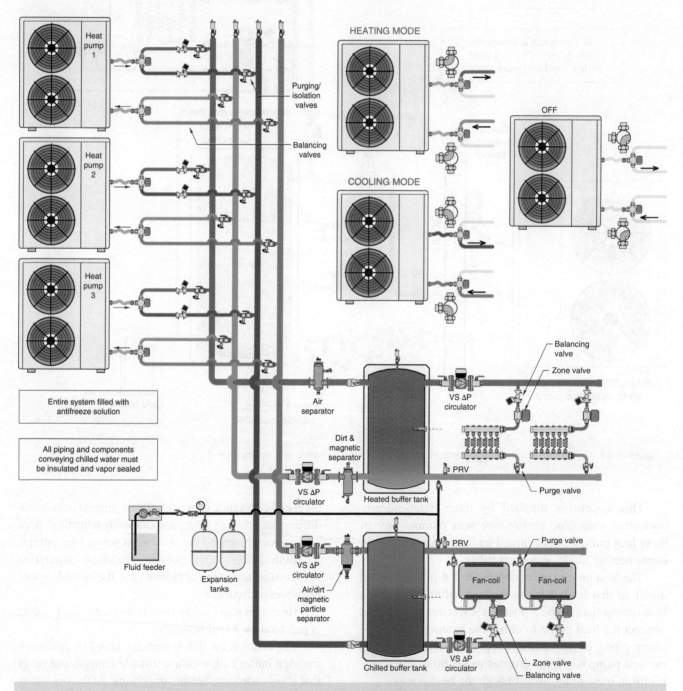

Figure 15-50 | Piping schematic for example system #3. Multiple air-to-water heat pumps supplying heated and chilled buffer tanks, allowing for simultaneous heating and cooling in building.

Summary

The global trend toward beneficial electrification will increase the use of hydronic heat pumps (e.g., air-to-water, or water-to-water). This will change the "relevancy" of hydronics technology in residential and light commercial buildings, from one that has been traditionally focused on heating, to one that enables both heating and cooling. This chapter was developed to demonstrate how to provide the latter.

Chilled-water cooling has been used for years in large buildings. It leverages the same superior thermal properties of water that allow hydronic heating to provide many benefits, such as small distribution piping and far higher distribution efficiency. Forward-thinking hydronic heating professionals should add hydronic cooling to the portfolio of services they offer. They should also invest in learning how to provide simple ducting design and installation, as well as design and installation of heat recovery ventilation systems. Doing so will open new market opportunities and enable them to meet client expectations with highly efficient heating, cooling, and ventilation solutions.

Key Terms

- 2-pipe air handler (or 2-pipe fan coil)
- 3-position motorized diverter valves
- 4-pipe air handler (or 4-pipe fan coil)
- absolute humidity
- ACCA Manual J
- air handler
- air-to-water heat pumps
- Arduino®
- automatic backwash filter
- carbon emissions
- chiller
- condensate
- condensate drain pan
- condensate P-trap
- condensation
- condenser unit
- console fan coils
- cooling capacity
- cooling COP
- cooling load
- decarbonization
- dedicated chillers
- dehumidification
- description of operation
- dewpoint
- dewpoint control
- dewpoint controller
- direct expansion ("DX") evaporator coils
- direct lake water cooling systems
- distribution efficiency
- dry bulb temperature
- effective coefficient of performance (COP_e)
- elastomeric foam insulation
- electric tankless water heater
- emergency shut off switch
- energy efficiency ratio (EER)
- energy recovery ventilator (ERV)
- external static pressure
- foil scrim kraft
- foot valve
- form-fitting insulation shell
- geothermal preconditioning coil
- grains
- heat of rejection
- heat recovery chiller
- heat recovery ventilator (HRV)
- heat tracing cable
- high wall fan coil
- inches of water column (inch w.g.)
- Indirect lake water cooling systems
- lake heat exchanger
- lake source cooling
- latent cooling load
- micro controller
- pre-insulated PEX
- process line
- proprietary controls
- psychrometric chart
- radiant panel cooling
- Raspberry Pi®
- relative humidity
- relative humidity controller
- reverse indirect water heater
- saturation condition
- seasonal energy efficiency ratio (SEER)
- secondary drain pan
- sensible cooling capacity
- sensible cooling load
- sensible heat ratio (SHR)
- sling psychrometer
- specific humidity
- superheat
- terminal units
- total cooling capacity
- total cooling load
- variable refrigerant flow (VRF)
- variable-speed pressure-regulated circulators
- water-to-water heat pumps
- wet bulb temperature

Questions and Exercises

1. Describe, in detail, three benefits or advantages of hydronic cooling.

2. Explain why electricity is becoming the preferred energy source in comparison to fossil fuels. Be specific.

3. Explain why piping carrying chilled-water for cooling can be much smaller in cross-section compared to ducting conveying cool air, when both streams provide the same rate of cooling.

4. The blower in a forced-air cooling system operated on 550 watts, while delivering cooling at a rate of 35,000 Btu/h. The circulator in a chilled-water cooling system operates on 35 watts while delivering the same rate of cooling. Calculate and compare the distribution efficiency for both systems.

5. Describe the difference between an air handler and a fan coil, as they would be applied in a chilled-water cooling system.

6. Describe three advantages that chilled-water cooling has relative to VRF systems.

7. What portions of a circulator and zone valve should *not* be insulated when these components are used in a chilled-water cooling system? Why?

8. (a) Determine the dewpoint of air at 85 °F and 40% relative humidity. How many grains of water would be contain in 3 pounds of water at this condition?

9. Use the psychrometric chart in Figure 15-10 to determine the relative humidity of air at 60 °F dry bulb temperature and 50 °F wet bulb temperature.

10. A person inside a house during warm weather adds to the:
    a. Sensible cooling load
    b. Latent cooling load
    c. Total cooling load
    d. Heating load

11. A "deep" cooling coil with multiple tube passes is preferred in which of the following situations:
    a. High sensible heat ratio
    b. Low sensible heat ratio
    c. A building in an arid climate with internal heat gain from electrical equipment

12. Explain why an air-to-water heat pump doesn't have a SEER rating.

13. As the chilled-water temperature produced by a heat pump increases,
    a. the cooling COP of the heat pump increases.
    b. the cooling capacity of the heat pump decreases.
    c. the heat pump's Btu/h cooling output per watt of electrical input remains constant.

14. Describe the difference between an air-to-water heat pump and a heat recovery chiller.

15. Why are the zone circuits in Figure 15-17 being supplied with water from the lower portion of the buffer tank?

16. What is the first step in evaluating the suitability of using lake water cooling, assuming the building is reasonably close to the lake shore?

17. Why is a heat exchanger required between the lake water and chilled-water distribution system for a direct lake water cooling system?

18. Explain the importance of a P-trap in the condensate drain piping from an air handler.

19. Does doubling the chilled-water flow rate through an air handler double its total cooling capacity? Explain why or why not.

20. Describe the difference between a standard diverter valve and a 3-position diverter valve.

21. Why is it necessary to monitor the dewpoint temperature of interior spaces when using radiant panel cooling?

22. Develop an electrical control system for the piping system shown in Figure 15-48. Use a ladder diagram to document that control system.

23. Develop a description of operation for the control system developed in question 22.

24. Why is it necessary to buffer the heating portion of the system shown in Figure 15-47, but not the cooling portion?

25. What is the primary reason to use a secondary drain pan under an air handler that contains a drip pan?

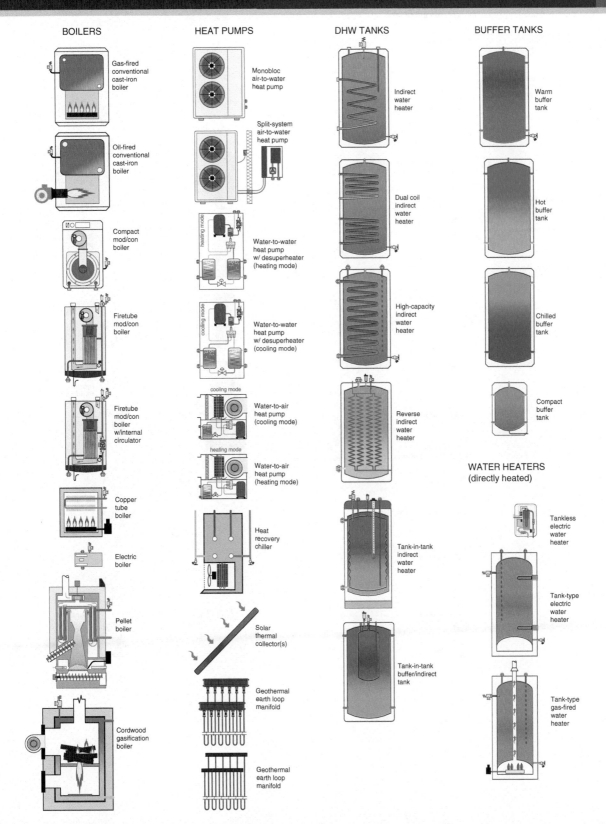

Appendix A

CIRCULATORS

- Wet rotor circulator
- Wet rotor circulator w/ internal check valve
- Wet rotor circulator w/ isolation flanges
- Wet rotor circulator w/ internal check valve & isolation flanges
- Variable-speed pressure-regulated circulator w/ isolation flanges
- Variable-speed pressure-regulated circulator w/ isolation flanges & internal check valve
- Wet rotor circulator w/ isolation unions
- Internal circulator
- Closely-coupled series circulators
- High capacity circulator
- High capacity circulator w/ isolation valves

VALVES

- Gate valve
- Globe valve
- Ball valve
- Drain valve
- Swing check valve
- Spring check valve
- Purge valve
- Dual port purge valve
- Thermostatic radiator valve (straight pattern)
- Thermostatic radiator valve (angle pattern)
- Thermostatic radiator valve (capillary tube actuator)
- Zone valve
- Diverter valve
- 3-position diverter valve
- Balancing valve
- ΔP valve
- Thermostatic mixing valve
- Pressure relief valve
- Temperature & pressure relief valve
- Pressure reducing valve
- Butterfly valve
- 3-way motorized mixing valve
- 4-way motorized mixing valve
- Motorized globe valve
- Motorized ball valve
- Flow switch

SEPARATORS

- Microbubble air separator
- Microbubble air separator (vertical pipe)
- Air & dirt separator
- Air & dirt separator w/magnet
- Y-strainer
- Low velocity zone dirt separator
- Low velocity zone dirt separator w/magnet
- Dirt separator (vertical pipe)
- Dirt separator w/ magnet (vertical pipe)
- Air separator (large / flanged)
- Dirt separator (large / flanged)
- Air & dirt separator (large / flanged)

MISC.

- Closely space tees
- Backflow preventer (testable)
- Backflow preventer
- Float air vent
- Tee
- Reducer tee
- Cross
- Union
- Brazed plate heat exchanger
- Brazed plate heat exchanger
- Brazed plate heat exchanger
- Plate & frame heat exchanger
- Expansion tank (small)
- Expansion tank (medium)
- Expansion tank (large)

SPECIALTY

- Fluid feeder
- Expansion compensator
- Expansion compensator
- Anti-condensation mixing valve (high capacity)
- Loading unit
- Dual isolation radiator valve
- Temperature sensor in well
- Thermometer
- Pressure gauge
- Hydraulic separator
- Hydraulic separator w/magnet
- Hydraulic separator (flanged)
- Hydraulic separator w/ magnet (flanged)

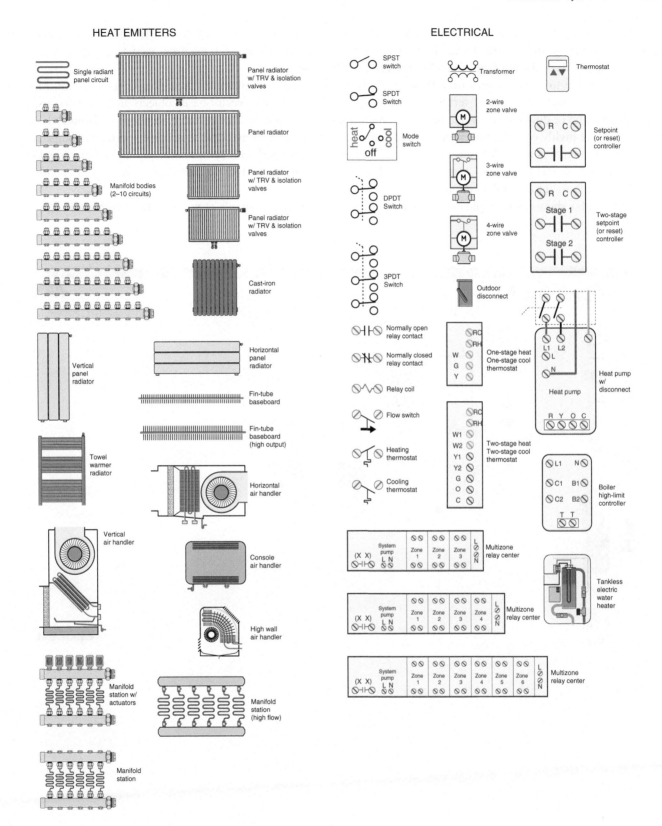

The symbols shown in Appendices A1, A2, and A3, as well as throughout this textbook, provide a broad palette for constructing piping and electrical schematics. Once these symbols are learned, it becomes easy to interpret schematic diagrams and reduce the amount of labelling needed on those diagrams.

It is possible to create images from these symbols using virtually any currently available CAD software. Most of these CAD softwares would create vector-based objects for each component. Vector-based objects are useful because they can be scaled up or down in size while retaining proportions, colors, line thickness, and resolution. Unfortunately, CAD systems, while highly capable, are expensive and require a significant learning curve to achieve speed and competence.

One alternative is a cloud-based software tool that is pre-populated with symbols for piping and electrical schematics. This tool enables simple "drag & drop" drawing construction techniques, and uses "intelligent" connectors to couple symbols into complete circuits.

Figure A4-1 shows an example of one such software tool, which is available at **www.hydrosketch.com**.

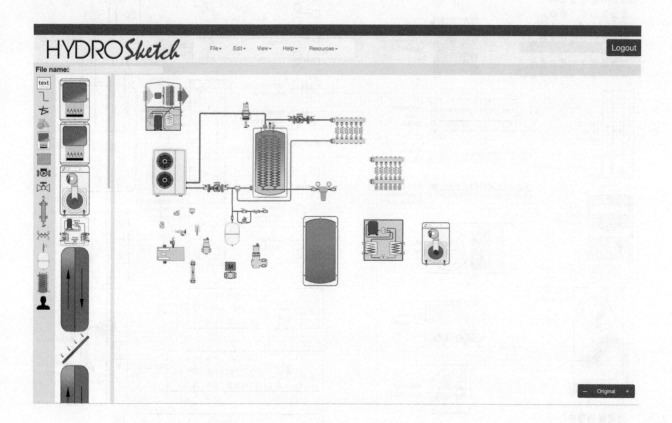

R-Values of Common Building Materials

Appendix B

Material	R-Value*	
INSULATIONS		
Fiberglass batts (standard density)	3.17	per inch
Fiberglass batts (high density)	3.5	per inch
Blown fiberglass	2.45	per inch
Blown cellulose fiber	3.1–3.7	per inch
Foam in place urethane	5.6–6.3	per inch
Expanded polystyrene panels (beadboard)	3.85	per inch
Extruded polystyrene panels	5.4	per inch
Polyisocyanurate panels (aged)	7.2	per inch
Phenolic foam panels (aged)	8.3	per inch
Vermiculite	2.1	per inch
MASONRY AND CONCRETE		
Concrete	0.10	per inch
8-in concrete block	1.11	for stated thickness
w/ vermiculite in cores	2.1	for stated thickness
10-in concrete block	1.20	for stated thickness
w/ vermiculite in cores	2.9	for stated thickness
12-in concrete block	1.28	for stated thickness
w/ vermiculite in cores	3.7	for stated thickness
Common brick	0.2–0.4	per inch
WOOD AND WOOD PANELS		
Softwoods	0.9–1.1	per inch
Hardwoods	0.8–0.94	per inch
Plywood	1.24	per inch
Waferboard or oriented strand board	1.59	per inch
FLOORING		
Carpet (1/4-in nylon level loop)	1.36	for stated thickness
Carpet (1/2-in polyester plush)	1.92	for stated thickness polyurethane
Foam padding (8 lb density)	4.4	per inch
Vinyl tile or sheet flooring (nominal 1/8 in)	0.21	for stated thickness
Ceramic tile	0.6	per inch

* The R-value for a specific thickness of a material may be obtained by multiplying the R-value per inch by the thickness in inches (or fractions of inches). The units on R-value are the standard U.S. units of °F $\times$ hr $\times$ ft^2/Btu.

Material (Continued)

MISCELLANEOUS

Material	R-Value*	
Drywall	0.9	per inch
Vinyl clapboard siding	0.61	for all thicknesses
Fiberboard sheathing	2.18	per inch
Building felt (15 lb/100 ft^2)	0.06	for stated thickness
Polyolefin housewrap	~0	for all thicknesses
Poly vapor barriers (6 mil)	~0	for stated thickness

The data in this table were taken from a number of sources including the ASHRAE Handbook of Fundamentals and literature from several material suppliers. It represents typical R-values for the various materials. In some cases a range of R-value is stated due to variability of the material. For more extensive data consult the ASHRAE Handbook of Fundamentals or contact the manufacturer of a specific product.

R-Values of Air Films

INSIDE AIR FILMS	R-VALUE*
Horizontal surface w/ upward heat flow (ceiling)	0.61
Horizontal surface w/ downward heat flow (floor)	0.92
Vertical surface w/ horizontal heat flow (wall)	0.68
45-degree sloped surface w/ upward heat flow	0.62

OUTSIDE AIR FILMS	
7.5 mph wind on any surface (summer condition)	0.25
15 mph wind on any surface (winter condition)	0.17

* The R-value for a specific thickness of a material may be obtained by multiplying the R-value per inch by the thickness in inches (or fractions of inches). The units on R-value are the standard U.S. units of °F $\times$ hr $\times$ ft^2/Btu.

Useful Conversion Factors and Data

Appendix C

LENGTH:

1 Foot (ft) = 12 inches (in) = 0.3048 meter (m)

AREA:

1 Square foot (ft^2) = 144 square inches (in^2) = 0.092903 square meters (m^2)

VOLUME:

1 Cubic foot (ft^3) = 7.49 gallons (gal) = 1728 cubic inches (in^3)
1 Gallon (gal) = 3.7853 liter (L)

FLOW RATE:

1 Gallon/minute (gpm) = 0.002225 ft^3/sec = 0.063008 L/sec = 0.227126 m^3/h.

VELOCITY:

1 m/sec = 3.2808 ft/sec = 2.2369 mph

PRESSURE:

1 bar = 14.7 (psi)
1 pound per square inch (psi) = 6894.76 N/m^2 = 6894.76 pascal (Pa)
Absolute pressure (psia) = [gauge pressure (psig) + 14.7]

ENERGY:

1 Kilowatt hour (kWhr) = 3,413 Btu
1 Therm = 100,000 Btu
1 MMBtu = 1,000,000 Btu
1 Joule (J) = 0.7376 ft·lb = 0.2388459 calorie (cal)
1 Kilojoule (KJ) = 0.9478 Btu
1 Btu (British Thermal Unit) = 777.98 ft·lb

POWER AND HEAT FLOW:

1 kilowatt (kW) = 3,413 Btu/hr.
1 Ton = 12,000 Btu/hr.
1 Horsepower (hp) = 0.746 kW = 2546 Btu/hr.

TEMPERATURE (SCALE):

°C = (°F − 32)/1.8
°F = (°C × 1.8) + 32

TEMPERATURE (DIFFERENCE):

1 °C = 1.8 °F
1 °F = 0.55555 °C

THERMAL CONDUCTIVITY:

1 Btu/(hr · ft · °F) = 1.7307 W/(m · °C)

HEAT TRANSFER COEFFICIENT:

1 Btu/(hr · ft^2 · °F) = 5.67859 W/(m^2 · °C)

SPECIFIC HEAT:

1 Btu/(lb · °F) = 4185.85 J/(kg · °C)

HEAT FLUX OR SOLAR RADIATION INTENSITY:

1 Btu/(hr · ft^2) = 3.15378 W/m^2

INLET FLUID PARAMETER:

1 (°F · hr. · ft^2/Btu) = 0.17615 (°C · m^2)/W

CHEMICAL ENERGY CONTENT OF COMMON FUELS:

1 Cubic foot of natural gas = 1020 Btu (chemical energy content)
1 Gallon #2 fuel oil = 140,000 Btu (chemical energy content)
1 Gallon propane = 91,200 Btu (chemical energy content)

Glossary

1-pipe valve (for panel radiators)   A special valve that allows multiple panel radiators to be connected in a sequence, but also allows independent flow control through each radiator.

2 to 10 VDC   A commonly used analog output signal from a controller for regulating devices such as motorized valves or the speed of a circulator.

2-pipe air handler (or 2-pipe fan-coil)   An air handler or fan-coil that only connects to two pipes.

2-pipe buffer tank   A piping configuration in which the load is connected between the heat source and the buffer tank.

4 to 20 mA   A commonly used analog output signal from a controller for regulating devices such as motorized valves or the speed of a circulator.

4-pipe air handler (or 4-pipe fan coil)   An air handler or fan-coil that contains two separate coil heat exchangers, one for cooling and the other for heating.

A

A-coil   A fluid to air heat exchanger consisting of aluminum fins and copper tubing. The overall shape of the coil resembles the sides of the letter A.

Above-floor tube and plate system   A radiant floor heating system in which the tubing and heat transfer plates are installed above the subflooring.

Absolute humidity   The number of pounds of water contained in a pound of air at some temperature. Also stated as the number of grains of water contained in a pound of air (7,000 gains = 1 pound).

Absolute pressure   The pressure of a liquid or gas relative to a pure vacuum. It is often expressed in units of psia (pounds per square inch absolute).

Absorber plate   The heat exchanger within a solar thermal collector that absorbs solar radiation and transfer a portion of that energy to a fluid passing through the collector.

ACCA Manual J   A manual, published by the Air Conditioning Contractors of America (ACCA), that provides procedures and data for estimating heating and cooling load.

Acceptance volume   The volume of fluid that can enter a closed expansion tank without the pressure of the air in the tank exceeding a set limit.

Active length   The portion of a fin-tube element that has fins on it.

Actuator   An electrically or pneumatically driven device that adjusts the position of a valve stem or damper based on the signal it receives from a controller.

Aerial boom   A specialized truck used for placing concrete in areas that are difficult or impossible to reach using other means.

Aggregate   The stone or sand used in concrete, but can also refer to other solid granular materials placed in concrete.

Air binding   The inability of a circulator to dislodge a large air pocket at a high point in a piping system.

Air change method   A traditional method of expressing the rate of air leakage into a building. One air change per hour means the entire volume of air in the building is replaced with outside air each hour.

Air film resistance   The thermal resistance of the air film along a wall, ceiling, or floor surface.

Air handler   A generic name for a device consisting of a blower and a finned-tube coil. It is used to either heat or cool air passing through it based on the temperature of the fluid circulating through the tubes of the coil.

Air infiltration   The process of outside air entering a building as inside air leaks out of the building.

Air purger (or air scoop, or air separator)   A device that separates air bubbles from flowing fluid and ejects them from the system.

Air-side pressurization   The pressure on the air side of a diaphragm-type expansion tank before fluid enters the tank. This pressure is adjusted by adding or releasing air through the Schrader valve on the tank.

Air temperature stratification   The natural tendency for warm air to rise to the upper portion of a room or interior space while cooler air settles near the floor.

Air-to-water heat pump   A heat pump that extract heat from outside air and transfers it to a stream of water (in heating mode), and vice versa in cooling mode.

Air vent   A manual or automatic device that releases air bubbles from a piping system.

Ampacity   The maximum current that can be carried by a conductor in a given application as determined by the National Electrical Code.

Analog control signal   An electrical voltage or current that varies over a predetermined range, and typically used to control a modulating device.

AND decision A digital logic function that has an output of (1) if all inputs are (1), otherwise the output is (0). Often used to represent the logical equivalent of two or more switches wired in series.

Angle valve A valve with its outlet port rotated 90 degrees from its inlet port. Also called an angle pattern valve.

Anion A negatively charged ion.

Annual fuel utilization efficiency (AFUE) An estimate of the seasonal efficiency of a heat source based on a federal government standard.

Approach temperature difference The difference between the temperature of the warmer fluid entering one side of a heat exchanger and the temperature of the cooler fluid leaving the other side.

Aquastat A device that measures the temperature of a liquid at some point in a system and opens or closes electrical contacts based on that temperature and its setpoint temperature.

Arduino® A low-cost single board microcontroller.

ASSE 1017-rated anti-scald tempering valve A special valve designed to prevent scalding hot domestic hot water from entering the piping distribution system in a building.

ASTM E3137/E3137M-17 An accuracy and testing standard used of heat meters.

Atmospheric: A term often used to describe gas burners that are designed to operate directly exposed to (unpressurized) room air.

Automatic air vents Air vents that can automatically eject air as it accumulates within them.

Automatic backwash filter A filter that automatically detects accumulated dirt and initiates a flushing cycle to clean itself.

Automatic flow balancing valve A special type of valve that attempts to hold a specified flow rates over a wide range of differential pressure.

Automatic glycol make-up system A device that automatically pumps glycol into a hydronic system upon a predetermined drop in system pressure.

Auxiliary loads Any additional heating load served by a hydronic space heating system.

Available floor area The floor area in a room that can be used to release heat as part of a floor heating system.

Average floor surface temperature The mean temperature of the upper surface of a heated floor.

Average flow velocity The average speed of a fluid particle across a specified cross-section in a piping component.

B

Backflow preventer A piping component that contains the functional equivalent of two check valves and a vent port. It prevents any fluid within a hydronic system from flowing backward and contaminating a fresh water source.

BACnet A programming language used in building automation systems.

Balance point temperature (a) The outdoor temperature at which the internal heat gains in a building equals the rate of heat loss from the building. (b) The outdoor temperature at which the heating capacity of a heat pump equals the building heating load.

Balance point temperature The outdoor temperature below which a building requires heat input from its space heating system.

Balancing The process of adjusting flow in a multiple branch piping system so that the proper amount of heat is delivered from each branch.

Balancing valve A valve operated at a partially open setting to achieve a desired flow rate at a given location in a hydronic system.

Ball valve A valve containing a rotating ball with a hole through it. It can be used for component isolation or limited flow regulation.

Band joist The edge (or perimeter) board on a wood-framed floor deck.

Bar A unit of pressure. 1 bar = 14.504 pounds per square inch (PSI).

Barefoot friendly floors A subjective term referring to warmed floors that feel very comfortable to a bare human foot.

Baseboard tee A special fitting that resembles a 90-degree elbow with an additional threaded tapping. This fitting is often used to mount an air vent at the outlet end of a finned tube baseboard element.

Basket strainer A component designed to separate dirt particle from a liquid flow stream and retain them within a fine mesh basket made of stainless steel or brass wire.

Bearing wall A wall within a building carrying structural loads downward from above.

Bell hanger A small copper device designed to support a tube or pipe a small distance away from a surface.

Bellows A closed metal container with pleated sidewalls that changes length when it experiences a change in internal pressure.

Below-floor tube and plate system A radiant floor heating system in which the tubing and metal heat transfer plates are attached to the underside of a subfloor.

Bend supports Preformed elbows or clips that allow flexible polymer tubing to be bent to a relatively small radius without kinking.

Best efficiency point (BEP): The point on a pump curve where the ratio of shaft input power divided by hydraulic output power (or electrical input power to a motor) is at its maximum value. Typically near the center of a pump curve.

Bimetal element An assembly composed of two dissimilar metal strips joined at their ends. It creates the movement of electrical contacts in a device such as electromechanical room thermostat.

Bin temperature data A means of tabulating the duration of ranges of outside temperature based on the number of hours in a year the temperature is within a given range. Typical bin temperature ranges are 5 °F or 1 °F. Bin temperature data is available for hundreds of locations in North American and globally.

Biomass Any plant material that can be combusted. In this textbook it refers to either cordwood or wood pellets.

Black ice A very thin film of ice that forms on pavement surfaces under certain conditions, and is very difficult to see.

Blower Different from a fan, a blower creates air motion by using an electric motor to spin a "squirrel cage" assembly of curved metal blades.

Blower door A testing instrument designed to measure the air leakage rate of a building by creating a measured pressure differential between the inside and outside of the building.

Blow-through fan-coils A configuration of a fan-coil (or air handler) where the fan (or blower) pushes air through a coil heat exchanger.

Bluetooth A communication protocol used for wireless communication between digital devices within relatively close distances of each other.

Boiler drain A valve that connects to a hose thread and can be placed at any point in a system requiring drainage.

Boiler feed water valve Another name for a pressure reducing valve that automatically allows water to flow into a hydronic system if the system pressure drops below a preset value.

Boiler jacket The outer sheet metal or polymer shell of a boiler.

Boiler protection A control action intended to prevent a conventional boiler from operating with sustained flue gas condensation.

Boiler reset control A control action that automatically changes the upper temperature limit of a boiler based on outdoor temperature.

Boiler rotation A control technique that varies the firing order of two or more boilers in a multiple boiler system with the goal of equalizing the operating hours of each boiler.

Boiler short-cycling An undesirable condition during which the burner of a boiler operates for a very short time. It leads to a high number of on/off cycles and premature wear of components.

Bond breaker A material used to prevent an adhesive bond between two materials.

Borehole A vertical hole in the earth created by a well drilling rig, and used to install a vertical piping assembly for a geothermal heat pump.

Boundary layer A thin layer of fluid that moves slowly along the inside wall of a pipe. The thicker the boundary layer is, the lower the rate of heat transfer between the bulk of the fluid and the tube wall.

Branch A segment of an overall piping system that is typically in parallel with another segment.

Brazed plate heat exchanger A compact heat exchanger consisting of parallel stainless steel plates brazed together at their edges and internal contact points.

Brine A generic term for a solution of water and antifreeze such as propylene glycol, ethylene glycol, or calcium chloride.

British thermal unit The amount of energy required to raise one pound of water by one degree Fahrenheit, also abbreviated Btu.

Brushless DC motor A motor with a permanent magnet rotor and electronically commutated stator coils. Also known as an ECM (electronically commutated motor).

BTU meter A device that senses both flow rate and temperature differential at some location in a hydronic system, and uses this data to calculate the rate of heat transfer and the total number of BTUs that have passed through that location.

BTU metering The process of measuring the thermal energy passing a given point in a hydronic system by measuring the flow rate and temperature drop of the fluid being circulated.

Bubble rise velocity The speed at which a bubble can rise through a liquid. Larger bubbles have greater rise velocities.

Buffer tank An insulated, water-filled storage tank that adds thermal mass to a hydronic distribution system. This tank can prevent the heat source from short cycling when the rate of heat input from the heat source is very different than the rate of heat output from the distribution system.

Building design heating load The total estimated rate of building heat loss under "design" conditions (based on the outdoor design temperature for a given location).

Building heat loss coefficient The building's design heat loss rate divided by the difference between inside and outside air temperature at which this rate of heat loss was determined.

Building heating load The rate at which heat must be added to a building to maintain a desired interior temperature.

Building mains A set of pipes that carry heated water from a mechanical room throughout a building, provide flow to several branch circuits.

Bulkhead fitting An assembly that allows a piping connection to be made through the sidewall of a tank, typically above the water level inside the tank.

Buoyancy The upward force that causes a less dense material to rise above a more dense fluid.

C

Candidate circulator A circulator that is being considered as a possible choice for a given flow rate and associated head loss situation.

Capacitance rate The product of mass flow rate and specific heat for a fluid. Can be expressed in units of Btu/h/°F.

Capacitance rate ratio The ratio of the smaller fluid capacitance rate through a heat exchanger divided by the larger capacitance rate.

Capillary tube   A very small diameter tube between a temperature sensing bulb and an electromechanical controller.

Carbon emissions   The carbon releases as the result of a chemical process such as combustion of a fuel.

Carbon-neutral fuel   A fuel such as wood that releases the same amount of carbon into the atmosphere if it is combusted or decomposes through natural processes.

Cations   A positively charged ion.

Cavitation   The formation of vapor pockets when the pressure on a liquid drops below its vapor pressure. Cavitation is very undesirable in circulators.

Cd factor   A correction factor used when estimating seasonal heating energy use based on degree days.

Central air separator   A device that separates air bubbles from the moving system fluid and ejects them from the system.

Centrifugal pump   An electrically driven device that adds mechanical energy (e.g., head) to a fluid using a rotating impeller. In this text, the word pump refers specifically to a centrifugal pump, and is used synonymously with the word circulator.

Characterized valve   A valve with a specifically shaped flow control element that creates an equal percentage flow characteristic.

Chase   A cavity within a building, usually in a vertical direction, through which mechanical or electrical services can be routed.

Check valve   A valve that limits flow to one direction.

Chemical energy content (of fuel)   The theoretical energy contained in a fuel prior to its combustion, and based on its chemical composition.

Chemical water quality   A general term referring to properties such as pH, or content of dissolved substances such as calcium, magnesium, or other minerals.

Chiller   A device that operates based on a refrigeration cycle to create chilled water—usually for building cooling purposes.

Circuit balancing valve   A valve used to adjust the flow rate to a desired value at a specific location within a hydronic system.

Circuit layout drawing   A drawing showing where tubing will be placed in floors, walls, or ceilings to create radiant panels.

Circuit simulator module   A software module with the Hydronics Design Studio that allows the flow and thermal performance of user-defined circuits to be analyzed.

Circulator   A centrifugal pump that creates flow in a hydronic circuit by adding head energy to the fluid. The term circulator is preferred to the word pump in the context of hydronic circuits.

Class (of SIM system)   One of three classes of snow and ice melting systems (SIMs) based on the rate of their ability to keep the pavement surface free of snow as it is falling.

Clevis hanger   A type of pipe support that is typically supported by a threaded rod from a ceiling structure.

Close-coupled   Bolting two or more inline circulators together end to end, in the same flow direction, to increase head.

Closed-loop control system   A control system configuration in which the controlled variable is sensed and the control action modified (if necessary) based on any deviation between the sensed value and target value of the controlled variable. Also known as a feedback control loop.

Closed-loop source   Usually refers to piping buried in the earth for the purpose of absorbing low-temperature heat for a geothermal heat pump.

Closed-loop system   A piping system that is sealed at all points from the atmosphere.

Closely spaced tees   Two tees that are installed immediately adjacent to each other to form a primary/secondary connection (e.g., hydraulic separation) between two piping circuits.

Closet flange   A fitting fastened to a flow for the drainage connector of a toilet.

Coalesce   A natural process in which molecules of gases such as oxygen and nitrogen, which are dissolved in water, join together under specific temperature and pressure conditions, to form microbubbles.

Coalescing media   A three-dimensional mesh designed to encourage dissolved gas molecules to coalesce (e.g., come together) to form microbubbles.

Cocurrent flow   When two fluids pass through a heat exchanger in the same direction.

Coefficient of linear expansion   A number that indicates the change in length of a specified material based on a change in temperature. Usually given in units of in/in/ °F.

Coefficient of performance (COP)   The ratio of the heat output rate of a heat pump divided by the rate of electrical energy input, where both rates are expressed in the same units. High COPs are desirable.

Coil   (a) A common name for a fluid to air heat exchanger mounted within an air handler or fan-coil. (b) The electromagnet that provides the motive force to move contacts in a relay or contactor.

Collector array   Two or more solar thermal collectors that operate as a group.

Combined air & dirt separator   A device designed to remove microbubbles of air as well as dirt from a liquid flowing through it.

Combi-system   Hydronic systems that provide both space heating and domestic hot water.

Combustion efficiency   The efficiency of a heat source in converting the chemical energy content of its fuel into heat, based on measurements of exhaust gas temperature and its carbon dioxide content.

Common piping The piping path in a multiple branch system that is in common with branch circuits and through which the full system flow passes.

Compact style panel A type of panel radiator made with preformed steel sheets welded together.

Compensated method A procedure used to balance parallel hydronic circuits by adjusting balancing valves in a prescribed sequence.

Component isolation The use of valves on all piping connections to a device that allow it to be isolated for service if necessary.

Composite tubing Tubing made of both metal and polymer materials. PEX-AL-PEX is an example of composite tubing.

Compressible fluids Fluids, usually gases, which can be compressed into smaller volumes when pressure is applied to them.

Compressive "donut" collar A mechanically-compressed collar designed to seal between the outer surface of a pipe and a hole through a concrete wall.

Compressive Load Rating The maximum load (often expressed in pounds per square inch) that should be placed on a foam insulation panel.

Compressor The device in which a refrigerant gas is compressed in volume, causing its temperature to increase.

Computer-aided drafting (CAD) The use of a computer and associated software for preparing technical drawings such as floor plans, and schematics.

Communicating thermostat An electronic thermostat that can send and receive digital information to other communicating thermostats or other communicating controllers within a networked control system.

Communication bus A set of wires (typically two), that link all devices in a networked control system together to enable digital communication.

Concrete sub-slab A slab of concrete that is poured but then covered with one or more layer(s) of other material(s).

Concrete thin-slab A layer of concrete in the range of 1 to 2 inches thick placed over hydronic tubing to create a radiant floor panel.

Condensate The liquid formed as a gas, such as water vapor, condenses.

Condensate drain pan A pan in some fan-coils and air handlers for catching water droplets falling from a chilled coil heat exchanger located above the pan.

Condensate P-trap A trap, similar to a plumbing P-trap, designed to prevent the negative air pressure inside an air handler from pulling air or sewer gases into the air passing through the air handler.

Condensation The process by which water vapor, or other vapors, condense into liquids.

Condenser The device in which a refrigerant gas releases heat, and in doing so changes from a vapor into a liquid.

Condenser unit The outside portion of an air conditioning or heat pump system used to exchange heat between a refrigerant and air.

Condensing mode Operating a combustion type heat source so that the water vapor produced during combustion condenses on the internal surfaces of the heat exchanger.

Conduction A natural process whereby heat is transferred by molecular vibrations through a solid or liquid material.

Confined space Any space not meeting the definition of unconfined space in the National Fuel Gas Code.

Conservation of energy A fundamental principle of physics that states that energy cannot be created or destroyed—only changed in form.

Conservation of mass A physical principle, that in the context of incompressible liquids, implies that the rate at which liquid enters a liquid-filled device (such as a tank, valve, or collection of several components) must equal the rate at which that liquid exits the device.

Console fan-coil A fan-coil for heating or cooling that is mounted low on a wall.

Constant differential pressure control A control algorithm that adjusts the speed of a circulator such that the pressure differential between its inlet and outlet ports remains at a set (fixed) value regardless of flow rate.

Constrained ΔT control A control algorithm for a circulator that automatically adjusts the circulator's speed, and thus the flow rate through it, in an attempt to maintain a preset temperature difference between two temperature sensor locations.

Contact adhesive An adhesive applied to both surfaces to be joined, that bonds those surfaces immediately upon contact.

Contactor Another name for a relay that is rated for relatively high current/line voltage loads.

Contacts The metal components within a switch or relay that come together or move apart to determine if electrical current will flow through a given pole of the switch or relay.

Continuous flow rating The rate at which a water heating device can supply heated water at a given temperature on a sustained basis. Often expressed in gallons per hour or gallons per minute.

Control algorithm The operating instructions or logic that a controller follows.

Control differential The intended difference in the sensed control variable from where the controller output is turned on to when it is turned off. For example, if a thermostat contact is expected to open at 72 °F and close at 68 °F the control differential is 4 °F.

Control joint A groove penetrating part way through a concrete slab for the purpose of forcing a tension crack to occur along a predetermined path.

Control state The current condition of the output of an on/off controller (e.g., the "ON" state or the "OFF" state).

Controlled device The hardware device that is being regulated by the controller.

Controlled variable The physical parameter that is being measured and controlled by the controller. Typical controlled variables in hydronic systems include temperature, pressure, and flow rate.

Controller The device that monitors the controlled variable and provides control outputs that direct operation of a controlled device.

Convection How heat is transferred between a solid surface and a fluid such as air or water.

Convection coefficient A number that, in part, establishes the rate of convective heat transfer between a fluid and a solid surface. The value of the convection coefficient depends on several factors such as the fluid's density, specific heat, viscosity, and the conditions under which it is flowing.

Convector A generic name for heat emitters that transfer the majority of their heat output by convection.

Conventional boiler A boiler that is intended to operate without sustained internal condensation of the fluid gases it creates during combustion.

Cooling capacity The rate at which a device absorbs heat from another material, typically measured in Btu/h.

Cooling COP The ratio of the cooling capacity of a device divided by the electrical power input to operate the device, where both rates are expressed in the same units.

Cooling load The rate at which a building requires cooling to maintain set conditions for indoor dry bulb temperature and relative humidity.

Coordinated zone management A control technique that seeks to operate zone valves or zone circulator in a manner that helps level the load on the heat source.

Copper water tube Copper tubing specifically manufactured for use in conveying water for domestic use, or within hydronic heating systems.

Correction factor A multiplying factor used to adjust a condition, such as heat output rate, for conditions other than those at which the condition is stated.

Counterflow When two fluids flow in opposite directions as they pass through a heat exchanger. This is generally desirable for maximum heat transfer rate.

Counterflow serpentine A shape for a radiant floor tubing circuit that is intended to evenly distribute heat within a room.

Counterflow spiral A shape for a radiant floor tubing circuit that is intended to evenly distribute heat within a room.

Coupling assembly The mechanical parts that couple a motor shot to the impeller shaft of a circulator.

Cover sheet A thin sheet of material (usually plywood or cement board) installed over the top of an above floor tube and plate radiant panel to serve as a smooth and stable substrate for finish flooring.

Cross-linked polyethylene Polyethylene in which long molecular chains have been bonded together to form a stronger and more temperature-resistant material. The acronym PEX is often used to represent cross-linked polyethylene.

Crossover bridge A piping segment that allows water to flow from a supply piping main to a return piping main. It usually contains a pair of closely spaced tees or a heat emitter.

C_v rating The number of gallons per minute (gpm) of 60 °F water that will cause a pressure drop of 1.0 psi through a valve when that valve is fully open.

C_v setting The adjustment of certain types of valves to set the number of gallons per minute of 60 °F water that will produce a pressure drop of 1.0 psi.

Cycle efficiency The net efficiency of a heat source accounting for heat losses during off-cycles. Cycle efficiency decreases as run fraction decreases.

D

Damper A device that closes to restrict the flow of air or other gases through a duct or vent.

Dead heading An undesirable operating condition in which a circulator is running, but all flow through it is blocked.

Dead loading The load placed on a structure such as a floor, due to the weight of the materials used to construct it.

Deaerated The condition of water in which it has been largely stripped of dissolved gases.

Decarbonization Any action intended to reduce the amount of carbon added to the earth's atmosphere.

Dedicated chillers A device that can only produce chilled water, or a chilled antifreeze solution, as its useful output.

Degree day The difference between the outside average temperature and 65 °F over 24 hours. Daily degree days can be added together to obtain monthly and annual total degree days.

Dehumidification Any process intended to remove water vapor from air.

Delay-on-break An operating mode for a time delay relay in which the contacts maintain their operating state until a set time has elapsed from when the input signal to the relay was turned off.

Delay-on-make An operating mode for a time delay relay in which the contacts maintain their normal (deenergized) state until a period of time has elapsed from when the input signal to the relay was turned on.

Delta T (ΔT) protection A control action used in snow-melting systems in which the controller regulates heat input so that the circuit temperature drop does not exceed a set value.

Demand charge A monetary charge utilities impose on certain commercial or industrial customers based on the peak power demand that occurs within each billing period.

Demand-fired A term describing a boiler that is only fired when there is a demand for heat (e.g., the boiler does not maintain temperature at all times).

Demineralization The process of removing mineral salts and ions from water, to reduce the potential for scaling or undesirable interactions with other materials inside a hydronic system.

Demineralizer column A vertical container in which water that is to be demineralized passes through chemically treated polymer resin beads. The beads strip the undesirable ions from the water and replace these ions with other ions that interact to form pure water.

Density In the context of this book, density refers to the weight of a substance divided by its volume. Some readers will recognize this as being the same as the specific weight of a substance. Common English units for density, as defined herein, are lb/ft^3 (pounds per cubic foot).

Description of operation A specific and detailed description of how a control system operates making reference to many or all of the devices involved in the control process.

Design dry bulb temperature (97.5%) The temperature the outside air is at, or above, 97.5% of the year. This is the outside temperature at which the design heating load is calculated.

Design heating load The heating load of a building or room when the outside air temperature is at its design dry bulb value.

Desuperheater A refrigerant to water heat exchanger that receives superheated refrigerant gas from the discharge of a compressor, and cools the refrigerant to near its vapor saturation temperature. The heat extracted from the refrigerant is typically used to heat domestic water.

Dewpoint control A control process that monitors the dry bulb temperature and relative humidity of an interior space, calculates the current dewpoint temperature based on these conditions, and produces an output to a controlled device.

Dewpoint temperature The temperature at which water vapor in a mixture of other gases begins to condense into liquid droplets. Other compounds in a mixture of gases also have an associated dewpoint temperature.

Diaphragm A flexible synthetic or metal sheet that changes shape based on the differential pressure across it.

Diaphragm-type expansion tank An expansion tank that contains a flexible diaphragm that separates the air and water within the tank.

Dielectric union A union made of steel on one side and brass on the other. A dielectric (electrically insulating) material separates the two metals to prevent galvanic corrosion.

Differential A range of temperature between the point where the electrical contacts of a controller open and when they close. The differential of some controllers is fixed, while on others it can be adjusted.

Differential pressure The difference in pressure between two points.

Differential pressure bypass valve A valve designed to allow flow through when the differential pressure across its ports reaches a specific value.

Differential pressure control A control algorithm that adjusts the speed of a circulator to maintain a constant differential pressure between its inlet and outlet regardless of the flow rate passing through it.

Differential pressure transducer A device for measuring the differential pressure between two points and generating an analog output signal based on that measurement.

Differential temperature controller A controller that monitors two temperature sensors and provides an output based on the difference between the temperatures at those sensors.

Diffuse The migration of molecules of one substance through the molecular structure of another material. Oxygen, for example, can slowly diffuse through the molecular structure of certain polymers.

Digital address A specific binary or hexadecimal sequence that identifies one digital device that is connected with other digital devices on a common communication bus.

Digital input A status of, on or off, 0 or 1, true or false, that is sensed as an input at a specific set of terminals on a controller.

Digitally encoded signal Information that exists in the form of binary numbers.

DIN rail An industry standard mounting rail for relays and other modular controls.

Direct digital control (DDC) A control network in which sensors, controllers, and controlled devices all communicate using digital signals sent along a common communication bus.

Direct expansion ("DX") evaporator coil A heat exchanger typically made of copper tubing and aluminum fins, used to exchange heat between a refrigerant and a fluid. That fluid could be air or a liquid such as water.

Direct-fired water heater A domestic water heater that includes an energy source such as an electric heating element or fuel burner.

Direct injection mixing A piping configuration in which hot water from a heat source is injected into a distribution system, and cool water from the return side of the distribution system flows back to be reheated by the heat source.

Direct lake water cooling system A cooling system that pumps water from a nearby lake to a building to be cooled.

Direct vent A method of exhausting combustion products from a boiler directly through an outside wall.

Direct-to-load heat transfer A buffer tank piping configuration that allow heat or chilled water to flow directly from a source device to a load circuit, when both are operating at the same time.

Dissolved air Molecules of oxygen, nitrogen, carbon dioxide, and other gases found in air that are present within liquid water.

Distribution efficiency The rate of heat delivery by a distribution system, under design load conditions, divided by the electrical wattage used to operate that distribution system. This index is valid for both hydronic and forced-air distribution systems.

Distribution system An assembly of piping, fittings, and valves that conveys heated water from a heat source to heat emitters. It may be a simple piping loop, or an elaborate, multibranch piping system.

Diverter tee A special fitting resembling a tee with an internal baffle. This fitting is used to force a portion of the flow entering the tee out through a branch circuit connected to its side port.

Diverter valve A valve (often electrically operated) that directs a single incoming fluid stream in one of two possible directions based on the position of the valve's flow element.

DOE heating capacity The heat output rating of boilers up to 300,000 Btu/h based on a federal government rating standard.

Domestic hot water (DHW) Heated water used for purposes of washing, cooking, or other uses. Domestic hot water is considered suitable for human consumption.

Domestic hot water usage profile A graph or bar chart that shows the number of gallons of domestic hot water required for each hour of a typical day.

Domestic water Water suitable for human consumption.

Double interpolation A mathematical method of estimating a value that is known to lie between two other values of each of two independent parameters.

Draft proving switch A device that measures the negative air pressure in an exhaust system and closes its electrical contacts whenever a safe negative pressure is maintained.

Drainback system A type of active solar energy system in which freezing is avoided by having all water drain out of the collectors and exposed piping into a holding tank at the end of each operating cycle.

Draindown system A type of active solar energy system in which freezing is avoided by having all fluid drain out of the collectors and exposed piping *to waste* at the end of each operating cycle.

Draw-through fan-coil A type of fan-coil or air handler where the fan or blower pulls air through the heat transfer coil.

Drip pan A metal or polymer pan located under the coil of an air handler or fan-coil, to catch condensate (e.g., liquid water) dripping from the coil. The condensate is then routed to an appropriate drain.

Driving ΔT The difference between the average water temperature in a heat emitter and the air surrounding the heat emitter.

Droop The tendency for a controller operating with only proportional logic to allow a temperature to stabilize below the desired target value when the rate of heat input to the load increases.

Dry base boiler A boiler design in which the boiler block rests on top of the combustion chamber.

Dry bulb temperature The temperature of air measured with a thermometer.

Dual isolation valve A special valve installed under a panel radiator that allows the radiator to be isolated from the supply and return piping.

Dumb mixing devices A mixing device such as a three-way valve that does not sense temperature or flow rate, and thus is incapable of reacting to changes in operating conditions.

Duty cycle The percentage of elapsed time in which the device is operating. A 50% duty cycle indicates the device is operating 50% of the elapsed time.

Dynamic pressure distribution The pressures at all locations within a piping system when the circulator is operating, and fluid is flowing.

E

Earth heat exchanger Piping buried in the soil for extracting or rejecting heat.

EEPROM An acronym for electrically erasable programmable read only memory.

Effective coefficient of performance (COP_e) Typically refers to the COP of a water-to-water heat pump in an application where both the heated and chilled fluid streams from the heat pump are useful.

Effective total *R*-value The average thermal resistance of a wall, ceiling, floor, etc., after compensating for the presence of framing members.

Effectiveness (of a heat exchanger) The ratio of the actual heat transfer across a heat exchanger to the maximum possible rate of heat transfer. Effectiveness is a unitless number between 0 and 1. Higher values are more desirable.

Efficiency The ratio of a desired output quantity or rate divided by the necessary input quantity or rate. Both quantities must have the same physical units.

Elastomeric foam insulation A type of flexible insulation used on piping and made from rubber compounds combined with PVC.

Electric thermal storage (ETS) system A system designed to use off-peak electrical energy to heat water for subsequent use in heating a building. The energy is stored in the form of heated water.

Electric valve actuators A device that when powered by electricity causes the stem of a valve to rotate or otherwise move which alters flow through the valve.

Electromagnetic radiation Energy that is transferred from one location to another by combined electric and magnetic fields. Examples include visible light, infrared light, microwaves, radio waves, and "radiant heat."

Electromagnetic spectrum The entire range of frequencies of electromagnetic radiation from low-frequency/long-wavelength radio waves to high-frequency/short-wavelength gamma rays.

Electromechanical controls Controls that use mechanical force generated by springs, bimetal elements, or pressure diaphragms to open and close electrical contacts.

Electronically commutated motor (ECM) Also known as a brushless DC motor. A motor with a permanent magnet rotor and electronically commutated stator coils. This type of motor must be run by a controller.

Element A name for the finned-tube component in a baseboard convector. The name also applies to the device that creates heat in an electric boiler or water heater.

Elevation head The head energy within a fluid due to its elevation. The higher the elevation, the greater its elevation head energy. It can be expressed as foot pounds of energy per pound of fluid (ft•lb/lb), which is often mathematically simplified to just ft.

Embedded controller A preassembled controller that performs a specific task, and is part of an overall control system.

Emissivity A number between 0 and 1 that indicates the ability of a surface to release thermal radiation. Surfaces with high emissivities release thermal radiation at higher rates compared to surfaces with low emissivities.

End suction circulator A circulator with an inlet connection perpendicular to the plane of its impeller.

End switch A switch built into a zone valve, or other electrical actuator, that closes its electrical contacts when the valve or actuator reaches its fully open or full travel position.

Energy efficiency ratio An index that expresses the efficiency of a cooling device as the cooling capacity in Btu/h divided by its input power in watts.

Energy rating (ER) A number indicating the relative energy use of circulator based on specific test standard.

Energy recovery ventilator (ERV) A device that extracts heat from an exhaust air stream and transfers it to an incoming fresh air stream.

EN442 standard A European standard for estimating the output of panel radiators over a wide range of operating conditions.

Entrained air Air bubbles that are carried along with the fluid flowing through a piping system.

EPDM Ethylene propylene diene monomer. A high-quality rubber compound resistant to many chemicals.

Equal percentage characteristic A valve characteristic in which the flow rate through the valve increases slowly as the stem starts to lift and increases faster the farther the stem lifts. This characteristic gives an equal percentage change over the previous flow for equal increments of stem movement. This characteristic is desirable when heat output is being controlled by varying the flow rate through the heat emitter.

Equilibrium A "balanced" operating condition in which energy input equals energy output.

Equivalent length The concept of replacing a piping component by a straight length of pipe, of the same diameter, that would yield the same hydraulic resistance as the component being replaced.

Equivalent resistance The result of mathematically combining two or more hydraulic resistances into a single hydraulic resistance that represents their combined effect.

Erosion corrosion A process where metal is removed from the inside surfaces of piping or piping components due to excessively high flow velocity.

Error The difference between a target value and the actual (measured) value of a controlled variable.

Escutcheon plate A decorative trim piece that slides over a pipe and covers the hole in the surface the pipe penetrates.

Evacuated tube A type of solar thermal collector with a glass tube that surrounds the absorber plate. Nearly all air has been evacuated from the glass tube to suppress convective heat loss.

Evacuated tube collector An assembly of several evacuated tubes, along with a manifold and structural supports.

Evaporator The portion of a refrigeration cycle where low-temperature heat is absorbed into a refrigerant. Also refers to the physical component where this action occurs.

EVOH (ethylene vinyl alcohol) A compound used to coat PEX tubing to reduce oxygen diffusion through tubing walls.

Expansion compensator A device installed into a piping path that absorbs expansion/contraction movement without inducing significant mechanical stress on the piping.

Expansion loop A loop formed within an otherwise straight run of tubing that can absorb the movement in the pipe due to thermal expansion or contraction.

Expansion offset An offset rectangular path within an otherwise straight run of tubing that can absorb the movement in the pipe due to thermal expansion/contraction.

Expansion tank A tank specifically designed and sized to accommodate the increased volume of system fluid when that fluid is heated.

Expansion tank sizer A module in the Hydronics Design Studio that assists in sizing diaphragm-type expansion tanks.

Exposed surface Any building surface exposed to outside air temperature.

External heat exchanger Any heat exchanger that is not contained inside another device, such as a tank.

External VA rating The volt-amps that can safely be supplied to an external electrical load from a controller with a built-in transformer.

Extruded polystyrene foam A type of closed-cell foam insulation board often used for insulation under a concrete slab on grade.

External heat exchanger A heat exchanger mounted outside (versus inside) a device such as a storage tank.

External static pressure The pressure differential developed by a fan or blower between its inlet and outlet.

Eye (of an impeller) The center opening where fluid flows into an impeller.

F

Fan Different from a blower. A fan is designed to move large volumes of air through a path with very low flow resistance. It consists of an electric motor and shaped, propeller-like blades.

Fan-coil A heat emitter that contains an internal blower or fan to force air through a finned-tube coil where it absorbs heat.

Fast fill mode A temporary condition in which water is rapidly added to a hydronic system to fill it and purge it of air.

Feed water valve A valve that automatically allows cold water to enter a closed hydronic system upon a drop in system pressure.

Feedback A signal sent from a sensor back to a controller that is regulating the controlled variable measured by the sensor.

Feet of head The common English units of expressing the head energy of a fluid. The units of feet represent the total mechanical energy content of each pound of fluid and are derived from simplifying the units of ft•lb/lb.

Female pipe threads (FPT) Tapered threads on the inside diameter of a fitting, valve, or other piping component.

Film-forming water stabilizer A chemical compound that is introduced into a hydronic piping system to form a very thin protective film on the surfaces of devices in contact with the system fluid, with the goal of protecting those surfaces from corrosion.

Finish floor resistance The thermal resistance or R-value of a finish floor material placed on top of the heated subfloor.

Finite element analysis (FEA) Use of specialized computer software to simulate and quantify heat flow.

Finned-tube A tube with added fins to enhance heat transfer.

Finned-tube baseboard convector A type of convector that consists of a copper tube with aluminum fins mounted in a steel enclosure located at the base of walls.

Fire-tube boiler A boiler design where hot combustion products passing through tubes heat water on the outside of the tubes.

Firmware Digital information stored in a special type of solid state memory that remains valid when power to the overall device is turned off. Sometimes called "nonvolatile" memory.

First hour rating The volume of domestic hot water that a given water heater can deliver within the first hour of a draw. It is based on a specified hot water supply temperature and rate of heat input.

Fittings Components used to attach straight lengths of pipe or tubing. Common fittings include elbows, tees, unions, couplings, reducer couplings, and bushings.

Fixed lead heat source An operating sequence for a multiple heat source system in which a designated heat source is always the first unit turned on upon a demand for heat.

Fixed point supports Supports for piping that do not allow the pipe to move.

Flanges (on a circulator) The common means of connecting a circulator to piping. The circulator's volute has integral cast flanges on its inlet and discharge ports. These bolt to matching flanges threaded onto the piping. An O-ring provides a pressure-tight seal between the flanges.

Flat plate heat exchanger A modern design for heat exchangers in which the fluids flow through alternating spaces between specially shaped stainless steel plates.

Flat plate solar collectors Devices that convert solar radiation to heat using a flat, fluid-cooled plate (e.g., the absorber plate) contained within an insulated and glazed housing.

Floating control A control output signal used to operate valves and dampers which always has one of three states: opening, closing, or holding the present position. It is also sometimes called "three-wire," or "tri-state" control.

Floating zone A specified range in which the error between the measured value of a controlled variable and the target value is small enough that no open or close signal is generated by the controller. It is sometimes called the "neutral zone."

Float-type air vents A type of automatic air vent that uses an internal float to open and close the air venting valve.

Flow-check valve A valve with a weighted plug that prevents backward flow and forward migration of hot water due to buoyancy.

Flow duration curve A graphical curve that shows the number of hours in a heating season during which flow through a hydronic system is at or above a given percentage of design load flow rate.

Flow element (or spool) The internal portion of a valve that moves to alter the flow rate through the valve.

Flow rate The volumetric rate of flow of a fluid. For liquids it is often expressed in units of gallons per minute (gpm). For gases, it is often expressed in units of cubic feet per minute (cfm).

Flow restrictor valve A valve used in some injection mixing applications to limit flow rate to improve the rangeability of the mixing hardware.

Flow switch A switch that closes its contacts when flow through it rises above a certain minimum value.

Flow transducer A device that senses flow rate and produces an electrical signal that can be interpreted by other control devices in the system.

Flow velocity The speed of an imaginary fluid particle at some point in a piping system. Common English units for flow velocity are feet per second.

Fluid A liquid or gas.

Fluid feeder A device that uses a pump to automatically add fluid to a hydronic system whenever the pressure at some location in the system drops below a preset value.

Fluid properties factor A number that combines the fluid properties of density and viscosity of a given fluid into a single index for use in determining hydraulic resistance.

Flux A grease-like paste that chemically cleans the surfaces of copper tubing and fittings being joined by soldering.

Foil scrim kraft A commonly used covering for fiberglass pipe insulation.

Foot-pound Abbreviated (ft•lb). A unit of energy. Often used to express mechanical energy. One foot-pound equals 0.0012850675 Btu.

Foot valve A type of check valve used on the submerged inlet pipe for a pump drawing water from a well or other body of water.

Forced convection Heat transfer between a solid surface and a moving fluid where the fluid's motion is created by a fan, blower, or circulator.

Forced-water purging The use of a forced-water stream to entrain air in a piping system and drive it out of a purging valve.

Form-fitting insulation shell A premolded foam shell that snaps in place over the volute of a circulator.

Fouling A term describing the undesirable accumulation of particulates, minerals, or other debris on the wetted surfaces of a heat exchanger.

Fouling factors Thermal resistances within heat exchangers cause by debris films that form on the metal surfaces separating the two fluids.

Four-way mixing valve A type of mixing valve designed for use with low flow resistance boilers.

Free area The unobstructed area for air to flow through a louvered panel or screen.

Free oxygen Oxygen molecules within water that are not part of the H_2O molecules and as such are free to react with other substances such as iron or steel in the system.

Free standing panel radiator A panel radiator that stands on support brackets secured to the floor rather than being supported by a wall.

Full port ball valve A ball valve in which the hole through the ball is approximately the same diameter as the pipe size of the valve.

Full reset control The ability of an outdoor reset control to reduce the water temperature supplied to the distribution system all the way from some maximum value at design load conditions down to room temperature at no load conditions.

Full storage systems A thermal storage system designed to store all the heat needed by the load, for a 24-hour period, during the preceding off-peak period.

Fuzzy logic Control logic that allows a controller to automatically modify the operating characteristics of the controller based on how the system is responding, and without human intervention.

G

Galvanic corrosion Corrosion between dissimilar metals caused by the flow of electrons from a less noble metal (on the galvanic chart) to a more noble metal.

Gaseous cavitation Cavitation caused by dissolved air or air bubbles entrained in the fluid rather than vapor pockets.

Gate valve A type of valve designed for component isolation purposes.

Gauge pressure The pressure of a liquid or gas measured relative to atmospheric pressure at sea level. Common English units for gauge pressure are psig (pounds per square inch gauge), or simply psi.

Gel-fired compression connectors Special connectors for small wires that prevent moisture from entering the electrical bond.

General purpose relay A type of relay that can be used for many different control applications.

Geological carbon Carbon found deep within the earth and as such not adding to the carbon content at the earth's surface unless it is brought to the surface by drilling or excavation.

Geothermal preconditioning coil A heat exchanger coil connected to a buried piping loop that is used to preheat or precool fresh air before that air passes into a heat recovery ventilator.

Globe valve A type of valve designed for flow regulation purposes.

Gooseneck piping Piping that is above the water level in an unpressurized thermal storage tank and as such under slight negative pressure relative to the atmosphere.

GPM (or gpm) Abbreviation for flow rate expressed in units of gallons per minute.

Grains 1/7,000 of a pound. A unit sometimes used to indicate the water content of air.

Gravity purging Filling a hydronic system from the bottom and allowing the air in the piping to rise up to, and out of, high point vents.

Grout A mixture of sand and bentonite clay used to fill voids after a piping assembly for an earth loop has been inserted into a borehole.

Gypsum thin slab A type of construction for radiant floor heating in which a slab of poured gypsum underlayment ranging from 1.25 to 2 inches thick is poured over tubing that has been fastened to a subfloor.

H

Half unions A portion of a union fitting often used on valves to allow quick disconnection from a device such as a radiator.

Hardwired logic Specific wiring connections between components that force a control system to operate in a predetermined manner.

Head The total mechanical energy content of a fluid at some point in a piping system.

Head loss The dissipation of head energy as fluid flows through any device.

Heat anticipator An adjustable resistor contained in an electromechanical room thermostat that preheats the bimetal element to minimize room temperature overshoot.

Heat capacity A material property indicating the amount of heat required to raise one cubic foot of the material by one degree Fahrenheit. Common English units for heat capacity are $Btu/ft^3/°F$.

Heat emitter A generic term for a device that releases heat from a circulating stream of heated fluid into the space to be heated. Examples include convectors, radiator, and fan-coils.

Heat exchanger A device that transfers heat from one fluid to another without allowing the two fluids to make physical contact.

Heat flux The rate of heat flow across a unit of area. The common English units for heat flux are $Btu/h/ft^2$.

Heat calculator unit The portion of a heat metering system that calculates the rate of heat transfer and total amount of heat transferred, based on temperature difference between two sensors and a measured flow rate.

Heat dump An assembly of components designed to release unneeded heat in a controlled and safe manner. Heat dumps are often used with cordwood gasification boilers.

Heat interface unit (HIU) An assembly of components that provides space heating and domestic hot to an occupied space such as an apartment by tapping into a distribution piping system from a central heat source.

Heat loss coefficient A value that indicates the rate of heat loss (usually for an entire building) per degree of temperature difference between the inside and outside air temperature. Common units are $Btu/h/°F$.

Heat meter A collection of sensing and processing hardware that measures the rate of heat transfer, and total amount of heat transferred over time, between two points in a piping system.

Heat migration Undesirable heat transfer within a system. Often caused by natural convection.

Heat motor actuator A device that moves the stem of a valve when supplied with a suitable voltage. The necessary movement is produced by the expansion/contraction of a wax-filled capsule that is heated by an internal resistor.

Heat of rejection Heat given off by a cooling system.

Heat output ratings One of several ratings describing the rate of heat output of a heat source such as a boiler.

Heat plant A general term for one or more boilers or other heat sources and their supporting equipment, most of which is located in the same room as the heat sources.

Heat pump A device that uses the refrigeration cycle to move heat from an area of lower temperature to one of higher temperature.

Heat purging Removing additional heat from a boiler by maintaining circulation through it after the burner has stopped firing.

Heat recovery chiller A special type of heat pump that can capture both the heat output and chilled water output from a refrigeration process, and transfer that heat and cooling effect into two separate fluid streams.

Heat recovery ventilator (HRV) A device for extracting heat from an exhaust air stream and transfer that heat into an incoming fresh air stream.

Heat sink A material or device that receives heat that is otherwise not needed.

Heat source A device that supplies heat to a space-heating distribution system.

Heat source reset A control process in which the temperature of water leaving a heat source is adjusted based on changes in outdoor temperature.

Heat source rotation A control process that automatically changes the start up and shut down order of each heat source in a multiple heat source system with the goal of establishing approximately equal run time on each heat source over a period of several days or weeks.

Heat transfer plate A metal plate that fits over heating tubing and provides lateral heat conduction away from the tubing.

Heating capacity The rate of heat output of a heat source, usually expressed in Btu/h.

Heating curve Another name for a reset line.

Heating effect factor An allowance of 15% extra heat output, beyond the laboratory tested heat output, allowed by the IBR rating standard for finned-tube baseboard convectors.

Hematite (Fe_2O_3) An oxide of iron that appears as a very fine dark gray sludge and is strongly attracted to magnetic fields.

Hi/Lo fire boiler A boiler that can produce two rates of heat output based on control input. Hi/Lo boilers typically have two independently controlled burner assemblies.

High limit setting An upper temperature limit set on a temperature controller.

High limit/switching relay controller   An electrical controller that limits the temperature of a boiler by turning off the burner.

High point vents   Devices located at the high point(s) of a piping system that can release air that may accumulate at these points.

High wall cassette   A type of fan-coil used for heating and cooling and mounted high on a wall, perhaps 6 to 8 inches below the ceiling.

Higher heating value (HHV)   The maximum amount of heat available from a fuel undergoing stoichiometric combustion with zero excess air, and including recovery of all latent heat through condensing of the water vapor produced during combustion. Normally stated in Btus per pound of fuel.

Home run distribution system   A distribution system in which each heat emitter has its own supply and return tube routed from a common manifold station.

Horizontal panel radiator   A panel radiator that is wider than it is tall, and has tubes oriented horizontally.

Hose bib   A common name for a valve that has threads for connecting a common garden hose.

Hunting   An undesirable response by a control system that prevents the system being controlled from stabilizing.

Hybrid distribution system   A distribution system containing more than one types of fundamental distribution piping topologies.

Hybrid staging/modulation   A method of controlling a multiple boiler system that regulates both the number of boilers operating and the rate of heat output from each boiler.

Hydraulic efficiency   The ratio of the hydraulic power output of a circulator divided by the rate of mechanical power input to the circulator's impeller shaft.

Hydraulic equilibrium   The condition where the head added by the system's circulator exactly equals the head dissipated by viscous friction of the fluid flowing through the system.

Hydraulic resistance   A means of representing the ability of piping components to remove head energy from a flowing fluid. The greater the hydraulic resistance of a component, the more head loss it creates for a given flow rate.

Hydraulic resistance diagrams   A drawing composed of hydraulic resistors that represents an assembly of piping components.

Hydraulic separation   A method of coupling two or more hydronic circuits so that the effect created by an operating circulator in one circuit does not influence the effect created by the operating circulator in another circuits.

Hydraulic separator   A device used to create hydraulic separation between two or more hydronic circuits.

Hydro-air distribution system   A distribution system in which a fluid stream carries heat to one or more air handlers or fan-coils, which dissipate that heat into interior air.

Hydronic Circuit Simulator   A module in the Hydronics Design Studio software for simulating the flow and heat transfer of a user-defined hydronic piping system.

Hydronic detergents   Commercially available chemical compounds formulated to clean debris and films from the wetted surfaces of components in a hydronic system.

Hydronic system stabilizer   Commercially available chemical compounds that create very thin protective films on the wetted surfaces of components in a hydronic system.

Hydroxide anion   An ion (OH$^-$) released by some of the resin beads in a demineralizer column. It combines with a hydrogen cation (H$^+$) to form pure water.

Hygroscopic air vent   A device used to vent air from a hydronic system that seals itself as water, rather than air, passes through it.

I

IBR gross output   An industry-standard rating for boiler heat output that includes the heat transferred to water flowing through the boiler but does not include heat loss from the boiler jacket. This output is expressed in Btu/h.

IBR method   A method of sizing fin-tube baseboard based on the average water temperature in the circuit.

IBR net output   The IBR gross output rating minus 15% for the assumed heat loss of distribution piping, and delay in heat delivery due to the thermal mass of the boiler. This output is expressed in Btu/h.

IBR testing and rating code   A standard for developing heat output ratings of fin-tube baseboard.

Idling (a melted pavement)   Maintaining sufficient heat flow to the slab so that its temperature is just above 32 °F at all times during the winter. Idling a slab allows it to reach melting conditions quickly when frozen precipitation occurs.

Impeller   A rotating disc with curved vanes contained within a centrifugal pump that adds head to a fluid as it is accelerated from its center toward its outer edge.

Implosion   A term describing the rapid and violent collapse of vapor pockets as the pressure around them rises above the vapor pressure of the fluid.

In solution   Referring to materials that are dissolved into water.

Inches of water column (inch w.g.)   A means of indicating the differential pressure developed by a blower or fan.

Incompressible   The ability of liquids to transmit large pressures without themselves being compressed into smaller volumes.

Indirect lake water cooling system   A method of cooling using a submerged heat exchanger placed on the bottom of a lake.

Indirect water heater   An insulated storage tank that contains an internal heat exchanger. Hot boiler water is circulated through one side of the heat exchanger, transferring heat to domestic water on the other side.

Indoor reset control   A control method in which the supply water temperature to heat emitters is adjusted based on changes in the indoor air temperature.

Inductive loads   Electrical load characteristic of certain motors and transformers that create a strong magnetic field around conductors that resist changes is electrical current.

Infiltration   The unintentional leakage of outside air into heated space.

Infrared thermal radiation   Another name for thermal radiation. It is technically electromagnetic radiation having wavelengths longer than can be seen by the eye.

Infrared thermograph   An image produced by a thermal camera that maps surface temperatures into a spectrum of colors that represent the overall range of temperatures present in the image.

Injection control device   A device such as a variable speed pump or modulating valve that regulates the rate at which hot water is injected into a circulating distribution system.

Injection mixing   A method of controlling the temperature of a hydronic distribution system by adding (injecting) hot water and simultaneously removing cooler water. The faster the injection rate, the warmer the distribution system becomes and the greater its heat output.

Inlet fluid parameter   The temperature difference between the fluid entering a solar collector and the outdoor air, divided by the intensity of solar radiation striking the collector. The thermal efficiency of a collector decreases in an approximately linear relationship with increasing inlet fluid parameter.

Inline circulator   A circulator constructed with its inlet and outlet ports on a common centerline.

Inside air film   The thermal resistance of a thin layer of air that clings to interior building surfaces.

Intelligent mixing device   A mixing device that senses fluid temperature and is able to adjust one or more flow rates based on how the sensed temperature compares to a set target value. Such devices may be electronic or use nonelectric thermostatic actuators.

Interface module   A controller that converts the digital control signals it receives into output signals that operate the device being controlled.

Internal heat exchanger   A heat exchanger contained in a storage tank and relying on natural convection for heat exchange with the water in the storage tank.

Internet accessibility   The ability of a control system to send or receive information from a remote site using the Internet.

Interpolation   A mathematical method of estimating a numerical value that is known to be between two other stated values. It is based on proportioning.

Interstage differential   A temperature range between two stages of a multistage controller in which there is no control action.

Inter-zone heat transfer   Heat movement across interior partitions, floors, and ceilings separating building areas that are kept at different temperatures.

Ion   An atom or molecule that has a net positive or net negative charge.

Ion-exchange process   A chemical process by which undesirable ions are exchanged for other (beneficial) ions.

Isolation flanges   Special flanges for circulators that contain built-in shutoff valves.

Isotherms   Lines connecting points having the same temperature. Isotherms are frequently used to display the results of finite element analysis of heat transfer through solid materials.

Iteration   A design process that entails making successively refined estimates until a mathematically stable operating condition is determined.

J

Jacket (of a boiler)   The insulation layer and metal or polymer enclosure surrounding the internal heat exchanger and other components of a boiler.

K

Kick-space heater   A small horizontal fan-coil convector that mounts in the recessed space at the base of a kitchen or bathroom cabinet.

KPa (Kilo-Pascals of pressure)   A pressure equal to 1,000 newtons of force distributed over 1 m^2 of area.

Kraft paper faced batt   A fiberglass insulation batt with a heavy paper facing.

L

Ladder diagram   Electrical schematic drawings that are used to design and document a control system.

Lake heat exchanger   A special stainless steel plate heat exchanger designed to be submerged in a lake, and either absorb or dissipate heat from or to lake water.

Lake source cooling   A method of cooling the uses the naturally cool water several feet below the surface of a lake.

Laminar flow   A classification of fluid flow where streamlines remain parallel as the fluid moves along the pipe. Laminar flow is more typical at very low flow velocities.

Latent cooling load   The portion of total cooling load that deals with removing moisture from interior air.

Latent heat   Heat that is added or removed from a substance without any temperature change. This occurs as a material changes phase from solid to liquid or from liquid to vapor.

Lead/lag control A control technique that automatically rotates the firing order of boilers in a multiple boiler system in an attempt to keep the operating hours of each boiler approximately equal.

Leaders The portion of a radiant panel tubing circuit that is not within the radiant panel, but instead connects the tubing within the radiant panel to a manifold station.

Least favored crossover A parallel piping branch that is farthest removed from the differential pressure created by a circulator.

Life cycle cost The total cost of owning and operating a device or system over a period of time considering material and installation cost, maintenance cost, and cost of capital.

Lifts Installing material in successive layers to allow adequate drying, curing, or compaction of one layer prior to covering it with another layer.

Lightweight concrete Concrete that uses a lightweight aggregate in place of dense stone aggregate. Lightweight aggregates include vermiculite, expanded shale, and polystyrene beads.

Line voltage A term that usually indicates a nominal 120 VAC relative to ground potential.

Line voltage section That portion of a ladder diagram containing line voltage devices.

Linear relationship When one variable changes in direct proportion to another variable.

Liquid A physical state of matter in which a material is pourable, and conforms to the shape of its container. Most liquids are also incompressible.

Load The heating requirement necessary to maintain a desired temperature in a building, room, or a volume of material such as a tank of water.

Lockshield valve A valve that mounts on the outlet of a heat emitter and serves as a combination isolation valve, flow balancing valve, and drain valve.

Log mean temperature difference (LMTD) The temperature difference at one end of a heat exchanger, minus the temperature difference at the other end, divide by the natural logarithm of the ratio of the two temperature differences. LMTD is an indication of the effective "average" temperature difference between the two fluids exchanging heat.

LONworks A digital communications protocol that allows multiple controllers to communicate with each other over a common two-wire cable that is connected to each controller.

Low-limit aquastat A controller that turns on the burner of a boiler when the water temperature within the boiler falls to a preset value.

Low loss header Any header with very low pressure drop along its length.

Low voltage For hydronic systems, most low voltage control circuits operate at 24 VAC.

Low water cutoff (LWCO) A safety control that opens electrical contacts to turn off the heat generating element or burner of a heat source if the water level in a system drops below the level of its probe.

Lower heating value (LHV) The maximum amount of heat available from a fuel undergoing a stoichiometric combustion process with zero excess air, but excluding the heat associated with water vapor produced during combustion. Normally stated in Btus per pound of fuel.

M

Magnetic dirt separator A device that uses a powerful permanent magnet to enhance the capture of ferrous metal particles entrained in a flow stream.

Magnetite (Fe_3O_4) An oxide of iron that appears as a dark gray sludge inside hydronic circuits, and it highly attracted to magnetic fields.

Mains Refers to the larger pipes in a hydronic system that carry flow to several branch circuits.

Make-up water assembly An assembly of components that automatically adds water to a hydronic system to make up for minor losses. It usually consists of a shutoff valve, feed water valve, and backflow preventer.

Male pipe threads (MPT) Tapered threads on the outside diameter of pipes or fittings.

Manifold A piping component that serves as a common beginning or ending point for two or more parallel piping circuits.

Manifold station A combination of a supply manifold and return manifold that services two or more circuits.

Manifold valve actuator An electrically operated device that attaches to the valves built into some types of manifolds. When supplied with the appropriate voltage, it opens the valve. When this voltage is removed, it closes the valve.

Manipulated variable The physical quantity that is being regulated by a control process.

Manometer A device used to measure the pressure difference between two points in a piping system.

Manual air vents Devices that, when manually opened, allow air to be released (vented) from a piping system.

Manual reset high limit (MRHL) A safety controller that opens an electrical contact when a set temperature is detected at its probe. A button on the device must be pushed to close the contacts once they have opened.

Master thermostat An electrical thermostat that controls operation of a circulator in a system using multiple nonelectric thermostatically operated valves on the heat emitters.

Maximum allowed upward heat flux The maximum rate of heat transfer from a radiant panel (often measured in Btu/h/ft^2), based on limiting the average surface temperature of the panel to a predetermined value.

Mean radiant temperature (MRT) The mathematical average of the temperatures of all wall, ceiling, and floor surfaces in a room. In some definitions of MRT, the temperature of each surface has a weighing factor based on the location in the room at which the MRT is being determined, and the area of each room surface.

Mechanical energy The total energy possessed by an object based on its height above the earth, its speed, and in the case of a fluid its pressure. The total mechanical energy content of a fluid is called head.

Metered balancing valves Specialized valves that allow the flow rate in a branch piping circuit to be adjusted to a desired value.

Microbubble air separator A type of air separator that captures microbubbles from the system's fluid. It lowers the dissolved air content of the water, enabling it to absorb any residual air in the system.

Microbubbles Very small air bubbles formed as gas molecules join together (e.g., coalesce). Individual microbubbles are often too small to be seen. Instead, they appear as a "cloud" within otherwise clear water.

Microcontroller A small self-contained controller that can be programmed for specific control actions.

Microprocessor An integrated circuit device capable of being electrically programmed to execute specific logical and mathematical operations.

Minimum supply temperature The lowest temperature allowed as the supply temperature from a boiler being operated by an outdoor reset control. This minimum is typically set to prevent a conventional boiler from operating with sustained flue gas condensation.

Minitube distribution system A system using one or more sets of minitubes between a common primary loop and one or more manifold stations.

Minitubes A pair of small diameter tubes running between a mechanical room and a remote manifold station for injection mixing control of the water temperature in the manifold circuits.

Mixing assembly A collection of hardware that constitutes a complete mixing system, and forms a "bridge" between a higher temperature boiler loop and a lower temperature distribution system.

Mixing device Any device in which two or more streams of fluid are intentionally mixed together to create a fluid temperature between the higher and lower entering fluid temperatures.

Mixing point Any location within a hydronic system where two or more streams of fluid come together.

Mixing reset control The process of automatically adjusting the supply temperature to a hydronic distribution system based on outdoor reset control, and accomplished by mixing hot water from a heat source with cooler return water from the distribution system.

Mixing valve A valve that blends two fluid streams entering at different temperatures to achieve a desired outlet temperature. The most common types in hydronic heating are three-way and four-way mixing valves.

MMbtu An abbreviation for one million (1,000,000) Btus.

Mod/con boiler An abbreviated name for a modulating/condensing boiler that can vary its high output rate (e.g., modulate) and is designed to operate with sustained flue gas condensation.

Modulating/condensing boiler A boiler designed to vary its firing rate in an attempt to match heat output to heating load. Modulating boilers are also capable of operating with sustained flue gas condensation, and typically achieve high thermal efficiency when doing so.

Modulation A process, such as rate of heat transfer, or fluid flow rate, that is continuously varied, as necessary, to meet a load.

Monobloc A self-contained air-to-water heat pump that is pre-charged with refrigerant.

Monoflo tee® A brand of diverter tees manufactured by the Bell and Gossett Corporation.

Motorized actuator An electrically operated device consisting of a motor and gear train, and used to rotate the shaft of a valve or a damper.

Multifunction time delay device A device that can be configured to provide several common time delay relay functions.

Multiload/multitemperature system Hydronic systems capable of simultaneously supplying two or more loads operating at two or more supply water temperatures.

Multiple boiler controller A controller capable of independently operating two or more boilers as required by the load.

Multistage heat production The ability to automatically turn on additional heat sources when necessary to reduce the error between a target supply water temperature and the measured supply water temperature.

Multistage thermostat A thermostat with two or more contacts to turn on additional heating or cooling sources when necessary to reduce the error between a target temperature and the measured temperature.

Multizone/multicirculator system A hydronic distribution system that uses a separate circulator for each zone circuit.

Multizone/zone valve system A hydronic distribution system that uses a separate zone valve for each zone circuit. A single circulator supplies the entire system.

Multizone relay center A specialized controller consisting of multiple relays and a transformer. This control is used in multizone hydronic systems.

Multizone systems A hydronic system capable of independently controlling heat input to two or more areas of a building.

N

National Fuel Gas Code A model code developed by the National Fire Protection Association that is widely recognized in the United States. It specifically covers the fuel supply, ventilation, safety, and exhaust requirements of combustion-type heat sources.

National pipe thread (NPT) The standard tapered threads used on piping, fittings, and valves in the United States.

Natural convection Heat transfer between a solid surface and a moving fluid when the fluid's motion is created only by natural processes such as the tendency of a warm fluid to rise (e.g., buoyancy).

Negative temperature coefficient (NTC) thermistor A solid state temperature sensor that decreases its resistance as its temperature increases, and in a very repeatable manner.

Net COP The rate of heat output from a heat pump divided by the sum of the electrical power input to the heat pump, plus the electrical power input to any circulators required to supply flow to the heat pump.

Net positive suction head available (NPSHA) The total head available to push water into a circulator at a given point in a piping system. This indicator is totally determined by the piping system and fluid it contains, and does not depend on the circulator. It is expressed in feet of head. To avoid cavitation, the NPSHA must be equal to or greater than the NPSHR of the selected circulator.

Net positive suction head required (NPSHR) The minimum total head required at the inlet of the circulator to prevent cavitation. This value is specified by pump manufacturers.

Networked control system A system of multiple controllers that are connected by a two-wire communication bus or linked as part of a wireless network.

Nominal inside diameter The *approximate* inside diameter of a pipe or tube.

Nonbarrier tubing PEX or other polymer tubing that does not have a barrier to limit oxygen diffusion through the tube wall.

Noncondensing mode An operating mode for a combustion-type heat source in which the water vapor produced during combustion do not condense (turn to liquid) within the heat source.

Normally closed contacts Electrical contacts in a relay that are closed when the relay coil is deenergized.

Normally open contacts Electrical contacts in a relay that are open when the relay coil is deenergized.

Number of transfer units (NTU) The UA value of a heat exchanger divided by the smaller of the two fluid capacitance rates. The higher the NTU value for a heat exchanger, the higher its effectiveness (E).

O

Off-cycle losses Heat losses from a heat source that occur when the heat source is not producing heat.

Off-peak periods Set times when electrical energy is sold at a reduced rate by the utility.

Offset A residual error between the target value and actual value of a controlled variable associated with proportional-only control.

On/off control Any control action that turns the device being controlled on and off in an attempt to control a process.

One-pipe system A hydronic distribution system that uses diverter tees to connect individual heat emitters to a main distribution piping circuit.

Open-loop source A source of ground water such as a pond, lake, or well, from which water is supplied to a geothermal heat pump.

Open-loop system A piping system that conveys fresh water, or is open to the atmosphere at any point.

Operating differential The actual variation in the controlled variable (often temperature) established by a process using on/off control.

Operating mode A specific condition or state of the controlled devices in a system that allows it to accomplish a specific objective. Hydronic heating systems can have several operating modes, including off, on, priority domestic water heating, heat purging, and so on.

Operating point The point on a graph where the pump curve intersects the system head loss curve. This point indicates the flow rate at which the system achieves hydraulic equilibrium.

Outdoor reset control A control method that increases the water temperature in a hydronic system as the outdoor temperature drops, and vice versa.

Outside air film A thin air film that clings to the exterior surfaces of buildings and provide a small thermal resistance to heat flow.

Overall heat transfer coefficient Usually referred to the "U" value, it is a number that includes the effects of convective and conductive heat transfer from one side of a heat exchanger to the other.

Overshoot The condition in which a controlled variable (often temperature) exceeds the desired temperature.

Over temperature/excess temperature A term often used in Europe to describe the difference between the average water temperature in a heat emitter and the room air temperature.

Oversurface percent A term used in computer selection of heat exchangers that gives the percentage of heat transfer area in a given heat exchanger model divided by the minimum required surface area to create a specified rate of heat transfer.

Oxides Chemical compounds formed when oxygen reacts with other materials. In hydronic system, oxides are typically formed when oxygen reacts with ferrous metals.

Oxygen diffusion A process by which oxygen molecules diffuse (travel) through certain materials due to a higher concentration of molecules on one side of that material. Travel is from higher to lower concentrations.

Oxygen diffusion barrier A layer of a special compound such as EVOH or aluminum on the outer surface or within the cross-section of tubing that significantly reduces the rate at which oxygen molecules can diffuse through the tube wall.

P

Packaged boiler Boilers that are traditionally supplied with factory-mounted components such as high limit control, circulator, pressure-relief valve, and oil or gas burners.

Packing gland A seal between the movable shaft of a valve and the valve's body.

Packing nut Part of a valve that is tightened to increase sealing pressure on a packing gland.

Panel radiator A metal heat emitter that mounts on a wall and releases heat through convection and thermal radiation.

Parallel direct-return piping A hydronic distribution system consisting of two or more branch piping paths that begin from a common supply pipe and end at a common return pipe. The branch *closest* to the circulator on the supply pipe is also *closest* to the circulator on the return pipe, and so forth with the other connected branches.

Parallel primary loop A primary loop in a primary/secondary system having two or more crossovers from a common supply to a common return. A pair of closely spaced tees and a balancing valve is installed in each crossover.

Parallel pumps Two or more pumps that are mounted to a common inlet header and discharge into a common outlet header.

Parallel reverse-return piping A hydronic distribution system consisting of two or more branch piping paths that begin from a common supply pipe and end at a common return pipe. The branch *closest* to the circulator on the supply pipe is the *farthest* from the circulator on the return pipe, and so forth with the other connected branches.

Parallel shifting Moving a reset line up or down without changing its slope to compensate for internal heat gain, or losses not proportional to the temperature difference between inside and output.

Parallel staging/modulation A control technique for a multiple mod/con boiler system where the firing rate (e.g., modulation) of all boilers is the same.

Partial reset Limiting the lowest target supply temperature for an outdoor reset controller to prevent a conventional boiler from operating with sustained flue gas condensation.

Partial storage systems A thermal storage system that stores only a portion of the energy needed by the load for a 24-hour period, during the preceding off-peak period.

Pellets Small cylindrical pieces of highly compressed wood fiber used as fuel in pellet-fueled boilers and also in pellet-burning stoves.

Performance envelope (of a mod/con boiler) The region between the boiler's maximum and minimum firing rate.

Perimeter circuits Radiant floor heating circuits that are routed near the exposed perimeter of a room.

Periodic flow reversal Any automatic means for reversing flow direction through a hydronic circuit at regular time intervals.

Permanent split capacitor (PSC) motor The type of electrical motor used in many small circulators, but being phased out over time and replaced by electronically commutated (EC) motors.

PEX tubing An acronym for cross-linked polyethylene tubing.

PEX-AL-PEX tubing A composite tubing having inner and outer layers of PEX, and a center layer of aluminum.

pH of a substance A logarithmic scale that indicates the relative acidity or alkalinity of a material. A pH of 7.0 is neutral. Any pH below 7.0 is acidic. Any pH above 7.0 is alkaline.

Photosynthesis A natural process by which plants convert carbon in the atmosphere into living vegetative matter.

Physical water quality A general term referring to the dirt and air content of water. High physical quality implies very low air and dirt content.

Pickup allowance A term used in describing the IBR net output rating of a boiler. It derates boiler heat output ratings by 15% assuming that 15% of the boiler's output is needed to cover the thermal losses and transient warm up of the distribution system that the boiler supplies.

Pipe size A term that, within the context of hydronics, and for pipes under 12 inches in diameter, refers to the *approximate inside diameter* of a pipe.

Pipe size coefficient A factor designated (c) used to describe all pipe geometry information when determining hydraulic resistance.

Piping topology The interconnected geometry formed by a specific arrangement of piping, fittings, valves, and other components. Examples of piping topology include two-pipe reverse return, single series circuit, home run, and so forth.

Plate & frame heat exchanger A heat exchanger constructed of multiple stainless steel plates that are held together by steel pressure plates and threaded rods. This type of heat exchanger can be disassembled if necessary, distinguishing it from a brazed plate heat exchanger.

Plateless staple-up system A method of installing radiant floor heating circuits in which tubing is directly stapled to the underside of a subfloor without use of metal heat transfer plates. *The author doesn't recommend this type of installation.*

Point of no pressure change (PONPC) The location where an expansion tank is attached to a hydronic circuit. The pressure at this point does not change when the circulator is turned on.

Poles (of a switch) The number of independent electrical signals that can be *simultaneously* passed through a switch or relay.

Polyethylene-raised temperature (PE-RT) A type of high-density polyethylene tubing that has been chemically structured so that it can be used at higher temperatures and pressures relative to typical high-density polyethylene tubing. PE-RT is not cross-linked like PEX tubing.

Polymer   A hydrocarbon substance consisting of long molecular chains. Many polymers are commonly called plastics. Common examples include polyethylene, polybutylene, and polypropylene.

Poured gypsum underlayment   A pourable slurry consisting of gypsum cement, sand, and other admixtures that cures into a hard subflooring. Often poured over hydronic tubing that has been stapled to the subfloor.

Power-venting   Using an electrically driven blower rather than a chimney to create the negative pressure required to safely vent combustion products from a heat source to outside air.

Premix combustion system   A method of mixing natural gas or propane with air in specific ratios prior to feeding that mixture to a burner.

Press-fit   A pipe joining method in which the socket of a special fitting is mechanically compressed against a tube wall (e.g., pressed) using a special tool. An elastomeric O-ring within the fitting forms a pressure-tight seal between the tube wall and fitting body.

Pressure head   The mechanical energy (e.g., head) possessed by a fluid due to its pressure.

Pressure-reducing valve   A special valve that reduces the pressure of water as it flows from a water supply system into a hydronic system. This valve is also called a feed-water valve.

Pressure-regulated circulator   A circulator that senses the differential pressure it is producing, and strives to maintain a specific value by adjusting its speed.

Pressure-relief valve   A spring-load valve that opens to release fluid from the system whenever the pressure at its location exceeds its rated opening pressure.

Primary circuit   A main distribution circuit that connects the system's heat source with several secondary circuits.

Primary combustion chamber   The upper chamber on a cordwood gasification boiler.

Primary side   The side of a transformer connected to line voltage. Also describes the side of a heat exchanger that operates at the higher fluid temperature.

Primary/secondary (P/S) piping   A method of interfacing individual (secondary) distribution circuits to a common (primary) circuit using closely spaced tees.

Primary/secondary interface   The connection of one hydronic piping loop to another using a pair of closely spaced tees.

Primary/secondary tees   A pair of closely spaced tees used to form a primary/secondary circuit interface.

Priming   Adding fluid to the volute of a circulator or pump when putting it into service.

Priority control   A control technique that allows lower priority loads to be turned off when necessary to ensure adequate heat deliver to designated load with the highest priority.

Priority load   A control strategy used in multiload systems in which some loads are temporarily turned off while the priority load is operating.

Priority override   A control technique to allows the highest priority load a specified maximum operating time before allowing lower priority loads to be turned back on.

Process line   A line on a psychrometric chart that connects two points, one of which represents the conditions (e.g., temperature and moisture content) of a return air stream, and the other representing the desired conditions of a supply air stream.

Processing algorithm   A specific set of programming instructions executed by a controller.

Processing algorithm   The encoded instructions the interpret one or more input conditions and generate an output.

Programmable thermostat   An electronic thermostat that can operate with multiple time schedules and different temperature setpoints for each schedule. Sometimes called a "setback thermostat."

Proportional band   The operating range of a controller in which the output signal or control action is proportional to the error between the target value and the measured value of the controlled variable.

Proportional control (P)   A control response where the value of the output signal is directly proportional to the error between the target value and the measured value of the controlled variable.

Proportional differential pressure control   An operating mode for a pressure-regulated circulator in which differential pressure decreases linearly with flow rate.

Proportional reset   A method of producing a reduced supply temperature in a hydronic distribution circuit by supplying a manually set mixing device with a "hot" fluid temperature that itself is reset by an intelligent mixing device.

Proportional-integral control (PI)   A control response where the value of the output signal is based on the error between the target value and the measured value of the controlled variable, *as well as how long that error has existed.*

Proportional-integral-derivative control (PID)   A control response where the value of the output signal is based on the error between the target value and the actual (measured) value of the controlled variable, how long that error has existed, and the time rate of change of error.

Proprietary controls   Physical controllers that contain components or operate on instruction sets that only one manufacturer has rights to, or the ability to distribute, service, or adjust.

Psychrometric chart   A complex chart originally developed by Willis Carrier that shows several thermodynamic properties of air.

Pulse width modulation (PWM)   A control algorithm in which the length of the on-cycle varies based on the error between the target value and the actual (measured) value of the controlled variable.

Pump (centrifugal) An electrically driven device that uses a rotating impeller to add mechanical energy (e.g., head) to a fluid. In this text, the word circulator refers specifically to a centrifugal pump.

Pump curve A graph indicating the head energy added to the fluid by the pump versus the flow rate through the pump. Such curves are provided by pump manufacturers.

Pump efficiency The ratio of the rate of head energy transferred to a flowing fluid divided by the rate of mechanical energy input to the shaft of the pump.

Purge cart An installation tool that uses a high capacity pump to purge air from a system, filter fluid, and help add antifreeze to a system.

Purge valves A specialized valve or combination of valves that allows forced-water purging of a hydronic system or a portion thereof.

Purging The process of filling a system with water and removing air from the system.

Q

Quick opening valve A valve characteristic that allows the flow rate through the valve to increase rapidly as the valve plug first moves away from the seat, and increase slowly as the valve plug nears its fully open position.

R

***R*-value (of a material)** A numerical value indicating the thermal resistance of the material. The greater the *R*-value, the slower heat will pass through the material, all other conditions being equal. Common North American units for *R*-value are $°F \cdot h \cdot ft^2/Btu$.

Radiant baseboard An extruded aluminum plate with integral tubes that is fastened to the base of walls, and releases the majority of its heat output as thermal radiation.

Radiant heating A natural process by which energy flows by thermal radiation from the surface of a heat emitter to the surfaces of cooler objects.

Radiant panel A heat emitter that releases 50% or more of its heat output as thermal radiation, and operates at surface temperatures under 300 °F.

Radiator A commonly used term for hydronic heat emitters that transfer some of their heat output by thermal radiation.

Rangeability The ability of a control system or device to maintain stable and accurate control under low loading conditions. A control system with high rangeability provides stable control under low loading conditions.

Raspberry Pi® A brand name for a programmable micro controller.

Real time inputs Information that is instantaneously supplied to a controller from sensors or other data sources.

Refrigerant Any substance that can change between liquid and vapor phase while undergoing a refrigeration cycle.

Refrigeration cycle The process of moving heat from a region of lower temperature to one of higher temperature by repeatedly expanding and compressing a compound called the refrigerant.

Relative humidity The amount of water vapor in the air as a percentage of the maximum amount of water vapor the air could retain at its current temperature.

Relay An electrically operated switch having a specific number of poles and throws.

Relay socket The base a relay plugs into, and to which external wires are connected.

Remote bulb aquastat An electromechanical temperature controller that uses a capillary tube several feet in length to connect its sensor bulb to its switch assembly.

Reset line A straight line having a specific slope on a graph of supply water temperature versus outdoor air temperature.

Reset ratio The ratio of the change in supply water temperature to the change in outdoor air temperature that a reset control will attempt to maintain. The reset ratio is also the mathematical slope of a reset line.

Resin beads Very small and porous polymer beads that have been chemically treated to absorb undesirable ions from water passing over them, while releasing other ions.

Resistance temperature detectors (RTDs) A type of solid state temperature sensor that varies its electrical resistance in a very precise manner as its temperature changes.

Reverse indirect water heater A tank in which domestic water is contained in one or more internal coils that are surrounded by water heated by the system's heat source.

Reverse injection mixing A piping configuration for injection mixing in which a portion of the fluid at the supply temperature of the distribution system is returned to the boiler loop to replace the hot water injected into the distribution system.

Reverse-return piping A piping configuration in which the branch piping circuit closest to the system's circulator on the supply main is farthest from the circulator on the return main. This arrangement helps to naturally balance flow rates through the branch circuits.

Reverse thermosiphon flow Undesirable flow from the upper portion of a tank containing heated water, through a piping path leading to the lower portion of the tank. If allowed to occur this flow increases heat loss from the tank. It can be prevented using a check valve or motorized valve to block the flow path when the circulator is off.

Reversing valve A valve used to change the direction of refrigerant flow in heat pumps, which allow the heat pump to operate as either a heat source or a chiller.

Reynold's number A unitless number that can predict whether a given flow situation will be laminar or turbulent.

Riser   A term used to describe vertical piping within a system, especially when that piping rises through multiple floor of a building.

Room air temperature profile   A graph showing how room air temperature varies from floor to ceiling.

Room heating load   The rate at which heat must be added to an individual room to maintain a desired indoor temperature.

Room side heat flux   The rate of heat output per unit of area from the room side of a radiant panel. It is usually expressed in Btu/h/ft^2.

Room temperature unit (RTU)   A device that measures room air temperature and communicates with a controller in another location.

Room thermostat   An adjustable, temperature-operated switch, mounted in a room, that turns heat delivery on and off.

Run fraction   The ratio of the on time of a heat source to the total elapsed time. In general, short run fractions are undesirable.

Rung   A horizontal path between vertical voltage lines on a ladder diagram. That path must contain a device that operates at the voltage between the two vertical lines.

S

Saddle fusion   A fabrication technique used with polyethylene and polypropylene pipe in which a special fitting is thermally fused to the outer surface of a pipe. After the saddle fusion is complete a hole is drilled through the pipe wall to create a tee.

Saturation condition   Typically refers to a vapor at a condition where any further cooling will result in condensation into a liquid.

Sealed combustion   A design for a combustion-type heat source in which all air required for combustion is drawn through sealed ducting from outside the building. All exhaust gases are also routed outside the building through sealed vent piping.

Seasonal energy efficiency ratio (SEER)   An indication of the seasonal average energy efficiency ratio (EER) of an air conditioner or heat pump, and established by a prescribed test procedure.

Secondary circuit   A piping circuit with its own circulator that receives heat from the building's primary circuit.

Secondary drain pan   A pan placed under an air handler to capture any condensate that may have leaked from the air handler's primary drip pan.

Secondary side   The side of a transformer connected to lower (e.g., control) voltage. Also refers to the "cooler" side of a heat exchanger.

Sectional boiler   A boiler assembled from cast-iron sections.

Sections   The cast-iron assemblies that are joined together to form a sectional boiler block.

Selective surface   An electroplated surface finish used on the absorber plates of solar collectors. The finish provides both high absorptivity and low emissivity.

Sensible cooling capacity   The sensible (rather than latent) heat transfer rate of an air-to-liquid heat exchange coil, typically within an air handler or fan-coil. This rate is based on the decrease in the dry bulb temperature of air passing through the coil, and the flow rate of that air. It is measured in Btu/h.

Sensible cooling load   The portion of the total cooling load created by heat gains to the interior of building that do not alter the moisture content of the interior air.

Sensible heat   Heat that, when added or removed from a material, is evidenced by a change in the temperature of that material.

Sensible heat ratio (SHR)   The ratio of the sensible cooling load to the total cooling load.

Sensing element   A probe used to detect a specific physical quantity such as temperature.

Sensor bulb   A fluid-filled capsule at the end of a capillary tube connected to an electromechanical temperature controller. The sensor bulb is fastened to a surface or inserted into a well at the location where the temperature is to be sensed.

Sensor well   A metal tube, open at one end and closed at the other, that is installed through the side wall of a tank or boiler. It serves as a sleeve through which a temperature sensor can be inserted. It holds the sensor in the desired location within a fluid-filled device.

Series baseboard simulator   A module in the Hydronics Design Studio software capable of simulating the thermal and hydraulic performance of a user-specified series baseboard circuit.

Series circulators   Two or more circulators installed in the same piping circuit, and having the same flow direction.

Series (piping circuit)   An assembly of pipe and piping components that are connected to form a single *closed* loop.

Series (piping path)   An assembly of pipe and piping components connected end to end to form a single path between two points.

Series primary loop   A primary loop in a primary/secondary piping system to which secondary circuits are connected in sequence, each using a pair of closely spaced tees.

Series staging/modulation   A control technique for a multiple mod/con boiler system in which each boiler is modulated up to it full output prior to firing the next boiler.

Setback   Reducing the temperature setting of a setpoint control, such as a room thermostat, to conserve energy.

Setpoint controller   A controller that attempts to maintain a preset temperature at some location in a system by controlling the operation of other devices. All setpoint controllers must operate with a differential.

Setpoint demand A demand for boiler operation from a load that needs to be supplied with water at a specific setpoint temperature.

Setpoint temperature A desired target temperature that is to be maintained if possible, and is set on some type of controller such as a thermostat or setpoint controller.

Shaft seal Metal or synthetic components of a pump that prevent fluid from leaking out where the impeller shaft penetrates the volute.

Shell & coil heat exchanger A heat exchanger constructed of an outer tank-like shell that surrounds an inner helical coil of tubing. One fluid passes through the coil while the other fluid passes through the space between the outer surface of the coil and the inner surface of the shell.

Shell & tube heat exchanger A heat exchanger in which one fluid passes through a group of parallel tubes while the other fluid passes through the spaces surrounding these tubes.

Shielded cable Electrical cable that has a metal foil sheathing wrapped around the insulated conductors. This shield is connected to an earth ground to prevent surrounding electromagnetic fields from inducing voltages into the insulated conductors.

Short-cycling An undesirable operating condition in which the duration of the heat source on cycle is considered too short for acceptable efficiency or equipment life. Short-cycling often occurs when a heat source is excessively oversized or connected to a highly zoned distribution system.

SIM system A system for melting snow and ice by circulating a warm antifreeze solution through tubing embedded in an external paving.

Single circuit/multizone (one-pipe) system A distribution piping system that uses diverter tees to connect each heat emitter branch circuit to the main piping circuit.

Sink (heat) A material into which heat is released.

Siphon The ability of a fluid to flow upward through a closed piping path, without use of a circulator, provided the fluid level at the outlet of that piping path is sufficiently lower than the inlet.

Slab-on-grade system A type of radiant floor panel in which tubing is cast into a concrete slab poured on stable soil.

Sleepers Wooden strips, usually arranged in parallel, and secured to a subfloor.

Sling psychrometer A device for simultaneously measuring the dry bulb temperature and wet bulb temperature of air.

Smooth pipe A generic term for tubing with relatively smooth inner walls. Examples include drawn copper tube, polybutylene tubing, and PEX tubing.

Snow switch A device that automatically turns on a SIM system when it detects the possibility of frozen precipitation.

Snow/ice pavement sensor A device that measure slab temperature as well as detects the possibility of frozen precipitation on the slab surface.

Socket fusion A type of thermal fusion for joining fittings and pipe made of thermoplastics such as polyethylene and polypropylene.

Soft temper tubing Copper tubing that is sold in coils, and can be bent with relative ease.

Solar circulation station A preassembled collection of hardware designed to interface a solar collector array to a storage tank.

Solar thermal collectors Devices for converting solar radiation into heat, and transferring that heat to a fluid stream passing through the collector.

Soldering A method of metallurgically joining copper or copper alloy pipe and components such as fittings and valves using a molten solder alloy fused to the base metal.

Solid fuel boiler A boiler capable of burning a solid fuel such as firewood, wood pellets, or coal.

Source (of a heat pump) The material from which the low-temperature heat is extracted.

Space-heat demand The control action whereby a heating system is signaled to deliver heat to a space-heating load.

Space heating The process of adding heat to a building or room to maintain indoor comfort.

Specific heat A material property indicating the amount of heat required to raise one pound of the material by one degree Fahrenheit, often expressed in Btu/lb/°F.

Specific humidity Another term for absolute humidity, usually expressed in pounds of water vapor per pound of air.

Spool (of a valve) An internal mechanism in a mixing valve that slides or rotates to regulate the amount of hot and cool fluid streams entering the valve.

Spring-loaded check valve A check valve that contains a spring to close the valve regardless of its mounting position.

Stacked home run system A piping system that uses a home run manifold assembly on each floor level of a multi-floor building. These manifolds are connected by a common set of supply and return riser pipes.

Stack effect The tendency of warm air to move upward in an enclosure, similar to how smoke rises through a chimney.

Stage (control) Using two or more stepped levels of heat input to a load.

Standard expansion tank An expansion tank that does not have a diaphragm.

Standard port ball valve A ball valve design where the orifice through the ball is slightly smaller than the diameter of the pipe to which the valve is connected.

Static pressure The pressure at some point in a piping system measured while the fluid is at rest. Static pressure increases from a minimum at the top of a system to a maximum at the bottom. Static pressure is created by the weight of the fluid itself, plus any extra pressurization applied at the top of the fluid.

Stationary air pockets Air that accumulates at the high points within a piping system.

Stator The portion of an electric motor surrounding the rotor and containing electromagnetic coils.

Steady state efficiency The efficiency of converting fuel energy into heat when a heat source operates continuously with steady input and output conditions.

Step down filtration A series arrangement of filters from course to fine particle capture size.

Stoichiometric combustion Combustion of a fuel under conditions where the exact amount of oxygen for complete combustion is present, with no excess air.

Straight pattern A term used to describe a valve that has inlet and outlet ports on a common centerline.

Strap-on aquastat An electromechanical controller that opens and closes electrical contacts in response to temperature, and mounts directly to a pipe.

Stratification The natural tendency for a warm fluid to rise above a cooler fluid due to density differences. This term is commonly used to describe warm air rising toward the ceiling of a room while cooler air descends toward the floor.

Street elbow An elbow fitting, either 90 degrees or 45 degrees, with a socket connection at one end and an insert (or ftg) connection at the other.

Striping The undesirable condition in which the surface of a radiant panel is significantly warmer directly above the tubing.

Subfloor A layer of wooden planks, plywood, or oriented strand board placed over and fastened to floor framing.

Submixing Mixing a warm fluid that has already passed through one mixing device with cooler fluid to achieve an even lower supply temperature.

Subsystem a collection of components that work together but are contained within a larger overall system.

Subgrade The soil under a concrete floor slab.

Substrate A solid layer of material that serves as a flat and stable mounting surface for another material.

Superheat The condition of a substance, such as water or a refrigerant, when it is at a temperature higher than its liquid/vapor saturation temperature. Superheat is measured in °F above the saturation temperature.

Superplasticizer An additive used to increase the slump of concrete without adding water.

Supply water temperature The temperature of water being supplied to a hydronic distribution system.

Suspended tube system A type of radiant floor heating system in which tubing is suspended in the air space between floor joists.

Sustained flue gas condensation An operating condition under which the water vapor produced during combustion is continuously condensing to liquid within a boiler. This condition is beneficial in boilers designed to operate with it, but very destructive to boilers not intended to operate with it.

Swing check valve A check valve with an internal disc that swings on a pivot.

Symbol palette A collection of symbols representing physical components in schematic drawings (piping and electrical).

System head loss curve A curve on a graph indicating the ability of a piping circuit to dissipate head energy as a function of the flow rate through it.

System turndown ratio The ratio of the maximum heat output from a group of multiple heat sources divided by the minimum stable heat output from that same group.

T

Tangential blower a blower with a wide but small diameter wheel.

Tankless coil A finned copper coil that is inserted into a boiler to provide domestic hot water.

Target flow rate A calculated flow rate through a piping system based on an expected rate of heat transport, and an assumed circuit temperature drop.

Target value The desired value of a controlled variable.

T-drill® A specialized tool for fabricating branch connections in copper tubing.

Temperature crossover An operating condition for a heat exchanger in which the outlet temperature of the fluid absorbing heat is higher than the outlet temperature of the fluid supplying.

Temperature drop The difference between the supply temperature and return temperature in an operating distribution circuit or part thereof.

Terminal unit A generic term for any device designed to create heat transfer between room air and chilled water.

Therm A quantity of energy equal to 100,000 Btus. The therm is the traditional unit in which natural gas is sold.

Thermal accumulator A special type of tank that accepts heat input from two or more heat sources, provides buffering for a zoned distribution system, and provides domestic hot water.

Thermal break The placement of insulation between two materials to reduce conduction heat transfer between them.

Thermal clutch An analogy between the clutch in an automobile and how a control system regulates heat transfer from a conventional boiler to a load, with the objective of avoiding sustained flue gas condensation.

Thermal conductivity A physical property of a material that indicates how fast heat can move through the material by conduction. The higher the thermal conductivity, the faster heat can move through the material.

Thermal energy Energy in the form of heat. The term thermal energy is used synonymously with the word heat.

Thermal envelope The shell formed by all building surfaces that separate heated space from unheated space.

Exterior walls, windows, and exposed ceilings are all parts of the thermal envelope of a building.

Thermal equilibrium The condition of a system when the rate of energy loss from the system is exactly the same as the rate of energy input to the system.

Thermal expansion The natural tendency of materials to expand as their temperature increases.

Thermal length A term used to describe the operating condition of a heat exchanger. It is the ratio of the temperature change of one fluid as it passes through the heat exchanger divided by the log mean temperature difference (LMTD) at which the heat exchanger is operating.

Thermal mass The ability of a material or object to store heat due to its mass and specific heat.

Thermal radiation Heat that is transferred in the infrared portion of the electromagnetic radiation. Thermal radiation cannot be seen by the human eye, but in many other ways behaves like visible light.

Thermal resistance The tendency of a material to resist heat flow. Insulation materials, for example, tend to have high thermal resistance. Other materials such as copper and glass have low thermal resistance. The term "R-value" is used to describe the thermal resistance of a material.

Thermal trap A piping assembly that discourages heat migration due to natural convection when the system circulators are off.

Thermistor A solid state temperature sensor that changes electrical resistance as its temperature changes.

Thermographic Image An image taken by an infrared sensing camera in which visible colors are used indicate the surface temperatures of objects in the image.

Thermoplastic A polymer that can be melted and reformed in shape.

Thermoset plastic A polymer that has been cross-linked or otherwise chemically altered so that it will not change back to a melted state.

Thermosiphoning The tendency of heated water to rise vertically within a hydronic piping system, even when the circulator is off.

Thermostatic mixing valve A mixing valve with an internal temperature control mechanism.

Thermostatic operator A nonelectric device that mounts to a radiator valve and opens and closes that valve in response to changes in room air temperature.

Thermostatic radiator valve (TRV) The assembly of a thermostatic operator and a radiator valve.

Three-piece circulator A circulator in which the wetted parts are separated from the motor by a coupling assembly.

Three-way diverter valve A valve designed to route entering flow to one of two possible outlet ports.

Three-way motorized mixing valves Mixing valves that blend hot water from a heat source with cooler water returning from a load to achieve a desired supply water temperature

to the load. These valves are automatically operated by some type of electrical controller.

Three-wire control Another name for floating control in which three wires are required between the controller and actuator.

Threshold setting (of a DPBV) The pressure differential at which a differential pressure bypass valve begins to open.

Throws The number of settings of a switch or relay in which an electrical current can pass through the switch or relay contacts.

Time delay relay A relay in which the contacts do not open or close at the same time coil voltage is applied or interrupted.

Time-of-use rates Electrical rates offered by some utilities that provided lower cost energy during specific times of day.

Ton (of capacity) A term representing a rate of energy flow of 12,000 Btu/h. It is frequently used to describe the capacity of heat pumps and air conditioners.

Total equivalent length The sum of the equivalent lengths of all piping, fittings, valves, and other components in a piping circuit.

Total cooling capacity The total of the sensible and latent cooling capacities of an air conditioner or a fan-coil or air handler operating at specific conditions.

Total cooling load The sum of the sensible cooling load and latent cooling load.

Total dissolved solids (TDS) An index used to describe the mineral content of water, usually expressed in parts per million (PPM).

Total head The total mechanical energy content of a fluid based on its pressure, elevation, and speed at some specified location in a piping system.

Total R-value of an assembly The overall total thermal resistance of an assembly of materials that are assembled together.

Towel warmer A specialty radiator designed to have towels draped over it for warming and drying.

Transformer An electrical component for reducing or increasing voltage in an AC circuit. Its common application in hydronic systems is to reduce line voltage (120 VAC) or (240 VAC) to control voltage (24 VAC).

Triac A solid state switching device used to turn an alternating current device on or off, or to operate it a variable speeds.

Triple action control A controller that provides high limit, low limit, and circulator control in hydronic systems using tankless water heaters.

Tri-state control Another name for floating control in which the actuator is either opening the valve, closing the valve, or holding the valve at its current position.

Tube and plate system A type of radiant panel construction in which tubing is saddled with or snaps into preformed metal plates that conduct heat away from the tubing.

Turbidity An indication of the suspended solids content of water based on its optical opacity. Water with high turbidity has low optical opacity.

Turndown ratio The ratio of the maximum heat output from a modulating heat source divided by the minimum stable heat output from that heat source.

Turbulators Twisted metal strips inserted in the firetubes of a boiler to create turbulent flow and enhance convective heat transfer.

Turbulent (flow) A classification of flow where streamlines repeatedly cross over each other as the fluid flows along the pipe. Turbulent flow is the dominant type of flow in hydronic heating systems.

Two-piece circulator A circulator with an air-cooled motor and having a direct connection of the motor shaft and impeller shaft.

Two-pipe system Another name describing a parallel direct-return or parallel reverse-return distribution piping system.

Two-wire digital bus A wiring method that allows communication between digital devices where each device on the bus has a unique digital address, and only two wires are required.

U

U-value The reciprocal of *R*-value. The higher the *U*-value of a material, or assembly of materials, the faster it conducts heat.

Uncoiler A device with a rotating platform that allows coiled tubing to be easily uncoiled without twisting.

Unconfined space As defined by the National Fuel Gas Code—space within a building that has a minimum of 50 ft^3 of volume per 1,000 Btu/h of gas input rating of all gas-fired equipment in that space.

Undershoot When the value of a controlled variable drops below the intended lower limit of operation.

Underside insulation Insulation placed under a framed floor deck.

Underslab insulation Insulation placed under a concrete slab.

Underslung thermal trap A piping assembly designed to discourage heat migration due to natural convection.

Unit heater A generic name for a fan-coil that is usually mounted near the ceiling of a room, with downward directed air flow.

Unsaturated state (of air solubility) When the amount of dissolved gases in the system water is low enough to enable the water to absorb residual air trapped in other parts of the system.

Usage profile A graph that shows the quantity of domestic hot water required at any hour of the day.

V

Valve actuator A device that physically moves the stem or shaft of a valve using electrical, thermal, or pneumatic input energy.

Valved manifold A manifold that has a flow regulating valve for each circuit connection.

Valveless manifold A manifold that does not have flow regulating valves for each circuit connection.

Vapor barrier A material that has low permeability to vapor flow.

Vapor pressure The minimum (absolute) pressure that must be maintained on a liquid to prevent it from changing to a vapor. The vapor pressure of most liquids increases as their temperature increases.

Variable-speed drive (VFD) An electronic device that controls the speed of an AC motor based on a control signal it receives.

Variable-speed injection mixing A method of controlling the temperature in a hydronic distribution system by using a variable speed pump to add hot water to a continuously circulating distribution system.

Variable-speed pressure-regulated circulators Circulators with electronically commutated motors that change speed automatically to maintain either constant differential pressure or proportional differential pressure between their inlet and outlet ports.

Velocity head The mechanical energy possessed by a fluid due to its velocity.

Velocity noise Undesirable sound created when the flow velocity through a pipe or piping component exceeds a certain value.

Velocity profile A drawing that shows how the velocity of the fluid varies from the centerline of a pipe out to the pipe wall.

Venturi fitting A tee that is designed to induce a differential pressure between its run and side ports whenever flow is passing through its run.

Venturi-type balancing valve A balancing valve that contains a venturi to create a differential pressure that correlates to the flow rate passing through the valve.

Vertical panel radiator A panel radiator that is taller than it is wide, and has tubes oriented in a vertical direction.

Viscosity A fluid property that infers the fluid's ability to resist flow. Fluids with high viscosities offer more resistance to flow.

Viscous friction Internal friction forces that develop within a fluid or between the fluid and a surface due to the viscosity of the fluid. Viscous friction is responsible for loss of head energy.

Volt-amp (VA) A multiplication of volts times amps. A common output rating for control transformers.

Volute The chamber that surrounds the impeller in a centrifugal pump.

W

Warm weather shut down (WWSD) A control mode in which the system's circulator and heat source are automatically turned off by a controller during warm weather.

Water hammer A loud noise created by a shock wave generated when a valve suddenly closes against a flowing fluid.

Water plate The portion of a steel panel radiator that contains fluid.

Water-reducing agent An additive that increases the slip of concrete without adding water to the mix.

Waterlogged A condition in which the captive air in an expansion tank is eventually replaced by water.

Water-to-water heat pump A heat pump that transfers heat from one stream of liquid to another stream of liquid. In hydronic applications, the liquid streams are typically water or an antifreeze solution.

Web enabled The ability of a control system to send and receive information over the Internet using the World Wide Web.

Wet base boiler A boiler design in which the boiler sections totally surround the combustion chamber.

Wet bulb temperature The temperature shown on a mercury bulb thermometer with a water-saturated wick over its bulb, and after the thermometer has been spun through the air several times as part of a sling psychrometer.

Wet rotor circulator A circulator in which the motor's armature and its impeller are an integral unit that is surrounded and cooled by the fluid passing through the circulator.

Wire-to-water efficiency The efficiency at which a circulator converts electrical energy input to head (mechanical energy) output.

Wood gasification A process in which wood is heated to the point where it releases pyrolytic gases that are then mixed with secondary air to produce a very hot and efficient combustion reaction.

Working pressure The allowable upper pressure limit at which a pipe or piping component can safely be used.

Y

Y-strainer A device designed to remove particles from a liquid stream passing through it. The particles are captured within a small basket made of fine stainless steel or brass wire.

Z

Zone An area within a building where space heating is controlled by a single thermostat or other temperature sensing device.

Zone valve Valves used to allow or prevent flow through individual zone distribution circuits.

Zoning Dividing a hydronic distribution system into two or more independently controlled distribution circuits.

Index

A

Above-floor tube and plate systems, 497, 503
  radiant floor panels, 476, 497–504
    installation procedure, 498–504
Absolute pressure, 146, 648
Absorber plate, 117
Acceptance volume, 640, 641
A-coils, 332
Active length, 312–313, 366
Activity level, effect of, 567
Actuator, 197, 203, 423, 441
Aerial boom, 486
AFUE. (*See* Annual Fuel Utilization Efficiency)
Aggregate, 487
Air
  binding, 655, 667
  change
    method, 33
    rates, 33
  dissolved within fluid, 147–148, 658–659
  entrained, 285, 657, 734
  entrapped, problems created by, 655–656
  film resistances, 24, 25
  flow rate, 339
  fluid, dissolved in, 658–659
  handler, 11, 327, 331
  infiltration, 33–36, 70
  pockets, stationary, 657
  problems, correcting chronic, 664–667
  purger, 663, 664. (*See also* Air, scoop)
  removal, 659–664
  requirements, combustion, 71–73
  scoop, 662. (*See also* Air, purger)
  sealing quality, 33
  separator, 17, 154
    central, 659
    microbubble, 663–664
  solubility, unsaturated state of, 663
  temperature stratification, 476, 667
  vent, 17, 552, 581
    automatic, 662
    float-type, 662
    manual, 660
Air-side pressurization, 640, 642
American Society of Mechanical Engineers, 67
Ampacity, 404, 409
Analog
  control signal, 83
  inputs, 370
AND decision, 403
Angle pattern, 194, 212, 214
Angle valve, 186–187, 572
Annual Fuel Utilization Efficiency (AFUE), 75, 78–79
Antifreeze-based systems, 122
Aquastat, 297
ASHRAE Handbook of Fundamentals, 24, 359
ASME. (*See* American Society of Mechanical Engineers)
Asphalt pavement, SIM installations in, 734–735
ASSE 1017-rated anti-scald tempering valve, 716
Atmospheric, 146
Automatic
  air vent, 660–662, 666
  flow balancing valves, 211
  flue damper, 57
Auxiliary
  loads, 688, 709
Available floor area, 534, 536, 540
Average
  floor surface temperature, 480, 493
  flow velocity, 222, 235
  surface temperature, 520

B

Backflow preventer, 15, 192, 672
BACNET, 462
Balance point temperature, 389, 390
Balancing, 211, 235
  introduction to, 763–776
  procedure, 765–770
  purpose of, 763–764
  reasons for, 763
  role of differential pressure control in, 770–771
  systems with manually setting balancing valves, 765
  theoretical approach to, 764–765
  using compensated method, 771–776
  valve, 211, 552, 599, 617, 750
Ball valve, 187–188, 262, 383
Band joist, 510
Barefoot friendly floors, 480, 540
Bars, 465, 597
Baseboard
  heating mode, 461
  tee, 179, 660
Basic loop systems, 13–14
Bearing walls, 482
Bell hangers, 160
Below-floor tube and plate system, 498, 504–508, 555
  installation procedure, 505–506
  radiant floor panels, 476

Bellows, 413
Bend supports, 484
Best efficiency point (BEP), 290
Bimetal element, 407, 409
Bin temperature data, 110
Biomass, 124–125
Black ice, 737, 739
Bleeders, 660
Blow-through fan-coils, 327
Blower, 10, 11, 327, 329
Bluetooth communication protocol, 299
Boiler
  basic control hardware, 426–435
  drain, 187, 671
  feed water valve, 671
  heating capacity, 74–75
  HI/LO fire, 378, 433
  high/low fired, 82–83
  jacket, 74, 81
  protection, 437–441, 442
  reset
    control, 504
    controllers, 431–432
  rotation, 435
  short-cycling, 381, 564
Boilers
  conventional, 203
  vs. condensing, 59–67
Borehole, 103
Bond breaker layer, 489
Boundary layer, 230, 691
Branch, 771
Brazed plate heat exchangers, 692–694, 695
Brushless DC motors. (*See* Electronically commutated motors (ECMs))
Brine, 2
British thermal units, 3, 9
BTU metering, 754
Btu/hr or Btuh. (*See* British thermal units)
Bubble's rise velocity, 657–658
Buffer tank, 56, 114–115, 381, 580, 743–749, 760
  design considerations for, 748–749
  piping, 743–746, 749
  sizing, 746
Building
  headers, 771
  heat loss coefficient, 129, 387, 389
  heating load, 21, 39
  mains, 760, 774
Building design heating load, 21
Bulkhead fitting, 93
Buoyancy, 2, 357

C

CAD. (*See* Computer-aided drafting)
Calculations, heating load, 21–22
Candidate circulators, 270, 569, 619
Capacitance rate (CR), 704, 705
 ratio, 704
Capacity of water, heat, 6
Capillary tube, 195, 413, 449
Carbon-neutral fuel, 125
Cast-iron
 circulators, 258
 sectional boilers, 54–55
Cavitation, 258, 292–297, 571
 correcting existing, 296
 gaseous, 295
 guidelines for avoiding, 295–296
 Net Positive Suction Head Available
  (NPSHA), 292–294
 Net Positive Suction Head Required
  (NPSHR), 294–295
Ceiling
 heated, thermal performance of,
  520–521
 panels, radiant, 518–521
Central air separators, 659, 660
Central heating plant, 760
Centrifugal pump, 258, 267
Characterized ball valve, 383
Chase, 67
Check valve, 188–189, 586–587, 593
Chemical energy content, 75
Circulators, 2, 8, 257–303, 351
 cavitation, 292–297
 connecting
  in parallel, 273–275
  to piping, 262
  in series, 272–273
 efficiency of, 285–287
 end suction, 258–259
 with flat pump curve, 595–596
 for hydronic systems, 258–263
 head
  explanation of, 267
  high vs. low, 271
 for injection mixing, 297
 with integral flow-checks, 299
 mounting considerations, 261
 multispeed, 271–272
 operating point of hydronic circuit,
  269–270
 performance, 266–275
  analytical methods for, 283–285
  curves, 268–269
 placement within the system,
  263–266
 primary, sizing the, 618–619
 push/pull arrangement, 273
 selecting, 302–303
 in series, 272–273
 smart, 275–283
  constant differential pressure
   control, 277–279
  energy savings achieved using,
   281–282
  proportional differential pressure
   control, 279–280
 special purpose, 297–302
 temperature-regulated, 297

three-piece, 260
two-piece, 260–261
wet-rotor, 259–260, 287
wire-to-water efficiency of, 287–288
zoning, 299
Circuit
 interior, 522
 layout drawings, 528–530, 532,
  533–534, 535
 perimeter, 522
 separation, 580
 single series, 569–572
Circuit-balancing valves, 548–549
Circuit-return temperature
   thermometers, 552
Class 1 SIM systems, 731, 736
Class 2 SIM systems, 731
Class 3 SIM systems, 731
Clean operation, 5–6
Clevis hangers, 160
Close-coupled, 130
Close coupling, 272–273
Closed-loop
 control system fundamentals,
  369–376
 sources, 101–103
 system, 13, 264, 275
Closely spaced tees, 579, 612, 614
Closet flange, 485
Coalescing media, 583, 663, 665
Coefficient of
 linear expansion, 182, 183
 performance (COP), 103–105, 116
Coil, 327, 399, 428
 tankless, 429, 710–711
 and shell heat exchangers, 708
Coin vents, 660
Cold start, 61
Collector array, 116, 122
Combined
 boiler/domestic host water tank
  assemblies, 57
 downward and edgewise heat loss, 29
 heat source reset and mixing reset,
  395–396
Combustion
 air requirements, 71–73
 efficiency, 76, 124
Comfort, 3
Common piping, 577, 580, 586
Communicating
 control systems, 462–465
 controllers, stand alone versus,
  462–463
 thermostats, 465
Communications bus, 462, 463
Compact style panel, 345
Compensated balancing method,
  771–776
Component isolation, 185, 188
Composite tubing, 172, 173
Compressible fluids, 222, 267
Compressive load rating, 733
Compressor, 100, 114
Computer-aided drafting (CAD), 214,
  528–530, 733

Concrete
 lightweight, 492–493
 pavement, SIM installation in,
  733–734
 placement, 486–489
 thin-slab radiant floor panels, 476,
  487–489
 installation procedure, 489–492
 sub-slab, 735–736
Condensate, 59
 neutralization, 65
Condenser, 99, 105
Condensing
 boilers, 62–64
 mode, 59, 60
Conduction, 3, 10, 24
 heat losses, 20, 22–26
Confined space, 71
Conservation of energy, 436
Conservation of mass, 436
Console fan-coil, 328
Constant circulation, 555, 557
Constant differential pressure control,
  277–279, 281
Contactor, 399
Contact adhesive, 515
Contacts, 399, 406
Continuous flow rating, 722
Control
 adaptive, 463
 algorithm, 463
 differential, 374, 391
 joints, 484–485, 525, 530
 processing algorithms, 370–371
 states, 373, 376
 systems, 368
  communicating, 462–465
  design principles, 457–459
  modern example, 459–462
  Web-enabled, 463
Controlled
 device, 370, 372
 shrinkage crack, 491
 variable, 370, 372
Controller, 370, 372
 embedded, 404
 output types, 373
 stand alone versus communicating,
  462–463
Convection, 3, 10, 12
 forced, 10, 11, 692
 natural, 10, 11, 308, 309, 692
Convectors, 308, 310
 finned-tube baseboard, 11–12, 309
 infloor, 357, 358
Conventional boilers, 59, 208, 297
 and mod/con boilers, multiple boiler
  systems using, 86–87
Conversion factors and data, useful,
  829
Coordinated zone management, 465
COP. (*See* Coefficient of, performance)
Copper water tube, 158–159
 boilers, 56–57
 heat loss from, 359–361
 mechanical joining of, 166–167
 supporting, 159–161
Correction factors, 313, 336, 337

Counterflow, 522
  arrangement, 698
  serpentine, 522, 523
  spiral, 522, 523
Coupling, 176
  assembly, 260
Cover sheet, 502, 545
Cross-linked polyethylene, 167, 472
Crossover bridge, 617, 619
Crossovers, 765, 771
Current, 231
C_v. (See Flow, coefficient)
C_v rating, 766
C_v setting, 766
Cycle efficiency, 76–78, 80
Cycle time, 374

D

Damper, 309, 312
Dead-heading, 604
Dead loading, 487, 497
Deaerated fluid, 655
Dedicated space heating device, 67
Degree days, 42–45
Delay-on-break, 402, 711
Delay-on-make, 402, 405
Delta T (ΔT) protection, 366
Demand charge, 90
Demand fired, 428
Demineralization, 150, 151
Demineralizer column, 152, 153
Density, 140, 141–142, 167
Depth stop attachment, 489
Description of operation, 405, 459
Design
  dry bulb temperature (97.5%), 21, 110
  flexibility, 5
  heating load, 21–22, 38
Desuperheater, 99–100
Dewpoint temperature, 59–62, 79
DHW. (See Domestic, hot water)
Diaphragm, flexible, 189
Diaphragm-type expansion tank, 15, 639–645, 646
  low-temperature systems, 641–642
  mounting, 643–645
  selecting, 643–645
  servicing, 643–645
  sizing, 641–642
Dielectric union, 180–181, 422
Differential, 14, 373, 408, 410
  control, 374
  operating, 374
  pressure, 209, 267
    bypass valves (DPBV), 209–211, 552, 596–597, 770–771
    control, 593, 770–771
    controller, 598
    threshold, 596
  temperature controller, 416–420
Diffuse, 174, 476, 487
Digital
  address, 462
  input, 370

Digitally encoded signal, 370
DIN
  4726 standard, 176
  rail, 401
Direct
  injection mixing, 447, 450
  through-the-wall openings, 71
  vent/sealed combustion, 70–71
Direct-fired water heater, 714, 720
Discharge port, 264, 265
Dissolved air, 147–148, 658, 663
Dissolved gases, 656
Distribution
  circuit, 572
  efficiency, 5, 622–625, 781
  piping systems, 563–632
  system, 1, 13, 21
District heating, 762
Diverter
  tee(s), 179–180, 572, 574
  valve, 629
Documentation, photo, 486
DOE heating capacity, 74, 76
Dollars per million Btu delivered ($/MMBtu), 45
Domestic
  hot water (DHW), 2, 405, 406, 709, 710
    heating, 709–724
    load estimates, 47
    mode, 461
    peak hourly demand for, 48–49
    tanks, 67–69
    usage patterns, 47–48
  water, 162, 189
Double
  interpolation, 335
  pole (DP) switches, 399
  pole/double throw (DPDT) toggle switch, 399
  throw (DT) switches, 399
DPBV. (See Differential pressure bypass valve)
Draft-proving switch, 69
Drain-back systems, 118, 122–123
Drain port, 16
Draw-through fan-coil, 331
Drawings
  circuit layout, 528–530, 531, 532, 533–534, 535
  layer, 529
Drip pans, 340, 802
Driving ΔT, 363, 365
Driving force, 232
Droop, 374, 409, 410
Dry-base boiler, 57, 59
Dual isolation valve, 346
Dual lockshield valve, 346, 347
Dual use device, 67
Dumb
  mixing devices, 442
  mixing valves, 199
  valve, 442
Duty cycle, 689
Dynamic pressure distribution, 647
Dynamic viscosity of water, 147

E

Earth heat exchanger, 101, 103
EEPROM (Electrically Erasable Programmable Read Only Memory), 275
Effective total R-value, 25, 37
Effectiveness, 704, 706
Efficiency, 74, 76
  of circulator, 285–287
    wire-to-water, 287–288
Elbows, 176
Electric
  boilers, 89–90
  radiant panel, 471
  thermal storage (ETS), 91–93
    layout, 92
    storage tanks for, 92–93
    system classification, 91–92
  valve actuators, 608–609
Electrical device model, 232
Electromagnetic
  energy, 11
  radiation, 116
  spectrum, 470
Electromechanical controls, 406–407, 415
Electronic controls, 372, 406–407
Electronically commutated motors (ECMs), 260, 598, 677
Element, 308, 310
Elevation head, 227
Embedded controllers, 404, 405
Emissivity, 12, 470
EN442 standard, 348
End
  suction circulator, 258–259
  switch, 209, 424, 554, 608
Energy
  delivered, 45
  Efficiency Ratios, 791, 793
  mechanical, 227, 258, 267
  savings, 3–5
    using smart circulators, 281–282
  usage, estimating annual heating, 41–46
Energy Rating (ER) label, 290
EnergyStar-certified, 35
Entrain, 191, 223
Entrained air, 657, 666
Entrapped air, 655–656
EPDM, 93, 166, 642
Equal percentage characteristic, 383, 384
Equilibrium, system, 567–568
Equivalent
  length, 233, 236, 238
  resistance, 233, 244, 246
  resistor, 240
Erosion corrosion, 250, 571
Error, 370, 371
Escutcheon plate, 311, 343
Estimating
  annual heating cost, 45–46
  the combined downward and edgewise heat loss, 31

Index

the infiltration heat loss, the air change method of, 33–36
Ethylene vinyl alcohol, 175
ETS. (*See* Electric, thermal storage)
Evacuated tube collectors, 116, 120
Evacuated tube solar collectors, 118, 119
Evaporation, 3
Evaporator, 101
EVOH. (*See* Ethylene vinyl alcohol)
Exhaust systems, power venting, 69–71
Expansion
  compensator, 183, 813
  loop, 183
  offset, 183
  system fluid, compatibility of the, 642–643
  tank, 14–15, 636, 639
    pressure and temperature ratings, 643
    standard, 636–639
  Tank Sizer, 636, 646, 647
Exposed surface, 9, 25
Extended surface suspended tube systems, 508–510
Exterior wall exposure, effect of, 567
External heat exchangers, 689, 692
External VA rating, 423, 459
Extruded
  aluminum heat transfer plates, 506
  polystyrene, 23, 481, 735
  underfloor plates, 506–507
Eye (of impeller), 258, 656

F

F. (*See* Female threads)
Fan, 327, 328
Fan coil, 11, 308, 327–334, 336
  advantages, 328
  cooling, using for, 339–340
  disadvantages, 328
  head loss of, 358
  heat output, 334–337
  overhead, 330–331
  performance principles, 336–339
  thermal performance, 334–340
  under-cabinet, 330
  wall-mounted, 328–330
Fan-enhanced panel radiator, 351
Fast-fill mode, 191
Fast-fill valve, 16
Feed water valve, 15, 189–191, 296
Feedback, 370, 764
Feet of head, 267, 270, 294, 768
Female threads, 176
Fibermesh®, 491
Filter fabric, 735
Finite element analysis (FEA), 26, 478, 517
Finish floor resistance, 548
Finned-tube, 764
  baseboard convector, 11, 308–312, 388
    head loss of, 357
    installing, 310–312
    placement considerations, 310
    sizing methods, 320–327

thermal ratings and performance of, 312–320
heat exchanger, 327. (*See also* Coil)
Fire-tube boiler, 56
Firmware, 275, 462
First hour rating, 48, 722
Fittings, 159
  common pipe, 176–179
  for hydronic systems, specialized, 179–182
  specifying, 176
  tips on using, 178–179
Fixed firing rate conventional boilers multiple boiler systems using, 80–89
Fixed lead boiler, 87
Fixed lead heat source, 433
Fixed point supports, 183
Flanges, 262, 663
Flashing, 292
Flat plate
  collectors, 116, 118
  heat exchanger, 695, 700, 708
  solar collectors, 116, 118
Flat pump curves, 277
  using circulator with, 595–596
Flat-tube panel radiators, 342–345
Float-type air vent, 662, 664
Floating
  control, 375–378
  zone, 376
Floor
  area, available, 536, 540
  covering considerations, 536
  heating mode, 461–462
  panels, tube placement considerations, 521–528
  surface temperature, 480
Flow
  classifications, 230
  coefficient (C_v), 235, 448
  duration curve, 281
  element (spool), 198
  laminar, 230
  meters, 552
  noise, 209
  rate, 143, 222
    on temperature drop and heat output, 539–540
  regulation, 185
  restrictor valve, 204
  switch, 717, 761
  turbulent, 230
  velocity, 222–223, 250
Flow-check valve, 16, 193, 586, 615
Fluid, 10, 22
  analyzing flow, 231–236
  compressible, 222
  feeder, 673
  flow
    in piping, 222–254
    rate, 338–339
  head of, 226–227
  incompressible, 198
  Properties Calculator module, 141, 234, 254, 638, 704
  properties factor (a), 234–235
  Reynold's number of the, 231
Flutes, 345
Fluted-channel panel radiator, 345–347

Fluted-steel panel radiators, 346
Flux, 163
Foot pound, 227
Forced
  air system, 378
  convection, 10, 11, 328, 690
Forced-water purging, 667–669
Fouling films, 695
Foundation heat loss, 27–33
4 to 20 mA, 372
Four-way motorized mixing valves, 201–202, 445–446
FPT. (*See* Female threads)
Framing members, effect of, 25–26
Free area, 71
Free oxygen, 655
Freestanding panel radiators, 344
French drain, 732
Ft.·lb. (foot pound), 227
Full
  port ball valve, 187, 188, 191
  reset control, 395
  storage ETS systems, 91
Fuzzy logic, 463

G

Gallons per minute, 222, 286
Galvanic corrosion, 149, 163, 181
Garage heating, intermittent, 724–726
Gas, 222
Gas- and oil-fired boiler
  designs, 54–59
  efficiency of, 75–79
Gaseous cavitation, 295, 656
Gate valves, 185–186, 262
General purpose relay, 400–401
Generic input/output ports, 464
Geotextile fabric, 733
Globe valves, 186
  gpm. (*See* Gallons per minute)
Graphite heat transfer plates, 507–508
Gravity
  purging, 667
Grout, 103
Gypsum-based underlayments
  installation procedure, 494–497
  radiant floor panels, 476, 493–497
Gypsum thin slab, 493

H

Half unions, 213, 262
Handwheel, 185–186
Hard
  drawn tubing, 159
  freeze, 114
Hard-wired logic, 403
Head, 227, 258, 267
  circulator, 267
  and differential pressure, converting between, 267–268
  loss, 230, 232, 250–251, 587
    due to viscous friction, 229

Index

relationship between pressure change and, 227–228
Heat, 8–10
  anticipator, 409
  capacity, 6, 140–141
  diffusion, 493
  dissipation line, 364
  dumps, 130
  emitters, 1, 10–11, 14, 307–366
    classification of, 308
    head loss of, 357–359
    output, controlling the, 382–384
  exchanger, 123, 689–709
    placement, pool, 729–730
    thermal performance of, 697–699
  flow, 23, 26–27
    rate of, 25–26
  flux, concept of, 532, 534
  gains from lighting/equipment, 567
  latent, 140
  load
    calculations, 21–22
    calculations, computer-aided, 40–41
    total building, 39
  loss
    coefficient, 387
    compensating for nonproportional, 389–390
    infiltration, 33–35
    rate of, 10
    through basement floors, 28–29
    through basement walls, 27–28
    through slab-on-grade floors, 29–32
    through windows, doors, and skylights, 32
  metering, 754
    hardware for, 756–759
    within heat sources, 762
    theory, 754–756
    migration, 206
    motor, 553
  migration, 206
  motor actuators, 423
  motors, 205
  output
    controlling, 377–382
    analytical model for baseboard, 314–315
    ratings, 314
    flow rate on temperature drop and, 539–540
  plant, 381
  pool, 5
  production, variable output rate, 380
  pump, 53, 93–98
    categories of, 94
  purging, 78
  quantity, 9, 142
  rate, 9
  sensible, 140
  sinks, 474
  source, 1, 2. (*See also* Heat source)
  specific, 140
  transfer
    modes of, 10–13
    plates, 498
    rate of, 23
Heat source, 7, 14
  controllers
    multiple, 433–435
    reset, 431–432

fixed lead, 433
reset
  for modulating heat sources, 392–393
  for on/off heat source, 391–392
rotation, 433
Heating
  capacity, 67
  curve, 475. (*See also* Reset line)
  district, 762
  effect factor, 314
  energy usage, estimating annual, 41–46
  garage, intermittent, 724–726
  load, estimate example, 36–39, 40, 44
  pool, 726–730
  spa, 726–730
  space, 5
Hematite (Fe_2O_3), 655
HI/LO fire boilers, 378, 434
High
  capacity water heating, hydronic solutions for, 719–724
  limit setting, 428, 429, 439, 461
  limit/switching relay control, 426–429
  point vents, 659–664
High/low-fire boilers, 82–83
High wall cassette, 329
Home run
  distribution system, 196, 199, 324, 606–610
    design procedure for, 610
    stacked, 609–610, 611
    zoning control options for, 608–609
Horizontal
  panel radiator, 343, 344
  supply air ducting, 72
Hose bib, 191
Hunting, 372
Hybrid distribution system, 625–632
Hybrid staging/modulation, 434
Hydraulic
  efficiency, 285
  equilibrium, 270, 567–568
  resistance, 231
    determining, 233–234
    diagrams, 240, 244
    of fittings, valves, and other devices, 236–238
  resistor, 240
  separation, 81, 577–579
    creating, 579–583
    mixing within, 583–585
    summary, 585
  separator, 81, 581
Hydraulic equilibrium, 270
Hydro-air distribution system, 333–334
Hydrogen cation, 151
Hydronic
  circuit, operating point of, 269–271
  Circuit Simulator module, 222, 249–250, 254, 285, 303, 326, 358, 548, 569, 571, 591, 595, 603, 764
  circulators. (*See* Circulators)

closed-loop system, 54, 142, 258, 636
control systems, 13
fan-coils, 334–340
heat
  emitters, 1, 11, 14, 307–366
  pump, 93–98
  sources, 52–136
heating
  benefits of, 2–8
  history of, 1
multizone systems, 193
radiant
  panel heating, 471
  panels, classification of, 476
solutions, for high-capacity water heating, 719–722
snow and ice melting systems (SIM), 731–743
  asphalt pavement installations, 734–735
  benefits of, 730–731
  boiler options, 737
  circuit temperature drop, 741, 742
  Class 1 SIM, 731
  Class 2 SIM, 731
  Class 3 SIM, 731
  design and control options, 736
  drainage provisions, 731–732
  estimating heat output, 741, 742
  flow requirements, 740–741
  freeze prevention options, 736–737
  installation in concrete pavement, 733–734
  on/off control options, 738–740
  pavement temperature control, 737–738
  for paver-covered surfaces, 735–736
  subsurface conditions, 732–733
  system classifications, 731
  tube spacing considerations, 741, 743
subassemblies, 557
subsystems, basic, 13–17
systems
  filling and purging, 667–672
  fluids in, 226
Hydronics Design Studio, 41, 141, 142, 235, 285, 303, 321, 323, 358, 361, 548, 569, 636, 641, 646, 764
Hydronic system stabilizer, 154
Hydroxide anion, 151
Hygroscopic air vent, 660, 661, 662
Hypocausts, 471

I

IBR (Institute of Boiler and Radiator Manufacturers)
  gross output, 74
  method, 320–321
  net output, 74–75
  Testing and Rating Code, 313
Ice
  black, 737
  and snow pavement sensor, 739
Idling (snow melt pavement), 733, 737

Index

Impeller, 258, 271
Implosion, 292
In solution, 658
Incompressible, 148–149, 222, 635
    fluids, 655
Indirect water
    heaters, 712–713
        benefits of, 713–714
        continuous flow rating, 722
        first hour rating, 722
        piping and settings for, 715–717
        sizing an, 723–724
Indoor reset control, 398
Inductive loads, 399
Infiltration, 33
    heat loss, 33–35
Infloor convectors, 357, 358
Infrared thermal radiation, 470
Infrared thermograph, 12
Injection control device, 449, 450
Injection mixing, 204, 446–447
    points on, 456–457
    using
        two-way motorized valve, 449–450
        two-way thermostatic valve, 447–449
        variable-speed pump (direct), 450–455
Inlet fluid parameter, 120
Inlet-temperature thermometer, 552
Inline circulators, 258, 259
Inside air film, 24
Inspection visit considerations, 34
Installation, noninvasive, 6–8
Instantaneous thermal efficiency, 119, 120
Institute of Boiler and Radiator Manufacturers. (*See* IBR)
Instruction set, 275
Integral flow-check valves, 299
Intelligent mixing device, 398
Interface module, 462
Interior circuits, 522
Intermittent garage heating, 724–726
Internal heat exchanger, 689–692
Internal heat gains, 389
Internet accessibility, 412
Interpolation, 334
    double, 335
Interstage differential, 410
Interzone heat transfer, 566
Ion exchange processes, 150
Isolation
    flange, 262
    valves, 552
Isotherms, 26, 477
Iteration, 284, 568
Iterative design, 568–569

J

Jacket (of boiler), 55

K

Kick space heater, 308, 330. (*See also* Fan coil, under-cabinet)
Kilo-Pascals, 597
Kinematic viscosity of water, 147
KPa, 597. (*See also* Kilo-Pascals)
Kraft paper faced batts, 492

L

Ladder diagram, 369, 403–404
    drawing, 404–405
    line voltage section, 403
Laminar, 230
    flow, 230
Latent heat, 140
Lead-based solders, 67
Leader length, 543
Leaders, 526, 609
Least favored crossover, 765
Life cycle operating cost, 289
Lightweight concrete, 492–493
Lift, 496
    hook, 487
Line voltage, 403, 404
    section, 403
Linear regression, 284
Linear relationship, 382–383
Liquid, 222
    crystal display (LCD), 410
    static pressure of, 223–226
Loads, 5
    auxiliary, 688
    estimates, daily DHW, 709
    priority, 689, 712
Lockshield valve, 212–213
Logic
    fuzzy, 463
    hard-wired, 403
LONworks, 462
Low
    loss headers, 581
    voltage section, 403
    water cutoff (LWCO), 430
Low-limit aquastat, 329
Low-velocity zone separator, 675–676

M

M. (*See* Male threads)
Magnetic dirt separator, 283
Magnetite (Fe_3O_4), 282, 655
Mains, 771
Make-up water
    assembly, 672
    system, 15–16, 672–673
Male threads, 176
Manifold, 502
    accessories, 550
    stations, 7, 482, 525, 548–550
    systems, multiple, 554
    valve actuators, 548, 552–554, 556
Manipulated variable, 370

Manometer, 211
Manual
    air vent, 660
    reset high limit (MRHL), 429–430
Manually set mixing valves, 441–442
Mass, 141
Master thermostat, 574
Maximum allowed upward heat flux, 542
Maximum deflection criteria, 487
Mean radiant temperature (MRT), 12
Measured value, 370
Mechanical energy, 227, 258, 267
Megawatt-hours (MWh) units, 756
Metabolism, 474
Meter rental, 46
Metered balancing valves, 751
Metering, BTU, 754
Microbubble, 658
    air separators, 663–664
Microfans, 351
Microprocessor, 369
Mineral oil, 490
Minimum supply temperature setting, 392
Minitube distribution system, 749–753
    benefits of, 750–752
    design details, 752–753
    multizone, 749
Minitubes, 749
Mixing
    assembly, 437, 457
    devices, 369
    within hydraulic separating devices, 583–585
    injection, 446–457
    points, 439
    purposes, 436
    reset control, 393–395
    strategies and hardware, 436–457
        four-way motorized valves, 445–446
        manually set valves, 441–442
        three-way motorized valves, 444–445
        three-way thermostatic valves, 442–444
    thermodynamics of, 436–437
    valve, 197–198
MMBtu, 44, 45
Modulating
    boilers, 64–65
    DC outputs, 376–377
    heat production, 380–381
    heat source, 381
Modulating/condensing boilers, 64–65, 380
    efficiency of, 79
    multiple boiler systems using, 83–86
Modulation, 64
Monoflo tees®, 179. (*See also* Diverter tee(s))
Most favored crossover, 765
Motorized actuators, 197
Mounting rail system, 160
MPT. (*See* Male threads)
Multicirculator systems, 575

hydraulic separation and, 577–579
  creating, 579–583
  mixing within, 583–585
  summary, 585
Multifunction tank, 633
Multifunction time delay device, 402
Multiload/multitemperature systems, 5
Multiple boiler
  controllers, 86
  system, 80–89, 88
    using conventional and mod/con boilers, 86–87
    using fixed firing rate conventional boilers, 80–82
Multiple heat source
  controllers
    for modulating heat sources, 434–435
    for on/off heat sources, 433–434
Multiple manifold systems, 554
Multispeed circulators, 271–272
Multistage heat production, 378–380
Multistage thermostat, 410
Multi-temperature radiant panel systems, 555–560
Multizone
  relay center, 424–426
  systems, 564, 585–593
    using zone valves, 594–595
    using zone circulators, design of, 587–593
Multizone/zone valve system, design procedure for, 593–599
MWh (megawatt-hours) units, 756

N

National Fenestration Rating Council (NFRC), 32
National Fire Protection Agency, 71
National Fuel Gas Code, 71
National pipe thread (NPT), 165, 174
Natural convection, 10, 11, 308, 309, 690
Negative temperature coefficient (NTC) thermistor, 410
Net COP, 105
Net Positive Suction Head Available (NPSHA), 292–294
Net Positive Suction Head Required (NPSHR), 294–295
Networked control system, 462
Networked heat metering systems, 756–757
NFPA. (*See* National Fire Protection Agency)
Nominal inside diameter, 158
Nonbarrier tubing, 176
Noncondensing mode, 60
Noninvasive installation, 6–7
Nonlinear, 383
Non-pressure-rated tanks, 93
Normally
  closed contacts, 400
  open contacts, 400

NPSHA. (*See* Net Positive Suction Head Available)
NPT. (*See* National pipe thread)
Nuclear radiation, 470

O

Off-cycle
  heat loss, 55
  losses, 55
Off-peak periods, 91
Offset, 371
Ohm's law, 232
Oil-fired condensing boilers, 64, 70
On/off
  control, 373
  output, 373–374
  of single-stage heat source, 377–378
180-degree return bend, 312
One-pipe
  systems, 180, 572
    design procedure for, 574–575, 576
1-Pipe valve (for panel radiators), 354
Open-loop
  sources, 101
  system, 13
Open protocols, 462
Operating
  analytical method for finding the, 284–285
  differential, 374
  cost, theoretical annual, 250–253
  modes, 399
  point, 270
  region, 275
Operation
  clean, 5
  quiet, 6
OR decision, 403
Outdoor reset control, 384–398
  theory of, 386–387
  implementation of, 390–393
Outdoor wood-fired furnaces, 126–127
Output, 74, 370
  signals, 372
Outside air film, 24
Over temperature/excess temperature, 347
Overshoot, 374, 409
Oxides, 655
Oxygen diffusion, 174–176
  barrier, 175

P

Packaged boilers, 54
Packing
  gland, 186
  nut, 186
Panel radiator, 307, 340–354
  benefits of, 340–342
  designer radiators, 348

fan-enhanced, 351, 354
flat-tube, 342–345
fluted-channel, 345, 348
freestanding, 344
head loss of, 358–359
piping for, 351–354
thermal performance, 348, 352–353
towel warmer, 348, 349
vertical, 342, 343
Parallel
  circulators in, 273–275
  direct return piping, 605
  direct-return system, 599–603
    manual design procedure for, 601–603
  distribution systems, 599
  hydraulic resistances, 243–247
  piping, 185
  primary circuit, 617
  reverse-return piping, 603
  reverse-return systems, 603–605
    manual design procedure for, 605
  shifting, 389
  staging/modulation, 434
Parallel-piped baseboard, 324–325
Partial
  load conditions, 55
  reset control, 392, 432
  storage ETS systems, 91
Partner valve, 773
PassivHaus standard, 35
Paste flux, 163–164
Paver-covered surfaces, SIM installations for, 735–736
Peak hourly demand, 29–30, 48–49
Pellet-fueled boilers, 131–133
Performance envelope, of boiler, 79
Perimeter circuits, 522
Periodic flow reversal, 742
Permanent split capacitor (PSC), 260, 275, 282, 450
PEX tubing, 167–169, 231, 343, 356, 472, 483, 487, 502, 606, 734
PEX-AL-PEX tubing, 169–171, 231, 343, 356, 483, 487, 502, 519, 606, 734
Photo documentation, 486
Pickup allowance, 74
Pipe adapters, 552
Pipe Heat Loss module, 361
Pipe size, 158, 243
  coefficient, 234
  considerations, 250–254
  selecting, 253
Piped in parallel, 244
Piping
  buffer tank, 743–749, 748
  common, 577
  components
    paths containing multiple pipe sizes, 243
    represented as series resistors, 240–243
    schematic symbols for, 213, 215, 823
  device model, 232, 233
  fluid flow in, 221–256
  installation, tips on, 214–219

Index

loops, fluid-filled, 229
materials, 158–176
for panel radiators, 351–352
parallel, 185
and settings for indirect water heaters, 715–717
standoff supports, 160
subassembly, 573
systems, reducing complex, 247–249
thermal expansion of, 182–185
topology, 610
Plate and frame heat exchanger, 694
Plateless staple-up systems, 510–511
radiant floor panels, 476
Point of no pressure change, 264, 647–650
Poles, 399
Polymer, 158
Polyethylene tubing (PEX), 167
Polyline, 529
Pool heat exchanger placement, 729–730
Pool/spa heating, 6, 726–730
Poured gypsum underlayments, 493
Power venting, 69
exhaust systems, 69–70
Precipitate, 150
Prefabricated
panels for suspended ceilings, 476
subfloor/underlayment radiant floor panels, 476, 511–512
Premix combustion system, 64
Preset method, 765
Press-fit joint, 166
Pressure
converting between head and differential, 267–268
drop charts, 235–236
head, 227, 258
relationship between head and, 227–229
testing, 485–486
Pressure-regulated circulator, 196
Pressure-rated steel tanks, 92
Pressure-reducing valve, 15, 189–190
Pressure-relief valve, 15, 192–193
Primary
circuit, 611
combustion chamber, 128
heat exchanger, 63
side, 403. (*See also* Transformer)
Primary/secondary
(P/S) piping, 610
systems, 596, 610–622
flow reversal in, 620–622
indirect water heaters in, 712
purging provisions for, 620
Priming, 258
Priority
control, 405
load, 402, 689, 712
override, 405
zone, 426, 593
Processing
algorithms, 370
derivative (D), 375
integral (I), 375

Programmable thermostats, 410
Proportional
band, 371
control (P), 371–372
differential pressure control, 279–281
reset, 396–398
limitations of, 398
Proportional-integral control (PI), 372
Proportional-integral-derivative control (PID), 372–373, 407
Proprietary protocols, 462
PSC. (*See* Permanent split capacitor)
Pulse width modulation (PWM), 374–375, 449
Pump curve, 268, 269
flat, using circulator with, 595–596
Pumps
centrifugal, 258
efficiency, 287
variable speed
injection mixing using, 450–455
Purge, 191
cart, 669, 670–671
valve, 191, 669
Purger, air, 663. (*See also* Air, scoop)
Purging, 16
gravity, 667
valves, 16, 191, 552
Push/pull arrangement, 273
PVC, 93
Pyrolytic gases, 126–128

Q

Quantity of heat, 142
Quick opening valve, 384
Quiet operation, 6

R

R-value, 21, 22
effective total, 25
of material, 23
of an assembly, total, 23
of common building materials, 827–828
Radiant
baseboard, 355–357
ceiling panels, 518–521
modular, 519
floor heating, 378
heat, 470. (*See also* Thermal radiation)
heating, 470–471
panel, 13, 471
benefits of, 473–474
ceiling panels, 476
circuit, thermal model of, 537–539
circuit sizing procedure, 530–531, 536
distribution systems, zoning option for, 554
history, 471–472
physiology, 474–476
surface temperature limitations, 536

system, multi-temperature, 555, 557–558
wall (tube and plate), 476, 513–518
Radiators, 308
Rangeability, 383, 750
Rate of heat
flow, 25–26
loss, 10
transfer, 22, 143
Rating, external VA, 423
Real time inputs, 463
Recessed fan-coils, 329
Recessed floor box, 357
Reference pressure, 773
Reference valve, 773
Refrigerant, 94
Refrigeration cycle, 94–95
Reinforced polypropylene (PP-R) tubing, 172–174
Relay, 398, 399–400
contacts
closed, 400
open, 400
general purpose, 400–401
socket, 400
time delay, 401–403
Remote bulb aquastat, 412
Renewable energy heat sources, 53
solar collectors, 116–118
evacuated tube collectors, 118, 119
flat plate collectors, 116, 117
performance of, 119–122
solar combisystems
antifreeze-based systems, 122–123
drain-back systems, 118, 122–123
solar energy, 116
solar thermal systems, 116
Reset
line, 387
ratio, 387–388
Resin beads, 150
Resistance, 232
loads, 399
Respiration, 5
Return main, 765
Return manifold, 548
Reverse injection mixing, 446–447
Reverse-return piping, 351
parallel, 603–606
Reversible hydronic heat pumps, 116
Reynold's number, 231
Rise velocity, of bubble, 658
Riser, 771
Rod clamp, 160
Room
air temperature profile, 474–476
-by-room
calculations, 38–39
flow control, 522
temperature control, 194
zoning, 473, 564
heating loads, 21
side heat flux, 532, 534
temperature unit (RTU), 388
thermostat, 14, 369

Rotor, 259
Rough openings, 489
Run fraction, 76
Rung, 403. (*See also* Ladder diagram)

S

Saddle fusion, 173
Saturation pressure, 656
Savings, energy, 3–5
Schematic symbols for
   electrical components, 825
   piping components, 213–214, 215, 823
Scoop, air, 662. (*See also* Air, purger)
Score and snap method, 502
Screed, 489
Sealed combustion, 64
   boilers, 70
Seasonal efficiency, 79
Secondary
   circuits, 611
   heat exchanger, 63
   side, 403. (*See also* Transformer)
Sectional boiler, 54
Sections, 54
Selective surface, 116
Self-balancing, 603
Sensible heat, 140
   quantity equation, 142–143
   rate equation, 143–145
Sensing element, 370
Sensor, 370
   bulb, 203
   snow/ice pavement, 739
   well, 204, 413
Series
   Baseboard Simulator, 320, 321, 570
   circuit, 240
   circulators in, 272
   piping
      circuit, 149
      path, 240
   primary circuits, 615–617
   staging/modulation, 434
Series-connected components, equivalent resistance of, 240–243
Serpentine patterns, 522
Setback, 280–281
Setpoint
   controller, 412–414
   demand, 433, 720
   single-stage electronic, 414
   temperature, 3, 373
   two-stage electronic, 415
Settings, 370
Shaft seals, 260
Shape memory, 168
Shell and tube heat exchanger, 56
Short-cycle, 743
Short cycling, 374
Shutoff valves, 15
SIM. (*See* Hydronic, snow and ice melting systems)

Single
   circuit/multizone (one-pipe) system, 572–575
   line drawing, 462
   pole (SP) switches, 399
   pole/double throw (SPDT) switches, 399, 413
   series circuit, 569–572
   throw (ST) switches, 399
Single-stage heat source, on/off control of, 377–378
Sink, 93
Siphon, 123
Sizing parallel-piped baseboards, 325–327
Slab-on-grade radiant floor panels, 477–487, 526
   installation procedure, 481–484
Sleepers, 489, 498
Sleeping comfort, desired, 567
Sleeve/roller supports, 183–184
Sliding scales, 45
Smoke pencil, 35
Smooth pipe, 231
Snow/ice pavement sensor, 739
Snow switch, 739
Snowmelting, 5
Socket fusion, 172
Soft soldering, 161
Soft temper tubing, 159
Software-assisted design, 548
Software-based circuit analysis, 249–250
Solar
   collectors, 118–119
      evacuated tube collectors, 118, 119
      flat plate collectors, 116–118
      performance of, 119–122
   combi-systems
      antifreeze-based systems, 122
      drain-back systems, 122–123
   energy, 116
   heat gain, effect of, 567
   radiation, 117
   thermal systems, 116
Solar Rating and Certification Corporation (SRCC), 120, 121
Soldering, 162
   joints, making good, 162
   tips on, 214–215
Solid-fuel boilers, 53, 436
   combination wood/oil boilers, 386
   outdoor wood-fired furnaces, 126
   pellet-fueled boilers, 131–136
   wood-burning boilers, efficiency of, 53
   wood-fueled boilers, 125
      types of, 125
   wood-gasification boilers, 137
Source, 93
   water, 101
Spa heating, 748
Space
   heat demand, 433
   heating, 5, 6
Specific heat, 140–141

Spool, 198
Spring-loaded check valve, 189, 586
Stack effect, 136
Stacked home run system, 609–610
Stage control, 415
Stages, 80, 378
Staging and modulation, combining, 381–382
Stagnation pressure, 122
Stand alone versus communicating controllers, 462–463
Standard
   expansion tank, 636
      sizing, 638–639
Standard-port ball valve, 187
Standoff piping supports, 160
Starting point line, 389
Starting torque, 275
Static pressure, 223–226
Stationary air pockets, 657
Stator, 259
Steady-state
   efficiency, 75–76
Steam flash, 146
Steel fire-tube boiler, 56
Steep pump curves, 271
Step-by-step circuit sizing procedure, 540–548
Storage tank, combining tankless coil with, 711–712
Strainer, 675
Strap-on aquastat, 413
Straight pattern, 194, 213, 214
Stratification, 4, 310
   air temperature, 476
Streamlines, 230
Street elbows, 176
Striping, 493, 511
Subassembly, piping, 573
Sub-atmospheric pressure, 667
Subfloor, 487
Subgrade, 481
Substrate, 498, 513
Subsystems, 13, 396
Superplasticizer, 491
Supply
   main, 765
   manifold, 548
   water temperature, 299, 351, 392
Suspended tube systems, 508–510
   radiant floor panels, 476
Sustained
   flue gas condensation, 55, 59, 61, 62
   water flow, 114–116
Swing-check valve, 188–189
Switches, 399
   flow, 717
   snow, 739
Symbol palette, 826
System
   curve, 239
   design, importance of, 17
   equilibrium, 567–568
   head loss curve, 239–240

piping and temperature control
  options, 548–560
turndown ratio, 83–86, 381, 724
volume, estimating, 645–646

T

T-bar ceiling systems, 517
T-Drill®, 613, 614
T-drilled connections, 116
Tank-type water heater, 69
Tankless coil, 429, 710–711
Tank-within-a-tank design, 712
Target
  flow rate, 571
  temperature, 391
  value, 370
Tees, 176
Temperature
  97.5 percent design dry bulb, 21
  controls, 14
  dewpoint, 59
  difference, 10
  drop, 539
  gradients, 510
  limit control, 361, 392
  mean radiant, 12
  setpoint, 14
Temperature-limiting controller, 14
Temperature-regulated circulators, 297
Thermal
  break, 489
  clutch, 439
  conductivity, 10, 22
  contraction, 184–185
  energy, 1, 270, 470
  envelope, 32–36
  equilibrium, 13, 361–363, 567
  expansion, 182, 635
    valve, 95
  load, 20
  mass, 55, 473
    effect of, 565–566
  migration prevention, in P/S systems, 613–615
  model, of radiant panel circuit, 537–539
  radiation, 3, 11, 308, 470, 474
  resistance, 23
  trap, 449
  wick, 477
Thermistor, 410
Thermodynamics
  first law of, 270
  of mixing, 436
Thermographic image, 470
Thermoplastics, 167
Thermos® bottle, 118
Thermoset plastic, 167
Thermosiphoning, 16, 586, 749
Thermostat, master, 574
Thermostat, room, 14, 407–409
  electronic, 410–411
  placement considerations, 409
  single-stage electromechanical, 407–409
  two-stage electromechanical, 410

Thermostatic
  actuator, 412, 628
  mixing valve, 630, 711
  motor, 203
  operator, 194
  radiator valves (TRV), 194–197, 344
Threaded adapters, 176
Three-piece circulator, 260
Three-wire control, 376. (See also Floating control)
Three-way diverter valves, 206, 207, 555
Three-way motorized mixing valves, 198–201, 444–445
Three-way thermostatic mixing valves, 203, 442–444
Threshold differential pressure, 596
Threshold setting (of DPBV), 596, 597
Throws, 399
Time delay relays, 401–403
Time-of-use rates, 53, 91
Ton (of capacity), 101
Total
  building heating load, 39
  equivalent length, 240, 241
  head, 227
  heat transferred, 754
  R-value of
    an assembly, 23
    room-by-room calculations, 38–39
    thermal envelope surfaces, 36–38
Towel warmer, 308, 348, 349
Transformer, 403
Triac, 450
Triple
  action control, 429
  pole (3P) switches, 399, 400
Tristate control, 376. (See also Floating control)
TRVs. (See Thermostatic radiator valves)
Tube
  depth within slab, 477–479
  placement considerations, 521–528
  and plate
    ceiling panels, 476
    system, 497
  size, effect of, 477
  spacing within slab, 480–481
  systems, extended surface suspended, 510
Tubing
  copper, heat loss from, 359–361
  layout drawing, 481
Turbulators, 56
Turbulent, 230
  flow, 230
Turndown ratio, 65
Turn-off differential, 467
Turn-on differential, 467
2 to 10 VDC, 372
Two-pipe
  circulators, 260–261
  direct return systems, 211
  reverse return piping configuration, 279
  systems, 599–601
Two-way motorized valve
  injection mixing using, 449–450

Two-way thermostatic valve, 203
  injection mixing using, 447–449
Two-wire digital bus, 462
TXV. (See Thermal expansion valve)
Tyvek®, 497, 504

U

U-value, 32
Uncoiler, 483
Unconfined space, 71
Undershoot, 374, 409
Underside insulation, 492
Underslab insulation, 477
Underslung thermal trap, 614
Unions, 176
Unit heater, 330–331. (See also Fan coils, overhead)
Unit U-value, 32
Unsaturated state of air solubility, 663
Usage profile, DHW, 710
User-definable I/O, 463–464

V

Valve
  actuators, 423, 552
  angle, 186–187
  balancing, 211–212, 617
  ball, 187–188, 262
  check, 188–189
    spring-loaded, 189, 586
  differential pressure bypass, 209
  diverter, 206
  dumb, 442
  dumb mixing, 199
  fast-fill, 16
  feed water, 15, 189–191
  flow-check, 16, 193, 586, 748
  four-way motorized mixing, 201–202, 445–446
  full port ball, 187, 191
  gate, 185–186, 262
  globe, 186
  integral flow-check, 299
  lockshield, 212–213
  metered balancing, 770
  mixing, 197–198
  pressure-reducing, 15
  pressure-relief, 15, 192–193
  purge, 15, 191
  purging, 191
  quick opening, 384
  shutoff, 15
  standard-port ball, 187
  thermal expansion, 95
  thermostatic
    mixing, 203
    radiator (TRV), 194–197
  three-way diverter, 206, 207–209
  three-way motorized mixing, 198–200, 444–445
  three-way thermostatic mixing, 203, 442–444
  two-way thermostatic, 203, 447–449
  venturi-type balancing, 211

Valved manifold
   station, 550
Valveless manifold, 550
Valves
   common types of, 185–189
   specialty for hydronic applications, 189–213
Vapor
   barrier, 481
   pressure, 145, 293
Variable
   controlled, 370
   firing rate heat production, 380
   flow rate control, 383–384
   frequency drive, 598
   manipulated, 370
   speed
      circulators, 271
      drive, 450
      injection mixing, 203
      pressure-regulated circulators, 598
   speed pump
      (direct) injection mixing using, 450–455
      (reverse) injection mixing using, 450–455
   water temperature control, 382–383
VAs. (*See* Volt-amps)
Velocity
   head, 227, 258
   noise, 209
   profile, 222
Vent ell, 179. (*See also* Baseboard tee)
Vents, high point, 660
   automatic, 666
   float-type, 662
   manual, 660

Venturi fitting. (*See* Monoflo tees®)
Venturi-type balancing valves, 211
Vertical
   panel radiator, 342, 343
   supply air ducting, 71
Viscosity, 146–147
Viscous friction, 229
Voltage, 231
Volt-amps (VA), 423
Volute, 258, 260

W

Wall-hung boilers, 58–59
Wall panels, radiant, 513–518
   installation procedure, 513–517
   thermal performance, 517–518
Warm weather shut down (WWSD), 716
Water
   hammer, 423
   physical properties of, 1, 139–156
   plate, 345
   temperature limitations, 114
Waterlogged, 637, 666
Water-reducing agent, 491
Water-tube boilers, 56–57
Web-enabled, 462
Welded wire fabric (WWF), 482, 733, 734
Wet-base boiler, 54, 57
Wet rotor circulator, 259–260, 287
Wind chill effect, 10, 474
Wireless thermostats, 411–412
Wire-to-water efficiency, 260, 285
   of circulator, 287–288

Wood-burning boilers, efficiency of, 131
Wood-fired
   boilers, efficiency of, 131
   furnaces, outdoor, 125
Wood-fueled boilers, 125
   types of, 125
Wood-gasification boilers, 125–131
Working pressure, 159
   rating, 161
WWF. (*See* Welded wire fabric)

Y

Youker system, 489–492

Z

Zone, 564
   circulators, 299
   coordination, 465
   priority, 426
   valve, 204–206, 205, 374, 423–424
      four-wire, 423, 424
      multizone systems using, 593–599
      systems, using variable-speed pressure-regulated circulators, 597–599
Zoned systems, occupant management of, 566–567
Zoning, 3
   of unoccupied areas, 567
   of transitory areas, 567

Hydronics Design Studio
PROFESSIONAL version 2.0
By Mario Restive & John Siegenthaler

A full set of powerful software tools to assist heating professionals in designing high performance hydronic heating systems

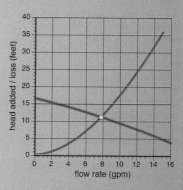

Here's what the tools do:

Visual Heat Load Estimator:
- Calculates the design heating load of user specified rooms
- Graphic displays of building components and material R-value database

Series Baseboard Simulator:
- Simulates thermal and hydraulic performance of circuits containing up to 12 baseboards
- Complete databases of baseboards, circulators, piping, fluids, and other components

Hydronic Circuit Simulator:
- Simulates thermal and hydraulic performance of extensive user defined circuits, including radiant panels
- Extensive databases of circulators, piping, fittings, valves, fluids, and other components

Expansion Tank Sizer:
- Calculates required expansion tank volume over a wide range of conditions
- Determines system pressurization, system volume, antifreeze required

Fluid Properties Calculator:
- Calculates density, specific heat, and viscosity of several common hydronic fluids
- Determines alpha valve of fluid

Equivalent Length Calculator:
- Determines the equivalent length of any user specified component based on flow rate and head loss characteristics

Hydraulic Resistance Calculator:
- Calculates hydraulic resistance of user defined components
- Determines inputs that can be used in Hydronic Circuit Simulator module

Pipe Sizer:
- Determines required size of copper, PEX, and PEX-AL-PEX tubing
- Sizes piping based on specified flow or required rate of heat transfer

Pipe Heat Loss Estimator:
- Uses highly accurate model to calculate heat loss of piping
- Allows comparison of insulated and uninsulated piping heat loss

Heating Cost Estimator:
- Estimates annual space heating cost for several fuels and efficiencies
- Includes weather data for hundreds of locations

Injection Mixing Simulator:
- Simulates operation of user-specified injection mixing system
- Can be used to simulate minitube distribution systems

Buffer Tank Simulator:
- Determine heat source on/off cycle based on tank size & loads
- Includes effect of standby heat loss

available for download at
www.hydronicpros.com